D1674609

Edited by
Ralf Riedel and I-Wei Chen

Ceramics Science and Technology

Related Titles

Ghosh, S. K. (ed.)

Self-healing Materials

Fundamentals, Design Strategies, and Applications

2009

ISBN: 978-3-527-31829-2

Krenkel, W. (ed.)

Ceramic Matrix Composites

Fiber Reinforced Ceramics and their Applications

2008

ISBN: 978-3-527-31361-7

Heimann, R. B.

Plasma Spray Coating

Principles and Applications

2008

ISBN: 978-3-527-32050-9

Riedel, R., Chen, I-W. (eds.)

Ceramics Science and Technology

Volume 1: Structures

2008

ISBN: 978-3-527-31155-2

Öchsner, A., Murch, G. E., de Lemos, M. J. S. (eds.)

Cellular and Porous Materials

Thermal Properties Simulation and Prediction

2008

ISBN: 978-3-527-31938-1

Li, K.

Ceramic Membranes for Separation and Reaction

2007

ISBN: 978-0-470-01440-0

Boch, P., Niepce, J.-C. (eds.)

Ceramic Materials

Processes, Properties, and Applications

2007

ISBN: 978-1-905209-23-1

Ghosh, S. K. (ed.)

Functional Coatings

by Polymer Microencapsulation

2006

ISBN: 978-3-527-31296-2

Scheffler, M., Colombo, P. (eds.)

Cellular Ceramics

Structure, Manufacturing, Properties and Applications

2005

ISBN: 978-3-527-31320-4

Raabe, D., Roters, F., Barlat, F., Chen, L.-Q. (eds.)

Continuum Scale Simulation of Engineering Materials

Fundamentals - Microstructures - Process Applications

2004

ISBN: 978-3-527-30760-9

Edited by
Ralf Riedel and I-Wei Chen

Ceramics Science and Technology

Volume 2: Properties

WILEY-VCH Verlag GmbH & Co. KGaA

The Editors

Prof. Dr. Ralf Riedel
TU Darmstadt
Institut für Materialwissenschaft
Petersenstr. 23
64287 Darmstadt
Germany

Prof. Dr. I-Wei Chen
University of Pennsylvania
School of Engineering
3231 Walnut Street
Philadelphia, PA 19104-6272
USA

Library of Congress Card No.: applied for

British Library Cataloguing-in-Publication Data
A catalogue record for this book is available from the British Library.

Bibliographic information published by the Deutsche Nationalbibliothek
The Deutsche Nationalbibliothek lists this publication in the Deutsche Nationalbibliografie; detailed bibliographic data are available on the Internet at http://dnb.d-nb.de.

Composition Thomson Digital, Noida, India
Printing and Bookbinding betz-druck GmbH, Darmstadt
Cover Design Schulz Grafik-Design, Fußgönheim

Printed in the Federal Republic of Germany
Printed on acid-free paper

ISBN: 978-3-527-31156-9

Contents

Preface *XXI*
List of Contributors *XXIII*

I **Ceramic Material Classes** *1*

1 **Ceramic Oxides** *3*
Dušan Galusek and Katarína Ghillányová
1.1 Introduction *3*
1.2 Aluminum Oxide *4*
1.2.1 Crystal Structure *4*
1.2.2 Natural Sources and Preparation of Powders *5*
1.2.2.1 High-Temperature/Flame/Laser Synthesis *6*
1.2.2.2 Chemical Methods *6*
1.2.2.3 Mechanically Assisted Synthesis *6*
1.2.3 Solid-State Sintered Alumina *7*
1.2.3.1 Submicrometer and Transparent Alumina *7*
1.2.4 Liquid-Phase Sintered (LPS) Aluminas *8*
1.2.5 Properties of Polycrystalline Alumina *11*
1.3 Magnesium Oxide *13*
1.3.1 Crystal Structure and Properties of Single-Crystal MgO *14*
1.3.2 Natural Sources and Production *14*
1.3.3 Polycrystalline Magnesia *15*
1.4 Zinc Oxide *15*
1.4.1 Crystal Structure and Properties of Single-Crystal ZnO *16*
1.4.2 Natural Sources and Production *16*
1.4.3 Properties *16*
1.4.4 Applications *17*
1.4.4.1 ZnO-Based Varistors *17*
1.4.4.2 Other Applications of ZnO Ceramics *19*
1.5 Titanium Dioxide *20*
1.5.1 Crystal Structure and Properties of Single-Crystal TiO_2 *20*

Ceramics Science and Technology Volume 2: Properties. Edited by Ralf Riedel and I-Wei Chen

ISBN: 978-3-527-31156-9

1.5.2 Natural Sources and Production 22
1.5.2.1 Synthesis of Anatase 22
1.5.2.2 Synthesis of Rutile 23
1.5.2.3 Synthesis of Brookite 23
1.5.3 Properties of TiO_2 Polymorphs 24
1.5.4 Polycrystalline Titania 24
1.5.5 Applications of TiO_2 25
1.6 Zirconium Oxide 27
1.6.1 Crystal Structure and Properties of Single Crystals 28
1.6.2 Natural Sources and Production 29
1.6.2.1 Phase Transformation of Zirconia 31
1.6.3 Partially Stabilized Zirconia 32
1.6.3.1 Mg-PSZ 32
1.6.3.2 Ca-PSZ 34
1.6.3.3 Y-PSZ 34
1.6.3.4 Ceria and Other Rare Earth-Stabilized Zirconias 34
1.6.4 Tetragonal Zirconia Polycrystals (TZP) 35
1.6.5 Zirconia-Toughened Alumina (ZTA) 37
1.6.6 Applications of Zirconia 38
1.6.6.1 Thermal Barrier Coatings 38
1.6.6.2 Solid Electrolytes 39
1.6.6.3 Fuel Cells 41
1.6.6.4 Bioceramics 42
1.7 Cerium Oxide 43
1.7.1 Crystal Structure and Properties of Single-Crystal CeO_2 43
1.7.2 Natural Sources and Production 44
1.7.3 Properties 45
1.7.4 Applications 46
1.7.4.1 Abrasives 46
1.7.4.2 Solid Electrolytes 46
1.7.4.3 Catalysts 47
1.8 Yttrium Oxide 48
1.8.1 Crystal Structure and Properties of Single Crystal Yttrium Oxide 48
1.8.2 Natural Sources and Preparation 48
1.8.3 Properties 49
1.8.4 Applications 50
References 51

2 Nitrides 59
Pavol Šajgalík, Zoltán Lenčéš, and Miroslav Hnatko
2.1 Silicon Nitride 59
2.1.1 Introduction 59
2.1.2 Amorphous Silicon Nitride 60
2.1.3 Silicon Nitride Single Crystals: Structure 60
2.1.3.1 α- and β- Si_3N_4 60

2.1.3.2 γ-Si_3N_4 61
2.1.4 Silicon Nitride Single Crystals: Mechanical Properties 63
2.1.4.1 α-Si_3N_4 63
2.1.4.2 β-Si_3N_4 63
2.1.4.3 γ-Si_3N_4 64
2.1.5 Silicon Nitride-Based Materials 64
2.1.6 Oxynitride Glasses 65
2.1.6.1 Properties of Oxynitride Glasses 66
2.1.7 Polycrystalline Si_3N_4 66
2.1.8 Lu-Doped Si_3N_4 Ceramics 67
2.1.9 SiAlON Ceramics 68
2.1.9.1 α- and β-SiAlONs 68
2.1.9.2 Si_2N_2O and O′-SiAlON 70
2.2 Boron Nitride 71
2.2.1 Crystallographic Structures 71
2.2.1.1 Hexagonal Boron Nitride 71
2.2.1.2 Cubic Boron Nitride 72
2.2.1.3 Wurtzitic Boron Nitride 72
2.2.2 Synthesis of BN 72
2.2.3 Properties of BN 73
2.2.3.1 h-BN 73
2.2.3.2 c-BN and w-BN 73
2.3 Aluminum Nitride 74
2.3.1 Structure 74
2.3.2 Synthesis 74
2.3.3 Properties 75
2.4 Titanium Nitride 75
2.4.1 Structure 75
2.4.2 Synthesis 75
2.4.3 Properties 76
2.5 Tantalum Nitride 77
2.6 Chromium Nitride 77
2.7 Ternary Nitrides 78
2.7.1 Ternary Silicon Nitrides 78
2.7.1.1 $MgSiN_2$ 78
2.7.1.2 Other Alkaline Earth Silicon Nitrides 79
2.7.1.3 $LaSi_3N_5$ 79
2.7.1.4 $LiSi_2N_3$ 80
2.8 Light-Emitting Nitride and Oxynitride Phosphors 80
References 81

3 Gallium Nitride and Oxonitrides 91
Isabel Kinski and Paul F. McMillan
3.1 Introduction 91
3.2 Gallium Nitrides 94

3.2.1 Phase Description *94*
3.2.2 Synthesis Routes to GaN *99*
3.2.2.1 Synthesis of Bulk Gallium Nitride *99*
3.2.2.2 Synthesis of Thin Film GaN and Epitaxial Growth Techniques *100*
3.2.2.3 Synthesis of GaN via Chemical Precursor Routes *101*
3.2.3 Properties of (Ga,Al,In)N Solid Solutions *101*
3.3 Gallium Oxides *102*
3.3.1 Phase Description and Properties *102*
3.3.2 Synthesis and Growth Techniques *106*
3.4 Gallium Oxonitrides *107*
3.4.1 Nomenclature Issues of Ga–O–N Materials *108*
3.4.2 Theoretical Predictions for Ga Oxonitride Compounds *108*
3.4.3 Literature Overview on Gallium Oxide Nitride Phases *110*
3.4.4 Synthesis and Growth Techniques *114*
3.4.4.1 Precursor Approach for Gallium Oxide Nitride Phases *115*
3.4.4.2 Crystalline Phases of Gallium Oxide Nitride Synthesized under High-Pressure/High-Temperature Conditions *118*
3.4.5 Potential Applications *122*
3.5 Outlook *124*
References *124*

4 Silicon Carbide- and Boron Carbide-Based Hard Materials *131*
Clemens Schmalzried and Karl A. Schwetz
4.1 Introduction *131*
4.2 Structure and Chemistry *131*
4.2.1 Silicon Carbide *131*
4.2.1.1 Phase Relations in the System Silicon–Carbon *132*
4.2.1.2 Structural Aspects *132*
4.2.2 Boron Carbide *134*
4.2.2.1 Phase Relations in the System Boron–Carbon *135*
4.2.2.2 Structural Aspects *135*
4.3 Production of Particles and Fibers *137*
4.3.1 Silicon Carbide *137*
4.3.1.1 Technical-Scale Production of α-Silicon Carbide *137*
4.3.1.2 β-Silicon Carbide Powder *138*
4.3.1.3 Silicon Carbide Whiskers *142*
4.3.1.4 Silicon Carbide Platelets *145*
4.3.1.5 Continuous Silicon Carbide Fibers *146*
4.3.1.6 Silicon Carbide Nanofibers *148*
4.3.1.7 Silicon Carbide Nanotubes (SiCNTs) *149*
4.3.2 Boron Carbide *149*
4.3.2.1 Technical-Scale Production *149*
4.3.2.2 Submicron B_4C Powders *150*
4.3.2.3 Boron Carbide-Based Nanostructured Particles *151*
4.4 Dense Ceramic Shapes *152*

4.4.1 Dense Silicon Carbide Shapes 152
4.4.1.1 Ceramically Bonded Silicon Carbide 152
4.4.1.2 Recrystallized Silicon Carbide 152
4.4.1.3 Reaction-Bonded Silicon Carbide 154
4.4.1.4 Sintered Silicon Carbide 154
4.4.1.5 Hot-Pressed Silicon Carbide 158
4.4.1.6 Chemical Vapor- and Physical Vapor-Deposited Silicon Carbide 159
4.4.1.7 Silicon Carbide Wafers 160
4.4.1.8 Silicon Carbide Nanoceramics 161
4.4.1.9 Silicon Carbide-Based Composites 162
4.4.1.10 Metal Matrix Composites (MMCs) 172
4.4.2 Dense Boron Carbide Shapes 174
4.4.2.1 Sintered Boron Carbide 175
4.4.2.2 Hot-Pressed and Hot Isostatic-Pressed Boron Carbide 178
4.4.2.3 Spark-Plasma-Sintered Boron Carbide 179
4.4.2.4 Boron Carbide-Based Composites 179
4.4.2.5 B_4C-Based MMCs 182
4.5 Properties of Silicon Carbide- and Boron Carbide-Based Materials 183
4.5.1 Silicon Carbide 183
4.5.1.1 Physical Properties 183
4.5.1.2 Chemical Properties 186
4.5.1.3 Tribological Properties 187
4.5.2 Boron Carbide 194
4.5.2.1 Physical Properties 194
4.5.2.2 Chemical Properties 200
4.6 Application of Carbides 202
4.6.1 Silicon Carbide 202
4.6.2 Boron Carbide 207
References 210

5 Complex Oxynitrides 229
Derek P. Thompson
5.1 Introduction 229
5.2 Principles of Silicon-Based Oxynitride Structures 230
5.3 Complex Si–Al–O–N Phases 231
5.3.1 Sialon X-Phase 231
5.3.2 The Sialon Polytypoid Phases 232
5.3.3 The Y-Si–O–N Oxynitrides 233
5.4 M–Si–Al–O–N Oxynitrides 238
5.4.1 α-SiAlON 238
5.4.2 JEM Phase 239
5.4.3 S-Phase 240
5.4.4 Lanthanum "New" Phase 242
5.4.5 $M_2(Si,Al)_5(O,N)_8$ Oxynitrides 242

5.4.6 $M(Si,Al)_3(O,N)_5$ Phases *243*
5.4.7 $M_3(Si,Al)_6(O,N)_{11}$ Phases *244*
5.4.8 Wurtzite Oxynitrides *245*
5.4.9 $MSi_2O_2N_2$ Oxynitrides *245*
5.4.10 More Complex Oxynitrides *246*
5.5 Oxynitride Glasses *246*
5.6 Oxynitride Glass Ceramics *248*
5.6.1 B-Phase *249*
5.6.2 Iw Phase *250*
5.6.3 U-Phase *250*
5.6.4 W-Phase *251*
5.6.5 Nitrogen Pyroxenes *251*
5.7 Conclusions *253*
References *254*

6 Perovskites *257*
Vladimir Fedorov
6.1 Introduction *257*
6.2 Crystal Structure *259*
6.2.1 Ideal Perovskite Structure *259*
6.2.2 Structural Distortions and Phase Transitions *260*
6.2.3 Other Perovskite-Related Structures *264*
6.2.3.1 Polytypes Consisting of Close-Packed Ordered AO_3 Layers *264*
6.2.3.2 Perovskite Intergrowth Structures *266*
6.2.4 Perovskite-Related Copper Oxide Structures *267*
6.2.5 Cation Ordering *269*
6.2.6 Nonstoichiometry *270*
6.2.6.1 A-Site Vacancies *270*
6.2.6.2 B-Site Vacancies *271*
6.2.6.3 Anion-Deficient Perovskites and Vacancy-Ordered Structures *272*
6.2.6.4 Anion-Excess Nonstoichiometry *274*
6.3 Physical Properties *274*
6.3.1 Electronic Properties *274*
6.3.1.1 The Colossal Magnetoresistance (CMR) Phenomenon *277*
6.3.2 Ferroelectricity and Related Phenomena *279*
6.3.3 Relaxor Ferroelectrics (Relaxors) *280*
6.3.4 Morphotropic Phase Boundary (MPB) Compositions *282*
6.3.5 Optical Properties *282*
6.3.6 Ion Conductivity *283*
6.3.6.1 Oxide-Ion Conductivity *283*
6.3.6.2 Proton Conductivity *285*
6.3.7 Cation Transport *287*
6.3.8 Computer Modeling of Ionic Transport in ABO_3 *288*
6.3.9 Applications *289*

6.4 Chemical and Catalytic Properties 289
6.4.1 Synthesis 292
6.5 Summary 292
References 293

7 The $M_{n+1}AX_n$ Phases and their Properties 299
Michel W. Barsoum
7.1 Introduction 299
7.2 Bonding and Structure 300
7.3 Elastic Properties 303
7.4 Electronic Transport 307
7.5 Thermal Properties 313
7.5.1 Thermal Conductivities 313
7.5.2 Thermal Expansion 316
7.5.3 Thermal Stability 318
7.5.4 Chemical Reactivity and Oxidation Resistance 318
7.6 Mechanical Properties 320
7.6.1 Introduction 320
7.6.2 Dislocations and their Arrangements 320
7.6.3 Plastic Anisotropy, Internal Stresses, and Deformation Mechanisms 321
7.6.4 Incipient Kink Band Microscale Model 325
7.6.5 Compression Behavior of Quasi-Single Crystals and Polycrystals 329
7.6.6 Hardness and Damage Tolerance 329
7.6.7 Thermal Shock Resistance 333
7.6.8 R-Curve Behavior and Fatigue 334
7.6.9 High-Temperature Properties 336
7.6.9.1 Compressive Properties 336
7.6.9.2 Tensile Properties 338
7.6.9.3 Creep 338
7.6.9.4 Solid-Solution Hardening and Softening 341
7.7 Tribological Properties and Machinability 342
7.8 Concluding Remarks 342
References 343

II Structures and Properties 349

8 Structure–Property Relations 351
Tatsuki Ohji
8.1 Introduction 351
8.2 Self-Reinforced Silicon Nitrides 352
8.3 Fibrous Grain-Aligned Silicon Nitrides (Large Grains) 355
8.4 Fibrous Grain-Aligned Silicon Nitrides (Small Grains) 361
8.5 Grain Boundary Phase Control 365
8.5.1 Fracture Resistance 365

8.5.2 Heat Resistance 368
8.6 Fibrous Grain-Aligned Porous Silicon Nitrides 370
8.6.1 Porous Silicon Nitride Through Tape-Casting 371
8.6.2 Porous Silicon Nitride Through Sinter-Forging 374
8.6.3 Comparison of Properties of Porous and Dense Silicon Nitrides 375
References 376

9 Dislocations in Ceramics *379*
Terence E. Mitchell
9.1 Introduction 379
9.2 The Critical Resolved Shear Stress 380
9.2.1 Experimental Observations 380
9.2.2 Kink Mechanism for Deformation 382
9.2.3 Modification of the Model for Kink Pair Nucleation on Point Defects, and Partial Dislocations 384
9.2.4 Comparison between Theory and Experiment 386
9.2.4.1 Sapphire and Stoichiometric Spinel 386
9.2.4.2 Nonstoichiometric Spinel 388
9.3 Crystallography of Slip 388
9.3.1 Crystal Structures 388
9.3.2 Slip Systems 390
9.3.3 Dislocation Dissociations 391
9.4 Dislocations in Particular Oxides 392
9.4.1 Magnesium Oxide and other Oxides with the Rock-Salt Structure 393
9.4.1.1 Magnesium Oxide 393
9.4.1.2 Transition Metal Oxides and the Effect of Stoichiometry 393
9.4.2 Beryllium Oxide and Oxides with the Wurtzite Structure 396
9.4.2.1 Beryllium Oxide 396
9.4.2.2 Dislocation Dissociation in BeO and Other Crystals with the Wurtzite Structure 397
9.4.2.3 Zinc Oxide 398
9.4.3 Zirconium Oxide and Other Oxides with the Fluorite Structure 398
9.4.3.1 Zirconia 398
9.4.3.2 Uranium Oxide 400
9.4.3.3 Dislocation–Dissociation in Oxides with the Fluorite Structure 401
9.4.4 Titanium Oxide and Oxides with the Rutile Structure 402
9.4.5 Silicon Oxide (Quartz) 402
9.4.5.1 Hydrolytic Weakening in Quartz 403
9.4.5.2 Dislocation–Dissociation in Quartz 404
9.4.6 Aluminum Oxide (Sapphire) 405
9.4.6.1 Dislocation–Dissociation in Sapphire 406
9.4.6.2 Basal Slip in Sapphire 406

9.4.6.3 Prism-Plane Slip in Sapphire *406*
9.4.6.4 Dipoles and Climb Dissociation in Sapphire *407*
9.4.6.5 Stacking Fault Energy of Sapphire *408*
9.4.6.6 Deformation Twinning in Sapphire *408*
9.4.7 $SrTiO_3$ and Oxides with the Perovskite Structure *410*
9.4.7.1 Inverse Brittle-to-Ductile Transition (BDT) in $SrTiO_3$ *410*
9.4.7.2 Dislocation–Dissociation and SFE in Strontium Titanate *411*
9.4.8 $MgO–Al_2O_3$ and other Spinels *413*
9.4.8.1 Slip Planes in Spinels *414*
9.4.8.2 Dislocation–Dissociations and the SFE in Magnesium Aluminate Spinel *415*
9.4.9 Mg_2SiO_4 (Forsterite) *417*
9.4.9.1 Water-Weakening and Dislocation–Dissociation in Olivine *418*
9.4.10 Other Oxides *419*
9.4.10.1 Oxides with the Cubic Rare-Earth Sesquioxide Structure *420*
9.4.10.2 YAG and other Oxides with the Garnet Structure *420*
9.4.11 SiC, Si_3N_4, and other Non-Oxide Ceramics *421*
9.4.11.1 Silicon Carbide *421*
9.4.11.2 Silicon Nitride *422*
9.4.12 Climb versus Glide Dissociation *422*
9.5 Work Hardening *423*
9.5.1 Work Hardening in MgO *423*
9.5.2 Work Hardening and Work Softening in Spinel *424*
9.5.3 Work Hardening in Sapphire *426*
9.5.3.1 Basal Plane Slip *426*
9.5.3.2 Prism Plane Slip *427*
9.6 Solution Hardening *428*
9.6.1 Isovalent Cations *428*
9.6.2 Aliovalent Cations *429*
9.7 Closing Remarks *430*
References *431*

10 Defect Structure, Nonstoichiometry, and Nonstoichiometry Relaxation of Complex Oxides *437*
Han-Ill Yoo
10.1 Introduction *437*
10.2 Defect Structure *438*
10.2.1 Pure Case *439*
10.2.2 Acceptor-Doped Case *443*
10.2.3 Donor-Doped Case *445*
10.2.4 Two-Dimensional Representations of Defect Concentrations *445*
10.2.5 A Further Complication: Hole Trapping *452*
10.2.6 Defect Structure and Reality *453*
10.3 Oxygen Nonstoichiometry *456*
10.3.1 Nonstoichiometry in General *457*

10.3.2 Experimental Reality 460
10.4 Nonstoichiometry Re-Equilibration 462
10.4.1 Undoped or Acceptor-Doped $BaTiO_3$ 464
10.4.1.1 Relaxation Behavior and Chemical Diffusion 464
10.4.1.2 Defect-Chemical Interpretation 466
10.4.2 Donor-Doped $BaTiO_3$ 468
10.4.2.1 Relaxation Behavior and Chemical Diffusion 468
10.4.2.2 Defect-Chemical Interpretation 469
10.4.3 Defect Diffusivities 475
References 477

11 Interfaces and Microstructures in Materials *479*
Wook Jo and Nong-Moon Hwang
11.1 Introduction 479
11.2 Interfaces in Materials 480
11.2.1 Surface Fundamentals 480
11.2.1.1 Surface Energy 480
11.2.1.2 Wulff Plot 486
11.2.1.3 Roughening Transition 495
11.2.1.4 Kinetics of Surface Migration 499
11.2.2 Solid/Liquid Interfaces 502
11.2.3 Solid/Solid Interfaces 506
11.2.3.1 Fundamentals 506
11.2.3.2 Structure and Energy 508
11.3 Practical Implications 513
11.4 Summary and Outlook 523
References 523

III Mechanical Properties *529*

12 Fracture of Ceramics *531*
Robert Danzer, Tanja Lube, Peter Supancic, and Rajiv Damani
12.1 Introduction 531
12.2 Appearance of Failure and Typical Failure Modes 532
12.2.1 Thermal Shock Failure 534
12.2.2 Contact Failure 536
12.3 A Short Overview of Damage Mechanisms 538
12.3.1 Sudden, Catastrophic Failure 539
12.3.2 Sub-Critical Crack Growth 539
12.3.3 Fatigue 540
12.3.4 Creep 540
12.3.5 Corrosion 540
12.4 Brittle Fracture 541
12.4.1 Some Basics in Fracture Mechanics 542
12.4.2 Tensile Strength of Ceramic Components, and Critical Crack Size 544

12.5 Probabilistic Aspects of Brittle Fracture *545*
12.5.1 Fracture Statistics and Weibull Statistics *545*
12.5.1.1 Weibull Distribution for Arbitrarily Oriented Cracks in a Homogeneous Uniaxial Stress Field *547*
12.5.1.2 Weibull Distribution for Arbitrarily Oriented Cracks in an Inhomogeneous Uniaxial Stress Field *548*
12.5.1.3 Weibull Distribution in a Multi-Axial Stress Field *549*
12.5.2 Application of the Weibull Distribution: Design Stress and Influence of Specimen Size *550*
12.5.3 Experimental and Sampling Uncertainties, the Inherent Scatter of Strength Data, and Can a Weibull Distribution be Distinguished from a Gaussian Distribution? *553*
12.5.4 Influence of Microstructure: Flaw Populations on Fracture Statistics *555*
12.5.5 Limits for the Application of Weibull Statistics in Brittle Materials *558*
12.6 Delayed Fracture *558*
12.6.1 Lifetime and Influence of SCCG on Strength *559*
12.6.1.1 Delayed Fracture under Constant Load *561*
12.6.1.2 Delayed Fracture under Increasing Load: Constant Stress Rate Tests *564*
12.6.1.3 Delayed Fracture under General Loading Conditions *565*
12.6.2 Influence of Fatigue Crack Growth on Strength *566*
12.6.3 Proof Testing *566*
12.7 Concluding Remarks *567*
References *569*

13 Creep Mechanisms in Commercial Grades of Silicon Nitride *577*
František Lofaj and Sheldon M. Wiederhorn
13.1 Introduction *577*
13.1.1 Motivation *577*
13.1.2 Creep of Silicon Nitride *578*
13.2 Material Characterization *580*
13.3 Discussion of Experimental Data *581*
13.3.1 Creep Behavior *582*
13.3.1.1 NT 154 *582*
13.3.1.2 SN 88 *583*
13.3.1.3 SN 281 *585*
13.3.2 TEM Characterization of Cavitation *587*
13.3.3 Density Change and Volume Fraction Cavities *588*
13.3.4 Size Distribution of Cavities Formed *590*
13.4 Models of Creep in Silicon Nitride *592*
13.4.1 Cavitation Creep Model in NT 154 and SN 88 *593*
13.4.2 Noncavitation Creep *595*
13.4.3 Role of Lutetium in the Viscosity of Glass *596*

13.5 Conclusions *596*
References *598*

14 Fracture Resistance of Ceramics *601*
Mark Hoffman
14.1 Introduction *601*
14.2 Theory of Fracture *601*
14.2.1 Stress Concentration Factors *608*
14.2.2 Crack Closure Concept and Superposition *609*
14.3 Toughened Ceramics *612*
14.3.1 Bridged Interface Methods for Increasing Fracture Resistance *613*
14.3.1.1 Grain Bridging *613*
14.3.1.2 Crack Growth Resistance Toughening *614*
14.3.1.3 Crack Bridging By a Second Phase *615*
14.3.1.4 Phase Transformation or Dilatant Zone Toughening *617*
14.3.1.5 Ferroelastic Toughening *619*
14.3.2 Toughening by Crack Tip Process Zone Effects *620*
14.4 Influence of Crack Growth Resistance Curve Upon Failure by Fracture *621*
14.5 Determination of Fracture Resistance *622*
14.5.1 Indentation Fracture Toughness *622*
14.5.2 Single-Edge Notched Beam *623*
14.5.3 Surface Crack in Flexure *625*
14.5.4 Determination of Intrinsic Toughness *625*
14.6 Fatigue *626*
14.6.1 Cyclic Fatigue Crack Propagation *626*
14.6.2 Contact Fatigue *627*
14.7 Concluding Remarks *629*
References *629*

15 Superplasticity in Ceramics: Accommodation-Controlling Mechanisms Revisited *633*
Arturo Domínguez-Rodríguez and Diego Gómez-García
15.1 Introduction *633*
15.2 Macroscopic and Microscopic Features of Superplasticity *634*
15.3 Nature of the Grain Boundaries *640*
15.4 Accommodation Processes in Superplasticity *643*
15.4.1 GBS Accommodated by Diffusional Flow *643*
15.4.2 GBS Accommodated by Dislocation Motion *648*
15.4.3 Solution–Precipitation Model for Creep *649*
15.4.4 Shear-Thickening Creep *655*
15.5 Applications of Superplasticity *656*
15.6 Future Prospective in the Field *659*
References *660*

IV Thermal, Electrical, and Magnetic Properties 665

16 Thermal Conductivity 667
Kiyoshi Hirao and You Zhou
16.1 Introduction 667
16.2 Thermal Conductivity of Dielectric Ceramics 668
16.2.1 Thermal Conductivity of Nonmetallic Crystals 668
16.2.2 High Thermal Conductivity in Adamantine Compounds 669
16.2.3 Estimate of Thermal Conductivity of β-Si_3N_4 671
16.3 High-Thermal Conductivity Nonoxide Ceramics 672
16.3.1 Thermal Conductivity of Composite Microstructures 672
16.3.1.1 Liquid-Phase Sintering of Nonoxide Ceramics 672
16.3.1.2 Effect of Secondary Phase on Thermal Conductivity of AlN Ceramic 674
16.3.1.3 Effect of Secondary Phase on Thermal Conductivity of Si_3N_4 Ceramic 676
16.3.1.4 Effect of Secondary Phase on Thermal Conductivity of SiC Ceramic 678
16.3.2 Improvement of Thermal Conductivity via Purification of Grains During Sintering 679
16.3.2.1 Aluminum Nitride Ceramics 679
16.3.2.2 Lattice Defects in β-Si_3N_4 Grains 680
16.3.2.3 Improvements in Thermal Conductivity for Silicon Nitride Ceramics 682
16.3.2.4 Lattice Defects in SiC Grains and Improvement of Thermal Conductivity 687
16.4 Mechanical Properties of High-Thermal Conductivity Si_3N_4 Ceramics 688
16.4.1 Harmonic Improvement of Thermal Conductivity and Mechanical Properties 688
16.4.2 Anisotropic Thermal Conductivity in Textured Si_3N_4 692
16.5 Concluding Remarks 693
References 694

17 Electrical Conduction in Nanostructured Ceramics 697
Harry L. Tuller, Scott J. Litzelman, and George C. Whitfield
17.1 Introduction 697
17.2 Space Charge Layers in Semiconducting Ceramic Materials 699
17.3 Effect of Space Charge Profiles on the Observed Conductivity 707
17.4 Influence of Nanostructure on Charge Carrier Distributions 708
17.5 Case Studies 710
17.5.1 Case Study: Nanostructured Sensor Films 710
17.5.2 Case Study: Interfaces in Ionic and Mixed Conducting Materials 714
17.5.3 Case study: Lithium-Ion Battery Materials 718

17.5.4 Case Study: Dye-Sensitized Solar Cells *719*
17.6 Conclusions and Observations *722*
References *724*

18 Ferroelectric Properties *729*
Doru C. Lupascu and Maxim I. Morozov
18.1 Introduction *729*
18.2 Intrinsic Properties: The Anisotropy of Properties *731*
18.3 Extrinsic Properties: Hard and Soft Ferroelectrics *739*
18.4 Textured Ferroelectric Materials *751*
18.4.1 OCAP *751*
18.4.2 TGG *752*
18.4.3 HTGG *753*
18.4.4 RTGG *753*
18.5 Ferroelectricity and Magnetism *755*
18.6 Fatigue in Ferroelectric Materials *764*
18.6.1 Macrocracking *765*
18.6.2 Microcracking *766*
18.6.3 Breakdown *767*
18.6.4 Frequency Effect *768*
18.6.5 Defect Agglomeration *769*
18.6.6 Electrode Effects *772*
18.6.7 Domain Nucleation or Wall Motion Inhibition *772*
18.6.8 Clamping in Thin Films *773*
18.6.9 Domain Splitting and Crystal Orientation Dependence *773*
18.6.10 Combined Loading *774*
18.6.11 Antiferroelectrics *775*
18.6.12 Fatigue-Free Systems *776*
References *777*

19 Magnetic Properties of Transition-Metal Oxides: From Bulk to Nano *791*
Polona Umek, Andrej Zorko, and Denis Arčon
19.1 Introduction *791*
19.2 Properties of Transition Metal 3d Orbitals *792*
19.3 Iron Oxides *793*
19.3.1 Iron Oxide Structures *793*
19.3.2 Magnetic Properties of Iron Oxides *794*
19.4 Ferrites *797*
19.5 Chromium Dioxide *800*
19.6 Manganese Oxide Phases *802*
19.6.1 Manganese Oxide Structures *804*
19.6.1.1 One-Dimensional Structures *804*
19.6.1.2 Layered Structures *807*
19.6.1.3 Three-Dimensional Structures *807*

19.6.2 Magnetic Properties of Selected Manganese Oxide Phases: The Manifestation of Magnetic Frustration *807*
19.6.2.1 Helical Order in Pyrolusite (β-MnO_2) *808*
19.6.2.2 Magnetic Properties of the Mixed-Valence Hollandite (α-MnO_2) *810*
19.6.2.3 Magneto-Elastic Coupling in the Layered α-$NaMnO_2$ Compounds *811*
19.6.2.4 Frustrated Magnetism of the "Defect" Spinel λ-MnO_2 *813*
19.6.3 Synthesis of MnO_2 Nanostructures *815*
19.6.3.1 Synthesis of α-MnO_2 Nanostructures *815*
19.6.3.2 Synthesis of β-MnO_2 Nanostructures *818*
19.6.3.3 Synthesis of γ-MnO_2 Nanostructures *819*
19.6.3.4 Synthesis of λ-MnO_2 Nanodiscs *820*
19.6.3.5 Synthesis of MnO_2 Nanostructures in other Crystallographic Phases *821*
19.6.4 Magnetic Properties of Manganese Dioxide Nanoparticles *821*
19.6.4.1 β-MnO_2 (Pyrolusite) *821*
19.6.4.2 α-MnO_2 (Hollandite) *824*
19.7 Concluding Remarks *828*
References *828*

Index *835*

Preface

Besides metals and polymers, advanced ceramics represent one of the most promising classes of materials for the key technologies of the twenty-first century. Recent developments in the field of ceramics have included a selection of synthesis, processing and sintering techniques applied to the production of novel structural and functional ceramics and ceramic composites. Significant progress has been made during the past two decades with respect to the production of novel multifunctional ceramics with a tailor-made microscale and/or nanoscale structures, reflecting the increasing technological importance of advanced ceramic materials.

The four-volume series of *Ceramics Science and Technology* covers various aspects of modern trends in advanced ceramics, and reflects the status quo of the latest achievements in ceramics science and developments. The contributions highlight the increasing technological significance of advanced ceramic materials, and present concepts for their production and application. Volume 1 covers the structural properties of ceramics by considering a broad spectrum of length scale, starting from the atomic level by discussing amorphous and crystalline solid-state structural features, and continuing with the microstructural level by commenting on microstructural design, mesoscopic and nanostructures, glass ceramics, cellular structures, thin films and multiphase (composite) structures. Volume 2 focuses first, on the distinct ceramic materials classes, namely oxides, carbides and nitrides, and second on the physical and mechanical properties of advanced ceramics. The series is continued with Volume 3, which will contain chapters related to the advanced synthesis and processing techniques used to produce engineering ceramics. The series will be completed by Volume 4, which will be devoted to the application of engineering and functional ceramics.

Quo vadis ceramics? The four-volume series also intends to provide comprehensive information relevant to the future direction of advanced or engineering ceramics. The present series evidences the technologically important trends related to the further development of this fascinating class of materials. The latest examples of technological achievements that already have achieved commercial status include piezoelectric ceramics based on PZT ($Pb(Zr,Ti)O_3$) and used, for example, in common-rail diesel engines, in Si_3N_4-based ball bearings, as glow plugs for diesel engines, as carbon fiber-reinforced silicon carbide (C/SiC) brake components in

Ceramics Science and Technology Volume 2: Properties. Edited by Ralf Riedel and I-Wei Chen

ISBN: 978-3-527-31156-9

vehicles, as luminescent ceramics based on sialon derivatives for LED applications, and as GaN-based ceramics for optoelectronics, among many others.

Furthermore, a variety of application fields is beginning to emerge in which novel ceramics are required, and these are expected to be established and commercialized in the near future. This technology-driven process requires a long-term alignment and a strong basis in continued fundamental research in ceramics science and technology. The intention is that this four-volume series will contribute to such development by providing the latest knowledge in ceramics science suitable not only for those students specializing in ceramics but also for university and industrial research groups.

We wish to thank all contributing authors for their great enthusiasm in compiling excellent manuscripts in their respective areas of expertise. We also acknowledge the support of Karen Böhling, who proofread each manuscript with due accuracy and patience. Last, but not least, we thank the Wiley-VCH editors, Bernadette Gmeiner and Martin Preuß, for their continuous encouragement to work on the book project.

Darmstadt and Philadelphia
February 2010

Ralf Riedel
I.-Wei Chen

List of Contributors

Denis Arčon
Institute Jožef Stefan
Jamova 39
1000 Ljubljana
Slovenia

and

University of Ljubljana
Faculty of Mathematics and Physics
Jadranska 19
1000 Ljubljana
Slovenia

Michel W. Barsoum
Drexel University
Department of Materials Science
and Engineering
Philadelphia, PA 19104
USA

Rajiv Damani
Sulzer Markets and Technology Ltd
Sulzer Innotec 1551
Sulzer-Allee 25, P.O. Box 65
8404 Winterthur
Switzerland

Robert Danzer
Montanuniversität Leoben
Institut für Struktur- und
Funktionskeramik
Franz-Josef-Straße 18
8700 Leoben
Austria

Arturo Domínguez-Rodríguez
Universidad de Sevilla
Departamento de Física de la Materia
Condensada
Avda. Reina Mercedes s/n
41012 Seville
Spain

Vladimir Fedorov
Russian Academy of Sciences
Nikolaev Institute of Inorganic
Chemistry, Siberian Branch
3, Akad. Lavrentiev prospect
Novosibirsk 630090
Russia

Dušan Galusek
VILA – Joint Glass Centre of the
Institute of Inorganic Chemistry
Slovak Academy of Sciences
Alexander Dubček University of Trenčín
and RONA, a.s.
911 50 Trenčín
Slovak Republic

Ceramics Science and Technology Volume 2: Properties. Edited by Ralf Riedel and I-Wei Chen

ISBN: 978-3-527-31156-9

Katarína Ghillányová
Slovak Academy of Sciences
Institute of Inorganic Chemistry
Dúbravská cesta 9
845 36 Bratislava
Slovak Republic

Diego Gómez-García
Universidad de Sevilla
Departamento de Física de la Materia
Condensada
Avda. Reina Mercedes s/n
41012 Seville
Spain

Kiyoshi Hirao
National Institute of Advanced
Industrial Science and Technology
(AIST)
2266-98 Shimo-Shidami, Moriyama-ku
Nagoya 463-8560
Japan

Miroslav Hnatko
Slovak Academy of Sciences
Institute of Inorganic Chemistry
Dúbravska cesta 9
84536 Bratislava 45
Slovakia

Mark Hoffman
The University of New South Wales
School of Materials Science and
Engineering
Sydney, New South Wales 2052
Australia

Nong-Moon Hwang
Seoul National University
School of Materials Science &
Engineering
Kwanak-gu
Seoul 151-744
Korea

Wook Jo
Nichtmetallisch-Anorganische
Werkstoffe
Material- und Geowissenschaften
Petersenstr. 23
64287 Darmstadt
Germany

Isabel Kinski
Fraunhofer Institut Keramische
Technologien und Systeme
Winterbergstraße 28
01277 Dresden
Germany

Zoltán Lenčéš
Slovak Academy of Sciences
Institute of Inorganic Chemistry
Dúbravska cesta 9
84536 Bratislava 45
Slovakia

Scott J. Litzelman
Massachusetts Institute of Technology
Department of Materials Science and
Engineering
Cambridge, MA 02139
USA

František Lofaj
Slovak Academy of Sciences
Institute of Materials Research
Watsonova 47
040 01 Košice
Slovakia

Tanja Lube
Montanuniversität Leoben
Institut für Struktur- und
Funktionskeramik
Franz-Josef-Straße 18
8700 Leoben
Austria

Doru C. Lupascu
Universität Duisburg-Essen
Institut für Matrialwissenschaft
Universitätsstr. 15
45141 Essen
Germany

Paul F. McMillan
University College London
Christopher Ingold Laboratories
Department of Chemistry and Materials
Chemistry Centre
20 Gordon Street
London WC1H 0AJ
UK

Terence E. Mitchell
Los Alamos National Laboratory
Structure–Property Relations Group
MST-8
Los Alamos, NM 87545
USA

Maxim I. Morozov
Karlsruhe Institute of Technology (KIT)
Institut für Keramik im Maschinenbau
Haid-und-Neu-Str. 7
76131 Karlsruhe
Germany

Tatsuki Ohji
National Institute of Advanced
Industrial Science and Technology
(AIST)
Advanced Manufacturing Research
Institute
Anagahora 2266-98, Shimo-shidami
Moriyama-ku
Nagoya 463-8560
Japan

Pavol Šajgalík
Slovak Academy of Sciences
Institute of Inorganic Chemistry
Dúbravska cesta 9
84536 Bratislava 45
Slovakia

Clemens Schmalzried
ESK Ceramics GmbH & Co. KG
Max-Schaidhauf-Str. 25
87437 Kempten
Germany

Karl A. Schwetz
retired ESK
Bergstr. 4
87477 Sulzberg
Germany

Peter Supancic
Montanuniversität Leoben
Institut für Struktur- und
Funktionskeramik
Franz-Josef-Straße 18
8700 Leoben
Austria

Derek P. Thompson
University of Newcastle upon Tyne
School of Chemical Engineering and
Advanced Materials
Advanced Materials Group
Newcastle upon Tyne NE1 7RU
UK

Harry L. Tuller
Massachusetts Institute of Technology
Department of Materials Science and
Engineering
Cambridge, MA 02139
USA

Polona Umek
Institute Jožef Stefan
Jamova 39
1000 Ljubljana
Slovenia

George C. Whitfield
Massachusetts Institute of Technology
Department of Materials Science and Engineering
Cambridge, MA 02139
USA

Sheldon M. Wiederhorn
National Institute of Standards and Technology
MSEL Bldg. 223
Gaithersburg, MD 20899-8500
USA

Han-Ill Yoo
Seoul National University
Department of Materials Science and Engineering
Seoul 151-744
Korea

You Zhou
National Institute of Advanced Industrial Science and Technology (AIST)
2266-98 Shimo-Shidami, Moriyama-ku
Nagoya 463-8560
Japan

Andrej Zorko
Institute Jožef Stefan
Jamova 39
1000 Ljubljana
Slovenia

I
Ceramic Material Classes

Ceramics Science and Technology Volume 2: Properties. Edited by Ralf Riedel and I-Wei Chen

ISBN: 978-3-527-31156-9

1 Ceramic Oxides

Dušan Galusek and Katarína Ghillányová

1.1 Introduction

Ceramic oxides represent the most extensive group of ceramic materials produced today. Traditionally, but rather artificially, the oxide ceramics are divided into "traditional" and "advanced" groups. The "traditional" ceramics include mostly silica-based products prepared from natural raw materials (clays), including building parts (bricks, tiles), pottery, sanitary ware, and porcelain, but also ceramics with other main components (e.g., alumina, magnesia), which are applied in the field of electroceramics (insulators), or industrial refractories.

"Advanced" ceramics require a much higher quality and purity of raw materials, as well as the careful control of processing conditions and of the materials' microstructure. They usually comprise oxides, which do not quite fall within the traditional understanding of the term "silicate" materials and ceramics. Oxides found in these ceramics include mostly oxides of metals such as aluminum, zirconium, titanium, and rare earth elements. Originally investigated mainly as materials for structural applications (especially alumina and zirconia), ceramic materials (and not only oxides) partly failed to meet the expectations, mainly due to problems with reliability and high production costs. In recent years, therefore, a significant shift has been observed in pursuing and utilizing the functional properties of ceramic materials, especially chemical (high inertness), optical, electrical, and magnetic properties. Another area of research which has been pursued in recent years is the refinement of microstructure to the nanolevel. It is widely anticipated that such microstructure refinement will not only improve the known properties of ceramics, but will also bring new properties to already known materials. The attempts to prepare nanostructure materials bring new challenges: from the synthesis of suitable nanopowders, through their handling, the rheology of nanosuspensions, and health and safety issues, to the development of sintering techniques that allow densification without any significant coarsening of the microstructure.

In the following sections, an attempt is made to address the questions of recent developments in the field of ceramic oxides. As this topic cannot be covered fully within the space available, oxides have been selected which are considered to be the

Ceramics Science and Technology Volume 2: Properties. Edited by Ralf Riedel and I-Wei Chen

ISBN: 978-3-527-31156-9

most important for the field (Al_2O_3, ZrO_2), as well as those that have recently become a subject of interest for the ceramic community due to their interesting properties, such as TiO_2, ZnO, CeO_2, and Y_2O_3. Those materials which are only used as minor components of ceramic materials, or ternary compounds such as titanates or spinels, have been excluded at this stage.

1.2 Aluminum Oxide

From the point of view of the volume of production, polycrystalline alumina is the material most frequently used as ceramics for structural applications. However, in comparison with for example, silicon nitride, where the influence of various additives on microstructure and properties has been well characterized and understood, and despite several decades of lasting research effort, alumina remains a material with many unknown factors yet to be revealed. Alumina-based materials can be divided roughly into three groups:

- **Solid-state sintered aluminas:** Here, research is focused on a better understanding of sintering processes with the aim of preparing nanocrystalline materials with superior mechanical properties (e.g., hardness and wear resistance), and possibly also transparency to visible light. The prerequisite for the successful preparation of submicrometer aluminas with desired properties are sufficiently fine-grained and reactive nanopowders of high purity. Their synthesis and characterization has, therefore, been intensively pursued during the past years.
- **Liquid-phase sintered (LPS) aluminas:** Despite the fact that LPS aluminas represent a substantial part of industrially produced alumina-based materials, and despite a tremendous amount of research work, many unknowns remain. Although sintering additives have a profound influence on mechanical properties (especially on hardness, creep, and wear resistance, and to a certain extent also on bending strength and fracture toughness), there remains some confusion as to how individual additives or their combinations influence the microstructure and behavior of alumina-based materials. The sintering additives used include mostly silica, alkali oxides, the oxides of alkali earth metals, and combinations thereof. Doping with rare earth oxides is studied with the aim of understanding and enhancing the creep resistance of polycrystalline aluminas.
- **Alumina-based composites:** These comprise especially zirconia-toughened alumina (ZTA), and alumina-based nanocomposites with non-oxide second phases such as SiC or TiC. However, the latter two, in particular, are beyond the scope of this chapter.

1.2.1 Crystal Structure

The only thermodynamically stable crystallographic modification of alumina is α-Al_2O_3, or corundum. *Corundum* has a hexagonal crystal lattice with the cell

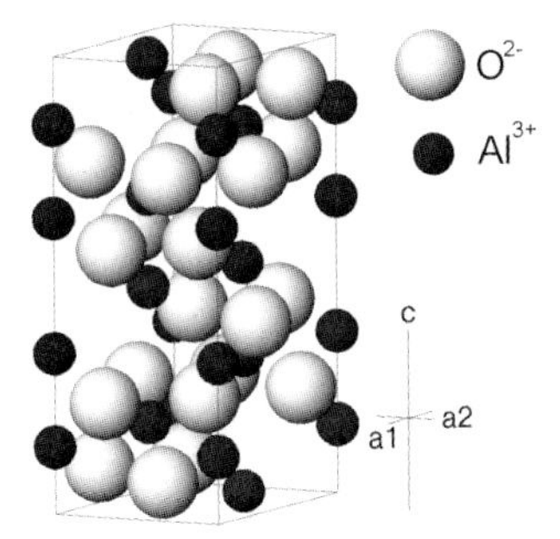

Figure 1.1 Crystal structure of α-Al_2O_3.

parameters a = 4.754 Å, and c = 12.99 Å. The ions O^{2-} are arranged in close hexagonal arrangement, with the cations Al^{3+} occupying two-thirds of the octahedral interstitial positions (Figure 1.1). Some selected materials properties of α-alumina single crystal are summarized in Table 1.1.

Except for the thermodynamically stable α modification, there exist also numerous metastable modifications, denoted γ, χ, η, ι, ε, δ, θ, and κ. These modifications are often used as supports for catalysts. All metastable modifications have a partially deformed closely packed hexagonal oxygen sublattice with various configurations of interstitial aluminum atoms. On approaching the equilibrium, the crystal lattice becomes more ordered until the stable α modification is formed. The type of metastable polymorph influences the morphology of the formed α-Al_2O_3 particles.

1.2.2
Natural Sources and Preparation of Powders

Aluminum is one of the most abundant elements on Earth and, in its oxidized form, is a constituent of most common minerals. Pure aluminum oxide is relatively rare, but may be found in the form of single crystal, when it is used as a gemstone in its colorless (sapphire) or red (ruby, due to the admixing of chromium) modifications.

Table 1.1 Some selected properties of single crystal α-alumina.

Property	Value	
Melting point	2053 °C	
Thermal conductivity	25 °C	40 W m K^{-1}
	1000 °C	10 W m K^{-1}
Thermal expansion coefficient (25–1000 °C)	Parallel with c	$8.8.10^{-6}$ K^{-1}
	Perpendicular to c	$7.9.10^{-6}$ K^{-1}
Density	3.98 g cm^{-3}	
Young's modulus	Parallel with c	435 GPa
	Perpendicular to c	386 GPa
Poisson ratio	0.27–0.30	

The most important raw material for the production of aluminum oxide is bauxite, which is a mixture of the minerals boehmite (α-AlO(OH)), diaspor (β-AlO(OH)), and gibbsite ($Al(OH)_3$), with a high content of various impurities such as Na_2O, SiO_2, TiO_2, and Fe_2O_3. Bauxite is refined using the Bayer process, which has been well described in many books dealing with the topic (e.g., Ref. [1]). Very pure commercial powders are prepared via the calcination of alum, $NH_4Al(SO_4)_2 \cdot 12\ H_2O$.

The preparation of submicrometer-grained aluminas requires well-defined pure nanopowders which, themselves, exhibit many exploitable characteristics, such as low-temperature sinterability, greater chemical reactivity, and enhanced plasticity. Thus, a range of methods has been developed for the preparation of nanopowders with desired properties. These can be roughly allocated to: (i) high-temperature/flame synthesis; (ii) chemical synthesis, including sol–gel; and (iii) mechanically assisted processes, such as high-energy milling.

1.2.2.1 High-Temperature/Flame/Laser Synthesis

The method usually comprises the injection of a suitable gaseous, or liquid aluminum-containing precursor into the source of intensive heat (e.g., laser [2], d.c. arc plasma [3–5], or acetylene, methane, or hydrogen flame [6, 7]), where the precursor decomposes and converts into the oxide. In most cases, transient aluminas are formed, and in order to obtain α-Al_2O_3 a further high-temperature treatment, usually accompanied by significant particle coarsening, is required. Possible precursors include metal–organic compounds such as trimethylaluminum or aluminum tri-*sec*-butoxide. Metastable alumina powders with particle sizes ranging from 5 to 70 nm can be prepared in this way.

1.2.2.2 Chemical Methods

These normally utilize the low- and medium-temperature decomposition of inorganic aluminum salts and hydroxides, or metal–organic compounds of aluminum. Typical precursors include aluminum nitrate and aluminum hydroxides. Hydrothermal conditions are often applied [8], but colloidal methods (sol–gel) have been extensively studied over the past three decades [9–11]. Recent efforts have been aimed at reducing the particle size of α-Al_2O_3, and decreasing the temperature of formation of α-Al_2O_3 from transient aluminas to $<1000\,^{\circ}C$ [12]. Results similar to those in sol–gel can be achieved with the use of metal ion–polymer-based precursor solutions. Here, the precursor solution (e.g., nitrate salt) is mixed with a water-soluble polymer, which provides a matrix for the dispersion of cations [13].

1.2.2.3 Mechanically Assisted Synthesis

These are based on the comminution of coarser-grained powders by high-energy milling or grinding. In this case, the minimum particle size is limited to approximately 40 nm [14], although further grinding may lead to severe aggregation and subsequent densification of the aggregates under mechanical stress. The high-energy ball-milling of γ-Al_2O_3 nanopowder with a small fraction of α-Al_2O_3 nanocrystalline seeds (both with an average particle size 50 nm) was reported to facilitate the transformation of milled powder to α-Al_2O_3 [15].

1.2.3
Solid-State Sintered Alumina

Strictly speaking, there are very few commercially available polycrystalline aluminas which could be defined as solid-state sintered. Even the materials prepared from ultrapure powders (purity >99.99%) develop a thin layer of intergranular amorphous film at the grain boundaries, as the result of the presence of trace impurities (e.g., Ca, Mg, Si, to mention the most important) which are always present in the starting powder [16]. Due to low lattice solubility of impurities in alumina, which usually do not exceed several ppm, the impurities segregate to the grain boundaries. Their concentration at interfaces increases as the grains grow during the course of densification, and the total area of interfaces (grain boundaries) per unit volume decreases. As soon as the preconcentration of the impurities exceeds a critical value, an amorphous grain boundary film is formed at the originally crystalline alumina–alumina interfaces. Hansen and Philips found that almost all the grain boundaries of a commercial 99.8% alumina were wetted by an amorphous film containing SiO_2 and CaO in addition to Al_2O_3 [17]. Harmer found that a commercial 99.98% alumina was sufficiently impure to contain a thin glassy film at the grain boundaries [18]. Processing (homogenization, pressing, the use of pressing additives, and sintering) represents another source of impurities. For example, the use of an electric furnace with $MoSi_2$ heating elements during sintering is known to contaminate the sintered materials with silicon, which is evaporated from the elements in the form of SiO, and then transported into the open porosity of the sintered body. Special precautions, such as the protection of specimens made from ultra-high-purity alumina (99.999%) in a closed sapphire tube, are then necessary to keep the grain boundaries clean [19]. The recently reported translucent or transparent aluminas with submicrometer microstructure represent one particular exception in this respect.

1.2.3.1 Submicrometer and Transparent Alumina

In submicrometer aluminas, the area of intergrain interfaces can be so high that the critical concentration of segregated impurities is not achieved and grain boundary glass is not formed. Submicrometer aluminas were reported to have increased hardness [20], a high mechanical strength [21–23], and a high wear resistance [24] in comparison with their coarser-grained counterparts, with possible application for cutting inserts [25].

Alumina is known to transmit infrared (IR) radiation and, if sintered to a high density (residual porosity <0.1%), it also transmits visible light. Possible applications then include high-pressure envelopes of sodium, or metal halide discharge lamps [26, 27], and impact layers of transparent armors with a high level of ballistic efficiency against armor-piercing ammunition (Figure 1.2). One potential use of alumina in the latter role has been already demonstrated with the use of sapphire single crystals in transparent armor systems [28].

At the present stage of knowledge, submicrometer-grained aluminas are in most cases only translucent or, at the best, are transparent with the linear transmission of

Figure 1.2 Comparison of an opaque tile (advanced ceramic armor, residual porosity 0.3%) and of a transparent window (residual porosity unmeasurable <0.03%) [35].

visible light at a level which is about 70% that of a sapphire single crystal of optical quality [29]. The visible light in polycrystalline alumina is scattered at grain boundaries as the result of alumina birefringence, and also due to light scattering at residual pores. It has been postulated that the linear transmission of visible light could be markedly increased by a further decrease in grain size to less than 340 nm – that is, below the wavelength of visible light [29], and by the complete elimination of any residual porosity (<0.1%). However, this goal is usually not attainable by normal sintering process, as the complete elimination of residual porosity requires long soaking times at high temperature, which is always accompanied by significant, and often abnormal, grain growth triggered by the presence of impurities and the formation of a grain boundary melt. The sintering trajectories of submicrometer alumina powders indicate that the majority of grain growth takes place above 90% relative density or, more specifically, at the very end of the sintering process when the last 3% of porosity is eliminated [30, 31]. As a result, a threefold or higher increase in grain size with respect to the original size of the alumina powder particles is observed. In order to solve this puzzle, special sintering techniques may be applied, such as microwave [32] or spark plasma sintering [33]; alternatively, pressure must be used, such as hot isostatic pressing (HIP) or sinter-HIP techniques [34], in order to prepare transparent polycrystalline alumina with a submicrometer grain size. Pressureless sintering to densities >99.9% at low temperatures (~1200 °C for minimum grain growth) is enabled by advanced processing approaches that yield green bodies with improved homogeneity and particle coordination [35].

1.2.4
Liquid-Phase Sintered (LPS) Aluminas

The glass present in alumina originates either from the materials' and processing impurities (Ca, Si, alkalis), or from substances added intentionally to influence

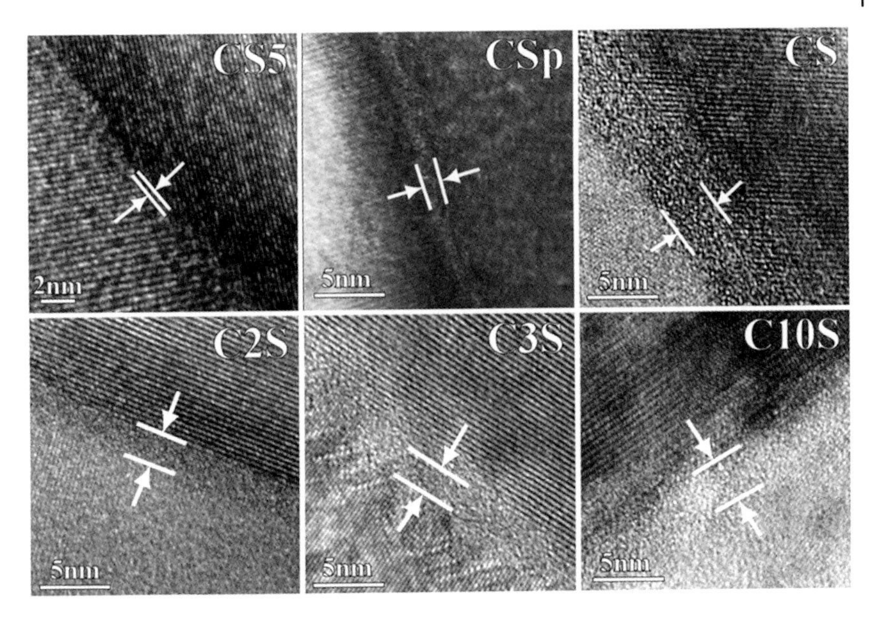

Figure 1.3 High-resolution transmission electron microscopy images of the grain boundaries of alumina samples sintered with 5 wt% calcium silicate additives. The numbers in the names of the specimens denote the molar ratio CaO/SiO_2 (e.g., CS5 = CaO·5 SiO_2). The change in grain boundary thickness with the composition of liquid forming additives is obvious [43].

sintering. Among these, the alumina–calcia–silica system is of considerable importance as it forms eutectics with melting temperatures in the range of 1200–1400 °C. The presence of SiO_2 and CaO results in the formation of a wetting aluminosilicate melt; sintering is then accelerated by dissolution–reprecipitation of the alumina particles in the melt. After cooling to room temperature, the melt solidifies to a glass, which forms several-nanometer-thick films at the intergrain faces. The equilibrium thickness of the film is defined by its chemical composition (Figure 1.3) [36, 37]. The co-doping of alumina with as little as 0.15 wt% CaO and 0.15 wt% SiO_2 ensures complete wetting of the grain boundaries with aluminosilicate glass [38]. The glass, which is "in excess" with respect to the formation of the film with equilibrium thickness, is accommodated in pockets at the grain boundary intersections, known as "triple pockets."

Both, silicon and calcium promote excessive growth along the basal planes of alumina crystals, known as abnormal grain growth (AGG), when present above critical concentrations that correspond roughly to their solubility limits [39]. As a result, platelike alumina grains grow. Abnormal platelike alumina grains are observed also in materials doped with $Na_2O + SiO_2$, $SrO + SiO_2$ and $BaO + SiO_2$ [40], TiO_2 [41], or SiO_2 and Y_2O_3 [42]. Although there is no general agreement on what is the driving force for AGG, it is usually attributed to the solubility anisotropy between growing and shrinking grains due to their curvature, and to the anisotropy of solid–liquid interfacial energies of various crystallographic faces of α-alumina crystals, and hence the differential wetting of various crystallographic faces of alumina crystals. However, others have suggested that AGG is in fact the consequence of an uneven distribution of impurities and sintering aids [43, 44].

Molecular dynamics (MD) simulations of grain boundaries wetted with calcium silicate glass indicate that cagelike structures form at the interfaces between alumina grains and amorphous silicate film, which can accommodate metal cations such as calcium or magnesium [45]. Up to a concentration of 12 atom% calcium atoms are segregated in these cagelike structures, decreasing the glass/grain interfacial energy (Figure 1.4) [46, 47]. As the content of calcia grows beyond 12 atom%, this is increasingly accommodated in the glassy phase, disrupts siloxane bonding, and weakens grain boundaries. If the content of calcium in grain boundary glass exceeds 30 atom%, then an abnormal grain growth is triggered [47]. The MD simulations also suggest that the presence of Ca in intergranular film affects the behavior of the (0 0 0 1) basal plane, but not of the (1 1 $\bar{2}$ 0) prism plane of alumina crystals [48, 49]. This different adsorption behavior is caused by the formation of specific bonds between intergranular film species and surface species that limit further adsorption. These simulation results are indicative of the preferential growth that is consistent

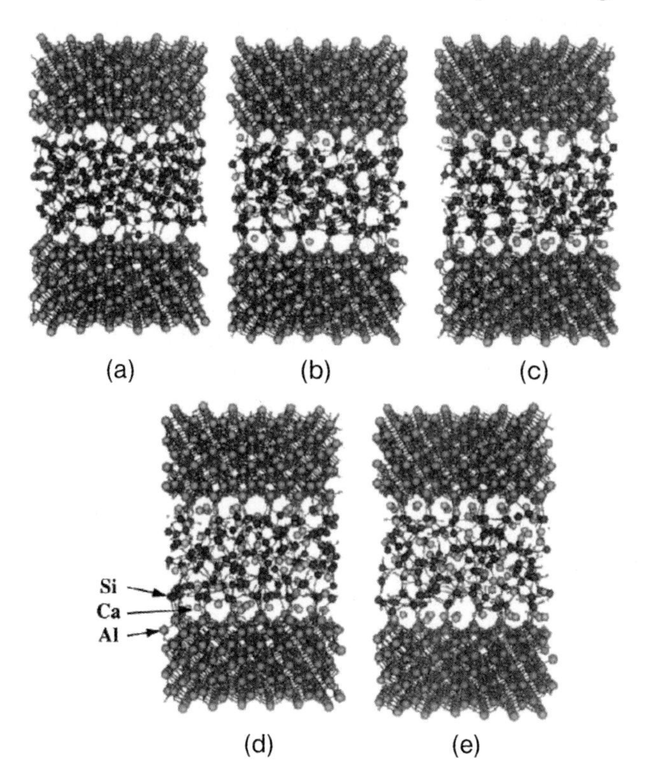

Figure 1.4 Examples of the final configurations of various compositions of the 600 atom calcium silicate intergranular films. The atom types are labeled in (d). As the number of Ca ions segregated to the interfaces increases, the interfaces become more ordered. (a) 0% CaO; (b) 12% CaO; (c) 22% CaO; (d) 32% CaO; (e) 41% CaO [64].

with the anisotropic growth of alumina crystals containing calcium aluminum silicate (CAS) intergranular films, as observed experimentally.

The addition of magnesium oxide is known to prevent the abnormal grain growth of alumina, both in very pure and liquid-phase-sintered materials. The microstructure then consists of equiaxed grains with narrow distribution of dihedral angles close to 120° [50]. There is no general agreement as to how MgO suppresses grain growth, but it is generally believed that the addition of MgO reduces the anisotropy of the solid–liquid interphase energies of alumina grains, changing the wetting of alumina with silicate liquids [50]. MgO also reduces alumina grain boundary mobility through solid solution pinning, thus protecting the material against abnormal grain growth arising from nonuniform liquid-phase distribution [51]. Most recent observations have indicated that MgO has a profound influence on the atomic structure of grain boundaries. Whilst the grain boundaries of an anorthite liquid containing aluminas are flat, atomically smooth and facetted, the addition of magnesia to an anorthite liquid results in curved, atomically rough grain boundaries [52]. Such a roughening effect is equivalent to a decrease in edge energy, so that the number of grains which can grow increases rapidly. As a consequence, many grains grow simultaneously and impinge upon each other such that AGG cannot occur [53].

Bae and Baik determined the minimum concentration of magnesia required to suppress the AGG in aluminas doped with ppm levels of silica and calcia [19]. The AGG does not take place if the amount of magnesia doping is approximately equal to the content of AGG-triggering impurities combined. Similar conclusions have been drawn from the investigation of commercial aluminas containing several vol% of glass. Typically, AGG is suppressed if the $MgO/(CaO + BaO + Na_2O + K_2O)$ ratio is >1 [54].

Partial crystallization often depletes the grain boundary glass of elements which are built into emerging crystalline phases. The structure, thermal expansion, and temperature dependence of viscosity of the grain boundary glass are then significantly altered. However, the crystallization of nanometer-thin glass grain boundary films is thermodynamically not favored, due to mechanical constraints from adjoining faces of alumina crystals [55]. Any volume change during crystallization gives rise to the strain energy which opposes the transformation, and under such circumstances crystallization will occur primarily in triple pockets. Consequently, a composition gradient will develop between the grain boundary and crystallized triple pocket. Phases such as anorthite, gehlenite, mullite, grossite, various spinels, and cordierite may crystallize in triple pockets [56, 57].

1.2.5
Properties of Polycrystalline Alumina

Polycrystalline alumina is especially praised for its great hardness, being the hardest of oxide ceramics. Although hardness ranges over a wide interval, and decreases with increasing content of the grain boundary glass, in ultrafine-grained solid-state sintered aluminas the hardness exceeds that of sapphire single crystal.

Table 1.2 Characteristic properties of various polycrystalline aluminas.

Property	**Al_2O_3 [wt%]**				
	86–94.5	**94.5–96.5**	**96.5–99**	**99–99.9**	**>99.9**
Density [g cm^{-3}]	3.4–3.7	3.7–3.9	3.73–3.8	3.89–3.96	3.97–3.99
RT hardness [GPa] HV 0.5	9.7–12	12–15.6	12.8–15	15–16	19.3
Young's modulus [GPa]	250–300	300	300–380	330–400	366–410
RT bending strength [MPa]	250–330	310–330	230–350	550	550–600

The other mechanical properties – especially fracture strength and fracture toughness – are usually inferior to those in other structural oxide ceramics such as zirconia. The presence of glass decreases the Young's modulus, and also the strength, hardness, and refractoriness of alumina ceramics. The typical properties of aluminas of various purities are listed in Table 1.2 [50].

Fracture strength varies over a broad range, and is often impaired by the thermal expansion mismatch of intergrain glass and alumina, and the extent of glass crystallization, due to volume changes and associated microcrack formation during the course of devitrification [58].

The presence of glass can either improve or impair fracture toughness, depending on its thermal expansion [59]. For example, if the thermal expansion of glass is higher than that of the matrix, the grain boundaries will be under tension whereas the grains will be in compression. Cracks may then propagate preferentially along the grain boundaries, potentially increasing the fracture toughness. In such a case the presence of platelike grains increases the fracture toughness by deviating the crack path and increasing the energy required for further propagation of the crack [60]. A glass with a lower thermal expansion will impose compressive stress into the grain boundaries, making them stronger and forcing the cracks to propagate intragranularly.

Direct evidence of grain boundary glass composition and crystallization on the magnitude and distribution of thermal residual stresses in calcium silicate sintered aluminas has been obtained recently [49]. Such studies have revealed the presence of residual fluctuating local stresses (i.e., very short-range stresses associated with dislocations, point defects, small inhomogeneities, etc.) between 420 and 460 MPa. This is an order of magnitude higher than the stresses in pure polycrystalline alumina that result from thermal expansion anisotropy of the alumina crystals along the c and a crystallographic axes (from 30 to 270 MPa, depending on grain size, and thermal history) [61–64].

The wear resistance of polycrystalline alumina is greater than that of sapphire single crystals, and increases with decreasing grain size [65–71]. The wear behavior of alumina cannot be simply related to its mechanical properties, such as hardness and fracture toughness [72, 73]. For pure alumina of mean grain size $<1\,\mu m$, the dominant wear mechanism is either plastic grooving or, in the presence of water, tribochemical wear, leading to polishing with very slow material removal rates. This is sometimes attributed to the formation of a layer of hydrated alumina, which accommodates the interfacial shear stresses [74]. For coarser-grain-sized aluminas (1–50 μm) the main wear mechanism is that of microfracture and crack interlinking, leading to grain detachment and the development of rough surfaces [71].

Little is known of the corrosion of polycrystalline aluminas, but in general it is considered to be very low, and even negligible (which explains why such little interest has been expressed in the study of the mechanism involved). However, the results of the few studies to be conducted have suggested that the corrosion of polycrystalline alumina is controlled by the corrosion resistance of the grain boundary glass.

The electric properties of alumina make it adaptable for use in many applications, ranging from electronic substrates to spark plug insulators to magnetohydrodynamic power generators. Alumina is a low-loss dielectric material, the dielectric constant placing it well within the insulating range with a value of 8.8 at 1 MHz [75]. The dielectric properties of sapphire are anisotropic, and there is a 20% difference in permittivity between the a- and c-axis directions of the sapphire single crystal [76]. The alumina single crystal has one of the lowest loss tangents known, at about 0.001 [77]. The electric properties of polycrystalline aluminas are influenced by the presence of grain boundary impurities, especially of alkalis, iron, and titanium [75]. Improvements in the insulating properties and a decrease in the loss tangent can be achieved by reducing the porosity [78], increasing the purity [79], or by eliminating any impurities with valence that is different from Al^{3+}, mainly H^{+}, Fe^{2+}, Mg^{2+}, Ca^{2+}, Si^{4+}, and Ti^{4+} [80]. Nevertheless, the presence of certain contaminants, such as TiO_2, can be advantageous in some cases, most likely due to the better final microstructure of Al_2O_3 [81].

1.3 Magnesium Oxide

Magnesia has been used traditionally for the fabrication of basic refractories mainly used in steelmaking, gradually developing from inexpensive doloma refractories through highly resistant magnesia linings to MgO refractories containing graphite, which decreases the wetting of bricks by molten slags. Recently, considerable attention has been paid to the fabrication of fully dense fine-grained, defect-free MgO, on the basis of its expected excellent mechanical, thermal, and optical properties [82]. Fully dense MgO exhibits a high transparency to both IR and visible light, from 300 nm to 7 μm, and therefore might potentially be used as a substitute for sapphire IR windows and protectors for sensors. Magnesium oxide (electrical grade) is used in the electrical heating industry due to its high dielectric strength and relatively high thermal conductivity. In addition, the combination of electrical and refractory properties facilitates its use in high-temperature crucibles, thermocouple tubes, kiln furniture, and insulators. Due to its high-temperature stability, porous magnesia is also a suitable candidate for the support of combustion

Table 1.3 Physical properties for single-crystal MgO.

Property	Value
Density	3.58 g cm^{-3}
Solubility	0.00 062 g in 100 g water
Thermal conductivity	42 W m^{-1} K^{-1} at 0 °C
Coefficient of thermal expansion	10.8×10^{-6}/K at 0 °C
Dielectric constant	9.65 at 1 MHz
Young's modulus	250 GPa
Shear modulus	155 GPa
Poisson's ratio	0.18

catalysts used in a high-temperature environment. Vanadium on a magnesium oxide support (V/MgO) is used as a selective catalyst for the oxidative dehydrogenation of alkanes.

1.3.1
Crystal Structure and Properties of Single-Crystal MgO

Magnesium oxide is a highly ionic crystal, with the Mg–O bonds having about 80% ionic character, and with a cubic face-centered crystal lattice (space group Fm3m). MgO has no polymorph transitions from room temperature to melting point at 3073 K. The physical properties of the MgO single crystal are listed in Table 1.3.

1.3.2
Natural Sources and Production

Magnesia (MgO) occurs in Nature largely as magnesite ($MgCO_3$) and dolomite (Mg, Ca)CO_3, and also rarely in oxide form as the mineral *periclase*. Magnesia refractories containing MgO can be obtained from high-purity magnesite ores simply by beneficiation and subsequent calcination at 500–700 °C. Sea water, brines, and deposits of MgO-rich salts represent other commercial sources of magnesia. Sea water contains about 1 kg of MgO per 500 l, in the form of magnesium chloride; the latter is reacted with an alkali source (commonly lime or slaked doloma) to form a precipitate of $Mg(OH)_2$ that is then washed, filter-pressed, dried, and calcined in large rotary kilns at 750–900 °C [50].

For the synthesis of high-purity, fine-grained MgO powders with excellent sinterability, as are required for the preparation of optical-grade magnesia, chemical techniques are used. The range of methods used includes the evaporative spray decomposition of a suitable precursor solution (e.g., magnesium acetate) [83], the thermal decomposition of $MgCl_2$ to MgO and HCl with subsequent washing, the reaction of $MgCl_2$ with citric acid in an excess of ammonia [84], and a vapor-phase oxidation process [85]. The sinterability of magnesia powders can be markedly improved by mechanical activation (e.g., high-energy milling).

1.3.3
Polycrystalline Magnesia

Due to the refractoriness of magnesia, either a high temperature or the application of pressure is usually required to attain a high density, although this generally leads to an undesirable microstructure coarsening and deterioration of the mechanical properties. Additives that lower the sintering temperature include $SiO_2 + B_2O_3$ [86], fluorides (LiF) [87], and V_2O_5. With the addition of approximately 0.5 cat%, the latter facilitates a complete densification by pressureless sintering at temperatures as low as 1250 °C. The final microstructure consists of 10 μm MgO grains with $Mg_3V_2O_8$ precipitates [88].

Translucent or transparent MgO is prepared from fine, active high-purity MgO powders with the use of pressure-assisted sintering techniques such as hot pressing (HP), HIP, or spark plasma sintering (SPS). Translucency is achieved under relatively moderate conditions (HIP 1100 °C/0.5 h, 99.7% theoretical density, average grain size 0.8 μm). Much higher temperatures are required to achieve full transparency (HIP 1500–1600 °C, relative density 99.9%, in-line transmission 65% of the MgO single crystal); however, the mechanical properties of the resultant material with a mean grain size of 132–199 μm are poor [89]. The hot pressing of a magnesia nanopowder (particle size 11 nm) at 790 °C and 150 MPa yields dense MgO ceramics with a relative density >99.5%, and with an average grain size of 73 nm [90]. SPS at temperatures between 700 and 825 °C and pressures of 100–150 MPa yields fully dense transparent nanocrystalline MgO with an average grain size of 52 nm, and with in-line transmissions of 40% and 60% of the MgO single crystal for the yellow and red light wavelengths, respectively [91].

It is generally believed that nanocrystalline MgO should possess superior mechanical properties, especially of strength and hardness, in comparison with its microcrystalline counterparts. If this hypothesis were true, it has never attracted any scientific evidence, and to the present authors' knowledge there is no information currently available on the mechanical properties of pure nanocrystalline MgO.

1.4
Zinc Oxide

Zinc oxide (ZnO) is perhaps best known as a good base for white pigments in paints, tableware, sanitary ware, tiles, and glasses. It also finds industrial use in the rubber industry, and as a component of opaque sunscreens. Recently, ZnO has attracted significant attention as a material for ultraviolet (UV) light-emitters, varistors, transparent high-power electronics, surface acoustic wave devices, piezoelectric transducers, and gas-sensors, and also as a window material for displays and solar cells. The quality and control of conductivity in bulk and epitaxial ZnO have raised interest in the use of this material for short-wavelength light emitters and transparent electronics [92, 93]. As a wide bandgap semiconductor ($E_g = 3.2$ eV), ZnO is a candidate host for solid-state blue to UV optoelectronics, including lasers. The applications include high-density data storage systems, solid-state lighting, secure communications, and biodetection. The most significant barrier to the widespread use of ZnO-related materials in electronic and photonic applications is the difficulty in carrier doping, particularly in achieving p-type material.

1.4.1
Crystal Structure and Properties of Single-Crystal ZnO

ZnO normally has the hexagonal (wurtzite) crystal structure with lattice parameters $a = 3.25$ Å and $c = 5.12$ Å (space group $P6_3mc$). The Zn atoms are tetrahedrally coordinated to four O atoms, where the Zn d-electrons hybridize with the oxygen p-electrons. Layers occupied by zinc atoms alternate with layers occupied by oxygen atoms [94]. Whilst a bond between the Zn and O atoms exhibits covalent characteristic in the *c*-direction, it is mostly ionic in the *a*-direction [95]; consequently, ZnO single crystals have highly anisotropic properties.

ZnO nanocrystals may have different structures, depending on the method of preparation. For example, nanoparticles formed by the oxidation of zinc vapor have the zinc-blende structure when smaller than 20 nm, and form tetrapod-like crystals on further growth [96]. ZnO particles prepared via the flash evaporation method have a cubic crystal structure [97].

1.4.2
Natural Sources and Production

The most important zinc ore is zinc sulfide, found as the mineral *sphalerite*. The majority of ZnO is produced by the so-called "French process," which has been utilized since 1844. For this, metallic zinc is melted at 419.5 °C in a graphite crucible and vaporized above 907 °C. The zinc vapor then reacts with oxygen in the air to form zinc oxide, which normally consists of agglomerated zinc oxide particles with sizes ranging from 0.1 μm to a few microns, and a purity of up to 99.9%. A modification of the French process, known as the catalyst-free combust-oxidized mesh (CFCOM) process, yields acicular ZnO nanostructures (rods, wires, tripods, tetrapods, plates). The so-called "active zinc oxide" which is used to prepare advanced ceramics, is produced by dissolving a zinc ore in hydrochloric acid, followed by alkali precipitation. The precipitated zinc hydroxide is then removed by filtration, calcined, and micronized to obtain the powdered form [98]. More sophisticated methods include the thermal decomposition of zinc oxalate dihydrate [$Zn(C_2O_4){\cdot}2\,H_2O$] [99], the microwave irradiation of a $Zn(NO_3)_2$ solution neutralized at pH ~8–12 [100], the homogeneous precipitation of $Zn(SO_4)$ or $Zn(NO_3)_2$ by urea at 100 °C [101], the direct conversion of $Zn(NO_3)_2{\cdot}6\,H_2O$-derived zinc hydroxide gel to crystalline product [102], or the combustion synthesis of zinc nitrate with glycine [103].

1.4.3
Properties

The physical properties of ZnO crystals depend heavily on the concentration of native defects caused by deviations from the stoichiometric composition. A review on this subject is available in Ref. [104].

Table 1.4 Basic physical properties of single-crystal ZnO.

Property	Value
Density	5.606 g cm^{-3}
Stable phase at 300 K	Wurtzite
Melting point	1975 °C
Thermal conductivity	a: 0.6 W m^{-1} K^{-1} c: 1–1.2 W m^{-1} K^{-1}
Electrical conductivity	a: 3 Ω cm c: 8.2 Ω cm
Coefficient of linear thermal expansion	a: 6.5×10^{-6} K^{-1} c: 3.0×10^{-6} K^{-1}
Refractive index	a: 2.008 c: 2.029
Energy gap	3.2 eV, direct

Although the basic physical characteristics of ZnO are summarized in Table 1.4, uncertainties persist with regards to some of these values. For example, the thermal conductivity exhibits a spread of values, most likely due to the presence of various crystal defects, such as dislocations [105].

Pure ZnO is an n-type semiconductor due to the incorporation of excess Zn, which causes a donor band conduction at low temperatures. A direct band gap of zinc oxide semiconductor is 3.2 eV (387 nm, deep violet/borderline UV). The bandgap can be altered by doping ZnO with divalent substitutes on the cation site: the addition of Cd reduces the bandgap to ~3.0 eV [106], while substitution of Zn by Mg increases the bandgap to ~4.0 eV.

The mechanical properties of polycrystalline ZnO ceramics are of special importance for their applications as varistors (see Section 1.4.4.1). When a varistor experiences a high-current pulse, the electrical energy is quickly converted to heat. The inertia of the material, which resists its thermal expansion, and the resonances of the resultant elastic waves in the block, may lead to microcracks and finally to mechanical failure [107]. Characteristic values of mechanical properties of ZnO ceramics are 1.2–1.4 MPa m$^{1/2}$ for fracture toughness, and 100–125 MPa for flexural strength.

1.4.4 Applications

1.4.4.1 ZnO-Based Varistors

Zinc oxide is best known for its use in varistors – that is, resistors with strongly nonlinear current–voltage characteristics. Such behavior was first reported in 1971 by Matsuoka, who observed the nonlinear electric properties of ZnO ceramics doped with Bi_2O_3, CoO, MnO, Cr_2O_3, and Sb_2O_3, and attributed such behavior to the presence of a bismuth-rich phase between the ZnO grains [108]. The nonlinear electrical characteristics of the nanometer-thick grain boundary phase is controlled by double Schottky barriers that result from electrons being trapped at the interface and screened by the ionized shallow and deep bulk defects [109–112]. As a result, the conductivity of ZnO-based varistors increases by several orders of magnitude when a characteristic voltage is exceeded. This switching property of varistors, which is not only reversible but also very fast, provides an opportunity for their use as surge arresters in power transmission, and for the protection of electronic devices.

The current–voltage curve of a varistor consists of three distinct regions [113]:

- Pre-switch, where the behavior is ohmic and the resistance is controlled by the low grain-boundary conductivity.
- Switch or breakdown, where the behavior is nonlinear.
- High-current ohmic, where the resistance is controlled by the grain conductivity.

Various dopants may either alter or improve the nonlinear behavior of ZnO varistors, the most important being Bi_2O_3 and Sb_2O_3. Today, a wide range of additives is used in commercial varistors, and these usually comprise up to 10 components including Bi_2O_3, Sb_2O_3, Co_3O_4, SnO_2, Cr_2O_3, MnO/MnO_2, Al_2O_3, and Ag_2O. To assess the exact role of any of these is virtually impossible, as many of the dopants have a high vapor pressure at the temperature of sintering, such that the composition often changes due to component vaporization. Uncontrolled Bi_2O_3 vaporization is often a critical parameter in the manufacture of commercial varistors [114, 115].

Bi_2O_3 is vital for the successful sintering of zinc oxide-based ceramics, due to the formation of a low-temperature eutectic with ZnO at 740 °C. The solubility limit of bismuth III-oxide in ZnO may be up to 25 mol% [116]. The bismuth-containing melt cools to form an amorphous phase (located mainly in triple grain boundary junctions and to a lesser extent at the grain boundaries), and a secondary crystalline pyrochlore phase of nominal composition $Zn_2Bi_3Sb_3O_{14}$. The segregation of bismuth and oxygen (and to a certain extent also of other dopants) at the grain boundaries is crucial to achieve specific varistor properties.

Sb_2O_3 is added for better microstructure control during the course of sintering. At low temperatures, Sb_2O_3 binds Bi_2O_3 and forms the pyrochlore phase, thus shifting the onset of liquid-phase sintering to a higher temperature. The pyrochlore then reacts with ZnO to yield an electrically insulating grain boundary pinning spinel phase ($Zn_7Sb_2O_{12}$) [117] and free Bi_2O_3 [116, 118]:

$$2ZnO + 3/2\,Sb_2O_3 + 3/2\,Bi_2O_3 + 3/2\,O_2 \xrightarrow{T<900\,^\circ C} Zn_2Bi_3Sb_3O_{14}\ (\text{pyrochlore}) \quad (1)$$

$$2Zn_2Bi_3Sb_3O_{14} + 17ZnO_3 \xrightarrow{900-1000\,^\circ C} Zn_7Sb_2O_{12} + 3Bi_2O_3\ (\text{liquid}) \quad (2)$$

Both, spinel and pyrochlore accommodate excess dopants, the concentrations of which exceed their solubility limits in ZnO, and therefore they concentrate at grain boundaries [113]. The microstructure of a ZnO varistor will then comprise ZnO grains, a bismuth-rich phase, and spinel grains, which can be located either inter- or intragranularly (Figure 1.5).

An inhomogeneous microstructure results in the formation of preferred current paths in a varistor, its local overheating, and mechanical failure. Microstructural

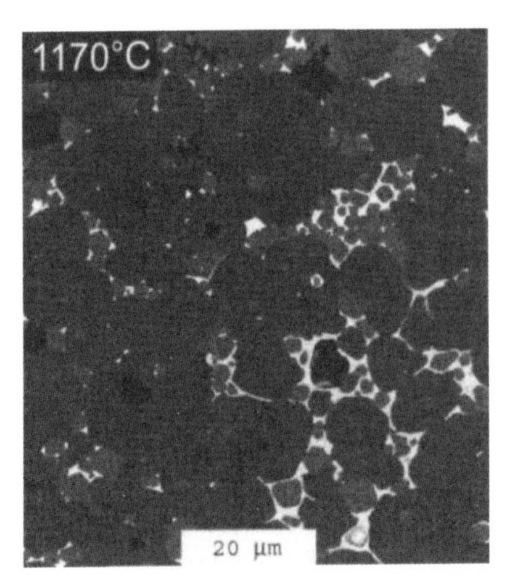

Figure 1.5 A backscattered-electron microscopy image of a polished ZnO varistor ceramics sintered at 1170 °C. The dark gray phase is ZnO, the light-gray is spinel, and the white is the bismuth-rich phase [131].

homogeneity is also important in order for a varistor to have a long service life; a continuous bismuth-rich network will contribute towards the leakage current that will provide the varistor with better pre-breakdown characteristics [119]. The leakage current in commercially formulated ZnO varistors is known to increase when Sb_2O_3 is replaced by SnO_2 [120].

The high volatility of Bi_2O_3 and the need for many minor additives in order to achieve a high performance, led to development of zinc oxide-based varistors with other varistor-forming oxides (VFOs). In Pr_6O_{11}-doped varistors, the most important two additives are Pr_6O_{11}, which gives rise to the nonlinear properties, and CoO, which enhances such properties [121–125]. Further improvement of the nonlinear properties by other minor additives (e.g., by Dy_2O_3) is not possible in the absence of cobalt oxide [126]. A simple three-oxide formulation, based on the addition of small amounts of vanadium and manganese oxides to zinc oxide, yields varistor behavior with a nonlinear coefficient in excess of 20 [127, 128]. Other VFOs which have been used include MnO_2 together with V_2O_5 [129], and a combination of NiO and CoO [130].

1.4.4.2 Other Applications of ZnO Ceramics

The majority of the ZnO that is produced is used in cosmetics, catalysis, and in the rubber industry, or as pigments in the production of tableware, sanitary ware, tiles, and glasses. The unique semiconducting properties of ZnO predestine its use in electronic devices. Notably, ZnO is interesting on the basis of its large exciton-binding energy (~60 meV), wide band gap and low lasing threshold. Consequently, ZnO one-dimensional semiconductor nanostructures (e.g., nanowires, nanorods or nanocolumns; see Figure 1.6) have become important fundamental building blocks for short-wavelength nanophotonic devices, and represent a substantial promise for integrated nanosystems [132]. Zinc oxide-based ceramics with nonlinear

Figure 1.6 Representative field emission scanning electron microscopy images of ZnO nanocolumns grown at 400 °C on a Si (001) substrate. (a) Grown for 30 min; (b) Grown for 50 min [132].

voltage–current characteristics (VCC) are used in gas sensors for NO_2 and volatile organic compounds (VOC), such as benzene, toluene, and xylene. ZnO-based diluted magnetic semiconductors (DMS) are used in spintronics applications.

1.5
Titanium Dioxide

Titanium dioxide (TiO_2) is probably best known, and most widely used, as a brilliant white pigment and component of sunscreens. Recently, it has attracted increasing attention in the electronics industry due to its high dielectric and semi-conducting properties, photocatalytic activity, and good biocompatibility.

1.5.1
Crystal Structure and Properties of Single-Crystal TiO_2

Titanium oxide forms three polymorphs: rutile, brookite, and anatase [133]:

- *Rutile* is a tetragonal mineral usually of prismatic habit, often twinned. The unit cell of rutile (space group $P4_2/mnm$, $a = 4.5845$ Å; $c = 2.9533$ Å, density 4.274 g cm^{-3}) contains titanium atoms at the corners and in the center. Each titanium atom is surrounded by an approximate octahedron of oxygen atoms, and each oxygen atom is surrounded by an approximate equilateral triangle of titanium atoms (Figure 1.7a).
- *Anatase* is a tetragonal mineral of octahedral habit. The unit cell has a space group $I4_1/amd$, with the parameters $a = 3.7842$ Å, $c = 9.5146$ Å, and a density of 3.895 g cm^{-3} (Figure 1.7b).
- *Brookite* is an orthorhombic mineral with the unit cell parameters $a = 9.184$ Å, $b = 5.447$ Å, $c = 5.145$ Å, the space group *Pbca*, and a density of 4.123 g cm^{-3} (Figure 1.7c).

Anatase and brookite are metastable phases, and their exothermic and irreversible conversion to rutile at high temperatures has been widely investigated. Both,

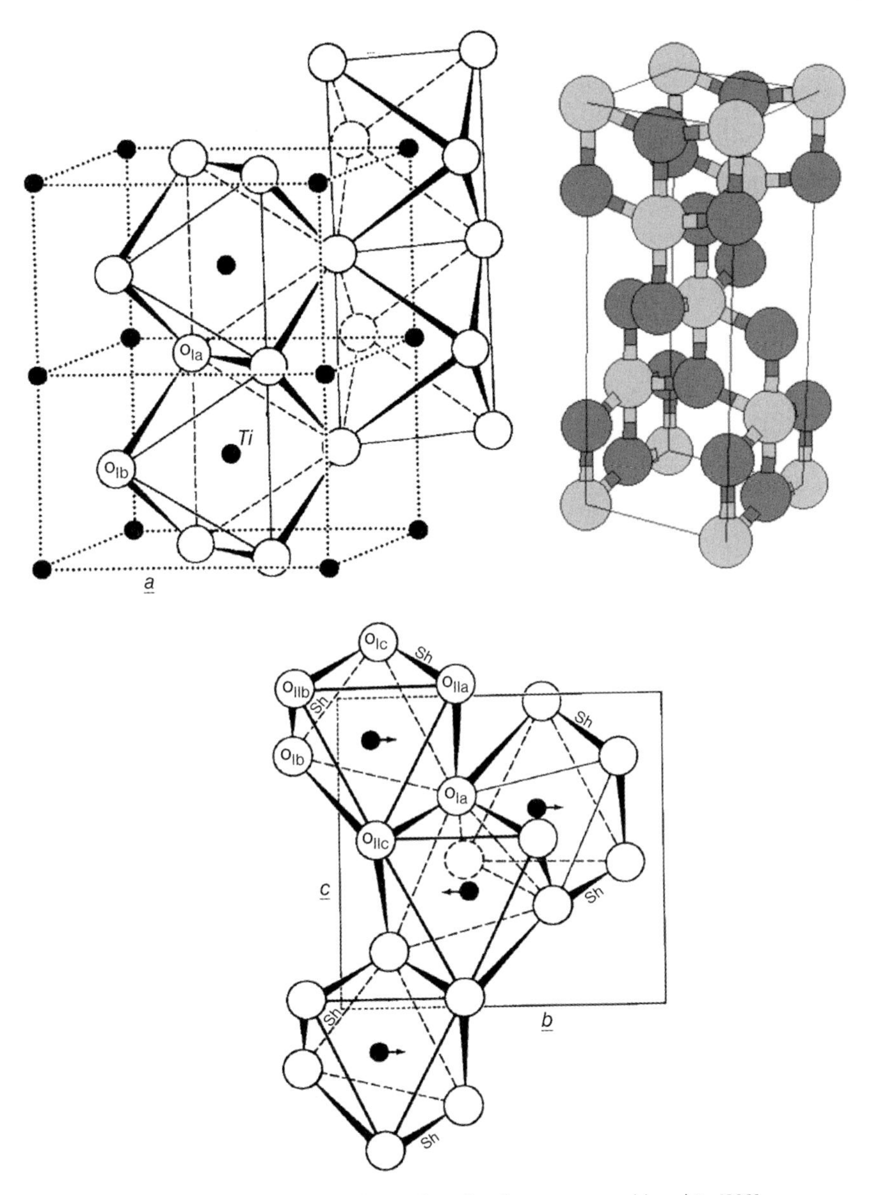

Figure 1.7 Crystal structures (from left to right) of rutile, anatase, and brookite [133].

ultrafine anatase and brookite transform upon coarsening to rutile when they reach a certain particle size [134, 135]. Once rutile has been formed, it grows much faster than anatase. The thermodynamic analysis of this phase stability indicates that anatase becomes more stable than rutile for particle sizes <14 nm. The transformation sequence and phase stability depend on the initial particle sizes of anatase and brookite [136–138].

1.5.2
Natural Sources and Production

The most important natural source of titania is iron titanate, known as the mineral *ilmenite*. Titanium oxide occurs also in its pure form, as the mineral rutile. Titanium dioxide is produced industrially via the sulfate process, whereby ilmenite ($FeTiO_3$) is hydrolyzed with sulfuric acid at >95 °C, after which TiO_2 is obtained by calcination at >800 °C. Rutile is purified by treatment with hydrochloric acid gas and conversion into titanium tetrachloride; TiO_2 is produced by treating the $TiCl_4$ with oxygen at >1000 °C.

The preparation of ultrafine-grained titania powders of specific phase composition has been described in several reports [139–141]; the starting materials and process conditions have a profound influence on the formation of TiO_2 nanocrystallites with a well-defined morphology (Figure 1.8.)

1.5.2.1 Synthesis of Anatase

Nanocrystalline anatase is usually prepared by hydrothermal synthesis or by sol–gel methods, using titanium alkoxides as the precursors. The TiO_2 precipitates can be produced by a reaction of $TiOCl_2$ and NH_4OH solutions, and subsequent treatment with NaOH or aging in boiling water [142]. A single-phase anatase powder is prepared by conventional or microwave hydrothermal (MH) synthesis from $TiOCl_2$ or a TiO_2 colloid [143, 144].

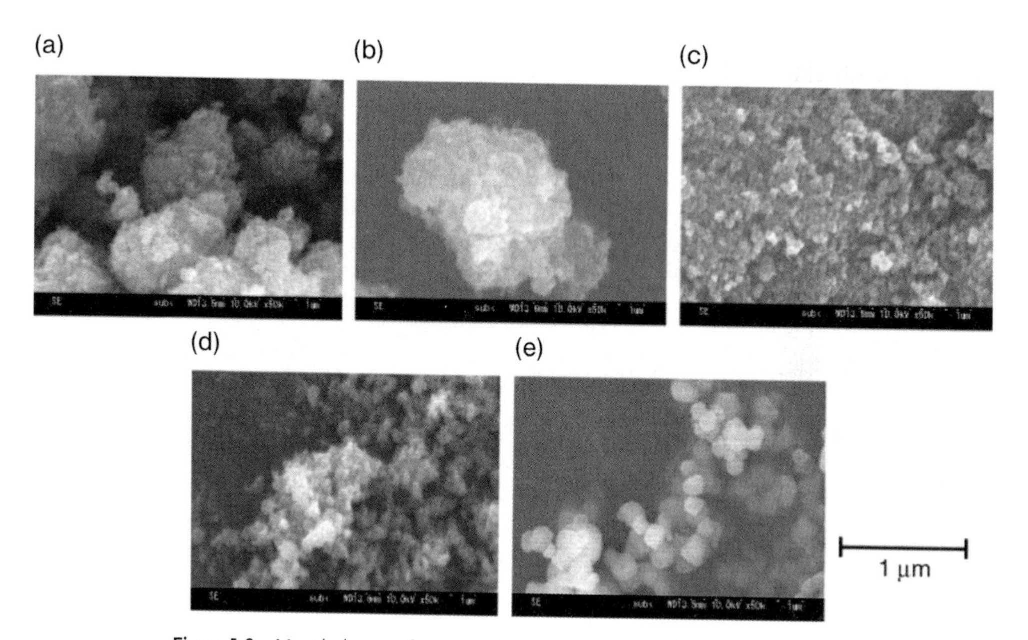

Figure 1.8 Morphologies of commercial anatase powders from various producers. (a) AMT-100; (b) AMT-600 (both Tayca Co., Osaka, Japan); (c) ST-01 (Ishihara Sangyo Kaisha Ltd., Osaka, Japan); (d) F-4 (Showa Denko K. K., Tokyo, Japan); (e) Sigma-Aldrich Co [147].

Anatase thin films are produced by dry processes such as sputtering and chemical vapor deposition (CVD), or by wet processes such as dip coating, sol–gel, spray-coating, and spin-coating. These methods require high temperatures (hundreds of °C) in order to achieve a fully crystalline anatase; the coating of surfaces with a lower thermal stability is therefore not possible. The development of procedures allowing near-ambient crystallization of anatase is therefore of profound importance. The available low-temperature syntheses utilize solutions of hazardous precursors, such as ammonium hexafluorotitanate [145], or titanium fluoride [146]. Crystalline anatase can be prepared under ambient conditions via the hydrolysis of tetraethylorthotitanate with acetylacetone, with added seeds of commercial anatase [147].

1.5.2.2 Synthesis of Rutile

The synthesis of pure rutile is difficult, as the crystallization normally yields mixtures of two, or even all three, polymorphs. Rutile is usually prepared via a hydrothermal synthesis from chlorides and oxychlorides of titanium seeded with rutile nanocrystals at temperatures below 250 °C. The addition of hydrochloric acid and aqueous alcohol solutions facilitates the preparation of rutile at temperatures between 40 and 90 °C [148]. Despite the risk of contamination, mineralizers (e.g., SnO_2, NH_4Cl or NaCl) are often used in order to reduce the size of rutile crystals. The reaction times of the hydrothermal synthesis of rutile can be significantly reduced by microwave irradiation [149]. A single-phase rutile with nanosized, well-dispersed particles prepared by a 2 h treatment of partially hydrolyzed 0.5 M $TiCl_4$ solution at 160 °C is shown in Figure 1.9.

1.5.2.3 Synthesis of Brookite

The synthesis of brookite is difficult since, as in the case of anatase, it normally yields a mixture of brookite and rutile and/or anatase. There is no general agreement as to

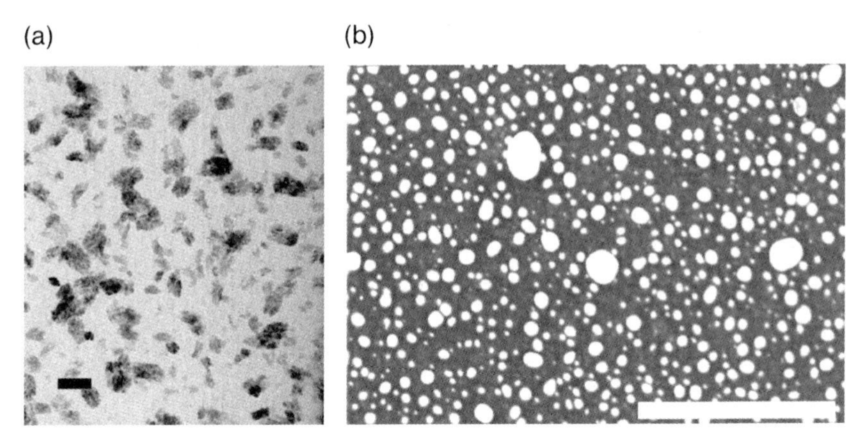

Figure 1.9 (a) Micrograph of rutile powder prepared by hydrothermal treatment of 0.5 M $TiOCl_2$ solution at 160 °C for 120 min (scale bar = 100 nm) [149]; (b) Micrograph of brookite nanoparticles with a mean size of ~30 nm, prepared by the method described in Ref. [152] (scale bar = 500 nm).

Table 1.5 Physical properties of various polymorphs of titania single crystals.

Property	Value		
	Anatase	Rutile	Brookite
Density [$g\,cm^{-3}$]	3.84	4.26	4.11
Melting point [°C]	—	1843	—
Refraction index	2.49	2.903	2.705 2.583
Electrical resistivity [Ω·cm, RT]	10^3–10^5		
Energy gap [eV]	3.23		

the factors responsible for the formation of brookite. Pottier *et al.* [150] have claimed that chloride ions are necessary in the reaction mixture to form brookite, whereas Kominami *et al.* [151] considered sodium salts, water and the organic titanium complex to be indispensable components for its successful synthesis. An example of 100% nanocrystalline brookite with an approximate particle size of 30 nm synthesized by the hydrolysis of $TiCl_4$ in acidic liquid media with isopropyl alcohol and water is shown in Figure 1.9b [152].

1.5.3 Properties of TiO_2 Polymorphs

Some physical properties of the three polymorphs of TiO_2 are listed in Table 1.5. Systematic data on titania single crystals are relatively scarce, and this applies especially to brookite, most likely due to difficulties with its preparation. The presence of impurities also results in a significant scatter of the property values.

1.5.4 Polycrystalline Titania

Polycrystalline TiO_2 is a functional ceramic of high technological importance that is used in the production of varistors, gas sensors, photovoltaic cells, dielectric resonators, and field-effect transistors as a gate dielectric. The electric properties of polycrystalline TiO_2 are heavily influenced by the presence of trace elements, impurities, as well as the oxygen partial pressure and temperature during the course of processing, all of which alter the bulk concentrations of the point and electronic defects. If sintering is allowed to occur in air or under a low oxygen partial pressure, a partial reduction of TiO_2 will take place. By using the Kröger–Vink notation, this reduction can be described in terms of the formation of oxygen vacancies [Eq. (3)], Ti^{4+} interstitials [Eq. (4)], Ti^{3+} interstitials [Eq. (5)], or oxygen vacancies and Ti^{3+} species in octahedral lattice sites [Eq. (6)] [153]:

$$2O_O^X \Leftrightarrow 2V_O^{\bullet\bullet} + 4e' + O_2 \quad (3)$$

$$\mathrm{Ti}_{\mathrm{Ti}}^{\mathrm{X}} + 2\mathrm{O}_{\mathrm{O}}^{\mathrm{X}} \Leftrightarrow \mathrm{Ti}_{i}^{\cdot\cdot\cdot\cdot} + 4e' + \mathrm{O}_2 \quad (4)$$

$$\mathrm{Ti}_{\mathrm{Ti}}^{\mathrm{X}} + 2\mathrm{O}_{\mathrm{O}}^{\mathrm{X}} \Leftrightarrow \mathrm{Ti}_{i}^{\cdot\cdot\cdot} + 3e' + \mathrm{O}_2 \quad (5)$$

$$2\mathrm{Ti}_{\mathrm{Ti}}^{\mathrm{X}} + \mathrm{O}_{\mathrm{O}}^{\mathrm{X}} \Leftrightarrow Ti'_{Ti} + \mathrm{V}_{\mathrm{O}}^{\cdot\cdot} + \frac{1}{2}\mathrm{O}_2 \quad (6)$$

The powders used for the preparation of polycrystalline ceramics usually comprise rutile, or a mixture of rutile with other titania polymorphs. Both, brookite and anatase convert irreversibly to rutile in the temperature range of 700–920 °C; sintered materials will therefore usually comprise polycrystalline rutile. Sintering temperatures of up to 1500 °C are required to attain dense samples for powders with particle sizes of 0.2–3 μm, although lower sintering temperatures are possible with finer powders. Sintering aids such as $SrCO_3$, Bi_2O_3, and SiO_2 are used, which act either through increasing the concentration of lattice defects through the formation of a solid solution, or by forming a liquid phase at the temperature of sintering [154]. The physical properties of polycrystalline titania are summarized in Table 1.6.

1.5.5 Applications of TiO_2

Due to its very high refractive index, TiO_2 is a brilliant white pigment and opacifier that is used in paints, plastics, papers, foods, and any other applications which require a bright white color. Titanium dioxide is an important component of sunscreens, as the nanoparticles of rutile are transparent to visible light but reflect UV light. Due to its high dielectric constant, TiO_2 deposited in thin films is used in optical coatings for dielectric mirrors and beam splitters.

Table 1.6 Physical properties of fully dense sintered polycrystalline titanium dioxide ceramics.

Property		Value
Density		4.0–4.2 g cm^{-3}
Poisson's ratio		0.27
Compressive strength		680 MPa
Fracture toughness		2.8–6.1 MPa m$^{-1/2}$
Young's modulus		230 GPa
Microhardness [HV 0.5]		880
Electrical resistivity	(25 °C)	10^{12} Ω · cm
	(700 °C)	2.5×10^4 Ω · cm
Dielectric constant [1 MHz]		85
Dielectric loss tanδ [25 °C]		6×10^{-5}
Thermal expansion [RT–1000 °C]		9×10^{-6} K^{-1}
Thermal conductivity [25 °C]		11.7 W m K^{-1}

Titanium dioxide is an n-type semiconductor that is used in thin-film oxygen and humidity sensors [155, 156]. Doping with other metal oxides (e.g., iron oxides) increases the sensitivity and selectivity of the titania oxygen sensors [157, 158]. In comparison to zirconia, titania sensors have a better resistance against lead poisoning.

Titania is known to exhibit varistor properties, with a lower breakdown voltage than ZnO-based varistors. The first reported (Nb, Ba)-doped TiO_2 varistors had a nonlinear exponent of about 3–4 [159], but today Ta- and Ba-doped TiO_2-based varistors with a nonlinear coefficient in the range of 20–30 are well known [160]. Those dopants with a +5 valence, such as Nb and Ta, and with an ionic radius similar to that of Ti^{4+}, dissolve in the TiO_2 lattice and reduce its resistivity by donating conductive electrons [161]. Some codopants, such as Ba and Bi, tend to exsolve during cooling and react with rutile to form secondary phases, $Ba_2Ti_9O_{20}$ and $Bi_2Ti_4O_{11}$. Subsequent slow cooling decreases the Bi^{+3} (Bi'_{Ti}) acceptor concentration in rutile grains in near-grain boundary regions, thus reducing the barrier height with a corresponding reduction in nonlinear exponent values (Figure 1.10) [162].

Titanium dioxide, and anatase in particular, has strong photocatalytic properties under UV light, with both electron and hole pairs being generated in TiO_2. These respectively reduce and oxidize any substances adsorbed onto the anatase surface, thus producing the radicals OH^- and O^{2-} that are capable of decomposing most organic compounds and killing bacteria; consequently, this represents a major potential for anatase crystals to be used in water and air purification, or as wastewater remediation [163]. The reaction activity of anatase crystals is orientation-dependent; for example, water reduction and photooxidation each take place at more negative potentials for the anatase (0 0 1) surface than for the anatase (1 0 1) surface [164]. The orientation of anatase nanocrystals with (0 0 1) preferred growth would, therefore, be

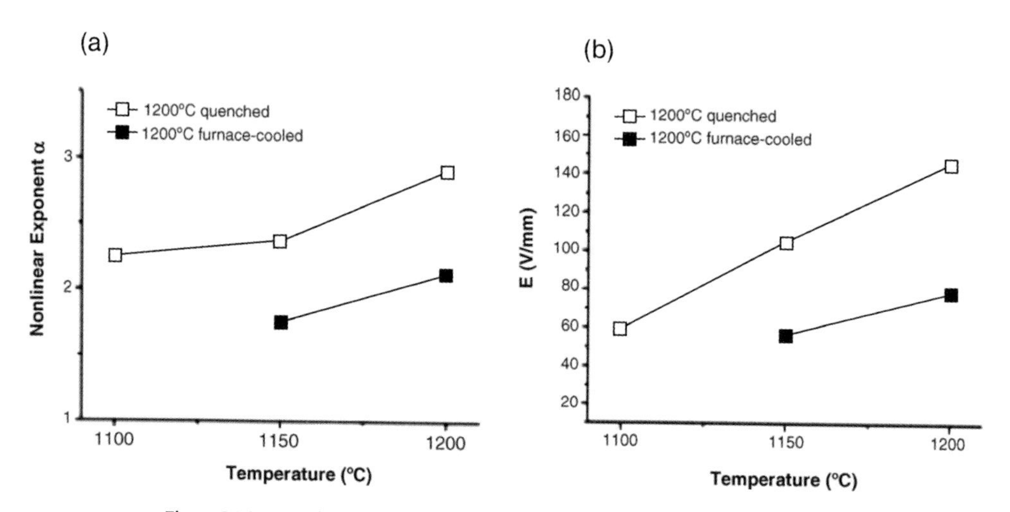

Figure 1.10 (a) The variation of nonlinear exponent and (b) breakdown voltage of Ba-doped titania varistors [162].

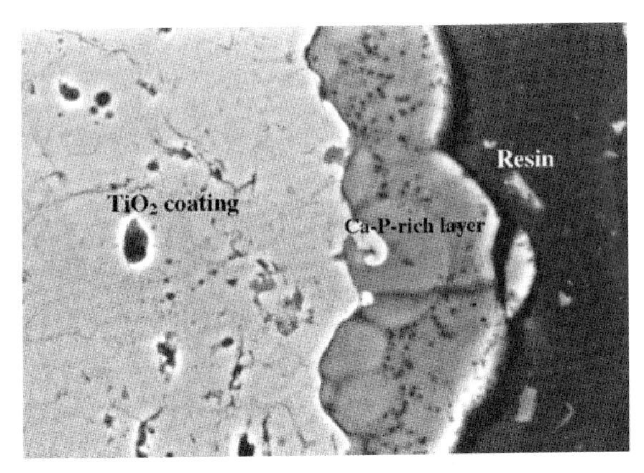

Figure 1.11 Cross-sectional view of hydroxyapatite layer formation on nano-TiO_2 coating after a four-week period soaking in a simulated body fluid [166].

expected to increase the charge conversion efficiency of photocatalysis. Anatase, when exposed to UV light, also becomes increasingly hydrophilic, providing a potential to produce windows with anti-fogging coatings or self-cleaning properties. Consequently, TiO_2 is added to paints, cements, windows, tiles, or other products in order to provide sterilizing, deodorizing, and anti-fouling properties. TiO_2 is also incorporated into outdoor building materials, where it will reduce the concentrations of airborne pollutants, such as VOC and NO_x.

A recently reported photocatalytic activity of brookite within the visible wavelength range also provides the opportunity for even broader applications of titania-based photocatalysts. For example, Showa Denko K.K. (Japan) have commercialized a new brookite nanoparticles-based photocatalyst that is responsive to visible light [165], although details of its synthesis are presently unavailable such that the company is the sole producer of brookite-based photocatalysts worldwide.

The biocompatibility of TiO_2 has been demonstrated by the formation of apatite on TiO_2 substrates in simulated body fluids [166–169] (Figure 1.11). As an example, plasma-sprayed TiO_2 coatings on Ti alloys have shown promising *in vivo* corrosion characteristics, and may act as a chemical barrier against the release of metal ions from medical implants [170].

1.6 Zirconium Oxide

Zirconia-based ceramics are characterized by a unique combination of high strength, toughness and chemical resistance, which allows their use in harsh environments under severe loading conditions. Typical applications include tools for cutting difficult materials such as Kevlar, magnetic tapes, plastic films, or paper items. It was shown long ago that wire-drawing dies and hot-extrusion dies made from zirconia could outperform their conventional counterparts [171]. Zirconia is considered as an

Table 1.7 Lattice parameters of zirconia polymorphs.

Crystal structure	Monoclinic	Tetragonal	Cubic
Space group	$P2_1/C$	$P4_2/nmc$	$Fm3m$
Unit cell parameters	a = 5.156 Å	a_t = 5.094 Å	a_c = 5.124 Å
	b = 5.191 Å	c_t = 5.177 Å	
	c = 5.304 Å		
	β = 98.9°		
Density [g cm^{-3}]	5.83	6.10 (calc.)	6.09 (calc.)

attractive matrix for nuclear applications, such as an inert matrix for the destruction of excess plutonium, or as a good host material for nuclear waste storage. Other applications include seals in valves, chemical and slurry pumps, thread guides, and bearings. Other important fields of application include biological implantation materials, examples being the replacement of worn or injured joints such as the hip or knee [172]. Functional applications of zirconia include (but are not limited to) thermal barrier coatings, solid electrolytes, oxygen sensors, and materials for fuel cells.

1.6.1
Crystal Structure and Properties of Single Crystals

Zirconia is known to exist as three, well-defined polymorphs, namely monoclinic, tetragonal, and cubic [173], although the existence of a high-pressure orthorhombic form has also been reported [174]. The lattice parameters of zirconia polymorphs are summarized in Table 1.7.

Monoclinic zirconia consists of a sevenfold coordinated Zr^{4+} cation, such that the oxygen ions with O_{11} coordination are almost tetrahedral, but with one angle in the structure differing significantly from the tetrahedral value (Figure 1.12a). Tetragonal

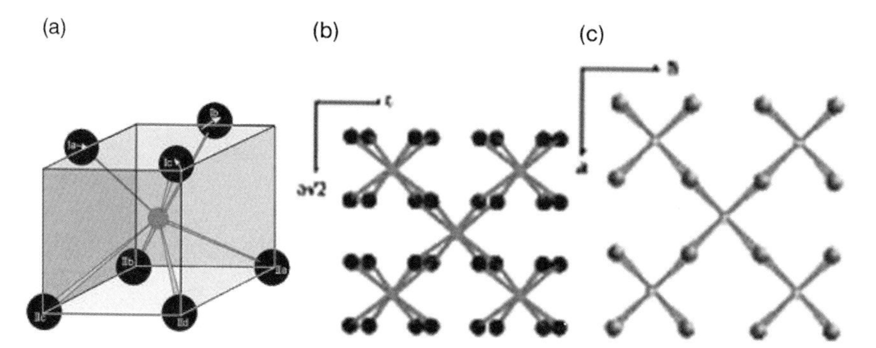

Figure 1.12 Crystal structures of (a) monoclinic, (b) tetragonal, and (c) high-temperature cubic ZrO_2 polymorphs.

Table 1.8 Some physical properties of single-crystal ZrO_2.

Property		Value
Transformation temperatures	m → t	950–1200 °C
	t → c	2370 °C
Melting point	Tetragonal	2677 °C
	Cubic	2500–2600 °C
Coefficient of thermal expansion		
Monoclinic	a	1.03×10^{-6}/K
	b	0.135×10^{-6}/K
	c	1.47×10^{-6}/K
Cubic		$7.05–13 \times 10^{-6}$/K (0–1000 °C)
Specific heat capacity [20 °C]		64.29 kJ mol^{-1} K^{-1}
Refractive index (cubic)		2.15–2.18
Young's modulus (tetragonal)		140–200 GPa
Hardness (cubic)		8–8.5 Mohs

zirconia contains the eightfold-coordinated Zr^{4+} cation with four oxygen ions placed at a distance of 2.065 Å in the form of a flattened tetrahedron, and four at 2.455 Å in an elongated tetrahedron rotated through 90 ° (Figure 1.12b). The high-temperature cubic polymorph has a face-centered CaF_2 structure with an eightfold-coordinated Zr^{4+} atom with oxygen ions arranged in two equal tetrahedra (Figure 1.12c).

Some physical properties of zirconia single crystals are summarized in Table 1.8. It should be noted, however, that these values are only for orientation; in the case of stabilized forms of the tetragonal and cubic polymorphs, the properties will be significantly influenced by the presence of stabilizing aids (see below).

1.6.2 Natural Sources and Production

Zirconia is found as the free oxide *baddeleyite*, always accompanied by hafnium oxide as the impurity; however, the most frequent source of zirconia is its compound oxide with silica, known as the mineral *zircon* ($ZrO_2 \cdot SiO_2$).

Pure zirconia is obtained via the chlorination and thermal decomposition of zirconia ores, their decomposition with alkali oxides, and lime fusion. The initial stage of the process is based on the chlorination of zircon in the presence of carbon at a temperature of 800–1200 °C in a shaft furnace:

$$ZrO_2 \cdot SiO_2 + C + 4Cl_2 \rightarrow ZrCl_4 + SiCl_4 + 4CO \quad (7)$$

On completion of the reaction the zirconium tetrachloride is distilled off, condensed, and then hydrolyzed with water to yield a solution of zirconium oxychloride, $ZrOCl_2$. The latter is then crystallized and the crystals calcined to produce hard granular ZrO_2.

The most common method for purifying zirconia ores is the breakdown of baddeleyite and zircon by reaction with sodium hydroxide at temperatures above 600 °C [175]:

$$ZrO_2 \cdot SiO_2 + 4NaOH \rightarrow Na_2ZrO_3 + Na_2SiO_3 + 2H_2O \quad (8)$$

Alternatively, sodium carbonate can be also used, at a temperature of approximately 1000 °C:

$$ZrO_2 \cdot SiO_2 + Na_2CO_3 \rightarrow Na_2ZrSiO_5 + CO_2 \quad (9)$$

$$ZrO_2 \cdot SiO_2 + 2Na_2CO_3 \rightarrow Na_2ZrO_3 + Na_2SiO_3 + CO_2 \quad (10)$$

The sodium silicate is removed by leaching in water, which at the same time hydrolyzes the zirconates to complex hydrated hydroxides of zirconia. These can be directly calcined to yield impure oxides, or further purified for example by treatment with sulfuric acid. The zirconyl sulfates formed are precipitated with a solution of ammonia to form the basic zirconium sulfate $Zr_5O_8(SO_4)_2{\cdot}x\,H_2O$, and then calcined.

Lime fusion is based on the reaction of zircon with calcia, or doloma, which yields calcium zirconium silicate, calcium zirconate, calcium silicate, zirconium oxide, calcium magnesium silicate and the mixtures thereof, according to the reaction conditions:

$$CaO + ZrO_2 \cdot SiO_2 \xrightarrow{1000\,^\circ C} CaZrSiO_5 \quad (11)$$

$$2CaO + ZrO_2 \cdot SiO_2 \xrightarrow{1600\,^\circ C} ZrO_2 + CaSiO_4 \quad (12)$$

The calcium silicate is removed by leaching with hydrochloric acid, and the remaining zirconia washed and dried.

Recent research efforts have been focused on the synthesis of doped, stabilized zirconia powders, especially with nanometer-sized particles. Nanocrystalline stabilized powders are vital for the preparation of nanocrystalline materials, as their properties are fundamentally different from those of conventional powders, due to the extremely small crystalline dimension, superior phase homogeneity and low-temperature sinterability that significantly determines the later-stage processing and sintering properties of ceramics.

A large variety of methods which include (but are not limited to) hydrothermal or solvothermal synthesis, spray-drying, air-plasma-spraying, combustion synthesis (spray pyrolysis), autoignition, coprecipitation, and polymerization/sol–gel have been used for the preparation of stabilized zirconia nanopowders. Typical precursors comprise $ZrOCl_2{\cdot}8\,H_2O$, $ZrO(NO_3)_2{\cdot}6\,H_2O$, and nitrates, chlorides or other inorganic salts of stabilizing metals, such as Mg, Ca, Y, Ce, and Pr. The application of microwaves represents an efficient means of enhancing the crystallinity and decreasing the processing time of doped zirconia nanopowders. Ultrafine particles may be synthesized using complexing process, whereby the metal ions are retained in homogeneous solution with the aid of complexion agents such as lactic acid, citric acid, or ethylenediaminetetra-acetic acid (EDTA). Precipitation of the particles occurs when the complex is broken and large amounts of nuclei are dumped into solution. At this point, almost all available energy is consumed in the formation of nuclei and, as a consequence, growth of the nuclei is limited such that ultrafine particles result from the process.

1.6.2.1 Phase Transformation of Zirconia

The phase transformation of zirconia is a process of major technological importance, notably the transformation from monoclinic (m) to tetragonal (t) zirconia that is associated with a volume decrease of approximately 3–5%. The transformation is generally described as a reversible, atermic, difusionless thermoelastic shear process, which proceeds at near-sonic velocities. However, recent investigations have suggested that the t $\rightarrow$ m transformation is a semi-thermoelastic rather than thermoelastic process, due to the presence of a large thermal hysteresis and a high critical driving force and reversible motion of the t–m interface, which can occur only under thermal stress [176].

A reverse t $\rightarrow$ m transformation and related volume increase during the course of cooling the zirconia parts from the processing temperature, which exceeds the reported temperature for unconstrained transformation of $1174 \pm 6\,^{\circ}$C [177], results in significant strains, which can be only accommodated by the formation of cracks. Thus, the fabrication of large parts of pure zirconia is not possible due to spontaneous failure on cooling.

Garvie *et al.* were the first to realize the potential for the phase transformation of zirconia in enhancing the mechanical properties of ceramics [178], and subsequently developed the concept of "transformation toughening." For this, the phase transformation of a metastable tetragonal zirconia particle can be induced by the stress field at the tip of a propagating crack. The volume change and shear strain associated with the transformation oppose the opening of the crack, thus increasing the resistance of the ceramic to crack propagation. One example of such a material is zirconia-toughened alumina (ZTA), in which the transformable tetragonal zirconia particles are embedded in a matrix with a high elastic modulus. This imposes elastic constraints that prevent transformation back to the monoclinic form. If a crack were to be extended under the stress, then large tensile stresses would be generated around the crack, especially ahead of the crack tip [179, 180]. These stresses would release the matrix constraints such that the transformable tetragonal inclusions would transform to monoclinic, and the resultant volume expansion ($\sim$3%) and shear strain ($\sim$1–7%) would lead to the generation of a compressive strain in the matrix. If this were to occur in the vicinity of the crack, then extra work would be required to move the crack farther.

There exists a size interval for zirconia particles, where the tetragonal particles can be transformed by stress. If the particles are less than critical size they will not transform, but if they are larger than the critical size then they will transform spontaneously. The spontaneous transformation of overcritical particles facilitates the additional toughening mechanism known as "microcracking." On cooling through the transformation temperature, the volume expansion of 3–5% is

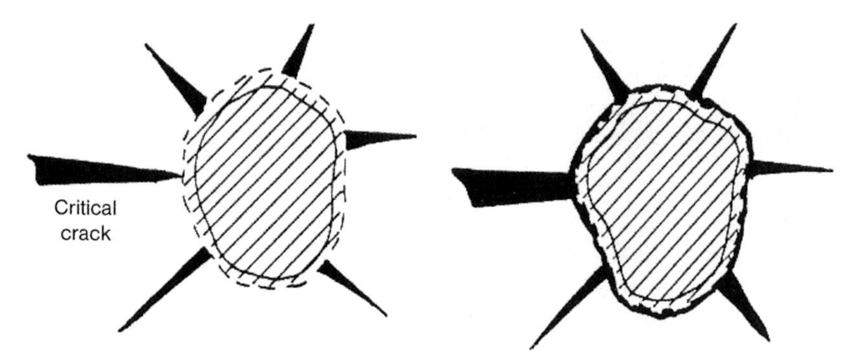

Figure 1.13 Microcrack formation around a transformed zirconia particle.

accommodated by the formation of radial microcracks around the zirconia particles. The fracture resistance is increased by the deviation of a propagating crack into the transformed particle, and bifurcation of the crack around it (Figure 1.13).

The critical size of transformable inclusions is significantly influenced by the presence of additives. If the content of stabilizing aids is sufficiently high, then the transformation can be suppressed entirely. This "stabilization" is in fact a kinetic stabilization of a solid solution in the cubic polymorph down to room temperature by alloying with an alkali earth oxide, or a rare earth oxide. A solid solution can be formed with any ion, provided that the ionic radius is within 40% of the ionic radius of Zr^{4+}. The term "full stabilization" refers to compositions which exhibit single-phase behavior over the whole range of temperatures from absolute zero to the melting temperature of zirconia. By avoiding the existence of a tetragonal phase at intermediate temperatures, the deleterious transformation back to a monoclinic polymorph is also avoided. Although the exact mechanism of stabilization is not clear, it has been suggested that alloying increases the ionic character of bonding, thus making the cubic structure more stable.

1.6.3 Partially Stabilized Zirconia

The addition of sufficient alloying component to facilitate a partial stabilization of the cubic phase leads to partially stabilized zirconia (PSZ), which exhibits an improved thermal shock resistance in comparison to fully stabilized zirconia, as well as an excellent fracture toughness. Today, four oxides – CaO, MgO, Y_2O_3, and CeO_2 – are commonly used to produce PSZ which, in fact, is a mixture of cubic and tetragonal/monoclinic phases that can be prepared by heat treatment of the cubic phase. This process is aimed at the development of a two-phase ceramic when the concentration of stabilizing agents is insufficient to produce full stabilization of the cubic structure.

1.6.3.1 Mg-PSZ

A typical Mg-PSZ contains approximately 8 mol% of MgO. The characteristic procedure when preparing Mg-PSZ involves sintering and heat treatment in the

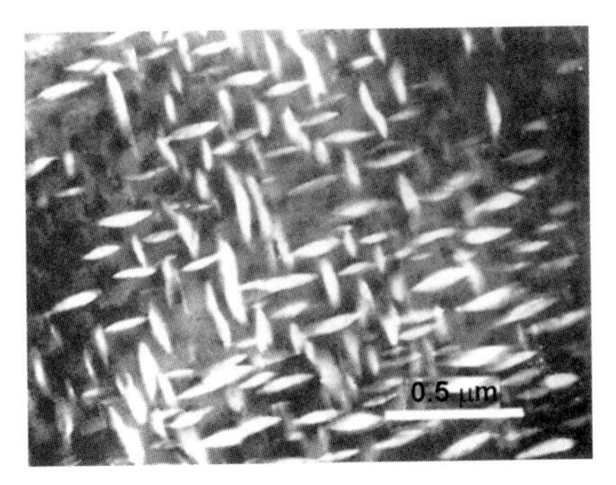

Figure 1.14 Oblate spheroids of tetragonal phase nucleated in a cubic ZrO_2 matrix.

cubic single-phase field (temperature >1750 °C), followed by a rapid cooling which does not allow the precipitation of equilibrium amounts of the tetragonal phase, but rather facilitates nucleation of the tetragonal phase in the form of nanometer-sized oblate spheroids in a cubic matrix (Figure 1.14). A subsequent subeutectoid heat treatment leads to growth of the tetragonal precipitates until they reach the size at which they can transform spontaneously to the monoclinic polymorph.

Normally, commercially available Mg-PSZ have rather complex microstructures, which consist of a coarse-grained cubic matrix (grains often on the order of tens of micrometers) with fine tetragonal and monoclinic grain boundary precipitates (Figure 1.15). The precipitates often transform during cooling to room temperature, and form the monoclinic phase; heterogeneously nucleated precipitates are often also formed within the grains. The fracture toughness of Mg-PSZ may be as high as 15 MPa m$^{1/2}$, but this will decrease with temperature. The fracture strength usually varies between 650 and 800 MPa.

Orthorhombic zirconia, which is also known as a "high-pressure polymorph," appears in Mg-PSZ when it is cooled to cryogenic temperatures [174]. In the nitrogen-quenched 9.4Mg-PSZ sample, the *ortho* phase is in the majority, at 46.6 ±

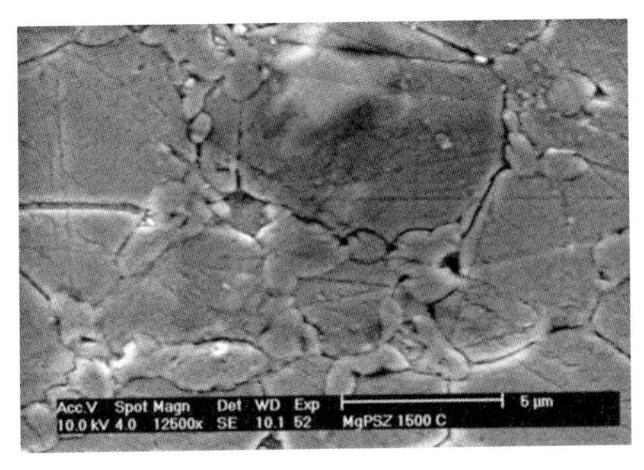

Figure 1.15 Microstructure of sintered and thermally aged Mg-PSZ consisting of large cubic grains and fine precipitates of tetragonal and monoclinic phases.

1.1 wt%. This material has a significantly higher Young's modulus (242 GPa, the second-highest of all zirconia-based materials, and the highest of all zirconia-based ceramics) than does the same material before cryogenic cooling; this is due to the very high modulus of the *ortho* phase, which is estimated to be ~285 GPa [181].

1.6.3.2 **Ca-PSZ**

Ca-PSZ is similar to Mg-PSZ, and develops similar microstructures during high-temperature aging. An important factor is the different critical size for unconstrained m → t transformation, which is approximately 6–10 nm for Ca-PSZ (compared to 25–30 nm for Mg-PSZ). An extensive systematic investigation of this material has been conducted by Garvie *et al.*, who achieved a considerable increase in the strength of the material, from 200 to 650 MPa, simply by ageing it at 1300 °C [182].

Various authors have reported different temperatures of eutectoid decomposition, ranging from 1000 to 1140 °C [183–185]. The eutectoid transformation is a rather slow process, and is therefore not seen in conventionally aged samples. However, a large cubic phase field also exists which, together with slow transformation, facilitates the existence of a fully cubic structure, providing the basis for calcia-stabilized zirconia solid electrolytes.

1.6.3.3 **Y-PSZ**

The addition of yttria to zirconia not only stabilizes the cubic or tetragonal form but also lowers the temperature of the t → m transformation. The practical consequence of this is that larger zirconia particles can be retained in the metastable tetragonal form, thus considerably easing any problems associated with the fabrication of a toughened ceramic, such as ZTA. One important feature of this system is the solubility of yttria in zirconia up to a concentration of approximately 2.5 mol% which, in conjunction with a low eutectoid temperature, will facilitate the formation of fully tetragonal ceramics which are referred to as "tetragonal zirconia polycrystals" (see Section 1.6.4).

A large cubic + tetragonal phase field in the Y_2O_3–ZrO_2 system permits the formation of a PSZ structure which is, in many respects, analogous to Mg-PSZ or Ca-PSZ, and consists of tetragonal precipitates embedded in a cubic phase matrix. The morphology of the precipitates depends on the conditions of ageing (e.g., time, temperature). For example, a rapid cooling will result in a displacive transformation and formation of the so-called t' phase, which has a lower c/a ratio than the normal tetragonal phase, and contains the same amount of yttria as the cubic phase [186].

1.6.3.4 **Ceria and Other Rare Earth-Stabilized Zirconias**

Ce-PSZ exhibits many similarities with Y-PSZ, having a very wide compositional range of formation of the solid solution of up to 18 mol% CeO_2. In the ZrO_2–CeO_2 system, the stabilization of t-ZrO_2 occurs over a wide composition range, from 12 to 20 mol% CeO_2, with a preferred composition of 12 mol% [187].

In order to obtain fully dense ceramics, it is necessary to use ultrafine powders that are normally prepared by the coprecipitation of precursors. The presence of liquid-forming additives, such as Ca and Si, is also often required. Although this material achieves a very high fracture toughness, with reported values as high as 30 MPa · $m^{1/2}$, Ce-TZP is rarely used as a monolithic material due to its relatively low level of hardness and strength.

The ZrO_2–$GdO_{1.5}$ system is important for the development of ceramics for thermal barrier coatings (TBC), other than the state-of-the-art yttria-stabilized zirconia (YSZ), which has a lower thermal conductivity and improved high-temperature performance and durability [188].

Scandia-stabilized zirconia (ScSZ) possesses the highest oxygen-ion conductivity among all zirconia-based oxides, and therefore represents a promising solid electrolyte for applications in electrochemical devices such as solid oxide fuel cells (SOFCs) and catalytic membrane reactors (further details are available in Section 1.6.6.3).

The possibility of PrO_x–ZrO_2 solid solution formation has been hypothesized (despite the phase diagram being unknown) on the basis of the lanthania–zirconia phase diagram, and by considering the ionic radii of Zr^{4+} (0.87 Å) and both Pr^{3+} (1.126 Å) and Pr^{4+} (0.90 Å). When, recently, 10 mol% Pr-doped zirconia powders were prepared using a microwave-assisted hydrothermal synthesis, the doped powder was shown to be a substitutional solid solution of praseodymium in tetragonal zirconia [189].

1.6.4 Tetragonal Zirconia Polycrystals (TZP)

As the toughening effect in partially stabilized zirconias increases linearly with the amount of retained tetragonal phase, the logical consequence is the development of ceramics that are wholly tetragonal. This achievement was first accomplished by Rieth [190], and later by Gupta *et al.* [191], who prepared the ceramic by sintering yttria and other rare earth-containing zirconia powders in the temperature range 1400–1500 °C. The resultant ceramic was fine-grained, with a strength of 600–700 MPa. Moreover, the critical grain size for the t → m transformation was found to depend on the amount of stabilizing aids, this being about 0.2 μm for 2 mol % Y_2O_3 and 1.0 μm for 3 mol% Y_2O_3, as well as on the presence or absence of mechanical constraints. The typical properties of TZP are summarized in Table 1.9.

The microstructures of commercial TZP ceramics are usually far from ideal, and often contain a significant amount of the cubic phase. TZP ceramics also often contain amorphous grain boundary phases, generally low-viscosity liquids from the systems Al_2O_3–Y_2O_3–SiO_2. The presence of a liquid and a small grain size result in a pseudosuperplasticity of TZPs with an extension in tension of over 100% at 1200 °C. High-tensile ductility is attributed to grain size stability during high-temperature deformation [192], which is achieved by suppressing the grain growth by adding various oxides, such as CuO [193–195], Sc_2O_3 [196], and Al_2O_3 [197]. An elongation to failure of 520% has been achieved in 5 wt% SiO_2-added 8 mol% Y_2O_3-stabilized c-ZrO_2 (8YCZ) at 1430 °C [198]. The limited grain growth was not always an indispensable condition for superplastic ceramics. Moreover, the addition of TiO_2 increased the grain size but reduced the flow stress and enhanced the elongation to

Table 1.9 Typical properties of TZP.

Property		Value
Melting point		2720 °C
Bulk density		6.05 g cm^{-3}
Bending strength	RT	900–1300 MPa
	800 °C	350 MPa
Young's modulus		140–200 GPa
Fracture toughness		5.5–11 MPa m$^{1/2}$
Hardness		~14 GPa
Thermal conductivity (RT)		1–2 W m^{-1} K^{-1}
Coefficient of thermal expansion	100 °C	8.3×10^{-6} K^{-1}
	800 °C	10.5×10^{-6} K^{-1}
Thermal shock resistance		ΔT = 360 °C

RT = room temperature.

failure [199]. It should be noted that this subject has been extensively reviewed by both Chokshi [200] and Jimenez-Melendo *et al.* [201].

One major obstacle against the full exploitation of TZP is the spontaneous surface t → m transformation that occurs when the ceramics are aged at temperatures of between 150 and 250 °C. The transformation is accompanied by a severe strength degradation, especially in a water vapor-containing environment, with the rate of degradation being governed by not only the number of oxygen vacancies but also the instability of *t*-ZrO_2. The aging rate was increased by about 50% in a 3Y-TZP dense sample sintered under reducing conditions, in comparison to a similar sample which had been sintered in air (Figure 1.16) [202]. Ageing is inhibited by the addition of liquid-forming additives, especially silica [203], and also by the addition of CeO_2 or

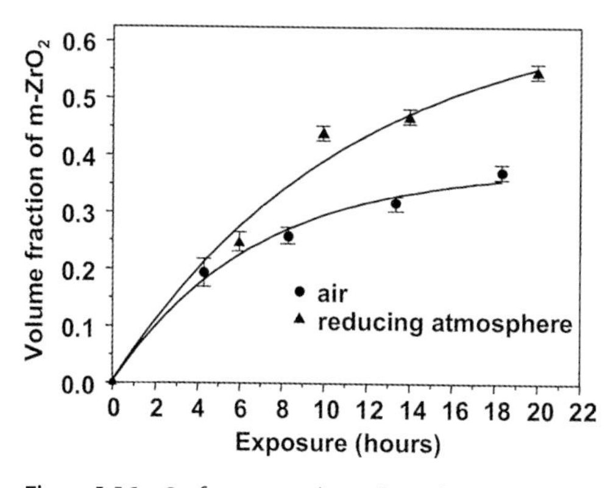

Figure 1.16 Surface monoclinic phase fraction versus exposure time at 140 °C in steam for the zirconia materials sintered in air and in a 90% Ar/10% H_2 atmosphere [202].

Figure 1.17 Scanning electron microscopy image of the 8 mol% CeO_2-stabilized ZrO_2 ceramic after spark plasma sintering [210].

Al_2O_3 [204, 205]. Al_2O_3 also raises the CeO_2 content in ZrO_2 grains, thus preserving a more tetragonal phase and enhancing the fracture toughness [206].

The applicability of Ce-TZP is limited also by its susceptibility to the reduction of Ce^{4+} to Ce^{3+} when sintering or using the ceramics in a nonoxidizing atmosphere. The reduction is accompanied by a change in color [207, 208]. Due to increases in the ionic radius of the cerium ions, from 0.101 nm for Ce^{4+} to 0.111 nm for Ce^{3+}, there is an approximate 40% mismatch in ionic radius with Zr^{4+}, and this results in a high elastic lattice strain and the segregation of Ce^{3+} to the grain boundaries [209]. Consequently, tetragonal ZrO_2 will be destabilized and cracking will occur in the bulk material. The sintering of Ce-TZP in nitrogen, or the hot isostatic pressing of presintered materials in a reducing environment with graphite heating elements, is not possible as the tetragonal phase is completely destabilized and transforms to monoclinic during cooling [207]. High heating rates and short soaking times, facilitated by SPS, partially overcomes this obstacle, such that Ce-PSZ consisting of monoclinic and tetragonal ZrO_2 with a volume ratio of 2 : 1, and a trace amount of Zr–Ce–O cubic solid solution, were prepared using this method (Figure 1.17) [210].

Both, Ce-TZP [211] and Ce–Y-TZP [212] exhibit martensitic transformation-associated shape memory. For example, the 8Ce–0.50Y-TZP exhibits a complete shape memory recovery under a recoverable strain of 1.2% at a relatively high operating temperature (>500 °C) [213].

1.6.5
Zirconia-Toughened Alumina (ZTA)

In the past, zirconia has been added routinely to a variety of ceramic materials in order to increase their toughness. Zirconia-toughened alumina was first developed by Claussen, who demonstrated a significant toughening effect of unstabilized, but transformable, tetragonal zirconia particles within the alumina matrix [214].

The retention of such particles is facilitated by mechanical constraints imposed by the surrounding alumina matrix with a high elastic modulus. Both, the transformation toughening and microcracking, contribute to increasing the fracture toughness of alumina. Unstabilized ZTAs can achieve the bending strength of up to 1200 MPa, and a fracture toughness of about 16 MPa $m^{1/2}$ at 15 vol% ZrO_2.

1.6.6 Applications of Zirconia

1.6.6.1 Thermal Barrier Coatings

The efficiency of gas turbine engines is dictated by the maximum temperature that the turbine rotors can sustain during continuous operation. If a thin coating of ceramic is supplied to a metal turbine blade, then the engine temperature can be increased by 50–200 °C, without increasing the temperature of metal. The desirable properties of a suitable ceramic thermal barrier coating include a high thermal expansion (close to that of metal), a low thermal conductivity, chemical stability in the gas turbine environment, and a high thermal shock resistance [215].

Plasma-sprayed zirconia coatings of PSZ composition (namely YSZ) deposited by electron beam physical vapor deposition (EB-PVD) and atmospheric plasma spraying (APS) have been investigated for this purpose. The microstructure of sprayed coatings is nonequilibrium and fine-grained, and contains both macrocracks and microcracks and a residual porosity. As a result, sintering through crack healing, accompanied by grain growth, takes place at elevated temperatures [216, 217]. Sintering affects the mechanical and physical properties, namely elastic modulus, strength and work of fracture, and also increases the thermal conductivity and impairs the strain-tolerant capability of TBCs. An increase in long-term phase, mechanical, and chemical stability under working conditions is therefore of primary importance. The chemical attack of coatings by the mineral constituents of fuel, especially Na, Mg, and S (in the form of liquid sulfates) and vanadium, deplete the YSZ coatings of yttrium, with subsequent destabilization and deterioration of the mechanical properties [218]. A review of mechanical behavior of YSZ-based thermal barrier coatings is provided in Ref. [219].

A decrease in thermal conductivity and an increase in thermal protection are achieved by adjusting the microstructure and porosity of ZrO_2–Y_2O_3 (7–8) wt% coatings [220–222]. The defect-cluster design approach, using high-stability, paired dopant oxides of distinctively different ionic sizes, produces lattice distortion in the oxide solid solutions, and also facilitates local ionic segregation and defect clustering. Oxide defect clusters with appropriate sizes attenuate and scatter the lattice and radiative phonon waves over a wide range of frequencies. The formation of thermodynamically stable, highly defective lattice structures with controlled defect-cluster sizes reduces oxide intrinsic lattice and radiation thermal conductivity. The influence of codoping with additional paired rare earth oxides Nd_2O_3–Yb_2O_3 or Gd_2O_3–Yb_2O_3 (i.e., ZrO_2–$(Y,Nd,Yb)_2O_3$ and ZrO_2–$(Y,Gd,Yb)_2O_3$) on thermal conductivity is shown in Figure 1.18 [215].

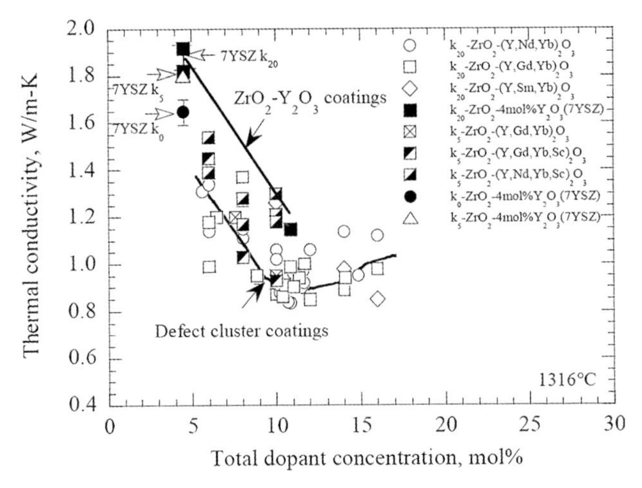

Figure 1.18 Thermal conductivity of various composition electron beam physical vapor deposition oxide defect-cluster coatings as a function of total dopant concentration [215].

The disadvantages of YSZ at high temperatures prompted an intense search for new TBC materials (for reviews on recent developments, see Refs [223, 224]). Interesting candidates for TBCs include zirconia-based materials with a pyrochlore structure and a high melting temperature, such as $La_2Zr_2O_7$, $Gd_2Zr_2O_7$, or $Nd_2Zr_2O_7$. Although these materials have a lower thermal conductivity and a higher thermal stability than YSZ, their thermal expansion is usually lower than that of YSZ, which leads to higher thermal stresses in the TBC. In addition, their toughness is lower due to an absence of toughening effects [225]. However, this problem can be solved by the use of layered topcoats; in this case, YSZ is used as a TBC material with a relatively high thermal expansion coefficient and high toughness. The YSZ layer is then coated with a new TBC material (e.g., $La_2Zr_2O_7$) which is then able to withstand higher temperatures (by about 100 °C) [226].

1.6.6.2 **Solid Electrolytes**

The use of zirconia as a solid electrolyte (and especially for oxygen-sensing devices) is facilitated by the fact that stabilized zirconia has a defect structure with a finite concentration of octahedral interstitial voids. The void space in the lattice is larger for the O^{2-} anions than for the Zr^{4+} cations, and the O^{2-} anions are therefore the rate-controlling species in the diffusion process.

The maximum ionic conductivity in ZrO_2-based systems is achieved when the concentration of acceptor-type dopant(s) is close to the minimum necessary for complete stabilization of the cubic fluorite structure [227, 228]. For example, the highest conductivity in $Zr_{1-x}Y_xO_{2-x/2}$ and $Zr_{1-x}Sc_xO_{2-x/2}$ ceramics is achieved at $x = 0.08$–0.11 and 0.09–0.11, respectively. Further additions decrease the ionic conductivity due to an increasing association of the oxygen vacancies and dopant cations into complex defects of low mobility [229]. This effect is more pronounced

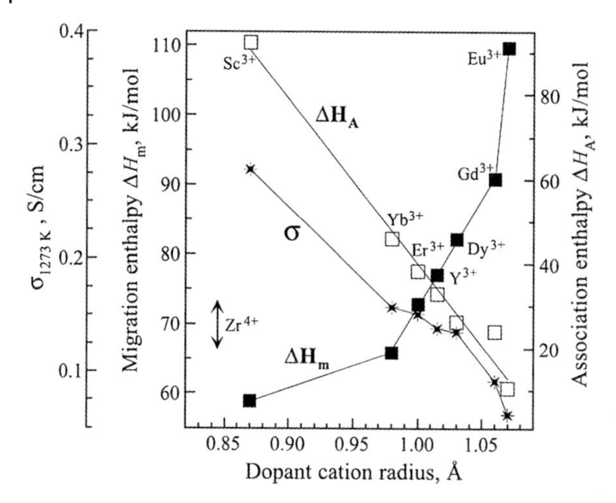

Figure 1.19 Maximum conductivity in the binary ZrO_2–Ln_2O_3 systems at 1000 °C, and the oxygen-ion migration and association enthalpies versus radius of Ln^{3+} cations [243].

if the mismatch between the host and dopant cation radii is larger [230, 231] (Figure 1.19). Because the Zr^{4+} ion is smaller than the trivalent rare earth cations, a maximum ionic transport is observed for Sc^{3+}. However, due to the high cost and problems with the ageing of Sc-FSZ at moderate temperatures, Y-FSZ is used for most practical applications.

The doping of ZrO_2 with alkaline earth metal cations (A^{2+}) is less effective due to a greater tendency to defect association and to a lower thermodynamic stability of the cubic fluorite-type solid solutions in ZrO_2–AO systems. To date, attempts to increase the stability of Sc-containing materials by codoping, or to reduce the cost of Ln^{3+}-stabilized phases by mixing them with cheaper alkaline earth dopants, have not yielded any worthwhile results [228].

Commercially available zirconia electrolytes often contain secondary phases, especially alumina- and silica-containing precipitates. The presence of a grain boundary glass causes a deterioration in the electrical properties due to its poor ionic conductivity [232], and a significant increase in grain boundary resistance [233–235]. In contrast, the minor addition of a highly dispersed Al_2O_3 decreases the grain boundary resistance by scavenging silica-rich impurities into new phases that do not wet the grain boundaries [236–238]. Moreover, the addition of alumina increases the mechanical strength by retarding grain growth. A similar effect was observed in Mg-PSZ, where MgO reacts with silica and forms discrete forsterite grains within the zirconia matrix [239]. Also, if the admixed MgO amount exceeds the solubility limit, then MgO will form a second phase in the zirconia matrix after sintering. Then, the MgO has a higher electrical conductivity than zirconia, and the conductivity of the composite electrolytes is therefore correspondingly higher [240].

The achievement of a high ionic conductivity in nanostructured ZrO_2 is questionable. In some studies, the overall conductivity of nanocrystalline YSZ ceramics was found to be comparable to that of their microcrystalline counterparts [241].

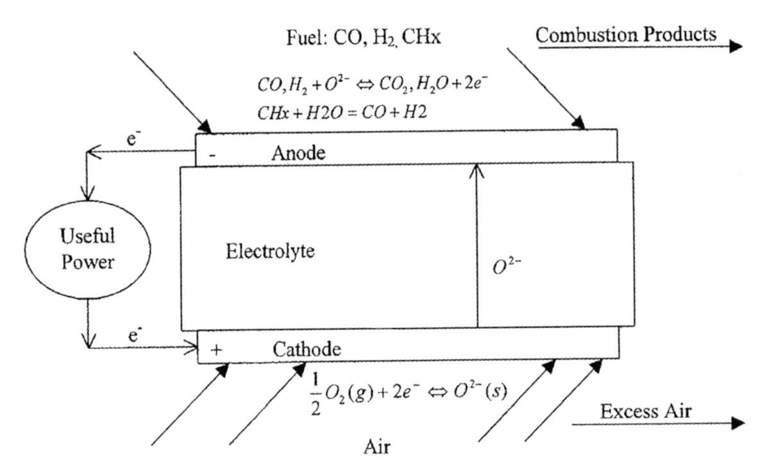

Figure 1.20 Schematic diagram of a fuel cell using a zirconia electrolyte [254].

In contrast, an increase in conductivity by about one order of magnitude was reported in YSZ nanostructured thin films [242] (for a review on this topic, see Ref. [229]).

1.6.6.3 **Fuel Cells**

The ionic conductivity of stabilized zirconia is utilized in solid oxide electrolyte fuel cells, as pioneered by Westinghouse in 1986. These cells function in reverse to a hydrogen generator, at temperatures approaching 1000 °C (Figure 1.20), with the diffusion of oxygen ions through a zirconia electrolyte. This is usually in the form of a tube, with air passing along one surface and the fuel (usually hydrogen) along another surface, which facilitates oxidation to proceed on the anode. The ionization of an oxygen molecule takes up to four electrons to the cathode, thus generating an electrical current in conjunction with oxidation of the fuel.

Current efforts are aimed at lowering the operating temperature of SOFCs from above 900 °C down to 500 °C, in order to improve both the longevity and cost of the peripheral materials and the electrical power generation efficiency [244]. Tetragonal zirconia is considered to be an electrolyte for intermediate-temperature solid oxide fuel cells (IT-SOFCs), due to better mechanical and electrical properties at lower temperatures in comparison with cubic zirconia [245]. Dense YSZ thin-film-based SOFCs can be operated at intermediate temperatures (650–800 °C), achieving power densities of between 0.35 and 1.9 W cm^{-2}, depending on the temperature and film thickness [246, 247]. Scandia-stabilized zirconia (ScSZ) has a higher ionic conductivity than YSZ, and a high mechanical strength and fracture toughness, which increases the reliability of ScSZ-based SOFCs [248–250]. ScSZ electrolyte sheets and electrolyte-supported-type cells have been designed for reduced-temperature operation (ca. 800 °C). However, for the working temperature range of the low-temperature SOFC (600–800 °C), the temperature dependence of electrical conductivity changes drastically across a composition boundary at approximately 10 mol% Sc_2O_3. At additive levels of between 3 and 8 mol%, the electrical conductivity changes smoothly with temperature, but at a content above 12 mol% a discontinuity of electrical conductivity appears at 650 °C, above which temperature a highly conductive cubic phase prevails. Below 650 °C, a less-conductive rhombohedral h-phase ($Zr_7Sc_2O_{17}$) is formed [251, 252].

One problem here is the low-temperature degradation and deterioration of mechanical properties under hydrothermal conditions, due to the t → m transformation of zirconia. This makes the application of t-ZrO_2 in IT-SOFC devices challenging, as water vapor is produced at the anode when the cell is in operation. The fuel cell must also be able to withstand thermal cycling, and to operate under pressures greater than atmospheric over the lifetime of the cell (>50 000 h) [253].

1.6.6.4 **Bioceramics**

Biomedical-grade zirconia was introduced 20 years ago to solve the problem of alumina brittleness, and the consequent potential failure of implants. The reason for this is that biomedical-grade zirconia exhibits the best mechanical properties of oxide ceramics as a consequence of transformation toughening, which increases its resistance to crack propagation. Likewise, partially stabilized zirconia shows excellent biocompatibility, and it has therefore been applied to orthopedic uses such as hip and knee joints [255].

The metastability of zirconia – and especially of Y-TZP, which is prone to ageing in the presence of water – represents a serious problem in biomedical applications [256]. Yttrium, as a trivalent ion, creates oxygen vacancies that aid hydroxyl group diffusion in the lattice, generating nucleation of the transformation via a stress corrosion-type mechanism [257]. The resultant degradation is characterized by surface roughening, microcracking at the surface, and the release of particles into the body. Although the manufacturers of zirconia claimed that this problem was limited under *in vivo* conditions, in the year 2001 approximately 400 implanted femoral heads constructed from zirconia failed within a very short period. Consequently, the aging and wear of zirconia has become a very important issue, the main aim being to renew the confidence of the medical community in zirconia-based biomaterials.

As shown in Figure 1.21, the aging-related nucleation and phase transformation leads to a cascade of events. The transformation of one grain, associated with a

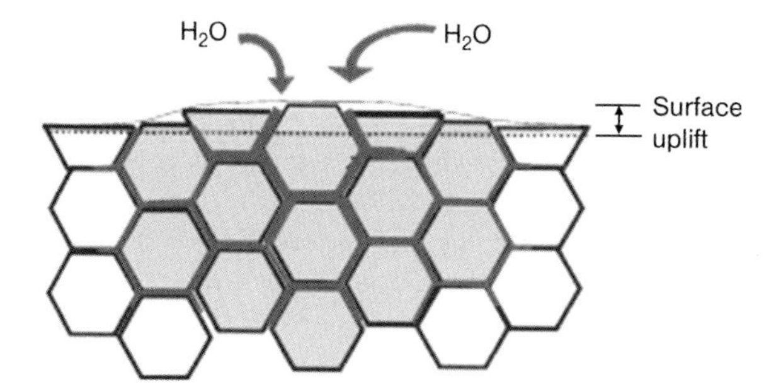

Figure 1.21 Schematic of the transformed zone in aged biomedical-grade zirconia, leading to extensive microcracking and surface roughening. The transformed grains are shown in gray. The dark gray path represents the penetration of water due to microcracking around the transformed grains [257].

volume increase, stresses the neighboring grains and results in the formation of microcracks; this, in turn, creates a pathway through which water can penetrate further into the material. The initial transformation of individual grains is process-related, and depends on their disequilibrium state – that is, the size of the grains and the content of the stabilizing aid [256], the specific orientation from the surface [258], the presence of residual stresses, and/or even the presence of a cubic phase [259]. The slowing down or even prevention of such aging is of the utmost importance, with proposed solutions including the addition of small amounts of silica [260] or the use of yttria-coated rather than coprecipitated powders [261]. Another possibility would be to use Ce-doped zirconia, as this material exhibits superior toughness (up to $20\,MPa \cdot m^{1/2}$) and negligible aging during the lifetime of an implant [257].

Due to the above-mentioned problems, the use of zirconia-based biomaterials in the surgery of large bones is currently restricted, and the manufacturers have responded by developing a series of toughened composites. These include an alumina-toughened zirconia (Bio-Hip®; Metoxit AG, Thayngen, Switzerland), which has a bending strength of up to 2000 MPa, and a zirconia-toughened alumina (BIOLOX® delta; Ceramtec AG, Plochingen, Germany) with a bending strength in excess of 1150 MPa and a fracture toughness of $\sim 8.5\,MPa \cdot m^{1/2}$.

1.7 Cerium Oxide

Cerium oxide (CeO_2) has found numerous applications as an electrolyte for SOFCs, as abrasive materials for chemical mechanical planarization (CMP), as a UV absorbent, as a material for oxygen pumps, and as an automotive exhaust promoter.

1.7.1 Crystal Structure and Properties of Single-Crystal CeO_2

Cerium is a common, naturally occurring element that is characterized chemically by having two valence states, +3 and +4. Ce^{4+} is the only nontrivalent rare-earth ion that is stable in an aqueous environment and it is, therefore, a strong oxidizing agent. The +3 state closely resembles the other trivalent rare earths.

Cerium oxide is a highly stable, nontoxic, refractory ceramic material with a melting point of 2600 °C and a density of $7.13\,g\,cm^{-3}$. Ceria has a fluorite face-centered cubic crystal structure with a lattice constant of 5.11 Å (Figure 1.22). CeO_2 easily transforms from the stoichiometric CeO_2 (+4) state to the Ce_2O_3 (+3) valence state via a relatively low-energy reaction, although even at a loss of considerable amounts of oxygen from the crystal lattice, and the formation of a large number of oxygen vacancies, its fluorite structure is retained. The suboxides thus formed may be readily reoxidized to CeO_2 in an oxidizing environment.

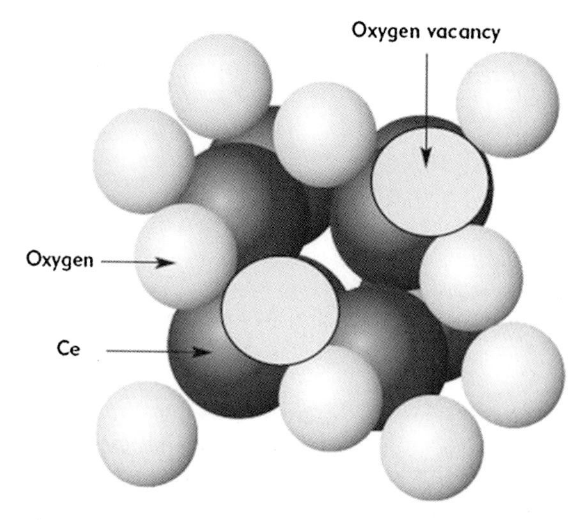

Figure 1.22 Fluorite structure of CeO_2.

1.7.2
Natural Sources and Production

The two largest sources of cerium and other rare earth elements are the minerals *bastnasite* and *monazite*. Bastnasite, which belongs to carbonate-fluoride minerals, exists as several types: bastnasite-(Ce) (Ce, La)CO_3F; bastnasite-(La) (La, Ce)CO_3F; and bastnasite-(Y) (Y, Ce)CO_3F. The most frequently occurring of these is bastnasite-(Ce), and cerium is by far the most common of the rare earths in this class of minerals.

Cerium oxide is extracted from bastnasite by roasting the ore with concentrated sulfuric acid, or with sodium carbonate [262]. Since, in the first process, HF forms as a byproduct, roasting with sodium carbonate is the preferred method. The rare earth elements contained in calcine are extracted by leaching with hydrochloric acid, and recovered from the leachate by precipitation with oxalic acid. During the course of roasting, cerium(III) is oxidized to cerium(IV). CeO_2 is insoluble in dilute hydrochloric acid; this is in contrast to other trivalent rare earth elements, which can be easily leached out, as can the impurities such as Fe, Ca, and Mg.

The present applications of ceria-based ceramics impose strict requirements on the quality and purity of the powders used. Several studies have described the synthesis of ceria nanopowders of high quality and with a well-defined morphology. Typical methods of preparation include hydrothermal synthesis [263, 264], the hydrolysis of an alkoxide solution (sol–gel) [265], chemical precipitation [266], mechanochemical processing [267], and gas-phase reaction [268]. *Emulsion techniques* can also be used, as these reduce not only the production costs of high-purity spherical powders but also the degree of aggregation. Thus, ceria powders with an average particle size <20 nm and a narrow particle size distribution can be prepared by reaction between two emulsions containing cerium nitrate and an alkaline precipitation agent, such as ammonium hydroxide or sodium hydroxide [269].

1.7.3 Properties

The mechanical properties of undoped ceria ceramics are usually rather poor; typically, these materials have room-temperature bending strengths of ~100 MPa and a fracture toughness of ~1.5 MPa · $m^{1/2}$ [270]. The fracture toughness of $Ce_{0.8}Gd_{0.2}O_{2-\delta}$ ceramics has been reported as 1.5 ± 0.2 MPa · $m^{1/2}$, this being independent of the crack length or grain size, within the range of 0.5 to 9.5 μm [271]. Such grain size-independence of fracture toughness is attributed to the almost 100% transgranular nature of the fracture of doped ceria ceramics. The bending strength is, in some cases, improved by the addition of other rare earth oxides [272], while the fracture strength is influenced by the method used to prepare the starting powder (Figure 1.23), the temperature of sintering, and by the concentration rather than the type of dopant (Figure 1.24). These ceramics fail transgranularly, with the proportion of the transgranular fraction ranging between 96% and 99%. The variation in fracture strength is attributed to the decrease in transgranular strength due to the generation of oxygen vacancies introduced by dopants, which distort the lattice and decrease the coulombic forces between ions [273].

As an ionic conductor (exhibiting a high mobility of oxygen ions), ceria represents a major candidate material for SOFCs. Microcrystalline ceria doped with various rare earth elements has a higher ionic conductivity than does stabilized ZrO_2, especially at lower temperatures. As ceria also exhibits electronic conductivity, it is in fact a mixed ionic–electronic conducting material, with ionic conductivity prevailing at temperatures above 500 °C. The high electronic conductivity of undoped nanocrystalline ceria is attributed to electronic conduction along the grain boundaries [274, 275].

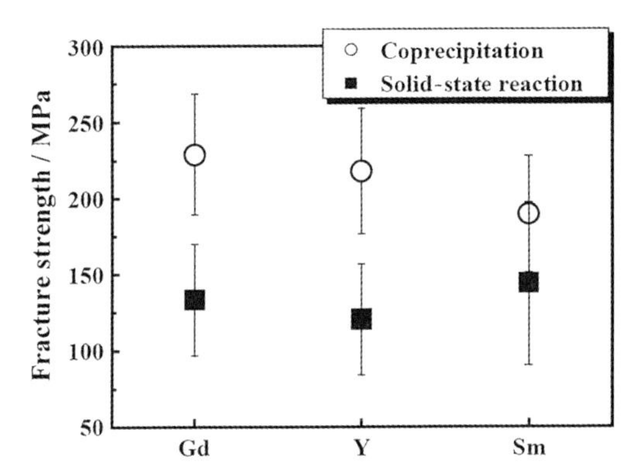

Figure 1.23 The fracture strength of $(CeO_2)_{0.80}(LnO_{1.5})_{0.20}$ (Ln = Y, Gd, and Sm) ceramics prepared by solid-state reaction of CeO_2 and the respective rare earth oxide powder and by an oxalate coprecipitation method [272].

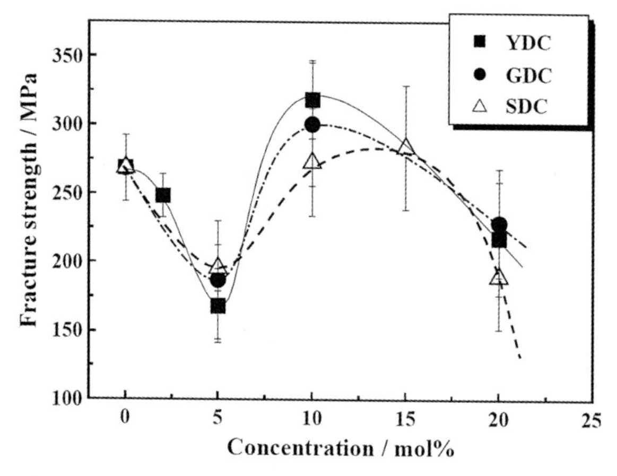

Figure 1.24 Dependence of room temperature fracture strength of doped CeO_2 ceramics on the concentration of rare earth-dopants: yttrium (YDC), gadolinium (GDC), and samarium (SDC). [272].

1.7.4 Applications

1.7.4.1 Abrasives

Because of its high chemical activity and unique crystal structure, cerium oxide is used as a polishing powder, the advantages of which include a long life, a high polishing efficiency, and a low residua. Today, ceria is gradually replacing traditional abrasives for some applications. Ceria polishing powders are classified as two types: (i) high-cerium abrasives with a ceria content >80%; and (ii) low-cerium abrasives with a ceria content of 48–50%. Currently, cerium oxide abrasive powders with nanosized spherical particles and a narrow particle size distribution are used for the chemical mechanical planarization of semiconductor devices, in order to reduce the scratching of wafers [276].

1.7.4.2 Solid Electrolytes

The main obstacle against the use of ceria in SOFCs is the partial reduction of Ce^{4+} to Ce^{3+} [277, 278]. Under the reducing conditions experienced on the anode side of the fuel cell, a large number of oxygen vacancies is formed within the ceria electrolyte, and consequently CeO_2 is reduced to Ce_2O_3, thus increasing electronic conductivity of the material. Finally, as a result of the oxygen vacancy formation, ceria undergoes a so-called "chemical expansion." The high ionic conductivity encountered at lower temperatures, coupled with problems related to the reduction, limit the application of ceria electrolytes at relatively low temperatures.

Due to the small association enthalpy between the dopant cation and oxygen vacancy in the fluorite lattice, the highest conductivity is achieved in ceria doped with Gd^{3+} or Sm^{3+} [279, 280]. The highest level of oxygen ionic transport is found in the solid solutions $Ce_{1-x}Ln_xO_{2-\delta}$, where Ln = Gd or Sm, and $x = 0.10–0.20$. The lattice ionic conductivity of Ln-doped ceria is about 0.01 S cm^{-1} at 500 °C; however, the substitution of a fraction of the ceria with Gd or Sm introduces vacancies without adding any electronic charge carriers. This has two main consequences: (i) it provides an n-type electronic conductivity which causes a partial internal electronic short circuit in a cell; and (ii) it generates nonstoichiometry with respect to normal valency in air and an expansion of the lattice, which can lead to mechanical failure. The effect of lattice expansion on mechanical integrity depends on the geometry of the cell, and the way in which the ceria is supported. In general, ceria electrolytes are considered to be mechanically unstable at temperatures above 700 °C [281].

The grain boundaries are of profound importance for the ionic conductivity, as they partially block ionic transport so that the total resistance will depend on the level of segregated impurities. The properties of CeO_2-based solid electrolytes have been reviewed elsewhere [229, 282–284].

Doped ceria is only viable for operating temperatures below 600 °C, and is therefore normally used as a supported thick film. In order to preserve a high activity and to maintain compatibility with metal supports, the processing temperature should be kept as low as possible (e.g., 1000 °C), although this may be difficult to achieve due to the high refractoriness of ceria. Consequently, small amounts (e.g., 1 mol%) of oxides of divalent transition metals (e.g., Co, Cu, Mn) are often used as sintering aids for Gd-doped ceria [285–287]. Unfortunately, CuO, CoO and MnO_2 partially impair the grain boundary conductivity by promoting the propagation of SiO_2 impurities at the grain boundaries [286, 287]. The addition of a small amount of Fe_2O_3 (~0.5 atom%) will reduce the sintering temperature by ~200 °C and also promote densification. Iron is also believed to promote the dissolution of Gd_2O_3 in CeO_2 at lower sintering temperatures, and to increase the grain boundary conductivity by scavenging the SiO_2. In general, the concentration of the transition metal sintering aids must be kept as low as possible in order to minimize their influence on the ionic or electronic performance of the electrolyte [285, 288].

Highly reactive nonagglomerated powders can be sintered to high density at low temperatures, without the need for additives. Fully dense ceramics with grain sizes between 0.15 and 0.75 μm were prepared via a pressureless sintering of Sm-doped nanopowders (14 nm) at 1000 °C [289].

1.7.4.3 **Catalysts**

Cerium oxide is used in exhaust three-way catalytic converters where the emissions from fuel burning are converted to harmless gases. The conversion includes the following reactions:

Hydrocarbon combustion:

$$(2x + y)\mathrm{CeO_2} + \mathrm{C}_x H_y \rightarrow [(2x + y)/2]\mathrm{Ce_2O_3} + x/2\mathrm{CO_2} + y/2\mathrm{H_2O} \tag{13}$$

Soot burning:

$$4\mathrm{CeO_2} + \mathrm{C_{soot}} \rightarrow 2\mathrm{Ce_2O_3} + \mathrm{CO_2} \tag{14}$$

NO_x reduction:

$$\mathrm{Ce_2O_3} + \mathrm{NO} \rightarrow 2\mathrm{CeO_2} + 1/2\mathrm{N_2} \tag{15}$$

The catalytic activity of cerium oxide depends on its particle size and surface area. As oxygen vacancy atomic point defects are formed more easily at the surface than in the bulk, high-surface-area materials will have a substantially higher catalytic activity [290]. The activation temperature of carbon combustion is reduced from approximately 700 °C for a micron-sized material to 300 °C, if the surface area of the material is increased by a factor of 20 [291].

1.8 Yttrium Oxide

Although yttrium oxide is rarely used as a ceramic material in its pure form, it is widely applied as an additive in various ceramics, where it is used as a component of sintering aids (e.g., in silicon nitride, or alumina), as a stabilizer in zirconia and in alumina-zirconia abrasives, in wear-resistant and corrosion-resistant cutting tools, seals and bearings, high-temperature refractories for continuous-casting nozzles, jet engine coatings, oxygen sensors in automobile engines, as a component of high-temperature superconductors of the Y–Ba–Cu–O composition, and in artificial gemstones. In electronics, yttrium–iron–garnets are used as components in microwave radars for the control of high-frequency signals. With aluminum oxide, yttrium forms yttrium–aluminum garnet, which is used in solid-state lasers. The use of yttrium oxide in its pure form, although less extensive, is described in greater detail in Section 1.8.4.

1.8.1 Crystal Structure and Properties of Single Crystal Yttrium Oxide

Yttrium oxide is a white refractory crystalline solid with a melting point of 2410 °C and a density of 5.03 g cm^{-3}. Yttrium oxide has a cubic, body-centered crystal lattice with the lattice parameter $a = 10.604$ Å, and with yttrium atoms octahedrally coordinated with six oxygen atoms. The cubic yttria undergoes a polymorphic phase cubic → hexagonal transition at about 2350 °C [292]. The H-type with space group D_{6h}^4 is a high-temperature hexagonal phase.

1.8.2 Natural Sources and Preparation

Yttrium is found together with other rare earth oxides in monazite sands [(Ce, La, etc.) PO_4] and in bastnasite [(Ce, La, etc.)(CO_3)F] (see Section 1.7.1). Yttrium is extracted together with other rare earth elements in a concentrated solution of sodium hydroxide at 140–150 °C; after cooling, the hydroxides of the rare earth elements are separated by filtration. Alternatively, bastnasite may be calcined to drive off CO_2 and fluorine, and then leached with hydrochloric acid to dissolve the trivalent rare earth elements. The rare earth hydroxides and chlorides obtained in this way are further processed to produce individual rare earth metal compounds

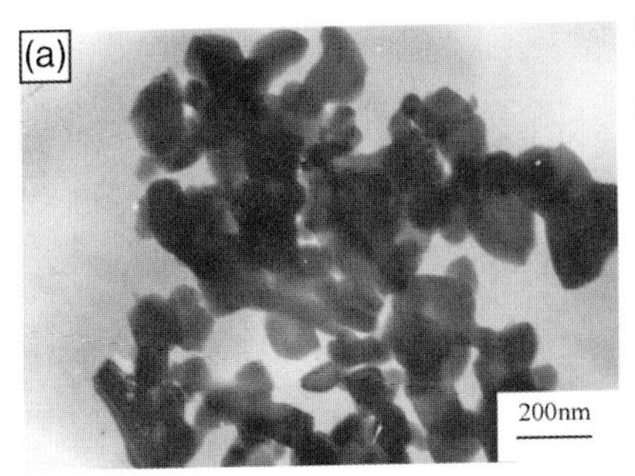

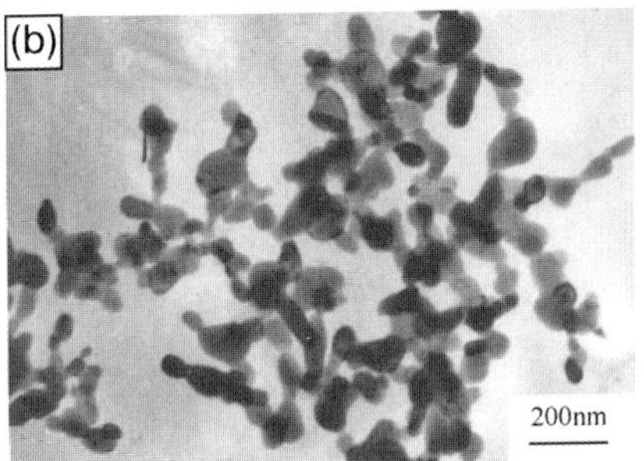

Figure 1.25 Morphology of yttria powder prepared by the precipitation of yttrium nitrate (a) without the addition of SO_4^{2-} and (b) with the addition of SO_4^{2-} [299].

such as fluorides, nitrates, carbonates, oxides, and pure metals for a variety of applications.

Ultra-fine-grained highly reactive yttria powders, suitable especially for the preparation of transparent ceramics, are prepared by various methods including combustion synthesis [293], precipitation [294, 295], hydrothermal synthesis [296], electrospray pyrolysis [297], and sol–gel [298]. In order to improve the dispersion and sinterability of yttria powders, seed crystals are often added [296]. A significant refinement of yttria powders prepared by precipitation from solution may be achieved by the addition of sulfate ions to the reaction mixture [299] (Figure 1.25).

Polycrystalline yttria ceramics are usually prepared by the conventional sintering of yttrium oxide powders [300, 301]. Due to its high refractoriness, yttrium oxide is rather difficult to sinter at ambient pressure. Even in a vacuum, the onset of densification of standard micrometer-sized powders occurs between 1400 and 1650 °C, while the achieved densities are low and the grain growth rapid (~65%) [302]. The use of sintering additives such as La_2O_3 [303], LiF [304] and ThO_2 [305], or the application of a high temperature (>2000 °C) [306] and pressure [307], are required for complete densification when fabricating fully dense or transparent yttria. The complete elimination of any residual porosity at lower temperatures, and refinement of the microstructure, can be achieved by hot isostatic pressing, which results in a material with a greater hardness, flexural strength, and thermal shock resistance (Figure 1.26) [308]. A fully dense nanocrystalline yttria with a grain size of 60 nm was recently prepared using a pressureless, two-stage sintering of a yttria nanopowder without additives at a temperature of about 1000 °C [309].

1.8.3 Properties

The use of yttria ceramics as standalone materials for structural applications is rather limited by their poor mechanical properties. For example, the commercially available 99.9% polycrystalline yttria Ceralloy® (Ceradyne Inc.) has a flexural strength of 99 MPa, a hardness of 5.85 GPa, and a fracture toughness of 1.4 MPa $m^{1/2}$. The hardness and fracture toughness of polycrystalline yttria are virtually grain-size

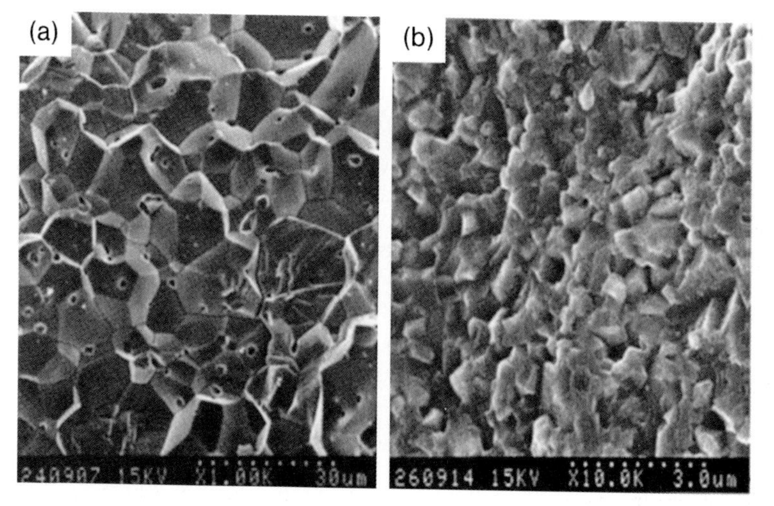

Figure 1.26 The microstructure of dense yttrium oxide prepared by (a) vacuum sintering and (b) hot isostatic pressing [308].

independent, with measured values being almost constant among all materials with grain sizes between 0.2 and 214 μm [310]. The fracture properties of yttria ceramics are influenced by the stoichiometry; a maximum fracture toughness of 3.5 $MPa \cdot m^{1/2}$ is achieved at stoichiometric composition, but this decreases to 2.3 $MPa \cdot m^{1/2}$ in an oxygen-deficient material. Such a change is reversible, however, with the initial value being restored after oxidation, and the hardness experiencing only a minor variation [311].

The Poisson ratio of yttria is 0.31, and the elastic modulus 170 GPa. The material has a relatively high thermal expansion coefficient ($9.1 \times 10^{-6} K^{-1}$), which is comparable to that of polycrystalline alumina, and a high thermal conductivity (14 $W mK^{-1}$), which is twice that of the other solid-state laser host material, $Y_3Al_5O_{12}$ (YAG). Yttria has also a very high electrical resistivity (10^{14} $\Omega \cdot cm$), is refractory by nature, is highly chemically and thermally stable, and is optically clear over a broad spectral region.

1.8.4 Applications

Three decades ago, Greskovich and Chernoch created a new field of application of yttria by producing a laser yttria host-based ceramic material [312]. Yttria is not only used as a solid-state laser material as a laser host crystal for trivalent lanthanide activators, such as Yb^{3+} and Nd^{3+}, but also shows significant potential for luminous pipes in high-intensity discharge lamps and heat-resistive windows.

As an optical ceramic, yttrium oxide transmits well in the IR range, from 1 to 8 μm wavelength. This high IR transmission, together with a good resistance to erosion and thermal shock, means that yttrium oxide would serve as an ideal material for protection domes for IR sensors [313].

Due to its high thermodynamic stability, yttria is used for the protective coatings of high-temperature containers or structural components intended for use in reactive environments (especially molten metals, such as titanium or uranium). In effect, the reactive material is in contact with the thin economical layer of Y_2O_3, while the container substrate is prevented from coming into direct contact with the molten metal. As a result, a double protection is achieved – that of the container against corrosion, and of the melt against container-originating impurities. Due to its high refractoriness and low neutron absorption, yttria may also be used as a structural material for nuclear reactors.

References

1 Cowley, J.D. and Lee, W.E. (2005) Materials Science and Technology; A Comprehensive Treatment; Chapter 2, Oxide Ceramics, in *Structure and Properties of Ceramics*, vol. 11 (ed. M. Swain), Wiley-VCH Verlag GmbH, Weinheim, Germany, pp. 87–91.

2 Borsella, E., Botti, S., Giorgi, R., Martelli, S., Turtù, S., and Zappa, G. (1993) *Appl. Phys. Lett.*, **63**, 1345.

3 Kumar, P.M., Borse, P., Rohatgi, V.K., Bhoraskar, S.V., Singh, P., and Sastry, M. (1994) *Mater. Chem. Phys.*, **36**, 354.

4 Hirayama, T. (1987) *J. Am. Ceram. Soc.*, **70**, C122.

5 Kumar, P.M., Balasubramanian, C., Sali, N.D., Bhoraskar, S.V., Rohatgi, V.K., and Badrinarayanan, S. (1999) *Mater. Sci. Eng.*, **B63**, 215.

6 Chen, Y., Glumac, N., Kear, B.H., and Skandan, G. (1997) *NanoStruct. Mater.*, **9**, 101.

7 Skandan, G., Chen, Y.-J., Glumac, N., and Kear, B.H. (1999) *NanoStruct. Mater.*, **11**, 149.

8 Almyasheva, O.V., Korytkova, E.N., Maslov, A.V., and Gusarov, V.V. (2005) *Inorg. Mater.*, **41**, 540.

9 Yoldas, B.E. (1975) *Am. Ceram. Soc. Bull.*, **54**, 289.

10 Ji, L., Lin, J., Tan, K.L., and Zeng, H.C. (2000) *Chem. Mater.*, **12**, 931.

11 Rao, G.V.R., Venkadesan, S., and Saraswati, V. (1989) *J. Non-Cryst. Solids*, **111**, 103.

12 Sharma, P.K., Varadan, V.V., and Varadan, V.K. (2003) *J. Eur. Ceram. Soc.*, **23**, 659.

13 Dhara, S. (2005) *J. Am. Ceram. Soc.*, **88**, 2003.

14 Karagedov, G.R. and Lyakhov, N.Z. (1999) *NanoStruct. Mater.*, **11**, 559.

15 Wang, Y., Suryanarayana, C., and An, L. (2005) *J. Am. Ceram. Soc.*, **88**, 780.

16 Bae, S.I. and Baik, S. (1993) *J. Mater. Sci.*, **28**, 4197.

17 Hansen, S.C. and Philips, D.S. (1983) *Philos. Mag. A*, **47**, 209.

18 Harmer, M. (1984) *Adv. Ceram.*, **10**, 679.

19 Bae, S.I. and Baik, S. (1994) *J. Am. Ceram. Soc.*, **77**, 2499.

20 Krell, A. and Blank, P. (1993) *J. Am. Ceram. Soc.*, **78**, 1118.

21 Morinaga, K., Torikai, T., Nakagawa, K., and Fujino, S. (2000) *Acta Mater.*, **48**, 4735.

22 Krell, A. and Blank, P. (1996) *J. Eur. Ceram. Soc.*, **16**, 1189.

23 O, Y.T., Koo, J., Hong, K.J., Park, J.S., and Shin, D.C. (2004) *Mater. Sci. Eng.*, **A374**, 191.

24 Krell, A. and Klaffke, D. (1996) *J. Am. Ceram. Soc.*, **79**, 1139.

25 Krell, A., Blank, P., Berger, L.M., and Richter, V. (1999) *Am. Ceram. Soc. Bull.*, **77**, 65.

26 Wei, G.C. (2005) *J. Phys. D: Appl. Phys.*, **38**, 3057.

27 Krell, A., Blank, P., Ma, H., Hutzler, T., van Bruggen, M.P.B., and Apetz, R. (2003) *J. Am. Ceram. Soc.*, **86**, 12.

28 Jones, C.D., Rioux, J.B., Lochem, J.W., Bates, H.E., Zanella, S.A., Pluen, V., and Mandelartz, M. (2006) *Am. Ceram. Soc. Bull.*, **3**, 24.

29 Apetz, R. and van Bruggen, M.P.B. (2003) *J. Am. Ceram. Soc.*, **86**, 480.

30 Lim, L.C., Wong, P.M., and Jan, M.A. (2000) *Acta Mater.*, **48**, 2263.
31 Nivot, C., Valdivieso, F., and Goeuriot, P. (2006) *J. Eur. Ceram. Soc.*, **26**, 9.
32 Cheng, J., Agrawal, D., Zhang, Y., and Roy, R. (2002) *Mater. Lett.*, **56**, 587.
33 Shen, Z., Peng, H., Liu, J., and Nygren, M. (2004) *J. Eur. Ceram. Soc.*, **24**, 3447.
34 Echeberria, J., Tarazona, J., He, J.Y., Butler, T., and Castro, F. (2002) *J. Eur. Ceram. Soc.*, **22**, 1801.
35 Krell, A. and Klimke, J. (2006) *J. Am. Ceram. Soc.*, **89**, 1985.
36 Raj, R. (1981) *J. Am. Ceram. Soc.*, **64**, 245.
37 Clarke, D.R. (1987) *J. Am. Ceram. Soc.*, **70**, 15.
38 Song, H. and Coble, R.L. (1990) *J. Am. Ceram. Soc.*, **73**, 2086.
39 Bae, S.I. and Baik, S. (1993) *J. Am. Ceram. Soc.*, **76**, 1065.
40 Song, H. and Coble, R.L. (1990) *J. Am. Ceram. Soc.*, **73**, 2077.
41 Chi, M., Gu, H., Wang, X., and Wang, P. (2003) *J. Am. Ceram. Soc.*, **86**, 1953.
42 McLaren, I., Cannon, R.W., Gülgün, A.M., Voytovych, R., Popescu-Pogrion, N., Scheu, C., Täfner, U., and Rühle, M. (2003) *J. Am. Ceram. Soc.*, **86**, 650.
43 Cho, S.-J., Lee, J.-Ch., Lee, H.-L., Sim, S.-M., and Yanagisawa, M. (2003) *J. Eur. Ceram. Soc.*, **23**, 2281.
44 Lee, S.-H., Kim, D.-Y., and Hwang, N.M. (2002) *J. Eur. Ceram. Soc.*, **22**, 317.
45 Blonski, S. and Garofalini, S. (1996) *J. Phys. Chem.*, **100**, 2201.
46 Zhang, S. and Garofalini, S.H. (2005) *J. Am. Ceram. Soc.*, **88**, 202.
47 Blonski, S. and Garofalini, S.H. (1997) *J. Am. Ceram. Soc.*, **80**, 1997.
48 Zhang, S. and Garofalini, S.H. (2005) *J. Am. Ceram. Soc.*, **88** (1), 202–209.
49 Švančárek, P., Galusek, D., Loughran, F., Brown, A., Brydson, R., Atkinson, A., and Riley, F. (2006) *Acta Mater.*, **54**, 4853.
50 Cahn, R.W., Haasen, P. and Kramer, E.J. (eds) (1995) Materials Science and Technology; A Comprehensive Treatment; Chapter 2, Oxide Ceramics, in *Structure and Properties of Ceramics*, vol. 11 (vol. ed. M. Swain), VCH Weinheim, Germany, p. 99.
51 Kaysser, W.A., Sprissler, M., Handwerker, C.A., and Blendell, J.E. (1987) *J. Am. Ceram. Soc.*, **70**, 339.
52 Kim, M.-J. and Yoon, D.-Y. (2003) *J. Am. Ceram. Soc.*, **86**, 630.
53 Kim, B.-K., Hong, S.-H., Lee, S.-H., Kim, D.-Y., and Hwang, D.M. (2003) *J. Am. Ceram. Soc.*, **86**, 634.
54 Goswami, A.P., Roy, S., Mitra, M.K., and Das, G.C. (2001) *J. Am. Ceram. Soc.*, **84**, 1620.
55 Raj, R. and Lange, F.F. (1981) *Acta Metall.*, **29**, 1993.
56 Powell-Dogan, C.A. and Heuer, A.H. (1990) *J. Am. Ceram. Soc.*, **73**, 3677.
57 Powell-Dogan, C.A. and Heuer, A.H. (1990) *J. Am. Ceram. Soc.*, **73**, 3684.
58 Powell-Dogan, C.A., Heuer, A.H., Ready, M.J., and Merriam, K. (1991) *J. Am. Ceram. Soc.*, **74**, 646.
59 Wu, Y.-Q., Zhang, Y.-F., Pezzotti, G., and Guo, J.-K. (2002) *J. Eur. Ceram. Soc.*, **22**, 159.
60 Becher, P.F. (1991) *J. Am. Ceram. Soc*, **74**, 255.
61 Blendell, J.E. and Coble, R.L. (1982) *J. Am. Ceram. Soc.*, **65**, 174.
62 Zimmermann, A., Fuller, E.W. Jr, and Rödel, J. (1999) *J. Am. Ceram. Soc.*, **82**, 3155.
63 Vedula, V.R., Glass, S.J., Saylor, D.M., Rohrer, G.S., Carter, W.C., Langer, S.A., and Fuller, E.R. Jr (2001) *J. Am. Ceram. Soc.*, **84**, 2947.
64 Litton, D.A. and Garofalini, S.H. (1999) *J. Mater. Res.*, **14**, 1418.
65 Mukhopaday, A.K. and Mai, Y.-W. (1993) *Wear*, **162–164**, 258.
66 He, C., Wang, Y.S., Wallace, J.S., and Hsu, S.M. (1993) *Wear*, **162–164**, 314.
67 Rice, R.W. and Speronello, B.K. (1976) *J. Am. Ceram. Soc.*, **59**, 330.
68 Wiederhorn, S.M. and Hockey, B.J. (1983) *J. Mater. Sci.*, **18**, 766.
69 Marshall, D.B., Lawn, B.R., and Cook, R.F. (1987) *J. Am. Ceram. Soc.*, **70**, C–139.
70 Gee, M.G. and Almond, E.A. (1990) *J. Mater. Sci.*, **25**, 296.
71 Miranda-Martinez, M., Davidge, R.W., and Riley, F.L. (1994) *Wear*, **172**, 41.
72 Him, S.S., Kato, H., Hokkirigawa, K., and Abe, H. (1986) *J. Tribol.*, **108**, 522.

73 Galusek, D., Brydson, R., Twigg, P.C., Riley, F.L., Atkinson, A., and Zhang, Y.-H. (2001) *J. Am. Ceram. Soc.*, **84**, 1767.

74 Kalin, M., Jahanmir, S., and Dražič, G. (2005) *J. Am. Ceram. Soc.*, **88**, 346.

75 Insley, R.H. (1990) Electrical Properties of Alumina Ceramics, in *Alumina Chemicals: Science and Technology Handbook* (ed. L.D. Hart), The American Ceramic Society, Westerville, OH, pp. 293–297.

76 Morrell, R. (1985) in *Handbook of Properties of Technical and Engineering Ceramics, Part I, An Introduction for Engineers and Designers*, National Physical Laboratory, Her Majesty's Stationary Office, London, pp. 162–167.

77 Krupka, J., Geyer, R.G., Kuhn, M., and Hinken, J.H. (1994) *IEEE T. Microw. Theory*, **42**, 1886.

78 Matsumoto, H. (1996) Proceedings, 1996 International Accelerator School in Japan, p. 1.

79 Woode, R.A., Ivanov, E.N., Tobar, M.E., and Blair, D.G. (1994) *Electronics Lett.*, **30**, 2120.

80 Vila, R., Gonzalez, M., Mola, J., and Ibarra, A. (1998) *J. Nucl. Mater.*, **253**, 141.

81 Alford, N.M. and Penn, S.J. (1996) *J. Appl. Phys.*, **80**, 5895.

82 Itatani, K., Tsujimoto, T., and Kishimoto, A. (2006) *J. Eur. Ceram. Soc.*, **26**, 639.

83 Gardner, T.J. and Messing, G.L. (1984) *Am. Ceram. Soc. Bull.*, **63**, 1498.

84 Jost, H., Braun, M., and Carius, C. (1997) *Solid State Ionics*, **101–103**, 221.

85 Itatani, K., Kishioka, A., and Kinoshita, M. (1993) *Gypsum & Lime*, **244**, 4.

86 Misawa, T., Moriyoshi, Y., Yajima, Y., Takenouchi, S., and Ikegami, T. (1999) *J. Ceram. Soc. Jpn*, **107**, 343.

87 Smethurst, E. and Budworth, D.W. (1972) *Trans. Br. Ceram. Soc.*, **71**, 45.

88 Köbel, S., Schneider, D., Schüler, C.Chr., and Gauckler, L.J. (2004) *J. Eur. Ceram. Soc.*, **24**, 2267.

89 Itatani, K., Yasuda, R., Howell, F.S., and Kishioka, A. (1997) *J. Mater. Sci.*, **32**, 2977.

90 Ehre, D., Gutmanas, E.Y., and Chaim, R. (2005) *J. Eur. Cerum. Soc.*, **25**, 3579.

91 Chaim, R., Shen, Z., and Nygren, M. (2004) *J. Mater. Res.*, **19**, 2527.

92 Wraback, M., Shen, H., Liang, S., Gorla, C.R., and Lu, Y. (1999) *Appl. Phys. Lett.*, **74**, 507.

93 Lee, J.-M., Kim, K.-K., Park, S.-J., and Choi, W.-K. (2001) *Appl. Phys. Lett.*, **78**, 2842.

94 Pearton, S.J., Bortin, D.P., Ip, K., Heo, Y.W., and Steiner, T. (2003) *Superlattice. Microst.*, **34**, 3.

95 (1957) *Zinc Oxide Rediscovered*, The New Jersey Zinc Company, New York.

96 Shiojiri, M. and Kaito, C. (1981) *J. Cryst. Growth*, **52**, 173.

97 Tanigaki, T., Komára, S., Tamura, N., and Kaito, C. (2002) *Jpn. J. Appl. Phys.*, **41**, 5529.

98 Kutty, T.R.N. and Padmini, P. (1992) *Mater. Res. Bull.*, **27**, 945.

99 Auffredic, J.P., Boultif, A., Langford, J.I., and Louer, D. (1995) *J. Am. Ceram. Soc.*, **78**, 323.

100 Komarneni, S., Bruno, M., and Mariani, E. (2000) *Mater. Res. Bull.*, **35**, 1843.

101 Tsuchida, T. and Kitajima, S. (1990) *Chem. Lett.*, **10**, 1769.

102 Dhage, S.R., Pasricha, R., and Ravi, V. (2005) *Mater. Lett.*, **59**, 779.

103 Hwang, C.-C. and Wu, T.-Y. (2004) *Mater. Sci. Eng. B*, **111**, 197.

104 Han, J., Mantas, P., and Senos, A.M.R. (2002) *J. Eur. Ceram. Soc.*, **22**, 49.

105 Florescu, D., Mourok, L.G., Pollack, F.H., Look, D.C., Cantwell, G., and Li, X. (2002) *J. Appl. Phys.*, **91**, 890.

106 Singh, L.K. and Mohan, H. (1975) *Indian J. Pure Appl. Phys.*, **13**, 486.

107 Vojta, A. and Clarke, D.R. (1997) *J. Am. Ceram. Soc.*, **80**, 2086.

108 Matsuoka, M. (1971) *Jpn. J. Appl. Phys.*, **10**, 736.

109 Pike, G.E., Kurtz, S.R., Gourley, P.L., Philipp, H.R., and Levinson, L.M. (1985) *J. Appl. Phys.*, **57**, 5521.

110 Gupta, T.K., Straub, W.D., Ramanachalam, M.S., Schaffer, J.P., and Rohatgi, A. (1989) *J. Appl. Phys.*, **66**, 6132.

111 Ulrich, S. and Hoffmann, B. (1984) *J. Appl. Phys.*, **57**, 5372.

112 Santos, J.D., Longo, E., Leite, E.R., and Varela, J.A. (1998) *J. Mater. Res.*, **13**, 1152.

113 Clarke, D.R. (1999) *J. Am. Ceram. Soc.*, **82**, 485.

114 Metz, R., Delalu, H., Vignalou, J.R., Achard, N., and Elkhatib, M. (2000) *Mater. Chem. Phys.*, **63**, 157.
115 Caballero, A.C., Valle, F.J., and Martín Rubí, J.A. (2001) *X-Ray Spectrom.*, **30**, 273.
116 Inada, M. (1980) *Jpn. J. Appl. Phys.*, **19**, 409.
117 Olsson, E., Dunlop, G., and Oslerlund, R. (1993) *J. Am. Ceram. Soc.*, **76**, 65.
118 Mergen, A. and Lee, W.E. (1997) *J. Eur. Ceram. Soc.*, **17**, 1049.
119 Huda, D., El Baradic, M.A., Hashmi, M.S.J., and Puyane, R. (1998) *J. Mater. Sci.*, **33**, 271.
120 Bernik, S. and Daneu, N. (2001) *J. Eur. Ceram. Soc.*, **21**, 1879.
121 Mukae, K. (1987) *Am. Ceram. Soc. Bull.*, **66**, 1329.
122 Alles, A.B. and Burdick, V.L. (1991) *J. Appl. Phys.*, **70**, 6883.
123 Lee, Y.-S., Liao, K.-S., and Tseng, T.-Y. (1996) *J. Am. Ceram. Soc.*, **79**, 2379.
124 Chun, S.-Y., Shinozaki, K., and Mizutani, N. (1999) *J. Am. Ceram. Soc.*, **82**, 3065.
125 Nahm, C.-W. (2001) *J. Eur. Ceram. Soc.*, **21**, 545.
126 Nahm, C.-W., Shin, B.-C., Park, J.-A., and Yoo, D.-H. (2006) *Mater. Lett.*, **60**, 164.
127 Kuo, C.T., Chen, C.S., and Lin, I.N. (1998) *J. Am. Ceram. Soc.*, **81**, 2949.
128 Hng, H.H. and Knowles, K.M. (2000) *J. Am. Ceram. Soc.*, **83**, 2455.
129 Pfeiffer, H. and Knowles, K.M. (2004) *J. Eur. Ceram. Soc.*, **24**, 1199.
130 Saleh, H.I. and El-Meliegy, E.M. (2004) *Br. Ceram. Trans.*, **103**, 268.
131 Balzer, B., Hagemeister, M., Kocher, P., and Gauckler, L.J. (2004) *J. Am. Ceram. Soc.*, **87**, 1932.
132 Qiu, D.J., Yu, P., and Wu, H.Z. (2005) *Solid State Commun.*, **134**, 735.
133 Meagher, E.P. and Lager, G.A. (1979) *Can. Mineral.*, **17**, 77.
134 Zhang, H.Z. and Banfield, J.F. (1998) *J. Mater. Chem.*, **8**, 2073.
135 Zhang, H.Z. and Banfield, J.F. (2000) *J. Phys. Chem.*, **B104**, 3481.
136 Hwu, Y., Yao, Y.D., Cheng, N.F., Tung, C.Y., and Lin, H.M. (1997) *Nanostruct. Mater.*, **9**, 355.
137 Ye, X.S., Sha, J., Jiao, Z.K., and Zhang, L.D. (1997) *Nanostruct. Mater.*, **8**, 919.
138 Arnal, P., Corriu, R.J., Leclerq, D., Mutin, P.H., and Viox, A. (1996) *J. Mater. Chem.*, **6**, 1925.
139 Cheng, H., Ma, J., Zhao, Z., and Qi, L. (1995) *Chem. Mater.*, **7**, 663.
140 Kim, S.J., Park, S.D., and Jeong, Y.H. (1999) *J. Am. Ceram. Soc.*, **8**, 927.
141 Dhage, S.R., Pasricha, R., and Ravi, V. (2003) *Mater. Res. Bull.*, **38**, 1623.
142 Seo, D.S., Lee, J.K., and Kim, H. (2001) *J. Cryst. Growth*, **233**, 298.
143 Wilson, G.J., Will, Frost, R.L., and Montgomery, S.A. (2002) *J. Mater. Chem*, **12**, 1787.
144 Murugan, A.V., Samuel, V., and Ravi, V. (2006) *Mater. Lett.*, **60**, 479.
145 Deki, S., Aoi, Y., Hiroi, O., and Kajinami, A. (1996) *Chem. Lett.*, **6**, 433.
146 Shimizu, K., Imai, H., Hirashima, H., and Tsukuma, K. (1999) *Thin Solid Films*, **351**, 220.
147 Funakoshi, K. and Nonami, T. (2006) *J. Am. Ceram. Soc.*, **89**, 2381.
148 Wang, W., Gu, B., Liang, L., Hamilton, W.A., and Wesolowski, D.J. (2004) *J. Phys. Chem.*, **B108**, 14789.
149 Baldassari, S., Komarneni, S., Mariani, E., and Villa, C. (2005) *Mater. Res. Bull.*, **40**, 2014.
150 Pottier, A., Chaneac, C., Tronc, E., Mazerolles, L., and Jolivet, J. (2001) *J. Mater. Chem.*, **11**, 1116.
151 Kominami, H., Kohno, M., and Kera, Y. (2000) *J. Mater. Chem.*, **10**, 1151.
152 Lee, B.I., Wang, X., Bhave, R., and Hu, M. (2006) *Mater. Lett.*, **60**, 1179.
153 Templeton, A., Wang, X., Penn, S.J., Webb, S.J., Cohen, L.F., and Alford, N.McN. (2000) *J. Am. Ceram. Soc.*, **83**, 95.
154 Meng, F. (2005) *Mater. Sci. Eng.*, **B117**, 77.
155 Li, M. and Chen, Y. (1996) *Sens. Actuators*, **B32**, 83.
156 Yamada, Y., Seno, Y., Masuoka, Y., Nakamura, T., and Yamashita, K. (2000) *Sens. Actuators*, **B66**, 164.
157 Kohl, D. (1991) *Sens. Actuators*, **B1**, 158.
158 Zhang, R.-B. (2005) *Mater. Res. Bull.*, **40**, 1584.
159 Yan, M.F. and Rhodes, W.W. (1982) *Appl. Phys. Lett.*, **40**, 536.
160 Gaikwad, A.B., Navale, S.C., and Ravi, V. (2005) *Mater. Sci. Eng.*, **B123**, 50.

161 Yang, S.L. and Wu, J.M. (1993) *J. Am. Ceram. Soc.*, **76**, 145.

162 Hsiang, H.-I. and Wang, S.-S. (2006) *Mater. Sci. Eng.*, **B128**, 25.

163 Kawai, T. and Sakata, T. (1980) *Nature*, **286**, 474.

164 Hengerer, R., Kavan, L., Krtil, P., and Grätzel, M. (2000) *J. Electrochem. Soc.*, **147**, 1467.

165 SDK News Release (2005) http://www.sdk.co.jp/contents_e/news05/05-02-04.htm.

166 Liu, X., Zhao, X., Fu, R.K.Y., Ho, J.P.Y., Ding, C., and Chu, P.K. (2005) *Biomaterials*, **26**, 6143.

167 Kasuga, T., Kondo, H., and Nogami, M. (2002) *J. Cryst. Growth*, **23**, 5235.

168 Keshmiri, M. and Troczynski, T. (2003) *J. Non-Cryst. Solids*, **324**, 289.

169 Uchida, M., Kim, H.M., Kokubo, T., Fujibayashi, S., and Nakamura, T. (2003) *J. Biomed. Mater. Res.*, **64A**, 164.

170 Kurzweg, H., Heimann, R.B., Troczynski, T., and Wayman, M.L. (1998) *Biomaterials*, **19**, 1507.

171 Gulami, S.T., Helfinstine, J.D., and Davis, A.D. (1980) *J. Am. Ceram. Soc.*, **59**, 211.

172 Garvie, R.C., Urbani, C., Kennedy, D.R., and McNeuer, J.C. (1984) *J. Mater. Sci.*, **19**, 3224.

173 Garvie, R.C. (1970) Oxides of Rare Earths, Titanium, Zirconium, Hafnium, Niobium, and Tantalum, in *High Temperature Oxides, Part II* (ed. A.M. Alper), Academic Press, p. 117.

174 Heuer, A.H. and Lenz, L.K. (1982) *J. Am. Ceram. Soc.*, **65**, 192.

175 Farnworth, F., Jones, S.L., and McAlpine, I. (1980) in *Speciality Inorganic Chemicals* (ed. R. Thompson), Royal Society of Chemistry, London, p. 249.

176 Zhang, Y.L., Jin, X.J., Rong, Y.H., Hsu, T.Y., Jiang, D.Y., and Shi, J.L. (2006) *Acta Mater.*, **54**, 1289.

177 Garvie, R.C. and Goss, M.F. (1986) *J. Mater. Sci.*, **21**, 1253.

178 Garvie, R.C., Hannink, R.H., and Pascoe, R.T. (1975) *Nature*, **258**, 703.

179 Evans, A.G. and Heuer, A.H. (1981) *J. Am. Ceram. Soc.*, **63**, 241.

180 McMeeking, R.M. and Evans, A.G. (1982) *J. Am. Ceram. Soc.*, **65**, 242.

181 Ma, Y. and Kisi, E.H. (2005) *J. Am. Ceram. Soc.*, **88**, 2510.

182 Garvie, R.C., Hughes, R.R., and Pascoe, R.T. (1978) *Materials Science Research*, vol. 11 (eds H. Palmour, R.F. Davis, and T.M. Hare), Plenum Press, New York.

183 Marder, J.M., Mitchell, T.E., and Heuer, A.H. (1983) *Acta Metall.*, **31**, 387.

184 Hellman, J.R. and Stubican, V.S. (1983) *J. Am. Ceram. Soc.*, **66**, 260.

185 Hellman, J.R. and Stubican, V.S. (1983) *J. Am. Ceram. Soc.*, **66**, 265.

186 Cowley, J.D. and Lee, W.E. (1995) Materials Science and Technology; A Comprehensive Treatment; Chapter 2, Oxide Ceramics, in *Structure and Properties of Ceramics*, vol. 11 (vol. ed. M. Swain), VCH Weinheim, Germany, p. 99.

187 Tsukuma, K. and Shimada, M. (1985) *J. Mater. Sci.*, **20**, 1178.

188 Levi, C.G. (2004) *Curr. Opin. Solid State Mater. Sci.*, **8**, 77.

189 Bondioli, F., Manfredini, C.T., Ferrari, A.M., Caracoche, M.C., Rivas, P.C., and Rodriguez, A.M. (2005) *J. Am. Ceram. Soc.*, **88**, 633.

190 Rieth, P.H., Reed, J.S., and Naumann, A.W. (1976) *Am. Ceram. Soc. Bull.*, **55**, 717.

191 Gupta, T.K., Bechtold, J.H., Kuznickie, R.C., Cadoff, L.H., and Rossing, B.R. (1977) *J. Mater. Sci.*, **12**, 2421.

192 Yoshizawa, Y. and Sakuma, T. (1992) *Acta Metall. Mater.*, **40**, 2943.

193 Hwang, C.M.J. and Chen, I.W. (1990) *J. Am. Ceram. Soc.*, **73**, 1626.

194 Hendrix, W., Kuypers, S., Vangrunderbeek, J., Luyten, J., and Vandermeulen, W. (1993) *Mater. Sci. Eng. A*, **168**, 45.

195 Seidensticker, J.R. and Mayo, M.J. (1994) *Scr. Metall. Mater.*, **31**, 1749.

196 Motohashi, Y., Akutsu, S., Kakita, S., and Maruyama, Y. (2004) *Mater. Sci. Forum*, **447–448**, 305.

197 Sharif, A.A. and Mecartney, M.L. (2004) *J. Eur. Ceram. Soc.*, **24**, 2041.

198 Dillon, R.P., Sosa, S.S., and Mecartney, M.L. (2004) *Scripta Mater.*, **50**, 1441.

199 Tsurui, K. and Sakuma, T. (1996) *Scripta Mater.*, **34**, 443.

200 Chokshi, A.H. (1993) *Mater. Sci. Eng.*, **A116**, 119.

201 Jimenez-Melendo, M., Dominguez-Rodriguez, A., and Bravo-Leon, A.J. (1998) *J. Am. Ceram. Soc.*, **81**, 2761.
202 Bartolome, J.F., Montero, I., Diaz, M., Lopez-Esteban, S., Moya, J.S., Deville, S., Gremillard, L., Chevalier, J., and Fantozzi, G. (2004) *J. Am. Ceram. Soc.*, **87**, 2282.
203 Gremillard, L., Chevalier, J., Epicier, T., and Fantozzi, G. (2002) *J. Am. Ceram. Soc.*, **85**, 401.
204 Sato, T. and Shimada, M. (1986) *J. Am. Ceram. Soc.*, **68**, 356.
205 Nettleship, I. and Stevens, R. (1987) *Int. J. High. Tech. Ceram.*, **3**, 1.
206 Fang, P., Gu, H., Wang, P., Van Landuyt, J., Vleugels, J., and Van der Biest, O. (2005) *J. Am. Ceram. Soc.*, **88**, 1929.
207 Heussner, K.-H. and Claussen, N. (1989) *J. Am. Ceram. Soc.*, **72**, 1044.
208 Theunissen, G.S.A.M., Winnubst, A.J.A., and Burggraaf, A.J. (1992) *J. Eur. Ceram. Soc.*, **9**, 251.
209 Hwang, S.L. and Chen, I.-W. (1990) *J. Am. Ceram. Soc.*, **73**, 3269.
210 Xu, T., Wang, P., Fang, P., Kann, Y., Chen, L., Vleugels, J., Van der Biest, O., and Van Landuyt, J. (2005) *J. Eur. Ceram. Soc.*, **25**, 3437.
211 Reyes-Morel, P.E. and Chen, I.W. (1988) *J. Am. Ceram. Soc.*, **71**, 343.
212 Jiang, B.H., Tu, J.B., Hsu, T.Y., and Qi, X. (1992) *Mater. Res. Soc. Symp. Proc.*, **246**, 213.
213 Zhang, Y.L., Jin, X.J., Hsu, T.Y., Zhang, Y.F., and Shi, J.L. (2002) *Mater. Sci. Forum*, **394–395**, 573.
214 Claussen, N. (1976) *J. Am. Ceram. Soc.*, **59**, 49.
215 Zhu, D. and Miller, R.A. (2004) *Int. J. Appl. Ceram. Technol.*, **1**, 86.
216 Eaton, H.E. and Novak, R.C. (1987) *Surf. Coat. Technol.*, **32**, 227.
217 Zhu, D. and Miller, R.A. (2000) *J. Thermal Spray Technol.*, **9**, 175.
218 Grot, A.S. and Martyn, J.K. (1981) *Am. Ceram. Soc. Bull.*, **60**, 807.
219 Choi, S.R., Zhu, D., and Miller, R.A. (2004) *Int. J. Appl. Ceram. Technol.*, **1**, 330.
220 Gu, S., Lu, T.J., Haas, D.D., and Wadley, H.N.G. (2001) *Acta Mater.*, **49**, 2539.
221 Lu, T.J., Levi, C.G., Wadley, H.N.G., and Evans, A.G. (2001) *J. Am. Ceram. Soc.*, **84**, 2937.
222 Nicholls, J.R., Lawson, K.J., Johnstone, A., and Rickerby, D.S. (2002) *Surf. Coat. Technol.*, **383–391**, 151.
223 Clarke, D.R. and Levi, C.G. (2003) *Annu. Rev. Mater. Res.*, **33**, 383.
224 Nicholls, J.R. (2003) *Mater. Res. Soc. Bull.*, **28**, 659.
225 Harmswoth, P.D. and Stevens, R. (1992) *J. Mater. Sci.*, **24**, 611.
226 Vaßen, R., Träger, F., and Stöver, D. (2004) *Int. J. Appl. Ceram. Technol.*, **1**, 351.
227 Etsell, T.H. and Flengas, S.N. (1970) *Chem. Rev.*, **70**, 339.
228 Kharton, V.V., Naumovich, E.N., and Vecher, A.A. (1999) *J. Solid State Electr.*, **3**, 61.
229 Kharton, V.V., Marques, F.M.B., and Atkinson, A. (2004) *Solid State Ionics*, **174**, 135.
230 Inaba, H. and Tagawa, H. (1996) *Solid State Ionics*, **83**, 1.
231 Mogensen, M., Sammes, N.M., and Tompsett, G.A. (2000) *Solid State Ionics*, **129**, 63.
232 Badwal, S.P.S. (1995) *Solid State Ionics*, **76**, 67.
233 Godickemeier, M., Michel, B., Orliukas, A., Bohac, P., Sasaki, K., Gauckler, L., Heinrich, H., Schwander, P., Kostorz, G., Hofmann, H., and Frei, O. (1994) *J. Mater. Res.*, **9**, 1228.
234 Appel, C.C. and Bonanos, N. (1999) *J. Eur. Ceram. Soc.*, **19**, 847.
235 Badwal, S.P.S. and Rajendran, S. (1994) *Solid State Ionics*, **70–71**, 83.
236 Drennan, J. and Auchterlonie, G. (2000) *Solid State Ionics*, **134**, 75.
237 Lee, J.-H., Mori, T., Li, J.-G., Ikegami, T., Komatsu, M., and Haneda, H. (2000) *J. Electrochem. Soc.*, **147**, 2822.
238 Yuzaki, A. and Kishimoto, A. (1999) *Solid State Ionics*, **116**, 47.
239 Leach, C.A. (1987) *Mater. Sci. Technol.*, **3**, 321.
240 Shiratori, Y., Tietz, F., Buchkremer, H.P., and Stover, D. (2003) *Solid State Ionics*, **164**, 27.
241 Mondal, P., Klein, A., Jägermann, W., and Hahn, H. (1999) *Solid State Ionics*, **118**, 331.
242 Kosacki, I., Suzuki, T., Petrovsky, V., and Anderson, H.U. (2000) *Solid State Ionics*, **136–137**, 1225.

243 Yamamoto, O., Arachi, Y., Sakai, H., Takeda, Y., Imanishi, N., Mizutani, Y., Kawai, M., and Nakamura, Y. (1998) *Ionics*, **4**, 403.

244 Haile, M. (2003) *Acta Mater.*, **51**, 5981.

245 Weppner, W. (1992) *Solid State Ionics*, **52**, 15.

246 de Souza, S., Visco, S.J., and de Jonghe, L.C. (1997) *Solid State Ionics*, **98**, 57.

247 Murray, E.P., Tsai, T., and Barnett, S.A. (1999) *Nature*, **400**, 649.

248 Tietz, F., Fischer, W., Hauber, T., and Mariotto, G. (1997) *Solid State Ionics*, **100**, 289.

249 Hirano, M., Watanabe, S., Kato, E., Mizutani, Y., Kawai, M., and Nakamura, Y. (1998) *Solid State Ionics*, **111**, 161.

250 Hirano, M., Inagaki, M., Mizutani, Y., Nomura, K., Kawai, M., and Nakamura, Y. (2000) *Solid State Ionics*, **133**, 1.

251 Ishii, T., Iwata, T., and Tajima, Y. (1992) *Solid State Ionics*, **57**, 153.

252 Ruh, R., Garrett, H.J., Domagala, R.F., and Patel, V.A. (1977) *J. Am. Ceram. Soc.*, **60**, 399.

253 Badwal, S.P.S. (2001) *Solid State Ionics*, **143**, 39.

254 Singh, P. and Minh, N.Q. (2004) *Int. J. Appl. Ceram. Technol.*, **1**, 5.

255 Hench, L.L. (1998) *J. Am. Ceram. Soc.*, **81**, 1705.

256 Lawson, S. (1995) *J. Eur. Ceram. Soc.*, **15**, 85.

257 Chevalier, J. (2006) *Biomaterials*, **27**, 535.

258 Deville, S., Guenin, G., and Chevalier, J. (2004) *Acta Mater.*, **52**, 5697.

259 Chevalier, J., Deville, S., Münch, E., Jullian, R., and Lair, F. (2004) *Biomaterials*, **25**, 5539.

260 Munoz-Saldana, J., Balmori-Ramirez, H., Jaramillo-Vigueras, D., Iga, T., and Schneider, G.A. (2003) *J. Mater. Res.*, **18**, 2415.

261 Gremillard, L., Chevalier, J., Fantozzi, G., and Epicier, T. (2002) *J. Am. Ceram. Soc.*, **85**, 401.

262 Chi, R., Li, Z., Peng, C., Gao, H., and Xu, Z. (2006) *Metall. Mater. Trans. B*, **37B**, 155.

263 Hirano, M., Fukuda, Y., Iwata, H., Hotta, Y., and Inagaki, M. (2000) *J. Am. Ceram. Soc.*, **83**, 1287.

264 Hirano, M. and Kato, E. (1999) *J. Am. Ceram. Soc.*, **82**, 786.

265 Rossignol, S., Gerard, F., and Duprez, D. (1999) *J. Mater. Chem.*, **9**, 1615.

266 Gu, Y., Li, G., Meng, G., and Peng, D. (2000) *Mater. Res. Bull.*, **35**, 297.

267 Tsuzuki, T. and McCormick, P.G. (2001) *J. Am. Ceram. Soc.*, **84**, 1453.

268 Bai, W., Choy, K.L., Stelzer, N.H.J., and Schoonman, J. (1999) *Solid State Ionics*, **116**, 225.

269 Lee, J.-S., Lee, J.-S., and Choi, S.-C. (2005) *Mater. Lett.*, **59**, 395.

270 Mashina, S., Shaizero, O., and Meriani, S. (1992) *J. Eur. Ceram. Soc.*, **9**, 127.

271 Zhang, T.S., Ma, J., Kong, L.B., Hing, P., and Kilner, J.A. (2004) *Solid State Ionics*, **167**, 191.

272 Ishida, T., Iguchi, F., Sato, K., Hashida, T., and Yugami, H. (2005) *Solid State Ionics*, **176**, 2417.

273 Hayashi, H., Sagawa, R., Inaba, H., and Kawamura, K. (2000) *Solid State Ionics*, **131**, 281.

274 Tuller, H.L. (2000) *Solid State Ionics*, **131**, 143.

275 Kim, S. and Maier, J. (2002) *J. Electrochem. Soc.*, **149**, J73.

276 Sigmund, W.M., Bell, N.S., and Bergstrom, L. (2000) *J. Am. Ceram. Soc.*, **83**, 1557.

277 Steele, B.C.H. (2000) *Solid State Ionics*, **129**, 95.

278 Goedickemeier, M. and Gauckler, L.J. (1998) *J. Electrochem. Soc.*, **145**, 414.

279 Gerhard-Anderson, R. and Nowick, A.S. (1981) *Solid State Ionics*, **5**, 547.

280 Kilner, J.A. (1983) *Solid State Ionics*, **8**, 201.

281 Atkinson, A. and Ramos, T. (2000) *Solid State Ionics*, **129**, 259.

282 Steele, B.C.H. (2001) *J. Mater. Sci.*, **36**, 1053.

283 Inaba, H. and Tagawa, H. (1996) *Solid State Ionics*, **83**, 1.

284 Mogensen, M., Sammes, N.M., and Tompsett, G.A. (2000) *Solid State Ionics*, **129**, 63.

285 Kleinvogel, C. and Gauckler, L.J. (2000) *Solid State Ionics*, **135**, 567.

286 Zhang, T.S., Kong, L.B., Zeng, Z.Q., Huang, H.T., Hing, P., Xia, Z.T., and Kilner, J.A. (2003) *J. Solid State Electr.*, **7**, 348.

287 Zhang, T.S., Ma, J., Kong, L.B., Hing, P., Leng, Y.J., Chan, S.H., and Kilner, J.A. (2003) *J. Power Sources*, **124**, 26.

288 Lewis, G.S., Atkinson, A., Steele, B.C.H., and Drennan, J. (2002) *Solid State Ionics*, **152–153**, 567.

289 Li, J.-G., Ikegami, T., and Mori, T. (2004) *Acta Mater.*, **52**, 2221.

290 Sayle, T., Parker, S.C., and Catlow, C.R.A. (1992) *J. Chem. Soc. Chem. Commun.*, **14**, 977.

291 Logothetidis, S., Patsalas, P., and Charitidis, C. (2003) *Mater. Sci. Eng. C*, **23**, 803.

292 Lopato, L.M., Shevchenko, A.V., Kushchevskii, A.E., and Tresvyatskii, S.G. (1974) *Izv. Akad. Nauk SSSR, Neorg. Mater.*, **10**, 1481.

293 Kim, W.J., Park, J.Y., Oh, S.J., Kim, Y.S., Hong, G.W., and Kuk, I.H. (1999) *J. Mater. Sci. Lett.*, **18**, 411.

294 Sordelet, D.J. and Akinc, M. (1988) *J. Am. Ceram. Soc.*, **71**, 1148.

295 Ikegami, T., Mori, T., Yajima, Y., Takenouchi, S., Misawa, T., and Moriyoshi, Y. (1999) *J. Ceram. Soc. Jpn, Int. Ed.*, **107**, 297.

296 Sharma, P.K., Jilavi, M.H., Nab, R., and Schmidt, H. (1998) *J. Mater. Sci. Lett.*, **17**, 823.

297 Rulison, A.J. and Flagan, R.C. (1994) *J. Am. Ceram. Soc.*, **77**, 3244.

298 Subramanian, R., Shankar, P., Kavithaa, S., Ramakrishnan, S.S., Angelo, P.C., and Venkataraman, H. (2001) *Mater. Lett.*, **48**, 342.

299 Wen, L., Sun, X.D., Xiu, Z., Chen, S.W., and Tsai, C.T. (2004) *J. Eur. Ceram. Soc.*, **24**, 2681.

300 Mogilenskii, V.I. and Polonskii, A.J. (1975) *Inorg. Mater.*, **11**, 218.

301 Richardson, K. and Akinc, M. (1988) *Ceram. Int.*, **14**, 101.

302 Dutta, S.K. and Gazda, G.A. (1975) US Patent No. 3 878 280.

303 Rhodes, W.H. (1981) *J. Am. Ceram. Soc.*, **64**, 13.

304 Lefever, R.A. and Matsho, J. (1967) *Mater. Res. Bull.*, **2**, 865.

305 Reetz, T., Haase, I., Ullmann, H., and Lang, H.J. (1989) *Solid State Ionics*, **36**, 193.

306 Tsukuda, Y. and Muta, A. (1976) *J. Ceram. Soc. Jpn, Int. Ed.*, **84**, 585.

307 Yeheskel, O. and Tevet, O. (1999) *J. Am. Ceram. Soc.*, **82**, 136.

308 Desmaison-Brut, M., Montintin, J., Valin, F., and Boncoeur, M. (1995) *J. Am. Ceram. Soc.*, **78**, 716.

309 Chen, I.-W. and Wang, X.-H. (2000) *Nature*, **404**, 168.

310 Tani, T., Miyamoto, Y., Koizumi, M., and Shimada, M. (1986) *Ceramurgia Int.*, **12**, 33.

311 Fantozzi, G., Orange, G., Liang, K., and Gautier, M. (1989) *J. Am. Ceram. Soc.*, **72**, 1562.

312 Greskovich, C. and Chernoch, J.P. (1973) *J. Appl. Phys.*, **44**, 4599.

313 Harris, D.C. (1999) *Materials for Infrared Windows and Domes*, SPIE-The International Society for Optical Engineering, Washington, USA.

2
Nitrides

Pavol Šajgalík, Zoltán Lenčéš, and Miroslav Hnatko

2.1
Silicon Nitride

2.1.1
Introduction

The proposal that silicon nitride-based ceramics represent one of the most promising synthetic materials for engineering applications has been made on a regular basis with regards to the creation of silicon nitride-based ceramics/composites. In fact, the last three decades of the twentieth century were full of expectation for the breakthrough of silicon nitride in the industry, such that silicon nitride-based ceramic composites were developed with bending strengths greater than 1 GPa, a fracture toughness of approximately 10 MPa $m^{1/2}$, and a Weibull modulus which ranged from 20 to 50. Moreover, between 70% and 80% of the room temperature strength was retained up to 1450 °C, and the creep rate remained consistently in the region of $10^{-10} s^{-1}$ at 1450 °C [1]. Yet, despite these clear achievements, the planned-for massive breakthrough into industry did not happen, no generally applicable silicon nitride with excellent properties was developed, and the original concept of replacing metals with silicon nitride was ultimately rejected. As a consequence, the direction of thought of many research groups was redirected towards developing "special" compositions for "special" applications.

Today, silicon nitride-based ceramics are considered to be living "on the edge": either they will be incorporated into a wide range of new technologies, with materials being tailored to particular applications, or they will be rejected, mainly because of their high price and lack of properties that renders them uncompetitive with metals. A pessimistic forecast might state that, as the huge amounts of money and intellectual capacity invested into the development of new ceramic materials has not brought about the desired industrial expectations, then future research in this area should be greatly restricted. On the other hand, a more optimistic forecast might predict that advanced ceramic materials should not only serve as "better metals," but also find their way into applications for completely new technologies – in much the

Ceramics Science and Technology Volume 2: Properties. Edited by Ralf Riedel and I-Wei Chen

ISBN: 978-3-527-31156-9

same way that plastics did during the 1950s. Alternatively, they might become economically competitive so as to become substitute materials within the classical domain of metals and alloys.

Silicon nitride-based ceramics are polycrystalline materials that consist at least of two phases, the silicon nitride grains and the grain boundary. In this chapter, an attempt will be made to describe, separately, the properties of both the grain and grain boundary phases. The effect of combining these phases, with regards to the mechanical properties of the materials, will also be discussed.

Silicon nitride exists in four structural phases: an amorphous α-Si_3N_4, and three crystalline phases, namely trigonal α-Si_3N_4, hexagonal β-Si_3N_4, and a high-pressure cubic γ-Si_3N_4 phase (or c-Si_3N_4).

2.1.2
Amorphous Silicon Nitride

Amorphous silicon nitride (α-Si_3N_4) is used extensively in the microelectronics industry in the form of thin films, which are prepared by both normal and reactive sputtering, chemical vapor deposition (CVD), ion-beam-assisted deposition, and ion plating [2, 3].

α-Si_3N_4 incorporates the basic SiN_4 and NSi_3 structural units of the crystalline forms, but lacks the longer-range ordering [4]. Its stoichiometry tends to depart from the 3 : 4 ratio, and for these nonstoichiometric SiN_x alloys the general formula a-SiN_x is used ($0 < x < 1.6$) [3, 5]. Depending on the type of precursor and the temperature of synthesis, these materials often incorporate other elements such as H, O, and C. a-SiN_x films deposited at a temperature of $>700\,°C$ tend to be reasonably close to stoichiometric Si_3N_4 ($x = 1.33$), although compositions with $1.20 < x < 1.46$ are the most important with regards to practical use. The a-Si_3N_4 powders are mostly produced from $Si(NH)_2$, and from $SiCl_4/NH_3$ in a N_2 plasma or by laser-induced CVD from SiH_4/NH_3 [6–8], although others have successfully synthetized high-purity amorphous Si_3N_4 nanopowder via gas-phase reactions of SiH_4 and NH_4 [9, 10].

The density of amorphous Si_3N_4 (up to $\sim 3\,g\,cm^{-3}$) is less than that of its crystalline forms. Rather, the most important properties of a-Si_3N_4 are its low permeability towards sodium, oxygen, H_2O, and hydrogen (because of the rigid covalent structure), its high electrical resistivity ($10^{12}\,\Omega\,m$), hardness, and its good resistance to chemical attack [2, 11, 12].

On heating, a-Si_3N_4 is converted to crystalline α-Si_3N_4; the time–temperature domain boundary for complete crystallization runs from about 10 min at 1400 °C to 1 h at 1250 °C, and to 4 h at 1100 °C [13].

2.1.3
Silicon Nitride Single Crystals: Structure

2.1.3.1 α- and β- Si_3N_4

Three crystallographic modifications of silicon nitride have been identified, namely α, β, and γ. In general, it is accepted that α-Si_3N_4 is the low-temperature

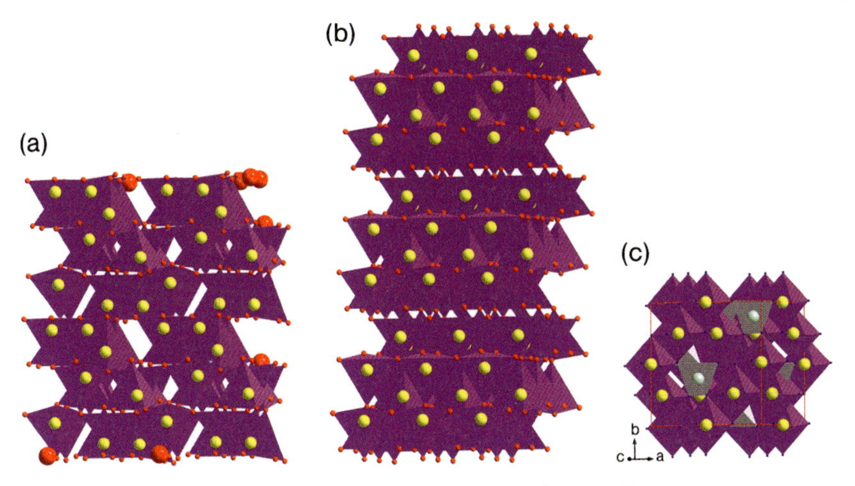

Figure 2.1 Crystal structures of: (a) α-Si_3N_4; (b) β-Si_3N_4; and (c) γ-Si_3N_4.

(metastable) modification, and β-Si_3N_4 the stable high-temperature modification at normal pressure [14]. The space group of α-Si_3N_4 is $P31c$ of trigonal symmetry [15–17], with each unit cell containing four Si_3N_4 units. The structure of β-Si_3N_4 has hexagonal symmetry with space group $P6_3$ (noncentrosymetric) [14, 18]. Earlier studies involved space group $P6_3/m$ (centrosymetric) [19–21]. The idealized crystal structures of both modifications (see Figure 2.1) consist of SiN_4 tetrahedra, in which the Si atom lies at the center of a tetrahedron, with four N atoms at each corner. The SiN_4 tetrahedra are joined by sharing corners in such a way that each N is common to three tetrahedra; thus each Si atom has four N atoms as nearest neighbors. α-Si_3N_4 consists essentially of alternate basal layers of β-Si_3N_4 and a mirror image of β, so that the c spacing of α-Si_3N_4 is almost twice that of β-Si_3N_4. The β-modification has open channels along the c-axis, while the α modification creates caves as a consequence of the mirror image of the A sequence. In the idealized forms, both α- and β-Si_3N_4 are geometrically related by $a_\alpha = a_\beta$, while $c_\alpha = 2c_\beta$. The unit cell dimensions of silicon nitride structures, as determined from the polycrystalline materials using X-ray diffraction (XRD), are listed in the Table 2.1.

2.1.3.2 γ-Si_3N_4

The synthesis of a novel γ-Si_3N_4 phase with a cubic spinel structure (see Figure 2.1c) was carried out under high pressure (15 GPa) and at temperatures above 1920 °C in a laser-heated diamond cell [22]. Today, many different processing techniques for γ-Si_3N_4 synthesis are available, the most common being the diamond anvil cell (DAC) synthesis (G. Serghiou, *et al.*, unpublished results), the multianvil pressure apparatus (MAP) synthesis [23], and the shock synthesis [24].

Cubic Si_3N_4 has the space group $Fd3m$ ($Z = 8$), and contains both tetrahedrally and octahedrally coordinated silicon atoms with a 1 : 2 ratio. Thus, the cubic phase may be

Table 2.1 Properties of Si_3N_4 measured on the polycrystalline ceramics [5, 68].

Property	α-Si_3N_4	β-Si_3N_4	γ-Si_3N_4
Lattice parameter a [nm]	0.7748	0.7608	0.7734
Lattice parameter c [nm]	0.5617	0.2911	
Density [g cm^{-3}]	3.192	3.198	4.012
Elastic modulus E_x [GPa]	340	310	
Volume modulus B_o [GPa]	248	256	300
Poisson number ν	0.3	0.27	
Hardness HV [GPa]	21	16	30
Strength σ [MPa]		800–1200	
Fracture toughness K_{IC} [MPa · m$^{1/2}$]	2–3	6–8	
Thermal expansion (0–1000 °C) [10^{-6} K^{-1}]	3.64	3.39	
Thermal conductivity λ [W m^{-1} K^{-1}]	110–150	90–180	
Resistance [Ω · cm]		10^{13}	
Refractive index n_o	2.03	2.02	

better described as $Si[Si_2N_4]$ rather than Si_3N_4. The unit cell measured by synchrotron radiation was $a = 0.77339 \pm 0.00001$ nm (Table 2.1), and the calculated density of the cubic phase was 3.75 ± 0.02 g cm^{-3} [22]. The bond length of Si(1)−N in the cubic phase was measured as about 1.786 Å, which was actually larger than the average Si−N bond length in both α-Si_3N_4 (1.738 Å) and β-Si_3N_4 (1.730 Å) [25]. The number of Si−N and Si−Si bonds in the cubic phase was seen to be higher than in both α- and β-Si_3N_4. γ-Si_3N_4 was shown to be stable at high temperatures and at pressures of between 15 and 30 GPa; at ambient pressure in air it persisted metastably to a temperature of at least 427 K.

Recently, a new δ-Si_3N_4 phase with a hexagonal structure and lattice parameters $a = 0.737$ nm and $c = 0.536$ nm, space group P62*c* [26, 27], was discovered. The phase was formed in the $Si_{3.0}B_{1.1}C_{5.3}N_{3.0}$ ceramics during annealing at 1800 °C for 3 h under a nitrogen pressure of 10 MPa. An abnormally large grain, which was found in the precursor-derived ceramics after crystallization, has been characterized by means of transmission electron microscopy (TEM) and electron energy-loss spectroscopy (EELS). The results of the latter analysis showed that this phase consisted of only silicon and nitrogen, with no other elements being detected [28].

In Nature, an additional mineral, *nierite* (Si_3N_4), was found in the perchloric acid-resistant residues of primitive meteorites (ordinary chondrites [29] and enstatite chondrites [30, 31]) [32]. This mineral was seen to occur in lathe-shaped grains that were 2 μm × 0.4 μm in size. An inspection of the mineral by using selected area electron diffraction (SAED) showed the d-spacings of nierite to be comparable to those of synthetic α-Si_3N_4. The unit-cell parameters of nierite, as calculated from measured d-spacings, were $a = 0.774 \pm 0.002$ nm, $c = 0.561 \pm 0.002$ nm, while the calculated density was 3.11 g cm^{-3} [33].

2.1.4
Silicon Nitride Single Crystals: Mechanical Properties

2.1.4.1 α-Si_3N_4

In the pioneering studies in this area, which were conducted by Remanis *et al.* [34], the Vickers indentation resulted in a hardness of 34 ± 5 GPa at room temperature and 22 ± 2.5 GPa at 1400 °C for the 1100 plane, and 25 ± 5 GPa at room temperature and 13 ± 2.5 GPa at 1400 °C, respectively. From these hardness measurements, the authors [34] were able to calculate also the values of fracture toughness, by using an average value of Young's modulus of 300 GPa for silicon nitride. The average fracture toughness (without regards to crystallographic orientation) ranged from 2.0 ± 0.51 MPa $m^{1/2}$ at room temperature to 1.5 ± 0.16 MPa $m^{1/2}$ at 1200 °C. The hardness of the prismatic plane was 40–55% higher compared to the basal level over the entire range of temperatures. The anisotropy of the thermal expansion coefficients for both planes of α-Si_3N_4 was also determined as $3.61 \times 10^{-6}\,K^{-1}$ and $3.70 \times 10^{-6}\,K^{-1}$, respectively, over the temperature range of 0 to 1000 °C [35].

Ogata *et al.* [36] performed an *ab initio* calculation of the "ideal" tensile strength of α-Si_3N_4 single crystal, and reported a value of ~51 GPa. From the calculated shear modulus, the same authors determined the hardness of 23 GPa for α-Si_3N_4, and found this calculated value to be in good agreement with the measured room temperature hardness of the basal plane of α-Si_3N_4 single crystal.

The density of the α-Si_3N_4 single crystal, as determined from crystallographic data, was 3.176 g cm^{-3} [37], which was comparable to an experimentally measured value of 3.19 ± 0.01 g cm^{-3} [38]. A list of available data relating to the densities of polycrystalline α-Si_3N_4 materials is provided in Ref. [39], with the majority of calculated and measured data being obtained from the interval between experimental and theoretical values, as noted above.

A molecular dynamics (MD) calculation of the ideal thermal conductivity of α-Si_3N_4 single crystal provided values of 105 and 225 W $m^{-1}\,K^{-1}$, respectively [40].

2.1.4.2 β-Si_3N_4

The hardness measurements of β-Si_3N_4 single crystals, as determined by several groups [41–43], all demonstrated the hardness anisotropy of the basal and prismatic planes. Similar to the α-Si_3N_4, the hardness of the prismatic plane was 60% higher than that of the basal plane [42], although the reported values were shown to depend on the applied load. In the case of the prismatic plane, the values ranged from 45 ± 6 GPa for a 1 g load to 22 ± 2 GPa for a 50 g load. In the case of the basal plane, the values ranged from 22 ± 2 GPa to 15 ± 1 GPa for the same applied loads.

The thermal expansion coefficient of β-Si_3N_4 depends on its crystallographic orientation, with values of $3.23 \times 10^{-6}\,K^{-1}$ being determined for the basal plane and $3.72 \times 10^{-6}\,K^{-1}$ for the prismatic plane, within a temperature range from 0 to 1000 °C [35]. Similar values were determined by Bruls *et al.* [44].

The "ideal" tensile strength of β-Si_3N_4 single crystal obtained by *ab initio* calculations [45, 46] was 57 GPa, with an estimated hardness of 20.4 GPa. The latter value was in good agreement with an experimentally determined basal plane hardness of β-Si_3N_4 grain, measured at a load of 50 g.

The density of β-Si_3N_4, as calculated from the structural data of the single crystal, was 3.214 g cm^{-3} [14], while that for polycrystalline β-Si_3N_4 was 3.202 g cm^{-3} [44].

The experimental thermal diffusivities were 0.32 cm^2 s^{-1} along the *a*-axis, and 0.84 cm^2 s^{-1} along the *c*-axis, which corresponded to thermal conductivities of 69 and 180 W m^{-1} K^{-1}, respectively [47]. The MD calculation of the ideal thermal conductivity of β-Si_3N_4 single crystal provided values of 170 and 450 W m^{-1} K^{-1}, respectively [40].

2.1.4.3 **γ-Si_3N_4**

The hardness of γ-Si_3N_4, as measured with the Vickers diamond indenter, was 35.3 GPa [48], whereas hardness estimated via *ab initio* calculations of the shear modulus of cubic Si_3N_4 were much higher, at ~47 GPa [49]. Zerr *et al.* [50] reported that the oxygen content of γ-Si_3N_4 was detrimental for the hardness of this modification, and showed that a low-oxygen-containing γ-Si_3N_4 (4 wt% O) had a hardness of 36 ± 8 GPa, and similar to that measured by others [48]. In comparison, a high-oxygen-containing γ-Si_3N_4 (14 wt% O) had a hardness of 30 ± 9 GPa.

Until now, no details have been reported of indentation fracture toughness (K_{IC}-ICL) measurements on pure γ-Si_3N_4. However, in the case of the spinel-sialon γ-Si_2AlON_3, mean K_{IC}-ICL values of ~4.6 MPa m$^{1/2}$ were reported at indentation loads of 0.5 and 1.0 kg [51].

The thermal expansion coefficient of γ-Si_3N_4, as determined by XRD measurements, can be described by the equation $\alpha = 2.716 \times 10^{-6}\,K^{-1} \pm 0.004 \times 10^{-6}\,T$. The linear thermal expansion coefficient at room temperature for the γ-Si_3N_4 was $\alpha = 3.89 \times 10^{-6}\,K^{-1}$ [52].

Values obtained by conducting experimental studies of the lattice parameters in the temperature range from 300 to 1000 K for γ-Si_3N_4, were increased from 3 to $6.5 \times 10^{-6}\,K^{-1}$ [53], which was significantly higher compared to the results for β-Si_3N_4 [44], which increased from 1 to $3 \times 10^{-6}\,K^{-1}$ within the same temperature range. A recently published theoretical study supported these results [54], with calculations resulting in $\alpha_{th}(c\text{-}Si_3N_4) = 3.3 \times 10^{-6}$ to $6.5 \times 10^{-6}\,K^{-1}$ for $T =$ 300–1000 K. The "ideal" tensile strength of γ-Si_3N_4 single crystals obtained by *ab initio* calculation [49] was ~45 GPa, and the "ideal" shear strength ~49 GPa. The density, as calculated from the structural data, was 3.93 ± 0.12 g cm^{-3} [22].

2.1.5
Silicon Nitride-Based Materials

The typical microstructure of dense silicon nitride-based ceramics is shown in Figure 2.2, where the dark phase in the form of elongated needle-like grains is silicon nitride. The individual silicon nitride grains in the microstructure are usually *single crystals* of one from two/three possible crystallographic modifications. The bright continuous phases are the grain boundary phases; these are normally amorphous or partly crystallized within the multigrain junctions, and are a conse-

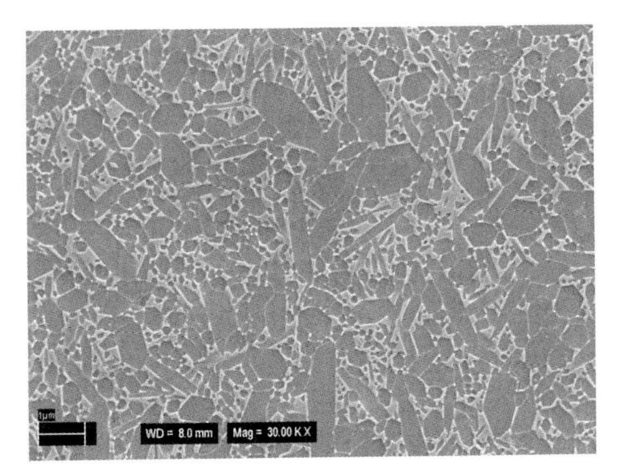

Figure 2.2 The microstructure of polycrystalline silicon nitride.

quence of the way in which the starting silicon nitride powder is densified (this is most often by a liquid-phase sintering). It is for this reason why silicon nitride-based materials should be considered to be a composite consisting of a major silicon nitride phase and a minor (usually silicate) phase and/or a silicate-rich oxynitride phase. As noted above, these phases may be either amorphous or crystallized, depending on the process used to create the silicon nitride materials. In order to understand the mechanical behavior of Si_3N_4-based ceramics, it would be beneficial to understand some of the properties of the silicate and/or silicate-rich oxynitride phases.

2.1.6 Oxynitride Glasses

The grain boundary/triple junction composition basically results from the oxide sintering additives that promote the densification process. These, with silica present as the impurity in all Si_3N_4 starting powders, create the liquid which enhances densification at the sintering temperature. The sintering temperature is usually controlled by the composition of the sintering additives, with different compositions having different eutectic temperatures, although normally such temperatures will range from 1750 to 2000 °C. Normally, the silicate phases will be formed, the chemical composition of which is dictated by the chemical composition of the sintering additives. Attempts have been made to simulate the crystallization behavior of these phases, for example in the sintering additive system SiO_2–Al_2O_3-Y_2O_3 [55]. In this case, a volume contraction and cavity formation were observed after crystallization of the glasses, together with an increased coefficient of thermal expansion with increasing glass density. In particular, these facts must be taken into account when designing silicon nitride-based ceramics for particular applications.

However, in the majority of cases the triple-point junctions and grain boundary composition is much more complicated, the reason being that in order to hinder the decomposition of Si_3N_4 during sintering at high temperatures, a nitrogen atmosphere must be used during densification. In addition to Si_3N_4, a proportion of the nitrogen will be dissolved in the liquid; consequently, not only the grain boundary films but also the triple junctions will be basically composed of a silicate-rich oxynitride glass.

2.1.6.1 Properties of Oxynitride Glasses

Recently, extensive studies of oxynitride glasses were conducted by various groups [56–67], the aim being to achieve a better understanding of the composite behavior of liquid-phase sintered Si_3N_4.

Among these groups, Lofaj *et al.* [56] showed that the density, hardness, thermal expansion coefficients and glass transition temperatures (T_g) of RE–Si–Mg–O–N glasses could be independently modified via rare earth additives and/or the nitrogen content. The addition of lanthanides (20–24 equiv% N) led to an increase of the hardness of these glasses by 11.4–14.4%, of the thermal expansion coefficient by 13%, and of the transition temperature by 31–35%. The lowest density (3.096 g cm^{-3}) was obtained for 13.4 Sc_2O_3–40.1 MgO–39.7 SiO_2–6.8 Si_3N_4 glass with 20 equiv% N, while the highest density (4.694 g cm^{-3}) was obtained for 13.7 Lu_2O_3–41.2 MgO–36.6 SiO_2–8.4 Si_3N_4 glass with 24 equiv% N. The other glasses, with Y_2O_3, La_2O_3, Nd_2O_3, Sm_2O_3, Gd_2O_3, and Yb_2O_3 had densities ranging between these two limits.

The lowest hardness of 7.93 GPa, as measured by the Vicker's hardness at a load of 0.5 kg, was obtained for 12.5 La_2O_3–37.5 MgO–46.7 SiO_2–3.2 Si_3N_4 glass with 10 equiv% N, while the highest value (of 10.53 GPa) was obtained for 13.7 Lu_2O_3–41.2 MgO–36.6 SiO_2–8.4 Si_3N_4 glass with 24 equiv% N.

The highest value of the thermal expansion coefficient, at $7.7 \times 10^{-6}\,K^{-1}$, was measured for 12.5 La_2O_3–37.5 MgO–46.7 SiO_2–3.2 (6.8) Si_3N_4 glass with 10 (20) equiv% N, and the lowest value, at $6 \times 10^{-6}\,K^{-1}$, was measured for 13.7 Lu_2O_3–41.2 MgO–36.6 SiO_2–8.4 Si_3N_4 glass with 24 equiv% N.

The lowest value of the T_g, at 809 °C, was measured for 12.5 La_2O_3–37.5 MgO–46.7 SiO_2–3.2 Si_3N_4 glass with 10 equiv% N, and the highest T_g was 882 °C for 13.7 Lu_2O_3–41.2 MgO–36.6 SiO_2–8.4 Si_3N_4 glass with 24 equiv% N.

Interestingly, the Lu_2O_3–MgO–SiO_2–N_2 glass with 24 equiv% N had the highest density, hardness and T_g, but the lowest thermal expansion coefficient.

The other glasses with Sc_2O_3, Y_2O_3, Nd_2O_3, Sm_2O_3, Gd_2O_3, and Yb_2O_3 additives each demonstrated physical properties which lay in between the given maximum and minimum limits.

The rare earth-doped oxynitride glasses demonstrated higher densities and thermal expansion coefficients, but a substantially lower hardness, compared to the α- and β-Si_3N_4 single crystals.

2.1.7 Polycrystalline Si_3N_4

Silicon nitride-based polycrystalline ceramic materials have the potential for structural applications at both room temperature and elevated temperatures, despite containing silicate/silicate-rich oxynitride grain boundary phases. These are generally much weaker compared to the silicon nitride single crystal grains of which the microstructure consists. The mechanical properties of some polycrystalline silicon nitrides are listed in Table 2.1. An existing synergistic microstructural effect leads to the properties of polycrystalline silicon nitride-based materials that are comparable, in some cases, to the silicon nitride single crystals. Si_3N_4-based ceramics have a relatively high temperature of dissociation (1840 °C at 0.1 MPa N_2) and a low density of 3.2 g cm^{-3}, with the exception of γ-Si_3N_4 (4.0 g cm^{-3}), and this may prove advantageous in comparison to refractory metals. The thermal shock resistance is good due to a low coefficient of thermal expansion (3.1 to 3.6×10^{-6} K^{-1}, from room temperature up to 1000 °C) and a higher thermal conductivity (60 to 110 W m^{-1} K^{-1} at 1000 °C). Whereas, α-Si_3N_4 has a greater hardness and good wear resistance, those ceramics composed of a β-modification have a higher strength and fracture toughness, owing to the presence of elongated β-Si_3N_4 grains.

Today, the predominant area of application for silicon nitride (Si_3N_4)-based ceramics is in engineering as structural components for room temperature/ high-temperature applications. As noted above, the room/high-temperature properties of Si_3N_4-based ceramics depend heavily on the oxide additives used for the densification, with the resultant amorphous phase present at the multigrain junctions and on the grain boundaries (1–2 nm thick intergranular glassy films) of Si_3N_4 grains dictating the room-temperature fracture behavior of ceramics, and also influencing the properties at elevated temperatures. The softening of glass will lead to a deformation of the ceramic body, mainly by grain-boundary sliding. In the past, many studies have been directed towards detecting the best combination of sintering additives with respect to the mechanical behavior of silicon nitride-based ceramics.

Recently, several theoretical and experimental studies were carried out to identify the role of sintering additives (mainly lanthanoides) with respect to the microstructure formation, and consequently, to the mechanical behavior of these materials [69–76]. The studies took into account also the properties of the silica-rich oxynitride glasses (as described above). The measured creep resistance of the current-generation Y- and Yb-doped silicon nitride ceramics was shown to be sufficient for long-term operation (>10 000 h) at temperatures up to 1325 °C. However, the development of Si_3N_4-based materials with a greater creep resistance is desirable, an example being that of small engines operating at higher temperatures. The general conclusion of the above-mentioned studies was that, for the high-temperature application of Si_3N_4 the most suitable additive would be Lu_2O_3. Notably, both an increase in lifetime and a decrease in creep rate (by two orders of magnitude) were observed for Lu_2O_3-containing Si_3N_4 compared to materials containing other additives.

2.1.8 Lu-Doped Si_3N_4 Ceramics

The creep resistance of silicon nitride-containing Lu-doped additives was shown to be three to five orders of magnitude greater than that of earlier grades containing Y- and/ or Yb-additives [74, 75]. This material has the potential for prolonged operation at temperatures up to 1470 °C, with the stress exponent, n, and the activation energy, Q, for creep being 5.3 ± 2.0 and 757 ± 117 kJ mol^{-1}, respectively. In particular, for 12.51 wt% Lu_2O_3-doped Si_3N_4, the flexural strengths at 1500 and 1600 °C were the highest yet to be reported, with the value at 1500 °C reaching ~700 MPa.

Others [76] first reduced the fraction of Lu_2O_3 in Si_3N_4 starting powder to 1.2 mol% and 4.8 mol%, after which the samples were hot-pressed at 1950 °C for 1 h under 20 MPa in 1 MPa N_2. Samples containing 1.2 mol% Lu_2O_3 had the highest strength (ca. 500 MPa at 1500 °C), as well as an excellent oxidation resistance (the weight gain during oxidation at 1500 °C for 1000 h was only 4 g m^{-2}). Moreover, the creep lifetime of this sample at 1500 °C, under a tensile stress of 137 MPa, exceeded 1678.5 h.

The specimen processed by sinter-forging with an anisotropic microstructure exhibited superior high-temperature strengths of 696 and 570 MPa at 1500 and 1600 °C, respectively, [77]. The starting Si_3N_4 powder contained 8 wt% Lu_2O_3 and 2 wt % SiO_2.

2.1.9
SiAlON Ceramics

The SiAlONs represent a group of ceramic materials that contain Si, Al, O, N (and other metals), with a very wide composition and crystal structure [78], and which could be generally described as solid solutions of Si_3N_4, SiO_2, Al_2O_3, and AlN. The basic SiAlON structures of α-SiAlON and β-SiAlON are isostructural with α-Si_3N_4 and β-Si_3N_4 respectively, and also O′-SiAlON which is based on the structure of Si_2N_2O. The positions of these SiAlON phases in the R–Si–Al–O N system are well illustrated by a so-called Jänecke prism, as shown in Figure 2.3 (here, the rare-earth element R = Y).

2.1.9.1 α- and β-SiAlONs

α-SiAlON is a solid solution based on the α-Si_3N_4 structure with the general formula $M_xSi_{12-(m-n)}Al_{(m+n)}O_nN_{(16-n)}$, where x is determined by the valence of stabilizing

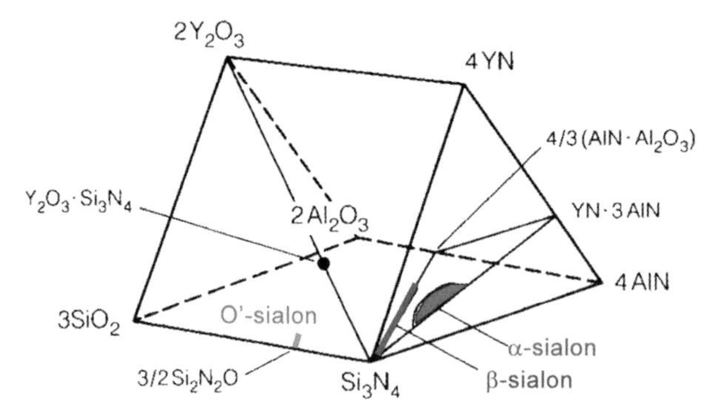

Figure 2.3 The Jänecke prism of Si_3N_4-4(AlN)-4(YN)-2(Y_2O_3)-2(Al_2O_3)-3(SiO_2) system, illustrating the positions of the α, β and O′-SiAlON phases.

cation M. The substitution of nitrogen by oxygen (n-value = Al–O for Si–N bonds) can be compensated either by the replacement of silicon by aluminum, or by the presence of a stabilizing cation M (M = Li, Mg, Ca, Y, rare-earth metals except for La, Ce, Pr, and Eu) [79] in interstitial positions (x) of the α-Si_3N_4 structure. The index, m, which expresses the amount of Si–N bonds replaced by longer Al–N bonds, depends on the valence of the M cation and on the x-value.

The α-sialon-forming area expands with decreasing size of the M ion in $M_xSi_{12\text{-}(m-n)}Al_{(m+n)}O_nN_{(16-n)}$. The maximum n-value is 1.0 for Nd, and approximately 1.2 for the other dopants. The m-value varies from 1.0 in all systems studied to a value of 2.75 in the Yb–α-sialon system, indicating that the substitution of Al–O units for Si–N in α-sialon is more restricted than the substitution of Al–N for Si–N.

The following empirical relationship between the sizes of the a and c lattice parameters axes and the m- and n-values of the α-sialon phase was obtained [80]:

$$a(\text{Å}) = 7.752 + 0436\text{m} + 0.02\text{n}$$

$$c(\text{Å}) = 5.620 + 0431\text{m} + 0.04\text{n}$$

Here, the mechanical properties of the α-SiAlONs are related to the α-Si_3N_4 structure, and are very similar. The Vickers hardness is in the range of 18 to 21 GPa [81–83] and the bending strength between 600 and 900 MPa [84, 85]. The fracture toughness of equiaxial α-SiAlONs ranges from 3 to 5.5 MPa·m$^{1/2}$; however, a significant improvement of K_{IC} was achieved by the development of α-sialons with elongated grains (see Figure 2.4) [86–92].

The β-SiAlON solid solutions are isostructural with β-Si_3N_4, and formed by the simultaneous equivalent substitution of Al–O for Si–N. The β-SiAlONs are most commonly described by the formula $Si_{6-z}Al_zO_zN_{8-z}$ $(0 < z < 4.2)$, retaining the 3 : 4 metal: nonmetal ratio, where z (Si–N) bonds are replaced by z (Al–O) bonds. As the

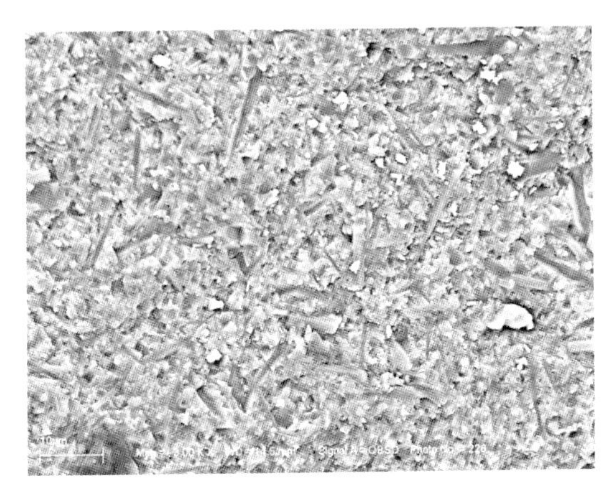

Figure 2.4 Fracture surface of α-SiAlON with elongated grains.

difference between both lengths (1.74 Å for Si–N and 1.75 Å for Al–O) is small, the lattice strain will also be small and the extent of replacement wide.

The β-SiAlON *z*-values can be determined from the measured lattice parameters using the mean of the following equations [93]:

$$a(\text{Å}) = 7.603 + 0.0297z$$

$$c(\text{Å}) = 2.907 + 0.0255z$$

In this case, those β-SiAlON materials with higher *z*-value can be densified by pressureless sintering, such that the resultant materials have high toughness (4–7 MPa $m^{1/2}$) due to their reinforcement with elongated β-SiAlON grains. The values of hardness of ~16 GPa correspond to the hardness of the original β-Si_3N_4. The bending strength of β-SiAlONs depends heavily on the *z*-value; normally, the strength is raised with a decreasing *z*-parameter (e.g., about 500 MPa for $z = 3$ [94] and 1.25 GPa for $z = 0.5$ [95]).

The mechanical properties of SiAlON ceramics can be tailored for specific applications by changing the starting composition; this will result in a different ratio of the α- and β-SiAlONs [96, 97]. Hardness will be increased markedly with an increasing α-SiAlON content, and the presence of an elongated β-SiAlON will result in a higher fracture toughness of the final composite.

2.1.9.2 Si_2N_2O and O′-SiAlON

Silicon oxynitride (Si_2N_2O) is a compound in the SiO_2–Si_3N_4 system with many interesting properties that provide it with promise for high-temperature applications [98–102]. For example, it can maintain an excellent oxidation resistance in air up to 1600 °C and a high flexural strength up to 1400 °C, without degradation [98–100]. Moreover, silicon oxynitride exhibits a very low density (2.81 g cm^{-3}), a high Vickers hardness (HV1 = 17–22 GPa) [100] that is comparable with that of α-Si_3N_4 and α-SiAlONs, a low thermal expansion coefficient (3.5×10^{-6} K^{-1}), good thermal shock resistance, and a high thermal stability (ca. 1800 °C) [101]. Moreover, the Si_2N_2O structure formed a narrow solid-solution range termed O′-SiAlON, by the same mechanism as β-SiAlON; that is, the Si–N bonds were replaced with Al–O bonds. O′-SiAlON can be represented by the formula $Si_{2-x}Al_xO_{1+x}N_{2-x}$, where the *x*-value varies from 0 to 0.2 [103].

Si_2N_2O can be synthesized either by the liquid-phase reaction sintering of an equimolar Si_3N_4 + SiO_2 mixture, or through a gas-phase nitridation of the mixture of Si and SiO_2. Dense bodies were prepared either by pressureless sintering with relatively large amounts of additives (MgO [104], Y_2O_3 and/or Al_2O_3 [104–106]), by hot pressing with Al_2O_3 [98, 107], CeO_2 [99b], or Li_2O additives [102], and also by hot isostatic pressing [100, 108]. The powders obtained by gas-phase synthesis have been densified by pressureless sintering with Y_2O_3 + Al_2O_3 sintering aids [109], by hot pressing with MgO, Y_2O_3 or CaO additives [110–112], or by spark plasma sintering (SPS) with Y_2O_3 [113] or CaO + Al_2O_3 aids [114]. The polymer-derived method of preparation can be also used for the synthesis of Si_2N_2O. As an example, Scheffler *et al.* prepared a SiC–Si_2N_2O microcomposite material with 13% porosity by the pyrolysis of poly(methylsilsequioxan) (this is known commercially as MK-polymer) at temperatures between 1250 and 1600 °C [115]. Subsequently, Riedel *et al.* [116] found that, during attrition milling of the pyrolyzed polysilazane-derived Si–C–N powder in isopropanol, the reactive amorphous Si_3N_4 powder took up oxygen by hydrolysis and alcoholysis in such amounts that a dense polycrystalline Si_2N_2O was formed after densification at 1600 °C in 0.1 MPa N_2, when Y_2O_3 + Al_2O_3 were added as sintering aids.

Recently, based on *ab initio* calculations, Ching [117] predicted that Si_2N_2O possessed a low dielectric constant, which is regarded as one of the most important physical properties for insulators. The good dielectric property of Si_2N_2O is mainly due to its bonding properties and special electronic structure, which results in a large band gap (5.97 eV) [118].

The potential applications of Si_2N_2O include roles as a high-temperature electric insulator [119, 120], as a nuclear-reactor moderator or reflector [121], and as materials for electrolytes [122]. O′-SiAlON has the potential for similar engineering applications.

2.2 Boron Nitride

Boron nitride (BN) is a chemical compound consisting of equal numbers of boron and nitrogen atoms. It is not found in Nature, and is therefore produced synthetically; the first synthesis of hexagonal BN was described in 1842 by Balmain [123].

2.2.1 Crystallographic Structures

Boron nitride exists in many different structures due to the special bonding behaviors of boron and nitrogen. Although the most well-defined crystallographic structures are hexagonal BN (h-BN), cubic BN (c-BN), and wurtzitic BN (w-BN), other crystalline structures, such as explosion boron nitride (e-BN) and ion beam-deposited boron nitride (i-BN) [124–135] and amorphous BN (a-BN) [136, 137] also exist.

2.2.1.1 Hexagonal Boron Nitride

Hexagonal BN crystallizes in similar fashion to graphite, in a hexagonal sheet layered structure, and thus is often referred to as "white graphite" (also called α-BN, or g-BN). Within each layer, the boron and nitrogen atoms are bound by strong covalent bonds (σ-bonding, sp^2-hybridization), whereas the layers are held together by weak van der Waals forces (π-bonding). As the planes are stacked on top of one another, without any horizontal displacement, this means that boron and nitrogen alternate along the c-axis. Hexagonal BN functions as an insulator because the π-electron is located at the nitrogen (due to the higher electronegativity of nitrogen) [138, 139].

2.2.1.2 Cubic Boron Nitride

Cubic BN is less stable than h-BN, but the conversion rate between these forms is negligible at room temperature. c-BN is prepared by annealing h-BN powder at higher temperatures, under pressures above 5 GPa. The cubic form has the sphalerite crystal structure (the B and N atoms are tetrahedrally coordinated), as in the diamond structure. Every boron atom is surrounded by four nitrogen atoms, and *vice versa*; in such an arrangement the boron and nitrogen atoms have sp^3 hybridization. c-BN is often also referred to as β-BN or sometimes z-BN (zinc-blende) [140, 141].

2.2.1.3 Wurtzitic Boron Nitride

Wurtzitic boron nitride (w-BN) has a similar structure to lonsdaleite, a rare hexagonal polymorph of carbon. In both c-BN and w-BN, the boron and nitrogen atoms are grouped into tetrahedra, although the angles between the neighboring tetrahedra are different [142]. The wurtzite-type of BN is a superhard hexagonal phase; this modification is a high-pressure phase and was first described by Bundy and Wentorf [143].

2.2.2 Synthesis of BN

The industrial synthesis of BN is based on a two-stage process. In the first stage, conducted at approximately 900 °C, a boron source such as B_2O_3 (boron oxide) or H_3BO_3 (boric acid) is converted in a reaction with a nitrogen source (mostly melamine or urea, sometimes even ammonia [NH_3]) to form amorphous BN:

$$B_2O_3 + 2NH_3 \rightarrow 2BN + 3H_2O \quad (T = 900\,^\circ C)$$

$$B_2O_3 + CO(NH_2)_2 \rightarrow 2BN + CO_2 + 2H_2O \quad (T > 1000\,^\circ C)$$

In the second step, the a-BN crystallizes at temperatures usually above 1600 °C in a N_2 atmosphere. Based on terminology borrowed from graphite processing, first a transition into the turbostratic phase, then into the mesographitic phase, and finally into the hexagonal phase of h-BN is observed. Control of the process parameters in the second annealing determines not only the degree of product purity, but also all the powder properties (particle size, crystallinity, specific surface area, etc.). h-BN sintered components are generally produced by pressure-assisted sintering (hot pressing/molding or hot isostatic pressing). The compaction mechanism can be described as a "glassy-phase-assisted compaction." Small quantities of B_2O_3 function as a liquid phase during sintering and as a binding phase for the h-BN platelets in the sintered body. In contrast to other nonoxide ceramics, no dissolution and reprecipitation during sintering has been reported.

Thin films of BN can be obtained by CVD from boron trichloride and nitrogen precursors [144]. The combustion of boron powder in nitrogen plasma at 5500 °C yields ultrafine BN which is for lubricants and toners [145].

Various composites, including SiC–BN, Si_3N_4–BN, AlN–BN, Sialon–BN, and AlON–BN, can be produced by the proposed *in situ* reactions. In this case, boron-bearing components such as B_4C and various metal borides such as AlB_2, SiB_4 and SiB_6 can be used as boron sources [160].

2.2.3 Properties of BN

2.2.3.1 h-BN

Major differences in bonding energies in the structure of h-BN – that is, strong covalent bonds within the basal planes and weak bonding between them – causes the high anisotropy of most properties of h-BN. For example, the hardness, electrical, and thermal conductivity are each much greater within the planes than perpendicular to them. Detailed information concerning the chemical and physical properties of h-BN are provided in Refs [146, 147]. Hexagonal BN may be used as a high-temperature solid lubricant, and it is a good thermal conductor and good electric insulator. It is stable in air up to 1000 °C, under vacuum up to 1400 °C, whilst in an inert atmosphere it can be used up to 2800 °C.

2.2.3.2 c-BN and w-BN

In contrast, the properties of c-BN and w-BN are more homogeneous. These materials are extremely hard; notably, the hardness of c-BN is slightly less than that of diamond, while that of w-BN is higher [148]. Because of much a better stability towards heat and metals, c-BN surpasses diamond in many mechanical applications [149]. The thermal conductivity of BN is among the highest of all electric insulators. The main properties of h-BN, c-BN, and w-BN are summarized in Table 2.2.

Boron nitride can be doped with Be (p-type) and with boron (n-type), sulfur, and silicon, or co-doped with carbon and nitrogen [153]. Both, h-BN and c-BN are wide-gap semiconductors with a band gap energy corresponding to the UV region. If a voltage is applied to h-BN [154, 155] or c-BN [156], then they emit UV light in the range 215–250 nm, and therefore can potentially be used as light-emitting diodes (LEDs) or lasers.

Table 2.2 Some properties of BN (⊥ – perpendicular to basal planes).

Property	h-BN	c-BN	w-BN
Mohs hardness	1–2	~10	~10
Young's modulus [$kg\,m^{-2}$]	3400–8700 [151]	4500 [150]	
Bulk modulus [GPa]	36	400	400
Thermal conductivity [$W\,cm^{-1}\,K^{-1}$]	0.6; (0.015 ⊥) [139]	13 [150]	—
Bandgap [eV]	5.2	6.4	4.5–5.5
Electrical resistivity [$\Omega\,cm$]	3.0×10^7; (3.0×10^9 ⊥) [139]	3.3×10^{13} [152]	
Density [$g\,cm^{-3}$]	~2.1	3.45	3.49

Further information concerning the synthesis and application of BN is available elsewhere [157–159].

2.3 Aluminum Nitride

Aluminum nitride (AlN) has interesting properties, such as a high thermal conductivity (70–210 $W\,m^{-1}\,K^{-1}$ for the polycrystalline material, and up to 285 $W\,m^{-1}\,K^{-1}$ for single crystals), a high volume resistance, and moderate dielectric properties. The thermal expansion coefficient of AlN is close to that of silicon, and it is one of the most mechanically strong and thermally stable ceramics. These excellent attributes make AlN a useful material for many applications [160, 161].

2.3.1 Structure

The wurtzite phase of AlN (w-AlN) is a wide band gap (6.2 eV) semiconductor material, which provides a potential application for deep UV optoelectronics. The space group is $C_{6v}{}^{4}$-$P6_3mc$, and the coordination geometry is tetrahedral.

2.3.2 Synthesis

AlN is synthesized by the carbothermal reduction of alumina, or the direct nitridation of aluminum. It is difficult to synthesize pure AlN powders due to their chemical instability [162]. The sintering behavior of AlN ceramics is greatly dependent on the size of the starting powders; nanometer-sized AlN particles have demonstrated a potential activity for low-temperature sintering [163–166]. The so-called low-temperature (~1500 °C) carbon thermal reduction is an effective method for synthesizing ultra-fine AlN powders [167, 168]. The general synthesizing temperature of AlN powders through carbon thermal reduction method is 1600–1800 °C [169, 170].

Owning to high covalent bonding and low self-diffusion coefficients of the constituent elements, the sintering of AlN ceramics is difficult. In general, AlN ceramics with complete densification and high thermal conductivity can be obtained at a temperature in excess of 1800 °C, with sintering additives [171, 172]. This not only increases the production costs of AlN ceramics but also promotes significant grain growth, which results in a deterioration of the mechanical properties. Consequently, recent investigations into AlN ceramics have been oriented towards achieving complete densification at a low sintering temperature [173]. Spark plasma sintering (SPS) is a newly developed technique that enables ceramic powders to be fully sintered at a relatively low temperature, in a very short time [174]. In the SPS process, a pulsed direct current is applied to the sintered powders, and the activation of powder particles is thought to be achieved as the application of electrical discharges [175].

2.3.3
Properties

Aluminum nitride may be used in composite structures containing aluminum for either structural or electronic applications, due to its attractive thermal, electronic, and mechanical properties [176–178]. AlN ceramics are also known to have a sufficiently high-temperature compatibility with refractory metals. Finally, AlN is an ecologically safe material. The structure of AlN as a ceramics layer of the multilayer Al/AlN composites has been investigated to only a limited degree [179].

The development of wide band-gap AlN semiconductors, and the emerging advances in a broad range of electronic, optical data storage, and optoelectronic devices, have witnessed dramatic success in recent years [180–186]. Group III-nitride technology which has developed steadily during the past decade has culminated in the commercial development of LEDs (which cover the violet–green range of the visible spectrum), laser diodes (LDs) [180], and modulation doped field effect transistors (MODFETs) [187, 188].

2.4
Titanium Nitride

Titanium nitride (TiN) is an extremely hard ceramic material which is often used as a coating on titanium alloys, steel, carbide, and aluminum components to improve the substrate's surface properties. When applied as a thin coating, TiN is used to harden and protect cutting and sliding surfaces, for decorative purposes, and as a nontoxic exterior for medical implants [189].

2.4.1
Structure

The crystal structure is cubic (cF8, space group *Fm3m*, No. 225), while the coordination geometry is octahedral with a lattice parameter of 4.240 Å [190]. The typical formation has a crystal structure of NaCl-type, with a roughly 1 : 1 stoichiometry; however TiN_x compounds with x ranging from 0.6 to 1.2 are thermodynamically stable [190–195].

The detailed structure and bonding character of the cubic TiN compounds have been extensively investigated [196].

2.4.2
Synthesis

The most common methods of TiN thin film creation are physical vapor deposition (PVD; usually sputter deposition, cathodic arc deposition, or electron beam heating) and CVD. For both methods, pure titanium is sublimated and reacted with nitrogen in a high-energy, vacuum environment. PVD is preferred when film-coating steel parts because the deposition temperatures lies beyond the austenitizing temperature of steel. PVD-applied TiN is also used for a variety of relatively higher melting point materials, such as stainless steels, titanium, and titanium alloys [197]. TiN coatings can also be deposited by thermal spraying, whereas TiN powders are produced by the nitridation of titanium with nitrogen or ammonia at 1200 °C [198].

Bulk ceramic objects can be fabricated by packing powdered metallic titanium into the desired shape, compressing it to the proper density, and then igniting it in an atmosphere of pure nitrogen. The heat released by the chemical reaction between the metal and gas is sufficient to sinter the nitride reaction product into a hard, finished item.

2.4.3 Properties

TiN will oxidize at 800 °C in a normal atmosphere. It is chemically stable at room temperature, and is attacked by hot concentrated acids [198]. TiN has excellent infrared (IR) reflectivity properties, creating a spectrum which is similar to that of elemental gold (Au) [199]. Depending on the substrate material and surface finish, TiN will have a coefficient of friction which ranges from 0.4 to 0.9 versus itself (nonlubricated) (Table 2.3).

Owing to its superior mechanical properties, TiN films are widely used in many industrial areas where high abrasion resistance, a low friction coefficient, a high temperature stability, and high hardness are required. The mechanical properties of TiN are strongly related to its preferred orientation [200, 201]; it has been reported [202, 203] that a TiN film with a (1 1 1) preferred orientation possesses the highest hardness. During the PVD deposition of a thin film, the packing density and preferred orientation of the film normally change with the increasing film thickness. Consequently, film thickness is an important parameter that affects the preferred orientation and hardness of the coating. It is also known that PVD-coated specimens will inevitably have residual stress when the process is complete. Such residual stress is a significant factor for influencing the preferred orientation [204, 205], adhesion, and hardness [206, 207] of the coating [208].

In general, the resistivity of TiN is low, and bulk values of about 25 $\mu\Omega \cdot$ cm are often quoted. The variation in resistivity for thin-film samples has been found to be wide, and to depend critically on the growth conditions [210–216].

Table 2.3 Characteristic properties of TiN.

Summary of characteristic [198, 209]	
Vickers hardness [GPa]	18–21
Modulus of elasticity [GPa]	251
Thermal conductivity [W m^{-1} K^{-1}]	19.2
Thermal expansion coefficient [10^{-6} K^{-1}]	9.35
Electrical resistivity [$\mu\Omega \cdot$ cm]	20

2.5 Tantalum Nitride

Tantalum mononitride (TaN), a representative refractory nitride, is widely used in many technological applications owing to its outstanding mechanical and chemical stability [190]. For example, TaN is used as a hard coating material or as a diffusion barrier in microelectronic devices. The electrical properties of TaN can vary between the metal and insulator, depending on the N_2 pressure in the growth condition. Recently, additional interest in TaN has been spurred by the high-density electronic devices that employ TaN as a metal gate to be compatible with high-dielectric constant insulators such as HfO_2 [217]. The thermal stability of TaN prevents the performance degradation after post-deposition annealing.

TaN is also known for its variety in compositions and metastable phases. The substoichiometric δ-TaN_x with cubic symmetry has been studied most extensively to date. This cubic phase can support a large density of vacancies up to 20%. On the other hand, three stable phases have been known for TaN close to the 1 : 1 stoichiometry. At ambient conditions, the hexagonal ε phase with the space group of $P\bar{6}2m$ is known to be stable [218]; this structure is close to the more symmetric CoSn ($P6/mmm$) type. (Some reports have indicated the CoSn structure as the ε phase [190, 219].) The ε phase transforms to the hexagonal θ phase (space group of $P\bar{6}m2$, WC-type) under high-pressure conditions [220], or to the cubic δ phase (space group of $Fm\bar{3}m$) with the rocksalt structure at high temperatures [221]. Compared to the cubic δ phase [222–225], the ε and θ phases have been investigated only minimally.

TaN_x is a widely used material for producing hard coatings, wear-resistant layers, thin film resistors, diffusion barriers in integrated circuits, and mask layers for X-ray lithography. Unlike the well-known hard coating material TiN (IVB–VA compound), the cubic TaN (VB–VA compound) is metastable. Whilst the Ti–N system is relatively simple, with only two compounds such as tetragonal Ti_2N and cubic TiN, [226], the Ta–N system has a variety of compounds and has not been extensively investigated. In addition to the equilibrium phases of bcc α-Ta, hexagonal γ-Ta_2N, hexagonal ε-TaN and many metastable phases have been reported [227–229]. These include tetragonal β-Ta, bcc β-Ta(N), hexagonal Ta_2N, hexagonal θ-TaN with the WC structure, δ-TaN with the cubic B1 NaCl-structure, hexagonal Ta_5N_6, tetragonal Ta_4N_5, and orthorhombic Ta_3N_5 [227, 228, 230].

2.6 Chromium Nitride

Chromium nitride (CrN) is an interstitial compound, with nitrogen atoms occupying the octahedral holes in the chromium lattice [231]; as such, it is not strictly a Cr(III) compound, nor does it contain nitride ions (N^{3-}). Chromium forms a second interstitial nitride, dichromium nitride, Cr_2N.

Among the transition metal nitrides, cubic CrN has been identified as a better coating and superhard material due to its hardness, excellent corrosion, oxidation and wear resistance [232–234]. It is well known that nanocrystallines have attracted considerable interest due to their unique physical and chemical properties. Recently, there was increasing interest in the preparation of nanocrystalline CrN. The prepared methods include high-energy reaction milling [235], ammonolysis of chromium chloride [236], benzene thermal synthesis [237], direct nitridation of nanosized Cr_2O_3 powder in N_2 atmosphere [238] and thermal nitridation of metal Cr with NH_4Cl [239]. Chromium (III) nitride can be prepared by the direct combination of chromium and nitrogen at 800 °C: $2\,Cr + N_2 \rightarrow 2\,CrN$. The chromium-based nitrides, Cr_2N and CrN, when prepared by plasma deposition, have been reported to have high hardness, enhanced wear- and corrosion-resistance, and can also be utilized as anti-wear and anti-corrosion coatings [240–244].

2.7 Ternary Nitrides

Ternary nitrides have recently been extensively studied, because they may display a wider variety of useful properties compared to the binary nitrides. The majority of reported ternary nitrides contain an electropositive element along with a transition metal or main-group metal [245]. The electropositive element (e.g., Ca, Sr, Ba, Mg) is included to increase the stability of the nitride, via the inductive effect [246].

2.7.1 Ternary Silicon Nitrides

Only a few ternary silicon nitrides have been prepared in a pure form and characterized. The isotypic compounds $MSiN_2$ (M = Be, Mg, Mn, Ca, Zn), with the same valence electron concentration of 4, can be considered as ternary substitution variants of AlN [247]. These compounds contain three-dimensional (3-D) infinite network structures, with SiN_4 tetrahedra linked through all four vertices by corner-sharing, which forms condensed $[Si_6N_6]$ twelve-membered rings in $MgSiN_2$ [248].

Silicon-based ternary nitrides such as $MgSiN_2$ and $LaSi_3N_5$ have been intensively studied as an alternative material for the substrates of integrated circuits [249–253], as host lattices for light-emitting phosphors [254–260], and as semiconductors.

2.7.1.1 $MgSiN_2$

$MgSiN_2$ has a good thermal conductivity 17–35 $W\,m\,K^{-1}$ [261–265] and a high electrical resistance at room temperature. These properties, combined with high strength, fracture toughness and hardness, make $MgSiN_2$ suitable for many industrial applications. Several synthetic procedures have been utilized to prepare $MgSiN_2$ powder, including a conventional synthesis starting with a Mg_3N_2 and Si_3N_4 mixture [261], or with a Mg and Si_3N_4 mixture [265], although in powders synthesized from Mg/Si_3N_4 mixtures some free Si metal was always present [266]. Single-phase $MgSiN_2$ was not obtained by using a processing route based on the direct nitridation of a Mg/Si mixture [266, 267], as (-Si_3N_4 and MgO were detected as minor phases. The synthesis of $MgSiN_2$ via the nitridation of Mg_2Si has also been carried out [267, 268] whereby, after heat-treatment at 1400 °C for 1 h, single-phase $MgSiN_2$ was obtained with an oxygen content of 0.5 wt% [267].

2.7.1.2 Other Alkaline Earth Silicon Nitrides

Other alkaline earth silicon nitrides, $MSiN_2$ (where M = Ca, Sr, Ba), were synthesized and characterized [269, 270]. $BaSiN_2$ crystallizes in space group *Cmca* with $a = 5.6046(1)$ Å, $b = 11.3605(3)$ Å, $c = 7.5851(2)$ Å, and $Z = 8$. The structure consists of pairs of SiN_4 tetrahedra edge-linked to form bow-tie-shaped Si_2N_6 dimers that share vertices to form layers, and has no analogue in oxide chemistry. $SrSiN_2$ has a distorted form of this structure ($SrSiN_2$: space group $P2_1/c$ (No. 14), $a = 5.9750(5)$ Å, $b = 7.2826(7)$ Å, $c = 5.4969(4)$ Å, $\beta = 113.496(4)°$, $Z = 4$). The structure of $CaSiN_2$ contains only vertex-sharing SiN_4 tetrahedra, linked to form a 3-D stuffed-cristobalite-type framework that is isostructural with $KGaO_2$ ($CaSiN_2$: space group *Pbca* (No. 61), $a = 5.1229(3)$ Å, $b = 10.2074(6)$ Å, $c = 14.8233(9)$ Å, $Z = 16$).

$Sr_2Si_5N_8$, $Ba_2Si_5N_8$ and $Ca_2Si_5N_8$ were synthesized by the reaction of silicon diimide with alkaline earth metals under a nitrogen atmosphere, and subsequently characterized [271–273].

2.7.1.3 $LaSi_3N_5$

Several synthetic procedures have been utilized to produce $LaSi_3N_5$ powder. The compound has been prepared by a reaction between Si_3N_4 and La_2O_3 at 2000 °C under 5 MPa nitrogen pressure [274] by hot pressing of Si_3N_4 and LaN at 1830 °C for 1 h under a 28 MPa load [275] by the direct reaction of Si_3N_4 and LaN at 1700 °C for 30 min under 14.7 MPa nitrogen pressure [276] by the direct reaction of $LaSi_3$ alloys with nitrogen at 1500 °C and 0.1 MPa N_2 pressure for one week [277], by the carbothermal reduction and nitridation (CRN) of La_2O_3 and SiO_2 with carbon [278], or by the CRN of lanthanum acetate and tetraethoxysilane with carbon [278, 279]. The oxygen content varied, depending on the method of preparation, between 1 and ~2.5 wt%.

Some of the properties of dense $MgSiN_2$ and $LaSi_3N_4$ are summarized in Table 2.4.

In contrast to binary nitrides (Mg_3N_2 and LaN), the ternary nitrides are stable in air for long periods of time. $LaSi_3N_5$ is resistant to aqua regia ($HNO_3 + 3HCl$) [277], and can be used at higher temperatures in a fluorine-containing environment compared to nickel alloys [275].

Table 2.4 Properties of dense ternary nitrides [280].

Ternary nitride	λ (W m^{-1} K^{-1})	*E* (GPa)	*HV1* (GPa)	ν	K_{1C} a) (MPa m$^{1/2}$)	σ_{4p} (MPa)
$MgSiN_2$	21.9	281	20.8 ± 0.4	0.237	3.5 ± 0.6	384 ± 32
$LaSi_3N_5$	4.1	246	18.3 ± 0.5	0.238	4.2 ± 0.5	426 ± 23

a) Indentation method.

2.7.1.4 $LiSi_2N_3$

Lithium silicon nitride ($LiSi_2N_3$) is used as anode material for lithium ion batteries [281]. It has a wurtzite structure (*Cmc2₁*) with lattice parameters $a = 9.2215(9)$, $b = 5.2964(8)$, $c = 4.7798(5)$ Å [282]. Other than $LiSi_2N_3$, other $Li_xSi_yN_z$ phases were synthesized: Li_2SiN_2, Li_5SiN_3, $Li_{18}Sin_3N_{10}$, $Li_{21}Sin_3N_{11}$, and Li_8SiN_4. The structure and lattice parameters of some of the phases are as follows: $LiSi_2N_3$; orthorhombic, $a = 9.198(3)$, $b = 5.307(2)$, $c = 4.779(2)$; Li_5SiN_3; cubic, $a = 4.7240(3)$; $Li_{18}Sin_3N_{10}$; tetragonal, $a = 14.168(4)$, $c = 14.353(8)$; $Li_{21}Sin_3N_{11}$; tetragonal, $a = 9.470(3)$, c 9.530(8); and Li_8SiN_4; tetragonal $a = 10.217(2)$, $c = 9.536(3)$ Å. All phases are lithium ion conductors. A new phase, Li_8SiN_4, has the highest lithium ion conductivity (5×10^{-2} S m^{-1} at 400 K) and the lowest activation energy (46 kJ mol^{-1}) among the present products [281]. Li_2SiN_2 has a conductivity of 2.7×10^{-3} S cm^{-1} at 770 K [283]. In addition to Li_2SiN_2, Li_3AlN_2 was also investigated as a lithium ion-conducting electrolyte [284, 285].

Ternary nitrides with a general formula $Si_{3-x}M_xN_4$ (M = Co, Fe, Ni) are newly identified anodes for lithium batteries. $Si_{3-x}M_xN_4$ exhibited improved specific capacity values in the range of 130 to 470 mA · h g^{-1}, with lesser capacity fade (<20%) after 50 cycles. More specifically, the $Si_{3-x}Fe_xN_4$ anode with a specific capacity of ~470 mA h g^{-1} may be exploited for practical applications [286, 287].

2.8
Light-Emitting Nitride and Oxynitride Phosphors

During the past decade, the luminescent properties of other Si-based ternary nitrides and oxynitrides have also been the subject of much interest, owing to their potential application in white LEDs and various types of display device. The LEDS can be created by: (i) using three individual monochromatic LEDs with blue, green, and red colors; (ii) combining a UV LED with blue, green, and red phosphors (Figure 2.5); and (iii) using a blue LED to pump yellow or green and red phosphors. Although the phosphors in LEDs should have a high absorption of UV or blue light,

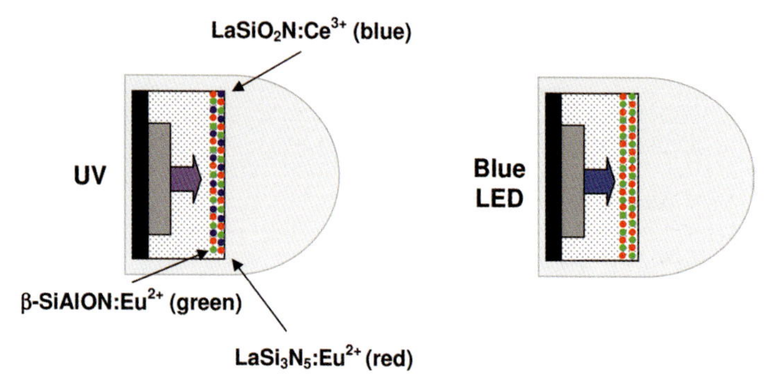

Figure 2.5 Schematic function of white light-emitting diode (LED).

most oxide-based phosphors have a low absorption in the visible-light spectrum, which makes it impossible for them to be coupled with blue LEDs. On the other hand, sulfide-based phosphors are thermally unstable and very sensitive to moisture.

Silicon-based oxynitride and nitride phosphors have encouraging luminescent properties (excitability by blue light, high conversion efficiency, and the possibility of full color emission), as well as a low thermal quenching, high chemical stability, and high potential for use in white LEDs [288, 289].

Compared with oxide-, boride-, sulfide-, or phosphate-based phosphors, the study of oxynitride and nitride phosphors is at a very early stage. However, a variety of oxynitride and nitride materials with promising luminescent properties have been discovered recently [290–292].

$Ca_2Si_5N_8$ doped with Eu^{2+} shows a strong emission band that peaks at 600 nm, and can be used as a light-emitting phosphor [290, 293]. The luminescence of Eu^{2+}-doped $Ba_2Si_5N_8$ was reported by Hoppe *et al.* [291], and that of Eu^{2+}-doped $M_2Si_5N_8$ (M = Ca, Sr, Ba) later described by Li *et al.* [294]. The peak emission wavelength was seen to shift upwards with increasing ionic size of the alkaline-earth metal, and shown to be 623, 640, and 650 nm for $Ca_2Si_5N_8$: Eu^{2+}, $Sr_2Si_5N_8$: Eu^{2+}, and $Ba_2Si_5N_8$: Eu^{2+}, respectively [288, 294].

Hirosaki *et al.* [289] reported the blue-emitting Ce^{3+}-doped $LaAl(Si_{6-z}Al_z)N_{10-z}O_z$ oxynitride phosphor. Blue emission also showed the Ce^{3+}-activated α-sialon.

Both, Hirosaki *et al.* [292, 295] and Zhou *et al.* [296] reported a green oxynitride phosphor based on Eu^{2+}-doped β-sialon. Other suitable host lattices for phosphors include $MSi_2O_2N_2$ compounds (M = Ca, Sr, and Ba); for example, $SrSi_2O_2N_2$: Eu^{2+} emits a green color, $BaSi_2O_2N_2$: Eu^{2+} yields blue–green emission, while $CaSi_2O_2N_2$: Eu^{2+} shows a yellowish emission [297, 298].

Eu^{2+}-doped Ca-α-sialon phosphor also shows a yellow emission [299]; the yellow emission of Ca-α-sialon: Eu^{2+} can be tuned by substituting Ca with other metals such as Li, Mg, and Y, and may even be adjusted by tailoring the composition of the α-sialon host lattice by altering the values of *m* and *n* in the chemical formula.

References

1 (a) Wiederhorn, S.M. and Ferber, M.K. (2001) *Curr. Opin. Solid State Mater. Sci.*, 5, 311–316; (b) Krause, R.F., Wiederhorn, S.M., and Li, C.-W. (2001) *J. Am. Ceram. Soc.*, **84**, 2394–2400.

2 Schaffer, P.S. and Swaroop, B. (1970) *Am. Ceram. Soc. Bull.*, **49**, 536–538.

3 Hockings, E.F. (1994) in *Gmelin Handbook of Inorganic and Organometallic Chemistry*, 8th edn, Springer-Verlag, Berlin, pp. 238–277.

4 Riley, F.L. (2000) *J. Am. Ceram. Soc.*, **83**, 245–265.

5 Sangster, R.C. (2005) *Materials Science Foundations*, Trans. Tech. Publ., Swizerland, pp. 22–24.

6 Li, Y., Liang, Y., and Zheng, F. (1994) *J. Mater. Sci. Lett.*, **13**, 1588–1590.

7 Sawhill, H.T. and Haggerty, J. (1982) *J. Am. Ceram. Soc.*, **65**, C131–C132.

8 Wang, T., Zhang, L., and Mo, C. (1994) *Nanostruct. Mater.*, **4**, 207–213.

9 Kavecký, Š., Janeková, B., and Šajgalík, P. (2000) *Key Eng. Mater.*, **175–176**, 49–56.

10 Halamka, M., Kavecký, Š., Dočekal, B., Madejová, J., and Šajgalík, P. (2003) *Ceram.-Silikáty*, **47**, 88–93.
11 Datton, J.V. and Drobeck, J. (1968) *J. Electrochem. Soc.*, **115**, 865–868.
12 Roenigk, K.F. and Jensen, K.F. (1987) *J. Electrochem. Soc.*, **134**, 1777–1785.
13 Alexandrov, L.N. and Edelman, F.L. (1977) Proceedings, 7th International Vacuum Congress, vol. 2, pp. 1619–1622.
14 Grün, R. (1979) *Acta Crystallogr.*, **B35**, 800–804.
15 Marchand, R., Laurent, Y., Lang, J., and Le Bihan, M.T. (1969) *Acta Crystallogr.*, **B25**, 2157–2160.
16 Kato, K., Inoue, Z., Kijima, K., Kawada, I., Tanaka, H., and Yamane, T. (1975) *J. Am. Ceram. Soc.*, **58**, 90–91.
17 Toraya, H. (2000) *J. Appl. Crystallogr.*, **33**, 95–102.
18 Wang, C.M., Pan, X., Rühle, M., Riley, F.L., and Mitomo, M. (1996) *J. Mater. Sci.*, **31**, 5281–5298.
19 Hardie, D. and Jack, K.H. (1957) *Nature*, **180**, 332–333.
20 Borgen, O. and Seip, H.M. (1961) *Acta Chem. Scand.*, **15**, 1789.
21 Wang, C.-M., Pan, X., Rühle, M., Riley, F.L., and Mitomo, M. (1996) *J. Mater. Sci.*, **31**, 5281–5298.
22 Zerr, A., Miehe, G., Serghiou, G., Schwarz, M., Kroke, E., Riedel, R., Fueß, H., Kroll, P., and Böhler, R. (1999) *Nature*, **400**, 340.
23 Kroke, E. (2002) *Angew. Chem., Int. Ed.*, **41**, 77.
24 Asay, J.R. and Shahinpoor, M. (1993) *High-Pressure Shock Compression of Solids*, Springer, New York.
25 Xu, Y.N. and Ching, W.Y. (1995) *Phys. Rev. B.*, **51**, 17379–17389.
26 Cai, Y., Zimmermann, A., Prinz, S., and Aldinger, F. (2002) *Physica Status Solidi*, **231**, R4–R6.
27 Cai, Y., Zimmermann, A., Prinz, S., and Aldinger, F. (2002) *J. Mater. Res.*, **17**, 2765–2767.
28 Cai, Y., Zimmermann, A., Prinz, S., Kramer, S., Philipp, F., Sigle, W., and Aldinger, F. (2002) *Philos. Mag. Lett.*, **82**, 553–558.
29 Alexander, C.M.O'D., Prombo, C.A., Swan, P.S., and Walker, R.M. (1991) *Lunar Planet. Sci.*, **22**, 5–6.
30 Alexander, C.M.O'D., Swan, P.S., and Prombo, C.A. (1994) *Meteoritics*, **29**, 79–84.
31 Stone, J. Hutcheon, I.D., Epstein, S., and Wasserburg, G.J. (1991) *Earth Planet. Sci. Lett.*, **107**, 570–581.
32 Russel, S.S., Lee, M.R., Arden, J.W., and Pillinger, C.T. (1995) *Meteoritics*, **30**, 399–404.
33 Lee, M.R., Russel, S.S., Arden, J.W., and Pillinger, C.T. (1995) *Meteoritics*, **30**, 387–398.
34 Reimanis, I.E., Suematsu, H., Petrovic, J.J., and Mitchel, T.E. (1996) *J. Am. Ceram. Soc.*, **79**, 2065–73.
35 Henderson, C.M.B. and Taylor, D. (1975) *Trans. J. Br. Ceram. Soc.*, **74**, 49.
36 Ogata, S., Hirosaki, N., Kocer, C., and Shibutani, Y. (2004) *Acta Mater.*, **52**, 233–238.
37 Marchand, R., Laurent, Y., and Lang, J. (1969) *Acta Crystallogr.*, **B25**, 2157.
38 Turkdogan, T., Bills, P.M., and Tippett, V.A. (1958) *J. Appl. Chem.*, **8**, 296–302.
39 Wang, Ch.-M. and Wang, T.-M. (1996) *J. Mater. Sci. Lett.*, **15**, 1805–1807.
40 Hirosaki, N., Ogata, S., and Kocer, C. (2002) *Phys. Rev.*, **B65**, 134110.
41 Chabrkaborty, D. and Mukerji, J. (1980) *J. Mater. Sci.*, **15**, 3051–3056.
42 Dusza, J., Eschner, T., and Rundgren, K. (1997) *J. Mater. Sci. Lett.*, **16**, 1664–1667.
43 Niihara, K. and Hirai, T. (1979) *J. Mater. Sci.*, **14**, 1952.
44 Bruls, R.J., Hintzen, H.T., de With, G., Metselaar, R., and van Miltenburg, J.C. (2001) *J. Phys. Chem. Solids*, **62**, 783.
45 Ogata, S., Li, J., and Yip, S. (2002) *Science*, **298**, 807.
46 Ogata, S., Hisrosaki, N., Kocer, C., and Shibutani, Y. (2003) *J. Mater. Res.*, **18**, 1168–1172.
47 Li, B., Pottier, L., Roger, J.P., Fournier, D., Watari, K., and Hirao, K. (1999) *J. Eur. Ceram. Soc.*, **19**, 1631–1639.
48 Jiang, J.Z., Kragh, F., Frost, D.J. *et al.* (2001) *J. Phys: Condens. Matter*, **13**, L515–L520.

49 Kocer, C., Hirosaki, N., and Ogata, S. (2003) *Phys. Rev.*, **B67**, 035210/1–035210/4.

50 Zerr, A., Kempf, M., Schwarz, M., Kroke, E., Göken, M., and Riedel, R. (2002) *J. Am. Ceram. Soc.*, **85**, 86.

51 Schwarz, M., Miehe, G., Zerr, A., Kroke, E., Heck, M., Thybusch, B., Poe, B.T., Chen, I.-W., and Riedel, R. (2002) *Angew. Chem.*, **114**, 804.

52 Jiang, J.Z., Lindelov, H., Gerward, L., Stahl, K. *et al.* (2002) *Phys. Rev.*, **B65**, 161202/1–161202/4.

53 Hintzen, H.T.J.M., Hendrix, M.M.R.M., Wondergem, H.J., Fang, C., Sekine, T., and de With, G. (2003) *J. Alloy. Compd.*, **351**, 40.

54 Fang, C.M., Wijs, G.A., Hintzen, H.T.J.M., and de With, G. (2003) *J. Appl. Phys.*, **93**, 5175.

55 Lichvár, P., Šajgalík, P., Liška, M., and Galusek, D. (2007) *J. Eur. Ceram. Soc.*, **27**, 429–436.

56 Lofaj, P., Satet, R., Hoffmann, M.J., and de Arellano López, A.R. (2004) *J. Eur. Ceram. Soc.*, **24**, 3377–3385.

57 Loehman, R.E. (1980) *J. Non-Cryst. Solids*, **42**, 433–446.

58 Wusirika, R.R. and Chyung, C.K. (1980) *J. Non-Cryst. Solids*, **28–39**, 39–44.

59 Messier, D.R. and Broz, A. (1982) *J. Am. Ceram. Soc.*, **65**, C123.

60 Hampshire, S., Drew, R.A.L., and Jack, K.H. (1984) *J. Am. Ceram. Soc.*, **67**, C46–C47.

61 Rouxel, T., Besson, J.-L., Rzepka, E., and Goursat, P. (1990) *J. Non-Cryst. Solids*, **122**, 298–304.

62 Becher, P., Waters, S.B., Westmoreland, C.G., and Riester, L. (2002) *J. Am. Ceram. Soc.*, **85**, 897–902.

63 Peterson, I.M. and Tien, T.-Y. (1995) *J. Am. Ceram. Soc.*, **78**, 1977–1979.

64 Rouxel, T., Huger, M., and Besson, J.-L. (1992) *J. Mater. Sci.*, **27**, 279–284.

65 Lofaj, F., Hvizdoš, P., Dorčáková, F., Satet, R., Hoffmann, M.J., and de Arellano López, A.R. (2003) *Mater. Sci. Eng. A*, **257**, 181–187.

66 Lofaj, F., Dorčáková, F., Kovalčík, J., Hoffmann, M.J., and de Arellano López, A.R. (2003) *Metall. Mater.*, **41**, 145–147.

67 Sun, E.Y., Becher, P.F., Hwang, S.L., Waters, S.B., Pharr, G.M., and Tsui, T.Y. (1996) *J. Non-Cryst. Solids*, **208**, 162–169.

68 Schneider, S.J. Jr(volume chairman) (1991) *Engineered Materials Handbook*, vol. 4, Ceramics and Glasses, ASM International.

69 (a) Becher, P.F. *et al.* (2006) *Mater. Sci. Eng. A*, **422**, 85–91; (b) Becher, P.F. *et al.* (2008) *J. Am. Ceram. Soc.*, **91**, 2328–2336.

70 Kitayama, M. *et al.* (2006) *J. Am. Ceram. Soc.*, **89**, 2612–2618.

71 (a) Satet, R.L. and Hoffmann, M.J. (2004) *J. Eur. Ceram. Soc.*, **24**, 3437–3445; (b) Satet, R.L. *et al.* (2006) *Mater. Sci. Eng. A*, **422**, 66–76.

72 (a) Shibata, N., Pennycook, S.J., Gosnell, T.R., Painter, G.S., Shelton, W.A., and Becher, P.F. (2004) *Nature*, **428**, 730–733; (b) Shibata, N. *et al.* (2005) *Phys. Rev. B*, **72**, 140101.

73 Wang, Ch.-M. *et al.* (1996) *J. Am. Ceram. Soc.*, **79**, 788–92.

74 Lofaj, F. *et al.* (2002) *J. Eur. Ceram. Soc.*, **22**, 2479–2487.

75 (a) Guo, S. *et al.* (2005) *Mater. Sci. Eng. A*, **408**, 9–18; (b) Guo, S., Hirosaki, N., Yamamoto, Y., Nishimura, T., and Mitomo, M. (2001) *Scripta Mater.*, **45**, 867–874.

76 Nishimura, T., Guo, S., Hirosaki, N., and Mitomo, M. (2006) *J. Ceram. Soc. Jpn*, **114**, 880–887.

77 Kondo, N., Asayama, M., Suzuki, Y., and Ohji, T. (2003) *J. Am. Ceram. Soc.*, **86**, 1430–32.

78 Jack, K.H. and Wilson, W.I. (1972) *Nature (London) Phys. Sci.*, **238**, 28–29.

79 Hampshire, S., Park, H.K., Thompson, D.P., and Jack, K.H. (1978) *Nature (London)*, **274**, 880–882.

80 Shen, Z. and Nygren, M. (1997) *J. Eur. Ceram. Soc.*, **17**, 1639–1645.

81 Santos, C., Strecker, K., Ribeiro, S., Suzuki, P.A., Kycia, S., Silva, O.M.M., and Silva, C.R.M. (2005) *Mater. Res. Bull.*, **40**, 1094–1103.

82 Zalite, I., Zilinska, N., and Kladler, G. (2008) *J. Eur. Ceram. Soc.*, **28**, 901–905.

83 Ye, F., Liu, C.-F., and Liu, L.-M. (2009) *Ceram. Int.*, **35**, 725–731.

84 (a) Mitomo, M., Tanaka, H., Muramatsu, K., Li, N., and Fujii, Y. (1980) *J. Mater. Sci.*,

15, 2661–2662; (b) Mitomo, M., Izumi, F., Bando, Y., and Sekikawa, Y. (1984) in *Ceramic Components for Engines*, KTK Science Publishers, Tokyo, Japan, pp. 377–386; (c) Mitomo, M. (1988) *Advanced Ceramics 11*, Elsevier, Barking, Essex, UK, pp. 147–161.

85 Bartek, A., Ekstrom, T., Herbertsson, H., and Johansson, T. (1992) *J. Am. Ceram. Soc.*, **75**, 432–439.

86 Chen, I.-W. and Rosenflanz, A. (1997) *Nature (London)*, **389**, 701–704.

87 Kim, J., Rosenflanz, A., and Chen, I.-W. (2000) *J. Am. Ceram. Soc.*, **83**, 1819–1821.

88 Zhang, C., Narimatsu, E., Komeya, K., Tatami, J., and Meguro, T. (2000) *J. Mater. Lett.*, **43**, 315–319.

89 (a) Zenotchkine, M., Shuba, R., Kim, J.-S., and Chen, I.-W. (2001) *J. Am. Ceram. Soc.*, **84**, 1651–1653; (b) Zenotchkine, M., Shuba, R., and Chen, I.-W. (2002) *J. Am. Ceram. Soc.*, **85**, 1254–1259.

90 (a) Rosenflanz, A. and Chen, I.-W. (1999) *J. Am. Ceram. Soc.*, **82**, 1025–1036;(b) Rosenflanz, A. (1997) α′-SiAlON: Phase stability, phase transformations and microstructural evolutions. PhD Dissertation, University of Michigan, Ann Arbor, MI;(c) Rosenflanz, A. and Chen, I.-W. (1999) *J. Eur. Ceram. Soc.*, **19**, 2325–2335;(d) Rosenflanz, A. and Chen, I.-W. (1999) *J. Eur. Ceram. Soc.*, **19**, 2337–2348.

91 Shuba, R. and Chen, I.-W. (2002) *J. Am. Ceram. Soc.*, **85**, 1260–1267.

92 Salamon, D., Šajgalík, P., Lenčéš, Z., and Křestan, J. (2005) *Mater. Lett.*, **59**, 3201–3204.

93 Shen, Z., Ekstrom, T., and Nygren, M. (1996) *J. Appl. Phys. D*, **29**, 893–904.

94 Kokmeijer, E., de Withand, G., and Metselaar, R. (1991) *J. Eur. Ceram. Soc.*, **8**, 71–80.

95 Kishi, K., Umebayashi, S., Tani, E., Shobu, K., and Zhou, Y. (2001) *J. Eur. Ceram. Soc.*, **21**, 1269–1272.

96 Zhang, C., Sun, W.Y., and Yan, D.S. (1999) *J. Eur. Ceram. Soc.*, **19**, 33–39.

97 Ye, F., Liu, L., Zhang, H., Zhou, Y., and Zhang, Z. (2009) *Scripta Mater.*, **60**, 471–474.

98 Huang, Z.K., Greil, P., and Petzow, G. (1984) *Ceram. Int.*, **10**, 14–17.

99 (a) Ohashi, M., Kanzaki, S., and Tabata, H. (1988) *J. Mater. Sci. Lett.*, **7**, 339–340; (b) Ohashi, M., Kanzaki, S., and Tabata, H. (1991) *J. Am. Ceram. Soc.*, **74**, 109–114; (c) Ohashi, M., Kanzaki, S., and Tabata, H. (1991) *J. Mater. Sci.*, **26**, 2608–2614.

100 Larker, R. (1992) *J. Am. Ceram. Soc.*, **75**, 62–66.

101 Rocabois, P., Chatillon, C., and Bernard, C. (1996) *J. Am. Ceram. Soc.*, **79**, 1361–1365.

102 Tong, Q., Wang, J. Li, Z., and Zhou, Y. (2007) *J. Eur. Ceram. Soc.*, **27**, 4767–4772.

103 Trigg, M.B. and Jack, K.H. (1987) *J. Mater. Sci. Lett.*, **6**, 407–408.

104 Bergmman, B. and Heping, H. (1990) *J. Eur. Ceram. Soc.*, **6**, 3–8.

105 Lewis, M.H., Reed, C.J., and Butler, N.D. (1985) *Mater. Sci. Eng.*, **71**, 87–94.

106 Mitomo, M., Ono, S., Asami, T., and Kang, S.-J.L. (1989) *Ceram. Int.*, **15**, 345–350.

107 Sekercioglu, I. and Wills, R.R. (1979) *J. Am. Ceram. Soc.*, **62**, 590–593.

108 Ekström, T., Ollson, P.-O., and Holmström, M. (1993) *J. Eur. Ceram. Soc.*, **12**, 165–176.

109 Sjöberg, J., Rundgren, K., Pompe, R., and Larsson, B. (1987) *High Technology Ceramics*, Elsevier Science Publishers, Amsterdam, Netherlands, pp. 535–543.

110 Boch, P. and Glandus, J.C. (1979) *J. Mater. Sci.*, **14**, 379–85.

111 Billy, M., Boch, P., Dumazeau, C., Glandus, J.C., and Goursat, P. (1981) *Ceram. Int.*, **7**, 13–18.

112 Siddiqi, A. and Hendry, A. (1986) *Br. Ceram. Soc. Proc.*, **37**, 1–13.

113 Barris, G.C., Shen, Z., Nygren, M., and Brown, I.W.M. (2003) *Key Eng. Mater.*, **237**, 37–42.

114 Radwan, M., Kashiwagi, T., and Miyamoto, Y. (2003) *J. Eur. Ceram. Soc.*, **23**, 2337–2341.

115 Scheffler, M., Pippel, E., Woltersdorf, J., and Greil, P. (2003) *Mater. Chem. Physics*, **80**, 565–572.

116 Riedel, R., Seher, M., Mayer, J., and Szabó, V. (1995) *J. Eur. Ceram. Soc.*, **15**, 703–715.

117 Ching, W.Y. (2004) *J. Am. Ceram. Soc.*, **87**, 1996–2013.

118 Ching, W.Y. and Ren, S.-Y. (1981) *Phys. Rev. B*, **24**, 5788–5795.

119 Brown, D.M., Gray, P.V., Heumann, F.K., Philipp, H.R., and Traft, E.A. (1968) *J. Electrochem. Soc.*, **115**, 311.

120 Frank, R.I. and Moberg, W.L. (1970) *J. Electrochem. Soc.*, **117**, 524.

121 Keiholtz, G.W. and Moore, R.E. (1972) *Nucl. Technol.*, **16**, 566.

122 Riley, F.L. (1977) *Nitrogen Ceramics*, NATO ASI Series E: Applied Science, No. 23, Noordhoff, Leyden, The Netherlands.

123 Balmain, W.H. (1842) *J. Prakt. Chem.*, **27**, 422.

124 Batzanov, S.S., Blokhina, G.E., and Deribas, A.A. (1965) *Struct. Chem.*, **6**, 209.

125 Akashi, T., Sawaoka, A., Saitoh, S., and Arahi, H. (1976) *Jpn. J. Appl. Phys.*, **15**, 891.

126 Sokolowska, A. and Olszyna, A. (1992) *J. Cryst. Growth*, **121**, 733–736.

127 Sokolowska, A. and Olszyna, A. (1992) *J. Cryst. Growth*, **116**, 507–710.

128 Michalski, A. and Olszyna, A. (1993) *Surf. Coat. Technol.*, **60**, 498–501.

129 Sokolowska, A. and Wronikowski, M. (1986) *J. Cryst. Growth*, **76**, 511–513.

130 Batsanov, S.S., Kopaneva, L.J., Lazareva, E.V., Kulikova, I.M., and Barinsky, R.L. (1993) *Propell. Explos. Pyrotech.*, **18**, 352–355.

131 Weissmantel, C., Bewilogua, K., Breuer, K., Dietrich, D., Ebersbach, U., Erler, H.J., Rau, B., and Reisse, G. (1982) *Thin Solid Films*, **91**, 31.

132 Murakawa, M. and Watanabe, S. (1990) *Surf. Coat. Technol.*, **43/44**, 128–136.

133 Rother, B., Zscheile, H.D., Weissmantel, C., Heiser, C., Holzhuter, G., Leonhardt, G., and Reich, P. (1986) *Thin Solid Films*, **142**, 83–99.

134 Halverson, W. and Quinto, D.T. (1985) *J. Vac. Sci. Technol.*, **A3**, 2141–2146.

135 Ikeda, T., Kawate, Y., and Hirai, Y. (1990) *J. Vac. Sci. Technol.*, **A4**, 3168–3174.

136 Hamilton, E.J.M., Dolan, S.E., Mann, C.M., Colijn, H.O., McDonald, C.A., and Shore, S.G. (1993) *Science*, **260**, 659–661.

137 Schmolla, W. and Hartnagel, H.L. (1983) *Solid State Electron.*, **26**, 931–939.

138 Mishima, O. and Era, K. (2000) in *Electric Refractory Materials* (ed. Y, Kumashiro), Marcel Dekker, New York Basel, pp. 495–549.

139 Pierson, H.O. (1975) *J. Compos. Mater.*, **9**, 228–240.

140 Chopra, K.L., Agarwal, V., Vankar, V.D., Deshpandey, C.V., and Bunshaw, R.F. (1984) *Thin Solid Films*, **126** (3–4), 307–312.

141 Wentzcovitch, R., Chang, K.J., and Cohen, M.L. (1986) *Phys. Rev*, **B34**, 1071–1079.

142 Silberberg, M.S. (2009) *Chemistry: The Molecular Nature of Matter and Change*, 5th edn, McGraw-Hill, New York, p. 483.

143 Bundy, F.P. and Wentorf, R.H. (1963) *J. Chem. Phys.*, **38**, 1144–1149.

144 Mirkarimi, P.B. *et al.* (1997) *Mater. Sci. Eng. Rep.*, **21**, 47–100.

145 Paine, R.T. and Narula, Ch.K. (1990) *Chem. Rev.*, **90**, 73–91.

146 Onak, T. and Barton, L. (eds) (1991) *Gmelin Handbook of Inorganic Organometallic Chemistry*, 8th edn, Boron Compounds, 4th Supplement, Vol. 3a, Boron and Nitrogen, Springer–Verlag, Berlin.

147 Schwetz, K.A. and Lipp, A. (1979) *Ber. Deutsch. Keram. Ges.*, **56**, 1.

148 Pan, Z. *et al.* (2009) *Phys. Rev. Lett.*, **102**, 140405.

149 Engler, M. *et al.* (2007) Hexagonal Boron Nitride (hBN) – Applications from metallurgy to cosmetics. *Cfi/Ber. DKG*, **84**, D25.

150 Demazeau, G. (1993) *Diam. Relat. Mater.*, **2**, 197–200.

151 Taylor, K.M. (1955) *Ind. Eng. Chem.*, **47**, 2506–2509.

152 Karim, M.Z., Cameron, D.C., and Hashmi, M.S.J. (1993) *Surf. Coat. Technol.*, **60**, 502–505.

153 Leichtfried, G. *et al.* (2002) *Landolt-Börnstein – Group VIII Advanced Materials and Technologies*, Powder Metallurgy Data. Refractory, Hard and Intermetallic Materials, vol. 2A2, Springer-Verlag, Berlin, pp. 118–139.

154 Kubota, Y. *et al.* (2007) *Science*, **317**, 932.

155 Watanabe, K., Taniguchi, T., and Kanda, H. (2004) *Nature Mater.*, **3**, 404.

156 Taniguchi, T. *et al.* (2002) *Appl. Phys. Lett.*, **81**, 4145.

157 Haubner, R., Wilhelm, M, Weissenbacher, R., and Lux, B. (2002) *High Performance Non-Oxide Ceramics II*, vol. 102, Springer-Verlag, Berlin.
158 Rogl, P. (2001) *Int. J. Inorg. Mater.*, **3**, 201–209.
159 Zhang, G.-J., Yang, J.-F., Ando, M., and Ohji, T. (2002) *J. Eur. Ceram. Soc.*, **22**, 2551–2554.
160 Sheppard, L.M. (1990) *Am. Ceram. Soc. Bull.*, **69**, 1801–1812.
161 Liu, F.S., Dong, H.W., Liu, Q.L. *et al.* (2006) *Opt. Mater.*, **28**, 1029–1036.
162 Slack, G.A., Tanzilli, R.A., and Pohl, R.O. (1987) *J. Phys. Chem. Solid*, **48**, 641–647.
163 Zheng, J., Song, X.B., Zhang, Y.H. *et al.* (2007) *J. Solid State Chem.*, **180**, 276–283.
164 Qiu, J.Y., Hotta, Y., Watari, K. *et al.* (2006) *J. Eur. Ceram. Soc.*, **26**, 385–390.
165 Lu, S.X., Tong, Y.H., Liu, Y.C. *et al.* (2005) *J. Phys. Chem. Solid*, **66**, 1609–1613.
166 Sardar, K. and Rao, C.N.R. (2005) *Solid State Sci.*, **7**, 217–220.
167 Kingsley, J.J. and Patil, K.C. (1988) *Mater. Lett.*, **11–12**, 427–432.
168 Qin, M.L., Qu, X.H., Lin, J.L. *et al.* (2002) *J. Inorg. Mater.*, **17**, 1054–1058.
169 Tsuge, A. (1990) *J. Mater. Sci.*, **25**, 2359–2361.
170 Silverman, L.D. (1988) *Adv. Ceram. Mater.*, **3**, 418–419.
171 Boey, F., Tok, A.I.L., Lam, Y.C. *et al.* (2002) *Mater. Sci. Eng. A*, **A335**, 281–289.
172 Yu, Y.D., Hundere, A.M., Høier, R. *et al.* (2002) *J. Eur. Ceram.*, **22**, 247–252.
173 Qiao, L., Zhou, H.P., Chen, K.X. *et al.* (2003) *J. Eur. Ceram. Soc.*, **23**, 1517–1524.
174 Ishiyama, M. (1993) Plasma activated sintering (PAS) system, in Proceedings of the 1993 Powder Metallurgy World Congress, Kyoto, (eds Y. Bando and K. Kosuge), Japanese Society of Powder and Powder Metallurgy, Japan, pp. 931–934.
175 Khor, K.A., Cheng, K.H., Yu, L.G. *et al.* (2003) *Mater. Sci. Eng. A*, **A347**, 300–305.
176 Cheng, H., Sun, Y., and Hing, P. (2003) *Surf. Coat. Technol.*, **166**, 231.
177 Kumar, A., Chan, H.L., Weimer, J.J., and Sanderson, L. (1997) *Thin Solid Films*, **308/309**, 406.
178 Jain, S.C., Willander, M., Narayan, J., and Van Overstraeten, R. (2000) *J. Appl. Phys.*, 87, 965.
179 Wang, X., Kolitsch, A., and Moller, W. (1997) *Appl. Phys. Lett.*, **71**, 1951.
180 Nakamura, S. (1997) *Solid State Commun.*, **102**, 237.
181 Ponce, F.A. and Bour, D.P. (1997) *Nature (London)*, **386**, 351.
182 Strite, S., Morkoc, H., and Vasc, J. (1997) *Sci. Technol.*, **B10**, 1237.
183 Davis, R.F., Ailey, K.S., Bremser, M.D., Carlson, E., Kem, R.S., Kester, D.J., Perry, W.G., Tanaka, S., and Weeks, T.W. (1996) in *Advances in Solid State Physics*, vol. 35, Vieweg, Braunschweig/Wiesbaden, p. 1.
184 Nakamura, S., Mukai, T., and Senoh, M. (1994) *Appl. Phys. Lett.*, **64**, 1687.
185 Nakamura, S., Senoh, M., Nagahara, S., Iwasa, N., Yamada, T., Matsushita, T., Kiyoku, H., and Sugimoto, Y. (1996) *Jpn. J. Appl. Phys.*, **35**, L74.
186 Khan, M.A., Sun, C.J., Yang, J.W., Chen, Q., Lim, B.W., Anwar, M.Z., Osisnsky, A., and Temkin, H. (1996) *Appl. Phys. Lett.*, **69**, 2418.
187 Khan, M.A., Shur, M.S., Chen, Q.C., and Kuznia, J.N. (1994) *Electron. Lett.*, **30**, 2175.
188 Aktas, O., Kim, W., Fan, Z., Botchkarev, A.E., Salvador, A., Mohammad, S.N., Sverdlov, B., and Morkoc, H. (1995) *Electron. Lett.*, **31**, 1389.
189 Sundgren, J.E. (1985) *Thin Solid Films*, **128**, 21–44.
190 Toth, L.E. (1971) *Transition Metal Carbides and Nitrides*, Academic Press, New York.
191 Sato, T., Tada, M., Huangand, Y.C., and Takei, H. (1978) *Thin Solid Films*, **54**, 61.
192 Yoshihara, H. and Moil, H. (1979) *J. Vac. Sci. Technol.*, **16**, 1007.
193 Sundgren, J.-E., Johansson, B.-O., Karlsson, S.-E., and Hentzell, H.T.G. (1983) *Thin Solid Films*, **105**, 367.
194 Chevallier, J., Chabert, J.P., and Spitz, J. (1981) Internal Rep. DMG 21/28, abstract in *Thin Solid Films*, **80**, 263.
195 Matthews, A. and Teer, D.G. (1980) *Thin Solid Films*, **73**, 367.
196 Dunand, A. Flack, H.D., and Yvon, K. (1985) *Phys. Rev.*, **B23**, 2299.
197 Molecular Metallurgy Inc. http://www.mmicoating.com/technology.html, Accessed on 25 June 2009.
198 Pierson, H.O. (1996) in *Handbook of Refractory Carbides and Nitrides: Properties,*

Characteristics, Processing, and Applications, William Andrew, p. 193.

199 Karlsson, B., Shimshoek, R.P., Seraphin, B.O., and Haygarth, J.C. (1982) *Phys. Scr.*, **25**, 775.

200 Ljungcrantz, H., Oden, M., Hultman, L., Greene, J.E., and Sundgren, J.-E. (1996) *J. Appl. Phys.*, **80**, 6725.

201 Johansson, B.O., Sundgren, J.-E., Greene, J.E., Rockett, A., and Barnett, S.A. (1985) *J. Vac. Sci. Technol.*, **A3**, 303.

202 Ljungcrantz, H., Oden, M., Hultman, L., Greene, J.E., and Sundgren, J.-E. (1996) *J. Appl. Phys.*, **80**, 6725.

203 Chen, C.T., Song, Y.C., Yu, G.-P., and Huang, J.-H. (1998) *J. Mater. Eng. Perform.*, **7**, 324.

204 Je, J.H., Noh, D.Y., Kim, H.K., and Liang, K.S. (1997) *J. Appl. Phys.*, **81**, 6126.

205 Pelleg, J., Zevin, L.Z., Lungo, S., and Croitora, N. (1991) *Thin Solid Films*, **197**, 117.

206 Hohl, F., Stock, H.-R., and Mayr, P. (1992) *Surf. Coat. Technol.*, **54/55**, 160.

207 LaFontaine, W.R., Paszkiet, C.A., Korhonen, M.A., and Li, C.-Y. (1991) *J. Mater. Res.*, **6**, 2084.

208 Valvoda, V. (1996) *Surf. Coat. Technol.*, **80**, 61–65.

209 Stone, D.S., Yoder, K.B., and Sproul, W.D. (1991) *J. Vac. Sci. Technol.*, **A9**, 2543–2547.

210 Noel, J.P., Houghton, D.C., Este, G., Shepherd, F.R., and Plattner, H. (1998) *J. Vac. Sci. Technol.*, **A2**, 284.

211 Igasaki, Y. and Mitsuhashi, H. (1980) *Thin Solid Films*, **70**, 17.

212 Posadowski, W., Krol-Stiepniewska, L., and Ziolowski, Z. (1979) *Thin Solid Films*, **62**, 347.

213 Ahn, K.Y., Wittmer, M., and Ting, C.Y. (1983) *Thin Solid Films*, **107**, 45.

214 Schiller, S., Beister, G., and Sieber, W. (1984) *Thin Solid Films*, **111**, 259.

215 Poitevin, J.M., Lemperiere, G., and Tardy, J. (1982) *Thin Solid Films*, **97**, 69.

216 Lemperiere, G. and Poitevin, J.M. (1984) *Thin Solid Films*, **111**, 339.

217 Kang, C.S., Cho, H.-J., Kim, Y.H., Choi, R., Onishi, K., Shariar, A., and Lee, J.C. (2003) *J. Vac. Sci. Technol.*, **B21**, 2026.

218 Christensen, A.N. and Lebech, B. (1978) *Acta Crystallogr.*, **B34**, 261.

219 Stampfl, C. and Freeman, A.J. (2005) *Phys. Rev. B*, **71**, 024111.

220 Tsvyashchenko, A.V., Popova, S.V., and Alekseev, E.S. (1980) *Phys. Stat. Solid.*, **B99**, 99.

221 Gatterer, J., Dufek, G., Ettmayer, P., and Kieffer, R. (1975) *Monatsch. Chem.*, **106**, 1137.

222 Sahnoun, M., Daul, C., Driz, M., Parlebas, J.C., and Demangeat, C. (2005) *Comp. Mater. Sci.*, **33**, 175.

223 Yu, L., Stampfl, C., Marshall, D., Eshrich, T., Narayanan, V., Rowell, J.M., Newman, N., and Freeman, A.J. (2002) *Phys. Rev. B*, **65**, 245110.

224 Stampfl, C. and Freeman, A.J. (2003) *Phys. Rev. B*, **67**, 064108.

225 Häglund, J., Fernández Guillermet, A., Grimvall, G., and Körling, M. (1993) *Phys. Rev. B*, **48**, 11685.

226 Sundgren, J.-E., Johansson, B.-O., Rockett, A., Barnett, S.A., and Greene, J.E. (1986) in *Physics and Chemistry of Protective Coatings*, vol. 149, American Institute of Physics, New York, p. 95.

227 Massalski, T.B. (1990) *Binary Alloy Phase Diagrams*, ASM International, Metal Park, Ohio, p. 2703.

228 Terao, N. (1971) *Jpn. J. Appl. Phys.*, **10**, 248.

229 Gerstenberg, D. and Calbick, C.J. (1964) *J. Appl. Phys.*, **35**, 402.

230 Lee, G.R., Kim, H., Choi, H.S., and Lee, J.J. (2007) *Surf. Coat. Technol.*, **201**, 5207–5210.

231 Greenwood, NN. and Earnshaw, A. (1984) *Chemistry of the Elements*, Pergamon, Oxford, p. 480.

232 Navinsek, B. and Panjan, P. (1993) *Thin Solid Films*, **223**, 4.

233 Ren, M., Yang, Z.G., and Shaw, L.L. (1999) *Nanostruct. Mater.*, **11**, 25.

234 Aizawa, T., Kuwahara, H., and Tamura, M. (2002) *J. Am. Ceram. Soc.*, **85**, 81.

235 Ren, R.M., Yang, Zh.G., and Shaw, L.L. (1999) *Nanostruct. Mater.*, **11**, 25.

236 Qian, X.F., Zhang, X.M., Wang, C., Tang, K.B., Xie, Y., and Qian, Y.T. (1999) *Mater. Res. Bull.*, **34**, 433.

237 Zhang, Z.D., Liu, R.M., and Qian, Y.T. (2002) *Mater. Res. Bull.*, **37**, 1005.

238 Li, Y.G., Gao, L., Li, J.G., and Yan, D.Sh. (2002) *J. Am. Ceram. Soc.*, **85**, 1295.

239 Yang, X.G., Li, C., Yang, B.J., Wang, W., and Qian, Y.T. (2004) *Mater. Res. Bull.*, **39**, 957.

240 Chiba, Y., Omura, T., and Ichimura, H.J. (1993) *Mater. Res.*, **8**, 1109.

241 Sikkens, M., Van Heereveld, A., and Volgelzang, E. (1983) *Thin Solid Films*, **108**, 229.

242 Engel, P., Schwarz, G., and Wolf, G.K. (1998) *Surf. Coat. Technol.*, **98**, 1002.

243 Lin, J.F., Liu, M.H., and Wu, J.D. (1996) *J. Eur. Ceram. Soc.*, **194**, 1–11.

244 Esaka, F., Furuya, K., Shimada, H., Imamura, M., Matsubayashi, N., Sato, H., Nishijima, A., Kawana, A., Ichimura, H., and Kikuchi, T. (1997) *J. Vac. Sci. Technol.*, **A15**, 2521.

245 Reckeweg, O. and DiSalvo, F.J. (2001) *Z. Anorg. Allg. Chem.*, **627**, 371–377.

246 Weller, M.T. (1977) *Inorganic Materials and Solid State Chemistry*, University of Southampton, pp. 1–2.

247 Schnick, W. (1993) *Angew. Chem., Int. Ed. Engl.*, **32**, 806–818.

248 Wintenberger, M., Tcheou, F., David, J., and Lang, J. (1980) *Z. Naturforsch. B*, **35**, 604.

249 Groen, W.A., Kraan, M.J., and deWith, G. (1993) *J. Eur. Ceram. Soc.*, **12**, 413–420.

250 Hintzen, H.T., Swaanen, P., Metselaar, R., Groen, W.A., and Kraan, M.J. (1994) *J. Mater. Sci. Lett.*, **13**, 1314–1316.

251 Davies, I.J., Uchida, H., Aizawa, M., and Itatani, K. (1999) *Inorg. Mater. Jpn*, **6**, 40–48.

252 Hayashi, H., Hirao, K., Toriyama, M., Kanzaki, S., and Itatani, K. (2001) *J. Am. Ceram. Soc.*, **84**, 3060–3062.

253 (a) Lenčéš, Z., Hirao, K., Kanzaki, S., Hoffmann, M.J., and Šajgalík, P. (2004) *J. Eur. Ceram. Soc.*, **24**, 3367–3375; (b) Lenčéš, Z., Hirao, K., Yamauchi, Y., and Kanzaki, S. (2003) *J. Am. Ceram. Soc.*, **86**, 1088–1093.

254 Gaido, G.K., Dubrovskij, G.P., and Zykov, A.M. (1974) *Izv. Akad. Nauk SSSR Neorg. Mater.*, **10**, 564–566.

255 Dubrovskij, G.P., Zykov, A.M., and Tchernovets, B.V. (1981) *Izv. Akad. Nauk SSSR Neorg. Mater.*, **17**, 1421–1425.

256 Uheda, K., Takizawa, H., Endo, T., Yamane, H., Shimada, M., and Wang, C.-M. (2000) *J. Lumin.*, **87–89**, 967–969.

257 Ueda, K., Takizawa, H., Endo, T., Yamane, H., and Shimada, M. (1999) *Kotai no Hannosei Toronkai Koen Yokoshu*, **10**, 65–68.

258 Ueda, K., Takizawa, H., Endo, T., Yamane, H., Shimada, M., and Mitomo, M. (2001) *Nippon Kagakkai Koen Yokoshu*, **79**, 251.

259 Ellens, A., Fries, T., Fiedler, T., and Huber, G. (26 March 2003) European Patent EP1296376.

260 Yoshimura, N., Suehiro, Y., Takahashi, Y., Ota, K., Mitomo, M., and Endo, T. (2005) United States Patent 20050001225.

261 Groen, W.A., Kraan, M.J., and de With, G. (1993) *J. Eur. Ceram. Soc.*, **12**, 413–420.

262 Davies, I.J., Shimazaki, T., Aizawa, M., Suemasu, H., Nozue, A., and Itatani, K. (1999) *Inorg. Mater. Jpn*, **6**, 276–84.

263 Davies, I.J., Uchida, H., Aizawa, M., and Itatani, K. (1999) *Inorg. Mater. Jpn*, **6**, 40–48.

264 Hintzen, H.T., Swaanen, P., Metselaar, R., Groen, W.A., and Kraan, M.J. (1994) *J. Mater. Sci. Lett.*, **13**, 1314–16.

265 Hintzen, H.T., Bruls, R., Kudyba, A., Groen, W.A., and Metselaar, R. (1995) Ceramic Transactions. vol. 51, International Conference on Ceramic Processing Science and Technology, Friedrichshafen, Germany, American Ceramic Society, Westerville, OH, pp. 585–89.

266 Bruls, R.J., Hintzen, H.T., and Metselaar, R. (1999) *J. Mater. Sci.*, **34**, 4519–31.

267 Uchida, H., Itatani, K., Aizawa, M., Howell, F.S., and Kishioka, A. (1997) *J. Ceram. Soc. Jpn*, **105**, 934–39.

268 David, J. and Lang, J. (1965) *Comptes Rendus Acad. Sci. Paris*, **261**, 1005–1007.

269 Gál, Z.A., Mallinson, P.M., Orchard, H.J., and Clarke, S.J. (2004) *Inorg. Chem.*, **43**, 3998–4006.

270 Groen, W.A., Kraan, M.J., and De With, G. (1994) *J. Mater. Sci.*, **29**, 3161–3166.

271 Schlieper, T. and Schnick, W. (2004) *Z. Anorg. Allg. Chem.*, **621**, 1037–1041.

272 Schlieper, T., Milius, W., and Schnick, W. (2004) *Z. Anorg. Allg. Chem.*, **621**, 1380–1384.

273 Fang, C.M., Hintzen, H.T., De With, G., and de Groot, R.A. (2001) *J. Phys. Condens. Matter*, **13**, 67–76.

274 Inoue, Z., Mitomo, M., and Li, N. (1980) *J. Mater. Sci.*, **15**, 2915–2920.
275 Holcombe, C.E. and Kovach, L. (1981) *Am. Ceram. Soc. Bull.*, **60**, 546–548.
276 Inoue, Z. (1985) *J. Mater. Sci. Lett.*, **4**, 656–658.
277 Woike, M. and Jeitschko, W. (1995) *Inorg. Chem.*, **34**, 5105–5108.
278 Fanelli, A.J., Solar, J.P., Wu, B.L., and Yamanis, J. (1994) United States Patent 5292489.
279 Hatfield, G.R., Li, B., Hammond, W.B., Reidinger, F., and Yamanis, J. (1990) *J. Mater. Sci.*, **25**, 4032–4035.
280 Kipsová, L., Lenčéš, Z., Zhou, Y., and Šajgalík, P. (2008) Proceedings of 4th Alternative Energy Recourses (ALER) Conference, Liptovský Mikuláš, pp. 46–52, ISBN 978-80-8070-912-9.
281 Yamane, H., Kikkawa, S., and Koizumi, M. (1987) *Solid State Ionics*, **25**, 183–191.
282 Orth, M. and Schnick, W. (1999) *Z. Anorg. Allg. Chem.*, **625**, 1426–1428.
283 Anderson, A.J., Blair, R.G., Hick, S.M., and Kaner, R.B. (2006) *J. Mater. Chem.*, **16**, 1318–1322.
284 Bharma, M.S. and Fray, D.J. (1995) *J. Mater. Sci.*, **30**, 5381–5388.
285 Kuriyama, K., Kaneko, Y., and Kushida, K. (2005) *J. Cryst. Growth*, **275**, 395–399.
286 Kalaiselvi, N., Doh, C.H., and SriKeerthi, P. (2007) *J. New Mater. Electrochem. Syst.*, **10**, 209–212.
287 Kalaiselvi, N. (2007) *Int. J. Electrochem. Sci.*, **2**, 478–487.
288 Xie, R.-J. *et al.* (2007) *Phosphor Handbook*, 2nd edn, CRC Press, Boca Raton, p. 331.
289 Hirosaki, N. *et al.* (2006) *Bull. Ceram. Soc. Jpn.*, **41**, 602.
290 Römer, S.R., Braun, C., Oeckler, O., Schmidt, P.J., Kroll, P., and Schnick, W. (2008) *Chemistry*, **14**, 7892–7902.
291 Hoppe, H.A. *et al.* (2000) *J. Phys. Chem. Solids*, **61**, 2001.
292 Hirosaki, N. *et al.* (2005) *Appl. Phys. Lett.*, **86**, 211905.
293 Piao, X. *et al.* (2006) *Chem. Lett.*, **35**, 334.
294 Li, Y.Q. *et al.* (2005) *Chem. Mater.*, **15**, 4492.
295 Xie, R.-J. *et al.* (2007) *J. Electrochem. Soc.*, **154**, J314.
296 Zhou, Y., Yoshizawa, Y., Hirao, K., Lenčéš, Z., and Šajgalík, P. (2008) *J. Am. Ceram. Soc.*, **91** (9), 3082–3085.
297 Bachmann, V. *et al.* (2006) *J. Lumin.*, **121**, 441.
298 Xie, R.-J. and Hirosaki, N. (2007) *Sci. Technol. Adv. Mater.*, **8**, 588–600.
299 Xie, R.-J. *et al.* (2006) *Appl. Phys. Lett.*, **89**, 241103.

3
Gallium Nitride and Oxonitrides

Isabel Kinski and Paul F. McMillan

3.1
Introduction

Nitride and oxonitride compounds provide several well-known families of useful technological materials, ranging from refractory ceramics and electronic substrates (AlN, AlO_xN_y, Si_3N_4) to ultrahard metallic phases and superconductors (TiN, NbN) [1–3]. Generally, the solid-state chemistry of nitride materials is less well developed than that of the corresponding oxides, but this situation is changing rapidly. A wide range of new nitride structures with unique properties has now been described, and further research is continuing [3–7]. Developing synthesis routes to the nitrides is generally hampered by the high stability of strongly bound molecular N_2. Although both, transition metal and covalently bonded main group nitrides are highly stable refractory materials, they are often nonstoichiometric and nitrogen-deficient, and low-temperature preparation routes often result in poorly crystalline materials. Annealing these materials under high-pressure/high-temperature conditions can lead to highly ordered compounds with improved electronic properties [8–10]; also, crystal growth under high N_2 pressures has been used to create single-crystalline AlN and GaN substrates for optoelectronic materials and devices [11, 12]. The field of high-pressure/high-temperature exploration has gained increasing importance with the synthesis of new nitrides of Pt and Os [13, 14], high-hardness cubic-structured Zr_3N_4 and Hf_3N_4 phases [15], and spinel-structured compounds based on Si_3N_4 and Ge_3N_4 [16–19]. The corresponding spinel-structured Sn_3N_4 compound was prepared for the first time using solid-state chemistry techniques, as a metastable phase [20], and later by a metathesis reaction carried out under high-pressure/high-temperature conditions [21]. The high-pressure/high-temperature synthesis studies on nitride spinels have been extended to oxonitride phases, including SiAlONs [22, 23]. Within the Al_2O_3–AlN system, several important $Al_xO_yN_z$ ceramic solid solutions and compounds are already well known [24]. At high AlN contents, layered phases based on hexagonal/cubic intergrowths are present; however, as the Al_2O_3 content is increased then cubic spinel-structured materials begin to appear. A large family of defect spinels (γ-Al_2O_3; $Al_xO_yN_z$) containing vacancies on both cation and anion sites are known. In this system, the

Ceramics Science and Technology Volume 2: Properties. Edited by Ralf Riedel and I-Wei Chen

ISBN: 978-3-527-31156-9

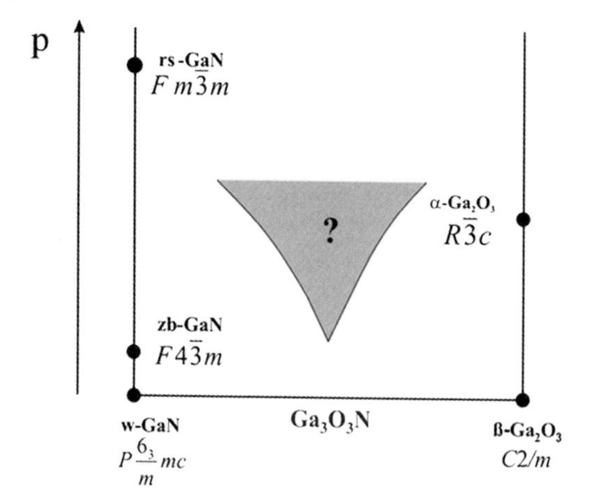

Figure 3.1 Schematic drawing of a possible stability field of a ternary compound in the binary phase diagram of the system GaN–Ga_2O_3 in function of pressure and molar ratio.

ideal stoichiometric oxide nitride spinel-structured compound would be at a composition Al_3O_3N, in which Al^{3+} ions are present on both octahedral and tetrahedral sites, and O^{2-} and N^{3-} occupy tetrahedral anion sites [25, 26]. It was predicted, by using *ab initio* techniques, that a corresponding Ga_3O_3N phase should exist and that it would have comparable optoelectronic properties to hexagonal wurtzite-structured GaN, and also the cubic spinel-structured phases of Si_3N_4 and Ge_3N_4 (see Figure 3.1) [27–31]. Insulating gallium oxide nitride layers are already established within the semiconductor community to provide an oxygen- and temperature-stable surface layer native to GaN for active surface portions in field effect transistor (FET) devices shielding the gate and semiconductor regions [32]. A bulk cubic phase with a composition approaching that of Ga_3O_3N has now been synthesized using high-pressure/high-temperature methods, either from starting mixtures of the oxides/nitrides (i.e., GaN + Ga_2O_3) [33] or using organometallic precursors [34, 35]. The precursor synthesis route could lead to the useful production of Ga–O–N films with the high-density structure at low-pressure conditions, and with variable chemical composition and optoelectronic properties. Interestingly, thin films of the related metastable high-hardness cubic phase of Zr_3N_4 first identified from high-pressure/high-temperature studies [36, 37] has recently been prepared under low-pressure conditions using physical vapor deposition (PVD) methods [38]. In this chapter, the solid-state structures, synthesis chemistry and physical properties of gallium nitrides and oxonitrides and their solid solutions with other Group 13 elements (Al^{3+}, In^{3+}) are reviewed. Details of the complex stable versus metastable polymorphism of Ga_2O_3 compounds that are just beginning to be developed for optoelectronic and catalysis applications are also described.

Since the seminal discovery of nitride-based blue light-emitting diodes (LEDs) by Maruska and Tietjen [39] in 1972, and of injection lasers [40] by Nakamura in 1995, the interest in Group 13 nitrides has increased rapidly, and has resulted in a broad usage for electrical and optical applications. Some thirty years after the first discovery of blue LEDs [41] based on gallium nitride (GaN) and its solid solutions formed with In and Al, the solid-state LEDs found entry into everyday life by replacing conventionally used lighting sources in traffic signs and signals, and for automotive lighting. In the fourth edition of the survey of worldwide markets and development activities, *Gallium Nitride-2005*, for gallium nitride-based devices, the commercial sales were noted as US$ 3.2 billion in 2004, and this is forecast to grow to US$ 7.2 billion by 2009 [42]. The major technology breakthroughs have taken place as applications in the fields of signage and signaling, automobiles, mobile phones, lighting, data storage and ultraviolet (UV) emitters for optoelectronics and in various military, aerospace, automotive, industrial and communication systems for electronics [42]. Although the achievements over the past three decades have established the great potential of this new class of materials based on the Group 13 nitrides, they have still not reached a point where their true performance or range of applications can be evaluated. There remain substantial experimental and developmental challenges to be met in order to fully exploit the variety of properties due to the large range of band gap energies available from 1.9 to 6.2 eV, from red to UV wavelengths, among the existing materials with known wide bandgap semiconducting and optoelectronic applications [43]. Current major research programs are directed at developing gallium nitride-based emitters that perform either at longer or shorter wavelengths than those of the blue to near-UV spectrum [44].

One of the critical issues in improving and developing efficient devices based on GaN is the absence of ideal substrates for growth of epitaxial thin films. In general, the Group 13 nitrides are grown on Al_2O_3 (sapphire) substrates that are widely available as single crystals with hexagonal symmetry, and that provide easy handling and cleaning, and stability at high temperatures. However, these lead to problems associated with the poor structural correspondence with the GaN phase and the wide difference in thermal conductivity. The lattice mismatch between Ga(In,Al)N and sapphire substrates results in films that contain threading dislocations with densities ranging from $\sim 10^8$ to $\sim 10^{10}\,cm^{-2}$, depending on the growth method and conditions [45]. Additionally, the nature, crystal quality and properties of the epitaxially-grown nitrides are strongly influenced by the substrate material and its crystalline orientation. Certain studies have proposed hexagonal 6H polytype of SiC (0001 plane) as a useful substrate for GaN deposition [46–48]. However, the best material is single crystalline GaN itself, grown by high-pressure/high-temperature methods in a nitrogen atmosphere [49]. Gallium nitride grown on planes with hexagonal symmetry [i.e., (0 0 0 1) of the sapphire or wurtzite-structured phases] results in the stable hexagonal polymorph with wurtzite structure; however, GaN grown on substrates with orientations directing formation of cubic symmetry, lead to a metastable zincblende-type structure. The cubic structure can also be synthesized by the use of certain precursors in vapor deposition experiments [50, 51]. A high-pressure treatment results in transformation into a rocksalt-structured phase (B1 phase) with octahedral coordination of the Ga and N atoms [52]. However, this phase is not recoverable to ambient conditions (see Table 3.2). All three polymorphs are predicted to have excellent optoelectronic properties for operation as LEDs or solid-state lasers in the blue-UV range; however, only the wurtzite-type structured w-GaN phase has been developed for technological industrial applications so far. The breakthrough for optoelectronic applications of the Group 13 nitrides is mainly due to the considerable achievement of doping GaN with Zn atoms, as found by Maruska and Tietjen, leading to materials that could emit either blue, green, yellow, or red light, depending on the concentration of zinc [39]. The research on gallium nitride doped with magnesium led to the first p-type conductor [53, 54], and this showed an increasing photoluminescence with radiation time during cathodoluminescence measurements carried out using scanning electron microscopy (SEM). The p-type conductivity of this Mg-doped GaN could be also achieved by annealing the material in nitrogen or vacuum at temperatures above 750 °C [55]. This constituted the fundamental discovery by Nakamura *et al.* that led to the rapid improvement of the electronic properties, and resulted in its past commercialization. The fact that LEDs are more energy-efficient, brighter, and more durable than tungsten-filament lighting has cleared the way for replacing the traditional incandescent bulb technology. The breakthroughs associated with this material was accelerated by introducing layers of solid solutions with the other Group 13 nitrides, AlN and InN, into the devices that enabled reaching even shorter wavelengths and opening up fields for application.

Not only the nitride but also the oxide of gallium (Ga_2O_3) is likewise a promising wide band gap semiconductor with potential optoelectronic device applications [56–65], and also for catalysis [66, 67]. Ga_2O_3 thin films can act as oxygen sensors [68, 69], and the crystalline material can be obtained as nanowires and various unusual shapes formed at the nanoscale that might have new applications [70, 71]. The material exhibits considerable polymorphism when prepared by various synthesis routes at ambient conditions [72]. The thermodynamically stable phase at ambient conditions has been determined as the β-Ga_2O_3 polymorph [73] that shows unique optical transparency in the deep-UV region due to its wide band gap of $\sim$4.9 eV. The properties of transparent insulating or conducting doped or defective gallium oxide can be tailored by using the Verneuil method with growth in oxidizing or reducing atmospheres, respectively [57]. For future technology applications in the areas of high-temperature electronics, optoelectronics and oxidation catalysis fields – including biotechnology and nanotechnology – the deep-UV region is most promising. Usefully, the transparent β-gallium oxide displays an electrical conductivity that reaches $\sim$40 S cm^{-1} [56, 57, 60–62]. When grown as an epitaxial film on α-Al_2O_3, the gallium oxide films can exhibit high transparency throughout the near infrared (IR)-deep-UV regions, and can be operated as a top gate for transparent FETs [63].

3.2 Gallium Nitrides

3.2.1 Phase Description

The Ga–N system contains a single composition that is nominally stoichiometric at the 1 : 1 composition first synthesized by Johnson *et al.* [74] by streaming ammonia over metallic gallium at temperatures between 900 and 1000 °C [74]. In these studies, an apparatus was used that had been described two years earlier by Tiede and Chomse [75]. Here, the stable polymorph under ambient pressure and temperature conditions was determined to be the hexagonal wurtzite-structured polymorph [76, 77]. However, there is also the known metastable cubic sphalerite-structured phase first mentioned by Pankove in 1973 as a private communication by Paff [78, 79]. At high temperature, GaN melts incongruently and decomposes above 900 °C to yield metallic Ga (which is liquid at the decomposition temperature) and N_2 [80]. Most studies on the investigation of GaN growth and the thermodynamic data have been carried out at the High Pressure Research Center of the Polish Academy of Sciences Warsaw, Poland. The vapor pressure (pN_2) reaches 1 atm below the thermodynamically predicted melting point. In order to melt GaN, and thus to carry out single-crystal growth experiments to obtain substrates for homoepitaxial deposition, it is necessary to apply a confining pressure of N_2 that exceeds the decomposition pressure at a given temperature. These parameters in the Ga–N system are now well established, so that the high-pressure growth of GaN single-crystal substrates has led to commercially available materials [81].

Both, nominally ionic and covalent binary AX compounds are found to crystallize in either hexagonal wurtzite or cubic zincblende (sphalerite) structure types, and the energy differences between the two are small [82]. It is often difficult to determine which polymorph is more thermodynamically stable under a given set of pressure and temperature conditions, and which is stabilized kinetically following a given synthesis route. In several cases, both structures have been found to coexist over wide ranges in temperature or pressure [83]. The stability of wurtzite versus zincblende structures is closely connected with deviations of the c/a ratio from the ideal value of 1.633, which in turn is linked to the ionicity of the compound. Based on the Phillips ionicity scale, van Vechten predicted phase transitions occurring among tetrahedrally coordinated semiconductors [84–87]. At high pressures, the wurtzite and zincblende compounds would transform into dense structures containing octahedrally coordinated atoms, that were often metallic. In the case of elemental Si and Ge, the high-pressure phases were of the β-Sn type, that is easily derived from the cubic diamond structure by a displacive transformation with a small energetic barrier. Continued densification yielded increasingly closely packed metallic phases. In the case of III–V (Group 13–15) semiconductors, the Phillips–van Vechten ionicity was seen to influence the high-pressure phases that were formed. More ionic Group 13–15 compounds favored the rocksalt (B1) structure at high density, whereas less ionic materials transformed into the β-Sn type structure under pressure. (Recent studies conducted by McMahon and Nelmes [88, 89] have indicated that several of the high-pressure structures may in fact be related to that of cinnabar, HgS.) The high ionicity of GaN (0.5) [90] indicates that it should undergo a phase transition from the wurtzite or zincblende structures into the rocksalt structure at high pressure, as verified experimentally by several groups [91–94].

The stable phase at ambient pressure and temperature conditions is the hexagonal wurtzite-type structure with space group $P6_3mc$ (no. 186). Figure 3.2a shows the unit cell of the wurtzite structure with alternating layers of gallium and nitrogen each

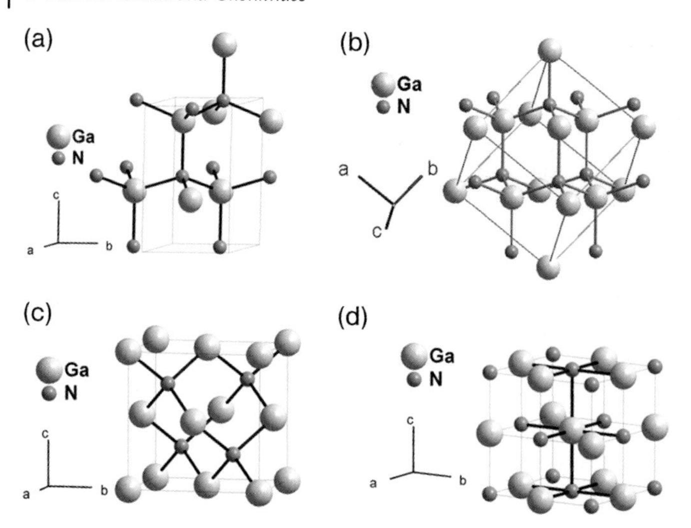

Figure 3.2 (a) Hexagonal wurtzite structure with space group $P6_3mc$ (no. 186); (b) Cubic zincblende structure tilted to display the analogy to the wurtzite structure; (c) Cubic zincblende structure; (d) Cubic rocksalt structure.

coordinated tetrahedrally by the other element. The atomic positions are for gallium atoms at 0, 0, 0 and 2/3, 1/3, 1/2, and for nitrogen atoms at 0, 0, u and 2/3, 1/3, 1/2 + u with u ~ 3/8. Paszkowicz *et al.* refined the structural parameters using Rietveld refinement on GaN powder data, and compared the crystallographic information with earlier determined structure parameters [95]. The lattice parameters for gallium nitride powder were refined as $a_0 = 3.18940(1)$ Å, $c_0 = 5.18614(2)$ Å, with $u = 0.3789(5)$ Å as the positional parameter for nitrogen.

Depending on the choice of substrate and growth conditions, as well as the precursor used, gallium nitride can be grown at ambient pressure in the closely related cubic zincblende structure [96–103]. The zincblende-type structured gallium nitride crystallizes in space group $F\bar{4}3m$ (no. 216) with lattice parameter $a = 4.528$ Å [104–107]. The atomic positions in the unit cell correspond to that of zincblende structure type sphalerite (ZnS), 0, 0, 0 for gallium and 1/4, 1/4, 1/4 for nitrogen atoms, forming again layers of gallium and nitrogen atoms coordinated tetrahedrally with each other, as shown in Figure 3.2c. The structure can be described by a cubic close-packing arrangement of gallium atoms, where the nitrogen atoms fill half of the tetrahedral sites, forming an interpenetrating cubic close-packed lattice.

Both structure types are based on tetrahedral coordination among the first nearest neighbors. However, differences between the two structures arise when the second nearest neighbors are considered. To emphasize the resemblance of both structures in Figure 3.2b, the zincblende structure is displayed with the [1 1 1] direction parallel to the main *c*-axis of wurtzite structure. Both structures form alternating wavy layers of gallium and nitrogen atoms. The wurtzite structure comprises two different layers which are rotated along the main axis by 180° with a hexagonal stacking sequence of ABAB, while the zincblende structure is built of a cubic stacking sequence along [1 1 1] the direction of ABCABC. At normal pressure the enthalpy difference between the stable hexagonal wurtzite-structured and metastable cubic sphalerite-structured polymorphs is low: it amounts to only 0.016 eV per formula unit as estimated by *ab initio* theoretical calculations (i.e., as low as 1–2 kJ mol^{-1}) [82]. The slightly higher energy for the zincblende stacking sequences makes the growth of GaN with the zincblende-type structure difficult, because the formation of stacking faults of wurtzite structure results in a lowered energy [106, 107].

As the zincblende (zb) phase has a smaller molar volume than the wurtzite polymorph, a high-pressure phase transition would be expected to occur between the two as the metastable zb-GaN was compressed. This metastable phase transition has been studied both theoretically and experimentally by several groups (Table 3.1). In this case, heteroepitaxially stabilized growth of zincblende-type GaN were used on various substrates in a variety of experimental studies, while Purdy *et al.* synthesized zb-GaN directly by an ammonothermal reaction of NH_4X with GaX_3 (X = Cl, Br, I) [108–110].

Table 3.1 Experimental and theoretical investigations on zincblende-structured GaN.

Heteroepitaxial on		Lattice parameter a_0 [Å]	Year	Reference
GaAs (1 0 0)	Precursor CVD	4.52–4.55	1986	[111]
β-SiC (1 0 0)	Modified MBE	4.54	1988	[112]
GaAs (0 0 1)	Gas-source MBE		1991	[113]
GaAs (1 0 0)	Plasma-assisted MBE	4.5	1991	[114]
Si (0 0 1)	Electron cyclotron resonance microwave-plasma-assisted MBE	4.52 ± 0.05	1992	[100]
MgO (0 0 1)	Reactive-ion MBE	4.531 ± 0.0005	1992	[115]
GaP (0 0 1)	Low-temperature modified MBE		1995	[116]
GaAs (0 0 1)	Plasma-assisted MBE		1996	[117]
GaAs (0 0 1)	Plasma-assisted MBE		1997	[118]
w-GaN on sapphire	Phosphorus-mediated MBE		1999	[119]
Bulk c-GaN	Ammonothermal reaction of NH_4X X = Cl, Br, I		1999	[108]
	Na-Ga-melt	4.5019(1)	2000	[120]
Calculated	Empirical pseudopotential method		2002	[121]
Si-doped GaAs (1 0 0)	MO VD	4.5008	2003	[122]
Bulk	Ammonothermal		2005	[109]
Calculated		4.558	2005	[98]

CVD = chemical vapor deposition; MBE = molecular beam epitaxy; MOVD = metal–organic chemical vapor deposition.

Table 3.2 Summary of the transformation of zb- or w-GaN under high-pressure conditions into rock salt (rs)-GaN (B1) phase.

Method	W-RS phase transition pressure [GPa]	Year	Reference
Experimental observation			
DAC	47–50 upstroke 30–20 downstroke	1993	[91]
DAC	37 beginning	1993	[123]
DAC	52.2 beginning	1994	[93]
Benzene thermal reaction of Li3N and GaCl3	5 MPa	1996	[124]
DAC	53.6 beginning	1997	[94]
Theoretical calculations			
Density functional theory	50 upstroke 30 downstroke	1991	[125]
First-principles nonlocal pseudopotential	55.1	1992	[126]
First-principles method	37–52	2000	[127]
Troullier–Martins pseudopotential	42.1 zb-rs		
Tersoff empirical potential model	42 zb-rs	2003	[128]
Molecular dynamics method with Buckingham potential	44	2005	[98]
Ab initio plane-wave pseudopotential density functional theory method	42.2 zb-rs	2005	[129]
Density functional theory VASP	40.4 w-rs	2005	[82]

In situ pressurization experiments in diamond anvil cells have been used to study the transition from w- and zb-GaN into the rock salt structure (B1) at high pressure (see Table 3.2 and Figure 3.2d). The observed transition pressure ranged from 47 to 54 GPa during the upstroke, with a back-transformation recorded at 20–30 GPa during decompression. The large hysteresis recorded in the experiments indicated that the first-order transition had a high activation enthalpy. In contrast to the wurtzite and zincblende polymorphs, B1-structured GaN no longer exhibited a direct band gap. This phase cannot be recovered to ambient pressure and temperature, and it has not yet been reported to be formed metastably at low-pressure conditions in any studies.

The color of the resulting GaN compound produced by different synthesis routes indicates the likely presence or absence of defects and deviations from the ideal composition:

- light-yellow to brown indicates nonstoichiometry present within the GaN phase;
- light-gray to dark gray indicates the beginning of decomposition, when metallic gallium is present;
- colorless indicates that an ideal 1 : 1 stoichiometry has been attained [130, 131].

3.2.2 Synthesis Routes to GaN

In the past, several reviews have summarized the different synthesis and growth techniques developed to improve the degree of crystallinity of GaN films or the bulk phase [43, 105, 130, 132–137]. The two main categories of preparation methods address different aims of applications: the first concerns the epitaxial deposition of highly ordered films on substrates for blue LEDs or lasers; the second other focuses on the growth of large single crystals for use as substrates or wafers. In both cases the main problem is the decomposition of GaN under elevated temperatures. The incongruent melting regime under 1 bar N_2 pressure is reported to occur at temperatures between 700 and 1157 °C, where GaN decomposes completely into a metallic gallium melt and nitrogen gas [138]. Considerable research has been carried out to determine the nitrogen pressure necessary to stabilize the congruent melting of GaN. Karpinski *et al.* have shown, experimentally, that GaN is stable up to temperatures of 2300 °C at a nitrogen pressure of 6 GPa (60 kbar) [139], and predicted that congruent melting would even take place at higher temperatures and pressures, because the system showed no evidence of recrystallization processes in the undecomposed specimen heated up to 2300 °C. Van Vechten calculated a much higher melting temperature of 2520 °C [85]. Utsumi *et al.* investigated the phase diagram of GaN under high pressure and temperature, and observed a rapidly increasing decomposition temperature as a function of pressure [140]. Congruent melting in the GaN system was found to occur at pressures higher than 6 GPa at temperatures around 2220 °C.

3.2.2.1 Synthesis of Bulk Gallium Nitride

Due to the decomposition of GaN at elevated temperatures of 1043 ± 20 °C under 1 bar flowing nitrogen pressure [138], the usual crystal growth techniques employed at $p = 1$ atm (e.g., Bridgman, Czochralski, Verneuil) cannot be applied. Although the synthesis of GaN has been known since 1932 [74], most bulk synthesis methods provide only powders or small crystals. Johnson *et al.* described the conversion of metallic gallium in an ammonia stream at 1100 °C into GaN, although at temperatures below 1100 °C the metallic gallium did not react with the ammonia. For their crystal structure analysis, Juzza and Hahn followed the synthesis description of Johnson *et al.*, but exchanged the porcelain container for an alumina boat, heated the metallic gallium for 2 h in an ammonia stream at 1100 °C, ground the light gray sample, and then reheated it [77]. They also adapted the synthesis method successfully used previously for indium nitride, and heated tris-ammoniumgalliumfluoride $(NH_4)_3GaF_6$ in an ammonia stream up to a temperature of 900 °C for 10 min; this led to the production of yellow-colored crystals [76, 141]. The initial synthesis processes and the names of the following deposition techniques leading to the injection laser diode are summarized in Table 3.3.

For the bulk crystal growth of GaN, the high-pressure nitrogen solution growth (HPNSG) method developed by Karpinski and Porowski has become the most useful technique [131, 139]. This high-pressure/high-temperature approach was initially

Table 3.3 Summary of the initial synthesis processes and deposition techniques leading to the injection laser diode.

Year	Reaction	Description	Reference
1932	$2\,Ga + 2\,NH_3 \rightarrow 2\,GaN + 3\,H_2$		[74]
1938	NH_3 flow over hot Ga	Needles and platelets	[76]
1940	$(NH_4)GaF6 + 4NH_3 \xrightarrow{1100\,°C} GaN + 6NH_4F$		[77]
1940	$(NH_4)InF6 + 4NH_3 \xrightarrow{630\,°C} InN + 6NH_4F$		[77]
1969	$GaCl + NH_3 \rightarrow GaN + HCl + H_2$	Most successful epitaxial growth technique	[39]
	HCl flow over metallic Ga	First single crystal GaN thin films	

Year	Growth technique	Description	Reference(s)
1972	CVD	First Zn-doped LED	[146]
1988	MBE	First p-type GaN	[54]
1995	MBE	First injection laser	[133, 147–158]

CVD = chemical vapor deposition; MBE = molecular beam epitaxy.

developed during investigations of the thermal stability of GaN as a function of high N_2-pressure. During annealing experiments on powder, the GaN started to sublime below the decomposition temperature, and this became the key issue for following crystallization experiments at high nitrogen pressure. By optimizing the pressure vessel and the sublimation–condensation process, crystals up to a maximum size of ~40 mm^2 could be synthesized [134].

Yamane *et al.* described a process based on a sodium flux leading to smaller crystals than HPNSG, but operating at a lower pressure and temperature [142–144]. Over the past ten years, the groups led by Yamane and DiSalvo have advanced, optimized and improved the technique leading to colorless transparent 2 mm × 1 mm GaN crystals at 5 MPa N_2 pressure prepared at 800 °C for 200 h. However, despite these advances, the growth of large samples of bulk stoichiometric GaN with a high degree of crystallinity still remains a considerable challenge, and new solutions to the problem would provide a welcomed solution to many obstacles associated with thin film growth occurring on non-native substrates.

3.2.2.2 Synthesis of Thin Film GaN and Epitaxial Growth Techniques

Most of the epitaxial growth techniques used today for the Group 13 nitrides have evolved from the early approaches to GaN growth by using a source material held in a zone heating furnace under flowing ammonia. The decomposition of GaN below its melting temperature, and the resulting high nitrogen pressures necessary to synthesize even small crystallites, mean that there are no GaN bulk single crystals available as substrates for thin film deposition, and so the technological development of GaN-based devices has relied on heteroepitaxial growth on various other substrates, including Al_2O_3 (sapphire) and SiC [145]. Maruska and Tietjen used their experience in the synthesis of other Group 13–15 semiconductors to adjust the growth technique to the demands of GaN single thin film synthesis [39]. During their early investigations on the epitaxial growth technique of Group 13 nitrides, Maruska and Tietjen used a vapor transport method where hydrochloric acid vapor was flowing over metallic gallium, which reacted at the substrate at a temperature of 825 °C in an ammonia atmosphere. Due to high growth rates of 0.5 μm min^{-1}, extremely thick films in the range of 50 to 150 μm could be deposited. Although films grown using this method were less influenced by lattice mismatch to the substrate or the thermal coefficient difference, the resulting GaN displayed very high background n-type carrier concentrations, approximately $\sim 6 \times 10^{19}\,cm^{-3}$.

3.2.2.3 Synthesis of GaN via Chemical Precursor Routes

Depending on the growth method employed, the requirements for the precursors used as starting materials in the synthesis of GaN vary considerably. For instance, precursors used for metal–organic chemical vapor deposition (MOCVD) should be sufficiently volatile and reactive to decompose thermally into the solid and easily removable gas. Neumayer *et al.* compiled a review on MOCVD precursors for the growth of Group 13 nitrides [132], while Monemar summarized the main growth techniques and the corresponding progress in the resulting materials' quality [134]. Parikh and Adomaitis developed a planetary radial-flow CVD method to characterize the growth chemistry of GaN, and thus created a process that provided a uniform film thickness [137]. In these studies, the gas phase and surface chemistry associated with GaN synthesis were summarized, and a model developed for epitaxial GaN growth using the precursor trimethylgallium ($(CH_3)_3Ga$) in an ammonia flow. The competing reaction pathways were scrutinized and the resultant model was applied to a GaN radial-flow CVD system with planetary wafer rotation, revealing the influence of reactor geometry on the deposition chemistry and the controllability of the competing mechanisms of GaN chemistry.

3.2.3 Properties of (Ga,Al,In)N Solid Solutions

The wurtzite-structured nitrides of the Group 13 elements (Al, Ga, In) form a series with wide direct band gaps ranging from 6.2 eV for AlN to 3.4 eV for GaN and ~0.9 eV for InN at room temperature. The formation of solid solutions between the nitrides leads to possibilities for band gap engineering, although the difference in lattice parameters between InN and the other two are large (Table 3.4). Both, AlN and GaN form continuous solid solutions, whereas the difference in lattice parameters causes miscibility gaps to appear in the GaN–InN and AlN–InN systems. The development of lattice strain accompanies the formation of $In_xGa_{1-x}N$ solid solutions, that provide the active material in injection laser diodes [159, 160]. Once formed, AlN and GaN are both chemically inert, resistant to radiation exposure, and stable up to high temperatures, which makes them attractive for use in chemically reactive environments as high-power devices. However, the focus lies more on their electronic properties with a high avalanche breakdown field, their high thermal conductivity, and their

Table 3.4 Listing of lattice parameters for hexagonal wurtzite-structured AlN, GaN, and InN.

Substance	Structure type	Lattice parameter a_0 [Å]	Lattice parameter c_0 [Å]	Reference
AlN	Wurtzite	3.11197(2)	4.98089(4)	[95]
GaN	Wurtzite	3.186(4)	5.176(5)	[76]
InN	Wurtzite	3.533(4)	5.693(4)	[76]

large high-field electron drift velocity. The layered architecture of wurtzite-structured Group 13 nitrides results in a polarity that is aligned with the growth direction. This polarity determines the orientation of spontaneous piezoelectric polarization inside the material [161].

3.3 Gallium Oxides

The stoichiometric compound of gallium and oxygen, Ga_2O_3, exists in a wide variety of polymorphic phases, including α, β, γ, δ, ε, and θ phases that so far have been identified using powder X-ray diffraction (XRD) [72]. The polymorphs α and β are usually prepared at ambient pressure from thermal decomposition of the nitrate ($Ga(NO_3)_3$), or via sol–gel synthesis routes followed by dehydration and recrystallization of the gel (Figure 3.3 adapted from [72]).

3.3.1 Phase Description and Properties

The monoclinic β-Ga_2O_3 ($C2/m$ no. 12) phase is the stable polymorph at ambient conditions [162]. It is isostructural with θ-Al_2O_3, that forms an intermediate phase during the metastable transformation of partly hydrated "transitional" aluminas as they evolve towards corundum (α-Al_2O_3). The O^{2-} anions form a slightly distorted face-centered cubic (*fcc*) lattice, and cations occupy both the tetrahedral and octahedral interstices (Figure 3.4). The α-Ga_2O_3 polymorph is isostructural with corundum; the O^{2-} ions form an hexagonal close packed (*hcp*) lattice, with two-thirds of the octahedral cation sites filled. Commercially available Ga_2O_3 usually consists of a mixture of the β- and α-phases: pure β-Ga_2O_3 can be obtained by heat treatment [72]. Several *ab initio* calculations have been carried out to determine the electronic properties, phonon dispersion, thermodynamic properties and relative stabilities of the β- and α-Ga_2O_3 polymorphs [28, 82, 163–165].

Beta-gallium oxide is a wide-band gap semiconductor ($E_g = 4.9$ eV) that exhibits promising properties for a range of optical and electronic applications. The wide gap leads to transparent semiconductors for optical windows into deep UV-wavelengths near 280 nm [65, 166]. The conductivity can be varied from insulating to conducting, depending on the preparation conditions and the presence of various dopants [167].

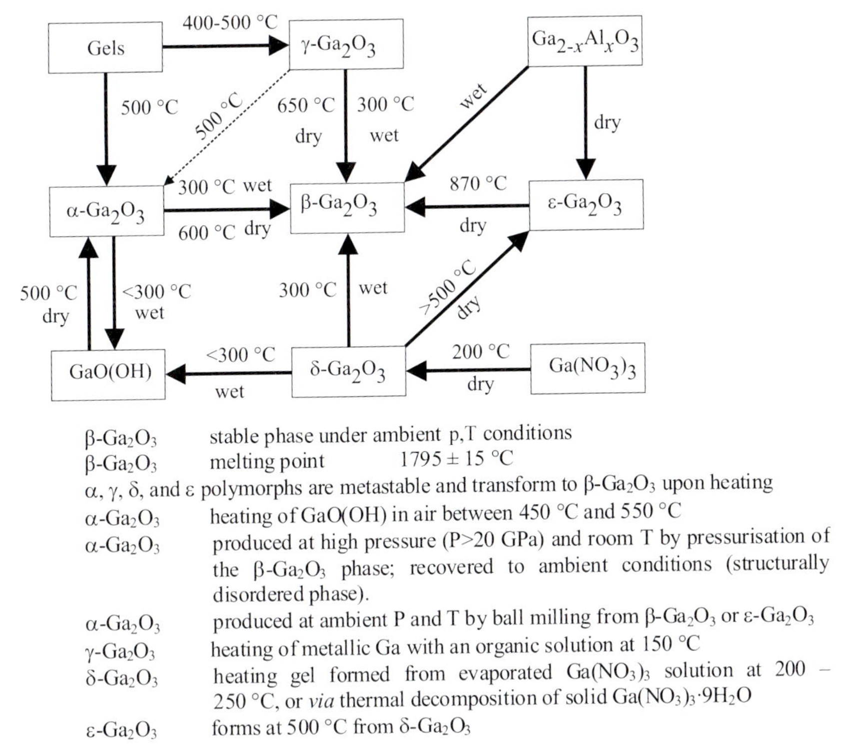

Figure 3.3 Phase transformations and synthesis temperatures in the gallium oxide system. Adapted from Roy *et al.* [72].

By doping Ga_2O_3 with Sn, high conductivities of up to $8.2\,S\,cm^{-1}$ can be reached [168]. Er^{3+} doping into β-Ga_2O_3 nanowires produced by deposition from the vapor phase results in visible cathodoluminescence, that may be partly due to $Er_3Ga_5O_{12}$ garnet particles in the nanoscale ceramics [169].

Ga_2O_3 is also being considered for use in high-temperature chemical gas sensors [68], as a magnetic memory material [170], and for dielectric thin films [171]. Recently, considerable effort has been devoted to the study of low-dimensional Ga_2O_3

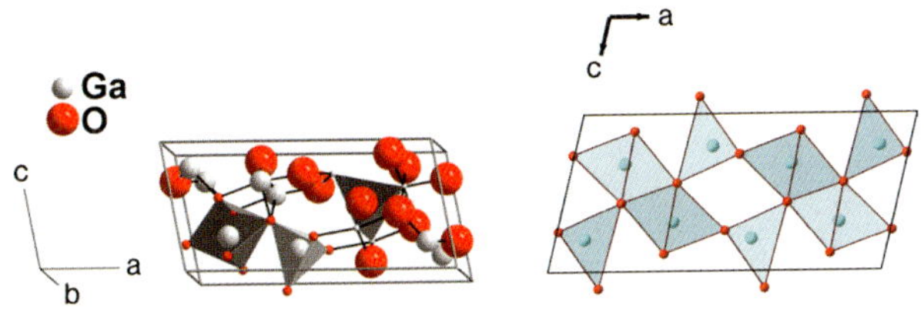

Figure 3.4 (a) Crystal structure of monoclinic β-Ga_2O_3; (b) Projection of the monoclinic structure of β-Ga_2O_3 onto the *a*–*c* plane with Ga^{3+} cations occupying both tetrahedral and octahedral interstices within the *ccp* lattice of O^{2-} ions.

materials, and β-Ga_2O_3 nanowires have been obtained through physical evaporation and arc-discharge methods [172]. β-Ga_2O_3 has also attracted recent interest as a phosphor host material for applications in electroluminescent displays [173, 174]. Due to its high chemical and thermal stability, β-Ga_2O_3 could become a useful alternative to sulfide-based phosphors [175]. The band gap of α-Ga_2O_3 is 2.41 eV, much narrower than that of β-Ga_2O_3 [176]. Ga_2O_3-based materials are also being explored for catalytic applications, both as a support material and as an active catalyst. Al_2O_3 is a well-known catalyst for nitric oxide (NO) reduction, for example by methanol [177]. Haneda *et al.* [66] have investigated the effect of SO_2 on the catalytic activity of Ga_2O_3–Al_2O_3 solid solutions and composite materials for the selective reduction of NO and propene to N_2 and CO + CO_2. Ga_2O_3 provides an active support for Bi–Mo–O catalysts for isobutane oxidation to methacrolein, where the hydrocarbon is initially adsorbed onto the Ga_2O_3 surface and then separates into H and a *tert*-butyl fragment, which then migrate to the $Bi_2Mo_3O_{12}$ surface where the methacrolein formation takes place [178]. Recently, Gebauer-Henke *et al.* have examined the selective hydrogenation of crotonaldehyde on Pt supported by α- and β-Ga_2O_3. The Ga oxide provides a partly reducible support material that helps to activate the carbonyl group prior to hydrogenation via the Pt surface [179].

Because of the density relationships between the two polymorphs (β-Ga_2O_3: 5.94 g cm^{-3}; α-Ga_2O_3: 6.48 g cm^{-3}), it is expected that a $\beta \rightarrow \alpha$ transformation should occur at high pressure. An early study in which synchrotron XRD was used in the diamond anvil cell, identified transformation into a tetragonal structure at pressure of approximately 13.3 GPa [180], although the starting materials in that study were impure and the reported transformation was not well characterized. Recently, Machon *et al.* [163] used a combination of XRD and Raman scattering with first-principles calculations to study the high-pressure behavior of β-Ga_2O_3. In this case, clear evidence was found of a $\beta \rightarrow \alpha$ transformation at a pressure of 20–22 GPa, but perhaps beginning as low as 18.5 GPa (Figure 3.5). However, the high-pressure phase was highly disordered, from the observed large width of the Raman bands along with a slight broadening in the X-ray reflections (Figure 3.6). The transformation involved a change in the O^{2-} packing from cubic to hexagonal, accompanied by a shift in Ga^{3+} ions between tetrahedral and octahedral sites. The resulting high-density structure had significant possibilities for disorder among the Ga^{3+} positions on octahedral sites within the *hcp* O^{2-} sublattice [163]. Transformations among the Ga_2O_3 polymorphs could also be achieved by using a "mechanochemistry" approach; for example, ball milling would result in a systematic decrease in particle size into the nanometer range, as a function of the milling time and temperature conditions. O'Dell *et al.* have examined the structural changes that occur in Ga_2O_3 polymorphs as a function of ball milling using solid-state ^{71}Ga NMR [181]. Recent results have suggested that the ball milling of β-Ga_2O_3 and α-Ga_2O_3 results in a transformation into the α-Ga_2O_3 polymorph as the particle size falls below ~50 nm (D. Machon *et al.*, unpublished results).

The structures of the other metastable polymorphs in the Ga_2O_3 system have not yet been fully characterized. The "γ-Ga_2O_3" phase is particularly problematic. It was first reported by Böhm [182] as a result of the rapid drying of a gel and heating to

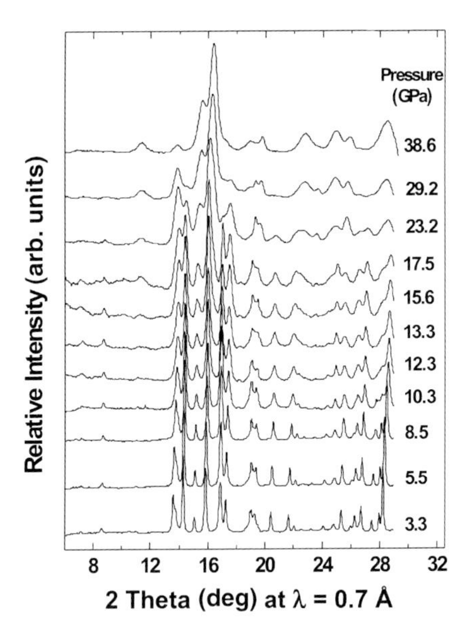

Figure 3.5 (a) X-ray diffraction patterns obtained during compression of the β-Ga_2O_3 phase at ambient temperature. The data were obtained in a diamond anvil cell at the Synchrotron Radiation Source (Daresbury, UK) and the ESRF (Grenoble, France). Reproduced with permission from Ref. [163]; © 2006, American Physical Society.

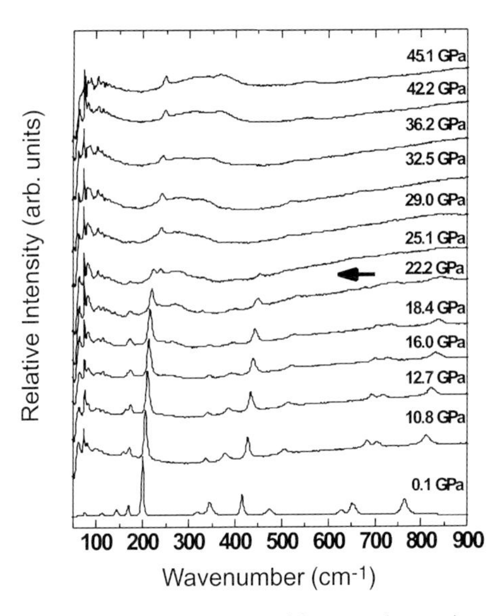

Figure 3.6 Raman spectra of β-Ga_2O_3 during slow compression. Reproduced with permission from Ref. [163]; © 2006, American Physical Society.

450–500 °C, who suggested that this polymorph has a defective cubic spinel structure analogous to γ-Al_2O_3. However, when Roy *et al.* [72] repeated the synthesis they obtained a material with only a few broad reflections in the XRD pattern that did not resemble those of γ-Al_2O_3. The δ-Ga_2O_3 polymorph appeared to have a structure which corresponded to that of the C-type rare earth sesquioxides (La_2O_3, etc.) or bixbyite. ε-Ga_2O_3, which is usually formed from δ-Ga_2O_3 by heating, has a characteristic XRD pattern, although no model for its structure could be suggested by Roy and coworkers. Recent studies with on Sn-doped Ga_2O_3 thin films synthesized using pulsed laser deposition have revived interest in the ε-Ga_2O_3 polymorph [168]. The XRD pattern matched that obtained by Roy *et al.* for ε-Ga_2O_3, and Orita *et al.* were able to index the pattern according to an orthorhombic space group $Pna2_1$ (no. 33). It is interesting that this symmetry is noncentrosymmetric. Kroll *et al.* [28, 82] carried out a theoretical investigation of possible Ga_2O_3 polymorphs, and suggested an orthorhombic equivalent of the ϰ-Al_2O_3 structure with the same space group. It has also been proposed that ϰ-Al_2O_3 and ε-Ga_2O_3 are not only isostructural, but also with phases such as $GaFeO_3$, $AlFeO_3$, and ε-Fe_2O_3. Yoshioka *et al.* [164] took the ϰ-Al_2O_3 structure as a starting point for their theoretical calculations, and optimized the ε-Ga_2O_3 structure within the $Pna2_1$ (no. 33). space group, and studied the phonon dispersion relations and thermodynamic properties of this phase. They also examined the relative stability of the five Ga_2O_3 polymorphs and examined the vacancy ordering within the defective spinel phase δ-Ga_2O_3. At this point, there appeared to be a good match between the experimental data and theoretical predictions for the ε-Ga_2O_3 polymorph, and this indeed corresponded to the ϰ-Al_2O_3 structure. However, recent investigations conducted by the group of one of the present authors, using a combination of synchrotron XRD, high-resolution transmission electron microscopy (HR-TEM) and diffraction, IR and Raman spectroscopy, and X-ray absorption spectroscopy (XRAS) and extended X-ray absorption fine structure (EXAFS) analysis, have begun to suggest that this structural solution is not the best in agreement with all the data. The initial refinement of the synchrotron XRD data indicated a good structure solution within the $Pna2_1$ (no. 33) space group, as suggested previously. Likewise, the IR and Raman data indicated that the structure was noncentrosymmetric, as was expected. However, the results of the transmission electron microscopic (TEM) studies, including those of selected area diffraction experiments (SAD), indicated that the structure must be body-centered. Clearly, the problem of resolving polymorphism within the Ga_2O_3 structures is far from complete, and this will continue to be an important point as the applications of these materials are developed, not only as semiconductors but also as catalysts (D. Machon *et al.*, unpublised work).

3.3.2 Synthesis and Growth Techniques

Various synthetic approaches have been applied to obtain stable and metastable Ga_2O_3 polymorphs, as either bulk crystalline phases or nanoparticles. Single crystals of the thermodynamically stable β-Ga_2O_3 phase can be obtained with and without dopants by using the Verneuil method, under oxidizing and reducing atmospheres [183, 184].

Today, the manufacturers of chemical compounds provide Ga_2O_3 at different levels of purity, produced by a variety of (usually unspecified) methods. Such commercially available samples often contain mixtures of polymorphs, especially of the β-Ga_2O_3 and α-Ga_2O_3 phases, although pure samples of β-Ga_2O_3 can be obtained by heating the mixture, for example, at a temperature of 800 °C for 10–12 h [163]. Yet, the different materials obtained from the same supplier can correspond to different polymorphs. For example, the Aldrich 99.99 + % sample studied by Machon *et al.* corresponded to a 70: 30 mixture of the β- and α-phases, whereas a sample purchased from the same supplier with 99.999 + % purity consisted mainly of the ε-Ga_2O_3 polymorph (D. Machon *et al.*, unpublised work).

Thin films of pure and doped Ga_2O_3 materials have been obtained by deposition on various substrates, including GaAs and Si, by using a range of physical vapor deposition (PVD) techniques including electron beam evaporation from a source such as crystalline $Gd_3Ga_5O_{12}$ garnet [166]. The resulting films often form with the high-density metastable α-Ga_2O_3 as well as stable β-Ga_2O_3 structure. An oriented epitaxial growth of β-Ga_2O_3 crystalline films has been achieved by pulsed laser deposition on a GaN (0 0 2)/Al_2O_3 (0 0 6) substrate [185]. Recently, the deposition of Ga_2O_3 and In_2O_3 thin films was achieved using single-source organometallic precursors [186–188]. A wide range of unusual and interesting nanoscale structures could be created from β-Ga_2O_3, including rods, wires, single and multiple tubes, ribbons, sheets, discs, belts, and "brushes" using methods that included arc-discharge or evaporation followed by condensation from the vapor phase onto a suitable substrate, hydrothermal synthesis or carbothermal reactions, or gas-phase reaction of evaporated Ga or GaN with traces of O_2 [172, 189–193]. In addition, an interesting "herringbone" nanostructure was observed for crystalline β-Ga_2O_3 nanowires formed by the thermal decomposition of GaAs mixed with Te in the presence of trace O_2 [194]. Ga_2O_3 quantum dots (QDs) with sizes ranging from 2 to 5 nm embedded in an SiO_2 matrix were produced using a sol–gel method [195].

3.4 Gallium Oxonitrides

It is clear from the discussions above that both GaN and Ga_2O_3 possess a wide range of useful and complementary electronic, optical, mechanical, and chemical properties. Some of these may be enhanced or optimized by preparing anion-substituted intermediate phases, solid solutions or new compounds within the Ga_2O_3–GaN system. An example that provides a template for new materials with rapid industrial breakthrough is represented by the analogous Al_2O_3–AlN system, where the formation of solid solutions led to new phases and crystal structures providing materials with properties that are located directly between those of both end members, or that are significantly enhanced relative to the end members, or that possess entirely new characteristics. Systematic investigations into the phase relationships and materials properties in the binary system Al_2O_3–AlN were first made during the 1970s, since which time the various solid solutions and a series of intermediate phases have been extensively characterized and several products have entered specialized markets for advanced ceramic materials [196–204]. To date, no corresponding systematic studies have been undertaken of the solid solutions and intermediate phases that might exist in the analogous system between Ga_2O_3 and GaN. Few reports have been published on the formation of crystalline gallium oxide/nitride phases formed as byproducts during the oxidation of GaN films, but the synthesis of the Ga–O–N phases has not been addressed directly. Thus, an overview of those materials that might currently exist within the Ga_2O_3–GaN system to provide an impetus for future investigations, is provided in the following sections.

3.4.1 Nomenclature Issues of Ga–O–N Materials

At this point, it may be worthwhile examining the nomenclature issues that currently surround $Ga_xO_yN_z$ solid-state compounds, in order to remove any potential confusion that is encountered and which may hamper future studies. In the binary system Al_2O_3–AlN, the mixed oxide–nitride or oxonitride phases in which Al is directly bonded to both oxygen and nitrogen have been formally designated as "alon" by IUPAC convention; this is consistent with the "sialon" solid solution system, where Al^{3+} cations are substitutionally replaced in the crystalline lattice by silicon (Si^{4+}) [205]. During a literature search for gallium oxide nitride compounds, several problems arose in relation to nomenclature that obscured the nature of the materials being examined, and might hamper future studies in this field. It was generally unclear, and indeed found to be an arbitrary choice, as to whether authors should use the designation "gallium oxynitride" or abbreviations such as GaN_xO_y to describe their compounds. In certain cases, both expressions were used for diverse and quite distinct materials, including gallium nitrates, in which no Ga–N bonds occurred, anion-substituted gallium nitrides (GaN: O) or corresponding oxides (Ga_2O_3: N) materials, simple mixtures of GaN and Ga_2O_3 phases [e.g., Refs [206–209]], and true solid solutions or intermediate compounds in the Ga_2O_3–GaN system. Even when inorganic substances, where anions are substituted by oxygen on the nitrogen positions (or *vice versa*) were widely described as "oxynitride" compounds, the authors decided to follow the previously established IUPAC conventions for the sialons and alons. In the discussions below, the terms "oxonitride" or "oxide nitrides" are used to describe solid solutions and intermediate compounds formed in the phase diagram of GaN–Ga_2O_3, as well as for oxygen/nitrogen anion-substituted GaN/Ga_2O_3 materials, that contain both Ga–N and Ga–O bonds. In the literature survey, all publications have been included where authors refer to "oxynitrides," except for those in which it is clear that nitrate or nitrite compounds were investigated, with no hint of the formation of any direct bond between nitrogen and the metal.

3.4.2 Theoretical Predictions for Ga Oxonitride Compounds

The theoretical search for new gallium oxonitride compounds was initiated following the observation of spinel-type structured materials that formed as side products during the preparation of epitaxial thin films of GaN using a CVD process. Puchinger *et al.* investigated these small, partially oxidized areas on GaN films by using TEM to selected area diffraction (SAD) technique [210]. The small particles were shown to possess a cubic spinel-type structure with a lattice parameter $a_0 = 8.20 \pm 0.07$ Å. The oxygen to nitrogen ratio determined in the TEM study indicated a composition of $Ga_{2.8}O_{3.5}N_{0.5}$, assuming completely filled anion sites. This first observation of a gallium oxide nitride phase triggered several theoretical studies aimed at establishing the existence of possible new spinel-structured solid solutions and intermediate compounds in the GaN–Ga_2O_3 system. The first computational study employed an ultrasoft pseudopotential within the local density approximation (LDA), including semicore d states being treated as valence states [27]. The Ga_3O_3N composition was assumed to have the ideal spinel structure, with space group $Fd\bar{3}m$ (no. 227) and 56 atoms in the unit cell. It was noted that such a model assumed that complete N/O disordering occurred over the anion sites, and that the tetrahedral and octahedral Ga^{3+} sites were assumed to be fully occupied. The calculated lattice parameter for this ideal oxonitride spinel structure was $a_0 = 8.22$ Å, only slightly larger than that determined experimentally for the nanosized particles of the $Ga_{2.8}O_{3.5}N_{0.5}$ solid solution phase [210]. Whilst it is known that LDA provides only an approximate view of the electronic structure, and that calculated band gaps are generally unreliable, the calculations provided a first suggestion that the Ga_3O_3N oxonitride spinel might have interesting and potentially useful electronic and optoelectronic properties. The band gap was predicted to be wide and direct, comparable with GaN, although an actual value was not reported in the first study. In fact, wide direct band gaps have also been predicted for the new nitride spinels γ-Si_3N_4 and γ-Ge_3N_4, using similar theoretical methods. Moreover, these results appear to be supported by experimental studies using X-ray absorption and emission measurements to determine the intrinsic gap [211, 212]. Shortly afterwards, Kroll *et al.* carried out a first-principles study using density functional theory (DFT) within the generalized gradient approximation (GGA), as implemented within the Vienna *ab initio* simulation package (VASP) [28, 82] to investigate the formation enthalpy of Ga_3O_3N spinel, as well as the defective spinel compositions $Ga_{23}O_{27}N_5$ and $Ga_{22}O_{30}N_2$, from Ga_2O_3 and GaN. In all cases the formation enthalpy were found to be positive, although the enthalpy change decreased with increasing pressure for N-rich compositions. Kroll *et al.* concluded that a substantial entropy contribution would be needed for a successful high-pressure/high-temperature synthesis of Ga_3O_3N spinel from its components. Such an entropy contribution could be gained from disordering O and N over the anion sites, or from Ga^{3+} vacancy distribution within cation-defective oxonitride spinel varieties, modeled on the γ-Ga_2O_3 structure [28]. Soignard *et al.* carried out a combined experimental and theoretical investigation of the synthesis and properties of Ga oxonitride spinel. For this, a mixture of Ga_2O_3 and GaN was treated at high pressure and temperature to obtain a spinel-structured material with the composition $Ga_{2.8}N_{0.64}O_{3.24}$ [33], and which contained vacancies on the octahedral Ga^{3+} sites, in agreement with the predictions of Kroll *et al.* [28, 82]. The theoretical study was implemented using DFT methods within the LDA and GGA, and included a statistical mechanical investigation of the entropy stabilization due to oxygen/

Table 3.5 Parameters of the third-order Birch–Murnaghan equations of states of a three unit-cell-based atomic models of the spinel-structured Ga_3O_3N that have the lowest energy calculated within the LDA[a)].

Model	E_0 (eV/Ga_3O_3N)	V_0 (Å^3/Ga_3O_3N)	a_0 (Å)	B_0 (GPa)	B_0
I	−48.109 (0.000)	69.5424	8.2246	210	4.13
II	−47.966 (0.143)	69.6928	8.2305	208	4.17
III	−47.854 (0.255)	69.6181	8.2275	209	4.14

a) Model I is found to be the minimal energy configuration, and models II and III occur at 143 and 255 meV/Ga_3O_3N (13.8 and 24.6 kJ mol^{-1}) higher energy, respectively, above the ground state. The three models have comparable equilibrium volumes and bulk moduli.

nitrogen (O/N) disorder among the anion sites of the Ga_3O_3N structure [33]. The band gap estimated within the LDA was 2.1 eV, but it was realized that this was likely to be far from the true value. In relation to comparable calculations for GaN and Ga_2O_3 phases, it was predicted that the likely band gap would reach 4 eV [27, 33]. The calculated structural parameters and compressibility values (K_o and K'_o assuming a third-order Birch–Murnaghan equation of state formalism) and the theoretical studies are summarized in Table 3.5.

The theoretical studies predict the formation of spinel-structured Ga_3O_3N from a mixture of GaN and Ga_2O_3 at pressures above approximately 6.6 GPa. Such a pressure range is comparable with that for the $\beta \rightarrow \alpha$ phase transition of Ga_2O_3. On the basis of the DFT calculations, the temperature required for a thermodynamically stabilized phase transition of GaN + Ga_2O_3 to Ga_3O_3N is predicted to be as high as ~2800 K. The theoretical studies indicated that the spinel phase formed at high temperature would be stabilized by O/N disorder on the anion sites and possible vacancy formation on the Ga^{3+} sites. These entropy-stabilizing effects would result in a lowering of the synthesis temperature at a given pressure. The results of the theoretical studies indicated that: (i) Ga oxonitride or oxide nitride compounds and solid solutions can be thermodynamically stabilized, and can be formed within the Ga_2O_3–GaN system, by either bulk synthesis or metastable thin film deposition or nanoparticle synthesis approaches; and (ii) that these oxide nitride materials can have interesting and useful electronic properties.

3.4.3
Literature Overview on Gallium Oxide Nitride Phases

During the past thirty years several reports have been published on the synthesis of gallium oxonitride phases. The first of these was mentioned by Verdier and Marchand in 1976 as a side product during the reaction of ammonia with gallium oxide, when a phase with the composition $Ga_{1-x/3}\square_{x/3}N_{1-x}O_x$ with $x \sim 0.1$ and $\square$ as vacancies was isolated under low-temperature conditions [213]. The authors did not report any detailed crystal structure data, but did note that the XRD pattern was consistent with a cubic zincblende-type structure with an a_0 lattice parameter of 5.50 Å [214]. In 1979, Grekov and Demidov attempted the first syntheses to prepare gallium oxide nitride phases [215]. For this, the solubility of oxygen in gallium nitride and the formation of solid solutions in the Ga–O–N system was investigated. By using the thermolysis of a $GaCl_3 \cdot NH_3$ complex and nitride/oxide deposition under controlled H_2O conditions, crystalline layers of GaN with variable contents of nitrogen and oxygen were produced. In those samples richest in oxygen, up to 25% nitrogen was replaced in the w-GaN structure with oxygen whereas, under conditions with almost equal amounts of nitrogen and oxygen atoms in the gaseous phase, amorphous layers composed of Ga, O, and N began to form. The amorphous layers synthesized at 900 °C displayed self-activated luminescence in the energy region of 1.75–1.8 eV. All of the synthesized films were found to be amorphous, with variable nitrogen/oxygen ratios. For films with a lower oxygen content, the luminescence maximum could be detected at 2.15–2.2 eV. During the next few years several reports and patents on gallium oxide nitrides with applications such as passivation films and insulating layers were published by Shiota and Nishizawa [216–220]. Although, neither the phases present within the gallium oxide nitride layers nor their crystal structures were clearly defined, the usefulness and effectiveness of gallium oxide nitride as an optical and electrical material, as well as the role of its manufacture, was firmly established. In order to improve the quality of the passivation layers in terms of achieving a higher resistivity and a better adherence to semiconducting gallium arsenide, small amounts of oxygen were incorporated into the pure gallium nitride films. Four different methods were suggested for this purpose, two of which led only to weakly adhering films and a poor reproducibility of oxygen incorporation into the films. Using pure oxygen gas mixed into an ammonia gas flow and an NO gas mixture thermal oxidation of the gallium nitride films as well as the reaction of a $GaBr_3 \cdot NH_3$ complex with adsorbed water resulted in improved properties of the passivation layers. In fact, the latter approach proved to be the best method for obtaining well-adhered films with a high resistivity. These films had an interface with the substrate that exhibited only a very low degree of oxidation, whereas the main body of the films was composed of the gallium oxide nitride material. Nevertheless, a detailed investigation of these films using Auger spectroscopy displayed only a mixture of GaN and Ga_2O_3 compounds that presumably formed a nanocrystalline composite containing small amounts of residual bromine. However, the simultaneous presence of both phases in the films resulted in an improved resistance to acid attack, especially when the N : O ratio was increased; all of the films could be dissolved with hot HCl and H_3PO_4.

Aleksandrov *et al.* investigated the kinetics of the formation of Ga oxide nitride layers on single-crystal silicon during pyrolysis [221]. $GaBr_3 \cdot NH_3$–H_2O was used as a precursor, which this resulted in GaN layers along with the growth of amorphous gallium oxide nitrides. These studies provided no further information regarding the detailed nature or structure of these oxide nitrides. During the 1990s, several attempts were made specifically to synthesize gallium oxide nitrides [222–227]. Some of these reports provided information on new crystalline phases present within the system. Merdrignac *et al.* claimed in a patent the synthesis of gallium nitride and oxide nitride for gas-sensing devices and applications. The formation of the gallium oxide nitride was achieved by reacting $NiGa_2O_4$ at low temperature (550–650 °C) in an ammonia atmosphere for 6 to 12 days. This reaction yielded a zincblende-type phase with the approximate composition $Ga_{1-x/3}\square_{x/3}N_{1-x}O_x$, in which $\square$ are vacancies on the tetrahedral sites and x ranges between $0 \leq x \leq 0.3$ [222]. This newly discovered phase exhibited useful semiconducting behavior, as well as sensing properties for gases such as NH and NH_2. Yoshida *et al.* filed a patent which detailed a method for forming and removing a selective growth mask on gallium arsenide [223]. In this case, a layer of gallium nitride was deposited onto gallium arsenide, and later irradiated with a halogen lamp light while being held in an oxygen atmosphere. During this process a gallium oxide nitride layer was formed that was subsequently removed by irradiating specific areas with an electron beam, while simultaneously applying a Cl-containing gas atmosphere [223]. Another route to the formation of gallium oxide nitrides in thin films was achieved by using low-temperature, plasma-assisted CVD [224]. Here, Kovalgin and Lermontova used what was then a new technique to deposit aluminum/gallium oxide nitride films. For this, $(Al,Ga)Cl_3 \cdot NH_3$ was used as a precursor and then reacted to form films in a nitrogen, oxygen, and argon gas atmosphere at temperatures between 300 and 400 °C, using a remote plasma-enhanced CVD process. The species present within the excited plasma were monitored directly, using optical emission spectrometry. While the aluminum oxide nitride films grown at temperatures below 400 °C were characterized by a low content of oxygen, the corresponding gallium oxide nitride films displayed essentially no influence of growth temperature on the composition of the layers, thus indicating the involvement of different processes for the formation of Al and Ga oxide nitrides. The IR absorption spectra of layers deposited at temperatures of 400 and 800 °C were comparable to the characteristic modes recorded for pure gallium oxide. These results raises the question of whether these layers really consisted of true oxide nitride compounds, or were simply nanocrystalline mixtures of the end member phases. In 1997, in their first report concerning the partial thermal oxidation of GaN layers treated with dry oxygen, Wolter *et al.* were unable to distinguish clearly between these possibilities, nor could they identify if the oxidation process of single crystal GaN in dry air underwent an initial transformation to an intermediate gallium oxide nitride phase [225]. Rather, the signals recorded in the X-ray photoelectron spectroscopy (XPS) spectra during the oxidation process were interpreted to be caused by the extreme roughness of the phase boundary, although it was suggested that an interfacial graded oxide nitride layer might also be formed. In 2000, the same group supplemented their XPS measurements with results from other experimental studies, to provide evidence for the formation of a true gallium oxide nitride phase with the general formula $Ga_{(x+2)}N_{3x}O_{3-3x}$ [228]. Tsuruoka *et al.* investigated the interaction of gallium nitride with oxygen over a temperature range of 300 to 700 K, using high-resolution electron energy loss spectroscopy (HREELS) and low-energy electron diffraction (LEED) methods [227]. After oxidation, a broad loss feature which appeared on the high energy side of the 702 cm^{-1} vibration mode was detected; however, with increasing temperatures of adsorption and annealing, this loss structure shifted to a higher energy. In addition, the shift of the band gap from 3.4 to 4.4 eV could indicate the formation of gallium oxide or of a gallium oxide nitride phase. Schmitz *et al.* described the formation of a monoclinic Ga_2O_3 phase that likely corresponded to β-Ga_2O_3, with some nitrogen substitution in the structure [226]. This homogeneous layer of monoclinic gallium oxide nitride was obtained by the adsorption of NO onto a CoGa (0 0 1) substrate at a temperature of 300 K, and subsequent annealing at 825 K. The band gap for this N-substituted β-Ga_2O_3 phase (ca. 4.1 ± 0.2 eV) lies intermediate between the values for the pure oxide and wurtzite-GaN (3.5 eV). During the following years, various intermediates at the interface between pure gallium oxide and gallium nitride or small oxidized regions on pure GaN layers were reported [229–231]. Nakano and Jimbo detected a broad intermediate gallium oxide nitride layer with graded composition at the interface of β-Ga_2O_3 and GaN by using secondary ion mass spectroscopy (SIMS) [230]. The interface had formed during the growth of a 100 nm β-Ga_2O_3 layer on n-type GaN using dry oxidation at a temperature of 880 °C for 5 h (see Figure 3.7). As no discrete interface traps could be measured by deep level transient spectroscopic measurements, Nakano and Jimbo concluded that the Fermi level was unpinned at the interfacial gallium oxide nitride layer. The SIMS profile displayed an effective thickness for the β-Ga_2O_3 of ~100 nm, where the GaN had been completely oxidized without any trace of nitrogen in the layer. The intermediate gallium oxide nitride layer detected in the depth region of 100–400 nm displayed a graded composition until the oxygen content was totally reduced and the GaN layer started.

In 2005 the first bulk gallium oxide nitride materials with the cubic spinel structure were reported as a result of high-pressure/high-temperature synthesis experiments conducted. First, in a collaboration between the groups of Riedel (Darmstadt, Germany) and Huppertz (Munich, Germany), an Ga oxonitride spinel phase was synthesized from organometallic precursors using a multi-anvil device [34]. In an

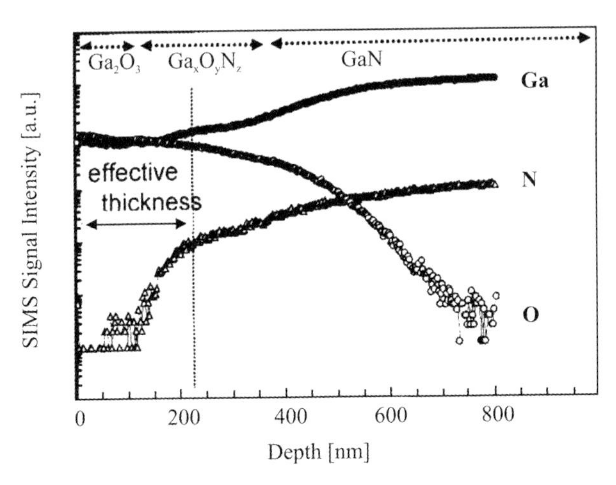

Figure 3.7 Secondary ion mass spectroscopy profiles of Ga, N, and O atoms in the thermally oxidized β-Ga_2O_3/GaN MOS structure. Reproduced with permission from Ref. [230]; © 2003, American Institute of Physics.

independent study, McMillan's group at University College London prepared a well-crystallized sample of $Ga_{2.8}N_{0.64}O_{3.24}$ spinel from Ga_2O_3 + GaN mixtures, by using a combination of laser-heated diamond anvil cell (DAC) and multi-anvil synthesis techniques [33]. These results are described later in the chapter.

Recently, additional patents have been filed relating to the properties of gallium oxide nitrides formed within thin films [208, 209, 232–235]. These describe the investigations on sensor properties of gallium oxide nitride films prepared by a synthesis route related to that described by the group of Marchand, in Rennes, France. Kerlau *et al.* used citric acid as chelating agent and nickel gallate as precursor, leading to crystalline gallium oxide nitride layers with mixed wurtzite- and zinc-blende-type structures. An elemental analysis resulted in 80 wt% of gallium and 13.8 wt% nitrogen, leading to a calculated formula of $GaO_{0.34}N_{0.86}$ [232]. The properties of this material, together with their potential applications, are described in Section 3.4.5. Koo *et al.* compared the photoconductivity of nanocrystalline gallium nitride along with amorphous gallium oxide nitride grown during ion-assisted deposition [234]. The oxygen concentration in films could be varied between 12 and 24 at%. The optoelectronic measurements for these materials are described below. In their patent, Peng *et al.* adopted ideas that had already been filed as a patent by Nishizawa *et al.* in 1982, in which suggestions for applications of the oxide nitride films and manufacturing method as either passivation film of 13–15 compound semiconductors or an insulating film shielding the gate in 13–15 compound semiconductors field effect transistors were claimed [235]. The method used to form the gate-insulating layer was improved by using a photoassisted electrochemical process, followed by a rapid thermal annealing process in an oxygen atmosphere at temperatures between 500 and 800 °C. Measurements using XPS revealed a graded composition within the thin films which varied from $GaO_{0.18}N_{0.82}$ to $GaO_{0.82}N_{0.18}$. Kikkawa *et al.* doped gallium oxide nitride with lithium [208] and magnesium [209], starting with nitrate mixtures as precursors and using citric acid as a gelling agent in aqueous solution. Structural investigations of the doped gallium oxide nitride were, in both cases, similar to the hexagonal structure of wurtzite-type GaN, as demonstrated using XRD [208, 209].

3.4.4 Synthesis and Growth Techniques

As noted above, most of the $Ga_xO_yN_z$ compounds or solid solutions reported to date have been obtained as side products during reactions designed to produce nitride materials, usually in thin film form, by using a variety of synthesis approaches. In most of these studies, the serendipitous or planned inclusion of O within nitride compounds (or, less frequently, of N incorporation in Ga oxides) has led to the investigation of properties of intermediate gallium oxide nitride phases or solid solutions identified during the studies. Reported studies of the directed syntheses of Ga oxonitride compounds are rare; these begin with the early investigations of Grekov *et al.* and Aleksandrov *et al.* [215, 221]. It was soon observed that highly crystalline intermediate gallium oxide nitride phases could not be obtained as readily as within the $AlN–Al_2O_3$ system, where the application of a high temperature is all that is necessary to achieve a series of new alon compounds and solid solution phases.

The first decisive investigations to establish the bulk synthesis of new Ga oxonitride phases used high-pressure/high-temperature synthesis techniques, and were related to the discoveries of the new dense spinel-structured nitrides γ-Si_3N_4 and γ-Ge_3N_4 [16–18, 31, 33, 34, 236–238].

3.4.4.1 Precursor Approach for Gallium Oxide Nitride Phases

The initial experiments conducted by the Darmstadt group to synthesize intermediate $Ga_xO_yN_z$ phases involved heating together mixtures of the end member compounds Ga_2O_3 and GaN, as applied to produce alons within the analogous $AlN–Al_2O_3$ system. However, none of these synthesis attempts was found to be successful. These initial negative results were rationalized by theoretical studies which showed that the formation of compounds such as Ga_3O_3N is highly endothermic at ambient conditions. Soignard *et al.* successfully developed this approach following an extensive series of high-pressure/high-temperature investigations to produce a spinel-structured compound $Ga_{2.8}N_{0.64}O_{3.24}$ [33]. In Darmstadt, an alternative approach was devised using chemical precursors that contained the required elements in suitable ratios [34].

Thermally Activated Transformation of Gallium Nitrate The complete or partial thermal decomposition of metal nitrates is commonly used for the preparation of oxides. The thermal treatment of gallium nitrate, $Ga(NO_3)_3 \cdot x\ H_2O$, leads to formation of the pure oxide in an inert gas atmosphere [239] and to the pure nitride when the reaction is carried out in a reducing gas atmosphere [231]. In order to explore the decomposition of $Ga(NO_3)_3 \cdot x\ H_2O$ as a function of temperature in a nitrogen gas atmosphere, Berbenni *et al.* combined thermal analysis with Fourier transform infrared (FTIR) spectroscopy, suggesting a mechanism that proceeds in three main steps, as displayed in Eqs. (1) to (3) [239].

At a temperature of 100 °C, a weight loss corresponding to 18% of the initial mass occurs and coincides with the reaction given in Eq. (1):

$$Ga(NO_3)_3 \cdot 8.07\ H_2O(s) \rightarrow Ga(OH)_2(NO_3)(s) + N_2O_5(g) + 7.07\ H_2O \tag{1}$$

During further heating, the $Ga(OH)_2(NO_3)$ intermediate decomposes via two steps at temperatures around 136 °C [see Eqs. (2) and (2′)]:

$$m Ga(OH)_2(NO_3)(s) \rightarrow (m/2)Ga(OH)_3(s) + (m/2)Ga(OH)O(s) + (m/2)N_2O_5(g)$$

The second slower mass loss step, that starts at a temperature of 150 °C and results in a mass decrease to 28.26% of the initial mass at 220 °C, corresponds to a reaction such as that proposed in Eq. (2′):

$$\begin{aligned}(1\text{-}m)Ga(OH)_2(NO_3)(s) &\rightarrow [(1\text{-}m)/2]Ga(OH)_3(s) \\ &+ [(1\text{-}m)/2]Ga(OH)O(s) + [(1\text{-}m)/2]N_2O_5(g)\end{aligned} \tag{2'}$$

The continuous mass loss process occurring up to 460 °C corresponds to the decomposition of $Ga(OH)_3$ and $Ga(OH)O$ to the oxide [Eq. (3)]:

$$(1/2)Ga(OH)_3(s) + (1/2)Ga(OH)O(s) \rightarrow (1/2)Ga_2O_3(s) + H_2O \tag{3}$$

Berbenni *et al.* proposed that the reaction in Eq. (2) is bypassed during rapid heating so that the starting material is directly decomposed into a phase mixture of $Ga(OH)_2(NO_3)$ and $Ga(NO_3)O$.

Jung chose the same starting material $Ga(NO_3)_3 \cdot x\ H_2O$ in order to synthesize pure gallium nitride, and monitored the conversion of the salt using XRD and ^{71}Ga magic angle spinning nuclear magnetic resonance (MAS-NMR) spectroscopy [23]. During the heating process in an ammonia gas atmosphere, the salt decomposed first to γ-Ga_2O_3, and then was transformed directly to gallium nitride, without any clear evidence of a phase transition from γ-Ga_2O_3 into another gallia polymorph. However, Jung concluded from the experimental results that the transformation from γ-Ga_2O_3 to w-GaN proceeds stepwise via amorphous gallium oxide nitride as an intermediate. Such an intermediate gallium oxide nitride would provide a useful precursor for the further synthesis of pure crystalline gallium oxide nitride. Additionally, it would prevent the formation of the pure thermodynamically stable phases gallium oxide and gallium nitride at the onset of the synthesis reaction. For these reasons, $Ga(NO_3)_3 \cdot x\ H_2O$ seemed to be a promising precursor to Ga oxonitride synthesis by using a more reactive gas atmosphere [39]. Therefore, the synthesis approach by Jung was slightly modified and $Ga(NO_3)_3 \cdot x\ H_2O$ was heated for 2 h in an ammonia atmosphere at 350 °C (see Figure 3.8a). In order to remove the water component, $Ga(NO_3)_3 \cdot x\ H_2O$ samples were dried for 12 h at 150 °C before the pyrolysis step. The

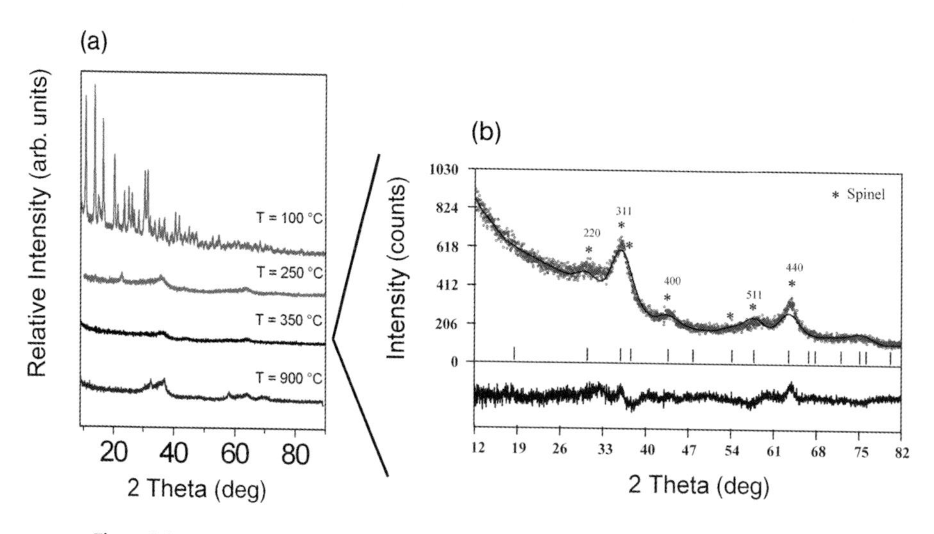

Figure 3.8 Monitoring the structural transformation with temperature and ammonia gas atmosphere starting from gallium nitrate reaching a spinel-type structured gallium oxonitride at 350 °C. At a temperature of 100 °C a crystalline phase of dried $Ga(NO_3)_3 \cdot x\ H_2O$ forms. Reproduced with permission from Ref. [34].

powder XRD pattern of the synthesized phase revealed a poor degree of crystallinity, but it could be indexed as a spinel-type structured gallium oxide nitride (see Figure 3.8b) [34]. Assuming completely filled anion sites, the chemical composition of this phase was calculated as $Ga_{2.7}N_{3.66}O_{0.34}$ from an elemental analysis for nitrogen and oxygen. However, further experiments to improve the degree of crystallinity failed, and led to phase separation into stable β-Ga_2O_3 and w-GaN upon prolonged annealing. At low temperatures, Ga_2O_3 also crystallizes with a defective spinel-type structure; therefore, it is not easy to ascertain if the reaction of the $Ga(NO_3)_3 \cdot x\,H_2O$ forms a γ-Ga_2O_3 phase with some solubility of nitrogen on the anion positions, or if it instead yields an anion-substituted gallium oxide nitride spinel.

The partial success of this precursor approach to produce the gallium oxide nitride phases has stimulated a search for further precursors that contain direct bonding of Ga to both N and O atoms within the precursor materials.

Pyrolysis of Molecular Precursors As Grekov *et al.* noted as early as 1979, the synthesis of gallium oxide nitride phases at ambient pressure conditions clearly requires the selection of a method resulting in the simultaneous formation of metal–nitrogen and metal–oxygen bonds [215]. If not, all of the synthesis procedures would result in mixtures of the stable Ga_2O_3 and GaN phases. The advantage of starting with molecular precursors derives from the presence of gallium bonded to predefined amounts of nitrogen and oxygen within the molecular compounds, and the fact that the precursor does not favor the formation of any special structure of either of the end members. Molecular precursors have been already used for the synthesis of the pure nitrides (see Section 3.2.2), where one of the appropriate precursors described was the bis,tris(dimethylamino)gallane, $[Ga(N(CH_3)_2)_3]_2$. In 2001, Valet and Hoffman described a modification of this precursor to $Ga(O^tBu)_3 \cdot HNMe_2$ as a product of the reaction between gallane and *tert*-butanol, following the main pathways expressed by Eqs. (4) and (5) [35]:

$$2\,GaCl_3 + 6\,LiNMe_2 \rightarrow [Ga(NMe_2)_3]_2 + 6\,LiCl \tag{4}$$

$$[Ga(NMe_2)_3]_2 + 6\,{}^tBuOH \rightarrow 2\,Ga(O^tBu)_3 \cdot HNMe_2 + 4\,HNMe_2 \tag{5}$$

The procedure leading to gallium tris(*tert*-butoxide) dimethylamine adduct gives various different substitution products as a function of the synthesis time and temperature, all of which can be used as precursors for the synthesis of gallium oxide nitride. The thermal conversion of these precursors was investigated using different gas atmospheres (including N_2, N_2/H_2 mixtures, argon, and NH_3) and optimizing the flow and heating rate and the final pyrolysis temperature. The products were examined by analyzing the final carbon, nitrogen, and oxygen contents, using elemental analysis [35]. The formation of crystalline phases, as well as the degree of crystallinity, was detected using XRD. The best results in terms of a low carbon content and well balanced oxygen to nitrogen ratios were achieved when $Ga(O^tBu)_3 \cdot HNMe_2$ was decomposed in an ammonia atmosphere at a maximum temperature of 350 °C. The XRD patterns of samples where the pyrolysis temperature exceeded 600 °C showed additional reflections, indicating the formation of a phase with a wurtzite-type structure. In order to investigate the microstructure and the homogeneity of the synthesized particles more accurately, the specimens were examined using TEM coupled with EELS. The high-resolution TEM images exhibited nanocrystalline particles throughout the sample areas examined, while the EELS analysis indicated an inhomogeneous nitrogen/oxygen distribution. Overall, the particles were found to correspond to a gallium oxide nitride material with a mean ratio of N/O about 0.84. There was no evidence for the presence of pure end member compounds, which indicated that no phase separation into GaN and Ga_2O_3 had occurred at low temperatures. The XRD patterns indicated the formation of a phase with a wurtzite structure, that could correspond to a gallium nitride phase with partially substituted oxygen on the anion sites. These results emphasized the need to use a high-pressure/high-temperature approach with appropriate starting materials to reach phases with high crystallinity in the system GaN–Ga_2O_3.

3.4.4.2 **Crystalline Phases of Gallium Oxide Nitride Synthesized under High-Pressure/High-Temperature Conditions**

Recently and independently, both the London and Darmstadt groups have applied high-pressure/high-temperature synthesis methods, combining large-volume synthesis and laser-heated diamond cell methods, to obtain bulk samples of $Ga_xO_yN_z$ spinel-structured phases, with compositions approaching that of the ideal spinel structure, Ga_3O_3N. The Darmstadt group used chemical precursors, whereas the London group followed a standard solid-state method, starting from w-GaN and β-Ga_2O_3 mixtures [33, 34]. In both studies, the recovered materials crystallized as single phases with the spinel structure, as determined by synchrotron and laboratory X-ray crystallography, as well as Raman spectroscopy and electron microprobe chemical analysis for all three elements. The determined lattice parameters were similar between the two studies: $a_0 = 8.264(1)$ Å [34] and 8.281(2) Å [33], respectively.

Soignard *et al.* used mixtures of pure β-Ga_2O_3 or α/β-Ga_2O_3 and w-GaN in approximately 1 : 1 molar ratios as a starting material in their attempts to realize the ideal structure Ga_3O_3N [33]. The high-pressure/high-temperature experiments were carried out using both a laser-heated diamond anvil cell (LH-DAC) and the multi-anvil technique. The high-pressure/high-temperature reaction was sluggish, and it was difficult to complete to obtain phase-pure products. In the LH-DAC experiments, the reaction and phase transformation into the gallium oxonitride spinel structure began at a pressure of 10.5 GPa and a temperature of 1700 °C. However, in the XRD patterns the reflections of both end member phases (α/β-Ga_2O_3 and w-GaN) could still be detected, and the results of the LH-DAC studies indicated that a full transformation to the cubic spinel-type gallium oxide nitride could only be completed at temperatures in excess of 2000 °C. In the multi-anvil experiments, the reaction and phase transformation was found to proceed to completion in a more controllable manner; however, small amounts of the starting materials could still be detected along with the gallium oxonitride spinel phase following synthesis runs at a pressure of 5 GPa and a temperature of 1700 °C (see Figure 3.9).

A detailed chemical analysis and X-ray structure refinements of the recovered samples indicated that the synthesized phase had a stoichiometry $Ga_{2.8}N_{0.64}O_{3.24}$,

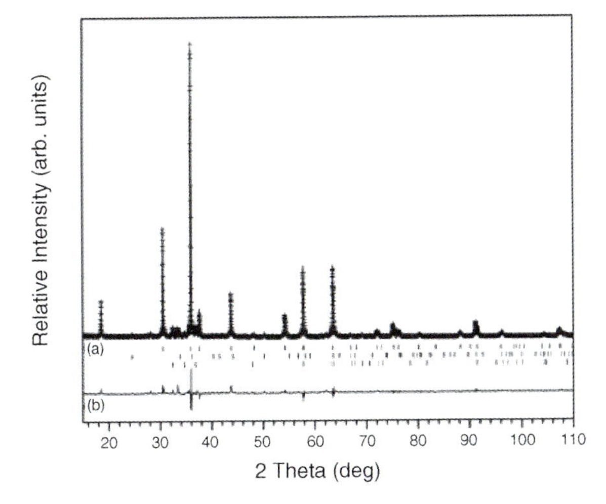

Figure 3.9 Rietveld refinement of spinel-type structured $Ga_{2.8}N_{0.64}O_{3.24}$ synthesized with a pressure of 5 GPa and a temperature of 1700 °C. The upper row of tick marks shows the reflections (2θ values) for the spinel phase; the sample also including 2% residual Ga_2O_3 is indicated by the middle row of tick marks; 14.5% GaN is represented by the lower row of tick marks [33].

and that it contained vacancies on the octahedrally-coordinated Ga^{3+} sites. The phase crystallized in cubic space group $Fd\bar{3}m$ (no. 227), and revealed a lattice parameter a_0 of 8.281 Å. The broad bands observed in the Raman spectrum indicated the presence of a substantial N/O disorder on the anion sites, along with the presence of cation site vacancies. The optical properties were investigated via photoluminescence studies excited using 335 nm laser illumination at room temperature. The results indicated a strong photoluminescence emission between wavenumbers 400–750 nm, that was likely to be dominated by the vacancy/defect structures and probably did not reflect the intrinsic band gap of the material (see Figure 3.10).

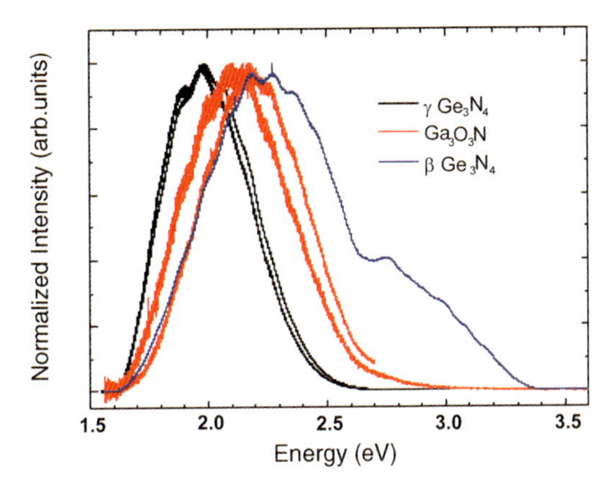

Figure 3.10 Photoluminescence measurement of gallium oxonitride in comparison to two polymorphs of germanium nitride [33].

Kinski *et al.* used as starting material a nanocrystalline gallium oxide nitride ceramic (see "Pyrolysis of Molecular Precursors," in Section 3.4.4.1) with a mean ratio of N/O of 0.84 for the high-pressure/high-temperature experiments. The ceramic was synthesized from $[Ga(O^tBu)_2NMe_2]_2$ as molecular precursor at a pyrolysis temperature of 350 °C in an ammonia gas atmosphere. A multi-anvil apparatus was used for the high-pressure/high-temperature synthesis experiments, with samples loaded into molybdenum capsules and compressed within 3 h at a pressure of 7 GPa and heated to a temperature of 1100 °C for 10 min. The retrieved transparent crystals displayed a light greenish color (the materials obtained by Soignard *et al.* were brown [33]). All crystallites of the new phase displayed a similar high degree of crystallinity and stability during the TEM analysis. The crystal structure was solved from the measured electron diffraction pattern, as well as by Rietveld refinement using powder XRD data recorded in the laboratory. The high-pressure/high-temperature phase crystallized in spinel-type structure with cubic space group $Fd\bar{3}m$ (no. 227) and a lattice parameter a_0 of 8.264 Å. Both methods affirmed that the anion positions did not deviate from the ideal x parameter of 0.25. For calculation of the chemical formula, the filled-anion model developed by McCauley for the corresponding aluminum oxide nitride was considered. This model presumes that vacancies can occur on the cation sites while maintaining complete occupation of the anion sites [201, 204]. The N/O ratio determined using EELS in TEM revealed that the average composition of individual crystals was chemically homogeneous, although scans taken from the core of the sample to its rim showed a small systematic variation. The chemical formula was calculated as $Ga_{2.81}O_{3.57}N_{0.43}$, assuming charge neutrality. The new phase contained less nitrogen than the starting material. It was also found that nitrogen leaving the main reaction center began to react with the capsule material (molybdenum) to form γ-Mo_2N. In addition to the new oxonitride phase, some gallium nitride was also formed under the experimental conditions, corresponding to the observations made by McMillan's group during experiments performed under high-pressure/high-temperature conditions using Ga_2O_3 and GaN as the starting materials. A main advantage of the precursor approach is that the crystalline gallium oxide nitride phase begins to form at lower temperatures (ca. 1100 °C), so that crystalline gallia is never detected.

In the case of high-pressure/high-temperature experiments starting with mixtures of gallia and gallium nitride, it was found that the β-Ga_2O_3 precursor underwent a transition into a disordered version of the high-pressure phase α-Ga_2O_3 [163] which then remained present in the samples along with hexagonal gallium nitride.

In estimating the phase diagram, Huppertz's group in Munich and Riedel's group in Darmstadt have continued their cooperation. In order to determine the stability fields, it is essential to start with mixtures of the under-ambient conditions thermodynamically stable phases β-Ga_2O_3 and w-GaN [240, 241]. As a result of a large number of high-pressure/high-temperature experiments, a wide range of stability of the spinel-structured gallium oxonitride could be detected; indeed, in one sample a single crystal could be isolated and its structure refined with a much higher nitrogen content: $Ga_{2.79}\square_{0.21}(O_{3.05}\square_{0.19})N_{0.76}$ with $\square$ as vacancies [242].

Soignard *et al.* [33] carried out detailed structural and chemical analysis of their sample, combining the results of the XRD/Rietveld analysis and an electron microprobe elemental analysis. The latter method provided an average composition $Ga_{43.6(8)}O_{46.9(5)}N_{9.5(8)}$, with the N : O ratio = 0.20(1). Such a formulation would imply vacancies present on the anion sites relative to the ideal spinel structure, and also a Ga oxidation state slightly below +3. However, as the electron probe analyses of such samples are known to underestimate the quantity of light elements present, the crystallographic data were used to provide additional constraints on the sample composition. The diffraction pattern of the $Ga_xO_yN_z$ phase was refined within the cubic spinel structure using the program GSAS [243]. To begin the refinement, the N : O ratio within the anion sites was initially set at the value determined from the electron microprobe analysis (i.e., 0.2). In a first stage of the refinement, the occupancies of both tetrahedral and octahedral Ga sites were refined freely; however, as occupancy of the tetrahedral site remained very close to unity, the occupancy of Ga atoms in that site was assumed to be equal to 1 in further refinement steps. The occupancy of Ga in the octahedral site was then further refined and determined as 0.90(4). If the Ga^{3+} oxidation state was assumed, then the chemical composition of the oxonitride spinel produced in the synthesis run was $Ga_{2.8}N_{0.64}O_{3.24}$, close to that determined for the material synthesized by Kinski *et al.* [34]. It is now believed that there are likely to be small variations in the O/N ratio of the Ga oxonitride spinels prepared by different routes and at different high-pressure/high-temperature synthesis conditions, that result in systematic variations in the lattice parameters, and likely also in the electronic properties. These variations must be investigated both experimentally and theoretically in future studies. Plotting the results available for the spinel-structured Ga oxonitrides studied to date shows an approximately linear dependence of the lattice parameter a_0 on the mol% of gallium nitride in the Ga_2O_3–GaN solid solutions (Figure 3.11), consistent with the theoretical study carried out by Kroll *et al.* [28].

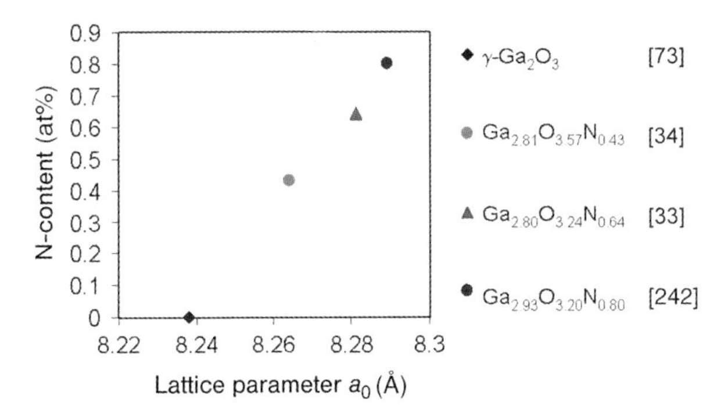

Figure 3.11 In spinel-type structured gallium oxonitrides the lattice parameter a_0 increases with higher nitrogen contents. The considered compositions are theoretically calculated regarding charge neutrality and considering a filled-anion model.

3.4.5
Potential Applications

Relatively few reports have been made concerning the possible applications for gallium oxide nitrides, although several are documented in patents [32, 220, 222, 223, 235]. Among these, the properties of the gallium oxide nitrides are mostly considered to be similar to those of the analogous alon systems, although there is no real evidence to support this assumption. It is suggested that the gallium oxonitrides represent an entirely new system with considerable future potential in several areas of wide bandgap semiconductor electronics, optoelectronics, gas sensors and ion conductors, catalytic science and biomedical applications that must be developed. On this basis, it is important first to examine the determined properties of gallium oxide nitride materials, before developing their possibilities and prospects. Since their first report on gallium oxide nitride in 1975, Marchand's group has dedicated their investigations to gallium oxide nitride phases [213, 222]. Kerlau *et al.* have reported the results of some preliminary investigations of gallium oxide nitride powders as gas sensor materials [232, 233]. Here, the synthesis of the gallium oxide nitride powder was related to that described in a patent from 1991, which utilized a nickel gallate $NiGa_2O_4$ as precursor and a citrate process [222]. Subsequent XRD studies of the yellowish powder showed the typical pattern for wurtzite-type structured material, along with some zincblende-type structured GaN. An elemental analysis of the powder gave 80 wt% gallium and 13.8 wt% nitrogen, from which data the chemical formula of the gallium oxide nitride was calculated as $GaO_{0.34}N_{0.86}$. The gallium oxide nitride was prepared as a thick film, and investigated to establish its thermal stability and gas-sensing properties as a function of influence of humidity and layer thickness, in comparison to a corresponding thick layer of gallium nitride. The gallium oxide nitride film showed a very high response to the presence of ethanol in the temperature range between 200 and 300 °C. According to Kerlau *et al.*, the thermal oxidation experiments exhibited an oxidation threshold at 500 °C, which indicated that the material could only be operated in a temperature range between 220 and 470 °C. In these preliminary studies, no clear correlation could be found between the sensing behavior, humidity, and layer thickness. Subsequent studies involved the detection of propane by the same gallium oxide nitride thick films, and also discussed the surface reactions involved [233]. In order to propose details for the sensing mechanism, the effect of water vapor as well as the oxygen concentration were investigated. The desorption products after exposure to propane were measured using temperature-programmed desorption mass spectrometry (TPD-MS), with the detected species including propene, propane, CO, CO_2, NH_3, and H_2O. Both, CO and CO_2 were among the products arising from the reaction between propane and the gallium oxide nitride in the absence of oxygen in the gaseous phase. These findings demonstrated that surface or lattice oxygen atoms or residual adsorbed oxygen-rich species must participate in the process. It is clear that newly developed gallium oxide nitride materials may play a role in the development of future catalysts for various oxidative processes, as is becoming recognized for Ga_2O_3-based materials [233].

Nakano and Jimbo investigated the interface electronic properties of thermally oxidized n-type GaN metal-oxide-semiconductor (MOS) capacitors [230]. The formation of an intermediate gallium oxide nitride layer with a graded composition could provide the origin of a small capacitance transient observed experimentally. The interface between β-Ga_2O_3 and GaN in a MOS capacitor has been investigated electrically using current–voltage (C–V) and capacitance transient techniques. Nakano and Jimbo were unable to detect any discrete interface traps, which was in reasonable agreement with the deep depletion feature and the low interface state density of $\sim 5.5 \times 10^{10}\,eV^{-1}\,cm^{-2}$. It was concluded by these authors that the surface Fermi level was most likely unpinned at the interface. Koo *et al.* developed an ion-assisted deposition procedure for the growth of GaN [234]; these films could be stabilized into an amorphous phase with incorporated 12 at% and 24 at% oxygen by using a controlled partial pressure of water vapor during deposition. For the most resistive films, the electric measurements exceeded the measurement limit of $10^{14}\,\Omega$ under a 9 V bias. Some of the films displayed very long-lived persistent photoconductivity (PPC) associated with carrier traps in the gap. As the PPC lasted for several days, the films had to be heated in order to return them to their dark conductivity. During absorption coefficient measurements, the amorphous gallium oxide nitride films displayed a blue shift of the band gap that was roughly proportional to the amount of oxygen incorporated in the films. Koo *et al.* concluded that the oxygen incorporation changed the intrinsic band structure of the nitride-based material, rather than introducing additional defect states into the gap. During these measurements the photoconductivity did not reach saturation after an hour of illumination. In summarizing their conclusions from these optical absorption, photoconductivity and thermally stimulated conductivity experiments, Koo *et al.* suggested the potential use of amorphous gallium oxide nitrides as visible-blind UV detectors. Without any attempts at optimization of their properties the gallium oxide nitride films achieved a noise equivalent power (NEP) of $10^{-8}\,W\,Hz^{-1}$ under 300 nm illumination using a simple photoresistor geometry and chopped light illumination at 10 Hz. Kikkawa *et al.* prepared gallium oxide nitride materials that were isostructural to hexagonal gallium nitride containing lithium [208] and magnesium [209] dopants, starting with nitrates as precursors and using citric acid in aqueous solution as the gelation agent; this resulted not only in substitution on the cation sites, but also an exchange of nitrogen with oxygen on the anion sites during synthesis. Further investigations did not provide any evidence of the expected room temperature ferromagnetism of Mn-doped gallium oxonitrides, that had been predicted previously [209]. However, the incorporation of Li^+ ions into the hexagonal gallium nitride structure at 750 °C increased the electrical resistivity of the semiconductor from 3×10^5 to $7 \times 10^5\,\Omega\,cm$. The Li-doped gallium oxonitride with hexagonal w-GaN structure displayed a yellow photoluminescence following excitation at 254 nm.

The results of studies conducted to date have indicated that O-substitution in wurtzite-structured GaN, and N-substitution in Ga_2O_3 phases, can be achieved effectively by using a variety of chemical and physical approaches, and that this has resulted in an important tuning of the optical and electronic properties. As yet, the oxygen and nitrogen substitution limits have not been determined, and the effects of

O/N substitution have not been fully characterized. It follows that many more investigations must be carried out to explore and establish the chemistry of these systems for their surface reactivity leading to catalysis, especially of oxidative processes, and optoelectronics materials development. Under high-pressure/high-temperature conditions, spinel-structured compounds close to the Ga_3O_3N composition have now been shown to exist, and these new materials can be recovered to ambient conditions. Similar materials with useful wide bandgap semiconductor and optoelectronic properties could be achievable in useful thin film forms by combining precursor synthesis and CVD/molecular beam epitaxy (MBE) approaches on appropriate substrates at low-pressure conditions. Consequently, further investigations are required to prepare and characterize the new gallium oxonitride ceramic compounds produced by high-pressure/high-temperature synthesis routes and thin film-deposition methods.

3.5
Outlook

Today, there is developing interest in $GaN–Ga_2O_3$ materials with the discovery of new compounds and solid solutions prepared under high-pressure/high-temperature conditions, or by metastable routes based on precursors. Interest is also centered on the recognition of the potential of these materials for catalysis and as gas sensors, as well as their valuable electronic and optoelectronic properties. Taken together, this is a completely new field of investigation, and synthesis studies are currently under way to prepare new phases and to characterize their properties both in bulk and in thin films. However, many more studies are required to establish the stable and metastable phase relationships over a wide range of pressure–temperature and synthetic conditions, as well as determining and optimizing the crystallinity and electronic properties of this new class of material.

References

1 Toth, L.E. (1971) *Transition Metal Carbides and Nitrides*, Academic Press, London, New York.
2 Oyoma, S.T. (1996) *Chemistry of Transition Metal Carbides and Nitrides*, Springer 1st edition.
3 Riedel, R. (2000) *Handbook of Ceramic Hard Materials*, Wiley-VCH, Weinheim.
4 Brese, N.E. and O'Keeffe, M. (1992) *Crystal Chemistry of Inorganic Nitrides*, vol. **79**, Springer-Verlag, p. 307.
5 Schnick, W. (1993) *Phosphorus Sulfur*, **76**, 443.
6 Niewa, R. and DiSalvo, F.J. (1998) *Chem. Mater.*, **10**, 2733.
7 Horvath-Bordon, E., Riedel, R., Zerr, A., McMillan, P.F., Auffermann, G., Prots, Y., Bronger, W., Kniep, R., and Kroll, P. (2006) *Chem. Soc. Rev.*, **35**, 987.
8 Bezinge, A., Yvon, K., Muller, J., Lengauer, W., and Ettmayer, P. (1987) *Solid State Commun.*, **63**, 141.
9 Lengauer, W. (1988) *J. Cryst. Growth*, **87**, 295.
10 Bull, C.L., McMillan, P.F., Soignard, E., and Leinenweber, K. (2004) *J. Solid State Chem.*, **177**, 1488.

11 Krukowski, S., Bockowski, M., Lucznik, B., Grzegory, I., Porowski, S., Suski, T., and Romanowski, Z. (2001) *J. Phys.-Condens. Matter*, **13**, 8881.

12 Bockowski, M., Wroblewski, M., Lucznik, B., and Grzegory, I. (2001) *Mater. Sci. Semicond. Proc.*, **4**, 543.

13 Gregoryanz, E., Sanloup, C., Somayazulu, M., Badro, J., Fiquet, G., Mao, H.K., and Hemley, R.J. (2004) *Nature Mater.*, **3**, 294.

14 Crowhurst, J.C., Goncharov, A.F., Sadigh, B., Evans, C.L., Morrall, P.G., Ferreira, J.L., and Nelson, A.J. (2006) *Science*, **311**, 1275.

15 Zerr, A., Miehe, G., and Riedel, R. (2003) *Nature Mater.*, **2**, 185.

16 Zerr, A., Miehe, G., Serghiou, G., Schwarz, M., Kroke, E., Riedel, R., Fue, H., Kroll, P., and Boehler, R. (1999) *Nature (London)*, **400**, 340.

17 Leinenweber, K., O'Keeffe, M., Somayazulu, M., Hubert, H., McMillan, P.F., and Wolf, G.H. (1999) *Chem. Eur. J.*, **5**, 3076.

18 Serghiou, G., Miehe, G., Tschauner, O., Zerr, A., and Boehler, R. (1999) *J. Chem. Phys.*, **111**, 4659.

19 Sekine, T., He, H.L., Kobayashi, T., Zhang, M., and Xu, F.F. (2000) *Appl. Phys. Lett.*, **76**, 3706.

20 Scotti, N., Kockelmann, W., Senker, J., Trassel, S., and Jacobs, H. (1999) *Z. Anorg. Allg. Chem.*, **625**, 1435.

21 Shemkunas, M.P., Wolf, G.H., Leinenweber, K., and Petuskey, W.T. (2002) *J. Am. Ceram. Soc.*, **85**, 101.

22 Sekine, T., He, H., Kobayashi, T., Tansho, M., and Kimoto, K. (2001) *Chem. Phys. Lett.*, **344**, 395.

23 Schwarz, M., Zerr, A., Kroke, E., Miehe, G., Chen, I.W., Heck, M., Thybusch, B., Poe, B.T., and Riedel, R. (2002) *Angew. Chem., Int. Ed.*, **41**, 789.

24 Corbin, N.D. (1989) *J. Eur. Ceram. Soc.*, **5**, 143.

25 Willems, H.X., de With, G., and Metselarr, R. (1993) *Tijdschr. Klei, Glas Keram.*, **14**, 43.

26 Dravid, V.P., Sutliff, J.A., Westwood, A.D., Notis, M.R., and Lyman, C.E. (1990) *Philos. Mag. A*, **61**, 417.

27 Lowther, J.E., Wagner, T., Kinski, I., and Riedel, R. (2004) *J. Alloys Compd.*, **376**, 1.

28 Kroll, P., Dronskowski, R., and Martin, M. (2005) *J. Mater. Chem.*, **15**, 3296.

29 Mo, S.D., Ouyang, L., Ching, W.Y., Tanaka, I., Koyama, Y., and Riedel, R. (1999) *Phys. Rev. Lett.*, **83**, 5046.

30 Dong, J., Tomfohr, J.K., Sankey, O.F., Leinenweber, K., Somayazulu, M., and McMillan, P.F. (2000) *Phys. Rev. B: Cond. Matter Mater. Phys.*, **62**, 14685.

31 Soignard, E., Somayazulu, M., Mao, H.K., Dong, J., Sankey, O.F., and McMillan, P.F. (2001) *Solid State Commun.*, **120**, 237.

32 Nishizawa, J. and Shiota, I. (1980) Patent 78-86573 ed., Edited by Semiconductor Research Foundation JP, p. -5.

33 Soignard, E., Machon, D., McMillan, P.F., Dong, J., Xu, B., and Leinenweber, K. (2005) *Chem. Mater.*, **17**, 5465.

34 Kinski, I., Miehe, G., Heymann, G., Theissmann, R., Riedel, R., and Huppertz, H. (2005) *Z. Naturforsch., B: Chem. Sci.*, **60**, 831.

35 Kinski, I., Scheiba, F., and Riedel, R. (2005) *Adv. Eng. Mater.*, **7**, 921.

36 Zerr, A., Miehe, G., and Riedel, R. (2003) *Nature Mater.*, **2**, 185.

37 Li, J., Dzivenko, D., Zerr, A., Fasel, C., Zhou, Y., and Riedel, R. (2005) *Z. Anorg. Allg. Chem.*, **631**, 1449.

38 Chhowalla, M. and Unalan, H.E. (2005) *Nature Mater.*, **4**, 317.

39 Maruska, H.P. and Tietjen, J.J. (1969) *Appl. Phys. Lett.*, **15**, 327.

40 Yamada, T. Mukai, T. Nakamura, S. (1995) Patent 94-22671 ed., Edited by J. Nichia Kagaku Kogyo Kk JP, p. -4.

41 Pankove, J.I., Maruska, H.P., and Berkeyheiser, J.E. (1970) *Appl. Phys. Lett.*, **17**, 197.

42 Steele, R. (2005) Strategies unlimited Report No. 4th.

43 Morkoc, H. (2001) *J. Mater. Sci. Mater. Electron.*, **12**, 677.

44 Naval Research (2008) National Compound Semiconductor Roadmap. Available at: http://www.onr.navy.mil/sci_tech/31/312/ncsr/materials/gan.asp.

45 Reshchikov, M.A. and Morkoc, H. (2005) *J. Appl. Phys.*, **97**, 061301.

46 Pavlovska, A., Bauer, E., Torres, V.M., Edwards, J.L., Doak, R.B., Tsong, I.S.T., Ramachandran, V., and Feenstra, R.M. (1998) *J. Cryst. Growth*, **189**, 310.
47 Pavlovska, A., Torres, V.M., Bauer, E., Doak, R.B., Tsong, I.S.T., Thomson, D.B., and Davis, R.F. (1999) *Appl. Phys. Lett.*, **75**, 989.
48 Torres, V.M., Edwards, J.L., Wilkens, B.J., Smith, D.J., Doak, R.B., and Tsong, I.S.T. (1999) *Appl. Phys. Lett.*, **74**, 985.
49 Krukowski, S., Bockowski, M., Lucznik, B., Grzegory, I., Porowski, S., Suski, T., and Romanowski, Z. (2001) *J. Phys. Condens. Matter*, **13**, 8881.
50 Carmalt, C.J. and Basharat, S. (2007) *Comprehensive Organometallic Chemistry III* (eds R.H. Crabtree and D.M.P. Mingos), Elsevier, Oxford, **12**, 1.
51 Haneda, M., Kintaichi, Y., and Hamada, H. (2001) *Appl. Catal. B - Environ.*, **31**, 251.
52 Boulfelfel, S.E., Zahn, D., Grin, Y., and Leoni, S. (2007) *Phys. Rev. Lett.*, **99**, 125505.
53 Amano, H., Kito, M., Hiramatsu, K., and Akasaki, I. (1989) *Jpn. J. Appl. Phys. 2*, **28**, L2112–L2114.
54 Amano, H., Akasaki, I., Kozawa, T., Hiramatsu, K., Sawaki, N., Ikeda, K., and Ishii, Y. (1988) *J. Lumin.*, **40–41**, 121.
55 Nakamura, S., Mukai, T., and Senoh, M. (1991) *Jpn. J. Appl. Phys. 2*, **30**, L1998–L2001.
56 Tippins, H.H. (1965) *Phys. Rev.*, **140**, 316.
57 Lorenz, M.R., Woods, J.F., and Gambino, R.J. (1967) *J. Phys. Chem. Solids*, **28**, 403.
58 Fleischer, M., Hanrieder, W., and Meixner, H. (1990) *Thin Solid Films*, **190**, 93.
59 Passlack, M., Hunt, N.E.J., Schubert, E.F., Zydzik, G.J., Hong, M., Mannaerts, J.P., Opila, R.L., and Fischer, R.J. (1994) *Appl. Phys. Lett.*, **64**, 2715.
60 Ueda, N., Hosono, H., Waseda, R., and Kawazoe, H. (1997) *Appl. Phys. Lett.*, **71**, 933.
61 Ueda, N., Hosono, H., Waseda, R., and Kawazoe, H. (1997) *Appl. Phys. Lett.*, **70**, 3561.
62 Matsuzaki, K., Yanagi, H., Kamiya, T., Hiramatsu, H., Nomura, K., Hirano, M., and Hosono, H. (2006) *Appl. Phys. Lett.*, **88**, 092106–092106/3.
63 Matsuzaki, K., Hiramatsu, H., Nomura, K., Yanagi, H., Kamiya, T., Hirano, M., and Hosono, H. (2006) *Thin Solid Films*, **496**, 37.
64 Song, Y.P., Zhang, H.Z., Lin, C., Zhu, Y.W., Li, G.H., Yang, F.H., and Yu, D.P. (2004) *Phys. Rev. B: Cond. Matter Mater. Phys.*, **69**, 075304.
65 Suzuki, K., Kuroki, Y., Okamoto, T., and Takata, M. (2007) *Adv. Technol. Mater. Mater. Proc. J.*, **9**, 77.
66 Haneda, M., Kintaichi, Y., and Hamada, H. (2001) *Appl. Catal. B - Environ.*, **31**, 251.
67 Mathew, T., Yamada, Y., Ueda, A., Shioyama, H., and Kobayashi, T. (2005) *Catal. Lett.*, **100**, 247.
68 Fleischer, M. and Meixner, H. (1991) *Sens. Actuators, B*, **B4**, 437.
69 Li, Y., Trinchi, A., Wlodarski, W., Galatsis, K., and Kalantar-Zadeh, K. (2003) *Sensor. Actuators B - Chemistry*, **B93**, 431.
70 Zhang, J., Jiang, F.H., Yang, Y.D., and Li, J.P. (2005) *J. Phys. Chem. B*, **109**, 13143.
71 Gundiah, G., Govindaraj, A., and Rao, C.N.R. (2002) *Chem. Phys. Lett.*, **351**, 189.
72 Roy, R., Hill, V.G., and Osborn, E.F. (1952) *J. Am. Chem. Soc.*, **74**, 719.
73 Zinkevich, M. and Aldinger, F. (2004) *J. Am. Ceram. Soc.*, **87**, 683.
74 Johnson, W.C., Parson, J.B., and Crew, M.C. (1932) *J. Phys. Chem.*, **36**, 2651.
75 Tiede, E. and Knoblauch, H.G. (1935) *Ber. Dtsch. Chem. Gesell. [Abteilung] B: Abhandlungen*, **68B**, 1149.
76 Juza, R. and Hahn, H. (1938) *Z. Anorg. Allg. Chem.*, **239**, 282.
77 Juza, R. and Hahn, H. (1940) *Z. Anorg. Allg. Chem.*, **244**, 133.
78 Pankove, J.I. (1973) *J. Lumin.*, **7**, 114.
79 Seifert, W. and Tempel, A. (1974) *Phys. Status. Solidi A*, **23**, K39–K40.
80 Karpinski, J. and Porowski, S. (1984) *J. Cryst. Growth*, **66**, 11.
81 Porowski, S., Grzegory, I., Krukowski, S., Leszczynski, M., Perlin, P., and Suski, T. (2004) *Europhysics News*, **35**, 69.
82 Kroll, P. (2005) *Phys. Rev. B: Condens. Matter Mater. Phys.*, **72**, 144407.
83 Lawaetz, P. (1972) *Phys. Rev. B*, **5**, 4039.
84 Philips, J.C. (1968) *Phys. Rev. Lett.*, **20**, 550.
85 Van Vechten, J.A. (1973) *Phys. Rev. B: Solid State*, **7**, 1479.

86 Van Vechten, J.A. (1974) *Phys. Rev. B: Solid State*, **10**, 4222.
87 Van Vechten, J.A. (1973) Phase Transitions. *Proceedings Conference on Phase Transitions and Their Applications in Materials Science* (eds. H.K. Henisch and R. Roy), Pergamon Press, New York, pp. 139–148.
88 Nelmes, R.J., McMahon, M.I., Wright, N.G., Allan, D.R., Liu, H., and Loveday, J.S. (1995) *J. Phys. Chem. Solids*, **56**, 539.
89 McMahon, M.I. and Nelmes, R.J. (1997) *Phys. Rev. Lett.*, **78**, 3697.
90 Van Vechten, J.A. (1969) *Phys. Rev.*, **187**, 1007.
91 Perlin, P., Gorczyca, I., Porowski, S., and Suski, T. (1993) *Jpn. J. Appl. Phys. Suppl.*, **32**, 334.
92 Xia, H., Xia, Q., and Ruoff, A.L. (1993) *Phys. Rev. B: Condens. Matter Mater. Phys.*, **47**, 12925.
93 Ueno, M., Yoshida, M., Onodera, A., Shimomura, O., and Takemura, K. (1994) *Phys. Rev. B: Condens. Matter Mater. Phys.*, **49**, 14.
94 Uehara, S., Masamoto, T., Onodera, A., Ueno, M., Shimomura, O., and Takemura, K. (1997) *J. Phys. Chem. Solids*, **58**, 2093.
95 Paszkowicz, W., Podsiadlo, S., and Minikayev, R. (2004) *J. Alloys Compd.*, **382**, 100.
96 Sun, X.W., Liu, Z.J., Song, T., Liu, X.B., Wang, C.W., and Chen, Q.F. (2007) *Chin. J. Chem. Phys.*, **20**, 233.
97 Sun, X.W., Chu, Y.D., Liu, Z.J., Liu, Y.X., Wang, C.W., and Liu, W.M. (2005) *Wuli Xuebao*, **54**, 5830.
98 Sun, X., Chen, Q., Chu, Y., and Wang, C. (2005) *Physica B*, **368**, 243.
99 Lei, T. and Moustakas, T.D. (1992) *Mater. Res. Soc. Symp. Proc.*, **242**, 433.
100 Lei, T., Moustakas, T.D., Graham, R.J., He, Y., and Berkowitz, S.J. (1992) *J. Appl. Phys.*, **71**, 4933.
101 Carmalt, C.J., Mileham, J.D., White, A.J.P., and Williams, D.J. (2003) *Dalton Trans.*, 4255.
102 Nakadaira, A. and Tanaka, H. (1996) *Proc. Int. Symp. Blue Laser and Light Emitting Diodes*, Ohmsha Ltd, Tokyo, 90.
103 Maruyama, T., Miyajima, Y., Hata, K., Cho, S.H., Akimoto, K., Okumura, H., Yoshida, S., and Kato, H. (1998) *J. Electron. Mater.*, **27**, 200.
104 Strite, S.C. III (1993) *Investigation of the Gallium Nitride, Aluminum Nitride and Indium Nitride Semiconductors; Structural, Optival, Electronic and Interfacial Properties*, UMI Company, 118.
105 Strite, S. and Morkoc, H. (1992) *J. Vacuum Sci. Technol., B: Microelectron. Nanometer Struct.*, **10**, 1237.
106 Strite, S., Ruan, J., Li, Z., Salvador, A., Chen, H., Smith, D.J., Choyke, W.J., and Morkoc, H. (1991) *J. Vacuum Sci. Technol., B: Microelectron. Nanometer Struct.*, **9**, 1924.
107 Wright, A.F. (1997) *Jpn. J. Appl. Phys.*, **82**, 2833.
108 Purdy, A. (1999) *Chem. Mater.*, **11**, 1648.
109 Purdy, A. (2005) *J. Cryst. Growth*, **281**, 355.
110 Purdy, A.P., Case, S., and Muratore, N. (2003) *J. Cryst. Growth*, **252**, 136.
111 Mizuta, M. (1986) *Jpn. J. Appl. Phys.*, **25**, L945–L948.
112 Paisley, M.J. (1989) *J. Vac. Sci. Technol., A*, **7**, 701.
113 Okumura, H., Misawa, S., and Yoshida, S. (1991) *Appl. Phys. Lett.*, **59**, 1058.
114 Strite, S., Ruan, J., Li, Z., Salvador, A., Chen, H., Smith, D.J., Choyke, W.J., and Morkoc, H. (1991) *J. Vac. Sci. Technol. B*, **9**, 1924.
115 Powell, R.C., Lee, N.E., Kim, Y.W., and Greene, J.E. (1992) *J. Appl. Phys.*, **73**, 189.
116 Cheng, T.S., Jenkins, L.C., Hooper, S.E., Foxon, C.T., Orton, J.W., and Lacklison, D.E. (1995) *Appl. Phys. Lett.*, **66**, 1509.
117 Yang, H., Brandt, O., Wassermeier, M., Behrend, J., Schönherr, H.P., and Ploog, K.H. (1996) *Appl. Phys. Lett.*, **68**, 244.
118 Ploog, K.H., Brandt, O., Yang, H., Menniger, J., and Klann, R. (1997) *Solid-State Electron.*, **41**, 235.
119 Zhao, Y., Tu, C.W., Bae, I.T., and Seong, T.Y. (1999) *Appl. Phys. Lett.*, **74**, 3182.
120 Yamane, H., Shimada, M., and DiSalvo, F.J. (2000) *Mater. Lett.*, **42**, 66.
121 Bouarissa, N. (2002) *Mater. Chem. Phys.*, **73**, 51.
122 Feng, Z.H., Yang, H., Zheng, X.H., Fu, Y., Sun, Y.P., Shen, X.M., and Wang, Y.T. (2003) *Appl. Phys. Lett.*, **82**, 206.
123 Xia, H., Xia, Q., and Ruoff, A.L. (1993) *Phys. Rev. B*, **47**, 12925.
124 Xie, Y., Qian, Y., Zhang, S., Wang, W., Liu, X., and Zhang, Y. (1996) *Appl. Phys. Lett.*, **69**, 334.

125 Munoz, A. and Kunc, K. (1991) *Phys. Rev. B*, **44**, 10372.
126 Van Camp, P.E., Van Doren, V.E., and Devreese, J.T. (1992) *Solid State Commun.*, **81**, 23.
127 Serrano, J., Rubio, A., Hernandez, E., Munoz, A., and Mujica, A. (2000) *Phys. Rev. B: Condens. Matter Mater. Phys.*, **62**, 16612.
128 Moon, W.H. and Hwang, H.J. (2003) *Phys. Lett. A*, **315**, 319.
129 Lu, L.Y., Chen, X.R., Cheng, Y., and Zhao, J.Z. (2005) *Solid State Communications*, **136**, 152.
130 Denis, A., Goglio, G., and Demazeau, G. (2006) *Mater. Sci. Eng., R*, **R50**, 167.
131 Karpinski, J., Jun, J., and Porowski, S. (1984) *J. Cryst. Growth*, **66**, 1.
132 Neumayer, D.A., Carmalt, C.J., Arendt, M.F., White, J.M., Cowley, A.H., Jones, R.A., and Ekerdt, J.G. (1996) *Mater. Res. Soc. Symp. Proc.*, **395**, 85.
133 Nakamura, S., Pearton, S., and Fasol, G. (2000) *The Blue Laser Diode - The Complete Story*, Springer-Verlag, Berlin.
134 Monemar, B. (1999) *J. Mater. Sci. - Mater. Electron.*, **10**, 227.
135 Morkoc, H. (1999) *Nitride Semiconductors and Devices*, Springer-Verlag, Berlin.
136 Liu, L. and Edgar, J.H. (2002) *Mater. Sci. Eng. R.*, **37**, 61.
137 Parikh, R.P. and Adomaitis, R.A. (2006) *J. Cryst. Growth*, **286**, 259.
138 Unland, J., Onderka, B., Davydov, A., and Schmid-Fetzer, R. (2003) *J. Cryst. Growth*, **256**, 33.
139 Karpinski, J., Porowski, S., and Miotkowska, S. (1982) *J. Cryst. Growth*, **56**, 77.
140 Utsumi, W., Saitoh, H., Kaneko, H., Watanuki, T., Aoki, K., and Shimomura, O. (2003) *Nature Mater.*, **2**, 735.
141 Hahn, H. and Juza, R. (1940) *Z. Anorg. Allg. Chem.*, **244**, 111.
142 Yamane, H., Shimada, M., Clarke, S.J., and DiSalvo, F.J. (1997) *Chem. Mater.*, **9**, 413.
143 Yamane, H., Kinno, D., Shimada, M., and DiSalvo, F.J. (1999) *J. Ceram. Soc. Jpn*, **107**, 925.
144 Yamane, H., Shimada, M., and DiSalvo, F.J. (2000) *Mater. Sci. Forum*, **325–326**, 21.
145 Beaumont, B., Vennegues, P., and Gibart, P. (2001) *Phys. Status Solidi B*, **227**, 1.
146 Pankove, J.I., Miller, E.A., and Berkeyheiser, J.E. (1972) *J. Lumin.*, **5**, 84.
147 Nakamura, S. (1992) Patent 91-116912 ed., Edited by L. Nichia Chemical Industries JP, p.-3.
148 Nakamura, S. (1995) *Kotai Butsuri*, **30**, 798.
149 Nakamura, S., Senoh, M., Iwasa, N., and Nagahama, S.i. (1995) *Appl. Phys. Lett.*, **67**, 1868.
150 Nakamura, S. and Yamada, M. (1995) Patent 93-288365 ed., Edited by J. Nichia Kagaku Kogyo Kk JP, 6.
151 Nakamura, S. and Yamada, M. (1995) Patent 93-300940 ed., Edited by J. Nichia Kagaku Kogyo Kk JP, p. 5.
152 Nakamura, S. and Iwasa, S. (1995) Patent 93-310533 ed., Edited by J. Nichia Kagaku Kogyo Kk JP, p. 5.
153 Nakamura, S. (1995) *Oyo Butsuri*, **64**, 717.
154 Nakamura, S. (1995) Patent 94-223739 ed., Edited by L. Nichia Chemical Industries US, p. -8.
155 Nakamura, S., Senoh, M., Iwasa, N., and Nagahama, S.i. (1995) *Jpn. J. Appl. Phys. 2*, **34**, 1797–1799.
156 Nakamura, S. and Mukai, T. (1995) Patent 93-157219 ed., Edited by J. Nichia Kagaku Kogyo Kk JP, p. 5.
157 Nakamura, S. (1995) *J. Vac. Sci. Technol. A*, **13**, 705.
158 Nakamura, S. and Mukai, T. (1995) Patent 93-146383 ed., Edited by J. Nichia Kagaku Kogyo Kk JP, p. 5.
159 Komaki, H., Nakamura, T., Katayama, R., Onabe, K., Ozeki, M., and Ikari, T. (2007) *J. Cryst. Growth*, **301–302**, 473.
160 Ho, I.h. and Stringfellow, G.B. (1996) *Appl. Phys. Lett.*, **69**, 2701.
161 Ambacher, O. *et al.* (1999) *Phys. Status Solidi B*, **216**, 381.
162 Foster, L.M. and Stumpf, H.C. (1951) *J. Am. Chem. Soc.*, **73**, 1590.
163 Machon, D., McMillan, P.F., Xu, B., and Dong, J. (2006) *Phys. Rev. B*, **73**, 094125.
164 Yoshioka, S., Hayashi, H., Kuwabara, A., Oba, F., Matsunaga, K., and Tanaka, I. (2007) *J. Phys.: Condens. Matter.*, **19**, 346211.
165 Liu, B., Gu, M., and Liu, X. (2007) *Appl. Phys. Lett.*, **91**, 172102–172102/3.
166 Passlack, M. *et al.* (1995) *Jpn. J. Appl. Phys.*, **77**, 686.

167 Fleischer, M. and Meixner, H. (1992) *J. Mater. Sci. Lett.*, **11**, 1728.

168 Orita, M., Hiramatsu, H., Ohta, H., Hirano, M., and Hosono, H. (2002) *Thin Solid Films*, **411**, 134.

169 Nogales, E., Mendez, B., and Piqueras, J. (2008) *Nanotechnology*, **19**, 035713/1.

170 Aubay, E. and Gourier, D. (1993) *Phys. Rev. B*, **47**, 15023.

171 Schubert, E., Passlack, M., Hong, M., Mannerts, J., Opila, R., Pfeiffer, L., West, K., Bethea, C., and Zydzik, G. (1994) *Appl. Phys. Lett.*, **64**, 2976.

172 Choi, Y.C. *et al.* (2000) *Adv. Mater.*, **12**, 746.

173 Miyata, T., Nakatani, T., and Minami, T. (2000) *J. Lumin.*, **87–9**, 1183.

174 Xiao, T., Kitai, A.H., Liu, G., Nakua, A., and Barbier, J. (1998) *Appl. Phys. Lett.*, **72**, 3356.

175 Hao, J.H. and Cocivera, M. (2002) *J. Phys. D - Appl. Phys.*, **35**, 433.

176 Kim, H.G. and Kim, W.T. (1987) *Jpn. J. Appl. Phys.*, **62**, 2000.

177 Tabata, M., Tsuchida, H., Miyamoto, K., Yoshinari, T., Yamazaki, H., Hamada, H., Kintaichi, Y., Sasaki, M., and Ito, T. (1995) *Appl. Catal. B - Environmental*, **6**, 169.

178 Obana, Y., Yashiki, K., Ito, M., Nishiguchi, H., Ishihara, T., and Takita, Y. (2003) *J. Jpn. Petrol. Inst.*, **46**, 53.

179 Gebauer-Henke, E., Grams, J., Szubiakiewicz, E., Farbotko, J., Touroude, R., and Rynkowski, J. (2007) *J. Catal.*, **250**, 195.

180 Tu, B.Z., Cui, Q.L., Xu, P., Wang, X., Gao, W., Wang, C.X., Liu, J., and Zou, G.T. (2002) *J. Phys. - Condens. Matter*, **14**, 10627.

181 O'Dell, L.A., Savin, S.L.P., Chadwick, A.V., and Smith, M.E. (2007) *Appl. Magn. Reson.*, **32**, 527.

182 Böhm, J. (1940) *Z. Angew. Chem.*, **53**, 131.

183 Remeika, J.P. (1963) I. Bell Telephone Laboratories Inc., US 3075831.

184 Chase, A.B. (1964) *J. Am. Ceram. Soc.*, **47**, 470.

185 Lee, S.A., Hwang, J.Y., Kim, J.P., Jeong, S.Y., and Cho, C.R. (2006) *Appl. Phys. Lett.*, **89**, 182906.

186 Basharat, S., Carmalt, C.J., King, S.J., Peters, E.S., and Tocher, D.A. (2004) *Dalton Trans.*, 3475.

187 Carmalt, C.J. and King, S.J. (2006) *Coordin. Chem. Rev.*, **250**, 682.

188 Basharat, S., Betchley, W., Carmalt, C.J., Barnett, S., Tocher, D.A., and Davies, H.O. (2007) *Organometallics*, **26**, 403.

189 Zhang, H.Z. *et al.* (1999) *Solid State Commun.*, **109**, 677.

190 Wu, X.C., Song, W.H., Huang, W.D., Pu, M.H., Zhao, B., Sun, Y.P., and Du, J.J. (2000) *Chem. Phys. Lett.*, **328**, 5.

191 Pan, Z.W., Dai, Z.R., and Wang, Z.L. (2001) *Science*, **291**, 1947.

192 Liang, C.H., Meng, G.W., Wang, G.Z., Wang, Y.W., Zhang, L.D., and Zhang, S.Y. (2001) *Appl. Phys. Lett.*, **78**, 3202.

193 Yang, Z.X., Wu, Y.J., Zhu, F., and Zhang, Y.F. (2004) *Phys. Status Solidi A*, **201**, 3051.

194 Jalilian, R., Yazdanpanah, M.M., Pradhan, B.K., and Sumanasekera, G.U. (2006) *Chem. Phys. Lett.*, **426**, 393.

195 Sinha, G., Ganguli, D., and Chaudhuri, S. (2006) *J. Phys.: Condens. Matter*, **18**, 11167.

196 Corbin, N. (1989) *J. Eur. Ceram. Soc.*, **5** (3), 143.

197 Corbin, N.D., Sundberg, G.J., Siebein, K.N., Wilkens, C.A., Pujari, V.K., Rossi, G.A., Hansen, J.S., Chang, C.L., and Hammarstrom, J.L. (1992) *Ceramic Technology Project*, 115.

198 Corbin, N.D. (1987) MTL-MS-87-3; Order No. AD-A184753, p. 21.

199 Corbin, N.D. and McCauley, J.W. (1981) *Proc. Soc. Photo-Optical Instrum. Eng.*, **297**, 19.

200 Corbin, N.D. and McCauley, J.W. (1980) *Am. Ceram. Soc. Bull.*, **59**, 373.

201 McCauley, J. (1978) *J. Am. Ceram. Soc.*, **61**, 372.

202 McCauley, J.W., Krishnan, K.M., Rai, R.S., Thomas, G., and Zangvill, A. (1988) Report, 22.

203 McCauley, J.W. and Corbin, N.D. (1979) *J. Am. Ceram. Soc.*, **62**, 476.

204 McCauley, J.W. and Corbin, N.D. (1979) *J. Am. Ceram. Soc.*, **62**, 476.

205 Metselaar, R. and Yan, D.S. (1999) *Pure Appl. Chem.*, **71**, 1765.

206 Janik, J.F., Drygas, M., Stelmakh, S., Grzanka, E., Palosz, B., and Paine, R.T. (2006) *Phys. Status Solidi A*, **203**, 1301.

207 Kamler, G., Weisbrod, G., and Podsiadlo, S. (2000) *J. Therm. Anal. Calorim.*, **61**, 873.

208 Kikkawa, S., Nagasaka, K., Takeda, T., Bailey, M., Sakurai, T., and Miyamoto, Y. (2007) *J. Solid State Chem.*, **180**, 1984.

209 Kikkawa, S., Ohtaki, S., Takeda, T., Yoshiasa, A., Sakurai, T., and Miyamoto, Y. (2008) *J. Alloys Compd.*, **450**, 152.

210 Puchinger, M., Kisailus, D.J., Lange, F.F., and Wagner, T. (2002) *J. Cryst. Growth*, **245**, 219.

211 Zerr, A., Miehe, G., Serghiou, G., Schwarz, M., Kroke, E., Riedel, R., Fuess, H., Kroll, P., and Boehler, R. (1999) *Nature*, **400**, 340.

212 Dong, J., Deslippe, J., Sankey, O.F., Soignard, E., and McMillan, P.F. (2003) *Phys. Rev. B*, **67**, 094104–094104/7.

213 Verdier, P. and Marchand, R. (1976) *Rev. Chim. Miner.*, **13**, 145.

214 Merdrignac, O., Guyader, J., Verdier, P., Colin, Y., Lent, Y., and Laurent, Y. (1990) Fr. Demande, 2638527.

215 Grekov, F.F., Demidov, D.M., and Zykov, A.M. (1979) *Zh. Prikl. Khim. (Leningrad)*, **52**, 1394.

216 Shiota, I. and Nishizawa, J. (1979) *Surf. Interf. Anal.*, **1**, 185.

217 Shiota, I., Miyamoto, N., and Nishizawa, J. (1979) *Surf. Sci.*, **86**, 272.

218 Shiota, I. and Nishizawa, J. (1979) *Surf. Interf. Anal.*, **1**, 185.

219 Nishizawa, J. and Shiota, I. (1980) *Conf. Ser. - Inst. Phys.*, **50**, 287.

220 Nishizawa, J. and Shiota, I. (1982) 80-215442 ed., Edited by J. Zaiden Hojin Handotai Kenkyu Shinkokai US, p. 14.

221 Aleksandrov, S.E., Zykov, A.M., Kryakin, V.A., and Tsoi, V.V. (1988) *Zh. Prikl. Khim.*, **61**, 1454.

222 Merdrignac, O., Guyader, J., Verdier, P., Colin, Y., and Laurent, Y. (1990) Patent 88-14270 ed., Edited by Fr. Centre National de la Recherche Scientifique and d. R. Universite, I FR, p. 16.

223 Yoshida, K. and Sasaki, M. (1995) 93-290193 ed., Edited by J. Hikari Gijutsu Kenkyu Kaihatsu JP, p. 5.

224 Kovalgin, A.Y. and Lermontova, N.A. (1997) *Proc. Electrochem. Soc.*, **97-25**, 701.

225 Wolter, S.D., Luther, B.P., Waltemyer, D.L., Onneby, C., Mohney, S.E., and Molnar, R.J. (1997) *Appl. Phys. Lett.*, **70**, 2156.

226 Schmitz, G., Eumann, M., Stapel, D., and Franchy, R. (1999) *Surf. Sci.*, **427–428**, 91.

227 Tsuruoka, T., Kawasaki, M., Ushioda, S., Franchy, R., Naoi, Y., Sugahara, T., Sakai, S., and Shintani, Y. (1999) *Surf. Sci.*, **427–428**, 257.

228 Wolter, S.D., DeLucca, J.M., Mohney, S.E., Kern, R.S., and Kuo, C.P. (2000) *Thin Solid Films*, **371**, 153.

229 Kim, H., Park, N.M., Jang, J.S., Park, S.J., and Hwang, H. (2001) *Electrochem. Solid-State Lett.*, **4**, G104–G106.

230 Nakano, Y. and Jimbo, T. (2003) *Appl. Phys. Lett.*, **82**, 218.

231 Jung, W.S. (2004) *Bull. Korean Chem. Soc.*, **25**, 51.

232 Kerlau, M., Merdrignac-Conanec, O., Reichel, P., Barsan, N., and Weimar, U. (2006) *Sens. Actuators, B*, **B115**, 4.

233 Kerlau, M., Reichel, P., Barsan, N., Weimar, U., sarte-Gueguen, S., and Merdrignac-Conanec, O. (2007) *Sens. Actuators, B*, **B122**, 14.

234 Koo, A., Budde, F., Ruck, B.J., Trodahl, H.J., Bittar, A., Preston, A., and Zeinert, A. (2006) *J. Appl. Phys.*, **99**, 034312–034312/7.

235 Peng, L., Wu, H., and Lin, J.Edited by TEKCORE CO LTD.

236 Riedel, R., Zerr, A., Kroke, E., and Schwarz, M. (2001) *Ceram. Trans.*, **112**, 119.

237 Soignard, E., Somayazulu, M., Dong, J., Sankey, O.F., and McMillan, P.F. (2001) *J. Phys. - Condens. Matter*, **13**, 557.

238 Soignard, E., McMillan, P.F., Hejny, C., and Leinenweber, K. (2004) *J. Solid State Chem.*, **177**, 299.

239 Berbenni, V., Milanese, C., Bruni, G., and Marini, A. (2005) *J. Therm. Anal. Calorim.*, **82**, 401.

240 Hering, S.A., Zvoriste, C.E., Riedel, R., Kinski, I., and Huppertz, H. (2009). *Z. Naturforsch. B*, **64** (10), 1115–1126.

241 Zvoriste, C.E., Dubrovinsky, L.S., Hering, S.A., Huppertz, H., Riedel, R., and Kinski, I. (2009) *High-Press. Res.*, **29** (3), 389–395.

242 Huppertz, H., Hering, S.A., Zvoriste, C.E., Lauterbach, S., Oeckler, O., Riedel, R., and Kinski, I. (2009) *Chem. Mater.*, **21** (10), 2101–2107.

243 Larson, A.C. and Von Dreele, R.B. (2000) Los Alamos National Laboratory Report 86.

4
Silicon Carbide- and Boron Carbide-Based Hard Materials

Clemens Schmalzried and Karl A. Schwetz

4.1
Introduction

Superhard compounds are obviously formed by a combination of the low atomic number elements boron, carbon, silicon, and nitrogen. Carbon–carbon as diamond, boron–nitrogen as cubic boron nitride, boron–carbon as boron carbide, and silicon–carbon as silicon carbide, belong to the hardest materials hitherto known. Because of their extreme properties and the variety of present and potential commercial applications, silicon carbide (SiC) and boron carbide (B_4C) are, besides tungsten carbide-based hard metals, considered by many as the most important carbide materials.

4.2
Structure and Chemistry

4.2.1
Silicon Carbide

Berzelius [1] first reported the formation of silicon carbide in 1810 and 1821, but it was later rediscovered during various electrochemical experiments, notably by Despretz [2], Schützenberger [3], and Moissan [4]. However, it was Acheson [5] who first realized the technical importance of silicon carbide as a hard material and, believing it to be a compound of carbon and corundum, he named the new substance "carborundum". By 1891, Acheson had managed to prepare silicon carbide on a large scale such that, today, it has become by far the most widely used nonoxide ceramic material.

Due to its great hardness, heat resistance, and oxidation resistance, silicon carbide has become firmly established as an abrasive as well as a raw material for producing refractories such as firebricks, setter tiles, and heating elements. Another major use of silicon carbide is as a siliconizing and carburizing agent in iron and steel metallurgy.

Ceramics Science and Technology Volume 2: Properties. Edited by Ralf Riedel and I-Wei Chen

ISBN: 978-3-527-31156-9

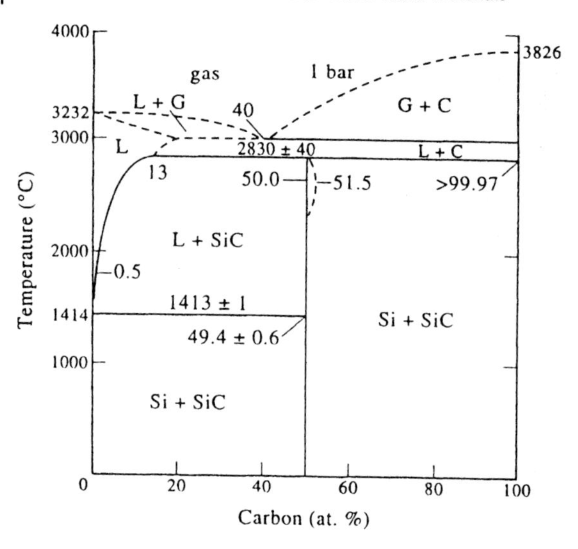

Figure 4.1 Phase diagram of the silicon–carbon system at 1 bar total pressure. After Ref. [6].

Self-bonded silicon carbide materials in molded form and of high SiC content have been available only from the late 1960s onwards. These are being increasingly used for structural components in mechanical engineering, and have proved to be highly successful for use under extreme abrasive, corrosive, and thermal conditions due to their excellent mechanical, chemical, and physical properties.

4.2.1.1 Phase Relations in the System Silicon–Carbon

As shown in the phase diagram in Figure 4.1, silicon carbide is the only binary phase in the silicon–carbon system, with the composition 70.05 wt% Si, 29.95 wt% C [6]. Silicon carbide does not have a congruent melting point, but in a closed system at a total pressure of 1 bar it decomposes into graphite and a silicon-rich melt at $2830 \pm 40\,°C$. This is the highest temperature at which silicon carbide crystals are formed. In an open system, silicon carbide begins to decompose at $\sim$2300 °C, with the formation of gaseous silicon and a residue of graphite. SiC and Si form a degenerate eutectic at 1413 °C and 0.02 atom% C. The solubility of carbon in liquid silicon is 13 atom% C at the peritectic temperature.

4.2.1.2 Structural Aspects

Silicon carbide exists in several modifications, being polymorphic and polytypical and crystallizing in a diamond lattice [7].

As silicon carbide exists predominantly in its beta form at temperatures below 2000 °C, this is referred to as the "low-temperature" modification. Cubic β-SiC is metastable and, in accordance with Ostwald's rule, is formed initially in SiC production from silicon dioxide and carbon. β-SiC can also be prepared at about 1450 °C from simple mixtures of silicon and carbon, or by the hydrogen reduction of organosilanes at temperatures below 2000 °C. Above 2000 °C, only the hexagonal and

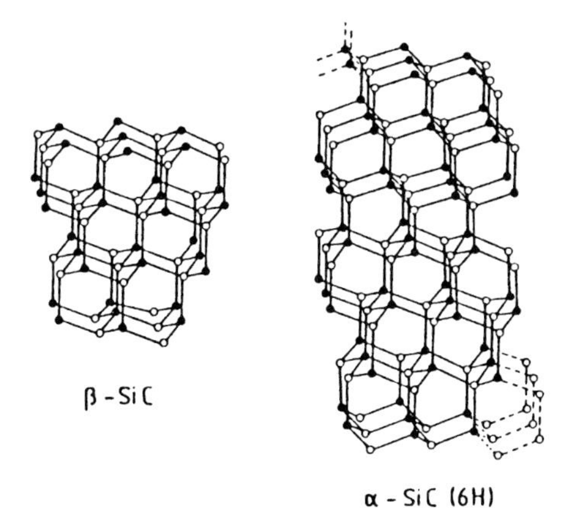

Figure 4.2 Crystal structures of the SiC polytypes 3C and 6H.

rhombohedral types are stable if there are no stabilizing influences which raise the transformation temperature [8, 9]. Thus, in a nitrogen atmosphere at high pressure, β-SiC is the stable form, so that above 2000 °C α-SiC transforms to β-SiC.

The basic element of the silicon carbide structure is the tetrahedron due to sp^3 hybridization of the atomic orbitals [10]. This tetrahedron consists of a silicon or a carbon atom at the spatial center, surrounded by four atoms of the other type. The Si–C bond is 88% covalent, and the tetrahedra are arranged in such a way that units of three silicon and three carbon atoms form angled hexagons which are arranged in parallel layers, as shown in Figure 4.2.

The layer sequences can repeat themselves in the cycles ABC, ABC . . . (zincblende, type 3C) or AB, AB . . . (wurtzite, type 2H), according to cubic or hexagonal close packing. In addition, numerous other stack sequences are formed in the case of silicon carbide, resulting in many similar polytypes.

Polycrystalline silicon carbide obtained by the Acheson process exhibits a large number of different polytypes, some of which dominate. More than 200 different polytypes are currently known; these can be classified into the cubic, hexagonal, and rhombohedral crystal system, and all have the same density of 3.21 g cm^{-3}. Written polytype nomenclature [11] indicates the number of layers in the repeating layer pack by a numeral, while the crystal system is denoted by the letters C, H, or R.

The most frequently encountered structures 3C, 4H, 6H, and 15R, are called "short-period" polytypes; "long-period" polytypes are much less common and consist of blocks of short-period polytypes which are broken by regularly occurring stacking faults.

The amounts of the polytypes 3C, 4H, 15R, and 6H can be quantitatively determined using X-ray diffraction (XRD) techniques [12], which can indicate the temperature of formation of the silicon carbide since the stability of a given polytype depends on, among other things, temperature. According to Inomata *et al.* [13], 2H-SiC seems to be stable below 1400 °C, 3C between 1400 and about 1600 °C, 4H between 1600 and 2100 °C, 6H above 2100 °C, and 15R above 2200 °C. There is no doubt, however, that impurities also play a part in the formation of the various

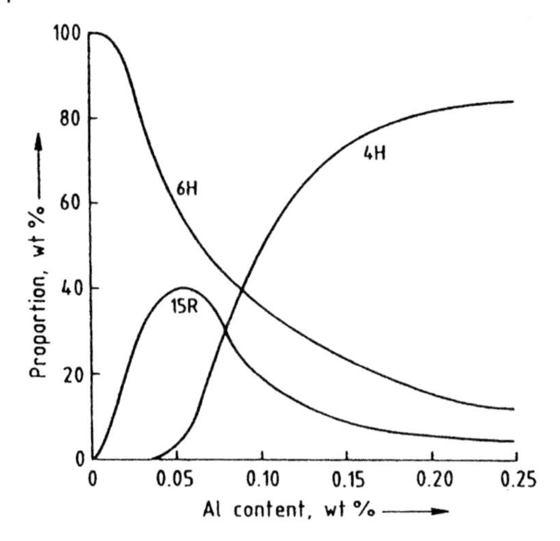

Figure 4.3 Proportions of polytypes in technical-grade SiC as a function of Al content. After Ref. [14].

polytypes, together with the surrounding gas atmosphere and growth-kinetic processes. Preferential stabilization of the 4H polytype is achieved with additions of aluminum up to concentrations of ~0.4 wt%, which is thought to substitute for silicon in the silicon carbide lattice [14–17]. According to Lundquist [14], pure α-SiC mainly crystallizes at 6H, with small amounts of aluminum 15R also occurring, whilst above 0.1 wt% aluminum, the 4H polytype predominates (see Figure 4.3). It is also well known today that nitrogen can be incorporated into the SiC lattice, whereby it stabilizes not only the cubic β-SiC form but also the wurtzite 2H polytype [8, 18]. Whereas, the nitrogen content obtained in 3C-SiC by gas/solid diffusion is low (e.g., <1 atom%), extremely high nitrogen contents of up to ~14 atom% were recently found for nitrogen-stabilized 2H SiC when formed from the carbothermal reduction of silicon nitride, Si_3N_4, in the presence of a liquid phase [19].

High-resolution transmission electron microscopy (HRTEM), using lattice imaging techniques, allows polytype analysis within single grains in the microstructure of dense SiC bodies [20].

4.2.2
Boron Carbide

Joly [21] reported the preparation of boron carbide in 1883, labeling the product as B_3C, whilst in 1899 Moisson [22] labeled the compound as B_6C. Yet, another 50 years passed until Ridgeway [23] suggested the stoichiometry to be 4 to 1. Today, it is well established that the composition of boron carbide has no exact stoichiometric composition but ranges from $B_{4.3}C$ to $B_{10.4}C$. The composition of commercially produced boron carbide, using the carbothermal reduction of boron oxide in arc furnaces, is usually close to B_4C, which corresponds to the stoichiometric limit on the high-carbon side.

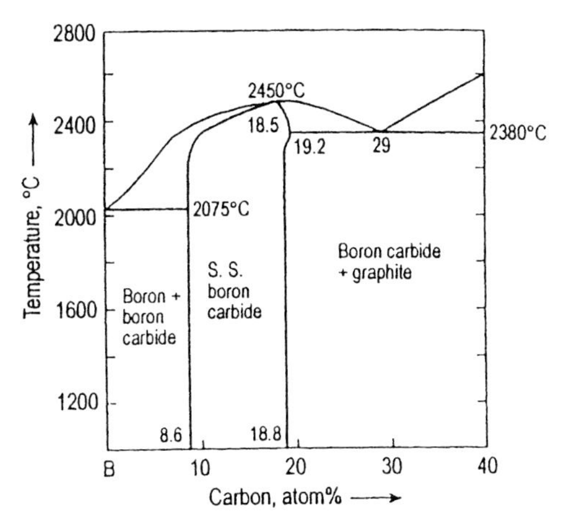

Figure 4.4 The boron–carbon phase diagram according to Schwetz and Karduck [25].

4.2.2.1 **Phase Relations in the System Boron–Carbon**

Today, it is generally accepted that only one binary phase $B_{13}C_{2\pm x}$ exists in the B–C system, with a homogeneity range of 8.8 to 20 atom% C, melting congruently at the composition $B_{13}C_2$ (18.5 atom% C) at 2450 °C [24] (see Figure 4.4). The maximum carbon content of 20% usually given corresponds to a stoichiometry of B_4C. Schwetz and Karduck [25] used quantitative microprobe analysis and found a maximum carbon content of fused boron carbide in equilibrium with graphite of 18.8 atom% ($B_{4.3}C$) in the temperature range from 2000 to 2350 °C; this was meanwhile confirmed by Werheit *et al.* [26]. This value differs significantly from those reported by Beauvy [27], who found a maximum carbon content of 21.6 atom% at "low temperature," and 24.3 atom% near the eutectic temperature. The eutectic temperature was detected to be at 2380 °C and 29–31 atom% C, which was in good accordance with thermodynamically assessed values [28, 29] and the values reported by Elliot [24]. In contrast to Bouchacourt and Thevenot [30], who suggested a peritectic reaction [Eq. (1)] at 2075 °C, assuming a melting point of boron at 2020 °C, the reaction should be eutectic in nature due to a generally accepted melting point of elemental boron slightly above 2075 °C. The solubility of carbon in β-boron is below 1 atom% [30].

$$(l) + B_4C \rightarrow \beta\text{-B} \tag{1}$$

The thermodynamic assessment of the binary phase diagram is in good agreement with experimental data [31, 32].

4.2.2.2 **Structural Aspects**

Boron carbide crystallizes in the trigonal–rhombohedral space group R3m; the unit cell is shown in Figure 4.5. The structure is usually described as an arrangement of distorted icosahedra located at the nodes of a rhombohedral Bravais lattice. Parallel to the space diagonal, which is the c-axis in hexagonal notation, a linear chain of three

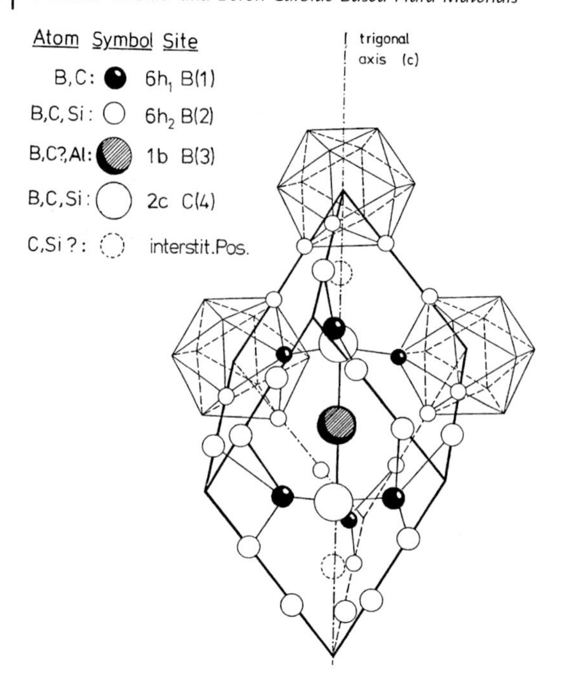

Figure 4.5 Crystal structure of boron carbide.

atoms connects two adjacent icosahedra. The unit cell is comprised of 12 icosahedral sites and three sites in the linear chain, while two inequivalent crystallographic sites exist in the icosahedron. The six atoms which form the top and bottom triangular faces of the icosahedron sit at the "polar" sites [B(2) in Figure 4.5], and are directly linked to atoms in neighboring icosahedra; the other six corners of the icosahedron form a puckered hexagon in the plane perpendicular to the [111] axis, and their symmetry-equivalent sites are termed "equatorial" [B(1) in Figure 4.5]. Each one of the six equatorial atoms is linked to an intericosahedral chain. Today, it is well established by structural investigations that, over the whole homogeneity range, the structure is composed of nearly isomorphous elementary cells with varying microstructures. None of the compositions exhibits a well-defined unit cell representing the structure. The B12 and B11C icosahedra, C–B–C, C–B–B, BϒB (ϒ, vacancy) chains form differently composed statistically distributed elementary cells. The concentrations of the structure elements in the corresponding structure formula [Eq. (2)] can be determined by evaluating the infrared (IR) and Raman active phonon spectra, and taking into account other reliable experimental results [33–37].

$$(B_{12})_n(B_{11}C)_{1-n}(CBC)_p(CBB)_q(B\Upsilon B)_{1-p-q} \tag{2}$$

Mauri *et al.* [38] postulated, according to analysis of nuclear magnetic resonance (NMR) spectra, a $B_{10}C_2$ structure with the two C atoms placed in antipodal polar sites.

Structural investigations in laser-assisted chemical vapor-deposited boron carbide thin films by Raman microspectroscopy and glancing incidence XRD, have provided evidence that the composition of the icosahedra and C–B–C chains are influenced by the overall carbon content [39]. The trends found are consistent with the micro-Raman spectra of the samples, and allow one to interpret the way in which carbon atoms are preferentially substituted by boron atoms in the lattice as the carbon content is reduced. As the carbon concentration begins to decrease, the C–B–C chains are the units preferentially affected, by replacement by C–B–B chains. Between about 17.5 and 13.5 atom% C, the most affected structures are the B11C icosahedra, leading to the formation of B12 structure units. Below 13.5 atom% C, the C–B–C and/or C–B–B diagonal chains are affected and probably disappear at low carbon concentrations. This seems to be consistent with the hypothesis of Yakel [40], which states that the main diagonal chains may be randomly replaced by a new linking component consisting of four boron atoms located at the center of the unit cell in a plane perpendicular to the c_h axis.

The existence of vacancies has been also reported by Kwei *et al.* [41], who observed that in the C–B–C chain, the central boron atom was relatively loosely held and that these locations could form vacancies. The increasing concentration of atomic defects has been attributed to weakly bound electrons, which explains the increasing electrical conductivity with increasing boron content [with a break at 13.3 mol% C; i.e., when (B11C) $\rightarrow$ (B12)].

The central position in the linear chain is considered to be occupied by larger atoms such as Si or Al, which form solid solutions with boron carbide [42–46].

4.3 Production of Particles and Fibers

4.3.1 Silicon Carbide

4.3.1.1 Technical-Scale Production of α-Silicon Carbide

Abrasive and refractory grade silicon carbide is produced industrially by the carbothermal reduction of silica, according to the reaction in Eq. (3), from high-purity quartz sand (99.5% SiO_2) and petroleum coke, in electrically heated resistance furnaces (Acheson furnace) [47]. The reaction is strongly endothermic: the heat of formation is 618 kJ mol^{-1} SiC, corresponding to 4.28 kWh kg^{-1} SiC.

$$SiO_2 + 3C \rightarrow SiC + 2CO \qquad (3)$$

The process takes place at temperatures of between 1600 and 2500 °C, and is far more complex than the above equation because of the intermediate reactions, some of which occur with participation of the gas phase [48, 49] (see Section 4.3.1.2).

Acheson SiC furnaces are up to 25 m long, 4 m wide, and 4 m high, consume up to 5000 kVA, and run for about 130 h. The furnace has a rectangular cross-section, and consists of two water-cooled graphite electrodes at the end and two movable side walls, the purpose of which is to retain the raw material mixture. A column of tamped graphite powder or a solid graphite rod is inserted into the reaction mixture core to

Figure 4.6 Silicon carbide roll at the ESD facility in Delfzijl, The Netherlands.

connect the electrodes. The silicon carbide is formed in the hot core in the form of a polycrystalline, compact cylinder – the so-called "roll". The carbon monoxide (CO) byproduct escapes through the surface of the reaction mixture and through slits in the side walls, where it is burned off. In contrast to classical Acheson furnaces, the ESK (Elektroschmelzwerk Kempten) process developed during the 1970s is ecologically safe [50]. In this case, the CO is collected and converted into electrical energy after burning, thus saving up to 20% of the energy necessary for the production of silicon carbide. The silicon carbide cylinders inside the linear or U-shaped resistance furnaces are up to 60 m long and weigh ~300 t after a run of 10 days (see Figure 4.6). A comparative review of some Acheson plants is given in Ref. [51].

The innermost SiC zone is of the highest quality, as the purest and largest crystals can form in this region. Moving outwards – that is, with decreasing temperature – the crystal growth rate decreases and the SiC roll becomes increasingly finely crystalline such that the outer zone is composed of very fine crystals of β-SiC.

The lumps of silicon carbide obtained when the furnace is dismantled are sorted, broken down, milled, sieved, air-classified and wet chemically purified so as to remove any adhering traces of elemental silicon, metals, metal compounds, graphite, dust, and silica [52–54].

4.3.1.2 **β-Silicon Carbide Powder**

Carbothermal Reduction of Silica The primary method to synthesize β-SiC is by carbothermal reduction reaction below 1900 °C; that is, carbon reacts with silica (SiO_2) to produce SiC and CO gas. The chemical equation for this reaction is commonly written as:

$$SiO_2 + 3C \rightarrow \beta\text{-}SiC + 2CO \tag{4}$$

According to the reaction mechanism proposed by Weimer [49], the reaction takes place in several steps [Eqs. (5) to (8)]:

$$SiO_2(s) + C(s) \rightarrow SiO(g) + CO(g) \tag{5}$$

$$SiO_2(s) + CO(g) \rightarrow SiO(g) + CO_2(g) \tag{6}$$

$$C(s) + CO_2(g) \rightarrow 2CO(g) \tag{7}$$

$$SiO(g) + 2C(s) \rightarrow SiC(s) + CO(g) \tag{8}$$

This reaction scheme is the most probable to occur according to the free energy functions of the reactions given above. The initial reaction, with the formation of gaseous SiO at points of direct contact between C and SiO_2, requires temperatures above ~1600 °C at atmospheric pressure (solid/solid reaction). This reaction has also been proposed by Blumenthal *et al.* [55] and Lee *et al.* [56]. The high rate of SiO(g) formation suggests that the further two gas/solid reactions make the final carbon reduction of the SiO to SiC possible once the C/SiO_2 contact points are consumed. This has been confirmed by quantitative investigations [57, 58]. The carbon crystallite diameter was found to have a substantial influence on the rate of reaction and the size of the synthesized SiC.

Various processing methods are available for obtaining carbon/silica mixtures with different scales of mixing. Depending on the nature of the reactants, these methods can usually be grouped into four categories:

- The reactants are physical mixtures of distinct powders of silica (or silica-bearing material) and carbon (or carbon-bearing material).

These powders are physically mixed and heat-treated to produce SiC. The mixing scale ranges from submicron to several tens or hundreds of microns. For example, Krstic [59] mixed carbon black (mean particle size <0.2 μm) and silica (mean particle size <40 μm) and obtained submicron SiC. Čerovič *et al.* [60] used colloidal SiO_2 (mean crystallite size 18 nm) and activated carbon (specific surface area 960 $m^2 g^{-1}$) to obtain submicron SiC. Seo *et al.* [61] obtained nanosized SiC (average size ~100 nm) by heat-treating a mixture of silica (mean particle size 0.8 μm) and carbon black (mean crystallite size 40–100 nm). Since 1981, a continuous version of the Acheson process has been operated commercially by the Superior Graphite Company [62]. For this, a mixture of silica sand and petroleum coke is fed continuously to the HSC (Hopkinsville Silicon Carbide) furnace and heated electrothermally to a temperature of 1900 °C. The run-off-furnace product is a free-flowing grain generally smaller 3 mm, comprising agglomerates of β-SiC crystallites and free carbon. The grain is friable and readily processed to produce sized microgrits and sintering powder by dry milling, wet grinding, classification, and wet chemical purification (removal of 5 wt% free carbon).

However, when the mixtures are prepared from raw materials with fine particle sizes, significant deviations from ideal narrow particle size distributions do occur. The scale of mixing may still be large due to agglomerates, even though fine particles are used. Additionally, the resultant powder may have a mixed morphology of equiaxed and acicular particles [60, 61]. The product may also contain a significant amount of unreacted carbon and/or silica due to the large scale of mixing and local compositional inhomogeneities in the mixture.

- The reactants are liquid mixtures in which both silicon-bearing and carbon-bearing materials are in liquid form, or are dissolved in a liquid.

Carbon-bearing materials (e.g., sugar, phenolic resin, etc.) and silica-bearing materials (e.g., tetraethoxysilane, TEOS; methyltrimethoxysilane, MTMS) are both dissolved in a solvent (usually ethanol) to form a homogeneous solution. The mixed solution is then processed to obtain a gel using methods such as solvent removal, freeze-drying, chemical gelation, and thermal gelation. The gel is then heat-treated at approximately 800–1100 °C in an inert atmosphere to remove the volatile species, after which the carbon/silica mixture is heat-treated to produce SiC by carbothermal reduction. Submicron or nanosized powders are obtained [63–74], depending on temperature and annealing time.

The major advantage of solution-based methods is that carbon-bearing and silica-bearing precursors are mixed on the molecular scale. Subsequent processing (solvent removal, pyrolysis, etc.) may result in some segregation of the components, such that the scale of mixing is no longer molecular. However, the silica/carbon mixtures produced by solution methods are generally more intimately mixed compared to batches prepared by the mixing of distinct powders. The resultant product usually has smaller crystallite sizes, and the powders may also be less aggregated and so will normally require less milling. Another advantage of solution-based methods is that high-purity SiC can be produced if the starting materials are of high purity. The major disadvantage of solution-based methods is that the processing costs tend to be high. Although many soluble carbon precursors are inexpensive (e.g., phenolic resin, sugar, etc.), solution-based silica sources are invariably much more expensive than powders, such as sand.

- Hybrid mixtures are used, in which one reactant is in particulate form while the other reactant is either a liquid or is dissolved to form a liquid precursor solution.

As an example, the particulate component may be colloidal silica, while the soluble component may be sugar or a carbon-bearing polymer (or oligomer). Wei *et al.* [75] prepared SiC by mixing a colloidal SiO_2 suspension (mean crystallite size 14 nm) and either a sucrose or a phenolic resin solution. The SiC product had a primary crystallite size of approximately 10–20 nm.

The method based on using particulate/liquid hybrid mixtures has some of the advantages of the mixed-solution method, such as a better mixing of the carbon and silica as compared to the mixing of distinct powders. In addition, SiC with fine crystallite sizes can be produced. However, as one component has a particulate form, the mixing is not expected to be as good as the method of mixed solutions.

- The reactants are fine-scaled silica/carbon mixtures formed as a decomposition product of silicon- and carbon-containing synthetic polymers or naturally occurring precursors.

This method is distinct from other methods in that both carbon and silica are single-sourced. Examples include synthetic polymers such as polycarbosilane (PCS) or naturally occurring precursors such as rice hulls. The precursors are usually pyrolyzed, whereupon the decomposition products (e.g., CH_4, H_2) are volatilized such that nanoscaled carbon/silica mixtures are formed. Heat treatment results in the formation of SiC via carbothermal reduction.

Yajima [76] was the first to study the preparation of silicon carbide fibers from carbosilanes. These and other SiC-containing polymers were used to produce SiC powders with a crystallite size as small as several nanometers [77, 78]. The advantage of the production route from liquid to solid to produce SiC has also attracted attention for SiC film production in microelectronics or as protection layers. In this way, amorphous, polycrystalline films of high purity produced by the dip-coating of substrates in PCS solutions and subsequent pyrolysis in an inert gas atmosphere, have been prepared [115].

The synthesis of SiC from rice hulls was investigated by Lee and Cutler [79], who pyrolyzed rice hulls (containing 18 wt% silica) in a nonoxidizing atmosphere to obtain a mixture of amorphous carbon and silica. The pyrolyzed material had a C/Si molar ratio of 5.6 to 1 (47 wt% silica, 53 wt% carbon). The pyrolyzed material was further heat-treated in argon to obtain submicron-sized SiC. Both, Krishnarao and Godkhindi [80] and Panigraphi *et al.* [81] also used rice hulls as the starting material to obtain submicron-sized SiC grains. High-purity β-SiC whiskers, 50–60 μm long and <1 μm in diameter, have been prepared by the pyrolysis of rice hulls in a reduced pressure of nitrogen at 1000–1500 °C by Raju and Verma [82].

The advantage of using SiC-containing synthetic polymers or rice hulls is that the fine-scale mixing of silicon and carbon is already present in the raw material. However, each type of precursor has some disadvantages. Notably, the synthetic polymers are very expensive, and the synthesis methods are complicated and require highly controlled processing conditions. In contrast, rice hulls are inexpensive and naturally occurring. However, the naturally occurring material contains large concentrations of undesirable impurity elements (including alkali metals). In addition, the silica distribution in the rice hulls is not uniform and the C/Si ratio also deviates far from the stoichiometric ratio (i.e., a C/Si molar ratio of 3). As a result, after the heat treatment there is a significant amount of excess carbon and/or unreacted silica remaining, which is difficult to separate from the SiC [80].

All of the above-described processes which result in ultrafine powders have in common that the reaction products have a high affinity to oxygen. If the handling of powders in an inert atmosphere is not appropriate for large-scale production, then particle size control by increasing the annealing temperature and time during the carbothermal reduction may be necessary to achieve surface areas in the range of $10–20\ m^2\ g^{-1}$.

Synthesis from the Elements Prochazka [73] synthesized submicron β-SiC powder with a surface area of $7\ m^2\ g^{-1}$ using a direct reaction of high-purity silicon powder with carbon black at temperatures between 1500 °C and 1650 °C, but saw little densification with boron and carbon additions. It was suggested that the free silicon was detrimental to densification.

Deposition from the Vapor Phase Submicron β-SiC can be continuously produced by decomposing gaseous or volatile compounds of silicon and carbon in inert or reducing atmospheres, at temperatures above 1400 °C [83]. The particle size and morphology will depend considerably on the reaction temperature and on the composition of the gas phase. A wide variety of reactants, and several methods of heating (e.g., dc arc jet plasma [84], high-frequency plasma [85], laser [86, 87] and thermal radiation [88]) are possible.

Small particle sizes are produced ranging from 0.01 to 0.5 μm, with greater than 70% efficiency, when using silicon tetrachloride ($SiCl_4$) and methane (CH_4) or methyltrichlorosilane (CH_3SiCl_3) as the gaseous precursors in the presence of hydrogen as the plasma gas [84].

If boron trichloride is used for the boron dopant, together with a small excess of the carbon feedstock, then sinterable SiC powders are obtained [85].

By using a CO_2 laser and a mixture of silane (SiH_4) and methane (CH_4) or acetylene (C_2H_2), β-SiC particles in the size range of from 5 to 200 nm have been obtained [86]. Najera *et al.*, using mixtures of silane with methane, ethane or acetylene, obtained nanometric powders (20 nm agglomerate size) by applying pulsed IR laser irradiation [87].

Fine, submicrometer and quasispherical SiC powders have been synthesized in a vertical tubular flow reactor by using the silane–acetylene–argon system [88].

Recently, chemical vapor deposition (CVD) methods based on single source raw materials have been used to synthesize nanosized powders. High-purity, low-oxygen-content nanosized SiC powders have been successfully produced by Huang *et al.* using CVD of dimethyldichlorosilane (DMS) at temperatures from 1100 to 1400 °C [89]. The powders grown at 1400 °C were SiC crystallites, while those grown below 1300 °C consisted of amorphous phases and SiC crystallites.

Li *et al.* deposited nanosized, partially crystallized SiC powders at temperatures as low as 900 °C in a hot-wall horizontal quartz tube reactor using liquid carbosilane as a precursor and hydrogen as the carrier gas [90]. Gupta *et al.* obtained β-SiC nanoparticles (10–30 nm) from a CVD process carried out in a hot-wall tube reactor using hexamethyldisilane as a single source for both silicon and carbon [91].

4.3.1.3 Silicon Carbide Whiskers

SiC whiskers are discontinuous, monocrystalline hair- or ribbon-shaped fibers in the size range of 0.1–5 μm diameter and 5–100 μm average length. Their aspect ratio ranges from 20 to several 100, and is an important parameter. SiC whiskers are all grown from the gas phase on solid or liquid substrates, under conditions that result in a small defect size (0.1–0.4 μm) in the whisker. Because they are nearly single crystals, the whiskers typically have very high tensile strengths (9–17 GPa) and elastic moduli (up to 550 GPa). Whiskers grown at temperatures below 1800 °C are composed of mostly β-SiC, while those grown above 2000 °C are mainly α-SiC.

According to Shaffer [92], the preparation methods for whiskers can be divided into three routes:

- Vapor–liquid–solid process (VLS)
- Vapor phase formation and condensation (VC)
- Vapor solid reaction (VS).

Vapor–Liquid–Solid Process In the VLS process, silicon- and carbon-rich vapors (usually CH_4, SiO, or $SiCl_4$) react at 1400 °C to SiC on a liquid alloy [Eq. (9)]. Microscopic particles of an alloy are first distributed on a substrate (graphite) and then exposed to silicon- and carbon-rich vapors. The presence of a liquid catalyst, such as a transition metal or usually an iron alloy, distinguishes this method from all other whisker preparation methods.

$$\text{Si (solution)} + \text{C (solution)} \rightarrow \text{SiC (whisker)} \qquad (9)$$

Although the VLS method yields near-perfect whiskers, and has been investigated since the early 1960s, it has only very recently become economically viable as a commercial scheme.

Figure 4.7 illustrates the VLS whisker growth sequence. At 1400 °C, the solid catalyst melts and forms the liquid catalyst ball. Carbon and silicon from the gas phase are dissolved in the liquid alloy, which soon becomes supersaturated such that solid SiC precipitates at the interface with the carbon substrate. Continued dissolution of the gas species into the liquid catalyst allows the whisker to grow, lifting the catalyst ball from the substrate as additional SiC precipitates. These VLS whiskers are typically larger in diameter (4–6 μm) than those formed by the vapor solid process (VS-whiskers); in fact, they frequently grow to a length of tens of millimeters.

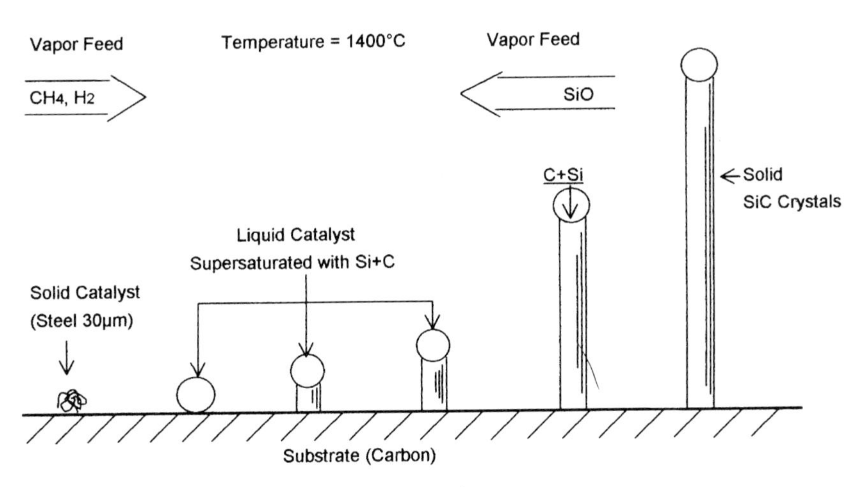

Figure 4.7 The VLS process for SiC whisker growth.

Vapor Phase Formation and Condensation In order to grow whiskers by vapor phase formation and condensation, the bulk SiC is first vaporized by heating to very high temperatures (>2200 °C), usually under reduced pressure. Upon cooling, α-SiC whiskers form on the nucleation sites. The addition of lanthanum, yttrium, neodymium, or zirconium leads to an increase in the growth rates. However, whiskers are no longer produced commercially using this sublimation process.

Vapor Solid Reaction At the present time, the principal commercial method of SiC whisker production is the carbothermic reduction of low-cost silica sources at temperatures of 1500–1700 °C. The reaction for the formation of VS-cubic β-SiC whisker occurs in two steps [Eqs. (10) and (11)]:

$$SiO_2(s) + C(s) \rightarrow SiO(g) + CO(g) \tag{10}$$

$$SiO(g) + 2C(s) \rightarrow \beta\text{-SiC (whisker)} + CO(g) \tag{11}$$

Many different VS-processes are known, depending on the starting raw materials and the catalysts species involved. In all cases, the use of a catalyst for whisker formation is favored over particulate SiC formation [93–96].

Commercially available VS-whisker grades are typically less than 1 μm in diameter (submicron) and up to 30 μm in length, and show strength/Young's modulus values of 1–5 GPa/400–500 GPa. During whisker handling and processing, rigorous protection must be provided as smaller whiskers (<3 μm diameter) can become lodged in the lungs and represent a health hazard. An inexpensive VS-process utilized calcined rice hulls, which contain both silica and carbon, as a precursor to SiC whiskers [97–99].

The characteristics of commercial SiC whiskers, as claimed by the manufacturers, are listed in Table 4.1.

β-SiC whiskers have been added to a wide variety of ceramics (Al_2O_3, Si_3N_4, AlN, $MoSi_2$, mullite, cordierite, and glass ceramics) in an effort to achieve increased toughness. A real success story in the advanced ceramic market was the introduction of hot-pressed alumina reinforced with 25–30 wt% SiC whisker. such that a composite with increased strength (by over 50%), fracture toughness (by 100%),

Table 4.1 Comparison of the properties of β-SiC whisker (manufacturers' data).

Property	Supplier						
	Tokai Carbon	Tateho	Shin Etsu	C-Axis Techn.	Millennium MATERIALS	ACM	ART
Phase	β	β	β	β	β	β	β
Length (μm)	20–50	5–200	3–20	—	30–200	5–80	—
Diameter (μm)	0.3–1.0	0.5–1.5	0.15–0.25	0.3–1.0	1–3	0.45–0.65	0.5–1.7
Aspect ratio	—	20–200	20–50	20–50	—	—	5–20
Density ($g\,cm^{-3}$)	3.2	3.18	—	3.21	3.21	3.2	3.21

increased fatigue and thermal shock resistance was obtained. Whisker pull-out and crack-bridging have been identified as the major toughening mechanisms.

These ceramic matrix composites (CMCs) have found an important niche in difficult-to-machine materials, and where there is a need for high machining rates. Reduced machining times of specific operations, ranging from 3 h to 18 s, can be achieved. Factors that allow the increase in productivity include not only increases in cutting speeds (by up to a factor of ten), but also lower wear rates and more predictable tool failure times. SiC_w/Al_2O_3 is also used as the wear components in ceramic/metal hybrid tooling and dies in the canning industry. Here, the key benefits include an increased tool life, a reduced down time, improved surface finishes, lower maintenance costs, and increased production rates. SiC_w/Al_2O_3 is presently used in cupping, drawing, ironing, and can-necking tools.

The use of whiskers as reinforcement in metallic matrix composites (MMCs) is described in Section 4.4.1.10.

4.3.1.4 **Silicon Carbide Platelets**

As reinforcement material for all matrix materials – ceramics, metals, and plastics – SiC platelets offer a similar potential as whiskers, but at lower cost and without any health hazard. Platelets are single-crystal, plate-like α-SiC particles with an aspect ratio of about 8–15. SiC platelets typically range in size from about 5–100 μm in diameter and 1–5 μm in thickness (see Table 4.2). They are produced commercially from inexpensive raw materials (silica and carbon, micron-sized β-SiC powders) at temperatures of 1900–2100 °C under an inert atmosphere [100]. Due to the presence of the boron and aluminum dopants added, platelet-shaped crystals are formed. Aluminum enhances the growth in the [0001] direction and decelerates the growth perpendicular to the [0001] direction. Boron enhances the growth notoriously perpendicular to the [0001] direction [101]. Aluminum (0.04–0.45 wt%), boron and nickel (each 0.4–0.8 wt%), and free silicon (0.3–3.6 wt%) were identified as impurity elements in SiC platelets produced by Millenium Materials [102].

When dense SiC platelet-reinforced SiC composites were fabricated by Lenk *et al.* [103], the highest sintered theoretical density (TD) achieved was 97–98%, with a 20% platelet content, using hot molding as the thermoplastic-forming technique [104]. Chou and Green [105–107] fabricated dense SiC platelet-reinforced

Table 4.2 Morphological characteristics of commercial SiC platelets.

Characteristic	Supplier		
	C-Axes Technology SF grade	Millenium Materials - 325 mesh	Carborundum Comp.
Maximum dimension (μm)	5–50	5–70	10–500
Thickness (μm)	2–5	0.5–5	1–5
Aspect ratio	8–10	8–15	10–15

alumina using hot pressing for consolidation. The platelets tended to lie parallel to each other after hot-pressing; however, the preferred orientation did not lead to any significant elastic anisotropy but rather to toughness anisotropy. The optimum properties were a Young's modulus of 421 GPa, fracture toughness of 7.1 MPa · $m^{1/2}$, and a flexural strength of 480 MPa at a reinforcement volume fraction of 0.3, using platelets with an average grain size of 12 μm. Crack deflection and grain bridging were both identified as possible toughening mechanisms. When Alexander *et al.* [108] investigated the influence of matrix grain size and platelet content on the mechanical properties and toughness mechanisms in SiC-toughened alumina, a threefold increase in toughness was observed which could be addressed to crack bridging and deflection.

In addition to reinforcing ceramics, SiC-platelets are also used to increase the strength, wear resistance and thermal shock resistance of aluminum matrices (see Section 4.4.1.10), and to enhance the properties of polymeric matrices.

4.3.1.5 **Continuous Silicon Carbide Fibers**

Together with boron filaments and carbon fibers, SiC fibers are important continuous inorganic reinforcement materials with a high modulus. Due to their high-temperature properties and resistance to oxidation, they are particularly well suited for structural parts where high stiffness at high temperatures is required. Today, three different types of SiC fiber are available: substrate-based fibers (CVD-fibers); polymer-pyrolysis-derived fibers (PP-fibers); and sintered powder-derived fibers (SP-fibers). The defect size in amorphous or polycrystalline fibers is very small, so that the strength can be high. The details of some commercially available SiC fibers are listed in Table 4.3, along with their mechanical characteristics.

Substrate-Based Fibers (CVD) β-SiC filaments of 100–150 μm thickness can be prepared by CVD onto tungsten or carbon monofilaments of 40 μm thickness, which act as hot substrates during heterogeneous nucleation. Various carbon-containing silanes have been used as reactants. Lackey *et al.* have produced small-diameter, high-strength, thermally stable SiC fibers by a continuous CVD technique [109] although,

Table 4.3 Mechanical characteristics of various SiC fibers.

Characteristic	Supplier				
	Textron	Nippon Carbon	UBE	COI Ceramics	Carborundum
Designation	SCS ULTRA	Hi-Nicalon Type S	Tyranno SA	Sylramic	Sintered-Alpha
Crystalline phase	β-SiC	β-SiC	β-SiC	β-SiC	α-SiC
Diameter (μm)	140	12	7,5	10 μm	25–100
Tensile strength (MPa)	6500	2600	2800	>2800	up to 1500
Young's modulus (GPa)	430	420	380	>310	400

owing to high process costs and the large diameters of the resultant fibers, this method is generally disadvantageous. Rather, it has been found that a faster SiC filament production is possible on a carbon filament substrate. For example, the US company Textron produces β-SiC fibers on a carbon-fiber core with a surface layer of pyrolytic carbon, which is itself coated with SiC. These fibers, which are designated SCS, have a surface carbon layer that provides a toughness-enhancing parting layer in composites having a brittle matrix.

Polymer Pyrolysis-Derived Silicon Carbide Fibers (PP-Fibers) As shown in 1976 by Yajima [76], β-SiC fibers with a smaller diameter (8–30 μm) and without a central core can be manufactured by the solid-state pyrolysis of a PCS precursor fiber.

This melt spinning–curing–pyrolyses route has been adopted by Nippon Carbon to produce a fiber called NICALON [110]. This has a composition which is almost identical to "Si_3C_4O"; thus, it is not pure SiC, but rather a Si–C–O composite, in which nanometer-sized β-SiC, SiO_2 and carbon are uniformly dispersed. These impurities affect both the thermal stability and strength of the fiber, which has been reported to degrade above 1200 °C. However, these fibers have been used successfully to reinforce aluminum, refractories and Li–Al–silicate glass ceramics. The coating of fibers, for example by carbon, allows a decrease in the interface bonding with the matrix, which in turn increases the strength and the toughness by favoring a pull-out mechanism. Recently, an electron beam curing method rather than oxidative curing has been used to synthesize NICALON fibers with a low oxygen content (Hi-Nicalon, 0.5 wt%) and oxygen-free SiC fibers (Hi-Nicalon S, 0.2 wt%, C/Si = 1.05) consisting mainly of crystallized β-SiC [111]. These types of fiber have a high degradation resistance up to 1800 °C, and excellent mechanical properties.

UBE Chemicals have synthesized amorphous Si–Ti–C–O fibers from the PCS–titanium alkoxide compound polymer. These so-called "Tyranno" fibers show excellent properties and can be spun thinner than the Nicalon fibers (see Table 4.3), although heating above 1000 °C results in their crystallization. Recent investigations have resulted in crystalline β-SiC fibers with a low oxygen content and the maintenance of high strength up to 1800 °C.

Since the mid 1990s, many research groups have concentrated on the development of fibers via the pyrolysis of appropriate pre-ceramic polymers [112]. In particular, the system Si–B–N–C has been of major interest due to the excellent high-temperature and oxidation stability of the resultant amorphous material [113]. In this case, the onset of crystallization may be as high as 1800 °C, while decomposition starts at 2000 °C in protective atmospheres.

Fibers produced from $SiBN_3C$, derived from the single source precursor $Cl_3Si(NH)BCl_2$, retain their high mechanical durability, as measured by creep resistance and tensile strength, up to a temperature of 1400 °C. This is related to the fact that these amorphous fibers do not show any grain boundaries and are virtually free from macroscopic pores. SiC, in contrast, is crystalline and suffers from grain boundary sliding and subcritical crack propagation at mechanical loading at elevated temperatures. A fiber based on this system has been developed by Bayer (SIBORAMIC) [114]. The feasibility of scaling up the synthesis of $SiBN_3C$ to technical dimensions has been demonstrated by processing 150 kg batches (of polymer) which have been spun and pyrolyzed to produce ceramic long fibers.

Sintered Powder-Derived α-Silicon Carbide Fibers (SP-Fibers) The Carborundum Company has developed a single-phase polycrystalline α-SiC fiber with a diameter of approximately 20–150 μm and density of over 96%. The fiber is prepared by first producing green filaments by melt extrusion or by the suspension spinning of plasticized mixtures of sinterable α-SiC powder with organic additives, such as polyethylene or polyvinylbutyral plus novolac, respectively. The green filaments are subsequently debindered during free fall from the extruder, and finally pressureless-sintered at 2100 °C in an argon atmosphere. These polycrystalline α-SiC fibers are stable up to 1600 °C in air and to 2250 °C in an inert gas, with such stability being far superior to most commercially available β-SiC-based fibers [100]. According to Prochazka [116], an increase in the strength of α-SiC fibers is achievable by using refined α-SiC powders, having a surface area greater than $20\,m^2\,g^{-1}$, a median SiC particle size ≤0.25 μm, and no particles larger than 1.5 μm.

4.3.1.6 **Silicon Carbide Nanofibers**

SiC nanofibers were synthesized by Honda *et al.* on Si substrates covered by Ni thin films using high-power microwave plasma CVD under hydrogen gas [117]. The resultant fibrous material was identified as β-SiC with a high crystallinity. SiC nanofibers were also fabricated by these authors using the VLS mechanism, in which SiC is precipitated from supersaturated liquid Ni nanoballs.

Huczko *et al.* synthesized β-SiC nanofibers via the dehalogenation of various alkyl- and acryl-halides with Si-containing compounds [118]. The combustion process resulted in SiC nanofibers and nanotubes 20–100 nm in diameter, with an aspect ratio higher than 1000. This method is simple and requires no catalyst or template.

The synthesis of SiC nanofibers (5–20 nm in diameter) by using sol–gel and polymer blend techniques was achieved by Raman *et al.* [119]. In this case, tetraethoxysilane and methyltriethoxysilane were used as a Si source, and polycarbonate as a carbon source. A polymer solution containing alkoxide, water, and a suitable solvent was stirred and freeze-dried to produce an intimate mixture of polycarbonate and sol–gel-derived silica. The dried precursor was then pyrolyzed at 1400 °C in an argon atmosphere to obtain nanostructured SiC.

More recently, Hao *et al.* investigated the influence of various metal catalysts on the morphologies of nanostructured SiC produced by a sol–gel route, using lanthanum nitrate as an additive to prepare a xerogel [120]. The xerogel was converted, via a VLS mechanism, into bamboo-like SiC nanofibers with diameters of 40–100 nm and lengths of hundreds of micrometers.

SiC nanofibers by melt-spinning of polymer blends have been prepared from PCS as a SiC ceramic precursor and a novolac-type phenol–formaldehyde resin [121]. These nanofibers were amorphous, about 100 nm in diameter, more than 100 μm long, and were rich in oxygen.

4.3.1.7 Silicon Carbide Nanotubes (SiCNTs)

Since the initial discovery of carbon nanotubes (CNTs) by Iijima *et al.* [122], much interest has been expressed in developing other inorganic nanostructured particles, due to the unique properties assigned to these nanostructures. Indeed, many potential applications have been proposed for silicon carbide nanostructures, including high-strength composites, nanosensors, and nanodevices.

SiC nanotubes (SiCNTs) have been synthesized by using CNTs as a template. For this, SiO gas is decomposed thermally to deposit Si on the CNT surface, and subsequent heat treatment converts the coated CNT into SiCNT [123]. Keller *et al.* prepared SiCNTs with an average outer diameter of 100 nm and lengths up to 1 μm by using a gas–solid reaction between CNTs and SiO vapor [124]. Sun *et al.* obtained SiCNTs and C–SiC coaxial nanotubes (CNTs coated with a SiC layer) by the reaction of CNTs with silicon powder [125]. Various shapes of nanostructured SiC were obtained by decomposing SiO on multiwalled carbon nanotubes (MWCNTs). In this case, β-SiC nanowires, biaxial SiC–SiO_x nanowires and a new type of multiwalled SiCNT were prepared with interlayer spacings of 3.5–4.5 Å. The outer layer of the MWCNT was transformed into a multiwalled SiCNT, while the inner part remained unchanged as a CNT. SiC was formed layer-by-layer via the surface diffusion of Si atoms into the CNT. A thermally induced synthesis was used by Rümmeli *et al.* [126] in which vaporized Si converted bundles of single-walled carbon nanotubes (SWCNTs) into nanostructured SiC. Various structures, including nanorods, nanotubes and nanocrystals, were obtained by changing the reaction time, temperature, carrier gas, and pressure. The NASA Glenn Research Center used the template method and SiCNT growth on a catalyst to synthesize SICNTs, with subsequent CVD on CNTs resulting in polycrystalline/amorphous tubes [127]. Cheng *et al.* prepared SiCNTs and bamboo-like nanofibers by using CVD or the infiltration of a liquid precursor into an alumina template [128]. SiCNTs prepared by the CVD method consist of nanocrystalline β-SiC or β-SiC/C, whereas the application of infiltration results in single-crystal SiC nanofibers and nanoparticles. However, further research will be required to overcome thermal instability, such as the transformation into nanoparticles, or excessive grain growth.

The potential applications of SiCNTs have been the subject of intense investigation by several groups. Hydrogen storage has been investigated by Mpourmpakis *et al.* [129], who concluded that SiCNTs would absorb more hydrogen than CNTs, particularly at low pressures around 1 MPa. Morisada *et al.* used SiC-coated CNTs as reinforcement for cemented carbides [130], and showed that the SiC coating could protect the nanotubes from reaction with molten metal. The nanotubes also showed potential use as "nanoreactors" or "nanotest tubes" at high temperatures [131].

4.3.2 Boron Carbide

4.3.2.1 Technical-Scale Production

The high-volume production of boron carbide powder was made available by the carbothermal reduction of boron oxide in huge electric arc or resistance furnaces, according to Eq. (12).

$$2B_2O_3 + 7C \rightarrow B_4C + 6CO \quad \Delta H = 1812\,\text{kJmol}^{-1} \tag{12}$$

The reaction takes place between 1500 and 2500 °C [132, 133], with large quantities of CO (2.3 $m^3\,kg^{-1}$) being generated. The overall reaction kinetics are more complicated than expressed by Eq. (12); the gaseous compounds boron oxide and CO contribute to the overall reaction, expressed by Eqs. (13–15):

$$B_2O_3 + 3CO \rightarrow 2B + 3CO_2 \tag{13}$$

$$2CO \leftrightarrow CO_2 + C \tag{14}$$

$$4B + C \rightarrow B_4C \tag{15}$$

In order to avoid huge losses of volatile boron oxide compounds, the furnace is cooled externally such that the outer shell remains unreacted. The core contains low-impurity boron carbide blocks (total metal impurities <0.5 wt%) with a stoichiometry of $B/C = 4.3$ and residual carbon [25]. The blocks are crushed, milled to the desired final grain size, and purified by chemical leaching.

High-purity fine boron carbide powder with an average particle size of 0.5–5 μm has been synthesized by the reaction of boric acid with a carbon-containing compound in a vented tube furnace [134].

Uniform submicron crystals of B_4C and boron-enriched boron carbide powders have been synthesized continuously by rapid carbothermal reduction at approximately 1950 °C in a pilot-scale graphite transport reactor. A unique reactor design allowed for the continuous feeding of a meltable boron oxide-containing precursor, rapid heating rates that completed the carbothermal reduction reaction in seconds, and an expanded cooling that allowed for the precipitation in space of volatile excess boron oxides. Rapid heating rates and minimized reaction times at high temperatures promoted nucleation with limited crystal growth [135].

As an alternative fabrication method, boron carbide powders may also be produced directly by the magnesiothermic reduction of boric oxide in the presence of carbon at 1000–1800 °C [136] [see Eq. (16)]. The MgO and unreacted Mg are removed by hydrochloric or sulfuric acid; moreover, the reaction is strongly exothermic and is carried out either directly by point ignition (thermite process) or in a carbon tube furnace under hydrogen. The product can be further purified by heating under vacuum at 1800 °C. Since the MgO acts as a particle growth inhibitor, ultrafine boron carbide particles in the range of 0.1–1.5 μm are obtained [137]. This process is unsuitable for producing highly pure boron carbide or boron carbide free of magnesium-containing impurities.

$$2B_2O_3(l) + 6Mg(s) + C(s) \rightarrow B_4C(s) + 6MgO(s) \tag{16}$$

4.3.2.2 Submicron B_4C Powders

Fine or nanometric boron carbide powders can be produced using ultrafine raw materials, plasma processing, elemental boron and carbon, CVD, or organic precursors. However, these production routes are costly and restricted to scientific applications.

Using nanometric carbon black and μm-sized boron oxide, nanometric boron carbide powder with an average particle size of 20 nm can be produced at temperatures as low as 1300 °C [138]. A nanometric-scale boron carbide powder (20–40 nm) has been produced in a plasma-assisted reaction process using boron oxide and a carbon source, whereby the reactants are vaporized in the plasma, thus leading to a reaction of atomized species [139].

Nanocrystalline B_4C was synthesized by an inexpensive carbothermal reduction method using carbon black and B_2O_3 as precursor; a full conversion was achieved at 1350 °C. The average particle size of the synthesized B_4C powder was 260 nm, but this was reduced to 70 nm after separation of the small particle fraction from the larger particles by sedimentation. The most likely reaction was the reduction of B_2O_3 vapor at the surfaces of the carbon particles after its vapor transport from the liquid B_2O_3 [140].

Crystalline boron carbide nanoparticles have been obtained by Chang *et al.* [141] by heating a mixture of amorphous carbon and amorphous boron at 1550 °C in a conventional high-temperature furnace reactor. The average size of B_4C particles was 200 nm.

Nanometric boron carbide particles can also be prepared by CVD. The reaction of boron trihalides with carbon or gaseous carbon-containing precursors using radiofrequency (RF) plasma [142, 143] or laser-assisted CVD [144] has been applied.

Nanocrystalline boron carbide (B_4C) with a particle size of approximately 15–40 nm was synthesized by Gu *et al.* [145] via a solvothermal reduction of carbon tetrachloride in the presence of amorphous boron powder at 600 °C in an autoclave, using metallic lithium as the reductant.

Recently, single-source organic precursors have been shown suitable for the production of nanosized boron carbide particles. The organic compounds polyhexenyldecaborane and 6,6′-$(CH_2)_6$-$(B_{10}H_{13})_2$ lead to B_4C and B_8C, respectively, upon heating [146].

The general problem associated with submicron powders is their high affinity to oxygen when exposed to air. Additionally, these highly sophisticated routes have poor yields and are too costly for technical-scale, high-volume production.

4.3.2.3 **Boron Carbide-Based Nanostructured Particles**

Recently, boron carbide nanostructures have attracted much attention as they have certain advantages over their bulk counterparts [147]. Nanoscale ceramic fibers, nanocylinders and nanoporous structures – as do their well-known carbon counterparts – have a tremendous number of potential applications, including uses as quantum electronic materials, structural reinforcements, and ceramic membranes for use as catalyst supports or in gas separations [148].

In order to synthesize boron carbide nanostructures, several techniques have been developed which include CNT template-mediated growth [149–152], CVD, and plasma-enhanced CVD (PECVD) [147, 153, 154], carbothermal reductions of boron oxides [155–159], a porous alumina templating technique [160], and electrostatic spinning [161]. The boron carbide nanostructures have diverse morphologies and structures. For example, boron carbide nanowires synthesized through CNT template-mediated growth can be either polycrystalline or crystalline, pure boron carbide nanowires, or boron carbide nanowires coated with boron nitride layers. The morphologies of PECVD boron carbide nanostructures may be cylindrical with smooth or rough and faceted surfaces, linear arrays of approximately equally spaced rhombohedral nanostructures, as well as nanosprings composed of helical amorphous boron carbide. Aligned boron carbide nanofibers and nanoscaled freestanding porous boron carbide ceramic-fiber matrices have been fabricated with the single-source precursors polyhexenyldecaborane and 6,6′-$(CH_2)_6$-$(B_{10}H_{13})_2$ by using a porous alumina templating method [160], and an electrostatic spinning method [161], respectively.

4.4 Dense Ceramic Shapes

4.4.1 Dense Silicon Carbide Shapes

SiC can be converted into more or less dense parts either by using various bonding phases to bind the SiC particles or grains together at elevated temperatures without any dimensional change of the initial compact; or by sintering – that is, by mechanisms of mass transport between particles which lead to porosity elimination together with shrinkage of the overall body.

4.4.1.1 Ceramically Bonded Silicon Carbide

This product group is based on oxide- or nitride-bonded SiC grains, and exhibits an apparent porosity of up to 20% by volume. A fracture surface of a silicon nitride bonded SiC is shown in Figure 4.8a.

Coarse SiC powder fractions are mixed with clay, or pure oxides such as SiO_2 and/or Al_2O_3, formed to "green" compacts by conventional ceramic methods, and fired at ~1400 °C. The fired clay or oxidic raw materials bind the SiC particles and, although they are only loosely bound, this is sufficient for many types of applications such as refractory furnace bricks and abrasive disks. Firing mixtures of SiC particles and elemental silicon in a nitrogen atmosphere, with or without the addition of air and possibly Al_2O_3, produces silicon nitride-, silicon oxynitride-, or SiAlON (silicon aluminum oxynitride) -bonded SiC products with enhanced high-temperature strengths [162–164].

4.4.1.2 Recrystallized Silicon Carbide

Silicon carbide shapes with a high green density made by slip casting or by press-molding SiC powders of bimodal particle size distribution are fired in an electric furnace at temperatures of up to 2500 °C, under the exclusion of air [165, 166]. Evaporation and condensation take place at temperatures above 2100 °C, and this

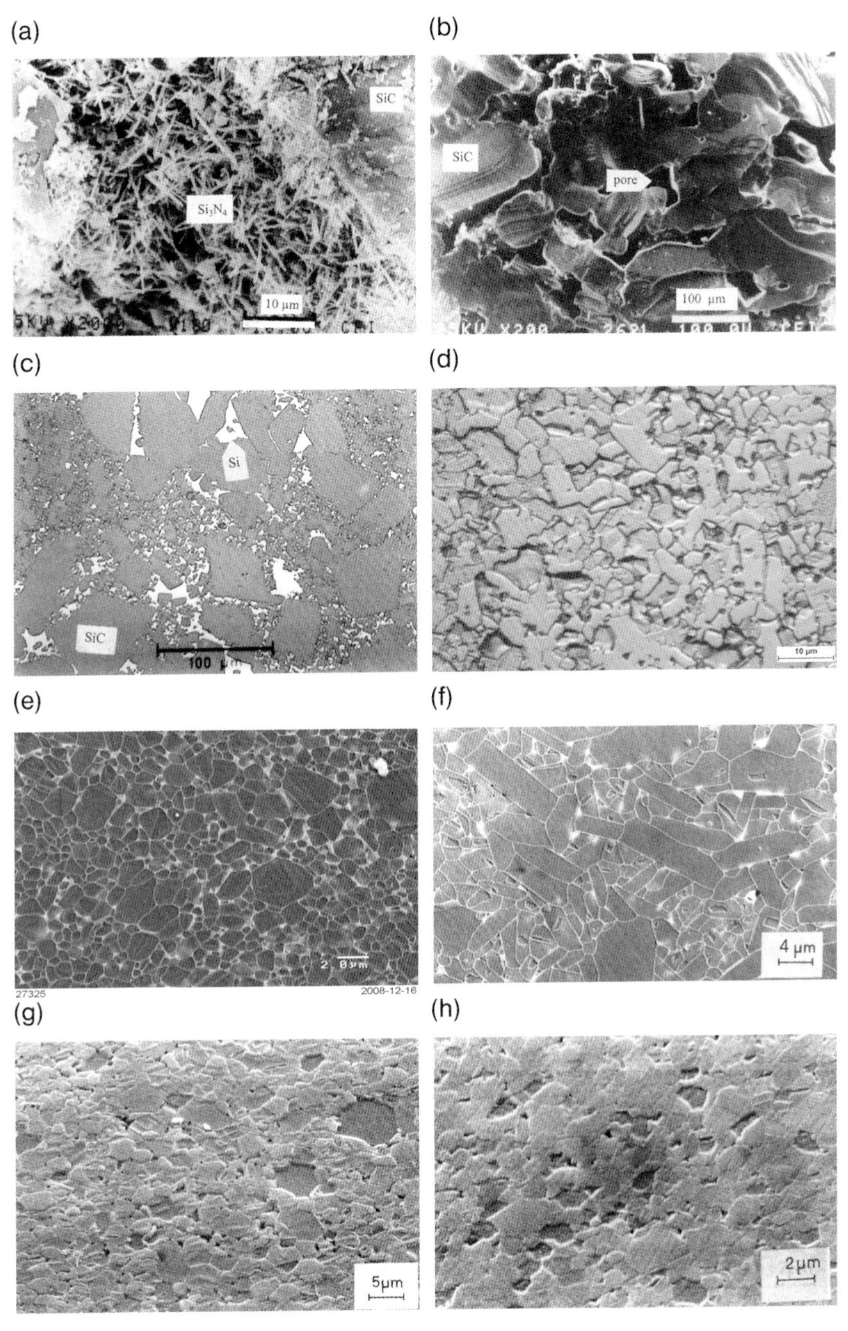

Figure 4.8 Microstructure of various SiC ceramics. (a) KSiC, nitride-bonded SiC (AnnaSicon 25); (b) RSiC, recrystallized SiC (AnnaNox CK); (c) SiSiC, silicon-infiltrated SiC (AnnaNox CD); (d) SSiC, sintered SiC (EKasic F); (e) LPS-SiC, liquid phase-sintered SiC (EKasic T); (f) LPS-SiC, platelike SiC grains (ESK); (g) HPSiC, Al-doped (ESK); (h) HIPSiC, undoped (ESK).

results in a selfbonded structure without shrinkage [167]. The initial (green) density and final density remain the same, and solid SiC bonds are formed between the crystals. The final RSiC product, which shows transgranular fracture (see Figure 4.8b), consists of 100% SiC and can have densities of up to 2.6 g cm^{-3}, with a porosity of approximately 20%.

4.4.1.3 **Reaction-Bonded Silicon Carbide**

In reaction sintering or reaction bonding, SiC is produced to some extent through chemical synthesis. A preform consisting of SiC, carbon, and a carbon-containing binder is first prepared using a conventional ceramic-shaping technique such as dry press-molding, slip-casting and extrusion- or injection-molding. The elemental carbon in the preform reacts with gaseous or liquid silicon upon heating, and the process can be controlled in such a way that a porous – but pure – SiC part is produced, or one in which the pores are filled with excess silicon [168–176]. As the latter form (SiSiC), is pore-free, dense (high density, showing no residual porosity; see Figure 4.8c), and thus much stronger, it is the one which is most frequently made, usually by infiltration with liquid silicon. Silicon infiltration can also be used to fill the pores of the recrystallized SiC moldings at a later stage [177].

The amount of free silicon in infiltrated SiC is usually between 10% and 15%, and it is difficult to reduce this value below 5%. Infiltration is relatively easy, because of the good wettability of the material, although problems may be encountered if the bodies are large or thick, because of the exothermic reaction and the expansion of silicon on freezing, which make perfect impregnation difficult.

Reaction sintering can, however, usually be controlled so that the change in volume is negligible, resulting in dimensionally accurate components. This is achieved by choosing the correct SiC and carbon contents in the initial mix, and ensuring correct porosity in the preform and by proper temperature control.

4.4.1.4 **Sintered Silicon Carbide**

Solid-State Sintered Silicon Carbide Although, initially, SiC was considered to be unsinterable, during the early 1970s it became possible to sinter SiC without applied pressure, thus achieving densities which were more than 95% of the theoretical density. This success was due largely to the pioneering studies of Prochazka [178] at the General Electric Corporation Research Center in Schenectady. The starting materials were submicrometer β-SiC [179] or later α-SiC [180], with the simultaneous addition of up to 2% carbon and boron.

A minimum of 0.3 wt% boron was required, although larger additions did not improve the density, which Prochazka attributed to the need to exceed the solubility limit of boron in SiC (0.2 wt% according to Shaffer [181]). Carbon is added in excess to ensure complete deoxidation; that is, removal of the SiO_2 film from the SiC grains. Additions of excess carbon are observed to inhibit grain growth due to the presence of carbon inclusions, and produce a more equiaxed grain structure [182], probably as a result of the energy required from the grain boundaries to overcome the carbon inclusions. Instead of boron and its compounds [183–186], it is possible also to use aluminum and aluminum compounds [187–193], or beryllium and its compounds [194], with the same net result. The sintering of Al/C-doped [195] or Al_2O_3-doped [193] SiC powders is believed to occur in two stages: (i) transient liquid phase sintering; and (ii) solid-state sintering during the final stage.

These findings led to the development of an inexpensive method of producing dense and complex parts consisting of pressureless pure silicon carbide (SSiC). The powders can be molded into a green body by using any of the methods used in ceramic molding [196, 197], depending on the shape required and the number of pieces involved. Sintering is carried out in an inert gas atmosphere or in a vacuum at temperatures between 1900 and 2200 °C.

Linear shrinkage exceeds 15% and depends on the green density of the molding. During the sintering process, there is SiC-polytype transformation, as well as grain growth, the extent of which depends on the type and amount of sintering additives, as well as the sintering temperature. Boron-doped β-SiC powders tend towards secondary recrystallization (exaggerated grain growth during 3C → 6H polytype transformation [198]), whereas, when starting with α-SiC powders, a homogeneous, fine-grained microstructure can be obtained (see Figure 4.8d). α-SiC powder, which consists predominantly of the 6H-type, usually shows a 6H → 4H polytype transformation during sintering with both aluminum-containing and boron-containing dopants. Aluminum promotes a 6H → 4H transformation at lower sintering temperatures (2050 °C), and boron most easily at higher temperatures (2200 °C). According to the investigations of Schwetz *et al.* [199], there appears to be no fundamental correlation between SiC-polytype transformation and microstructure development during the sintering of SiC.

Spark-Plasma-Sintered Silicon Carbide Spark plasma sintering (SPS) represents a newly developed rapid sintering technique with great potential for achieving fast densification results with minimal grain growth in a short sintering time [206]. The effects that accelerate densification have been studied in detail [207, 208], it having been proved by experimental data that the enhanced sinterability of a powder subjected to SPS is mainly associated with particle surface activation and increased diffusion rates on the contact zones caused by the applied pulse current. During the past decade, many investigations have been carried out on the consolidation of various materials [209]. Notably, ultrafine powders which are especially difficult to sinter can be easily consolidated by SPS. Low-temperature mass transport by surface diffusion during the SPS of nanoparticles can lead to rapid densification kinetics, with negligible grain growth.

The consolidation of a SiC nanopowder synthesized by the mechanical alloying method was subsequently accomplished by SPS at 1700 °C for 10 min under an applied pressure of 40 MPa, without using any sintering additives. The SiC sintered compact with a relative density of 98% consisted of nanosized particles smaller than 100 nm. The mechanical properties of sintered compacts were close to those of reference samples sintered with boron and carbon as the sintering additives [210].

Likewise, 99% dense samples with a submicron-sized microstructure have been obtained by Tokita, by the consolidation of a submicron starting powder without sintering additives [211]. In this case, a high hardness of 29 GPa and a fracture toughness of 4.7 MPa·m$^{1/2}$ were achieved.

Tamari *et al.* [212] achieved 99% dense samples within a 5 min dwell time at 1800 °C when using Al_2O_3 and Y_2O_3 as sintering additives; these samples showed a high bending strength of 900 MPa.

Liquid-Phase-Sintered Silicon Carbide Liquid-phase-sintered SiC (LPS-SiC) is a dual-phase SiC ceramic, with typically ~95% SiC content and ultrafine microstructure.

The mechanisms of liquid-phase sintering processes have been described in detail by Kingery [213, 214], Petzow [215], and German [216]. The sintering process can be treated as a sequence of three stages: particle rearrangement; solution–reprecipitation; and skeleton sintering.

Significant densification results from particle rearrangement at an early stage, as soon as the liquid is formed and wetting occurs; such rearrangement is driven by capillary forces. Parameters such as the quantity of liquid, green density, sintering temperature, and wetting strongly affect this stage, which proceeds rather rapidly. If enough liquid phase is formed and the wetting is sufficient, then a full densification can be achieved almost instantaneously.

The dissolution and reprecipitation of the solid phase dominate the microstructural development, including densification, shape accommodation, and grain growth. It is the chemical potential gradient that induces the dissolution of grains with a small surface radius, and their reprecipitation on those with larger radii during microstructure development. Because the diffusivities in a liquid are orders of magnitude higher than those in a solid, the kinetics of densification are much faster than in the case of solid-state sintering. It is worth noting that, in addition to densification, coarsening due to Ostwald ripening will occur simultaneously during the solution–reprecipitation process.

Once a rigid skeleton of contacting particles has been formed, solid-state sintering which is rate-controlled by the grain boundary diffusion begins. The overall shrinkage or densification rates are significantly reduced, and the microstructure approaches stability. Although the three stages overlap, particle rearrangement is always the fastest compared to the other two.

As early as 1975, SiC was hot-pressed to 99% relative density by using Al_2O_3 as an additive [217]. During sintering, the aluminum oxide was thought to form an oxidic melt, with silica always present on the surface of fine SiC particles. This fact was supported by a report that completely dense bodies were obtained at a temperature of 1800 °C, which is less than the melting point of pure alumina. The mechanism was shown to be a rapid rearrangement during the initial period of pressure application and densification by a solution–reprecipitation process. Suzuki [193] optimized this process by using pressureless sintering, and obtained a sintered body with a high density which exceeded 97% of the theoretical value by adding at least 2 wt% Al_2O_3. The densification proceeded via a liquid-phase sintering mechanism, and was accompanied by a β- to α-SiC phase transformation and a reduction in the amount of liquid phase.

However, as mentioned by Mulla [218], and thermodynamically and experimentally assessed by Thevenot [219, 220], the reaction between SiC and Al_2O_3 according to

Eq. (17) is a major problem associated with sintering in this system, because of the formation of volatile components which retard densification.

$$Al_2O_3(s) + 2SiC(s) + Al_2O(g) \rightarrow 4Al(g) + 2CO(g) + 2SiO(g) \quad (17)$$

Since the innovative approach of liquid-phase sintering (LPS) by Omori and Takei [221], who showed that a wide variety of rare-earth oxides, in combination with Al_2O_3 and/or boron compounds, could be used as sintering additives to promote the densification of SiC, extensive studies have been conducted in order to obtain fine-grained dense SiC ceramics with improved mechanical properties. The major inspiration for this densification strategy came from the sintering procedures applied to silicon nitride using compositions within the system Al_2O_3–Y_2O_3 as additives [222–224]. By adopting these, a superior self-reinforced SiC was successfully fabricated [225, 226].

The advantage of using a liquid phase is that SiC can be sintered to high densities at temperatures well below 2000 °C, with much shorter cycle times compared to conventional solid-state sintering [227]. Another advantage is that SiC powder with a higher oxygen content (SiO_2) can be densified. The relevant quasiternary system Al_2O_3–Y_2O_3–SiO_2 revealed the formation of a liquid phase as low as 1660 K [228]. Pressureless sintering, hot pressing, gas pressure sintering and sinter-HIPing are used for achieving high densities. The microstructure of LPS-SiC consists of equiaxed grains (grain size ~1 μm; see Figure 4.8e) or heterogeneous mixtures of equiaxed and elongated grains (Figure 4.8f), embedded into a secondary grain boundary phase that consists of more or less crystallized oxidic or oxynitridic compounds, depending on the starting powders [200] and sintering conditions. Stable amorphous grain boundary films in the dimensional range of a few nanometers have been observed by several authors [201, 229–236], although clean interfaces have also been reported [237–240]. The difference in sintering additives or sintering processes might affect the equilibrium film thickness or the absence of the films.

Toughness values of 6–7 MPa $m^{1/2}$ have been reported for heterogeneous microstructures revealing anisotropic grain growth of 4H-polytype. Four-point bending strengths of up to 800 MPa have been achieved in fine-grained LPS SiC materials (see Table 4.5).

Although LPS SiC may not be suitable as a universal tribology material such as SSiC, the combination of properties suggests that it will be suitable for high-stress engineering components, and also in many wear as well as armor applications where strength, reliability, and toughness are demanded [202, 203].

The detailed features of the different additive systems include:

Oxide additives: Sigl and Kleebe [243] investigated the densification mechanism of silicon carbide using yttrium–aluminum–garnet (YAG) powder. The core and rim structure in this system suggested that Ostwald ripening by solution and reprecipitation controls the sintering process. Grande *et al.* [244] found that SiC containing 7.5 wt% Y_2O_3–Al_2O_3 formed an eutectic, and densification occurred by LPS. Much lower sintering temperatures (1700–1800 °C) for SiC were observed by Mulla and Kristic [227], if a small amount of SiO_2 was present with the Y_2O_3–Al_2O_3

system. Nader [243] also studied oxide additives (SiO_2, Y_2O_3, and Al_2O_3) and found that the reaction of SiC with SiO_2 impeded densification. The previously described LPS materials have in common that α-SiC has been used as the starting powder, resulting in fine microstructures with equiaxed SiC grains. When the fracture toughness of the sintered body is concerned, however, the use of β-SiC is preferred as the starting powder because the β- to α-transformation at high temperatures results in a microstructure with interlocking plate-like grains [225, 244]. Platelet-like shaped grains have been shown to increase the fracture toughness of SiC by crack bridging [222, 226] or crack deflection [224]. The transformation can be accelerated by adding some α-SiC as transformation seeds in the starting powders [245, 246], but retarded by incorporating N into the intergranular glass phases [247, 248].

Nitride additives: The densification of SiC using AlN–Y_2O_3 was first attempted by Chia *et al.* [249]. Nader [243] studied extensively the sintering behavior of SiC with the addition of AlN–Y_2O_3, and found the beneficial effect of nitrogen atmospheres on densification. Jun *et al.* [247], using β-SiC as the starting powder, reported that the amount of nitride or a nitrogen atmosphere had a strong influence in terms of densification and grain growth behavior of the SiC sintered bodies. N_2 significantly retarded both the β- to α-transformation and the anisotropic grain growth. Apart from improved densification behavior, the addition of AlN to LPS-SiC even could improve the beneficial effects of yttria and/or alumina dopants. An optimized composition of 10 vol% of additive with 60 mol% AlN and 40 mol% Y_2O_3 was found by Nader [243].

Al–B–C-based additives: Several authors have reported the occurrence of LPS using aluminum- with boron- and/or carbon-bearing compounds. The methods of addition have included direct mixing of aluminum, boron, and carbon with SiC powders [233, 236, 250, 251, 253], formation of an Al solid-solution in the SiC powder [254], and the use of AlB_2 [255] or Al_4C_3 and B_4C compounds [185, 256]. The addition of aluminum not only initiates densification at substantially lower temperatures than boron and carbon alone, but also enhances the β- to α-phase transformation [257, 258], as well as the transformation of 6H-SiC to 4H-SiC [259, 260] that leads to heterogeneous microstructures with elongated grains that enhance toughness.

4.4.1.5 **Hot-Pressed Silicon Carbide**

Dense parts made by hot-pressing are generally considered to have a very fine microstructure (see Figure 4.8g) and the best mechanical properties. Completely pure SiC does not exhibit any noticeable plastic behavior up to its decomposition temperature, and can therefore be densified only under diamond synthesis conditions (35 kbar, 2300 °C) to 100% of the theoretical density [261]. In axial hot-pressing with graphite tools it is, however, necessary to use small amounts of sintering aids; typically, 0.5–3% boron, aluminum, aluminum oxide, beryllium oxide, yttrium oxide, or tungsten carbide are added to fine α-SiC powder before molding [19, 217, 262–268]. The molding temperature is between 1900 and 2000 °C, depending on the particle size of the powder and the amount of sintering additives used. The pressure that can be used is limited by the strength of the graphite, the maximum being

50 MPa. The conventional method requires large amounts of energy and mold material, and can normally be used only for parts with simple, uncomplicated shapes. Due to these limitations, precision parts can only be produced by machining with diamond tools [269], and these, of course, are expensive.

Hot Isostatic-Pressed SiC A more convenient (but even more expensive) method than conventional hot-pressing is hot isostatic pressing (HIP). During the development of this technique [270–272] it proved possible to prepare high-purity SiC products with SiC contents of more than 99.5% by the hot isostatic pressing of SiC powder or SiC preforms in vacuum-sealed casings, to a final density which was exactly equivalent to the theoretical density, and with a uniform, fine-grained microstructure (see Figure 4.8h). Due to the higher isostatic pressure of 2 kbar (argon gas), compared to hot pressing, no sintering aids need to be added; that is, the thin film of silica on the SiC particles (which represents residual oxygen impurity of the SiC powder) is sufficient to achieve complete consolidation.

The post-densification of SSiC by HIP is less complicated because no gastight encapsulation is necessary. Hence, not only is it possible to achieve more than 99% of the theoretical density, but the variation of density and strength can also be reduced. Pressureless sintering and HIP can now be performed within the same cycle, in a process termed sinter-HIP, and this offers clear economic benefits [273–275].

Although the importance of hot-pressed SiC has decreased considerably since the introduction of pressureless sintering, it is currently still the most suitable method for obtaining the best mechanical properties for pure, monolithic SiC.

4.4.1.6 Chemical Vapor- and Physical Vapor-Deposited Silicon Carbide

CVD-SiC is obtained by the chemical reaction of volatile silicon- and carbon-containing compounds in the presence of hydrogen in the temperature range of 1000–1800 °C, as shown in Eq. (18):

$$CH_3SiCl_3(g) \rightarrow SiC(s) + 3HCl(g) \tag{18}$$

Besides the formation of SiC powders (see Section 4.3.1.2), SiC fibers (see Section 4.3.1.5), and SiC thin or thick films [276, 277], monolithic bodies in sections up to 1.5 m diameter and 25 mm thickness can now be prepared using this method.

CVD-SiC is of high purity (99.9995%), cubic β-modification, and very fined grained. Coatings of CVD-SiC were initially developed during the 1960s for nuclear fuel particles that were used to reduce the diffusive release of metallic fission products from the fuel kernel. With its high resistance to wear and abrasion, CVD-SiC is an extremely durable, non-particle-generating material that is ideal for the ultraclean environment of semiconductor manufacturing facilities. With their resistance to corrosion, oxidation and chemical erosion, the components stand up to the plasmas and acids used in semiconductor processing and cleaning. With its outstanding thermal properties (thermal conductivity $\geq$250 W mK^{-1} at 20 °C, thermal expansion 4.5×10^{-6} °C from 20–400 °C), CVD-SiC is also used in rapid thermal processing, where a wafer may be heated from 20 to 1100 °C in 6–7 s and then cooled just as rapidly. Bulk cubic phase CVD-SiC is additionally used in applications such as electronic packaging, kiln furniture for diode manufacture, laser optics for high temperatures, and sputtering targets and substrates for computer storage media [278, 279]. It can be polished to variations of less than 0.3 nm.

Today, SiC thin films (<1 μm thickness) can also be produced by physical vapor deposition (PVD), for example, sputtering. This method allows lower substrate temperatures, but operates more slowly. Electrically conductive B/N-doped sintered α-SiC with up to 9 wt% free carbon has been developed as a target material [280]. Novel applications for PVD-SiC include films for computer storage media, protective coating for lenses, and microwaveable packaging for food.

4.4.1.7 **Silicon Carbide Wafers**

In 1955, a laboratory simulation process for growing SiC crystals was developed by J. A. Lely [281], in which the nucleation of individual crystals was uncontrolled and the resulting crystals were randomly sized, hexagonal α-SiC platelets. In 1978, Tairov and Tsvetkov [282] introduced the growth of SiC single crystals by the vapor transport process, while in 1983 Ziegler [283] introduced the modified sublimation process for growing SiC single crystals. In 1987, the research group of R.F. Davis, at North Carolina State University (NCSU), announced a modification to the original Lely sublimation process [284] in which only one large crystal was grown that consisted of a single polytype. In this process, SiC powder or lumps of SiC were placed inside a cylindrical graphite crucible that was closed with a graphite lid onto which a seed crystal had been attached. The crucible was heated to approximately 2200 °C, normally in an argon atmosphere, at a pressure below atmospheric. A temperature gradient was applied over the length of the crucible in such a way that the SiC powder at the bottom of the crucible was at a higher temperature than the seed crystal. The SiC powder sublimed at the high temperature, such that the volume inside the crucible was filled with a vapor of different carbon and silicon molecules, such as Si_2C, SiC_2, Si_2, and Si. As the temperature gradient was chosen such that the coldest part of the crucible was the position of the seed, the vapor could condense on it and the crystal would grow.

In this way, boules of SiC may be grown, which may be further processed into wafers. The most severe problems encountered are with the micropipes; these are small holes approximately 0.1–5 μm in diameter which penetrate the substrates. The density of the micropipes is in the order of 10^2–$10^3\,cm^{-2}$, from the commercial material obtained from Cree Research Inc. Recently, Cree Inc. announced the availability of n-type 4H SiC-wafers with zero micropipe density, and today 50 mm 6H SiC- and 100 mm 4H SiC-wafers are available (from Cree, Inc.). Based on the growth of epitaxial thin films of single crystal polytypes on boule-grown substrates, electronic devices are now available commercially. Due to the high thermal conductivity, high breakdown electric field, high saturated electron drift velocity and high power density, SiC is today the material of choice for high-power and high-frequency devices. Notably, it is also the material of choice for blue light-emitting diodes (LEDs).

As a spin-off, currently colorless SiC gemstones, of one-half to one carat in size, and cut from 6H-SiC wafers, are entering the jewelry market for about 10–15% of the price of diamonds.

4.4.1.8 Silicon Carbide Nanoceramics

Sintering nanosized powders to achieve nanosized microstructures (grain size $\leq$100 nm), with the aim of flaw size reduction or flaw avoidance, is recognized as a promising means of improving the mechanical properties and reliability of SiC ceramics. However, nanosized SiC powders are not easy to process (see Section 4.3.1.2), and several difficulties must be overcome. Notably, the powder flows badly and has a low filling and compaction density if unconditioned, it exhibits a low oxidation resistance [74], and it is currently too expensive for large-scale use.

Fabrication by Solid-State Sintering In 1991, Vassen *et al.* [285] showed that polycrystalline β-SiC bodies with a density of at least 95% of the theoretical density, and a fine grain size of 150 nm, could be prepared by an encapsulated HIP of B/C-doped laser-synthesized powders with particle sizes below 20 nm. The required HIP temperature was 1500 °C, which is 250 °C below the temperature needed to densify conventional submicron powders. This fabrication process was later optimized [286] by preheating the shaped SiC bodies in a vacuum during an annealing step before encapsulation. Owing to this additional annealing step, the residual oxygen contents were minimized and a mean grain size of only 60 nm could be obtained in the HIPed SiC shapes. It was found that the reduced final grain size had a strong effect on mechanical properties; typically, as a consequence of grain size reduction from 1 μm to 150 nm, the Vickers Hardness increased from 2000 HV10 to 2500 HV10, whereas the fracture toughness decreased from 4 to 3 MPa, respectively [287]. However, by the adaption of a bimodal grain-sized distribution – that is, with the introduction of larger-sized SiC grains into a nanosized matrix – an increase in fracture toughness to 6 MPa $m^{1/2}$ was achieved [288, 289]. For a fine-grained HIP-SiC (300 nm), even under a stress of 100 MPa at 1600 °C very moderate creep rates of $1 \times 10^{-6}\,s^{-1}$ were measured [290].

Recently, the fabrication of dense SiC bodies with nanosized microstructures has been demonstrated successfully by Zhou *et al.* [291] and Pan *et al.* [292], using SPS.

Fabrication by Liquid-Phase Sintering The preparation of β-SiC nanoceramics with an average grain size of 110 nm by LPS has been demonstrated by Mitomo *et al.* [293], who were subsequently awarded a patent on "Superplastic SiC Sintered Body" [294].

For this, an ultrafine β-SiC powder with an average particle size of 90 nm was axially hot-pressed with additions of Al_2O_3, Y_2O_3 and CaO at 1750 °C. The SiC nanoceramic showed large deformation with a high strain rate of $5.0 \times 10^{-4}\,s^{-1}$ at 1700 °C. On the basis of these results [301], the maximum temperature and the minimum deformation rate for nanosized β-SiC might be defined as 1800 °C and $10^{-4}\,s^{-1}$, respectively. The superplastic deformation at temperatures as low as 1700 °C is based on the fine-grained microstructure and the presence of a glassy phase at grain boundaries. Thus, a new technology has been developed whereby nano-SiC parts can be subjected to plastic deformation as in the case of metals, and can be made into complicated shapes with near net-shape quality – that is, without the need for expensive post-processing such as diamond machining.

Lee *et al.* [295, 296] achieved a 99% dense LPS-SiC with an average grain size of 40 nm by using a two-step sintering process. Here, nanocrystalline SiC powder with a mean particle size of 20 nm was axially hot pressed with addition of Al_2O_3, Y_2O_3 and CaO. The powder compacts were heated to a temperature of 1750 °C and immediately annealed for 8 h at 1550 °C.

Hotta and Hojo [297] recently fabricated LPS-SiC by SPS with the addition of ALN and Y_2O_3. After an annealing period of 10 min at 1900 °C, a nanosized microstructure (50–100 nm) was obtained which revealed a bending strength as high as 1200 MPa.

4.4.1.9 Silicon Carbide-Based Composites

The alloying of SiC is, and has been, carried out basically for two reasons: (i) to improve the properties (toughness, wear, etc.) by the formation of tailored composites/solid solutions; or (ii) to improve the processing.

Improvements in processing can occur:

- In solid-state sintering, where the second phase acts simultaneously as a sintering aid for SiC, accelerating material transport by grain boundary and/or lattice diffusion.
- In reactive liquid sintering, due to reduced sintering temperatures ("transient liquid phase sintering").

In the latter case, the SiC and/or additions are reacted to an intermediate liquid which not only provides densification at reduced temperatures but also, as it is consumed in the reaction, yields a SiC-based material without grain boundary films.

The fabrication of SiC composites by second-phase dispersion is widely applied to improve material toughness. The various toughening mechanisms [298] that have high potential to reduce crack extension in SiC-composite materials are crack deflection, microcrack formation, crack bridging by reinforcement with metallic ligaments (e.g., TiC, TiB_2), and crack bridging and pull-out by platelet- or fiber-reinforcement.

Tensile fracture in SiC-based composites will only occur after a large enough load is applied to exceed the compressive stress in the process zone formed along the crack path. In order to achieve an increased crack deflection and crack–wake interaction in SiC, the microstructure can be modified in various ways:

- by the addition of discontinuous reinforcements; that is, particulates, whiskers or platelets (SiC-based composites)
- by continuous reinforcements (C-fiber/SiC or SiC-fiber/SiC)
- by inducing the growth of elongated SiC grains (see Section 4.4.1.4, *in situ*-toughening of LPS-SiC).

For optimum toughening by crack deflection, Telle *et al.* [299] pointed out that geometric factors such as grain size, volume fraction, orientation, and morphology of the added or "*in situ*-grown" phases, as well as the grain boundary strength, must be considered. The current state of the art with regards to some more or less important SiC-nonoxide composites is reviewed in the following sections.

SiC–TiC Very promising composites have been developed in the SiC–TiC system with SiC as the matrix phase [300–302]. Dispersed TiC particles significantly improve both the strength and toughness. Although, the addition of TiC does not reduce the densification temperature significantly below 2100 °C, the coarsening of SiC is completely retarded, which in turn raises the strength to 700–800 MPa [300–303]. The increase in K_{IC} to 6.5–7.5 is attributed to the misfit of the thermal expansion coefficients of TiC and SiC, introducing considerable radial tensile stresses at the phase boundaries and hoop compressive stresses in the matrix. These stresses enable crack deflection, crack branching, and microcracking above a critical particle size of ~3 μm. The optimum volume content of TiC ranges between 20 and 30 vol.%.

SiC–TiB$_2$ SiC-based composites with transition metal diboride (TiB_2, ZrB_2, etc.) particulates have been developed for electroconductive applications such as heating elements and igniters [304, 305], and also as wear-resistant structural parts for high temperatures such as valve-train components and rocker arm pads in super-hot-running engines [301]. These composites combine the high thermal and electrical conductivity of TiB_2 and ZrB_2 with the oxidation resistance of SiC. Additionally, due to thermal mismatch stresses of the order of 2 GPa, toughening mechanisms such as crack deflection and stress-induced microcracking with a pronounced process zone as well as flank friction have been proven to occur. Both, Cai *et al.* [306] and Faber *et al.* [307] have presented a detailed analysis of the contributions of the particular mechanisms to the total fracture toughness, stating that stress-induced microcracking is operational in a process zone of approximately 150 μm width.

The typical conditions for densification by axial hot-pressing are a temperature of 2000–2100 °C, at a pressure of 20–60 MPa for 30–60 min, and this results in 96–99.8% density. The particle sizes of the matrix and dispersed phases range between 1–5 and 4–8 μm, respectively. An optimum volume fraction of reinforcing particles of 25–30 vol% has been reported, yielding a flexural strength of 710 MPa and a fracture toughness of 5.0–5.7 MPa $m^{1/2}$ [303]. Composites with a lower TiB_2 content of 15 vol% exhibit a mean strength of 485 MPa, combined with a K_{IC} of 4.5 MPa [301].

The strength of SiC-based materials with 50 vol% ZrB_2, HfB_2, NbB_2 or TaB_2 particles also ranges between 400 and 500 MPa [304]. Similar strength values (480 MPa), combined with an exceptionally higher fracture toughness of 7–9 MPa $m^{1/2}$, have been reported for large-scale lots of pressureless sintered 16 vol% TiB_2 composites [305]. Since the sintering was carried out at temperatures exceeding 2000 °C, yielding 98–99% of the theoretical density and an average particle size of 2.0 μm, it is obvious that the reinforcing phase also acts as a grain growth inhibitor for SiC. The high-temperature strength of the SiC–TiB_2 and SiC–ZrB_2 composites was found to remain nearly constant at 480 MPa up to 1200 °C, and is hence superior to that of many sialons [304, 305].

Tani and Wada [308] fabricated optimized SiC–TiB_2 composites by reactive sintering, starting from an intimate mixture of SiC, TiO_2, B_4C, and C powders. The mixture was either hot-pressed or pressureless sintered and post-HIPed at temperatures of >1900 °C. Titanium diboride was formed *in situ* according to Eq. (19) during an intermediate heating step at 1400–1500 °C in vacuum or in an argon atmosphere.

Overstoichiometric amounts of B_4C and C (1–2 wt% each) can be adjusted to aid sintering. The primary advantages claimed for this reaction sintering process are the use of water in powder processing due to the disuse of highly reactive, preformed TiB_2 powders, and the very small size of reinforcing TiB_2 particles formed *in situ* due to the use of ultrafine TiO_2, B_4C, and C starting powders.

$$TiO_2 + 0.5B_4C + 1.5C \rightarrow TiB_2 + 2CO \quad (19)$$

An effective reactive pressureless sintering of SiC–TiB_2 composites was reported by Blanc, Thevenot and Treheux [309], who also studied the tribological behavior using a pin on flat configuration (flat: SiC, pin: SiC–TiB_2). Under dry conditions, the composites showed less wear resistance than monolithic SiC, although when water was used as the lubricant the opposite case applied.

In the study of Kuo and Kriven [310], indentation-strength tests were used to determine the retained strength, flaw tolerance, and toughness-curve characteristics of two types of SiC–TiB_2 composite. Here, β-SiC–TiB_2 composites which had been hot-pressed with an Al_2O_3 sintering aid, were compared with the well-studied α-SiC–TiB_2 composites, which were pressureless sintered with boron and carbon additives. The addition of TiB_2 (15 vol%) to the B- and C-doped α-SiC only increased the retained strength, without any significant improvement in toughening. On the other hand, the addition of TiB_2 (30 vol%), along with the effect of Al_2O_3 sintering aid for the β-SiC–TiB_2 composite, caused a great improvement in the properties, with a higher retained strength in the long crack regions, a better flaw-tolerance behavior, and a sharply rising toughness versus crack size curve. These different toughening behaviors for α- and β-SiC–TiB_2 were related to the weak nature of the SiC–SiC and SiC–TiB_2 interfaces, as well as to the fraction and size of TiB_2.

Cho *et al.* [311] fabricated tough SiC–TiB_2 composites by hot pressing and subsequent annealing using β-SiC, TiB_2, Y_2O_3 and Al_2O_3 as the starting powders. During the annealing step the α-SiC-grains showed anisotropic grain growth leading to an elongated grain morphology. The toughness of the SiC–50 wt% TiB_2 composites was 7.3 MPa $m^{1/2}$, which was approximately 60% higher than that of hot-pressed composites. Bridging and crack deflection by the elongated α-SiC grains and coarse TiB_2 grains appeared to account for the increased toughness of the composites.

Chu *et al.* [312] investigated the ballistic performance of SiC–TiB_2 composites, by consolidating various compositions by pressureless sintering to near-theoretical density. The addition of TiB_2 ($\leq$10% by volume) increased the density of the material by less than 3% over that of pure SiC. TiB_2 additions also hindered SiC grain growth and the formation of elongated grains during high-temperature pressureless sintering. The SiC–TiB_2 composites demonstrated improved ballistic properties in depth-of-penetration (DOP) tests over pure, pressureless-sintered SiC material; in fact, the DOP values approached those of SiC produced by hot-pressing.

SiC–B_4C During the mid 1980s, pressureless sintering and post-hipping were developed by Schwetz *et al.* [313] to produce 100% dense SiC–B_4C composite materials having SiC : B_4C weight ratios within the range of from 90: 10 to 10: 90, and a free carbon content of 4–5 wt%.

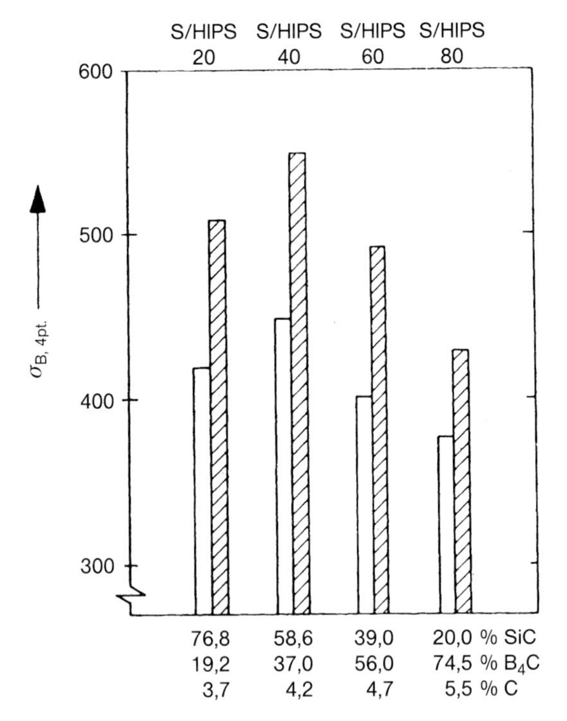

Figure 4.9 Flexural strength of pressureless sintered (S, unhatched bars) and post-HIPed (HIPS, hatched bars) SiC–B_4C particulate composites. After Ref. [313].

These composites combine the good thermal shock resistance and oxidation resistance of SiC with the hardness, wear resistance and low specific gravity of boron carbide. In this way, a maximum strength of 550 MPa (four-point bend) was achieved for a composite of 59 wt% SiC/37 wt% B_4C/4 wt% C (see Figure 4.9). The composite, which can be used in oxidizing atmospheres up to 1200 °C, has a microstructure that is characterized by equiaxed B_4C and graphite grains of $<1\,\mu m$ diameter, which were embedded in a matrix of SiC grains with an average grain size of 1.5 μm. However, no improvement in fracture toughness was achieved, as the fracture mode was almost 100% transgranular. Similar results with sintered SiC–B_4C composites were obtained by Thevenot [314] and later by Tomlinson *et al.* [315], who observed a 20% increase in strength when 25 vol% B_4C was added to SiC.

Excellent tribological properties for SiC–B_4C–C composite materials were reported by Kevorkiijan *et al.* [316], who prepared SiC–B_4C–C sealing rings for magnetic pumps by pressureless sintering; the rings were then characterized using a pin-on-disc method (medium: water, pressure: 16–25 MPa, speed: 35–75 ms^{-1}). The introduction of a lower level of B_4C particles (5–20 wt%) into the SiC matrix resulted in an almost linear decrease in the wear rate. Moreover, the further addition of B_4C (20–40 wt%) led to an almost parabolic wear rate response. For example, with 40 wt% B_4C, a decrease in the relative wear volume of the sealing rings by more than 55% was achieved. These results recommended the use of wear-resistant SiC–B_4C composites for heavily loaded mechanical face seals in the pairing hard/hard against themselves. They may, likewise, be suitable for the production of shaft protection sleeves and components for sliding bearings, the wear resistance of which must be improved.

SiC– AlN A series of solid solutions between SiC and AlN over the whole composition range was concurrently discovered at the Universities of Utah and Newcastle upon Tyne [317], and has since received considerable attention [318–321].

A 2H wurtzite-type structure is formed by the carbothermal reduction of fine SiO_2 and Al_2O_3 with a carbon source under a nitrogen atmosphere at 1600 °C [see Eq. (20)]. Kinetically favored is the carbothermal reduction of α′-SiAlONs or α′-SiAlON–precursor mixtures (3 Si_3N_4 + 3 AlN + CaO) at 1800 °C [see Eq. (21)]. Because the diffusion coefficients in covalent solids are extremely small, a solid solution was thought unlikely to be obtained by heating and annealing of the powdered solid components.

$$3SiO_2 + 0.5Al_2O_3 + 4.5C + N_2 \rightarrow 3SiC \cdot AlN + 1.5CO + 0.5N_2 \quad (20)$$

$$CaSi_9Al_3ON_{15} + 10C \rightarrow 3(3SiC \cdot AlN) + Ca + CO + 6N_2 \quad (21)$$

Zangvil and Ruh [322] however, obtained SiC–AlN solid solutions by hot-pressing powder mixtures. Li [323] obtained completed SiC–AlN solid solution formation by the heat treatment of high-energy, ball-milled powder mixtures. The ball milling of AlN and SiC powders at ambient temperature resulted in the mixing of AlN and SiC on a nanometer scale. After annealing at 2100 °C for 1 h, the resultant product was composed of 2H and 4H type SiC–AlN solid solution. Shirouzu *et al.* [324] recently consolidated powder mixtures of β-SiC (0.03 μm) and AlN (1.1 μm) by SPS. In this case, annealing the powder compact at 2100 °C for 30 min led to the production of dense, homogeneous 2H SiC–AlN solid solutions. Energy dispersive spectroscopy (EDS) imaging revealed oxidic triple junction pockets, thus emphasizing the enhancement of densification by LPS.

Alternatively, a process based on the combustion synthesis starting with the elements initiated by electric field activation has been used by Xue and Munir [325]. Likewise, homogeneous single-phase AlN–SiC solid solution powders with well-crystallized hexagonal morphologies and a fine particle size of 3 μm have been obtained by Mei and Li through the combustion reaction of aluminum, carbon, and Si_3N_4 powder mixtures in air [326].

Kimura *et al.* [327] used a thermally activated CVD process to form AlN–β-SiC powder mixtures. Here, by the reaction of $AlCl_3$ with a N_2–NH_3 mixture, AlN is deposited on submicron SiC particles in a fluidized powder bed.

Czekaj *et al.* [328] used the pyrolysis of organic precursors starting with dialkylaluminumamide and polyvinylsilane or polycarbosilane to synthesize ultrafine powders. Paine *et al.* [329] pyrolyzed a single-source organic precursor to obtain solid solution powders. However, both methods had low yields of 50%.

Recently, Liu *et al.* [330] synthesized SiC–AlN solid solution whiskers by heating a mixture of α-Si_3N_4, Al_2O_3, MgO, and Y_2O_3 in a graphite crucible in a N_2 atmosphere. The as-synthesized whiskers were well-crystallized, with widths ranging from 0.5 to

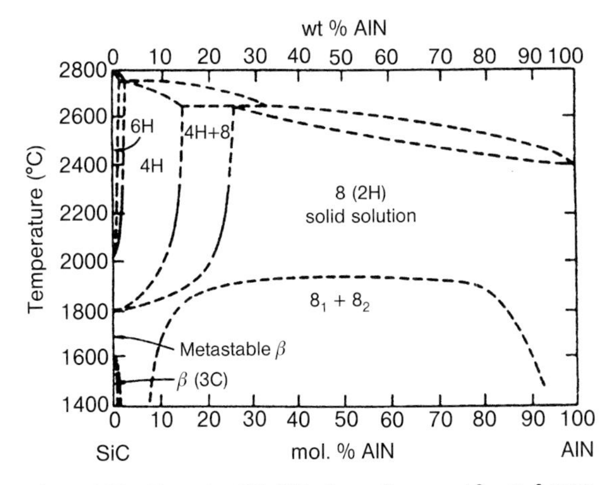

Figure 4.10 Tentative SiC–AlN phase diagram. After Ref. [322].

5 μm and lengths ranging from 0.1 to 1 mm. It was proposed that the whiskers were produced via the VS process.

According to the phase diagram proposed by Zangvil and Ruh [322] (see Figure 4.10), at temperatures above 2000 °C, a 6H-4H-2H series of solid solutions appears with increasing amounts of AlN. AlN strongly stabilizes specific polytypes (4H and 2H) at certain composition ranges, enabling the engineering of single-phase SiC materials with discrete physical properties. At temperatures below about 1900 °C, a miscibility gap was first proposed by Rafaniello *et al.* [319], a suggestion later supported by several studies. For example, Xu *et al.* [321], Chen *et al.* [331] and Li *et al.* [332] each obtained strong SiC–AlN materials with flexure strengths up to 1 GPa. Several mechanisms of grain refinement resulting from SiC-polytype transformations into a wurtzite (2H) solid solution were also reported. Kuo *et al.* [333] reported the formation of modulated structures within the miscibility gap, while Lee and Wei [334] described pressureless sintering with 2 wt% Y_2O_3 as sintering aid at 2050 °C to produce a duplex structure composed of large $(SiC)_x(AlN)_y$ grains and small SiC grains. A further solid solution treatment (>2225 °C), followed by annealing within the miscibility gap (1860 °C), resulted in a spinoidal decomposition, yielding various duplex/modulated structures with improved fractured toughness of the alloys as high as 5.5 MPa·m$^{1/2}$.

In addition to their possible use as high-temperature structural ceramics, materials in the SiC–AlN system have demonstrated potential as wide band-gap semiconductors and for opto-electronic applications [322]. The band gap can be easily adjusted in the range from 3.3 eV (pure SiC) to 6.0 eV (pure AlN) [327, 335, 336]. Likewise, the thermal and electric conductivities can be modulated over a wide range [318, 328].

SiC–Al_2OC Extensive SiC–Al_2OC solid solutions have been identified by Cutler *et al.* [317] for between 1% and 100% Al_2OC. Moreover, it was shown that wurtzite 2H-SiC could incorporate substantial amounts of AlN and Al_2OC in solid solution; hence, the acronym "SiCAlON" was coined to describe these materials, in analogy to "SiAlON" ceramics.

Jackson *et al.* [337] have sintered SiC at temperatures between 1850 and 1950 °C, using a transient liquid phase produced by the carbothermal reduction of Al_2O_3 by Al_4C_3. The resultant ceramic was fine-grained (average grain size <5 μm), and consisted of SiC (starting polytypes) and Al_2OC as the two major phases. The properties of the hot-pressed ceramics varied with the amount of Al_2OC but, at an optimum composition of about 5–10 wt% Al_2OC, the strength (660 MPa) and fracture toughness ($K_{IC} = 3.1\,MPa{\cdot}m^{1/2}$) obtained were comparable with or superior to the corresponding properties of commercial-grade sintered SiC. Huang *et al.* [338] found encapsulation to be necessary for effective sintering with additions of Al_2O_3 and Al_4C_3, the densification occurring above ~1860 °C. These authors attributed such densification to a transient liquid phase in the system Al_2O_3–Al_4O_4C in the route to forming Al_2OC with a eutectic temperature of ~1840 °C. The strength decreased with Al_2OC content in hot-pressed samples, from ~600 MPa at 10 wt% Al_2OC to around 250 MPa at ~50 wt% Al_2O_3. Most significantly, the fracture toughness of some SiCAlON compositions appeared to be higher than that of SiC (~4.2, compared to 3.0 $MPa{\cdot}m^{1/2}$), using an indentation technique. Lihrmann and Tirlocq [339] proposed the fabrication of sintered or hot-pressed SiC-based composites containing SiC, as well as 5–30 wt% solid solutions composed of Al_2OC and AlN. Starting from SiC–AlN–Al_4C_3–Al_2O_3 powder mixtures, the densification was greatly enhanced by the occurrence of a transient liquid phase that originated in the Al_2O_3–Al_4C_3 system at temperatures above 1800 °C. The composites with 5–10% (Al_2OC–AlN) solid solution exhibited a mean grain size <2 μm and a mean strength of 620–670 MPa, combined with a K_{IC} of 5.1–6.8 $MPa{\cdot}m^{1/2}$. Both, strength and fracture toughness retained their values up to 1400 °C, before weakening and thus demonstrating the highly refractory nature of the Al_2OC–AlN second phase.

SiC– SiC and SiC–C (Continuous Fiber-Reinforced SiC Matrix Composites) Three different processes are commonly used to manufacture carbon fiber-reinforced SiC materials: (i) chemical vapor infiltration (CVI) [340]; (ii) liquid polymer infiltration (LPI; also termed polymer infiltration and pyrolysis, PIP) [341]); and (iii) melt infiltration or liquid silicon infiltration (MI/LSI) [342].

Depending on the method, the SiC composites vary in their properties as well as in their manufacturing costs. The advantages of the MI processes are a short manufacturing time and the use of low-cost raw materials. The unique thermal and mechanical properties of these MI ceramic matrix composites (CMCs) has opened up a wide field of new applications beyond aerospace. The data relating to these different materials are listed in Table 4.4.

Fabrication by CVI: This is performed by the deposition of SiC vapor inside a porous preform (40–60% porosity) made from high-strength C- or SiC- fibers, and results in a composite with ~10% residual porosity. The fracture mode is noncatastrophic, and typical flexural strength of 300–400 MPa and toughness values of over 20 $MPa\,m^{1/2}$ are obtained. Using the "forced flow thermal-gradient CVI process" developed at Oak

Table 4.4 Mechanical and physical data for fiber-reinforced SiC matrix composites. After Heidenreich [345].

		CVI		LPI	LSI		
	Manufacturer SI-unit	C/SiC SPS (SNECMA)	C/SiC MT Aerospace	C/SiC EADS	C/C-SiC DLR	C/C-SiC SKT	C/SiC SGL[i]
Density	$g\,cm^{-3}$	2.1	2.1–2.2	1.8	1.9–2.0	>1.8	2/2.4
Porosity %	%	10	10–15	10	2–5	—	2/<1
Tensile strength	MPa	350	300–320	250	80–190	—	110/20–30
Strain to failure	%	0.9	0.6–0.9	0.5	0.15–0.35	0.23–0.3	0.3
Young's modulus	GPa	90–100	90–100	65	50–70	—	65/20–30
Compression strength	MPa	580–700	450–550	590	210–320	—	470/250
Flexural strength	MPa	500–700	450–500	500	160–300	130–240	190/50
ILSS[j]	MPa	35	45–48	10	28–33	14–20	—
Fiber content	Vol.%	45	42–47	46	55–65	—	—
Coefficient thermal expansion ‖	$10^{-6}\,K^{-1}$	3[a]	3	1.16[d]	−1 −0.5[b]	0.8–1.5[d]	−0.3/1.8[e]
⊥		5[a]	5	4.06[d]	2.5–7[b]	5.5–6.5[d]	0.03–1.36[f]/3[g]
Thermal conductivity ‖	$Wm^{-1}\,K^{-1}$	14.3–20.6[a]	14	11.3–12.6[b]	17.0–22.6[c]	12–22	23–12[h]/40–20[h]
⊥		6.5–5.9[a]	7	5.3–5.5[b]	7.5–10.3[c]	28–35	
Specific heat	$J\,kg^{-1}\,K$	620–1400	—	900–1600[b]	690–1550	—	—

a) ‖ and ⊥ = Fiber orientation; [a] RT–1000 °C; [b] RT–1500 °C; [c] 200–1650 °C; [d] = RT–700 °C; [e] 1200 °C; [f] 200–1200 °C; [g] 300–1200 °C; [h] 20–1200 °C; [i] values for fabric/short fiber-reinforced material; [j] interlaminar shear strength.

Ridge National Laboratory/USA, the infiltration time is reduced from weeks (isothermal CVI at SEP/France) to less than 24 h (ORNL), and final densities for composites of >90% TD [343] are obtained.

It is now well established that a fiber coating must be deposited on the fiber prior to infiltration of the matrix, in order to control the fiber-matrix bonding and the mechanical behavior of the composite. Pyrocarbon (PyC), boron nitride or (PyC-SiC)- and (BN-PyC)-multilayers, with an overall thickness ranging from about 0.1 μm to about 1 μm, and displaying a layered crystal structure (PyC, BN) or a layered microstructure (multilayers), are the most common interphase materials in non-oxide CMCs. The main role of the interphase is to deflect the microcracks which form in the matrix under loading, and hence to protect the fiber from notch effect.

Finally, pressure-pulsed-CVI (P-CVI) has recently been presented as a means of engineering, at the micrometer (or even nanometer) scale, either the interphase or the matrix. Based on this technique, multilayered selfhealing interphases and matrices (combining crack-arrester layers and glass-former layers) have been designed and produced, through a proper selection of chemical composition of the layers [539].

Fabrication by LPI: A carbon-fiber preform infiltrated with resin (e.g., PCS) is pyrolyzed to form SiC. Infiltration and pyrolyzation are repeated a number of times until the pores are narrow enough so that further infiltration ceases. Finally, the body is heated to temperatures between 1000 and 1500 °C for crystallization of the SiC matrix. Densification takes less time and costs less than densification by CVI.

Fabrication by LSI: The LSI/MI processes can generally be subdivided in three main steps:

(1) A carbon fiber-reinforced plastic (CFRP) preform is made using two-dimensional (2-D) fiber fabrics (cut fibers and filament-wound or braided preforms or high- carbon-yield precursors).
(2) The CFRP is pyrolyzed and transformed into a porous C/C preform in inert gas atmosphere at $T > 900\,^{\circ}C$.
(3) Molten silicon is infiltrated in a vacuum process; the silicon then reacts with carbon to build up the SiC matrix.

In order to protect the carbon fiber from reaction with molten Si, and to ensure a weak embedding of the brittle fibers in the brittle matrix, fiber embedding in the carbon matrix via PIP, fiber coating via CVI, and *in situ* fiber embedding in the carbon matrix are each applied. Fiber protection via PIP is used widely to manufacture short-fiber-reinforced C/SiC brake disks, for example leading to the so-called Sigrasic materials from SGL Carbon AG. Thereby, endless fiber bundles are impregnated with phenolic resin, which is cured and paralyzed, thus embedding the fiber filaments in a dense carbon matrix and resulting in a C/C-like raw material. When using CVI for fiber coating, a thin layer of pyrolithic carbon is deposited on each fiber filament, resulting in a C/SiC material with the filaments mainly embedded individually in the SiC matrix. This method is used by EADS Astrium for the manufacture of so-called Sictex materials. These time-consuming and costly fiber coatings are not necessary at all if particularly suitable precursors, which offer a strong fiber matrix bonding, are used for the manufacture of the CFRP preform [344], leading to a segmentation of each fiber bundle into dense C/C bundles during pyrolysis. This cost-efficient method is the basis of the LSI process, which has been developed at DLR German Aerospace center [345]. In this way, C/C-SiC parts can be made in a near net shape technique with almost no restrictions to size, wall thickness, and geometry

Automotive brake disks represent the first large-scale application of CMC materials. For the first time ever, the upscaling from a single part or small series manufacture to the viable industrial production of several thousand safety-relevant components per year was achieved successfully by SGL Brakes GmbH, Meitingen. Compared to cast iron, the C/SiC brake disks offer weight savings of up to 50%; moreover, their very high abrasion resistance leads to brake disks that will last for the lifetime of the vehicle, with a possible total service life of up to 300 000 km.

Biomorphous Silicon Carbide Composites The use of biostructures such as wood for the preparation of microcellular-designed ceramic materials has become a matter of increasing interest in recent years. Various attempts have been made to utilize native as well as preprocessed biological plant materials (lignocellulosics) for the preparation of carbon monoliths [346], as well as for biomorphous oxide or SiC-based ceramic materials [347, 348].

The morphology of natural plants is characterized by different cellular and microporous structures ranging from the nanometer to the millimeter scale [349]. The major biopolymeric constituents of wood are cellulose, hemicellulose, and lignin. Cellulose chains are bundled to fibrils and form a bioorganic, cellulose fiber-reinforced composite material in a matrix of lignin and hemicellulose. The average elemental composition of wood is approximately 50 wt% C, 43.4 wt% O, 6.1 wt% H, 0.2 wt% N, and 0.3 wt% inorganic materials.

Since the study of Byrne and Nagle [351], biomorphous SiSiC-ceramics derived from wood have been studied intensively over the past few years [350, 352, 353]. The fabrication involves the preparation of porous biocarbon monoliths by pyrolysis of the native plant materials, which are subsequently subjected to the liquid Si-infiltration. The processing yields biomorphous, dense or less-porous SiSiC-ceramics, depending on the type of wood/species used (Figure 4.11). The LSI processing of plant-derived biocarbon templates is capable of reproducing the micro-morphology of the native tissue with high precision in the SiSiC-ceramic composite, thus representing a net-shape process on the micrometer scale [347]. This process yields SiSiC-ceramic materials with a unidirectional microstructure and anisotropic structural and mechanical properties.

In contrast to the Si-melt infiltration, highly porous, single-phase, biomorphous SiC-ceramics can be manufactured by the reactive infiltration of gaseous Si-containing reactants such as Si/SiO-vapor or silanes [355, 356]. Based on rapid fluid infiltration into the accessible cellular template structure, a variety of chemical processing routes offer a wide range of chemical compositions and structural modifications in the ceramic reaction products.

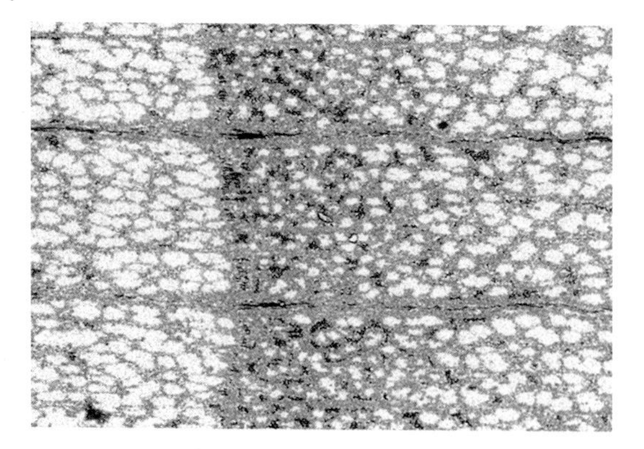

Figure 4.11 SEM image (BSE mode) overview of axial section of biomorphous SiSiC ceramics prepared from beech wood-derived biocarbon templates by LSI processing. The infiltration/reaction time was 60 min at 1550 °C. After Ref. [354].

The cellular micro- and macrostructure pseudomorphs with naturally grown wood tissue show a complex mechanical behavior, which is governed by the unique arrangement of cells. In some aspects, the fracture behavior in biomorphous ceramics is similar to that of fibrous monolithic ceramics, as well as that of laminate composite ceramics showing a noncatastrophic stress–strain behavior. A pronounced anisotropy of fracture behavior is a characteristic feature which depends on the loading conditions with respect to the orientation of the cell-packing structure [357].

Due to their anisotropic morphology with a unidirected pore structure on the micrometer level, biomorphous SiC- and SiSiC-ceramic composites represent interesting candidates for applications such as liquid metal filters, high-temperature catalyst supports or nozzle structures, and micro-reactor devices.

4.4.1.10 Metal Matrix Composites (MMCs)

One of the main advantages of adding a reinforcement to light metals such as aluminum, titanium or magnesium, is an increase in the specific strength of the metal. The reinforcement of light metals opens up the possibility of applying these materials to areas where a low component mass has a high priority. The addition of a reinforcement can result in:

- an increase in yield strength and tensile strength at room temperature and above, whilst maintaining acceptable toughness;
- an increase in creep resistance at higher temperatures compared to that of conventional alloys;
- an increase in fatigue strength, especially at higher temperatures;
- an increase in Young's modulus;
- a controlled coefficient of thermal expansion matching that of directly mounted integrated circuits; and
- a high thermal conductivity.

Discontinuously Reinforced MMCs Discontinuously reinforced MMCs (particulates, whisker, platelets) are produced by either liquid- or solid-state processing [358–360]. Three different processes are used commercially used for liquid-state processing:

Stir casting (Duralcan™ process): In this process, liquid aluminum and ceramic particles are mixed in a vacuum [361]. The melt is stirred at a temperature slightly above the liquidus temperature of the alloy so as to distribute the ceramic particles homogeneously. The Duralcan process of making particulate composites by a liquid metal casting route involves the use of 8–12 μm particulate reinforcement. For particles that are much smaller (2–3 μm), the result is a very large interface region and, thus, a very viscous melt during MMC manufacture, which leads to a poor quality product. In foundry-grade MMCs, high-Si-aluminum base alloys are used as these prevent formation of the brittle compound Al_4C_3 in the interfacial reaction layer between Al and SiC. Al_4C_3 can be extremely detrimental to the mechanical properties of the MMC. After casting, the solidified ingot may undergo secondary processing by extrusion, forging, or rolling.

Preform infiltration: This process usually involves the infiltration of a preform of fibers or particulate reinforcement with a liquid metal. Depending on the wetting characteristics and the type of alloy, either pressureless or pressure-assisted infiltration can be used.

The pressureless liquid metal infiltration process of making MMCs is known as the Primex process (Lanxide). This process can only be used with certain reactive metal alloys, such as Al–Si–Mg, to infiltrate ceramic performs. The magnesium reacts with the nitrogen atmosphere to build up an easily wettable coating of Mg_3N_2 on SiC. Silicon must also be present in the alloy to suppress the formation of aluminum carbide by the reaction of SiC with Al. During infiltration, the nitride reacts with aluminum to AlN, which can be found as small precipitates and as a thin film on the SiC. The particle loading can be as high as 75 vol%, depending on the particle shape and size.

Pressure-assisted production routes such as squeeze casting or gas pressure-assisted infiltration force liquid metal into an evacuated freestanding preheated fibrous or particulate preform at high pressure [362–364]. The pressure is applied continuously until the solidification is complete. The use of a high applied pressure is especially useful when poor wetting occurs. Composites manufactured using this method also have the advantage of minimal reaction between the reinforcement and molten metal, since the casting and solidification times are both short. MMCs with a range of reinforcement levels ranging from 30% to >70% have been successfully fabricated.

Spray deposition: The spray deposition process, which was developed by Osprey Ltd during the late 1970s for producing monolithic alloys [365], has been adapted by several manufacturers to produce particulate-reinforced MMC billets with a residual porosity of 5% [366]. The porosity is eliminated by a secondary processing, such as extrusion or rolling. A spray gun is used to produce an atomized stream of aluminum alloy, into which heated SiC particles are injected. An optimum particle size is required for efficient transfer into the atomized melt stream. High melt feed rates can be used, allowing ingots to be produced in tonnage quantities with relatively short run times (0.2–2 $kg s^{-1}$). Spray deposition is essentially a liquid metal process, but deleterious reaction products at the particle–matrix interface are generally avoided as the contact times in the liquid state are very short. There is a practical limit of ceramic reinforcement of 25% due to high loss rates at high feed ratios.

The solid-state processing of MMCs is used primarily to produce particulate-reinforced MMCs, in which process the gas-atomized matrix alloy and reinforcement powders are blended, followed by cold compaction to roughly 80% density [367–369]. Before high-temperature consolidation by vacuum hot pressing, extrusion or hot isostatic pressing, any volatile components must be removed by degassing. The resultant fully dense components can be subsequently processed by extrusion, rolling, and forging.

Whisker-reinforced composites can also be manufactured by this process, although a homogeneous composite structure can often be difficult to achieve with whiskers as they can form tangled agglomerates during the blending process.

Continuously Reinforced MMCs For the production of continuous fiber-reinforced MMCs, the above-described methods of preform infiltration also apply.

Another applied manufacturing route is vapor phase processing by PVD, which typically involves the deposition of a relatively thick layer of evaporated matrix material onto the surface of a fiber, in high vacuum [370]. The matrix alloy vapor is produced by directing a high-power electron beam onto the solid bar feedstock of the matrix alloy. The alloy composition can be tailored, as differences in the evaporation rate between different solutes in the matrix alloy become compensated by changes in composition of the molten pool formed on the end of the bar, until a steady state is reached in which the alloy content of the deposit is the same as that of the feedstock. Typical deposition rates for the fiber coating are 5–10 $\mu m\ min^{-1}$. After fiber coating, the MMC is fabricated by bundling the coated fibers and consolidating them using hot pressing or HIP. A very uniform distribution of fibers or monofilaments can be produced by this process, with fiber contents of up to 80%.

Further methods have been developed for Al–SiC and Ti–SiC MMCs which make use of diffusion bonding/solid-state processing of stacks of plasma sprayed foils or stacks of alternating metal foils and fiber fabrics (foil–fiber–foil).

4.4.2
Dense Boron Carbide Shapes

The sintering of boron carbide, with its high fraction of covalent bonding (>90%), is difficult due to low diffusion coefficients. Additionally, naturally occurring oxide layers on the surface of fine powder particles promote evaporation–condensation processes that do not contribute to densification but rather enhance grain growth. Extreme high vapor pressures of gaseous boron oxide compounds at temperatures exceeding 1500 °C lead to reactions such as that in Eq. (22). Thus, boron carbide is transported from regions of high curvature (convex) to regions of low curvature (concave); that is, the grains coarsen with no shrinkage at all [371].

$$5B_2O_3(g) + B_4C(s) \rightarrow 14BO(g) + CO(g) \quad (22)$$

According to Greskovich and Rosolovski [372] and Prochazka [373], the specific surface area during the initial state of sintering is mainly consumed by a coarsening of the particles and pores, which in turn reduces the driving force for densification (local chemical potential). Densification comes to an end before pore closure is obtained. This so-called "terminated density" has been observed for pure boron carbide, SiC and silicon nitride, when no sintering aids are present.

4.4.2.1 Sintered Boron Carbide

Due to the high fraction of covalent bonding of boron carbide, transport mechanisms such as volume and grain boundary diffusion which contribute to the overall shrinkage become effective at temperatures above 2000 °C – that is, close to the melting point (2450 °C). According to Grabchuk and Kislyi [374], transport along surfaces is predominant in the temperature range from 1500 to 1800 °C, whereas sublimation predominates at temperatures exceeding 1800 °C. A high resistance to plastic deformation and grain boundary sliding, combined with low surface energies, constrains any considerable particle rearrangement and shape accommodation before densification by the transport of matter occurs. Therefore, for the densification of stoichiometric boron carbide, the use of very fine powders low in oxygen content is a prerequisite condition; in addition, temperatures in excess of 2250 °C are required. Because excessive grain growth occurs above 2000 °C, residual porosity is entrapped inside the boron carbide grains.

Adlassing [375], for example, achieved densities >90% at a temperature close to 2450 °C yet, by using finer powders densities >99% have been obtained by Grabchuk and Kislyi [376] annealing at 2300 °C. These samples showed low residual porosities that ranged from 0.5 to 1%, and grain sizes of at least 10–15 μm, although the strength of such samples is approximately 15–20% less than that of similar samples produced by hot-pressing.

In order to enhance pressureless densification at lower temperatures, several different methods apply:

- Increasing the density of line- and planar-defects by high-energy milling [377].
- Increasing the point defect concentration by doping with aluminum [378] or aluminum-containing compounds (e.g., Al_4C_3, Al_2O_3, AlF_3 [378–380] or boron [381]), which substitute for carbon in the CBC chain, thus introducing vacancies and electron holes.
- Adding carbon or carbon-containing compounds such as aluminum carbide, SiC or related compounds to remove any boron oxide layers from the boron carbide particle surface. This in turn will increase the surface energy and inhibit exaggerated grain growth due to evaporation–condensation [382].

The addition of Be_2C resulted in densities of 94% when sintering at temperatures between 2200 and 2280 °C [383]. The addition of transition metal diborides inhibits grain growth by the pinning of grain boundaries or, as in the case of W_2B_5, by activating LPS [384]. Because boron carbide reacts with most metals except for Cu, Zn, Sn, Ag, Pb -forming borides and free carbon- the use of metals to enhance densification is not appropriate, for example Mg, Cr, Co, and Ni [379, 385, 386]. Stibbs *et al.* [387] added up to 10 wt% Al, Mg, or TiB_2 to achieve densities >99% when sintering at temperatures in the range of 2150–2280 °C.

The only technical relevant sintering aid to consolidate boron carbide to almost theoretical density is carbon [388–390]. Schwetz and Grellner [391] used a phenolic resin as a carbon-containing compound (1–3% carbon) to obtain >90% density at 2150 °C. Sintering is promoted by a carbon layer on the submicron boron carbide powder resulting from pyrolysis of the phenolic resin. Residual carbon at grain boundaries not consumed by reaction with boron oxide may control the surface diffusion, evaporation and grain boundary movement, thus limiting excessive grain growth [382, 392]. By adding 6 wt% carbon and sintering at 2220 °C, a fine-grained microstructure with 1–5 μm grain size is obtained, and exaggerated grain growth starts above 2235 °C. The formation of grains exceeding 0.5 mm was explained by the presence of impurities with low melting points. Bourgoin *et al.* used Novolac-type resins as a carbon precursor [393].

Because of the complex handling of powders, in addition to environmental problems during the pyrolysis of precursors, a route for uniform doping with carbon black was developed by Matje, Sigl and Schwetz [394, 395]. Nearly full density was achieved at a temperature of 2150 °C, leaving 5–7% of free carbon in the sintered compact (see Figure 4.12). The fine-grained microstructure was very uniform, but some grains showed a discontinuous growth accompanied by stress-induced twinning, which could be attributed to the thermal expansion anisotropy of boron carbide. Above approximately 3 wt%, the density was independent of the amount of added carbon (see Figure 4.13).

Figure 4.12 Microstructure of boron carbide pressurelessly sintered with carbon black. After Ref. [395].

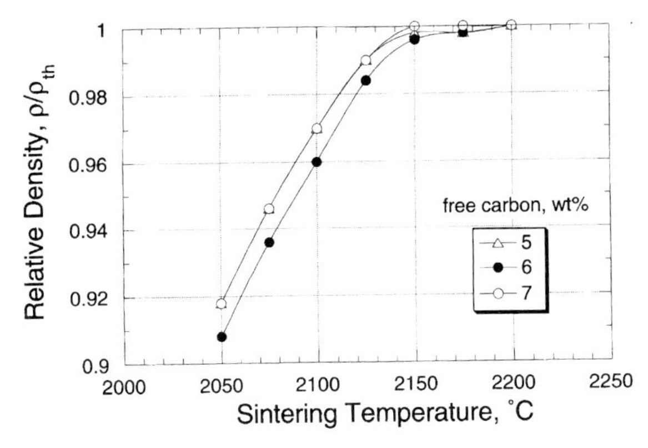

Figure 4.13 Densification of boron carbide with addition of carbon. After Ref. [395].

Speyer *et al.* [396] recently developed a process for pressureless consolidation obtaining fine-grained boron carbide without the use of sintering aids. The amount of boron oxide was reduced during an annealing step in vacuum or hydrogen, followed by heating up to sintering temperatures in the range of 2300–2400 °C, using very high heating rates of 100–150 °C min^{-1}. According to these authors, the extraction rate of B_2O_3 was accelerated by rapid heating through the range of about 1870–1950 °C, which left less time for oxide-facilitated particle coarsening to take place. Additionally, by rapid heating through the range of about 2010–2140 °C, the time over which coarsening could occur by evaporation and condensation of B_4C was minimized. Rapid heating brought comparatively small, high-surface-energy particles to a temperature range at which LPS or activated sintering was rapid relative to the coarsening. Densities of up to 97% were achieved in this way. Post-HIPed samples showed a greater hardness and strength than hot-pressed samples, due to less free carbon and a smaller grain size.

The combined addition of deoxidizing and diffusion enhancing agents and/or grain growth inhibitors has been utilized by numerous authors to synthesize boron carbide-based ceramic materials, for example, B + C, SiC + Al, TiB_2 + C [393, 397, 398]. Schwetz *et al.* [399] obtained densities exceeding 97% by using 9–10 wt% SiC and 1–3 wt% C, and sintering at 2000–2100 °C. A post-HIP process was used to remove any residual porosity. Yet, instead of using inorganic additives, organic precursors may be introduced, these being converted to SiC and carbon upon pyrolysis [400–402].

Boron carbide was reaction-sintered by Sigl [403] using boron carbide and TiC, leading to a composite material with TiB_2 as the reinforcing secondary phase and free carbon.

The densification of boron carbide using silicon or aluminum is difficult to obtain due to the high vapor pressures of liquid silicon or aluminum at reasonable sintering temperatures, and which may hinder densification due to the formation of degassing channels and entrapment of residual gas in the closed pores. Telle and Petzow [404, 405] have demonstrated the existence of a binary equilibrium between a $B_{12}(B,C,Si)_3$ solid solution and boron-rich liquid silicon above 1560 °C, which might be beneficial for compaction by LPS of boron carbide. A $B_{12}(B,C,Al)_3$ solid solution, in equilibrium with a boron-rich Al melt in the temperature range of 1000–1800 °C, has been proposed by Lukas [406]. The use of these additives is nevertheless of interest when preparing metal matrix composites using the production routes described in Section 4.4.1.10.

Complex shaped parts of boron carbide may be formed using classical green state-forming procedures, including slip casting [407, 408] and injection molding [409].

4.4.2.2 Hot-Pressed and Hot Isostatic-Pressed Boron Carbide

In order to obtain dense shapes and fine microstructures, hot pressing has been applied to induce grain boundary sliding, twinning, creep, and bulk diffusion accompanied by recrystallization [410–412].

The main prerequisite conditions for achieving high densities are the use of submicron powders, temperatures in the range of 2100–2200 °C, pressures of 25–40 MPa, a dwell time of 15–20 min, and a vacuum or argon atmosphere. In order to provide carbon as a sintering aid and to resist the high applied pressures at temperatures exceeding 2000 °C, carbon dies are commonly used.

Sintering aids such as boron [413, 414], carbon [396], or metals (Mg, Al, Si, Ti, V, Cr, Fe, Ni, Cu) [385–387, 405, 413] have been used to enhance densification and to decrease the sintering temperature. According to Telle and Petzow [415], the combined addition of boron and silicon or boron, silicon and titanium, enhances densification due to liquid film formation and reduces grain growth by pinning of grain boundaries due to the formation of fine-grained SiC or TiB_2. Additionally, compounds such as alumina, sodium silicate with $Mg(NO_3)_2$, Fe_2O_3, MgF_2, AlF_3 or ethyl silicate have been used to lower the sintering temperature and improve mechanical properties and toughness [379, 416, 417].

Recently, Mikijelj *et al.* [418] fabricated dense boron carbide shapes by hot pressing at temperatures as low as 1825 °C, using additive mixtures of the quasiternary system Al_2O_3–AlN–Y_2O_3. Starting with a submicron boron carbide powder (average particle size 0.6 μm), bending strength values in the range of 700–820 MPa, a maximum fracture toughness of 3.9 $MPa{\cdot}m^{1/2}$, and a hardness (HK0.3) of 25.1–26.8 GPa, were achieved. The high strength could be assigned to the fine-grained microstructure (1–2 μm average grain size). All of these values were superior to those achieved with dense hot-pressed shapes when using the same boron carbide starting powder.

Hot isostatic pressing (HIP) may be used to densify boron carbide, as well as various other difficult-to-densify materials, in order to achieve near-100% theoretical density. By using a diffusion barrier and a special boron oxide glass as a canning material, Larker *et al.* [419] produced dense boron carbide shapes without using sinter aids. As almost no grain growth occurred, strength values of 714 MPa could be obtained in a material with a 3 μm median grain size.

The post-HIP of materials with closed porosity – that is, >93% density after pressureless or uniaxial pressure-assisted heat treatment – allows for the closure of entrapped porosity, thus improving the mechanical properties. The four-point bending strength of 485 MPa for samples with 2% carbon content could be improved to 580 MPa by HIP at 2000 °C with 200 MPa [409, 420].

4.4.2.3 Spark-Plasma-Sintered Boron Carbide

Recently, SPS has attracted interest as a consolidation method for boron carbide, because the sintering temperatures can be lowered without using additives that would reduce grain growth and achieve enhanced mechanical properties compared to classical sintering routes. Kim *et al.* [421] densified commercial submicron powders at 1950 °C and obtained >99% density. Washing the material in methanol before sintering lowered the final grain size from 4.5 to 3.5 μm. A Vickers Hardness of 34.5 GPa and a toughness (ICL) of 4.7 MPa·m$^{1/2}$ (3.5 for unwashed powder) have been reported. Anselmi-Tamburini and Munir [422] obtained 95% dense submicron structured samples at sintering temperatures as low as 1700 °C, starting with elemental powders. The rapid densification rates obtained at temperatures as low as 1600 °C could be addressed to excessive crystallographic stacking faults in the grains formed below 1200 °C. A rapid densification is achieved by means of a disorder–order transformation.

4.4.2.4 Boron Carbide-Based Composites

Among ceramic matrix composites with boron carbide as matrix phase, SiC and TiB_2 are the most cited discontinuous reinforcements. Thus, the following sections will provide a review on these two classes of reinforced material.

B_4C–SiC The presence of particulate SiC limits grain growth in sintered B_4C–SiC compacts, thus leading to fine-grained materials with improved strength compared to the single-phase boron carbide materials [313, 423].

Since the thermal expansion mismatch is comparably small, there is no large toughening effect observable in B_4C–SiC particulate composites. Nevertheless, the combination of hardness and wear resistance of boron carbide with the oxidation and thermal shock resistance of SiC may be of interest for certain applications.

Different processing routes have been applied to synthesize this class of composite material. Besides the classical method of powder mixing using submicron SiC, B_4C and carbon or carbon-containing precursors, followed by forming and subsequent heat treatment [313], the infiltration of boron carbide preforms with organic precursors such as PCS, followed by heat treatment [400, 401] or the pressure-assisted reaction sintering of silicon borides with elemental carbon [423], have been used.

The microstructure of the composites fabricated by powder mixing consists of a fine-grained boron carbide matrix with isolated SiC grains, whereas the materials starting with organic precursors [425] or with silicon borides [423] exhibit a continuous silicon carbide phase located in the grain boundaries of the matrix phase.

Strength values of 380 MPa and 400 MPa have been obtained for pressureless sintered composites of the composition 75 wt% B_4C, 20 wt% SiC, 5 wt% C and 56 wt% B_4C, 39 wt% SiC, 5 wt% C, respectively. Post-HIP improved the bending strengths by 15% and 25%, respectively.

The oxidation kinetics of the composite materials compared to single-phase boron carbide has been verified by Telle [425] and Narushima *et al.* [426]. The combined monitoring of weight change and X-ray photoelectron spectroscopy (XPS) analysis [425] or *in situ* Raman spectroscopy [426] showed that, in dry systems, the oxidation is initially governed by the build-up of a boron oxide layer, whereas in humid air no stable boron oxide layer is formed due to the formation of boric acid which evaporates immediately, thus showing an increased weight loss in the initial stage. The weight loss in dry air at temperatures exceeding 1000 °C can be ascribed to the formation of gaseous boron oxide and boron suboxides. In the case of isolated SiC grains, the oxidation of boron carbide is not retarded until a liquid passivating borosilica glass is formed on the surface above 1200 °C in ambient dry air, releasing gaseous B_2O_3. In contrast, a continuous layer is formed in materials fabricated from SiC precursors, thus passivating the oxidation in the early stage of oxidation. At temperatures above 1400 °C, gaseous silicon monoxide may be released from the melt, whereas SiC is formed at the oxidation front by reaction of SiO_2 with evolving CO. The rate of oxidation in the stage of passivated oxidation obeys parabolic oxidation kinetics, which may be attributed to diffusion-controlled oxidation governed by the diffusive motion of oxygen through the liquid oxidic layer.

B_4C–TiB_2 Boron carbide-based composites with TiB_2 as the discontinuous reinforcing phase have been studied for cutting tools and wear parts by various authors [415, 427–433].

Regardless of their fabrication method, these composites exhibit improved mechanical properties due to grain refinement by pinning of grain boundaries and crack deflection and/or microcrack toughening due to stress fields related to thermal mismatch of boron carbide and transition metal diboride. Strength values of 600–870 MPa and toughness values of 3.2–7.3 MPa·m$^{1/2}$ have been reported for boron carbide/transition metal diboride composites [434, 436].

In the past, different routes have been utilized to produce dense B_4C–TiB_2 composite materials, differing in starting powder mixtures. Two main methods can be distinguished: (i) to use B_4C and TiB_2; and (ii) to make use of the reactions building up the final components of boron carbide and titanium diboride.

As B_4C and TiB_2 are thermodynamically stable up to 2300 °C, composites can be fabricated according to method (i). Nishijama and Umekawa [437] fabricated fully dense shapes by the pressureless sintering of compositions in the range 20–60 vol% TiB_2 at 2100 °C in vacuum. As a consequence, they obtained a maximum three-point bending strength of 620 MPa on composites with 35 vol% TiB_2, and a hardness HR_A of 93.8 was measured. Dense samples have been also obtained by pressureless sintering mixtures with Fe as additive at 2175 °C [437]. Composites with the addition of 1 wt% Fe exhibited a bending strength of 420 MPa at a volume fraction of 20% TiB_2. Sigl and Schwetz [438] sintered specimens at 2175 °C, followed by HIP at 2050 °C, starting with 20 and 40 vol% TiB_2 and carbon. The dense composites revealed a fracture toughness of 3.0 and 3.6 MPa·m$^{1/2}$, respectively.

Skorokhod and Krstic [435] used powder mixtures of submicron boron carbide, titanium oxide, and carbon to produce dense shapes with 15 vol% TiB_2 (reaction hot pressing at 2000 °C). As a result, they measured bending strengths of up to 630 MPa, and toughness values in the range of 6–7 MPa·m$^{1/2}$. Interfacial cracks and weak interfaces have been observed that clearly enhance toughness. Yamada *et al.* [436]

used reaction hot pressing, starting with submicron boron carbide powder (average size 0.5 μm), nanometric TiO_2 and carbon black, in order to achieve submicron-structured fully dense samples. Composites with 20 mol% TiB_2 exhibited a flexural strength of 866 MPa and toughness of 3.2 $MPa{\cdot}m^{1/2}$, compared to 582 MPa and 2.4 $MPa{\cdot}m^{1/2}$ obtained for monolithic boron carbide using the same starting powder. The increased toughness has been addressed to crack deflection and microcrack toughening.

Another method of consolidation makes use of the reaction hot pressing of TiC–B mixtures according to Eq. (23). The rapid reaction velocity due to very high local temperatures during the extremely exothermic conversion may be controlled by the addition of TiB_2 or B_4C [439]. The latter addition is beneficial due to the involvement in the overall reaction scheme, leading to a grain refinement of the involved powder. The resultant microstructure consists of micron-sized grains of both reaction products.

$$TiC + 6B \rightarrow TiB_2 + B_4C \qquad (23)$$

Another possible route of self-propagating high-temperature synthesis (SHS) starts with the elemental powders. Dudina *et al.* [440] recently produced composites by SPS, starting with elemental powders. A bulk composite material of near-full density, a fine uniform microstructure, and an increased fracture toughness was obtained by SPS at 1700 °C. Here, a fine-grained microstructure was observed with grain sizes of boron carbide and titanium diboride of 5–7 μm and 1–2 μm, respectively. Heian *et al.* [441] used the same SPS method, starting with mechanically activated elemental powders, and achieved relative densities of up to 95%. The product was composed of a mixture of nanometric crystallites and grains in the micrometer range, while transmission electron microscopy (TEM) images showed a highly defective structure containing a high density of twins.

Another promising development of B_4C–transition metal boride composite materials is the *in situ* formation of boron carbide, titanium diboride and tungsten boride during the reaction hot pressing of B_4C–B–Si–WC–TiC–Co mixtures [415, 430, 442]. Both, WC and TiC reacted with B and B_4C to form W_2B_5 and TiB_2, respectively; these reactions were strongly exothermic. Cobalt, silicon, and boron built up a boron-rich melt that was stable as a transient liquid phase up to 1600 °C, and which subsequently segregated and dissolved in the transition metal borides. The starting composition had been adjusted to end in a final composition of 72 vol% boron carbide, 20 vol% titanium diboride, and 8 vol% ditungsten pentaboride. A fine-grained microstructure of 97% density was obtained with a medium grain size of approximately 0.8–1 μm. The composites, when hot-pressed at 1820 °C, exhibited a bending strength of 830 MPa combined with a fracture toughness of 2.5 $MPa{\cdot}m^{1/2}$ and a Vickers hardness of 32 GPa (HV_{10}). The toughness could be increased at the expense of hardness and strength to values in the range of 4.2–5 MPa $m^{1/2}$ by hot pressing at temperatures exceeding 1900 °C, which in turn led to a coarsening of the microstructure. Samples undergoing HIP at 1700 °C (99.8% density) using a Si content of 7.5 wt%, exhibited remarkable mechanical properties, with a bending strength of 1130 MPa, fracture toughness of 5.2 $MPa{\cdot}m^{1/2}$, and hardness of 29 MPa having been reported. Crack deflection caused by internal stresses was identified as the toughening mechanism.

Sigl [403] utilized the reaction of boron carbide with titanium carbide to obtain high strength–high toughness composite materials. The emerging free carbon is beneficial for the reduction of oxide layers, and also enhances toughness. Starting with submicron boron carbide powder, the reaction was completed at 1250 °C, while subsequent sintering at temperatures above 2100 °C led to dense samples with closed porosity. Subsequent HIP at 2050 °C consolidated the composites to densities >99%. In this way, samples with up to 6% free carbon have been fabricated, with the range of strength and toughness values being comparable to those of boron carbide sintered on the resin route. A maximum toughness of 4.2 MPa·$m^{1/2}$, combined with a bending strength of 290 MPa, has been achieved for a material with 1.5% residual carbon. A maximum strength of 500 MPa, combined with a toughness of 2.9 MPa·$m^{1/2}$, has been measured for a maximum free carbon content of 6%, corresponding to a minimum grain size of 3 μm. Both, TiB_2 and graphite precipitates are effective as grain growth inhibitors. Indeed, a strong relationship between fracture toughness and grain size was observed to follow the characteristic dependence of increasing toughness with increasing grain size. The fall in toughness at grain sizes exceeding 10 μm can be addressed to a switch from an intergranular to a transgranular fracture.

Recently, TEM images of sintered and post-HIPed composites, starting with B_4C, TiB_2 and carbon, clearly showed the occurrence of thin, nanosized carbon interlayers at the B_4C–TiB_2 interfaces [438, 443]. This weakening of grain boundaries was accompanied by microcracking, which released stresses above a threshold value. In addition to particle toughening by crack deflection, a further increase in toughness due to the formation of microcracks has been achieved in composites with 20 and 40 vol% TiB_2. Maximum toughness values of 6 MPa have been reported for both materials.

4.4.2.5 B_4C-Based MMCs

The advantage of MMCs among metallic alloys without reinforcement has been described in detail in Section 4.4.1.10.

Liquid-phase sintering cannot be utilized to densify powder mixtures of boron carbide and aluminum, because of the very high vapor pressure of the liquid metal at reasonable temperatures above 1000 °C, where aluminum is in equilibrium with an Al-saturated $B_{12}(B,C,Al)_3$-solid solution [406]. Nevertheless, MMCs with particulate boron carbide as a dispersed phase have been obtained by infiltration of presintered or compacted boron carbide preforms with liquid metal [444], although aluminum reacts with boron carbide to form carbides, borides, and borocarbides according to the equilibrium state [406]. At temperatures in the range of 1000–1200 °C, a suitable wetting behavior is achieved within minutes of annealing. The final dense shape consists of a continuous network of aluminum alloy. Toughness values of 5–16 MPa·$m^{1/2}$ and flexural strengths of 200–680 MPa have been achieved, depending on the volume fraction of the metal. Upon subsequent heat treatment, the matrix can be hardened due to the precipitation of carbides and borides. Thus, the Vickers hardness of a composite with 31 vol% Al could be raised from 15.7 to 19.4 GPa. The reactivity of boron carbide preforms with aluminum could be lowered by a passivation step – that is, annealing of the compacted preform at temperatures in the range 1250–1800 °C. Thus, wetting is improved and undesirable reactions are suppressed [445].

It has been shown that the powder metal route can be used to produce dense, extruded shapes of composites that exhibit properties for use in structural and neutron- absorbing applications [446]. When the particles have been sufficiently mixed, they are directed first into a die, and then into a cylindrical container where the particulates are subjected to extremely high pressures that transform the elements into a solid ingot. These ingots are annealed at 625 °C to achieve the final extrudable shape. Such extrusion is carried out at temperatures in the range of 450–500 °C, and the resultant advanced metal matrix composites are 60% lighter, 30% stronger, 40–45% stiffer, and 50% higher in fatigue strength than any of the top-of-the-line 7000 series aluminum alloy materials. Depending on the ratio of boron carbide to aluminum, and also depending on the particular aluminum alloy used as the base material metal, the resultant material will have a tensile strength of 427 to 745 MPa, a yield strength of 400 to 670 MPa, a modulus of elasticity of 98 to 100 GPa, and is extremely fracture-resistant. Furthermore, the resulting material has a hardness which is comparable to that of titanium and chromoly steel, but a density which is roughly one-third that of steel and roughly 60% that of titanium. Another material mainly used in neutron-absorbing applications is commercially available as Boral®. This is produced by mixing boron carbide granules and aluminum powder inside an aluminum box, heating the box and its content to form an ingot, and then hot-rolling the ingot to form a plate consisting of a coarse core of B_4C–Al composite material bonded between two thin sheets of aluminum cladding [447, 448].

4.5 Properties of Silicon Carbide- and Boron Carbide-Based Materials

4.5.1 Silicon Carbide

4.5.1.1 Physical Properties

Color Whereas, pure α-SiC is colorless, the cubic β-modification is yellow. The only other elements that can be included in the SiC crystal lattice in amounts >1 ppm are N, Al, and B. Nitrogen gives a green color to 3C and 6H, and a yellow color to 4H and 15R. The presence of the trivalent elements boron and aluminum gives all the modifications and polytypes a blue–black color [449].

Optical Properties α-SiC is birefringent due to its crystal structure: for 6H, $n_e = 2.6669$, $n_o = 2.6273$; for 15R, $n_e = 2.6720$, $n_o = 2.6293$; for 4H, $n_e = 2.6874$, $n_o = 2.6365$ (Li 671 nm) [450]. For β-SiC, a refractive index of ~2.63 (Li 671 nm) has been reported [450].

Electrical Properties Silicon carbide is a semiconductor. The most important electronic properties of SiC are its wide energy band gap of 3.26 eV for 4H-SiC and 3.03 eV for 6H-SiC, a high breakdown electric field of $2.2 \times 10^6\,\mathrm{V\,cm^{-1}}$ for 100 V operation, and high saturated electron drift velocity of $2 \times 10^7\,\mathrm{cm\,s^{-1}}$. Doping with the trivalent elements aluminum and boron gives the SiC p-character, while the pentavalent element nitrogen produces n-character when incorporated into the SiC crystal lattice. The resistivity can be varied between 0.1 Ω·cm and 10^{12} Ω·cm, depending on the concentration of the dopings [451].

Whereas compact, homogeneous SiC obeys Ohm's law, aggregates of SiC grains show nonlinear current–voltage behavior. At low applied voltages, they behave as insulators; however, when the applied voltage is increased above a certain value the current increases exponentially. Thus, the points of contact between the grains cause the electrical resistance to be voltage-dependent, a typical effect in grain contacts of semiconducting materials [451, 452]. This nonlinear behavior is used in the electronic devices called the "varistor" (variable resistor). The current–voltage characteristic is described by Eq. (24), with typical values for the exponential factor α ranging from 5 to 7. Over a wide current range, the voltage remains within a very narrow band for a specific device, and can be referred to as the "varistor voltage" for that device. The nonlinear electrical characteristic makes the device useful in voltage regulation applications, and in particular for limiting surges and transient voltages that may appear on power lines.

$$\mathrm{I} = \pm \mathrm{K}\,(\mathrm{U})^{\alpha} \tag{24}$$

Thermal and Calorific Properties For a ceramic material, SiC has an unusually high thermal conductivity of $150\,\mathrm{W\,m^{-1}\,K^{-1}}$ at 20 °C, and $54\,\mathrm{W\,m^{-1}\,K^{-1}}$ at 1400 °C [453]. The high thermal conductivity and low thermal expansion ($4.7 \times 10^{-6}\,\mathrm{K^{-1}}$ for 20–1400 °C) explain why the material has such good resistance to thermal shock.

The specific heat capacity of SiC is $26.8\,\mathrm{J\,mol^{-1}\,K^{-1}}$ at room temperature, and $50.2\,\mathrm{J\,mol^{-1}\,K^{-1}}$ at 1000 °C. The standard enthalpy of formation ΔH^0_{298K} is $-71.6 \pm 6.3\,\mathrm{kJ\,mol^{-1}}$, and the entropy S^0_{298K} is $16.50 \pm 0.13\,\mathrm{J\,mol^{-1}\,K^{-1}}$ [454].

Mechanical Properties Silicon carbide is noted for its extreme hardness [204, 455–457], its high abrasive power, high modulus of elasticity (410 GPa), high temperature resistance up to above 1500 °C, as well as a high resistance to abrasion. The industrial importance of SiC is mainly due to its extreme hardness of 9.5–9.75 on the Mohs scale; only diamond, cubic boron nitride, and boron carbide are harder. The Knoop microhardness, HK0.1, is ~26GPa (20GPa for α-Al_2O_3, 30GPa for B_4C, 47GPa for cubic BN, and 70GPa–80GPa for diamond). Silicon carbide is very brittle, and can therefore be crushed comparatively easily, in spite of its great hardness. Some typical physical properties of the SiC ceramics are summarized in Table 4.5.

As the grain size, pore content, and chemical composition of the various ceramic products differ considerably, it follows that the properties are also different.

Recrystallized SiC is much stronger than ceramically bonded material, but its high residual porosity imposes limits as far as mechanical strength is concerned [458].

Table 4.5 Physical properties of various SiC ceramics.

Material type	SiC content (%)	Density ($g\,cm^{-3}$)	Porosity (%)	Young's modulus (GPa)	Thermal expansion 30–1000 °C ($10^{-6}\,K^{-1}$)	Thermal conductivity 30–100 °C ($W\,m^{-1}\,K^{-1}$)	Flexural strength 20 °C (MPa)
Nitride bonded SiC (N–SiC)	75–85	2.6–2.83	10–15	150–240	4.5–5	14–17	180–200
Recrystallized SiC (RSiC)	100	2.6–2.8	15–20	230–280	4.8–5	25–30	80–120
Infiltrated SiC (SiSiC)	80–90	3–3.12	0	270–400	4–4.8	100–160	180–450
Sintered and hot pressed SiC (SSIC, HPSiC, HIP-SiC, HIPSSiC)	>98	3.15–3.21	<2	350–450	4–4.8	40–150	260–640
Liquid phase- sintered SiC (LPS-SiC)	>95	3.20–3.24	<1	420–460	4–4.5	90–100	500–800

Reaction-sintered SiSiC is still stronger, but only up to 1400 °C, the softening point of the accompanying silicon phase [459].

The best mechanical strength is exhibited by pressureless sintered SiC, hot-pressed [460, 461] and liquid-phase-sintered materials (see Sections 4.4.1.4, 4.4.1.5).

Solid-state-sintered, hot-pressed, and isostatically hot-pressed materials offer considerable advantages over all other ceramic materials in plastic deformation under a sustained load (creep), because of the low content or almost complete absence of sintering aids [462, 463].

4.5.1.2 **Chemical Properties**

One of the outstanding characteristics of SiC is its oxidation resistance, which is due to the high affinity of silicon for oxygen. The reaction of silicon with oxygen in an aqueous medium causes passivity and, if exposed to thermal oxidation (e.g., in air), this leads to the formation of glassy silica films [464–466]. The oxidation of pure SiC begins at around 600 °C, forming a coating of SiO_2 on the surface that prevents further oxidation [467] [see Eq. (25)]. The reaction rate varies with time according to a parabolic law [464]. The kinetics are determined by the diffusion of oxygen through the SiO_2 layer, and the temperature-dependence of oxidation is of an Arrhenius type.

$$SiC + 2O_2 \rightarrow SiO_2 + CO_2 \qquad (25)$$

The "active" oxidation of SiC is distinguished from the "passive" oxidation reaction described in Eq. (26). Active oxidation takes place under conditions of oxygen deficiency above 1000 °C, and leads to decomposition of the SiC and the formation of silicon monoxide [465, 467]. A review on the oxidation of SiC and the conditions for the boundary has been recently provided by Nickel and Prescher [468]. A likely high-temperature boundary for SiC is ~1700–1800 °C, where a secondary active-to-passive transition by bubble formation and spallation occurs.

$$SiC + O_2 \rightarrow SiO + CO \qquad (26)$$

Thus, SiC is attacked and decomposed by oxidizing agents (e.g., $Na_2CO_3 + Na_2O_2$ or $Na_2CO_3 + KNO_3$) if the protective layer of SiO_2 is removed, thereby enabling the reaction to proceed unhindered.

Pure SiC is twice as oxidation resistant, even at 1500 °C, than the best current superalloys at their maximum service temperature of 1200 °C.

Silicon carbide is resistant to most chemicals, resisting acids and alkalis, and even aqua regia and fuming nitric acid [205].

A mixture of hydrofluoric acid, nitric acid, and sulfuric acid slowly attacks SiC, with β-SiC being somewhat more reactive than α-SiC [469]. However, a complete dissolution takes place only if the SiC is very finely divided, under pressure, and at an elevated temperature (e.g., 250 °C for 16 h). Alkali melts will attack SiC in the presence of oxidizing agents [470, 471]; indeed, oxides, molten metals, and water vapor all have destructive effects on SiC at temperatures exceeding 1000 °C. Chlorine reacts exothermally with SiC above 800 °C, with the formation of silicon tetrachloride ($SiCl_4$) and carbon.

Silicon carbide behaves in various ways towards molten metals. It is not attacked by molten zinc or zinc vapor [472], but molten aluminum will attack SiC slowly to form Al_4C_3 and silicon; however, as the silicon content increases the reaction will eventually cease because an equilibrium is established [473]. Molten iron dissolves SiC to form iron carbide and iron silicide.

4.5.1.3 Tribological Properties

The term "tribology" is derived from the Greek τϱίβω ("tribo") meaning "I rub". Tribology is the science and technology of interacting surfaces in relative motion. It includes the study and application of the principles of friction, lubrication, and wear on the macroscopic and microscopic scale.

Review on Tribology of Sintered α-SiC Ceramics It was during the early 1980s that the advantage of sintered SiC as a seal face material was first realized [474, 475], and since then the mechanical engineering industry has focused attention on this material. Notably, numerous tribological studies have been carried out to better understand those physical and chemical properties of SSiC that will affect its behavior when in contact with itself, and with other ceramics or metals. Model investigations involving SiC have shown the coefficients of friction against various materials – and even against itself – to be a function of contact stress [476], and have documented the anisotropic wear behavior of monocrystalline SiC crystals [477]. Lashway *et al.* [478] found, that a controlled amount of porosity improved the ability of sintered-SiC to retain a hydrodynamic film with a lower friction. Seal tests have also indicated a lower power dissipation at varying pressure–velocity values for combinations with sintered-SiC seal face materials.

Consistently positive practical experience gathered by Knoch *et al.* on sintered-SiC seal rings [479, 480] has helped the material to gain popularity rapidly for use in situations involving wear problems.

Excellent results have been achieved in developing sliding bearings for hermetically sealed pumps [481]. All of these are absolutely leak-proof, whether in operation or shut down, and they are therefore of great value environmentally. Moreover, they all share the same design requirement – that the sliding bearing of the pump shaft must be flushed and lubricated by the pumped medium. Whilst traditional seal materials were unable to cope with the harsh conditions of the chemical industry, and were rapidly destroyed by both corrosion and abrasion, sintered SiC proved to be an excellent solution to the problem. Moreover, liquid-containing abrasive particles do not restrict the use of sintered SiC.

Many sliding-wear problems that occur in the field are attributable to the interruption of ideal (i.e., properly lubricated) running conditions. In such cases, the sliding faces of the bearing or seal in question make contact with each other, giving rise to a solid-state or dry friction that is marked by a pronounced increase in the coefficients of friction. Local frictional heat leads to peak thermal stresses that may be of such intensity as to cause a breakout of the microstructural constituents. Then, when lubrication (and cooling) is restored, the material is in danger of cracking or fracturing due to thermal shock. Sintered SiC, however, is better able to cope with such situations than other ceramic materials, because it is stronger and has a lower thermal expansion coefficient, and also a higher thermal conductivity. Consequently, SiC can even survive brief periods of dry running.

The nonlubricated wear behavior of sintered SiC under unidirectional sliding at room temperature was studied by Derby *et al.* [482], as a function of load and sliding time. At low loads, polishing and plowing mechanisms were observed, and the microstructure revealed an etched appearance. With increased loads and sliding times, microcracking at the grain boundaries occurred, leading to subsequent grain-pullout. In the study of Cramner [483] it was shown that, for sintered SiC sliding against itself, surface plastic deformation, plowing and cracking were also operative as wear mechanisms. Miyoshi *et al.* [484] carefully studied the friction and wear of the SiC/iron pair in a vacuum environment at temperatures up to 1500 °C. By using X-ray photoelectron spectroscopy, the surface chemistry of sintered SiC (graphite and SiO_2), as well as changes in the chemistry due to the increased temperature, were shown to be highly influential and to dramatically change the friction coefficient. Breznak *et al.* [485] quantified the role of the initial surface roughness on the frictional behavior of sintered SiC seals rubbing against each other in a cyclic oscillatory motion. As a result of the consequent surface polishing, the coefficient of friction was shown to decrease to 0.25, from an initial value of 0.40. Further studies by Breval *et al.* [486] indicated that the addition of graphite to SiC reduced the wear of the SiC/SiC couples, but that improving their initial surface finishes had the opposite effect and led to a greater loss of material during the "running in" stage. The wear debris of SiC/SiC couples exhibited a bimodal particle size distribution; some particles were micrometer-sized, while others ranged from 5 nm to 50 nm. By using pin-on-disk and abrasion wheel testing, Wu *et al.* [487] reported that the amount of wear would increase with increasing grain size for sintered SiC.

Smythe and Richerson [488] conducted experiments to study the dynamic sliding contact behavior at temperatures greater than 1000 °C, and found that surface film formation at higher temperatures governed the frictional behavior. It is believed that SiO_2 or SiO_2 modified by inherent impurities or from the environment acts as a lubricating film, thus substantially decreasing the friction coefficient [489].

When Tomizawa *et al.* [490] performed pin-on-disk experiments to determine the friction of SSiC against itself in water at room temperature, they measured a friction coefficient of 0.26, and also noted that the wear of sintered SiC occurred by a combination of tribochemical dissolution and the formation of pits by fracture of the SiC grains. The amount of material removal varied from one SiC grain to the other, due to a strong dependence of tribochemical wear on the crystallographic orientation of the SiC grains.

Knoch and Kracker [491] observed that such anisotropic tribological behavior was pronounced with sintered SiC, which has a bimodal grain structure – that is, about 30 vol% of larger platelets of hexagonal α-grains (ca. 100 μm length) and about 70 vol% of smaller α-grains (ca. 10 μm length). This material showed a superior performance, particularly when paired against a softer carbon material. A relief structure developed in the water-lubricated surface, while the depressions in such a textured surface clearly served as reservoirs for lubricant, thus improving the emergency running properties of the components. Boch *et al.* [492] studied the dry friction behavior of SiC/SiC couples (fixed-ball rotating disk) at temperatures ranging from 20 to 1000 °C, and observed a rapid decrease in the coefficient of friction from 0.45 to 0.16 when the temperature exceeded 400 °C. Between 400 and 700 °C, the debris agglomerated in the form of SiO_2 rolls, which arranged themselves perpendicular to the sliding direction. These rolls subsequently acted as minute roller-bearings, decreasing the coefficient of friction and preventing the formation of cracks in the wear track.

Habig and Woydt [493] studied the unlubricated friction and wear of self-mated SSiC sliding couples at temperatures between 22 and 1000 °C, using a stationary pin and a rotating disk assembly. At room temperature, the coefficient of friction decreased with increasing sliding velocity, from 0.8 to 0.6, but at temperatures of 400 °C and over a scatter of friction coefficients between 0.2 and 0.7 was observed. By using small-spot-ESCA (electron spectroscopy for chemical analysis), the low friction coefficients of SiC/SiC were shown to be due to the formation of thin oxide layers formed by tribo-oxidation, while the higher friction coefficients accompanied by higher wear coefficients were related to thicker oxide layers. These results were in excellent agreement with those of Yamamoto *et al.* [494], who observed a decrease in the coefficient of friction after heating SiC/SiC couples for 1 h at 1000 °C, with the formation of a thin oxidation layer. At a higher oxidation level, however, the coefficient of friction increased rapidly. A low coefficient of friction for the SiC/SiC couple was also measured by Martin *et al.* [495], who monitored friction under an oxygen partial pressure of 50 MPa, which should permit the formation of only a very thin oxidation layer.

When Sasaki [496] studied the influence of humidity on friction and wear behavior of SiC/SiC couples, a humid atmosphere was seen to reduce the coefficient of friction from 0.5 (dry air) to 0.2 (wet air), whilst a simultaneous decrease in the coefficient of wear was observed, from $10^{-5}\,mm^3\,N^{-1}\,m^{-1}$ in dry air to $10^{-6}\,mm^3\,N^{-1}\,m^{-1}$ in humid air.

Denape and Cannon [497] considered the important contribution of wear particles towards the wear mechanisms, and showed that polishing at low loads (5 N) was due to fine (<1 μm) individual wear particles circulating in the sliding interface, whereas abrasion and grain pull-out (at high loads of 20 N) were associated with an accumulation of large particles adherent to the sliding phases. Subsequent wear tests conducted under water also showed an anisotropic wear of the individual SiC crystallites. The authors concluded that the circulation of wear debris in the sliding interface controlled both the wear rate and the friction response.

In a study on the friction and wear behavior of lubricated ceramic journal bearings, Maurin-Perrier *et al.* [498] showed that both SiC/SiC and SiC/Si_3N_4 couples provided a significantly better behavior in terms of film stability compared to the classical materials used for water- or hydrocarbon-lubricated bearings. A self-improvement of surface roughness during the running-in period by a tribochemical reaction significantly increased the range of stability of the film; consequently it was considered unnecessary to carry out highly expensive surface finishing to increase the performance of the SiC/SiC couple. The lifetime of the components is determined mainly by the behavior during the start and stop phases, where boundary lubrication occurs. Sliding wear on sintered SiC leads to a very smooth surface between the residual porosity. It was also found [499] that, in dry run situations, a sintered SiC of bimodal grain structure (EKasic D) would outperform other SiC materials. The bimodal EKasic D showed the lowest development of heat, and thus it had the lowest coefficient of friction. In a major US study on "Tribological Fundamentals of Solid Lubricated Ceramics" [500], the bimodal grain size/shape distribution was judged best in terms of wear resistance, when compared to other sintered α-SiC materials.

Löffelbein *et al.* [501] tested self-mated sliding couples of SSi_3N_4, HIP-$RBSi_3N_4$, SSiC, SiSiC, MgO–ZrO_2 and Al_2O_3 in different aqueous solutions (H_2O, NaOH, KOH, NH_4OH, HNO_3, H_2SO_4, H_3PO_4, CH_3COOH, HCl, $HClO_4$) under conditions of boundary or mixed lubrication. The best frictional behavior was observed with couples of SSiC, with steady-state friction coefficients of 0.05. The lowest wear coefficients were measured for couples of the two types Si_3N_4, SSiC and Al_2O_3, with values of approximately $10^{-7}\,mm^3\,N^{-1}\,m^{-1}$. If low friction and low wear are required, then couples of SSiC appear to be the best option in most aqueous solutions. In a study conducted by Anderson and Blomberg [502], based on tests with sintered SiC sliding unlubricated on itself in point (pin-on-disc), line (journal bearing), and plane (mechanical face seal) contacts, tribo-oxidation and surface fracture were identified as the dominant deterioration mechanisms. The oxidation products formed were SiO_2 and, within narrow operational regimes, SiO. The highest wear rates occurred in the pin-on-disc configuration, and the lowest rates in the journal bearing tests.

Kitaoka *et al.* [503] investigated the effects of temperature and sliding speed on the tribological behavior of sintered SiC (0.1 wt% B/1.0 wt% C), by sliding on the same material in deoxygenated water from room temperature to 300 °C under high vapor pressures (120 °C/2 bar, 300 °C/85 bar). The wear mechanism appeared to consist of the dissolution of reaction products such as silica and the hydrothermal oxidation of SiC, according to Eqs. (27–30). Fine, mirrorlike, worn surfaces were observed without wear debris under all sliding conditions.

$$SiC + 2H_2O \rightarrow SiO_2 + CH_4 \tag{27}$$

$$SiC + 4H_2O \rightarrow SiO_2 + CO_2 + 4H_2 \tag{28}$$

$$SiC + 2H_2O \rightarrow SiO_2 + 1/2\,C_2H_6 + 1/2\,H_2 \tag{29}$$

$$SiC + 2H_2O \rightarrow SiO_2 + C + 2H_2 \tag{30}$$

Sintered SiC: Material Development in Sliding Wear Application The coefficients of friction of hard and wear-resistant ceramic materials, and of most surface coatings, are always greater than 0.1 under dry running conditions. However, this value is far too high for the design of dry-running bearings since, if frictional heat were to develop and the loads were high, then very high temperatures would be generated.

Although these would not affect the sintered SiC, the housing and overall structure might well be distorted and damaged, leading to bearing failure [504]. As materials research has shown it impossible to develop hard ceramic materials with a coefficient of friction <0.01 (which is the typical value for a lubricated bearing), then the target for product development must include:

- stabilizing the hydrodynamic lubrication film;
- preventing dry running; and
- reducing friction and wear if the lubrication film breaks down.

There are clearly two routes towards stabilizing the hydrodynamic film. The first route is to optimize the design of the component, including the most suitable bearing surface characteristics. In other words, the structure of a hydrodynamic load-bearing film can be positively influenced by the selective introduction of surface texture, thus offering a means of reducing the frictional forces and wear rates. Potential applications include particularly those tribological contacts with higher relative velocities. The second route is to optimize the SiC material itself by tailoring the microstructure of the SSiC.

The latter has been achieved by introducing a bimodal coarse-grained microstructure, thus improving corrosion resistance and shifting disintegration of the grains towards higher stress parameters, or by introducing isolated pores [505] which would serve as lubricant stores in a period of low lubricant supply, and increase the applicable tribological load (= surface pressure·sliding velocity, p·v). An alternative would be to introduce graphite particles, to serve as a dry lubricant.

Figure 4.14 shows the microstructure of the materials EKasic C, EKasic P, and EKasic G. The microstructure of EKasic C consists of SiC grains in the range of 10 to 1500 μm, while EKasic P exhibits a fine-grained microstructure (typically <5 μm) with isolated pores in the range of 50 to 200 μm (median diameter 70–80 μm). EKasic G shows a bimodal grain size distribution in the range of 10 to 1000 μm, with homogeneously distributed graphite particles (2 wt%) in the range of 50 to 120 μm. The properties of these SiC materials are listed in Table 4.6.

EKasic G exhibits a reduced coefficient of friction and an improved resistance to wear in the event of boundary lubrication. This allows for higher p·v-values in a graphite-loaded material when compared to porosity-included, fine-grained, and coarse-grained materials.

In applications with hot water, where the chipping out of fine grains in hot spots, the tribochemical reaction with water and the formation of damaging SiO_2 layers on the sliding surfaces reduces the time before failure substantially, the coarse-grained bimodal microstructures are the materials of choice [506]. Any catastrophic failure of components would be avoided as the large SiC platelets close to the surface are anchored to a depth at which there is no grain-boundary corrosion.

The effectiveness of the coarser microstructure in improving corrosion resistance has been clearly demonstrated in practical tests on a mechanical seal test rig when, even after a 500 h test session (deionized water, 60 °C, 6 bar, hard/hard couples), no SiO_2 layer had formed on the functional surface. Coarse-grained SSiC microstructures have also been shown to improve service life in the regime of mixed lubrication,

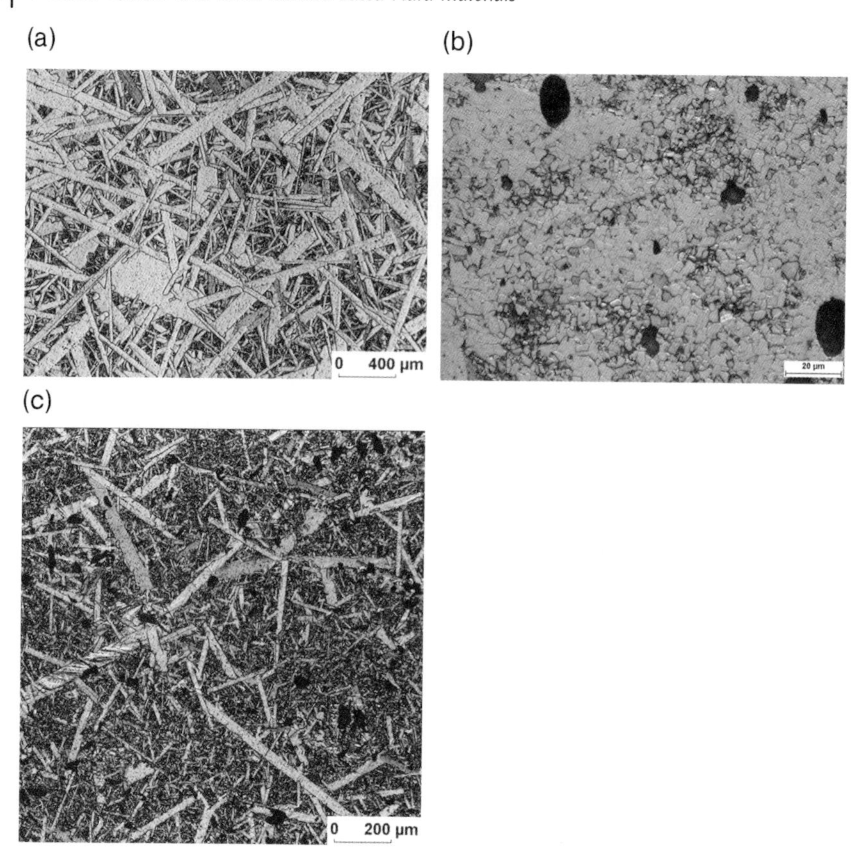

Figure 4.14 Microstructure of different SiC ceramics developed for tribological applications. (a) EKasic C; (b) EKasic P; (c) EKasic G.

when compared to fine-grained SSiC. This can be explained by the relatively high depth of roughness under steady-state conditions, which acts as a reservoir for residual liquid in the seal gap in case of any breakdown of the hydrodynamic film (see Figure 4.15).

Table 4.6 Physical properties of SiC ceramics used in tribological applications.

Material type	Density ($g\,cm^{-3}$)	Porosity (%)	Young's modulus (GPa)	Thermal expansion 20–500 °C [$10^{-6}\,K^{-1}$]	Thermal conductivity 20 °C ($W\,m^{-1}\,K^{-1}$)	Flexural strength 20 °C (MPa)	Knoop hardness HK0.1 (GPa)
EKasic C	>3.15	0.5–0.6	410	4.0	110	400–430	25
EKasic G	>3.0	<3	410	4.1	110	250–300	23
EKasic P	2.70–2.90	10–14	350	4.0	90	210–230	22

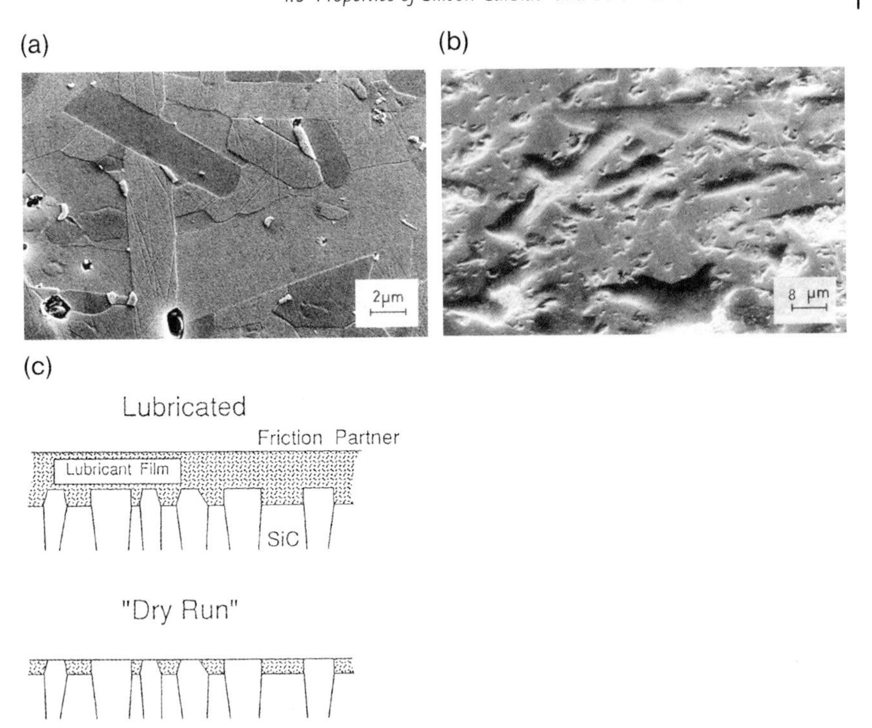

Figure 4.15 (a) SEM image of a polished and plasma-etched section of sintered SiC with coarse bimodal grained microstructure (EKasic); (b) SEM image of sliding face after ~ 4000 h of successful service as a mechanical seal face (SiC/carbon pair, lubricated with water, maximum pressure 50 bar); (c) Schematic of the operational condition of a relief-structured sliding surface in the lubricated state (above), and after breakdown of the lubricant film (below).

An outstanding mechanical strength of more than 550 MPa has been demonstrated by liquid-phase-sintered EKasic T (mean grain size <2 μm), and its high fracture toughness, fine microstructure and edge stability make it the material of choice for gas seal rings. Under operating conditions, gas seal rings must be separated by the gas film at a sliding velocity of 0.5 m s^{-1}, and such seal rings typically operate at 200 m s^{-1}. Whilst common gas film thicknesses range from 3 to 5 μm, the gap between the seal rings must not exceed 5 μm. Consequently, the ceramic material must have a low surface roughness, and therefore fine-grained LPS SiC is the material of choice.

The introduction of a surface texture by laser ablation represents an effective design measure for developing more efficient sealing systems [507]. The application of a laser allows the preparation of either open (especially for sliding bearings) or closed structures (mainly for mechanical seals), with dimensions down to the micrometer range. Tribological tests on a test rig [60 °C deionized water, pressure (p) = 48 bar, velocity (v) = 3.8 m s^{-1}] with self-paired components, revealed still-intact structures and no detectable wear after 16 h for the textured component (circular structures of 200 μm diameter). In contrast, the untextured surface showed an increased number of sliding traces and deposits under similar test conditions. Additionally, a reduction of the long-term energy consumption by a factor of two to three has been observed [506].

A somewhat different approach to improve the mixed-friction behavior of hard/hard-pairs is the application of tribologically active coating systems. For example, a DLC (diamond-like-carbon) coating leads to lower coefficients of friction and less friction-induced heating. Consequently, the service life in the event of dry running is increased. A second benefit is a comparably low wear rate. However, because the layer thickness is limited to maximal 15 μm by virtue of the coating process, DLC coating is primarily beneficial during the initial running-in phase and cannot offer a long-term solution for dry running [508].

Compared to DLC coatings, a carbon coating based on the *in situ* formation of carbon from the SSiC substrate material [509], which is adjustable in thickness (10–50 μm) and may be texturized by means of laser beam in a reactive gas atmosphere [510], exhibited an increase in life-time (the time to reach a frictional-induced temperature of 160 °C) by a factor of 10 ($v = 1.3\,m\,s^{-1}$; $p = 0.2\,N\,mm^{-2}$; 45–55% relative humidity). However, as increasing the p·v-values decreases the lifespan of the integral carbon coating, polycrystalline diamond coatings are currently tested on diverse SSiC materials and under varying testing conditions. Yet, the preliminary results achieved with self-paired coarse-grained SSiC (10 μm layer thickness) have shown much promise [506]. Under soft testing conditions ($v = 1.3\,m\,s^{-1}$; $p = 0.2\,N\,mm^{-2}$) and a very dry atmosphere (1.8% relative humidity), the coefficient of friction was reduced to 0.6 compared to 0.85 for uncoated, self-paired coarse-grained SSiC with the same surface quality. As a consequence, the 160 °C temperature limit of the test rig was reached after 20 min compared to 5 min for the uncoated components. Moreover, the diamond layer showed no signs of damage. Under hard test conditions ($v = 3.9\,m\,s^{-1}$; $p = 0.4\,N\,mm^{-2}$), the DLC- and carbon coatings were both worn after only a few minutes. However, the polycrystalline diamond coatings did not show any decrease in thickness; rather, they survived the initial running-in phase and thus offered a high potential for long-term use in practice. Burgmann Industries GmbH & Co. KG has recently announced the commercialization of diamond-faced SiC seals that offer outstanding protection against wear during dry-running in applications with mixed friction, and under exposure to abrasive media.

4.5.2 Boron Carbide

4.5.2.1 Physical Properties

Optical Properties Ion beam-deposited boron carbide, as well as ion beam-deposited SiC, are very attractive as coatings on optical components for instruments for space astronomy and earth sciences operating in the extreme-UV (EUV) spectral region, because of their high reflectivity, which is significantly higher than any conventional coating below 105 nm. Boron carbide shows a normal incidence reflectance of 30–40% in the EUV region (70–200 nm) [526].

The complex refractive index in the EUV region has been determined by Larruquert and Keski-Kuha [527] for hot-pressed boron carbide and ion beam-deposited B_4C, respectively. The refractive index and extinction coefficient for hot-pressed boron carbide increased from 0.5 to 1.77 and from 0.41 to 2.05, respectively (wavelength range 49–121 nm). Ion beam-deposited boron carbide showed similar values, the extinction coefficient being slightly lowered (from 0.45 to 1.72).

Electrical Properties Icosahedral boron-rich solids contain high concentrations of well-defined intrinsic structural defects. These defects include missing or incomplete occupation of atomic sites, the statistical occupation of equivalent sites, or antisite defects.

Theoretical band structure calculations performed to date have disregarded the fact that the high defect concentrations in the icosahedral boron-rich structures far exceed a negligibly weak disorder. Rather, they predict a metallic character in consequence of high valence electron deficiencies, yet experimental results have demonstrated a semiconducting behavior (see Table 4.7).

This apparent contradiction has been clarified, however. At boron carbide (homogeneity range $B_{4.3}C$ to $B_{\sim 11}C$), there is a quantitative correlation between intrinsic point defects and electron deficiencies [511]. Point defects in semiconductors generate split-off valence states in the band gap. Accordingly, the agreement of the concentration of experimentally proved point defects with calculated electron deficiencies implies that, in this way, the unoccupied valence states of the idealized structures are exactly compensated in the real solids. Hence, the real valence bands are completely filled, and the solids are semiconductors in accordance with experiment. This implies that the electron deficiency evokes the generation of compensating intrinsic defects, probably for energetic reasons [511, 512, 526].

Such high concentrations of gap states attached to the valence band essentially affect the electronic charge transport; in particular, they are responsible for the p-type character and the very low electrical conductivity. Aside from the electric conductivity in extended band states, a hopping-type conduction must be expected in localized gap states. The electronic properties of boron carbide can be consistently described by a band scheme, which highlights deep energy levels in the band gap (2.09 eV) at 0.065, 0.18, 0.47, 0.77, 0.92 and 1.2 eV (values based on optical measurements), related to the valence band edge. This allows the largely consistent description of all reliable experimental results [537].

The electrical conductivity of boron carbide can be satisfactorily described by Mott's law of variable range hopping in a large range of temperature. This transport clearly takes place within the high-density gap states. At high temperatures, thermally activated carriers remarkably contribute to the charge transport. This superposition of hopping and band conduction was separately proved at lower temperatures by analyzing the dynamic transport of boron carbide determining the Fourier transform infrared (FTIR) spectra [513, 514]. The conductivity was shown to range from approximately $1\ S\ m^{-1}$ at room temperature to $600\ S\ m^{-1}$ at 1700 °C.

The doping of semiconductors is important for tailoring their electronic properties in applications. Boron carbide is p-type because of its high concentration of largely

Table 4.7 Boron carbide: Comparison between theoretical electronic properties, experimental characterization and intrinsic point defects determined experimentally. After Werheit [516].

	Idealized structure				Real structure		
Structure	**Valence states [(unit cell)$^{-1}$]**	**Reference**	**Valence electrons [(unit cell)$^{-1}$]**	**Electronic character: theoretical**	**Electronic character: experimental**	**Intrinsic point defects [(unit cell)$^{-1}$]**	**Reference(s)**
B13C2 B12(CBC)	48	[522]	47	Metal	Semiconductor	0.97(5)	[533, 534]
B4.3C B11C(CBC)	48	[522]	47,83	Metal	Semiconductor	0.19(1)	[533, 534]

unoccupied gap states close to the valence band. Therefore, an overcompensation to n-type requires a high concentration of doping elements. The systematic doping of boron carbide has become known for few elements only [512]. However, whilst n-type conductivity was not realized, the Seebeck coefficient occasionally became considerably higher. In fact, because of the high concentrations of unoccupied gap states ($\sim 10^{21}\,\mathrm{cm}^{-3}$), it is questionable as to whether the n-doping of boron carbide is possible at all. Systematic investigations on enhancing the p-type character of icosahedral boron-rich structures, for example, by the substitution of regular B atoms by Be, have been missing to date. In the case of boron carbide, attempts at doping have been reported for numerous elements (e.g., H, He, Mg, C, Si, N, P, O, Al, Cr, Fe; see Ref. [526], and for Fe, V, P; see Ref. [530]).

The electronic application of boron-rich solids requires their properties to be highlighted in comparison with classical semiconductors; an example is the high Seebeck coefficient of boron carbide, which increases monotonously up to more than 2000 K. Unfortunately, as mentioned above, the high concentrations of unoccupied gap states prevent compensation to n-type in boron carbide. A more detailed review of this topic is available in Ref. [516].

The suitability of materials for thermoelectric devices is to be checked on the basis of their theoretical efficiency Eqs. (31)–(33). To optimize the factor $T_{high} - T_{low}/T_{high}$, which is the theoretical efficiency of a Carnot machine, materials allowing application at very high temperatures are required. The factor $(M-1)/M + T_{high}/T_{low})$ essentially depends on z containing the transport properties, which are relevant for thermoelectrical applications. Particular attention should be given to the fact that, towards high temperatures, semiconductors usually become intrinsic; this means that their charge transport is essentially determined by electrons and holes, both thermally excited across the band gap. In this temperature range doping becomes ineffective, the Seebeck effects of electrons and hole largely compensate each other, and hence the resulting efficiency becomes rather small.

$$\begin{aligned}\eta &= [(T_{high} - T_{low})/T_{high}] \quad [(M-1)/(M + T_{high}/T_{low})]\, M \\ &= [1 + z\,(T_{high} - T_{low})/2]^{1/2}\end{aligned} \tag{31}$$

$$M = [1 + z\,(T_{high} - T_{low})/2]^{1/2} \tag{32}$$

$$z = S^2\,\sigma/\varkappa \tag{33}$$

The dependence of Seebeck coefficients on temperature and composition has been recently summarized by Werheit [517]. The essential result is that S weakly varies between 100 and 300 $\mu\mathrm{V\,K^{-1}}$ within the homogeneity range, the minimum close to $B_{13}C_2$, and monotonously increases up to at least 2000 K. There is no remarkable indication of intrinsic behavior in this range. The reason is the extraordinary high concentration of gap states generated by structural defects [518] and pinning the Fermi level up to very high temperatures.

Some of the doping elements hitherto investigated increase S considerably. Both, Si [519, 520] and Al [521] are accommodated in the chain-free (B⌴B) elementary cells of boron carbide, and therefore their maximum concentration is limited. For Al, S varies from 250 to 450 μV K^{-1} in the compositional range 0 to 1.5 wt% Al.

Thermophysical Properties Over the whole homogeneity range, the thermal conductivity is very low in the range of from10 to 30 W m^{-1} K^{-1} at $B_{10}C$ and $B_{4.3}C$, respectively, and it remains low up to the limit of the hitherto available measurements at about 1100 K [523, 524] (also H. Werheit and N. Kuss, unpublished results). This is not surprising, when the enormous density of structural defects is taken into account.

The thermal expansion coefficient for sintered $B_{4.3}C$ is 4.5–4.6 × 10^{-6} K^{-1} (25–800 °C).

Calorific Properties The specific heat capacity of B_4C is 26.8 J mol^{-1} K^{-1} at room temperature, and 121.82 J mol^{-1} K^{-1} at 1000 °C. The standard enthalpy of formation $\Delta H^0_{298\,K}$ is −62.0 kJ mol^{-1}, and the entropy $S^0_{298\,K}$ is 27.11 J mol^{-1} K^{-1} [525].

Neutron-Absorbing Capability Boron carbide is a neutron-absorbing material used to control the reactivity of nuclear reactors by taking advantage of the nuclear reactions shown in Eqs. (34) and (35). The abundance of ^{10}B in naturally occurring Boron is ~20%. The neutron capture cross-section for thermally activated neutrons (0.025 eV) is 3850 barn (1 barn = 10^{-28} m^2).

$$^{10}B + {}^{1}n \rightarrow {}^{7}Li\ (0.84\ MeV) + {}^{4}He\ (1.47\ MeV) + \gamma(0.48\ MeV)\ 94\% \quad (34)$$

$$^{10}B + {}^{1}n \rightarrow {}^{7}Li\ (1.02\ MeV) + {}^{4}He\ (1.78\ MeV)\ 6\% \quad (35)$$

Mechanical Properties The variation of mechanical properties cited in the literature can be strongly attributed to microstructural variations in the carbon content, grain size, inhomogeneities, and residual porosity. The values given here are taken from reviews given by Lipp [132], Thevenot [137, 528, 529], and Greim and Schwetz [530].

The room-temperature elastic constants of single crystals with the stoichiometry $B_{5.6}C$ synthesized by the optical floating zone method have been measured using a resonant ultrasound spectroscopy technique [531]. The single crystal elastic constants indicate the strong anisotropy of the structure: $c_{11} = 542.81$, $c_{33} = 534.54$, $c_{13} = 63.51$, $c_{12} = 130.59$, and $c_{44} = 164.79$ GPa, respectively.

The room temperature isotropic elastic moduli of boron carbide show that its bulk, shear and Young's moduli are substantially higher than those of most solids. A theoretical value of Young's modulus of 483 GPa has been calculated by Fransevitch *et al.* [532]. Sintered carbon-rich boron carbide exhibits a Young's modulus of 440–460 GPa at room temperature, which decreases only slightly with temperature (400 MPa at 1000 °C). Schwetz *et al.* [409] reported on the influence of carbon content on the mechanical properties of post-HIPed boron carbide materials. The Young's modulus decreased from 444 to 412 GPa by increasing the free carbon content from 4 to 6.5 wt%.

The anisotropy in Knoop hardness has been determined by Werheit *et al.* using single crystals grown by the floating zone method ($B_{4.3}C$). The hardness values HK0.1 of the surfaces perpendicular and parallel to the c direction are 36.7 and 33.8 GPa, respectively. These substantially higher values compared to polycrystalline samples have been attributed to the perfect structure free of graphite precipitates and twinning. The hardness of polycrystalline boron carbide depends heavily on indentation load, grain size, porosity, and free carbon content. Using small indentation loads, the hardness values represent single grain values; hardness values HK0.01 of 48 GPa have been reported by Thevenot *et al.* [137]. With increasing load, the hardness approaches a constant value, representing an indentation integrating across the average microstructure. A Knoop hardness HK0.1 of 29 GPa has been reported by Schwetz *et al.* [409] for post-HIPed samples with an average grain size of 3.9 μm (doped with 2 wt% carbon black). The hardness continuously decreases with increasing carbon content towards 26.8 GPa at 5 wt% carbon black.

Low fracture toughness values that increase continuously with increasing grain size, from 2 MPa·m$^{1/2}$ (2 μm) to 4 MPa·m$^{1/2}$ (10 μm) have been reported by Schwetz *et al.* [409].

Schwetz *et al.* [409] investigated the influence of carbon content and sintering temperature on flexural strength of sintered and post-HIPed boron carbide shapes. A bending strength of 480 MPa for dense shapes sintered at 2175 °C (3.6 wt% free carbon, 98% of theoretical density) has been achieved. By post-HIP, the strength could be raised to 580 MPa. The strength varies strongly with carbon content and sintering temperature (Figure 4.16). Discontinuous grain growth has been observed

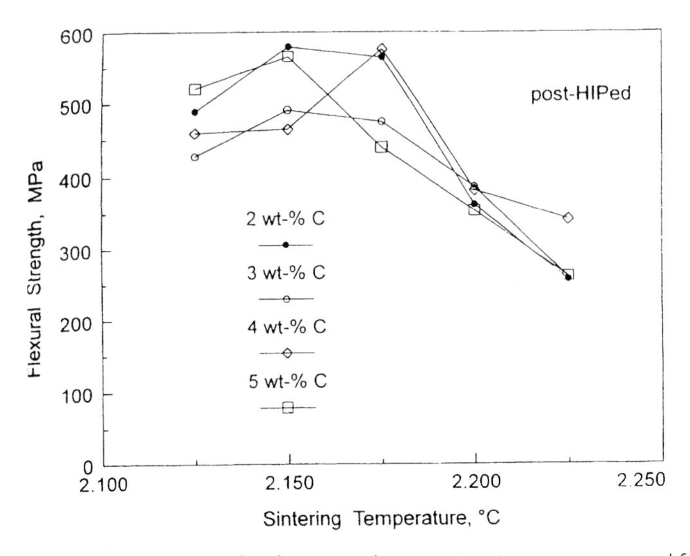

Figure 4.16 Four-point bending strength versus sintering temperature and free carbon content. After Ref. [409].

at sintering temperatures above 2225 °C, thus detrimentally decreasing the strength below 300 MPa by the formation of microcracks.

Poisson's ratio of boron carbide is significantly lower than that of most solids (0.15).

Recently, the variation of mechanical properties with stoichiometry has been investigated. An increase in fracture toughness from ~3 MPa·m$^{1/2}$ at B_4C to ~5–5.5 MPa·m$^{1/2}$ approaching the composition of $B_{13}C_2$ has been observed by Chheda *et al.* [533]. Whilst Young's modulus slightly decreased, the modulus of rupture (MOR) increased from 387 to 557 MPa.

The analysis of existing shock compression data on boron carbides, irrespective of stoichiometry, impurities, and processing technique, indicates that boron carbides suffer a loss of shear strength when shocked above their Hugoniot elastic limit (HEL) – that is, 17–20 GPa [534]. This is a significant feature which must be considered when choosing ceramics as armor materials to defeat ballistic threats. Recently, results from a number of studies have supported this effect by means of microscopic (TEM) and spectroscopic (Raman) studies. Chen *et al.* [535], using TEM, revealed the presence of amorphous bands of nanometric width that occurred parallel to specific crystallographic planes and contiguously with apparent cleaved fracture surfaces. The amorphization of B_4C was previously suggested (though not concluded) by Domnich *et al.* [536], based on the results of a nanoindentation/ Raman spectroscopy study on B_4C. The change in intensity of the amorphous carbon peaks in the Raman spectrum taken from within the indent was indicative of structural change. A TEM examination by Ge *et al.* [537] of *in situ* indents in B_4C confirmed the presence of shear amorphization regions, as well as comminution at the nanoscale. The energetic driving force for the formation of these amorphous bands was recently studied by Fanchini *et al.* [538]. Based on a Gibbs free-energy analysis of B_4C polytypes with different possible boron and carbon arrangements, it was concluded that the B12(CCC) polytype (i.e., B12-icosahedra connected by C–C–C chains) possessed the lowest mechanical stability when subjected to high pressures. The collapse of this crystal structure led to the formation of 2–3 nm narrow bands along the (113) lattice plane, which are composed of B_{12} icosahedra and amorphous C.

4.5.2.2 **Chemical Properties**

Boron carbide is rather stable in acids or alkali liquids. Due to its low reactivity with HF-H_2SO_4 or HF-HNO_3 mixtures, metal impurities or other boron compounds can be readily removed from the as-received raw powder by leaching. Boron carbide is instable on contact with metals and metal melts at high temperatures, and forms the corresponding borides and carbides/free carbon. The reaction of boron carbide with transition metal oxides is used in the large-scale production of transition metal borides, such as titanium diboride. Boron carbide is unstable in the presence of gaseous hydrogen and nitrogen, forming borane/methane (>1200 °C) and boron nitride (>1800 °C), respectively. Boron halides are evolved by reaction with gaseous chlorine (>600 °C) and bromine (>800 °C).

Boron carbide starts to react with oxygen or water vapor at 500–600 °C, and the reaction accelerates significantly above 800 °C [540, 541]. According to Telle [425], the oxidation in dry air may be described by formation of a stable liquid boron oxide layer at low temperatures and the volatilization of boron oxide and suboxides at temperatures exceeding 1000 °C [Eqs. (36)–(39)]. These two competitive reactions govern the overall kinetics of oxidation, with the rate-controlling mechanism clearly being the diffusion of oxygen through the oxidic surface layer.

$$B_4C + 4O_2 \rightarrow B_2O_3(l) + CO_2 \quad (36)$$

$$B_2O_3 \rightarrow B_2O_2(g) + 0.5O_2 \quad (37)$$

$$B_2O_3(l) \rightarrow B_2O(g) + O_2 \quad (38)$$

$$B_2O_3(l) \rightarrow B_2O_3(g) \quad (39)$$

The oxidation of boron carbide in humid air has been studied in detail by Steinbrück *et al.* [542], using quantitative mass spectrometry in order to measure the reaction kinetics continuously. The chemical reactions in Eqs. (40)–(42) are thought to play a role during the oxidation of boron carbide. Surplus steam then reacts with liquid boron oxide to form more volatile boric acid [Eqs. (43) and (44)]. At higher temperatures, the evaporation of liquid boron oxide also takes place [Eq. (45)]. According to thermochemical calculations, a significant methane release is obtained only at temperatures below 700 °C. In the temperature range between 800 °C and 1600 °C, the kinetics are supposed to be determined by at least two processes: (i) the formation of liquid boron oxide, which covers the surface and thus acts as a diffusion barrier for the starting materials and products of the reaction; and (ii) the evaporation of boron oxide and the reaction products (mainly boric acid) with steam. The former process follows a parabolic rate law, whereas the latter is thought to be of linear kinetics, resulting altogether in paralinear oxidation kinetics. The two processes lead to an equilibrium thickness of the oxidic layer.

$$B_4C + 7H_2O \rightarrow 2B_2O_3(l) + CO + 7H_2 \quad (40)$$

$$B_4C + 8H_2O \rightarrow 2B_2O_3(l) + CO_2 + 8H_2 \quad (41)$$

$$B_4C + 6H_2O \rightarrow 2B_2O_3(l) + CH_4 + 4H_2 \quad (42)$$

$$B_2O_3(l) + H_2O \rightarrow 2HBO_2 \quad (43)$$

$$B_2O_3(l) + 3H_2O \rightarrow 2H_3BO_3 \quad (44)$$

$$B_2O_3(l) \rightarrow B_2O_3(g) \quad (45)$$

A somewhat different approach favors the surface reaction and mass transport in the gas phase as the rate-determining steps [543].

4.6
Application of Carbides

4.6.1
Silicon Carbide

Of the ~700 000 tons of SiC produced each year, about 33% is used in metallurgy as a deoxidizing plus alloying agent, and about 50% in the abrasives industry [544]. The remainder is used in the refractory and structural ceramics industries and, to a small extent, also in the electrical and electronics industries as heating elements, thermistors, varistors, light-emitting diodes, high-power and high-frequency devices, and as an attenuator material for microwave devices.

In its loose granular form, SiC is used for cutting and grinding precious and semiprecious stones, and for the fine grinding and lapping of metals and optical glasses [545–547]. When bound with synthetic resins and ceramic binders, SiC grits are used in grinding wheels, whetstones, hones, abrasive cutting-off wheels, and as monofiles for the machining of metals, ceramics, plastics, and coal-based materials [548].

Coated abrasives include abrasive paper and cloth in sheet or band form. These are produced by strewing the SiC grains onto a substrate coated with glue or bonding resin, and then covering with a second layer of bonding agent [544].

The addition of SiC during the melting of cast iron aids carburization and siliconization, and improves the quality of the cast iron as a result of its seeding action [549]. In the production of steel in an arc furnace, SiC acts as deoxidant and helps in slag melting.

The need to control thermal expansion and to increase the tensile strength, fatigue strength and stiffness of aluminum alloys, has led to the development of Al–SiC composites alloys containing up to 50 vol% SiC particles [550]. Recently, developmental products have become available commercially; for example, DWA Aluminum Composites manufactures an extruded discontinuously reinforced aluminum (DRA) for fan exit guide vanes for Pratt & Whitney's high-bypass gas turbine engines on the Boeing 777 airliner, as well as thin-gage sheet for the ventral fins and fuel access covers of the F-16 fighter plane. Other applications include aircraft hydraulic components for the F-18 E/F, brake fins for Walt Disney World thrill rides, and global positioning system (GPS) satellite electronic packaging chip carriers with 50% SiC particulate reinforcement. Other potential applications for DRA include sporting goods such as golf equipment, and bicycle frames and components. For reinforcement applications, SiC is also used in the form of whiskers [551], platelets, and fibers.

The resistance of ceramically bonded and recrystallized SiC to thermal shock, oxidation and corrosion is utilized in its use as a refractory construction material, for example, in the linings and skid rails for furnaces and hot cyclones, and as a kiln furniture, especially in saggars [164, 552–556]. The good electrical conductivity of the material at high temperatures, coupled with its outstanding oxidation resistance, led to its early use in the electric heating industry [557–559], which markets its products mainly in the form of rods and tubes that operate up to 1500 °C. Recrystallized SiC igniters are used in home gas appliances, replacing pilot lights. High-purity SiC shapes are used in the electronic industry as furnace components for the processing of silicon wafers. The thermoelectric properties of SiC suggest the use of sintered SiC rods as high-temperature thermoelements [560], and also as Seebeck elements [561] for high-temperature thermoelectric energy conversion. Voltage-dependent resistors (varistors) consist of ceramic- or polymer-bonded SiC, and are used in overvoltage protection equipment.

Silicon carbide is an outstanding material for the construction of electronic equipment. Blue light-emitting diodes having an improved 470 nm peak wavelength are being produced and marketed as the first commercial SiC semiconductor device.

The continual development of the deposition of SiC thin films and of large-diameter single crystal SiC wavers, and the associated technologies of doping, etching and electrical contacts, have culminated in a host of new solid-state devices that includes field effect transistors (FETs) capable of operation up to 650 °C [562].

High-density, high-strength SiSiC, SSiC, HP-, and HIPSiC materials, which have been available commercially only since the 1970s, have opened up a new field of application [563], namely, in mechanical and high-temperature engineering. Today, SiSiC and, especially SSiC, are displacing the chemically less-resistant tungsten carbide (hard metal) and the thermal-shock-sensitive aluminum oxide in modern mechanical seals, where they are used in the form of slide rings. The excellent wear resistance of sintered SiC, combined with its excellent chemical resistance and outstanding tribological characteristics, ensure that mechanical seals made from this material last longer, which in turn results in much-reduced maintenance and production costs for pump-dependent processes in the chemical industry [564].

SiSiC seal rings – in contrast to those made from pure SSiC – can only be used in acid media because of the accompanying silicon phase, which is attacked by alkalis.

For similar requirements involving radial loads, sliding bearings (see Figure 4.17) are manufactured from SSiC. The erosion and chemical resistance of SiC enable the designer to position the bearings in the medium to be transported; that is, to eliminate lubrication and sealing problems. Other components include shaft protection sleeves for waste gas exhaust fans, and precision spheres for dosing and regulating valves.

The tailored properties of liquid phase-sintered silicon carbide (LPS-SiC) allow it to be used as dewatering elements in the paper machinery, as armor plates [202] and as rings for highly stressed gas seals. It is a price-competitive alternative to silicon nitride materials, and outperforms both alumina and tungsten carbide materials. In addition, LPS-SiC has been proposed as a neutral matrix in ceramic matrix composites containing plutonium to burn the world's stockpiles of military plutonium in thermal or fast reactors [565].

Hot-pressed SiC is the preferred material to replace oxide ceramics for rods, fixtures, and punches in high-temperature strength-testing equipment. In view of the low level of plasma contamination, the low induced radioactivity, and excellent high-temperature resistance of SiC, it is the ideal material – especially in its isostatically hot-pressed form – for use in fusion reactors [566].

(a)

(b)

Figure 4.17 EKasic silicon carbide laser-structured sliding bearings used for example, in highly loaded chemical pumps, in magnetic couplings for hermetically sealed pumps and in stirrers for chemical and pharmaceutical processes. (a) A radial bearing; (b) A thrust bearing. Illustrations courtesy of ESK Ceramics GmbH & Co. KG.

Combustion tubes made from slip-cast SiSiC [567] have better resistance to corrosion, to high-temperatures, and to thermal shock, so that they will last far longer than, for example, tubes made from heat-resistant steel. Recently, the heat-treatment industry has begun to use SSiC radiant tubes in indirect gas-fired heat-treating operations. In such systems, the tubes are internally heated by combustion burners and radiate heat to some external work load, such as an ingot of alloy, which is isolated from the combustion atmosphere [568].

Due to fine soot particulates that evolve from diesel engines being classified as hazardous substances, diesel particulate filters (DPFs) have been developed during the past two decades [569]. Currently, exhaust gases are treated by filtering the particulates onto the walls of a wall-flow ceramic honeycomb, with the entire exhaust gas stream being forced through porous thin walls made from a ceramic material (usually SiC, cordierite, or aluminum titanate) with open-pore channels. The soot particles accumulate on the filter surface and later are burned off catalytically at temperatures of at least 500 °C before the pressure drop becomes excessive. Although these wall-flow devices provide a trapping efficiency of >95%, a regeneration step is then required to combust the collected carbon particulates. Consequently, today's developments concentrate increasingly on catalyzed DPFs and, by placing a catalyzed washcoat on the filter wall, the temperature of combustion can be significantly lowered. Recent on-road studies conducted with catalyzed DPFs have shown that the temperatures in the driving cycle are high enough for the combustion reaction to occur on the catalyst surface, so that no regeneration step is necessary.

Both, SiSiC and SSiC are destined for use in heat-exchanger systems because of their high thermal conductivity and corrosion resistance [570–574]. Recently, SSiC (EKasic) based heat exchangers have been successfully commercialized (by ESK Ceramics GmbH & Co. KG), in which diffusion bonding has been applied

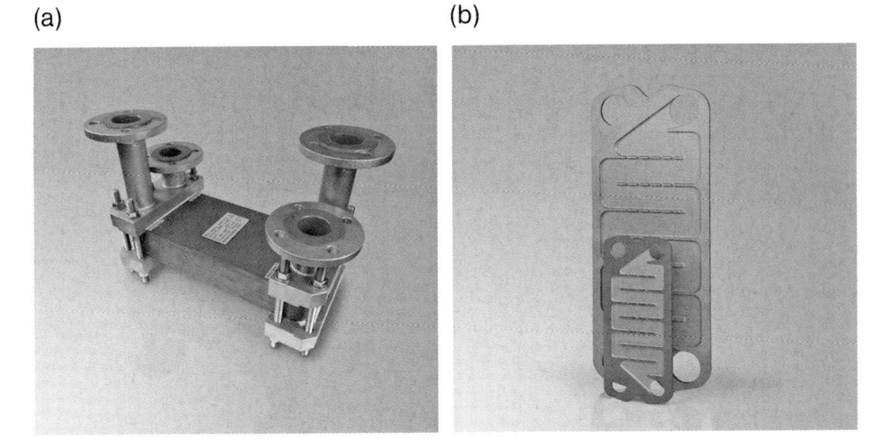

Figure 4.18 (a) Monolithic SiC-plate heat exchanger made from (b) flat EKasic plates with embedded channels, by diffusion welding. Illustrations courtesy of ESK Ceramics GmbH & Co. KG.

as an appropriate method to produce seamless large monolithic shapes (Figure 4.18) [575]. The special plate design, high thermal efficiency, high corrosion and wear resistance and high capability to resist pressure differences of up to 16 bar, also allows for potential applications in the chemical, steel, pharmaceutical, and semiconductor industries. In addition, the high resistance to acids is a prerequisite for use in the field of ore winning and fertilizers. Gasketed and semi-welded designs are also commercially available.

Sintered SiC can be used as a ceramic armor to defeat ballistic threats [576]; the properties of ceramic armors are listed in Table 4.8.

Sintered and isostatically hot-pressed SiC materials, as well as silicon nitride (Si_3N_4), have recently been playing an important role in the research and development of ceramic components for motor vehicle engines, and for higher-temperature hot-section components in gas turbines [577–582]. Although the latter role has been ongoing for about 50 years, these materials have not been widely adopted due primarily to their poor fracture toughness and poor resistance to damage in aggressive turbine environments, in interfacing with adjacent metallic components, and in their ability to be scaled-up to large parts [583, 584]. Over the past decade, however, a variety of ceramic matrix composite (CMC) systems reinforced by continuous-length ceramic fibers have been developed showing performance benefits over monolithics for hot-section components. Indeed, recent field tests using SiC/SiC liners with multilayer environmental barrier coatings (EBCs) have demonstrated accumulated service times of more than 12 000 h at up to 1260 °C, and a calculated stress load of about 76 MPa [585]. Current research is focused on SiC/SiC materials with a Si-free matrix, reliable coating and lifetime calculation models to improve service in highly corrosive combustor environments (high water vapor pressure, $T > 1200\,°C$, gas velocity up to 90 m s^{-1}). Further developments for the application of SiC-based CMCs have concentrated on hot components in military and space technology, including jet vanes for thrust vector control in military rockets, and hot gas ducts, combustion

Table 4.8 Physical properties of SiC ceramics used in armor applications. After Karandikar *et al.* [576].

Material	Designation	Manufacturer	Density ($g\,cm^{-3}$)	Grain size (μm)	Young's modulus E (GPa)	Flexural strength (MPa)	Fracture toughness ($MPa{\cdot}m^{1/2}$)	Hardness (GPa)
SiC	SiC-N	Cercom	3.22	2–5	453	486	4.0	18.7 (HK 2 kg)
	Ceralloy 146-3E	Ceradyne	3.20	—	450	634	4.3	22.6 (HV 0.3 kg)
	Hexoloy	Saint Gobain	3.13	3–50	410	380	4.6	27.5 (HK 0.1 kg)
	Purebide 5000	Morgan AM&T	3.10	3–50	420	455	—	18.9 (HK 2 kg)
	SC-DS	Coors Tek	3.15	3–50	410	480	3–4	27.5 (HK 1 kg)
	MCT SSS	M Cubed Technologies	3.12	3–50	424	351	4	19.3 (HK 2 kg)
	MCT LPS	M Cubed Technologies	3.24	1–3	425	372	5.7	18.4 (HK 2 kg)
	Ekasic-T	Ceradyne	3.25	1–3	453	612	6.4	18.9 (HK 2 kg)
RBSiC	SSC-902	M Cubed Technologies	3.12	45	407	260	4	15.1 (HK 2 kg)

Properties for manufacturer's materials are taken from their respective websites (except for M Cubed Technologies), except details of grain size.

chambers and thermal protection systems for space re-entry vehicles. Due to the high thermal shock capabilities of this class of materials, and to the high fracture toughness of some of their composites, a number of civil applications have entered the market, including highly loaded brake discs in high-speed trains and sports cars, and highly loaded journal bearings. In both of these applications, conventional materials (metals for brakes, monolithic ceramics for bearings) were shown to be unsuitable because of the thermal loads and the brittle failure mode, respectively.

Today's high-speed laser scanning systems are quickly being adopted for printing, cutting, welding, drilling, and various other applications that require scanning with a laser beam. The main performance drivers for these scanning systems are the scan speed and laser output power. After having provided a laser of sufficient power, the scan speed is determined primarily by the material and design of the scan mirror. Lightweight SiC mirrors offer clear advantages over alternatives such as glass or silicon mirrors, including high stiffness and high thermal conductivity, these being ideal for dynamic optomechanical systems. The key overall benefit of using SiC mirrors is the ability to increase the scan rate and consequent productivity of the tool in which the scanner is used.

In keeping with the launch capability and cost, space optics should be compact and light. In particular, large mirrors should be stiff and stable, and the mass should be kept to a minimum. Reducing the mass of the mirror can have a major impact on reducing not only its self-deflection but also of its mass, and the cost of the entire system. Traditional optical materials include beryllium, fused silica, zerodur, and aluminum, while SiC serves as a type of neotype light-mass optical material. The two main features in evaluating the properties of optical materials are the specific rigidity (E/ϱ) and the thermal distortion ratio (λ/α). SiC exhibits the best specific stiffness except for beryllium, and the best thermal stability of all the materials mentioned above. In addition, SiC offers other features and advantages, such as low-cost machining and short manufacturing times. Hence, SiC is the best choice as substrate material for mirrors in space applications [590].

4.6.2 Boron Carbide

Due to its high hardness and wear resistance, loose boron carbide grains are readily used for the grinding and lapping of hard metals and ceramics.

Another widespread application of boron carbide grains is the *boriding* of steel (also known as "boronizing") – a surface-hardening process that involves the diffusion of boron into the metal surface at high temperature. In a number of applications, boriding has replaced processes such as carburizing, nitriding, nitrocarburizing and hard chromium plating. The hardness of the formed iron boride layer (FeB and/or Fe_2B) far exceeds that of other hardening processes, and is equivalent to that of tungsten carbide [605]. In addition to an enhancement of wear resistance, the corrosion resistance in nonoxidizing acids, alkali media and molten metals, as well as the fatigue life and service performance under oxidizing and corrosive environments, are all improved.

Boron carbide is widely used as an antioxidant for carbon-bonded refractories. The carbon in refractory materials improves their resistance to thermal shock, and also protects them against wetting by metals and slag. However, it is essential that the carbon is protected against oxidation, and boron carbide has proved particularly effective for this purpose. When B_4C is oxidized, an interaction occurs with the matrix material to form liquid and/or gaseous phases that protect the carbon from oxidation, and this prolongs the service life of the carbon-bonded refractory materials. The effects of boron carbide addition on the oxidation resistance of magnesia–carbon bricks was investigated by Gokce *et al.* [587], at temperatures in the range of 1300–1500 °C. Magnesium borate, which exists in a liquid state above 1360 °C, had a major effect on the oxidation resistance of the bricks by filling the open pores and forming a protective layer on the surface.

Due to its neutron-absorbing efficiency, boron carbide is attractive as a neutron absorber material, and is used both in powdered and solid forms to control the rate of fission in nuclear reactors (Figure 4.19b)[530]. B_4C mixed with other materials, such as aluminum metal or polyethylene plastic, is applied to protect it against oxidation in the reactor environment. Al–B_4C metal–matrix composite plates (e.g., Boral, Bortec) have wide applications as isolators in spent fuel element racks, in the inner sections of reactor shields as shutdown control rods and neutron curtains, as shutters for thermal columns, and as shipping containers.

The extremely high wear resistance of boron carbide is a prerequisite condition for its application as blast or spray nozzles (Figure 4.19a), and also as dressing sticks (tools used to dress the surface of a grinding wheel).

The hitherto only known application of boron carbide with respect to its semiconducting properties is as a graphite/boron carbide thermocouple [588].

Over the past decade, the use of ceramics in personal armor has increased exponentially. Both, boron carbide and SiC are attractive armor materials as they

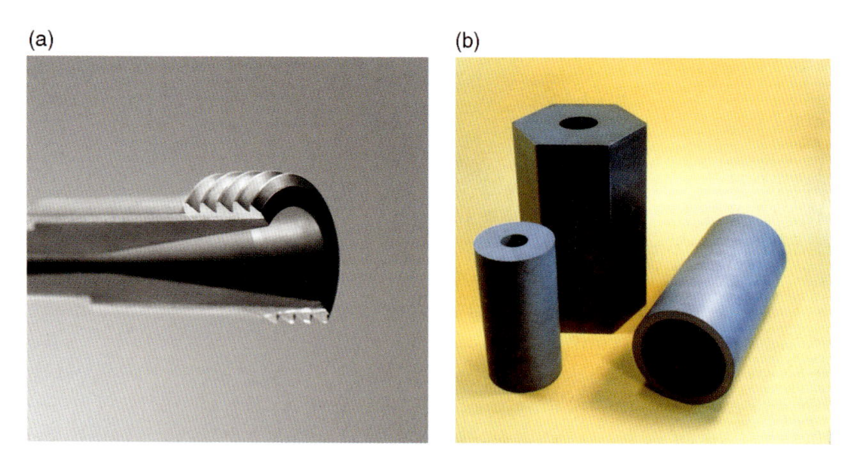

Figure 4.19 (a) Cross-sectional area of a venturi blast nozzle, TETRABOR boron carbide. (b) boron carbide shield and control components for nuclear application. Illustrations courtesy of ESK Ceramics GmbH & Co. KG.

Table 4.9 Physical properties of boron carbide ceramics used in armor applications. After Karandikar *et al.* [576].

Material	Designation	Manufacturer	Density ($g\,cm^{-3}$)	Grain size (μm)	Young's modulus E (GPa)	Flexural strength (MPa)	Fracture toughness ($MPa{\cdot}m^{1/2}$)	Hardness (GPa)
B4C	Norbide	Saint Gobain	2.51	10–15	440	425	3.1	27.5 (HK 0.1 kg)
	Ceralloy-546 4E	Ceradyne	2.50	10–15	460	410	2.5	31.4 (HV 0.3 kg)

Properties for manufacturer's materials are taken from their respective websites, except for grain size.

are more weight-efficient than traditional monolithic armors, such as armored steel (density 7.8 g cm^{-3}) or aluminum oxide (3.9 g cm^{-3}), against a variety of threats [576–589]. Additional properties that make ceramics efficient to defeat ballistic threats include their extreme hardness, high compressive strength, and high elastic modulus. Unfortunately, the low tensile strength typical of ceramics may severely limit their performance, and thus standalone ceramic armors can be inferior to traditional steel armors. The damage of armor occurs by an activation of pre-existing defects caused by hydrostatic, shear, and tensile stresses which, during a ballistic event, interact with any pre-existing defects. Thus, an understanding of how to reduce any defects stemming from the powder preparation, green-forming and sintering processes, or from the final machining and surface-finishing operations, represents an essential step towards increasing such efficiency. Today's enhanced performance is made possible by delaying penetration and/or by increasing the "comminuted ceramics" erosion efficiency; an example is the use of a modest lateral confinement to constrain any broken ceramic pieces. The properties of boron carbide ceramics, as used in armor applications, are listed in Table 4.9.

Acknowledgments

Some parts of this chapter were included in an earlier review by K.A. Schwetz (1994), Silicon Carbide, in: *Encyclopedia of Advanced Materials*, 1994, pp. 2455–2461. Elsevier Science Ltd, Kidlington, OX5 16B, UK is kindly thanked for permission to use these.

References

1 Berzelius, J.J. (1824) *Amer. Phys. u. Chem.*, **1**, 169–230.
2 Despretz, C.M. (1849) *Compt. Rend.*, **29**, 709–724.
3 Schützenberger, P. and Colson, A. (1881) *Compt. Rend. Akad. Sci.*, **92**, 1508.
4 Moissan, H. (1900) *Der elektrische Ofen*, Verlag M. Krayer, Berlin.
5 Acheson, E.G. Patent (1893) US 492 767; Patent (1894) DE 76629; Patent (1896) DE 85195.
6 Kleykamp, H. and Schumacher, G. (1993) *Ber. Bunsenges. Phys. Chem.*, **97** (6), 799–805.
7 Dietzel, A., Jagodzinski, H., and Scholze, H. (1960) *Ber. Dtsch. Keram. Ges.*, **37**, 524–537.
8 Kieffer, R., Gugel, E., Ettmayer, P., and Schmidt, A. (1966) *Ber. Dtsch. Keram. Ges.*, **43**, 621–623.
9 Jeeps, N.W. and Page, T.F. (1981) *J. Am. Ceram. Soc.*, **64**, C177–C178.
10 (a) Ott, H. (1925) *Z. Krist.*, **61**, 515–532;(b) Ott, H. (1925) *Z. Krist.*, **62**, 210–218;(c) Ott, H. (1926) *Z. Krist.*, **63**, 1–19.
11 Ramsdell, L.S. (1947) *Am. Mineral.*, **32**, 64–82.
12 Ruska, J., Gauckler, L.J., Lorenz, J., and Rexer, H.U. (1979) *J. Mater. Sci.*, **14**, 2013–2017.
13 Inomata, Y., Inoue, Z., Mitomo, M., and Suzuki, H. (1968) *Yogyo-Kyokai-Shi*, **76**, 313–319.
14 Lundquist, D. (1948) *Acta Chem. Scand.*, **2**, 177–191.
15 Shaffer, P.T.B. (1969) *Mater. Res. Bull.*, **4**, 213–220.
16 Mitomo, M., Inomata, Y., and Kumanomido, M. (1970) *Yogyo-Kyokai-Shi*, **78**, 365–369.

17 Mitomo, M., Inomata, Y., and Tanaka, H. (1971) *Mater. Res. Bull.*, **6**, 759–764.

18 Patience, M.M. (1983) *Silicon Carbide Alloys*, University of Newcastle upon Tyne.

19 Foster, D. (1996) Densification of silicon carbide with mixed oxide additives, Ph.D. Thesis, University of Newcastle upon Tyne.

20 (a) Jeeps, N.W. and Page, T.F. (1979) *J. Microsc.*, **116**, 159–171;(b) Jeeps, N.W. and Page, T.F. (1980) *J. Microsc.*, **119**, 177.

21 Joly, A. (1883) *C. R. Acad. Sci.*, **97**, 456.

22 Moisson, H. (1894) *C. R. Acad. Sci.*, **118**, 536.

23 Ridgeway, R.R. (1934) *Trans. Electrochem. Soc.*, **66**, 117.

24 Elliot, R.P. (1965) *Constitution of Binary Alloys*, (1st suppl.), McGraw-Hill, New York.

25 Schwetz, K.A. and Karduck, P. (1991) *J. Less-Common Met.*, **175**, 1.

26 Werheit, H. (2000) Boron Compounds, in *Landolt-Börnstein, Numerical Data and Functional Relationships in Science and Technology*, New Series, Group III, vol. **41D** (ed. O. Madelung), Springer, Berlin, pp. 1–491.

27 Beauvy, M. (1984) in *Conference Abstracts, 8th International Symposium on Boron, Borides, and Related Compounds* (ed. G.V. Tsagareishvili), Acad. Sci. Georg. S.S.R., Tbilisi, p. 25.

28 Kasper, B. (1996) Phase Equilibria in the B-C-N-Si System (in German), Thesis, University of Stuttgart.

29 Lukas, H.L. (1990) in *Constitution of Ternary Alloys 3* (eds G. Petzow and G. Effenberg), VCH, Weinheim, pp. 140–146.

30 Bouchacourt, M. and Thevenot, F. (1979) *J. Less-Common Met.*, **67**, 327.

31 Lim, S.K. and Lukas, H.L. (1995) in *Deutsche Forschungsgemeinschaft, Hochleistungskeramik, Herstellung, Aufbau und Eigenschaften* (eds G. Petzow, J. Tobolski, and R. Telle), VCH, Weinheim, pp. 605–616.

32 Seifert, H.J., Kasper, B., Lukas, R.L., and Aldinger, F. (2001) Thermodynamic Assessment of the B-C System, in Journées d'Etude des Equilibres Entre Phases XXVII JEEP, March 22–23, Montpellier (eds R.M. Marin-Ayral and M.C. Record), pp. 123–126.

33 Kuhlmann, U. and Werheit, H. (1992) *Solid-State Commun.*, **83**, 849.

34 Kuhlmann, U., Werheit, H., and Schwetz, K.A. (1992) *J. Alloys Compd.*, **189**, 249.

35 Werheit, H., Au, T., Schmechel, R., Shalamberidze, S.O., Kalandadze, G.I., and Eristavi, A.M. (2000) *J. Solid-State Chem.*, **154**, 79.

36 Werheit, H., Rotter, H.W., Meyer, F.D., Hillebrecht, H., Shalamberidze, S.O., Abzianidze, T.G., and Esadze, E.G. (2004) *J. Solid State Chem.*, **177**, 569–574.

37 Werheit, .H. (2006) *J. Phys. Condens. Matter*, **18**, 10655–10662.

38 Mauri, F., Vast, N., and Pickard, C.J. (2001) *Phys. Rev. Letters*, **87** (8), 085506.

39 Conde, O., Silvestre, A.J., and Oliveira, J.C. (2000) *Surf. Coat. Technol.*, **125**, 141.

40 Yakel, H.L. (1975) *Acta Crystallogr.*, **B31**, 1791.

41 Kwei, G.H. and Morosin, B. (1996) *J. Phys. Chem.*, **100**, 8031–8039.

42 La Placa, S. and Post, B. (1961) *Planseeberichte für Pulvermetallurgie*, **9**, 109.

43 Lipp, A. (1966) *Ber. Dtsch. Keram. Ges.*, **43** (1), 60.

44 Lipp, A. and Röder, M. (1966) *Z. An. Allg. Chem.*, **343**, 1.

45 Lipp, A. and Röder, M. (1966) *Z. An. Allg. Chem.*, **344**, 225.

46 Matkovich, Y.I. and Economy, J. (1977) in *Boron and Refractory Borides* (ed. Y.I. Matkovich), Springer, Berlin, pp. 98–106.

47 Anonym (1954) *Ind. Heating*, 992–1004.

48 Poch, W. and Dietzel, A. (1962) *Ber. Dtsch. Keram. Ges.*, **39**, 413–426.

49 Weimer, A.W. (1997) Carbothermal reduction synthesis processes, in *Carbide, Nitride and Boride Materials Synthesis and Processing* (ed. A.W. Weimer), Chapman & Hall, London-Madras, pp. 75–180.

50 Liethschmidt, K. (1982) Siliciumcarbid, in *Chemische Technologie, Band 2* (eds K. Winnacker and L. Küchler), Anorg. Techn., 1, 4. Aufl., Hanser-Verlag, pp. 626–629.

51 Mehrwald, K.H. (1992) *Ber. Dtsch. Keram. Ges.*, **69**, 72–81.

52 Boecker, W., Landfermann, H., and Hausner, H. (1981) *Powder Metall. Int.*, **13**, 37–39.
53 Prochazka, S. (1986) Techn. Report 86-CRD-158 General Electric Corp.
54 Matje, P. and Schwetz, K.A. (1988) in *Proceedings 1st International Conference on Ceramic Powder Processing Science, November 1987, Orlando FL*, vol. 1 (ed. G. Messing *et al.*), Ceramic-Transactions Am. Ceram. Soc., pp. 460–468.
55 Blumenthal, J.L., Santy, M.J., and Burns, E.A. (1966) *AIAA J.*, **4** (6), 1053–1057.
56 Lee, J.G., Miller, P.D., and Cutler, I.B. (1977) in *Reactivity of Solids* (eds J. Wood, O. Linquist, C. Helgesson, and N.G. Vannerberg), Plenum, New York, pp. 707–711.
57 Viscomi, F. and Himmel, L. (1978) *J. Metall.*, **6**, 21–24.
58 Benaissa, M., Werckmann, J., Ehert, G., Peschiera, E., Guille, J., and Ledoux, M.J. (1994) *J. Mater. Sci.*, **29**, 4700–4707.
59 Krstic, V.D. (1992) *J. Am. Ceram. Soc.*, **75** (1), 170–174.
60 Čerovič, L., Milonjič, S.K., and Zec, S.P. (1995) *Ceram. Int.*, **21**, 271–276.
61 Seo, W.S., Koumoto, K., and Arai, S. (1998) *J. Am. Ceram. Soc.*, **81** (5), 1255–1261.
62 Goldberger, W.M., Reed, A.K., and Morse, R. (1989) in *SiC '87 Ceramics Transactions, Vol. 2* (ed. J.D. Cawley), American Ceramics Society, Westerville, OH, pp. 93–104.
63 Tanaka, H. and Kurachi, Y. (1988) *Ceram. Int.*, **14**, 109–115.
64 Hasegawa, I., Nakamura, T., Motojima, S., and Kajiwara, M. (1995) *J. Mater. Chem.*, **5** (1), 193–194.
65 Raman, V., Bahl, O.P., and Dhawan, U. (1995) *J. Mater. Sci.*, **30**, 2686–2693.
66 Ono, K. and Kurachi, Y. (1991) *J. Mater. Sci.*, **26**, 388–392.
67 Huang, D., Ikuhara, Y., Narisawa, M., and Okamura, K. (1998) *J. Am. Ceram. Soc.*, **81** (12), 3173–3176.
68 Narisawa, M., Okabe, Y., Iguchi, M., Okamura, K., and Kurachi, Y. (1998) *J. Sol-Gel Sci. Technol.*, **12**, 143–152.
69 Li, J., Tian, J., and Dong, L. (2000) *J. Eur. Ceram. Soc.*, **77**, 1853–1857.
70 Meng, G.W., Cui, Z., Zhang, L.D., and Phillipp, F. (2000) *J. Cryst. Growth*, **209**, 801–806.
71 Prener, J.S. (1960) US Patent 3 085 863.
72 O'Connor, T.L. and McRae, W.A. (1963) US Patent 3 236 673.
73 Prochazka, S. (1972) Final Report SRD 72-171, General Electric.
74 Schwetz, K.A. and Lipp, A. (1978) *Radex-Rundschau*, (2), 479–498.
75 Wei, G.C., Kennedy, C.R., and Harris, L.A. (1984) *Ceram. Bull.*, **63** (8), 1054–1061.
76 Yajima, S., Hayashi, J., Omori, M., and Okamura, K. (1976) *Nature*, **261**, 683–685.
77 Soraru, G.D., Babonneau, F., and Mackenzie, J.D. (1990) *J. Mater. Sci.*, **25**, 3886–3893.
78 Bouillon, E., Langlais, F., and Pailler, R. (1991) *J. Mater. Sci.*, **26**, 1333–1345.
79 Lee, J.G. and Cutler, I.B. (1975) *Ceram. Bull.*, **54** (2), 195–198.
80 Krishnarao, R. and Godkhindi, M.M. (1992) *Ceram. Int.*, **18**, 243–249.
81 Panigraphi, B.B., Roy, G.G., and Godkhindi, M.M. (2001) *Br. Ceram. Trans.*, **100** (1), 29–34.
82 Raju, C.B. and Verma, S. (1997) *Br. Ceram. Trans.*, **96** (3), 112–115.
83 Hahn, F., Rudakoff, G., and Tiller, H.J. (1990) *Hermsdorfer Techn. Mitt.*, **79**, 2546–2550.
84 Baumgartner, H.R. and Rossing, B.R. (1989) in *SiC '87 Ceramics Transactions, Vol. 2* (ed. J.D. Cawley), American Ceramics Society, Westerville, OH, pp. 3–16.
85 Stroke, F.G. (1981) (PPG) US Patent 4 295 890.
86 Cannon, W.R., Danforth, S.C., Flint, J.H., Haggerty, J.S., and Marra, R.A. (1982) *J. Am. Ceram. Soc.*, **65** (7), 324–330.
87 Nájera, J.J., Cáceres, J.O., Ferrero, J.C., and Lane, S.I. (2002) *J. Eur. Ceram. Soc.*, **22** (13), 2371–2378.
88 Kavecky, S., Janekova, B., Madejova, J., and Sajgalik, P. (2000) *J. Eur. Ceram. Soc.*, **20**, 1939–1946.
89 Huang, Z.R., Liang, B., Jiang, D.L., and Tan, S.H. (1996) *J. Mater. Sci.*, **31** (16), 4327–4332.
90 Li, B., Zhang, C., Hu, H., and Qi, G. (2007) *Front. Mater. Sci. China*, **1** (3), 309–311.

91 Gupta, A. and Jacob, C. (2006) *Mater. Sci. Forum*, **527–529** (1), 767–770.

92 Shaffer, P.T.B. (1994) *Handbook of Advanced Ceramic Materials*, Advanced Refractory Technologies Inc., Buffalo, NY.

93 Komaki, K. (1974) Japanese Patent Pub. SHO 49-32.719.

94 Tanaka, M., Kawabe, T., and Kobune, M. (1987) Japanese Patent 1.361.044.

95 Yamamoto, A. (1983) Japanese Patent Prov. Pub. SHO 58-145.700.

96 Milewski, J.V., Gac, F.D., Petrovic, J.J., and Skaggs, S.R. (1985) *J. Mater. Sci.*, **20**, 1160–1166.

97 Cutler, I.B. (1973) US Patent 3.754.076.

98 Sharma, N.K., Williams, W.S., and Zangvil, A. (1984) *J. Am. Ceram. Soc.*, **67** (11), 715–720.

99 Krishnarao, R.V., Mahajan, Y.R., and Kumar, T.J. (1998) *J. Eur. Ceram. Soc.*, **18**, 147–152.

100 Böcker, W.D.G., Chwastiak, S., Frechette, F., and Lau, S.K. (1989) in *SiC '87 Ceramics Transactions, Vol. 2* (ed. J.D. Cawley). American Ceramics Society, Westerville, OH, pp. 407–420.

101 Kistler-De Coppi, P.A. and Richarz, W. (1986) *Int. J. High Technol. Ceram.*, **2**, 99–113.

102 Meier, B., Hamminger, R., and Nold, E. (1990) *Microchim. Acta*, **2**, 195–205.

103 Lenk, R. and Adler, J. (1995) in *Fourth Euro-Ceramics Vol. 2, Part II, BaSiC Science* (ed. C. Galassi), Faenca Editrice SpA, Italy, pp. 407–414.

104 Lenk, R., Kriwoschepov, A.F., and Große, K. (1995) *Sprechsaal: Ceramics & Materials*, **128** (1), 17–20.

105 Chou, Y.S. and Green, D.J. (1992) *J. Am. Ceram. Soc.*, **75** (12), 3346–3352.

106 Chou, Y.S. and Green, D.J. (1993) *J. Am. Ceram. Soc.*, **76** (6), 1452–1458.

107 Chou, Y.S. and Green, D.J. (1993) *J. Am. Ceram. Soc.*, **76** (8), 1985–1992.

108 Alexander, K.B., Becher, P.F., and Waters, S. (1990) in *Proceedings of Twelfth International Congress for Electron Microscopy*, San Francisco Press, San Francisco, CA, pp. 106–107.

109 Lackey, W.J., Fohn, A., Hanigofsky, G.B., Freeman, R.D., Hardin, R.D., and Prasad, A. (1995) *J. Am. Ceram. Soc.*, **78** (6), 1564–1570.

110 Ishikawa, T. (1991) in *SiC Ceramics-2* (eds S. Somiya and Y. Inomata), Elsevier, pp. 81–98.

111 Ichikawa, H. and Takeda, M. (1999) Advanced Structural Fiber Composites, in *Proceedings of the 9th CIMTEC-World Ceramics Congress and Forum on New Materials*, Florence, Italy, 14–19 June, 1998 (ed. P. Vincenzini), Faenza, Techna.

112 Riedel, R., Mera, G., Hauser, R., and Klonczynski, A. (2006) *J. Ceram. Soc. Jap.*, **114** (6), 425–444.

113 Jansen, M., Jäschke, B., and Jäschke, T. (2002) in *High Performance Non-Oxide Ceramics* (ed. M. Jansen), Springer-Verlag, Berlin, Heidelberg, New York, pp. 137–193.

114 Baldus, H. P., and Yansen, M. (1997) *Angew. Chemie*, **109**, 338–354.

115 Heimann, D., Wagner, T., Bill, J., Aldinger, F., and Lange, F.F. (1997) *J. Mater. Res.*, **12** (11), 3099–3101.

116 Prochazka, S. (1997) US Patent US 566 8068.

117 Honda, S., Baek, Y., Ikuno, T., Kohara, H., Katayama, M., Oura, K., and Hirao, T. (2003) *Appl. Surf. Sci.*, **212–213**, 378–382.

118 Huczko, A., Bystrzejewski, M., Lange, H., Fabianowska, A., Cudzilo, S., Panas, A., and Szala, M. (2005) *J. Phys. Chem. B*, **109**, 16244–16251.

119 Raman, V., Bathia, G., Bhardwaj, S., Srivastva, A., and Sood, K. (2005) *J. Mater. Sci.*, **40**, 1521–1527.

120 Hao, Y., Jin, G., Han, X., and Guo, X. (2006) *Mater. Lett.*, **60**, 1334–1337.

121 Correa, Z., Murata, H., Tomizawa, T., Tenmoku, K., and Oya, A. (2006) Disclosing Materials at the Nanoscale, in Proceedings, 11th International Ceramics Congress and 4th Forum on New Materials (eds P. Vincenzini and G. Marletta), TransTech Publications, Switzerland.

122 Iijima, S. (1991) *Nature*, **354**, 56–58.

123 Pham-Huu, C., Keller, N., Ehret, G., and Ledoux, M. (2001) *J. Catal.*, **200**, 400–410.

124 Keller, N., Pham-Huu, C., Ehret, G., Keller, V., and Ledoux, M. (2003) *Carbon*, **41**, 2131–2139.

125 Sun, X., Li, C., Wong, W., Wong, N., Lee, C., Lee, S., and Teo, B. (2002) *J. Am. Chem. Soc.*, **124**, 14464–14471.

126 Rümmeli, M., Borowiak-Palen, E., Gemming, T., Knupfer, M., Biedermann, K., Kalenczuk, R., and Pichler, T. (2005) *Appl. Phys. A*, **80**, 1653–1656.

127 Lienhard, M.A. and Larkin, D.J. (2003) Silicon Carbide Nanotube Synthesized, Technical report, Nasa Glenn research Center. Available at: http://www/rt/rt2002/5000/5510lienhard.html.

128 Cheng, Q., Interrante, L., Lienhard, M., Shen, Q., and Wu, Z. (2005) *J. Eur. Ceram. Soc.*, **25**, 233–241.

129 Mpourmpakis, G., Froudakis, G., Lithoxoos, G., and Samios, J. (2006) *Nano Lett.*, **6**, 1581–1583.

130 Morisada, Y. and Miyamoto, Y. (2004) *Mater. Sci. Eng. A*, **381**, 57–61.

131 Hulteen, J. and Martin, C. (1997) *J. Mater. Chem.*, **7**, 1075–1087.

132 (a) Lipp, A. (1965) *Tech. Rundschau*, **57**, (14), 5;(b) Lipp, A. (1965) *Tech. Rundschau*, **57**, (28), 19;(c) Lipp, A. (1965) *Tech. Rundschau*, **57**, (33), 5.

133 Lipp, A. (1966) *Tech. Rundschau*, **58** (7), 5.

134 Duffek, G., Wross, W., Vendl, A., and Kieffer, R. (1976) *Planseeberichte Pulvermet.*, **24**, 280.

135 Weimer, A.W., Roach, R.P., Haney, C.N., Moore, W.G., and Rafaniello, W. (1991) *AIChE J.*, **37** (5), 759–768.

136 Gray, E.G. (1980) European Patent Appl. 1.152.428.

137 Thevenot, F.J. (1990) *J. Eur. Ceram. Soc.*, **6**, 205.

138 Pradhan, B., Tandon, D., Taylor, R.L., and Hoffman, P.B. (2006) US Patent Appl. 20060051281.

139 Campbell, J., Klusewitz, M., LaSalvia, J. *et al.* (2008) Novel processing of boron carbide; plasma synthesized nano powders and pressureless sintering forming of complex shapes. Proceedings 26th Army Science Conference, December 2008, Orlando, Florida.

140 Herth, S., Joost, W.J., Doremus, R.H., and Siegel, R.W. (2006) *J. Nanosci. Nanotechnol.*, **6** (4), 954–959.

141 Chang, B., Gersten, B.L., Szewczyk, S.T., and Adams, J.W. (2007) *Appl. Phys. A*, **86** (1), 83–87.

142 Ploog, K. (1974) *J. Less-Common Met.*, **25**, 115.

143 McKinnon, I.M. and Reuben, B.G. (1975) *J. Electrochem. Soc.*, **122**, 806.

144 Knudsen, A.K. (1987) *Adv. Ceram.*, **21**, 237.

145 Gu, Y., Chen, L., Qian, Y., Zhang, W., and Ma, J. (2005) *J. Am. Ceram. Soc.*, **88** (1), 225–227.

146 Sneddon, L.G. (2006) Final Report AFRL-SR-AR-TR-06-0355, Air Force Office of Scientific Research.

147 Zhang, D., McIlroy, D.N., Geng, Y., and Norton, M.G. (1999) *J. Mater. Sci. Lett.*, **18**, 349.

148 U.S. Department of Energy (1999) Nanoscale Science, Engineering and Technology; Research Directions Report of the BaSiC Energy Sciences Nanoscience/Nanotechnology Group. Available at: http://www.er.doe.gov/bes/reports/archives.html

149 Dai, H.J., Wong, E.W., Lu, Y.Z., Fan, S.S., and Lieber, C.M. (1995) *Nature*, **375**, 769.

150 Han, W., Bando, Y., Kurashima, K., and Sato, T. (1999) *Chem. Phys. Lett.*, **299**, 368.

151 Han, W., Kohler-Redlich, P., Ernst, F., and Röhle, M. (1999) *Chem. Mater.*, **11**, 362.

152 Wei, J.Q., Jiang, B., Li, Y.H., Xu, C.L., Wu, D.H., and Wei, B.Q. (2002) *J. Mater. Chem.*, **12**, 3121.

153 McIlroy, D.N., Zhang, D., Kranov, Y., and Norton, M.G. (2001) *Appl. Phys. Lett.*, **79**, 1540.

154 Ma, R., Bando, Y., Sato, T., and Kurashima, K. (2001) *Chem. Phys. Lett.*, **350**, 434.

155 Xu, F.F. and Bando, Y. (2004) *J. Phys. Chem. B*, **108**, 7651.

156 Ma, R. and Bando, Y. (2002) *Chem. Mater.*, **14**, 4403.

157 Ma, R. and Bando, Y. (2002) *Chem. Phys. Lett.*, **364**, 314.

158 Carlsson, M., Garcia, F.J., and Johnsson, M. (2002) *J. Cryst. Growth*, **236**, 466.

159 Johnsson, M. (2004) *Solid State Ionics*, **172**, 365.

160 Pender, M.J. and Sneddon, L. (2000) *Chem. Mater.*, **12**, 280.

161 Welna, D.T., Bender, J.D., Wei, X., Sneddon, L.G., and Allcock, H.R. (2005) *Adv. Mater.*, **17**, 859.

162 Kappmeyer, K.K., Hubble, D.H., and Powers, H.W. (1966) *Am. Ceram. Soc. Bull.*, **45** (12), 1060–1064.

163 Washburn, M.E. and Love, R.W. (1962) *Am. Ceram. Soc. Bull.*, **41** (7), 447–449.

164 Fickel, A. (1997) in *Refractory Materials - Pocket Manual* (ed. G. Routschka), Vulkan-Verlag Essen, pp. 74–80.

165 van der Beck, R. and O'Connor, J. (1957) *Ceram. Ind.*, **3**, 96–98.

166 Alliegro, R.A. (1974) in *Ceramics for High Performance Applications* (eds J.J. Burke*et al.*), Metals and Ceramics Information Center, Columbus, OH, pp. 253–263.

167 (a) Kriegesmann, J. (1988) *Interceram*, **2**, 27–30;(b) Kriegesmann, J. (1986) *Powder Metall. Int.*, **18**, 341–343.

168 Popper, P. and Davies, D.G.S. (1961) *Powder Metall.*, **8**, 113.

169 (a) Taylor, K.M. (1956) *Materials and Methods*, 92–95;(b) Taylor, K.M. (1956) US Patent 3.189 472.

170 Kieffer, R., Gugel, E., Schmidt, A., (1965; 1967) Patent DE-OS 1 671 092.

171 Kennedy, P. and Shennan, J.V. (1974) in *Silicon Carbide 1973* (eds R.C. Marhall*et al.*), University of South Carolina Press, Columbia, pp. 359–366.

172 Weaver, G.Q., Baumgartner, H.R., and Torti, M.L. (1975) in *Special Ceramics 6* (ed. P. Popper), British Ceramics Research Association, pp. 261–281.

173 Hillig, W.B., Mehan, R.L., Morelock, C.R., De Carlo, V.J., and Laskow, W. (1975) *Am. Ceram. Soc. Bull.*, **54**, 1054–1056.

174 Hase, T. *et al.* (1976) *J. Nucl. Mater.*, **59**, 42–48.

175 Anonymous (1982) *Interceram*, **1**, 50–51.

176 Willmann, G. and Heider, W. (1983) *Werkstofftechnik*, **14**, 158.

177 Taylor, K. (1962) US Patent 3.205.043.

178 Prochazka, S. (1981) Technical Report 81-CRD-314, General Electric Company, p. 16.

179 Prochazka, S. (1974) in *Ceramic for High Performance Applications* (eds J.J. Burke*et al.*), Metals and Ceramic Information Center, Columbus, OH, pp. 239–252.

180 Coppola, J.A. and McMurtry, G.H. (1976) Substitution of ceramics for ductile materials in design. National Symposium on Ceramics in the Service of Man, Carnegie Institution, Washington, D.C.

181 Shaffer, P.T.B. (1969) *Mater. Res. Bull.*, **4** (3), 213–220.

182 Hamminger, R. (1989) *J. Am. Ceram. Soc.*, **72** (9), 1741–1744.

183 Murata, Y. and Smoak, R.H. (1979) in *Proceedings International Symposium on Densification and Sintering* (eds S. Somiya and S. Saito), Gakujutsu Bunken, Tukyu-Kai/Tokyo, pp. 382–399.

184 Boecker, W. and Hausner, H. (1978) *Powder Metall. Int.*, **10**, 87–89.

185 Stutz, D.R., Prochazka, S., and Lorentz, J. (1985) *J. Am. Ceram. Soc.*, **68** (9), 479–482.

186 Fetahagic, T. and Kolar, D. (1990) *Ceram. Acta*, **2** (2), 31–37.

187 Schwetz, K.A. and Lipp, A. (1980) in *Science of Ceramics 10* (ed. H. Hausner), Verlag DKG, pp. 149–158.

188 Grellner, W., Schwetz, K.A., and Lipp, A. (1981) in *Proceedings 7th Symposium on Special Ceramics* (eds D. Taylor and P. Popper), British Ceramics Research Association, Stoke on Trent, pp. 27–36.

189 Boecker, W., Landfermann, H., and Hausner, H. (1979) *Powder Metall. Int.*, **11**, 83–85.

190 Hausner, H. (1980) in *Proceedings 4th CIMTEC, Energy and Ceramics* (ed. P. Vincenzini), Elsevier, Oxford, New York, pp. 582–595.

191 Tanaka, H., Inomata, Y., Hara, K., and Hasegawa, H. (1985) *J. Mater. Sci. Lett.*, **4**, 315–317.

192 Suzuki, K. (1986) in Pressureless Sintering of SiC with Addition of AL2O3, vol. 36, Research Laboratory of the Asahi Glass Co., Ltd, pp. 25–36.

193 Suzuki, K. (1991) in *SiC Ceramics - 2* (eds S. Somiya and Y. Inomata), Elsevier, London, New York, pp. 163–182.

194 Smoak, R.H. German Patent -DE-OS 27.51.851.

195 Mohr, A. (1989) Untersuchungen zur Minimierung der Additivgehalte für die drucklose Sinterung von α-SiC, Diplomarbeit, Institut für Keramik im Maschinenbau, Universität Karlsruhe.

196 Richerson, D.W. (1992) *Modern Ceramic Engineering*, Marcel Dekker Inc., New York, Basel, Hong Kong.

197 Thümmler, F. and Oberacker, R. (1993) *Introduction to Powder Metallurgy*, The

Institute of Materials Series, University Press, Cambridge.

198 Johnson, C.A. and Prochazka, S. (1977) in *Ceramic Microstructures 1976* (eds R.M. Fullrath and J.P. Pask), Westview Press, Boulder, CO, pp. 366–378.

199 Schwetz, K.A., Isemann, F., and Lipp, A. (1984) in *Proceedings 1st International Symposium on Ceramic Components For Engines* (eds S. Somiya *et al.*), KZK Scientific Publ., Tokyo, Boston, Lancaster, pp. 583–594.

200 Schwetz, K.A., Schäfter, E., and Telle. R., (2003) *Cfi/Ber. DKG 80*, No. 3, E 40–E 45.

201 Schwetz, K.A., Werheit, H., and Nold. E., (2003) *Cfi/Ber. DKG 80*, No. 12, E 37–E 44.

202 Schwetz, K.A., Kempf, T., Saldsleder, D., and Telle, R.(2004) Ceramic Eng. and Science Proceedings **25** [3], 579–588.

203 Schwetz, K.A., Sigl, L., Kempf, T., and Victor, G. (2001) European Patent EP 1070686, Schwetz, K.A., Sigl, L., Kempf, T., and Victor, G., (2003) Patent US 6531423.

204 Rendtel, A., Mössner, B., and Schwetz, K.A., (2005) Ceramic Eng. and Science Proceedings 26, 161–168.

205 (a) Schwetz, K.A., and Hassler, J., (2002) *Cfi/Ber. DKG 79*. No. 10 D 15–D 18. (b) Schwetz, K.A., and Hassler, J., (2002) *Cfi/Ber. DKG 79*. No. 11 D 14–D 19.

206 Tokita, M. (1999) *Mater. Sci. Forum*, **83–88**, 308.

207 Munir, Z.A. and Tamburini, U.A. (2006) *J. Mater. Sci. A*, **41**, 763.

208 Omori, M. (2000) *Mater. Sci. Eng. A*, **287**, 193.

209 Mamedov, V. (2002) *Powder Metall.*, **45** (4), 322–328.

210 Yamamoto, T.A., Kondou, T., Kodera, Y., Ishii, T., Ohyanagi, M., and Munir, Z.A. (2005) *J. Mater. Eng. Perform.*, **14** (4), 460–466.

211 Tokita, M. (1997) *Nyn Seramikkasu*, **10**, 43–53.

212 Tamari, N., Tanaka, T., Tanaka, K., Kawahara, K., and Tokita, M. (1995) *J. Ceram. Soc. Jap.*, **103**, 740–742.

213 Kingery, W.D. (1959) *J. Appl. Phys.*, **30** (3), 301–306.

214 Kingery, W.D. (1959) *J. Appl. Phys.*, **30** (3), 307–310.

215 Petzow, G. and Huppmann, W.J. (1976) *Z. Metallkde*, **67** (9), 579–590.

216 German, R.M. (1985) *Liquid Phase Sintering*, Plenum Press, New York.

217 Lange, F.F. (1975) *J. Mater. Sci.*, **10**, 314–320.

218 Mulla, M.A. and Krstic, V.D. (1994) *Acta Metall. Mater.*, **42** (1), 303–308.

219 Bauda, S., Thevenot, F., Pisch, A., and Chatillon, C. (2003) *J. Eur. Ceram. Soc.*, **23**, 1–8.

220 Bauda, S., Thevenot, F., and Chatillon, C. (2003) *J. Eur. Ceram. Soc.*, **23**, 9–18.

221 (a) Omori, M. and Takei, H. (1982) *J. Am. Ceram. Soc.*, **65**, C–92;(b) Omori, M. and Takei, H. (1995) US Patent 5.439.853.

222 Padture, N.P. and Lawn, B.R. (1992) *J. Am. Ceram. Soc.*, **77**, 2518.

223 Mulla, M.A. and Kristic, V.D. (1994) *J. Mater. Sci.*, **29**, 934.

224 Mulla, M.A. and Kristic, V.D. (1994) *Acta Metall.*, **42**, 303.

225 Padture, N.P. (1994) *J. Am. Ceram. Soc.*, **77**, 519.

226 Kim, Y.W., Mitomo, M., and Hirotsuru, H. (1995) *J. Am. Ceram. Soc.*, **78**, 3145.

227 Mulla, M.A. and Kristic, V.D. (1991) *Am. Ceram. Soc. Bull.*, **70**, 439.

228 Gröbner, J. (1994) Konstitutionsberechnungen im System Y-Al-Si-C-O. Doctoral Thesis, University of Stuttgart.

229 Falk, L.K.L. (1997) *J. Eur. Ceram. Soc.*, **17**, 983–994.

230 Falk, L.K.L. (1998) *J. Eur. Ceram. Soc.*, **18**, 2263–2279.

231 Wang, C., Mitomo, M., and Emoto, H. (1997) *J. Mater. Res.*, **12** (12), 3266–3270.

232 Kaneko, K., Yoshiya, M., Tanaka, I., and Tsurekawa, S. (1999) *Acta Mater.*, **47** (4), 1281–1287.

233 Moberlychan, W.J., Cannon, R.M., Chan, L.H., CaO, J.J., Gilbert, C.J., Ritchie, R.O., and De Jonghe, L.C. (1996) *Mater. Res. Soc. Symp. Proc.*, **410**, 257–262.

234 Moberlychan, W.J. and De Jonghe, L.C. (1998) *Acta Metall.*, **46**, 2471.

235 Nagano, T. and Kaneko, K. (2000) *J. Am. Ceram. Soc.*, **83** (10), 2497–2502.

236 Ca,O, J.J., Moberlychan, W.J., De Jonghe, L.C., Gilbert, C.J., and Ritchie, R.O. (1996) *J. Am. Ceram. Soc.*, **79** (2), 461–469.

237 Volz, E., Roosen, A., Wang, S.-C., and Wei, W.-C.J. (2004) *J. Mater. Sci.*, **39**, 4095–4101.

238 Carpenter, R.W., Braue, W., and Cutler, R.A. (1991) *J. Mater. Res.*, **6** (9), 1937.

239 Hamminger, R., Krüner, H., and Böcker, W. (1992) *J. Hard Mater.*, **3** (2), 93.

240 Kleebe, H.J. (2002) *J. Am. Ceram. Soc.*, **85** (1), 43.

241 Sigl, L.S. and Kleebe, H.J. (1993) *J. Am. Ceram. Soc.*, **76**, 773.

242 Grande, T., Sommerset, H., Hagen, E., Wiik, K., and Einarsrud, M.A. (1997) *J. Am. Ceram. Soc.*, **80**, 1047.

243 Nader, M. (1995) Untersuchung der Kornwachstumsphänomene an flüssigphasengesinterten SiC-Keramiken und ihre Möglichkeit zur Gefügeveränderung. Doctoral Thesis, University of Stuttgart.

244 Lenk, R. and Adler, J. (1997) *J. Eur. Ceram. Soc.*, **17**, 197–202.

245 Kim, Y.-W., Mitomo, M., Emoto, H., and Lee, J.-G. (1998) *J. Am. Ceram. Soc.*, **81** (12), 3136–3140.

246 Lee, S.K. and Kim, C.H. (1994) *J. Am. Ceram. Soc.*, **77**, 1655–1658.

247 Jun, H.W., Lee, H.W., Kim, G.H., Song, H.S., and Kim, B.H. (1997) *Ceram. Eng. Sci. Proc.*, **18**, 487.

248 Kim, Y.-W., Lee, Y.-I., Mitomo, M., Choi, H.-J., and Lee, J.-G. (1999) *J. Am. Ceram. Soc.*, **82** (4), 1058–1060.

249 Chia, K.Y., Böcker, W.D.G., and Storm, R.S. (1994) US Patent 5.298.470.

250 Lin, B.W., Imai, M., Yano, T., and Iseki, T. (1986) *J. Am. Ceram. Soc.*, **69**, C–67.

251 Duval-Riviere, M.-L. and Vicens, J. (1993) *J. de Physique IV France*, **03** (C7), 1417–1421.

252 Moberlychan, W.J., Cannon, R.M., Chan, L.H., Cao, J.J., Gilbert, C.J., Ritchie, R.O., and De Jonghe, L.C. (1996) *Mater. Res. Soc. Symp. Proc.*, **410**, 257–262.

253 Suzuki, K. (1986) *Bull. Ceram. Soc. Jpn*, **21**, 590.

254 Tanaka, H., Inomata, Y., Hara, K., and Hasegawa, H. (1985) *J. Mater. Sci. Lett.*, **4**, 315–317.

255 Coppola, J.A. *et al.*, Japanese Patent Application, 53-121810.

256 Zhou, Y., Tanaka, H., Otani, S., and Bando, Y. (1999) *J. Am. Ceram. Soc.*, **82** (8), 1959–1964.

257 Williams, R.M., Juterbock, B.N., Shinozaki, S.S., Peters, C.R., and Whalen, T.J. (1985) *Am. Ceram. Soc. Bull.*, **64** (10), 1385–1389.

258 Shinozaki, S., Williams, R.M., Juterbock, B.N., Donlon, W.T., Hangas, J., and Peters, C.R. (1985) *Am. Ceram. Soc. Bull.*, **64** (10), 1389–1393.

259 Zhang, X.F., Yang, Q., and de Jonghe, L.C. (2003) *Acta Mater.*, **51** (13), 3849–3860.

260 Tanaka, H., Yoshimura, H.N., and Otani, S. (2000) *J. Am. Ceram. Soc.*, **83** (1), 226–228.

261 Nadeau, J.S. (1973) *Ceram. Bull.*, **52** (2), 170–174.

262 Alliegro, R.A., Coffin, L.B., and Tinklepaugh, J.R. (1956) *J. Am. Ceram. Soc.*, **39**, 386–389.

263 Prochazka, S. and Charles, R.J. (1973) *Am. Ceram. Soc. Bull.*, **52**, 885–891.

264 Kriegesmann, J. (1978) *Ber. Dtsch. Keram. Ges.*, **55**, 391–397.

265 Bind, J.M. and Biggers, J.V. (1976) *J. Appl. Phys.*, **47**, 5171–5174.

266 Broussaud, D. (1976) in *Ceramic Microstructures* (eds R.M. Fullrath and J.A. Pask), Westview Press, Boulder, Colorado, pp. 679–688.

267 Iseki, T., Arakawa, K., Matsuzaki, H., and Suzuki, H. (1983) *Yogyo-Kyokai-Shi*, **91**, 349.

268 Nakamura, K. and Asai, O. (1982) *Kagaku Kogyo*, **33**, 977.

269 Kessel, H. and Gugel, E. (1978) *Industrie Diamanten Rundschau*, **12** (3), 180–185.

270 Kriegesmann, J., Hunold, K., Lipp, A., Reinmuth, K., and Schwetz, K.A. (1981) European Patent 71241.

271 Whalen, T.J., Williams, R.M., and Juterbock, B.N. (1985) 10th Plansee Seminar Proceedings, Reutte, Tirol, p. 783.

272 (a) Hunold, K. (1984) *Powder Metall. Int.*, **16** (5), 236–238;(b) Hunold, K. (1985) *Powder Metall. Int.*, **17**, (2), 91–93.

273 Hunold, K. (1988) Proceedings, Advanced Materials Technology Ceramic Workshop No. 4, Advances in Materials, Processing and Manufacturing Science, International Committee for Advanced Materials Technology, Nagoya, Japan, pp. 49–62.

274 Oberacker, R., Kühne, A., and Thümmler, F. (1987) *Powder Metall. Int.*, **19** (6), 43–50.

275 Fetahagic, F., Oberacker, R., and Thümmler, F. (1989) Emerging materials by advanced processing, in Proceedings of IXth German-Yugoslav Meeting on Materials Science Development (eds W.A. Kaysser and J. Weber-Bock), Berichte der Kernforschungsanlage Jülich, pp. 313–325.

276 Powell, J.A. and Matus, L.G. (1988) *Springer Proc. Phys.*, **34**, 2–12.

277 Davis, R.F. *et al.* (1991) *Proc. IEEE*, **79** (5), 677–700.

278 Hirai, T. and Sasaki, M. (1991) in *SiC Ceramics - 1*, (eds S. Somiya and Y. Inomata), Elsevier, London, New York, pp. 77–98.

279 Pickering, M.A. and Haigis, W. (1993) *Am. Ceram. Soc. Bull.*, **72** (3), 74–78.

280 Tenhover, M., Ruppel, I., Lyle, S.S., and Pilione, L.J. (1993) Proceedings 36th Annual Technical Conference of the Society of Vacuum Coaters, pp. 362–365.

281 Lely, J.A. (1955) *Ber. Dt. Keram. Ges.*, **32**, 229–231.

282 Tairov, Y.M. and Tsvetkov, V.F. (1978) *J. Cryst. Growth*, **43**, 209–212.

283 Ziegler, G., Lanig, P., Theis, D., and Weyrich, C. (1983) *IEEE Trans. Electron Devices*, **30**, 277–281.

284 Carter, C.H. Jr, Tang, L., and Davis, R.F. (1987) Proceedings 4th National Review Meeting on the Growth and Characterization of Silicon Carbide, Raleigh, NC, USA, pp. 15–28.

285 Vassen, R., Stöver, D., and Uhlenbusch, J. (1993) Sintering and grain growth of ultrafine amorphous SiC-Si-Powder mixtures, in Proceedings, 2nd European Ceramic Society Conference, 11–14 September 1991, Augsburg. *Euro-Ceramics II, Vol. 2, Structural Ceramics and Composites* (eds G. Ziegler and H. Hausner), Deutsche Keramische Gesellschaft e.V., Cologne, Germany, p. 791.

286 Vassen, R., Buchkremer, H.P., and Stöver, D. (1996) Verfahren zum Herstellen feinkristalliner Siliciumkarbidkörper, German Patent DE196.42. 753.3-45.

287 Vassen, R. and Stöver, D. (1997) *Philos. Mag. B*, **76** (4), 585.

288 Förster, J., Vassen, R., and Stöver, D. (1995) *J. Mater. Sci. Lett.*, **14**, 214–216.

289 Vassen, R., Förster, J., and Stöver, D. (1995) *Nanostruct. Mater.*, **6**, 889–892.

290 Kaiser, A., Vassen, R., Stöver, T., and Buchkremer, H. (1997) *Nanostruct. Mater.*, **8** (4), 489–497.

291 Zhou, Y., Hirao, K., and Toriyama, M. (2000) *J. Am. Ceram. Soc.*, **83** (3), 654–656.

292 Pan, Z., Pan, W., and Li, R. (2002) *Key Eng. Mater.*, **224–226**, 713–716.

293 Mitomo, M., Kim, Y.-W., and Hirotsuru, H. (1996) *J. Mater. Res.*, **11** (7), 1601–1604.

294 Mitomo, M., Hirotsuru, H., and Kim, Y.-W. (1997) US 5.591.685.

295 Lee, Y.-I., Kim, Y.-W., Mitomo, M., and Kim, D.-Y. (2003) *J. Am. Ceram. Soc.*, **86** (10), 1803–1805.

296 Lee, Y.-I., Kim, Y.-W., and Mitomo, M. (2004) *J. Mater. Sci.*, **39**, 3801–3803.

297 Hotta, M. and Hojo, J. (2006) *Mater. Sci. Forum*, **510–511**, 1022–1025.

298 Becher, P.F. (1991) *J. Am. Ceram. Soc.*, **74** (2), 255–269.

299 Telle, R., Brook, R.J., and Petzow, G. (1991) *J. Hard Mater.*, **2** (1–2), 79–114.

300 Wei, G.C. and Becher, P.F. (1984) *J. Am. Ceram. Soc.*, **67** (8), 571.

301 Janney, M.A. (1987) *Am. Ceram. Soc. Bull.*, **66** (2), 322.

302 Jiang, D.J., Wang, J.H., Li, Y.L., and Ma, L.T. (1989) *Mater. Sci. Eng.*, **A 109**, 401.

303 Ly Ngoc, D. (1989) Gefügeverstärkung von SiC Keramiken, Thesis, University of Stuttgart.

304 Jimbou, R., Takahashi, K., and Matsushita, Y. (1986) *Adv. Ceram. Mater.*, **1**, 341.

305 McMurtry, C.H., Böcker, W.D.G., Seshadri, S.G., Zanghi, J.S., and Garnier, J.E. (1986) *Am. Ceram. Soc. Bull.*, **66**, 325.

306 Cai, H., Gu, W.H., and Faber, K.T. (1990) Proceedings American Ceramics Society Composites, 5th Technical Conference on Composite Materials, pp. 892–901.

307 Faber, K.T., Gu, W.H., Cai, H., Winholtz, R.A., and Magleg, D.J. (1991) in *Toughening Mechanisms in Quasi-Brittle Materials* (ed. S.P. Shah), NATO ASI series. Series E, Applied sciences no. 195, Kluwer, Dordrecht, pp. 3–17.

308 Tani, T. and Wada, S. (1988 European Patent EP 303192.

309 Blanc, C., Thevenot, F., and Treheux, D. (1997) *Key Eng. Mater.*, **132–136**, 968–971.

310 Kuo, D.H. and Kriven, W.M. (1998) *J. Eur. Ceram. Soc.*, **18**, 51–57.

311 Cho, K.S., Choi, H.J., Lee, J.G., and Kim, Y.W. (1998) *J. Mater. Sci.*, **33** (1), 211–214.

312 Chu, H., Lillo, T.M., Merkle, B., Bailey, D.W., and Harrison, M. (2006) Advances in Ceramic Armor: A Collection of Papers Presented at the 29th International Conference on Advanced Ceramics and Composites, January 23–28, 2005, Cocoa Beach, Florida, in *Ceramic Engineering and Science Proceedings*, vol. **26** (No. 7), (eds J.J. Swab, D. Zhu, and W.M. Kriven), Wiley-VCH, Weinheim, pp. 279–286.

313 Schwetz, K.A., Reinmuth, K., and Lipp, A. (1985) *World Ceram.*, **2**, 70–84.

314 Thevenot, F. (1987) Proceedings 9th International Symposium on Boron Borides and Related Compounds (ed. H. Werheit), pp. 246–256.

315 Tomlinson, W.J. and Whitney, J.C. (1992) *Ceram. Int.*, **18**, 207–211.

316 Kevorkiijan, V., Bizjak, A., Vizintin, J., Thevenot, F., Interdonato, G., and Reimondi, C. (1995) 4th Euro Ceramics Vol. 4, BaSiC Science (ed. A. Bellosi), pp. 209–216.

317 Cutler, I.B., Miller, P.D., Rafaniello, W., Park, H.K., Thompson, D.P., and Jack, K.H. (1978) *Nature*, **275**, 434–435.

318 Rafaniello, W., Cho, K., and Virkar, A.V. (1981) *J. Mater. Sci.*, **16**, 3479–3488.

319 Rafaniello, W., Plichta, M.R., and Virkar, A.V. (1983) *J. Am. Ceram. Soc.*, **66** (4), 772–776.

320 Ruh, R. and Zangvil, A. (1982) *J. Am. Ceram. Soc.*, **65** (5), 260–265.

321 Xu, Y., Zangvil, A., Landon, M., and Thevenot, F. (1992) *J. Am. Ceram. Soc.*, **75** (2), 325–333.

322 Zangvil, A. and Ruh, R. (1989) in *Silicon Carbide' 87* (eds J.D. Cawley and L.E. Semler), American Ceramics Society, Westerville, OH, pp. 63–82.

323 Li, J.-L. (2002) *Mater. Sci. Technol.*, **18** (12), 1589–1592.

324 Shirouzu, K., Ohkusa, T., Kawamoto, T., Enomoto, N., and Hojo, J. (2008) *J. Ceram. Soc. Jap.*, **116** (7), 781–785.

325 Xue, H. and Munir, Z.A. (1997) *J. Eur. Ceram. Soc.*, **17**, 1787–1792.

326 Mei, L. and Li, J.T. (2008) *Acta Mater.*, **56** (14), 3543–3549.

327 Kimura, I., Hotta, N., Niwano, H., and Tanaka, M. (1991) *Powder Technol.*, **68**, 153–158.

328 Czekaj, C.L., Hackney, M.L.J., Hurley, W.J. Jr, Interrante, L.V., Sigel, G.A., Schields, P.J., and Slack, G.A. (1990) *J. Am. Ceram. Soc.*, **73**, 352–357.

329 Paine, R.T., Janik, J.F., and Narula, C. (1988) *Mater. Res. Soc. Symp. Proc.*, **121**, 461–464.

330 Liu, G., Chen, K., Li, J., Zhou, H., and Mei, L. (2008) *Cryst. Growth Des.*, **8** (8), 2834–2837.

331 Chen, Z., Tan, S., Zhang, Z., and Jiang, D. (1997) *J. Mater. Sci. Technol.*, **13**, 342–344.

332 Li, J.-F., Sugimori, M., and Watanabe, R. (1999) *Key Eng. Mater.*, **159–160**, 83–88.

333 Kuo, S.Y., Virkar, A.V., and Rafaniello, W. (1987) *J. Am. Ceram. Soc.*, **70**, C–125.

334 Lee, R.R. and Wei, W.C. (1990) *Ceram. Eng. Sci. Proc.*, **11** (7–8), 1094–1121.

335 Ohyanagi, M., Shirai, K., Balandina, N., and Hisa, M. (2000) *J. Am. Ceram. Soc.*, **83**, 1108–1112.

336 Carrillo-Heian, E.M., Xue, H., Ohyanagi, M., and Munir, Z.A. (2000) *J. Am. Ceram. Soc.*, **83**, 1103–1107.

337 Jackson, T.B., Hurford, A.C., Brunner, S.L., and Cutler, R.A. (1998) in *Silicon Carbide '87* (eds J.W. Cawley and C. Semler), American Ceramics Society, Columbus, Ohio, pp. 227–240.

338 Huang, J.L., Herford, A.C., Cutler, R.A., and Virkar, A.V. (1986) *J. Mater. Sci.*, **21**, 1448–1456.

339 Lihrmann, J.M. and Tirlocq, J. (1997) World Patent, PCT WO 97/06119.

340 Leuchs, M. (2008) in *Ceramic Matrix Composites, Fiber Reinforced Ceramics and their Applications* (ed. W. Krenkel), Wiley-VCH, Weinheim, pp. 141–161.

341 Motz, G., Schmidt, S., and Beyer, S. (2008) in *Ceramic Matrix Composites, Fiber Reinforced Ceramics and their Applications* (ed. W. Krenkel), Wiley-VCH, Weinheim, pp. 165–184.

342 Heidenreich, B. (2008) in *Ceramic Matrix Composites, Fiber Reinforced Ceramics and their Applications* (ed. W. Krenkel), Wiley-VCH, Weinheim, pp. 113–137.

343 Stinton, D.P., Lowdon, R.A., and Krabill, R.H. (1990) Mechanical Property Characterization of Fiber-Reinforced SiC Matrix Composites. Oak Ridge National Laboratory Report ORNL/TM-11524 April.
344 Krenkel, W. and Fabig, J. (1995) Proceedings of the10th International Conference on Composite Materials ICCM-10, 4, pp. 601–609.
345 Heidenreich, B. Proceedings Sixth International Conference on High Temperature Ceramic Matrix Composites, HTCMC-6, New Delhi, India.
346 Byrne, C.E. and Nagle, D.E. (1997) *Carbon*, **35**, 26.
347 Greil, P. (2001) *J. Eur Ceram. Soc.*, **21**, 105.
348 Ota, T., Takahashi, M., Hibi, T., Ozawa, T., Suzuki, S., Hikichi, Y., and Suzuki, H. (1995) *J. Am. Ceram. Soc.*, **78**, 3409.
349 Gibson, L.J. (1992) *Met. Mater.*, **8**, 333.
350 Byrne, C.E. and Nagle, D.E. (1996) US Patent 6051096; (1998) Patent US6124028.
351 Byrne, C.E. and Nagle, D.E. (1997) *Mater. Res. Innovat.*, **1**, 137.
352 Greil, P., Lifka, T., and Kaindl, A. (1998) *J. Eur. Ceram. Soc.*, **18**, 1961.
353 Guanjun, Q., Rong, M., Ning, C., Chunguang, Z., and Zhihao, J. (2002) *J. Mater. Process. Technol.*, **120**, 107.
354 Zollfrank, C. and Sieber, H. (2005) *J. Am. Ceram. Soc.*, **88** (1), 51–58.
355 Vogli, E., Mukerji, J., Hoffman, C., Kladny, R., Sieber, H., and Greil, P. (2001) *J. Am. Ceram. Soc.*, **84**, 1236.
356 Sieber, H., Vogli, E., Mueller, F., Greil, P., Popovska, N., Gerhard, H., and Emig, G. (2002) *Key Eng. Mater.*, **206–213**, 2013.
357 Greil, P., Vogli, E., Fey, T., Bezold, A., Popovska, N., Gerhard, H., and Sieber, H. (2002) *J. Eur. Ceram. Soc.*, **22**, 2697–2707.
358 Miracle, D.B. (2001) *Composites, ASM Handbook*, vol. **21**, ASM International, pp. 579–587.
359 Chawla, N. and Chawla, R.(eds) (2006) *Metal Matrix Composites*, Springer Science + Business Media, Inc., p. 399.
360 Kainer, K.U.(ed.) (2006) *Metal Matrix Composites*, Wiley-VCH, Weinheim, p. 266.
361 Skibo, M.D., Schuster, D.M., and Bruski, R.S. (1996) US Patent 5531425.
362 Blucher, J.T. (1992) *J. Mater. Process. Technol.*, **30** (2), 381–390.
363 Mortensen, A., Michand, V.J., and Flemings, M.C. (1993) *JOM*, **45** (1), 36–43.
364 Cook, A.J. and Werner, P.S. (1991) *Mater. Sci. Eng. A*, **A144**, 189–206.
365 Evans, R.W., Letham, A.G., and Brooks, R.G. (1985) *Powder Metall.*, **28** (1), 13–19.
366 Willis, T.C. (1988) *Met. Mater.*, **4**, 485–488.
367 Huda, D., El Baradie, M.A., and Hashmi, M.S.J. (1993) *J. Mater. Process. Technol.*, **37** (1–4), 513–518.
368 Ghosh, A. (1993) in *Fundamentals of Metal Matrix Composites* (eds S. Suresh, A. Mortensen, and A. Needleman), Butterworth-Heinemann, Newton, MA, pp. 23–41.
369 Ferguson, B.L. and Smith, O.D. (1984) in *Powder Metallurgy*, vol. 7, Metals Handbook, 9th edn, ASM International, p. 537.
370 Ward-Close, C.M. and Partridge, P.G. (1990) *J. Mater. Sci.*, **25**, 4315–4323.
371 Shaw, N.J. (1989) *Powder Metall. Int.*, **21** (16), 31.
372 Greskovich, C. and Rosolowski, J.H. (1976) *J. Am. Ceram. Soc.*, **59**, 336.
373 Prochazka, S. (1989) GE Corp. Res. and Dev. Center Technical Information Series, Report No. 89CRD025, Schenectady, New York.
374 Grabchuk, B.L. and Kislyi, P.S. (1976) *Sov. Powder Metall. Met. Ceram.*, **15**, 675.
375 Adlassing, K. (1958) *Planseeberichte Pulvermetallurgie*, **6**, 92.
376 Grabchuk, B.L. and Kislyi, P.S. (1974) *Poroshkovaya Metallurgiya*, **8**, 11.
377 Tkachenko, Y.G., Kovtun, V.I., Britun, V.F., Yurchenko, D.Z., and Bovkun, G.A. (2005) *Powd. Metall. Met. Ceram.*, **44** (7–8), 4–7.
378 Kriegesmann, J. (1989) German Patent DE 37.11.871.C2.
379 Lange, R.G., Munir, Z.A., and Holt, J.B. (1980) *Mater. Sci. Res.*, **13**, 311.
380 Kanno, Y., Kawase, K., and Nakano, K. (1987) *J. Ceram. Soc. Jpn*, **95**, 1137.
381 Kislyi, P.S. and Grabchuk, B.L. (1975) Proceedings 4th European Symposium on Powder Metallurgy, Société Francaise Metallurgie, Grenoble, pp. 10–21.

382 Dole, S.L. and Prochazka, S. (1985) in *Ceramic Engineering and Science Proceedings*, vol. **6** (7/8), (ed. W.J. Smothers), American Ceramics Society, Westerville, OH, pp. 1151–1160.
383 Prochazka, S. U.S. Patent US 4005235.
384 Zhakariev, Z. and Radev, D. (1977) *J. Mater. Sci. Lett.*, **7**, (1988) 695.
385 Glasson, D.R. and Jones, J.A. (1969) *J. Appl. Chem.*, **19** (5), 125.
386 Janes, S. and Nixdorf, J. (1966) *Ber. Dtsch. Keram. Ges.*, **43**, 136.
387 Stibbs, D., Brown, C.G., and Thompson, R. (1973) US Patent US 3749571.
388 (a) Schwetz, K.A. and Vogt, G. (1977) German Patent DE 27.51.998;(b) Schwetz, K.A. and Vogt, G. (1980) US Patent US 4.195.066.
389 Henney, J.W. and Jones, J.W.S. (1978) British Patent Appl. 2014193A.
390 Suzuki, H. and Hase, T. (1979) Proceedings, Conference on Factors in Densification of Oxide and Nonoxide Ceramics Japan (eds S. Somiya and S. Saito), p. 345.
391 Schwetz, K.A. and Grellner, W.J. (1981) *J. Less-Common Met.*, **82**, 37.
392 Dole, S.L., Prochazka, S., and Doremus, R.H. (1989) *J. Am. Ceram. Soc.*, **72**, 958.
393 Bougoin, M., Thevenot, F., Dubois, F., and Fantozzi, G. (1985) *J. Less-Common Met.*, **114**, 257.
394 Matje, P. (1990) Internal Report, Elektroschmelzwerk Kempten GmbH, Germany.
395 Sigl, L.S. and Schwetz, K.A. (1991) *Euro-Ceramics II*, (1), 517.
396 Speyer, R.F., Lee, H., and Bao, Z. (2004) US Patent Appl. Serial No. 60/638,007; European Pat. EP0755161 (2004); World Patent WO 110685 (2004).
397 Grabchuk, B.L. and Kislyi, P.S. (1975) *Sov. Powder Metall. Met. Ceram.*, **14**, 338.
398 Ob, J.H., Orr, K.K., Lee, C.K., Kim, D.K., and Lee, J.K. (1985) *J. Korean Ceram. Soc.*, **22**, 60.
399 Schwetz, K.A., Reinmuth, K., and Lipp, A. (1983) *Sprechsaal*, **116**, 1063.
400 Telle, R., Brook, R.J., and Petzow, G.J. (1991) *J. Hard Mater.*, **2**, 79.
401 Lörcher, R., Strecker, K., Riedel, R., Telle, R., and Petzow, G. (1990) *Solid State Phenomena 8 & 9, Proceedings, International Conference on Sintering of Multiphase Metal and Ceramic Systems* (ed. G.S. Upadhyaya), Sci-Tech Publications, Vaduz, Gower Publ. Co., Brookfield VT, pp. 479–492.
402 Bougoin, M. and Thevenot, F. (1987) *J. Mater. Sci.*, **22**, 109.
403 Sigl, L.S. (1998) *J. Eur. Ceram. Soc.*, **18** (11), 1521.
404 Telle, R. (1990) *The Physics and Chemistry of Carbides, Nitrides and Borides*, vol. **185** (ed. R. Freer), NATO ASI Series E, Kluwer, Dordrecht, The Netherlands, pp. 249–268.
405 Telle, R. and Petzow, G. (1987) *High Tech Ceramics* (ed. P. Vincenzini), Material Science Monographs, Elsevier, Amsterdam, pp. 961–973.
406 Lukas, H.L. (1990) *Constitution of Ternary Alloys 3* (eds G. Petzow and G. Effenberg), VCH, Weinheim, pp. 140–146.
407 Williams, P.D. and Hawn, D.C. (1991) *J. Am. Ceram. Soc.*, **74**, 1614.
408 Matsumoto, A., Goto, T., and Kawakami, A. (2004) *J. Ceram. Soc. Jpn*, **112** (1305), 399–402.
409 Schwetz, K.A., Sigl, L.S., and Pfau, L. (1997) *J. Solid-State Chem.*, **133**, 68.
410 Kuzenkova, P.S., Kislyi, P.S., Grabchuk, B.L., and Bodnaruk, N.I. (1979) *J. Less-Common Met.*, **67**, 217.
411 Ostapenko, J.T., Slezov, V.V., Tarasov, R.V., Kartsev, N.F., and Podtykan, V.P. (1979) *Sov. Powder Metall. Met. Ceram.*, **19**, 312.
412 Brodhag, C., Bouchacourt, M., and Thevenot, F. (1983) in *Ceramic Powders* (ed. P. Vincenzini), Material Science Monographs, vol. **16**, Elsevier, Amsterdam, pp. 881–890.
413 Ekbom, L.B. and Amundin, C.O. (1980) *Sci. Ceram.*, **10**, 237.
414 Champagne, B. and Angers, R. (1979) *J. Am. Ceram. Soc.*, **62**, 149.
415 Telle, R. and Petzow, G. (1988) *Mater. Sci. Eng.*, **A105/106**, 97.
416 Vasilos, T. and Dutta, S.K. (1974) *Ceram. Bull.*, **53**, 453.
417 Furukawa, M. and Kitahara, T. (1979) *Nippon Tungsten Rev.*, **12**, 55.
418 Mikijelj, B., Victor, G., and Schwetz, K.A. (2007) US Patent 7309672.
419 Larker, H.T., Hermansson, L., and Adlerborn, J., (1988) *J. Ind. Ceram.*, **8**, 17.

420 Schwetz, K.A., Grellner, W., and Lipp, A. (1986) in *Proceedings 2nd International Conference on Science of Hard Materials* (eds E.A. Almond, C.A. Brookes, and R. Warren), Institute of Physics Conference Series, Elsevier, London, pp. 415–426.
421 Kim, K.H., Chae, J.H., Park, J.S., Kim, D.K., Shim, K.B., and Lee, B.H. (2007) *J. Ceram. Process. Res.*, **8** (4), 238–242.
422 Anselmi-Tamburini, U. and Munir, Z.A. (2005) *J. Am. Ceram. Soc.*, **88** (6), 1382–1387.
423 Talmy, I.G. and Zavskoski, J.A. (2000) US Patent 6069101.
424 Bougoin, M. and Thevenot, F. (1987) *J. Mater. Sci.*, **22**, 109.
425 Telle, R. (1991) in *Boron-Rich Solids, 10th International Symposium on Boron, Borides, and Related Compounds*, AIP Conference Proceedings 231 (eds D. Emin, T.L. Aselage, A.C. Swickendick, B. Morosin, and C.L. Beekel), American Institute of Physics, New York, pp. 553–560.
426 Narushima, T., Goto, T., Maruyama, M., Arashi, H., and Iguchi, Y. (2003) *Mater. Trans.*, **44** (3), 401–406.
427 Nishiyama, K., Mitra, I., Momozawa, N., Watanabe, T., Abe, M., and Telle, R. (1997) *J. Jpn. Res. Inst. Mater. Technol.*, **15**, 292.
428 Nowotny, H., Benesovsky, F., and Brukl, C. (1961) *Monatsch. Chem.*, **92**, 403.
429 Holleck, H., Leiste, H., and Schneider, W. (1987) in *High Tech Ceramics, Proceedings 6th CIMTEC* (ed. P. Vincenzini), Elsevier, Amsterdam, pp. 2609–2622.
430 Hofmann, H. and Petzow, G. (1986) *J. Less-Common Met.*, **117**, 121.
431 Telle, R. and Petzow, G. (1986) in *Horizons of Powder Metallurgy II, Proceedings International Powder Metallurgy Conference Exhibit* (eds W.A. Kaysser and W.J. Huppmann), Verlag Schmid, Freiburg, pp. 1155–1158.
432 Telle, R., Meyer, S., Petzow, G., and Franz, E.D. (1988) *Mater. Sci. Eng.*, **A105/106**, 125.
433 Kim, D.K. and Kim, C.H. (1988) *Adv. Ceram. Mater.*, **3**, 52.
434 Lange, D. and Holleck, H. (1985) Proceedings 11th International Plansee-Conference Planseewerke, Reutte, HM50, p. 747.
435 Skorokhod, V. and Krstic, V.D. (2000) *J. Mater. Sci. Lett.*, **19**, 237–239.
436 Yamada, S., Hirao, K., Yamauchi, Y., and Kanzaki, S. (2003) *J. Eur. Ceram. Soc.*, **23**, 1123–1130.
437 Nishiyama, K. and Umekawa, S. (1985) *Trans. Jpn. Soc. Ceram. Mater.*, **11** (2), 53.
438 Sigl, L.S. and Schwetz, K.A. (1994) Proceedings 11th International Symposium on Boron, Borides and Related Compounds, Tsukuba, Japan, August 22–26, 1993 (eds R. Uno and I. Higashi), Japanese Journal of Applied Physics Series, vol. 10, p. 224.
439 Mogilevski, P., Gutmanas, E.Y., Gotman, I., and Telle, R. (1995) *J. Eur. Ceram. Soc.*, **15**, 527.
440 Dudina, D.V., Hulbert, D.M., Dongtao, J., Unuvar, C., Cytron, S.J., and Mukherjee, A.K. (2008) *J. Mater. Sci.*, **43** (10), 3569–3576.
441 Heian, E.M., Khalsa, S.K., Lee, J.W., Munir, Z.A., Yamamoto, T., and Ohyanagi, M. (2004) *J. Am. Ceram. Soc.*, **87** (5), 779–783.
442 Telle, R. (1997) *Ceramics for High-Tech Applications, Materials Science, European Concerted Action COST 503 Powder Metallurgy – Powder Based Materials* (ed. T. Valente), The European Communities, Brussels.
443 Sigl, L.S. and Kleebe, H.J. (1994) *J. Am. Ceram. Soc.*, **78** (9), 2374.
444 Halverson, D.C., Pyzik, A.J., Aksay, J.A., and Snowden, W.E.J. (1989) *J. Am. Ceram. Soc.*, **72**, 775.
445 Pyzik, A.J., Deshmukh, U.V., Dunmead, S.D., Ott, J.J., Allen, T.L., and Rossow, H.E. (1996) US Patent 5521016.
446 Carden, R.A. (1999) US Patent 5.980.602.
447 Kinney, V.L. and Rockwell, T. (1949) Boral: A New Thermal Neutron Shield, Oak Ridge National Laboratory Report ORNL 242.
448 Kitzes, A.S. and Hullings, W.Q. (1953) Oak Ridge National Laboratory Report ORNL 981.
449 Greim, J. and Schwetz, K.A. (2002) Boron Carbide, Boron Nitride, and Metal Borides, in *Ullmann's Encyclopedia of*

Technical Chemistry (1982), 4th edn, vol. **21**, Verlag Chemie, 431–438.
450 Shaffer, P.T.B. (1971) *Appl. Opt.*, **10** (5), 1034–1036.
451 Zückler, K. (1956) in *Halbleiterprobleme III, Advances in Solid State Physics Bd. HP3* (ed. W. Schottky), Springer-Verlag, Berlin, Heidelberg, pp. 207–229.
452 Hagen, S.H. (1970) *Ber. Dtsch. Keram. Ges.*, **47**, 630–634.
453 Gugel, E., Schuster, P., and Senftleben, G. (1972) *Stahl Eisen*, **92**, 144–149.
454 (1971) *JANAF Thermochemical Tables*, 2nd edn, NSRDS-NBS37, Washington, D.C.
455 Niihara, K. (1984) *Am. Ceram. Soc. Bull.*, **63**, 1160–1164.
456 Page, T.F. (1978) *Proc. Br. Ceram. Soc.*, **26**, 193–208.
457 Kollenberg, W., Mössner, B., and Schwetz, K.A. (1990) *VDI-Berichte*, **804**, 347–358.
458 Washburn, M.E. and Coblenz, W.S. (1988) *Ceram. Bull.*, **67**, 356–363.
459 Thümmler, F. (1980) *Sintering Processes* (ed. G.C. Kuczynski), Plenum Publ. Corp., pp. 247–277.
460 Kriegesmann, J., Lipp, A., Reinmuth, K., and Schwetz, K.A. (1984) in *Ceramics for High Performance Appl. III* (eds E.M. Lenoe*et al.*), Plenum Publ. Corp., pp. 737–751.
461 Hamminger, R., Grathwohl, R.G., and Thümmler, F. (1986) in *Proceedings 2nd International Conference on the Science of Hard Materials, Rhodes, September 1984, Institute of Physics Conference Series, No. 75*, Hilger, Bristol, pp. 279–292.
462 Grathwohl, G., Reets, Th., and Thümmler, F. (1981) *Sci. Ceram.*, **11**, 425–431.
463 Chermant, J.L., Moussa, R., and Osterstock, F. (1981) *Rev. Int. Hautes Temper. Refract. Fr.*, **18**, 5–55.
464 Wiebke, G. (1960) *Ber. Dtsch. Keram. Ges.*, **37**, 219–226.
465 Schlichting, J. (1979) *Ber. Dtsch. Keram. Ges.*, **56**, 196–199 and 256–261.
466 Schlichting, J. and Schwetz, K. (1982) *High Temp.-High Press.*, **14**, 219–223.
467 Frisch, B., Thiele, W.R., Drumm, R., and Münnich, B. (1988) *Ber. Dtsch. Keram. Ges.*, **65**, 277–284.
468 Presser, V. and Nickel, K.G. (2008) *Crit. Rev. Solid State*, **33** (1), 1–99.
469 Konopicky, K., Patzak, I., and Dohr, H. (1972) *Glas Email Keramo. Tech.*, **23**, 81–87.
470 McKee, D.W. and Chatterji, D. (1976) *J. Am. Ceram. Soc.*, **59**, 441–444.
471 Jepps, N.W. and Page, T.F. (1981) *J. Microsc.*, **127**, 227–237.
472 Shaffer, P.T.B. (1966) *Ceram. Age*, **82**, 42–44.
473 Iseki, T., Kameda, T., and Maruyama, T. (1984) *J. Mater. Sci.*, **19**, 1692–1698.
474 Labus, T. (1981) *Lubr. Eng.*, **37** (7), 387–394.
475 Eisner, J.H. (1982) *Chemie-Anlagen + Verfahren*, **46**, 51 54.
476 Richerson, D.W. (1981) *Mater. Sci. Res.*, **14**, 661–676.
477 Miyoshi, K. (1981) in *Proceedings, International Conference on the Wear of Materials* (eds S.K. Rhee, A.W. Ruff, and K.C. Ludema), ASME, New York.
478 Lashway, R.W., Seshadri, S.G., and Srinivasan, M. (1983) Various forms of SiC and their effects on seal performance. 38th Annual Meeting of the American Society of Lubrication Engineers, Houston, Texas.
479 Knoch, H., Kracker, J., and Schelken, A. (1983) *Chemie-Anlagen + Verfahren*, **2**, 28–30.
480 Knoch, H., Kracker, J., and Schelken, A. (1985) *Chemie-Anlagen + Verfahren*, **3**, 101–104.
481 Knoch, H., Kracker, J., and Schelken, A. (1985) *World Ceram.*, **2**, 96–98.
482 Derby, J., Seshadri, S.G., and Srinivasan, M. (1986) in *Fracture Mechanics of Ceramics*, vol. **8** (ed. R.C. Bradt*et al.*), Plenum, New York, pp. 113–125.
483 Cramner, D.C. (1985) *J. Mater. Sci.*, **20**, 2029–2037.
484 Miyoshi, K., Buckley, D.H., and Srinivasan, M. (1983) *Ceram. Bull.*, **62**, 494–500.
485 Breznak, J., Breval, E., and Macmillan, N.H. (1985) *J. Mater. Sci.*, **20**, 4657–4680.
486 Breval, E., Breznak, J., and Macmillan, N.H. (1986) *J. Mater. Sci.*, **21**, 931–935.
487 Wu, C.C., Rice, R.W., Platt, B.A., and Carrir, S. (1985) *Ceram. Sci. Eng. Proc.*, **6**, 1023.
488 Smythe, J.R. and Richerson, D.W. (1983) *Ceram. Sci. Eng. Proc.*, **4**, 663–673.

489 Lindberg, L.J. and Richerson, D.W. (1985) *Ceram. Sci. Eng. Proc.*, **6**, 1059–1066.

490 Tomizawa, H. and Fisher, T.E. (1987) *ASLE Trans.*, **30**, **1**, 41–46.

491 Knoch, H. and Kracker, J. (1987) *cfi/Ber. DKG*, **64**, 159–163.

492 Boch, P., Platon, F., and Kapelski, G. (1989) *J. Eur. Ceram. Soc.*, **5**, 223–228.

493 Habig, K.H. and Woydt, M. (1989) Proceedings 5th International Congress on Tribology, Vol. 3, Espoo, Finland (eds K. Holmberg and I. Nieminen), p. 106.

494 Yamamoto, Y., Okamoto, K., and Ura, A. (1989) Proceedings 5th International Congress on Tribology, Vol. 3, Espoo, Finland (eds K. Holmberg and I. Nieminen), p. 138.

495 Martin, J.M., LeMogne, T., Montes, H., and Gardos, N.N. (1989) Proceedings 5th International Congress on Tribology, Vol. 3, Espoo, Finland (eds K. Holmberg and I. Nieminen), p. 132.

496 Sasaki, S. (1989) *Wear*, **134**, 185–200.

497 Denape, J. and Lamon, J. (1990) *J. Mater. Sci.*, **25**, 3592–3604.

498 Maurin-Perrier, P., Farjandon, J.P., and Cartier, M. (1991) *Wear Mater.*, **8** (2), 585–588.

499 Maurin-Perrier, P. (1992) Final Technical Report BRITE-EURAM Project, Proposal P-2231, Contract No. RJ-1b-295.

500 Gardos, N.N. (1990) Determination of the Tribological Fundamentals of Solid/ Lubricated Ceramics, WRDC-TR-90-4096. Hughes Aircraft, El Segundo, CA 90245.

501 Löffelbein, B., Woydt, M., and Habig, K.H. (1993) *Wear*, **162–164**, 220–228.

502 Andersson, P. and Blomberg, A. (1994) *Wear*, **174**, 1–7.

503 Kitaoka, S., Tsuji, T., Katoh, T., Yamaguchi, Y., and Kashiwagi, K. (1994) *J. Am. Ceram. Soc.*, **77** (7), 1851–1856.

504 Knoch, H., Sigl, L., and Long, W.D. (1990) Product development with pressureless sintered SiC. 37th Sagamore Army Materials Research Conference. AMTL-Watertown, Massachusetts, 1 October 1990.

505 Knoch, H. and Fundus, M. (1995) *Ceram. Tech. Int.*, 59–63.

506 Wildhack, S., Kracker, J., and Rendtel, A. (2008) New technology for SiC components for mechanical seals and sealless pumps. Proceedings VDMA International Rotating Equipment Conference, Düsseldorf, Germany.

507 Etsion, I. and Halperin, G. (2003) *Sealing Technol.*, **3**, 6–10.

508 Fundus, M. and Knoch, H. (1997) Diamond-like carbon coatings – tribological possibilities and limitations in applications on sintered silicon carbide bearing and seal faces. Proceedings of the 14th International Pump Users Symposium, Houston, USA, pp. 93–98.

509 (a) Thaler, H., Schwetz, K.-A., and Kayser, A. (2004) Patent US 6.777.076;(b) Thaler, H., Schwetz, K.-A., and Kayser, A. (2002) European Patent EP 1.188.942.B2.

510 (a) Meschke, F. (2004) German Patent DE 102.35.269.A1;(b) Meschke, F. (2004) World Patent WO 2004/013505 A1.

511 Schmechel, R. and Werheit, H. (1999) *J. Phys. Condens. Matter*, **11**, 6803–6813.

512 Werheit, H. and Schmechel, R. (1998) Boron, in *Landolt-Börnstein, Numerical Data and functional relationships in Science and Technology Group III, Vol. 41C* (ed. O. Madelung), Springer, pp. Berlin, pp. 3–148.

513 Schmechel, R. and Werheit, H. (1997) *J. Solid-State Chem.*, **133**, 335–341.

514 Schmechel, R. and Werheit, H. (1998) *J. Mater. Process. Manufact. Sci.*, **6**, 329–337.

515 Aoki, Y., Miyazaki, Y., Okino, Okinaka, N., Yatu, S., Kayukawa, N., Takahashi, H., Hirai, T., Omori, M., and Abe, Y. Proceedings 33rd Plasmadynamics and Laser Conference, 20–23 May 2002, Maui, Hawaii, AIAA-2002-2213.

516 Werheit, H. (2009) *Proceedings, 16th International Symposium on Boron, Borides and Related Materials, September 7–12, Matsue, Shimane, Japan*, Journal of Physics Conference Series, 176 012037 (ed. T. Tanaka), IOP Publishing Ltd.

517 Werheit, H. (2006) *Proceedings 25th International Conference on Thermoelectrics, August 2006, Vienna, Austria* (ed. P. Rogl), IEEE, Piscataway, NJ, pp. 159–163.

518 Schmechel, R. and Werheit, H. (1999) *J. Phys. Condens. Matter*, **11**, 6803–6813.

519 Werheit, H., Kuhlmann, U., Laux, M., and Telle, R. (1993) *Proceedings 11th International Conference on Boron, Borides and Related Compounds* (eds R. Uno and I. Higashi), Japanese Journal of Applied Physics Series, vol. **10**, pp. 86–87.

520 Werheit, H., Kuhlmann, U., Laux, M., and Telle, R. (1994) *J. Alloys Compd.*, **209**, 181.

521 Schmechel, R., Werheit, H., and Robberding, K. (1997) *J. Solid-State Chem.*, **133**, 254–259.

522 Armstrong, D.R., Bolland, J., Perkins, P., Will, G., and Kirfel, A. (1983) *Acta Crystallogr. B*, **39**, 324.

523 Gosset, D., Guery, M., and Kryger, B. (1991) Proceedings, 10th International Conference on Boron, Borides and Related Compounds, Albuquerque, New Mexico, 1990, in *AIP Conference Proceedings 231* (eds D. Emin, T.L. Aselage, A.C. Switendick, B. Morosin, and C.L. Beckel), pp. 380–383.

524 Kuhlmann, U. and Werheit, H. (1993) in *Proceedings 11th International Conference on Boron, Borides and Related Compounds, Tsukuba, Japan* (eds R. Uno and I. Higashi), Japanese Journal of Applied Physics Series, vol. **10**, pp. 84–85.

525 Kosolapova, T.Y. (1990) *Handbook of High-Temperature Compounds*, Hemisphere Publishing Corporation, 225, ISBN 0-89116-849-4.

526 Keski-Kuba, R.A.M., Larruquert, J.I., Gum, J.S., and Fleetwood, C.M. (1999) in *Ultraviolet-Optical Space Astronomy Beyond HST* (eds J.A. Morse, J.M. Shull and A.L. Kinney), ASP Conference Series, **164**, 406–419.

527 Larruquert, J.I. and Keski-Kuha, R.A.M. (2000) *Appl. Opt.*, **39** (10), 1537–1540.

528 Thevenot, F.J. (1990) The Physics and Chemistry of Carbides, Nitrides, and Borides, in *NATO ASI Series E*, vol. **185**, Kluwer, pp. Dordrecht, NL, pp. 87–96.

529 Thevenot, F. and Bouchacourt, M.L. (1979) *L'Industrie Ceramique*, **732**, 655.

530 Greim, J. and Schwetz, K.A. (2009) Boron carbide, boron nitride, and metal borides, in *Ullmann's Encyclopedia of Industrial Chemistry, (7th edn on CD ROM)*, Wiley-VCH, Weinheim, Germany.

531 McClellan, K.J., Chu, F., Roper, J.M., and Shindo, I. (2001) *J. Mater. Sci.*, **36** (14), 3403–3407.

532 Frantsevitch, I.N., Gnesin, G.G., and Kudyumov, A.V. (1980) *Superhard Materials*, Naukova Dumka, Kiev, p. 296.

533 Chheda, M., Normandia, J., and Shih, J. (2006) *Ceram. Ind.*, 124–126.

534 Dandekar, D.P. (2001) Shock Response of Boron Carbide, Technical Report U.S. Army Research Laboratory, ARL-TR-2456.

535 Chen, M., McCauley, J.W., and Hemker, K.J. (2003) *Science*, **299** (5612), 1563–1566.

536 Domnich, V., Gogotsi, Y., Trenary, M., and Tanaka, T. (2002) *Appl. Phys. Lett.*, **81** (20), 3783–3785.

537 Ge, D., Domnich, V., Juliano, T., Stach, E.A., and Gogotsi, Y. (2004) *Acta Mater.*, **52**, 3921–3927.

538 Franchini, G., McCauley, J.W., and Chhowalla, M. (2006) *Phys. Rev. Lett.*, **97** (3), 035502.

539 Naslain, R.R., Pailler, R., Bourat, X., Bertrand, S., Heurtevent, F., Dupel, P., and Lamouroux, F. (2001) *Solid State Ionics*, **141–142**, 541–548.

540 Matje, P. and Schwetz, K.A. (1989) *Proceedings 2nd International Conference on Ceramic Powder Processing Science* (eds H. Haussner, G.L. Messing, and S. Hirano), DKG, Köln, pp. 377–384.

541 Heuberger, M., Telle, R., and Petzow, G. (1992) *Powder Metall.*, **35**, 125.

542 Steinbrück, M., Meier, A., Stegmeier, U., and Steinbock, L. (2004) Experiments on the oxidation of boron carbide at high temperatures. Scientific report, Forschungszentrum Karlsruhe, Wissenschaftliche Berichte FZKA 6979.

543 Veshchunov, M.S., Berdyshev, A.V., Boldyrev, A.V., Palagin, A.V., Shestak, V.E., Steinbrück, M., and Stuckert, J. (2005), Modelling of B4C oxidation by steam at high temperature based on separate-effects tests and its application to the bundle experiment QUENCH-07. Scientific report, Forschungszentrum Karlsruhe, Wissenschaftliche Berichte FZKA 7118.

544 Coes, L. Jr (1971) Abrasives, in *Applied Mineralogy*, vol. **1**, pp. Springer-Verlag, Wien.

545 Martin, K. (1972) *Fachberichte Oberflächentechnik*, **10** (6), 197–202.

546 Spur, G. and Sabotka, I. (1987) *Z. Wirtschaft. Fertig.*, **82** (7), 275–380.

547 Spur, G. and Simpfendörfer, D. (1988) *Z. Wirtschaft. Fertig.*, **83** (4), 207–212.

548 Britsch, H.B. (1976) *Ber. Dtsch. Keram. Ges.*, **53** (5), 143–149.

549 Benecke, T., Venkateswaran, S., Schubert, W.D., and Lux, B. (1993) *Gießerei*, **80** (19), 256–662.

550 Nair, S.V., Tien, J.K., and Bates, R.C. (1985) *Int. Met. Rev.*, **30** (6), 275–290.

551 Lipp, A. (1970) *Feinwerktechnik*, **74** (4), 150–154.

552 Gugel, E. (1966) *Ber. Dtsch. Keram. Ges.*, **43**, 354–359.

553 (a) Fickel, A.F. (1980) *Sprechsaal*, **113** (7), 517–531;(b) Fickel, A.F. (1980) *Sprechsaal*, **113** (10), 737–747.

554 Rasch, R. and Maatz, H. (1978) *Maschinenschaden*, **51**, 145–147.

555 Stavric, Z. and Hue, M. (1975) *Keram. Z.*, **27**, 125–128.

556 Bierbauer, G. (1972) *Keram. Z.*, **24**, 142–145.

557 Rubisch, O. and Schmitt, R. (1966) *Ber. Dtsch. Keram. Ges.*, **43**, 173–179.

558 Buchner, E. and Rubisch, O. (1974) in *Silicon Carbide – 1973* (eds R.C. Marshall*et al.*), University of South Caroline Press, Columbia, pp. 428–434.

559 Nakamura, Y. and Yajima, S. (1982) *Am. Ceram. Soc. Bull.*, **61**, 572–573.

560 Huether, W. (1982) Thermoelement zur Temperaturmessung und Verfahren zur Herstellung desselben. European Patent EP. 72 430.

561 Kuomoto, K., Shimohigashi, M., Takeda, S., and Yanagida, H. (1987) *J. Mater. Sci. Lett.*, **6**, 1453–1455.

562 Davis, R. (1990) in *The Physics and Chemistry of Carbides, Nitrides and Borides* (ed. R. Freer), Kluwer Academic Publ., Dordrecht, pp. 589–623.

563 Srinivasan, M. (1989) in *Structural Ceramics - Treatise on Materials Science and Technology*, vol. **29** (ed. J.B. Wachtman), Academic Press, San Diego, CA, pp. 100–159.

564 Zeus, D. (1991) *cfi/Ber. DKG*, **68** (1/2), 36–45.

565 Krstic, V.D., Vlajic, M.D., and Verrall, R.A. (1996) *Advanced Ceramic Materials* (ed. H. Mostaghaci), Trans Tech. Publications, pp. 387–396.

566 Porz, F., Grathwohl, G., and Hamminger, R. (1984) *J. Nucl. Mater.*, **124**, 195–214.

567 Heuschmann, G. and Willmann, G. (1986) *Interceram*, **1**, 24–29.

568 Butt, D.P., Tressler, R.E., and Spear, K.E. (1994) in *Corrosion of Advanced Ceramics* (ed. K.G. Nickel), Kluwer Academic Publishers, Dordrecht, pp. 153–164.

569 Heck, R.M., Farrauto, R.J., and Gulati, S.T. (2002) in *Catalytic Air Pollution Control*, 3rd edn, Wiley-VCH, Weinheim, pp. 156–357.

570 Penty, R.A. and Bjerklie, J.W. (1982) *Ceram. Eng. Sci. Process.*, **3** (1–2), 120–127.

571 Heinrich, J., Huber, J., Foster, S., and Quell, P. (1987) in *High-Tech Ceramics* (ed. P. Vincenzini), Elsevier, Amsterdam, pp. 2427–2440.

572 Kerr, M.C. (1987) in *High-Tech Ceramics* (ed. P. Vincenzini), Elsevier, Amsterdam, pp. 2441–2449.

573 Heider, W. (1987) *Proceedings, International Conference-Exhibition for Technical Ceramics and Innovative Materials: The future of technical ceramics in practical application up to the year 2000*, Wiesbaden, Germany, 16–18 November (eds S. Schnabel and J. Kriegesmann), Demat Exposition Managing, pp. 18.01–18.19.

574 Hof, W. (1991) *Chemie-Technik*, **20** (12), 18–22.

575 (a) Meschke, F., Kayser, U., and Rendtel, A. (2006) German Patent DE 044.942.2 A1;(b) Meschke, F., Kayser, U., and Rendtel, A. (2006) World Patent WO/2006/029741 A1;(c) Meschke, F., Kayser, U., and Rendtel, A. (2006) European Patent EP 1.637.271.

576 Karandikar, P.G., Evans, G., Wong, S., and Aghajanian, M.K. (2009) Advances in Ceramic Armor IV, in *Proceedings 32nd International Conference on Advanced Ceramics and Composites, Symposium 4, Ceramic Armor, 27 January–1 February 2008, Daytona Beach, Florida*. Ceramic Engineering and Science Proceedings, vol. **29** (6) (ed. L. Prokurat Franks), John

Wiley & Sons, Inc., Hoboken, New Jersey, p. 163–178.

577 Storm, R.S., Ohnsorg, R.W., and Frechette, F.J. (1982) *J. Eng. Power: Trans ASME*, **104**, 76.

578 Schwetz, K.A., Grellner, W., Hunold, K., Lipp, A., and Langer, M. (1986) *Proceedings 2nd International Symposium Ceramic Materials and Components for Engines*, (eds W. Bunk and H. Hausner), Verlag DKG, pp. 1051–1062.

579 Storm, R.S., Boecker, W.D.G., McMurtry, C.H., and Srinivasan, M. (1985) Sintered alpha silicon carbide ceramics for high temperature structural application - Status review and recent developments, NASA Technical Reports, ASME PAPER 85-IGT-127.

580 Westerheide, R., Hollstein, T., and Schwetz, K.A. (1998) Proceedings 6th International Symposium on Ceramic Materials and Components for Engines, October 1997, Arita, Japan (eds K. Niihara *et al.*), pp. 253–258.

581 Gutmann, Ch., Schulz, A., and Wittig, S. (1996) A New Approach for a Low-Cooled Ceramic Nozzle Vane. ASME-Paper 96-GT-232.

582 Dilzer, M., Gutmann, Ch., Schulz, A., and Wittig, S. (1999) International Gas Turbine and Aeroengine Congress and Exhibition Stockholm 1998. *J. Eng. Gas Turbine Power*, **121** (2), 254–258.

583 Andrees, G. (1990) Entwicklung eines keram. Werkstoffes zur Auskleidung thermisch hochbeanspruchter Brennräume und Heißgasführungen. Final Report, Project BMFT 03M–2028, MTU Motoren- und Turbinen-Union München GmbH.

584 Schwetz, K.A., and Sigl, L.S. (1999) Fabrication and properties of HIP-treated sintered SiC for combustor liners of stationary gas turbines, in *Proceedings, Werkstoffwoche 98*, Vol. 3 (eds A. Kranzmann and U. Gramberg), Wiley-VCH, Weinheim, NY, pp. 15–24

585 van Rohde, M. *et al.* (2007) *J. Eng. Gas Turbine Power*, **129**, 21–30.

586 Davis, J.R.(ed.) (2002) *Surface hardening of steels: understanding the basics*, ASM International, Materials Park, OH, pp. 213–223.

587 Gokce, A.S., Gurcan, C., Ozgen, S., and Aydin, S. (2008) *Ceram. Int.*, **34** (2), 323–330.

588 Hunold, K., Lipp, A., Reinmuth, K., and Arnold, P. (1988) Patent 4732620.

589 Normandia, M.J., LaSalvia, J.C., and Gooch, W.A. Jr (2004) *AMPTIAC Q.*, **8** (4), 21–27.

590 Robichaud, J. (2003) Proceedings, Future EUV/UV and Visible Space Astrophysics Missions and Instrumentation, Proceedings of SPIE, vol. 4854 (eds J.C. Blades and O.H.W. Siegmund), SPIE, pp. 39–49.

5
Complex Oxynitrides

Derek P. Thompson

5.1
Introduction

The title of this chapter clearly invites a definition of the term "complex," and how a dividing line can be drawn between those materials which are deemed "complex" as compared to others which are of a "simpler" nature. In providing such a definition, it is appropriate to refer to the mineral silicates which, in many ways, demonstrate similarities with their complex oxynitride counterparts. Because silicate structures can be described as assemblies of [SiO_4] (or more often [(Si,Al)O_4]) tetrahedra, the polymeric nature of these crystal structure types automatically imparts a complexity, which also shows itself in the large number of different Si : O (or Si + Al : O) ratios observed for silicates. More complex ratios of atoms in a formula unit automatically correlate with larger unit cells of lower symmetry, which would then be described as "crystallographically complex." Most minerals only contain oxygen as the nonmetal species; however, occasionally OH groups, and other nonmetals (e.g., fluorine) or nonmetallic groups (e.g., carbonate, sulfate) do occur. In Nature, nitrogen is not found in silicates; however, when it is present in synthetic materials it is deliberately incorporated into the starting mix, and this provides scope for further complexity in chemical formulation, which in turn usually promotes a further increase in structural or crystallographic complexity.

Historically, the development of nitrides proceeded slowly until the emergence of silicon nitride during the late 1950s. From that time onwards, the intricacies of the relationship between the two forms of silicon nitride, combined with the observation that, when hot-pressed or sintered into dense products, silicon nitride displayed excellent mechanical properties at temperatures in excess of 1000 °C, propelled silicon nitride into a subject area of intense interest. As a result, it now resides in a well-defined niche as a structural material for wear parts and related applications. At the same time, this provided a catalyst for the development of other nitrides, both of a binary and ternary character. Increasing complexity was provided by the observation that, just as with mineral silicates, aluminum could replace silicon in

Ceramics Science and Technology Volume 2: Properties. Edited by Ralf Riedel and I-Wei Chen

ISBN: 978-3-527-31156-9

$[SiN_4]$ tetrahedra. This was combined with the other important observation, that if oxygen replaced nitrogen at the same time in equal amounts, then there is no overall valency imbalance. Equally significantly, as the Al–O bondlength is approximately the same as the Si–N bondlength, then the simultaneous replacement of Si–N by Al–O can take place very simply from a geometric point of view, without introducing any additional strain or charge imbalance into the lattice. The resulting β-sialon ceramics, of composition $Si_{6-z}Al_zO_zN_{8-z}$ have been the subject of intense study during the past thirty-five years, and in this chapter will be treated as noncomplex oxynitrides and hence excluded. In contrast, the α-sialon group of sialons will be treated as complex, partly because of the incorporation of an additional metallic element, and partly because these are a more recent group of materials, for which there have been some important developments in the past twenty years. Also, compositionally they are not significantly different to many other nitrogen-rich M–Si–Al–O–N oxynitrides which occur with different structures in metal (M)–sialon systems. Ternary nitrides will only be described if they also display a significant substitution of Si–N by Al–O, whereas other four-component (two metal and two nonmetal) plus five-component and higher oxynitrides will form the essential substance of the chapter, with additional comments included on more complex structures. For convenience, the chapter proceeds in the order of decreasing nitrogen content of the phases considered.

Those readers interested in the general field of oxynitride structures may also wish to consult previous reviews in this area, as provided by Lang [1], Thompson [2], Marchand [3], and Thompson [4].

5.2
Principles of Silicon-Based Oxynitride Structures

Many nitride- and oxynitride-based structures are broadly similar to those of mineral silicates, although the different valency of nitrogen allows different compensation by cations which, in some cases, allows similar structures to be observed (e.g., $YSiO_2N$ with the α-wollastonite form of $CaSiO_3$) but in other cases permits different structures to occur that are not found in the mineral kingdom. The Q^n terminology introduced to describe silicate structures has provided a simpler method than the use of nonsystematic long names to describe the various structural possibilities that arise when $[SiO_4]$ tetrahedra are joined together. In the silicate field, the range extends from Q° (orthosilicates) to the three-dimensional (3-D) tectosilicates characterized by the feldspars and the various polymorphs of silica. However, in nitrides, the additional electron provided by the nitrogen atom participates in sp^2 bonding and allows the nitrogen to bond to three silicon neighbors in a plane at 120° to one another. In contrast to silicates, where oxygen commonly only links to two silicon atoms, this additional possibility means that each nitrogen atom at the corner of an $[SiN_4]$ tetrahedron can be in contact with another two silicon (or aluminum) atoms; hence, Q numbers up to Q^8 are possible, the latter being exemplified by silicon nitride itself. Clearly the higher the *n*-value, the more tightly bonded is the structure, and therefore $n = 7$–8-type materials are the likely candidates for high-temperature materials with good mechanical properties.

The similarity between oxynitrides and mineral silicates raises the question of whether all the silicate structure types can be reproduced in the oxynitride field. First, it is important to realize that the total number of oxynitrides observed so far is notably fewer (typically not many more than 200) than the number of mineral silicate structures and, in some cases, there is only one clear representative in one particular group. Thus for example for the Q° type, the apatite family – of general formula $M_{10}(SiO_4)_6N_2$ (M = Y, Ln) – is the primary oxynitride example observed to date. The Q^1 structure is well illustrated by the J-phases (see Section 5.3.3), where the 2 : 7 structural unit contains a maximum of two nitrogen atoms. The nitrogen pyroxene family (see Section 5.6.5), which are of the type $MgMSi_2O_5N$ (M = Y, Ln) [5], are examples of Q^2 chain structures, and the Q^2 ring type of structure is exemplified by the nitrogen wollastonites, $MSiO_2N$ (M = Y, Ln). A pure Q^3 arrangement in the traditional clay-type mineral structures has not been reproduced by any oxynitride so far, possibly because most of the mineral derivatives contain OH, and the simultaneous presence of N and OH is not possible; however, the melilite structure discussed below does show the presence of some Q^3 in the tetrahedral distribution. The Q^4 arrangement, exemplified in the mineral kingdom by the polymorphic forms of silica and also by the feldspars, has not been observed in oxynitride analogues. This may represent a lack of research in those systems most likely to yield framework structures, but it is also a reflection that many likely compositions tend to form liquids at reaction temperatures, which on cooling then form glasses rather than crystalline modifications. This discussion highlights certain difficulties that occur when working with oxynitrides, namely that higher temperatures are needed with covalently bonded starting reagents, and this can result in unwanted side reactions (e. g., with crucible materials) and a loss of volatiles. Moreover, as the growing of single crystals is also more difficult, materials may not always be produced that can be used for detailed crystal structure determinations.

The question of O/N ordering, which originally was neglected but then more recently was studied using neutron diffraction, has today become more tractable since the arrival of magic angle spinning nuclear magnetic resonance (MAS-NMR). The basic principles were summarized by Thompson [6], and these still remain broadly correct today. The way in which this has helped to complete structural information is provided for individual structures in the following sections.

5.3 Complex Si–Al–O–N Phases

5.3.1 Sialon X-Phase

The X-phase was observed as a secondary crystalline phase in many of the early preparations of β-sialon ceramics (Figure 5.1). Further studies have shown that the

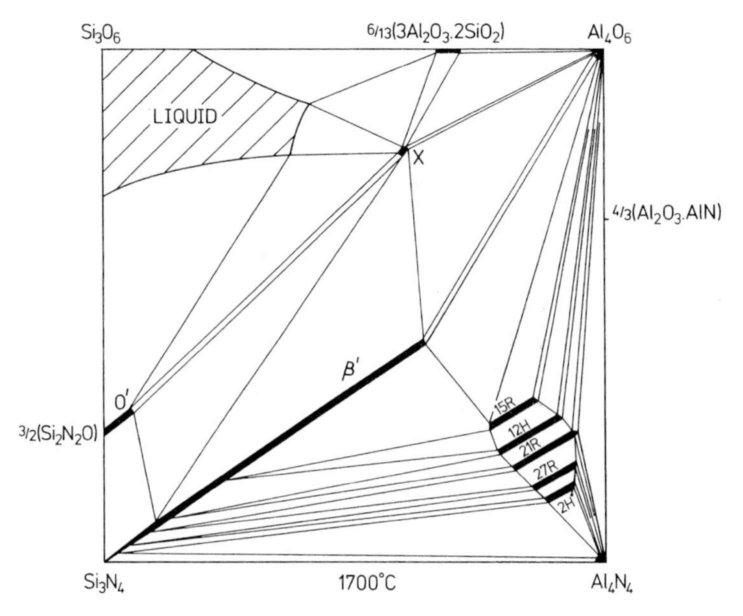

Figure 5.1 Phase relationships in the Si–Al–O–N system at 1700 °C.

composition of this phase was close to the maximum Al and N solubility limit in the liquid phase in the Si–Al–O–N system. However, the melting point of X, at just over 1700 °C, corresponds to a typical preparation temperature for sialon ceramics. Moreover, as the viscosity of the liquid phases in this system (in the absence of additional metallic species) was extremely high, the crystallinity of X-phase samples was often poor and many of the lines on the X-ray diffraction (XRD) pattern were not clearly defined. Heat treatment at lower temperatures sharpened up the pattern and, for this reason, two forms referred to as "high-" and "low-" X were distinguished, with low-X being a superlattice of the high-X form. The triclinic unit cell dimensions measured for low-X give values of $a = 0.968$, $b = 0.855$, $c = 1.120$ nm; $\alpha = 90.0°$, $\beta = 124.4°$, $\gamma = 98.5°$, with high-X having almost identical parameters except that $b_{low} = 3b_{high}$.

Single crystal data for X-phase were collected by Korgul and Thompson [7], and refinement showed the structure to be very similar to mullite (Figure 5.2), with chains of octahedra extending along the b-axis and linked together by arrangements of tetrahedra. It is believed that the difference between the high- and low-X forms is due to the ordering of Si and Al in tetrahedra, which takes place if X-phase is heated for extended times at temperatures below its melting point.

5.3.2
The Sialon Polytypoid Phases

At the corner of the Si–Al–O–N phase diagram closest to aluminum nitride (see Figure 5.1), there occur a series of phases, characterized by extended ranges of

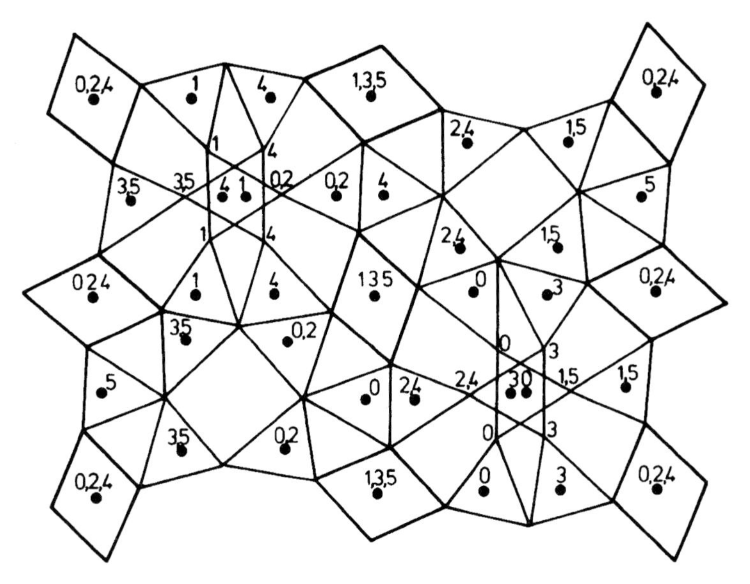

Figure 5.2 Projection of the crystal structure of sialon X-phase down [0 1 0]. The numbers indicate y-coordinates in sixths.

composition (due to Si–N by Al–O substitution) along lines of constant metal (M): nonmetal (X) ratio of the type 4 : 5, 5 : 6, 6 : 7, 7 : 8, 9 : 10, and approximately 11 : 12, to which Ramsdell symbols 8H, 15R, 12H, 21R, 27R, and $2H^{\delta}$ (the latter being a disordered variant) have been attributed. These are a series of polytypoid phases, with structures similar to that of AlN, but in which layers of octahedrally coordinated aluminum have been inserted at regular intervals, the frequency of these layers being determined by the composition. The basic structures are now well known, and the reader is directed to the excellent reports of Westwood and coworkers [8–10] for more detail. An interesting feature of these structures is that in order for them to have M : X ratios less than unity, at some point in between the layers of octahedra there must be a change in the orientation of the tetrahedra from "pointing upwards" to "pointing downwards," and at this point the tetrahedra must share bases. This is observed, but only one metal atom is shared between each pair of base-sharing tetrahedra. An interesting observation emerging from careful transmission electron microscopy (TEM) measurements [10] is that, in contrast to the octahedral layers which are perfectly flat, the layers of base-shared tetrahedra are undulating (see Figure 5.3). Other cations can be incorporated into these polytypoid structures, the most common of which is magnesium [11].

5.3.3
The Y-Si–O–N Oxynitrides

Investigations conducted during the 1970s established the existence of 4 Y–Si–O–N oxynitrides (see Figure 5.4): apatite ($Y_{10}(SiO_4)_6N_2$); wollastonite ($YSiO_2N$); melilite

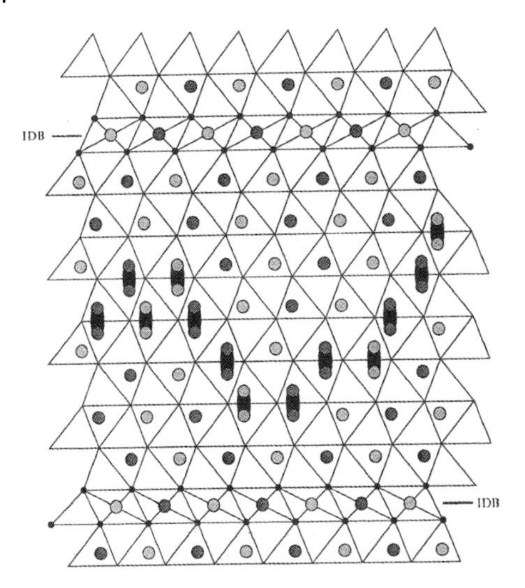

Figure 5.3 Undulating layers of base-shared tetrahedra in sialon polytypoid phases [10].

($Y_2Si_3O_3N_4$); and J-phase ($Y_4Si_2O_7N_2$). Previous reviews have provided details of the basic crystal structures of these materials, and their similarities with mineral analogues. In general, oxygen always occupies nonbridging sites (as in $YSiO_2N$), or sites not coordinated to silicon (as in $Y_{10}(SiO_4)_6N_2$) and J-phase). Only a few comments relating to additional more recent work will be included in this chapter.

An excellent example of the use of NMR to determine O/N order is provided by the studies of Koroglu *et al.* [12] on the nitrogen melilite phase ($Y_2Si_3O_3N_4$). Only one peak is observed on the ^{29}Si spectra of this phase (Figure 5.5a), and this led

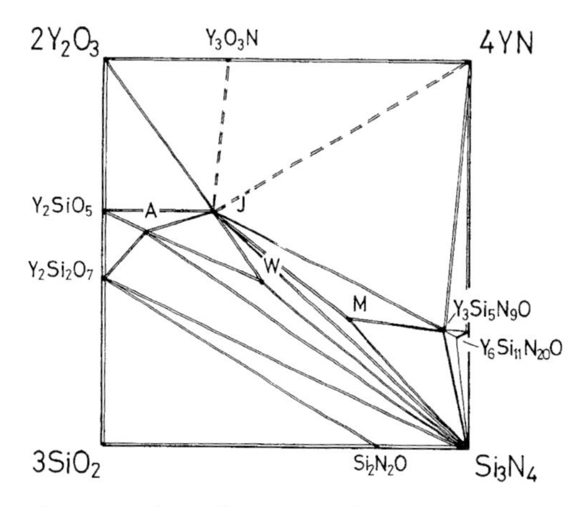

Figure 5.4 Phase relationships in the Y–Si–O–N system at 1700 °C, showing W (wollastonite), A (apatite), M (melilite), and J-phase.

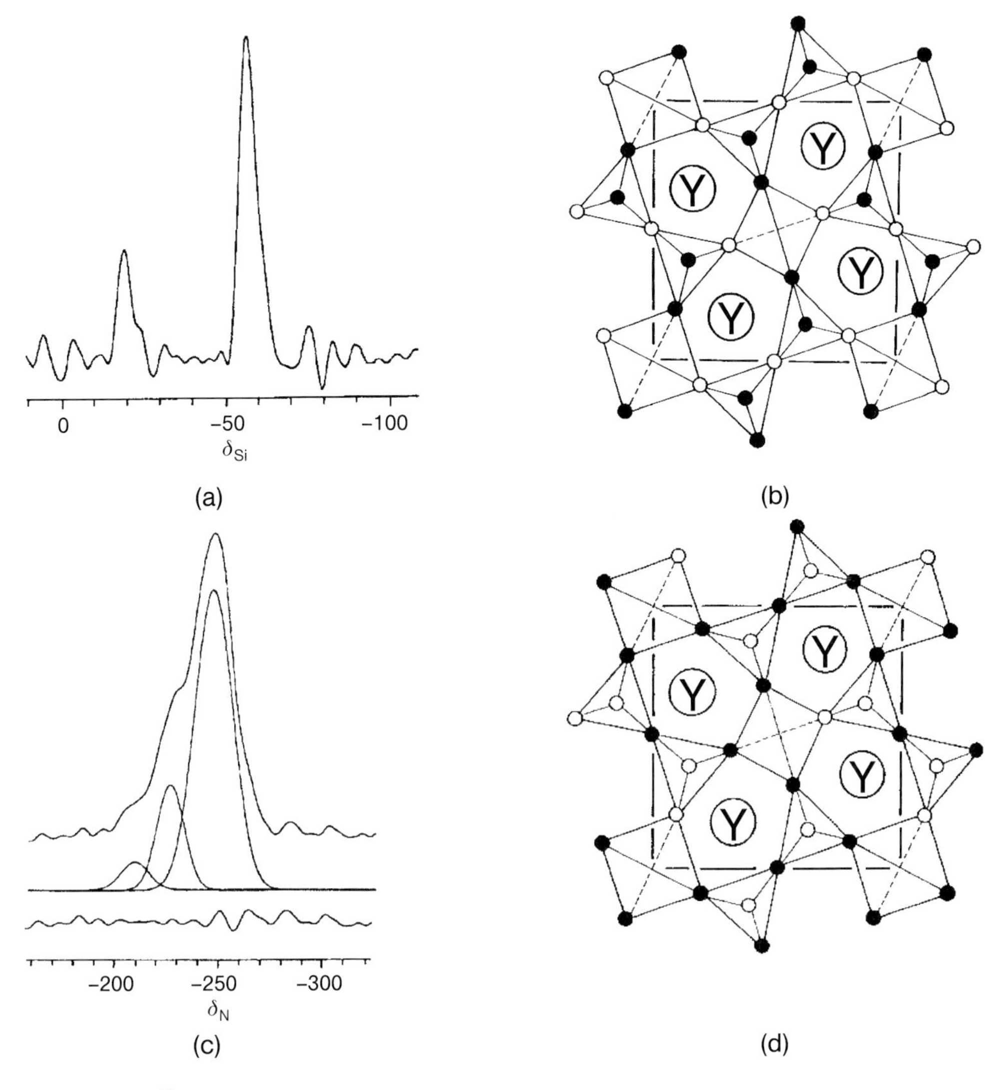

Figure 5.5 (a) ^{29}Si MAS-NMR spectrum of $Y_2Si_3O_3N_4$. The peak at $\delta = -19$ ppm arises from SiC impurity; (b) O/N ordering scheme assuming a single ^{29}Si peak; (c) ^{15}N MAS-NMR spectrum for $Y_2Si_3O_3N_4$; (d) O/N ordering scheme based on the 1 : 2 : 5 intensity ratio observed on the ^{15}N spectrum. In panels (b) and (d), O is represented by open circles, and N by filled circles.

Dupree *et al.* [13] to predict the O/N distribution shown in Figure 5.5b. However, the siting of nitrogen atoms in nonbridging sites is not observed in any other oxynitride phases, and moreover is not consistent with Pauling's rules; therefore it was necessary to obtain the ^{15}N NMR spectrum (Figure 5.5c) to see the full picture. This shows three peaks, in the intensity ratio 1 : 2 : 5, corresponding to N atoms linking two [SiO_2N_2] tetrahedra, one [SiO_2N_2] tetrahedron and one [$SiON_3$]

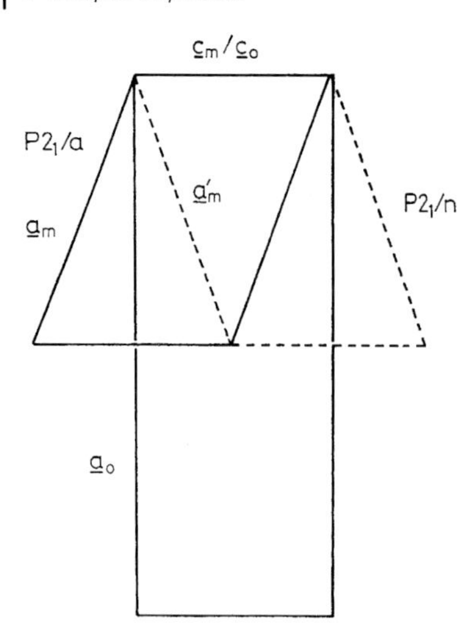

Figure 5.6 The two equivalent (and almost identical) monoclinic cells for J-phase.

tetrahedron, and two $[SiON_3]$ tetrahedra, respectively; this can be explained exactly (and uniquely) by the O/N arrangement shown in Figure 5.5d. It is interesting that subsequent modeling studies confirmed this arrangement, but with only a marginally more negative ΔG value for the preferred arrangement; this showed that, in practice, some disorder would also be expected to be present.

Further studies have also been carried out on the family of J-phases. In addition to the yttrium compound, this cuspidine-like phase occurs in all rare earth oxynitride systems. Because of the pseudo-orthorhombic nature of the cuspidine unit cell, X-ray powder data can be indexed on either of two very similar monoclinic cells (in space groups $P2_1/a$ or $P2_1/n$ respectively; see Figure 5.6), and this caused considerable confusion in the early indexings of oxynitride J-phases. However, by using consistent indexing procedures, a plot of unit cell dimensions versus the ionic radius of the rare earth cation (Figure 5.7) showed that there are in fact two very similar types of rare earth J-phase structure [14], with elements La–Dy showing the first type, and Ho–Lu the second. Careful investigations by Takahashi and colleagues [15, 16] showed that this was due to shifts in $[Si_2(O,N)_7]$ groups relative to each other (Figure 5.8)

In addition, because of the existence of cuspidene-like aluminates of general formula $M_4Ln_2O_9$, there exists a pseudo-solid solution all the way between the J-phase, $M_4Si_2O_7N_2$, and its corresponding $M_4Al_2O_9$ aluminate. However, recent studies [14] have shown that in fact this is not an exact solid solution and, just as in the Al-free compositions (where there are slight differences in structure) so too in the Al-containing phases there appears to be a similar structure for the end member regions. However, J-phases of slightly different unit cell dimensions are observed centered about the mid-point composition.

Early studies on the $YN–Si_3N_4$ system have identified three ternary compounds [17] which, on the basis of very poor-quality multiphase samples, were tentatively attributed to compositions $Y_6Si_3N_{10}$, $Y_2Si_3N_6$, and YSi_3N_5. Subsequent detailed

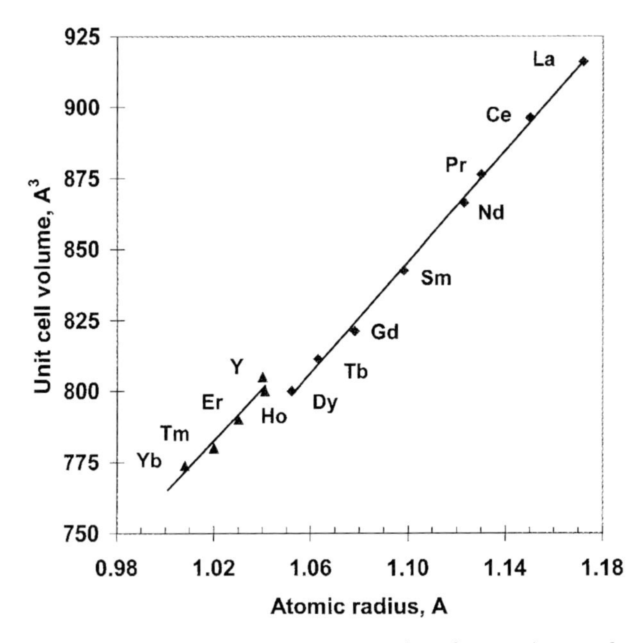

Figure 5.7 Unit cell volume versus ionic radius of rare earth cation for J-phase.

studies showed that the "$Y_6Si_3N_{10}$" phase was in fact a carbonitride of composition $Y_2Si_4N_6C$ (Figure 5.9), and is a member of a quite extensive family of 2 : 4 : 7 nitrides and carbonitrides [18]. The compound YSi_3N_5 is now believed to have a lower silicon content than was previously thought, because the XRD pattern is very similar to that of the compound identified in the Er–Si–O–N system as $Er_6Si_{11}N_{20}O$. Although a structure determination was published for this latter compound [19], there were

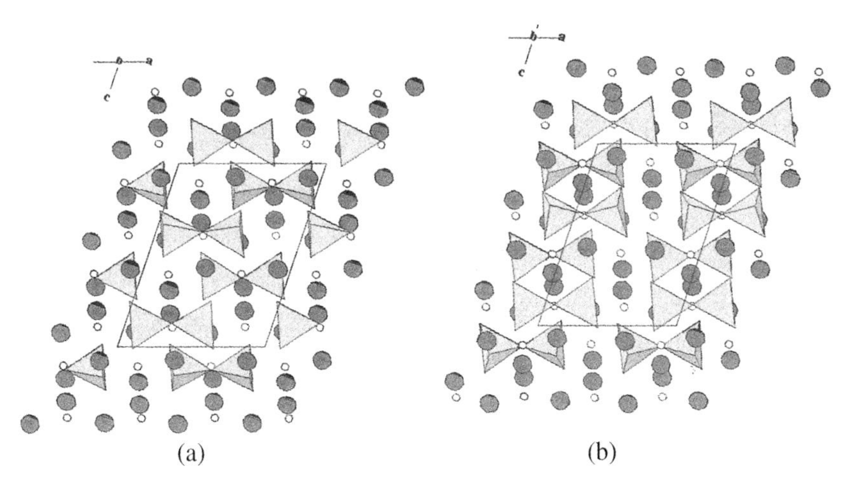

Figure 5.8 [0 1 0] projections of the crystal structures of (a) $La_4Si_2O_7N_2$ and (b) $Lu_4Si_2O_7N_2$. Reproduced from Refs [15, 16].

Figure 5.9 Atomic arrangement in $Y_2Si_4N_6C$. Reproduced from Ref. [18].

features of the structure which were not completely convincing. Hence, further studies on better quality samples are recommended before it can be said that this structure is in fact correct.

However, when the third Y–Si–O–N compound, which originally was thought to be $Y_2Si_3N_6$, was investigated further [20], the diffraction pattern was found to index on an orthorhombic unit cell of dimensions $a = 0.498$, $b = 1.615$, $c = 1.065$ nm. A single-crystal structure determination carried out on this compound gave the structure shown in Figure 5.10. As with the La–Si–Al–O–N "new" phase (see Section 5.4.4), it is clear that some of the nitrogen atoms are 3-coordinated by silicon, while most are coordinated by two; this leaves some nonmetals (presumably oxygen) to be coordinated by only one silicon.

5.4 M–Si–Al–O–N Oxynitrides

5.4.1 α-SiAlON

Since their original discovery in 1978 [21], there have been many crystal structure determinations on a variety of different α-sialons. α-Sialon can incorporate the additional large cations (M) required for the stabilization of its structure, because the

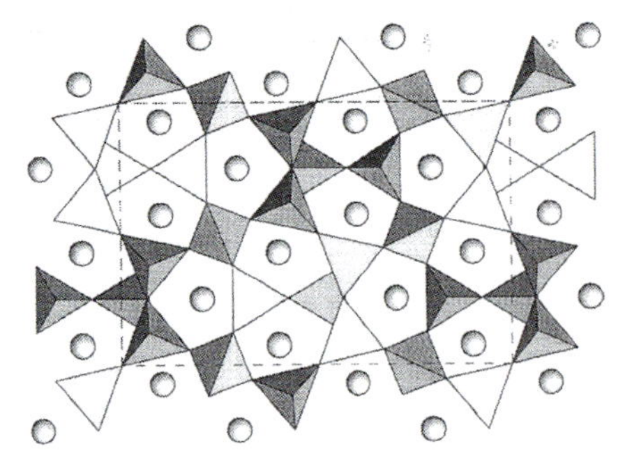

Figure 5.10 Atomic arrangement in $Y_3Si_5N_9O$. Reproduced from Ref. [20].

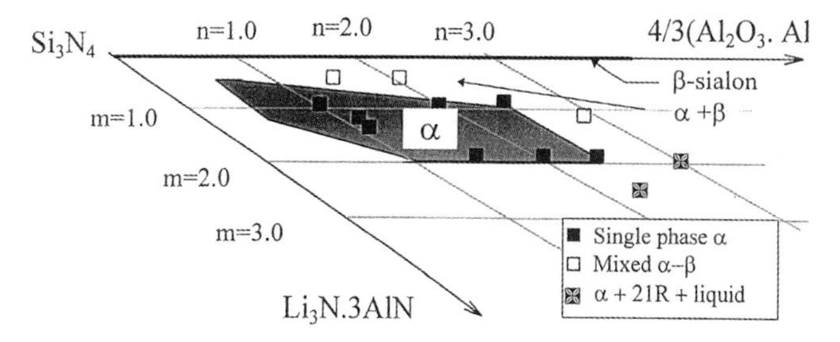

Figure 5.11 Region of α-sialon stability in the lithium sialon system. Reproduced from Ref. [22].

long channels parallel to the *c*-axis (which are a characteristic feature of the structures of β-Si_3N_4 and β-sialon) are broken up by the different stacking sequence of Si–N layers in the α structure; this leaves large holes which are of an appropriate size to accommodate large cations. An additional significant characteristic is that, whereas in the β structure the surrounding Si(Al) and N(O) atoms are all equidistant from the centre of the channel, in the α structure the nonmetal atoms surrounding the large interstice are drawn in towards its center, such that a metal cation placed in this site is coordinated by nonmetal atoms and not by Si(Al) atoms. In the $M_xSi_{12-m-n}Al_{m+n}N_{16-n}O_n$ unit cell-based formula for α-sialon, the maximum value of x should be 2, which is consistent with the two interstices per unit cell. However, the maximum x-value observed in different systems will differ greatly, with a tendency to decrease with increasing valency of M. Thus, for Li α-sialons, the $x = 2$ compositions (Figure 5.11) are relatively easy to prepare [22], whereas in the Ca-sialon system, $x = 2$ compositions originally could not be prepared. However, more recent studies using CaH_2 as a more active Ca-containing ingredient have shown that almost the full $x = 2$ composition can be realized [23]. For trivalent metals (Y and Ln), the maximum x-value observed is closer to 1. The actual shape of the interstice in the α-sialon structure is not spherical, but rather is elliptical, with the long axis of the ellipse elongated along the *c*-axis. In total, the interstice is coordinated by 11 nonmetal atoms, with one at the top and bottom, and three layers of three in a triangular arrangement at different heights. It is, therefore, impossible for any cation to be coordinated by all eleven nonmetals, and in practice most cations link either to the top or the bottom nonmetal, as well as to two of the three other layers, giving a sevenfold coordination (Figure 5.12). In the case of lithium, it is believed that the lithium coordinates to only one layer of three nonmetals, giving an overall 4-coordination [24]. In this latter case, it was questioned whether there was room for two lithium cations in the interstices – one at the top and the other at the bottom. However, no evidence has yet been obtained for this, especially as it would then be possible to incorporate more atoms in the unit cell than the two indicated in the above discussion.

5.4.2
JEM Phase

This phase was first identified by the Stockholm group, and the name coined for it was an acronym of the initial authors [26]. This phase occurs closest to the

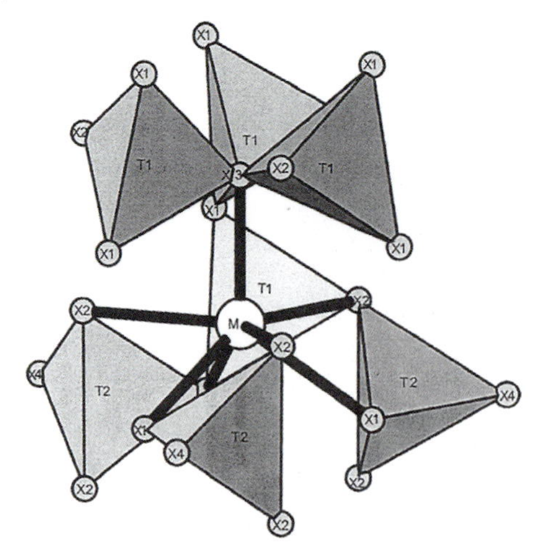

Figure 5.12 Coordination of the M cation in a $(Ca,La)_{0.3}(Si,Al)_{12}(O,N)_{16}$ α-sialon [25].

composition $LnSi_5Al_2N_9O$, and has an orthorhombic unit cell of dimensions $a = 0.943$, $b = 0.977$, and $c = -0.894$ nm. The crystal structure (Figure 5.13) shows a 3-D interlinked arrangement of $[(Si,Al)(O,N)_4]$ tetrahedra, with the large cations distributed statistically between two equivalent sets of sites. As is similar for many of these structures, the JEM phase has only been observed for the first three rare earth elements. Figure 5.14 shows the interesting way in which the JEM phase grains tend to occur in "clumps" when present as a second phase with α-sialon [27].

5.4.3
S-Phase

Sialon S-phase was first observed by Hwang *et al.* [28] as a second phase in the preparation of α-sialons, with strontium as the densifying additive. The general

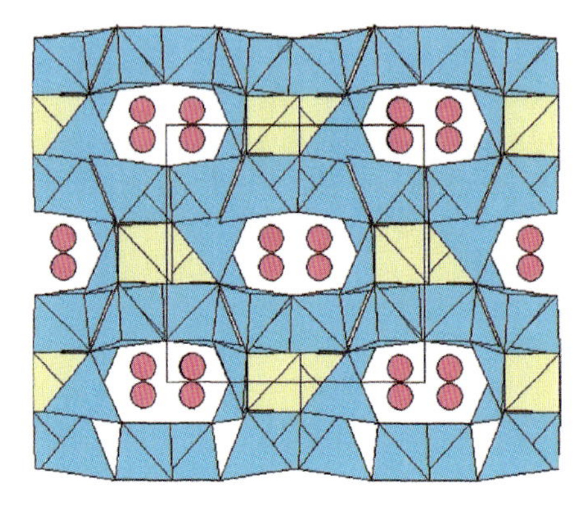

Figure 5.13 Crystal structure of JEM phase. Reproduced from Ref. [26].

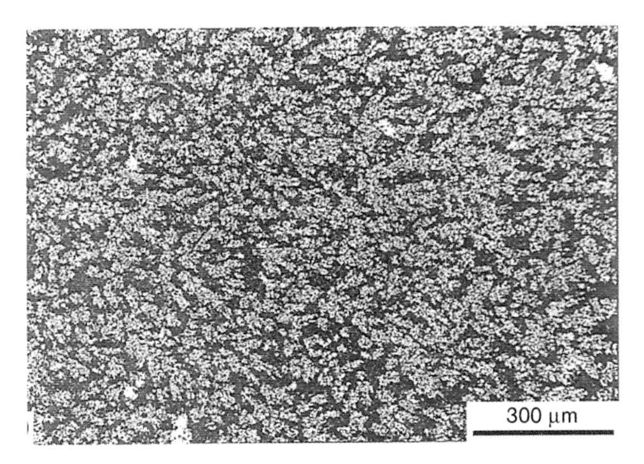

Figure 5.14 A typical microstructure of JEM phase occurring with α-sialon [27].

formula given by Esmaeilzadeh *et al.* [29] was $M_2Al_xSi_{12-x}N_{16-x}O_{2+x}$, where M can be either Ba or Sr. In preparative studies conducted by Lewis *et al.* [30], the composition $BaSi_5AlO_2N_7$ was used (i.e., $x = 2$ in the above formula), which was convenient for the production of a dense ceramic because this composition could be prepared using mixes of Si_3N_4, AlN and Al_2O_3 powders, with $BaCO_3$ as the starting barium compound. The unit cell is orthorhombic, with the dimensions of the above Ba derivative of $a = 0.823$, $b = 0.966$ and $c = 0.491$ nm, and a structure (see Figure 5.15) consisting of a 3-D interlinked assembly of (Si,Al)-centered tetrahedra with the large cations in holes in this arrangement. Esmaeilzadeh *et al.* [29] identified a larger unit of ten tetrahedra (Figure 5.16) of overall composition $[(Si,Al)_{10}(O,N)_{25}]$, which link together to define the complete structural framework. Lewis *et al.* [30] were aiming to prepare dense samples of S-phase to compare the mechanical properties with those of α- and β-sialons;

Figure 5.15 Atomic arrangement for the orthorhombic structure of S-phase [29].

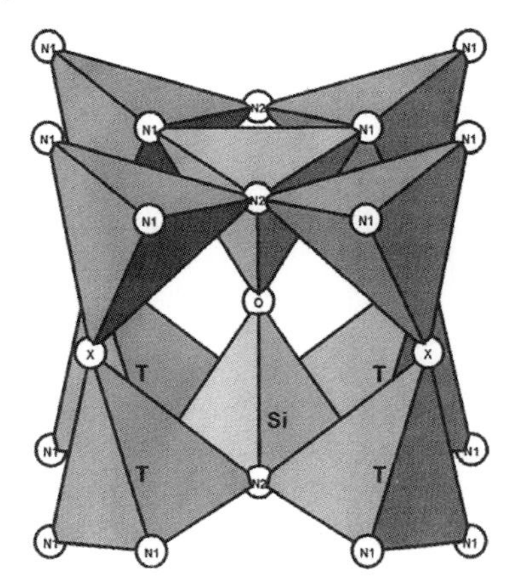

Figure 5.16 The $[(Si,Al)_{10}(O,N)_{25}]$ structural unit in S-phase.

however, the results were disappointing in that the morphology of the S-phase grains was plate-shaped, in contrast to the acicular morphology observed in α- and β-sialons (which is an important microstructural requisite for promoting good toughness).

5.4.4
Lanthanum "New" Phase

The title of "new" phase was originally conferred on the compound observed close to the composition $La_2O_3{\cdot}2\ Si_3N_4$ in the La–Si–O–N system [31, 32], with equivalent compounds also being observed in the Ce and Nd sialon systems. After considerable uncertainty, both about indexing of the powder X-ray data and also the correct composition, this was finally resolved when the crystal structure was determined by Grins *et al.* [33], who confirmed the composition as being $Ln_3Si_8N_{11}O_4$. These authors also showed that a considerable amount of Si–N by Al–O substitution could occur, to the extent of at least 1.5 Si–N units in the above formula unit. The unit cell is monoclinic and the dimensions of this substituted derivative in the La system are $a = 1.585$, $b = 0.490$, $c = 1.804$ nm, with $\beta = 114.8°$; however some new phase compositions can be indexed perfectly satisfactorily with a c-dimension of half this value. Figure 5.17 shows a representation of this structure on the x-z plane, demonstrating the 3-D interlinking of tetrahedra; however the structure is of interest in that it has certain regions where the nitrogen atoms are common to three tetrahedra (just as in silicon nitride), whereas elsewhere Figure 5.17 shows $[(Si,Al)_2(O,N)_7]$-type units where there are nonbridging oxygens linked to only one silicon atom plus three large rare earth atoms.

5.4.5
$M_2(Si,Al)_5(O,N)_8$ Oxynitrides

These derivatives are based on the ternary compounds of the type $M_2Si_5N_8$, where M = Ba, Sr, Ca). To the present author's knowledge, the exact extent of the solid

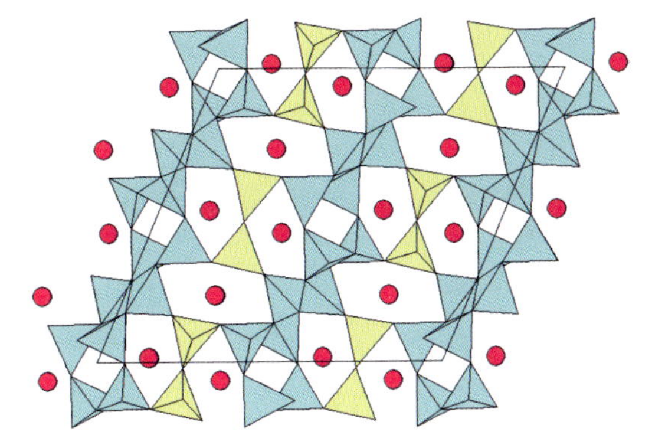

Figure 5.17 Crystal structure of the La "new" phase. Reproduced from Ref. [33].

solution range has not been determined, but is probably of the order of one Al + O in the above formula unit. Structures of these compounds have been determined for each of the three alkaline earth cations mentioned; Figure 5.18 shows the atomic arrangement for the barium derivative. These compounds are of particular importance because of their properties as phosphors when activated with small amounts of Eu^{2+}.

5.4.6
$M(Si,Al)_3(O,N)_5$ Phases

The compound $LaSi_3N_5$ was first reported by Inoue *et al.* [35], who prepared it at high pressures at 2000 °C. The unit cell of $LaSi_3N_5$ is orthorhombic, with $a = 0.784$, $b = 1.124$ and $c = 0.481$ nm, while the structure (Figure 5.19) is another variant of 3-D- linked $[SiN_4]$ tetrahedra, where some of the nitrogens are 3-coordinated by

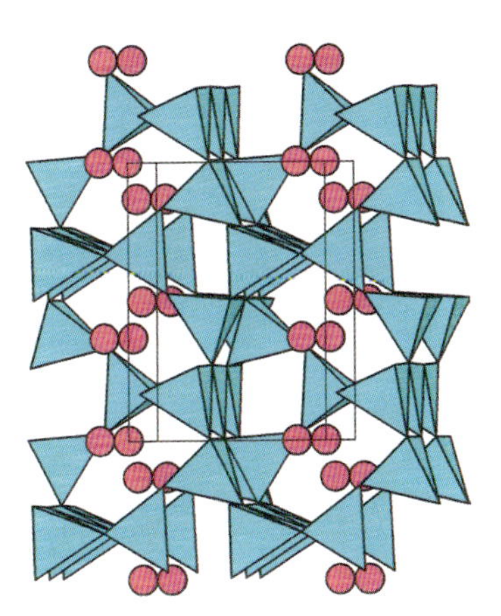

Figure 5.18 Atomic arrangement in $Ba_2Si_5N_8$. Reproduced from Ref. [34].

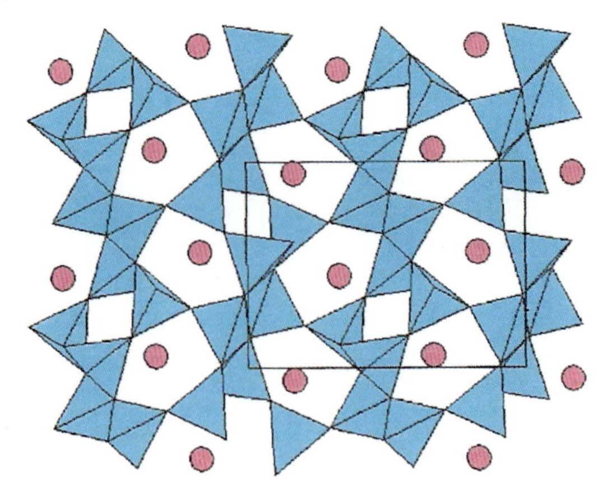

Figure 5.19 Atomic arrangement in $LaSi_3N_5$. Reproduced from Ref. [35].

silicon (just as in silicon nitride) and others are only 2-coordinated. This type of structure is observed for only the first few members of the rare earth series.

5.4.7 $M_3(Si,Al)_6(O,N)_{11}$ Phases

These compounds are observed for the first few members of the rare earth series, and are of interest in showing tetragonal symmetry, the unit cell of the cerium end-member having dimensions $a = 1.014$ and $c = 0.484$ nm. The 3-D linked arrangement of tetrahedra, with the rare earth atoms in the interstices (Figure 5.20), has more or less every nitrogen linked to two silicon atoms, as would be expected from the almost 1 : 2 ratio of Si : N.

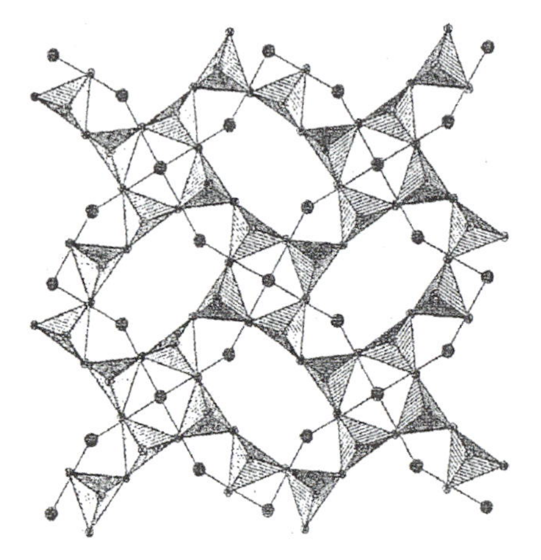

Figure 5.20 Atomic arrangement in $Sm_3Si_6N_{11}$. Reproduced from Ref. [36].

5.4.8
Wurtzite Oxynitrides

During the 1960s, the crystal structure of $MgSiN_2$ [37] was established as being analogous to that of aluminum nitride (wurtzite-type), but with total ordering of the cations, in such a way as to give an orthorhombic unit cell, as compared with the hexagonal AlN. Analogous compounds were established such as $MnSiN_2$ [38]. Subsequent studies with aluminum-containing systems showed that more complex nitrides such as $MgAlSiN_3$ [39], $MnAlSiN_3$ [40] and $CaAlSiN_3$ [41] existed, apparently with $LiSi_2N_3$-type ordering schemes based on an AlN wurtzite arrangement. However, the early studies failed to result in successful crystal structure refinements, mainly because of difficulties in excluding oxygen from the final products. Nonetheless, more recent studies with $CaAlSiN_3$ [42] have resulted in the successful formation of pure $CaAlSiN_3$ (Figure 5.21), and this material has shown considerable promise in optical applications. Previous samples, which showed a disparity in unit cell dimensions as compared with pure $CaAlSiN_3$, most likely contained oxygen; such compositions very likely lie along a line of compositions that extends from $CaAlSiN_3$ to Si_2N_2O.

5.4.9
$MSi_2O_2N_2$ Oxynitrides

Another series of oxynitrides which, in contrast to most of the preceding nitrogen-rich compounds, has more or less equal amounts of the two nonmetals is the $MSi_2O_2N_2$ series, where M is a divalent cation (typically Ca, Sr, or Ba) [43]. All of these compounds are monoclinic, and would appear to be essentially point compositions. No data have yet been reported on the further substitution of Al–O in place of Si–N, even though this would be expected to a limited extent. Studies conducted by

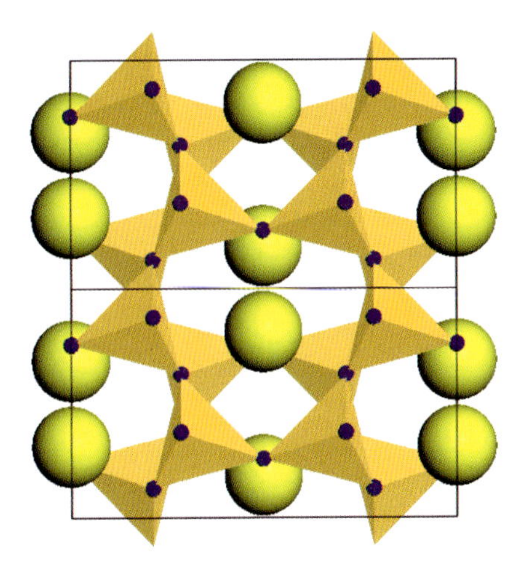

Figure 5.21 Atomic arrangement in $CaAlSiN_3$ [42].

Figure 5.22 Schematic representation of the structural units in $Sr_3Ln_{10}Si_{18}Al_{12}O_{18}N_{36}$. Reproduced from Ref. [46].

Li [44] have strongly indicated that these materials will prove promising as light-emitting diode (LED) conversion phosphors.

5.4.10
More Complex Oxynitrides

The investigations carried out by Schnick and coworkers at the University of Munich at the end of the twentieth century must be singled out as being significant in opening up new and complex oxynitride structures, and pointing the way towards further novel possibilities. Schnick and colleagues reported the first nitrogen zeolite [45], of composition $Ba_2Nd_7Si_{11}N_{23}$, while their synthesis and characterization of the six-component series of compounds of the type $Sr_3Ln_{10}Si_{18}Al_{12}O_{18}N_{36}$ [46] (see Figure 5.22) shows that, in systems of increased complexity, there are clearly new compounds waiting to be discovered. This indeed provides further fulfilment for the predictions by Jack [47], made some forty years ago, that the extent of the field of oxynitride ceramics may ultimately be comparable to that of the mineral silicates.

5.5
Oxynitride Glasses

Oxynitride glasses also fit into the category of "complex oxynitrides" because they typically contain three or more cations, in addition to oxygen and nitrogen. Oxynitride glasses have been well known since the late 1960s, when it first became apparent that the densification of silicon nitride using oxide sintering additives could be facilitated by reacting the latter with other starting constituents to form liquid phases. Densification could then be achieved by using a liquid phase sintering mechanism. Whereas, originally it was thought that the liquid was an oxide eutectic, it soon became clear that nitrogen was also present in this liquid phase, and on cooling to room temperature the liquid would remain in the grain boundaries as a glass. When single-oxide additives were added to silicon nitride, the resultant liquid phase (and hence the glass-forming region) was relatively limited in composition, whereas the presence of aluminum in the starting mix (which resulted in the formation of an M–Si–Al–O–N oxynitride liquid, where M is the sintering additive) significantly enlarged the glass-forming region. It also provided the added advantage of lowering the eutectic temperature, thereby stabilizing the liquid phase compared to the M–Si–O–N liquid phases which, at the higher temperatures required for melting, could easily lose volatile species such as silicon monoxide. Although the early studies were conducted using Mg as the metal, this was soon expanded to include Ca and Y; a further expansion included Li, Sc, Ba, and the rare earths (e.g., see Ref. [48]). In these studies, it has been common to express compositions in terms of equivalent % (eq%) of both the metals and nonmetals. For example, in the Y–Si–Al–O–N system the glass of highest nitrogen content occurred at a composition close to $Y_3Si_3Al_2O_9N_3$, which would be expressed as 33.3 eq% yttrium, 44.4 eq% silicon, and 22.3 eq% aluminum for the metal contents, and as 33.3 eq% nitrogen and 66.7 eq% oxygen for the nonmetal contents. Initially, it was found that the maximum nitrogen content that could be incorporated into these glasses tended to increase with the valency of the M cation. However, in practice defining the upper limit of nitrogen solubility is determined more by the procedure used for glass melting, because the viscosity of these oxynitride glasses increases with increasing nitrogen content. Hence, the higher the viscosity the more difficult it is for the glass to dissolve yet more nitrogen, when it is provided in the form of either silicon nitride or aluminum nitride. Recent studies conducted by a research group at Stockholm [49] have shown that much more nitrogen can be incorporated into these glasses if more reactive starting materials are used. The easiest way to achieve this is to add the M metal in the form of a finely divided metal powder, rather than as the oxide (which was the conventional approach). Alternatively, it was shown that in other cases (e.g., calcium), it is convenient for the metal to be added in the form of a hydride, which then decomposes at a relatively low temperature to give a finely divided dispersion of reactive metal powder in the starting glass mix. Many of the investigations carried out at Stockholm involved the use of lanthanum as the metal additive, and in the La–Si–O–N and La–Si–Al–O–N systems glasses have been produced which contain up to 70 eq% of nitrogen. This has led to important questions being asked concerning the structure of these glasses, because the increased covalency of nitrogen would be expected to promote more of a tendency to form crystalline products in glasses with such high levels of nitrogen. Despite NMR studies having been conducted [50], the actual coordination of cations and anions in these glasses is still being explored.

One common feature of all oxynitride glasses is that they are black. This blackness is not an intrinsic phenomenon of the materials themselves, but rather is consistent with the similar dark color of silicon nitride materials, especially those which have been produced at higher temperatures, and also in a carbon-rich environment. It is now known that the dark color is due to a fine dispersion of metallic particles within the glass [51], which form fundamentally as a result of trace metallic impurities (typically iron) in the starting materials. At temperatures above ~1200 °C, the metallic impurities react with silicon nitride in the starting mix to form metal silicides (e.g., $FeSi_2$); these are molten at such temperatures and further catalyze the decomposition of silicon nitride grains, as a result of which the final dark particles consist mainly of silicon and surround the original metal silicide nucleus. As a higher nitrogen content glass will require a higher content of silicon nitride in the starting material, there is a trend towards an increasing blackness with increasing nitrogen content. In principle, it should be possible to avoid such discoloration, first by using very pure starting powders, or second by using either a low glass-melting temperature or less-reducing melting conditions. However, whereas reasonably transparent oxynitride glass samples have been produced in thin sections, the production of bulk samples of transparent materials remains a challenge for the future.

It is well known [48] that many physical properties of oxynitride glasses increase systematically with their nitrogen content. Such properties include the glass transition temperature (T_g), the refractive index, elastic modulus, viscosity, and hardness. Data relating to these characteristics have been provided over a wide range of M–Si–Al–O–N systems. This behavior can generally be understood by the fact that, as soon as nitrogen is incorporated into the glass, a certain proportion of its atoms will become coordinated with three silicon atoms, rather than with two (as in oxide glasses). The covalent bonding associated with these [NSi_3] units then confers a greater rigidity to the glass network.

Clearly, although it is possible for more combinations of cations to be used for oxynitride glass formation, it is unlikely that they will result in significant benefits in performance compared to those already determined for five-component glasses. However, it is possible for alternative anions to be used, and the use of fluorine (in addition to oxygen and nitrogen) in calcium-containing glasses has demonstrated some interesting benefits. In fact, the incorporation of fluorine results in a significant expansion of the glass-forming region in the Ca–Si–Al–O–N system towards Ca-rich compositions, while the addition of even very small amounts of fluorine (e.g., 1 eq%) resulted in an increase (from 26% to 40%) in the maximum incorporation of nitrogen into this system [52]. These effects most likely arise from a reduction of the liquidus temperature and a subsequent lowering of viscosity as a result of fluorine addition. However, there are significant benefits with regards to the possible future commercial production of nitrogen glasses.

5.6 Oxynitride Glass Ceramics

As with oxide glasses, heat treatment at temperatures above T_g but below the eutectic temperature in the system results in the formation of crystalline products. For most oxynitride glasses, the results of heat treatment will be a mixture of oxide and oxynitride crystalline products, with the exact phases produced differing significantly with the temperature(s) used. In oxide systems, where the aim of the heat treatment is to produce a single-phase crystalline product of controlled (small) grain size, it is quite normal either to add a nucleating agent in the form of a very finely divided powder, or to perform an initial nucleation treatment aimed at forming a uniform dispersion of nuclei; the temperature is then raised to allow crystallization. Although, in the case of oxynitride glasses, there is always a fine dispersion of black particles, many research groups tend to carry out an initial nucleation stage so as to facilitate crystallization. Unfortunately, the product can only be single phase if the composition of the starting glass is the same as that of the final desired end-product. Because of the relatively restricted size of the glass-forming regions in oxynitride glass systems, only a small number of single-phase oxynitride glass ceramics can be prepared in this way. A number of crystalline oxynitride phases which readily form during the nucleation and crystallization of oxynitride glass ceramics will be discussed in the next section. Although, inevitably, these compositions are oxygen-rich, samples with quite high nitrogen contents can be produced, with perhaps the most nitrogen-rich being the aluminum-substituted melilite phase that occurs in low-Z rare earth sialon systems. A composition such as $Ln_2Si_2AlO_4N_3$ is close to the limit of Al-substitution in the melilite structure, and in the low-Z rare earth systems. This composition is also close to the limit of nitrogen substitution in the liquid (and hence glass-forming) region in these systems.

5.6.1 B-Phase

This phase was originally observed in the Y–Si–Al–O–N system, and the simplicity of the X-ray powder pattern allowed immediate indexing in terms of a hexagonal unit cell of approximate dimensions $a = 0.384$ nm, $c = 0.975$ nm. The low-temperature form of $YAlO_3$ [53] has dimensions $a = 0.368$ nm, $c = 1.052$ nm, while the oxynitride $YSiO_2N$ has a monoclinic unit cell which is pseudohexagonal with dimensions of the pseudohexagonal sub-cell $a = 0.405$ nm, $c = 0.910$ nm. Therefore, since the dimensions of the B-phase cell were almost exactly halfway between those of these two phases, the composition was thought to be close to Y_2SiAlO_5N. Moreover, as both the above-mentioned compounds had structures based on alternate layers of yttrium atoms and three-membered silicate or silicon oxynitride rings, it was assumed that this type of structure would also be the same in B-phase. Whereas, the initial studies focused entirely on B-phase produced in the Y–Si–Al–O–N system, more recent investigations have examined the extent to which B-phase occurs in rare-earth sialon systems. In fact, it appears to have a quite restricted range, only forming for rare earths with atomic numbers higher than about holmium (Ho).

More recent studies have involved the formation, composition, and microstructure of B-phase in much more detail [54] – especially in the Y- and Er-sialon systems. The composition is variable, both in cation and O : N ratios, depending to some extent on the starting composition, with the simplest whole number M : Si : Al ratio being 4 : 3 : 2, and a nitrogen content of approximately 10–15 eq%. It is apparent that the B-phase forms most readily from glasses close to this composition when heat-treated at low temperatures (1000–1100 °C). Although additional structural studies have been conducted on this phase, and appear to imply that B-phase has a chain [55] rather than a ring structure, the results have not been convincing, and there is clear scope for further structural studies to be carried out on this phase.

5.6.2 Iw Phase

This compound was first noted in the Y–Si–Al–O–N system by Leng-Ward and Lewis [56], who found it to have a Y : Si : Al composition close to 3 : 2 : 1. Later studies showed that, just as with B-phase, the composition could be variable, depending on the starting cation ratio, but with a mean value quite close to 3 : 2 : 1. An indexing of the XRD pattern has been proposed by Liddell and Thompson [57], who attributed to it a monoclinic unit cell of dimensions : $a = 1.128$ nm, $b = 1.007$ nm, $c = 1.002$ nm, and $\beta = 101.0°$. There is general agreement that the structure of the Iw phase is related to that of the B-phase, and Liddell and Thompson considered the c-axes to be similar, and the x and y dimensions of Iw to both be three times the value of B-phase, though with some distortion due to rotation within the tetrahedral network. Further evidence for the relationship between B and Iw was provided by from high-temperature XRD studies [58], which showed that glasses giving B-phase as the product of crystallization at temperatures just above T_g, also give Iw at higher temperatures, this being consistent with the idea of a degree of ordering among the tetrahedra. Although further structural studies are required, the main difficulty here is to prepare samples that contain high proportions of the Iw phase.

5.6.3 U-Phase

This phase is one of the best examples of a nitrogen glass ceramic, and is also the best documented as regards composition and structure. It occurs at compositions of the type $Ln_3Si_3Al_3O_{12}N_2$, and even though it has been prepared in the Y–Si–Al–O–N system, it is much more readily prepared in rare earth systems of low atomic number (typically La, Ce, and Nd). The structure is well established [59]; Figure 5.23 shows the

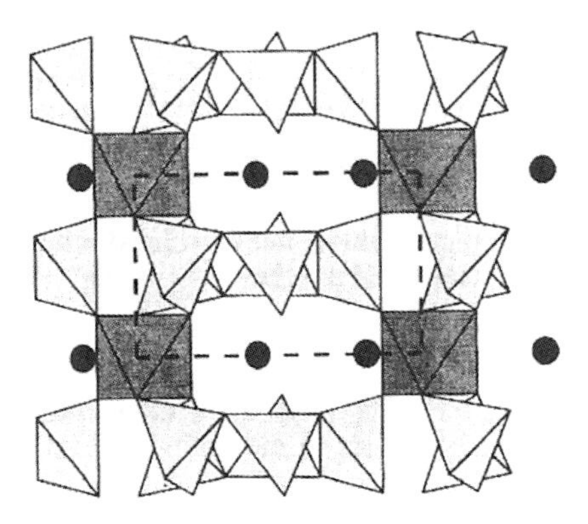

Figure 5.23 [1 1 0] projection of sialon U-phase. Reproduced from Ref. [59].

origin-sited aluminum atoms in 6-coordination, with the remaining cations in 4-coordination. The NMR spectra provide no convincing evidence for any ordering of (Si,Al) in the tetrahedra, nor for any nitrogen/oxygen ordering, apart from the nonbridging nonmetals which must be oxygen, based on a consideration of Pauling's rules.

5.6.4 W-Phase

W-phase was first reported by Fernie *et al.* [60] as one of several new phases prepared in the Nd–Si–Al–O–N system. Further preparative studies have confirmed the existence of a phase with the same XRD pattern as that described by Fernie *et al.*, but with a range of compositions, centered about a mean of $Nd_4Si_9Al_5O_{30}N$. It was also shown that the best way to prepare this phase was via a glass ceramics route; it was also clear that this phase formed more easily in the La- and Ce-sialon systems. To date, it has not been possible to prepare single crystals of this phase, and materials prepared via the glass ceramic route show a polycrystalline, needle-shaped microstructure (Figure 5.24). The diffraction pattern of the La derivative can be indexed on the basis of a monoclinic unit cell of dimensions: $a = 1.206$, $b = 0.508$, $c = 1.081$ nm, $\beta = 106°$.

Preliminary attempts to find mineral analogues showed some similarities with the mineral latiumite ($KCa_3(Si,Al)_5O_{11}[(SO_4,CO_3)]$), but further studies are required to determine the details of this structure.

5.6.5 Nitrogen Pyroxenes

Mineral pyroxenes occur at compositions of the type $MSiO_3$, and these structures are characterized by the presence of chains of SiO_4 tetrahedra, joined at two corners.

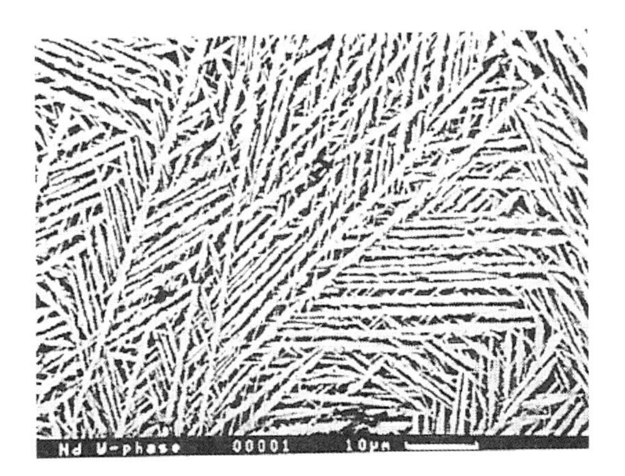

Figure 5.24 Acicular microstructure of sialon W-phase. Reproduced from Ref. [61].

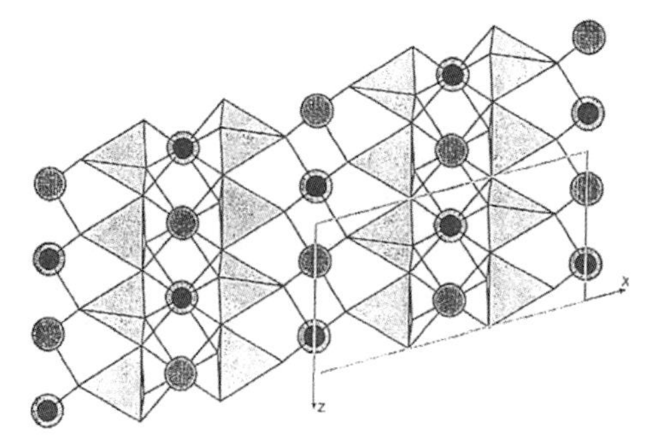

Figure 5.25 Atomic arrangement in nitrogen pyroxene $MgYSi_2O_5N$. Reproduced from Ref. [5].

The first oxynitride compounds having this type of structure were reported by Patel and Thompson [5], who observed them as secondary phases in samples of silicon nitride densified with mixtures of yttrium and magnesium. The XRD pattern of this phase indexed on a monoclinic unit cell of dimensions: $a = 0.973$, $b = 0.875$, $c = 0.531$ nm; $\beta = 105.6^\circ$, which compared very favorably with those of the mineral diopside $CaMgSi_2O_6$. An energy-dispersive X-ray (EDX) analysis confirmed that the overall Mg : Y : Si cation ratio was close to 1 : 1 : 2, and by balancing the valencies whilst at the same time keeping the characteristic metal: nonmetal ratio of a pyroxene, the formula of the oxynitride emerged as $MgYSi_2O_5N$. The structure of this compound is shown in Figure 5.25. Further studies of other systems showed that more nitrogen-rich pyroxenes could be prepared if Mg was substituted by rare earths. In addition, more nitrogen could be incorporated if another trivalent metal was used; Figure 5.26 shows the excellent fine-grained microstructure of a $YScSi_2O_4N_2$ glass ceramic prepared by the careful heat-treatment of a glass of this composition.

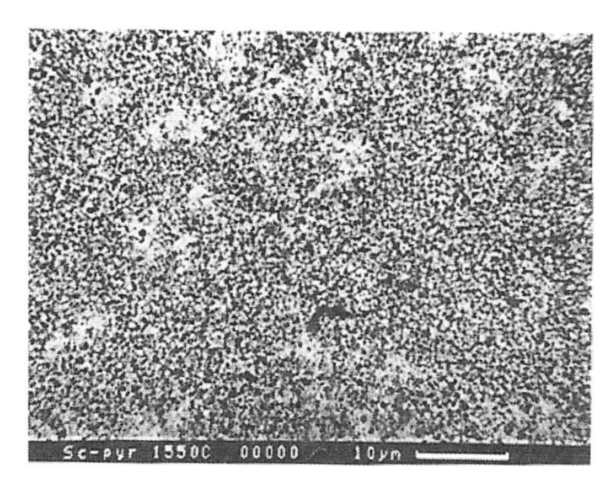

Figure 5.26 Fine-grained microstructure of a $ScYSi_2O_4N_2$ glass ceramic.

5.7 Conclusions

The resurgence of interest in nitrides that occurred during the 1950s was based on the realization that these materials had significant potential, based mainly on their high melting points. At the time, the concept of oxynitrides was thought unimportant, as oxygen was considered to be an element that caused problems for nitrides as a result of either oxidation or hydrolysis. It was only during the early 1970s that sialons emerged as a significant new development, when it was realized that oxygen and nitrogen could coexist side by side, and in materials that were substantially covalently bonded. Moreover, the ease with which it was possible to substitute aluminum plus oxygen in place of silicon plus nitrogen, without having to arrange any additional substitution among the cation distribution so as to balance valencies, enabled the identification of solid solution series for many silicon-based nitrides, and this immediately offered the major advantages of easier densification and the optimization of microstructures. Yet, it was the need of silicon nitride-based materials for a densification aid, and the immediate success of adding metal oxides such as MgO and Y_2O_3, that broadened the range of new compounds in oxynitride systems, notably because more cations were used. Unfortunately, the investigations carried out during the 1970s were limited, mainly because attention was focused on producing good quality silicon nitride and sialons for turbine applications, whilst the additional M–Si–O–N and M–Si–Al–O–N phases that appeared at the grain boundaries of silicon nitride and sialon ceramics were seen as phases that simply diluted the properties of the main phase. Consequently, it has been only during the past thirty years that attention has centered on studying four- and five-component phases in their own right, with increased funding being directed towards other types of ceramics, rather than nitrides. Nonetheless, interest from chemists rather than ceramists has increased, to a point where the investigations conducted by groups in Stockholm since the mid-1980s, and in Munich since the 1990s, have resulted in the establishment of a wide new range of materials, with more interesting structures. In addition, the ability to devise methods to produce single crystals of these materials has resulted in high-quality crystal structure determinations.

During the early investigations with nitrides and oxynitrides, attention was focused mainly on materials with good mechanical properties that could be used at high temperatures. Such focus persisted until the 1980s, since when increasing numbers of studies have been conducted with oxynitrides, examining their functional applications. As an example, among the materials described in this chapter, nitrides of the $M_2Si_5N_8$ and $MAlSiN_3$ series are currently being explored as phosphor materials. In fact, one significant advantage of these materials is that they not only have a good thermal stability over the working range of temperatures, but can also carry out substitutions at both M sites (and also by aluminum + oxygen for silicon + nitrogen) which, in turn, can provide a distinct ability to tailor specific electrical properties.

Recently, there has also been a considerable thrust in the direction of producing transparent oxynitride ceramics, the aim being to overcome the longstanding problem of conventional oxynitride glasses of gray or black discolorations. In this respect, recent studies conducted in Stockholm have shown that much more nitrogen can be incorporated than had been previously thought, thereby opening a new field of interest for these materials. Studies with fluorine-containing oxynitride glasses have also demonstrated a major potential for development, on this occasion as a biomaterial, as well as highlighting the ability of fluorine to expand the glass-forming field, thus facilitating glass formation.

In summary, the development of complex oxynitrides during the past thirty years has been slow, but systematic. These are not easy materials to produce, and the cost and associated high temperatures required may act as deterrents against their immediate industrial use. However, an impressive array of these materials is available today, displaying an impressive range of properties. Clearly, as technologies continue to improve, these materials will find appropriate niche applications in the future.

Acknowledgments

The author gratefully acknowledges the help of numerous colleagues, both past and present, who have been involved in the Newcastle contribution to the studies described in this chapter. Particular thanks are extended to Dr Kath Liddell, who carried out many of the X-ray studies, and to Professor Hasan Mandal, for his significant contribution to the characterization of many of the phases mentioned.

References

1 Lang, J. (1983) Progress in Nitrogen Ceramics, in *Proceedings NATO Advanced Study Institute* (ed. F.L. Riley), Nijhoff, The Hague, pp. 23–45.

2 Thompson, D.P. (1989) The crystal chemistry of nitrogen ceramics. *Mater. Sci. Forum*, **47**, 21–42.

3 Marchand, R. (1998) Ternary and high order nitride materials, *Handbook on the Physics and Chemistry of Rare Earths*, North Holland, vol. **25**, **ch. 166**, pp. 51–99.

4 Liddell, K. and Thompson, D.P. (2003) The future for multicomponent sialon ceramics. *Key Eng. Mater.*, **237**, 1–10.

5 Mandal, H., Patel, J.K., and Thompson, D.P. (1993) Nitrogen pyroxenes, in *Euroceramics III*, vol. **3** (eds P. Duran and J.F. Fernandez), Faenze Editrice Iberica S.L., pp. 1163–1168.

6 Thompson, D.P. (1993) O/N ordering in oxynitride ceramics, in *Euroceramics II* (eds G. Ziegler and H. Hausner), DKG, Köln, pp. 75–79.

7 Thompson, D.P. and Korgul, P. (1983) Progress in Nitrogen Ceramics, in *Proceedings NATO Advanced Study Institute* (ed. F.L. Riley), Nijhoff, The Hague, pp. 375–380.

8 Westwood, A.D., Youngman, R.A., McCartney, M.R., Cormack, A.N., and Notis, M.R. (1995) *J. Mater. Res.*, **10** (5), 1270–1286.

9 Westwood, A.D., Youngman, R.A., McCartney, M.R., Cormack, A.N., and Notis, M.R. (1995) *J. Mater. Res.*, **10** (5), 1287–1300.

10 Westwood, A.D., Youngman, R.A., McCartney, M.R., Cormack, A.N., and Notis, M.R. (1995) *J. Mater. Res.*, **10** (5), 2573–2585.

11 Thompson, D.P., Korgul, P., and Hendry, A. (1983) Progress in Nitrogen Ceramics,

in *Proceedings NATO Advanced Study Institute* (ed. F.L. Riley), Nijhoff, The Hague, pp. 61–74.

12 Koroglu, A., Apperley, D.C., Harris, R.K., and Thompson, D.P. (1996) *J. Mater. Chem.*, **6**, 1031–1034.

13 Dupree, R., Lewis, M.H., and Smith, M.E. (1988) *J. Am. Chem. Soc.*, **110**, 1083–1091.

14 Liddell, K., Thompson, D.P., Wang, P.L., Sun, W.Y., Gao, L., and Yan, D.S. (1998) *J. Eur. Ceram. Soc.*, **18**, 1479–1492.

15 Takahashi, J., Yamane, H., Hirosaki, N., Yamamoto, Y., Suehiro, T., Kamiyama, T., and Shimada, M. (2003) *Chem. Mater.*, **15**, 1099–1104.

16 Takahashi, J., Yamane, H., Shimada, M., Yamamoto, Y., Hirosaki, N., Mitomo, M., Oikawa, K., Torii, S., and Kamiyama, T. (2002) *J. Am. Ceram. Soc.*, **85**, 2072–2077.

17 Jameel, N.S. (1984) Structural chemistry of some yttrium sialons. Ph.D. Thesis, University of Newcastle upon Tyne.

18 Liddell, K., Thompson, D.P., Bräuniger, T., and Harris, R.K. (2005) *J. Eur. Ceram. Soc.*, **25**, 37–47.

19 Woike, M. and Jeitschko, W. (1997) *J. Solid State Chem.*, **129**, 312–318.

20 Liddell, K. and Thompson, D.P. (2001) *J. Mater. Chem.*, **11**, 507–512.

21 Hampshire, S., Park, H.K., Thompson, D.P., and Jack, K.H. (1978) *Nature*, **274**, 880–881.

22 Yu, Z.B., Thompson, D.P., and Bhatti, A.R. (2000) *J. Eur. Ceram. Soc.*, **20**, 1815–1828.

23 Cai, Y., Shen, Z., Grins, J., Esmaeilzadeh, S., and Höche, T. (2007) *J. Am. Ceram. Soc.*, **90**, 608–613.

24 Kempgens, P.F.M., King, I.J., Harris, R.K., Yu, Z.B., and Thompson, D.P. (2001) *J. Mater. Chem.*, **11**, 2507–2512.

25 Grins, J., Esmaeilzadeh, S., and Shen, Z. (2003) *J. Am. Ceram. Soc.*, **86**, 727–730.

26 Grins, J., Shen, Z., Nygren, M., and Ekström, T. (1999) *J. Mater. Chem.*, **9**, 1019–1022.

27 Mandal, H. and Thompson, D.P. (1996) *J. Mater. Sci. Lett.*, **15**, 1435–1438.

28 Hwang, C.J., Susnitzky, D.W., and Beaman, D.R. (1995) *J. Am. Ceram Soc.*, **78**, 588–592.

29 Esmaeilzadeh, S., Grins, J., Shen, Z., Eden, M., and Thiaux, M. (2004) *Chem. Mater.*, **16**, 2113–2120.

30 Lewis, M.H., Basu, B., Smith, M.E., Bunyard, M., and Kemp, T. (2004) *Silic. Indus.*, **69** (7–8), 225–231.

31 Mah, T., Mazdiyasni, K.S., and Ruh, R. (1979) *J. Am. Ceram. Soc.*, **62**, 12–16.

32 Leach, M.J. (1990) Synthesis and multinuclear magnetic resonance studies of some nitrogen-containing ceramic phases. Ph.D. Thesis, University of Durham.

33 Grins, J., Shen, Z., Esmaeilzadeh, S., and Berastegui, P. (2001) *J. Mater. Chem.*, **11**, 2358–2362.

34 Schlieper, T., Milius, W., and Schnick, W. (1995) *Z. Anorg. Allg. Chem.*, **621**, 1380–1388.

35 Inoue, Z., Mitomo, M., and Ii, N. (1980) *J. Mater. Sci.*, **15**, 2915–2922.

36 Schlieper, T. and Schnick, W. (1995) *Z. Anorg. Allg. Chem.*, **621**, 1535–1542.

37 David, J. and Lang, J. (1967) *C.R. Acad. Sci. (Paris)*, **265**, 581–586.

38 Maunaye, M., Marchand, R., Guyader, J., Laurent, Y., and Lang, J. (1971) *Bull. Soc. Fr. Miner. Crist.*, **94**, 561–566.

39 Perera, D.S. (1976) Magnesium sialons. Ph.D. Thesis University of Newcastle.

40 Siddiqi, S.A. (1980) The structural characterisation of some metal sialons. M.Sc. Thesis, University of Newcastle.

41 Patience, M.M. (1983) Silicon carbide alloys. Ph.D. Thesis, University of Newcastle.

42 Mikami, M., Uheda, K., and Kijima, N. (2006) *Phys. Status Solidi*, **203**, 2705–2711.

43 Huang, Z.K., Sun, W.Y., and Yan, D.S. (1985) *J. Mater. Sci. Lett.*, **4**, 255–258.

44 Li, Y.Q. (2005) Structural and luminescence properties of novel rare-earth doped silicon nitride based materials. Ph.D. Thesis, Technical University of Eindhoven.

45 Huppertz, H. and Schnick, W. (1997) *Angew. Chem., Int. Engl. Ed.*, **36**, 2651–2655.

46 Lauterbach, R., Irran, E., Henry, P.F., Weller, M.T., and Schnick, W. (2000) *J. Mater. Chem.*, **10**, 1357–1363.

47 Jack, K.H. (1973) *Proc. Br. Ceram. Soc.*, **22**, 376–392.

48 Hampshire, S. and Pomeroy, M.J. (2008) *Int. J. Appl. Ceram. Technol.*, **5**, 155–163.

49 Hakeem, A.S., Grins, J., and Esmaeilzadeh, S. (2007) *J. Eur. Ceram. Soc.*, **16**, 4773–4781.

50 Leonova, E., Hakeem, A.S., Jansson, K., Stevensson, B., Shen, Z., Grins, J., Esmaeilzadeh, S., and Eden, M. (2008) *J. Non-Cryst. Solids*, **354**, 49–60.

51 Korgul, P. and Thompson, D.P. (1993) *J. Mater. Sci.*, **28**, 506–512.

52 Hanifi, A.R., Genson, A., Pomeroy, M.J., and Hampshire, S. (2007) *Mater. Sci. Forum*, **554**, 17–23.

53 Bertaut, F. and Mareschal, J. (1963) *C.R. Acad. Sci. (Paris)*, **257**, 867–873.

54 Maclaren, I., Falk, L.K.L., Menke, Y., Diaz, A., Hampshire, S., Parmentier, J., and Thompson, D.P. (1999) *Inst. Phys. Conf. Ser.*, **161** (4), 153–156.

55 Gonon, M.F., Descamps, J.-C., Cambier, F., and Thompson, D.P. (2000) *Proceedings, 2nd International Symposium on Nitrides, Limerick, 1998* (eds S. Hampshire and M.J. Pomeroy), Trans Tech, Zurich, pp. 325–334.

56 Leng-Ward, G. and Lewis, M.H. (1985) *Mater. Sci. Eng.*, **71**, 101–105.

57 Liddell, K. and Thompson, D.P. (1998) *Br. Ceram. Trans.*, **97**, 155–161.

58 Dolekcekic, E. (2004) Properties and crystallisation of M-Si-Al-O-N glasses: effects of composition on the stability of B-phase. Ph.D. Thesis, University of Limerick.

59 Grins, J., Käll, P.O., Liddell, K., Korgul, P., and Thompson, D.P. (1990) *Inst. Phys. Conf. Ser.*, **111**, 427–434.

60 Fernie, J.A., Leng-Ward, G., and Lewis, M.H. (1989) *J. Mater. Sci. Lett.*, **9**, 29–35.

61 Thompson, D.P. (1993) Silicon nitride ceramics: science and technological advances, in *Research Society Symposium K*, vol. **287** (ed. I-.W. Chen), Materials Research Society, pp. 79–91.

6
Perovskites

Vladimir Fedorov

6.1
Introduction

Although ceramics, as processed inorganic materials, are widely used today by humankind, there are only a half-dozen specific ceramic materials that dominate the different applications. Among similar materials, the perovskites take special place since, with some skilled chemical manipulation, they are able to produce an incredibly wide array of phases with totally different functions. In fact, the perovskites have been dubbed "inorganic chameleons," as they display a rich diversity of chemical compositions and properties.

Archetypal perovskite is a mineral having a composition of $CaTiO_3$. During the 1830s, the German geologist Gustav Rose described this mineral and named it after the Russian mineralogist, Count Lev Perovskii [1]. The ideal perovskite, ABO_3, has a cubic structure with the corner-sharing octahedra BO_6 that form the skeleton of the structure in which the center position is occupied by the A cation (Figure 6.1). The surprising point here is that although the naturally occurring compound $CaTiO_3$ was originally thought to be cubic, its true symmetry was later shown to be orthorhombic [2].

The most numerous compounds with the perovskite structure are complex oxides, although some halides, nitrides, carbides and hydrides may also crystallize in this structure. In general, the complex compounds with perovskite-type structure are today referred to as "perovskites."

Although the perovskites were first investigated by Goldschmidt back in the 1920s [3], widespread studies of these materials first began twenty years later during World War II, when the extraordinary high dielectric constant in $BaTiO_3$ was detected at the Institute of Physics (Russian Academy of Sciences, Moscow) [4] and, at the same time, in the TAM laboratories in Cleveland, USA [5]. The discovery of a high dielectric constant in $BaTiO_3$ was the key event in the history of the emergence of the perovskite structure as the preeminent "high-tech" ceramic material. Very quickly, it was realized that $BaTiO_3$ was the most promising insulator for capacitors, and in quick succession the reason for the extraordinary value of the

Ceramics Science and Technology Volume 2: Properties. Edited by Ralf Riedel and I-Wei Chen

ISBN: 978-3-527-31156-99

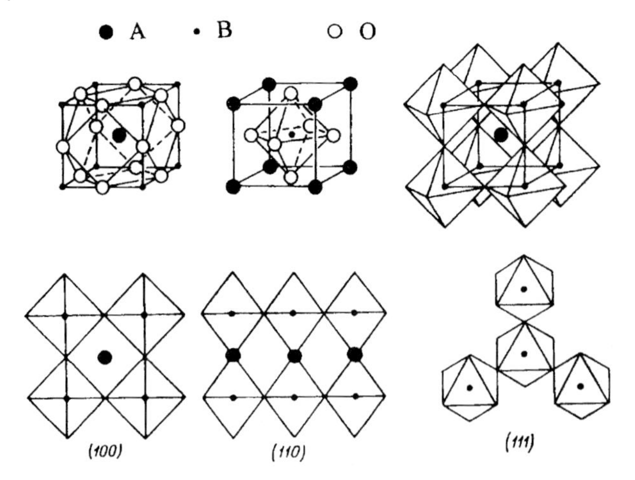

Figure 6.1 Different presentations of the ideal perovskite ABO_3 structure and the projections of structure along [1 0 0], [1 1 0], and [1 1 1].

dielectric constant recognized in $BaTiO_3$ was understood [6, 7]. The theory of domain structure opened the link to *ferroelectricity*. Since some important properties were discovered in perovskites, the studies of these materials have grown exponentially, with the perovskites having attracted considerable attention for many years, owing to both the fundamental phenomena of crystalline solids and their wide range of applications. The ABO_3 perovskites display several interesting physical properties that are useful for different types of electronic devices, including as ferroelectricity ($BaTiO_3$), piezoelectricity ($PbZr_{1-x}Ti_xO_3$), strong relaxor ferroelectricity ($PbMg_{1/3}Nb_{2/3}O_3$), ferromagnetism ($SrRuO_3$), weak ferromagnetism ($LaFeO_3$), large thermal conductivity due to exciton transport ($LaCoO_3$), mixed ionic-electronic conductivity suited for cathode materials in solid oxide fuel cells (SOFCs) ($La_{1-x}Sr_xMnO_3$), insulator-to-metallic transitions of interest for thermistor applications ($LaCoO_3$), oxygen ion conductivity which is superior to that of conventional zirconia-based electrolytes at moderate temperatures (doped $LaGaO_3$), colossal magnetoresistance ($Ln_{1-x}A_xMnO_3$, where Ln = rare earth, A = Ca or Sr), fluorescence compatible with laser action ($LaAlO_3$: Nd), nonlinear optical properties ($LiNbO_3$, $LiTaO_3$, $NaBa_2Nb_5O_{15}$), transport properties of interest for high-temperature thermoelectric power (La_2CuO_4), high proton conductivity with potential applications in hydrogen fuel cells and sensor technologies (particularly cerates and zirconates), ultrahigh melting point ($BaMg_{1/3}Ta_{2/3}O_3$), and a fascinating and widely studied high T_c superconductivity ($BaPb_{1-x}Bi_xO_3$ and perovskite-based cuprates $YBa_2Cu_3O_7$). All of these various properties and possible applications have attracted considerable attention towards the perovskites, and continue to stimulate widespread interest in the study of these materials.

Today, a great deal of material, including original articles, comprehensive reviews and excellent books, have been devoted to the perovskites (e.g., Refs [8–37]). In addition, the specialist journal *Ferroelectrics* has been published since 1971.

6.2 Crystal Structure

6.2.1 Ideal Perovskite Structure

The different ways of presenting the crystal structure of ABO_3 perovskite are shown in Figure 6.1. The ideal perovskite structure is cubic with space group $Pm\bar{3}m$-O_h^1, and is typified not by $CaTiO_3$ but by $SrTiO_3$ with $a = 3.905$ Å, $Z = 1$. The structure has a very simple arrangement of ions. In perovskite ABO_3, the smaller B cations lie at the center of an octahedron of anions, such that each BO_6 octahedron shares every corner with other BO_6 units, leading to a three-dimensional (3-D) network. The A cations then sit at the center of a cube formed by eight corner-sharing BO_6 octahedra, with cubo-octahedral coordination to twelve anions.

In the almost ideal perovskite structure of $SrTiO_3$, the TiO_6 octahedra are perfect with 90° angles and six equal Ti–O bonds at 1.952 Å. Each Sr atom is surrounded by twelve equidistant oxygens at 2.761 Å [2].

The perovskite structure is thus a superstructure with a ReO_3-type framework built up by the incorporation of A cations into the BO_6 octahedra. The significance and role of the ReO_3-type framework as a host structure for deriving numerous structures of metal oxides has been emphasized by Raveau [38].

The crystal chemistry of perovskites was first developed by Goldschmidt [3], who prepared and studied not only a large number of the synthetic perovskites with different compositions (including $BaTiO_3$) but also went on to establish the principles for the synthesis of similar compounds, which remain valid to the present day.

In the ideal perovskite structure, the A and O atoms touch one another and form the cubic ABCABC stacking of close-packed AO_3 layers along the cubic [1 1 1] direction; B is contained in the octahedral interstices coordinated by six O^{2-} anions. In the ideal structure ABO_3, there is the following ratio: $r_A = r_O$ (1.40 Å) and $r_B = 0.414$, $r_O = 0.58$ Å; the B–O distance is equal to $a/2$ (where a is the cubic unit cell parameter), while the A–O distance is $a/\sqrt{2}$ and the following relationship between the ionic radii holds $(r_A + r_O) = \sqrt{2}(r_B + r_O)$. However, it was found that the cubic structure was still retained in ABO_3 compounds, even though these equations were not exactly obeyed. As a measure of the deviation from the ideal situation, Goldschmidt introduced a tolerance factor t, defined by the equation:

$$t = (r_A + r_O)/\sqrt{2}(r_B + r_O) \tag{1}$$

For an ideal perovskite, t is unity. In practice, perovskite-structured compounds are found to adopt the ideal cubic arrangement for $\approx 0.97 < t < \approx 1.03$, but a great many elements can form perovskite structures for lower t-values. The concept of the "tolerance factor" for the perovskite arrangement of interpenetrating dodecahedra and octahedra indicates how far from ideal packing the ionic sizes can move but still be "tolerated" by the perovskite structure. The $t = 1$ condition represents the "close-packing" in the perovskite structure. For $t < 1$, the size of the unit cell is governed by the B-site ion, and as a result the A-site ions have too much room for vibration.

For $t > 1$, just the opposite situation occurs in the unit cell; that is, in this case the B-site ions have too much room to vibrate. For larger values of t, hexagonal structures are preferred in which there is some degree of edge-sharing of the BO_6 octahedra. It is important to mention here that all of these calculations hold only when the radii of the ions used for calculating "t" are taken constantly from within the same tables (e.g., Goldschmidt, or Pauling, or Shannon/Prewitt, or Roy/Müller). Moreover, it should be emphasized that the tabular ionic radii are obtained from X-ray data taken at room temperature and ambient pressure. Since the A–O and B–O bonds have different thermal expansions and compressibilities, t is a temperature- and pressure-dependent parameter, and is unity only at a single temperature for a given pressure. A larger thermal expansion of the A–O bonds makes $dt/dT > 0$, except where a spin-state transition occurs on the B cation. Normally, the A–O bonds are more compressible than the B–O bonds, which makes $dt/dP < 0$; however, where transition metal B atoms have 3d electrons that approach or enter the crossover from localized to itinerant electronic behavior, the B–O bond may be more compressible and give $dt/dP > 0$ [39].

As a matter of fact, the tolerance factor is a rather complex crystallo-chemical parameter, which can reflect the structural distortion, force constants of binding, rotation and tilt of the BO_6 octahedrons. These in turn affect the dielectric properties, transition temperature, temperature coefficient of the dielectric constant of material, and even the dielectric loss behavior in a perovskite dielectric.

At a later time, it was found that those compounds with a perovskite-type structure can be stable if $r_A \geq 0.90$, and $0.51\ \text{Å} \leq r_B \leq 1.1\ \text{Å}$ [40–42]. If $r_A > r_O$, then the oxygen ions in layers AO_3 are separated and the size of octahedral voids are increased, so that cations with $r_B > 0.41\ r_O$ can be placed. If $r_A < r_O$, the ions A are found smaller than the voids with $Z_A = 12$, and a parameter of the cubic unit cell is defined primarily by the value of $2(r_B + r_O)$. Thus, in a perovskite structure the sizes of positions with $Z_A = 12$ and $Z_B = 6$ are interconnected. Therefore, the ABO_3 perovskite structure has a great versatility to adapt to unmatched equilibrium A–O and B–O bond lengths.

Many ternary oxides with compositions of $A^{+}B^{5+}O_3$, $A^{2+}B^{4+}O_3$, $A^{3+}B^{3+}O_3$, $A^{4+}B^{2+}O_3$, and an abundance of compounds with more complex compositions, are crystallized in perovskite structure. The perovskite structure is very flexible, allowing not only the substitution of different cations in positions A and B over a wide range of compositions $A_{1-x}A'_xB_{1-x}B'_xO_3$, but also the introduction of vacancies or substitutions on the anion sublattice. It is for this reason that about 90% of the metallic elements of the Periodic Table are known to be stable in a perovskite-type oxide structure.

6.2.2
Structural Distortions and Phase Transitions

Most perovskites adopt a cubic structure at high temperature. Although it would seem that the perovskite structure is very simple, it does in fact have many degrees of freedom that lead to different distortions of the structure. As a rule, the useful properties of perovskites appear in distorted compounds. The nature of distortions is due to the features of nuclear or electron configurations [43]. The former type of distortion arises when there is a mismatch between the sizes of ions and holes occupied by these ions. To increase the stability of the structure, the anionic polyhedra are forced to deform in such a way that the interatomic distances are reduced to a minimum. The second type of distortion is due to a spontaneous deformation through unsymmetrical covalent bonds, with the distortions being caused by a presence of Jahn–Teller cations in BO_6 octahedra ($B = Cu^{2+}$, Mn^{3+} and others). Under normal conditions, more than 80% of known perovskites are characterized by distorted structures. In many cases, small distortions produce essential changes in physical properties.

The BO_6 octahedra can not only deform but may also tilt and rotate along their fourfold or twofold axes, giving rise to different superstructures or modulated structures. Besides, there is a strong dependence of structural symmetry on temperature: at lower temperatures, numerous modifications or structural distortions from the ideal perovskite structure exist [43–45], and all of these causes lower the symmetry of the structure from cubic to tetragonal, orthorhombic, rhombohedral, or monoclinic. A lowering in symmetry will introduce different orientation variants (twins) and translation variants (antiphase boundaries). Structures with a lower symmetry, derived from the cubic structure by tilting and/or deformation of the BO_6 octahedra, become stable such that one (or several) phase transformation(s) may take place.

For example, barium titanate ($BaTiO_3$), as a more representative member of the perovskites, demonstrates similar transformations distinctly (Figure 6.2). $BaTiO_3$ has an ideal cubic structure, space group $Pm\bar{3}m$, at a temperature above 120 °C; by lowering the temperature, $BaTiO_3$ is subsequently transformed into the tetragonal phase with the space group *P4mm*. Here, the oxygen octahedra have a minor

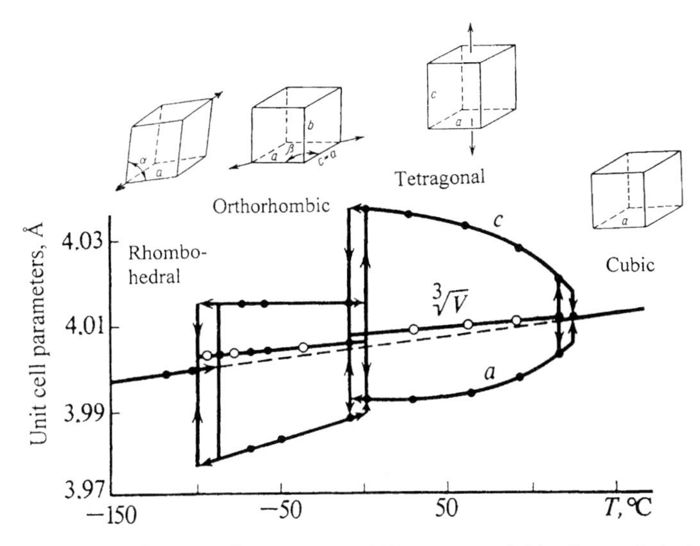

Figure 6.2 Phase transformations in $BaTiO_3$ accompanied by changes in lattice parameters.

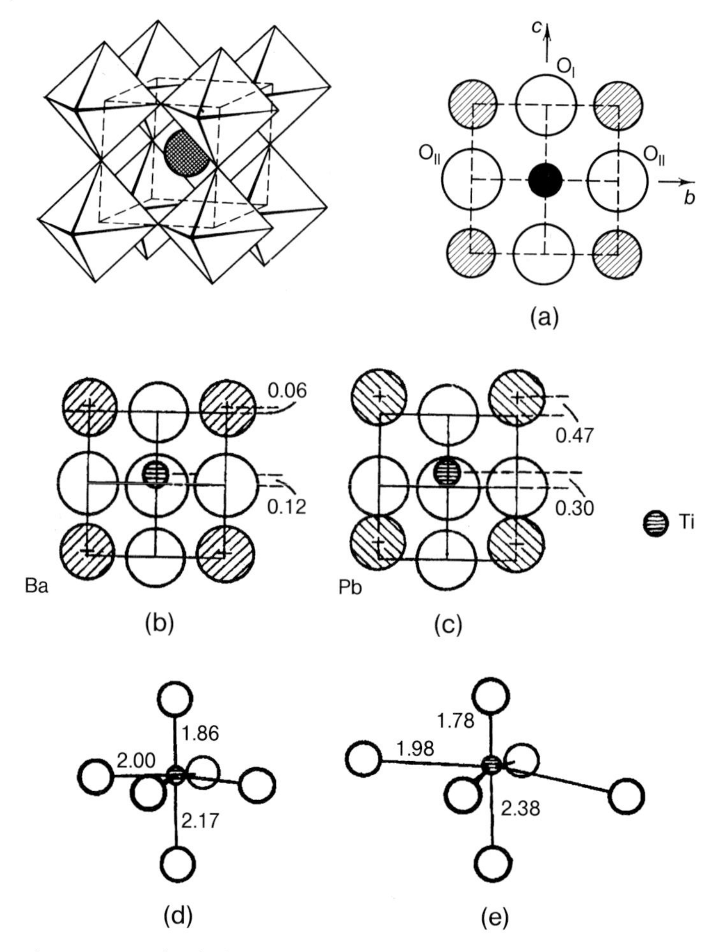

Figure 6.3 (a) The ideal perovskite ABO_3 structure; (b–e) Changes of the environment of Ti^{4+} ions in tetragonal modifications of $BaTiO_3$ (b,d) and $PbTiO_3$ (c,e).

distortion: a displacement of O_I with respect to O_{II} is small (Figure 6.3a). The displacement of Ti atoms from the centers of corresponding O_6 octahedra is equal to 0.12 Å (Figure 6.3b); as a result, two bonds O_{II}–Ti which form the angle different from 180° (171°28′) (Figure 6.3d) are displaced from their normal lattice positions in the same direction on 0.06 Å (Figure 6.3b) [46]. (Note that in tetragonal $PbTiO_3$ the corresponding displacements are equal to 0.30 Å and 0.47 Å, respectively [47]; Figure 6.3c and e). This distortion brings about the remarkable difference in Ti–O distances (Figure 6.3d and e).

Upon further cooling, the tetragonal $BaTiO_3$ undergoes a second phase transition and becomes orthorhombic, space group *Bmm*2. This structure can again be interpreted as a slight distorted cubic form that is obtained by stretching in the direction of one face diagonal, and compression of the other face; these diagonals then become orthorhombic axes. The third axis saves its initial direction, but the lattice spacing along this direction is changed slightly. On approaching the temperature $-70 \div -90\,^{\circ}C$, $BaTiO_3$ undergoes a third phase transition to become a rhombohedral below this temperature (space group R3*m*).

With regards to similar distortions, it should be emphasized that displacements of the A and B cations from the centers of corresponding polyhedra which lead to dipole moments are remarkable, as these produce ferroelectricity and related famous phenomena. Perovskites containing high-charge B^{n+} cations such as Ti^{4+}, Zr^{4+}, Nb^{5+}, Ta^{5+} represent an important group of ferroics.

In similar compounds the size of the B^{n+} cations is smaller than the occupied octahedral voids; thus, B^{n+} cations located in a highly polarizable oxygen environment will be displaced from a center of octahedron to one of main direction. If ferroic displacements of the B^{n+} cations in all the octahedral sites are parallel, and occur along the [0 0 1], [1 1 0] and [1 1 1] directions of the initial phase, then the distortional structures will belong to polarized space groups – tetragonal, orthorhombic or rhombohedral symmetry, respectively. $BaTiO_3$ is a unique compound because it has five crystal modifications (in addition to the four phases described above there is a high-temperature hexagonal phase which is realized at 1460–1612 °C), three of which are polarized and ferroic.

As noted above, the perovskite structure is inclined to different distortions. The two types of distortion with lowering symmetry, derived from the perovskite cubic structure by tilting the BO_6 octahedra, have undergone extensive investigation. For example, in the case where $t < 1$, the B–O bonds are situated under compression and the A–O bonds under tension, and the structure alleviates these stresses by a cooperative rotation of the $MO_{6/2}$ octahedra. Glazer [48] identified fifteen tilt systems that are possible, together with their space groups, and supposed that during such rotations all BO_6 octahedra in the distorted phase remained regular, while tilting of the octahedra around different axes of a parent unit cell was mutually independent and displacements of A ions were disregarded. Glazer also provided a notation for their description, namely that if neighboring octahedra along any direction of type [0 0 1] of this initial crystal turn around this axis in phase, then such rotations would be labeled by an upper index " + ". However, if such tilts were out of phase then they would be labeled by the mark "–". For example, if similar distortions take place around one axis *z*, the correct notations are $(a^0 a^0 c^+)$ and $(a^0 a^0 c^-)$, respectively. Both types of tilt may occur in distorted structures around different axes of the initial cubic phase. A group theoretical analysis of the domain fragmentation in ABO_3 with orthorhombic, rhombohedral or monoclinic space groups was conducted by Van Tendeloo and Amelinckx [49] and by Portier and Gratias [50]. For example, the rhombohedral $R\bar{3}c$ phase was seen to be related to the cubic parent phase, and could be described as $a^-a^-a^-$, with the c_R axis being parallel to one of the $[1\ 1\ 1]_c$ directions. There are, therefore, four different orientation variants. The orthorhombic *Pnma* phase can be described as $a^+a^-a^-$ (Figure 6.4b), and can be formed from the cubic phase in six different orientation variants. The singular long axis b_o ($\approx 2a_c$) can adopt any of the three $[1\ 0\ 0]_c$ cube directions; for each choice of b_o, the a_o and c_o axes can be interchanged.

Similar descriptions were made by Aleksandrov using another notation [33, 51].

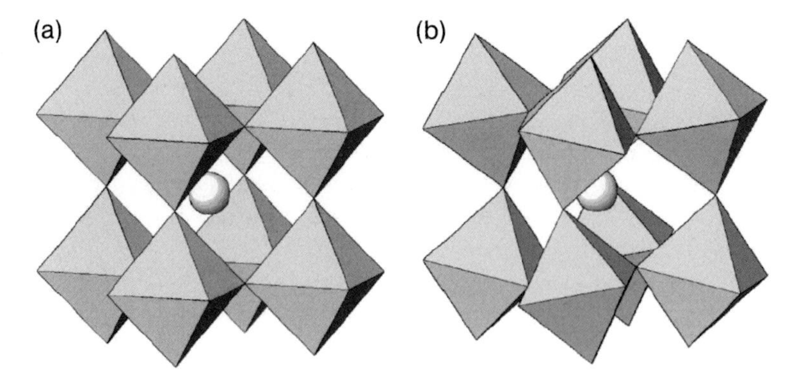

Figure 6.4 Transformation of cubic perovskite structure (a) in orthorhombic (b) by tilting the BO_6 octahedra.

6.2.3
Other Perovskite-Related Structures

6.2.3.1 Polytypes Consisting of Close-Packed Ordered AO_3 Layers

The stacking of AO_3 layers in the structure may be either cubic (*c*) or hexagonal (*h*) with respect to two adjacent layers. If the stacking is entirely cubic, the BO_6 octahedra share only corners in three dimensions to form the normal perovskite 3C structure (Figure 6.5a). If the stacking is all-hexagonal, the BO_6 octahedra share opposite faces and form chains along the *c*-axis, as in $BaNiO_3$ (2H) (Figure 6.5b). In between these two extremes there may be several polytypic structures consisting of mixed cubic and hexagonal stacking of AO_3 layers. Thus, the structural family of perovskites includes numerous stacking varieties, with the c and *h* layers in different sequences. Both the cubic and the hexagonal perovskite structure can be generated from the stacking of close-packed AO_3 layers and subsequent filling of the generated octahedral sites by the B-cation, where an ABC-type stacking results in the cubic structure, and an AB

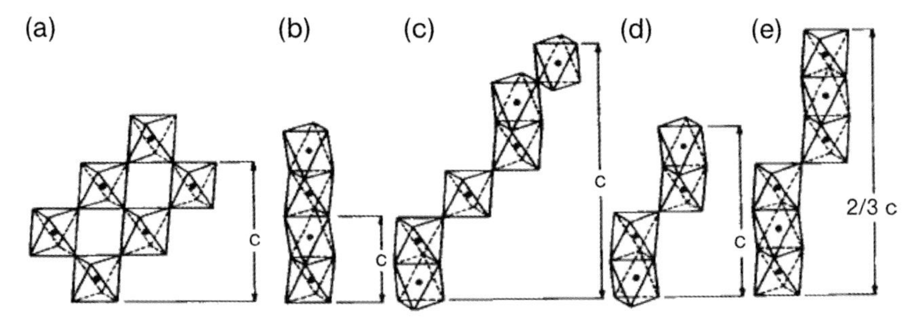

Figure 6.5 Network of $BO_{6/2}$ octahedra in the ABO_3 perovskite polytypes. (a) Cubic 3C; (b) Hexagonal 2H; (c) Hexagonal 6H; (d) Hexagonal 4H; (e) Rhombohedral 9R [34].

stacking results in the hexagonal perovskite structure. For example, in a sequence such as *chhc*, there is a group of three octahedra that share faces at the *h* layers; this group is connected with other octahedra by vertex-sharing at the *c* layers. The size of the groups of face-sharing octahedra depends on the nature of the metal atoms in the octahedra, and especially on the ionic radius ratios. For example, if $t > 1$, the B–O bonds will be situated under tension, and the A–O bonds under compression. In this case, the structure may alleviate the tensile stresses by introducing hexagonal ABAB stacking, with the octahedral sites coordinated by six anions sharing common octahedral-site faces in isolated columns rather than common octahedral-site corners, as in cubic stacking. Large A-site cations can be accommodated between columns of face-shared octahedra. Goodenough [34, 52] has identified the energy reasons for the hexagonal stackings and formation of the series of polytypes presented in Figure 6.5.

The nomenclature for a polytype gives the number of AO_3 close-packed planes in a unit cell, followed by a letter to designate the symmetry as cubic, hexagonal, or rhombohedral; thus, the series goes from 3C to 6H to 4H to 9R to 2H as the fraction of hexagonal stackings increases. The loss of Madelung energy is reduced in the 6H and 4H polytypes by displacements of the B^{m+} cations away from their shared octahedral-site face; and in the 9R polytype by displacements of the two outer B^{m+} cations of a triple-B unit away from their shared octahedral-site face. It has been demonstrated for the system $Ba_{1-x}Sr_xRuO_3$ with $dt/dP < 0$, that under pressure a 9R polytype can be transformed to 4H, a 4H to 6H, and a 6H to a cubic 3C perovskite (Figure 6.6) [34, 52].

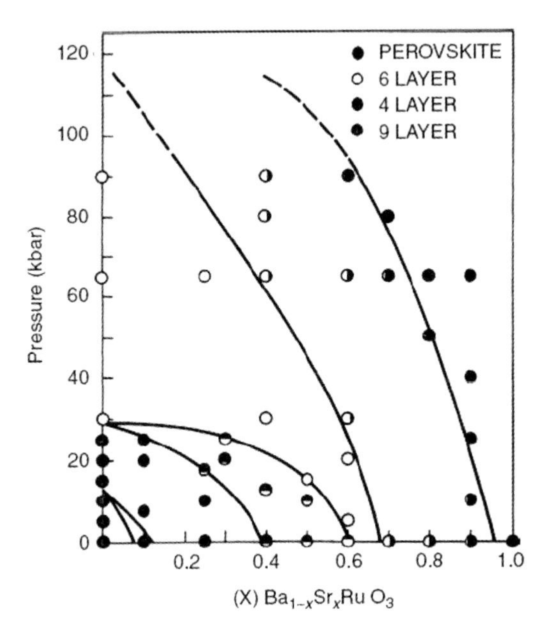

Figure 6.6 Room-temperature pressure versus composition phase diagram of $Ba_{1-x}Sr_xRuO_3$ [34].

6.2.3.2 Perovskite Intergrowth Structures

Ruddlesden–Popper (RP) Phases The cubic perovskite structure may be described as alternating AO and BO_2 (0 0 1) planes on traversing the [0 0 1] axis, and this architecture lends itself to the formation of intergrowths. Ruddlesden and Popper [53] designed a series of homologous compounds (RP-type phases) with the general formula $AO(ABO_3)_n$, where AO represents the rock salt structure layer separating the blocks of perovskite layers characterized by $n = 1,2,3 \ldots \infty$. The first member of this series can be described as the K_2NiF_4 structure. As n increases, the phases tend to be more perovskite in nature, and finally reach the pure perovskite structure for $n = \infty$ (Figure 6.7).

Such phases exhibit properties that range from a typical insulator or ferroelectric to a conductor or superconductor. Several single-crystal phases of the $(SrO)(SrTiO_3)_n$ series have been synthesized with different n-values, and their dielectric constants measured [54]. The $(LaO)(LaNiO_3)_n$ compounds have also been prepared successfully [55]. Some other RP-type phases studied by various groups are detailed in Refs [5, 56].

Aurivillius Phases The Aurivillius bismuth compounds, which have the structural formula $(Bi_2O_2)^{2+}(A_{n-1}B_nO_{3n+1})^{2-}$, form intergrowths of the perovskite structure and $Bi{-}O_2{-}Bi$ sheets that consist of edge-shared BiO_2 square pyramids [57]. The $n = 3$ compound $Bi_4Ti_3O_{12} = (Bi_2O_2)^{2+}(Bi_2Ti_3O_{10})^{2-}$ is illustrated in Figure 6.8. The Aurivillius phases represent a large family of oxides and oxyfluorides, where

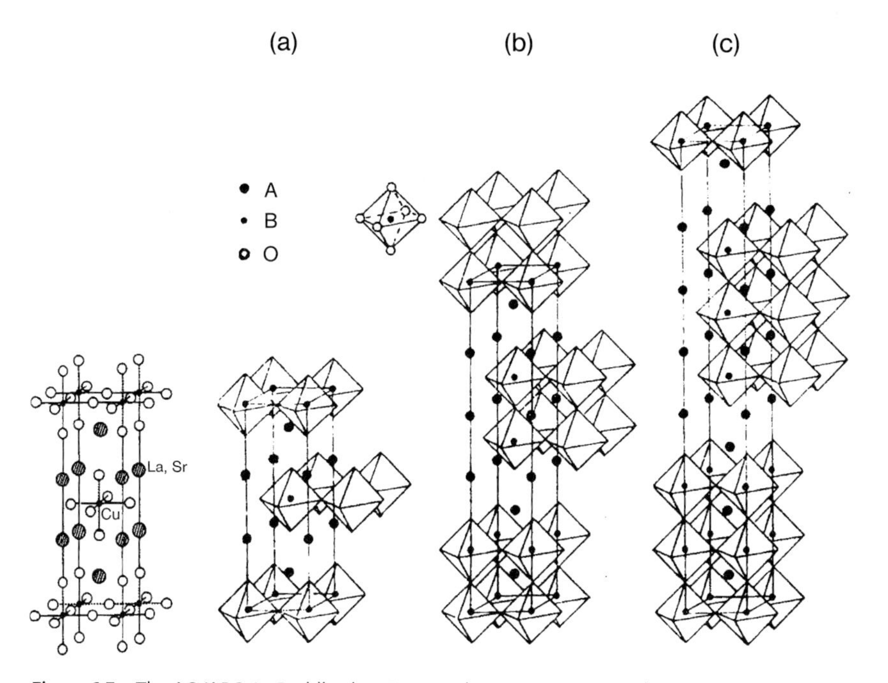

Figure 6.7 The $AO(ABO_3)_n$ Ruddlesden–Popper phase. (a) T-tetragonal La_2CuO_4 ($n = 1$); (b,c) Compounds with stoichiometry $A_3B_2O_7$ ($n = 2$) (b) and $A_4B_3O_{10}$ ($n = 3$) (c).

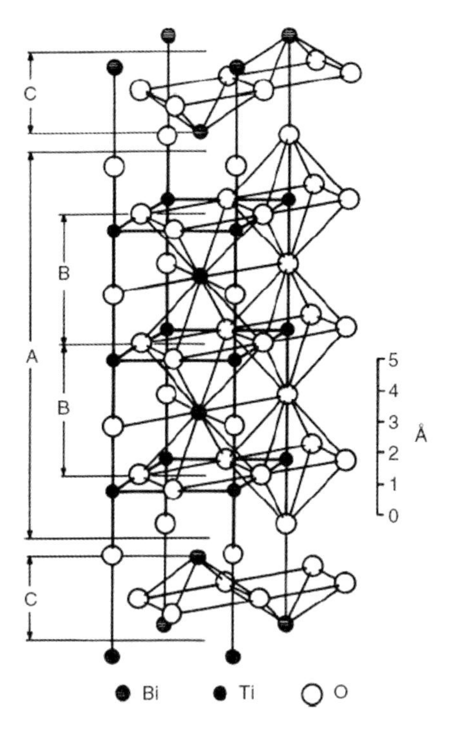

Figure 6.8 One half of the pseudo-tetragonal unit cell of $Bi_4Ti_3O_{12}$ [34].

B = Ti, V, Nb, Ta, Mo, and W, and the oxygen/fluorine ratio depends on the valencies of the B and A cations.

The bismuth–molybdenum/tungsten compounds $(Bi_2O_2)^{2+}(MO_4)^{2-}$ (M = Mo^{6+} or W^{6+}) is a representative of Aurivillius phase with $n = 1$ (Figure 6.9) [58, 59].

Closely related oxygen-deficient compounds can be generated by replacing the M^{6+} with lower-valence cations, such as V^{5+}, to form $Bi_4V_2O_{11}$ [60]. In the case of a high-temperature (>570 °C) γ-tetragonal phase of $Bi_4V_2O_{11}$, the perovskite-like layers of composition $VO_{3.5}$ contain a random distribution of oxygen vacancies. Numerous attempts to stabilize this high-temperature phase at ambient temperature have been made by replacing some of the V^{5+} with other metal cations. This generates the so-called BIMEVOX family of oxide-ion conductors, where ME represents the dopant metal (e.g., BICUVOX) for Cu substitution. Over 35 cation species have been incorporated into $Bi_4V_2O_{11}$ [61]. Many of the BIMEVOX compounds form extremely complex structures, with ordering of the anion vacancies leading to incommensurate superstructures.

6.2.4 Perovskite-Related Copper Oxide Structures

As the copper ions are stable in square-pyramidal as well as square-coplanar and octahedral oxygen coordination, this allows the layered copper oxides to form a large

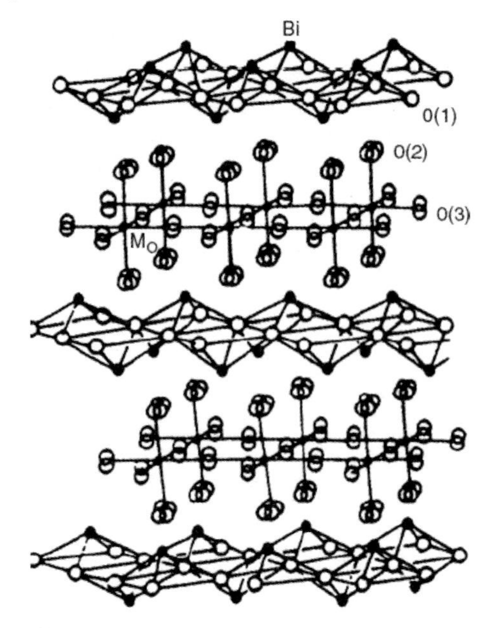

Figure 6.9 The ideal structure of Bi_2MoO_6 [58].

family of perovskite-related structures. The T′-tetragonal structure of Nd_2CuO_4 is shown in Figure 6.10a, where Cu^{2+} cations have a square-coplanar coordination [62]. If two A-site cations of quite different size are present, they may order in alternate intergrowths to give a T*-tetragonal phase with alternating rock-salt and fluorite layers, as illustrated for $Nd_{2-y-z}Ce_ySr_zCuO_4$ (Figure 6.10b) [63].

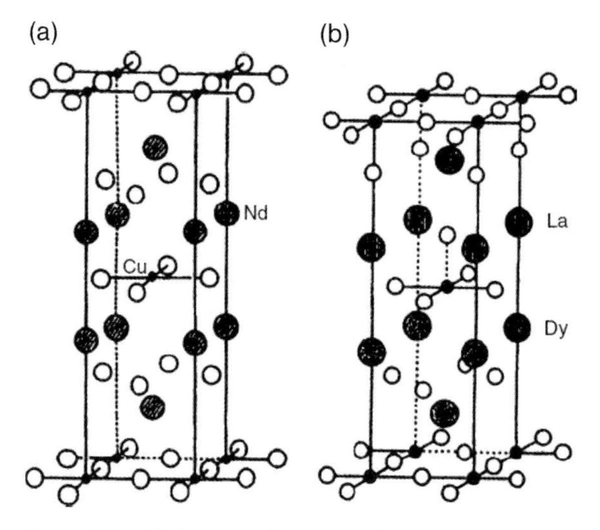

Figure 6.10 (a) T′-tetragonal structure of Nd_2CuO_4; (b) T*-tetragonal structure of $Nd_{2-y-z}Ce_ySr_zCuO_4$ [34].

In the structure $Nd_{2-y-z}Ce_ySr_zCuO_4$, the Cu^{2+} ions have a fivefold oxygen coordination. However, with a smaller A′ cation that is stable in eightfold oxygen coordination, and a larger A cation in an AO layer, the family of copper oxide intergrowth structures that are related to the perovskite structure becomes extensive. Similar compounds may be represented by their sequence of successive (0 0 1) planes as follows:

- $AO \cdot CuO_2 \cdot AO$
- $\phi \cdot AO \cdot CuO_2 \cdot AO \cdot \phi$
- $\phi \cdot AO \cdot CuO_2 \cdot A'CuO_2 \cdot AO \cdot \phi$
- $\phi \cdot AO \cdot CuO_2 \cdot A' \cdot CuO_2 \cdot A' \cdot CuO_2 \cdot AO \cdot \phi$
- $\phi \cdot AO \cdot CuO_2 \cdot A' \cdot CuO_2 \cdot A' \cdot CuO_2 \cdot A' \cdot CuO_2 \cdot AO \cdot \phi$

where ϕ represents a more complex layer inserted between two AO sheets. In Cu containing high-T_c superconductors, the ϕ layer may act as a "charge reservoir" relative to the superconductive CuO_2 sheets, which have Cu in all sixfold or all fivefold oxygen coordinations.

6.2.5
Cation Ordering

The perovskite structure is very flexible, and the substitution of cationic positions A and B occurs over a wide range of compositions $A_{1-x}A'_xB_{1-x}B'O_3$. These compounds have a special position because of their remarkable features, which include: diffuse phase transitions and relaxor properties; spontaneous phase transitions; and ferroelectric-to-antiferroelectric transformations. Often, the complex compounds remain disordered solid solutions having a regular $Pm\bar{3}m$ structure. However, under prolonged high-temperature heating they can be transformed into other phases with partial (local) or full ordering. Hence, cation ordering in $A_{1-x}A'_xB_{1-x}B'O_3$ can take place on both the A sites and the B sites; moreover, the x-value can vary widely and also take on a value of rational numbers such as 1/2, 1/3, and 1/4.

Today, there are many examples of double perovskites $A_2BB'O_6$, in which the B and B′ atoms are ordered because of a charge difference and/or a large size and electronegativity difference. For some compositions that have simple B: B′ ratios, such as 1 : 1 and 1 : 2, the cation ordering of B and B′ ions will occur (Figure 6.11).

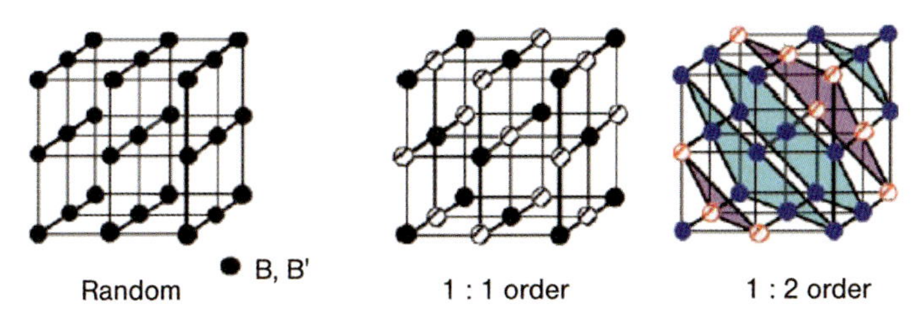

Figure 6.11 Cation arrangements on octahedral site of perovskite-type structures of types $A_2BB'O_6$ (1 : 1 ordering) and $A_3BB'_2O_9$ (1 : 2 ordering) [36].

Cation ordering may allow a stabilization in an octahedral site of a cation that is too large to be normally found in such a site. Some examples of 1 : 1 ordered perovskites are Ba_2LnNbO_6 (Ln = rare earth), La_2MgRuO_6, $LaSrCoTaO_6$, Sr_2MMoO_6, (M = Fe, Co, Mn), Sr_2FeWO_6, Sr_2FeReO_6, and many others. The 1 : 2 ordered $A_3BB'_2O_9$ perovskites have been found as in $Ba_3MgTa_2O_9$, $Ba_3ZnTa_2O_9$, $Sr_3CoTa_2O_9$, and others. The wide series of ordered $Pb_2BB'O_6$ perovskites has been discussed in detail [5, 18, 30].

Similarly, the ordering of A-site cations in $AA'B_2O_6$ or $AA'_3B_4O_{12}$ perovskites has been demonstrated. Of particular interest is the situation where smaller ions such as Cu^{2+} or Mn^{3+} that are stable in a square-coplanar coordination, induce a cooperative rotation of the $BO_{6/2}$ octahedra so as to provide four coplanar nearest neighbors from the 12-fold oxygen coordination of an A site. Examples of this include $(CaCu_3)B_4O_{12}$ (B = Ge, Mn, Ti, Ru) and $(NaMn_3)Mn_4O_{12}$.

6.2.6 Nonstoichiometry

The perovskites can tolerate vacancies on any of the atomic sites. In addition, ordering of the atomic vacancies occurs in perovskite-related structures.

6.2.6.1 A-Site Vacancies

Since the BO_3 array in the perovskite structure forms a stable network, the large A cations can be missing either partly or wholly. Cation vacancies on the cubo-octahedral A sites can be generated by increasing the valence of the octahedral B cation species; that is, $A^{3+}B^{3+}O_3 \rightarrow A^{3+}{}_{2/3}B^{4+}O_3 \rightarrow A^{3+}{}_{1/3}B^{5+}O_3 \rightarrow B^{6+}O_3$. This sequence is illustrated by, for example, the compounds $LaGaO_3 \rightarrow La_{2/3}TiO_3 \rightarrow La_{1/3}NbO_3 \rightarrow ReO_3$ [64–66].

$\square ReO_3$ is striking a representative of A cation vacancy perovskite. Strong Re–O–Re interactions stabilize BO_3 array sufficiently to allow all the A-site cations to be removed without its collapse whilst, at the same time, the structure of the ReO_3 remains cubic. Nonetheless, a partial or complete collapse of the array is found in many compounds. For example, WO_3 occurs in a wider variety of modifications, all of which are distorted forms of the ReO_3 type (with W atoms shifted from the octahedron centers and with varying the W–O bond lengths). The channels of the structure can be occupied by alkali metal ions in varying amounts, and this results in compositions A_xWO_3 that are termed *tungsten bronzes*. Cubic tungsten bronzes have the ReO_3 structure with partial occupation of the voids by Li^+ or Na^+; they are intermediate between the ReO_3 type and perovskite structure. It is interesting to note that an ordering occurs at a larger x, and this is accompanied by a collapse of the W–O–W bond angle in WO_2 planes perpendicular to a unique axis to give the tetragonal and hexagonal structures illustrated in Figure 6.12. Tetragonal tungsten bronzes are similar to the hexagonal bronzes, but contain three distinguishable tunnels; the larger pentagonal tunnels may be occupied by K^+ ions, while Na^+ ions may occupy both pentagonal and square tunnels, and the Li^+ ions can occupy the smaller, triangular tunnels.

Figure 6.12 Structures of A_xWO_3 bronzes. (a) Tetragonal bronze for $x \geq 0.4$; (b) Hexagonal bronze for $x \geq 0.7$.

Tungsten bronzes are metallic conductors, and have a metallic luster and colors that go from gold to black depending on composition. They are very resistant chemically, and serve as industrial catalysts and as pigments in "bronze colors."

Besides bronzes, A-site-defective perovskite oxides are known to be formed when B = Ti, Nb, Ta, and so on. One interesting example of a similar phase is $Cu_{0.5}TaO_3$, where copper atoms are ordered at the A sites; these compounds will exhibit metallic properties if the B atom occurs in a low oxidation state.

6.2.6.2 **B-Site Vacancies**

B-site vacancies in perovskite oxides are energetically not favored because of the large formal charge and the small size of the B-site cations. If vacancies are to occur at the B-sites, there must be other compensating factors such as B–O covalency and B–B interaction. Whilst the covalency of the B—O bond would increase with increasing charge and decreasing size of the B cation, the B–B interaction would be favored by a hexagonal rather than a cubic stacking of the AO_3 layers. Thus, B-site vacancies would be expected to occur more frequently in the perovskites of highly charged B cations that possess hexagonal polytypic structures. Many well-characterized B-site vacancy hexagonal perovskites are indeed known, and the B-site vacancies in these compounds are ordered between *h–h* layers, where the BO_6 octahedra share faces. For example, $Ba_5Ta_4O_{15}$ adopts a five-layer (*ccchh*) sequence where the octahedral site between the *h–h* layers is vacant [67] (Figure 6.13).

The structure of $Ba_3Re_2O_9$ is similar, consisting of a $(hhc)_3$ layer sequence where the ReO_6 octahedra share corners, with the vacancy occurring at octahedral sites between the *h–h* layers [68]. Another two compounds, $Ba_3W_2O_9$ and $Ba_3Te_2O_9$, each adopt an all-hexagonal BaO_3 layer sequence; however, in $Ba_3W_2O_9$ two-thirds of the octahedral sites in each layer are filled by tungsten, whereas in $Ba_3Te_2O_9$ the octahedral sites in two adjacent layers are fully occupied, with every third layer being completely vacant [69].

Oxide perovskites consisting of ordered B-site vacancies exhibit interesting luminescent properties. For example, $Ba_3W_2O_9$ shows an efficient blue (460 nm) photoluminescence below 150 K [6], while other compounds show different color emissions when they are doped with various rare-earth activators.

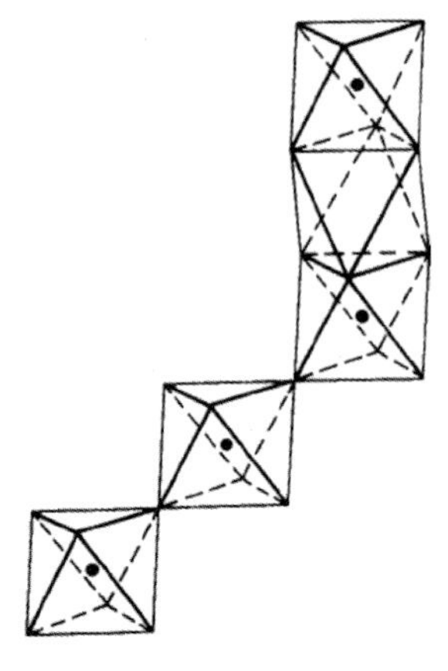

Figure 6.13 Schematic (1 1 0) projection of the $Ba_5Ta_4O_{15}$ structure. The horizontal lines refer to BaO_3 close-packed planes.

6.2.6.3 Anion-Deficient Perovskites and Vacancy-Ordered Structures

Anion vacancy nonstoichiometry in perovskite oxides is more common than that involving vacancies in the cation sublattice. Anion-deficient perovskites ABO_{3-x} are of technical interest as oxide-ion conductors, which is achieved by the hopping of vacancies. In many of the ABO_{3-x} perovskites, nonstoichiometry spans the composition range $0 \leq x \leq 0.5$. For example, the systems $SrTiO_{3-x}$ and $SrVO_{3-x}$ each have a complete solid solution of oxygen vacancies over the range $0 \leq x \leq 0.5$, in which the oxygen vacancies remain random at room temperature [71]. It is noteworthy that both these materials exhibit metallic properties, thus confirming the decisive role of d-electrons when associated with B cations. However, strong electrostatic forces between the vacancies usually create ordered phases in which the vacancies are not mobile. Even where the vacancies remain random, the short-range order may trap out the mobile vacancies. An oxygen vacancy reduces the anion coordination at its two neighboring B cations; B cations stable in fivefold oxygen coordination will have a vacancy ordering that leaves the cations with no more than one nearest-neighbor oxygen vacancy. However, if the B cation is more stable in fourfold oxygen coordination, then the oxygen-vacancy ordering will be organized so as to leave the B cations in sixfold and fourfold oxygen coordination.

Anion-deficient $ABO_{3-\delta}$-type perovskites, with $\delta = 0.5$, correspond to the stoichiometry $A_2B_2O_5$. There are many possible models of vacancy ordering when some of the BO_6 octahedra are replaced by BO_5 square pyramids ($Ca_2Mn_2O_5$), BO_4 square planar units ($La_2Ni_2O_5$), BO_4 tetrahedra ($Ca_2Fe_2O_5$), or other combinations [72]. The formation of BO_4 tetrahedra leads to the so-called *brownmillerite* structure type exhibited by $Ca_2Fe_2O_5$ and Ca_2FeAlO_5 [71, 73]. The compositions could be considered as anion-deficient perovskites with one-sixth of the anion sites being vacant. In the [0 1 0] direction of brownmillerite structure there is an alternating sequence of corner-sharing octahedra and tetrahedra, with alternate rows of the latter having different orientations. Thus, oxygen vacancies are ordered in alternate (0 0 1) BO_2 planes of the cubic perovskite structure (Figure 6.14).

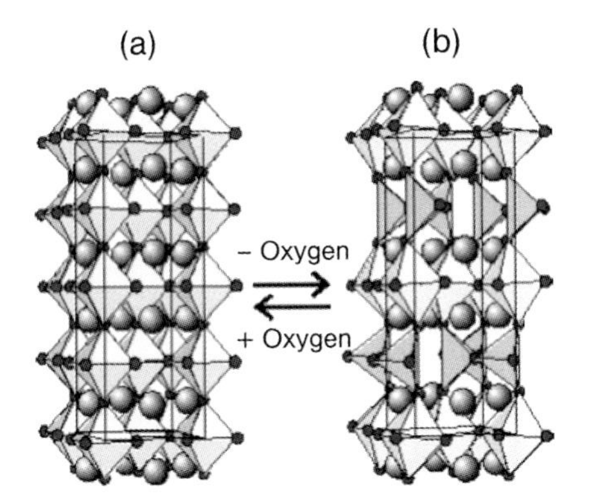

Figure 6.14 Cubic perovskite ABO_3 (a) and brownmillerite structure $ABO_{2.5}$ with vacancy ordering in the *ab* plane (b).

The structures of $Ca_2Mn_2O_5$ and $Ca_2Co_2O_5$, which have compositions similar to brownmillerite, involve different modes of vacancy ordering [72, 74, 75] (Figure 6.15). Oxygen vacancies in $Ca_2Mn_2O_5$ are ordered in every (0 0 1) BO_2 plane of the cubic perovskite, such that one-half of the oxygen atoms in alternate [1 1 0] rows are removed. The vacancy ordering in $Ca_2Co_2O_5$ is similar to that in $Ca_2Mn_2O_5$, but the doubling of the *c*-axis is likely to be due to the ordering scheme shown in Figure 6.15. Under ambient conditions the brownmillerite-structured compound $Ba_2In_2O_5$ possesses *Icmm* symmetry and contains disorder of the InO_4 tetrahedra over their two different orientations. This leads to a rather complex microstructure [76].

In complex systems with a 50 : 50 mixture of a larger A cation and a smaller A′ cation, the two cations may order into alternate planes so as to retain the larger A cations in the AO (0 0 1) planes, where they retain 12-fold oxygen coordination; in this case, oxygen is only removed from the $A'O_{1-x}$ planes, as is shown for the ordered $BaTbCo_2O_{5.5}$ structure [77]. The removal of all oxygen from the $A'O_{1-x}$ planes would leave the A′ cations in eightfold oxygen coordination, as occurs in $RBaFe_2O_5$ [78]. Many other relevant examples of anion-deficient perovskites have been discussed elsewhere, by Rao, Smyth, Poeppelmeier, Grenier, Reller, and others [20, 21, 65–82].

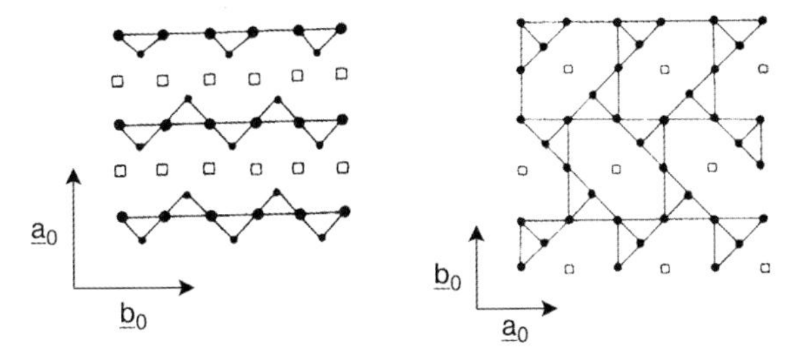

Figure 6.15 Vacancy ordering in the *ab* plane of $Ca_2Mn_2O_5$ (a) and $Ca_2Mn_2O_5$ (b).

Finally, it should be noted that there is a series of copper-containing complex compounds with superstructures based on anion-deficient perovskite type. These have been the subject of intense research in recent years, because of their remarkable high-temperature superconducting properties, and many appropriate reviews have been produced.

6.2.6.4 Anion-Excess Nonstoichiometry

Anion-excess nonstoichiometry in perovskite oxides is less common than anion-deficient nonstoichiometry, most likely because the introduction of interstitial oxygen into a perovskite structure is thermodynamically unfavorable. For example, careful studies of system $LaMnO_{3+x}$ (including neutron diffraction) have revealed that an oxygen excess is accommodated by vacancies at the A- and B-sites, and in reality the compound with the proposed composition of $LaMnO_{3.12}$ should be described as $La_{0.94}\,\square_{0.06}Mn_{0.98}\,\square_{0.02}O_3$ [83].

6.3 Physical Properties

The physical properties of perovskites are manifold, interesting, and useful. Perovskite oxides can demonstrate, among others, high dielectric constants, ferroelectricity, piezoelectricity, strong relaxor ferroelectricity, ferromagnetism, large thermal conductivity, oxygen ion conductivity, high proton conductivity, mixed ionic-electronic conductivity, insulator-to-metallic transitions, colossal magnetoresistance, fluorescence, nonlinear optical properties, and superconductivity. All of these various properties and possible applications have attracted – and continue to attract – considerable attention to the perovskites and stimulate further widespread interest in the study of these materials [84–89].

6.3.1 Electronic Properties

In the perovskite structures, the direct cation–cation interaction is remote because of the large intercation distance along the cube–face diagonal. Therefore, the electronic properties of perovskites are determined by the cation–anion B–O, and the cation–anion–cation interactions of octahedral site cations along the cube edge, B–O–B ($180°-\phi$). The character of these interactions depends on the electronic energy of the cations.

The perovskite oxides are ionic compounds with an electrostatic Madelung energy that is large enough to raise most cation outer s and p electronic energies well above the top of the O^{2-} anion p^6 bands, where they remain essentially unoccupied at ordinary temperatures. The exceptions are some A-site cations with $5s^2$ or $6s^2$ cores, such as Sn^{2+}, Tl^+, Pb^{2+}, or Bi^{3+}. However, the empty 5s or 6s bands of cations Sn^{4+}, Tl^{3+}, Pb^{4+} or Bi^{5+}, located on the B-sites, are at low enough energies to accept electrons.

Although lanthanide cations normally occupy A sites, they can occupy B sites in ordered double A_2BLnO_6 (Ln = rare earth) perovskites having a large A-site cation such as Ba^{2+}. The lanthanides introduce empty 5d bands that overlap their s and p bands; the bottom of these conduction bands have primarily a 5d character. The $4f^n$ configurations at the lanthanide cations are localized and the intra-atomic electron–electron coulomb energies separate successive $4f^n$ and $4f^{n+1}$ configurations by an energy that is larger than the energy gap between the bottom of the 5d band and the top of the anion p^6 bands. Therefore, the lanthanide cations usually have only a single valence state unless a $4f^n$ level falls in a gap between the filled and empty states, in which case they may have two valence states.

The five d orbitals of a transition metal atom B are degenerate; however, with more than one electron in the d manifold, the spin degeneracy is removed by the ferromagnetic direct-exchange interaction between electron spins in atomic orthogonal orbitals. Transition metal B cations usually introduce filled and/or empty d states within the gap between the anion p^6 bands and lanthanide 5d bands, which lowers the probability that the lanthanide ion can have two valence states in a transition metal perovskite.

Since in the perovskite structure the interactions between the neighboring A-site cations and between A-site and B-site cations are weak, the particular interest is the interatomic interactions between *d* electrons on neighboring transition metal B-ions. The dominant interactions between B-cation 3*d* electrons are normally the (180°–ϕ) B–O–B interactions. These spin–spin interactions are between nonorthogonal orbitals, and therefore involve different types of virtual charge transfer (superexchange or antiferromagnetic interactions between half-filled orbitals, and semicovalent exchange or ferromagnetic interactions between half-filled and empty or full orbitals) and real electron transfers, namely, double exchanges (the so-called *Zener* double exchange and *de Gennes* double exchange), indirect exchange and others [34, 90–93]. A transition from localized to itinerant behavior of the B-*d* electrons occurs where the interatomic interactions become greater than the intra-atomic interactions. All of these interactions determine the many physical properties of the perovskites. In the approach to itinerant-electron behavior from the localized-electron side, itinerant electron bags may be described by molecular orbitals in a localized-electron matrix. This situation is found in the mixed-valent colossal magnetoresistive (CMR) manganites and the underdoped high-T_c copper oxide superconductors, where the itinerant-electron bags are hole-rich.

The electrical conductivity of perovskites shows wide variations. Some compounds exhibit high dielectric properties, most are semiconductors, while others show metallic conductivity. The group of perovskites containing cations with empty s- and d-states ($CaTiO_3$, $SrTiO_3$, $BaTiO_3$, $CaZrO_3$, $SrZrO_3$, $BaZrO_3$, $CaSnO_3$, $SrSnO_3$, $BaSnO_3$, $LiNbO_3$, $NaNbO_3$, $KNbO_3$, $NaTaO_3$, $KTaO_3$, etc.) is dielectric, and they are characterized by a high resistivity ($\sim 10^{14}$ Ω·m), a high relative dielectric constant (~20–10 000), and they also exhibit ferroelectric behavior [11, 17, 21, 89].

Several perovskites exhibit metallic conductivity, typical examples being $LaTiO_3$, $LaNiO_3$, $SrVO_3$, $AMoO_3$ (A = Ca, Sr, Ba), ReO_3, and A_xWO_3. Metallic conductivity in similar perovskite oxides is due to a strong cation–anion–cation interaction [21, 24, 34].

The perovskites containing rare earth and transition metal ions often show widely differing electrical properties. For example, $LaTiO_3$ and $LaNiO_3$, both of which contain Ni^{3+} in the low-spin state, show a metallic conductivity and Pauli paramagnetism. $LaCrO_3$ is a semiconductor, while $LaMnO_3$ exhibits an abrupt change in both conductivity and magnetic susceptibility as a function of temperature close to 720 K. $LaCoO_3$ displays more complex behavior; it is a semiconductor up to about 400 K, after which its conductivity increases much more rapidly up to 823 K. Between 823 to 1200 K the conductivity passes through a broad, flat maximum, whilst finally, above 1200 K, it shows a metal-like behavior. The conductivity can be enhanced considerably by a partial substitution of the lanthanide by a divalent ion. Thus, for $La_{1-x}Sr_xMnO_3$ the manganese ions are Mn^{3+} for $x = 0$. However, an increase in x results in the creation of Mn^{4+} holes, thereby increasing conductivity. For substitutions $0.2 < x < 0.4$, the system becomes ferromagnetic and shows a metal-semiconductor transformation. A similar type of behavior was observed in Sr-substituted cobaltites.

There is a strong correlation between the electrical and magnetic properties of perovskites, because they both depend on the outermost electrons localized at specific atomic sites. In the ideal cubic perovskite structure, each oxygen is shared by two B ions, forming a B–O–B angle of 180°. Such a configuration is favorable for superexchange interactions between magnetic B^{3+} cations. The interactions between 5s or 6s orbitals on neighboring B sites give rise to a narrow conduction band that is accessible to reduction; however, the electrons in these bands usually segregate into localized s^2 cores, since the on-site electron–electron coulomb energies between spin-paired s^2 electrons is relatively small. The exchange interaction usually results in an antiparallel coupling of nearest-neighbor magnetic moments. When the B^{3+} ions are in two sublattices ($A_2BB'O_6$), other spin arrangements are possible. If B′ is a diamagnetic ion, the B^{3+} ions are aligned antiferromagnetically, and the most important exchange mechanism is believed to be a longer range superexchange interaction through two oxygens of the type B–O–B′–O–B. The B–B separation is now considerably longer than the 4 Å separation found in the ideal perovskite.

Although most perovskites with main-group B cations are insulators, electrons can be introduced into empty 5s or 6s bands, as in $Sn(IV)O_3$, $Pb(IV)O_3$, or $Bi(V)O_3$ arrays. Special attention has been paid to the superconductivity found in the systems $BaPb_{1-x}Bi_xO_3$ and $Ba_{1-x}K_xBiO_3$, and this in turn has stimulated considerable fundamental interest. In the orthorhombic $BaPbO_3$, the Pb-6s and O-2p bands overlap so as to make $BaPbO_3$ a semi-metal. $BaBiO_3$ is a semiconductor because the half-filled 6s band is unstable relative to a disproportionation reaction:

$$2Bi^{4+} \rightarrow Bi_{I}^{(4-d)} + Bi_{II}^{(4+d)}$$

The reason for these semiconductive properties of $BaBiO_3$ is the creation of two distinguishable Bi sites that change the translational symmetry of the crystal and split the 6s band into two. The substitution of Pb for Bi in $BaBi_{1-x}Pb_xO_3$ suppresses the disproportionation reaction, and introduces a metallic behavior for $x > 0.65$; the system also becomes superconductive in the range $0.65 < x < 0.95$ [69], as illustrated

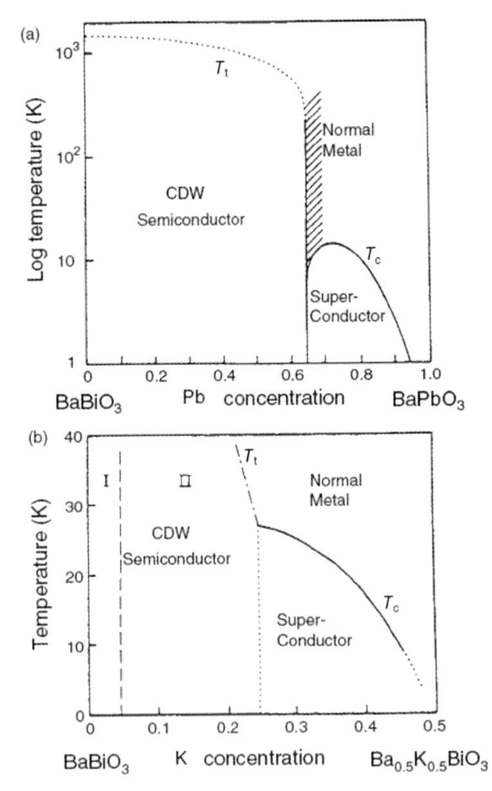

Figure 6.16 Phase diagrams for the systems. (a) $BaBi_{1-x}Pb_xO_3$; (b) $Ba_{1-x}K_xBiO_3$ [34].

in the phase diagram of Figure 6.16a. The related effect is also observed by a substitution of K^+ for Ba^{2+} in $Ba_{1-x}K_xBiO_3$ (Figure 6.16b). Then, at $x > 0.25$, $Ba_{1-x}K_xBiO_3$ becomes superconductive.

This is only a fleeting glance at a variety of electronic behavior of perovskites, and a comprehensive discussion of the electrical and magnetic properties of many perovskite compounds may be found in a series of reviews [34, 90–95].

6.3.1.1 The Colossal Magnetoresistance (CMR) Phenomenon

The CMR effect was discovered in distorted perovskite manganites [or manganates (III, IV) according to chemical nomenclature rules] of the type $Ln_{1-x}A'_xMnO_3$ [96]. The electric and magnetic properties of the manganites $(Ln_{1-x}A'_x)MnO_3$ exhibiting CMR properties depend heavily on the composition (Ln, A′, x), including the exact oxygen content, the structural and microstructural features, and the charge ordering [34, 35, 97, 98]. A comprehensive discussion of this phenomenon and proper CMR materials has been made by van Tendeloo and colleagues [35].

The composition $(La_{1-y}Pr_y)_{0.7}Ca_{0.3}MnO_3$ represents a model system for understanding the CMR phenomenon and adjusting its parameters [99]. The parent compounds $LaMnO_3$ and $PrMnO_3$ are A-type antiferromagnets, due to the antiferromagnetic (AFM) superexchange interaction; the electrons of HS Mn^{3+} (t^3e^1 configuration) is localized. By introducing A^{2+} cations (Ca^{2+}, Sr^{2+}, Ba^{2+}, Pb^{2+}) – that is, by introducing Mn^{4+} ions – these materials become ferromagnetic and metallic below the Curie temperature (T_C). If the MnO_3 array is Mn(IV)/Mn(III) mixed-valent, then the crossover from localized to itinerant electronic behavior in the orbitally disordered state becomes possible, and at higher temperatures these two-manganese polarons $Mn^{3+}-O-Mn^{4+} = Mn^{4+}-O-Mn^{3+}$ ("Zener polarons") can be formed according to the Zener double-exchange mechanisms involving fast electron transfer within two-M-atoms system M (d^{n+1})-O-M (d^n) [100].

In the mixed-valent systems $(La_{1-y}Ln_y)_{0.7}Ca_{0.3}MnO_3$ [99], the ratio Mn(IV)/Mn 0.3 is held constant and the average Mn–O–Mn bond angle θ_{av} is reduced by substituting a smaller lanthanide ion Ln^{3+} for La^{3+}. In these systems, a crossover of the Mn electrons from localized to itinerant electronic behavior is displaced to larger Mn(IV)/Mn ratios as θ_{av} decreases with the Ln^{3+}-ion concentration, y. At crossover, two-manganese Zener polarons are formed at higher temperatures, but at a lower temperature there is a dynamic spinodal phase segregation into a Mn(IV)-rich ferromagnetic phase and a polaronic, paramagnetic matrix having the cooperative JT distortion of the parent $LnMnO_3$ phase. On lowering the temperature, the Zener polarons are progressively trapped out into the conductive, ferromagnetic phase. Where the T_C-value of the ferromagnetic phase is greater than the magnetic ordering temperature (T_N) of the type-A AFM matrix, a CMR is found in an interval $T_m < T < T_C$; a first-order transition on cooling through $T_m \geq T_N$ signals a discontinuous growth of the orbitally disordered ferromagnetic phase to beyond percolation to give the "bad-metal" conduction. The application of a magnetic field in the interval $T_m < T < T_C$ stabilizes the conductive ferromagnetic phase relative to the paramagnetic, polaronic matrix, and the CMR phenomenon results where the ferromagnetic, conductive phase grows to beyond percolation in the paramagnetic matrix.

The CMR phenomenon can be illustrated with Figure 6.17, which shows the change in resistivity $(\varrho_{0T} - \varrho_{5T})/\varrho_{5T}$, a logarithmic scale) as a function of temperature in the system $(La_{1-x}Pr_x)_{0.7}Ca_{0.3}MnO_3$. Although, the transition from polaronic to "bad-metal" (vibronic) conductivity is first-order, ϱ_{0T} changes smoothly as the more conductive phase grows to beyond percolation below T_m. There are two important features here: first, the extraordinary decrease in T_m with the isovalent substitution of Pr^{3+} for La^{3+}, falling from 250 K at $x = 0$ to about 80 K at $x = 0.6$; and second, the dramatic increase in the negative magnetoresistance just above T_m as T_m decreases. In these perovskites, the regions of short-range ferromagnetic order above T_m are not conventional magnetic polarons; rather, they are the result of a spinodal phase segregation. The hole-rich, more conductive minority phase is orbitally disordered, while the orbitally ordered polaronic matrix is hole-poor. As θ_{av} increases, the relative stability of the orbital ordering decreases, and ferromagnetic vibronic spin–spin interactions can stabilize an orbitally disordered ferromagnetic phase at a higher T_m. Moreover, as T_m decreases, more Zener polarons are trapped out in the hole-rich minority phase to increase the resistivity of the matrix and the volume fraction of the minority phase that grows to the percolation threshold in a magnetic field of 5 T.

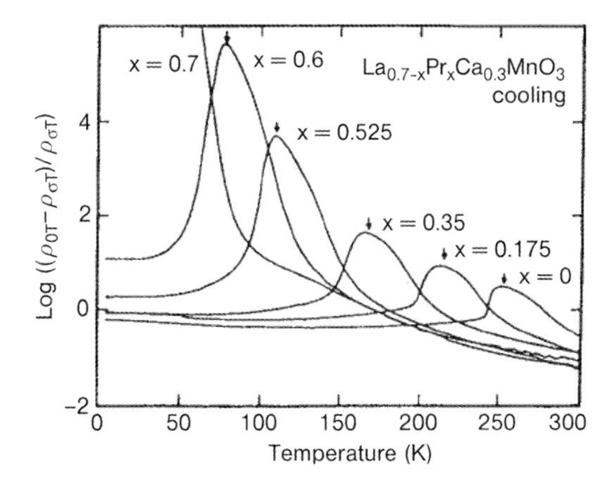

Figure 6.17 The change in normalized resistivity $(\varrho_{0T} - \varrho_{5T})/\varrho_{5T}$ as a function of temperature for different values of x in $(La_{1-x}Pr_x)_{0.7}Ca_{0.3}MnO_3$. ϱ_{0T} and ϱ_{5T} represent the resistivity in a zero magnetic field and in a magnetic field of 5 T, respectively. The arrows indicate the Curie temperature, T_c [99].

Thus, not only T_C but also the magnitude of the magnetoresistance effect are strongly affected by the average size of the $(Ln_{1-x}A'_x)$ cations.

In the $Ln_{1-x}A'_xMnO_{3-y}$ systems, the manganese valence can be modified not only by varying x, but also by replacing A′ by a monovalent cation or vacancies, by decreasing the oxygen content, or by doping the Mn site. All of these modifications will induce a mixed valency of the manganese (i.e., the hole carrier density). Several related compounds which exhibit a CMR effect, including $(La,Sr)CoO_3$, $LaNiO_3$, and Sr_2FeMoO_6, have been studied.

In some manganites, "charge ordering" is observed below a temperature T_{CO} [35, 97]. This rather highly complex phenomenon cannot at present be explained by double exchange alone, because charge ordering and double exchange are competing interactions, responsible for the fascinating properties of these materials.

6.3.2
Ferroelectricity and Related Phenomena

Some crystal modifications of $BaTiO_3$ and many other perovskites possess a high dielectric susceptibility. The origin of a high dielectric constant in $BaTiO_3$ is a cooperative displacement of the Ti^{4+} ions from the center of symmetry of their B sites below a critical temperature, T_c. These spontaneous displacements help to create local dipoles that are aligned parallel to one another, giving a permanent dipole moment. The Ti^{4+}-ion displacements may equally well be in the opposite direction, which allows switching of the polarization $\boldsymbol{P}$ of the crystal in an applied electric field $\boldsymbol{E}$, with a P–E hysteresis loop analogous to the M–H hysteresis loop of a ferromagnet. Similar materials are referred to as *ferroelectrics* (above T_c, ferroelectric materials transform to the *paraelectric* state, where dipoles are randomly oriented). The crystals with an antiparallel arrangement of the permanent dipoles are called *antiferroelectrics*;

crystals exhibiting ferroic displacements are also termed *ferroelastic*. A ferroelastic crystal contains two or several stable orientations of the ferroic axis when there is no mechanical stress. It is possible to change from one orientation to another by applying a stress in a given direction. Reversing the stress from tensile to compressive gives rise to a stress–strain elastic hysteresis. The coupling between the ferroelectric and ferroelastic properties allows $\boldsymbol{P}$ to be modified by an applied stress (*piezoelectric effect*) or by modifying the strain by an applied electric field (*converse piezoelectric and electrostrictive effects*). In nonconducting *pyroelectric* crystals, a change in polarization can be observed by a change in temperature.

Primary ferroics are characterized by large values of the Curie constant C ($\sim 10^5$ K), and contain either small, highly charged B-ions (Ti^{4+}, Nb^{5+}, W^{6+}) or lone-pair ions Pb^{2+}, Bi^{3+}, and Sb^{3+}. Owing to their electronic structure, these ions promote distortions that give rise to spontaneously polarized structures. *Secondary ferroics*, such as an incipient ferroelectric $SrTiO_3$ and antiferroelectric $NaNbO_3$, may be considered to be ferrobielectric: the dielectric anisotropy in antiferroelectric domains can give rise to high values of induced polarization that vary orientationally in different domains. Thus, they give rise to domain rearrangement under applied fields.

Besides ferroelectricity, a large number of perovskites exhibit paired properties which have applied values; examples include $KNbO_3$ (ferroelectric–ferroelastic), $YMnO_3$, $HoMnO_3$ (ferroelectric–antiferromagnetic), $BiFeO_3$ (antiferroelectric–antiferromagnetic), $SrTiO_3$, $YMnO_3$ (ferroelectric–semiconducting), $SrTiO_3$ (ferroelectric–superconducting), $LiNbO_3$, $LiTaO_3$, $Ba_2NaNb_5O_{15}$, and $Pb(Zr_{1-x}Ti_x)O_3$ (ferroelectric–electro-optic).

6.3.3 Relaxor Ferroelectrics (Relaxors)

Many perovskites demonstrate relaxational properties. For example, the simple doped ferroelectrics $KTaO_3$ and $SrTiO_3$ exhibit relaxational effects, but these are relatively weak and occur only at low temperatures. Much stronger relaxational effects that occur at much higher temperatures are prominent in disordered perovskites with mixed compositions, among which there is a special family of lead-based complex compounds with the formula $PbB'B''O_3$, $PbA'B'B''O_3$ where $A' = La^{3+}$, etc., $B' = Fe^{2+}$, Mg^{2+}, Zn^{2+}, In^{3+}, Sc^{3+}, and $B'' = Nb^{5+}$, Ta^{5+}, W^{6+}. Similar compounds have been called *relaxor ferroelectrics* or *relaxors* [31, 32, 101, 102]. Structurally, most perovskite- based relaxors, due to the presence of a slight lattice distortion, are in the rhombohedral symmetry. They are characterized by extraordinarily high dielectric constants, and the existence of nanoscale/microscale ordered polar domains in a disordered matrix. When these dipolar entities possess more than one equivalent orientation, they may undergo dielectric relaxation in an applied alternating current field. Cooperative ferroelectric distortions of varying orientation exist within a paraelectric matrix above T_{on}. In the very dilute limit (<0.1 at%), each polar domain behaves as a noninteracting dipolar entity with a single relaxation time. Yet, at higher concentrations of disorder the polar domains can interact, leading to a more complex relaxational behavior with distributions of the relaxation times. A statistical fluctuation of the size and composition of these microregions produces large differences in their ferroelectric T_C-values. The disordering leads to the transformation of a sharp ferroelectric phase transition to a diffuse ferroelectric phase transition. The crystals in the highly disordered state are characterized by diffuse ferroelectric phase transition and so acquire relaxor properties. Thus, relaxor behavior in perovskites requires some essential component parts, namely, the lattice disorder and the existence of polar nanodomains at temperatures much higher than T_m (the temperature at maximum ε) that must exist as islands in a highly polarizable host lattice. Relaxor materials have some distinct characteristics which normal perovskite ferroelectrics do not exhibit (Figure 6.18), and have shown great promise in most ferroelectric-related application arenas, including pyroelectrics ($PbSc_{1/2}Ta_{1/2}O_3$), capacitors/dielectrics ($PbMg_{1/3}Nb_{2/3}O_3$), electrostriction/actuators ($PbMg_{1/3}Nb_{2/3}O_3$, $PbZn_{1/2}Nb_{1/2}O_3$), medical ultrasound/high-efficiency transducers ($Pb[Zn_{1/3}Nb_{2/3}]_{1-x}Ti_x]O_3$, $Pb[Sc_{1/2}Nb_{1/2}]_{1-x}Ti_x]O_3$), piezoelectrics

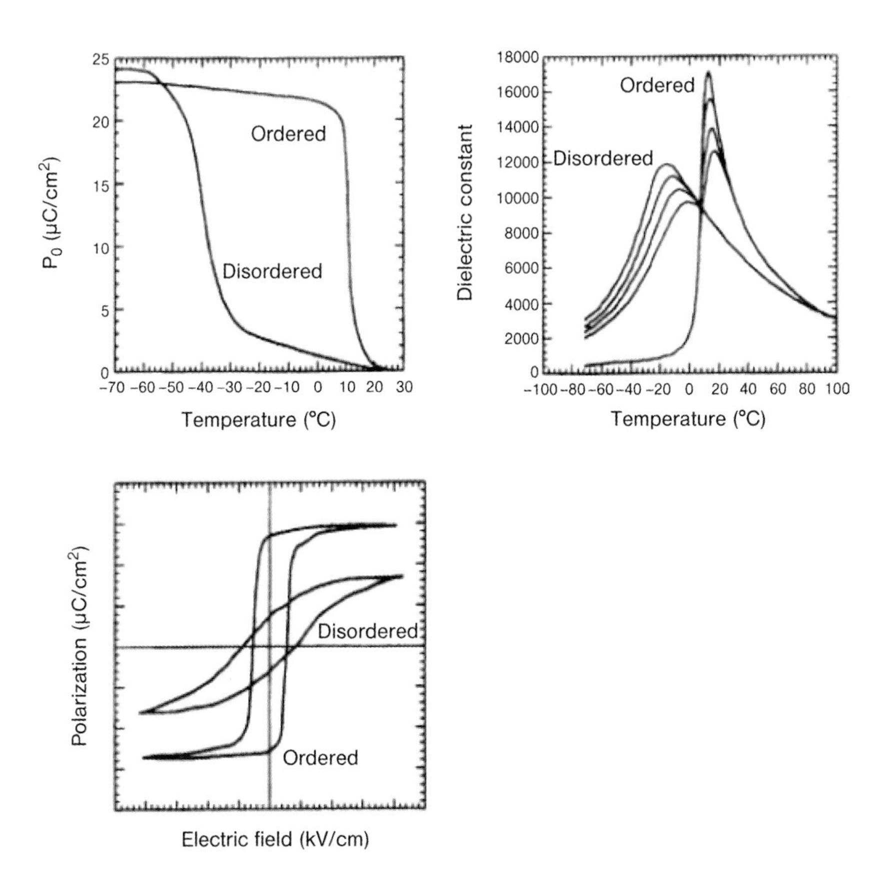

Figure 6.18 Some physical properties observed in the perovskites with ordered and disordered structures [5].

($PbZr_{1-x}Ti_xO_3$, $Pb[Zn_{1/3}Nb_{2/3}]_{1-x}Ti_x]O_3$, $Pb[Sc_{1/2}Nb_{1/2}]_{1-x}Ti_x]O_3$), and electro-optics ($Pb_{1-x}La_{2x/3}Zr_{1-y}Ti_yO_3$).

6.3.4
Morphotropic Phase Boundary (MPB) Compositions

In most cases of solid solutions between various classes of perovskite ferroelectrics and relaxors, a morphotropic phase boundary*(MPB)* occurs at which point several ferroelectric properties change drastically in favor of the device applications [5, 102–104]. The MPB represents an abrupt structural change within a solid solution, with variations in composition that are almost independent of temperature. Usually, this occurs because of the instability of one phase against the other at a critical composition, where two phases are energetically very similar but elastically quite different. At the same time, the dielectric constant, piezoelectric and electromechanical characteristics, spontaneous polarization and pyroelectric behavior each attain maxima, whereas the elastic constants tend to be softer in the vicinity of the MPB. These high property coefficients and unique structural characteristics of the MPB compositions give rise to an increased interest in these materials for use in various device applications, especially for electromechanical actuations. For example, $Pb(Zn_{1/3}Nb_{2/3})O_3$-$PbTiO_3$ and $Pb(Mg_{1/3}Nb_{2/3})O_3$-$PbTiO_3$ MPB materials have been found to exhibit remarkably large piezoelectric constants, while $Pb(Zr_{0.52}Ti_{0.48})O_3$ (PZT) shows a maximum piezoelectric response at compositions close to the temperature-independent compositional phase boundary. [105–107]. Bhalla *et al.* [5] have recently compiled an extensive amount of information on structure–property diagrams of the relaxor MPB systems.

6.3.5
Optical Properties

In the past, perovskites have been used as model systems for spectroscopic studies in the infrared (IR), visible, and ultraviolet (UV) ranges. For the $SrTiO_3$ perovskite containing an empty d-shell transition ion (Ti^{4+}), the emission spectrum shows luminescence with a band maximum at about 500 nm, which is independent of the sample's history and any contamination [108]. The optical properties of metal ions with a partially filled d-shell containing perovskites have also been studied. In this case, the Cr^{3+} (d^3) ion in $ATiO_3$ (A = Ca, Sr, Ba) and ABO_3 (A = La, Gd, Y; B = Al, Ga) may show two completely different types of emission, namely a narrow-line and spin-forbidden emission ($^2E \rightarrow {}^4A_2$) and a broad-band and spin-allowed emission ($^4T_2 \rightarrow {}^4A_2$), with the latter occurring for relatively weak fields. The transition metal-doped titanates became of interest soon after the discovery of photoelectrochemical water splitting by titanates [109], when it appeared possible to sensitize an $SrTiO_3$ electrode for visible light by doping with transition metal ions.

Several of the perovskite crystals, including $LiNbO_3$, $KNbO_3$, $KTaO_3$, $KNb_{1-x}Ta_xO_3$, and $Ba_2NaNb_5O_{15}$, possess nonlinear optical properties that can be used in electro-optical devices as modulators of laser radiation; for example, $KNbO_3$ is currently the best available second harmonica generator [110].

6.3.6 Ion Conductivity

6.3.6.1 Oxide-Ion Conductivity

The majority of the perovskite oxides are acceptor-doped with low-valent cations, giving rise to the formation of extrinsic oxygen vacancies (and/or electronic species) as charge-compensating defects. The series of compounds based on $LaBO_3$ ($LaCoO_3$, $LaMnO_3$, $LaGaO_3$) represent some of the most fascinating members of the perovskite family, as they exhibit high oxide ion conductivity when suitably doped with lower-valent metal ions. These materials have attracted considerable attention owing to both their range of applications, such as electrolytes and components in the composite electrodes in SOFCs, in solid oxide electrolyzer cells (SOECs), in oxygen sensors (e.g., lambda sensors in automobile engines), in heterogeneous catalysts, and in the fundamental fascination of ionic transport in crystalline solids [24, 25, 111]

Figure 6.19 shows the temperature dependence of the O^{2-}-ion conductivity in an oxygen-deficient brownmillerite-type perovskite $Ba_2In_2O_5$ that exhibits a first-order order–disorder transition at 930 °C. The Arrhenius plot of O^{2-}-ion conductivity shows that, above T_t, a considerable short-range order persists; nevertheless, the oxide-ion conduction in this field can be competitive with that in the commercial electrolyte yttria-stabilized zirconia (YSZ).

Some selection criteria have been proposed as a means of identifying promising candidates for similar materials [111, 112]. First, they should possess a high ionic conductivity, σ_i, coupled with a low electrical conductivity, σ_e. The other requirements based largely on steric considerations and "chemical sense," can be summarized as follows: (i) a high concentration of mobile charge carriers (i.e., oxide ion vacancies); (ii) the material should possess low metal–oxygen binding energies to allow the O^{2-} to "break free" and diffuse through the lattice; (iii) B-site cations with multiple valence states should be avoided, as they can give rise to an unwanted contribution from electronic conduction; and (iv) the Goldschmidt tolerance factor t should be close to unity, so that the material adopts the ideal cubic perovskite structure. This enhances σ_i by making all the oxygen sites crystallographically – and therefore also energetically – equivalent (though this is not the case for the distorted *Pnma* perovskite structure). There are some additional requirements connected with crystallochemical parameters of perovskite structure, namely that a cation and the lattice free volume V_f (defined as the unit cell volume minus the volumes of all constituent ions) should be as large as possible, and that open paths should exist between the oxide-ion sites.

Although it has been commonly assumed that the migrating O^{2-}-ion within the perovskite lattice takes a direct linear path along the $\langle 1\ 1\ 0 \rangle$ edge of the BO_6 octahedron into a neighboring vacancy, calculations of the potential energy surface [25] have shown that a small deviation from the direct path for migration channel is revealed (Figure 6.20). The curved route with the saddle-point away from the adjacent B-site cation is due to a significantly lower energy barrier.

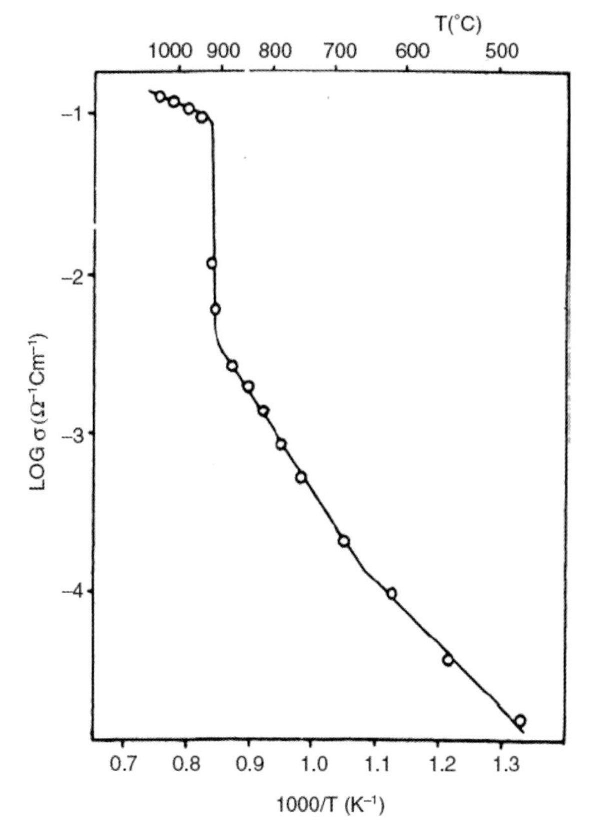

Figure 6.19 Arrhenius plot of O^{2-}-ion conductivity for $Ba_2In_2O_5$ under $P_{O2} = 10^{-6}$ atm [34].

In the case of $A^{3+}B^{3+}O_3$ perovskites, the largest suitable trivalent cation is La^{3+}, and the requirement for $t \approx 0.96$ makes Ga^{3+} the most appropriate B-site cation. $LaGaO_3$ and its derivatives are indeed excellent oxide-ion-conducting perovskites [113–115]. The highest oxide-ion conductivity in perovskites (without significant electronic conductivity) was found in Sr- and Mg-doped $LaGaO_3$: the optimized

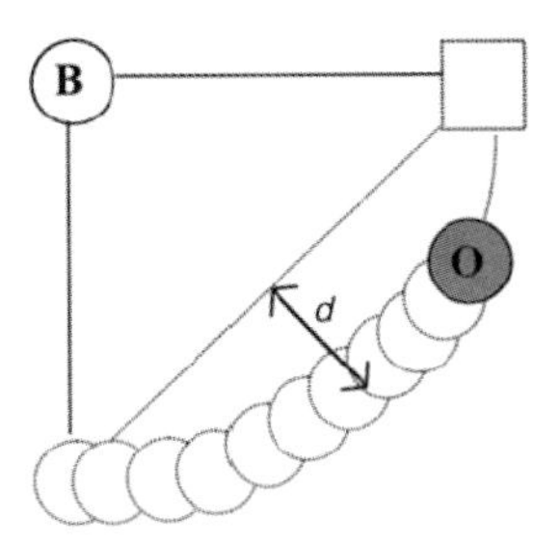

Figure 6.20 Curved path for oxygen vacancy migration between adjacent anion sites [25].

composition $La_{0.9}Sr_{0.1}Ga_{0.8}Mg_{0.2}O_{2.85}$ has an oxide-ion conductivity $\sigma_O > 10^{-2}$ S cm^{-1} at 600 °C with a transport number $t_O \equiv \sigma_O/\sigma \approx 1$ over the oxygen partial pressure range $10^{-20} < P_{O2} < 0.4$ atm.

Mogensen *et al.* [111] have observed the following trends in ionic conductivity considering the lattice stresses: a stress-free lattice gives the maximum oxide-ion conductivity. In the cubic closest packing of the ideal perovskite, the A-site ions should be equal in size to the oxide ion, implying that the matching A-site cation radius, $r_{m,A}$, is 1.40 Å. This explains why La^{3+} (r = 1.36 Å in 12-fold coordination) and Sr^{2+} (r = 1.44 Å) usually are the best A-site ions for maximizing the conductivity.

The importance of the size of the B-site ion is further evidenced by the results for the rare earth gallates. Even when the A-site cation is considerably smaller than 1.40 Å, as in $PrGaO_3$ (r_{Pr3+} = 1.31 Å) or $NdGaO_3$ (r_{Nd3+} = 1.27 Å), the oxide-ion conductivity is still relatively high. Because the oxide ion interacts strongly with the B cation, the optimum oxide ion conductor should be found among the $A^{3+}B^{3+}O_3$-type of perovskites; interaction between the B-site ion and the oxide ion is expected to increase in $A^{2+}B^{4+}O^{2-}{}_3$, and even more in $A^{1+}B^{5+}O^{2-}{}_3$.

Thus, based on reported experimental data, it can be argued that lattice distortion (lattice stress and deviation from cubic symmetry) due to ion radii mismatch determines the ionic conductivity to a very large extent, and that lattice distortion is of much greater importance than many other proposed parameters. At the same time, the fully symmetric and stress-free lattice, in which the oxide vacancy has the highest mobility and the lowest activation energy for the vacancy hopping, is a more proper candidate for high-ion conductors.

6.3.6.2 **Proton Conductivity**

In addition to oxygen ion conduction, perovskite oxides have attracted considerable attention as high-temperature proton conductors, with promising applications in fuel cells, hydrogen sensors, and steam electrolysers [116].

Some perovskites can absorb water from the atmosphere below 400 °C via the reaction

$$(H_2O)_g + \square_s + O_s^{2-} \rightarrow 2(OH^-)_s$$

where $\square_s$ represents an oxygen vacancy in the solid. A similar process (the enthalpy of which is exothermic) can be described by a water incorporation reaction, which indicates that proton uptake in such materials increases with decreasing temperature. In other words, these materials will be dominated by oxygen vacancies at high temperatures, and by protons at low temperatures. For example, during this process, oxygen-deficient brownmillerite-structured $Ba_2In_2O_5$ with the oxide-ion conductivity transfers in $Ba_2In_2O_4(OH)_2$, which becomes a cubic perovskite and a good proton conductor at 400 °C [117]. Iwahara [118] first identified $BaCe_{1-x}Ln_xO_{3-0.5x}$ (Ln = rare earth) perovskites as proton- rather than oxide-ion-conductors at low temperatures as a result of water absorption. Liu and Nowick [119] have shown that the conductivities of hydrated $BaCe_{1-x}Nd_xO_{3-0.5x}$, $SrCe_{1-x}Yb_xO_{3-0.5x}$, and $KTa_{1-x}Fe_xO_{3-x}$ all conduct bare protons.

Relevant samples which exhibit high H^+ ion conductivity in water vapor, and also in an atmosphere containing hydrogen, are substituted $A^{2+}B^{4+}{}_{(1-x)}B^{3+}{}_xO_3$ perovskites. When these are exposed to water vapor, the oxygen vacancies are replaced by hydroxy groups OH^- in which the very small proton H^+ is not in a "free" state but rather is closely associated with an oxygen ion O^{2-} located near the B^{3+} ion. At high temperatures, this can result in the H^+ ion "jumping" to a position near to other O^{2-} ions.

Most attention has focused on $A^{2+}B^{4+}O_3$ perovskites such as $ACeO_3$ (A = Sr, Ba) [120, 121] and $AZrO_3$ (A = Ca, Sr) [122], with very few studies having been conducted with $A^{3+}B^{3+}O_3$ oxides [123]. More recently, a high proton conductivity was discovered in complex systems of the type $A_3B'B''_2O_9$ ($Ba_3Ca_{1.18}Nb_{1.82}O_9$).

Several mechanisms have been proposed for the "jumping" of H^+ ions in similar materials [124–129]. It has been established that the origin of the H^+-ion motion is based on the vibration amplitude of host lattice, and especially on the OH^- vibration mode. At the same time, the characteristics of crystal structures make a significant impact on host lattice vibrations and the conduction ions. With regards to the concentration-dependence of proton conductivity, there appears to be a maximum value close to 10% substituted atoms. This was interpreted by noting the density of the substituted ions around the O-ion, in addition to the effects of vibration amplitude [128]. This conduction mechanism is due to proton "hopping" between adjacent oxygen ions, rather than by hydroxyl ion migration (Figure 6.21). According to an isotope effect (H^+/D^+) analysis, the proton jump involves quantum effects (tunneling) and cooperative motions of the structure (lattice phonons), both of which lead to modulations of the O–O distance [129].

The binding force of movable H^+ ions was found to be weakened with the increasing atomic mass of the constituent cations for proton conductors; at the same time, the E_{ac} values decreased and the conductivity increased linearly with increasing cation mass (Figure 6.22). A strong influence of the cations on the motion of the H^+ ions revealed a significant role of long-range interionic coulombic forces which were dominant for proton mobility.

The current state of proton conductivity in perovskites has been discussed recently in detail by Wakamura [128].

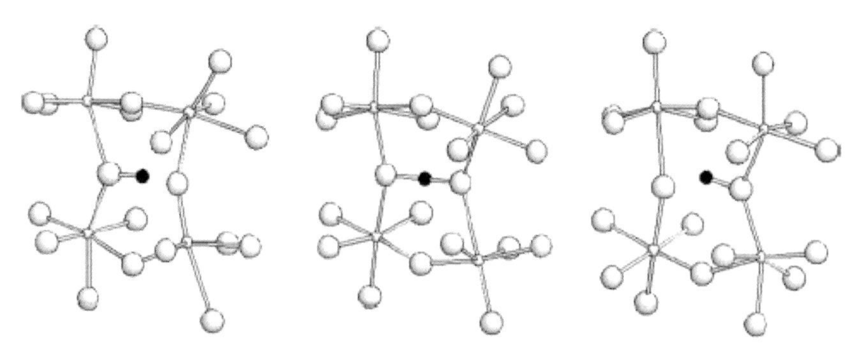

Figure 6.21 Sequence of three snapshots from dynamical simulations showing inter-octahedra proton hopping in orthorhombic $CaZrO_3$ [25].

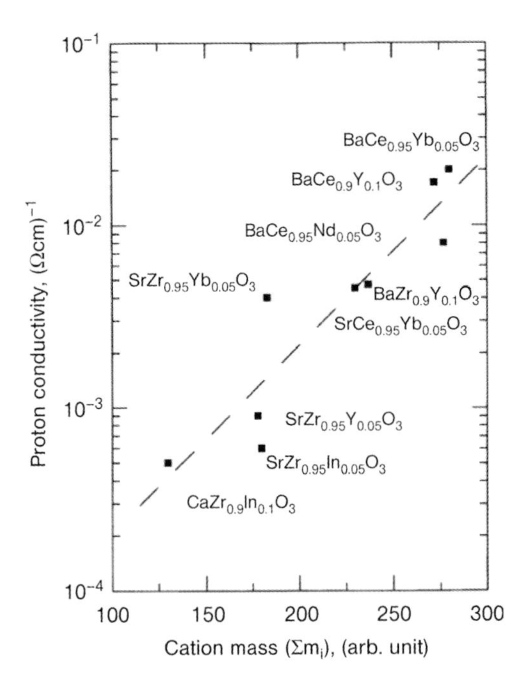

Figure 6.22 Total masses of constituent atoms in a unit cell versus the H^+-ion conductivity at 600 °C for several perovskite-type proton conductors. The compounds have almost the same concentration of substituted B-ions that predicts a near-equal concentration of doped H^+-ions. The line is drawn through the squares indicating the different compounds, with formulas adjacent [128].

6.3.7
Cation Transport

Despite the extensive investigations devoted to oxygen transport, very little effort has been devoted to the study of cation diffusion. Meanwhile, cation mobility plays an important role not only in the processing steps, but also in their operation and degradation. It was proposed that lanthanum diffusion in manganites would most likely take place by vacancy migration between neighboring sites. The calculated values of the activation energies for this La vacancy mechanism indicated a value of 3.93 eV for the cubic structure (this value was consistent with an experimental activation energy of 4.98 eV for La diffusion in $LaCrO_3$ [130] and a predicted value of 4.7 eV for $LaGaO_3$ [131]), with a clear trend towards higher energies as the perovskite lattice became more distorted from the cubic structure [25]. This fact was associated with the Mn–O–Mn bending and MnO_6 octahedra tilting. The calculations revealed a correlation between the shortest O–O separation across the octahedral interstice and the activation energy for La migration, with the activation energy increasing as the O–O distance decreased.

At first sight, the obvious pathway for the migration of B ions in a perovskite lattice is between the diagonal B sites in the $\langle 1\,1\,0\rangle_{cubic}$ directions, as shown in Figure 6.23.

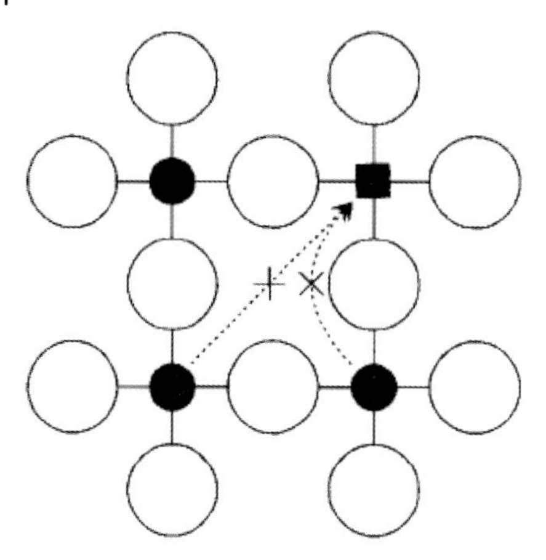

Figure 6.23 Mn vacancy migration along diagonal and curved paths in $LaMnO_3$ [25].

However, the calculated migration energies for this pathway for three structures of $LaMnO_3$ (cubic, rhombohedral, and orthorhombic) are extremely high (ca. 14 eV) [25]. Nonetheless, much lower activation energies were found for an alternative curved path between adjacent Mn sites along the $\langle 1\ 0\ 0\rangle_{cubic}$ direction, for which the ion trajectory lies approximately in the $\{011\}_{cubic}$ plane when the migrating Mn cation moves up and around the oxygen ion.

By modeling A-cation transport in Ca- and Sr-substituted manganites $La_{1-x}A_xMnO_3$, in which the divalent dopant cations are compensated by electron holes (Mn^{4+} sites), it was found that the migration energies decreased in the order La > Sr > Ca (although the ionic radius of La^{3+} lies between that of Sr^{2+} and Ca^{2+}). This unexpected trend could be explained by assuming that the electrostatic and ion polarizibility factors, as well as steric hindrance at the saddle point, are important.

6.3.8 Computer Modeling of Ionic Transport in ABO_3

The study of diffusion mechanisms and defect phenomena is very important, not only for understanding macroscopic transport behavior and being able to predict transport parameters in solid materials, but also for real technological tasks that include the application of oxide electrodes for SOFCs, and for ceramic membranes designed to generate oxygen by its separation from air. Unfortunately, the study of diffusion or conductivity in solids encounters certain difficulties in extracting sufficient information to identify the atomistic mechanisms (or local defect structures) that control the ionic transport. In order to overcome these problems, computer modeling techniques have been used with great success; information relating to the computer modeling of ionic transport in perovskites have been reviewed recently by Islam [25].

6.3.9 Applications

Perovskites of different compositions represent prime device materials in a wide variety of applications, including [5]:

- $BaTiO_3$, for capacitors
- $Pb(Mg_{1/3}Nb_{2/3})O_3$, for capacitors and dielectrics
- $Ba(Zn_{1/3}Nb_{2/3})O_3$, $Ba(Mg_{1/3}Ta_{2/3})O_3$, for microwave dielectrics
- $BaRuO_3$, for resistors
- $PbTiO_3$, for pyroelectric detectors and hydrophones
- $LiTaO_3$, for pyroelectric sensors
- $Pb(Sc_{1/2}Ta_{1/2})O_3$, $(Ba_{0.60}Sr_{0.40})TiO_3$, for pyroelectrics
- $Pb(Zr_{1-x}Ti_x)O_3$, for piezoelectric applications of all types
- $Pb(Mg_{1/3}Nb_{2/3})O_3$, $Pb(Zn_{1/3}Nb_{2/3})O_3$, $Pb[(Mg_{1/3}Nb_{2/3})_{1-x}Ti_x]O_3$, for electrostriction and actuators
- $Pb[(Zn_{1/3}Nb_{2/3})_{1-x}Ti_x]O_3$, $Pb[(Sc_{1/2}Nb_{1/2})_{1-x}Ti_x]O_3$, for medical ultrasound and high-efficiency transducers
- $Pb[(Zn_{1/3}Nb_{2/3})_{1-x}Ti_x]O_3$, $Pb[(Sc_{1/2}Nb_{1/2})_{1-x}Ti_x]O_3$, for piezoelectrics
- $SrRuO_3$, $LaCoO_3$, for conducting electrodes
- $BaZrO_3$, for dielectric resonators
- $SrTiO_3$, ($KTaO_3$), for tunable microwave devices
- $GdFeO_3$, for magnetic bubbles
- $(Ca,La)MnO_3$, for ferromagnetics
- $LaFeO_3$, for NO_x sensors
- $YAlO_3$, for laser hosts
- Doped Rh: $BaTiO_3$, $KNbO_3$, Fe: $LiNbO_3$, for photorefractive materials
- $KNbO_3$, for best second-harmonic generators
- $(Pb_{1-x}La_{2x/3})(Zr_{1-y}Ti_y)O_3$, for electro-optics
- YBCO, BiSCO, $Ba(Pb,Bi)O_3$, for superconductors
- $LaAlO_3$, $NdGaO_3$, $Sr(Al_{0.5}Ta_{0.5})O_3$, $Sr(Al_{0.5}Nb_{0.5})O_3$, for microwave dielectric substrates for high-temperature superconductors.

6.4 Chemical and Catalytic Properties

Because the majority of industrial heterogeneous catalysts are based on mixed-metal oxides, it is reasonable that perovskites have also been examined. The preparation of specific, tailor-made mixed oxides able to perform complex physico-chemical functions is one of the main topics of research in the field of catalysis. To achieve this goal, ample information on the physical and solid-state chemical properties of the catalytic materials should be accumulated.

The perovskite families are characterized by the great flexibility of their crystal structures so as to accommodate wide cation substitutions or anion vacancies, and many relevant examples of this were considered above. Besides cationic substitutions and the introduction of defects into the oxygen sublattice, the exchange of oxygen by other anions such as halides or nitrides is well known. Similar substitutions can significantly influence the properties of the compound. As noted above, the oxygen-deficient $Ba_2In_2O_5$, having an oxide-ion conductivity, transforms in $Ba_2In_2O_4(OH)_2$ by reaction with water, such that it becomes not only a cubic perovskite but also a good proton conductor:

$$Ba_2In_2O_5 + 2H_2O \rightarrow Ba_2In_2O_4(OH)_2$$

The oxygen–nitrogen exchange in titanates is achieved via an ammonolysis reaction [132]:

$$La_2Ti_2O_7 + 2NH_3 \rightarrow 2LaTiO_2N + 3H_2O$$

The properties of the oxynitride are also changed dramatically, with the absorption of colorless titanate being shifted into the region of the visible light in a new compound, such that it acquires an intense bright color [133]. The compound also shows significantly better transport properties due to a reduction in the band gap. As a consequence of these effects, the oxynitride perovskites represent promising candidates as redox catalysis and photocatalysts for the solar water splitting reaction [134].

The catalytic activity for redox reactions relies on the oxygen adsorption possibility and reducibility of the metal oxides, although the reactivity of the applied perovskites was found to vary drastically. Oxygen adsorption on perovskites has been studied mainly because of the importance of these compounds as oxidation–reduction catalysts. As a general rule, oxygen adsorption on the perovskite oxide surface is a complex process. Oxygen adsorption on $LaBO_3$ (B = Cr, Mn, Fe, Co, Ni) has displayed maxima for Mn and Co, the oxygen adsorption being dominated by a molecular form according to a rapid process followed by much slower, activated adsorption kinetics, associated to the formation of oxygen-charged species ($O_2 \rightarrow 2O^{2-}$).

The thermal stability of perovskites ABO_3 is determined by the sum of the energies of the A–O and B–O bonds, which in turn depends on the nature of cations at positions A and B. For example, the extent of H_2-reduction of cobalt in $LnCoO_3$ increased from Ln = La to Eu – that is, with the decreasing ionic radius of the lanthanide element, while the ease of reduction increased with decreasing Ln–O bond energy. On the other hand, the stability in hydrogen of $LaFeO_3$ and $LaRhO_3$ was found to be greater than that of the corresponding yttrium perovskites. Partial substitution of the A ion by another ion of a lower oxidation state, such as Sr ($La_{1-x}Sr_xCoO_3$), may also induce important changes in stability. With increasing values of x, the concentration of unstable Co^{4+} and/or of oxygen vacancies is also increased, favoring the diffusion of lattice oxygen from the bulk to surface, as charge compensators. Thus, increasing the strontium content in the defective structure of $La_{1-x}Sr_xCoO_3$ would account for the instability of the lattice in a reducing environment [135]. In $La_{1-x}Ce_xCoO_3$, where a tetravalent Ce ion is introduced into an A-site, vacancies arise at the A-site due to the limit of solid solution, whilst at the same time tetravalent ion Co^{4+} is also produced. In the case of $LaNiO_3$, since the trivalent La is stable the nickel ion at a B-site exists not in its usual bivalent state but rather in a trivalent state. Thus, perovskite-type oxides are highly flexible materials that are able to incorporate various metal ions into their stable structures, and can give rise to an abnormal valency as well as a vacancy by the substitution of constituent elements.

The thermal stability of perovskite oxides was found to depend also on the type of cation at the B position. Structural changes of a series of $LaBO_3$ (B = V, Cr, Mn, Fe, Co, Ni) oxides induced by H_2-reduction at 1273 K as a function of the oxygen partial pressure were studied [136]. The following order of stability in terms of the $-\log P_{O2}$ values can be established: $LaNiO_3 < LaCoO_3 < LaMnO_3 < LaFeO_3 < LaCrO_3 \sim LaVO_3$. This order parallels the Madellung constant, except for $LaNiO_3$ and $LaCoO_3$, which show higher constants.

Historically, the initial interest of perovskites during the mid-1970s was focused on their application as catalysts to remove noxious materials from the exhaust gases of automobile engines. However, the low resistance of perovskites to poisoning by sulfur dioxide, as compared to catalysts based on noble metals, has led to a reduction in research activity in this field. Nonetheless, several very interesting systems have been discovered, including perovskites incorporated by noble metals (Pt, Pd, Rh) that were highly active for exhaust treatments, while the incorporation of a small amount of Rh into the $LaMnO_{3.15}$ led to a high three-way catalytic activity in a synthetic $CO + NO + C_3H_6$ mixture.

Later, several reactions were studied in which perovskite oxides have been used as catalysts. The strategy of designing perovskite catalysts to enhance their catalytic performance was based on: (i) the selection of B-site elements, valency and vacancy controls; (ii) synergistic effects mainly of B-site elements; (iii) the enhancement of surface area by forming fine particles or dispersing on supports; and (iv) the addition of precious metals with their regeneration method. Studies of the catalytic properties of pure bulk perovskites or supported perovskites, conducted since the 1980s [26, 27, 137–144], have shown that transition and rare earth metal perovskites appear best suited to high-temperature processes such as catalytic combustion, methane reforming, CO and ammonia oxidation, NO decomposition, and sulfur dioxide reduction. Such efficacy is based on their well-known thermal stability over a broad range of oxygen partial pressures, and also to their resistance against catalytic poisons such as Cl-, F-, and S-containing compounds.

Some perovskites have displayed effective performances as cathode materials in high-temperature fuel cells (SOFCs), although inherent barriers (e.g., electrolyte ohmic resistance and electrode overpotentials) to the reduction of operating temperatures for SOFCs must first be overcome. Some Sr-substituted cobaltites can act as mixed oxide-ion and electronic conductors, affording excellent cathode performance with very low overpotentials.

Although perovskites have not yet found application as commercial catalysts, they remain excellent models for the study of catalytic reactions. Moreover, they also represent promising candidates for polyfunctional catalysts, because they are characterized by a great flexibility of the crystal structure to accommodate cation substitutions, or to provide anion vacancies, both of which are extremely useful. For example, a combination of two different ions at the B-site in the system

$LaMn_{1-x}Cu_xO_3$ brings about synergistic effects that can serve as a powerful tool for catalyst design. Such synergism was presumed to be due to a bifunctional catalysis of the Mn and Cu, and several successful examples of this have been reported [27]. Fine-tuning of the properties can be achieved by suitable cation- and anion-substitutions, and also by adjusting the morphology of the compounds. A comprehensive review of the relevance of perovskites in heterogeneous catalysis and surface chemistry has been provided in Ref. [26].

6.4.1 Synthesis

Oxide perovskites are often synthesized using standard solid-state ceramic reactions, with the most frequently used scheme including the preparation of mixture oxides (or carbonates and oxides) and firing at high temperatures (ca. two-thirds of the melting point) for periods of up to 10 h. In some processes, however, the rate of formation of the target phases is very low. Problems may also occur if one of the oxides (especially a toxic oxide, such as Pb) partly vaporizes during the long reaction times. The application of a microwave field is very effective for solid-state reactions; indeed, on the basis of its simplicity and speed, microwave sintering is especially attractive for Pb-containing perovskites as it minimizes any Pb loss.

Additional synthetic approaches include:

- Sol–gel techniques: these are widely used for producing thin films at low temperatures, or for preparing highly oriented perovskites.
- A templated grain growth method was applied for the fabrication of textured ceramics with single crystal-like properties.
- Both hydrothermal and hydrothermal–electrochemical methods have been used for the synthesis of $BaTiO_3$ at very low temperatures.
- Single crystals can be obtained using the methods of growth processes from flux.
- Many of the methods originally developed for use in electronics have also been applied successfully for the preparation of thin films.

6.5 Summary

The extraordinary versatility of the perovskite structure has led to the creation of an essential family of materials with different functional properties that can be used for a wide variety of present-day technologies, and is likely to remain an integral part of future developments. Clearly, the perovskites represent important commercial materials, with their remarkable structures providing a wealth of new compounds with various and, often, unique properties. By employing the vast crystal chemistry resources and principles available (e.g., ionic radii, valence, tolerance factor, etc.), innumerable perovskite compounds with a wide variety of properties can be designed to provide structural identities that will facilitate the development of superior integrated devices. The diversity of perovskite compounds that can be synthesized covers extremes of electrical, magnetic, optical and mechanical properties over a wide temperature range. By using currently available novel approaches for the design and fabrication of different materials, perovskites ranging from macroscale to nanoscale can be engineered that exhibit desirable properties for new device concepts.

References

1 At the invitation of the Russian Czar, Nicholas I (1796–1855), three German scientists undertook a scientific expedition into the remote Siberian reaches of the Russian Empire in 1837. Their travels covered nearly 16 000 km, reaching as far east as the Altai Mountains, just north of the Mongolian border. Gustav Rose, the mineralogical expert of the group, recorded the account of their investigations on the famous gem deposits of Siberian regions (*Humboldt's Travels in Siberia* (1837–1842) – *The Gemstones* - By Gustav Rose).

2 (a) Megaw, H.D. (1946) *Proc. Phys. Soc.*, **58**, 133; (b) Naray-Szabo, J. (1943) *Naturwissenschaften*, **31**, 202.

3 (a) Goldschmidt, V.M., Barth, T., Lunde, S. *et al.* (1926) *J. Math.-Nat. Klasse.*, **N2**, 97; (b) Goldschmidt, V.M. (1927) *Geochemisce Verterlungsgesetze der Elemente*, Norske Videnskap, Oslo.

4 Vul, B.M. and Goldman, I.M. (1945) *Dokl. Akad. Nauk SSSR*, **46**, 154 (in Russian).

5 Bhalla, A.S., Guo, R., and Roy, R. (2000) *Mater. Res. Innovat.*, **4**, 3.

6 (a) Ginsburg, V.L. (1945) *J. Expt. Theor. Phys. SSSR*, **15**, 739; (b) Ginsburg, V.L. (1949) *J. Expt. Theor. Phys. SSSR*, **19**, 36 (in Russian).

7 von Hippel, E. (1959) *A Molecular Science and Molecular Engineering*, MIT Press, Cambridge, MA.

8 Megaw, H.D. (1957) *Ferroelectricity in Crystals*, Methuen, London.

9 Smolensky, G.A. (1970) *J. Phys. Soc. Jpn*, **28** (Suppl.), 26.

10 Smolensky, G.A., Bokov, V.A., Isupov, V.A., Krainik, N.N., Pasynkov, R.E., and Shur, M.S. (1971) *Ferroelectrics and Antiferroelectrics*, Nauka, Leningrad (in Russian).

11 Fesenko, E.G. (1972) *Family of Perovskite and Ferroelectricity*, Atomizdat, Moscow (in Russian).

12 Galasso, F.S. (1969) *Structure, Properties and Preparation of Perovskite-type Compounds*, Pergamon Press, Oxford.

13 Academy of Sciences of USSR (1973) *Barium Titanate*, Science, Moscow (in Russian).

14 Vainshtein, B.K., Fridkin, V.M., and Indenbom, V.L. (eds) (1979) *Modern Crystallography*, vol. 4, Physical Properties of Crystals, Nauka, Moscow.

15 Aleksandrov, K.S., Anistratov, A.T., Beznosikov, B.V., and Fedoseeva, N.V. (1981) *Phase Transitions in the Crystals of Halide Compounds* ABX_3, Science, Novosibirsk (in Russian).

16 Bazuev, G.V. and Shveikin, G.P. (1985) *Complex Oxides of Elements with Completing d- and f-Shells*, Nauka, Moscow (in Russian).

17 Venevtsev, Yu.N., Politova, E.D., and Ivanov, S.A. (1985) *Ferro- and Antiferroelectrics of Barium Titanate Family*, Chemistry, Moscow (in Russian).

18 Smolensky, G.A., Bokov, V.A., Isupov, V.A., Krainik, N.N., Pasynkov, R.E., Sokolov, A.I., and Yushin, N.K. (1985) *Physics of Ferroic Phenomena*, Nauka, Leningrad (in Russian).

19 Aleksandrov, K.S. and Beznosikov, B.V. (1981) *Perovskite Crystals*, Science, Novosibirsk (in Russian).

20 Rao, C.N.R., Gopalakrishnan, J., and Vidyasagar, K. (1984) *Indian J. Chem.*, **23A**, 265.

21 Rao, C.N.R. and Gopalakrishnan, J. (1986) *New Directions in Solid State Chemistry*, Cambridge University Press.

22 Palguev, S.F., Gilderman, V.K., and Zemtsov, V.I. (1990) *High-temperature Oxide Electronic Conductors for Electrochemical Devices*, Nauka, Moscow (in Russian).
23 Ganguly, P. and Shah, N. (1993) *Physica C*, **208**, 307.
24 Kharton, V.V., Yaremchenko, A.A., and Naumovich, E.N. (1999) *J. Solid State Electrochem.*, **3**, 303.
25 Islam, M.S. (2000) *J. Mater. Chem.*, **10**, 1027.
26 Pena, M.A. and Fierro, J.L.G. (2001) *Chem. Rev.*, **101**, 1981.
27 Tanaka, H. and Misono, M. (2001) *Curr. Opin. Solid State Mater. Sci.*, **5**, 381.
28 Stitzer, K.E., Darriet, J., and zur Loye, H.-C. (2001) *Curr. Opin. Solid State Mater. Sci.*, **5**, 535.
29 Mitchel, R.H. (2002) *Perovskites. Modern and Ancient*, Almaz Press, Inc., Ontario, Canada.
30 Schaak, R.E. and Mallouk, T.E. (2002) *Chem. Mater.*, **14**, 1455.
31 Samara, G.A. (2003) *J. Phys.: Condens. Matter*, **15**, R367.
32 Isupov, V.A. (2003) *Ferroelectrics*, **289**, 131.
33 Aleksandrov, K.S. and Beznosikov, B.V. (2004) *Perovskites. The Present and Future*, Science, Novosibirsk (in Russian).
34 Goodenough, J.B. (2004) *Rep. Prog. Phys.*, **67**, 1915.
35 van Tendeloo, G., Lebedev, O.I., Hervieu, M., and Raveau, B. (2004) *Rep. Prog. Phys.*, **67**, 1315.
36 Kato, S., Ogasawara, M., Sugai, M., and Nakata, S. (2004) *Catal. Surv. Asia*, **8**, 27.
37 Demazeau, G. (2005) *Z. Anorg. Allg. Chem.*, **631**, 556.
38 Raveau, B. (1986) *Proc. Indian Natl Sci. Acad. Part A*, **52**, 67.
39 (a) Katrusiak, A. and Ratuszna, A. (1992) *Solid State Commun.*, **84**, 435; (b) Goodenough, J.B. (2001) *Struct. Bond.*, **98**, 1; (c) Goodenough, J.B. and Zhou, J.-S. (2001) *Struct. Bond.*, **98**, 17.
40 Wainer, E. and Wentworth, C. (1952) *J. Am. Chem. Soc.*, **35**, 207.
41 Wood, E.A. (1951) *Acta Crystallogr.*, **4**, 358.
42 Roth, R.S.J. (1957) *Res. Nat. Bur. Standards*, **58**, 75.
43 (a) Miller, O. and Roy, R. (1974) *The Major Ternary Structural Families*, Springer-Verlag, Berlin; (b) Hyde, B.G. and Andersson, S. (1989) *Inorganic Crystal Structures*, John Wiley & Sons Inc., New York.
44 Woodward, P.M. (1997) *Acta Crystallogr. B*, **53**, 32.
45 Rao, C.N.R. and Rao, K.J. (1978) *Phase Transitions in Solids*, McGraw-Hill, New York.
46 Frazer, B.C. and Danner, H.R. (1955) *Phys. Chem.*, **100**, 745.
47 Shirane, G. and Pepinsky, R. (1956) *Acta Crystallogr.*, **9**, 131.
48 (a) Glazer, A.M. (1972) *Acta Crystallogr. B*, **28**, 3384; (b) Glazer, A.M. (1975) *Acta Crystallogr. A*, **31**, 756.
49 Van Tendeloo, G. and Amelinckx, S. (1974) *Acta Crystallogr. A*, **30**, 431.
50 Portier, R. and Gratias, D. (1982) *J. Phys.*, **43**, C4.
51 Aleksandrov, K.S. (1976) *Ferroelectrics*, **14**, 801.
52 Goodenough, J.B., Kafalas, J.A., and Longo, J.M. (1972) Chapter 1, *Preparative Methods in Solid State Chemistry* (ed. P. Hagenmuller), Academic, New York.
53 (a) Ruddlesden, S.N. and Popper, P. (1957) *Acta Crystallogr.*, **10**, 538; (b) Ruddlesden, S.N. and Popper, P. (1958) *Acta Crystallogr.*, **11**, 54.
54 Hawley, M.E. (1998) *Ceram. Transac.*, **86**, 41.
55 Rao, C.N.R. and Thomas, J.M. (1985) *Acc. Chem. Res.*, **18**, 113.
56 Sharma, J.B. and Singh, D. (1998) *Bull. Mater. Sci.*, **21**, 363.
57 (a) Aurivillius, B. (1949) *Arkiv. Kemi.*, **1**, 463; (b) Aurivillius, B. (1950) *Arkiv. Kemi.*, **2**, 499; (c) Aurivillius, B. (1950) *Arkiv. Kemi.*, **2**, 519.
58 Teller, R.G., Brazdil, J.F., Grasselli, R.K., and Jorgensen, J.D. (1984) *Acta Crystallogr. C*, **40**, 2001.
59 Knight, K.S. (1992) *Miner. Mag.*, **56**, 399.
60 Abraham, F., Boivin, J.C., Mairesse, G., and Nowogrocki, G. (1990) *Solid State Ionics*, **40–41**, 934.
61 (a) Kendall, K.R., Navas, C., Thomas, J.K., and zur Loye, H.-C. (1996) *Chem. Mater.*, **8**, 642; (b) Abraham, F., Debreuille-Gresse, M.F., Mairesse, G., and Nowogrocki, G. (1988) *Solid State Ionics*, **28–30**, 529.

62 Müller-Buschbaum, H. and Wallschlager, W. (1975) *Z. Anorg. Allg. Chem.*, **414**, 76.

63 (a) Sawa, H., Obara, K., Akimitsu, J., Matsui, Y., and Horiuchi, H. (1989) *J. Phys. Soc. Jpn*, **58**, 2253; (b) Izumi, F., Kito, H., Sawa, M., Akimitsu, J., and Asano, H. (1989) *Physica C*, **160**, 235.

64 Abe, M. and Uchino, K. (1974) *Mater. Res. Bull.*, **9**, 147.

65 Nadiri, A., le Flem, G., and Dalmas, C. (1988) *J. Solid State Chem.*, **73**, 338.

66 Howard, C.J., Luca, V., and Knight, K.S. (2002) *J. Phys.: Condens. Matter*, **14**, 377.

67 Shannon, J. and Katz, L. (1970) *Acta Crystallogr. B*, **26**, 102.

68 Calvo, C.N.N. and Chamberland, B.L. (1978) *Inorg. Chem.*, **17**, 699.

69 Jacobson, A.J., Scanlon, J.C., Poeppelmeier, K.R., Longo, J.M., and Cox, D.E. (1982) *Mater. Res. Bull.*, **16**, 359.

70 Blasse, G. and Dirksen, G.J. (1981) *J. Solid State Chem.*, **36**, 124.

71 Wadsley, A.D. (1964) *Non-Stoichiometric Compounds* (ed. L. Mandelcorn), Academic, New York, p. 134.

72 (a) Poeppelmeier, K.R., Leonowicz, M.E., Scanlon, J.C., and Longo, J.M. (1982) *J. Solid State Chem.*, **45**, 71; (b) Anderson, M.T., Vaughey, J.T., and Poeppelmeier, K.R. (1993) *Chem. Mater.*, **5**, 151.

73 Grenier, J.C., Pouchard, M., and Hagenmuller, P. (1981) *Struct. Bond.*, **47**, 1.

74 Vidyasagar, K., Gopalakrishnan, J., and Rao, C.N.R. (1984) *Inorg. Chem.*, **23**, 1206.

75 Vidyasagar, K., Reller, A., Gopalakrishnan, J., and Rao, C.N.R. (1985) *J. Chem. Soc., Chem. Commun.*, 7.

76 Berastegui, P., Hull, S., Garcıa-Garcıa, F.J., and Eriksson, S.-G. (2002) *J. Solid State Chem.*, **164**, 119.

77 Soda, M., Yasui, Y., Fujita, T., Miyashita, T., Sato, M., and Kakurai, K. (2003) *J. Phys. Soc. Jpn*, **72**, 1729.

78 Woodward, P.M., Suard, E., and Karen, P. (2003) *J. Am. Chem. Soc.*, **125**, 8889.

79 Smyth, D.M. (1985) *Annu. Rev. Mater. Sci.*, **15**, 329.

80 Smyth, D.M. (1993) in *Properties and Applications of Perovskite-Type Oxides* (eds L.G. Tejuca and J.L.G. Fierro), Marcel Dekker, New York, p. 47.

81 Roth, R.S. and Stephenson, N.C. (1970) in *The Chemistry of Extended Defects in Nonmetallic Solids* (eds L. Eyring and M. O'Keefe), North-Holland, Amsterdam.

82 Reller, A., Thomas, J.M., Jefferson, D.A., and Uppal, M.K. (1984) *Proc. Roy. Soc. London*, **A394**, 223.

83 Tofield, B.C. and Scott, W.R. (1974) *J. Solid State Chem.*, **10**, 183.

84 Tejuca, L.J. and Fierro, J.L.G. (eds) (1993) *Properties and Applications of Perovskite Type Oxides*, Marcel Dekker, New York.

85 Takenafa, T. and Nagata, H. (2005) *J. Eur. Ceram. Soc.*, **25**, 2693.

86 Hoffmann, M.J. and Kungl, H. (2004) *Curr. Opin. Solid State Mater. Sci.*, **8**, 51.

87 Haertling, G. (1999) *J. Am. Ceram. Soc.*, **82**, 797.

88 Xu, Y. (1991) *Ferroelectric Materials and their Application*, Elsevier Scientific Publications, Amsterdam, The Netherlands.

89 Vereschagin, V.I., Pletnev, P.M., Surgikov, A.P., Fedorov, V.E., and Rogov, I.I. (2004) *Functional Ceramics*, Siberian Branch, Russian Academy of Sciences, Novosibirsk.

90 Goodenough, J.B. (1963) *Magnetism and the Chemical Bond*, John Wiley & Sons, Inc., New York.

91 Goodenough, J.B. and Longo, J.M. (1970) *Landolt–Börnstein Tabellen, New Series III/4a*, Springer-Verlag, Berlin.

92 Goodenough, J.B. (1971) *Prog. Solid State Chem.*, **5**, 149.

93 Goodenough, J.B. (1974) in *Solid State Chemistry* (ed. C.N.R. Rao), Marcel Dekker, New York.

94 Sarma, D.D. (2001) *Curr. Opin. Solid State Mater. Sci.*, **5**, 261.

95 Zhou, J.-S., Goodenough, J.B., Dabrowski, B., Klamut, P.W., and Bukowski, Z. (2000) *Phys. Rev. Lett.*, **84**, 526.

96 von Helmolt, R., Wecker, J., Holzapfel, B., Schultz, L., and Samwer, K. (1993) *Phys. Rev. Lett.*, **71**, 2331.

97 Rao, C.N.R. and Raveau, B. (eds) (1998) *Colossal Magnetoresistance, Charge Ordering and Related Properties of Manganese Oxides*, World Scientific, Singapore.

98 Coey, J.M.D., Viret, M., and von Molnar, S. (1999) *Adv. Phys.*, **48**, 167.
99 Hwang, H.Y., Cheong, S.-W., Radaelli, P.G., Marezio, M., and Batlogg, B. (1995) *Phys. Rev. Lett.*, **75**, 914.
100 Zener, C. (1951) *Phys. Rev.*, **82**, 403.
101 Smolenski, G.A., Isupov, V.A., and Agranovskaya, A.I. (1959) *Sov. Phys. Sol. State*, **1**, 909.
102 Cross, L.E. (1987) *Ferroelectrics*, **76**, 241.
103 Randall, C.A. and Bhalla, A.S. (1990) *Jpn. J. Appl. Phys.*, **29**, 327.
104 Jaffe, B., Cook, W.R. Jr, and Jaffe, H. (1971) *Piezoelectric Ceramics*, Academic Press, London.
105 Bhattacharya, K. and Ravichandran, G. (2003) *Acta Mater.*, **51**, 5941.
106 Messing, G.L., Trolier-McKinstry, S., Sabolsky, E.M., Duran, C., Kwon, S., Brahmaroutu, B., Park, P., Yilmaz, H., Rehrig, P.W., Eitel, K.B., Suvaci, E., Seabaugh, M., and Oh, K.S. (2004) *Crit. Rev. Solid State Mater. Sci.*, **29**, 45.
107 Park, S.E. and Shrout, T.R. (1997) *J. Appl. Phys.*, **82**, 1804.
108 Blasse, G. (1988) *Prog. Solid State Chem.*, **18**, 79.
109 Butler, M.A. and Ginley, D.S. (1980) *J. Mater. Sci.*, **51**, 1.
110 Kuzminov, Yu.S. (1982) *Ferroelectric Crystals for Control of Laser Radiation*, Science, Moscow.
111 Mogensen, M., Lybye, D., Bonanos, N., Hendriksen, P.V., and Poulsen, F.W. (2004) *Solid State Ionics*, **174**, 279.
112 Pouchard, M. and Hagenmuller, P. (1978) in *Solid Electrolytes* (eds P. Hagenmuller and W. van Gool), Academic Press, New York.
113 Ishihara, T., Matsuda, H., and Takita, Y. (1994) *J. Am. Chem. Soc.*, **116**, 3801.
114 Goodenough, J.B. (2003) *Annu. Rev. Mater. Res.*, **33**, 91.
115 Ishihara, T., Sammels, N.M., and Yamamoto, O. (2003) in *High Temperature Solid Oxide Fuel Cells: Fundamentals, Design and Applications* (eds S.C. Singhal and K. Kendall), Elsevier, p. 83.
116 Norby, T. (1999) *Solid State Ionics*, **125**, 1.
117 Goodenough, J.B., Manthiram, A., and Kuo, J.-F. (1993) *Mater. Chem. Phys.*, **35**, 221.
118 Iwahara, M. (1988) *Solid State Ionics*, **28–30**, 573.
119 Liu, J.F. and Novick, A.S. (1992) *Solid State Ionics*, **50**, 131.
120 Shima, D. and Haile, S.M. (1997) *Solid State Ionics*, **97**, 443.
121 Guan, J., Morris, S.E., Balachandran, U., and Liu, M. (1998) *Solid State Ionics*, **110**, 303.
122 Huang, P. and Petric, A. (1995) *J. Mater. Chem.*, **5**, 5.
123 Ruiz-Trejo, E. and Kilner, J.A. (1997) *Solid State Ionics*, **97**, 33.
124 Colomban, P. (1999) *Ann. Chim. Sci. Mater.*, **24**, 1.
125 Kreuer, K.D. (1996) *Chem. Mater.*, **8**, 610.
126 Samgin, A.L. (2000) *Solid State Ionics*, **136–137**, 291.
127 Wakamura, K. (2004) *Solid State Ionics*, **171**, 229.
128 Wakamura, K. (2005) *J. Phys. Chem. Solids*, **66**, 133.
129 Cherry, M., Islam, M.S., and Catlow, C.R.A. (1995) *J. Solid State Chem.*, **138**, 125.
130 Akashi, T., Nanko, M., Maruyama, T., Shiraishi, Y., and Tanabe, J. (1998) *J. Electrochem. Soc.*, **145**, 2090.
131 Khan, M.S., Islam, M.S., and Bates, D.R. (1998) *J. Phys. Chem. B*, **102**, 3099.
132 Marchand, R., Pors, F., and Laurent, Y. (1986) *Rev. Int. Hautes Temp.*, **23**, 11.
133 Jansen, M. and Letschert, H.P. (2000) *Nature*, **404**, 980.
134 Kasahara, A., Nukumizu, K., Takata, T., Kondo, J.N., Nara, M., Kobayashi, H., and Domen, K. (2003) *J. Phys. Chem. B*, **107**, 791.
135 Nakamura, T., Misono, M., and Yoneda, Y. (1982) *Bull. Chem. Soc. Jpn*, **55**, 394.
136 Nakamura, T., Petzow, G., and Gauckler, L.J. (1979) *Mater. Res. Bull.*, **14**, 649.
137 Sorenson, S.C., Wronkiewicz, J.A., Sis, L.B., and Wirtz, G.P. (1974) *Ceram. Bull.*, **53**, 446.
138 Guilhaume, N. and Primet, M. (1997) *J. Catal.*, **165**, 197.
139 Trimm, D.L. (1983) *Appl. Catal.*, **7**, 249.
140 Tejuca, L.G., Fierro, J.L.G., and Tascon, J.M.D. (1989) *Adv. Catal.*, **36**, 237.
141 Isupova, L.A., Sadykov, V.A., Tikhov, S.F., Kimkhai, O.N., Kovalenko, O.N.,

Kustova, G.N., Ovsyannikova, I.A., Dovbii, Z.A., Kryukova, G.N., Rozovskii, A.Y., Tretyakov, V.F., and Lumin, V.V. (1994) *React. Kinet. Catal. Lett.*, **53**, 223.

142 Baran, E.T. (1990) *Catal. Today*, **8**, 133.

143 Yamaxoe, V. and Tekaoka, J. (1990) *Catal. Today*, **8**, 175.

144 Pauli, I.A., Avvakumov, E.G., Isupova, L.A., Poluboyarov, V.A., and Sadykov, V.A. (1992) *Si.b Khim. Zhurn.*, **3**, 133 (in Russian).

7
The $M_{n+1}AX_n$ Phases and their Properties

Michel W. Barsoum

7.1
Introduction

Today, it has been fairly well established that the layered ternary carbides and nitrides with the general formula $M_{n+1}AX_n$, and where $n = 1$, 2 or 3, M is an early transition metal, A is an A-group element (mostly group IIIA or IVA), and X is either C or N, represent a new class of solids [1, 2]. These phases are layered, with $M_{n+1}X_n$ layers interleaved with pure A-group element layers.

The $M_{n+1}AX_n$ phases combine an unusual, and sometimes unique, set of properties [1]. Like their corresponding binary carbides and nitrides, they are elastically stiff, good thermal and electrical conductors, resistant to chemical attack, and have relatively low thermal expansion coefficients. Mechanically, however, they cannot be more different: They are relatively soft (1–5 GPa) and most readily machinable, thermal shock-resistant, and damage-tolerant [3, 4]. Moreover, some are also fatigue- and creep- and oxidation-resistant. At higher temperatures, they go through a plastic-to-brittle transition, whilst at room temperature they can be compressed to stresses as high as 1 GPa and fully recover upon removal of the load, while dissipating 25% of the mechanical energy [5].

In this chapter, the physical and mechanical properties of bulk $M_{n+1}AX_n$ phases are summarized. The chapter is subdivided into six sections, in the first two of which the structure and bonding (including theoretical) characteristics of these materials are reviewed. Their elastic properties are summarized in Section 7.3., while in Sections 7.4 and 7.5 their electrical properties and thermal properties are reviewed, respectively. The mechanical properties are dealt with in Section 7.6. What is not detailed in this chapter are the MAX-phase thin films, for which several excellent groups, especially in Europe [6–9], have provided details. Another topic that has encouraged much activity, but which will not be reviewed here, is that of the processing of the MAX phases; rather, at this point emphasis is placed on the properties of bulk materials.

Ceramics Science and Technology Volume 2: Properties. Edited by Ralf Riedel and I-Wei Chen

ISBN: 978-3-527-31156-9

7.2
Bonding and Structure

The hexagonal $M_{n+1}AX_n$ unit cells – space group $P63/mmc$ – have two formula units per unit cell. There are approximately 50 M_2AX phases [10]; five M_3AX_2, namely Ti_3SiC_2 [11], Ti_3GeC_2 [12], Ti_3AlC_2, [13] Ti_3SnC_2 [14], and Ta_3AlC_2 [15]. The number of M_4AX_3 phases is also growing since that structure was first established in Ti_3AlN_4 [16, 17]. More recently, the following 413 phases have been discovered: Ta_4AlC_3 [15, 18, 19], Nb_4AlC_3 [20], and V_4AlC_{3-x} [21]. However, neither Ti_3SnC_2 [14], Ta_3AlC_2 [15], nor V_4AlC_{3-x} [21] have been synthesized in a predominantly pure form, and consequently their characterization awaits a processing breakthrough.

The 413 phases exist in two polymorphs, α and β [16]. The layering in the α-polymorph is: ABABACBCBC ...; that of the β-polymorph is: ABABABABA. With the exception of Ta_4AlC_3, which exists in bulk form in both polymorphs [15, 18, 19, 22], to date the remainder of the 413 phases are known to crystallize in bulk form in the α-polymorph [16, 17, 21].

The 211 phases (Figure 7.1a), for which $n = 1$, are by far the most prevalent (Table 7.1). The unit cell (Figure 7.1a) is characterized by near close-packed M layers interleaved with layers of pure group A-element, with the X-atoms filling the octahedral sites between the former. The M_6X octahedra are identical to those found in the rock salt structure of the corresponding binary MX carbides. The A-group elements are located at the centers of trigonal prisms that are slightly larger, and thus better able to accommodate the larger A-atoms, than the octahedral sites [23–26]. When $n = 2$, for example, Ti_3SiC_2, the A-layers are separated by two M-layers (Figure 7.1b), whereas for $n = 3$ (e.g., Ti_4AlN_3) there are four M-layers (Figure 7.1c).

Table 7.1 lists most of the known MAX phases – most of which were discovered by Nowotny *et al.* [10] during the 1960's – together with their lattice parameters and theoretical densities. The A-group elements are mostly IIIA and IVA, and all but five compounds are 211s. The most versatile element is Al, as it forms nine compounds, including two nitrides, one 312 phase, and four 413 phases. Ga also forms nine 211 phases, six of which are carbides and three are nitrides.

Given the close chemical and structural similarities of the MAX and MX phases, much can be learned about the former from what is known about the latter. For example, for the most part the M–M distances in the MAX phases are strongly correlated to, and almost equal to, the same distance in the MX phases [1, 10]. Like the MX compounds [27, 28], it is useful to consider the ternaries to be interstitial compounds in which the *A*- and *X*-atoms fill the interstitial sites between the M atoms. In such a scheme, the c-parameter of the 211 phases – comprised of four M-layers per unit cell – should be approximately four times the a-parameter. Similar arguments for the 312 and 413 phases, with six and eight M layers per unit cell, respectively, predict ratios of about 6 and about 8 [29]. The actual c/a ratios of approximately 4, 5.8–6, and 7.8, are consistent with this simple structural notion.

Currently, much effort is being expended in theoretical modeling of the MAX phases [30–37]. Not unlike the MX phases, the bonding in the MAX phases is

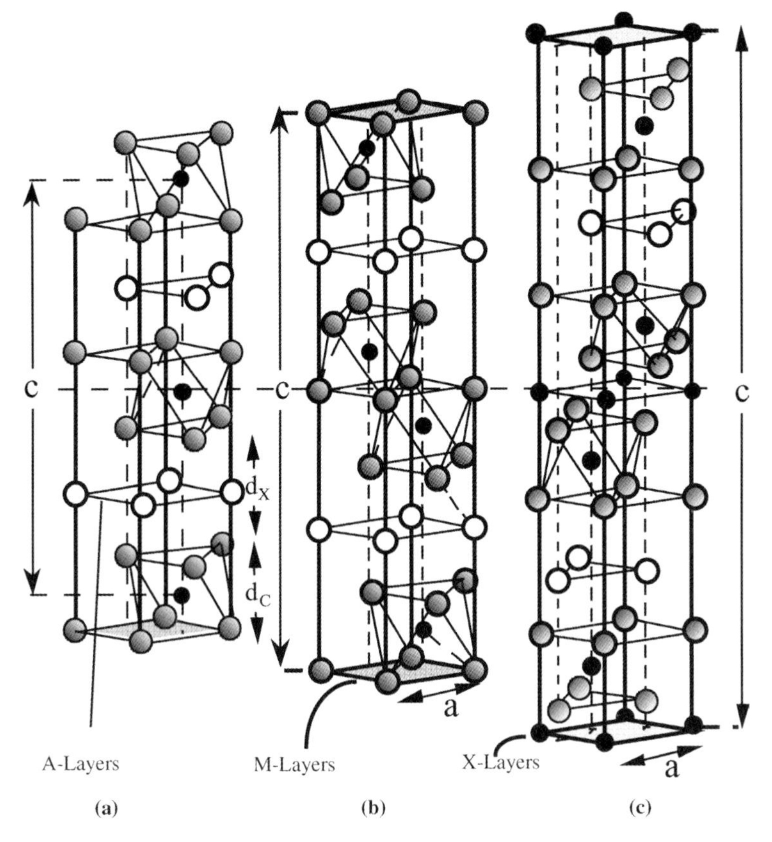

Figure 7.1 MAX phase unit cells. (a) 211; (b) 312; (c) 413.

a combination of metallic, covalent, and ionic [30, 38]. The following conclusions of the theoretical studies are noteworthy:

- As in the MX compounds, there is a strong overlap between the p levels of the X elements and the d levels of the M atoms, leading to strong covalent bonds that are comparable to those in the MX binaries [30].
- The density of states at the Fermi level, $N(E_F)$ – like in many MX binaries, but notably *not* TiC – is substantial (Figure 7.2).
- The p-orbitals of the A atoms overlap the d-orbitals of the M atoms (Figure 7.2b).
- The electronic states at the Fermi level are mostly d-d M orbitals [30–33, 38, 39].
- In the M_2AlC phases, there is a net transfer of charge from the A-group element to the X-atoms [32]. Whether this is true of other MAX phases as well awaits further study.

Lastly, the number of possible solid solution permutations and combinations is clearly quite large. It is possible to form solid solutions on the M sites, the A-sites and the X sites, and combinations thereof. A continuous series of solid solutions,

Table 7.1 List of $M_{n+1}AX_n$ phases known to date to exist. The theoretical density ($Mg\,m^{-3}$) is shown in bold text. The *a*- and *c*-lattice parameters (Å) are shown in brackets. Most of this list appeared in a 1970 review paper [10]. In this and other tables, the 312s are highlighted yellow, and the 413s gray. The list does not include solid solutions.

	Group		
	IVA	VA	VIA
Al	Si	P	S
Ti_2AlC, 4.11 (3.04, 13.60)	**$Ti_3SiC_2$4.52** (3.0665, 17.671)	**V_2PC 5.38** (3.077, 10.91)	**Ti_2SC, 4.62** (3.216, 11.22)
V_2AlC, 4.82 (2.914, 13.19)	Al	**Nb_2PC 7.09** (3.28,11.5)	**Zr_2SC, 6.20** (3.40, 12.13)
Cr_2AlC, 5.24 (2.86, 12.8)	**Nb_4AlC_3, 7.09** (3.123, 24.109)		**$Nb_2SC_{0.4}$,** (3.27,11.4)
Nb_2AlC, 6.50 (3.10, 13.8)	**$V_4AlC_{3-1/3}$, 5.16** (2.9302, 22.745)		**Hf_2SC,** (3.36, 11.99)
Ta_2AlC, 11.82 (3.07, 13.8)			
Ti_2AlN, 4.31 (2.989, 13.614)			
Ti_3AlC_2, 4.5 (3.075, 18.578)			
Ti_4AlN_3, 4.76 (2.988, 23.372)			
α-Ta_4AlC_3, 12.9 (3.11, 24.12)			
β-Ta_4AlC_3, 13.2 (3.087, 23.70)			
Ga	Ge	As	Se
Ti_2GaC, 5.53 (3.07, 13.52)	**Ti_2GeC, 5.68** (3.07, 12.93)	**V_2AsC, 6.63** (3.11, 11.3)	
V_2GaC,6.39 (2.93, 12.84)	**V_2GeC, 6.49** (3.00, 12.25)	**Nb_2AsC, 8.025** (3.31, 11.9)	
Cr_2GaC, 6.81 (2.88, 12.61)	**Cr_2GeC, 6.88** (2.95, 12.08)		
Nb_2GaC, 7.73 (3.13, 13.56)	**Ti_3GeC_2, 5.55** (3.07, 17.76)		
Mo_2GaC, 8.79 (3.01, 13.18)			
Ta_2GaC, 13.05 (3.10, 13.57)			
Ti_2GaN, 5.75 (3.00, 13.3)			
Cr_2GaN, 6.82 (2.875, 12.77)			
V_2GaN, 5.94 (3.00, 13.3)			

Table 7.1 (*Continued*)

	Group		
	IVA	VA	VIA
In	Sn		
Sc_2InC (?)	Ti_2SnC, 6.36 (3.163, 13.679)		
Ti_2InC, 6.2 (3.13, 14.06)	Zr_2SnC, 7.16 (3.3576, 14.57)		
Zr_2InC, 7.1 (3.34, 14.91)	Nb_2SnC, 8.4 (3.241, 13.802)		
Nb_2InC, 8.3 (3.17, 14.37)	Hf_2SnC, 11.8 (3.320, 14.388)		
Hf_2InC, 11.57 (3.30, 14.73)	Hf_2SnN, 7.72 (3.31, 14.3)		
Ti_2InN, 6.54 (3.07, 13.97)			
Zr_2InN, 7.53 (3.27, 14.83)			
Tl	Pb		
Ti_2TlC, 8.63 (3.15, 13.98)	Ti_2PbC, 8.55 (3.20, 13.81)		
Zr_2TlC, 9.17 (3.36, 14.78)	Zr_2PbC, 9.2 (3.38, 14.66)		
Hf_2TlC13.65 (3.32, 14.62)	Hf_2PbC, 12.13 (3.55, 14.46)		
Zr_2TlN, 9.60 (3.3, 14.71)			

$Ti_2AlC_{0.8-x}N_x$, where $x = 0$ to ≈ 0.8, occurs at 1490 °C [41]. Recently, the existence of this solid solution was confirmed; $Ti_3Al(C_{0.5},N_{0.5})_2$ and $Ti_3Al(C_{0.5},N_{0.5})_2$ were also synthesized and characterized [42–44]. Similarly, a continuous solid solution, $Ti_3Si_xGe_{1-x}C_2$, $x = 0$ to 1 exists [45]. On the M-sites the following solids solutions are known to exist: $(Nb,Zr)_2AlC$, $(Ti,V)_2AlC$, $(Ti,Nb)_2AlC$, $(Ti,Cr)_2AlC$, $(Ti,Ta)_2AlC$, $(V,Nb)_2AlC$, $(V,Ta)_2AlC$, $(V,Cr)_2AlC$ [46, 47], $(Ti,Hf)_2InC$ [48] and $(Ti,V)_2SC$ [49]; moreover, this list is by no means exhaustive.

7.3 Elastic Properties

As shown in Table 7.2 and Figure 7.3, the $M_{n+1}AX_n$ phases are, for the most part, elastically quite stiff. When combined with the fact that the densities of some of MAX phases are relatively low (ca. 4.1–5.0 g cm^{-3}; see Table 7.1) their specific stiffness

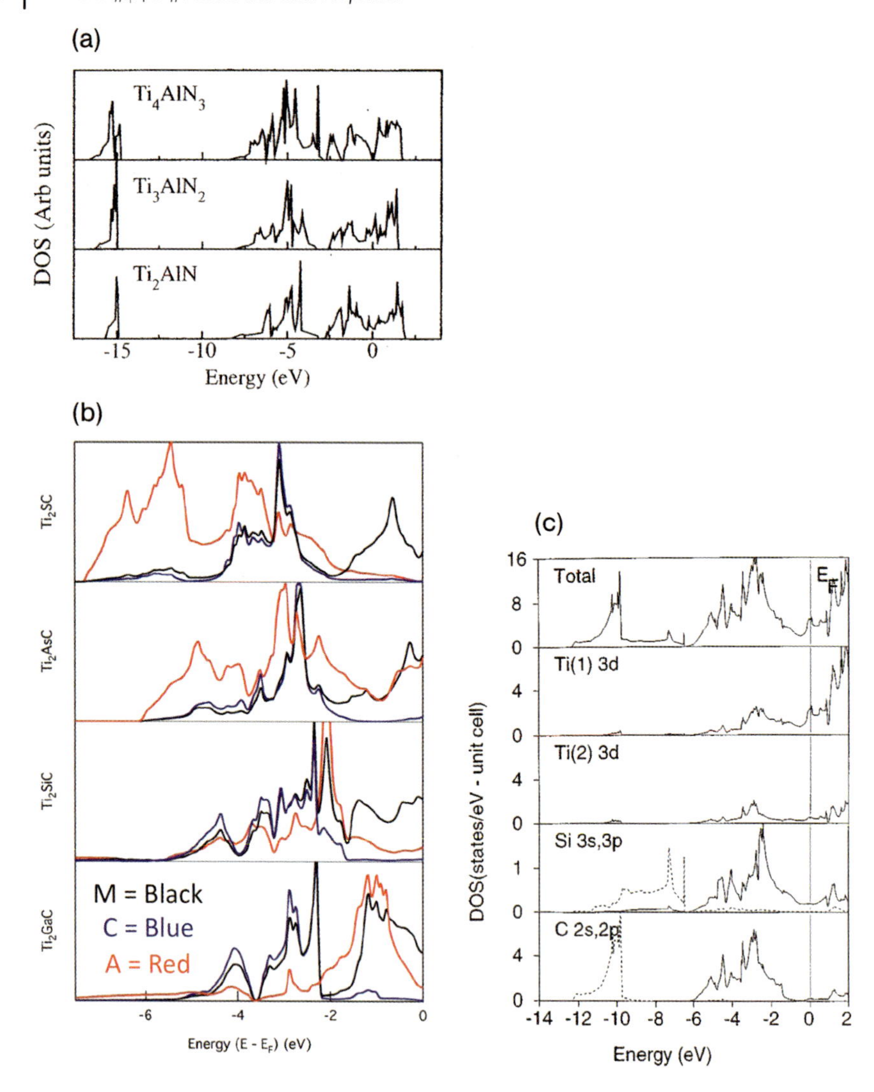

Figure 7.2 Typical total and partial DOS at E_F for select MAX phases. (a) Total DOS for Ti_2AlN, Ti_3AlN_2 (which does not exist) and Ti_4AlN_3 [38]; (b) Partial density of states of Ti, A and C atoms in Ti_2AC compounds with increasing p electron concentrations [40]. The DOS are color-coded as indicated in the bottom panel; (c) Partial and total DOS for Ti_3SiC_2 [30].

values can be high. For example, the specific stiffness of Ti_3SiC_2 is comparable to that of Si_3N_4, and approximately three times that of Ti.

Until the discovery of the MAX phases, the price paid for high specific stiffness values has been a lack, or at least a difficulty, of machinability. It is important, therefore, to note that one of the most characteristic properties of the MAX phases is the ease with which they can be machined, with nothing more sophisticated than

Table 7.2 Young's modulus (*E*), shear modulus (*G*), and Poisson ratio (ν) of select MAX phases. Also listed are the bulk moduli values (B^*), measured directly in an anvil cell, and $B\ddagger$, calculated from the shear and longitudinal sound velocities. The references and values are color-coded; the 312s are highlighted yellow, and 413s gray.

Solid	Density (g cm^{-3})	*G* (GPa)	*E* (GPa)	ν	*B*‡ (GPa)	*B** (GPa)	Reference(s)
Ti_2AlC	4.1	118	277	0.19	144	186	[50][51]
$Ti_2AlC_{0.5}N_{0.5}$	4.2	123	290	0.18			
V_2AlC	4.81	116	235	0.20	152	201	[43]
Cr_2AlC	5.24	102	245	0.20	138	166	[52][53][54]
	5.1	116	288				
Nb_2AlC	6.34	117	286	0.21	165	208	[52][53]
Ta_2AlC	11.46	121	292			251	[55][53]
Ti_3SiC_2	4.52	139	343–339	0.20	190	206	[50][56]
Ti_3GeC_2	5.02	142	340–347	0.19	169	179	[50][57]
$Ti_3(Si,Ge)C_2$	4.35	136.8	322	0.18	166	183 ± 4	[50][58]
Ti_3AlC_2	4.2	124	297	0.20	165	226 ± 3	[50][42]
Ti_3AlCN	4.5	137	330	0.21		219 ± 4	[43][42]
Cr_2GeC	6.88	80	245	0.29	165	182 ± 2	[59][60]
						169 ± 2	[61]
V_2GeC	6.5					165 ± 2	[60]
Ti_2SC	4.62	125	290	0.16	145	191 ± 3	[62][63]
Ti_2SnC	6.36					152 ± 3	[64]
Nb_2SnC	8.4		216			180 ± 5	[65][64]
Zr_2SnC	7.16		178				[65]
Hf_2SnC	14.39		237			169 ± 4	[64][65]
Nb_4AlC_3	6.98	127	306				[66]
β-Ta_4AlC_3	13.2	132	324			261 ± 2	[18][19]
Ti_4AlN_3	4.7	127	310	0.22	185	216	[67][68]
$TiC_{0.96}$	4.93	205	≈500	0.19	272		

a manual hacksaw. Their excellent electrical conductivities (see below) also allows them to be readily electron-discharge machined.

Poisson's ratios for most of the MAX phases hover around 0.2, which is lower than the 0.3 of Ti, and closer to the 0.19 of near-stoichiometric TiC.

In general, the In-, Pb- and Sn-containing MAX phases are less stiff than those composed of lighter A-elements. For example, at 178, 216 and 237 GPa, the Young's moduli (*E*) of Zr_2SnC, Nb_2SnC and Hf_2SnC [65], respectively, are all lower than any of the Al-containing ternaries or Ti_3SiC_2 (Figure 7.3). At 127 GPa, the bulk modulus (*B*) of Zr_2InC [69] is one of the lowest reported for a MAX phase. At the other extreme is Ta_4AlC_3, the *B*-value of which, at 260 GPa, is the highest experimentally reported to date [19].

As shown in Figure 7.3, for the most part, the agreement between the measured and calculated values of *E* and *B* are acceptable, but there are exceptions. The theoretical calculations overestimate the elastic properties of Cr_2AlC although,

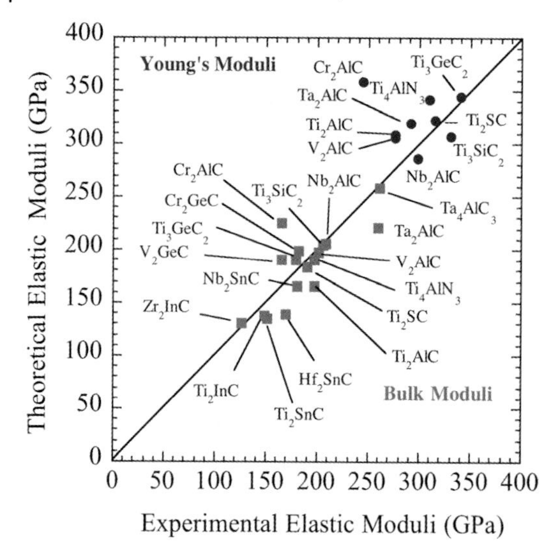

Figure 7.3 Comparison of experimental and theoretical bulk (B) and Young's (E) moduli of select MAX phases.

as discussed below, this may be due to the presence of vacancies in the measured samples. They also appear to underestimate the bulk modulus of Ta_2AlC, but for reasons that are not clear.

Several theoretical reports have shown a one-to-one correspondence between the bulk moduli of the binary MX and ternary MAX compounds – a not too-surprising result given that the latter are comprised of blocks of the former [40, 70, 71].

An important, but subtle, factor influencing the B-values of the MAX phases is their stoichiometry, and more specifically their vacancy concentrations. This effect is best seen in the B-values of Ti_2AlN, where theory and experiment show a *decrease* in lattice parameters as C is substituted for by N. Given that the lattice parameters shrink, it is not surprising that theory predicts that this substitution should increase the B-value, when, in fact, experimentally it decreases with increasing N-content [42]. This paradox is resolved when it is appreciated that B is a strong function of vacancies, and that the addition of N results in the formation of vacancies on the Al and/or N sites. As discussed below, the presence of these defects also influence other properties [43, 44].

Another subtle effect on B is the puckering or corrugation of the basal planes [32]. Results obtained with electron energy loss spectroscopy (EELS), together with *ab initio* calculations, have shown that in the solid solution, TiNbAlC, the basal planes are corrugated. This corrugation leads, in turn, to a larger decease in B – due to a softening along the c-axis – than might otherwise be anticipated [73].

In contradistinction to other layered solids such as graphite and mica (which are elastically quite anisotropic), the MAX phases are mildly so. For example, Holm *et al.* predicted that c_{33} and c_{11} for Ti_3SiC_2 would be almost equal [31], a prediction that was later confirmed experimentally [74]. The same is true of some of the M_2AlC

phases [75, 76]; with a c_{11} of 308 GPa and a c_{33} of 270 GPa, Ti_2AlC is slightly more anisotropic [35]. For Ti_2SC, one of the stiffest 211 phases known to date, c_{11} and c_{33} are, respectively, predicted to be approximately 338 and 348 GPa [77, 78], although values predicted by others were 10% higher [79]. The corresponding experimental values – *viz.* 334 and 336 GPa – are in good agreement with predictions.

The Al-containing MAX phases and Ti_3SiC_2 have another useful attribute, namely that their elastic properties are not a strong function of temperature. For example, at 1273 K the shear and Young's moduli of Ti_3AlC_2 are about 88% of their room-temperature values [50, 67]. In that respect, their resemblance to the MX binaries is notable.

Finally, it should be noted that the Raman modes for the MAX phases have been deciphered and shown to be, for the most part, in excellent agreement with experimental results (Figure 7.4) [80–83]. There are essentially two types of vibration; low-energy ($<300\,cm^{-1}$) shear modes (along the a-direction) involving the A and M atoms; and higher-energy modes involving vibrations along the c-axis. The low-energy modes are a manifestation of the weakness of the M–A bonds in shear *relative* to the M–X bonds. This comment notwithstanding, it is incorrect to conclude from this statement that any of the bonds in the MAX phases are necessarily weak. For the Al-containing MAX phases, for example, in some cases, the M–Al bonds are stronger – at least in the c-direction – than the M–X bonds (see Section 7.5.2).

7.4 Electronic Transport

The electrical conductivities of the MAX phases, like those of their M and MX counterparts, are metallic-like in that the resistivity, ϱ, increases linearly with increasing temperature, T (Figure 7.5a). The temperature dependencies of $(\varrho - \varrho_0)$, where ϱ_0 is the residual resistivity at 0 K, show clearly that $d\varrho/dT$ depends on the transition metal in the MAX phase (Figure 7.5a). The actual room temperature resistivities, ϱ_{RT}, are listed in Table 7.3, together with their residual resistivity ratios (RRRs), defined as $\varrho_{RT}/\varrho_{4K}$, where ϱ_{4K} is the resistivity at 4 K. The RRR values are a measure of the quality of a given material, with higher RRR values indicating less defective solids.

In order to understand the electronic transport of a solid, it is necessary to know its charge carrier densities and mobilities. For most solids, the Hall coefficient (R_H) is used to determine the concentration and sign of the majority charge carriers. Once known, the mobility is determined from the conductivity values, σ. The MAX phases, however, are unlike most other metallic conductors in that their Hall and Seebeck coefficients are quite small – in some cases vanishingly small – and a weak function of temperature [52, 84–87]. Furthermore, the magnetoresistance (MR) ($\Delta\varrho/\varrho = [\varrho(B) - \varrho(B=0)/\varrho(B=0)]$), where B, the applied magnetic field intensity, is positive, parabolic, and nonsaturating. Said otherwise, the MAX phases are compensated conductors, and a two-band conduction model is needed to understand their electronic transport. In the low-field, B, limit of the two-band model, the

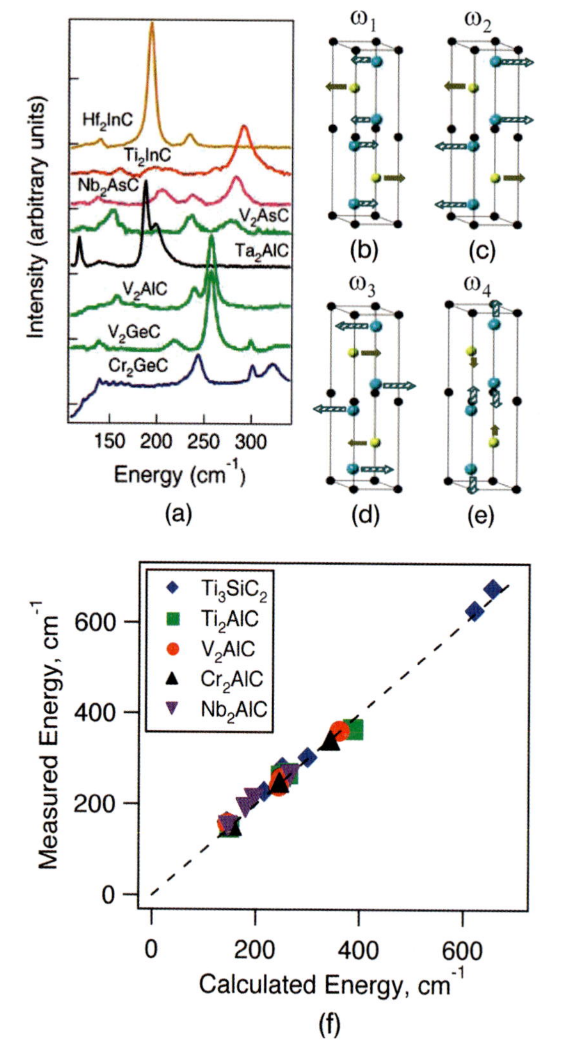

Figure 7.4 (a) Measured 300 K Raman spectra, from Ta_2AlC, V_2AlC, Hf_2InC, Ti_2InC, V_2GeC, Cr_2GeC, V_2AsC, and Nb_2AsC polycrystalline samples; (b–e) Lattice displacements are shown for each of the four Raman-active modes for 211 phases; (f) Comparison of calculated one-phonon, Raman-active Γ-point phonon energies with corresponding measured peaks, collected at 300 K, from eleven 211 phases [83].

following applies:

$$\sigma = \frac{1}{\varrho} = e(n\mu_n + p\mu_p) \tag{1}$$

$$\frac{\Delta\varrho}{\varrho_o} = \alpha B^2 = \frac{\mu_n\mu_p np(\mu_n + \mu_p)^2}{(\mu_n n + \mu_p p)^2} B^2 \tag{2}$$

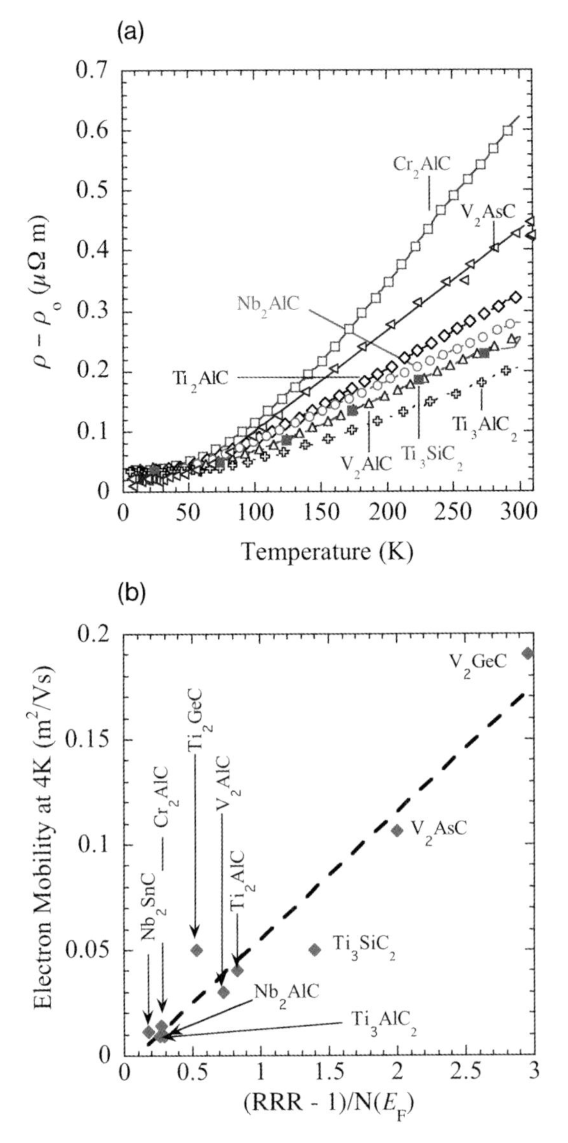

Figure 7.5 Temperature-dependence of $(\varrho - \varrho_o)$ for select MAX phases, where ϱ_o is resistivity at 4 K; (b) Functional dependence of electronic charge mobilities at 4 K on RRR and the density of states at the Fermi level, $N(E_f)$.

$$R_H = \frac{(\mu_p^2 p - \mu_n^2 n)}{e(\mu_p p + \mu_n n)^2} \qquad (3)$$

There are four unknowns: the concentration of electrons and holes, n, p, and their mobilities, μ_n and μ_p, respectively. Given the small R_H and Seebeck coefficient values,

Table 7.3 Summary of electrical transport parameters calculated from the resistivity (ϱ), Hall coefficient (R_H), and magnetoresistance coefficient (α), for select MAX phases. Unless otherwise noted, $\mu_n = \mu_p = \sqrt{\alpha}$ is assumed. Note that this approach can, and does, result in slightly different values than assuming $n = p$. The residual resistivity ratio is listed in column 4.

Composition	T (K)	ϱ ($\mu\Omega \cdot m$)	RRR	R_H ($\times 10^{11}$) ($m^3 C^{-1}$)	α (T^{-2}) ($m^4 V^{-2} s^{-2}$)	μ_n ($m^2 Vs^{-1}$)	μ_p ($m^2 Vs^{-1}$)	n (10^{27} m^{-3})	p (10^{27} m^{-3})	Reference
$Ti_4AlN_{2.9}$	300	2.61	1.1	90	3×10^{-7}	0.00 055	0.00 055	0.8	3.51	[44]
	300	2.61		90 ± 5	3×10^{-7}	—	0.00 034	—	7.0	[85][b]
Ti_3SiC_2	300	0.22	8.33	38	0.29×10^{-4}	0.0054	0.0054	2.0	3.2	[72]
	4	0.03		30	0.0021	0.045	0.045	1.2	1.6	
Ti_3GeC_2	300	0.28	5.6	−18	0.00015	0.002	0.002	1.00	0.80	[65]
	4	0.05		−2.5	0.002	0.045	0.045	1.4	1.3	
Ti_3AlC_2	300	0.353	1.95	−1.2	0.37×10^{-4}	0.0063	0.0063	1.41	1.4	[44]
	4	1.81		1.0	1.1×10^{-4}	0.01	0.01	1.71	1.73	
	300	0.387	—	−1.2	4×10^{-5}	0.0046–0.0042	0.0054–0.003	1.5–1.6	2–4	[52]
Ti_3AlCN	300	0.40	1.5	17.4	6.5×10^{-6}	0.0025	0.0025	2.5	3.5	[44]
	4	0.27		33	0.33×10^{-4}	0.0057	0.0057	1.8	2	
Ti_2AlC	300	0.36	4.9	−27	0.26×10^{-4}	0.0051	0.0051	1.39	1.2	
	4	0.073		−8	4.5×10^{-4}	0.021	0.021	2.1	1.9	
	300	0.36	4.8	−28	2.0×10^{-4}	0.009	0.0082	≈1	≈1	[52][a]
$Ti_2Al(C_{0.5}N_{0.5})$	300	0.36	2.86	45.6	3.5×10^{-5}	0.0059	0.0059	1.1	1.8	[44]
	4	0.126		60	2.2×10^{-4}	0.015	0.015	1.2	2.4	
Ti_2AlN-a	300	0.25	8.5	−3.9	1.7×10^{-4}	0.012	0.012	1.02	1.05	
	4	0.029		6.1	68.7×10^{-4}	0.083	0.083	1.24	1.3	
Ti_2AlN-b	300	0.343	2.8	−7	0.48×10^{-4}	0.0063	0.0063	1.46	1.4	
	4	0.123		16	2×10^{-4}	0.014	0.014	1.6	2.0	
Ti_2SC	300	0.52	2.3	160	2.3×10^{-4}	≈0.015	≈0.015	0.45	0.31	[62]

Cr_2AlC	300	0.74	4.9	−12.3	0.75×10^{-5}	0.0083	0.0083	1.5	1.7	[52]
V_2AlC	300	0.25	7.1	−10	2×10^{-5}	0.0045	0.0045	3	2.5	
Nb_2AlC	300	0.39	2.8	−25	0.9×10^{-5}	0.003	0.003	2.8	2.67	
Ti_2GeC	300	0.30	3.2	27	5×10^{-5}	0.0071	0.0071	1.3	1.6	[88]
	4	0.094		16	7.3×10^{-4}	0.027	0.027	1.1	1.4	

a) Assuming $n = p$.
b) Assuming a single band model.

and the nonsaturating and parabolic MR, it is reasonable to assume either $n \approx p$ or $\mu_n = \mu_p$. With that assumption, it is possible to solve for the four unknowns.

The transport parameters for select MAX phases are summarized in Table 7.3. Based on these results, it is apparent that:

- Most of the room-temperature resistivities fall in the relatively narrow range of 0.2 to 0.7 $\mu\Omega \cdot m$ [1]. One notable exception, discussed below, is that of Ti_4AlN_3.
- For the most part, $n \approx p$ and $\mu_n \approx \mu_p$. The densities of electronic carriers fall into the relatively narrow range of 0.3 to 3×10^{27} m^{-3}. Unless otherwise noted, the values listed in Table 7.3 were calculated assuming $\mu_n \approx \mu_p = \sqrt{\alpha}$. The advantages and limitations of using this approximation are discussed in Ref. [44]. In previous reports, the assumption $n \approx p$ was made instead, which leads to slightly different values for the electronic parameters. Whichever assumption is made, however, does not change the fundamental notion that many of the MAX phases are compensated conductors with $n \approx p$ and $\mu_n \approx \mu_p$, and a two-band model is needed to explain their electronic transport parameters.
- At 4 K the less-defective samples, as measured by the RRR, have higher mobilities. The 4 K mobilities (Figure 7.5b) are also inversely proportional to the density of states at the Fermi level, $N(E_F)$.
- Note that while n and p are *not* related to $N(E_F)$, $d\varrho/dT$ (not shown) is related. The idea that n and p are not related to $N(E_F)$ is *only* valid when that density is substantial. As noted above, the absence of levels at the Fermi level will clearly influence the resistivity values. What is *not* true, however, as several theoretical reports have claimed, is that compounds with higher values of DOS at E_F would be more conductive. As shown here, higher values actually lead to reduced mobilities (Figure 7.5b). It follows that caution must be taken in making any conclusions about conductivity from the values of the DOS at E_F.

To better understand some of the trends observed in Figure 7.5 and Table 7.3, it is important to realize that since the electronic properties of the MAX phases are dominated by the d-d M-orbitals [30, 32, 89, 90], it follows that their behavior should be similar to that of their respective transition metal, M. It is thus useful to briefly summarize what is known about the latter. The conductivities of transition metals are inversely proportional to $N(E_F)$ [91, 92]. The RRR is also important. Accordingly, the electron mobilities at 4 K (μ_{4k}) should be *inversely* proportional to $N(E_F)$ and *directly* proportional to (RRR – 1). That such a correlation exists is shown in Figure 7.5b.

Similarly, $d\varrho/dT$ scales with $N(E_F)$, and the electron/phonon coupling factor [52]. It must be re-emphasized, that no correlation exists between n and p and $N(E_f)$ for the MAX phases.

The ternary Ti_4AlN_3 is somewhat unique, in that not only is it nonstoichiometric (its actual chemistry is $Ti_4AlN_{2.9}$), but it also behaves more as a semi-metal than a metal [85]. Consistent with these results are *ab initio* predictions which show that, if stoichiometric, the Fermi level of this compound falls in a small gap (see Figure 7.2a) [38]. Interestingly, when the resistivity of two nominally identical samples were measured, a significant difference between their room-temperature resistivities was found; these variations were attributed to slight variations in stoichiometry [85]. The high defect concentration clearly leads to significantly reduced mobilities (Table 7.3). The same trend can be seen in Ti_2AlN (Table 7.3), where the 4 K mobility of sample "a" is approximately sevenfold higher than that of a nominally identical sample "b"; the RRR of the former is about sixfold that of the latter. This difference was ascribed to the presence of electron-scattering defects, most probably vacancies [44].

Solid solutions also result in reduced mobilities, but not equally. Substitutions on the A sites appear to have little effect on ϱ [86], whereas substitutions on the X-sites have an effect that is only observed if the concentration of defects – presumably vacancies and displaced atoms – in the end-members are low. Consistent with the fact that the Fermi level is dominated by d-d orbitals of the M-sites, substitution on these, however, can have a more dramatic effect on increasing resistivities above those of the end members [93].

One of the unique properties the Si- and Ge-containing MAX phases is their small and temperature independent Seebeck coefficients [86, 87]. Using *ab initio* calculations, Chaput *et al.* [94] calculated the thermopower to be negative along the c axis and positive in the basal planes. The small value experimentally observed was thus ascribed to a compensation between the thermopowers of the two nonequivalent crystallographic axes. Yet, while certainly reasonable, this prediction awaits experimental verification on epitaxial thin films or large single crystals.

Interestingly, solids with essentially zero thermopower can, in principle, be used as leads to measure the *absolute* thermopower of other solids. In other words, they could be used as reference materials in thermoelectric measurements.

7.5 Thermal Properties

7.5.1 Thermal Conductivities

Typically the total thermal conductivity of a solid, k_{th}, can be considered to be the sum of the electronic, k_e and phonon, k_{ph}, contributions; that is:

$$k_{th} = k_e + k_{ph} \tag{4}$$

The former can be estimated from the Wiedmann–Franz law:

$$k_e = \frac{L_o T}{\varrho} \tag{5}$$

where L_o is the classic Lorenz number, $2.45 \times 10^{-8}\,W \cdot \Omega\,K^2$. Thus, knowing ϱ as a function of T it is straightforward to calculate k_{th} as a function of temperature. The room-temperature values of k_{th}, k_e and k_{ph} of about 20 MAX phases, including some solid solutions, are summarized in Table 7.4.

The MAX phases are good room-temperature thermal conductors. Figure 7.6a plots the values of k_{th} as a function of temperature for a select number of

Table 7.4 Summary of total (k_{th}), phonon (k_{ph}) and electron (k_e) thermal conductivities ($W\,mK^{-1}$) for a number of MAX phases, near-stoichiometric TiC_x and NbC_x. Color-coding as Tables 7.1 and 7.2.

Compound		k_{th}	k_e	k_{ph} (%)	Reference
Ti_3SiC_2		34	33 (97)	≈1 (3)	[97]
		40	36.2 (90)	3.8 (10)	[86]
$Ti_3Si_{0.5}Ge_{0.5}C_2$		39	38 (97)	1 (3)	[86]
Ti_3GeC_2		38	38 (100)	—	[86]
Ti_3AlC_2		40	21 (52)	19 (42)	[44]
Ti_2AlC		33	20.5 (62)	12.5 (38)	[52]
		46	20 (43)	26 (57)	[98]
V_2AlC		48	29 (61)	19 (39)	[52]
Cr_2AlC		23	9 (39)	14 (61)	[52]
Nb_2AlC		29	19 (66)	10 (34)	[52]
		23	23 (100)	—	[93]
TiNbAlC		16.6	9.4 (56)	7.2 (43)	[93]
Nb_2SnC		17.5	17.5 (100)	—	[99]
Ti_2InC		≈26.5	26.5 (100)	—	[48]
TiHfInC		≈20	20 (100)	—	
Hf_2InC		≈26.5	26.5 (100)	—	
Ta_2AlC		28.4	28.3 (100)	—	[55]
Ti_2SC		60	30 (50)	30 (50)	[62]
Ti_2AlN	a	60	29 (49)	31 (51)	[44]
	b	34	23 (67)	11 (33)	
Ti_3AlCN		53.4	18.3 (34)	36 (66)	[44]
$Ti_2AlC_{0.5}N_{0.5}$		29.3	16.9 (58)	12.4 (42)	[44]
Nb_4AlC_3		13.5	9.6 (70)	3.9 (30)	[66]
$Ti_4AlN_{2.9}$		12	2.8 (23)	9.2 (77)	[100]
Ta_4AlC_3		38.4	19 (50)	19 (50)	[18]
TiC_x		33.5	12 (36)	21.5 (64)	[101]
$TiC_{0.96}$		14.4	7.35 (50)	7.05 (50)	[102]
NbC_x		14	21[a]		[27]

a) Implies $L_0 < 2.45 \times 10^{-8}\,W\Omega\,K^{-2}$.

Al-containing MAX phases and Ti_3SiC_2; the corresponding values of k_{ph} are plotted in Figure 7.6b.

The room-temperature results are listed in Table 7.4, together with the corresponding parameters for near-stoichiometric TiC, TiC_x and NbC_x for comparison. From these results it is reasonable to conclude that:

- The MAX phases are good thermal conductors because they are good electrical conductors.
- In general, for the non S- or Al-containing MAX phases, $k_{ph} \ll k_e$ (Table 7.4).
- The S- and Al-containing MAX phases are decent phonon conductors (Table 7.4). At $36\,W\,mK^{-1}$, the k_{ph} of Ti_3AlCN is the highest reported for a MAX phase to date.
- With a few notable exceptions (as discussed below), a correlation exists between the quality of the crystal, as measured by RRR, and k_{ph} (Figure 7.6c).

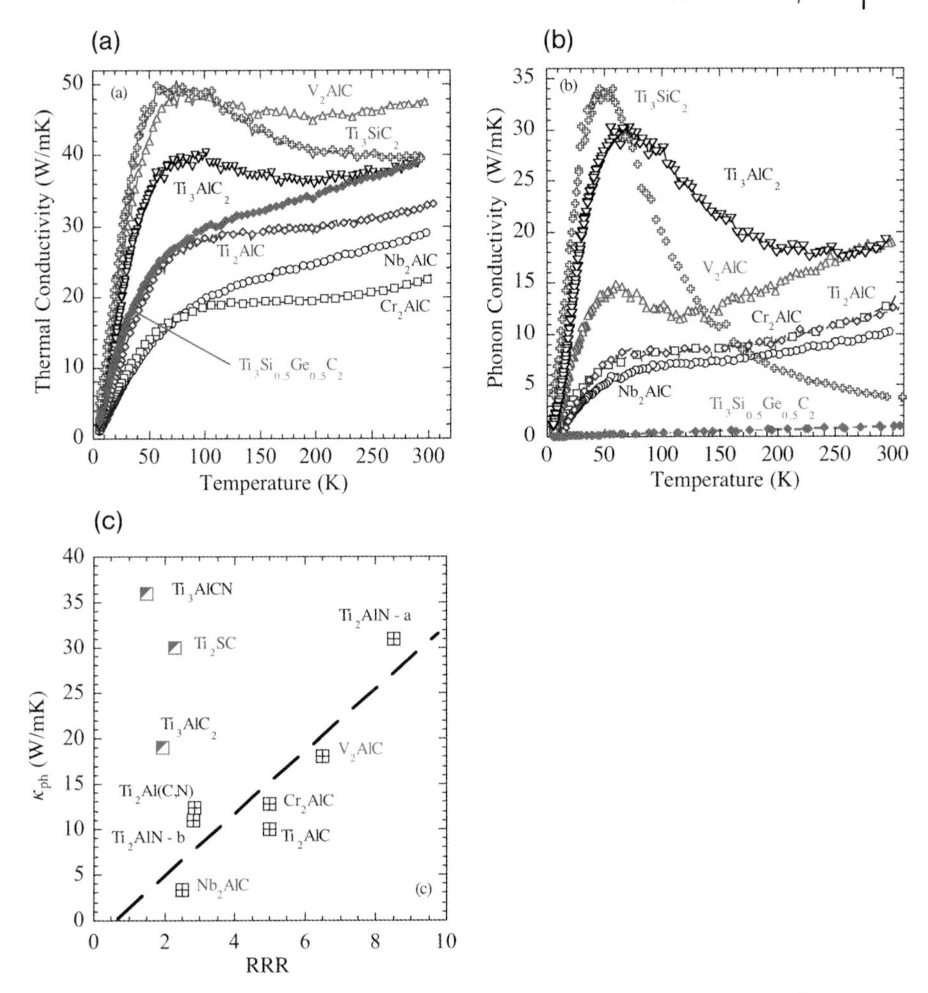

Figure 7.6 Temperature-dependence of thermal conductivities of select MAX phases. (a) k_{th}; (b) k_{ph}; (c) Effect of RRR on k_{ph}.

Stiff, lightweight solids with high Debye temperatures are typically good phonon conductors. Given the rigidity of some MAX phases (Table 7.2), the fact that k_{ph} is suppressed is somewhat surprising. As discussed in more detail elsewhere [1], this result can be attributed to the scattering efficiency of the A-group atoms that tend to play the role of a "rattler" in these structures. Rattlers are atoms that vibrate about their equilibrium position more than other atoms [95, 96].

Probably the most convincing evidence for the rattler conjecture is to compare k_{ph} for the isostructural compounds, Ti_3SiC_2 and Ti_3AlC_2. Based on their RRR values (which is higher for Ti_3SiC_2) and the peak heights of the curves shown in Figure 7.6a and b, it is reasonable to conclude that the Ti_3SiC_2 sample is less defective than its Ti_3AlC_2 counterpart. Yet, at room temperature, k_{ph} for the latter is about fivefold higher than for the former (Table 7.4). Given the similarities of their elastic properties, molecular weights and Debye temperatures, it is clear that Si is a much more potent phonon scatterer than Al. An analysis of high-temperature (up to 1200 °C) neutron diffraction spectra has shown that Si is indeed a rattler in Ti_3SiC_2 [1, 97]. The same is presumably true of most MAX phases with A-group elements with atomic numbers greater than that of Zn (e.g., compare the k_{ph} of Ti_2InC with that of Ti_2AlC). Along the same lines, k_{ph} for Ti_3GeC_2 is essentially zero (Table 7.4), which suggests that Ge is even more of a rattler than Si.

The situation for Al is more ambiguous. The results for $Ti_4AlN_{2.9}$, over the same temperature range, have shown that while the vibrational amplitudes of the Al atoms are greater than those of Ti or N [100], the differences are not as large as in the case of Si. In general, Al is better bound, and is thus less of a rattler, which partially explains why k_{ph} is not negligible in these phases.

Another exception to this rule is Ti_2SC which, because of its low c/a ratio (Table 7.1), was anticipated to have exceptionally strong Ti–S bonds [1]. Consequently, and despite a low RRR value, at 30 W mK^{-1} its k_{ph} is one of the highest reported to date (Table 7.4), the only higher value being that of Ti_3AlCN [43]. It should be noted that Ti_2SC, Ti_3AlCN, and possibly also Ti_3AlC_2, are apparent outliers in Figure 7.6c; the main reason for this is believed to be the strengths of the M–A bonds in these phases [44, 62].

Similar to the binary MX compounds [101, 102], k_{ph} in the MAX phases are sensitive to the presence of vacancies. A good example can be found by examining the k_{ph} of two Ti_2AlN samples – Ti_2AlN "a" and "b" in Table 7.4 – with RRR values of 8.5 and 2.8; the corresponding k_{ph} values are 31 and 11 W mK^{-1} [43]. The same is true of $Ti_4AlN_{2.9}$; despite it being quite stiff, the presence of vacancies – presumably on the N-sites – resulted in quite low RRR and k_{ph} values (Table 7.4). In other words, k_{ph} is a function of the quality of the crystal (Figure 7.6c). At the present time, however, it is not clear on which sublattice these phonon-scattering vacancies reside.

7.5.2
Thermal Expansion

The thermal expansion coefficients (TECs) of the MAX phases fall in the range of ~5 to 13 × 10^{-6} K^{-1} (Table 7.5). With two exceptions, Cr_2GeC and Nb_2AsC, the agreement between the TCE values measured dilatometrically and those measured using high-temperature diffraction (both X-ray and neutron) are reasonable. However, why Cr_2GeC and Nb_2AsC are such outliers is not clear at the present time.

A correlation exists between the TECs of the ternaries and the corresponding MX binaries [1, 65]. For example, the TECs of the Hf-containing MAX phases are lower than those containing Ti, which in turn are lower than Cr_2AlC. For comparison, the TECs of HfC, TiC and Cr_3C_2 are 6.6, 7.4 and 10.5 × 10^{-6} K^{-1}, respectively [27].

The anisotropies in thermal expansions along the a and c directions for the 211 phases were found to be a function of the A-group element (Figure 7.7a) [77]. For example, the highest anisotropies were observed for the As-containing ternaries. Interestingly, the anisotropies in thermal expansion were also found to correlate with c_{13} of the 211 phases (Figure 7.7b).

Table 7.5 Summary of TECs for various MAX phases determined from both high-temperature diffraction and dilatometry. Numbers in parentheses are estimated standard deviations in the last significant figure of the refined parameter.

Compound	α_a (10^{-6} °C^{-1})	α_c (10^{-6} °C^{-1})	Anisotropy (α_c/α_a)	α_{av} (10^{-6} °C^{-1})	α_{dila} (10^{-6} °C^{-1})	Reference(s)
Ti_3SiC_2	8.9(1)	10.0(2)	1.12(2)	9.3(2)	9.1(5)	[77]
	8.4(1)††	9.3(10)††	1.12	8.6(4)††	9.1(2)	[103]
	8.6(1)†	9.7(1)†	1.13	9.1(2)†	9.1(2)	[97]
Ti_3AlCN	6.0(2)	11.3(2)	1.89(2)	7.8(2)	7.5(5)	[77]
Ti_3GeC_2	8.1(2)	9.7(2)	1.20(3)	8.6(2)	7.8(2)	[77]
$Ti_3Si_{0.25}Ge_{0.75}C_2$	8.8(6)	11.1(3)	1.27(6)	9.6(5)		[77]
Ti_3AlC_2	8.3(1)	11.1(1)	1.33(1)	9.2(1)		[77]
					9.0(2)	[104]
Ti_2AlCN	8.4(1)	8.8(1)	1.05(2)	8.5(1)	7.9(5)	[77]
Ti_2AlN	10.6(2)	9.75(2)	0.92(3)	10.3(2)		[77]
	8.6(2)††	7.0(5)††	0.82	8.1(5)††		[1]
					8.8(2)	[98]
Ti_2AlC	7.1(3)††	10.0(5)††		8.1(5)††	8.8(2)	[1]
Ti_2SC	8.6(1)	8.7(2)	1.01(2)	8.7(1)	9.3(6)	[77]
	8.5(5)††	8.8(2)††	1.04	8.6(6)††		[105]
V_2AlC	9.1(2)	10.0(7)	1.10(16)	9.4(10)	9.4(5)	[77]
	9.3(5)††	9.5(4)††	1.0	9.4(5)††		[106]
V_2GeC	6.9(1)	15.8(3)	2.27(1)	9.9 (2)	9.4(6)	[77]
V_2AsC	7.2(1)	14.0(1)	1.92(1)	9.5(1)		[77]
Cr_2GeC	12.9(1)	17.6(2)	1.37(1)	14.5(2)	9.5(5)	[77]
Cr_2AlC	12.8(3)	12.1(1)	0.94(3)	12.6(2)	12.8(5)	[77]
					13.3, 13	[107], [108]
Nb_2SnC	6.6(4)	14.5(2)	2.17(2)	9.3(3)		[77]
					7.8(2)	[65]
Nb_2AlC	8.8(2)	6.8(3)	0.78(6)	8.1(2)		[77]
					7.5(2), 8.7(2)	[1], [93]
Nb_2AsC	2.9(1)	10.6(1)	2.57(1)	5.5(1)	7.3(5)	[77]
Hf_2InC	7.2(1)	7.6(2)	1.05(3)	7.3(2)		[77]
					7.6(2)	[48]
Ti_2InC					9.5	[48]
Ti_2SnC					10	[65]
Hf_2SnC					8.1	[65]
Zr_2SnC					8.3	[65]
Zr_2PbC					8.2	[65]
Nb_2SnC					7.8	[65]
Hf_2PbC					8.3	[65]
Ta_2AlC	9.2(2)	6.4(2)	1.43(3)	8.3(2)	7.2(6)	[77]
					8	[55]
Ti_4AlN_3	8.3(2)	8.3(9)	1.00(1)	8.3(5)		[77]
	9.6(1)	8.8(1)	0.92	9.4(1)†	9.7(2)	[109]
Nb_4AlC_3					7.2	[66]
Ta_4AlC_3					8.2 ± 0.3	[18]

† Neutron diffraction; †† X-ray diffraction

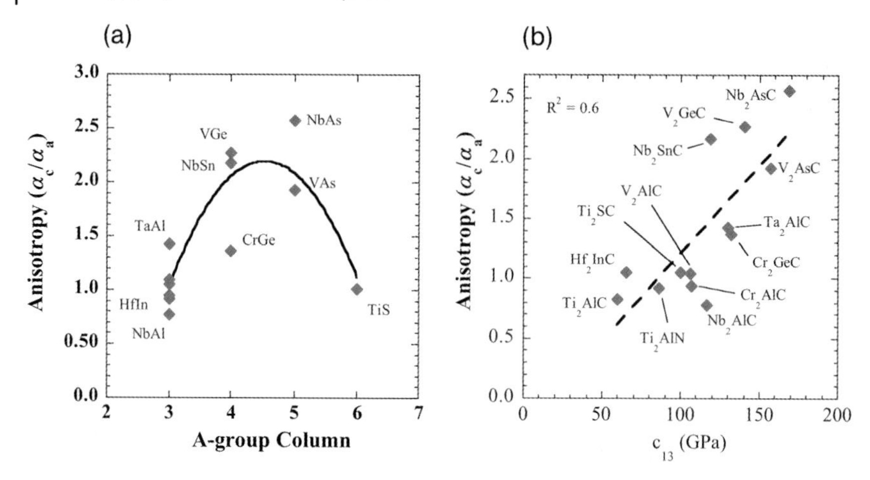

Figure 7.7 Correlation between thermal expansion anisotropy, as measured by the ratio of the thermal expansion along the a-axis, α_a, to that along the c-axis, α_c and (a) A-group element; (b) c_{13}.

7.5.3
Thermal Stability

The MAX phases do not melt congruently, but rather decompose peritectically according to the following reaction (I):

$$M_{n+1}AX_n \Rightarrow M_{n+1}X_n + A \tag{I}$$

Given the chemical stability of the $M_{n+1}X_n$ blocks, and the fact that the A-layers are relatively loosely held, this result is not too surprising. The decomposition temperatures vary over a wide range; from ~850 °C for Cr_2GaN [110] to above 2300 °C for Ti_3SiC_2 [111]. The decomposition temperatures of the Sn-containing ternaries range from 1200 to 1400 °C [65].

It is important to note here that the stability of the MAX phases, for the most part, is *kinetic* in origin, since in most gaseous environments the activity of the A-group element is negligible. Under those circumstances the free energy change for reaction I will always be *negative*. It is also for this reason that the decomposition temperatures are a function of many variables, the most important of which is oxygen contamination and other impurities [112].

7.5.4
Chemical Reactivity and Oxidation Resistance

The $M_{n+1}X_n$ layers are chemically quite stable. By comparison, because the A-group atoms are relatively weakly bound, they are the most reactive species. For example, heating Ti_3SiC_2 in a C-rich atmosphere results in the loss of Si and the formation of TiC_x [113]. When the same compound is placed in molted cryolite [114], or molten

Al [115] essentially the same reaction occurs, namely that the Si escapes and TiC_x forms.

In some cases, such as Ti_2InC, simply heating in vacuum at ~800 °C results in a loss of the A-group element and the formation of TiC_x [48].

Given their excellent high-temperature mechanical properties (see below), some of the MAX phases are currently being considered for a number of high-temperature applications, both structural and nonstructural. To be used in air, however, their oxidation resistance is of paramount importance. Early, short-term oxidation results on Ti_3SiC_3 suggested that it was oxidation-resistant to temperatures as high as 1400 °C [116]. However, a more recent long-term oxidation study [117] on the same compound indicated that, as with other rutile-forming MAX-phases [118], the oxidation kinetics start out parabolic, but then switch to linear. Based on this study, it now appears that the highest temperature at which Ti_3SiC_3 can be used continuously in air is ~900 °C [117].

The most promising MAX phase to date, with superb oxidation resistance, is Ti_2AlC [119]. After 10 000 cycles from 1350 °C to room temperature, a thin, adherent protective 15 μm-thick layer was found [119]. The formation of Al_2O_3 is key to high-temperature oxidation protection. The formation of Al_2O_3 is another example of the reactivity of the A-group element *vis-à-vis* the $M_{n+1}X_n$ blocks. It should be noted that alumina forms despite the fact that the Al concentration is half that of Ti, another reactive metal. Wang and Zhou [120] also reported the formation of alumina layers in Ti_3AlC_2, where the Al concentration is one-third that of Ti.

The oxidation resistance of Cr_2AlC is also quite good [108, 121–123]. At 1000 and 1100 °C, the oxidation resistance is excellent because of the formation of a thin Al_2O_3 oxide layer, with a narrow Cr_7C_3 underlayer. At 1200 and 1300 °C, an outer (Al_2O_3, Cr_2O_3) mixed-oxide layer, an intermediate Cr_2O_3 oxide layer, an inner Al_2O_3 oxide layer, and a Cr_7C_3 underlayer formed on the surface. At 1200 °C, scale cracking and spalling was observed, and at 1300 °C the cyclic oxidation resistance deteriorated owing to the formation of voids and scale spallation [121]. It follows that, despite its excellent oxidation resistance, it is unlikely that Cr_2AlC can be used at temperatures much higher than 1100 °C or even 1000 °C, because of this propensity for spallation that can be traced to the relatively high thermal expansion of this compound (Table 7.5).

In general, the oxidation of the MAX phases occurs according to reaction (II):

$$M_{n+1}AX_n + b\,O_2 = (n+1)\,M\,O_{x/n+1} + AO_y + X_nO_{2b\text{-}x\text{-}y} \qquad \text{(II)}$$

The oxidation of Ti_3SiC_2, for example, results in the formation of an outer pure rutile (TiO_2) layer and an inner layer comprised of rutile and SiO_2. Even in the case of $Ti_{n+1}AlC$, the formation of a continuous alumina layer, is a function of the purity of the samples. Impure samples, or those with high contents of TiC, tend to form Al_2O_3 and rutile, rather than pure Al_2O_3 layers; Ti_2InC forms TiO_2 and In_2O_3, the Sn-containing ternaries, SnO_2 [124]; Ti_2SC, forms TiO_2 [105], Ta_2AlC forms Ta_2O_5, $TaAlO_4$ and an X-ray diffraction (XRD) amorphous phase [125], and so on.

7.6
Mechanical Properties

7.6.1
Introduction

Given the close structural and chemical similarities between the MAX and their corresponding MX phases, it is not surprising (as discussed above) that these two classes of compounds share some common attributes and properties. For example, they are both metal-like conductors dominated by d-d bonding; their phonon conductivities are both susceptible to the presence of vacancies; and the TCEs of the ternaries track those of the binaries, to name a few.

However, in sharp contradistinction, the mechanical properties of the MAX phases cannot be more different than those of their binary cousins. The mechanical properties of the MAX phases are dominated by the fact that basal-plane dislocations multiply and are mobile at temperatures as low as 77 K and higher. The presence of basal slip is thus crucial to understanding their response to stress. This is true despite the fact that the number of independent slip systems is less than the five needed for ductility. In typical ceramics at room temperature, the number of independent slip systems is essentially zero. The MAX phases, thus occupy an interesting middle ground, in which in constrained deformation modes, highly oriented microstructures, and/or at higher temperatures they are pseudo-ductile. In unconstrained deformation, and especially in tension at lower temperatures, they behave in a brittle fashion.

As noted above, basal plane – and only basal plane – dislocations are responsible for how the MAX phases respond to stress. There are no credible reports that twins and/or nonbasal dislocations participate in any meaningful way in their deformation. It follows that at all times the number of slip systems active is less than the five are needed for polycrystalline ductility. As will become apparent shortly, most of the present understanding on the deformation of the MAX phases is based on early work carried out on Ti_3SiC_2, which is the most extensively MAX phase studied and best understood to date. However, there is little doubt – as confirmed by more recent studies – that what applies to Ti_3SiC_2 also applies to other MAX phases.

In order to understand the response of the $M_{n+1}AX_n$ phases to stress, it is imperative to understand the nature of their dislocations and, equally important, how they assemble. The dislocations and their arrangement are described in Section 7.6.2, while the implications of having fewer than the necessary five independent slip systems is discussed in Section 7.6.3. The most important deformation mode of kinking – both incipient and regular – is described in Section 7.6.4, while Section 7.6.5 deals with the room-temperature mechanical properties. The final section is devoted to the mechanical response at elevated temperatures.

7.6.2
Dislocations and their Arrangements

Both, regular and high-resolution transmission electron microscopy (TEM) [16, 126, 127] studies of Ti_3SiC_2 revealed the presence of only perfect dislocations lying in the basal planes, with a Burgers vector equal to the a-lattice parameter or $\mathbf{b} = 1/3 \langle 11\bar{2}0 \rangle$ [126]. Every dislocation is of a mixed nature, with an edge and screw component [126, 127]. It should be noted that since nonbasal dislocations would have Burgers vectors > c (i.e., >11–23 Å), their presence is highly unlikely and even if present – as a result of growth for example – would not play any role in the deformation.

The dislocations arrange themselves either in pile-ups or arrays on the same basal planes, or in walls, as low- or high-angle grain boundaries, normal to the basal planes (Figure 7.8a and b). The walls have both tilt and twist components [126, 127], and to account for both the boundary was interpreted to be composed of parallel, *alternating*, mixed perfect dislocations with two different Burgers vectors lying in the basal plane at an angle of 120° relative to one another (Figure 7.9a). The excess of one type of dislocation accounts for the twist (Figure 7.9b–d) [126, 127].

Dislocation interactions, other than orthogonal, have not been reported to date. As discussed later, this fact has far-reaching ramifications in that substantial deformation can now occur without work hardening in the classic sense.

7.6.3
Plastic Anisotropy, Internal Stresses, and Deformation Mechanisms

The MAX phases, ice and graphite, and other layered minerals such as mica, are plastically anisotropic. This plastic anisotropy, combined with the fact that they lack the five independent slip systems needed for ductility, quickly lead to a very uneven states of stress when a polycrystalline sample is loaded [129]. The glide of basal plane dislocations takes place only in favorably oriented or soft grains, which rapidly transfer the load to hard grains – that is, those not favorably oriented to the applied stress. Needless to say, this leads to high internal stresses.

Most important – and as discussed at some length herein – the plastic anisotropy leads to kink band formation, typical micrographs of which are shown in Figure 7.10. Note that elastic anisotropy is not a necessary requirement for kinking; several of the MAX phases kink, despite being elastically quite isotropic [31].

All MAX phases tested to date pass though a brittle-to-plastic transition. The most important clue as to what is occurring above the transition temperature, and in contradistinction to almost all other crystalline solids, is the fact that the fracture toughness, K_{1C}, actually *decreases* [132]. Such a decrease not only rules out the activation of other slip systems, but also – and as importantly – indicates that the backpressure from dislocation arrays and mobile dislocation walls at the crack tip is reduced at higher temperatures. The most likely mechanism is that delaminations and/or grain boundary decohesion is occurring. In other words, one assumption needed to understand the high-temperature response of the MAX phases is a temperature-dependent grain boundary decohesion and/or delamination stress.

At this juncture, all of the elements needed to understand the response of the MAX phases to stress are in place.

As noted above and discussed below, the key micro-mechanism in the deformation of the MAX phases is the kink band (KB). The KBs in crystalline solids were first observed in Cd single crystals loaded parallel to their basal planes by Orowan, who

(a)

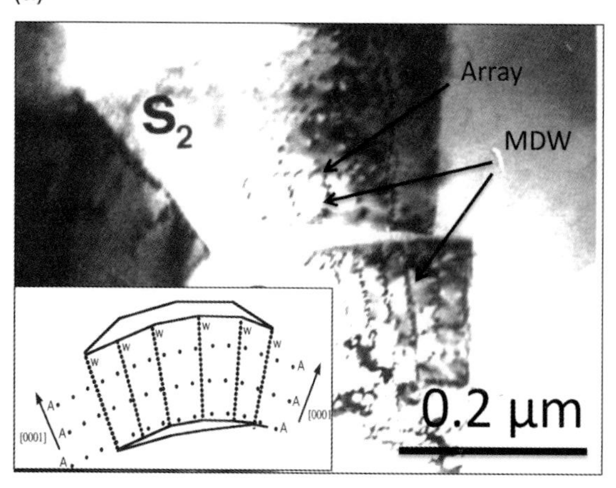

(b)

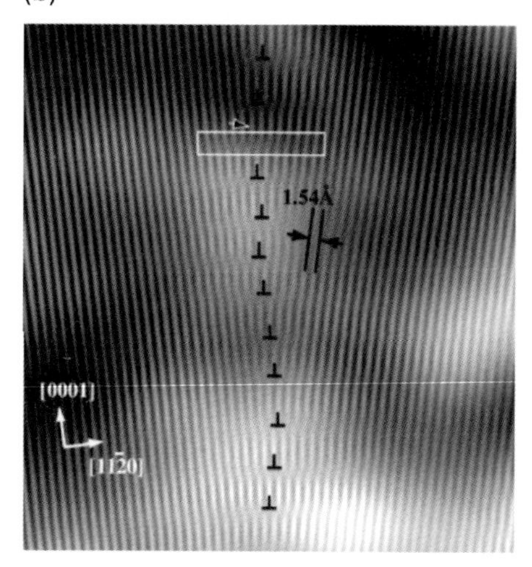

Figure 7.8 (a) Bright-field image of a bent region delaminated in three slices (only one labeled here as S_2 is shown) and containing separated walls and arrays. The inset shows a schematic of S_2, showing walls (W) and arrays (A) [128]. Herein, such separated walls, because they are mobile are termed mobile dislocation walls (MDWs); (b) Bright-field image of a low-angle boundary or MDW-filtered in 11$\bar{2}$1 reflection using fast Fourier transform. An example of the Burgers circuit drawn around one of the dislocations is shown by the rectangle; the termination of the extra plane is indicated by the arrow. The locations of all other dislocations are designated by the inverted T symbol [126].

labeled it a new type of deformation mechanism in metals [133]. Later, Hess and Barrett [134] proposed a model to explain KB formation by the regular glide of dislocations. In their model, above a critical value of shear stress a pair of dislocation walls of opposite signs nucleate and move in opposite directions, within the volume that is eventually to become the KB (Figure 7.11c). The end result is two regions of

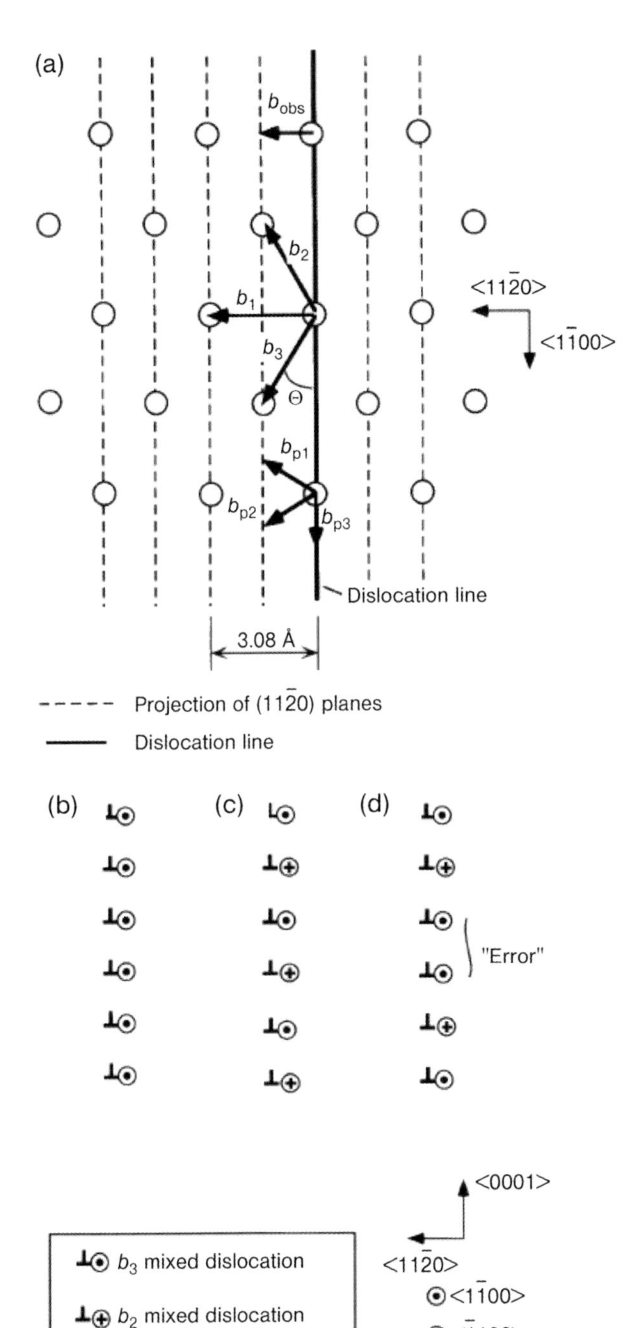

Figure 7.9 (a) Schematic representation of the 0001 projection showing the dislocation line, and projections of observed and possible Burgers vectors. Possible arrangements of mixed dislocations b_2 and b_3 in a dislocation wall: (b) Screw components are all in the same direction; (c) Alternating screw components; (d) Same as panel (c), but showing a stacking error in the screw component which gives rise to twist; panel (b) is unstable, whereas panels (c) and (d) are stable [126].

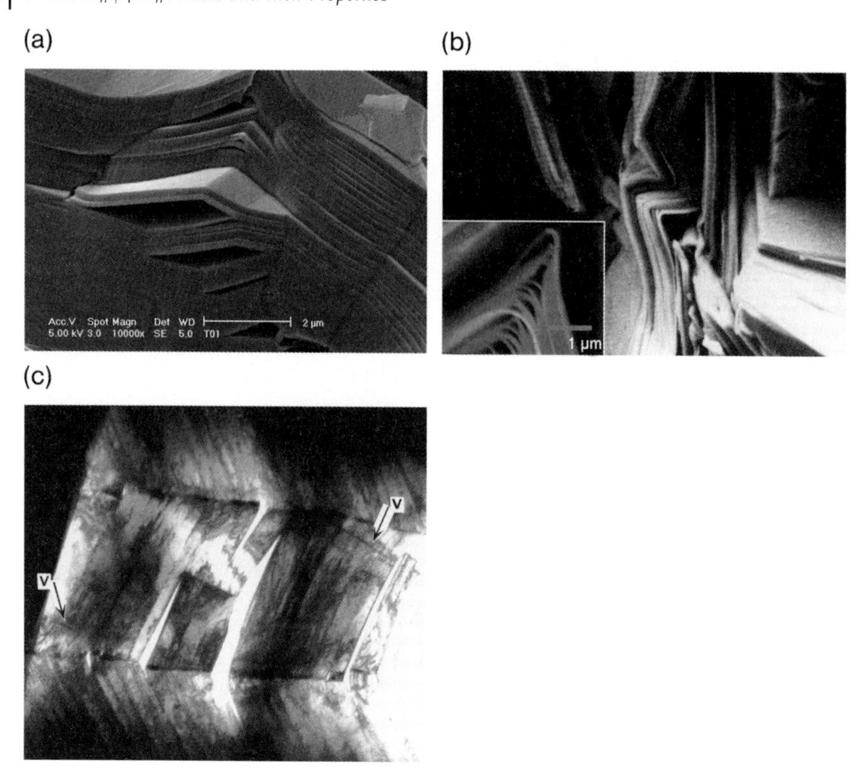

Figure 7.10 Typical examples of the nanolaminate nature of the MAX phases at different lengths scales. (a) SEM images of a porous Ti_3SiC_2 sample that failed in compression [130], where delamination and kink bands and boundaries are clearly seen; (b) Typical kinking and delaminations observed on fractured surfaces of Ti_3SiC_2 [131]; (c) Bright-field image of a kink band with a high misorientation angle ("stove-pipe") containing delamination cracks. Separate dislocation walls, which are generally normal to the basal planes, but are inclined to the kink boundaries, are denoted by V [128].

lattice curvature, separated from each other and from the unkinked crystal by well-defined kink boundaries, BC and DE (Figure 7.11d).

It should be noted in passing that a $[11\bar{2}1]$ twin is nothing but a special case of a KB, where a basal plane dislocation is nucleated every c-lattice parameter [136]. The fundamental difference between a KB and a twin is in the shear angle: for the latter, it is crystallographic, but for the former it is not. What determines the angle of a kink boundary is the number of mobile dislocation walls that end up in that boundary.

The model of Hess and Barrett was qualitative; a few years later Frank and Stroh [135] proposed a more quantitative model in which they considered the energetics of the process that is the starting point for the microscale model (as discussed in the next section) that is currently used to qualitatively and quantitatively explain the typical response of the MAX phases to cyclic compressive and tensile stresses at room temperature (Figure 7.12). In Figure 7.12a are plotted typical cyclic compressive stress–strain curves for Ti_3SiC_2 with two different grain sizes. Also

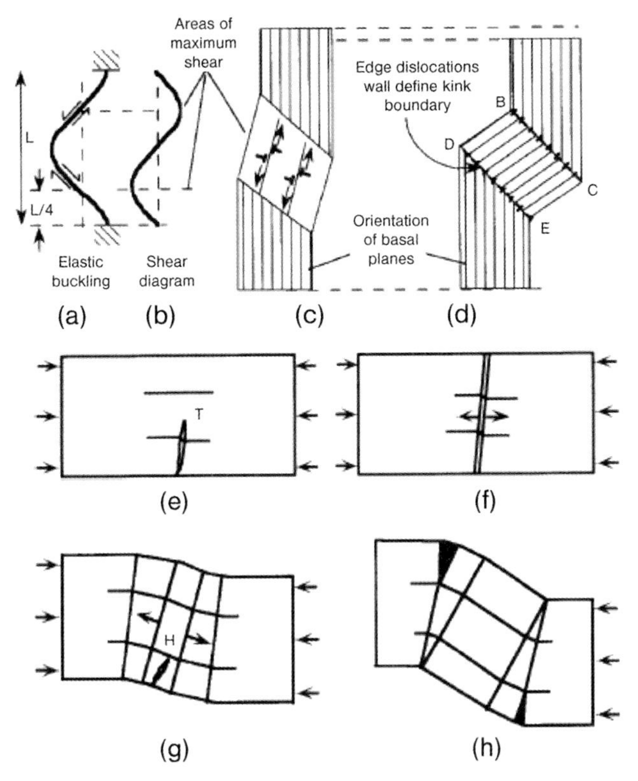

Figure 7.11 Scheme of kink-band formation. (a–d) According to Ref. [134]; (e–h) According to Frank and Stroh [135]. (a) Elastic buckling; (b) Corresponding shear diagram; (c) Initiation of pairs of dislocations in areas of maximum shear; (d) Kink band and kink boundaries comprised of edge dislocations of one sign, giving rise to the classic stove-pipe shape; (e) Initiation of kink band at tip of narrow kink, *T*; (f) Intersection of *T* with free surface, removes the attractive energy between the walls and allows them to separate and move in opposite directions; (g, h) Repeat of the same process to create more dislocation walls and kink boundaries.

shown in Figure 7.12a are the stress–strain curves for two linear elastic solids, Al and alumina. In Figure 7.12b are plotted typical cyclic compressive stress–strain curves for Ti_2AlC with different densities. It is important to note the nested loops and the presence of a unique loading or unloading trajectory in case of increasing or decreasing loadings, respectively.

7.6.4
Incipient Kink Band Microscale Model [135, 137–140]

The model presented herein is a simplified version of the actual model; a more detailed exposition can be found elsewhere [139, 140]. Frank and Stroh [135] considered an elliptic subcritical KB with a length of 2α and a width of 2β, such that $\alpha > \beta$ (Figure 7.13a), and showed that the remote shear stress (τ_c) or normal

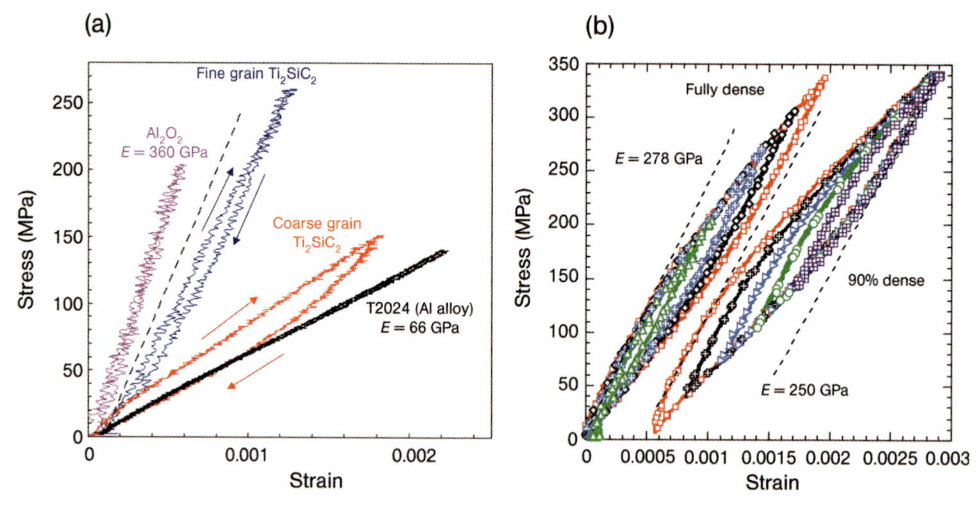

Figure 7.12 Typical compressive cyclic stress–strain curves. (a) Fine-grained and coarse-grained Ti_3SiC_2. The dotted line is a linear elastic response expected for Ti_3SiC_2 had kinking *not* occurred. Also included are the results on Al_2O_3 and Al for comparison [5]; (b) T_2AlC_2 samples with comparable grain sizes but slightly different densities. The 10 vol% porous sample, shifted to the right for clarity, dissipated *more* energy on an absolute basis than its fully dense counterpart. Note the nested loops and the presence of only one loading trajectory on the left, and one unloading trajectory on the right [137].

stress (σ_c) needed to render such a subcritical KB unstable – that is, have it grow spontaneously – is given by [135]:

$$\tau_c \approx \frac{\sigma_c}{2} = \sqrt{\frac{2G^2 b\gamma_c}{2\alpha}} \tag{6}$$

where b is the Burgers vector and γ_c is the critical kinking angle, given by [135]:

$$\gamma_c = \frac{b}{D} \approx \frac{3\sqrt{3}(1-\nu)\tau_{loc}}{2G} \tag{7}$$

where ν is Poisson ratio and D is the distance between dislocations along the c-axis (Figure 7.13a). In studies conducted to date, mostly on the MAX phases, it has been shown that 2α is equal to the grain dimension normal to the direction of easy slip [135, 138, 141].

In the Frank and Stroh analysis, it was assumed that once a KB was nucleated, it would immediately extend to a free surface, thus eliminating the mutual attraction and resulting in two parallel, mobile noninteracting dislocations walls (Figure 7.11f). It is the repetition of this process that results in the generation of new dislocation walls, the coalescence of which forms the kink boundaries (Figure 7.11h).

To explain many of our observations (see below), it was necessary to invoke the idea of an IKB – a KB that does *not* dissociate into mobile dislocation walls (MDWs) (Figure 7.13b). Because of its shape, the IKB shrinks when the load is removed and is thus, by definition, fully and spontaneously reversible. An IKB consists of multiple

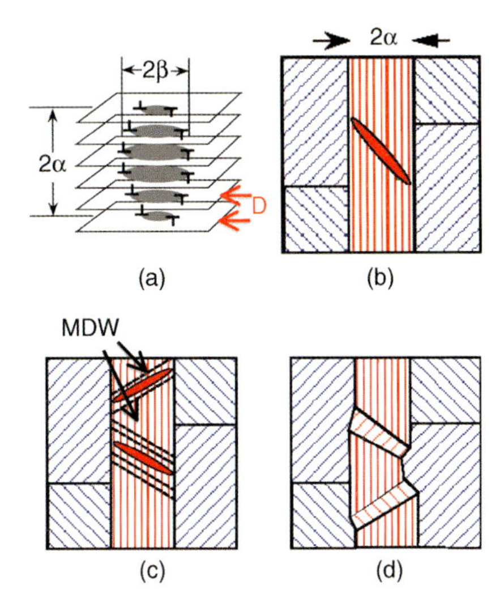

Figure 7.13 (a) Schematic representation of a thin elliptic cylinder with axes 2α and 2β, and $\alpha \gg \beta$. The sides are comprised of two dislocation walls (shown in red) of opposite sign, and uniform spacing D; (b) Formation of an IKB in a hard (red) grains adjacent to soft (blue) grains. The lines in the grains denote basal planes. At this stage it is fully reversible; (c) Multiple mobile dislocation walls (MDWs) in a large grain. The dotted lines denote walls that have separated from the source and are moving away from it. This only happens at higher temperatures and/or stresses; (d) Same as panel (c), but now the MDWs are all subsumed into the kink boundaries. At this stage, the "effective" grain size is smaller than the original grain size.

parallel dislocation loops (Figure 7.13a). As a first approximation, each loop can be assumed to be a circle with radius, 2β, that is related to τ and 2α by: [135]

$$2\beta \approx \frac{2\alpha(1-\nu)}{G\gamma_c}\tau \tag{8}$$

The formation of an IKB can be divided into two stages, namely *nucleation* and *growth* [139]. The nucleation process is not well understood, but occasionally small plastic strains are required to nucleate them [140]. The present model considers IKB growth only from $2\beta_c$ to 2β; the dislocation segment with radius $2\beta_c$ is presumed to pre-exist, or to be nucleated during pre-straining. The values of $2\beta_c$ are typically estimated from Eq. (6), assuming $\sigma = \sigma_t$, where σ_t is a threshold stress. The latter is in turn obtained experimentally from W_d versus σ^2 plots [140].

It follows that for $\sigma > \sigma_t$, the IKB nuclei grow and the IKBs-induced axial strain resulting from their growth is assumed to be given by [139]:

$$\varepsilon_{IKB} = \frac{\Delta V N_k \gamma_c}{k_2} = \frac{\pi(1-\nu)N_k\alpha^3}{12k_2G^2\gamma_c}(\sigma^2-\sigma_t^2) = m_1(\sigma^2-\sigma_t^2) \tag{9}$$

where m_1 is the coefficient before the term in brackets in the second term, and N_k is the number of IKBs per unit volume. The product $\Delta V N_k$ is thus the volume fraction of the material that is kinked. The factor k_2 relates the volumetric strain due to the IKBs to the axial strain along the loading direction. In general, k_2 is a function of texture; for random-oriented MAX phase microstructures, k_2 is assumed to be 2 [137].

The area enclosed by the hysteretic loops (Figure 7.12) represents the energy dissipated per unit volume per cycle, W_d, resulting from the growth of the IKBs from β_c to β is given by:

$$W_d = 3k_2 \frac{\Omega}{b} m_1 (\sigma^2 - \sigma_t^2) \tag{10}$$

where Ω is the energy dissipated by a dislocation line sweeping a unit area. When combining Eqs (9) and (10), it follows that:

$$W_d = 3k_2 \frac{\Omega}{b} \varepsilon_{IKB} \tag{11}$$

Thus, Ω/b should be proportional, if not equal, to the critical resolved shear stress (CRSS) of an IKB dislocation loop. In the present authors' studies to date, this has been shown repeatedly to be the case [137–139].

Armed with this model, it is now possible to understand and quantify the response of MAX phases to compressive (Figure 7.12), and tensile (not shown, but almost identical to the compressive loops; see for example Ref. [142]) stresses at room temperature. Before delving into the details of these results, it is crucial to appreciate the following. In Figure 7.12b, it is clear that the 10% porous solid dissipates *more* energy per unit volume on an absolute scale per cycle than its fully dense counterpart. This result is probably the strongest evidence to date that the IKB model is valid, for the simple reason that it eliminates all mechanisms, such as dislocation pileups and/or twinning, that scale directly to the volume of the material tested. It is, however, in full agreement with the IKB model in that kinking is a form of plastic instability or buckling, and thus a less-rigid solid is more prone to kinking that a fully dense one [137]. Similar conclusions were reached for porous Ti_3SiC_2 samples, where the enhancements in W_d could be accounted for by a reduction in G [130, 141]. The strong effect of grain size on the shape and size of the reversible loops can be traced back to Eq. (6), and can be explained by equating 2α to the grain dimension along the [0001] direction. In other words, kinking is easier and occurs at a lower threshold stress in a large-grained solid than in its fine-grained counterparts.

According to the model, plots of ε_{NL} versus σ^2, W_d versus σ^2 and W_d versus ε_{NL} should all yield straight lines (not shown), which they do [59, 137, 143]. In all cases, and in agreement with the present model, the correlation coefficients, or R^2 values, are typically >0.99. Of equal importance, the slopes of the lines yield quite reasonable numbers for the values of Ω/b, which are comparable to the only value of CRSS of basal plane dislocations measured in the MAX phases, which is of the order of 35 MPa [144].

In summary, the fully reversible cyclic stress–strain loops are due to the growth and shrinkage of basal plane dislocation loops associated with the IKBs. The values of W_d achieved for the MAX phases are some of the highest ever reported for crystalline solids, and are comparable to those of some woods. It should also be noted that W_d increases as σ^2 such that, the higher the stress, the higher the energy dissipated.

7.6.5
Compression Behavior of Quasi-Single Crystals and Polycrystals

When highly oriented, macrograined (~2 mm diameter) Ti_3SiC_2 samples were tested in compression the response was quite anisotropic [144]. When the basal planes are oriented in such a way that allows for slip (x-direction in inset in Figure 7.14a), the samples yield at ~200 MPa (Figure 7.14a) and the deformation occurs by the formation of a classic shear bands (not shown). By contrast, when the slip planes are parallel to the applied load (z-direction in inset in Figure 7.14a), and deformation by ordinary dislocation glide is not easy the stress–strain curves show clear maxima at stresses between 230 and 290 MPa, followed by a region of strain-softening and, finally, recovery (Figure 7.14a). Here, the deformation occurs by a combination of KB formation at the corners of the tested cubes, delaminations within individual grains, and ultimately shear-band formation (Figure 7.14b). It should be noted that the KB visible on the bottom left-hand side of the cube in Figure 7.14b did *not* result in the total delamination of the outermost grain in which it was initiated. This observation was taken to be compelling microstructural evidence that KBs are potent suppressers of delaminations, and is one important reason why the MAX phases are as damage-tolerant as they are [144].

Not all MAX phases fail suddenly; some, especially coarse-grained 211 phases, exhibit graceful-failure characteristics in that the stress–strain curves are closer to an inverted shallow V, than to a sharp drop at a maximum stress [98, 104]. The failure mode is still shear failure across a plane inclined 45° to the loading axis, but presumably enough ligaments reach across the plane to result in a less sudden loss in load-bearing capability. This tendency increases with increasing grain size and reduced loading rates.

The ultimate compressive stresses of the MAX phases are a function of grain size, with larger grains being weaker. The compressive strengths range from about 300 MPa to close to 2 GPa [18, 55, 145–147]. For any microstructure, the flexural strengths are typically about 50% lower than the compressive strengths. Lastly, and akin to other structural ceramics, the ultimate tensile strengths are the lowest and range between <100 and ~300 MPa [142, 148]. In all cases, larger grains, result in weaker solids.

7.6.6
Hardness and Damage Tolerance

The MAX phases, unlike the MX carbides, are relatively soft and exceedingly damage-tolerant. The Vickers hardness values of polycrystalline MAX phases fall in the range of 2–8 GPa; thus, they are softer than most structural ceramics, but harder than most metals. The low hardness persists – at least in Ti_3SiC_2 – even at temperatures as low as 77 K [149].

(a)

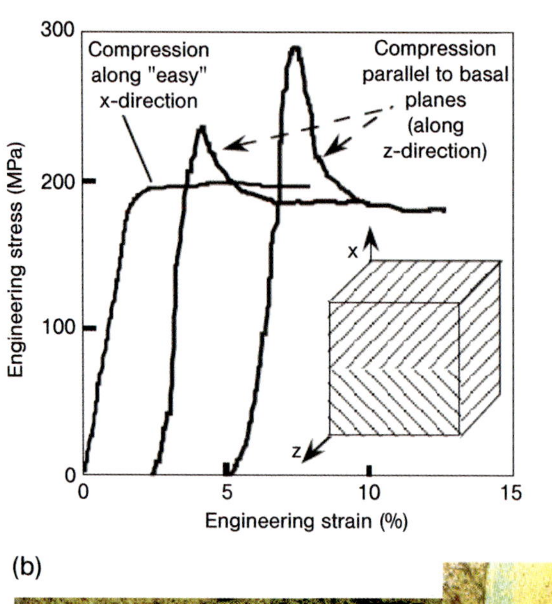

(b)

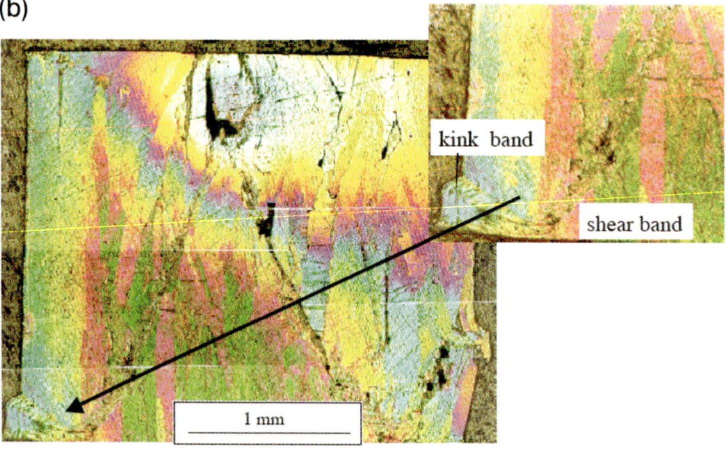

Figure 7.14 (a) Engineering stress–strain curves of 2 mm cubes of highly oriented samples of Ti_3SiC_2. The inset shows a schematic of the cube and basal plane orientations; (b) Optical microscope micrograph of polished and etched sample after deformation parallel to the basal planes. Note the kinking at the corners. The inset shows a higher-magnification view of the lower left corner, emphasizing the kink band and, equally important, the genesis of the shear band that cuts across the cube face [144].

When working with chemical vapor deposition (CVD) single crystals, Nickl *et al.* [150] were the first to note that the hardness of Ti_3SiC_2 was anisotropic, and higher when loaded along the c-direction. This was later confirmed by Goto and Hirai [151], who were the first to show that the hardness was a function of indentation load. Both observations are characteristic of the MAX phases [1, 152]. With decreasing

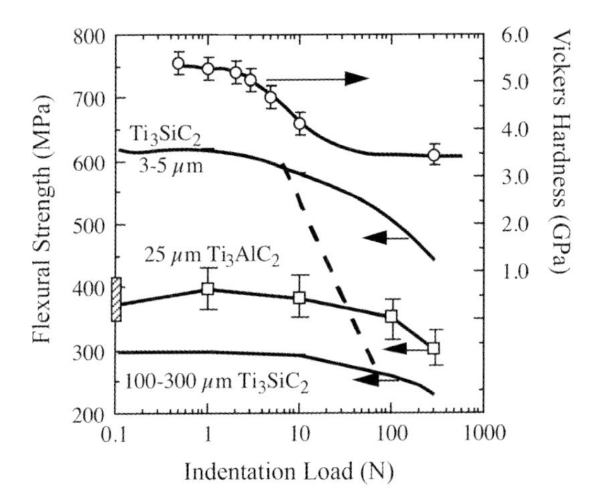

Figure 7.15 Vickers hardness versus log of indentation load (top curve); four-point flexural strength versus log of indentation loads for fine-grained and coarse-grained Ti_3SiC_2 and Ti_3AlC_2 with a grain size of ~25 μm. The inclined dashed line shows the expected behavior for brittle solids [104].

load, the hardness increases (right y-axis in Figure 7.15), and below a certain load it is not measurable, as no trace of the indentations is found. These observations were not understood until recent nanoindentation studies in which the reversible nature of the KBs was elucidated [74].

Not surprisingly, given their plastic anisotropy, the response to nanoindentations is anisotropic; when the basal planes are parallel to the surface, the extent of plastic deformation is higher and the hardness is lower, than if the basal planes are loaded edge-on (compare Figure 7.16a and b) [74, 127]. In the former case, it is easier to form KBs because the top surface is unconstrained (upper inset in Figure 7.16b). The lower inset in Figure 7.16a is a schematic of how the pileup around the indentation shown in the upper inset in Figure 7.16b, probably developed. To form the pileup, both delaminations and kinking are required. A cross-sectional TEM image of an epitaxial Ti_3SiC_2 film is shown in the lower inset of Figure 7.16b, and demonstrates the phenomenon even more directly and convincingly [7]. It should be noted that when the basal planes are loaded edge-on, they tend to delaminate *without* a pileup (upper inset in Figure 7.16a).

The MAX phases are quite damage-tolerant. The damage tolerance of Ti_3SiC_2 [145, 152] and Ti_3AlC_2 [104] are best exemplified in Figure 7.15 (lower three curves), in which the functional dependencies of post-indentation flexural strengths are plotted versus the log of the Vickers indentation loads. The post-indentation flexural strengths are considerably less dependent on the indentation loads than typical brittle ceramics (dashed line in Figure 7.15). Similar to ceramics, however, the damage tolerance of coarse-grained microstructures is superior to those of finer-grained samples. This damage tolerance has since been shown to be an important characteristic of the MAX phases [18, 45, 55, 59, 98, 146, 147, 153].

Typically, Vickers indentations in brittle solids result in sharp cracks that extend from corners of the indents and result in sharp reductions in strength. As first

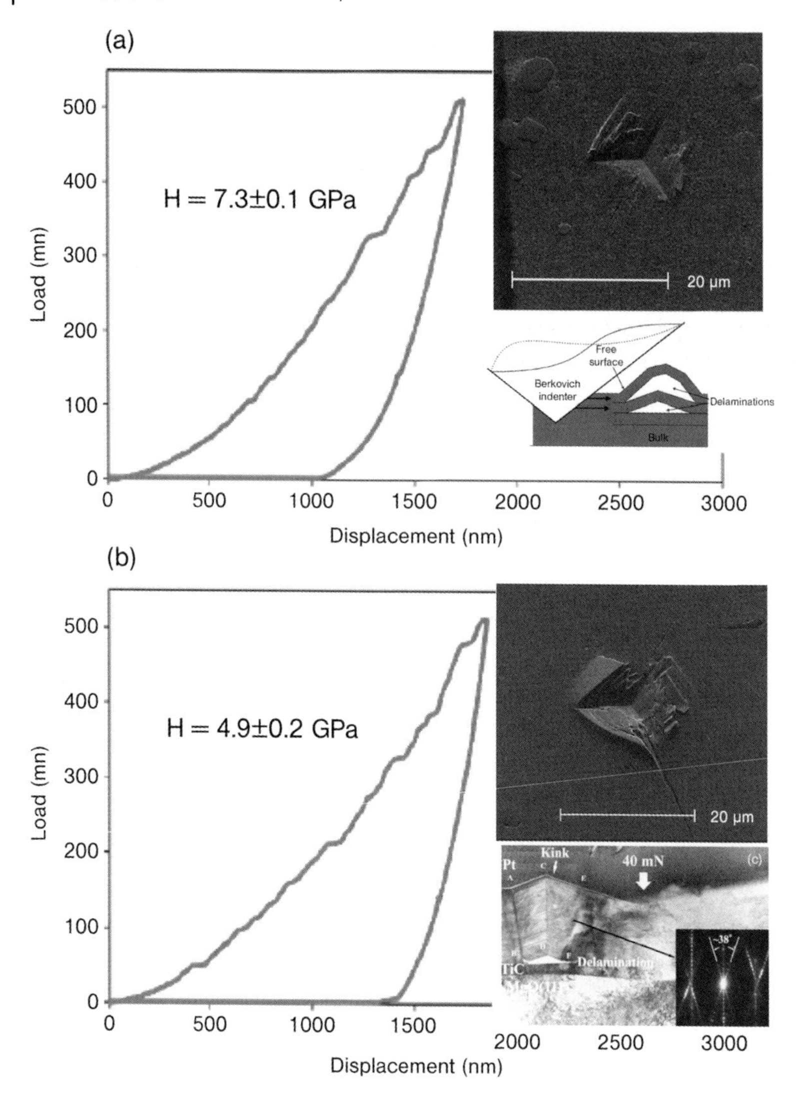

Figure 7.16 (a) Load-displacement curve of an indentation in a Ti_3SiC_2 grain with the basal plane (a) perpendicular and (b) parallel to the surface [127]. The upper inset is SEM image of the indentation in a Ti_3SiC_2 grain with the basal plane perpendicular to the surface. The lower inset in a is a schematic of delamination and kink-band formation at the surface for Ti_3SiC_2 with the basal plane parallel to the surface caused by the nanoindentations [127]. The upper inset in panel (b) is a SEM image of the indentation in a Ti_3SiC_2 grain with the basal plane parallel to the surface. The lower inset is a cross-sectional TEM image of an indentation in a thin epitaxial Ti_3SiC_2 film, showing kink band formation and delaminations [7].

reported by Pampuch *et al.* [154, 155], it is quite difficult to induce cracks from the corners of Vickers indentations in most MAX phases. Instead of the formation of cracks, delaminations, the kinking of individual grains, grain push-outs and pull-outs are observed in the area around indentations [104, 146, 152]. In short, the main reason for this damage tolerance is the ability of the MAX phases to contain and confine the extent of damage to a small area around the indentations by plastic deformation.

The importance of such high damage tolerance cannot be overemphasized, as it implies that the MAX phases are much more tolerant to processing and service flaws that typically are quite detrimental to the mechanical properties of brittle solids. This, in turn, should also greatly increase manufacturing yields, as the need for full density is relaxed.

7.6.7
Thermal Shock Resistance

The MAX phases are also quite thermal shock-resistant [4]. The response of Ti_3SiC_2 to thermal shock depends on the grain size [145]; the post-quench flexural strengths of coarse-grained samples are not a function of quench temperature and actually become slightly stronger when quenched from 1400 °C (not shown). The response of

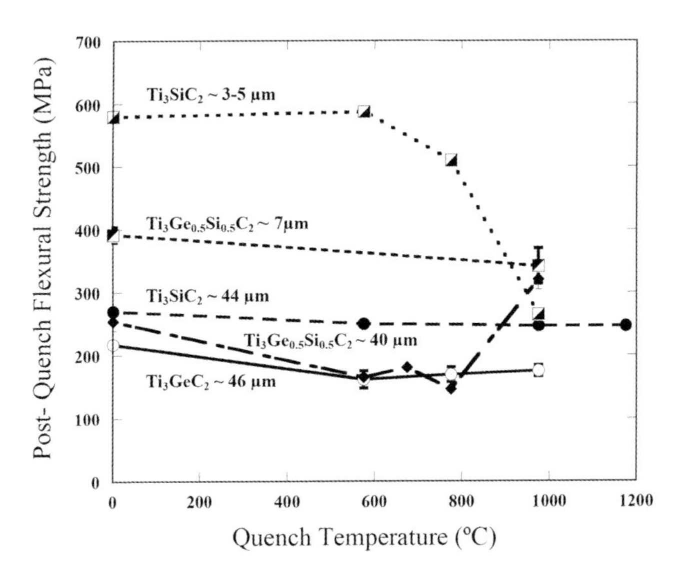

Figure 7.17 Post-quench four-point flexural strength versus quench temperature of coarse-grained Ti_3GeC_2 samples and fine- and coarse-grained $Ti_3Si_{0.5}Ge_{0.5}C_2$ and Ti_3SiC_2 samples. The numbers in the figure denote the average grain size of the quenched samples. The quench was into ambient temperature water. Each point is the average of at least three separate tests. Note that the flexural strength of the 40 μm $Ti_3Si_{0.5}Ge_{0.5}C_2$ sample was *higher* than the as-fabricated, unquenched sample with the same composition [45].

fine-grained Ti_3SiC_2 samples is different; instead of exhibiting a critical quenching temperature above which the strength is greatly reduced, as is typical for ceramics, their post-quenching strengths gradually decrease over a 500 °C range (Figure 7.17). The same nonsusceptibility to thermal shock is also exhibited by $Ti_3(Si_xGe_{1-x})C_2$ solid solutions [45]. In one case, the post-quench flexural strengths of a CG $Ti_3(Si_{0.5},Ge_{0.5})C_2$ sample were actually about 20% *higher* than the as-received CG samples (line labeled $Ti_3Si_{0.5}Ge_{0.5}C_2$ 40 μm in Figure 7.17). The reasons for this quench hardening is not entirely clear at the present time, but they are most likely related to the formation of smaller domains as a result of thermal residual stresses.

7.6.8
R-Curve Behavior and Fatigue

Gilbert *et al.* [156] reported on the fracture and cyclic fatigue-crack growth behavior of Ti_3SiC_2. At room temperature, both fine- and coarse-grained Ti_3SiC_2 microstructures exhibited substantial R-curve behavior, where the stress needed to extend a crack increased with crack length (Figure 7.18a). The fine-grained samples initiated at a $K_{1c} \approx 8\ \mathrm{MPa}\sqrt{m}$, rising to $\approx 9.5\ \mathrm{MPa}\sqrt{m}$ after 1.5 mm of crack extension. The coarse-grained samples exhibited a substantially stronger R-curve behavior, with crack growth initiation between 8.5 to 11 $\mathrm{MPa}\sqrt{m}$, and peaking at 14-16 $\mathrm{MPa}\sqrt{m}$ after 2.7–4 mm crack extension (Figure 7.18a). The latter value is believed to be one of the highest K_{1c} values reported for monolithic, single-phase, nontransforming ceramics.

The cyclic-fatigue study of coarse- and fine-grained Ti_3SiC_2 samples [156] confirmed a high dependence of cyclic-fatigue crack growth rates, da/dN, on the applied stress intensity range (ΔK; not shown) that is typical of ceramics. However, the fatigue crack growth thresholds, ΔK_{th}, are comparatively higher than those for typical ceramics and some metals (e.g., 300 M alloy steel) tested under the same conditions. Here again, the fatigue threshold of the coarse-grained samples (9 $\mathrm{MPa}\sqrt{m}$) was one of the highest fatigue thresholds ever observed in monolithic, nontransforming ceramics. For the fine-grained structure, the fatigue threshold fell to $\approx 6.5\ \mathrm{MPa}\sqrt{m}$.

With the increase in testing temperature, the fatigue thresholds for both microstructures decrease slightly up to 1100 °C. At 1200 °C, which is above the brittle-to-plastic transition (see below), the da/dN versus ΔK curves showed three distinctive regions that were more pronounced for the fine-grained samples [132].

Evidence for elastic-ligaments bridging and frictional pull-outs similar to those observed in other structural ceramics were found in all samples after fracture toughness and cyclic-fatigue testing. However, the excellent K_{1c} and K_{th} values can be mostly attributed to the unique KB-induced lamellae (Figure 7.18b) that serve as very tenacious crack bridges, somewhat reminiscent of wood.

Zhang *et al.* [157] investigated the short crack fatigue behavior of Ti_3SiC_2 polycrystalline samples and reported that, for a relatively short crack lengths, the crack growth rates often *decreased* with the number of cycles, reflecting a faster increase in crack growth resistance through grain bridging than the applied crack driving force.

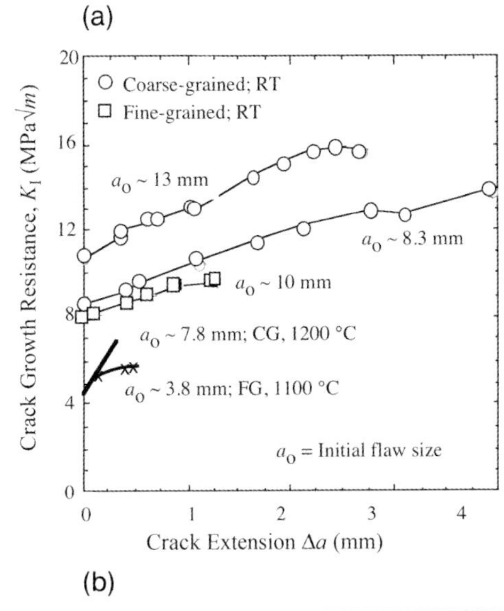

Figure 7.18 (a) Crack-growth resistance, K_R, plotted as a function of crack extension, Δa, for both the fine- and coarse-grained Ti_3SiC_2 microstructures at ambient and elevated temperatures. The initial flaw size, a_0, is indicated for each measured *R*-curve. Note that K_R *drops* above the brittle-to-plastic transition temperature [132]; (b) Field-emission SEM image of a bridged crack, which is growing from left to right, in the coarse-grained Ti_3SiC_2 microstructure [156]. Heavily deformed lamella bridge the crack, and significant amounts of delamination and bending are observed. Such processes must, at least partially, account for the extremely high plateau K_{1C}-values.

It was concluded that the high fatigue resistance originated from the laminated nature of Ti_3SiC_2, that in turn led to extensive energy dissipation through micro-faults, crack deflection, branching, and intact grain bridging. It is also more likely than not, that IKBs played an important role in this enhanced fatigue resistance. The same authors also confirmed that the long crack behavior was similar to that reported by Gilbert *et al.*, with a quite steep dependence of da/dN on ΔK [156].

7.6.9 High-Temperature Properties

As noted above, all MAX phases tested to date pass through a brittle-to-plastic transition (BPT) [98, 145, 158, 159]. The temperature of the transition varies from phase to phase, but for many Al-containing phases and Ti_3SiC_2, it is between 1000 and 1100 °C. The fact that K_{1c} *drops* above the BPT temperature (Figure 7.18a) [132], categorically rules out the activation of additional slip systems, as some have suggested, and this is why it is more accurate to label the transition as a BPT transition, rather than as the more common ductile-to-brittle transition.

All of these conclusions have recently been confirmed by others [158, 159]. Wan *et al.* [159] however, have ascribed the drop in K_{1c} to a dramatic decrease in modulus at higher temperatures. In the present authors' studies, however, no correlation was found between the plastic-to-brittle transition and a drop in modulus, for the simple reason that the modulus does not drop precipitously, but rather linearly with the same slope as below the transition temperature [50]. In measuring the modulus, care must be taken not to load the sample, as this can trigger the formation of IKBs that will result in apparently lower modulus values.

7.6.9.1 Compressive Properties

Further compelling evidence for the KB-based model discussed above can be found in the high-temperature response of Ti_3SiC_2 to compressive cyclic loadings. Up to ~900 °C, and consistent with an athermal model for the formation of IKBs, minimal changes occur to the shapes or areas of the cyclic stress–strain loops in compression [5]. However, at temperatures >1000 °C, the stress–strain loops are open (inset in Figure 7.19a) and the response becomes strongly strain rate-dependent [5, 160]. Cyclic hardening at temperatures as high as 1200 °C has been observed for both fine-grained (inset in Figure 7.19a) and coarse-grained samples, although the effect on the latter is more dramatic [5]. The hardening is manifested two ways: (i) the areas enclosed by the loops become smaller (Figure 7.19a); and (ii) the *initial* slopes of the stress–strain curves approach the linear elastic limit (as measured by ultrasound) as the number of cycles increases (Figure 7.19a). The latter finding clearly implies that cycling results in microdomains that are harder to kink than the initial grains [i.e., a reduction of 2α in Eq. (6)].

Along the same lines, and for the same reasons, after cycling at high temperature the response of the coarse-grained samples becomes comparable to that of fine-grained samples, and the latter approach their linear elastic limit. This is best exemplified in Figure 7.19b, where the room temperature stress–strain curves of a coarse-grained Ti_3SiC_2 sample *before and after* a 2% deformation at 1300 °C (inset in Figure 7.19b) are compared. After deformation, the stress–strain loops are distinctly steeper [5]. From these results, it is clear that deformation at high temperatures effectively reduces the grain size.

Of equal importance, the first loop after deformation (traced in red in Figure 7.19b) is open, but subsequent loops (shown in blue) are closed [160]. The simplest interpretation for this observation is that, during the first cycle (red cycle in

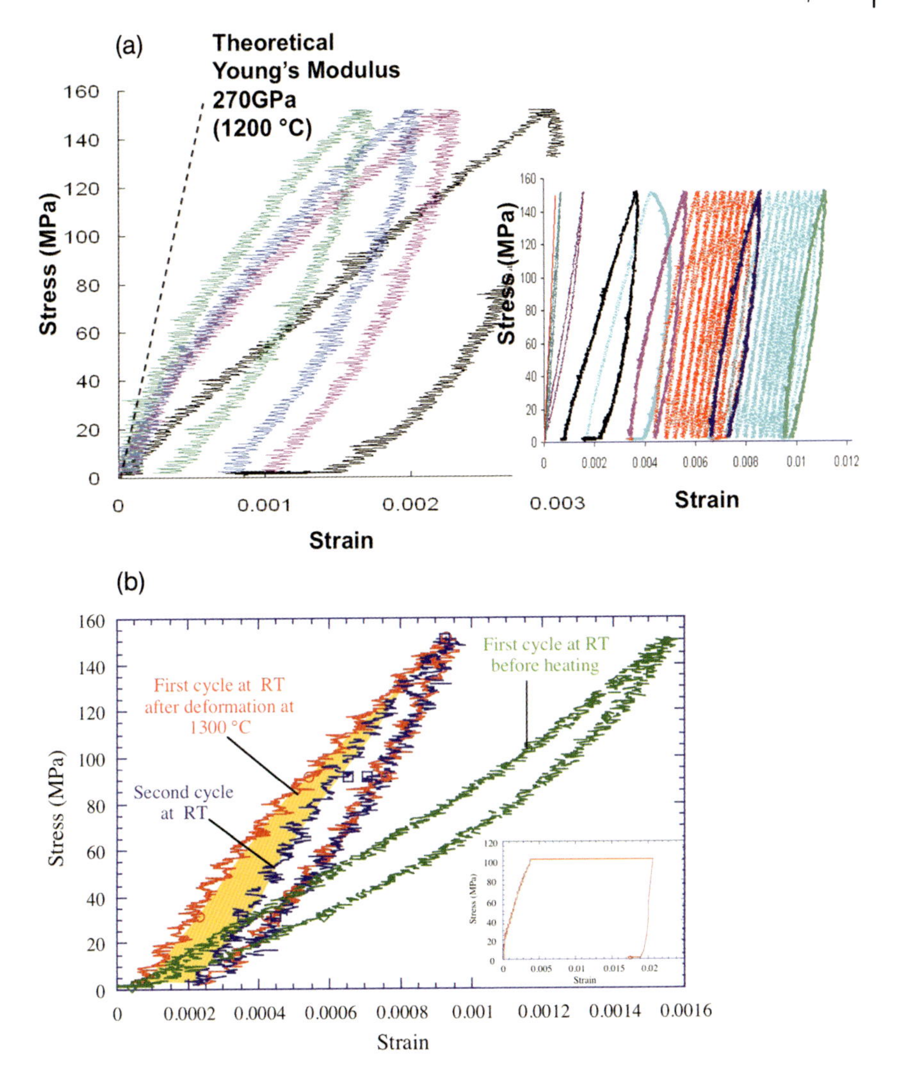

Figure 7.19 (a) Fine-grain sample of Ti_3SiC_2 cycled in compression at 1200 °C [5]. The inset shows all the cycles. Select, color-coded, cycles are extracted from the inset and plotted starting at zero strain for comparison. Note the significant cyclic hardening manifested by both reduction in dissipated energy, W_d, and stiffening at low loads. The dashed line is the response predicted for purely elastic deformations; (b) Comparison of compressive stress–strain curves of a coarse-grained Ti_3SiC_2 sample tested before and after a 2% deformation at 1300 °C (inset). The latter was loaded twice; the first loading resulted in an open loop, and the second and subsequent loading in only closed loops [160].

Figure 7.19b) the mobile dislocation walls (Figure 7.8a) generated at high temperature are swept into the kink boundaries, immobilizing them. On subsequent loading (blue cycle in Figure 7.19b), only the IKBs are activated such that the loop is fully reversible and reproducible.

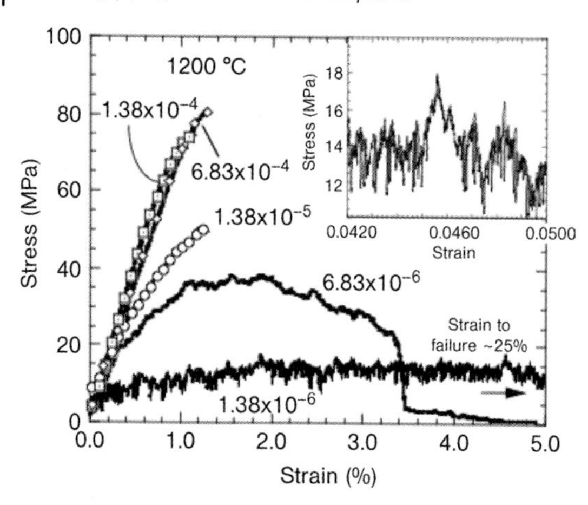

Figure 7.20 Typical tensile stress–strain curves for coarse-grained Ti_3SiC_2 tested in air at 1200 °C as a function of the strain rate. Note the strong strain rate sensitivity; at high strain rates, the samples were brittle. At low rates, deformations of 25% were possible. Inset shows extent of stress fluctuations [161].

7.6.9.2 **Tensile Properties**

The response of Ti_3SiC_2 to tensile stresses is a strong function of temperature and strain rate. Below the BPT, they are brittle, but above they can be quite plastic, with strains to failure up to 25% in some cases, especially at low strain rates (Figure 7.20) [161]. The deformation occurs *without* necking, and the majority of the failure strain is due to damage accumulation in the form of cavitations, pores, and microcracks [142, 148, 162, 163].

At ~0.5, the strain rate sensitivity for both coarse- and fine-grained microstructures of Ti_3SiC_2 is quite high [142, 148] and is more characteristic of super-plastic solids than of typical metals or ceramics. This does not imply that the deformation mechanisms of the MAX phases are in any way comparable to those of superplastic solids. The latter have comparable strain rate sensitivities, but only at grain sizes that are at least two orders of magnitude smaller than those of the CG Ti_3SiC_2 samples.

7.6.9.3 **Creep**

In the 1000–1300 °C temperature range, Ti_3SiC_2 loaded in tension or compression exhibits primary, secondary, and tertiary creep [162–164]. A short primary creep stage, where the deformation rate decreases rapidly, is followed by a secondary creep regime during which the creep rate, although not truly reaching a constant level, does not change significantly over a significant time period.

This stage is characterized by a minimum creep rate, $\dot{\varepsilon}_{min}$, which during both tension and compression for the fine-grained and coarse-grained samples can be described by [162–164]:

$$\dot{\varepsilon}_{min} = \dot{\varepsilon}_o A \left(\frac{\sigma}{\sigma_o}\right)^n \exp\left(-\frac{Q}{RT}\right) \quad (12)$$

Table 7.6 Summary of creep parameters for Ti_3SiC_2.

Creep	Grain	ln *A*	*n*	*Q* (kJ mol^{-1})	$\dot{\varepsilon}_{min}$ (s)$^{-1}$ at 1200 °C and 50 MPa	Reference
Compression	CG	22	2	585 ± 20	2.3×10^{-7}	[164]
	FG	14	1.5	768 ± 20	2.2×10^{-7}	[164]
Tensile	CG	17	2	458 ± 12	3.5×10^{-6}	[162]
	FG	19	1.5	445 ± 10	1.0×10^{-5}	[163]

CG = coarse-grained; FG = fine-grained.

where *A*, *n*, and *Q* are, respectively, a stress-independent constant, the stress exponent, and the activation energy for creep. Here, both *R* and *T* have their usual meanings, and $\dot{\varepsilon}_0 = 1\,s^{-1}$ and $\sigma_o = 1$ MPa. The various creep parameters are summarized in Table 7.6.

From these, and other, results it can be concluded that:

- Based on the similarities in the values of *n*, especially at low stresses (see Table 7.6 or Figure 7.21), it is fair to conclude that the same mechanism is responsible for the creep of both fine-grained and coarse-grained microstructures. This claim is made despite the differences in activation energies. Why the activation energies are a function of the type of loading, or even what their origins are, is unclear. Hence, further studies are required.
- In all cases, at any given stress, $\dot{\varepsilon}_{min}$ is at least an order of magnitude lower in compression than in tension (Table 7.6).
- When large internal stresses develop during testing [163–165], their most likely sources are dislocation pileups and/or MDWs.
- In tension, $\dot{\varepsilon}_{min}$ is a weak function of grain size, with grain size exponents <1. The same is true in compression, but only at low stresses, which implies the dominant creep mechanism is most likely dislocation creep, despite the fact that *n* is less than the exponents (3–7) typically associated with dislocation creep [163, 165].

In compression, and especially at relatively high stresses and/or temperatures, $\dot{\varepsilon}_{min}$ can actually be *lower* for the fine-grained than the coarse-grained microstructures (not shown) [164]. In that regime, the stress exponents also become quite large. The reasons for this unusual phenomenon are not totally understood at present, but post-testing microstructural examinations of samples in that regime strongly suggest a change in mechanism, from dislocation creep to subcritical crack growth. This regime is more difficult to document in tension because the stress state is such as to quickly cause fast fracture. These comments notwithstanding, further studies are required to better understand what is occurring.

At lower stresses, the times to failure measured during tensile creep are considerably longer for the coarse-grained than for fine-grained microstructures [163, 165]. This was attributed to the higher damage tolerance of the coarse-grained microstructure – that is, to the ability of the microstructure to sustain higher bridging

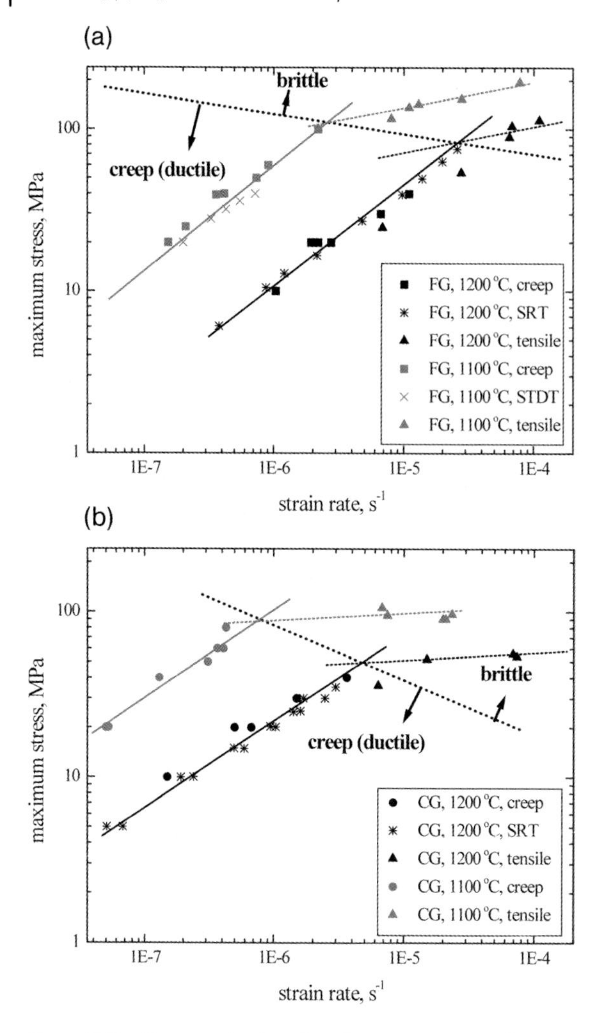

Figure 7.21 Summary log-log plot of stress versus strain rate results obtained from tensile tests at constant cross-head displacement (CHD) rate, relaxation tests, creep tests and strain transient dip tests (STDTs). The plotted lines are only guides for the eye, for (a) a coarse-grained microstructure and (b) a fine-grained microstructure. The gray symbols represent results at 1100 °C, and black symbols at 1200 °C [142]. SRT = stress relaxation test; STD = stress transient dip.

stresses than the finer-grained counterparts. In this case, two different bridging mechanisms were observed:

- A mechanism which begins with delaminations that are initiated at opposite ends of a single grain, but on different basal planes. With further deformation, the torque separates the lamellae between the two delamination cracks to ultimately form crack bridges not unlike those shown in Figure 7.18b.

- A mechanism in which grains where the basal planes are parallel to the applied load serve as classic crack-bridges (not shown) [162].

The simplest explanation for these observations, as well as to the fact that K_{1c} actually *decreases* above the PBT, is the following. As in ice [161], and for the same reason (viz., the paucity of operative slip systems), the glide of basal plane dislocations takes place only in favorably oriented grains or soft grains. During the short primary creep regime, basal plane dislocations glide and form pile-ups and/or MDWs, which results in large internal stresses. As these internal stresses increase, the creep rate decreases, which leads to a quasi-steady state. If the rate at which these stresses accrue is significantly higher than the rate at which they can relax, failure (as shown in Figure 7.21) is brittle insensitive to strain rate or grain size [142]. At the lowest strain rates and/or highest temperatures, the internal stresses can be accommodated by interlaminar decohesion, grain bending or kinking, and/or grain boundary decohesion and sliding. As the number of cavities and cracks increase and coalesce, this leads naturally to an increase in $\dot{\varepsilon}_{min}$ and a transition from secondary to tertiary creep. In other words, dislocation creep gives way to subcritical crack growth [164, 165].

To put it most succinctly, the response of the MAX phases to stress at elevated temperatures is directly related to the rate at which the internal stresses are relaxed.

Lastly, it should be noted that, despite the potential of using the MAX phases as structural materials at high temperatures, and the clear importance of tensile creep in those applications, it is somewhat surprising that so few creep studies have been conducted.

7.6.9.4 **Solid-Solution Hardening and Softening**

As noted above, since substitutions can be made on either of the three sites in the MAX phases, the number of solid solutions possible is quite high. It is thus not unreasonable to assume, that once the effects of such substitutions have on the mechanical properties are understood, then major improvements should accrue.

Few studies have been conducted on MAX phase solid solutions, but the few carried out have provided some intriguing results. Based on a limited data set, it appears that *only* substitutions on the X-sites have a *dramatic* (see below) effect on the mechanical properties. Substitutions on the M-sites [e.g., $(Ti_{0.5},Nb_{0.5})_2AlC$, [47]] or the A-sites [e.g., $Ti_3(Si_{0.5},Ge_{0.5})C_2$ [45]], if they result in solid-solution hardening at all, is more modest. The hardening also appears to be a function of MAX phase explored and the solute atom chosen. For example, Meng *et al.* [166] showed that the Vickers hardness, at 10 N, of Ti_2AlC doped with 15 at% V increased from 3.5 to 4.5 GPa as a result. The flexural and compressive strengths were also increased. The same group showed that substituting 25 at% of the Al in Ti_3AlC_2 by Si, increased the Vickers hardness, at 10 N, from ~3 to 4 GPa [167]. Here again, the flexural and compressive strengths were increased. Intriguingly, substituting Si for Ge in Ti_3GeC_2, did not result in hardening, but rather a decrease in compressive strengths. If there was an increase in hardness, it was not obvious [45]. It is worth noting that the bulk moduli of Ti_3SiC_2 and Ti_3AlC_2 [57] are both higher than that of the $Ti_3Si_{0.5}Ge_{0.5}C_2$ solid solution [58].

In contradistinction, $Ti_2AlC_{0.5}N_{0.5}$, is harder, more brittle and significantly stronger, than the end members Ti_2AlC and Ti_2AlN, which can be deformed plastically up to 5% strain, even at room temperature [98]. At a temperature of ~1200 °C, a solid solution softening effect is observed, the reasons for which are still unclear, although they may be related to the aforementioned interlaminar decohesions.

7.7
Tribological Properties and Machinability

As early as 1996, it was noted that Ti_3SiC_2 shavings had a graphitic feel to them [4], and this was later confirmed [168] when the friction coefficient, μ, of Ti_3SiC_2 basal planes was measured using lateral force microscopy and shown to be one of the lowest ever reported, at $2–5 \times 10^{-3}$. Not only was this value exceptionally low, but it remained very low during up to six months' exposure to the atmosphere.

These low μ-values do not translate to low friction in polycrystalline samples [169]. More recent results [170], however, have suggested that it might be possible to obtain surfaces with μ-values against steel and Si_3N_4 as low as 0.15, especially at low stresses. However, further investigations are required to better understand, and ultimately exploit, this important property.

Finally, it would be remiss not to mention probably the most characteristic trait of the MAX phases, and one that truly sets them apart from other structural ceramics or high-temperature alloys – their ease of machinability [4, 72, 131, 171]. The MAX phases are quite readily machinable with regular high-speed tool steels or even a manual hacksaw. Machining does not occur by plastic deformation, as in the case of metals, but rather by the breaking off of tiny microscopic flakes. In that respect, they are not unlike other machinable ceramics such as Maycor™. The analogy with ice is also apt [161]; the MAX compounds do not machine as one scoops ice cream (as in metals), but rather as in shaving ice. The MAX phases have some of the highest specific stiffnesses for readily machinable solids, with the exception of Be and Be-alloys.

7.8
Concluding Remarks

Some of the physical properties of the MAX phases, such as thermal expansion, elastic properties and thermal conductivity, have much in common with their respective MX binaries. However, their electronic structure and transport properties are more akin to those of the transition metals themselves.

The unique combination of properties possessed by the MAX phases – their ease of machinability, low friction, thermal and structural stability, good thermal and electrical conductivities–render them attractive for many applications such as rotating electrical contacts and bearings, heating elements, nozzles, heat exchangers, tools for die pressing, among many others. Many of these applications are currently being field-tested and are at various stages of development. The hurdles to commercialization are several, and include the relative high costs of the powders and sintered components as compared to other structural ceramics such SiC and alumina. Another hurdle is the newness of these solids and the reluctance of some manufacturers to use them instead of solids with which they have been working for many years, and are more familiar. The lack, until recently, of −325 mesh, pressureless, sinterable powders has also slowed the development of these materials.

At the present time, it appears that the best – and probably the first – MAX phase to be adopted for high-temperature applications will be Ti_2AlC, for two reasons: (i) because it is an alumina former and thus has superb oxidation resistance; and (ii) perhaps most importantly, that it is most likely the cheapest of the MAX phases. Currently, −325 mesh Ti_2AlC and Ti_3SiC_2 powders are available commercially from Kanthal, Sweden; these can be pressureless sintered to full density [137, 172] or thermal sprayed [173]. Ti_3SiC_2 targets are also currently being used to fabricate Ti–Si–C thin films [174] by Impact Coatings in Sweden. However, given the remarkable set of properties exhibited by the MAX phases, it is simply a matter of time before many applications will be identified.

References

1 Barsoum, M.W. (2000) *Prog. Solid State Chem.*, **28**, 201.

2 Barsoum, M.W. and El-Raghy, T. (2000) *Am. Sci.*, **89**, 336.

3 Barsoum, M.W. and Radovic, M. (2004) Mechanical Properties of the MAX Phases, in *Encyclopedia of Materials Science and Technology* (eds R.W.C.K.H.J. Buschow, M.C. Flemings, E.J. Kramer, S. Mahajan, and P. Veyssiere), Elsevier, Amsterdam.

4 Barsoum, M.W. and El-Raghy, T. (1996) *J. Am. Ceram. Soc.*, **79**, 1953.

5 Barsoum, M.W., Zhen, T., Kalidindi, S.R., Radovic, M., and Murugahiah, A. (2003) *Nature Mater.*, **2**, 107.

6 Palmquist, J. *et al.* (2002) *Appl. Phys. Lett.*, **81**, 835.

7 Molina-Aldareguia, J.M., Emmerlich, J., Palmquist, J., Jansson, U., and Hultman, L. (2003) *Scripta Mater.*, **49**, 155.

8 Wilhelmsson, O., Palmquist, J.-P., Nyberg, T., and Jansson, U. (2004) *Appl. Phys. Lett.*, **85**, 1066.

9 Schneider, J.M., Sun, Z., Mertens, R., Uestel, F., and Ahuja, R. (2004) *Solid State Commun.*, **130**, 445.

10 Nowotny, H. (1970) *Prog. Solid State Chem.*, **2**, 27.

11 Jeitschko, W. and Nowotny, H. (1967) *Monatsh. Chem.*, **98**, 329.

12 Wolfsgruber, H., Nowotny, H., and Benesovsky, F. (1967) *Monatsh. Chem.*, **98**, 2401.

13 Pietzka, M.A. and Schuster, J. (1994) *J. Phase Equilib.*, **15**, 392.

14 Dubois, S., Cabioch, T., Chartier, P., Gauthier, V., and Jaouen, M. (2007) *J. Am. Ceram. Soc.*, **90**, 2642.

15 Etzkorn, J., Ade, M., and Hillebrecht, H. (2007) *Inorg. Chem.*, **46**, 1410.

16 Barsoum, M.W. *et al.* (1999) *J. Am. Ceram. Soc.*, **82**, 2545.

17 Rawn, C.J. *et al.* (2000) *Mater. Res. Bull.*, **35**, 1785.

18 Hu, C. *et al.* (2007) *J. Am. Ceram. Soc.*, **90**, 2542.

19 Manoun, B., Saxena, S.K., El-Raghy, T., and Barsoum, M.W. (2006) *Appl. Phys. Lett.*, **88**, 201902.

20 Hu, C. *et al.* (2007) *Scripta Mater.*, **57**, 893.

21 Etzkorn, J., Ade, M., and Hillebrecht, H. (2007) *Inorg. Chem.*, **46**, 7646.

22 Eklund, P. *et al.* (2007) *Acta Mater.*, **55**, 4723.

23 Jeitschko, W., Nowotny, H., and Benesovsky, F. (1963) *Monatsh. Chem.*, **94**, 1198.

24 Jeitschko, W., Nowotny, H., and Benesovsky, F. (1964) *Monatsh. Chem.*, **95**, 178.
25 Jeitschko, W., Nowotny, H., and Benesovsky, F. (1963) *Monatsh. Chem.*, **94**, 844.
26 Jeitschko, W., Nowotny, H., and Benesovsky, F. (1963) *Monatsh. Chem.*, **94**, 672.
27 Pierson, H.O. (1996) *Handbook of Refractory Carbides and Nitrides*, Noyes Publications, Westwood, NJ.
28 Cottrell, A. (1995) *Chemical Bonding in Transition Metal Carbides*, Institute of Materials, London.
29 Barsoum, M.W. and Schuster, J.C. (1998) *J. Am. Ceram. Soc.*, **81**, 785.
30 Medvedeva, N., Novikov, D., Ivanovsky, A., Kuznetsov, M., and Freeman, A. (1998) *Phys. Rev. B*, **58**, 16042.
31 Holm, B., Ahuja, R., and Johansson, B. (2001) *Appl. Phys. Lett.*, **79**, 1450.
32 Hug, G., Jaoun, M., and Barsoum, M.W. (2005) *Phys. Rev. B*, **71**, 24105.
33 Hug, G. and Frie, E. (2002) *Phys. Rev. B*, **65**, 113104.
34 Sun, Z.M. and Zhou, Y.C. (1999) *Phys. Rev. B*, **60**, 1441.
35 Zhou, Y.C. and Sun, Z.M. (2000) *Phys. Rev. B*, **61**, 12570.
36 Zhou, Y.C., Sun, Z.M., Wang, X.H., and Chen, S.Q. (2001) *J. Phys. Condens. Matter*, **13**, 10001.
37 Sun, Z.M. and Zhou, Y.C. (2002) *J. Phys. Soc. Jpn*, **71**, 1313.
38 Holm, B., Ahuja, R., Li, S., and Johansson, B. (2002) *J. Appl. Phys.*, **91**, 9874.
39 Sun, Z., Ahuja, R., Li, S., and Schneider, J.M. (2003) *Appl. Phys. Lett.*, **83**, 899.
40 Hug, G. (2006) *Phys. Rev. B*, **74**, 184113.
41 Pietzka, M.A. and Schuster, J.C. (1996) *J. Am. Ceram. Soc.*, **79**, 2321.
42 Manoun, B. *et al.* (2007) *J. Appl. Phys.*, **101**, 113523.
43 Radovic, M., Ganguly, A., and Barsoum, M.W. (2008) *J. Mater. Res.*, **23**, 1517.
44 Scabarozi, T. *et al.* (2008) *J. Appl. Phys.*, **104**, 073713.
45 Ganguly, A., Zhen, T., and Barsoum, M.W. (2004) *J. Alloys Compd.*, **376**, 287.
46 Schuster, J.C., Nowotny, H., and Vaccaro, C. (1980) *J. Solid State Chem.*, **32**, 213.
47 Salama, I., El-Raghy, T., and Barsoum, M.W. (2002) *J. Alloys Compd.*, **347**, 271.
48 Barsoum, M.W., Golczewski, J., Siefert, H.J., and Aldinger, F. (2002) *J. Alloys Compd.*, **340**, 173.
49 Nowotny, H., Schuster, J.C., and Rogl, P. (1982) *J. Solid State Chem.*, **44**, 126.
50 Radovic, M. *et al.* (2006) *Acta Mater.*, **54**, 2757.
51 Manoun, B., Saxena, S.K., Barsoum, M.W., and El-Raghy, T. (2006) *J. Phys. Chem. Solids*, **67**, 2091.
52 Hettinger, J.D. *et al.* (2005) *Phys. Rev. B*, **72**, 115120.
53 Manoun, B. *et al.* (2006) *Phys. Rev. B*, **73**, 024110.
54 Tian, W., Wang, P., Zhang, G., Kan, Y., and Li, Y. (2007) *J. Am. Ceram. Soc.*, **90**, 1663.
55 Hu, C. *et al.* (2008) *J. Eur. Ceram. Soc.*, **28**, 1679.
56 Onodera, A., Hirano, H., Yuasa, T., Gao, N.F., and Miyamoto, Y. (1999) *Appl. Phys. Lett.*, **74**, 3782.
57 Manoun, B. *et al.* (2007) *J. Alloys Compd.*, **433**, 265.
58 Manoun, B. *et al.* (2004) *Appl. Phys. Lett.*, **84**, 2799.
59 Amini, S. *et al.* (2008) *J. Mater. Res.*, **23**, 2157.
60 Manoun, B., Amini, S., Gupta, S., Saxena, S.K., and Barsoum, M.W. (2007) *J. Phys. Condens. Matter*, **19**, 456218.
61 Phatak, N.A. *et al.* (2008) *J. Alloys Compd.*, **463**, 220.
62 Scabarozi, T.H. *et al.* (2008) *J. Appl. Phys.*, **104**, 033502.
63 Kulkarni, S.R. *et al.* (2008) *J. Alloys Compd.*, **448**, L1.
64 Manoun, B. *et al.* (2009) *Solid State Commun.*, **149**, 1978.
65 El-Raghy, T., Chakraborty, S., and Barsoum, M.W. (2000) *J. Eur. Ceram. Soc.*, **20**, 2619.
66 Hu, C. *et al.* (2008) *J. Am. Ceram. Soc.*, **91**, 2258.
67 Finkel, P., Barsoum, M.W., and El-Raghy, T. (2000) *J. Appl. Phys.*, **87**, 1701.
68 Manoun, B., Saxena, S.K., and Barsoum, M.W. (2005) *Appl. Phys. Lett.*, **86**, 101906.

69 Manoun, B. *et al.* (2004) *Appl. Phys. Lett.*, **85**.
70 Fang, C.M. *et al.* (2006) *Phys. Rev. B*, **74**, 054106.
71 Du, Y.L., Sun, Z.M., Hashimoto, H., and Tian, W.B. (2009) *Phys. Status Solidi.*, **246**, 1039–1043.
72 Finkel, P., *et al.* (2001) *Phys. Rev. B*, **65**, 35113.
73 Manoun, B., Zhang, F., Saxena, S.K., Gupta, S., and Barsoum, M.W. (2007) *J. Phys. Condens. Matter*, **19**, 246215.
74 Murugaiah, A., Barsoum, M.W., Kalidindi, S.R., and Zhen, T. (2004) *J. Mater. Res.*, **19**, 1139.
75 Sun, Z., Li, S., Ahuja, R., and Schneider, J.M. (2004) *Solid State Commun.*, **129**, 589.
76 Wang, J. and Zhou, Y. (2004) *Phys. Rev. B*, **69**, 214111.
77 Scabarozi, T.H. *et al.* (2009) *J. Appl. Phys.*, **105**, 013543.
78 Du, Y.L., Sun, Z.M., Hashimoto, H., and Tiana, W.B. (2008) *Phys. Lett. A*, **372**, 5220.
79 Bouhemadou, A. and Khenata, R. (2008) *Phys. Lett. A*, **372**, 6448.
80 Spanier, J.E., Gupta, S., Amer, M., and Barsoum, M.W. (2005) *Phys. Rev. B*, **71**, 12103.
81 Amer, M. *et al.* (1998) *J. Appl. Phys.*, **84**, 5817.
82 Wang, J., Zhou, Y., Lin, Z., Meng, F., and Li, F. (2005) *Appl. Phys. Lett.*, **86**, 101902.
83 Leaffer, O.D., Gupta, S., Barsoum, M.W., and Spanier, J.E. (2007) *J. Mater. Res.*, **22**, 2651.
84 Barsoum, M.W., Yoo, H.I., and El-Raghy, T. (2000) *Phys. Rev. B*, **62**, 10194.
85 Finkel, P., Barsoum, M.W., Hettinger, J.D., Lofland, S.E., and Yoo, H.I. (2003) *Phys. Rev. B*, **67**, 235108.
86 Finkel, P. *et al.* (2004) *Phys. Rev. B*, **70**, 085104.
87 Yoo, H.I., Barsoum, M.W., and El-Raghy, T. (2000) *Nature*, **407**, 581.
88 Scabarozi, T.H. *et al.* (2008) *Solid State Commun.*, **146**, 498.
89 Lofland, S.E. *et al.* (2004) *Appl. Phys. Letts.*, **84**, 508.
90 Drulis, M.K., Drulis, H., Gupta, S., and Barsoum, M.W. (2006) *J. Appl. Phys.*, **99**, 093502.
91 Mott, N.F. and Jones, H. (1936) *The Theory of the Properties of Metals and Alloys*, Dover Publications, New York.
92 Ashcroft, N.W. and Mermin, N.D. (1976) *Solid State Physics*, Saunders College Publishing, Philadelphia.
93 Barsoum, M.W. *et al.* (2002) *Metall. Mater. Trans.*, **33a**, 2779.
94 Chaput, L., Hug, G., Pecheur, P., and Scherrer, H. (2005) *Phys. Rev. B*, **71**, 121104(R).
95 Keppens, V. *et al.* (1998) *Nature*, **395**, 876.
96 Sales, B.C., Chakoumakos, B.C., Mandrus, D., and Sharp, J.W. (1999) *J. Solid State Chem.*, **146**, 528.
97 Barsoum, M.W. *et al.* (1999) *J. Phys. Chem. Solids*, **60**, 429.
98 Barsoum, M.W., Ali, M., and El-Raghy, T. (2000) *Metall. Mater. Trans.*, **31A**, 1857.
99 Barsoum, M.W., El-Raghy, T., and Chakraborty, S. (2000) *J. Appl. Phys.*, **88**, 6313.
100 Barsoum, M.W. *et al.* (2000) *J. Appl. Phys.*, **83**, 825.
101 Taylor, R.E. (1961) *J. Am. Ceram. Soc.*, **44**, 525.
102 Lengauer, W. *et al.* (1995) *J. Alloys Compd.*, **217**, 137.
103 Manoun, B., Saxena, S.K., Liermann, H.-P., and Barsoum, M.W. (2005) *J. Am. Ceram. Soc.*, **88**, 3489.
104 Tzenov, N. and Barsoum, M.W. (2000) *J. Am. Ceram. Soc.*, **83**, 825–832.
105 Kulkarni, S. *et al.* (2009) *J. Alloys Compd.*, **469**, 395.
106 Kulkarni, S.R. *et al.* (2007) *J. Am. Ceram. Soc.*, **90**, 3013.
107 Tian, W. *et al.* (2006) *Scripta Mater.*, **54**, 841.
108 Lin, Z.J., Zhou, Y.C., and Li, M.S. (2007) *J. Mater. Sci. Technol.*, **23**, 721.
109 Rawn, C.J., Barsoum, M.W., El-Raghy, T., Procopio, A.T., and Hoffman, C.M. (2000) *Mater. Res. Bull.*, **35**, 1785.
110 Farber, L. and Barsoum, M.W. (1999) *J. Mater. Res.*, **14**, 2560.
111 Du, Y., Schuster, J., Seifert, H., and Aldinger, F. (2000) *J. Am. Ceram. Soc.*, **83**, 197.
112 Tzenov, N., Barsoum, M.W., and El-Raghy, T. (2000) *J. Eur. Ceram. Soc.*, **20**, 801.

113 El-Raghy, T. and Barsoum, M.W. (1998) *J. Appl. Phys.*, **83**, 112.
114 Barsoum, M.W. *et al.* (1999) *J. Electrochem. Soc.*, **146**, 3919–3923.
115 El-Raghy, T., Barsoum, M.W., and Sika, M. (2001) *Mater. Sci. Eng. A*, **298**, 174.
116 Barsoum, M.W., El-Raghy, T., and Ogbuji, L. (1997) *J. Electrochem. Soc.*, **144**, 2508.
117 Barsoum, M.W., Ho-Duc, L.H., Radovic, M., and El-Raghy, T. (2003) *J. Electrochem. Soc.*, **150**, B166.
118 Salama, I., El-Raghy, T., and Barsoum, M.W. (2003) *J. Electrochem. Soc.*, **150**, C152–C158.
119 Sundberg, M., Malmqvist, G., Magnusson, A., and El-Raghy, T. (2004) *Ceram. Int.*, **30**, 1899.
120 Wang, X.H. and Zhou, Y.C. (2003) *Corros. Sci.*, **45**, 891.
121 Lee, D.B. and Nguyen, T.D. (2008) *J. Alloys Compd.*, **464**, 434.
122 Lee, D.B., Nguyen, T.D., Han, J.H., and Park, S.W. (2007) *Corros. Sci.*, **49**, 3926.
123 Tian, W., Wang, P., Kan, Y., and Zhang, G. (2008) *J. Mater. Sci.*, **43**, 2785.
124 Chakraborty, S., El-Raghy, T., and Barsoum, M.W. (2003) *Oxid. Met.*, **59**, 83.
125 Gupta, S., Filimonov, D., and Barsoum, M.W. (2006) *J. Am. Ceram. Soc.*, **89**, 2974.
126 Farber, L., Levin, I., and Barsoum, M.W. (1999) *Philos. Mag. Lett.*, **79**, 163.
127 Kooi, B.J., Poppen, R.J., Carvalho, N.J.M., De Hosson, J.Th.M., and Barsoum, M.W. (2003) *Acta Mater.*, **51**, 2859.
128 Barsoum, M.W., Farber, L., El-Raghy, T., and Levin, I. (1999) *Metall. Mater. Trans.*, **30A**, 1727.
129 Duval, P., Ashby, M.F., and Andermant, I. (1983) *J. Phys. Chem.*, **87**, 4066.
130 Sun, Z.-M., Murugaiah, A., Zhen, T., Zhou, A., and Barsoum, M.W. (2005) *Acta Mater.*, **53**, 4359.
131 Barsoum, M.W., Brodkin, D., and El-Raghy, T. (1997) *Scripta Metall. Mater.*, **36**, 535.
132 Chen, D., Shirato, K., Barsoum, M.W., El-Raghy, T., and Ritchie, R.O. (2001) *J. Am. Ceram. Soc.*, **84**, 2914.
133 Orowan, E. (1942) *Nature*, **149**, 463.
134 Hess, J.B. and Barrett, C.S. (1949) *Trans. AIME*, **185**, 599.
135 Frank, F.C. and Stroh, A.N. (1952) *Proc. Phys. Soc.*, **65**, 811.
136 Freise, E.J. and Kelly, A. (1961) *Proc. Phys. Soc. A*, **264**, 269.
137 Zhou, A.G., Barsoum, M.W., Basu, S., Kalidindi, S.R., and El-Raghy, T. (2006) *Acta Mater.*, **54**, 1631.
138 Barsoum, M.W., Zhen, T., Zhou, A., Basu, S., and Kalidindi, S.R. (2005) *Phys. Rev. B*, **71**, 134101.
139 Zhou, A.G., Basu, S., and Barsoum, M.W. (2008) *Acta Mater.*, **56**, 60.
140 Zhou, A. and Barsoum, M.W. (2009) *Metall. Mater. Trans.*, **40A**, 1741–1756.
141 Fraczkiewicz, M., Zhou, A.G., and Barsoum, M.W. (2006) *Acta Mater.*, **54**, 5261.
142 Radovic, M., Barsoum, M.W., El-Raghy, T., Wiederhorn, S.M., and Luecke, W.E. (2002) *Acta Mater.*, **50**, 1297.
143 Barosum, M.W. and Basu, S., (2010) Kinking Nonlinear Elastic Solids, in Encyclopedia of Materials Science and Technology (Eds. R.W.C.K.H.J. Buschow, M.C. Flemings, E. J. Kramer, S. Mahajan and P. Veyssiere) *Elsevier*, Amsterdam.
144 Barsoum, M.W. and El-Raghy, T. (1999) *Metall. Mater. Trans.*, **30A**, 363.
145 El-Raghy, T., Barsoum, M.W., Zavaliangos, A., and Kalidindi, S.R. (1999) *J. Am. Ceram. Soc.*, **82**, 2855.
146 Procopio, A., Barsoum, M.W., and El-Raghy, T. (2000) *Metall. Mater. Trans.*, **31A**, 333.
147 Wang, X.H. and Zhou, Y.C. (2002) *Acta Mater.*, **50**, 3141.
148 Radovic, M., Barsoum, M.W., El-Raghy, T., Seidensticker, J., and Wiederhorn, S.M. (2000) *Acta Mater.*, **48**, 453.
149 Kuroda, Y., Low, I.M., Barsoum, M.W., and El-Raghy, T. (2001) *J. Aust. Ceram. Soc.*, **37**, 95.
150 Nickl, J.J., Schweitzer, K.K., and Luxenberg, P. (1972) *J. Less-Common Met.*, **26**, 335.
151 Goto, T. and Hirai, T. (1987) *Mater. Res. Bull.*, **22**, 2295.
152 El-Raghy, T., Zavaliangos, A., Barsoum, M.W., and Kalidindi, S.R. (1997) *J. Am. Ceram. Soc.*, **80**, 513.
153 Amini, S., Barsoum, M.W., and El-Raghy, T. (2007) *J. Am. Ceram. Soc.*, **90**, 3953.

154 Lis, J., Pampuch, P., Piekarczyk, J., and Stobierski, L. (1993) *Ceramics Inter.*, **19**, 219.
155 Lis, J. *et al.* (1995) *Material Letter*, **22**, 163.
156 Gilbert, C.J. *et al.* (2000) *Scripta Mater.*, **238**, 761.
157 Zhang, H., Wang, Z.G., Zang, Q.S., Zhang, Z.F., and Sun, Z.M. (2003) *Scripta Mater.*, **49**, 87.
158 Tian, W., Sun, Z.-M., Hashimoto, H., and Du, Y. (2009) *J. Mater. Sci.*, **44**, 102.
159 Wan, D.T. *et al.* (2008) *J. Eur. Ceram. Soc.*, **28**, 663.
160 Zhen, T., Barsoum, M.W., and Kalidindi, S.R. (2005) *Acta Mater.*, **53**, 4163.
161 Barsoum, M.W., Radovic, M., Finkel, P., and El-Raghy, T. (2001) *Appl. Phys. Lett.*, **79**, 479–481.
162 Radovic, M., Barsoum, M.W., El-Raghy, T., and Wiederhorn, S.M. (2003) *J. Alloys Compd.*, **361**, 299.
163 Radovic, M., Barsoum, M.W., El-Raghy, T., and Wiederhorn, S.M. (2001) *Acta Mater.*, **49**, 4103.
164 Zhen, T. *et al.* (2005) *Acta Mater.*, **53**, 4963.
165 Radovic, M., Barsoum, M.W., El-Raghy, T., and Wiederhorn, S.M. (2003) *J. Alloys Compd.*, **361**, 299.
166 Meng, F.L., Zhou, Y., and Wang, J.Y. (2005) *Scripta Mater.*, **53**, 1369.
167 Zhou, Y.C., Chen, J.X., and Wang, J.Y. (2006) *Acta Mater.*, **54**, 1317.
168 Myhra, S., Summers, J.W.B., and Kisi, E.H. (1999) *Mater. Lett.*, **39**, 6.
169 El-Raghy, T., Blau, P., and Barsoum, M.W. (2000) *Wear*, **42**, 761.
170 Souchet, A. *et al.* (2005) *Tribol. Lett.*, **18**, 341–352.
171 Barsoum, M.W. and El-Raghy, T. (1997) *J. Mater. Synth. Process.*, **5**, 197.
172 Murugaiah, A., Souchet, A., El-Raghy, T., Radovic, M., and Barsoum, M.W. (2004) *J. Am. Ceram. Soc.*, **87**, 550.
173 Frodelius, J. *et al.* (2008) *Surf. Coat. Technol.*, **202**, 5976.
174 Hogberg, H. *et al.* (2005) *Surf. Coat. Technol.*, **193**, 6.

II
Structures and Properties

Ceramics Science and Technology Volume 2: Properties. Edited by Ralf Riedel and I-Wei Chen

ISBN: 978-3-527-31156-9

8
Structure–Property Relations

Tatsuki Ohji

8.1
Introduction

As the properties of materials are closely related to their structures, the realization of a material with certain properties can be achieved through structure control. The structure of ceramic materials, however, consists of many types of microstructural elements such as particles, grains, pores, defects, fibers, layers, and interfaces. These microstructural elements can be classified by size into four scale levels: (i) atomic-molecular scale; (ii) nanoscale; (iii) microscale, and (iv) macroscale, as shown in Figure 8.1. Besides their physical and chemical nature, other features of these elements including their morphology, configuration, distribution, and orientation are also important. It is obvious, therefore, that there are many factors which can control the structure of materials – not only the types of the elements, but also their features.

During the past few decades, the properties of advanced ceramics have been greatly improved through the well-organized control of these structural elements. In the past, research groups have attempted to improve the properties of ceramic materials by controlling their structural elements at a single-scale level. Whilst with this approach, it is possible to improve a specific property, the other properties may very well be compromised because the microstructural elements generally extend over more than one scale level. A recent trend, therefore, has been simultaneously to control the morphology and distribution of these diverse structural elements at multiple size-scales, in order to optimize the competing properties and/or introduce new functions. For example, in the field of ceramics, it has been recognized that a homogeneous and fine-grained microstructure yields materials with a high strength but a low fracture toughness, whereas materials with a coarse microstructure may exhibit a high toughness but a low strength. However, it has become possible to realize both high strength and high fracture toughness through controlling the size, morphology, and alignment of the grains simultaneously. At the same time, controlling the crystal phase of grain boundaries gives rise to further significant improvements in other properties such as heat-resistance and thermal conductivity. In fact, by controlling the morphology and orientation of the pores as well as the

Ceramics Science and Technology Volume 2: Properties. Edited by Ralf Riedel and I-Wei Chen

ISBN: 978-3-527-31156-9

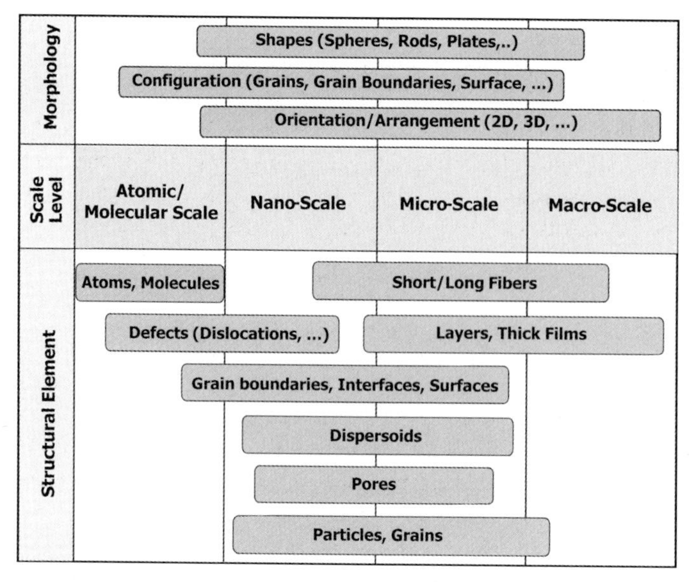

Figure 8.1 The structural elements of ceramics.Reprinted with permission from Ref. [20]; © 1997, Elsevier Ltd.

grains, even porous ceramics have been fabricated that are stronger and tougher than dense ceramics.

The research efforts made to tailor ceramic microstructure so as to improve properties are described in this chapter. Although many material properties are affected by the microstructure, the emphasis here is placed on the mechanical properties, especially strength and fracture toughness. In particular, attention is focused on silicon nitride ceramics, the mechanical properties of which have been improved substantially through microstructure tailoring during the past two decades.

8.2 Self-Reinforced Silicon Nitrides

Silicon nitride represents one of the most important engineering ceramics, because of its high fracture toughness and high strength. In the past, many studies have been conducted to further improve fracture toughness, with microstructures such as grain size and morphology, grain alignment, and boundary chemistry, each having been controlled to enhance a variety of crack-wake toughening mechanisms, including grain bridging, grain sliding, grain rotation, and frictional grain pull-out. These toughening mechanisms come into operation when a crack propagates from an initial flaw, and result in a pronounced increase in fracture resistance with increasing crack length, which is referred to as the *R*-curve. The *R*-curves were first reported in about 1990 for β-silicon nitride with a coarsened microstructure of fibrous (or elongated) grains [1–4]. For example, Li and Yamanis [1] demonstrated that silicon nitrides containing large (>1 μm-diameter) fibrous grains exhibited plateau toughness values close to 10 MPa $m^{1/2}$ for long cracks, whereas submicrometer grain-sized materials had toughness values of about 3 MPa $m^{1/2}$. The bridging of cracks by intact fibrous grains and grain pull-out was observed in the region behind the crack tips, just as with whisker-reinforced ceramics. The plateau fracture toughness of silicon nitrides increased with the diameter of the larger fibrous grains, which is consistent with the prediction for frictional bridging and pull-out processes [5, 6]. These materials have been referred to as "*in situ*-toughened" or "self-reinforced silicon nitrides."

These large fibrous grains are hexagonal prisms of β-silicon nitride, which are grown anisotropically with the *c*-axis growth rates larger than those normal to the prism faces. Such grain growth can be controlled by the ratio of α- to β-phase in the starting powders, as well as the sintering temperatures, pressure, and hold time [7–10]. For example, the addition of some β-phase into mostly α-phase powders enhances the formation of large fibrous β-grains in a matrix of fine β-grains.

However, the fracture strength of the self-reinforced silicon nitrides decreases with the size of the larger fibrous grains, despite the increasing plateau (or long crack) toughness [3, 11–13]. This phenomenon has been frequently observed in ceramics materials, namely that fracture strength generally decreases with increasing grain size. This antagonistic relationship between fracture toughness and strength suggests that the high value of plateau toughness is associated with long cracks which are much larger than the inherent flaw size, and thus it is not necessarily relevant for catastrophic fracture. In fact, *R*-curve measurements for the self-reinforced silicon nitrides have indicated that a coarsening of the overall microstructure of both the finer matrix grains and the larger reinforcing grains increases the plateau toughness, but also decreases the initial slope, which is important for the stability of a short (natural) crack. Thus, a more judicious control of microstructure is required to enhance the overall *R*-curve response; that is, an increase in the initial slope to improve fracture strength, and an increase in the plateau toughness to improve long crack stability.

The use of fibrous β-seed crystals [14–18] is a refined approach to enhance the formation of the larger fibrous grains, and to precisely control the size, morphology, and fraction of these grains. The seed crystals are usually prepared by heating a powder mixture of α-Si_3N_4 and oxides as sintering additives. By selecting the types of Si_3N_4 starting powders and oxides, different sizes and shapes of seed crystals can be obtained. As an example, Figure 8.2 shows seed crystals of a high aspect ratio, which were obtained by heating fine and homogeneous α-silicon nitride powders, the specific surface area of which was 11.0 $m^2 g^{-1}$, with 5 mol% Y_2O_3 as sintering additives under 0.5 MPa N_2 pressure at 1850 °C for 2 h [18].

Hirao and coworkers [15] have succeeded in realizing both a high fracture toughness and a high fracture strength, using seeds of morphologically regulated β-silicon nitride crystals. These crystals have features that are important for seeds – that is, single crystals, large cross-section, and short length. They were grown by heating the powder mixture of α-Si_3N_4, 5 mol% Y_2O_3 and 10 mol% SiO_2 in a silicon

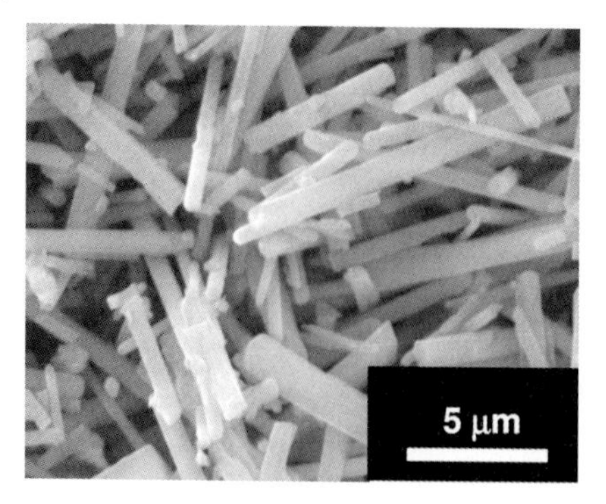

Figure 8.2 Fibrous crystals of silicon nitride seeds. Illustration courtesy of Yoshiaki Inagaki *et al.*

nitride crucible at 1850 °C for 2 h under a nitrogen pressure of 0.5 MPa. The obtained particles were screened through a 100-mesh sieve, and then subjected to an acid rinse treatment to remove any residual glassy phases. This procedure resulted in fibrous crystals with typical diameters of 1 μm and lengths of 4 μm. The seed crystals, the fraction of which was varied from 0% to 5% (by volume), were dispersed into a well-milled methanol slurry of α-Si_3N_4, 5 wt% Y_2O_3 and 2 wt% Al_2O_3 powders, in order to avoid particle fracture. The powder mixture was sintered at 1850 °C in a relatively low nitrogen gas pressure of 0.9 MPa, and a bimodal distribution of silicon nitride grains was obtained owing to the epitaxial grain growth of the dispersed seed crystals with the faster rate along the *c*-axis (as shown in Figure 8.3). The grain growth followed the empirical relationship, $D^n - D_0^n = Kt$, with growth exponents of 3 and 5 for the *a*- and *c*-axis directions, respectively, where D is the grain size, D_0 is the initial size, K is a constant, and t is time. The fracture toughness, measured using the single-edge precracked-beam (SEPB) method, was improved from $\sim$6 MPa·m$^{1/2}$ at no seed addition to $\sim$9 MPa·m$^{1/2}$ at 5 vol%, while the fracture strength (as determined by four-point flexure tests) remained at a high level of about 1 GPa (as shown in Figure 8.4). The crack path observation for seeded materials revealed a substantial

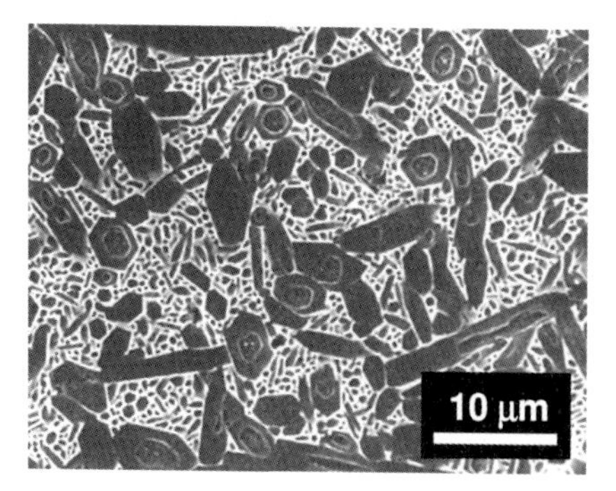

Figure 8.3 Microstructure of the silicon nitride where the 2 vol% of β-seed crystals was dispersed into the raw powders of α-phase and sintering additives before sintering. Note the bimodal distribution of silicon nitride grains. Illustration courtesy of Kiyoshi Hirao, *et al.*

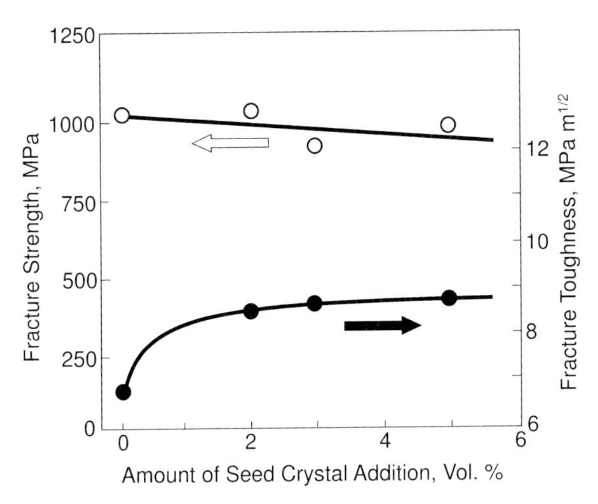

Figure 8.4 Fracture toughness and strength of the seeded silicon nitride as a function of amount of β-seed crystal addition. Reprinted with permission from Ref. [15]; © 1994, Blackwell Publishing, Inc.

deflection at the larger fibrous grains, compared to an almost flat crack propagation in unseeded materials. A synergistic improvement of high fracture toughness and high strength could be attributed to the well-controlled size, fraction, and distribution of the larger fibrous grains. The success in tailoring the microstructures of silicon nitrides with fibrous β-Si_3N_4 seed crystals resulted in a material with ideal material mechanical properties.

8.3
Fibrous Grain-Aligned Silicon Nitrides (Large Grains)

Single-crystal seeding not only provides a novel approach for improving fracture toughness while retaining high strength [15], but also serves as an effective method to control the orientation in addition to the size, content, and morphology of the fibrous β-silicon nitride grains [19–21]. In whisker-reinforced ceramic composites, it is well-known that the orientation of the whiskers plays an important role in their mechanical properties. Moreover, the whisker orientation can be controlled by many techniques that include tape-casting, slip-casting, hot-pressing, and extrusion. The fracture toughness measured for a crack perpendicular to the whisker orientation in an oriented composite is higher than that of the corresponding composites with a random distribution of whiskers.

By combining seeding and tape-casting techniques, Hirao and colleagues [19, 20] have attempted to regulate the orientation as well as the size and morphology of the fibrous grains in silicon nitrides. For this, they added various amounts of the above-described seed crystals into a slurry of α-Si_3N_4, 5 wt% Y_2O_3, and 2 wt% Al_2O_3 powder mixture, with 3 wt% dispersing agent and a mixing medium composed of toluene and *n*-butanol. The slurry was tape-cast to form green sheets by using a doctor-blade

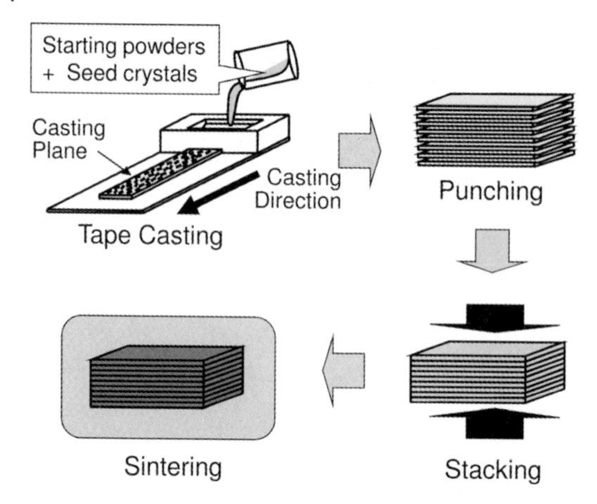

Figure 8.5 Schematic illustration of fabrication procedure for silicon nitride with controlled grain orientation through seeding and tape casting. Reprinted with permission from Ref. [19]; © 1995, Blackwell Publishing, Inc.

equipment. The width of the resultant sheets was 100 mm, and their thickness was controlled to range from 130 to 150 μm by adjusting the blade height and casting rate. Subsequently, the green sheets were punched into rectangular shapes and stacked together at 130 °C under a pressure of 70 MPa aligned in the casting direction (Figure 8.5). The stacked sheets were then calcined at 600 °C to remove the organic components. After cold isostatic pressing at 500 MPa, the compact was sintered in a boron nitride-coated carbon crucible at 1850 °C for 6 h under a nitrogen pressure of 1 MPa.

Figure 8.6 shows the microstructure of the seeded (2 vol%) and tape-cast silicon nitride. Here, the large fibrous grains with a diameter of a few micrometers and a

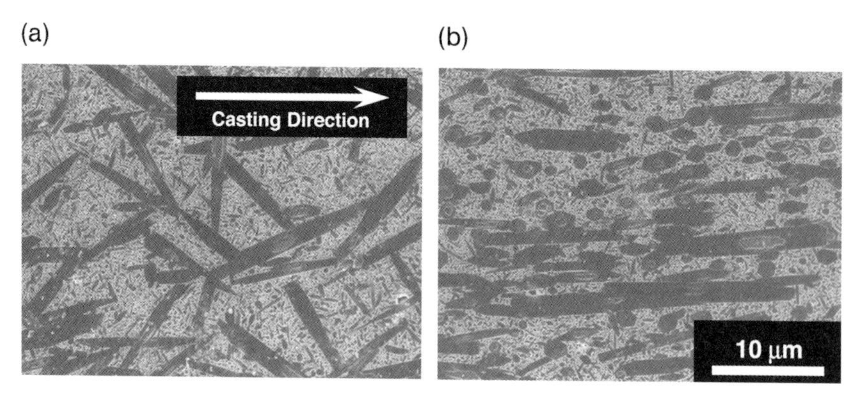

Figure 8.6 Microstructure of the silicon nitride fabricated through dispersing 2 vol% of β-seed crystals into the raw powder slurry of α-phase and sintering additives and tape-casting it (a) parallel to and (b) perpendicular to the casting plane. The fibrous grains elongated from the seed crystals were somewhat randomly oriented, but always stayed within the casting plane. Illustration courtesy of Kiyoshi Hirao *et al.*

length of about 20 μm, embedded in a matrix of fine grains, tend to be aligned parallel to the casting direction. The polished surface parallel to the casting direction exhibits fibrous grains that are somewhat randomly oriented but still stay within the casting plane. It appears that the seed crystals had aligned themselves during the casting process, since fibrous grains were grown epitaxially from the seed crystals. However, it should be noted that the seed crystals were not completely aligned during this process due to their relatively small aspect ratio. Yet, this incomplete alignment sometimes resulted in better mechanical properties rather than a complete alignment, owing to the wedge effects of the inclined fibrous grains (this is discussed later in the chapter).

The fracture toughness and strength, determined using the SEPB method and four-point flexure testing, respectively, as a function of amount of seed crystals, are illustrated graphically in Figure 8.7. When the stress was applied in directions parallel to the grain alignment, it was clear that both the fracture toughness and strength increased with increasing amounts of the seed crystals.

Another feature of this material is its narrow distribution of the fracture strength. When the strength distribution was expressed in Weibull statistics, the Weibull modulus was 46 – substantially higher than the value of 26 obtained for a conventional, self-reinforced silicon nitride. Thus, the seeded and tape-cast silicon nitride showed a synergistic improvement in all of the important fracture attributes, such as fracture strength, fracture toughness, and strength stability (Weibull modulus).

In the seeded and tape-cast silicon nitride, a greater number of fibrous grains were involved with the crack-wake toughening mechanism, so that the toughening mechanism worked more effectively even at short crack extensions. Ohji *et al.* [22] investigated the fracture resistance behavior when a crack propagated in the direction normal to the grain alignment. In this case, the *R*-curves were determined by

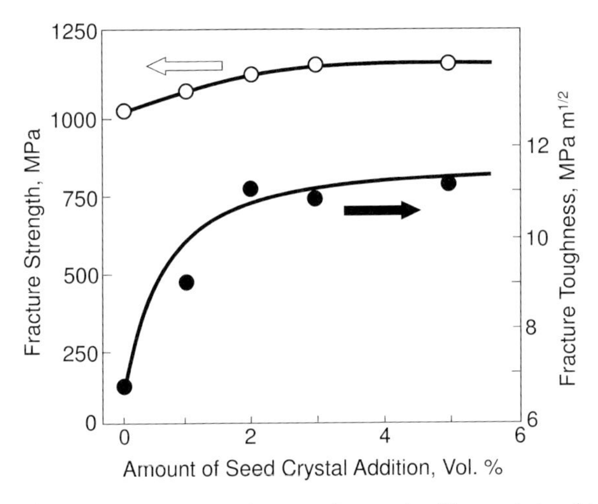

Figure 8.7 Fracture toughness and strength of the seeded and tape-cast silicon nitride as a function of amount of β-seed crystal addition. Reprinted with permission from Ref. [20]; © 1997, Elsevier Ltd.

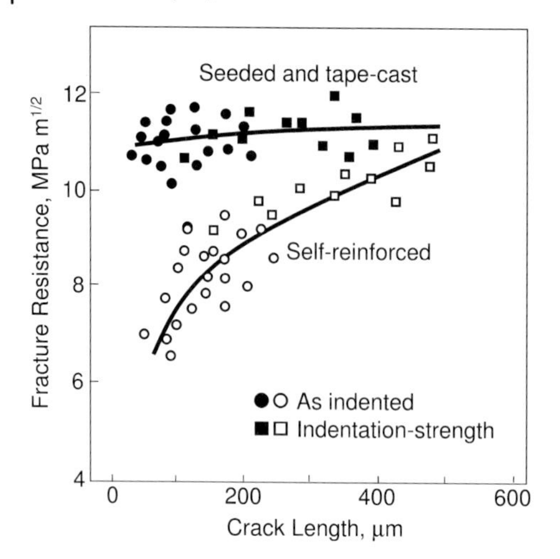

Figure 8.8 *R*-curve behaviors of the self-reinforced (open symbols) and the seeded and tape-cast (closed symbols) materials, determined by as-indented crack lengths (circles) and indentation-strength measurements (squares). Reprinted with permission from Ref. [22]; © 1995, Blackwell Publishing, Inc.

measurements of the initial (as indented) and the fracture instability (indentation-strength measurement) crack lengths, covering a crack length range of up to 500 μm. Flexural test specimens were cut from the sintered billets so that the tensile axis and tensile surface were parallel to the casting direction and the green sheet plane, respectively. Vickers indentations were made so that one diagonal of the Vickers impression and the emanating surface crack were perpendicular to the grain growth direction (to the tensile axis of the flexural test specimens). As shown in Figure 8.8, the fracture resistance of the conventional self-reinforced material, with randomly oriented fibrous grains, increased continuously as a crack propagated, reaching 11 $MPa{\cdot}m^{1/2}$ at about 500 μm, similar to other measurements for self-reinforced silicon nitrides [1, 6, 23–25]. On the other hand, the seeded and tape-cast material exhibited a high initial fracture resistance above 10 $MPa{\cdot}m^{1/2}$, and the fracture resistance remained almost constant in the following crack extension, when a crack propagated in the direction normal to the fiber axis (transverse crack propagation). Thus, although the rising fracture resistances determined from the above method are affected by many experimental factors that are difficult to quantify rigorously, the relative merit of the *R*-curve behavior of the seeded and tape-cast material was apparent and consistent with its superior fracture strength and plateau fracture toughness.

In general, the fibrous grains that span the cracked surfaces behind the crack tip induce two important crack-wake toughening mechanisms, namely elastic grain bridging and frictional grain pull-out. It is claimed that the fibrous grain alignment produces at least two benefits: the first benefit is that a greater number of the fibrous grains are involved with the toughening, while the second benefit is that the grains work more effectively because they stand normal to the crack plane. If the fibrous grains were to be inclined to the crack plane, bending may cause premature grain fracture and thus substantially reduce the toughening effects [25]. These benefits give rise to the above unique *R*-curve behavior of the seeded and tape-cast material in the transverse crack propagation. Because of these benefits, it is speculated that the fracture resistance should rise steeply from the intrinsic material toughness (or toughness before crack extension, which is assumed to be between 2 and 5 $MPa{\cdot}m^{1/2}$ [26]), to about 10 $MPa{\cdot}m^{1/2}$. It is generally known that crack bridging (or elastic bridging) is responsible for short crack toughening, while pull-out is responsible for long crack toughening. This suggests that crack bridging is very effective in the seeded and tape-cast material due to fibrous grain alignment.

The rising *R*-curve behavior generally observed in the conventional self-reinforced material has little effect on catastrophic fracture, as long crack extension (unrealizable in catastrophic fracture from natural flaws) is required to derive such a substantial increase in fracture resistance. On the contrary, the steep rise in the fracture resistance in the seeded and tape-cast materials should be relevant to catastrophic failure. Therefore, high strength and high fracture toughness are compatible in this material, as revealed by Hirao *et al.* [19]. Another benefit brought about by this steep *R*-curve is the narrow strength distribution, as reflected in the higher Weibull modulus. It is known that if the flaws involved in strength determination exhibit rising *R*-curve behavior, then the Weibull modulus will be substantially increased [27–29]. Since even natural flaws that are short can benefit from the very steep *R*-curve and the high initial fracture resistance, this will lead to a high Weibull modulus of 46, as stated above.

Becher and coworkers [21] have also undertaken a systematic investigation of the effects of microstructure on the rise in fracture resistance with crack extension (*R*-curve) and fracture strength in seeded and unseeded silicon nitrides. For this, samples of fibrous β-reinforcing grains of both different sizes and size distributions were prepared. Four representative microstructures of samples created using a mixture of 6.25 wt% Y_2O_3 + 1 wt% Al_2O_3 as sintering additives are shown in Figure 8.9. In the seeded and tape-cast samples, the intentionally added large β-seed crystals (2 vol%) dominated the generation of large fibrous grains, while the unseeded and hot-pressed samples prepared from α-phase powders contained the fibrous grains, due to an $\alpha \rightarrow \beta$ transformation. With increasing hold time during the hot pressing, the fraction of large fibrous grains was increased, and this was accompanied by significant grain growth. The grains in the sample fabricated from the β-phase powders were essentially equiaxed.

The *R*-curve response of the composites was measured using an applied moment double cantilever beam (AMDCB) specimen (10 mm wide × 2.5 mm thick × 35 mm long), which contains a 1 mm-deep and 1.5 mm-wide centerline groove parallel to the length of the sample on one wide surface [30]. In the seeded and tape-cast samples, the AMDCB specimens were prepared with the crack plane perpendicular to the tape-casting direction, in which the large fibrous grains tended to be oriented. The results for the samples of Figure 8.9 are shown in Figure 8.10. The nearly equiaxed material (d) showed the long crack toughness of only 3.5 $MPa{\cdot}m^{1/2}$ and a fracture resistance below 2 $MPa{\cdot}m^{1/2}$ for short crack lengths of <100 μm. For the unseeded and hot-

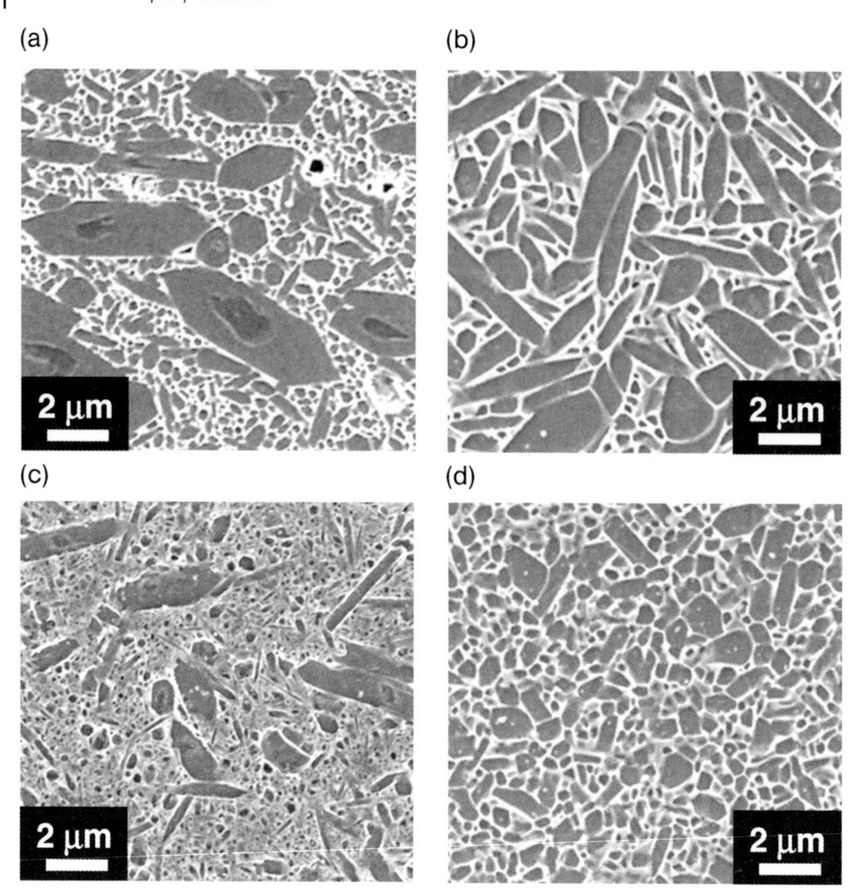

Figure 8.9 Microstructures of silicon nitrides. (a) α-powder with 2 wt% β-seed crystals, tape-cast and gas-pressure-sintered at 1850 °C for 6 h; (b) α-powder without β-seed crystals, hot-pressed at 1750 °C for 2 h; (c) α-powder without β-seed crystals, hot-pressed at 1750 °C for 0.33 h; (d) β-powder without β-seed crystals, hot-pressed at 1750 °C for 2 h. Reprinted with permission from Ref. [21]; © 1998, Blackwell Publishing, Inc.

pressed samples (b and c), the toughness became higher with a longer sintering time, as both the overall grain size distributions and the number of larger fibrous grains were increased. On the other hand, in the seeded and tape-cast sample (a), the fracture resistance was markedly increased, even compared to those of unseeded and hot-pressed materials. The increased fracture resistance was attributable to the larger diameter of the fibrous grains and their alignment. For the tougher materials, the initial portion of the *R*-curve appeared to be extremely steep, which was in agreement with the results of fracture resistance measurements made by Vickers indentation. The incorporation of large β-seed crystals to generate the large fibrous grains, combined with a fine matrix grain size, represents the most effective method for enhancing the *R*-curve response.

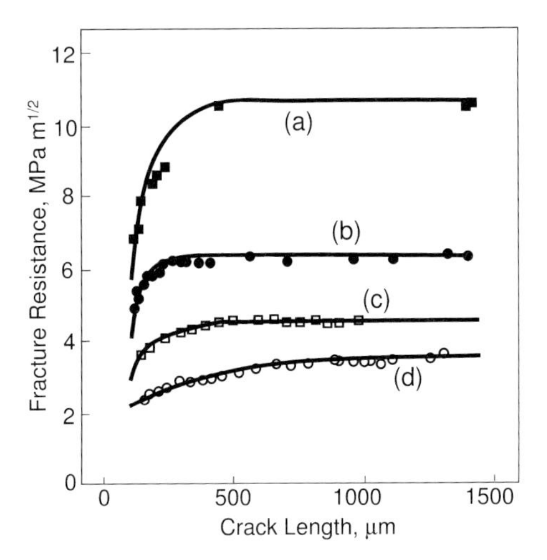

Figure 8.10 *R*-curve behaviors determined by the AMDCB techniques for the silicon nitrides (a), (b), (c) and (d) in Figure 8.9. Reprinted with permission from Ref. [21]; © 1998, Blackwell Publishing, Inc.

8.4
Fibrous Grain-Aligned Silicon Nitrides (Small Grains)

Besides the seeding technique, several other approaches have been developed to align the fibrous grains in silicon nitrides. Typical examples are superplastic forging and superplastic sinter-forging techniques; in the former approach, a deformation stress is applied to a sintered body, whereas in the latter approach the stress is applied to a formed body. Schematic illustrations of these techniques are shown in Figure 8.11. By using either technique, the fibrous grains will be aligned along the tensile direction, or in the direction perpendicular to the compression. The feature of these approaches is an alignment of relatively small fibrous grains, compared to that of the above-described seeding technique.

The superplastic deformation of silicon nitride occurs due to the grain boundary sliding when the grain size is small, even if the grain shape is fibrous. In the case of fibrous grains, these are aligned along the tensile direction in the tensile deformation [31]. By use of this phenomenon, Kondo *et al.* [32, 33] prepared a superplastic plane–strain compressive deformation on silicon nitrides of fine and fibrous grains, and investigated their strength and fracture toughness in comparison to those of the original material. Silicon nitride prepared by gas-pressure sintering with additives of Y_2O_3 and Al_2O_3 was plane–strain-deformed superplastically in a graphite channel die, with a pressure of 49 kN for 3 h at 1750 °C. This deformation resulted in 50% height reduction. A subsequent X-ray diffraction (XRD) analysis confirmed that both the original and deformed specimens consisted of β-silicon nitride grains, without any trace of the α-phase. The original and deformed specimens and their microstructures are shown in Figure 8.12. Whilst the original

Superplastic forging

Original Sinter

Forged Body

Fibrous Grain Alignment

High-Temperature & Compression

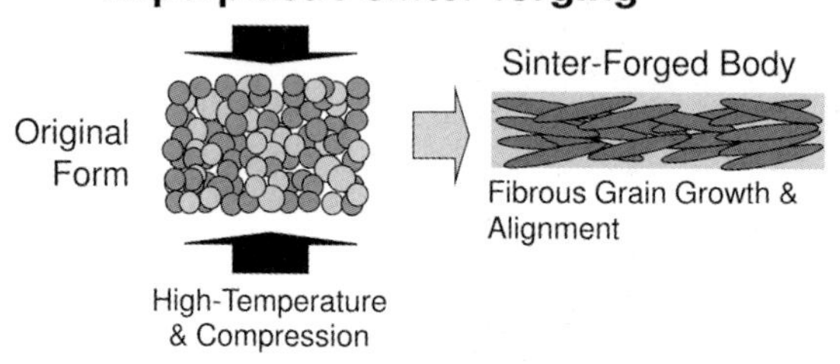

Figure 8.11 Schematic illustration of superplastic forging and superplastic sinter-forging techniques for fabricating fibrous-grain-aligned silicon nitrides.

specimen consisted of randomly oriented fibrous β-silicon nitride grains, its deformed counterpart showed the fibrous grains to be aligned perpendicularly to the pressing direction. Although these tended to be aligned along the extruding direction, the degree of alignment was small compared to that perpendicular to the pressing direction. The fracture strength determined by three-point flexure tests for the original specimen was ~1.1 GPa, and this increased to ~1.7 GPa for the deformed specimen when the stress was applied along the extruding direction. The fracture toughness, as measured using a single-edge-V-notched-beam (SEVNB) method, was also increased from 8.5 to 12 $MPa \cdot m^{1/2}$. The silicon nitride was also reported to show remarkably high fracture energies, from 200 to 630 $J\,m^{-2}$, particularly at high temperatures [34].

This superplastically deformed silicon nitride also showed substantially improved creep resistance at high temperatures, when the stress was applied along the extruding direction. For example, the creep rates of the deformed body in tensile creep tests conducted at 1200 °C was found to be about one order of magnitude lower than that of the original body [35].

Superplastic sinter-forging is a process where sintering and forging (or shaping) are conducted concurrently [36]. This process is advantageous in at least two ways when compared to sintering and forging being conducted individually. The first benefit is that there is no need to prepare a sintered silicon nitride as the starting material; rather, the product of required shape can be obtained directly from the compacted green body. The second benefit is that the deformation can be conducted at a stage

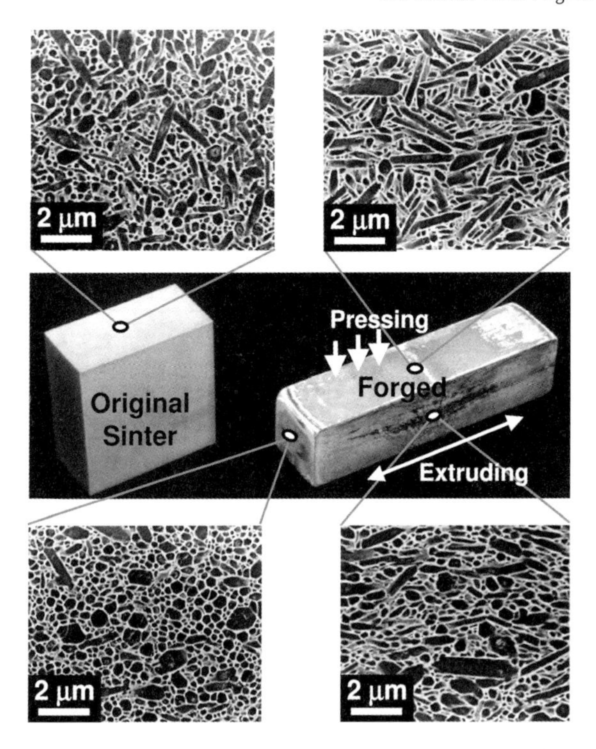

Figure 8.12 Original and superplastically deformed silicon nitride specimens and their microstructures. In the deformed specimen, many of the fibrous grains are aligned along the extruding direction, and perpendicular to the pressing direction. Reprinted with permission from Ref. [32]; © 1998, Blackwell Publishing, Inc.

when the grains are still fine and equiaxed, during the sintering process of silicon nitride. Indeed, a microstructure with fine and equiaxed grains is generally believed to be advantageous for superplastic forging.

A compacted powder mixture of α-Si_3N_4, 5 wt% Y_2O_3 and 3 wt% Al_2O_3 was sinter-forged with a pressure of 49 kN at 1750 °C for 3 h in 0.1 MPa of nitrogen. The superplastically sinter-forged body had approximate dimensions of 80 mm × 20 mm 7.5 mm, while the original formed body had dimensions of 40 mm × 20 mm × 28 mm (see Figure 8.13). The microscopic study revealed that the obtained samples had similar microstructures to the superplastically deformed samples, but that the grain size was relatively small in comparison. When the stress was applied perpendicular to the compressive direction, the fracture strength, as determined by three-point flexure tests, was as high as 2.1 GPa, and the SEVNB fracture toughness was 8.3 $MPa \cdot m^{1/2}$.

As stated above, it has been considered that the fracture toughness of silicon nitride often varies inversely with the fracture strength. The fracture toughness and strength data for various silicon nitrides studied to date (including these materials) are plotted in Figure 8.14. It is known that the antagonistic relationship can be applied to the

Figure 8.13 Original formed silicon nitride body and superplastically sinter-forged body. Reprinted with permission from Ref. [36]; © 1999, Blackwell Publishing, Inc.

silicon nitrides fabricated by the two techniques described here, although both properties are substantially improved compared to those of other silicon nitrides. The superplastically sinter-forged specimen shows a higher fracture strength but a lower fracture toughness compared to those of the superplastically deformed counterpart [32], due to a substantially small grain size. The former also exhibits a higher strength than that reported by Yoshimura *et al.* [37], for silicon nitride with a smaller grain size (~50 nm). This is very likely due to the effects of the steep *R*-curve behavior, as well as to the improved fracture toughness, caused by the grain alignment, and to the reduced sizes of flaws during the deformation.

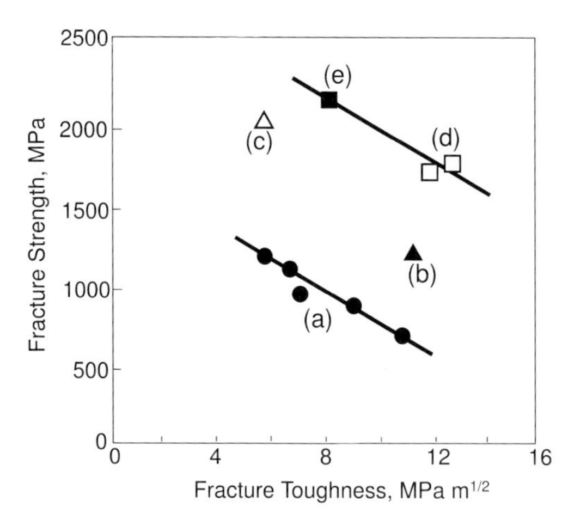

Figure 8.14 Fracture toughness and strength for various silicon nitrides studied so far. papers. The filled circles (a) denote the self-reinforced material [3, 22, 38]; the filled triangle (b) denotes the seeded and tape-cast material [19]; the open triangle (c) denotes the nanosized material [37]; the open square (d) denotes the superplastically forged material [32]; the filled square (e) denotes the superplastically sinter-forged material [36]. Reprinted with permission from Ref. [36]; © 1999, Blackwell Publishing, Inc.

8.5
Grain Boundary Phase Control

8.5.1
Fracture Resistance

Significant improvements in the fracture toughness of silicon nitrides can be obtained also by tailoring the chemistry of the intergranular amorphous phase. Debonding should occur at the interface between the β-silicon nitride grains and the intergranular glassy phase, so that the fibrous β-silicon nitride grains can effectively contribute to the toughening mechanisms such as crack deflection and crack bridging. The interfacial debonding in silicon nitrides can be controlled by the composition of densification additives which ultimately alter the composition of the intergranular amorphous phase [39–42]. It has been suggested that chemical bonding across the interface determines the strength of the interface, while the residual thermal expansion mismatch stress imposed on the interface alters the debonding length. Sun *et al.* [42] described improvements in the fracture resistance of seeded and tape-cast silicon nitride ceramics by controlling the chemistry of the intergranular glassy phase. These authors prepared the seeded and tape-cast silicon nitrides with different yttrium–aluminum ratios, and investigated their *R*-curve behaviors in relation to the microstructural features, using an applied moment double cantilever beam (AMDCB) geometry similar to that of Becher *et al.* [21]. Although, the different sintering additives generally result in microstructures with different grain morphologies and sizes, the seeding method is effective in regulating the grain morphology and size, and consequently represents an idealistic approach for investigating the effect of sintering additives on fracture behavior.

The different yttrium–aluminum ratios in the sintering additives lead to different compositions of the intergranular glass. Examples are shown in Figure 8.15, where spectra acquired using energy-dispersive X-ray spectrometry (EDS) were obtained in transmission electron microscopy (TEM) from glass pockets at triple grain junctions in two samples with additives of 6.25 wt% Y_2O_3–1.0 wt% Al_2O_3 and 4.0 wt% Y_2O_3–2.8 wt% Al_2O_3. The aluminum concentration was increased, while the yttrium concentration was decreased from the former to the latter. The fracture resistance, as determined using the AMDCB techniques, also varied as a function of the yttrium–aluminum ratio in the sintering additives, as shown in Figure 8.16. The stress was applied in a direction parallel to the fibrous grain alignment, whereupon all of the behaviors obtained for different compositions exhibited steeply rising *R*-curves; that is, the fracture resistance of the materials was increased rapidly in line with the initial crack extension. It should be noted, however, that the long crack (or plateau) toughness values increased systematically as the yttrium–aluminum ratios were increased.

When Sun *et al.* [42] used *in-situ*scanning electron microscopy (SEM) to observe the interaction between the cracks and the microstructures, the interfacial debonding behavior between the large fibrous grains and the intergranular glass was seen to vary

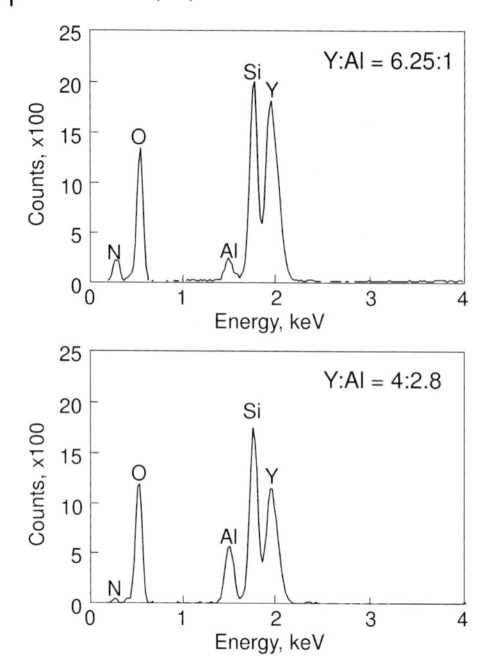

Figure 8.15 Energy-dispersive X-ray spectra collected from glass pockets at triple grain junctions in two samples with additives of 6.25 wt% Y_2O_3–1.0 wt% Al_2O_3 and 4.0 wt% Y_2O_3–2.8 wt% Al_2O_3. The Al concentration was increased, while the Y concentration decreased from the former to the latter. Reprinted with permission from Ref. [42]; © 1998, Blackwell Publishing, Inc.

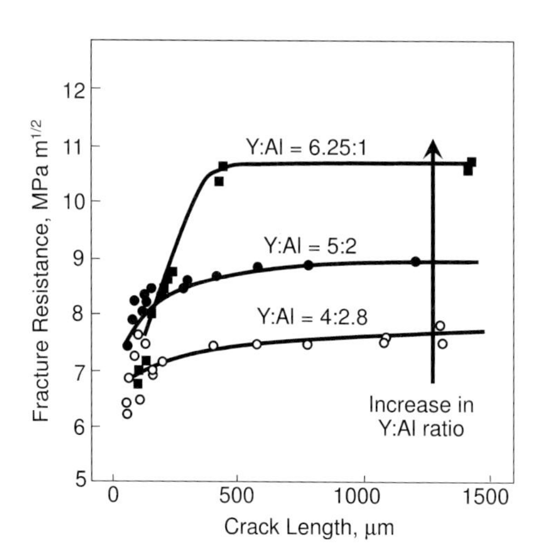

Figure 8.16 *R*-curve behaviors determined by the AMDCB techniques for the seeded and tape-cast silicon nitrides with additives of 6.25 wt% Y_2O_3–1.0 wt% Al_2O_3, 5 wt% Y_2O_3–2.0 wt% Al_2O_3, and 4.0 wt% Y_2O_3–2.8 wt% Al_2O_3. Reprinted with permission from Ref. [42]; © 1998, Blackwell Publishing, Inc.

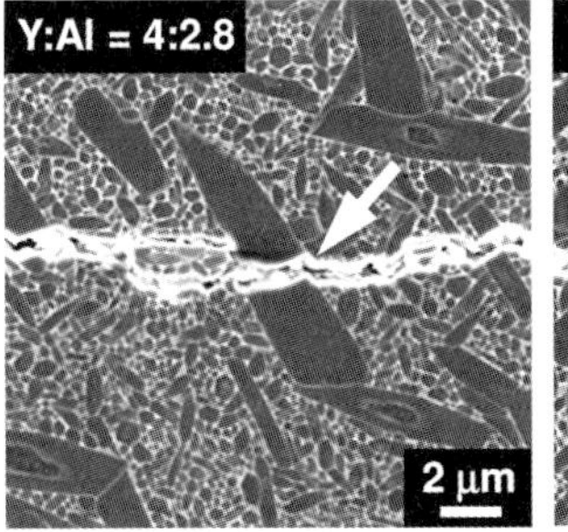

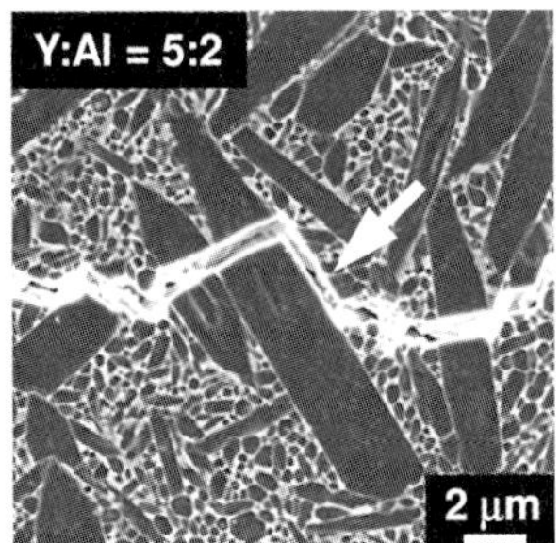

Figure 8.17 *In situ* observations of crack interacting with the large fibrous grains. In spite of the similar angles of incidence, the fibrous grain in the sample of 4.0 wt% Y_2O_3–2.8 wt% Al_2O_3 failed transgranularly, while crack deflection and interfacial debonding occurred in that of 5.0 wt% Y_2O_3–2.0 wt% Al_2O_3. Reprinted with permission from Ref. [42]; © 1998, Blackwell Publishing, Inc.

according to the sintering additives employed. As shown in Figure 8.17, despite similar angles of incidence, the fibrous grain in the sample of 4.0 wt% Y_2O_3–2.8 wt% Al_2O_3 failed transgranularly, whereas crack deflection and interfacial debonding occurred in the 5.0 wt% Y_2O_3–2.0 wt% Al_2O_3 sample. Although, for the smaller angles of incidence, interfacial debonding occurred in both samples, the latter sample showed longer debonding lengths. The critical debonding angle, beyond which no crack deflection could occur at the interface, was also estimated at different yttrium–aluminum ratios. The debonding angle was found to increase upon increasing the yttrium–aluminum ratio, and was about 60°, 70°, and 75° at ratios of 4 : 2.8, 5 : 2, and 6.25 : 1, respectively. This indicates that, in samples with additives of a higher yttrium–aluminum ratio, the interface between the large fibrous grains and the surrounding glass had a lower interfacial debonding energy, and was more readily debonded. Upon increasing the yttrium–aluminum ratio, the amount of large fibrous grains involved with the toughening process, where the crack intersects at the smaller angles, should be significantly increased. This is consistent with results which indicated significant improvements in the *R*-curve behavior and long crack fracture toughness values with high yttrium–aluminum ratios, and that the overall extent of crack deflection and crack bridging during crack extension would be increased upon increasing the same ratios in *in-situ* SEM observations.

Sun *et al.* [43] further investigated the influence of intergranular glass on the debonding behavior of the interface between the prismatic faces of β-silicon nitride whiskers and oxynitride glasses, using model systems based on various Si-(Al)-Y(Ln)-O-N (Ln: rare-earth) oxynitride glasses. It was revealed that the critical debonding angle decreased and the interfacial debonding strength increased on raising the Al and O concentrations in the β′-SiAlON layer formed epitaxially on the β-silicon nitride whiskers. Becher *et al.* [44] also revealed that the incorporation of fluorine into the intergranular film of silicon nitrides allows the crack to circumvent the grains. Atomic cluster analyses have shown the strong Si—O and Al—O bonding across the glass/crystalline interface with the SiAlON layer and the weakened amorphous network of the intergranular film in the presence of fluorine.

8.5.2 Heat Resistance

It is essentially important to control the chemistry of the intergranular amorphous phase in silicon nitrides, not only to improve the fracture toughness but also to produce supreme heat or creep resistance. Although ceramics are expected to be used for heat application, the glassy phases present at the grain boundaries of many ceramics (including silicon nitrides) may become softened at high temperatures, leading to a severe degradation of their mechanical reliabilities. The highly refractory phase is then required for the interface when such ceramics materials are used for high-temperature applications.

During recent decades, extensive efforts have been made to control the grain boundary phase and to improve the heat resistance of silicon nitride, and this has led to significant improvements in high-temperature mechanical reliability. For example, some grades of commercial silicon nitrides have shown excellent creep resistance, even at temperatures above 1400 °C [45, 46]. Subsequent XRD analyses of these materials have revealed $Lu_2Si_2O_7$ and $Lu_4Si_2N_2O_7$ as secondary phases.

Zeng *et al.* [47] attempted to produce a highly refractory phase at grain boundaries in seeded and tape-cast silicon nitrides, by using Lu_2O_3-SiO_2 sintering additive systems to provide rigid grain boundaries. The slurry of α Si_3N_4 powder, 3 wt% β Si_3N_4 seed crystals, 1 wt% SiO_2, and 9 wt% Lu_2O_3 was tape-cast, with sintering being carried out in a 0.9 MPa N_2 atmosphere at 1950 °C for 6 h. A typical example of the obtained microstructure is shown in Figure 8.18. Because of the higher sintering temperature than in previous investigations [19–21], very large fibrous grains, the lengths and diameters of which were typically several tens of micrometers and ∼5 μm, respectively, were grown from fibrous β-Si_3N_4 seed crystals and tended

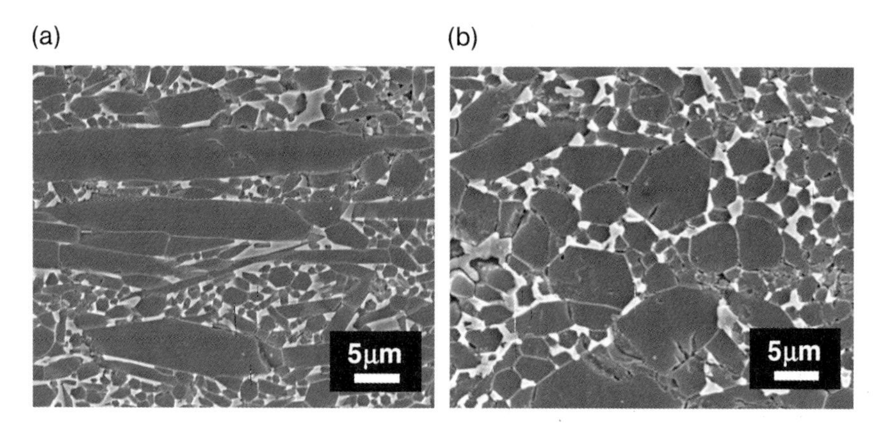

Figure 8.18 Microstructure of the silicon nitride fabricated through dispersing 3 wt% of β-seed crystals into the raw powder slurry of α-phase and Lu_2O_3–SiO_2 sintering additives and subsequent tape-casting. (a) Plane parallel to the casting plane; (b) Plane perpendicular to the casting direction. Note the large fibrous grains elongated from the seed crystals. Illustration courtesy of Yu-Ping Zeng *et al.*

to align along the casting direction. Again, it should be noted that many of these were not completely aligned but rather were substantially inclined in the tape planes.

In these studies, the fracture energy was employed to evaluate the fracture toughness at high temperatures, as the critical stress intensity factor does not clearly describe it when a rising *R*-curve behavior is observed at high temperatures. The chevron-notched beam (CNB) technique was employed for measuring the fracture energy γ [48], which is defined as follows:

$$\gamma = W_{\mathrm{WOF}}/2A \qquad (8.1)$$

where W_{WOF} is the total energy until final failure, obtained from the load–displacement curves, and A is the area of the specimen web portion.

Figure 8.19 shows the fracture strength (determined by three-point flexure tests) and fracture energies measured at room temperature and 1500 °C, in comparison with those of the unseeded material, which was obtained in the same procedures, except for not adding β-Si_3N_4 seed crystal. The stress was applied in two directions parallel (strong direction) and perpendicular (weak direction) to the fibrous grain alignment. The strength measured in the strong direction at 1500 °C was equivalent to the room temperature strength, while the unseeded material showed a substantial strength degradation. The strength in the weak direction was low, but comparable to

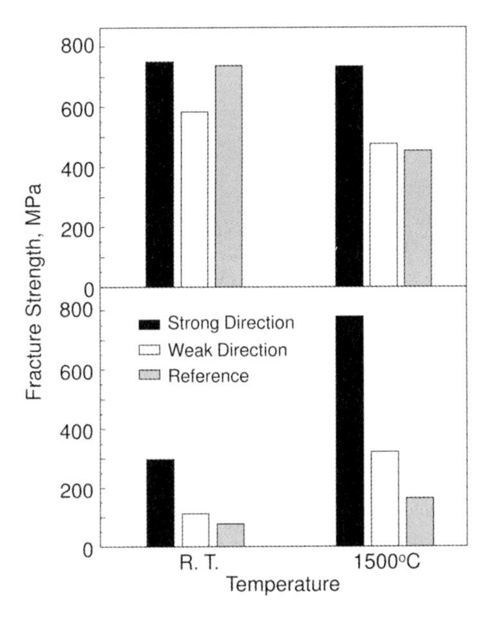

Figure 8.19 Fracture strength and fracture energies at room temperature (R.T.) and 1500 °C of the seeded and tape-cast silicon nitride with Lu_2O_3–SiO_2 sintering additives and reference material (unseeded but otherwise fabricated in the same procedures). The stress was applied in two directions: parallel (strong direction); and perpendicular (weak direction) to the fibrous grain alignment. Illustration courtesy of Yu-Ping Zeng *et al.*

that of the unseeded material at 1500 °C. The excellent fracture strength at high temperatures was very likely attributable to the improved facture resistance. As shown in Figure 8.19, the fracture energy of the unseeded material was about 100 and 450 $J m^{-2}$, while that of the seeded material in the strong direction was about 300 and 800 $J m^{-2}$, at room temperature and 1500 °C, respectively. These large fracture energies apparently were attributable to enhanced crack shielding effects of the bridging and frictional pull-out of the very large fibrous grains. It should be noted however that, even in the weak direction, the seeded material showed substantially high fracture energies, particularly at the higher temperature. This was presumably due to the large number of the substantially inclined fibrous grains, which gave rise to grain bridging and frictional pull-out effects, even in the weak direction. The inclined fibrous grains are also beneficial to toughening in the strong direction due to so-called "wedge effects" [49], as shown in Figure 8.20. Compared to the completely aligned grains, further energy is required for the grains to be pulled out, owing to the additional bending applied to the grains. However, this mechanism only works effectively when the grains are strong enough, or when the surrounding phases are weak enough, for the inclined grains not to be fractured during crack opening by such additional bending forces. Thus, a large fracture energy can be obtained in seeded materials which contain large fibrous grains, and at high temperatures where the grain boundary phases are softened.

8.6 Fibrous Grain-Aligned Porous Silicon Nitrides

Although, in structural materials, the pores are generally believed to cause a deterioration in mechanical reliability, this is not always true. Rather, the pores may cause an improved or even unique mechanical performance of the material which cannot be attained in the dense counterpart, especially when the porous microstructure is carefully tailored. Two examples of porous silicon nitrides are described in the

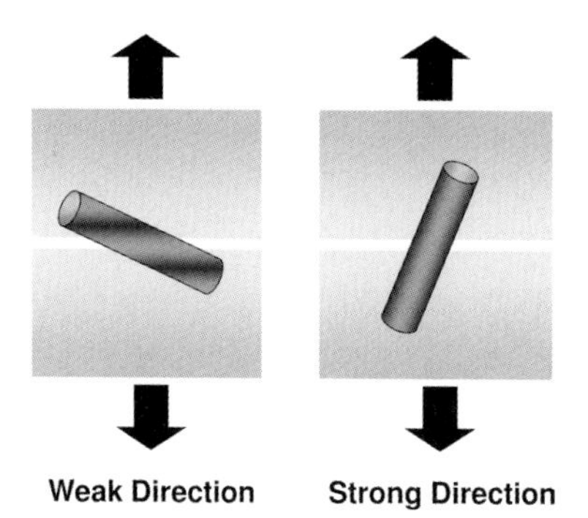

Figure 8.20 Schematic of alignment with tilted fibrous grains. Because of the wedge effects, tilted fibrous grains give rise to improved properties both in the strong (parallel to the alignment) and weak (perpendicular to the alignment) directions.

following sections. The first example is the porous silicon nitride fabricated by tape-casting whiskers [50, 51], where the relatively large fibrous grains have been aligned uniaxially and show a high fracture strength in excess of 1 GPa, as well as a high damage tolerance, depending on the porosity. The fracture energy was several-fold greater than that of dense silicon nitride, and primarily attributable to a grain "pull-out" mechanism enhanced by the pores. The second example is a porous silicon nitride with 25% porosity, fabricated by using a sinter-forging technique [52]. This material contained aligned fibrous grains of a relatively small size, and exhibited not only an excellent strain tolerance but also a 25% reduction in weight.

8.6.1 Porous Silicon Nitride Through Tape-Casting

For the tape-cast porous material, β-Si_3N_4 seed crystals were mixed with 5 wt% Y_2O_3 and 2 wt% Al_2O_3 as starting powders. The green sheets formed by tape-casting were stacked and bonded under pressure, and sintering was performed at 1850 °C under a nitrogen pressure of 1 MPa. The texture of porous silicon nitride with a porosity of 14% is shown in Figure 8.21. The fibrous grains of silicon nitride are seen to be well-aligned towards the casting direction, while the pores – the shapes of which are mostly plate-like along the same direction – exist among the grains.

Figure 8.22 shows the porosity dependencies of fracture strength and fracture energy when the stress is applied in the parallel direction to the grain alignment of the porous silicon nitride. The strength was measured by the three-point flexural test, while the fracture energy was determined by the CNB technique. The fracture strength became larger as the porosity decreased such that, in the porosity range below 5%, the strength attained was above 1.5 GPa. This value was almost comparable to that of fibrous-grain-aligned dense silicon nitride fabricated through superplastic forging, as described above [33]. When the fibrous grains were aligned, the pores around the grains promoted debonding between the interlocking fibrous

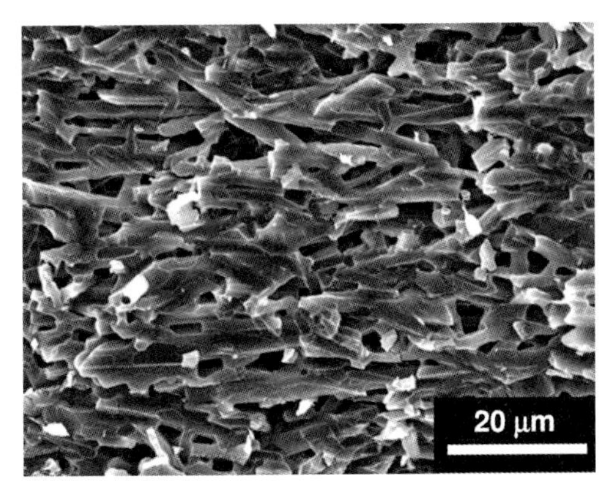

Figure 8.21 Microstructures of porous silicon nitride with large fibrous grain alignment (porosity 14%). Reprinted with permission from Ref. [50]; © 2000, Blackwell Publishing, Inc.

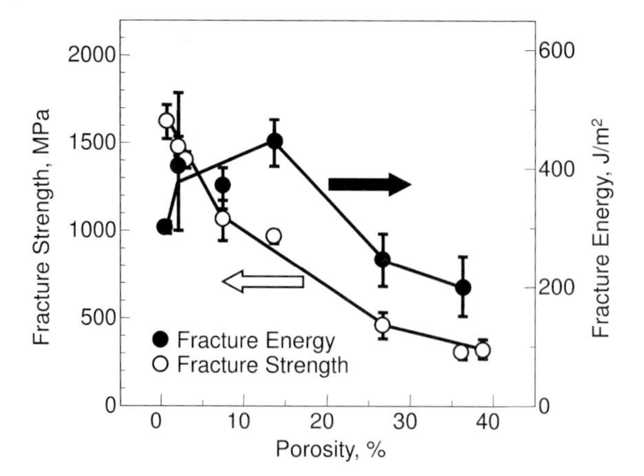

Figure 8.22 Porosity dependence of fracture strength and fracture energy of porous silicon nitride with large fibrous grain alignment. Stress is applied parallel to the alignment direction. The vertical bars indicate standard deviations. Reprinted with permission from Ref. [51]; © 2004, Elsevier Ltd.

grains, without breaking. Consequently, even a small amount of pores could enhance the grain bridging so as to improve the strength. On the other hand, the fracture energy of the porous silicon nitride ranged from 300 to 500 J m^{-2} in the porosity range below 20%; these values were considerably higher than were found in other studies [34, 48]. The large fracture energy was mainly due to the effect of crack-bridging by aligned fibrous grains and/or a pull-out of the grains. Debonding was promoted by the existence of pores, and the aligned grains bridging the crack or interlocking with each other were drawn apart without breaking, which in turn resulted in an increase of the sliding resistance. Figure 8.23 shows the SEM image of

Figure 8.23 Fracture surface of porous silicon nitride with large fibrous grain alignment (porosity 14%). The large fracture energy of this porous silicon nitride is attributable to sliding resistance associated with the drawn-out fibrous grains enhanced by pores. Reprinted with permission from Ref. [50]; © 2000, Blackwell Publishing, Inc.

the ligament area of the fractured surface; clearly, the protruding grains and holes from which the fibrous grains have been pulled out, thus supporting these speculations. As the porosity approached 15% the fracture energy became larger, but then decreased monotonously when the porosity exceeded 15%. This drop in fracture energy was presumed due to reductions in substantial bridging and/or the pull-out area.

It is possible to estimate fracture toughness from fracture energy using the following relationship:

$$K_{IC} = (2E'\gamma)^{1/2} \tag{8.2}$$

where E' is the Young's modulus for the plane–strain condition, which is given by $E' = E/(1-\nu^2)$, where ν is Poisson's ratio. The obtained fracture toughness is an averaged value when the toughness varies with crack extension (R-curve). Figure 8.24 shows the porosity dependence of the obtained fracture toughness compared to that of the fracture strength. A high fracture toughness above 17 MPa·m$^{1/2}$, as well as a high strength above 1.5 GPa, was attained in the porosity range below 5%. In particular, the fracture toughness of a specimen with a very small porosity (<5%) exceeded that of a fully dense specimen.

Figure 8.25 shows the dependencies of fracture strength and fracture energy on porosity in the perpendicular direction to the grain alignment of the porous silicon nitride. The values of fracture strength and fracture energy in the porosity range from 0% to 35% indicated lower values than in the parallel direction. Nonetheless, the fracture energy was relatively high compared to that reported for an isotropic dense silicon nitride, of about 70 J m^{-2} [50]. Consequently, owing to the existence of tilted fibrous grains, a wedge effect was also caused in the perpendicular direction to the grain alignment.

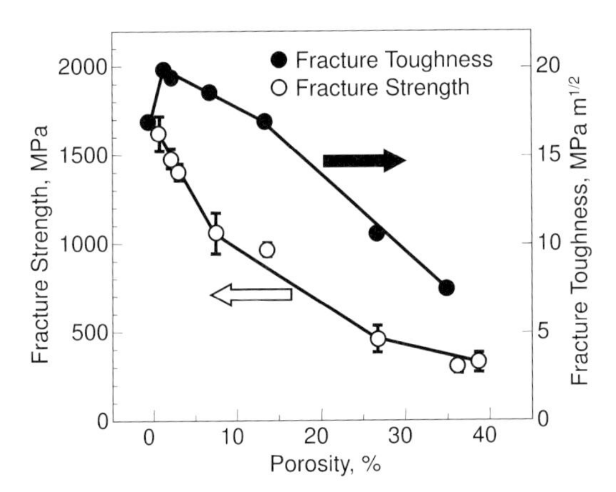

Figure 8.24 Porosity dependence of fracture strength and fracture toughness, K_{IC} of porous silicon nitride with large fibrous grain alignment. Reprinted with permission from Ref. [51]; © 2004, Elsevier Ltd.

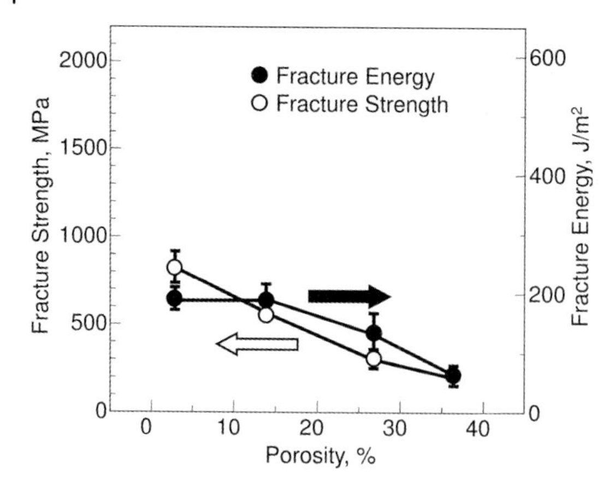

Figure 8.25 Porosity dependence of fracture strength and fracture energy of porous silicon nitride with large fibrous grain alignment. Stress is applied perpendicularly to the alignment direction. The vertical bars indicate standard deviations. Reprinted with permission from Ref. [51]; © 2004, Elsevier Ltd.

8.6.2
Porous Silicon Nitride Through Sinter-Forging

The sinter-forging technique described above has been applied for fabricating the porous silicon nitride with aligned fibrous grains. α-Si_3N_4 powder was used as the starting powder, and mixed with 5 wt% Y_2O_3 as the sintering additive. The initial α-phase powders with equiaxed shape are fibrous through phase transformation to β-phase and Ostwald ripening during soaking at elevated temperatures. In this process, the deformation pressure was applied after grain elongation until the desired porosity was reached; this in turn led to a porous microstructure with aligned silicon nitride fibrous grains. The partial sinter-forging was conducted using a hot-press furnace. After soaking for 30 min at 1850 °C, a uniaxial mechanical compressive force of 60 kN (the pressure averaged 30 MPa during forging) was applied for 150 min at the same temperature. The piston transfer was controlled so that the porosity reached 25%, and the resultant porosity was 24%. It should be noted that during the first soaking without applying the force, some α- to β-phase transformation should occur. The microstructural observation revealed that the fibrous grains, the size of which was substantially small compared to that of the porous material through tape-casting, tended to align in the direction (see Figure 8.26). The relative density of the green compact before sintering was approximately 45~50%; therefore, a height reduction of 25~30% could be applied by the partial sinter-forging, resulting in the anisotropic microstructure.

The elastic modulus, measured using the pulse echo method, was about 180 GPa, and the fracture strength determined by the three-point flexure test was about 800 MPa in the grain-alignment direction. This strength, which is relatively high for the large porosity of 25%, is almost equivalent to that of commercially available

Figure 8.26 Fracture surfaces of porous silicon nitride with fine fibrous grain alignment fabricated through a sinter-forging technique. The porosity is 24%. Illustration courtesy of Naoki Kondo *et al.*

dense silicon nitrides. This high strength of porous materials is very likely due to the reduced flaw size of fracture origin, in addition to the toughening effects of fibrous grain alignment. Even if large defects exist in the green compact, their sizes can be substantially reduced by forging [32], as the green compact shrinks in the pressing direction.

8.6.3 Comparison of Properties of Porous and Dense Silicon Nitrides

The fracture energy, strength, fracture strain and lightweight nature of the porous silicon nitrides introduced here, compared to two conventional dense materials, are shown graphically in Figure 8.27. Here, PSN-A and -B are porous materials fabricated

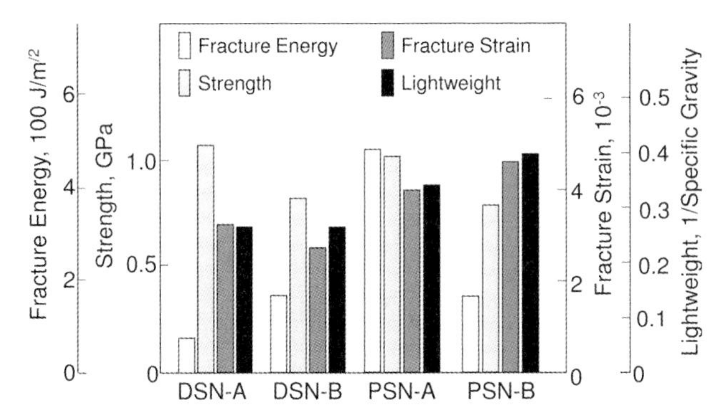

Figure 8.27 Comparison of fracture energy, strength, fracture strain, and lightweight nature of various porous and dense silicon nitrides. PSN-A and -B are porous materials fabricated through tape-casting (14% porosity) and sinter-forging (24% porosity), respectively; DSN-A and -B are the fine-grained and coarse grained (*in situ*-toughened) dense materials. Reprinted with permission from Ref. [53]; © 2002, Elsevier Ltd.

through tape-casting (14% porosity) and sinter-forging (24% porosity), respectively, while DSN-A and -B are fine-grained and coarse grained (*in situ*-toughened) dense materials [53]. For each of these materials the fracture energy was measured using the CNB technique. The fracture strain can be calculated given by dividing the strength by the elastic modulus, and the lightweight nature is expressed as a reciprocal of the specific gravity. When applying a stress parallel to the grain alignment, the porous materials show almost equivalent strength to the dense materials. However, with regards to fracture energy or damage tolerance, the porous materials show greater advantages than the dense materials. A typical case is that of PSN-A, where the coarse fibrous grains are aligned. In contrast, in PSN-B (fine fibrous grain alignment) it is possible to introduce pores with only a minimal sacrifice of strength. In this case, the elastic modulus is decreased by almost half at 24% porosity, while the fracture strain or strain tolerance become substantially larger. An increased porosity also further improves the light weight, which is an intrinsic characteristic of ceramic materials; typically, 24% porosity will lead to a 24% reduction in weight. Consequently, the existence of pores does not necessarily lead to a degradation of the properties for ceramics. Rather, by controlling the size, shape, and alignment of the pores and the matrix grains, it is possible to provide excellent and/or unique performances.

In brittle ceramic materials, any strains generated undesirably cannot be relaxed by plastic deformation, and this may often lead to an abrupt fracture under even a small strain; this is particularly the case for high-elastic-modulus materials. However, if the elastic modulus can be reduced while retaining the strength, then the strain-to-failure ratio can be increased, thus ensuring structural reliability.

References

1 Li, C.W. and Yamanis, J. (1989) *Ceram. Eng. Sci. Proc.*, **10**, 632–645.
2 Matsuhiro, K. and Takahashi, T. (1989) *Ceram. Eng. Sci. Proc.*, **10**, 807–816.
3 Kawashima, T., Okamoto, H., Yamamoto, H., and Kitamura, A. (1991) *J. Ceram. Soc. Jpn*, **99**, 320–323.
4 Mitomo, M. and Uenosono, S. (1992) *J. Am. Ceram. Soc.*, **75**, 103–108.
5 Becher, P.F. (1991) *J. Am. Ceram. Soc.*, **74**, 255–269.
6 Steinbrech, R.W. (1992) *J. Eur. Ceram. Soc.*, **12**, 131–142.
7 Lee, D.D., Kang, S.J.L., Petzow, G., and Yoon, D.N. (1990) *J. Am. Ceram. Soc.*, **73**, 767–769.
8 Krämer, M. and Hoffmann, M.J. (1993) *J. Am. Ceram. Soc.*, **76**, 2778–2784.
9 Han, S.-M. and Kang, S.-J.L. (1995) *MRS Bull.*, **20**, 33–37.
10 Dressler, W., Kleebe, H.J., Hoffmann, M.J., Rühle, M., and Petzow, G. (1996) *J. Eur. Ceram. Soc.*, **16**, 3–14.
11 Tajima, Y. and Urashima, K. (1994) Improvement of Strength and Toughness of Silicon Nitride Ceramics, in *Tailoring of Mechanical Properties of Si3N4 Ceramics* (eds M.J. Hoffmann and G. Petzow), Kluwer Academic Publishers, Dordrecht, The Netherlands, pp. 101–109.
12 Becher, P.F., Lin, H.T., Hwang, S.L., Hoffmann, M.J., and Chen, I.-W. (1993) The Influence of Microstructure on the Mechanical Behavior of Silicon Nitride Ceramics, in *Silicon Nitride Ceramics* (eds I-.W. Chen, P.F. Becher, M. Mitomo, G. Petzow, and T.S. Yen), Materials Research Society, Pittsburgh, PA, pp. 147–158.
13 Hirosaki, N., Akimune, Y., and Mitomo, M. (1993) *J. Am. Ceram. Soc.*, **76**, 1892–1894.

14 Wittmer, D.E., Doshi, D., and Paulson, T.E. (1992) *Ceram. Eng. Sci. Proc.*, **13**, 907–917.

15 Hirao, K., Nagaoka, T., Brito, M.E., and Kanzaki, S. (1994) *J. Am. Ceram. Soc.*, **77**, 1857–1862.

16 Emoto, H. and Mitomo, M. (1997) *J. Eur. Ceram. Soc.*, **17**, 797–804.

17 Hirao, K., Tsuge, A., Brito, M.E., and Kanzaki, S. (1993) *J. Ceram. Soc. Jpn*, **101**, 1071–1073.

18 Inagaki, Y., Ando, M., and Ohji, T. (2001) *J. Ceram. Soc. Jpn*, **109**, 978–980.

19 Hirao, K., Ohashi, M., Brito, M.E., and Kanzaki, S. (1995) *J. Am. Ceram. Soc.*, **78**, 1687–1690.

20 Kanzaki, S., Brito, M.E., Valecillos, M.C., Hirao, K., and Toriyama, M. (1997) *J. Eur. Ceram. Soc.*, **17**, 1841–1847.

21 Becher, P.F., Sun, E.Y., Plucknett, K.P., Alexander, K.B., Hsueh, C.H., Lin, H.T., Waters, S.B., Westmoreland, C.G., Kang, E.S., Hirao, K., and Brito, M.E. (1998) *J. Am. Ceram. Soc.*, **81**, 2821–2830.

22 Ohji, T., Hirao, K., and Kanzaki, S. (1995) *J. Am. Ceram. Soc.*, **78**, 3125–3128.

23 Ramachandran, N. and Shetty, D.K. (1991) *J. Am. Ceram. Soc.*, **74**, 2634–2641.

24 Li, C.-W., Lee, D.-J., and Lui, S.-C. (1992) *J. Am. Ceram. Soc.*, **75**, 1777–1785.

25 Becher, P.F., Hwang, S.L., Lin, H.T., and Tiegs, T.N. (1994) Microstructural Contribution to the Fracture Resistance of Silicon Nitride Ceramics, in *Tailoring of Mechanical Properties of Si3N4 Ceramics* (eds M.J. Hoffmann and G. Petzow), Kluwer Academic Publishers, Dordrecht, The Netherlands, pp. 87–100.

26 Tanaka, I., Pezzotti, G., Okamaoto, T., and Miyamoto, Y. (1989) *J. Am. Ceram. Soc.*, **72**, 1656–1660.

27 Kendall, K., Alford, N.McN., Tan, S.R., and Birchall, J.D. (1986) *J. Mater. Res.*, **1**, 120–123.

28 Cook, R.F. and Clarke, D.R. (1988) *Acta Metall.*, **36**, 555–562.

29 Shetty, D.K. and Wang, J.-S. (1989) *J. Am. Ceram. Soc.*, **72**, 1158–1162.

30 Freiman, S.W., Mulville, D.R., and Mast, P.W. (1973) *J. Mater. Sci.*, **8**, 1527–1533.

31 Kondo, N., Sato, E., and Wakai, F. (1998) *J. Am. Ceram. Soc.*, **81**, 3221–3227.

32 Kondo, N., Ohji, T., and Wakai, F. (1998) *J. Am. Ceram. Soc.*, **81**, 713–716.

33 Kondo, N., Ohji, T., and Wakai, F. (1998) *J. Mater. Sci. Lett.*, **17**, 45–47.

34 Kondo, N., Inagaki, Y., Suzuki, Y., and Ohji, T. (2001) *J. Am. Ceram. Soc.*, **84**, 1791–1796.

35 Kondo, N., Suzuki, Y., Brito, M.E., and Ohji, T. (2001) *J. Mater. Res.*, **16**, 2182–2185.

36 Kondo, N., Suzuki, Y., and Ohji, T. (1999) *J. Am. Ceram. Soc.*, **82**, 1067–1069.

37 Yoshimura, M., Nishioka, T., Yamakawa, A., and Miyake, M. (1995) *J. Ceram. Soc. Jpn*, **103**, 407–408.

38 Kawashima, T., Okamoto, H., Yamamoto, H., and Kitamura, A. (1990) Characteristic Variety of Silicon Nitride Made Through Microstructural Control, in *Silicon Nitride Ceramics II* (eds M. Mitomo and S. Somiya), Uchida Rokakuho Publishing, Tokyo, Japan, pp. 135–146.

39 Tajima, Y. (1993) *Mater. Res. Soc. Symp. Proc.*, **287**, 189–196.

40 Urashima, K., Tajima, Y., and Watanabe, M. (1992) *R*-curve and Fatigue behavior of Gas Pressure Sintered Silicon Nitride, in *Fracture Mechanics of Ceramics*, vol. **9** (ed. R.C. Bradt), Plenum Press, New York, USA, pp. 235–310.

41 Wötting, G. and Ziegler, G. (1984) *Ceram. Int.*, **10**, 18–22.

42 Sun, E.Y., Becher, P.F., Plucknett, K.P., Hsueh, C.H., Alexander, K.B., Waters, S.B., Westmoreland, C.G., Hirao, K., and Brito, M.E. (1998) *J. Am. Ceram. Soc.*, **81**, 2831–2840.

43 Sun, E.Y., Becher, P.F., Hsueh, C.H., Painter, G.S., Waters, S.B., Hwang, S.L., and Hoffmann, M.J. (1999) *Acta Mater.*, **47**, 2777–2785.

44 Becher, P.F., Painter, G.S., Sun, E.Y., Hsueh, C.H., and Lance, M.J. (2000) *Acta Mater.*, **48**, 4493–4499.

45 Ohji, T. (2001) *Ceram. Sci. Eng. Proc.*, **22**, 159–166.

46 Lofaj, F., Wiederhorn, S.M., Long, G.G., Jemian, P.R., and Ferber, M.K. (2001) *Ceram. Sci. Eng. Proc.*, **22**, 166–173.

47 Zeng, Y.P., Yang, J.F., Kondo, N., Ohji, T., Kita, H., and Kanzaki, S. (2005) *J. Am. Ceram. Soc.*, **88**, 1622–1624.

48 Ohji, T., Goto, Y., and Tsuge, A. (1991) *J. Am. Ceram. Soc.*, **74**, 739–745.

49 Kovalev, S., Miyajima, T., Yamauchi, Y., and Sakai, M. (2000) *J. Am. Ceram. Soc.*, **83**, 817–824.

50 Inagaki, Y., Ohji, T., Kanzaki, S., and Shigegaki, Y. (2000) *J. Am. Ceram. Soc.*, **83**, 1807–1809.

51 Inagaki, Y., Shigegaki, Y., Ando, M., and Ohji, T. (2004) *J. Eur. Ceram. Soc.*, **24**, 197–200.

52 Kondo, N., Suzuki, Y., and Ohji, T. (2001) *J. Mater. Res.*, **16**, 32–34.

53 Inagaki, Y., Kondo, N., and Ohji, T. (2002) *J. Eur. Ceram. Soc.*, **22**, 2489–2494.

9
Dislocations in Ceramics

Terence E. Mitchell

9.1
Introduction

The concept of the dislocation was first introduced in 1934 to explain why pure metals are soft and easy to deform plastically. Ceramics, by comparison, were seen as hard and usually brittle. In fact, a great deal of early work had been done by 1934 on the deformation of single crystals of alkali halides and natural minerals. In their classic book, Schmid and Boas [1] devoted a chapter to the plasticity and strength of ionic crystals. They presented a table listing the glide and twin elements of 30 different minerals, which was essentially a condensation of a much longer listing from the 1927 Landolt–Börnstein tables [2]. Much of these investigations were conducted by Otto Mügge, whose pioneering studies included the determination of not only the glide elements of copper, silver and gold [3], but also the glide and twinning elements of literally dozens of different minerals.

Mügge and others found that the only minerals that could easily be deformed under ambient conditions were the alkali halides and a few sulfides and carbonates. An exception to this was periclase (MgO), which deformed by $\{1\bar{1}0\}\langle 110\rangle$ dodecahedral glide in the same way as halite (NaCl). A more recently discovered exception is $SrTiO_3$ with the cubic perovskite structure, which can be deformed plastically at ambient and high temperatures but is brittle at intermediate temperatures (see Section 9.4.7). Other oxides and silicate minerals either cleaved or twinned when attempts were made to deform them at normal temperatures and pressures [1].

The application of dislocation theory to ceramics developed slowly over the years, with the exception of the alkali halides [4, 5]. The main reason for this was the difficulty of inducing plastic flow in many ceramics, as well as the scarcity of available crystals. This experimental barrier was overcome by Wachtman and Maxwell [6], who developed high-temperature mechanical testing techniques and were able to deform single crystals of sapphire (α-Al_2O_3) at temperatures above 900 °C, periclase (MgO) at temperatures above 1100 °C, and rutile (TiO_2) at temperatures above 600 °C. This opened up the field of dislocations and plastic deformation of ceramics, and was followed by the seminal reports of Kronberg [7] on dislocations in sapphire, and by Hornstra [8] on dislocations in spinel ($MgAl_2O_4$). The 1970s and 1980s saw much

Ceramics Science and Technology Volume 2: Properties. Edited by Ralf Riedel and I-Wei Chen

ISBN: 978-3-527-31156-9

activity in this field, especially in the application of electron microscopy to the study of dislocations in ceramics, and there was the usual mixture of predicted and unpredicted observations, stimulating the need for more research. "Ductile" ceramics have remained elusive, however, and perhaps that is the reason for the present quiescent stage of research. "Toughened" ceramics based on zirconia and silicon nitride emerged as a reality [9]; however, discussions of this topic are beyond the scope of this chapter.

In this chapter, attention will be focused on oxide ceramics. Related reviews are available on ionic crystals by Haasen [5] and by Sprackling [4], on covalent crystals by Hirsch [10] and by Alexander [11], and on silicate minerals by Paterson [12]. Previous reviews on oxides by Wachtman [13], Terwillinger and Radford [14], Bretheau *et al.* [15], and Mitchell and Heuer [16] are also available.

9.2 The Critical Resolved Shear Stress

9.2.1 Experimental Observations

Most ceramics are brittle at low and medium temperatures, and can be deformed plastically above the brittle-to-ductile transition temperature. The critical resolved shear stress (CRSS) then decreases rapidly with increasing temperature. In many cases there is a linear relationship between log(CRSS) and temperature, as first shown by Castaing for semiconductor crystals [17]. Examples are shown in Figures 9.1–9.4. For MgO in Figure 9.1 [5], the relationship is well obeyed for both easy slip on the

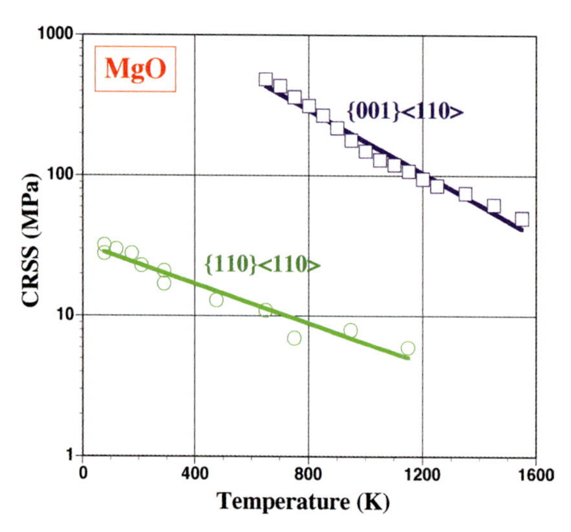

Figure 9.1 Log(CRSS) versus temperature for the $\{110\}\langle 110\rangle$ and $\{001\}\langle 110\rangle$ slip systems in MgO. Data from Ref. [5].

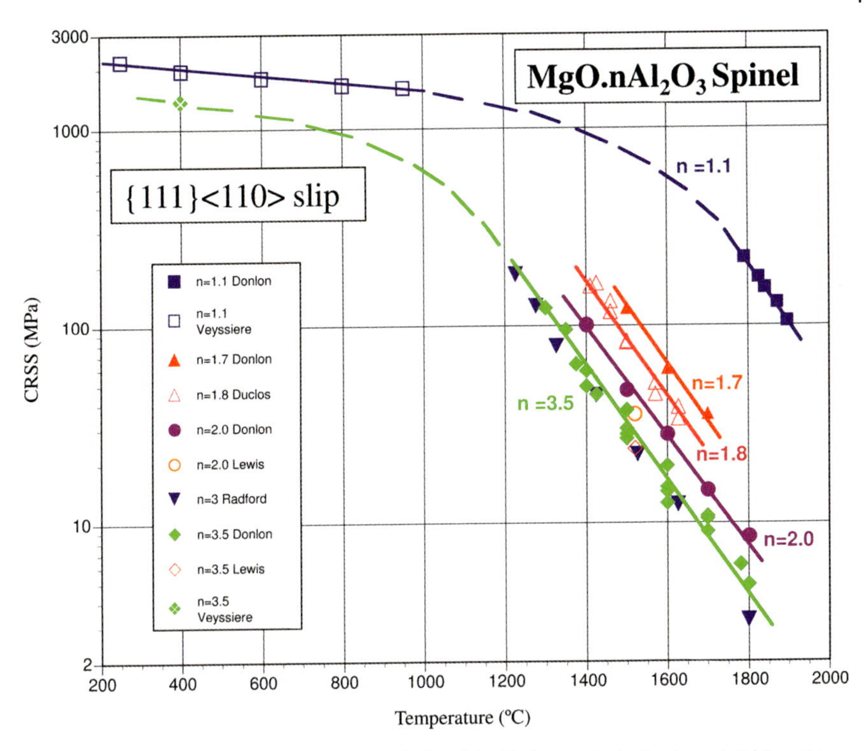

Figure 9.2 Log(CRSS) versus temperature for {111}⟨110⟩ slip in spinel of various stoichiometries. The low-temperature data represent either high-pressure confinement experiments or hardness tests [18].

{110} plane and hard slip on the {001} plane; the CRSS is approximately 40-fold higher for the latter compared to the former at 800 K. MgO can be deformed at very low temperatures on the {110} slip plane. For MgO·$n$$Al_2O_3$ spinel in Figures 9.2 and 9.3 [18], the behavior is more complicated. The log(CRSS)–T relationship is obeyed at very high temperatures for both {111}⟨110⟩ and {110}⟨110⟩ slip. The CRSS values are about the same for the two systems for stoichiometric spinel ($n = 1$), but lower on the {110} plane for nonstoichiometric compositions. It should be noted that the CRSS is almost two orders of magnitude lower for $n = 3.5$ crystals than for $n = 1.1$ crystals. The data at low temperatures were obtained either under a high confining pressure [19, 20] or under a hardness indent [21]. The interpolated lines between the high- and low-temperature data show a sharp change in slope for {111}⟨110⟩ slip (Figure 9.2), and a yield stress anomaly with a maximum at ~1200 °C for {110}⟨110⟩ slip. Dissociated screw dislocations cross-slip from {110} to {001} planes at low temperatures [19], thereby becoming immobilized, and possibly this is the reason for the yield stress anomaly [18]. The softening of nonstoichiometric crystals has been ascribed to cation vacancies assisting kink diffusion at high temperatures or kink nucleation on partials (as described in Section 9.2.3). The behavior of sapphire in Figure 9.4 is different. The log(CRSS)–T

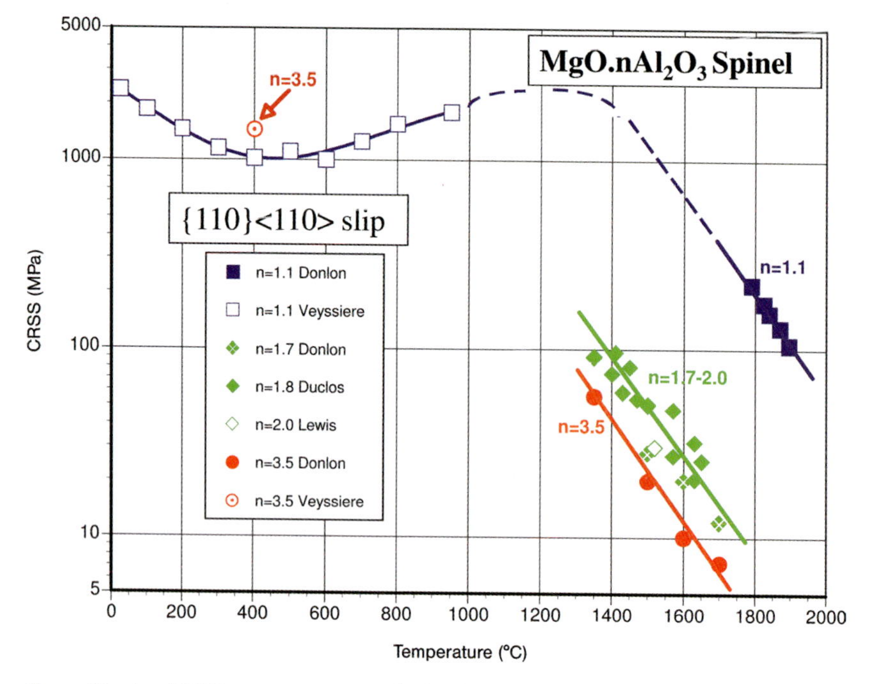

Figure 9.3 Log(CRSS) versus temperature for {110}⟨110⟩ slip in spinel of various stoichiometries. The low-temperature data represent either high-pressure confinement experiments or hardness tests. Note the yield stress anomaly for $n = 1.1$ crystals [18].

law is obeyed approximately for both basal and prism plane slip. However, the slopes are different so that the curves cross at ~900 K, with prism plane slip becoming easier at lower temperatures. The curves in Figure 9.4 are not straight lines but were fitted from Eq. (1), as described in Section 9.2.4. The mechanical behavior of other ceramics is described in Section 9.4.

9.2.2
Kink Mechanism for Deformation

The high strength of most ceramics is due to the difficulty of moving dislocations through the lattice; that is, most ceramics have a high Peierls stress. Dislocations move from one Peierls valley to the next by the nucleation of a kink pair, under the action of applied stress and temperature. The kinks are abrupt and their further motion is controlled by a secondary Peierls barrier. Mitchell, Peralta and Hirth [22] have adapted the standard treatment of Hirth and Lothe [23] to show that the resulting strain-rate is given by:

$$\dot{\varepsilon} = \dot{\varepsilon}_o \frac{\sigma/\mu}{\alpha^{1/2} kT/\mu b^3} \exp\left(-\frac{Q_D + 2F_k}{kT}\right) \exp\left(\frac{\mu (bh)^{3/2}}{kT\sqrt{2\pi}}\left(\frac{\sigma}{\mu}\right)\right), \quad \dot{\varepsilon}_o = \rho h^2 L\upsilon/b \tag{1}$$

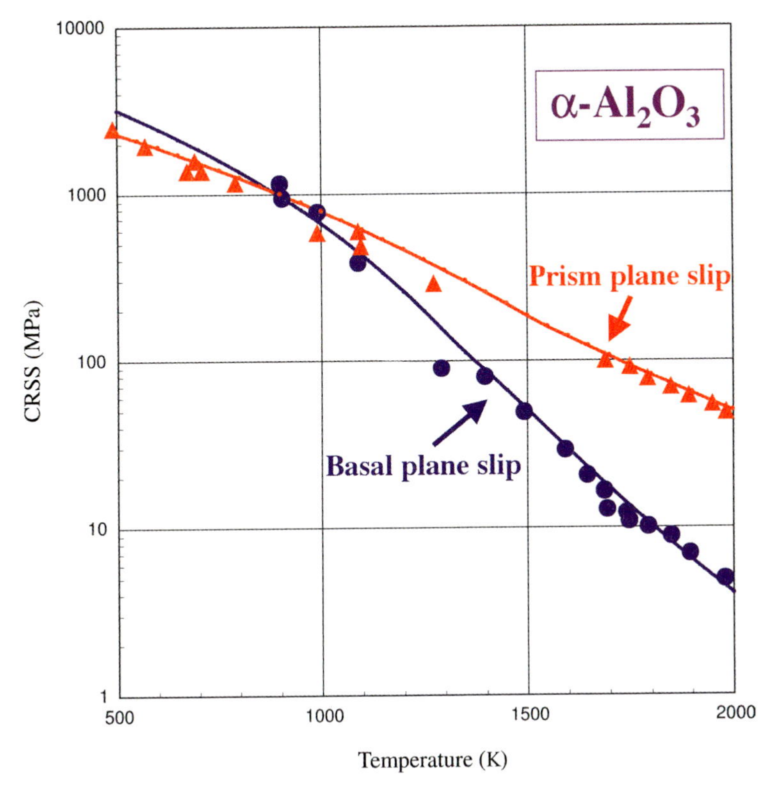

Figure 9.4 Plots of log(CRSS) versus temperature for basal and prism plane slip in sapphire showing both experimental data and curves fitted from Eq. (1) [22].

where σ is the stress acting on the dislocation, μ is the shear modulus, T is the temperature, b is the Burgers vector of the dislocation (which may be partial or perfect), α is a factor-dependent on σ and T [22], h is the periodicity of the Peierls barriers in the glide plane, a is the periodicity for the kinks along the dislocation line, Q_D is the activation energy for a kink to overcome its own Peierls barrier, F_k is the free energy of formation of a single kink, ρ is the mobile dislocation density, ν is the attempt frequency, and L is the length of a dislocation segment (the distance to which the kink pairs will expand). Equation (1) has been re-examined by Rodriguez *et al.* [24] using somewhat different assumptions, namely the so-called *long-segment* limit of the kink pair model rather than the *short-segment* limit assumed in Ref. [22]. Rodriguez *et al.* also assumed that the mobile dislocation density ρ increases as σ^2 (much as occurs with the total dislocation density in metals during work hardening). This is discussed further in Section 9.2.4.

Because stress and temperature occur in several places inside and outside the exponents, Eq. (1) must be solved numerically. The activation energy, $Q = 2F_k + Q_D$, can be determined approximately from the slope of an Arrhenius plot or the slope of a plot of log σ against T. However, the apparent activation energy determined in this way is up to ~20% lower than the actual activation energy [25]. In general, both the activation energy and the pre-exponential factor, $\dot{\varepsilon}_o = \rho b L \upsilon$, must be determined numerically from experimental curves of CRSS against temperature [22]. The stress exponent (determined by assuming that the strain rate is proportional to some power of stress) also varies continuously, from 10 or more at high stresses and low temperatures to unity at low stresses and high temperatures [25].

The relative contributions of the kink formation energy ($2F_k$) and the kink diffusion energy (Q_D) are unknown. However, F_k should be of the form $\alpha' \mu b_k^2 h$, where α is a constant of the order 0.1 for an abrupt kink (a kink with a width of atomic dimensions), and an order of magnitude smaller for a relaxed kink (a kink with a width much greater than atomic dimensions) [23], so that:

$$Q = 2\alpha' \mu b_k^2 h + Q_D \tag{2}$$

where b_k is the Burgers vector associated with the kink. Kinks should be abrupt in high-Peierls stress materials, such as ceramics and intermetallics, and relaxed for low-Peierls stress materials, such as fcc metals. How this applies to particular oxides will be examined in Section 9.4.

9.2.3
Modification of the Model for Kink Pair Nucleation on Point Defects, and Partial Dislocations

In $MgO \cdot nAl_2O_3$ ($n > 1$), the CRSS decreases dramatically with increasing deviation from stoichiometry; that is, with increasing n. In fact, the CRSS was found [26] to be proportional to $[V_c]^{-2}$, where $[V_c]$ is the concentration of charge-compensating cation vacancies which is related to n by

$$[V_c] = \frac{n-1}{3(3n+1)} \tag{3}$$

The dependence on T and n is illustrated in Figure 9.5, where $[V_c]\sigma^{1/2}$ is plotted logarithmically against linear temperature for slip on both {111} and {110} planes. The data fall on two separate straight lines, with a fair amount of scatter but clearly show that {110} slip has a lower CRSS (σ) over almost the entire temperature range, and has a smaller slope. The proportionality between σ and $[V_c]^{-2}$ suggests strongly that cation vacancies are catalysts for kink nucleation and also possibly for kink diffusion. In fact, it has been shown [22] that an attractive interaction between a point defect and a kink leads to the observed proportionality between σ and $[V_c]^{-2}$, whether the kinks nucleate separately at two independent vacancies or at a single divacancy [22], as illustrated in Figure 9.6. In spinel, it is also likely that, at the high temperatures of deformation, the cation vacancies can move rapidly with the kinks by pipe diffusion along the dislocation lines, effectively reducing the activation energy.

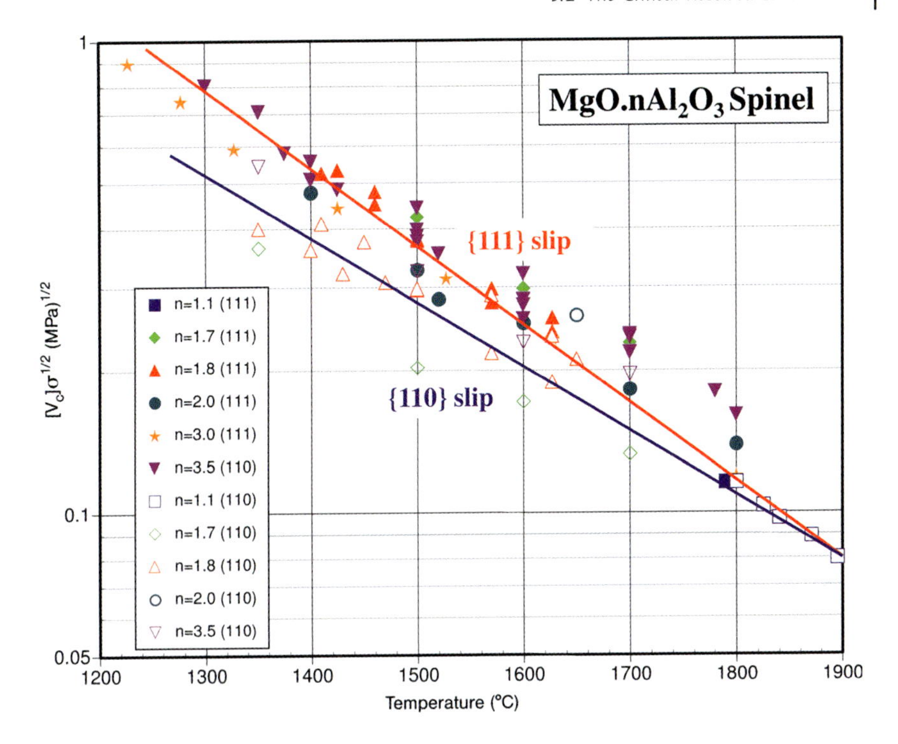

Figure 9.5 $[V_c]\sigma^{1/2}$ plotted logarithmically against linear temperature for slip on both {111} and {110} planes in $MgO{\cdot}nAl_2O_3$ spinel for a range of values of n. $[V_c]$ is the concentration of cation vacancies, and σ is the CRSS. Only the high-temperature data of Figures 9.2 and 9.3 have been used [18].

Kink nucleation on partial dislocations has also been considered [27] because of the experimental observation that the critical resolved shear stress for $\{110\}\langle 111\rangle$ slip in molybdenum disilicide decreases when substitutional alloying elements are added that decrease the stacking fault energy, and increases when substitutional elements are added that increase the stacking fault energy. This modification may apply not only to other intermetallics but also to ceramics such as spinel, where increasing

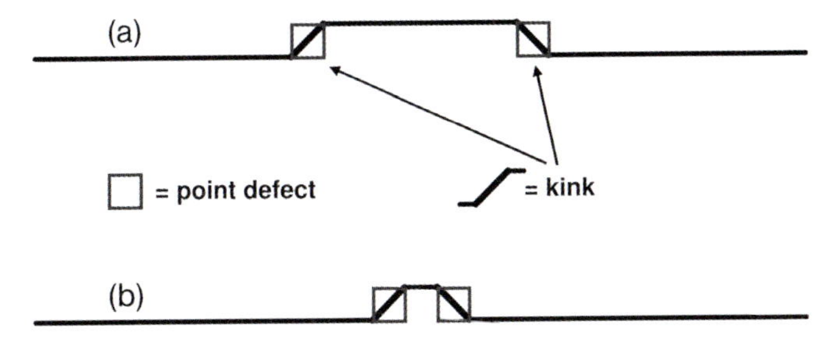

Figure 9.6 Diagrams showing kink pair nucleation. (a) At two separate vacancies; (b) At a divacancy [22].

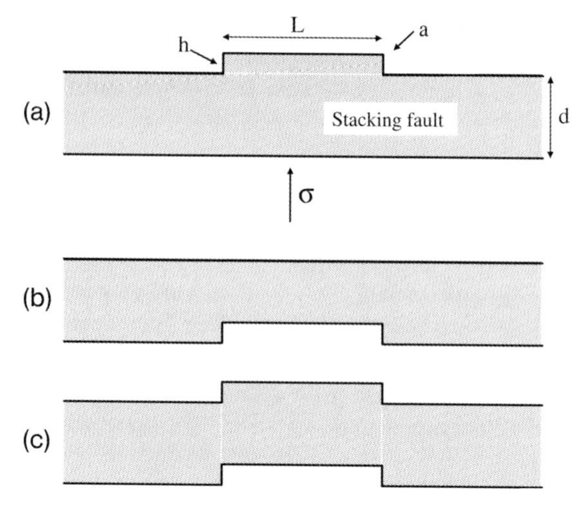

Figure 9.7 (a) Kink pair nucleation on the leading partial of a dissociated dislocation; (b) Kink pair nucleation on the trailing partial; (c) Simultaneous nucleation on both partials [22, 27].

deviation from nonstoichiometry both increases the concentration of cation vacancies and decreases the stacking fault energy.

The modification to the kink nucleation model for the case of extended dislocations is illustrated in Figure 9.7. Kink pair nucleation can occur on either the leading or trailing partial (Figure 9.7a and b), and it turns out that these are equally likely, if the partials are initially at their equilibrium separation. Kink pair nucleation will occur simultaneously on both partials (Figure 9.7c), if the stacking fault energy is high, so that the ribbon width is a few Burgers vectors or less; in this case there is no modification to the theory given in Section 9.2.2. If the stacking fault energy is low enough that kink pair nucleation occurs independently on the partials, Mitchell *et al.* [27] show that the stress σ in Eq. (1) is replaced by an effective stress σ' given by

$$\sigma' = \sigma - \frac{\pi\gamma^2 h}{\mu b_1^3} \tag{4}$$

where γ is the stacking fault energy and b_1 is the partial Burgers vector. The net result is that the stress is predicted to increase with γ^2, as observed experimentally in $MoSi_2$. It is not clear whether the reduction in CRSS in nonstoichiometric spinel is due to the increase in cation vacancy concentration, or to the reduction in stacking fault energy.

9.2.4
Comparison between Theory and Experiment

9.2.4.1 Sapphire and Stoichiometric Spinel

This section will be confined to discussions of sapphire and spinel, as in the original analysis of the application of kink pair nucleation theory to ceramics [22]. The applicability to other ceramics will be considered in Section 9.4. The experimental and fitted curves [by numerical solution of Eq. (1)] of CRSS plotted logarithmically against linear temperature are shown in Figure 9.4 for basal and prismatic slip in sapphire. Parameters and curve-fitting results for sapphire and stoichiometric spinel are given in Ref. [22]. The Burgers vectors of the partial dislocations have been used in the calculations, since the activation energy Q scales with b^2 [Eq. (2)]; using the Burgers vectors of the perfect dislocations gives unreasonably large values of Q.

The fitted curves for stoichiometric spinel (not shown) are as good as those for sapphire, at least in the high-temperature regime. The results can be summarized as follows:

- The value of Q is higher for basal slip (1.9 eV) than prism plane slip (1.5 eV) in sapphire, as expected from the stronger temperature dependence of the CRSS for the former.
- The value of Q is much higher for stoichiometric spinel (4.7 eV), which has an even stronger temperature dependence. Q has the same value for $\{111\}$ and $\{110\}$ slip because the CRSS is reported to be the same for stoichiometric spinel [26]. However, in nonstoichiometric spinel (Figure 9.5), the CRSS values are lower for $\{110\}$ slip and have a lower temperature dependence (lower Q).
- Q is given by Eq. (4), and it is uncertain how much the kink formation energy and activation energy contribute to Q. For sapphire, if a value of $\alpha' = 0.04$ is selected in Eq. (4), then $2F_k = 1.6$ eV and Q_D would be near zero for prism plane slip and small but finite for basal slip. On the other hand, taking $\alpha' = 0.1$ for spinel gives $2F_k = 2.9$ eV, so that $Q_D = 1.8$ eV. Other values of α' would give different answers, but the point is that the behavior of sapphire and spinel on their various slip systems can be explained by ascribing different contributions from the kink formation and diffusion energies.
- The estimated value of $\dot{\varepsilon}_o = \rho b L \upsilon$ is smaller for prism plane slip (6.7) than basal slip (1.8×10^3) in sapphire, but much larger in spinel (3.4×10^7). A simple estimate with $\rho = 10^8\,\text{m}^{-2}$, $L = \rho^{-1/2}$ and $\upsilon = 10^{13}\,\text{s}^{-1}$ gives $\dot{\varepsilon}_o \sim 3 \times 10^7$, so that the spinel value is reasonable. The lower values for sapphire can only be explained by lower values of ρ, the mobile dislocation density, and/or L, the mean distance of movement of a kink along a dislocation line. A possible reason for this is that dislocations in sapphire are trapped to form dipoles which quickly break up by self-climb into strings of loops (see Figures 9.15 and 9.16 in Section 9.4.6), thereby reducing the pool of mobile dislocations and limiting the slip distance.

Rodriguez *et al.* [24] point out, correctly, that basal dislocations are undissociated by glide in sapphire, so that the perfect Burgers vector should be used in Eq. (1). This results in a larger value of Q. Rodriguez *et al.* applied both the short-segment and long-segment theories to basal and prismatic slip in sapphire, and whilst finding better fits with the long-segment model, they needed to introduce a stress-dependent mobile dislocation density (as described in Section 9.2.1). It was found, for example, that $\rho = \sim 10^{16}\,\text{m}^{-2}$ at 973 K for basal slip, which is a large number at high strains let alone at the yield point. Although the original papers should be referred to for details, it is clear that atomistic calculations are needed to determine kink energies and kink barriers in order to resolve some of these uncertainties.

9.2.4.2 Nonstoichiometric Spinel

The CRSS values in Figure 9.5 for nonstoichiometric spinel were analyzed by Mitchell *et al.* [22] using Eq. (1). For $\{111\}\langle 110\rangle$ slip, they found that the activation energy was constant at 2.8 eV (lower than the 4.7 eV for stoichiometric spinel), while the pre-exponential factor $\dot{\varepsilon}_o$ was increased by an order of magnitude; they ascribed the latter effect to the increasing cation vacancy concentration given by Eq. (3), and the former effect to a lower kink migration energy due to the presence of cation vacancies. For $\{110\}\langle 110\rangle$ slip, the activation energy was slightly lower (~2 eV) due to a lower kink migration energy on $\{110\}$ than on $\{111\}$ planes. The lower kink diffusion energy is likely because, at ambient temperature where vacancies would be immobile, nonstoichiometric crystals are actually *harder* [18]. The continued softening with increasing n at high temperatures must then be due to the increasing $[V_c]$ – that is, increasing $\dot{\varepsilon}_o$ from Eq. (3).

It is also possible that the softening observed in nonstoichiometric spinel is due to a reduction in the stacking fault energy, using the modification described in the previous section. The experimental and calculated softening due to the reduction of stacking fault energy with increasing values of n in nonstoichiometric $MgO{\cdot}nAl_2O_3$ spinel are in remarkably good agreement [27]. Unfortunately, there is no way of deciding whether softening is due to kink nucleation on point defects or kink nucleation on partials, although the latter may be found to be the more satisfying. But, at least there are now two possible (and plausible) explanations for solution softening in nonstoichiometric spinel, whereas for many years there was none!

9.3 Crystallography of Slip

In this section, the crystal structures of ceramics (confined to oxides for simplicity) and the implications of crystal geometry to dislocation behavior will be examined.

9.3.1 Crystal Structures

A number of oxides can be described in terms of fcc or hcp packing of the oxygen anions, with the cations occupying octahedral or tetrahedral interstitial sites. These are listed in Table 9.1, along with some other oxides that cannot be described in this way. Care must be taken when describing the structures as close-packed – the analogy with metals can be misleading, as the following examples will show that:

- the anions in spinel are not in perfect fcc packing (see Figure 9.8);
- the anion basal planes in sapphire are flat but irregular hexagons, while the cations are in puckered layers in between the anion layers (see Figure 9.9);

Table 9.1 Crystal structures of some oxides.

Oxide	Space group	Anion packing	Cation occupancy		Other oxides
			Octahedral	Tetrahedral	
MgO	$Fm\bar{3}m$	fcc	1	0	NiO, CoO, etc.
BeO	$P6_3mc$	hcp	0	1/2	ZnO
UO_2	$Fm\bar{3}m$	simple cubic	—	—	ThO_2, c-ZrO_2
TiO_2	$P4_2/mnm$	~hcp	1/2	0	—
SiO_2	$P3_121$	—	—	—	(quartz form)
Al_2O_3	$R\bar{3}c$	hcp	2/3	0	Cr_2O_3, Fe_2O_3
Y_2O_3	$Ia\bar{3}$	—	—	—	Rare-earth oxides
$SrTiO_3$	$Pm\bar{3}m$	—	—	—	Many perovskites
$MgAl_2O_4$	$Fm\bar{3}m$	fcc	1/2	1/8	Many spinels
Mg_2SiO_4	$Pbnm$	~hcp	1/2	1/8	Fe_2SiO_4, etc.
$Y_3Al_5O_{12}$	$Ia\bar{3}d$	—	—	—	Many garnets

- rutile is tetragonal, but the anions can be thought of being in a distorted hcp array, which is useful for describing the "crystallographic shear" structures in reduced rutile;
- forsterite (Mg_2SiO_4) is orthorhombic but the anion packing is close to hcp, giving a structure which is the hexagonal analogue of cubic spinel;

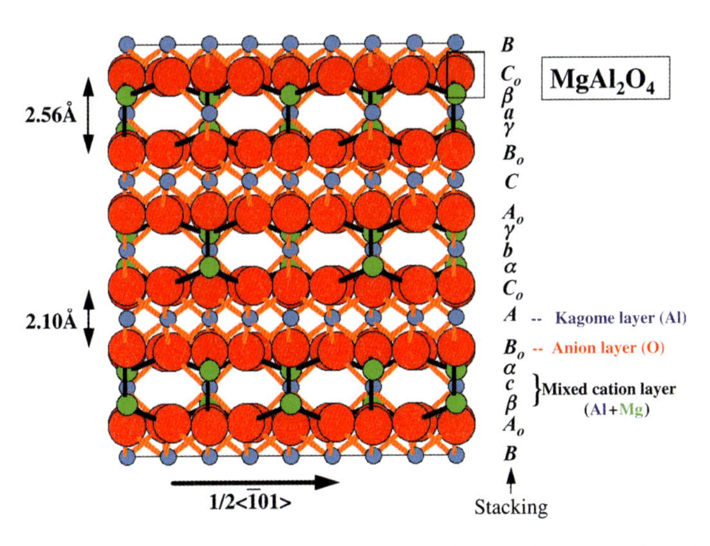

Figure 9.8 Stacking of {111} planes in spinel viewed along ⟨211⟩. $A_OB_OC_O$ refers to oxygen stacking, ABC to Al stacking in the Kagomé layers, abc to Al stacking in the mixed cation layers, and αβγ to stacking of Mg cations. Note that the oxygen fcc packing is imperfect, and that the {111} anion layers are alternately separated by 0.210 nm and 0.256 nm [18].

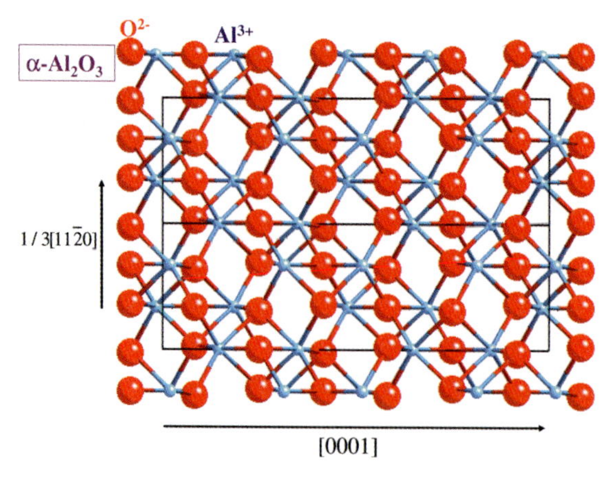

Figure 9.9 Structure of sapphire viewed along $\langle 1\bar{1}00 \rangle$. The oxygen anion planes are flat, but consist of linked larger and smaller triangles. The aluminum cation planes are puckered. The unit cell is outlined.

- the cations in fluorite-structured crystals (e.g., UO_2) are in fcc array but the anions are in simple cubic packing.

The other oxides listed in Table 9.1 are included because their mechanical properties have received some attention (especially quartz). Quartz is by no means close-packed because of the stability of the $[SiO_4]$ tetrahedron. Yttria and the rare-earth sesquioxides and the mixed oxide garnet structures are compact, in spite of their complexity, but cannot be described as close-packed in the traditional sense. In fact, the C-type M_2O_3 structure of yttria can be thought of as having the fluorite structure of urania with one-quarter of the anion sites vacant.

9.3.2
Slip Systems

Crystals generally slip on close-packed planes in close-packed directions, such that the Burgers vector is equal to the shortest lattice vector. This is true in metals and is expected from Peierls stress arguments. It is also true in ceramics, but there are exceptions to the rule. Observed slip systems are summarized in Table 9.2. In general, the Burgers vectors are well behaved: in crystals with fcc lattices, $\mathbf{b} = 1/2\langle 110 \rangle$; in crystals indexed on a hexagonal unit cell, $\mathbf{b} = 1/3\langle 11\bar{2}0 \rangle$; in both cases, $\mathbf{b}$ is the shortest lattice vector. However, the magnitude of the Burgers vector varies widely according to the size of the unit cell. In terms of the distance, d_O, between oxygen anions, $\mathbf{b}$ is equal to d_O for MgO and BeO, and is equal to $2d_O$ for $MgAl_2O_4$, TiO_2, and Mg_2SiO_4. For prism plane slip in α-Al_2O_3, $\mathbf{b}$ is $\langle 10\bar{1}0 \rangle$, which has a magnitude of $3d_O$ ($= 0.822$ nm). Even larger Burgers vectors are encountered in the garnets and cubic rare-earth sesquioxides (see Table 9.2).

Table 9.2 Slip systems in oxides.

Oxide	Slip system	Burgers vector (nm)	Other slip systems
MgO	$\{1\bar{1}0\}\langle 110\rangle$	$1/2\langle 110\rangle = 0.298$	$\{001\}\langle 110\rangle$
BeO	$(0001)\langle 11\bar{2}0\rangle$	$1/3\langle 11\bar{2}0\rangle = 0.270$	$\{1\bar{1}00\}\langle 11\bar{2}0\rangle$, $(0001)\langle 10\bar{1}0\rangle$
UO_2	$\{001\}\langle 110\rangle$	$1/2\langle 110\rangle = 0.387$	$\{1\bar{1}1\}\langle 110\rangle$
TiO_2	$\{100\}\langle 011\rangle$	$\langle 011\rangle = 0.546$	$\{110\}[001]$
SiO_2	$(0001)\langle 11\bar{2}0\rangle$	$1/3\langle 11\bar{2}0\rangle = 0.491$	$\{11\bar{2}0\}[0001]$, $\{10\bar{1}0\}[0001]$ (see Section 9.4.5)
Al_2O_3	$(0001)\langle 11\bar{2}0\rangle$	$1/3\langle 11\bar{2}0\rangle = 0.476$	$\{1\bar{2}10\}\langle 10\bar{1}0\rangle$, $\{01\bar{1}2\}\langle 0\bar{1}11\rangle$
Y_2O_3	$\{110\}\langle 1\bar{1}1\rangle$	$1/2\langle 111\rangle = 0.918$	See Section 9.4.10
$SrTiO_3$	$\{1\bar{1}0\}\langle 110\rangle$	$\langle 110\rangle = 0.552$	See Section 9.4.7
$MgAl_2O_4$	$\{1\bar{1}1\}\langle 110\rangle$	$1/2\langle 110\rangle = 0.594$	$\{1\bar{1}0\}\langle 110\rangle$
Mg_2SiO_4	$(100)[001]$,$\{110\}[001]$	$[001] = 0.599$	$(100)[010]$, $\{0kl\}[100]$
$Y_3Al_5O_{12}$	$\{110\}\langle 1\bar{1}1\rangle$	$1/2\langle 111\rangle = 1.040$	$\{112\}\langle 1\bar{1}1\rangle$, $\{132\}\langle 1\bar{1}1\rangle$

Slip plane behavior is less consistent. In MgO, the separate anion and cation closest-packed planes are {111}, but the combined anion and cation closest-packed planes are {100}; the observed {110} slip plane is explained in terms of the lower repulsion between like ions during slip on this plane compared with {100} [28]. In more complex structures, it becomes difficult to decide which is the closest-packed plane, and the simple-minded approach loses its value. In $MgAl_2O_4$, for example, stoichiometric crystals prefer the close-packed {111} anion planes (see Figure 9.1), whereas nonstoichiometric crystals ($MgO{\cdot}nAl_2O_3$, $n > 1$) slip on {110} planes, as in rock salt. On the other hand, garnets and rare-earth sesquioxides, in spite of the complexity of their structures, follow the dictates of the underlying bcc lattice and slip on the $\{110\}\langle 111\rangle$ system, as in bcc metals.

9.3.3 Dislocation Dissociations

As pointed out in Section 9.1, Kronberg [7] and Hornstra [8] produced seminal reports on the dissociation of dislocations in ceramics, particularly sapphire and spinel. Kronberg suggested that basal dislocations in sapphire should dissociate according to

$$1/3\langle 11\bar{2}0\rangle \rightarrow 1/3\langle 10\bar{1}0\rangle + 1/3\langle 01\bar{1}0\rangle \quad (5)$$

creating a fault in the cation sublattice but leaving the anion sublattice unfaulted. Kronberg also suggested a further dissociation into Shockley-like quarter partials, with the creation also of a fault in the anion sublattice. Kronberg pointed out that the motion of the quarter partials requires the anions and cations to move along different paths in a process known as "synchroshear." However, he based his

Table 9.3 Dissociation of dislocations in oxides.

	Dissociation reaction	Reference(s)
MgO	None reported	[15, 16]
BeO	$1/3\langle 11\bar{2}0\rangle \rightarrow 1/3\langle 10\bar{1}0\rangle + 1/3\langle 01\bar{1}0\rangle$	[30]
UO_2	None reported	[15, 16]
TiO_2	$\langle 101\rangle \rightarrow 1/2\langle 101\rangle + 1/2\langle 101\rangle$ (see Section 9.4.4)	[31]
SiO_2	$1/3\langle 11\bar{2}0\rangle \rightarrow 1/6\langle 11\bar{2}0\rangle + 1/6\langle 11\bar{2}0\rangle$ (see Section 9.4.5)	[32, 33]
Al_2O_3	$1/3\langle 11\bar{2}0\rangle \rightarrow 1/3\langle 10\bar{1}0\rangle + 1/3\langle 01\bar{1}0\rangle$ $\langle 10\bar{1}0\rangle \rightarrow 1/3\langle 10\bar{1}0\rangle + 1/3\langle 10\bar{1}0\rangle + 1/3\langle 10\bar{1}0\rangle$	[15, 16]
Y_2O_3	$\langle 100\rangle \rightarrow 1/2\langle 100\rangle + 1/2\langle 100\rangle$	[34]
Er_2O_3	$1/2\langle 111\rangle \rightarrow 1/4\langle 110\rangle + 1/4[112]$	[35]
$SrTiO_3$	$\langle 110\rangle \rightarrow 1/2\langle 110\rangle + 1/2\langle 110\rangle$	[36, 37]
$MgAl_2O_4$	$1/2\langle 110\rangle \rightarrow 1/4\langle 110\rangle + 1/4\langle 110\rangle$	[15, 16]
Mg_2SiO_4	$[001] \rightarrow 1/2[001] + 1/2[001]$ [010] undissociated [100] undissociated	[12, 38]
$Y_3Al_5O_{12}$	$1/2\langle 111\rangle \rightarrow 1/4\langle 111\rangle + 1/4\langle 111\rangle$	[39]

analysis on the idealized structure of sapphire, rather than on the puckered structure shown in Figure 9.9. In fact, as pointed out by Bilde-Sørensen *et al.* [29], dislocations should be able to move more easily through the puckered layers in Figure 9.9 on an electrically neutral plane. Such slip via motion of $1/3\langle 10\bar{1}0\rangle$ partials only involves half of the cations in a given puckered layer; synchroshear is then unnecessary.

For spinel, Hornstra suggested that $1/2\langle 110\rangle$ dislocations should dissociate into collinear half-partials:

$$1/2\langle 110\rangle \rightarrow 1/4\langle 110\rangle + 1/4\langle 110\rangle \qquad (6)$$

again with the formation of a cation fault only. The further dissociation into the equivalent of Shockley partials in the fcc anion sublattice (shown later in Figure 9.19) creates an anion fault in addition to the cation fault. Again, the second dissociation has never been observed. As in sapphire, anion fault energies in spinel must be very large. However, the dissociations given by Eqs (5) and (6) are commonly observed, albeit often by climb, as described below. Dissociations which have been observed in these and other oxides are summarized in Table 9.3. Particular oxides are discussed in the following section.

9.4
Dislocations in Particular Oxides

The following sections relate to monoxides (Section 4.1 and 4.2), followed by dioxides (Sections 4.1–4.5), sesquioxides (Section 4.6), mixed oxides (Sections 4.7–4.9), other oxides (Section 4.10), and non-oxides (Section 4.11).

9.4.1 Magnesium Oxide and other Oxides with the Rock-Salt Structure

The rock-salt structure is frequently found among the oxides of the alkaline earth metals (MgO, CaO, etc.) and the transition metals (MnO, FeO, CoO, NiO). In the latter case, the oxides are nonstoichiometric and should be written $M_{1-x}O$, due to the existence of trivalent ions and their charge-compensating cation vacancies. In all cases, the easy slip system is $\{110\}\langle 110\rangle$ (as described in Sections 9.2.1 and 9.3.2). Slip on the $\{001\}$ plane is much more difficult, and slip on $\{111\}$ has never been reported. This is discussed below for MgO, followed by a discussion of the effect of nonstoichiometry in the transition metal oxides.

9.4.1.1 Magnesium Oxide

Slip on the $\{110\}\langle 110\rangle$ system occurs at low stresses, and plastic deformation is possible at low temperatures [5, 40]. In fact, the CRSS extrapolated to 0 K is only 30 MPa, as shown in Figure 9.1. The CRSS for $\{001\}\langle 110\rangle$ slip is more than an order of magnitude higher, as is also shown in Figure 9.1. The activation energy for $\{110\}$ $\langle 110\rangle$ slip can be estimated from Eq. (1) to be ~0.1 eV. Since the activation energy is the sum of the elastic energy of the kink pair and the migration energy [Eq. (2)], both must be very small for the easy slip system. In fact, since dislocations in MgO are undissociated, the appropriate value of b in Eq. (2) is 0.298 nm (Table 9.2), and so $\mu b_k^2 h$ is calculated to be about 20 eV. This means that either the kinks are diffuse (α' is small) and the secondary Peierls barrier is very low, or that the Peierls stress is so low that the kink mechanism does not apply. The activation energy for $\{001\}\langle 110\rangle$ slip can be estimated from the slope of the CRSS curve in Figure 9.1 to be ~1 eV. This is lower than for sapphire or spinel (Section 9.2.4), but again the perfect Burgers vector should be used in Eq. (2) and so α' would have to be on the order of 0.01 – that is, a diffuse kink with a low secondary Peierls barrier.

Why slip on the $\{001\}$ plane in MgO is so much more difficult than on the $\{110\}$ plane is an interesting question. Gilman [28] pointed that, in the former case, each ion is required to move past a pair of ions of like sign at the half-slipped position; in the $\{110\}$ case, a pair of ions of like sign is again present at the half-slipped position, but a single ion of *opposite* sign is present at a slightly closer distance. This is believed to alleviate the charge problem, even though the $\{110\}$ planes are less close-packed than the $\{001\}$ planes. Atomistic calculations of $1/2\langle 110\rangle$ dislocations in MgO have been performed by Woo and Puls [41] and, later, by Watson *et al.* [42]; unfortunately, these authors did not tackle the problem of the slip plane. However, Woo and Puls calculated the Peierls stress for slip on the $\{110\}$ plane and found it to be quite low at approximately 70 MPa. This is about a factor of two higher than the CRSS extrapolated to 0 K, but should be considered to be in satisfactory agreement in terms of the computing techniques of the time.

9.4.1.2 Transition Metal Oxides and the Effect of Stoichiometry

Plastic deformation of single crystals of the transition metal oxides $M_{1-x}O$ with the rock-salt structure has been studied for M = Mn [43], Fe [44], Co [45–48], and

Ni [49, 50]. The maximum value of x varies from 0.001 for Ni and 0.013 for Co to 0.15 for Fe and Mn. $Fe_{1-x}O$ is unstable below ~570 °C, and cannot therefore be deformed plastically at low temperatures. There have been no reports of low-temperature deformation of $Mn_{1-x}O$, but there is no reason to believe that this should not be possible. On the other hand, plastic deformation on the $\{110\}\langle 110\rangle$ slip system occurs readily in $Co_{1-x}O$ and $Ni_{1-x}O$ at low temperatures [46, 48–50]; in fact $Co_{1-x}O$ has even been deformed at 4.2 K [48]. Care must be taken in the case of $Co_{1-x}O$ to cool the crystals in a low-p_{O_2} atmosphere so as to avoid the precipitation of Co_3O_4, which causes hardening [48]. The resultant crystals have a CRSS of ~100 MPa at 4.2 K, comparable with that of MgO on the same system (Figure 9.1). The range of stoichiometry at low temperatures is small and so there is no effect on the CRSS, except through the possibility of precipitation.

It is a different story at high temperatures, where there have been a number of studies of the influence of stoichiometry on the creep rate [43–45, 47, 49] or the flow stress [43, 47]. Typically, for a given applied stress, the creep rate increases with increasing p_{O_2}, that is, as the crystals become more stoichiometric (i.e., as x decreases). However, the behavior is less straightforward in $Mn_{1-x}O$; at temperatures between 1000 and 1400 °C, the flow stress first increases with increasing p_{O_2} and then decreases with further increase in p_{O_2}, by a factor of about 2 in each case.

The creep behavior is often analyzed in terms of the phenomenological equation

$$\dot{\varepsilon} = A\sigma^n p_{O_2}^m \exp\left(-\frac{Q}{kT}\right) \tag{7}$$

where n and m are exponents and Q is an activation energy. Equation (7) is interpreted in terms of climb-controlled glide. Thus, Q is the activation energy for diffusion of the rate-controlling species; the value of m is used to identify the diffusing species (see Table 9.4), and the value of n to identify the climb mechanism. For example, Jolles and Monty [44] studied wüstite ($Fe_{1-x}O$) at temperatures between 850 and 1150 °C and determined that $Q \sim 3$ eV, independent of temperature; they interpreted this as the activation energy for oxygen self-diffusion (the slowest moving species). The value of m (0.03–0.11) indicated that the rate-controlling species was doubly-charged oxygen interstitials, O''_i, with a contribution from singly-charged interstitials, O'_i, at the highest temperature. Although the value of n was unusually high (~9), they still

Table 9.4 Values of m [the pO_2 exponent in the creep rate of Eq. (7)] for various oxygen defects in the transition metal oxides with the rock-salt structure [44].

Defect	Symbol	m
Neutral oxygen vacancy	V_O^x	−0.56
Singly-charged oxygen vacancy	$V_O^{\bullet}$	−0.33
Doubly-charged oxygen vacancy	$V_O^{\bullet\bullet}$	−0.11
Neutral oxygen interstitial	O_i^x	+0.42
Singly-charged oxygen interstitial	O'_i	+0.20
Doubly-charged oxygen interstitial	O''_i	+0.03

suggested that the mechanism was a recovery-controlled process involving dislocation climb. The dominant charge-compensating defects are cation vacancies, V''_{Fe}, but these are thought to be highly mobile and so do not affect dislocation motion. Oxygen interstitials are presumably minority charge-compensating defects, and so are the defects which control diffusive processes.

The behavior of $Co_{1-x}O$ makes an interesting contrast. Dominguez-Rodriguez *et al.* [45] also analyzed creep data at temperatures from 1000 to 1450 °C using Eq. (7), and found that Q increased from 2.5 eV at low temperatures and high p_{O_2} to 5 eV at high temperatures and high p_{O_2}. Their m values increased from 0.5 to 0.1 with increasing temperature, independent of p_{O_2}, while the n values decreased from 8.5 to 6.5 with increasing temperature. These results were also interpreted in terms of recovery controlled by oxygen diffusion, with the rate-controlling species being *neutral* oxygen interstitials at low temperature and high p_{O_2}, and oxygen vacancies at high T and high p_{O_2}. The interstitials have a lower activation energy for diffusion than the vacancies.

The constant strain-rate compression tests performed by Routbort [47] on $Co_{1-x}O$ and by Goretta and Routbort [43] on $Mn_{1-x}O$ single crystals, provide different insights into the deformation mechanisms. In both cases, yield points are observed followed by steady-state flow. For $Mn_{1-x}O$, the steady-state flow stresses were analyzed in terms of Eq. (7) over a temperature range from 900 to 1400 °C and a p_{O_2} range from 10^{-11} to 10^2 Pa; the m values were positive (~0.6 to 0.25) at low p_{O_2} and negative (~ −0.2 to −0.5) at high p_{O_2}. The primary rate-controlling species are then singly-charged oxygen vacancies at low p_{O_2} and neutral oxygen interstitials at high p_{O_2}. For $Co_{1-x}O$ compressed over a similar range of temperature and p_{O_2}, constant strain-rate tests give similar results to the creep tests described above; for small x, oxygen vacancies are rate-controlling and for larger x, the oxygen interstitials control the deformation rate. However, there are complications; serrated yielding is observed at 1000 °C for small x, suggesting that impurities are causing dynamic strain-aging (the Portevin–Le Chatelier effect). In addition, there is a prominent yield-point, and the upper yield stress has a different dependence on x than the steady-state flow stress. In both cases, the yield stress first increases with increasing x (increasing p_{O_2}), and then decreases with any further increase in x. As it is unlikely that climb can control yielding, the initial increase may represent solution hardening due to the cation vacancies, while the subsequent decrease in yield stress may represent solution softening from the same cation vacancies.

Caution should be exercised in using Eq. (7) to determine the exact rate-controlling mechanism. For high-Peierls stress materials, serious errors occur in the use of such empirical expressions when determining activation energies and stress exponents [25]. For low-Peierls stress materials, as in the present case, a well-grounded theoretical equation must be developed for the particular diffusion-controlled model which is thought to be extant. Only then can the empirical values of m, n, and Q be used to imply mechanisms with confidence. Another point is that the high values of n reported above sometimes imply that σ in Eq. (7) should be replaced by $(\sigma - \sigma_o)$, where σ_o is an internal stress or friction stress; the value of n is then reduced.

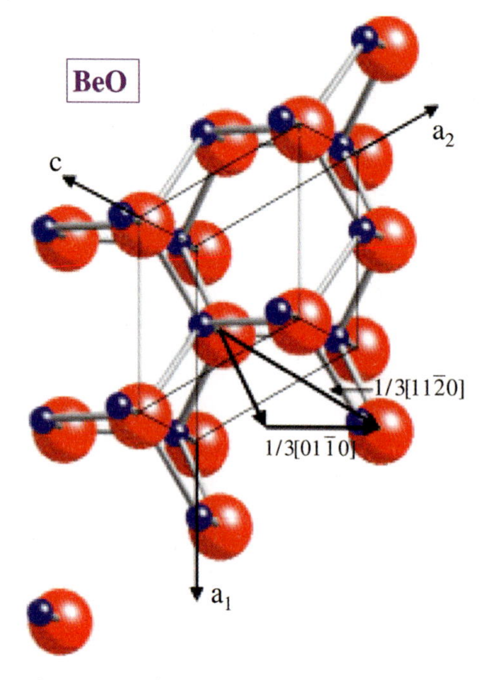

Figure 9.10 The wurtzite structure of BeO, showing the small cations and large anions in tetrahedral coordination. A hexagonal unit cell is outlined. The 1/3 [11$\bar{2}$0] Burgers vector is shown with its dissociation into two 1/3 ⟨10$\bar{1}$0⟩ partials.

9.4.2
Beryllium Oxide and Oxides with the Wurtzite Structure

Both beryllium oxide (BeO) and zinc oxide (ZnO) are technologically the most important tetrahedrally coordinated oxides that crystallize in the hexagonal wurtzite structure, space group $P6_3/mmc$. This structure involves hcp packing of the oxygen sublattice, with the cations occupying half of the tetrahedral interstices, as illustrated in Figure 9.10. Curiously, in marked contrast to sulfides, no oxide examples are known that crystallize with the zincblende or sphalerite structure, in which oxygen anions are in fcc stacking with cations in half of the tetrahedral interstices. In neither case is any extensive literature available on single crystal deformation.

9.4.2.1 Beryllium Oxide

Early studies on BeO single crystals by Bentle and Miller [51] identified four slip systems; basal slip, (0001)⟨11$\bar{2}$0⟩; prismatic slip, {1$\bar{1}$00}⟨11$\bar{2}$0⟩ and {1$\bar{1}$00}[0001]; and pyramidal slip, {$\bar{1}\bar{1}$22}⟨11$\bar{2}$3⟩. Not surprisingly, significant plastic anisotropy was found. At 1000 °C, the yield stresses for these systems were as follows: 35 MPa for basal slip; 49 and 110 MPa for prismatic slip along ⟨11$\bar{2}$0⟩ and [0001], respectively; and >250 MPa for pyramidal slip. More recent studies on creep deformation in BeO single crystals by Corman [52] confirmed the anisotropic plastic deformation behavior; these studies also revealed that basal slip in BeO could be activated at surprisingly low temperatures. For example, at 50 MPa and using crystals with a high Schmid factor for basal slip (compression direction along $[\bar{1}101]$), a creep rate of $1.3 \times 10^{-7}\,s^{-1}$ was found at 650 °C, which increased to $7.3 \times 10^{-7}\,s^{-1}$ for a stress of 70 MPa.

By way of contrast, temperatures of 1650 to 1850 °C had to be used for samples compressed along $[1\bar{1}00]$ and [0001]. Although most of the $[1\bar{1}00]$ samples buckled, it was possible to determine a creep rate of $7.2 \times 10^{-9}\,s^{-1}$ for one sample at a stress of 50 MPa at 1650 °C, and this increased to $1.6 \times 10^{-7}\,s^{-1}$ for a stress of 100 MPa. For a [0001]-oriented sample, an upper bound for the creep rate of $1.7 \times 10^{-8}\,s^{-1}$ at 1750 °C and 100 MPa could be determined.

9.4.2.2 **Dislocation Dissociation in BeO and Other Crystals with the Wurtzite Structure**

Suzuki *et al.* [30] studied dislocations in a number of crystals with the wurtzite structure, including ZnO, BeO, AlN, GaN, and InN. The dislocations were introduced at ambient temperature by pulverization during specimen preparation. Most of the dislocations were screw in character, with a $1/3\,\langle 11\bar{2}0\rangle$ Burgers vector. Suzuki *et al.* used weak-beam and high-resolution transmission electron microscopy (TEM) to show that they were dissociated into a pair of $1/3\,\langle 10\bar{1}0\rangle$ Shockley-type partials (shown schematically in Figure 9.10). The stacking fault energies determined from the ribbon width are shown in Table 9.5. Also included in Table 9.5 are values for CdS, CdSe, and SiC; in the last case, it was assumed that the stacking fault energy (SFE) for 2H-SiC (with the wurtzite structure) was between that for 4H- and 6H-SiC [53]. Suzuki *et al.* found there to be a good correlation between the normalized SFE (the energy per bond across the stacking fault between anions and cations in the basal plane) and the c/a ratio. This is shown in Table 9.5 and plotted in Figure 9.11. This correlation was interpreted in terms of the second-nearest-neighbor π-electron interaction, which is an attractive coulombic force between unlike ions. The stronger this interaction, the shorter would be the c-axis; hence the inverse correlation between the normalized SFE and c/a ratio in Figure 9.11. Suzuki *et al.* also pointed out that the wurtzite structure was more stable than the zinc blende structure in crystals with a high ionicity, and that a stacking fault in wurtzite represented a small

Table 9.5 Stacking fault energy and c/a ratio for a number of crystals with the wurtzite structure.

Crystal	SFE ($mJ\,m^{-2}$)	Reduced SFE (meV per bond)	c/a ratio	Reference
CdSe	10	10	1.635	[30]
SiC	3–15	3–15	1.633	[53]
CdS	10	9	1.632	[30]
BeO	41	16	1.623	[30]
InN	41	27	1.611	[30]
ZnO	100	57	1.603	[30]
AlN	220	115	1.600	[30]

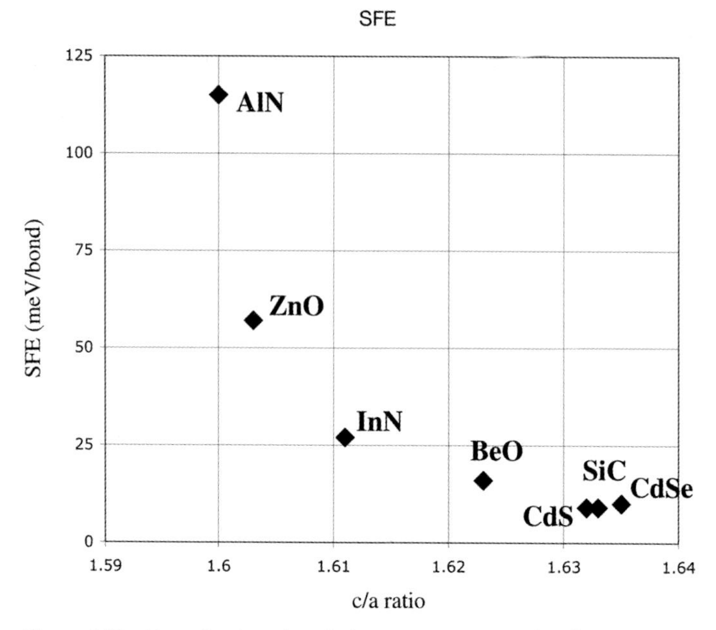

Figure 9.11 Normalized stacking fault energy (energy per bond across the basal fault plane) versus c/a ratio for various crystals with the wurtzite structure [30, 53].

region of zinc blende. It is certainly true in Figure 9.11 that CdS and CdSe not only have a low SFE but also occur in both the wurtzite and zinc blende structures.

9.4.2.3 Zinc Oxide

Information relating to dislocations and plasticity in ZnO is surprisingly sparse, especially given that sizable ZnO single crystals are now available. Detailed studies on ZnO would be most welcome.

9.4.3 Zirconium Oxide and Other Oxides with the Fluorite Structure

Simple oxides with large cations – such as Zr^{4+}, Th^{4+}, and U^{4+} – crystallize with the fluorite structure (the mineral fluorite itself is CaF_2). Both, ZrO_2 and UO_2 have been well studied and will be emphasized in this chapter, though neither story is simple.

9.4.3.1 Zirconia

The cubic form of pure ZrO_2 is only stable from its melting point (~2750 °C) to ~2400 °C, where it undergoes a distortion to tetragonal symmetry. A further transformation to monoclinic symmetry occurs below ~1200 °C, which is martensitic in character. Solid solution additions of various "stabilizers" such as CaO, Y_2O_3, and CeO_2 reduce the temperature of these transformations. While the tetragonal → monoclinic transformation forms the basis of the technologically important phenomenon of "transformation toughening" [54, 55], which has led to a commercially very useful class of advanced ZrO_2-containing structural ceramics, the issue of transformation toughening is beyond the scope of this chapter.

Single crystals of cubic Y_2O_3-stabilized ZrO_2 are readily available [56]; in fact, crystals with the nominal 9.4 m/o composition are widely sold as imitation diamonds, as the refractive index and dispersion of Y_2O_3-stabilized cubic ZrO_2 are very similar to those of diamond. Thus, many investigations have been undertaken in the study of cubic Y_2O_3-stabilized ZrO_2 with solute contents between 9.4 and 20 mol% Y_2O_3. In addition to the intrinsic scientific interest, very high flow stresses of 300–400 MPa are found for higher solute content crystals at elevated temperatures (1400 °C) [57, 58], suggesting the possibility of some technological exploitation of this material.

The "early" studies on plastic deformation of cubic Y_2O_3-stabilized ZrO_2 were performed by Heuer, Dominguez-Rodriguez, and their collaborators [59–63], and were conducted mostly at elevated temperatures of 1400 °C and higher. As a consequence, $\{001\}\langle 110\rangle$ was identified as the easy slip system, although some slip was also noted on $\{111\}$. On the other hand, if cube slip was suppressed, for example by deforming along $\langle 100\rangle$ [63], then $\{110\}$ slip could also occur [64]. Messerschmidt and colleagues [65] later showed that, by careful alignment, a deformation of cubic ZrO_2 was possible along the soft $\langle 112\rangle$ direction at temperatures down to 400 °C. Moreover, under hydrostatic confirming pressures, deformation down to 250 °C was possible [66], as shown in the plot of yield stress against temperature in Figure 9.12.

The wide temperature range employed by Messerschmidt's group (400–1400 °C; see Figure 9.12 [58]) has revealed that cubic ZrO_2 displays essentially textbook behavior of dislocation plasticity. At all temperatures, part of the flow stress is of an athermal nature, due to long range dislocation–dislocation interactions, and to the back stress of dislocation segments bowing out between large jogs. This type of athermal deformation dominates crystals containing 10 m/o Y_2O_3 deforming by easy slip between ~1000 and ~1250 °C; the flow stresses decrease at higher temperatures and increase at lower temperatures. At these lower temperatures, the kink mechanism causes a large increase in the flow stress, to 1 GPa or more at 400 °C [64].

At the highest temperatures – perhaps 1400 °C and above – deformation is controlled by recovery [65, 67, 68]. One consequence of this is a reduction in the plastic anisotropy, as shown in Figure 9.12. A marked apparent solution hardening at these elevated temperatures is most likely explained by the strong effect of Y^{3+} concentration on bulk diffusion kinetics; the diffusivity of the rate-controlling (slower) species decreases by an order of magnitude in going from 9.6 to 20 mol % Y_2O_3 [69, 70]. Yet, at lower temperatures, perhaps below 1000 °C, the size misfit between Zr^{4+} and Y^{3+} can cause a more "classical" solid-solution hardening.

Although Y^{3+} ions present only weak obstacles to dislocation motion [58], they are present in high concentrations and could, in theory, yield a large contribution to the flow stress. However, the crystals presently available apparently contain very small precipitates of ZrN [71] which provide stronger obstacles to slip than do unassociated Y^{3+} ions [58, 72, 73]. Nonetheless, these unassociated Y^{3+} ions do cause plastic instabilities, such as dynamic strain aging or the Portevin–Le Châtelier effect in

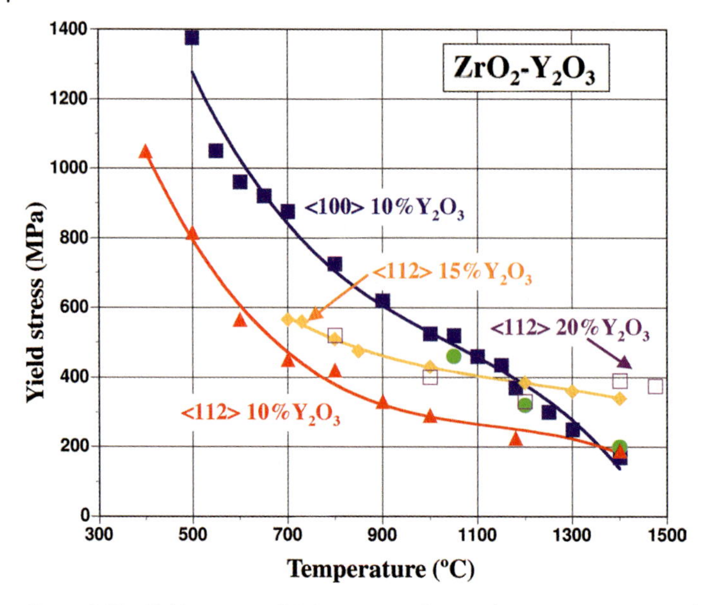

Figure 9.12 Yield stresses of ZrO_2–Y_2O_3 single crystals versus temperature. ⟨100⟩ and ⟨112⟩ in the legend refer to the compression axis; the percentages are the mol% of Y_2O_3. Strain rates are variable (see Ref. [58]).

high-solute content crystals deformed over a broad range of temperatures between 800 and 1400 °C [72, 73].

9.4.3.2 Uranium Oxide

The other fluorite-structured material that has been well studied is UO_{2+x}; in this case, the nonstoichiometry is an issue, as the value of x can extend from zero (stoichiometric UO_2) to 10^{-2}, or even larger. The most comprehensive study was conducted by Keller *et al.* [74], to which the reader is referred for reference to the earlier literature. In following these authors, deformation at low and high temperatures will be considered separately.

UO_2 has a surprisingly low brittle–ductile transformation. The only observed slip system at low temperatures is $\{111\}\langle 110\rangle$, and this does not depend on stoichiometry. Sources of mobile dislocation are an issue, however, and in order to achieve deformation at temperatures below 600 °C the crystals must be pre-deformed at 600 °C. With such pre-deformed crystals, deformation to plastic strains >1% is possible with modest yield stresses, typically 80 MPa at 450 °C, 110 MPa at 400 °C, and 120 MPa at 250 °C. Attempts to deform these crystals at room temperature were not successful, although perhaps a more careful alignment of the load train might have allowed plastic deformation at temperatures below 250 °C. Microindentation at room temperature is always possible, however, and the Knoop hardness anisotropy at room temperature is also consistent with $\{111\}$ slip [74]. The yield stress at 600 °C was variable, but surprisingly was not a function of the O/U ratio; the plastic deformation

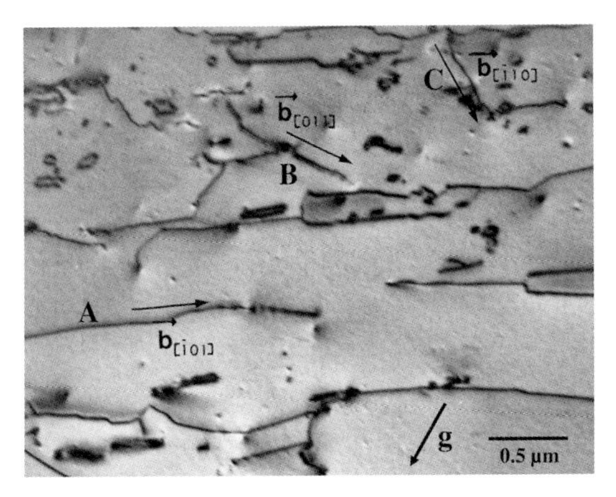

Figure 9.13 Dislocation structure in UO_{2+x} ($x = 0.0001$) deformed at 600 °C. Primary screw dislocations are marked A and secondary dislocations are marked B and C. Debris in the form of loops and dipoles are also seen. Foil cut parallel to the (111) slip plane. Data from Ref. [74].

appeared to be controlled by the interaction of mobile screw dislocations with unknown but extrinsic impurity defects. Screw dislocation debris dominated the dislocation substructures (as shown in Figure 9.13), indicating that screw dislocations were less mobile than edge dislocations at these low temperatures. Although the Knoop hardness did increase with increasing nonstoichiometry at room temperature, this may have been due to the precipitation of U_4O_{9-y} on cooling from elevated temperatures; samples with $x = 0.01$ or greater are in the two-phase field (UO_{2+x} + U_4O_{9-y}) at room temperature.

The situation at higher temperatures (800–1400 °C) is more complex. The primary glide plane depends on orientation, temperature, and the O/U ratio. Typically, $\{001\}$ $\langle 110 \rangle$ slip is most common at low O/U ratios and higher temperatures, while $\{111\}$ $\langle 110 \rangle$ slip dominates at high O/U ratios and lower temperatures. The CRSS for $\{001\}$ slip is almost independent of the O/U ratio, while that for $\{111\}$ glide decreases with increasing x. At low values of x, the flow stress for slip is controlled by extrinsic impurity defects as at low temperatures, whereas at high O/U ratios it appears that oxygen interstitial defects ("Willis defects" [75]), which are necessarily present in crystals with high O/U ratios) actually reduce the hardening effect of these extrinsic impurities. Keller *et al.* [74] suggested, somewhat speculatively, that the core structure of the glide dislocations would be changed by the presence of Willis defects in such a way that dislocations on $\{111\}$ could overcome the hardening effect of impurity defects, whereas those on $\{100\}$ could not. Experiments conducted with more pure crystals than those available to Keller *et al.* [74] would be desirable to understand the plasticity of UO_{2+x} crystals, particularly with regards to issues related to the nonstoichiometry.

9.4.3.3 Dislocation–Dissociation in Oxides with the Fluorite Structure

Dislocation–dissociation has never been reported in any of the oxides with the fluorite structure, although this is not surprising as the $1/2\langle 110 \rangle$ Burgers vector joins neighboring anions and neighboring cations. Dissociation into, say, Shockley-type partials would create a high-energy fault.

9.4.4
Titanium Oxide and Oxides with the Rutile Structure

Although binary oxides with a relatively large quadrivalent cation in octahedral coordination can occur in a variety of crystal structures (the anatase form of TiO_2, the Ramsdellite form of MnO_2, etc.), the rutile structure is generally the stable form and is found in a number of 3d transition metal oxides (TiO_2, VO_2, CrO_2, MnO_2, etc.). Of these, only rutile itself – TiO_2 – has received much attention with regard to dislocation activity; hence, attention will be focused on the behavior of rutile.

In the seminal report of Wachtman and Maxwell [6] that began the modern era of studies of dislocation plasticity in oxides, the brittle $\rightarrow$ ductile transition of TiO_2 was reported to be 600 °C, lower than was found for either Al_2O_3 (900 °C) or MgO (1100 °C). The first systematic studies of crystal plasticity were carried out by Ashbee and Smallman [76], who identified the active slip systems as $\{101\}\langle\bar{1}01\rangle$ and $\{110\}$ $[001]$. Blanchin *et al.* [77, 78] confirmed $\{101\}\langle\bar{1}01\rangle$ as the easy slip system, and deformed [001] oriented crystals at temperatures as low as 600 °C; at this temperature, yielding occurred at a CRSS of about 45 MPa. Neither extended faults on $\{101\}$ nor dissociation of $\langle 101\rangle$ dislocations were observed [79], in agreement with elastic energy calculations [80], but not with the results of some early experiments [76, 81]. However, Bursill and Blanchin [79], using high-resolution (HR) TEM, identified dissociation close to the surface of foils. More recently, Suzuki *et al.* [31] found a 1 nm dissociation into half-partials of the $\langle 101\rangle$ dislocation, while Blanchin *et al.* [82] observed a much larger dissociation in Cr_2O_3-doped rutile. The secondary $\{110\}[001]$ system could be activated above 900 °C, but the total plastic strain achieved prior to fracture was limited [82]. Reduced rutile and Cr- and Al-doped TiO_2 have also been studied [82]; significant solution hardening was seen to occur that involved glissile dislocations interacting with both point defects and extended defects. In some cases, this gave rise to dynamic strain aging [78].

9.4.5
Silicon Oxide (Quartz)

α-Quartz is the polymorph of silicon oxide (SiO_2) that is stable at ambient temperature and pressure, and is one of the most important of the rock-forming minerals. Plastic deformation of quartz-rich rocks occurs whenever there is tectonic activity, and over the years many studies of quartz deformation have been reported in the earth sciences literature. α-Quartz is trigonal, space group $P3_121$, and transforms displacively to β-quartz, space group $P6_222$, at 573 °C. The finite dilatation associated with the $\alpha \rightarrow \beta$ transformation often involves cracking, and this has hampered studies of plastic deformation of quartz at elevated temperatures. It has, accordingly, not proven possible to deform quartz in the laboratory under ambient pressure.

Furthermore, if the quartz is "dry" – that is, if it contains little or no dissolved H_2O or other water-related species (H^+, OH^-, etc.) – then any deformation is difficult even under hydrostatic-confining pressures. For example, under a confining pressure of 1.5 GPa, Griggs and Blacic [83, 84] found the CRSS for basal slip of dry quartz to be ~2.2 GPa at 300 °C and ~1.6 GPa at 700 °C, which is about 5% of the shear modulus of 40 GPa. This intrinsic resistance to plastic deformation of quartz remains high to at least 1300 °C (~0.8 T_m), where the CRSS is still in excess of 0.5 GPa [12, 85]. Where deformation has been induced under hydrostatic-confining pressures, various slip planes – (0001), $\{10\bar{1}0\}$, $\{11\bar{2}0\}$, $\{10\bar{1}1\}$, $\{01\bar{1}\bar{1}\}$ – and dislocations with various Burgers vectors – $1/3\langle 11\bar{2}0\rangle$, [0001], and $1/3\langle 11\bar{2}3\rangle$ – have often been found. These Burgers vectors correspond to the three shortest lattice translations in quartz, namely 0.491, 0.541, and 0.730 nm, respectively.

9.4.5.1 Hydrolytic Weakening in Quartz

In the presence of even small quantities of water, plastic deformation in the laboratory becomes much easier, albeit still necessitating hydrostatic-confining pressures. Moreover, pressure expands the stability field of α-quartz to higher temperatures (e.g., ~1300 °C at 3.4 GPa confining pressure [86]). This "hydrolytic weakening" was discovered by Griggs and Blacic [83, 84], and has resulted in a very large literature devoted to understanding the softening effect of water. The dramatic effect of water content is illustrated in Figure 9.14, which shows the stress–strain curves obtained by Ord and Hobbs [87] under various buffered assemblies at 800 °C under a confining pressure. As the buffering issue is beyond the scope of the present chapter, attention will be focused here only on the "dry" and "H_2O only" curves of Figure 9.14. The following (albeit still tentative) summary on the effect of water on the deformation of quartz is taken from Paterson [85].

The ability to deform quartz to large strains is determined primarily by dislocation mobility rather than mobile dislocation density. There is evidence, however, that water can influence nucleation of potentially mobile dislocations in both synthetic and natural quartz [88, 89], as discussed by McLaren [38]. Water apparently influences dislocation mobility by: (i) facilitating the nucleation and diffusion of kinks; and (ii) facilitating dislocation climb [90]. Atomistically, the water appears to accelerate the self-diffusion of oxygen and/or silicon by mechanisms that are still obscure. This suggests that three regimes of deformation in quartz can be recognized [90]: (i) the *intrinsic* regime at relatively low temperatures and controlled by the kink pair nucleation and diffusion; (ii) a *hydrolytic glide control* regime involving kink nucleation and/or diffusion; and (iii) a *recovery control* regime at high temperatures.

In the hydrolytic glide regime, dislocation activity and the mechanism of water weakening depend on the water content of the quartz. High-quality synthetic quartz can be relatively dry (~100 at. ppm [H]/[Si]), and in these samples the water is accommodated in the quartz lattice by point defects, specifically a $(4H)_{Si}$ species (i.e., one Si is formally replaced by 4H [91].) Crystals with, say, 1000 at. ppm [H]/[Si] contain aggregated molecular water in the form of bubbles. These can be fluid inclusions which have no long-range strain fields associated with them, or water

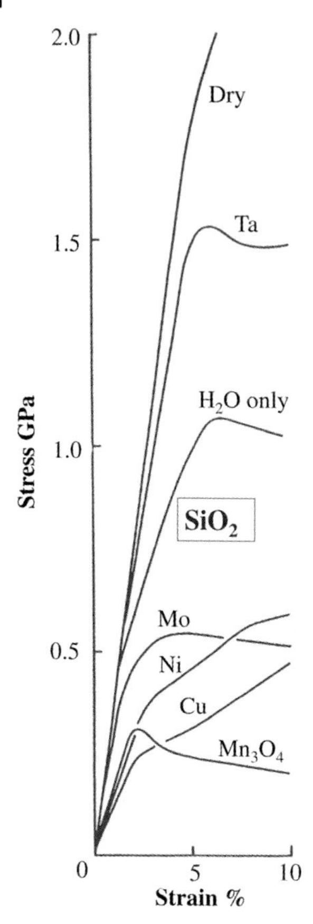

Figure 9.14 Stress–strain curves of quartz deformed in compression under a confining pressure of 1.6 GPa in various buffered assemblies at 800 °C and a strain rate of $10^{-5}\,s^{-1}$. The effective water content increases from the top to the bottom curve. Data from Ref. [87].

clusters under a high internal pressure and showing strain contrast [92]; on heating, these clusters evolve into strain-free bubbles and associated prismatic dislocation loops with $\mathbf{b} = 1/3\,\langle 11\bar{2}0 \rangle$.

9.4.5.2 **Dislocation–Dissociation in Quartz**

Dislocation–dissociation in quartz was first observed by McLaren *et al.* [32] in a natural dry quartz deformed at 500 °C, and has been studied most recently by Cordier and Doukhan [33] in a synthetic crystal containing ~100 at. ppm [H]/[Si]. By using a crystal oriented to have high Schmid factors for $(0001)1/3\,\langle 11\bar{2}0 \rangle$basal slip, $\{11\bar{2}0\}[0001]$ prism plane slip, and $\{10\bar{1}0\}\ 1/3\,\langle \bar{1}2\bar{1}3 \rangle$ pyramidal slip, and deformed under a hydrostatic pressure of 1.0–1.1 GPa, Cordier and Doukhan found a flow stress at 500 °C of almost 3 GPa, which decreased to ~2 GPa at 900 °C. These high flow stresses correspond to a significant fraction of the shear modulus and reflect the fact that the quartz is relatively dry. Slip at 500 °C was heterogeneous (in slip bands), but conformed to the basal slip system. Some of the basal dislocations were dissociated into collinear partials by as much as 0.5 to 5 μm by the reaction

$$1/3\,\langle 11\bar{2}0\rangle \rightarrow 1/6\,\langle 11\bar{2}0\rangle + 1/6\,\langle 11\bar{2}0\rangle \quad (8)$$

while others appeared undissociated. At 700 °C and a flow stress of ~2.5 GPa, similar evidence for basal slip was found, although pyramidal slip had also occurred. The pyramidal dislocations were also widely dissociated into collinear partials via the reaction

$$1/3\,\langle 11\bar{2}3\rangle \rightarrow 1/6\,\langle 11\bar{2}3\rangle + 1/6\,\langle 11\bar{2}3\rangle \quad (9)$$

At 900 °C, in the β-quartz phase field, pyramidal glide dominated, the dislocations again being widely dissociated. In a few regions, which the authors suggested had a higher water content than the average, dislocation microstructures typical of recovery were observed. Thus, these authors have provided direct evidence of dissolved water enhancing both the glide and climb of dislocations. No attempt was made to derive stacking fault energies for quartz, as the variable – but very widely spaced – stacking faults clearly indicated nonequilibrium structures, undoubtedly due to the very high flow stresses at which the dislocations were gliding. The atomic structure of stacking faults in quartz has been considered by Doukhan and Trepied [90]; low-energy faults on (0001) and on $\{10\bar{1}0\}$, as observed experimentally by Cordier and Doukhan [33], have all the Si and O atoms in the planes adjacent to the fault-forming $[SiO_4]$ tetrahedra which are bonded to neighboring tetrahedra at all four corners; furthermore, no dangling bonds, nor stretched or distorted Si–O–Si bonds, were present.

9.4.6
Aluminum Oxide (Sapphire)

Sapphire (α-Al_2O_3) was the first synthetic gemstone to be grown in the laboratory by Verneuil over a century ago [93], and indeed Verneuil crystals were widely used as the raw material for the "jewels" in fine Swiss watches during the first half of the twentieth century. The ready availability of undoped and Cr-doped Al_2O_3 (ruby) single crystals led to the first demonstration of laser action in synthetic ruby by Maiman in 1960 [94], and sapphire wafers up to 20 cm in diameter have been substituted for Si wafers in certain demanding electronic integrated circuit applications.

The availability of sizable single crystals has led to a significant literature on the deformation of sapphire of various orientations, and at various temperatures. As already noted, the first such study was by Wachtman and Maxwell in 1954 [6], who activated (0001) $1/3\,\langle 11\bar{2}0\rangle$ basal slip at 900 °C via creep deformation. Since that time, it has become clear that basal slip is the preferred slip system at high temperatures, but that prism plane slip, $\{11\bar{2}0\}\,\langle 1\bar{1}00\rangle$, can also be activated and becomes the preferred slip system at temperatures below ~600 °C. Additional slip systems, say on the pyramidal plane $\{\bar{1}012\}\ 1/3\,\langle 10\bar{1}1\rangle$, have very high CRSSs and are thus difficult to activate. Both, basal and rhombohedral deformation twinning systems, are also important in Al_2O_3 (these are discussed later in the chapter).

The yield stress for both basal and prism plane slip as a function of temperature have been determined over a wide temperature range (Figure 9.4). The "Castaing Law" – ln(CRSS) decreasing linearly with temperature – has been explained by Mitchell *et al.* [22] in terms of conventional kink pair nucleation and kink diffusion (as discussed in Section 9.2).

9.4.6.1 Dislocation–Dissociation in Sapphire

Dislocation–dissociation has been an important issue in studies of basal deformation in sapphire since the seminal studies of Kronberg [7], who suggested that basal dislocations would dissociate into half-partials via the reaction:

$$1/3\,\langle 11\bar{2}0\rangle \rightarrow 1/3\,\langle 10\bar{1}0\rangle + 1/3\,\langle 01\bar{1}0\rangle \qquad (10)$$

with each half partial further dissociating into quarter partials via the reaction:

$$1/3\,\langle 10\bar{1}0\rangle \rightarrow 1/6\,\langle 2\bar{1}\bar{1}\rangle + 1/6\,\langle 11\bar{2}0\rangle \qquad (11)$$

The reaction in Eq. (10) involves faulting only in the cation sublattice, as $1/3\,\langle 10\bar{1}0\rangle$ is a translation vector of the anion sublattice. Equation (11) involves faulting of both sublattices, and must lead to a very high stacking fault energy, as no reliable evidence for quarter partial dislocations has ever been found.[1] Kronberg utilized an idealized version of the corundum structure of Al_2O_3 – a perfect hcp oxygen sublattice and flat as opposed to puckered cation layers – in considering the dissociations in Eqs (10) and (11) and imagined that the *motion* plane for the dislocations was between the cation and anion layers.

9.4.6.2 Basal Slip in Sapphire

Bilde-Sørensen *et al.* [29] presented a new model for basal slip, in which the motion plane is within the puckered cation layer. Each $1/3\,\langle 10\bar{1}0\rangle$ partial transports only half of the cations as it moves to the half-slipped position; motion of the trailing partial restores perfect lattice without any charge transport. Because of the rhombohedral symmetry of α-Al_2O_3, the $\langle 10\bar{1}0\rangle$ and $\langle 01\bar{1}0\rangle$ directions are not crystallographically equivalent. Based on these considerations, the new model of basal slip led to a unique basal stacking fault (one of four possible basal stacking faults that can be imagined in Al_2O_3). *Ab initio* calculations performed by Marinopolous and Elsässer [96] showed that this fault had the lowest basal SFE in Al_2O_3, at 1.49 J m^{-2}; it was also the basis for a new model of basal twinning (see below).

9.4.6.3 Prism-Plane Slip in Sapphire

Prism-plane slip occurs by the motion of $\langle 10\bar{1}0\rangle$ dislocations on the $\{1\bar{2}10\}$ plane, rather than $1/3\,\langle 1\bar{2}10\rangle$ dislocations in the $\{10\bar{1}0\}$ plane, in spite of the unusually large magnitude of the Burgers vector of $\langle 10\bar{1}0\rangle$ dislocations. Such dislocations

1) Kronberg's *synchroshear* model of basal dislocation motion required these quarter partials; they were also involved in his model of basal twinning. As first emphasized by Bilde-Sørensen *et al.* [29], and as further discussed by Heuer *et al.* [95], this requires that dislocation motion involves charge transport, a significant impediment to slip in a material with such strongly ionic bonding as Al_2O_3.

might be expected to be unstable and *decompose* according to the reaction:

$$\langle 10\bar{1}0\rangle \rightarrow 1/3\,\langle 2\bar{1}\bar{1}0\rangle + 1/3\,\langle 11\bar{2}0\rangle \qquad (12)$$

This reaction can occur by glide in the basal plane only for screw dislocations, and is thought to be the mechanism for the formation of a dislocation network in crystals undergoing deformation by prism plane slip [97]. Alternatively, the dislocation can lower its energy by dissociating into three collinear partials according to the reaction (see also Table 9.3):

$$\langle 10\bar{1}0\rangle \rightarrow 1/3\,\langle 10\bar{1}0\rangle + 1/3\,\langle 10\bar{1}0\rangle + 1/3\,\langle 10\bar{1}0\rangle \qquad (13)$$

This dissociation has been observed to occur by climb for high-temperature deformation, but can also occur by glide.

9.4.6.4 Dipoles and Climb Dissociation in Sapphire

Dislocation substructures formed after basal glide are similar to those found in hcp metals, in that a high density of dislocation dipoles is seen, as shown in Figure 9.15. However, because of the high temperatures required for deformation (>1200 °C), dislocation climb is relatively easy and the dipoles tend to break up into strings of loops. This climb process occurs by a redistribution of point defects by pipe diffusion along the dislocation – that is, the climb occurs conservatively (self-climb). Narrow dipoles form faulted dipoles by a process of climb dissociation of the individual dislocations and annihilation of the inner partials, as illustrated in Figure 9.16, and again, this can occur by conservative climb. The faulted dipole can break up into a string of faulted loops, just as a perfect dipole can break up into a string of perfect loops.

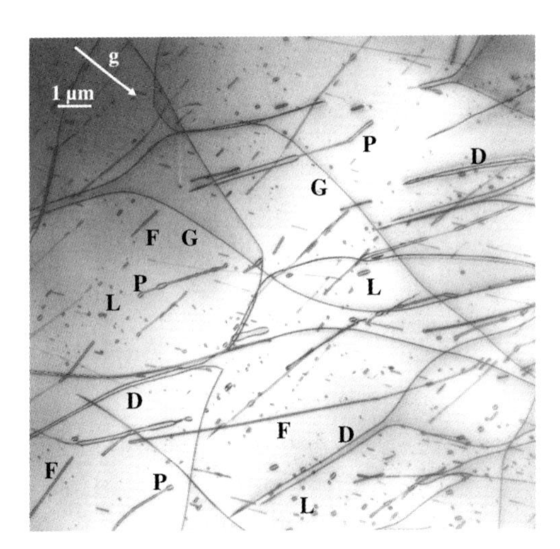

Figure 9.15 Dislocations in sapphire deformed on the basal plane to 3.6% shear strain at 1400 °C. Examples of glide dislocations (G), regular dipoles (D), faulted dipoles (F), isolated loops (L) and loops in the process of pinching off (P) are indicated. Foil parallel to (0001), bright-field image, $g = 03\bar{3}0$.

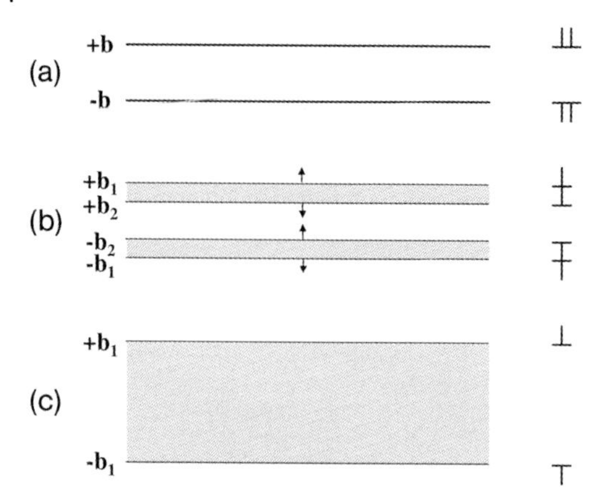

Figure 9.16 Formation of a faulted dipole in sapphire. (a) Original perfect dislocation dipole; (b) Dipole with each partial undergoing conservative climb dissociation; (c) Faulted dipole formed by the annihilation of the inner partials.

Climb dissociation initially occurs on $\{11\bar{2}0\}$ planes, as expected for the edge dipoles. Subsequently, the $1/3\ \langle 10\bar{1}0\rangle$ faulted dipoles rotate into the $\{10\bar{1}0\}$ plane so as to bring them into the edge orientation. The whole process of dipole break-up has been explained in terms of elasticity theory [98, 99]. The faulted dipoles form via the Kronberg dissociation [Eq. (10)], although Kronberg envisioned the dissociation occurring by glide rather than climb.

9.4.6.5 **Stacking Fault Energy of Sapphire**

These various glide and climb configurations have been used to estimate stacking fault energies in sapphire; these are summarized in Table 9.6 and compared with atomistic calculations. In general, the SFE on the prismatic planes is lower than that on the basal plane, both for experimental and calculated values. The exception is the value on the basal plane measured from the diameter of faulted loops (produced by radiation) at which the loops unfault. This is probably a nonequilibrium situation, and a value of about 1.5 J m^{-2} for the SFE in the basal plane is most likely. Such a value would provide only a core dissociation for glide dislocations in the basal plane, in agreement with the fact that dislocations dissociated in the basal plane have never been observed.

9.4.6.6 **Deformation Twinning in Sapphire**

Among the oxides described in this chapter, deformation twinning is most important in the case of sapphire. Two twinning systems are known, on the rhombohedral and basal planes, and dislocation models for each have been suggested and confirmed using TEM.

Rhombohedral Twinning Rhombohedral twinning is the deformation mode with the lowest CRSS in sapphire [105], and can occur at temperatures as low as −196 °C [106].

Table 9.6 Experimental and theoretical fault energies for sapphire on various planes.

Fault vector	Fault plane	Fault energy (J m^{-2})	Method
$1/3\langle 10\bar{1}0\rangle$	$\{10\bar{1}0\}$	0.4	TEM study of faulted $1/3\langle 1\bar{2}10\rangle$ dipole [98]
$1/3\langle 10\bar{1}0\rangle$	$\{10\bar{1}0\}$	0.23, 0.16	TEM study of elongated $\langle 10\bar{1}0\rangle$ loop [100]
$1/3\langle 10\bar{1}0\rangle$	$\{10\bar{1}0\}$	0.64	TEM study of faulted loop produced by radiation [101]
$1/3\,[0001] \equiv 1/3\,\langle 10\bar{1}0\rangle$	$\{10\bar{1}0\}$	0.5, 0.5	Atomistic calculation [102]
$1/3\langle 10\bar{1}0\rangle$	$\{11\bar{2}0\}$	0.15	TEM study of faulted $1/3\langle 1\bar{2}10\rangle$ dipole [100]
$1/3\langle 10\bar{1}0\rangle$	$\{1\bar{2}10\}$	0.3	TEM study of $\langle 10\bar{1}0\rangle$ glide dissociation [103]
$1/3\langle 10\bar{1}0\rangle$	$\{\bar{2}110\}$	0.4	TEM study of $\langle 10\bar{1}0\rangle$ climb dissociation [104]
$1/3\langle 10\bar{1}0\rangle$	$\{1\bar{2}10\}$	~0.1	TEM study of elongated $\langle 10\bar{1}0\rangle$ loop [100]
$1/3\,[0001] \equiv 1/3\,\langle 10\bar{1}0\rangle$	$\{11\bar{2}0\}$	0.5	Atomistic calculation [102]
$1/3\langle 10\bar{1}0\rangle$	$\{10\bar{1}2\}$	0.25	TEM study of $\langle 10\bar{1}0\rangle$ climb dissociation [100]
$1/3\,[0001] \equiv 1/3\,\langle 10\bar{1}0\rangle$	(0001)	0.27	TEM study of faulted loop produced by radiation [101]
$1/3\langle 10\bar{1}0\rangle$	(0001)	1.2, 2.3	Atomistic calculation [102]

The twin law is as follows: $K_1 = \{01\bar{1}2\}$; $\eta_1 = \langle 01\bar{1}1\rangle$; $K_2 = \{0\bar{1}14\}$; $\eta_2 = \langle 02\bar{2}1\rangle$; and $s = 0.202$ [106]. The dislocation model suggested for rhombohedral twinning [107] is based on the double cross-slip mechanism proposed by Pirouz [108] for silicon (see also Lagerlöf *et al.* [109]). Briefly, twin nucleation involves a pinned zonal dislocation undergoing the following dissociation:

$$1/3\,\langle 01\bar{1}1\rangle \rightarrow 1/21.9\,\langle 0\bar{1}11\rangle + [1/3\,\langle 01\bar{1}1\rangle - 1/21.9\,\langle 0\bar{1}11\rangle] \tag{14}$$

The leading dislocation, with $b_t = 1/21.9K\langle 0\bar{1}11\rangle$, bows out, in the same manner as a Frank–Read source, to form a loop of faulted crystal. Because b_t is a partial dislocation, it cannot bow out again on the same plane but can, if in screw orientation, cross-slip onto an adjacent plane and bow out again. Continued growth of the faulted loop on each plane and cross-slip onto neighboring planes allow for twin growth. The unusual magnitude of b_t arises from the notion that twinning involves shear of every neighboring $(01\bar{1}2)$ plane, by a partial dislocation with $b_t = 0.07$ nm. The $(01\bar{1}2)$ interplanar spacing is 0.348 nm; a shear of $1/21.9\,\langle 0\bar{1}11\rangle$ causes the observed twinning shear of 0.202 ($0.0703/0.348 = 0.202$) [107]. This model predicts that the rhombohedral twin is a screw twin – with η_1 being a twofold screw axis – rather than a type I or type II twin. The proposed structure of the twin interface has been confirmed using HRTEM [110]. Moreover, *ab initio* calculations performed by Marinopoulos and Elsässer [111] have shown that the energy of the twin interface predicted by this model ($0.63\,\mathrm{J\,m^{-2}}$) is much lower than that of two other possible structural models of a rhombohedral twin boundary, at 1.35 and $3.34\,\mathrm{J\,m^{-2}}$.

Basal Twinning Basal twinning can also occur in sapphire [112], and involves the following twin law: $K_1 = (0001)$; $\eta_1 = \langle 10\bar{1}0\rangle$; $K_2 = \{10\bar{1}1\}$; $\eta_2 = \langle\bar{1}012\rangle$; and $s = 0.635$. A cross-slip model has also been employed for this case [95, 113, 114], and involves a $1/3\,\langle 10\bar{1}0\rangle$ twinning dislocations produced via either reaction Eq. (10) or Eq. (13). The model predicts that the twin is type II (η_1 is a twofold rotation axis), and theoretical calculations [115] indicate a twin boundary energy of $0.73\,\mathrm{J\,m^{-2}}$, much lower than other possible basal twin boundaries with either mirror or glide symmetry, which have energies of 1.99 and $2.63\,\mathrm{J\,m^{-2}}$, respectively. Persuasive TEM evidence supporting this model, and involving a nucleating dislocation produced by the reaction in Eq. (13), has been found [114]. In particular, twin boundaries containing three $1/3\,\langle 10\bar{1}0\rangle$ partials have been imaged, as well as the presence of a threading dislocation within the twin; the latter is the "fingerprint" of the double cross-slip twinning model. Furthermore, Castaing *et al.* [115] were able to activate basal twinning in suitably oriented sapphire crystals by compressing under a hydrostatic confining pressure at temperatures down to 600 °C. They found that the temperature dependence of the CRSS for basal slip and basal twinning were virtually the same, and concluded that both were controlled by kink pair nucleation and diffusion on $1/3\,\langle 10\bar{1}0\rangle$ partial dislocations.

9.4.7
$SrTiO_3$ and Oxides with the Perovskite Structure

The oxide perovskites comprise a technologically important family of ceramics, as all practical inorganic ferroelectrics and piezoelectrics crystallize in this structure. Furthermore, the structure of the so-called high-T_c superconductors is closely related, and perovskite single crystals such as $SrTiO_3$ are widely used as substrates for thin film deposition of the high-T_c materials and other useful functional oxides. Lastly, it is widely believed that olivines, $(Mg, Fe)_2SiO_4$, the principal constituent of the Earth's upper mantle, transform to magnesio-wüstite (Mg, Fe)O and a $(Mg, Fe)SiO_3$ perovskite in the lower mantle. Thus, knowledge of the high-temperature rheology of perovskites is of crucial importance for understanding mantle convection. Unfortunately, it is not possible to produce such silicate perovskites in the laboratory, and it has been necessary to study nonsilicate surrogates such as $BaTiO_3$, $CaTiO_3$, and $SrTiO_3$ to understand the tectonic processes of the Earth. The structure of cubic perovskite is shown in Figure 9.17.

9.4.7.1 Inverse Brittle-to-Ductile Transition (BDT) in $SrTiO_3$

High-temperature creep studies identified $\{110\}\,\langle 1\bar{1}0\rangle$ as the dominant slip system for both $BaTiO_3$ and $SrTiO_3$ [116, 117]. Evidence also exists for $\langle 100\rangle$ Burgers

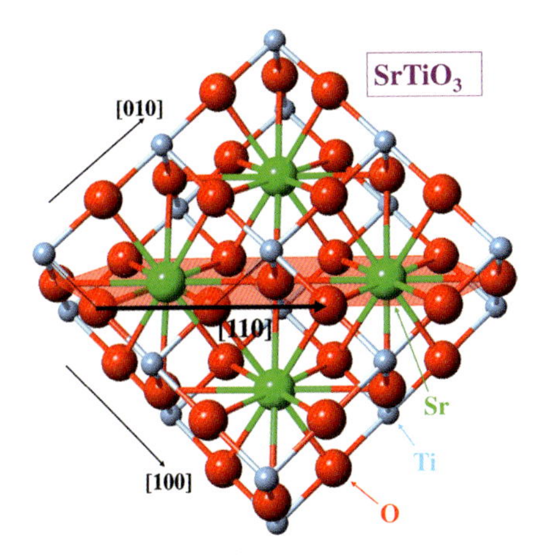

Figure 9.17 The crystal structure of cubic $SrTiO_3$ (four unit cells), viewed in perspective approximately along the [001] direction. The [110] Burgers vector is shown. Splitting into two collinear 1/2 [110] partials obviously maintains the oxygen anion packing, but places Sr cations in juxtaposition across the {110} fault plane (shown shaded).

vectors [118, 119] and {100} slip planes [36, 118]. More recently, Brunner *et al.* [120] and Gumbsch *et al.* [121] found unexpected low-temperature plasticity and an "inverse" BDT – one with *decreasing* temperature – in $SrTiO_3$ deformed along $\langle 100\rangle$ or $\langle 110\rangle$, and involving $\{110\}\ \langle 1\bar{1}0\rangle$ slip, as shown in the yield stress versus temperature curves in Figure 9.18. Such behavior is similar to the yield stress anomaly found in some ceramics such as $MgO\text{-}Al_2O_3$ spinels [18], in many intermetallics such as $MoSi_2$ (see Mitchell *et al.* [122]), and in a number of LI_2 and B2 compounds (see Caillard [123]). The difference is that, for the yield stress anomalies, the materials remain plastic in the intermediate temperature range where the yield stress increases with increasing temperature, whereas $SrTiO_3$ is brittle in the intermediate temperature range and its fracture stress increases with increasing temperature. As shown in Figure 9.18, at temperatures below about 900 K, the critical resolved shear stress for $\{110\}\langle 1\bar{1}0\rangle$ slip is low (ca. 50 MPa) and only weakly dependent on temperature; below 300 K, it begins to rise, but at 100 K, it is still only ~300 MPa.

9.4.7.2 Dislocation–Dissociation and SFE in Strontium Titanate

The $\langle 110\rangle$ Burgers vector for $SrTiO_3$ is quite long (0.552 nm), and its dominance over the shorter $\langle 100\rangle$ Burgers vector (0.391 nm) is, at first sight, surprising. However evidence for the dissociation of $\langle 110\rangle$ dislocations has been obtained in samples deformed at room temperature, which implies a low SFE on the {110} plane of $SrTiO_3$. Mao and Knowles [36] studied the dissociation of lattice dislocations in sintered $SrTiO_3$, and found that the reaction

$$\langle 110\rangle \rightarrow 1/2\,\langle 110\rangle + 1/2\,\langle 110\rangle \tag{15}$$

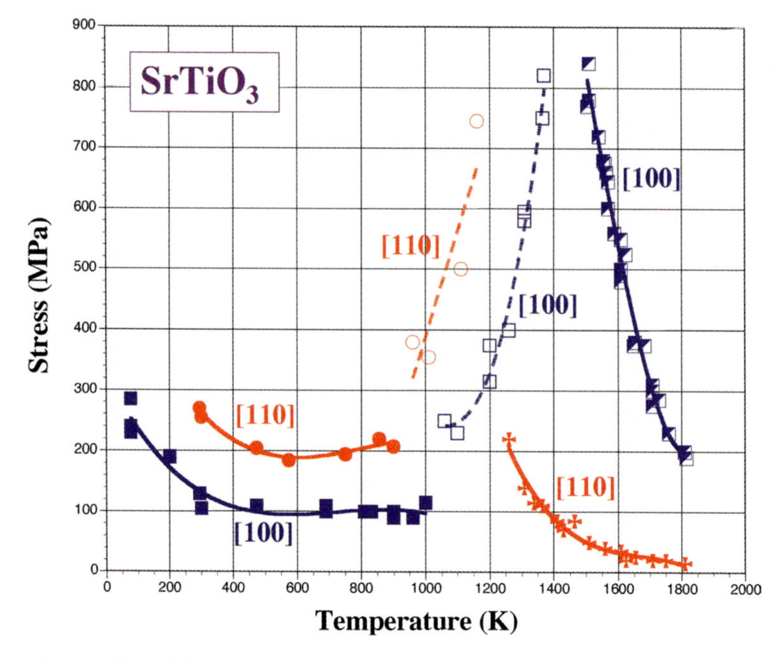

Figure 9.18 Yield stress versus temperature for $SrTiO_3$ crystals compressed along ⟨100⟩ and ⟨110⟩. No measurable plastic flow occurs in the intermediate temperature range; the fracture stress is indicated by the open symbols. Data from Refs [120] and [121].

occurred for mixed ⟨110⟩ dislocations lying along the ⟨100⟩ direction. The dissociation is illustrated in Figure 9.17, and shows that the fault juxtaposes the large Sr cations across the fault plane but maintains the anion sublattice. Mao and Knowles calculated a SFE of 145 mJ m^{-2} for glide dissociation on (001) and 245 mJ m^{-2} for climb dissociation on (010). The different SFEs found for the glide and climb cases was surprising, as the fault planes (001) and (010), respectively, were crystallographically equivalent. On the other hand, Matsunaga and Saka [37] found that ⟨110⟩ dislocations introduced by Vickers indentation at room temperature were dissociated by glide into two 1/2⟨110⟩ partials on $\{1\bar{1}0\}$ planes and, from the ribbon width, deduced a SFE of 136 mJ m^{-2}. Sigle *et al.* [124] found the same double contrast in ⟨110⟩ dislocations produced by deformation at low temperatures, but showed that the contrast was due to dislocation dipoles. They concluded that, at high temperatures, there was no glide dissociation but possibly climb dissociation.

The inverse BDT for $SrTiO_3$ deformed along ⟨100⟩ or ⟨110⟩ is thought to be due to transformation of the dislocation core structure [121]; Gumbsch *et al.* [121] suggested a possible climb dissociation (more properly a *decomposition*) via the reaction

$$\langle 110\rangle \rightarrow \langle 100\rangle + \langle 010\rangle \tag{16}$$

However, the ⟨100⟩ product dislocations in Eq. (16) are both perfect dislocations and are perpendicular to one another. Frank's rule indicates that, to first order, such a reaction would not lead to reduction in the dislocation line energy, whether it occurs by glide or climb. Of course, if the screw/edge character of the ⟨110⟩ dislocation is taken into account, then Eq. (16) will be favorable for an edge dislocation and unfavorable for a screw dislocation.

On the other hand, as discussed later in Section 4.12, barring a pathological anisotropy of the SFE, the climb configuration for Eq. (15) should always have a lower energy than a glide configuration. It is likely that the structural transformation at the heart of the inverse BDT in $SrTiO_3$ is the climb of dissociated ⟨110⟩ dislocations into a sessile configuration, and that the onset temperature of the inverse BDT (~750 °C) can be correlated with the temperature at which pipe diffusion kinetics permit climb along the dislocation cores to effect the core transformation. In fact, climb-dissociated ⟨110⟩ dislocations have been found by Zhang *et al.* [125] in a hot-pressed $SrTiO_3$ bicrystal containing a 5° (110)[001] tilt boundary [126]; however, the SFE indicated by the separation of the partials was high (720 mJ m^{-2}) compared to the value obtained from glide dissociation in one of the studies [36]. Although the dissociation is by climb rather than by glide, the stacking fault plane is also {110}. It is possible that the partials had not achieved their equilibrium separation by climb, and this rather puzzling situation needs to be resolved.

9.4.8
$MgO–Al_2O_3$ and other Spinels

The mechanical properties of $MgO–Al_2O_3$ spinel single crystals have been studied extensively by both creep and constant strain-rate compression tests at high temperature [18–20, 26, 127–148], as well as by indentation testing [21, 142, 149]. Studies have also been conducted on other oxides with the spinel structure, such as magnetite (Fe_3O_4) [150], manganese zinc ferrite ($Mn_{0.7}Zn_{0.2}Fe_{2.1}O_4$) [151, 152], manganese aluminum spinel ($Mn_{1.8}Al_{1.2}O_4$) [153], nickel ferrite ($Ni_{0.7}Fe_{2.3}O_4$) [154–156], and chromite ($FeCr_2O_4$) [157]. Perhaps the most unusual mechanical behavior exhibited by $MgO{\cdot}nAl_2O_3$ spinel is the compositional softening (as discussed in Section 9.2.5), whereby the critical resolved yield stress decreases by almost two orders of magnitude when the stoichiometry is changed from $n=1$ (stoichiometric) to $n=3.5$ [20, 26, 128, 130, 139, 144, 148]. Stoichiometric crystals can be deformed plastically only at temperatures above ~1700 °C, while nonstoichiometric crystals can be deformed at temperatures down to about 1200 °C. (Actually, this refers to ambient pressure conditions; under a high confining pressure, spinel can be deformed plastically down to ambient temperatures [20, 139].) The temperature dependence of the CRSS has already been discussed in Section 9.2.4 in terms of kink pair nucleation and kink migration on dislocations. Compositional softening is due either to kink pair nucleation on cation vacancies, or to kink pair nucleation on partial dislocations and the decrease in SFE with increasing deviation from stoichiometry.

Table 9.7 Slip planes in oxides with the spinel structure.

Spinel oxide	Slip planes	Reference(s)
$MgO{\cdot}nAl_2O_3$	{111}, {110}, {100}	[20, 26, 127–130, 132–135, 140, 141]
Fe_3O_4	{111}, {100}	[150]
$Mn_{0.7}Zn_{0.2}Fe_{2.1}O_4$	{111}, {110}	[151, 152]
$Mn_{1.8}Al_{1.2}O_4$	{110}	[153]
$Ni_{0.7}Fe_{2.3}O_4$	{111}, {110}, {100}	[155, 156]
$FeCr_2O_4$	{111}, {110}, {100}	[157]

9.4.8.1 Slip Planes in Spinels

The slip direction in spinel is always parallel to the $\langle 110\rangle$ close-packed direction of the fcc lattice; that is, $1/2\langle 110\rangle$ is the shortest perfect Burgers vector. In terms of the approximate fcc packing of the oxygen anion sublattice (Figure 9.8), $1/2\langle 110\rangle$ represents two interatomic distances. On the other hand, the observed slip plane varies, as shown in Table 9.7 for Mg–Al spinel and a few other oxide spinels that have been tested. The most commonly observed slip planes are {111} and {110}, although {100} has been reported in magnetite, nickel ferrite, and chromite. Slip on {100} planes has also been observed in Mg–Al spinel, but this was at temperatures below 1000 °C under confining pressure for crystals compressed along the $\langle 111\rangle$ axis [20]. The Schmid factor for all the $\{110\}\langle 110\rangle$ slip systems is zero for the $\langle 111\rangle$ compression axis, while that for $\{111\}\langle 110\rangle$ slip is small (0.27) and that for $\{100\}\langle 110\rangle$ slip is relatively large (0.47).

Slip in spinel can be understood to some extent by examining the crystal structure. As shown in Figure 9.8, stacking of the {111} planes can be written in the following 18-layer sequence:

$$\ldots A_O \beta c \alpha B_O A C_O \alpha b \gamma A_O C B_O \gamma a \beta C_O B \ldots \qquad (17)$$

where $A_O B_O C_O$ refers to the approximate fcc stacking of the oxygen anions, ABC refers to octahedral Al cations in the two-thirds occupied Kagomé layers (shown in Figure 9.19), abc refers to the octahedral Al cations in the one-third occupied mixed layers, and $\alpha\beta\gamma$ refers to the tetrahedral Mg cations in the one-quarter occupied mixed layers. Several interesting features emerge from Figure 9.8:

- The anion planes on either side of the mixed $\beta c \alpha$ layer have a spacing of 0.256 nm, whereas those surrounding the Kagomé layer have a significantly smaller spacing of 0.210 nm; the difference is due to the fact that the tetrahedral sites have to expand in order to accommodate the larger Mg^{2+} cations.
- The individual layers are all charged, but the Kagomé layers, along with half of the anions on either side, have a neutral composition of Al_2O_3. The mixed layers have a neutral composition of $Mg_3Al_2O_6$.
- The $1/2\langle 110\rangle$ slip vector corresponds to two inter-anion distances, but the anions are not evenly spaced, because of the nonideality of the fcc packing (see Figure 9.19).

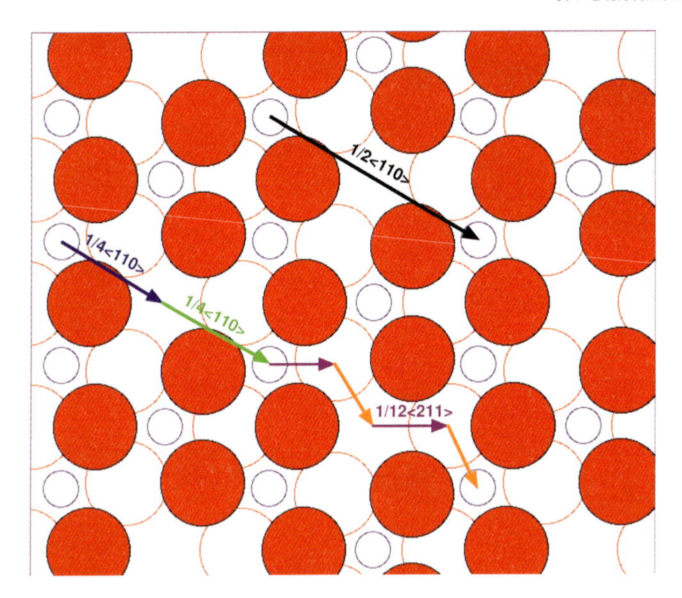

Figure 9.19 The {111} Kagomé layer in spinel. The large open circles represent the lower anion layer, the large filled circles the upper anion layer, and the small open circles the middle cation layer. The perfect 1/2⟨110⟩ Burgers vector is shown, as well as possible dissociation schemes. Note the larger anion octahedra surrounding the empty cation sites and the imperfect fcc stacking.

The {110} planes can be analyzed in a similar way [18]. The stacking sequence is given as ...*ABCD*... although it would be ...*ABAB*... but for the nonideality of the fcc anion packing. The compositions of the *A* and *B* planes are $MgAlO_2$ and Al_2O_4 respectively, and so edge dislocations on the {110} planes would be charged, although not so much as dislocations on the {111} plane. The spacing between the {110} planes is 0.143 nm compared with 0.105 nm for the spacing between the {111} anion and cation planes in the Kagomé layer. This may explain why {110} slip is easier. The spacing of the {100} planes (0.101 nm) is even smaller than the {111} planes, so that slip on {100} planes should only be observed under favorable circumstances, as described above.

9.4.8.2 Dislocation–Dissociations and the SFE in Magnesium Aluminate Spinel

The Hornstra dissociation into collinear half-partials [Eq. (6)] is commonly observed, most often by climb. However, evidence has been found for combinations of glide and climb dissociation on both rational and irrational planes. The {100} fault plane predominates, regardless of stoichiometry, followed by {110}, {112}, {113}, and {111} [26]. Occasionally, pure glide dissociation is observed, for example, of screw dislocations on the {110} plane of stoichiometric crystals [26]. Although faults in spinel exhibit a clear preference for the {100} plane and other low-index planes, they are also observed to be wavy and to lie on irrational planes (see Figure 9.20).

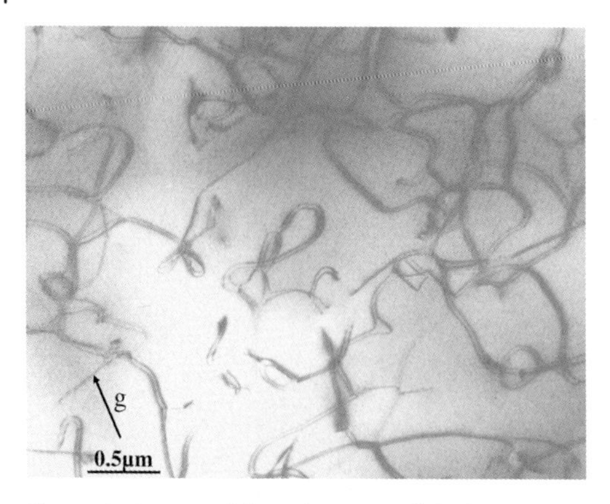

Figure 9.20 Typical three-dimensional climb structure in a $n = 3.5$ spinel deformed 5% along the ⟨111⟩ axis at 1500 °C. Note the widely extended, and appropriately named, ribbons. TEM bright-field image, $\mathbf{g} = 440$.

Apparently, the fault energy is relatively isotropic, in much the same way as anti-phase domain boundaries in ordered alloys. Estimates of the fault energy have been made from ribbon widths. Values of γ from the literature [18, 26, 129, 131, 133, 137] are plotted against n in Figure 9.21. The stacking fault energy is seen to decrease by about a factor of ten with increasing nonstoichiometry as n rises from 1 to 3.5. There is fair agreement between different investigations, the most serious discrepancy being for the stoichiometric spinel, where Welsch *et al.* [135] determined $\gamma = 180\,\text{mJ m}^{-2}$ for edge dislocations on the {111} plane produced by deformation at 1800 °C; in contrast, Veyssiére and Carter [19] determined $\gamma = 530\,\text{mJ m}^{-2}$ for screw disloca-

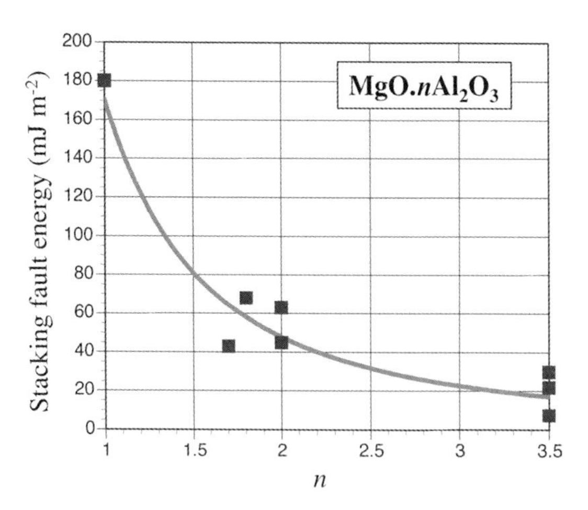

Figure 9.21 Stacking fault energy of $MgO{\cdot}nAl_2O_3$ versus n [18].

tions on the {001} plane produced by deformation at 400 °C under a confining pressure (the latter value is not shown on Figure 9.21). It is not known if such a difference is due to the different dissociation planes, or to the different deformation temperatures. Various fault planes are represented in Figure 9.21, and there is no consistency about the variation of γ from one plane to another. It is also possible that the segregation of vacancies or impurities to stacking faults could occur at high temperatures, giving a lower fault energy than would otherwise be the case.

9.4.9 Mg_2SiO_4 (Forsterite)

Olivine, the mineral name give to the solid-solution series between forsterite (Mg_2SiO_4) and fayalite (Fe_2SiO_4), is thought to be the dominant component in the Earth's upper mantle. Thus, its deformation behavior has aroused considerable geophysical interest. The structure of forsterite is illustrated in Figure 9.22; it can be considered the hcp equivalent of spinel, that is, the oxygens are in almost perfect hcp packing, with the larger divalent ions occupying half of the octahedral interstices, and the smaller Si ions occupying one-eighth of the tetrahedral interstices. A small distortion of the hcp oxygen sublattice (similar to the small departure from perfect fcc packing in the oxygen sublattice in spinel) leads to orthorhombic symmetry, space

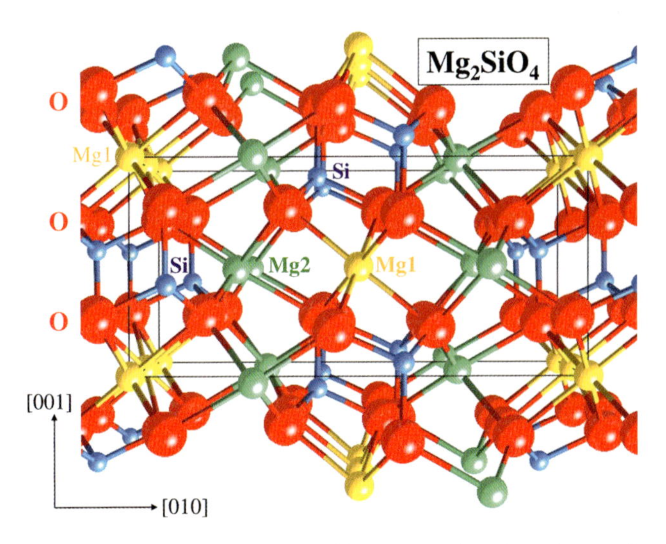

Figure 9.22 The crystal structure of forsterite, Mg_2SiO_4, viewed in perspective along the [100] direction. The isolated SiO_4 tetrahedra alternately point up and down in rows along the **a** axis. The Mg1 octahedra are at the corners of the indicated unit cell and half-way along the **a** axis in rows. The Mg2 octahedra are more distorted. The irregular close-packed anion planes are seen edge-on, parallel to (001).

group *Pbnm*,[2)] with lattice parameters (for pure forsterite) of $a = 0.476$ nm, $b = 1.023$ nm and $c = 0.599$ nm. There are two types of octahedra, one more distorted than the other, and the $[SiO_4]$ tetrahedra point alternately up and down along the *a* axis. Dislocations with the shorter [100] and [001] Burgers vectors dominate the deformation, as expected [158–160]. (In fact, [100] and [001] actually correspond to the [0001] and $\langle 11\bar{2}0\rangle$ directions in the pseudohexagonal oxygen sublattice of olivine [[161]).) Furthermore, slip in these directions can occur without disruption of the SiO_4 tetrahedra, as the core structure of the dislocations need not involve breaking of covalent Si−O bonds but only rearrangement of the more ionic Mg(Fe)−O bonding, as discussed by Paterson [12, 85].

High-temperature creep deformation of synthetic forsterite crystals, as studied by Darot and Gueguen [162], occurred by (010)[100] slip, which is apparently the preferred slip system. This same slip system was detected around Vickers indents produced at temperatures of 600 °C and above by Gaboriaud *et al.* [163, 164], who used natural olivine single crystals from San Carlos, Arizona; (110)[001] slip was also activated. There have been many other extensive deformation studies of the olivine minerals (e.g., see Kohlstedt and Ricoult [165] and Poumellec and Jaoul [166]).

9.4.9.1 Water-Weakening and Dislocation–Dissociation in Olivine

"Water weakening" can occur in olivine, as in quartz [167–170]. It has been suggested by Mackwell *et al.* [171] that water enhances climb mobility, although more detailed explanations are still lacking. Another possibility is that the point defects associated with the addition of water cause enhanced kink nucleation and/or kink diffusion (as described in Section 9.2.3).

Dislocation–dissociation has also been an issue in olivine since the early studies of Poirier *et al.* [161, 172]. Evidence for such dissociation, using TEM, was first reported by Van der Sande and Kohlstedt [173], and more recently by Smith *et al.* [174] and Drury [34]. The dissociation [Eqs (18) and (19)] suggested by Poirier assumes that the oxygen ions can be considered as hard spheres:

$$[100] \rightarrow \left[\tfrac{1}{6}\,\tfrac{\bar{1}}{36}\,\tfrac{\bar{1}}{4}\right] + \left[\tfrac{2}{3}00\right] + \left[\tfrac{1}{6}\,\tfrac{1}{36}\,\tfrac{1}{4}\right] \tag{18}$$

for dislocations on (010)and

$$[100] \rightarrow \left[\tfrac{1}{6}\,\tfrac{\bar{1}}{9}\,\tfrac{1}{6}\right] + \left[\tfrac{2}{3}00\right] + \left[\tfrac{1}{6}\,\tfrac{1}{9}\,\tfrac{\bar{1}}{6}\right] \tag{19}$$

for dislocations on (001). Subsequent TEM evidence for Eqs (18) and (19) was reported by Van der Sande and Kohlstedt [173], as well as for the dissociation of a [001] dislocation lying on (100). In all of these cases, the extent of the dislocation–dissociation was small – less than three Burgers vectors in glide, and less than ten Burgers vectors in climb [171].

Hydration-induced climb dissociation of dislocations in naturally deformed mantle olivine has been reported by Drury [171]. Here, the dissociation is pro-

2) The "official" space group in the International Tables of X-ray crystallography is *Pnma* (no. 62), but the historical axes given above mean that the space group permutes to *Pbnm*.

nounced and involves the climb of [001] dislocations (rather than [100] dislocations) on (001) and {021} planes via the reaction

$$[001] \rightarrow \left[0 \tfrac{3}{11} \tfrac{1}{4}\right] + \left[0 \tfrac{\bar{3}}{11} \tfrac{1}{4}\right] + \left[0 \tfrac{\bar{3}}{11} \tfrac{1}{4}\right] + \left[0 \tfrac{3}{11} \tfrac{1}{4}\right] \tag{20}$$

The partial Burgers vectors of this dissociation reaction are close to, but not exactly parallel to, $\langle 011 \rangle$. Two of the planar defects produced by Eq. (20) are cation-deficient and probably stabilized by H^+ to produce layers with composition (Mg, Fe)$(OH)_2$, isostructural with the OH-rich layer of the humite group of minerals. (The humite group of minerals consists of (Mg, Fe)(OH) layers separated by n olivine (002) layers, where n ranges from 1 to 4, as described by White and Hyde [175].) The central planar defect is formed by removing (conceptually) a stoichiometric layer of olivine, and hence does not involve any local chemical change. Drury has suggested that the climb-dissociated dislocations may be involved in the olivine $\rightarrow$ humite transformation, which occurs in mafic rocks.

The lack of significant glide dissociation in dry olivine suggests that glide is controlled by kink pair nucleation on perfect dislocations. For the case of hydrated olivine, it is not known whether the wider dissociation applies to glide as well as to climb dissociation. If it does, then it could well be that kink nucleation is occurring on partials for hydrated olivine and on perfect dislocations for dry olivine, and this would explain the water weakening. As described in Section 9.4.12, the climb-dissociated configuration has a lower energy than the glide-dissociated configuration (provided that the stacking fault energy is fairly isotropic), so that the observation of the former does not necessarily mean that climb is rate-controlling. In materials with a high Peierls stress, kink pair nucleation and propagation should always be rate-controlling, even when extensive climb occurs.

9.4.10
Other Oxides

Slip systems in other oxides are summarized in Table 9.2, and observed dislocation–dissociation in Table 9.3. None of the other oxides appear to exhibit the diversity of behavior of sapphire and spinel with regards to dislocation substructures and dissociations, although this may simply be because they have not received the same attention. Given that the stacking fault energy is likely to be high (particularly if anion faults are involved), it is not surprising that dissociation has not been observed in the rock salt oxides, as described in Section 9.4.1, and in oxides with fairly small unit cells (such as BeO and TiO_2). Similarly, no dissociation has been reported in the fluorite-structured oxides ZrO_2–Y_2O_3 and UO_2 [74], or in the anti-fluorite oxide, Cu_2O [176]. In all of these cases, a high-energy cation plus anion fault would be created.

Two other oxide structures are listed in Tables 9.1–9.3 that deserve further discussion, namely the rare-earth sesquioxides (M_2O_3) and oxides with the garnet structure. Both are compact oxides, but not close-packed in the traditional sense, with a large unit cell based on a body-centered cubic I lattice. The shortest lattice vector, $1/2\langle 111 \rangle$, is very large in both cases (0.92 and 1.04 nm, respectively) and plastic deformation is possible only at very high temperatures.

9.4.10.1 **Oxides with the Cubic Rare-Earth Sesquioxide Structure**

Plastic deformation has been studied in single crystals of Y_2O_3 [34, 177, 178] and Er_2O_3 [35, 179]. For Y_2O_3, Gaboriaud *et al.* [34] reported the presence of $1/2\langle 111\rangle$, $\langle 100\rangle$ and $\langle 110\rangle$ dislocations after creep deformation at temperatures from 1550 to 1800 °C and stresses from 20 to 140 MPa. It is not clear whether the $\langle 110\rangle$ dislocations (with a 1.5 nm Burgers vector!) result from glide on $\{001\}$ planes or from reactions between $1/2\langle 111\rangle$ dislocations. The $\langle 100\rangle$ dislocations are important in the high temperature range and are dissociated by climb into $1/2\langle 100\rangle$ collinear partials. The $1/2\langle 111\rangle$ dislocations are dissociated by glide in the $\{110\}$ plane according to:

$$1/2\,\langle 111\rangle \rightarrow 1/4\,\langle 110\rangle + 1/4\,\langle 112\rangle \tag{21}$$

The $\langle 110\rangle$ dislocations are dissociated into four $1/4\langle 110\rangle$ collinear partials. The SFE of the middle fault is surprisingly low ($\sim$80 mJ m^{-2}), and it is suggested that this is because only the fourth plane–plane interactions are changed by the fault [178]. For Er_2O_3 single crystals [35, 179], temperatures in excess of 1600 °C and stresses of $\sim$100 MPa were required for plastic deformation at a strain rate of $\sim 10^{-4}\,s^{-1}$. Only $\langle 111\rangle\{110\}$ slip was observed. The $1/2\langle 111\rangle$ dislocations were found to be dissociated according to Eq. (21), and the authors pointed out that the partial Burgers vectors connect cation positions in the structure.

9.4.10.2 **YAG and other Oxides with the Garnet Structure**

Synthetic garnets such as YAG (yttrium–aluminum–garnet; $Y_3Al_5O_{12}$), YIG (yttrium–iron–garnet) and GGG (gadolinium–gallium–garnet) are of interest for use as substrates for bubble memory and other electronic devices. In contrast, natural silicate garnets are of interest because they are the dominant minerals in the transition zone of the Earth's mantle, and hence may control its rheology [180]. For the synthetic garnets, YIG single crystals can be deformed plastically at temperatures as low as 1200 °C [181, 182], GGG can be deformed at temperatures above $\sim$1450 C [39, 183, 184], while YAG requires temperatures as high as 1600 °C [185, 186]. As noted by Karato *et al.* [180], the stress–temperature behavior of the various garnets (including natural minerals) scales as the elastic modulus and melting temperature. In all cases, the slip direction is $\langle 111\rangle$, as expected for the bcc lattice; the most commonly observed slip plane is $\{110\}$, but $\{112\}$ and $\{123\}$ slip planes are also frequently seen. Such behavior is reminiscent of bcc metals. Indeed, Rabier and Garem [39] found that GGG slips on $\{011\}$ planes when compressed along [001], and $\{112\}$ slip planes (in the twinning sense) when compressed along [110]; such behavior is similar to the slip asymmetry observed in bcc metals. These authors suggested that the core structure of screw dislocations in garnets is also asymmetrical, as all of the synthetic garnets exhibited a dissociation of $1/2\langle 111\rangle$ dislocations into two collinear partials [39, 181, 182, 185]:

$$1/2\,\langle 111\rangle \rightarrow 1/4\,\langle 111\rangle + 1/4\,\langle 111\rangle \tag{22}$$

The ribbon width is variable and involves both glide and climb, so that it may be nonequilibrium. The $1/4\langle 111\rangle$ fault vector causes initially empty sites in the perfect structure to be filled by tetrahedral and dodecahedral cations, so that the SFE should be high. Equation (22) could not be detected in dislocations introduced into silicate garnets by compression under a high confining pressure at temperatures of 1000 °C and above [187], and this indeed suggested that the SFE in all garnets would be high.

9.4.11
SiC, Si_3N_4, and other Non-Oxide Ceramics

Although this chapter has concentrated on oxide-ceramics, a considerable amount of data has been produced on non-oxide ceramics – principally borides, carbides, and nitrides – that would in fact be worthy of a second chapter. This section will be confined to a few remarks about two important structural ceramics, silicon carbide and silicon nitride. However, it is worth mentioning the so-called refractory hard metals, as typified by the carbides and nitrides of the Group IVA and VA transition metals. These have the rock-salt structure but, unlike the alkali halides, single crystals are not ductile at ambient temperatures and temperatures greater than ~900 °C are required for plastic deformation, in spite of the fact that the $1/2\langle 110\rangle$ dislocations, as in fcc metals, are dissociated into Shockley partials on the $\{111\}$ plane [188]. Most of these refractory compounds have a wide range of stoichiometry. In TiC_{1-x}, for example, Tsurekawa *et al.* [189] observed that the CRSS decreased by ~50% as x was increased from 0.05 to 0.14; that is, compositional softening occurred. The CRSS then increased again when x was increased to 0.25, a phenomenon that the authors attributed to ordering. Solution softening also occurred when Mo was substituted for Ti. It is believed that these results should be re-evaluated in terms of the kink model presented in Section 9.2, including the possible role of SFE and point defects.

9.4.11.1 **Silicon Carbide**

As noted in Section 9.4.2, SiC has a relatively low SFE, which is reflected in the fact that SiC occurs in numerous polytypes with different stacking of the Si–C tetrahedra. The β-phase consists of a single 3C cubic polytype with the sphalerite structure, while the α-phase consists of a mixture of hexagonal or rhombohedral polytypes. Only the hexagonal 6H polytype is available in sufficiently large single crystal form to conduct plastic deformation studies. Pirouz [190], in summarizing the results of such studies, reported that 6H-SiC is highly anisotropic and deforms at high temperature on the basal plane. If crystals are compressed parallel to the basal plane, then kinking occurs, but only at stresses of ~6 GPa and temperatures >1000 °C [191]. For crystals oriented for basal slip, plastic deformation occurs at much lower stresses (down to ~20 MPa) in the temperature range of 800–1400 °C [190]. In the higher temperature range, the $1/3\ \langle 11\bar{2}0\rangle$ dislocations are dissociated into Shockley partials with a separation corresponding to a SFE of ~2 mJ m^{-2}. Within the lower temperature range, only Shockley partials are observed; in heavily deformed regions the motion of the partials is such as to induce the $\alpha \rightarrow \beta$ transformation. Pirouz pointed out that the core structures of the two types of partials are quite different, one with C–C bonds and the other with Si–Si bonds, so that the activation energies for kink pair formation and/or kink migration are also quite different. This possibility was not allowed for in the analysis of kink behavior on partial dislocations in Section 9.2.4, but certainly could be.

9.4.11.2 **Silicon Nitride**

There are two phases of silicon nitride, α-Si_3N_4 and β-Si_3N_4. The β form is hexagonal ($P6_3/m$) with $a = 0.761$ nm and $c = 0.291$ nm, while the α form is trigonal (P31c) with roughly the same dimensions for the a axis and approximately double the dimensions for the c axis. Synthesized powder consists generally of α-Si_3N_4, and subsequent sintering converts the structure to the β form. Most interest in silicon nitride has been in using sintering additives to control the grain shape and the grain boundary phase so as to produce a polycrystalline material with high toughness and high-temperature creep resistance. Plastic deformation at high temperatures is due to grain boundary sliding, with little contribution from dislocation motion [192]. Single crystals of α-Si_3N_4 suitable for plastic deformation studies have been grown using chemical vapor deposition (CVD). These crystals could be deformed plastically at a yield stress of ~250 MPa at 1760 °C and ~50 MPa at 1820 °C. Such a steep drop in yield stress indicates a high activation energy, as would be expected for kink pair nucleation with a large Burgers vector [193]. In fact, TEM observations have revealed $1/3\,\langle 11\bar{2}0\rangle$ dislocations (0.78 nm Burgers vector) with no discernible dissociation [193], and the slip plane was $\{1\bar{1}01\}$. Knoop hardness anisotropy indicated the same slip system at room temperature [194]. Although there have been no comparable studies of β-Si_3N_4, dislocations with Burgers vectors $1/3\,\langle 11\bar{2}0\rangle$, [0001] and $1/3\,\langle 11\bar{2}3\rangle$ occur in sintered material [192]. The same Burgers vectors are found in as-grown α-Si_3N_4 single crystals [195]. It should be noted that the [0001] Burgers vector in β-Si_3N_4 is relatively short (0.291 nm), and this would be expected to be the most active dislocation in plastically deformed single crystals.

9.4.12
Climb versus Glide Dissociation

One phenomenon of interest that became apparent during the early application of electron microscopy to the study of dislocations in ceramics was the predominance of climb dissociation in many oxides during high-temperature deformation. This was first discovered in sapphire and spinel, but has since become a commonplace observation in other ceramics such as perovskite and forsterite, and also in many intermetallics deformed at high temperatures. Climb dissociation involves a redistribution of matter between the half-planes of the partials by short-circuit diffusion along the fault plane (see Figure 9.16). It is easily shown that the climb configuration has the lowest energy, barring some pathological anisotropy in the fault energy. This is shown by writing down the sum of the interaction energy and the fault energy per unit length:

$$\frac{W_{12}}{L} + \frac{W_{SF}}{L} = \frac{\mu b_1 b_2}{2\pi(1-\upsilon)}\left[-\ell n\frac{r}{\rho} - \sin^2\theta\right] + \gamma(\theta)r \tag{23}$$

where b_1 and b_2 are the Burgers vectors of the two partials (assumed here for simplicity to be collinear), ρ is the core radius, $\gamma(\theta)$ is the fault energy, r is the spacing of the partials, and θ is the angle between the fault plane and the slip plane. The differentiation of Eq. (23) gives the equilibrium separation:

$$r_{eq} = \frac{\mu b_1 b_2}{2\pi(1-\upsilon)\gamma(\theta)} \tag{24}$$

which is independent of θ except through γ. The difference in energy between the glide ($\theta = 0°$) and climb ($\theta = 90°$) configurations is

$$\frac{\Delta W}{L} = \frac{\mu b_1 b_2}{2\pi(1-\upsilon)} \left[\ell n \frac{\gamma(90)}{\gamma(0)} - 1 \right] \tag{25}$$

which is positive provided $\gamma(90) > 2.72\gamma(0)$; that is, the climb configuration should have the lowest energy unless the fault energy in the glide plane is very small. The data in Table 9.6 show that, as the fault energy in sapphire is not very anisotropic, the climb configuration should always dominate.

Glide dissociation should always make glide easier by reducing the kink energy, whereas climb dissociation would make glide more difficult. However, at elevated temperatures, where short-circuit diffusion between the partials is rapid, it is possible for a dislocation to change quickly from a climb-dissociated to a glide-dissociated configuration as the dislocation begins to move under stress, and to adopt the climb-dissociated configuration when stationary. This can be accomplished by the simultaneous glide of one partial and the climb of the other partial onto the glide plane. At one time, several mechanisms were proposed to explain how climb dissociation could control the flow stress [16]; however, it is now thought that the stress is controlled by a kink mechanism (as described in Section 9.2). Nevertheless, climb dissociation has been a fascinating side-show.

9.5 Work Hardening

Although oxide crystals exhibit a variety of work-hardening behavior, only three systems will be discussed in this section, namely MgO, spinel, and sapphire.

9.5.1 Work Hardening in MgO

MgO crystals are generally found to show two-stage hardening: a region of low hardening rate after yielding, followed by a linear region with a work hardening rate of approximately $\mu/300$ [40]. The latter value is the standard rate observed in stage II of fcc crystals. As MgO crystals can only be deformed in compression to a strain $\sim$10%, it is possible that a third stage is never reached. The dominant features of the deformation microstructure of MgO in stage I are dipoles and elongated loops [196], much as in stage I of fcc metals. The hardening mechanisms are likely the same, namely that the dipoles are formed by the trapping of dislocations of opposite signs on parallel slip planes, and this controls the hardening [197]. At higher temperatures.

the dipoles tend to break up into rows of loops, as in sapphire (see Figure 9.15). It can be recalled that, in the rock-salt structure, the $\{110\}\langle 1\bar{1}0\rangle$ and $\{1\bar{1}0\}\langle 110\rangle$ slip systems have the same Schmid factor and are both activated. Apparently, these two systems do not intersect in stage I but rather are forced to do so in stage II. At high temperatures, where climb can occur, a cell structure is formed but the climb also contributes to recovery, such that the work-hardening rate decreases. The hardening rate in stage II of $\sim\mu/300$ occurs in the temperature range of 200–500 K [40], where no cell structure forms and the most significant microstructural features are the elongated loops and dipoles. In the absence of a cell structure, the mechanism for the stage II work-hardening rate in MgO must be quite different from that in fcc metals (see, for example Hirsch and Mitchell [198]). The likely mechanism for MgO is that slip on one system must cut through the debris (dipoles, elongated loops) formed on the other system. In the alkali halides, a small amount of slip is observed on the oblique $\{110\}\langle 110\rangle$ systems, and this may also contribute to work hardening [4]. The alkali halides also exhibit a stage III with a decreasing hardening rate, which is ascribed to recovery by cross-slip on the $\{001\}$ plane [4].

9.5.2
Work Hardening and Work Softening in Spinel

$MgO{\cdot}nAl_2O_3$ spinel reveals a variety of stress–strain behaviors, depending on stoichiometry, orientation, and temperature. Examples are shown in Figure 9.23 [26, 142]., and some general observations are as follows:

- There is sometimes a yield-point, but this is never as sharp as has been reported in, for example, sapphire [199, 200].
- Sometimes, the yield-point is followed by an extensive region of work-softening, especially in stoichiometric spinel.
- Sometimes, the yield-point is followed by a region of easy glide (low work-hardening rate), but it is never as extensive or obvious as in sapphire [201] or MgO [40].
- By 10% shear strain or before, the flow stress has reached a steady-state value; in other words, the specimen is creeping at the imposed strain rate under the resulting flow stress – that is, the work-hardening and recovery rates are equal.
- Stress–strain curves for $\{111\}\langle 110\rangle$ and $\{110\}\langle 110\rangle$ slip show similar features, although work softening appears to be more dramatic for $\{110\}\langle 110\rangle$ slip in stoichiometric spinel (see Figure 9.23d).

The work-hardening rate in the first percent or so of strain is very high, with values given by:

$$\theta = \frac{d\tau}{d\gamma} \approx 2000 \text{ to } 3000\ \text{MPa} \approx \frac{\mu}{40} \text{ to } \frac{\mu}{60} \qquad (26)$$

where τ and γ are the shear stress and shear strain, respectively. This should be compared with the value of $\theta \sim \mu/300$ for stage II hardening in fcc metals. The high

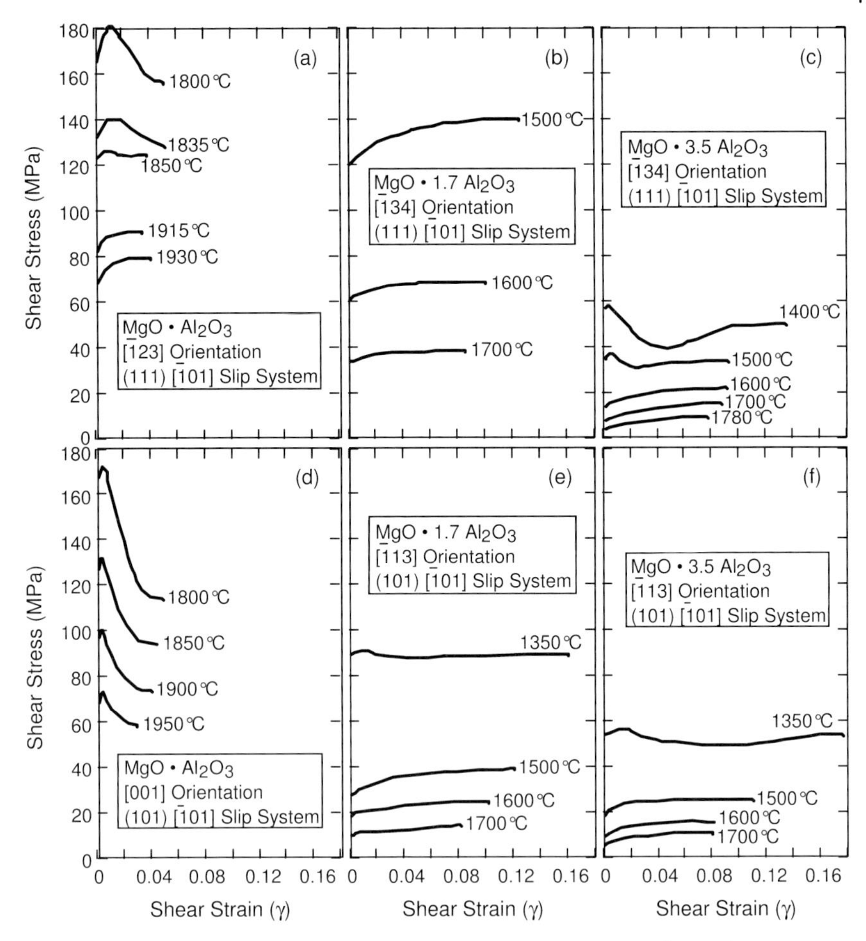

Figure 9.23 Stress–strain curves of spinel crystals at various temperatures for stoichiometry $n = 1$ (panels a and d), $n = 1.7$ (panels b and e), and $n = 3.5$ (panels c and f) for slip on {111} planes (panels a, b and c) and {110} planes (panels d, e, and f) [18].

initial hardening rate in spinel has been ascribed to a rapid build-up in the dislocation density, to a value of about $10^{13}\ \mathrm{m}^{-2}$ at a strain of less than 1% [142]. Such rapid build-up is due to the small slip distance, which in turn is due to the formation of a three-dimensional dislocation network by climb. Apparently, the very first dislocation sources emit dislocations which travel far enough to give the observed slip lines on the surface. Dislocations which remain in the crystal are able to climb into low-energy dislocation networks, which then form effective barriers to further dislocation glide, reducing the slip distance and increasing the dislocation density. Donlon *et al.* [142] have estimated the slip distance at the maximum in the shear stress to be approximately 2 μm in stoichiometric spinel.

The work-softening phenomenon is due then to a *reduction* in dislocation density with increasing strain. This was shown clearly by Donlon *et al.* [142], who measured the dislocation density, ρ, by using TEM. In this case, the dislocation density was measured as a function of stress for stoichiometric spinel deformed at various stresses and temperatures, to give:

$$\tau = \tau'_o + \alpha\mu b\rho^{1/2} \tag{27}$$

where $\tau'_0 \sim 30$ MPa is a constant frictional stress, and α is a constant estimated to be $\sim$0.7 from the data. This value of α is at the upper end of the range from 0.1 to 1.0 typically found in other materials. However, it should be appreciated that most of these data were obtained in the work-softening regime, whereas Eq. (27) is usually used to describe work hardening! For example, for the 1800 °C data, it was found that the dislocation density decreased as the strain increased from 1.5% to 30%. Normally, yield-points are ascribed to dislocation multiplication, and to the concomitant increase in mobile dislocation density and decrease in average dislocation velocity. In spinel, the initial dislocation multiplication leads to work hardening, while subsequent dislocation annihilation by climb leads to work softening. The role of the mobile dislocation density is not at all clear, but it is possible that climb effectively immobilizes the majority of the dislocations so that the available mobile dislocation density (for glide) remains approximately constant. Of course, both glide and climb of dislocations contribute to the strain rate; however, Donlon *et al.* [142] have estimated that glide is approximately twice as important as climb in controlling the strain rate.

9.5.3 Work Hardening in Sapphire

Stress–strain curves in sapphire exhibit a three-stage hardening both for basal and prism plane slip [97, 201]. The mechanisms for each are quite different from each other, and also from those described above.

9.5.3.1 Basal Plane Slip

Typical basal slip stress–strain curves are shown in Figure 9.24 [201]. A yield point is followed by a short region of easy glide, then by a region of high work hardening (termed stage A), after which the work-hardening rate decreases (stage B) until a plateau is reached (stage C). Dislocation density measurements have provided results consistent with Eq. (27). The region after the yield point is characterized by the accumulation of dipoles and elongated loops, as in fcc metals, although, even at this early stage, the dipoles have begun to break up into small loops – both faulted and unfaulted (see Figure 9.15). Stage A work hardening is thought to be due to the interaction of glide dislocations with loops, which have accumulated to the point that they can block a glide dislocation all around its perimeter [201]. Recovery in stage B is due to the annihilation of loops by climb; in stage C, the rate of accumulation of loops by dipole break-up is balanced by their rate of annihilation by climb. That work hardening occurs at all is because dipole break-up is controlled by pipe diffusion, while annihilation is controlled by self-diffusion. The interaction cross-section is related to the product of the length of a loop, l, and its height, h; the latter is inversely proportional to the stress. The rate of accumulation of loops is calculated from the number of dipoles formed and the wavelength, λ, for the break-up of the dipoles. Pletka *et al.* [201] showed that the stage A work hardening rate is

$$\theta_A = \frac{\mu l}{48\pi(1-\upsilon)\lambda} = \frac{\mu}{96\pi(1-\upsilon)} \quad (28)$$

assuming that $l/\lambda \approx 1/2$. Equation (28) yields a work-hardening rate of about $\mu/225$, which is in good agreement with the experimental value of $\mu/300$ in stage A. Stages B and C are easily modeled by taking into account standard expressions for the shrinkage rate of the loops [202]. Diffusion rates calculated from the plateau stress are in good agreement with experiment. Overall, the model for work hardening and recovery, based on loop accumulation by dipole break-up and loop annihilation by climb, is very satisfactory.

9.5.3.2 **Prism Plane Slip**

Stress–strain curves similar to those in Figure 9.24 are observed for prism plane slip, except that the hardening rate in region A is much higher, at about $\mu/40$ [97]. Although many loops are observed in deformed crystals, the most important feature of the dislocation substructure is thought to be the three-dimensional network that forms. The network is mostly made up from $1/3\,\langle 11\bar{2}0\rangle$ dislocations, and it is suggested that these arise from the decomposition of the $\langle 10\bar{1}0\rangle$ prism plane dislocations that have cross-slipped onto the basal plane via the decomposition reaction into $1/3\,\langle 11\bar{2}0\rangle$ dislocations [Eq. (12)]. The network acts as a "forest," threading the prism slip plane and impeding slip on this plane. A forest theory of hardening is inapplicable to metals because the model is too homogeneous; it does not take into account the observed cell structure, and gives a hardening rate that is too high [197]. By contrast, the forest theory is well suited for explaining the high work-hardening rate in sapphire deforming by prism plane slip, because the observed

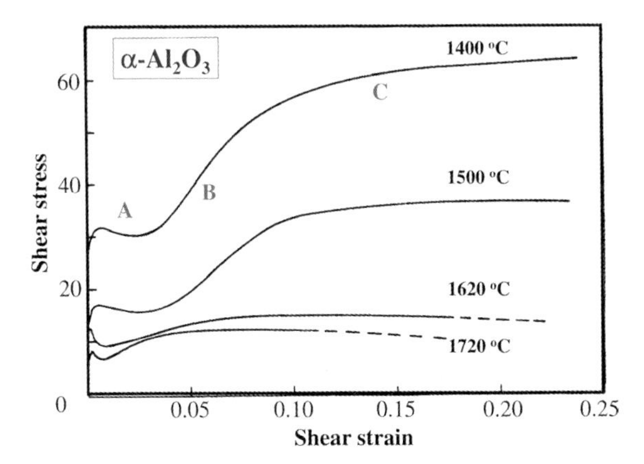

Figure 9.24 Stress–strain curves of sapphire deformed in basal glide at various temperatures [201].

network is relatively homogeneous and the slip distance is just a few times the network spacing.

9.6 Solution Hardening

Solution hardening in oxides can occur by the substitution of either isovalent or aliovalent cations (substitutional anions have not been studied); the latter cations produce much greater hardening rates. Solution hardening theory basically follows along the lines developed for metals. Solution hardening due to isovalent cations results from a combination of size and elastic modulus misfit. Solution hardening due to aliovalent cations is the result of an asymmetric distortion around defect complexes formed with charge-compensating defects. Solution hardening occurs more rapidly at low temperatures, and there is usually a plateau in the temperature dependence at high temperatures. However, at low temperatures and low concentrations, where the solutes act as point obstacles, Friedel–Fleischer statistics are followed and the yield stress increases as $c^{1/2}$. In the plateau region and at higher concentrations, where the solutes act as diffuse obstacles, Mott–Labusch statistics are followed and the yield stress increases as $c^{2/3}$. However, even in metals, for which much more data are available, these simple rules are not always obeyed very well. Solution hardening due to isovalent substitutionals will be discussed first.

9.6.1 Isovalent Cations

Very few systematic studies of solution hardening by isovalent solutes in oxides have been performed compared to substitutional solutes in fcc metals. For sapphire, Pletka *et al.* [203] have compared the effects of Cr^{3+} and Ti^{3+} on the yield strength at high temperatures. Their results are shown in Figure 9.25; the curves through the data correspond to a $c^{2/3}$ dependence, which is a slightly better fit than the $c^{1/2}$ dependence for the Cr^{3+} data, while the Ti^{3+} data are too sparse to make any distinction between them. The hardening rate due to Ti^{3+} is about a factor of two higher than that due to Cr^{3+}. Yield stress curves at other temperatures were found to be parallel to those in Figure 9.25, which suggests that the stress due to solution hardening is additive with the yield stress of pure sapphire, and that the data are in the plateau regime of solution-hardening behavior.

The only data for MgO appear to be for isovalent Ni^{2+} additions [204]. The hardening rate at ambient temperature is somewhat higher than that for Ti^{3+} in sapphire, and the data best fit a $c^{1/2}$ law. The plateau regime begins at about 400 °C for MgO (at least for Fe^{3+} additions [205]), and so the yield stress behavior is in the low-temperature region where the solution-hardening rate is temperature-dependent.

Pletka *et al.* [203] analyzed the solution-hardening rate in sapphire using the Labusch model [202], where the increment in yield stress due to solution hardening

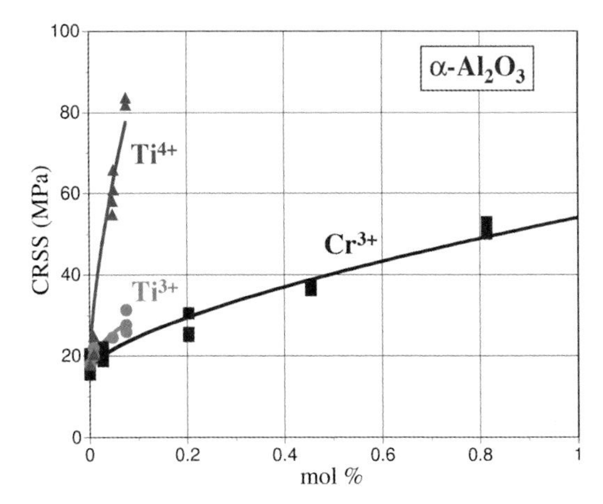

Figure 9.25 Critical resolved shear stress versus mol% of oxide solute (for cations in the form of Cr^{3+}, Ti^{3+} and Ti^{4+}) in sapphire deformed at 1500 °C. The curves through the data correspond to a $c^{2/3}$ dependence. Data from Ref. [203].

in the plateau regime results from a combination of the size misfit and modulus misfit. It was shown that, since mostly edge dislocations are observed at high temperatures in sapphire, the size misfit term dominated. Pletka *et al.* [203] showed that the misfit for Ti^{3+} additions was about 1.6-fold larger than for Cr^{3+} additions, which explains quantitatively the difference in the hardening rates in Figure 9.25. Rodriguez *et al.* [206] have also studied solution hardening due to Cr^{3+} in sapphire, but over a wider temperature range (900–1500 °C). These authors also found that the increment in stress due to solution hardening was independent of temperature, so that the temperature range was in the plateau region. They then applied the kink pair mechanism (see Section 9.2) by subtracting the solution-hardening increment from the applied stress, and found a good agreement between theory and experiment. Rodriguez *et al.* [206] use the so-called "long-segment limit" and the stress-dependent mobile dislocation density (see Section 9.2.4); even though these assumptions were debatable, they showed the way in which the intrinsic strength due to the kink pair mechanism could be additive with the extrinsic strength due to solution hardening.

9.6.2
Aliovalent Cations

The most extensively studied systems are Ti^{4+} in sapphire (Figure 9.25) and Fe^{3+} in MgO (Figure 9.26). Figure 9.25 shows that the hardening rate due to Ti^{4+} is about sixfold that due to Ti^{3+} [203]. There are no equivalent data for Fe^{2+} compared to Fe^{3+} in MgO, but it is certainly true that the hardening rate due to Fe^{3+} is also about sixfold that due to Ni^{2+} in MgO [204]. As stated above, this is due to the formation of charge-compensating defect complexes – three Ti^{4+} substitutionals with one Al cation vacancy (or a titanium–oxygen interstitial defect [207]) in the sapphire case and

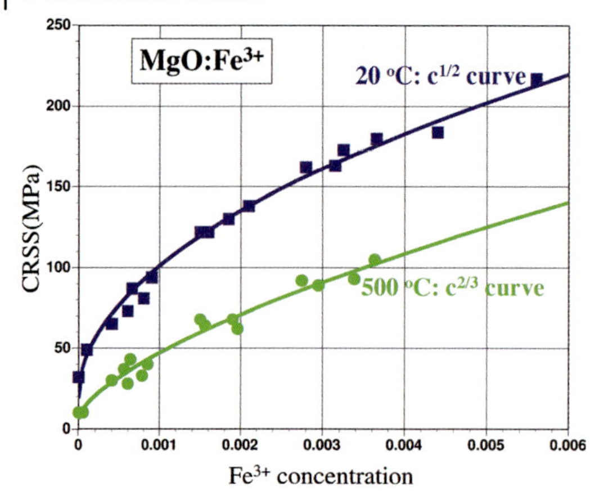

Figure 9.26 Yield stress versus concentration of Fe^{3+} dopant in MgO. Data from Ref. [205].

two Fe^{3+} substitutionals with one Mg cation vacancy in the MgO case. The elastic strains around such complexes are large and asymmetric (the so-called "tetragonal" distortion), and they interact more strongly with dislocations than do isovalent cations. Figure 9.26 shows that the hardening due to Fe^{3+} in MgO followed a $c^{1/2}$ dependence at low temperatures, and a $c^{2/3}$ dependence in the plateau regime at high temperatures [205]. Unfortunately, the data for Ti^{4+} in Al_2O_3 were too sparse to distinguish between the two laws.

The $c^{1/2}$ law at low temperatures implies Friedel–Fleischer statistics for the increment of yield stress due to solution hardening:

$$\Delta\sigma_y = f_m^{3/2} c_p^{1/2} / (2\Gamma)^{1/2} b \tag{29}$$

where f_m is maximum force between defect and dislocation, c_p is the planar concentration of defects in the glide plane of the dislocation, and Γ is the line tension. Each of the terms in Eq. (29) can be evaluated [208] and, in particular, $f_m \cong \frac{1}{3}\mu\Delta\varepsilon b^2$ where $\Delta\varepsilon$ is the tetragonality (the difference in strain along two orthogonal directions). From Figure 9.25, $\Delta\varepsilon$ due to Fe^{3+} in MgO could be calculated as 0.12 and, from Fig.ure 9.26, $\Delta\varepsilon$ due to Ti^{4+} in Al_2O_3 was found to be ~0.05 [208]. Of course, as the defect clusters are in no sense tetragonal, the value of $\Delta\varepsilon$ represents the strength of the cluster.

9.7
Closing Remarks

In this chapter, the commonly held belief that the terms "ceramic" and "brittle" are synonymous has been shown not to be exactly tenable. Oxide single crystals with the rock-salt structure can be deformed plastically at liquid helium temperatures, while strontium titanate shows an inverse BDT and can be deformed plastically at ambient temperatures and below. Oxides with the fluorite structure can be deformed plastically at temperatures of a few hundred degrees above ambient, especially if great care is taken to use crystals that are free from defects and well-aligned. There is also the issue of crystal purity; it is believed that many lessons could be learned from experience with bcc metals which, in the early days before techniques were developed to reduce interstitial contents to the level of a few ppm, were thought to be brittle. It is unclear how crystals such as BeO and ZnO would behave if such levels of purity could be achieved. Of course, it is unlikely that oxides with large unit cells such as garnets, spinels and rare-earth sesquioxides would lower their BDT temperatures significantly. However, with the surprising behavior of the perovskite strontium titanate, a renewed effort would be worthwhile in both the experimental and modeling arenas.

For those ceramics with a high Peierls stress, the CRSS can be understood consistently in terms of a model of kink pair nucleation and motion on dislocations. The steep temperature dependence is governed by an activation energy that is the sum of the elastic energy for kink pair formation and the energy for the kinks to overcome their secondary Peierls barriers.

The kink model has been modified to take into account kink nucleation on point defects, and kink nucleation on partial dislocations. Kink nucleation on partial dislocations is easier because the elastic energy is so much less, and this may indeed be the reason for plasticity at low temperatures in $SrTiO_3$, and possibly also for the remarkable softening observed in nonstoichiometric $MgO–Al_2O_3$ spinel as the SFE is reduced. As most ceramic oxides have a high SFE, kink nucleation is forced to occur on perfect dislocations at much higher stresses. If the SFE of a given oxide could be reduced by alloying, then the BDT temperature would also be reduced. With regards to kink nucleation on point defects, it is a viable mechanism for explaining the water-weakening phenomenon that is well known in quartz, and also occurs in olivine and alumina. This may also represent a possible mechanism for softening in spinel, in competition with the SFE effect.

Acknowledgments

Over the years, this research on the plastic deformation of ceramics has been supported by the Department of Energy, Office of Basic Energy Sciences. The author has also enjoyed collaborations with many students and colleagues, such as the late J. Cadoz, J. Castaing, A. Dominguez-Rodriguez, W.T. Donlon, A.H. Heuer, J.P. Hirth, R.J. Keller, K.P.D. Lagerlöf, K.J. McClellan, P. Peralta, the late D.S. Phillips, P. Pirouz, and B.J. Pletka, to name but a few.

References

1 Schmid, E. and Boas, W. (1935) *Kristallplastizität*, Springer, Berlin.
2 Landolt, H. and Bornstein, R. (1927) *Physikalische-Chemische Tabellen*, Springer, Berlin.
3 Mügge, O. (1899) *Gotting. Nachr.*, 56.
4 Sprackling, M.T. (1976) *The Plastic Deformation of Simple Ionic Crystals*, Academic Press, London.

5 Haasen, P. (1985) *Dislocations and Properties of Real Materials* (ed. M.H. Loretto), The Institute of Metals, London, p. 312.
6 Wachtman, J.B. and Maxwell, L.H. (1954) *J. Am. Ceram. Soc.*, **37**, 291.
7 Kronberg, M.L. (1957) *Acta Metall.*, **37**, 291.
8 Hornstra, J. (1960) *J. Phys. Chem. Solids*, **15**, 311.
9 Evans, A.G. and Heuer, A.H. (1980) *J. Am. Ceram. Soc.*, **63**, 241.
10 Hirsch, P.B. (1985) *Dislocations and Properties of Real Materials* (ed. M.H. Loretto), Institute of Metals, London, p. 333.
11 Alexander, H. (1986) *Dislocations in Solids* (ed. F.R.N. Nabarro), North Holland, Amsterdam, p. 113.
12 Paterson, M.S. (1985) *Dislocations and Properties of Real Materials* (ed. M.H. Loretto), Institute of Metals, London, p. 359.
13 Wachtman, J.B. (1967) *Am. Ceram. Soc. Bull.*, **46**, 756.
14 Terwillinger, G.R. and Radford, K.C. (1974) *Am. Ceram. Soc. Bull.*, **53**, 172.
15 Bretheau, T., Castaing, J., Rabier, J., and Veyssière, P. (1979) *Adv. Phys.*, **28**, 835.
16 Mitchell, T.E. and Heuer, A.H. (2004) *Dislocations in Solids*, **14**, 339.
17 Castaing, J., Veyssière, P., Kubin, L.P., and Rabier, J. (1981) *Philos. Mag. A*, **44**, 1407.
18 Mitchell, T.E. (1999) *J. Am. Ceram. Soc.*, **82**, 3305.
19 Veyssière, P. and Carter, C.B. (1988) *Philos. Mag. Lett.*, **57**, 211.
20 Veyssière, P., Kirby, S.H., and Rabier, J. (1980) *J. Phys. (Paris)*, **41**, 175.
21 Westbrook, J.H. (1966) *Rev. Hautes Tempér. et Réfract.*, **3**, 47.
22 Mitchell, T.E., Peralta, P., and Hirth, J.P. (1999) *Acta Mater.*, **47**, 3687.
23 Hirth, J.P. and Lothe, J. (1992) *Theory of Dislocations*, Krieger, Melbourne, FL.
24 Rodriguez, M.C., Castaing, J., Munoz, A., Veyssiere, P., and Rodriguez, A.D. (2008) *J. Am. Ceram. Soc.*, **91**, 1612.
25 Mitchell, T.E., Hirth, J.P., and Misra, A. (2002) *Acta Mater.*, **50**, 1087.
26 Donlon, W.T., Heuer, A.H., and Mitchell, T.E. (1998) *Philos. Mag. A*, **78**, 615.
27 Mitchell, T.E., Anderson, P.M., Baskes, M.I., Chen, S.P., Hoagland, R.G., and Misra, A. (2003) *Philos. Mag. A*, **83**, 1329.
28 Gilman, J.J. (1959) *Acta Metall.*, **7**, 608.
29 Bilde-Sørenson, J.B., Lawlor, B.F., Geipel, T., Pirouz, P., Heuer, A.H., and Lagerlöf, K.P.D. (1996) *Acta Mater.*, **44**, 2145.
30 Suzuki, K., Ichihara, M., and Takeuchi, S. (1994) *Jpn. J. Appl. Phys.*, **33**, 1114.
31 Suzuki, K., Ichihara, M., and Takeuchi, S. (1991) *Philos. Mag. A*, **63**, 657.
32 McLaren, A.C., Retchford, J.A., Griggs, D.T., and Christie, J.M. (1967) *Phys. Status Solidi*, **19**, 631.
33 Cordier, P. and Doukhan, J.C. (1995) *Philos. Mag. A*, **72**, 497.
34 Gaboriaud, R.J., Denatot, M.F., Boisson, M., and Grihlé, J. (1978) *Phys. Status Solidi A*, **46**, 387.
35 Sharif, A.A., Misra, A., Petrovic, J.J., and Mitchell, T.E. (2000) *Mater. Sci. Eng.*, **A290**, 164.
36 Mao, Z. and Knowles, K.M. (1996) *Philos. Mag. A*, **73**, 699.
37 Matsunaga, T. and Saka, H. (2000) *Philos. Mag. Lett.*, **80**, 597.
38 McLaren, A.C. (1991) *Transmission Electron Microscopy of Minerals and Rocks*, Cambridge University Press, Cambridge.
39 Rabier, J. and Garem, H. (1984) *Deformation of Ceramic Materials II* (eds R.E. Tressler and R.C. Bradt), Plenum, New York, p. 187.
40 Srinavasan, M. and Stoebe, T.G. (1974) *J. Mater. Sci.*, **9**, 121.
41 Woo, C.H. and Puls, M.P. (1977) *Philos. Mag.*, **35**, 1641.
42 Watson, G.W., Kelsey, E.T., and Parker, S.C. (1999) *Philos. Mag. A*, **79**, 527.
43 Goretta, K.C. and Routbort, J.L. (1987) *Acta Metall.*, **35**, 1047.
44 Jolles, E. and Monty, C. (1991) *Philos. Mag. A*, **64**, 765.
45 Dominguez-Rodriguez, A., Sanchez, M., Marquez, R., Castaing, J., Monty, C., and Philibert, J. (1982) *Philos. Mag. A*, **46**, 411.
46 Castaing, J., Spendel, M., Philibert, J., Rodriguez, A.D., and Marquez, R. (1980) *Revue Phys. Appl.*, **15**, 277.
47 Routbort, J.L. (1982) *Acta Metall.*, **30**, 663.
48 Dominguez-Rodriguez, A., Castaing, J., Koisumi, H., and Suzuki, T. (1988) *Revue Phys. Appl.*, **23**, 1361.

49 Cabrero-Cano, J., Dominguez-Rodriguez, A., Marquez, R., Castaing, J., and Philibert, J. (1982) *Philos. Mag. A*, **46**, 397.

50 Dominguez-Rodriguez, A., Castaing, J., and Philibert, J. (1977) *Mater. Sci. Eng.*, **27**, 217.

51 Bentle, G.G. and Miller, K.T. (1965) *J. Appl. Phys.*, **38**, 4247.

52 Corman, G.S. (1992) *J. Am. Ceram. Soc.*, **75**, 71.

53 Hong, M.H., Samant, A.V., and Pirouz, P. (2000) *Philos. Mag. A*, **80**, 919.

54 Heuer, A.H. (1987) *J. Am. Ceram. Soc.*, **70**, 689.

55 Green, D.J., Hannink, R.H.J., and Swain, M.W. (1989) *Transformation Toughening of Ceramics*, CRC Press, Boca Raton.

56 Wenckus, J.F. (1993) *J. Crystal Growth*, **128**, 13.

57 Dominguez-Rodriguez, A., Lagerlöf, K.P.D., and Heuer, A.H. (1986) *J. Am. Ceram. Soc.*, **69**, 281.

58 Tikhonovsky, A., Bartsch, M., and Messerschmidt, U. (2003) *Phys. Status Solidi A*, **201**, 26.

59 Heuer, A.H., Lanteri, V., and Dominguez-Rodriguez, A. (1989) *Acta Metall.*, **37**, 559.

60 Martinez-Fernandez, J., Jimenez-Melendo, M., Dominguez-Rodriguez, A., Cordier, P., Lagerlöf, K.P.D., and Heuer, A.H. (1995) *Acta Metall. Mater.*, **43**, 1917.

61 Cheong, D.S., Dominguez-Rodriguez, A., and Heuer, A.H. (1989) *Philos. Mag. A*, **60**, 107.

62 Cheong, D.S., Dominguez-Rodriguez, A., and Heuer, A.H. (1991) *Philos. Mag. A*, **63**, 377.

63 Dominguez-Rodriguez, A., Cheong, D.S., and Heuer, A.H. (1991) *Philos. Mag. A*, **64**, 923.

64 Baufeld, B., Betukhov, B.V., Bartsch, M., and Messerschmidt, U. (1998) *Acta Mater.*, **46**, 3077.

65 Baufeld, B., Baither, D., Barsch, M., and Messerschmidt, U. (1998) *Phys. Status Solidi A*, **166**, 127.

66 Teracher, P., Garem, H., and Rabier, J. (1991) *Strength of Metals and Alloys* (eds D.G. Brandon, R. Chaim, and A. Rosen), Freund Publishers, London, p. 217.

67 Gomez-Garcia, D., Martinez-Fernandez, J., Dominguez-Rodriguez, A., Eveno, P., and Castaing, J. (1996) *Acta Mater.*, **44**, 991.

68 Messerschmidt, U., Baufeld, B., McClellan, K.J., and Heuer, A.H. (1995) *Acta Metall. Mater.*, **43**, 1917.

69 Kilo, M., Borchadt, G., Weber, S., Scherrer, S., and Tinschert, K. (1997) *Ber. Bunsen-Ges. Phys. Chem.*, **101**, 1361.

70 Chien, F.R. and Heuer, A.H. (1996) *Philos. Mag. A*, **73**, 681.

71 Gomez-Garcia, D., Martinez-Fernandez, J., Dominguez-Rodriguez, A., and Westmacott, K.H. (1996) *J. Am. Ceram. Soc.*, **79**, 487.

72 Bartsch, M., Tikhonovsky, A., and Messerschmidt, U. (2002) *Phys. Status Solidi A*, **201**, 46.

73 McClellan, K.J., Heuer, A.H., and Kubin, L.P. (1996) *Acta Mater.*, **44**, 2651.

74 Keller, R.J., Mitchell, T.E., and Heuer, A.H. (1988) *Acta Metall.*, **36**, 1061.

75 Willis, B.T.M. (1978) *Acta Crystallogr. A*, **34**, 88.

76 Ashbee, K.H.G. and Smallman, R.E. (1963) *Proc. Roy. Soc. Lond. A*, **274**, 195.

77 Blanchin, M.G. and Faisant, P. (1979) *Rev. Phys. Appl.*, **14**, 619.

78 Blanchin, M.G., Fontaine, G., and Kubin, L.P. (1980) *Philos. Mag. A*, **41**, 261.

79 Bursill, L.A. and Blanchin, M.G. (1984) *Philos. Mag. A*, **49**, 365.

80 Motohashi, Y., Blanchin, M.G., Vicario, E., Fontaine, G., and Otake, S. (1979) *Phys. Status Solidi A*, **54**, 355.

81 Ashbee, K.H.G., Smallman, R.E., and Williamson, G.K. (1964) *Proc. Roy. Soc. Lond. A*, **216**, 54.

82 Blanchin, M.G., Bursill, L.A., and Lafage, C. (1990) *Proc. Roy. Soc.*, **A429**, 175.

83 Griggs, D.T. and Blacic, J.D. (1964) *Science*, **147**, 292.

84 Griggs, D.T. and Blacic, J.D. (1964) *Trans. Am. Geophys. Union*, **45**, 102.

85 Paterson, M.S. (1989) *Rheology of Solids and of the Earth* (eds S. Karato and M. Torium), Oxford University Press, London, p. 107.

86 Frondel, C. (1962) *The System of Mineralogy of J. D. and E. S. Dana*, John Wiley & Sons, Inc., New York.

87 Ord, A. and Hobbs, B.E. (1986) *Mineral and Rock Deformation: Laboratory Studies* (eds B.E. Hobbs and H.C. Heard),

American Geophysical Union, Washington, D.C., p. 51.
88 McLaren, A.C., Fitzgerald, J.D., and Gerretson, J. (1989) *Phys. Chem. Miner.*, **16**, 465.
89 Fitzgerald, J.D., Boland, J.N., McLaren, A.C., Ord, A., and Hobbs, B.E. (1991) *J. Geophys. Res.*, **96**, 2139.
90 Doukhan, J.C. and Trepied, L. (1985) *Bull. Mineral.*, **108**, 97.
91 Cordier, P., Weil, J.A., Howarth, D.F., and Doukhan, J.C. (1994) *Eur. J. Mineral.*, **6**, 17.
92 McLaren, A.C., Cook, R.F., Hyde, S.T., and Tobin, R.C. (1983) *Phys. Chem. Miner.*, **9**, 79.
93 Verneuil, A. (1887) *Compt. Rend. Paris*, **104**, 501.
94 Maiman, T.H. (1960) *Nature*, **187**, 493.
95 Heuer, A.H., Lagerlöf, K.P.D., and Castaing, J. (1998) *Philos. Mag. A*, **78**, 747.
96 Marinopoulos, A.G. and Elsässer, C. (2001) *Philos. Mag. Lett.*, **81**, 329.
97 Cadoz, J., Castaing, J., Phillips, D.S., Heuer, A.H., and Mitchell, T.E. (1982) *Acta Metall.*, **30**, 2205.
98 Mitchell, T.E., Pletka, B.J., Phillips, D.S., and Heuer, A.H. (1976) *Philos. Mag.*, **34**, 441.
99 Phillips, D.S., Pletka, B.J., Heuer, A.H., and Mitchell, T.E. (1982) *Acta Metall.*, **30**, 491.
100 Lagerlöf, K.D.P., Mitchell, T.E., Heuer, A.H., Riviére, J.P., Cadoz, J., Castaing, J., and Phillips, D.S. (1984) *Acta Metall.*, **32**, 97.
101 Howitt, D.G. and Mitchell, T.E. (1981) *Philos. Mag.*, **44**, 229.
102 Kenway, P.R. (1993) *Philos. Mag. B*, **68**, 171.
103 Bilde-Sørensen, J.B., Tholen, A.R., Gooch, D.J., and Groves, G.W. (1976) *Philos. Mag.*, **33**, 877.
104 Phillips, D.S. and Cadoz, J.L. (1982) *Philos. Mag. A*, **46**, 583.
105 Scott, W.D. and Orr, K.K. (1983) *J. Am. Ceram. Soc.*, **66**, 27.
106 Heuer, A.H. (1966) *Philos. Mag.*, **13**, 379.
107 Geipel, T., Lagerlöf, K.P.D., Pirouz, P., and Heuer, A.H. (1994) *Acta Metall. Mater.*, **42**, 1367.
108 Pirouz, P. (1987) *Scripta Metall. Mater.*, **21**, 1463.
109 Lagerlöf, K.P.D., Castaing, J., and Heuer, A.H. (2002) *Philos. Mag. A*, **82**, 2841.
110 Nufer, S., Marinopoulos, A.G., Gemming, T., Elsässer, C., Kurtz, W., Köstmeier, S., and Rühle, M. (2001) *Phys. Rev. Lett.*, **86**, 5066.
111 Marinopoulos, A.G. and Elsässer, C. (2000) *Acta Mater.*, **48**, 4375.
112 Castaing, J., He, A., Lagerlof, K.P.D., and Heuer, A.H. (2004) *Philos. Mag. A*, **84**, 1113.
113 Pirouz, P. (1996) *Acta Metall. Mater.*, **44**, 2153.
114 He, A., Lagerlöf, K.P.D., Castaing, J., and Heuer, A.H. (2002) *Philos. Mag. A*, **82**, 2855.
115 Marinopoulos, A.G., Nufer, S., and Elsässer, C. (2001) *Phys. Rev. B*, **63**, 165112.
116 Nishigaki, J., Kuroda, K., and Saka, H. (1991) *Phys. Status Solidi A*, **128**, 319.
117 Doukhan, N. and Doukhan, J.C. (1986) *Phys. Chem. Miner.*, **13**, 403.
118 Takeuchi, S., Suzuki, K., Ichihara, M., and Suzuki, T. (1989) *Jpn. J. Appl. Phys., Part 2*, **28**, 17.
119 Poirier, J.P., Peyronneau, J., Gesland, J.Y., and Brebec, G. (1983) *Phys. Earth Planet. Inter.*, **33**, 273.
120 Brunner, D., Taeri-Baghbadrani, S., Cigle, W., and Rühle, M. (2001) *J. Am. Ceram. Soc.*, **84**, 1161.
121 Gumbsch, P., Taeri-Baghbadra, S., Brunner, D., Sigle, W., and Rühle, M. (2001) *Phys. Rev. Lett.*, **87**, 85505.
122 Mitchell, T.E., Baskes, M.I., Chen, S.P., Hirth, J.P., and Hoagland, R.G. (2001) *Philos. Mag. A*, **81**, 1079.
123 Caillard, D. and Couret, A. (1996) *Dislocations in Solids* (eds F.R.N. Nabarro and M.S. Duesbery), Elsevier, p. 69.
124 Sigle, W., Sarbu, C., Brunner, D., and Ruhle, M. (2006) *Philos. Mag. A*, **86**, 4809.
125 Zhang, Z., Sigle, W., Kurtz, W., and Rühle, M. (2003) *Phys. Rev. B*, **66**, 214112.
126 Zhang, Z., Sigle, W., and Rühle, M. (2002) *Phys. Rev. B*, **66**, 94108.
127 Radford, K.C. and Newey, C.W.A. (1967) *Proc. Br. Ceram. Soc.*, **9**, 131.
128 Newey, C.W.A. and Radford, K.C. (1968) *Anisotropy in Single Crystal Refractory Compounds* (eds F.W. Vahldiek and S.A. Mersol), Plenum Press, New York, p. 321.

129 Lewis, M.H. (1968) *Philos. Mag.*, **17**, 481.

130 Doukhan, N., Duclos, R., and Escaig, B. (1976) *J. Phys. (Paris)*, **37**, 566.

131 Welsch, G., Hwang, L., Heuer, A.H., and Mitchell, T.E. (1974) *Philos. Mag.*, **29**, 1374.

132 Hwang, L., Heuer, A.H., and Mitchell, T.E. (1975) *Deformation of Ceramic Materials* (eds R.C. Bradt and R.E. Tressler), Plenum Press, New York, p. 257.

133 Mitchell, T.E., Hwang, L., and Heuer, A.H. (1976) *J. Mater. Sci.*, **11**, 264.

134 Doukhan, N., Duclos, R., and Escaig, B. (1973) *J. Phys. (Paris)*, **34**, 397.

135 Duclos, R., Doukhan, N., and Escaig, B. (1978) *J. Mater. Sci.*, **13**, 1740.

136 Veyssière, P., Rabier, J., and Garem, H. (1979) *Philos. Mag. A*, **39**, 815.

137 Donlon, W.T., Mitchell, T.E., and Heuer, A.H. (1979) *Philos. Mag. A*, **40**, 351.

138 Doukhan, N. (1979) *J. Phys. Lett.*, **40**, L603.

139 Kirby, S.H. and Veyssière, P. (1980) *Philos. Mag. A*, **41**, 129.

140 Duclos, R., Doukhan, N., and Escaig, B. (1981) *Philos. Mag.*, **43**, 1595.

141 Duclos, R. (1981) *J. Phys. (Paris)*, **42**, 49.

142 Donlon, W.T., Mitchell, T.E., and Heuer, A.H. (1982) *Philos. Mag. A*, **45**, 1013.

143 Doukhan, N., Duclos, R., and Escaig, B. (1982) *J. Phys. (Paris)*, **43**, 1149.

144 Duclos, R., Doukhan, N., and Escaig, B. (1982) *Acta Metall.*, **30**, 1381.

145 Suematsu, H., Suzuki, T., Iseki, T., and Mori, T. (1989) *J. Am. Ceram. Soc.*, **72**, 306.

146 Corman, G.S. (1991) *Ceram. Eng. Sci. Proc.*, **12**, 1745.

147 Corman, G.S. (1992) *J. Mater. Sci. Lett.*, **11**, 1657.

148 Mitchell, T.E., Donlon, W.T., and Heuer, A.H. (1998) *Scripta Mater.*, **39**, 537.

149 Yano, T., Ikari, M., Iseki, T., Farnum, E.H., Clinard, F.W., and Mitchell, T.E. (1995) *J. Am. Ceram. Soc.*, **78**, 1469.

150 Charpentier, P., Rabbe, P., and Manenc, J. (1968) *Mater. Res. Bull.*, **3**, 69.

151 Ito, T. (1971) *J. Am. Ceram. Soc.*, **54**, 24.

152 Callahan, S.L., Tressler, R.E., Johnson, D.W., and Reece, M.J. (1984) *Deformation of Ceramic Materials II* (eds R.E. Tressler and R.C. Bradt), Plenum Press, New York, p. 177.

153 Dekker, E.H.L. and Rieck, G.D. (1974) *J. Mater. Sci.*, **9**, 1839.

154 Veyssière, P., Rabier, J., Garem, H., and Grihlé, J. (1978) *Philos. Mag. A*, **38**, 61.

155 Veyssière, P., Rabier, J., Garem, H., and Grihlé, J. (1976) *Philos. Mag.*, **33**, 143.

156 Veyssière, P., Rabier, J., and Garem, H. (1979) *Philos. Mag.*, **39**, 815.

157 Doukhan, N., Doukhan, J.-C., Nicholas, A., and Secher, D. (1984) *Bull. Minéral.*, **107**, 777.

158 Carter, N.L. and AvéLallemant, H.G. (1970) *Geol. Soc. Am. Bull*, **81**, 2181.

159 Phakey, P.P., Dollinger, G., and Christie, J.M. (1972) *Geophysical Monograph Series* (eds H.C. Heard I.Y. Borg N.L. Carter, and C.B. Raleigh), American Geophysical Union, Washington, p. 117.

160 Raleigh, C.B. (1968) *Science*, **150**, 739.

161 Poirier, J.P. (1975) *J. Geophys. Res.*, **80**, 4059.

162 Darot, M. and Gueguen, Y. (1981) *Bull. Mineral.*, **104**, 261.

163 Gaboriaud, R.J. (1986) *Bull. Mineral.*, **109**, 185.

164 Gaboriaud, R.J., Darot, M., Gueguen, Y., and Woirgard, J. (1981) *Phys. Chem. Minerals*, **7**, 100.

165 Kohlstedt, D.L. and Ricoult, D.L. (1984) *Deformation of Ceramic Materials II* (eds R.E. Tressler and R.C. Bradt), Plenum Press, New York, p. 251.

166 Poumellec, B. and Jaoul, O. (1984) *Deformation of Ceramic Materials II* (eds R.E. Tressler and R.C. Bradt), Plenum Press, New York, p. 281.

167 Karato, S. (1989) *Rheology of Solids and of the Earth* (eds S. Karato and M. Torium), Oxford University Press, Oxford, p. 176.

168 Karato, S., Paterson, M.S., and Fitzgerald, J.D. (1986) *J. Geophys. Res.*, **91**, 8151.

169 Blacic, J.D. (1972) *Geophysical Monograph Series* (eds H.C. Heard, I.Y. Borg, N.L. Carter, and C.B. Raleigh), American Geophysical Union, Washington, p. 109.

170 Mackwell, S.J., Kohlstedt, D.L., and Paterson, M.S. (1985) *J. Geophys. Res.*, **88**, 3122.

171 Drury, M.R. (1991) *Phys. Chem. Miner.*, **18**, 106.

172 Poirier, J.P. and Vergrobbi, B. (1978) *Phys. Earth Planet. Inter.*, **16**, 370.

173 VanderSande, J.B. and Kohlstedt, D.L. (1976) *Philos. Mag.*, **34**, 653.

174 Smith, B.K., Allen, F.M., and Buseck, P.R. (1987) *EOS Trans. Am. Geophys. Union*, **68**, 426.
175 White, T.J. and Hyde, B.G. (1982) *Phys. Chem. Miner.*, **8**, 55.
176 Audouard, A., Castaing, J., Rivière, J.P., and Sieber, B. (1981) *Acta Metall.*, **29**, 1385.
177 Gaboriaud, R.J. (1981) *Philos. Mag. A*, **44**, 561.
178 Boisson, M. and Gaboriaud, R.J. (1981) *J. Mater. Sci.*, **16**, 3452.
179 Sharif, A.A., Misra, A., Petrovic, J.J., and Mitchell, T.E. (2000) *Key Eng. Mater.*, **171–174**, 801.
180 Karato, S., Wang, Z., Liu, B., and Fujino, K. (1995) *Earth Planet. Sci. Lett.*, **130**, 13.
181 Rabier, J., Veyssière, P., and Garem, H. (1981) *Philos. Mag. A*, **44**, 1363.
182 Rabier, J., Veyssière, P., Garem, H., and Grihlé, J. (1979) *Philos. Mag. A*, **39**, 693.
183 Wang, Z., Karato, S., and Fujino, K. (1996) *Phys. Chem. Miner.*, **23**, 73.
184 Garem, H., Rabier, J., and Veyssière, P. (1982) *J. Mater. Sci.*, **17**, 878.
185 Karato, S., Wang, Z., and Fujino, K. (1994) *J. Mater. Sci.*, **29**, 6458.
186 Blumenthal, W.R. and Phillips, D.S. (1996) *J. Am. Ceram. Soc.*, **79**, 1047.
187 Voegele, V., Ando, J.I., Cordier, P., and Liebermann, R.C. (1998) *Phys. Earth Planet. Inter.*, **108**, 305.
188 Hollox, G.E. (1968/69) *Mater. Sci. Eng.*, **3**, 121.
189 Tsurekawa, S., Kurishita, H., and Yoshinaga, H. (1989) *Lattice Defects in Ceramics* (eds T. Suzuki and S. Takeuchi), Japanese Journal of Applied Physics, Tokyo, p. 47.
190 Pirouz, P. (2002) *Understanding Materials* (ed. C.J. Humphreys), Institute of Materials, London, p. 225.
191 Suematsu, H., Suzuki, T., Iseki, T., and Mori, T. (1991) *J. Am. Ceram. Soc.*, **74**, 173.
192 Wang, C.M., Pan, X., Rühle, M., Riley, F.L., and Mitomo, M. (1996) *J. Mater. Sci.*, **31**, 5281.
193 Suematsu, H., Petrovic, J.J., and Mitchell, T.E. (1993) *Mater. Res. Soc. Symp. Proc.*, **287**, 449.
194 Suematsu, H., Petrovic, J.J., and Mitchell, T.E. (1997) *J. Ceram. Soc. Jpn*, **105**, 842.
195 Zhou, D.S. and Mitchell, T.E. (1995) *J. Am. Ceram. Soc.*, **78**, 3133.
196 Groves, G.W. and Kelly, A. (1963) *Proc. Roy. Soc. Lond.*, **A275**, 233.
197 Hirsch, P.B. and Mitchell, T.E. (1968) *Work Hardening* (eds J.P. Hirth and J. Weertman), Gordon and Breach, New York, p. 365.
198 Hirsch, P.B. and Mitchell, T.E. (1967) *Can. J. Phys.*, **45**, 663.
199 Conrad, H. (1965) *J. Am. Ceram. Soc.*, **48**, 195.
200 Kronberg, M.L. (1962) *J. Am. Ceram. Soc.*, **45**, 274.
201 Pletka, B.J., Heuer, A.H., and Mitchell, T.E. (1977) *Acta Metall.*, **25**, 25.
202 Labusch, R. (1970) *Phys. Status Solidi*, **41**, 659.
203 Pletka, B.J., Mitchell, T.E., and Heuer, A.H. (1977) *Phys. Status Solidi A*, **39**, 301.
204 Srinivasan, M. and Stoebe, T.G. (1970) *J. Appl. Phys.*, **41**, 3726.
205 Reppich, B. (1976) *Mater. Sci. Eng.*, **22**, 71.
206 Rodriguez, M., Munoz, A., Castaing, J., Veyssiere, P., and Rodriguez, A.D. (2007) *J. Eur. Ceram. Soc.*, **27**, 3317.
207 Phillips, D.S., Mitchell, T.E., and Heuer, A.H. (1980) *Philos. Mag. A*, **42**, 417.
208 Mitchell, T.E. and Heuer, A.H. (1977) *Mater. Sci. Eng.*, **28**, 81.

10
Defect Structure, Nonstoichiometry, and Nonstoichiometry Relaxation of Complex Oxides

Han-Ill Yoo

10.1
Introduction

Crystalline oxide materials with optical, electrical, magnetic, thermal, chemical, and electrochemical applications have been of great technological interest for many years. They often comprise no less than three components in no less than three sublattices, and tend to be even more complex in terms of the number of components and sublattices. The simplest examples may be $BaTiO_3$, with a perovskite structure, and $MnFe_2O_4$ with a spinel structure. These are quintessential ingredients of dielectrics and soft ferrites, respectively, and currently take the lion's share in the worldwide electroceramics market.

For all applications of complex oxides, a knowledge of – and an ability to control – their defect structures, especially the types, concentrations and spatial distribution mainly of point defects, is essential in order to endow these materials with necessary functions, as well as to design and/or optimize the processing routes of the devices ultimately produced. For example, in order to produce a positive-temperature coefficient resistor (PTCR) with $BaTiO_3$, the interior of the $BaTiO_3$ grains should be made into n-type semiconductors by doping with donor impurities, and grain boundaries into electrical insulation by employing oxidation [1]. The thickness of the insulation must be carefully controlled in order to achieve optimum PTCR performance, by appropriately adjusting the magnitude and distribution of oxygen nonstoichiometry during the cooling period that follows sintering. Another example may be a $BaTiO_3$-based multilayer ceramic capacitors (MLCC) employing base metal electrodes (e.g., Ni). This should be sintered in a reducing atmosphere so as to avoid oxidation of the base metal, while suppressing the reduction-generated electrons and oxygen vacancies so as to retard degradation of the dielectric properties and insulation resistance while the device is in service. In a way, controlling and/or tailoring the defect structures is like blowing the "soul" into the body – the art of making oxides alive in a functional manner.

The basic principle behind the control or tailoring of defect structure is that point defects are thermodynamically stable; hence, not only their concentrations but also

Ceramics Science and Technology Volume 2: Properties. Edited by Ralf Riedel and I-Wei Chen

ISBN: 978-3-527-31156-9

their thermodynamic equilibrium properties can be uniquely determined by the independent thermodynamic variables such as temperature, pressure, and the compositions or the conjugate chemical potentials of the system. This is a basic thermodynamic postulate, that might even be called the "first law of thermodynamics." Thus, by adjusting the temporal and spatial distribution of these thermodynamic variables it is possible, at least in principle, to adjust the temporal and spatial distribution of the defects.

Defects are not a conserved entity: they can be annihilated or generated normally via solid-state diffusion from or to the repeatable growth sites such as surfaces, grain boundaries, and dislocations. Solid-state diffusion is a time- and energy-consuming process, but it is possible to kinetically cheat the defect structure by adjusting the time rate of the thermodynamic variables. The latter is quite often taken advantage of in actual processes to freeze-in a nonequilibrium defect structure for property control purposes.

In this chapter, the thermodynamic and kinetic aspects of defect structure in complex oxides will be considered, with applications in mind. As a prototype of complex oxide, reference will be made specifically to $BaTiO_3$ among others because, to the best of the present author's knowledge, the related experimental data have been the most extensively and consistently documented. The intention is not to provide an exhaustive literature review, but rather to convey the thermodynamic and kinetic behaviors of defect structures in complex oxides. Although most of the treatments here relate to $BaTiO_3$, they may be easily modified to other complex oxide systems, with only minor modifications.

The chapter will be structured as follows. In Section 10.2, a description is provided of how to calculate the equilibrium defect structure of a prototype complex oxide as a function of independent thermodynamic variables. When a batch composition in terms of the component oxides (e.g., BaO and TiO_2) is formulated, the only way to control defect structure during processing is normally via the exchange of a volatile component (e.g., oxygen) during heat treatment. In this case, the oxygen nonstoichiometry can be adjusted so as eventually to govern the concentration of charge carriers, oxygen vacancies, electrons and holes that often drastically affect the performance of oxide-based devices. In Section 10.3, the way in which oxygen nonstoichiometry, as a measure of the concentration of electronic carriers, varies with the thermodynamic variables is examined, along with its defect-chemical implications. The adjustment of oxygen nonstoichiometry is largely possible by utilizing chemical diffusion processes under an oxygen potential gradient imposed during the processing of an oxide, or of the device itself. In Section 10.4, details are provided of how oxygen nonstoichiometry may be relaxed in an oxygen potential gradient in relation to defect structure.

10.2
Defect Structure

The logical framework to calculate defect structure of a complex oxide was originally proposed by Wagner and Schmalzried [2, 3]. The basic concept was to solve the simultaneous equations involving concentrations of the structure elements – be they regular or irregular – that are based on a series of constraints of internal (thermal) equilibria, external (particle exchange) equilibria, crystal structure preservation, charge neutrality, and mass conservation (when doped). Whereas the internal equilibria are determined by temperature only under ambient pressure ($P = 1$ atm), the external equilibria are determined by the independent activities of the chemical components, in addition to the temperature. Hence, the concentration will become a function of the intensive thermodynamic variables of the given system – that is, the temperature and independent activities of the components. This concept will be applied to calculate the defect structures of $BaTiO_3$ for pure, acceptor-doped, and donor-doped cases, respectively.

10.2.1 Pure Case

First, we consider pure $BaTiO_3$. For a ternary system, there are two composition variables, for example, the mole fractions of Ti and O. When there are three sublattices, however, it is more convenient to choose the molecular ratio of the component oxides, BaO and TiO_2 and the equivalent ratio of the nonmetallic component to the metallic components [3, 4] or

$$1+\eta \equiv \frac{[\mathrm{Ti}]_t}{[\mathrm{Ba}]_t}; \qquad 1-\frac{1}{3}\delta \equiv \frac{2[\mathrm{O}]_t}{2[\mathrm{Ba}]_t+4[\mathrm{Ti}]_t} \tag{1}$$

where $[\;]_t$ represents the total concentration of the component therein. The lattice molecule may then be represented more appropriately as

$$\mathrm{BaTi}_{1+\eta}\mathrm{O}_{3+2\eta-\delta} \tag{2}$$

where η is termed the deviation from the molecularity ($\eta = 0$) or nonmolecularity, and δ is the deviation from the stoichiometry ($\delta = 0$) or nonstoichiometry of the compound. Any thermodynamic equilibrium property of the system is given as a function of independent thermodynamic variables of the system under the atmospheric pressure, temperature (T), the activity of a component oxide, such as TiO_2 ($a_{\mathrm{TiO_2}}$), and the activity of oxygen ($a_{\mathrm{O_2}}$) or equivalently by their conjugate variables η and δ, considering the Gibbs–Duhem equation for the system.

One may start by conjecturing the possible defects of the oxide from its structural and energetic considerations. For the system of perovskite structure, the interstitial defects may be ruled out and hence, the structure elements, the concentrations of which need to be known, may be the irregular structure elements: V''_{Ba}, V''''_{Ti}, $V^{\bullet\bullet}_{\mathrm{O}}$, e', $h^{\bullet}$; in addition to the regular structure elements: $\mathrm{Ba}^{\mathrm{x}}_{\mathrm{Ba}}$, $\mathrm{Ti}^{\mathrm{x}}_{\mathrm{Ti}}$, and $\mathrm{O}^{\mathrm{x}}_{\mathrm{O}}$; in terms of the Kröger–Vink notation.

Letting $[S]$ denote the concentration (in number cm^{-3}) of the structure element S ($[e'] \equiv n$, $[h^{\bullet}] \equiv p$), it is possible to formulate all of the constraints, assuming an ideal dilute solution behavior of defects as:

- Internal equilibria:

$$0 = e' + h^{\bullet}; \quad K_i = np \tag{3}$$

$$0 = V''_{Ba} + V''''_{Ti} + 3V_O^{\bullet\bullet}; \quad K_S = [V''_{Ba}][V''''_{Ti}][V_O^{\bullet\bullet}]^3 \tag{4}$$

- External equilibria:

$$O_O^{\times} = V_O^{\bullet\bullet} + 2e' + \frac{1}{2}O_2(g); \quad K_{Re} = [V_O^{\bullet\bullet}]n^2 a_{O_2}^{1/2} \tag{5}$$

$$TiO_2 = Ti_{Ti}^{x} + 2O_O^{\times} + V''_{Ba} + V_O^{\bullet\bullet}; \quad K_T = \frac{[V''_{Ba}][V_O^{\bullet\bullet}]}{a_{TiO_2}} \tag{6}$$

- Charge neutrality:

$$n + 2[V''_{Ba}] + 4[V''''_{Ti}] = p + 2[V_O^{\bullet\bullet}] \tag{7}$$

- Structure preservation:

$$1\beta = [Ba_{Ba}^{x}] + [V''_{Ba}] \tag{8}$$

$$1\beta = [Ti_{Ti}^{x}] + [V''''_{Ti}] \tag{9}$$

$$3\beta = [O_O^{x}] + [V_O^{\bullet\bullet}] \tag{10}$$

with $\beta = N_A/V_m$, where N_A and V_m are the Avogadro number and the molar volume of the system, respectively.

Here, the mass-action law constants are denoted as K_j ($j = Re, T, S, i$), which may be represented as

$$K_j = K_j^o \exp\left(-\frac{\Delta H_j}{kT}\right) \tag{11}$$

where K_j^o is the pre-exponential factor and ΔH_j is the enthalpy change of the associated reaction $j (= i, S, Re, T)$.

It is noted that there are exactly eight equations [Eqs. (3)–(10)] for eight unknowns (V''_{Ba}, V''''_{Ti}, $V_O^{\bullet\bullet}$, e′, $h^{\bullet}$, Ba_{Ba}^{x}, Ti_{Ti}^{x}, and O_O^{x}). As the concentrations of irregular structure elements are normally much smaller than those of regular structure elements, Eqs. (8)–(10) become trivial (i.e., $[Ba_{Ba}^{x}] \approx [Ti_{Ti}^{x}] \approx [O_O^{x}]/3 \approx 1\beta$) and hence, it is possible to delete the regular structure elements from the list of unknowns. It should be noted that, when starting with "q" irregular structure elements in a c-component system, there would always be (c – 1) external equilibrium conditions and (q – c) internal equilibrium conditions and one charge neutrality condition to determine all of those n unknowns completely.

Table 10.1 Matrix of majority disorder types in the systems of $BaTiO_3$. The top left rectangle demarcated by thick solid lines is for the pure case; this rectangle plus the rightmost column for the acceptor-doped case; and the rectangle plus the bottom-most row for the donor-doped case. A pair of signs out of +,0,− at each element are for η and δ: e.g., (+,0) is for $\eta > 0$ and $\delta \approx 0$.

− / +	n	$2[V''_{Ba}]$	$4[V''''_{Ti}]$	$[A'_C]$
p	$n = p$ (0; 0)	$p = 2[V''_{Ba}]$ (+; −)	$p = 4[V''''_{Ti}]$ (−; −)	$p = [A'_C]$ (+; −)
$2[V_O^{\bullet\bullet}]$	$2[V_O^{\bullet\bullet}] = n$ (0; +)	$[V_O^{\bullet\bullet}] = [V''_{Ba}]$ (+; 0)	$[V_O^{\bullet\bullet}] = 2[V''''_{Ti}]$ (−; 0)	$2[V_O^{\bullet\bullet}] = [A'_C]$ (+; 0)
$[D_C^{\bullet}]$	$[D_C^{\bullet}] = n$ (−; +)	$[D_C^{\bullet}] = 2[V''_{Ba}]$ (−; 0)	$[D_C^{\bullet}] = 4[V''''_{Ti}]$ (−; 0)	

In principle, this set of equations can be solved simultaneously for each defect concentration in terms of $K_j(T)$, a_{TiO_2}, and a_{O_2} or

$$[S] = f(K_j, a_{TiO_2}, a_{O_2}) \tag{12}$$

However, this is usually prohibitively messy, the problem being that the algebraic structure of the charge neutrality condition, Eq. (7), is different from the remainder [Eqs. (3)–(6)]. Thus, the normal practice is to approximate the charge neutrality condition by a limiting condition or in terms of an oppositely charged pair of disorders in the majority (Brouwer approximation [4]), depending on the thermodynamic conditions.

All of the possible types of majority disorder may be distinguished by constructing a matrix with the positively charged disorders as a row, and the negatively charged disorders as a column (or *vice versa*, of course), as shown in Table 10.1 [3, 4]. In the case of pure $BaTiO_3$, there can be 2×3 elements or six possible majority disorder types.

By using the limiting charge neutrality condition in terms of a majority disorder pair in Table 10.1, it is possible to solve, with absolutely no algebraic awkwardness, each defect concentration in the form of [3, 4]:

$$[S] = a_{O_2}^m a_{TiO_2}^n \prod_j K_j^s \tag{13}$$

The numerical values for the exponents "m" and "n" [not to be confused with the electron concentration in Eq. (3)] in each majority disorder regime are listed (in the form of m; n at each element) in Table 10.2.

In order to construct a complete picture of defect structure against the thermodynamic variables, and particularly of a_{TiO_2} and a_{O_2} over their entire ranges, these piecewise solutions should be combined, Eq. (13), in appropriate order. For this purpose, it is necessary first to know the distribution of the majority disorder types in the thermodynamic configuration space of $\log a_{O_2}$ versus $\log a_{TiO_2}$ at a fixed temperature. A simple method to identify the configuration of the majority disorder types is as follows [3, 4].

For the present system [Eq. (2)], the nonmolecularity may be represented in terms of the irregular structure elements, due to Eqs. (1) and (8)–(10) as:

$$\beta\eta \approx [V''_{Ba}] - [V''''_{Ti}] \tag{14}$$

Table 10.2 Numerical values for m and n such that $[S] = a_{O_2}^m a_{TiO_2}^n \prod_j K_j^s$ in each majority disorder regime of Table 10.1.

	$n=p$	$p=4[V_{Ti}'''']$	$2[V_{Ti}'''']=[V_O^{\bullet\bullet}]$	$2[V_O^{\bullet\bullet}]=n$	$[V_O^{\bullet\bullet}]=[V_{Ba}'']$	$2[V_{Ba}'']=p$
n	0; 0	−1/5; 1/5	−1/4; 1/6	−1/6; 0	−1/4; −1/4	−1/6; −1/3
p	0; 0	1/5; −1/5	1/4; −1/6	1/6; 0	1/4; 1/4	1/6; 1/3
$[V_{Ba}'']$	1/2; 1	1/10; 7/5	0; 4/3	1/6; 1	0; 1/2	1/6; 1/3
$[V_{Ti}'''']$	1; −1	1/5; −1/5	0; −1/3	1/3; −1	0; −2	1/3; −7/3
$[V_O^{\bullet\bullet}]$	−1/2; 0	−1/10; −2/5	0; −1/3	−1/6; 0	0; 1/2	−1/6; 2/3
$[A_C']$	−; −	−; −	−; −	0; 0	−; −	−; −
$[A_C^\times]$	−; −	−; −	−; −	1/6; 0	−; −	−; −

	$p=[A'_C]$	$2[V_O^{\bullet\bullet}]=[A'_C]$		$[D_C^\bullet]=4[V_{Ti}'''']$	$[D_C^\bullet]=2[V_{Ba}'''']$	$n=[D_C^\bullet]$
		$[A'_C]\cong[A]_t$	$[A_C^\times]\cong[A]_t$			
n	0; 0	−1/4; 0	−1/6; 0	−1/4; 1/4	−1/4; −1/2	0; 0
p	0; 0	1/4; 0	1/6; 0	1/4; −1/4	1/4; 1/2	0; 0
$[V_{Ba}'']$	1/2; 1	0; 1	1/6; 1	0; 3/2	0; 0	1/2; 1
$[V_{Ti}'''']$	1; −1	0; −1	1/3; −1	0; 0	0; −3	1; −1
$[V_O^{\bullet\bullet}]$	−1/2; 0	0; 0	−1/6; 0	0; 1/2	0; 1	−1/2; 0
$[A_C']$	0; 0	0; 0	−1/6; 0			
$[A_C^\times]$	0; 0	1/4; 0	0; 0			

as normally $[S]/\beta \ll 1$ if S is irregular. The nonstoichiometry, on the other hand, may be written as:

$$\beta\delta \approx [V_O^{\bullet\bullet}] - [V_{Ba}''] - 2[V_{Ti}''''] = \frac{1}{2}(n-p) \tag{15}$$

for the same reason. The second equality is due to the charge neutrality condition [Eq. (7)].

For each of the majority disorder types in Table 10.1 it can then be determined, by using Eqs. (14) and (15), whether η and δ are respectively larger than 0 (denoted as +), close to 0 (denoted as 0), or smaller than 0 (denoted as −). The results (as pairs out of −, 0, and +) are given in the form of (−; 0), for example, to each element in Table 10.1, where the first symbol is for η and the second for δ. In this identification, it is important to be aware that for the majority disorder type for which η can be both + and − as, for example, for $n \approx p$ or $n \approx 2[V_O^{\bullet\bullet}]$, it is possible to set $\eta \approx 0$.

There may be three regions along the axis of log a_{O_2}: $\delta > 0$ (oxygen deficit); ≈ 0 (near stoichiometry); and <0 (oxygen excess) with increasing a_{O_2}. There are also three regions along the axis of log a_{TiO_2}: $\eta < 0$ (TiO_2-deficit); ≈ 0 (near molecularity); and >0 (TiO_2-excess) with increasing a_{TiO_2}. The configuration plane may, therefore, be divided into nine regions, with each majority disorder type being assigned to the region corresponding to its sign combination for (η; δ) in Table 10.1. Quite often, multiple types fall into one region, usually as the number of majority disorder types increases above nine. Again, old wisdom [5] helps prevent getting lost in finding the appropriate sequence of the majority disorder types. These are:

- **Rule 1. Near molecularity and stoichiometry region (0; 0)**: It is to be occupied by an intrinsic majority disorder type, either electronic or ionic disorder pair, depending on which is energetically more favorable. In the present pure case, it depends on whether $K_i^{1/2} > K_S^{1/5}$ or $K_i^{1/2} < K_S^{1/5}$. Consequently, intrinsic electronic and ionic disorder pairs normally do not fall congruently.
- **Rule 2. Other regions**: The sequence in a given region outside the region (0; 0) is to be determined by the continuity principle. That is, any two neighboring regimes of majority disorder should have one defect of the defect pairs in common.
- **Rule 3. Majority disorder regime boundaries**: Once the majority disorder types are located sequentially in the configuration plane of log a_{O_2} versus log a_{TiO_2}, the two neighboring regimes of majority disorder meet in line when one defect of the two pairs is in common and otherwise at point. In the three-dimensional space of log a_{O_2}−log $a_{TiO_2}$$-1/T$, the former would be a plane and the latter a line.

These are termed the "allocation rules," and by applying them it is possible to construct the configuration planes, as shown in Figure 10.1a and b for the case of $K_i^{1/2} > K_S^{1/5}$ and $K_i^{1/2} < K_S^{1/5}$, respectively. Until now, however, there appears to be no experimental evidence for $BaTiO_3$ that might even hint at any possibility of $n \approx p$ being the majority disorder under any thermodynamic condition.

10.2.2
Acceptor-Doped Case

Next, we will consider $BaTiO_3$ doped with a fixed amount (x) of acceptor impurities A on, for example, cation sites (generically denoted as A'_C). Specifically, these are assumed to be trivalent substituting Ti or A'_{Ti}, for example, Al'_{Ti}. The lattice molecule may then be written as:

$$BaTi_{1-x+\eta}A_xO_{3-x/2+2\eta-\delta} \tag{16}$$

As one more defect species A'_{Ti} is added, there is a need for one more constraint, in addition to those for the pure case: that is, mass conservation or

$$[A'_{Ti}] = \beta x \tag{17}$$

The charge neutrality condition [Eq. (7)] and the site conservation conditions [Eq. (9)] should be accordingly modified, respectively, as:

$$n + 2[V''_{Ba}] + 4[V''''_{Ti}] + [A'_{Ti}] = p + 2[V_O^{\bullet\bullet}] \tag{18}$$

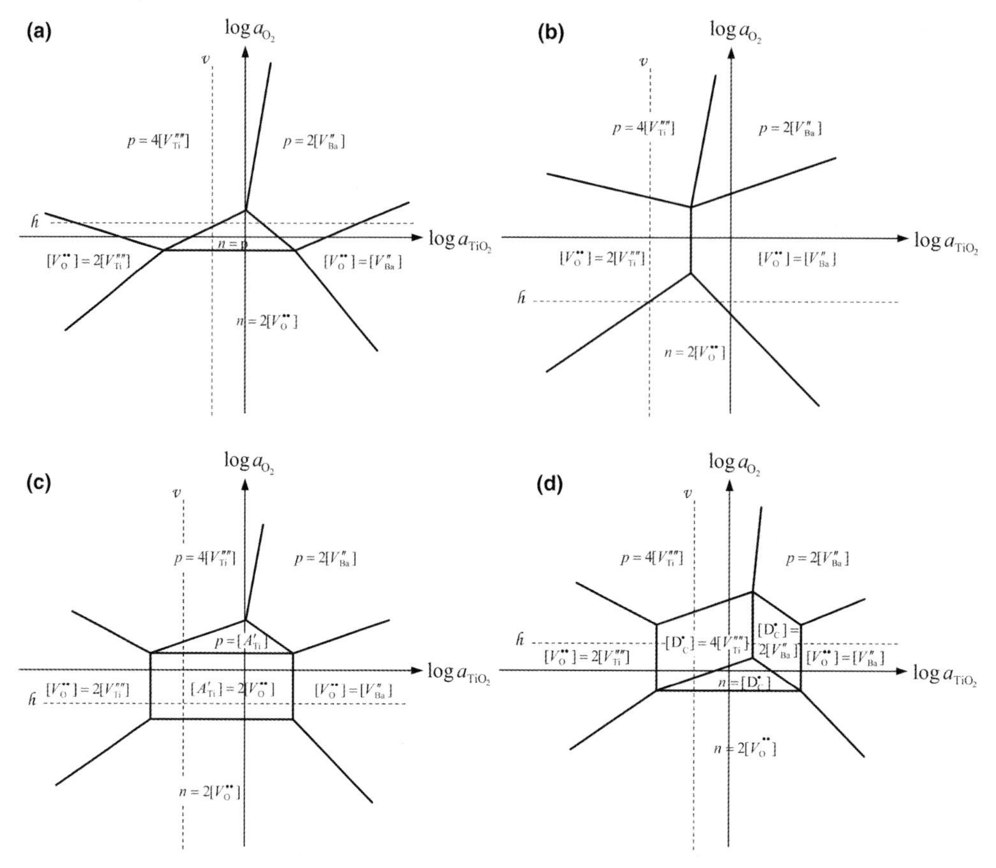

Figure 10.1 Configuration of the majority disorder types on a plane of log a_{O_2} versus log a_{TiO_2} at fixed temperature: (a) $K_i^{1/2} \gg K_S^{1/5}, [A'_C], [D^{\bullet}_C]$; (b) $K_S^{1/5} \gg K_i^{1/2}, [A'_C], [D^{\bullet}_C]$; (c) $[A'_C] \gg K_i^{1/2}, K_S^{1/5}, [D^{\bullet}_C]$; (d) $[D^{\bullet}_C] \gg K_i^{1/2}, K_S^{1/5}, [A'_C]$.

$$1\beta = [\mathrm{Ti}^{\mathrm{x}}_{\mathrm{Ti}}] + [V''''_{\mathrm{Ti}}] + [A'_{\mathrm{Ti}}] \tag{19}$$

In addition, while the nonmolecularity is still the same as Eq. (14), the nonstoichiometry should be modified due to Eq. (1) as:

$$\beta\delta \approx [V^{\bullet\bullet}_{\mathrm{O}}] - [V''_{\mathrm{Ba}}] - 2[V''''_{\mathrm{Ti}}] - \frac{1}{2}[A'_{\mathrm{Ti}}] = \frac{1}{2}(n-p) \tag{20}$$

From Eq. (18), or from the matrix in Table 10.1, it is possible to immediately distinguish 2×4 limiting conditions or eight possible majority disorder types. However, it should be noted that any intrinsic disorder type ($n = p$ in the present case) cannot occupy the near molecularity–stoichiometry region (0; 0), because obviously $[A'_{Ti}] \gg K_i^{1/2}, K_S^{1/5}$, or it would otherwise not be extrinsic due to the doped acceptors.

Again, following Rules 1–3, it is possible to construct the configuration plane of the majority disorder types, as in Figure 10.1c; in each regime, the values for the exponent m and n of Eq. (13) are given in Table 10.2.

10.2.3
Donor-Doped Case

It can now be assumed that the $BaTiO_3$ is doped with a fixed amount (γ) of donor impurities D on, for example, the cation sites ($D_C^{\bullet}$). They are again assumed to be fixed-valent, but this time to be occupying Ba-site instead, or $D_{Ba}^{\bullet}$, for example, $La_{Ba}^{\bullet}$. The lattice molecule may then be represented as:

$$Ba_{1-\gamma}D_{\gamma}Ti_{1+\eta}O_{3+\gamma/2+2\eta-\delta} \tag{21}$$

As one more unknown $[D_{Ba}^{\bullet}]$ is added similarly to the acceptor-doped case, one mass-conservation equation for the dopant is added to Eqs. (3)–(6) as:

$$[D_{Ba}^{\bullet}] = \beta\gamma \tag{22}$$

In addition, the charge neutrality and site conservation conditions [Eqs. (7) and (8)] are modified, respectively, to:

$$n + 2[V_{Ba}''] + 4[V_{Ti}''''] = p + 2[V_O^{\bullet\bullet}] + [D_{Ba}^{\bullet}] \tag{23}$$

$$1\beta = [Ba_{Ba}^{x}] + [V_{Ba}''] + [D_{Ba}^{\bullet}] \tag{24}$$

The nonstoichiometry subsequently takes the form:

$$\beta\delta \approx \frac{1}{2}(n-p) = [V_O^{\bullet\bullet}] - [V_{Ba}''] - 2[V_{Ti}''''] + \frac{1}{2}[D_{Ba}^{\bullet}] \tag{25}$$

while the nonmolecularity still remains the same as Eq. (14).

From Eq. (23) or Table 10.1, it is possible to distinguish 3×3 possible limiting conditions or possible majority disorder types. However, the intrinsic disorder type $n = p$ is again ruled out from the near molecularity–stoichiometry region (0; 0) in the configuration plane and hence, one is left with eight majority disorder types. The configuration map of the majority disorder types can be constructed following Rules 1–3, as in Figure 10.1d. The reader may wish to check from the maps in Figure 10.1 how those rules are working.

10.2.4
Two-Dimensional Representations of Defect Concentrations

Equation (13) is basically a four-dimensional representation (under a fixed total pressure) of defect structure, that is not easy to visualize with ordinary vision. The normal practice is, thus, to represent in two dimensions: log $[S]$ versus log a_{TiO_2} (at fixed a_{O_2} and T), log $[S]$ versus log a_{O_2} (at fixed a_{TiO_2} and T), or log $[S]$ versus $1/T$ (at fixed a_{TiO_2} and a_{O_2}).

By combining the piecewise solutions, Eq. (13) with m and n values as given in Table 10.2 in accord with the configuration of the majority disorder types in Figure 10.1, it is easy to draw log $[S]$ versus log a_{TiO_2} (at fixed T and a_{O_2}) or versus log a_{O_2} (at fixed a_{TiO_2} and T). The cross-section along a horizontal dotted line in each of Figures 10.1a–d (designated as "h"), for example, is shown in Figures 10.2a–d, which are simply log [S] versus log a_{TiO_2} (at fixed T and a_{O_2}). Similarly, log [S] versus log a_{O_2} (at given T and a_{TiO_2}) are shown in Figures 10.3a–d, which are the cross-sections along the vertical dotted lines denoted as "v" in Figures 10.1a–d, respectively.

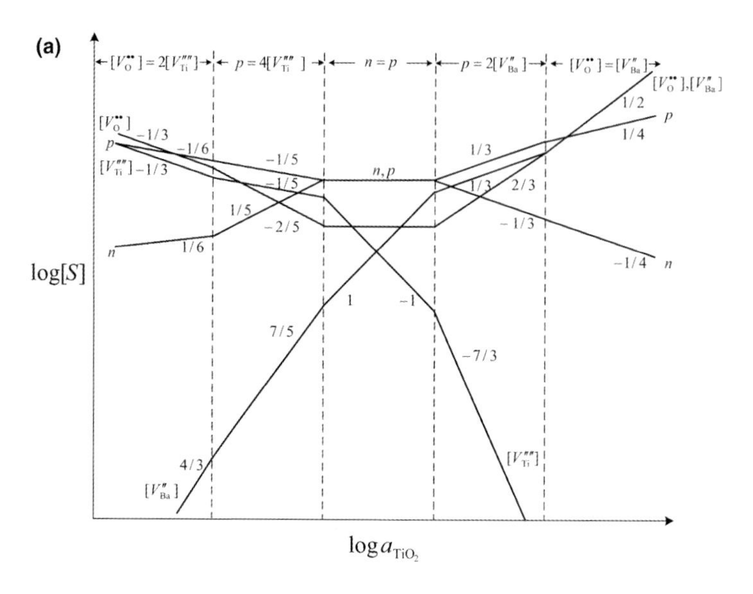

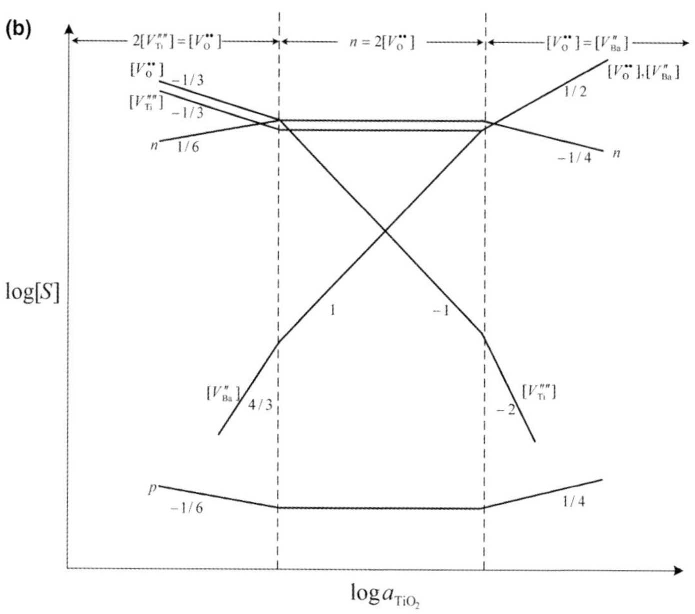

Figure 10.2 Log [S] versus log a_{TiO_2} (a–d), corresponding to the horizontal dotted lines designated as "h" (fixed a_{O_2}) in Figures 10.1a–d, respectively.

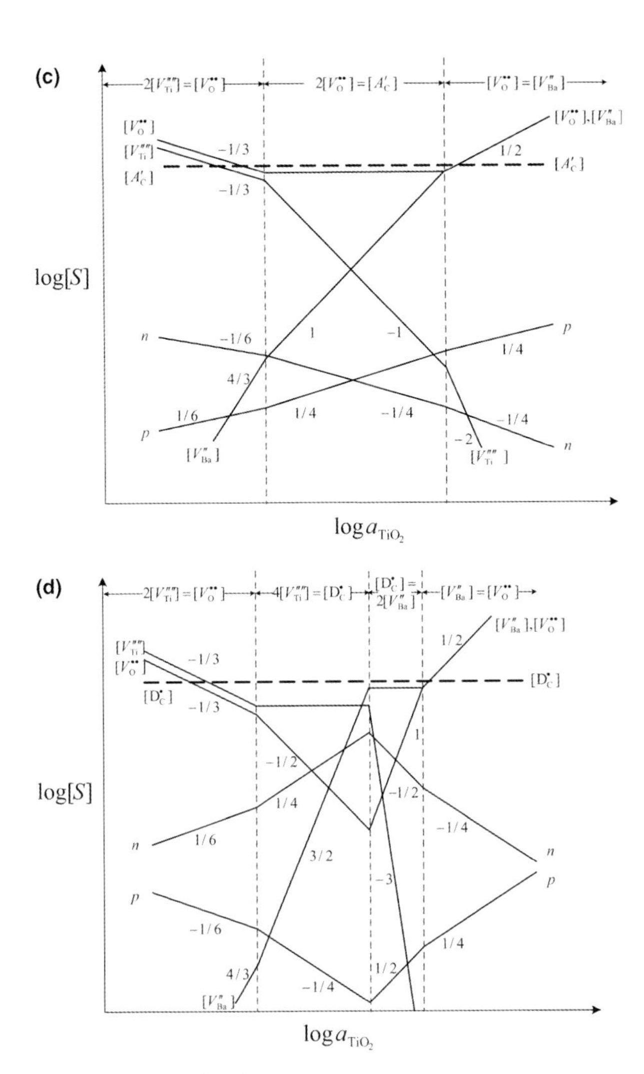

Figure 10.2 *(Continued)*

The reader is reminded that the representations, Figures 10.2 and 10.3, are for the systems in equilibrium internally [Eqs. (3) and (4)] as well as externally [Eqs. (5) and (6)]. In many cases, however, the external equilibria are often suspected for kinetic reasons. For $BaTiO_3$ and the like, for example, the chemical diffusion of oxygen is normally much faster than for the metallic components; hence, whilst the oxygen exchange equilibrium [Eq. (5)] can be readily achieved, the metallic component exchange equilibrium [Eq. (6)] can often be barely achieved. If this is the case, then a_{TiO_2} may not be held fixed while a_{O_2} varies. The extreme situation may be the case in which the system is closed with respect to the metallic components exchange, so that the nonmolecularity η, instead of a_{TiO_2}, remains fixed by the initial batch composition across the entire range of $\log a_{O_2}$. Depending largely on the temper-

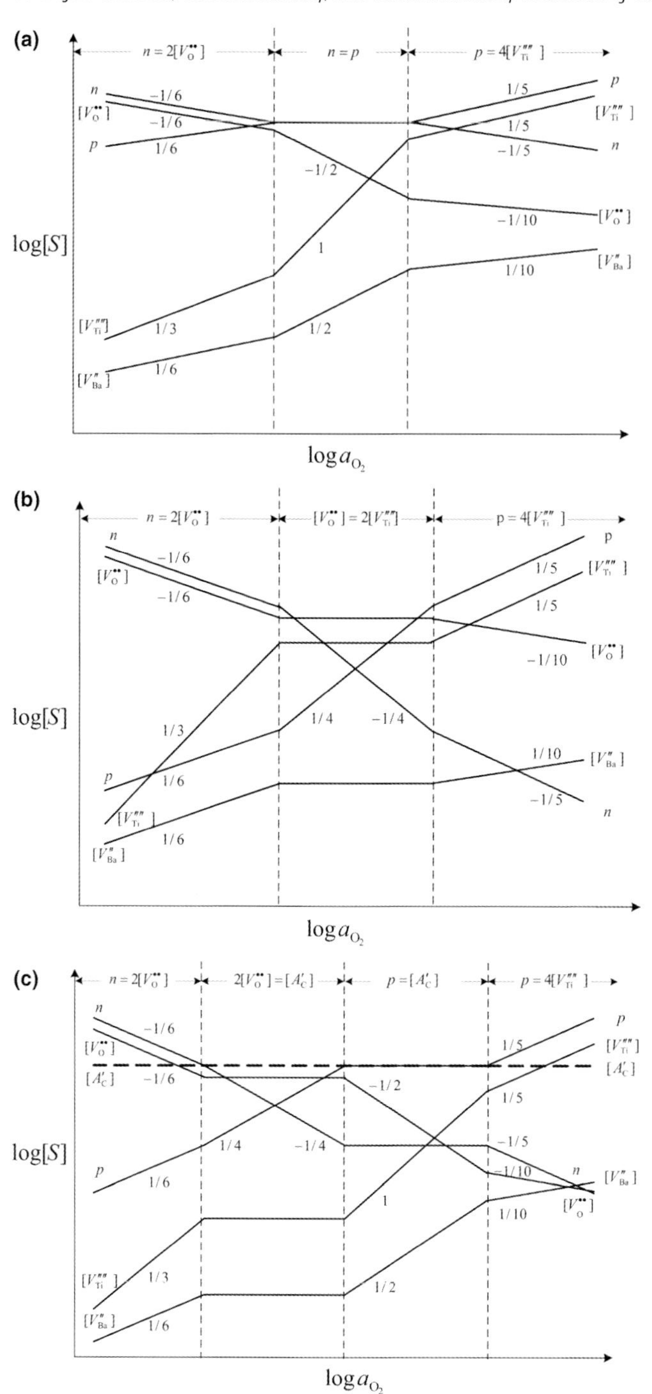

Figure 10.3 Log [S] versus log a_{O_2} (a–d), corresponding to the vertical dotted lines designated as "ν" (fixed a_{TiO_2}) in Figures 10.1a–d, respectively.

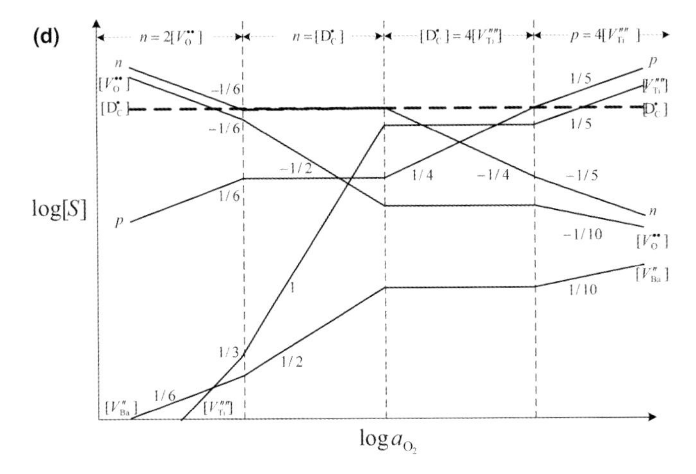

Figure 10.3 *(Continued)*

ature, this situation may be more realistic than the complete thermodynamic equilibrium. Thus, there is often a need to know how the defect structure varies against a_{O_2} while η, instead of a_{TiO_2}, is held constant. Wisdom [6] has already considered this problem. Now, log [S] versus log a_{O_2} at fixed a_{TiO_2} will be transformed (in Figure 10.3) to those at fixed η to determine how they appear.

To make the situation simpler, it can be assumed that the initial batch composition, whether doped or not, is with a_{TiO_2}-deficit ($\eta < 0$) such that:

$$-\beta\eta \approx [V_{Ti}^{''''}] \gg [V_{Ba}^{''}] \tag{26}$$

It is then possible to immediately dictate the variation of $-\eta$ versus a_{O_2} at fixed a_{TiO_2}, by following the variation of $[V_{Ti}^{''''}]$ in Figures 10.3a–d. It should be noted here that, even if a_{TiO_2} is held fixed, the nonmolecularity is changing with oxygen activity, which means that material must be exchanged with the surroundings in order to keep the system in the external equilibrium.

In each of the majority disorder regimes, it is possible to obtain from Eq. (13) for $[S] = [V_{Ti}^{''''}] \approx -\beta\eta$,

$$a_{TiO_2} = a_{O_2}^{-m/n}(-\beta\eta)^{1/n}\prod_j K_j^{-s/n} \tag{27}$$

Each defect concentration in Eq. (3) can then be transformed, by replacing a_{TiO_2} with Eq. (27), into a form

$$[S] = a_{O_2}^{p}(-\beta\eta)^{q}\prod_j K_j^{r} \quad \text{for} \quad S \neq V_{Ti}^{''''} \tag{28}$$

By combining these piecewise solutions in accord with the appropriate sequence of the majority disorder types it is possible, in principle, to transform Figure 10.3 to those for fixed η. The sequence of the majority disorder types for fixed η will remain the same as that for fixed a_{TiO_2}, due to Le Châtlier's principle:

$$\left(\frac{\partial(-\beta\eta)}{\partial \ln a_{TiO_2}}\right)_{T, \ln a_{O_2}} > 0 \tag{29}$$

Now, let us suppose that the fixed nonmolecularity is such that

$$K_i^{1/2}, K_S^{1/5} \ll -\beta\eta \ll [A'_{Ti}], [D^{\bullet}_{Ba}] \tag{30}$$

Then, for the pure undoped case in Figure 10.3a, the regime of $n = p$ cannot be seen and the sequence of majority disorder types subsequently turn the same as that in Figure 10.3b, whether $K_i^{1/2} \gg K_S^{1/5}$ or $K_i^{1/2} \ll K_S^{1/5}$. The latter is transformed as in Figure 10.4a, where it is noted that $[V_{Ti}^{''''}] \approx -\beta\eta$ is flat against log a_{O_2}. For the acceptor-doped case (Figure 10.3c), the majority disorder regime of $p = 4[V_{Ti}^{''''}]$ is impossible because $[V_{Ti}^{''''}] \approx -\beta\eta$ is fixed. Consequently, the transformed plot should be as in Figure 10.4b. For the donor-doped case of Figure 10.3d, the majority disorder regime $[D^{\bullet}_{Ba}] = 4[V_{Ti}^{''''}]$ cannot exist due to the assumption on η [Eq. (30)], and the regime $p = 4[V_{Ti}^{''''}]$ either for the same reason as in the acceptor-doped case. Figure 10.3d is, thus, transformed as in Figure 10.4c.

Actually, any shift of disorder regime with the component activities is only made possible by exchanging the chemical components, say O and Ti (or Ba), with the surrounding. Otherwise, the shift itself would be impossible. For example, when the donor-doped $BaTiO_3$ undergoes a shift from $[D^{\bullet}_{Ba}] \approx 4[V_{Ti}^{''''}]$ to $n \approx [D^{\bullet}_{Ba}]$ at a fixed a_{TiO_2} (see Figure 10.3d), the lattice molecule changes nominally from $Ba_{1-y}D_yTi_{1-y/4}O_3$ to $Ba_{1-y}D_yTiO_3$; that is, η changes from $-y/4$ to 0, indicating that Ti is supplied from (or Ba is drained to) the surroundings. This is why there are limitations on the availability of the majority disorder regimes for the partially closed cases (η = constant) in Figure 10.4b and c. Advantage may be taken of this fact to tell whether the system is closed with respect to metallic component exchange. If a donor-doped $BaTiO_3$ exhibits an a_{O_2}-region where $n \propto a_{O_2}^{-1/4}$ is in agreement with Figure 10.3d, for example, then it can be said that the system remains open to keep a_{TiO_2}, rather than η, constant, indicating the presence of a second phase [11].

Finally, there may be interest in the variation of a_{TiO_2} with a_{O_2} while η is held fixed. This can be calculated via Eq. (27) or Eq. (6), or more easily in the present case, as

$$a_{TiO_2} = K'_T[-\beta\eta]^{-1}[V_O^{\bullet\bullet}]^{-2} \tag{31}$$

due to the external equilibrium

$$Ti_{Ti}^{x} + 2O_O^{x} = V_{Ti}^{''''} + 2V_O^{\bullet\bullet} + TiO_2 \tag{32}$$

that is another aspect of (thus redundant with) Eq. (6) or $K_T K'_T = K_S$. Now, it is possible to draw in Figures 10.4a–c the variations of log a_{TiO_2}, simply by following that of $[V_O^{\bullet\bullet}]$, but with twice its slope in the opposite sense.

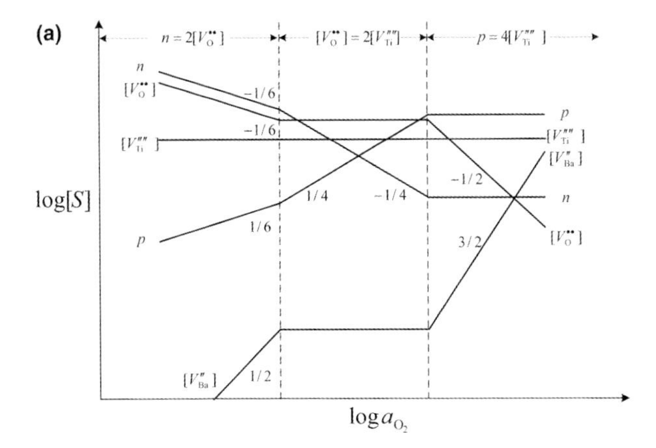

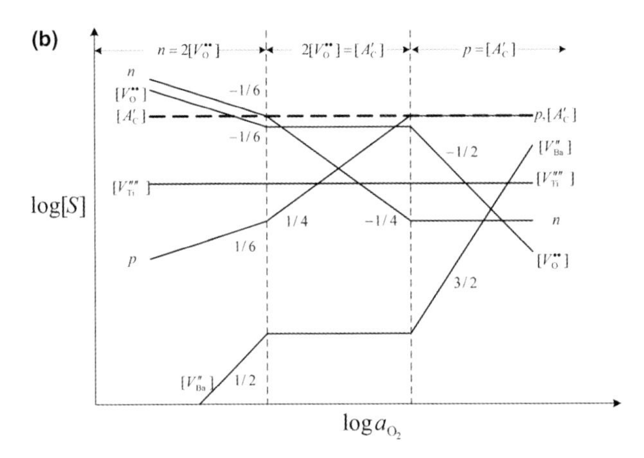

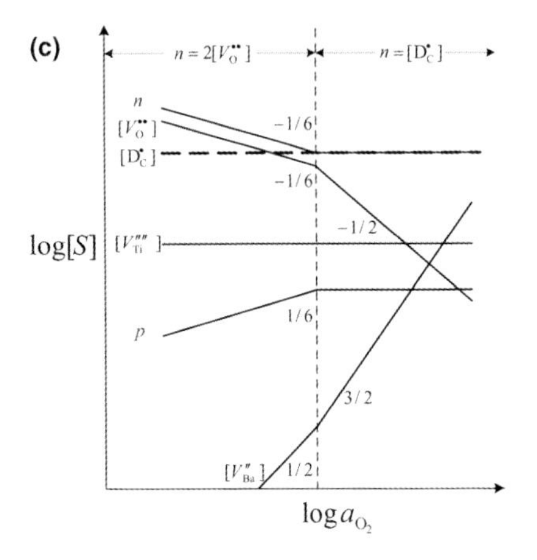

Figure 10.4 Log [S] versus log a_{O_2} at fixed nonmolecularity η, (a–c) corresponding to Figures 10.3b–d, respectively.

10.2.5
A Further Complication: Hole Trapping

Until now, it has been considered that all donors or acceptors, whether intrinsic or extrinsic, are fully ionized. As the temperature is lowered, however, those centers particularly of extrinsic origin tend to trap electrons or holes. Once electronic carriers are fully trapped, the defect-sensitive properties are often altered appreciably, and at low temperatures in particular [7, 8]. This trapping effect may be utilized ingeniously when designing the composition of actual material for special functions.

For example, when considering hole trapping by fixed-valent acceptor impurities (e.g., A'_{Ti}), when hole trapping is no longer negligible – and hence, the concentration of the trapped holes (e.g., $[A^{x}_{C}]$) is of concern, then one more equilibrium condition can be added to Eqs. (3)–(6). This is the internal equilibrium condition with respect to hole trapping or ionization equilibrium condition of the acceptors,

$$A^{x}_{C} = A'_{C} + h^{\bullet}; \qquad K_a = \frac{[A'_{C}]p}{[A^{x}_{C}]} \tag{33}$$

The mass conservation constraint, Eq. (17) is accordingly modified as

$$[A'_{C}] + [A^{x}_{C}] = [A]_t, \tag{34}$$

The charge neutrality equation, however, remains the same or as Eq. (18), because the trapped holes make the acceptor centers only neutral (A^{x}_{C}) in the present case. Otherwise, it would have to be modified to include the acceptors with different effective charges due to hole trapping.

It is possible to define the trapping factor χ^{-1} as

$$\chi^{-1} \equiv \frac{[A^{x}_{C}]}{p} = \frac{[A'_{C}]}{K_a} \tag{35}$$

Two extreme cases may be distinguished. When $\chi^{-1} \ll 1$, the concentration hierarchy will be such that

$$[A'_{C}] \approx [A]_t > p \gg [A^{x}_{C}] \tag{36}$$

and the defect structure is essentially the same as in Figure 10.3c or Figure 10.4b, only with $[A^{x}_{Ti}]$ added as a minority disorder if preferred. On the other hand, when $\chi^{-1} \gg 1$, most of holes are trapped and the concentration hierarchy in the near-stoichiometry regime ($2[V^{\bullet\bullet}_{O}] \approx [A'_{C}]$) turns to

$$[A'_{C}] \approx [A]_t > [A^{x}_{C}] \gg p \tag{37}$$

As a_{O_2} increases further, the trapped centers come to overwhelm the acceptor impurities, or

$$[A^{x}_{C}] \approx [A]_t \gg [A'_{C}] > p \tag{38}$$

As a_{O_2} increases even further, the charged disorders in majority will finally become $p = [A'_{C}]$. The defect structure is as shown in Figure 10.5, in which the portion

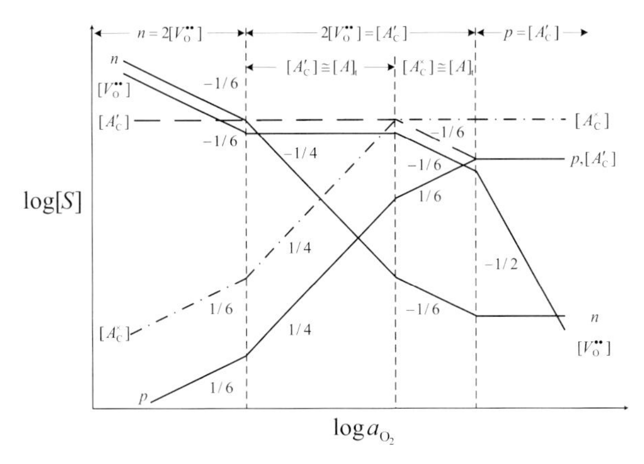

Figure 10.5 Defect structure for the acceptor-doped case (corresponding to Figure 10.3c or Figure 10.4b), with holes trapped fully ($\chi^{-1} \gg 1$).

corresponding to the concentration hierarchy of Eq. (37) is essentially the same as that given by Waser [7] for $SrTiO_3$.

10.2.6
Defect Structure and Reality

Experimental studies on defect structure of complex oxides are very much limited compared to those on binary systems. It is, thus, fair to say that the present understanding of the defect structure of the complex oxides is still far from complete. For example, although $BaTiO_3$ is one of the most studied systems, it is still not known whether donor impurities, such as $La_{Ba}^{\bullet}$, are compensated by $V_{Ti}^{''''}$ or $V_{Ba}^{''}$ or both, in the ion-compensation regime. Furthermore, it has been believed from a structural point of view that Ba is more mobile than Ti, although the results of a more recent study [9] have shown that Sr substituting Ba and Zr substituting Ti are all comparatively mobile at elevated temperatures, leaving the defect structure even more puzzling.

Information on the defect structure has been obtained usually from the observations of defect-structure-sensitive properties against the component activities, say a_{TiO_2}, and a_{O_2}, in particular. The simplest and most straightforward properties may be electrical conductivity and (self- or impurity-) tracer diffusivities among others, because they are directly proportional to the concentrations of relevant defects in the majority. In the case of the system of $BaTiO_3$, for example, a multitude of studies has been conducted on the electrical conductivity against a_{O_2} (see, e.g., Ref. [8]), but never against a_{TiO_2}; however, the latter may be experimentally extremely difficult to control, if not impossible. Cation tracer diffusion studies against oxygen activity have recently commenced [9], although even for the conductivity studies it is not always clear whether a_{TiO_2} or η is held fixed during the measurement. The system under examination is, therefore, most likely ill-defined thermodynamically in the strictest sense.

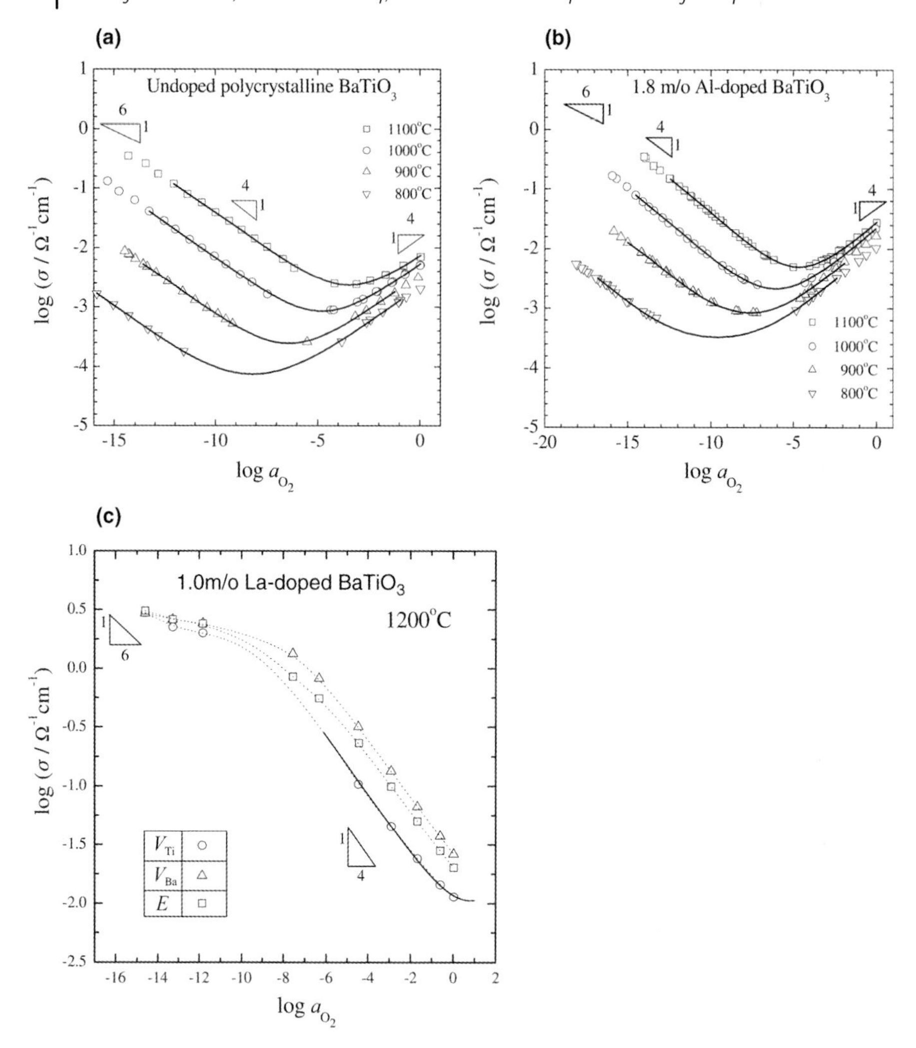

Figure 10.6 Equilibrium total electrical conductivity (σ) versus log a_{O_2} of $BaTiO_3$. (a) Undoped, polycrystalline $BaTiO_3$ [10]; (b) Acceptor (1.8 mol% Al)-doped single crystalline $BaTiO_3$ [28]; (c) Donor (1.0 mol% La)-doped polycrystalline $BaTiO_3$ [11].

For the system of $BaTiO_3$, the equilibrium conductivity has been the most extensively documented (see, e.g., Refs. [8, 10, 11]) over the experimentally viable range of $-20 < \log a_{O_2} \leq 0$ at elevated temperatures, although the remaining variable, a_{TiO_2} or η has not been explicitly specified. Figure 10.6a–c shows the typical results for the pure [10, 12], acceptor (Al'_{Ti})-doped [13], and donor ($La^{\bullet}_{Ba}$)-doped cases [11], respectively.

The conductivity is mostly attributed to electronic carriers, thus proportional to n or p, depending on which is in the majority against a_{O_2} and hence, $\sigma_{el} \propto a_{O_2}^{m}$ due to

Eq. (13) or (28) in each disorder regime. It has always been found for the undoped and acceptor-doped cases alike that, over the entire range of oxygen activity examined, the oxygen exponent "m" takes a value close to $-1/6$, $-1/4$, and $+1/4$ in sequence with increasing a_{O_2} [8, 10]. For the acceptor-doped case, this sequential variation is no doubt in agreement with the sequential shift of the majority disorder types from $n \approx 2[V_O^{\bullet\bullet}]$ to $2[V_O^{\bullet\bullet}] \approx [Al'_{Ti}]$ with increasing a_{O_2}, as shown in Figure 10.3c or Figure 10.4b. For the undoped case, however, the interpretation of the same m-sequence is less straightforward because, depending on a_{TiO_2} or η, there may be, next to the $n \approx 2[V_O^{\bullet\bullet}]$ regime where $n \propto a_{O_2}^{-1/6}$, two possible ionic disorder regimes, $[V_O^{\bullet\bullet}] \approx [V''_{Ba}]$ or $[V_O^{\bullet\bullet}] \approx 2[V''''_{Ti}]$ where $n \propto a_{O_2}^{-1/4}$ and $p \propto a_{O_2}^{+1/4}$ (see Figure 10.1b and Table 10.2 or Figure 10.3b). It is not yet clear for the system of "pure" $BaTiO_3$, whether the experimental finding of $\sigma_{el} \propto a_{O_2}^{\pm 1/4}$ is due to $[V_O^{\bullet\bullet}] \approx [V''_{Ba}]$ or $[V_O^{\bullet\bullet}] \approx 2[V''''_{Ti}]$, or even due to the background impurity acceptors [10]. This issue may be elucidated by observing the conductivity variations on pure specimens with different nonmolecularity η. However, the experiment may prove to be problematic because of the extremely limited range of η of the single-phase $BaTiO_3$ [14, 15]. Furthermore, none of the exclusively p-type regimes, ($h^\bullet$, A'_C), ($h^\bullet$, V''_{Ba}) and ($h^\bullet$, V''''_{Ti}), has ever revealed itself up to $a_{O_2} = 1$ for the system of $BaTiO_3$.

For the donor-doped case, on the other hand, the oxygen exponent of the conductivity takes values $m \approx -1/6$, 0, $-1/4$ in sequence as a_{O_2} increases up to 1 [8] (see Figure 10.6c). This sequence reflects the shift of the majority disorder types from $n \approx 2[V_O^{\bullet\bullet}]$ ($m = -1/6$) to $n \approx [La_{Ba}^\bullet]$ ($m = 0$), and finally to either $[La_{Ba}^\bullet] \approx 2[V''_{Ba}]$ or $[La_{Ba}^\bullet] \approx 4[V''''_{Ti}]$ ($m = -1/4$) (see Figure 10.1d and Table 10.2 for m-values). There seems to be no doubt about the majority disorder types for the former two regimes ($m = -1/6$ and 0), but the last, ionic compensation regime is not clear with regards to whether $[La_{Ba}^\bullet] \approx 2[V''_{Ba}]$ or $[La_{Ba}^\bullet] \approx 4[V''''_{Ti}]$. An attempt was once made to elucidate this issue, when Yoo *et al.* [11] measured the equilibrium conductivity on the three donor-doped specimens with nominal compositions $Ba_{1-y}La_yTi_{1-y/4}O_3$ (designated as V_{Ti}), $Ba_{1-3y/2}La_yTiO_3$ (as V_{Ba}) and $Ba_{1-y}La_yTiO_3$ (as E), all with $y = 0.01$ for the purpose of sorting out one out of V''''_{Ti}, V''_{Ba} and e' that can compensate the impurity donors $La_{Ba}^\bullet$ in an air atmosphere. The results are as shown in Figure 10.6c. There found to be a minority second phase for all three cases, while the p-type conductivity increased up to 16% of the total electronic conductivity at $a_{O_2} = 1$. Consequently, no conclusion could be drawn as to the identity of the ionic disorder compensating the doped donor impurities, but it has been concluded that a_{TiO_2}, rather than η, is held fixed (as in Figure 10.3d). This was because the region where $n \propto a_{O_2}^{-1/4}$ and $p \propto a_{O_2}^{+1/4}$ is observed at a high oxygen activity region that otherwise would not be seen (see Figure 10.4c).

For the quantitative, defect-chemical analyses of these conductivities in Figure 10.6, the reader is referred to Refs [10, 11, 13, 16]. It should be mentioned in passing that, for the case of Al-doped $BaTiO_3$ (Figure 10.6b), practically all of the holes were trapped by Al-acceptors at the measurement temperatures [17]. The responsible defect structure is, therefore, as given in Figure 10.5, and the p-type conductivity ($m = 1/4$) in Figure 10.6b is essentially due to the trapped holes [17].

10.3
Oxygen Nonstoichiometry

Equations (15), (20) and (25) indicate that, whether the system is undoped or doped, its oxygen nonstoichiometry is a measure of the concentrations of the electronic charge carriers, n and p:

$$\beta\delta = \frac{1}{2}(n - p) \tag{39}$$

Figure 10.7 shows how the nonstoichiometry alters the trend of defect-sensitive properties, such as the electronic electrical conductivity, thermoelectric power, and oxygen chemical diffusivity [10]. Thus, the control of oxygen nonstoichiometry during processing is very often crucial to ensure the required properties of an oxide, or functions of the device thereof.

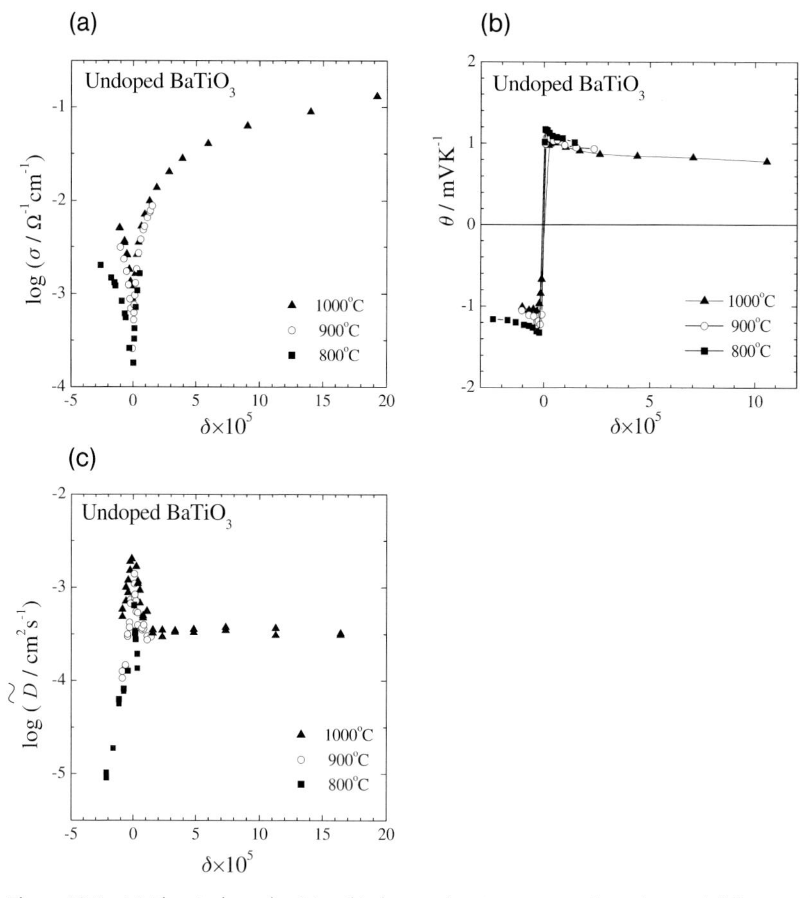

Figure 10.7 (a) Electrical conductivity, (b) thermoelectric power, and (c) chemical diffusivity of undoped $BaTiO_{3-\delta}$ versus oxygen nonstoichiometry, δ. Note that the stoichiometric composition is the demarcation point between n-type and p-type behavior. Data from Ref. [10].

As might be expected from Figure 10.7, in either the far oxygen-deficit ($\delta \gg 0$) or oxygen excess region ($\delta \ll 0$), the nonstoichiometry variation is rather trivial because $n \gg p$ or $n \ll p$. In the near stoichiometry region ($\delta \approx 0$), on the other hand, the nonstoichiometry variation is less trivial because there are both electrons and holes. If electronic carriers are trapped by, for example, doped impurities, then the nonstoichiometry variation is even further informative. The most general case will be considered here, namely, nonstoichiometry in the near-stoichiometry region of the acceptor-doped $BaTiO_3$ with the holes trapped.

10.3.1
Nonstoichiometry in General

For the acceptor-doped $BaTiO_3$ with the lattice molecular formula of Eq. (16), the nonstoichiometry has been given as Eq. (20) or

$$\beta\delta = [V_O^{\bullet\bullet}] - [V_{Ba}''] - 2[V_{Ti}''''] - \frac{1}{2}[A_{Ti}]_t \tag{40}$$

When the concentration of trapped holes is no longer negligible compared to that of the free holes, it is possible to rewrite Eq. (40), due to the mass conservation constraint, Eq. (34), and the charge neutrality constraint, Eq. (18), as:

$$\beta\delta = \frac{1}{2}(n - p - [A_{Ti}^{x}]) = \frac{1}{2}[n - p(1 + \chi^{-1})] \tag{41}$$

where χ^{-1} is the trapping factor defined as the concentration ratio of trapped holes to free holes in Eq. (35). It should be noted that the stoichiometric composition $\delta = 0$ falls, in general, at $n = p(1 + \chi^{-1})$, not at $n = p$. Only when $\chi^{-1} \to 0$, would the latter be the case.

The solutions for the defects in Eq. (41) are given in the majority disorder regime of $2[V_O^{\bullet\bullet}] = [A_C']$ (see Table 10.2) as:

$$p = \chi[A_C^{x}] \tag{42}$$

$$n = K_i \chi^{-1} [A_C^{x}]^{-1} \tag{43}$$

$$[A_C^{x}] \approx 2^{-1/2} [A]_t^{3/2} K_{Ox}'^{1/2} a_{O_2}^{1/4} \tag{44}$$

Equation (41) then takes the form [17]:

$$\beta\delta = -\sqrt{K_i'} \sinh\left(\frac{1}{4} \ln \frac{a_{O_2}}{a_{O_2}^{o}}\right) \tag{45}$$

where

$$K_i' \equiv K_i(1 + \chi^{-1}) \tag{46}$$

$$a_{O_2}^{o} = 4(1 + \chi^{-1})^{-2} K_i^{-2} K_{Re}^{2} [A]_t^{-2} \tag{47}$$

The latter $a^o_{O_2}$ is the oxygen activity corresponding to $\delta = 0$ or $n = p(1+\chi^{-1})$ (see Figure 10.5). At this specific oxygen activity $a^o_{O_2}$, the variation $\beta\delta$ as a function of $\ln a_{O_2}$ exhibits an inflection or

$$\left(\frac{\partial^2(\beta\delta)}{\partial\,(\ln a^2_{O_2})^2}\right)_T = 0 \tag{48}$$

This fact actually provides a means to locate the stoichiometric point on a nonstoichiometry isotherm that is measured [18].

Equation (46), in association with Eq. (3), indicates that $K_i' = n(p + [A_C^x])$; thus, the latter may be termed the "pseudo-equilibrium constant for intrinsic electronic excitation." It is noted that only when $\chi^{-1} \to 0$, $K_i' \to K_i$ and Eq. (45) takes the conventional, familiar form:

$$\beta\delta = -\sqrt{K_i}\,\sinh\left(\frac{1}{4}\ln\frac{a_{O_2}}{a^o_{O_2}}\right) \tag{49}$$

This may be the case if the system is pure enough and/or temperature is high enough [see Eq. (35)]. Attempts have often been made to determine K_i from the nonstoichiometry isotherms via Eq. (49), even for a system which bears possible trapping centers (e.g., acceptor impurities), whether they are intentionally doped, or not. In any case, it is necessary to be aware that what is actually being determined is not the true K_i, but rather a K_i' in general that is dependent on the type and concentration of the trap centers. One way of evaluating K_i may be to determine K_i' as a function of $[A]_t$ via Eq. (45) at a fixed temperature or $K_i' = K_i(1 + [A]_t/K_a)$, due to Eqs (8), (35), and (36). The limiting value may then be taken as $[A]_t \to 0$ for the true K_i. However, this approach has not yet been implemented experimentally.

Once the nonstoichiometry has been measured as a function of oxygen activity at different temperatures, the partial molar enthalpy of component oxygen relative to gas oxygen at the standard state, $\Delta\bar{H}_O(= \bar{H}_O - H^o_{O_2}/2)$, will become of some interest. By noting that the relative partial molar Gibbs free energy of the component oxygen $\Delta\bar{G}_O = \mu_O - \mu^o_{O_2}/2 = RT\ln a^{1/2}_{O_2}$ in Eq. (45), it is possible to obtain, via the Gibbs–Helmholtz equation:

$$\Delta\bar{H}_O(\delta) = \frac{1}{2}\left[\frac{\partial \ln a_{O_2}}{\partial(1/RT)}\right]_\delta = \Delta\bar{H}_O(\delta = 0) - \frac{\beta\delta}{\sqrt{(\beta\delta)^2 + K_i'}}\Delta H_i' \tag{50}$$

with

$$\Delta H_i' \equiv -\frac{\partial \ln K_i'}{\partial(1/RT)} = \Delta H_i - \frac{\partial \ln(1+\chi^{-1})}{\partial(1/RT)} \tag{51}$$

where $\Delta\bar{H}_O(\delta = 0)$ denotes the relative partial molar enthalpy of component oxygen at the stoichiometric composition ($\delta = 0$). It should be noted that as $\chi^{-1} \to 0$, then Eq. (50) will converge to that of the pure case corresponding to Eq. (49). The variation of $\Delta\bar{H}_O(\delta)$ against δ is illustrated graphically in Figure 10.8 for $\chi^{-1} \to 0$ (no trapping) and for $\chi^{-1} \to \infty$ (full trapping), respectively.

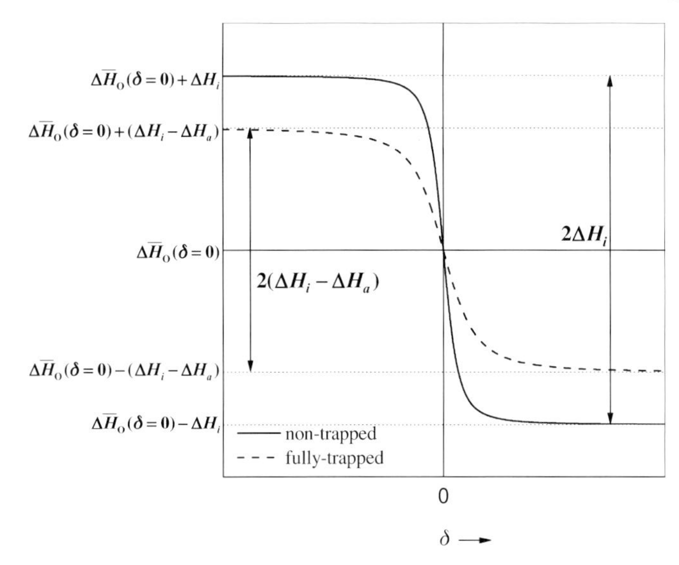

Figure 10.8 Variation of the relative partial molar enthalpy of component oxygen versus oxygen nonstoichiometry for nontrapped case ($\chi^{-1}\to 0$; solid curve) and fully trapped case ($\chi^{-1}\to\infty$; dashed curve).

Here, it can be seen that $\Delta\overline{H}_O(\delta)$ are bounded by $\Delta\overline{H}_O(\delta=0)+\Delta H_i'$ and $\Delta\overline{H}_O(\delta=0)-\Delta H_i'$ as $\delta \ll -\sqrt{K_i'}/\beta$ and $\delta \gg \sqrt{K_i'}/\beta$, respectively, and vary anti-symmetrically crossing $\delta=0$.

The sigmoid variation of $\Delta\overline{H}_O(\delta)$ crossing $\delta=0$ in Figure 10.8 may be understood from a slightly different viewpoint [19]. For simplicity's sake, consider the pure $BaTiO_3$ with no traps. Then, oxidation or oxygen incorporation reaction may take place simultaneously in two ways: (i) by producing the free holes; and (ii) by consuming free electrons or

$$\frac{1}{2}O_2(g)+V_O^{\bullet\bullet}=O_O^{\times}+2h^{\bullet} \tag{52}$$

$$\frac{1}{2}O_2(g)+V_O^{\bullet\bullet}+2e'=O_O^{\times} \tag{53}$$

Whilst these are indistinguishable thermodynamically, they are distinguishable defect-chemically. Allowing K_p and K_n to denote the equilibrium constant for these reactions, respectively, they are:

$$K_p=\frac{p^2}{[V_O^{\bullet\bullet}]a_{O_2}^{1/2}};\qquad K_n=\frac{1}{[V_O^{\bullet\bullet}]n^2a_{O_2}^{1/2}} \tag{54}$$

Obviously, $K_n=K_{Re}^{-1}=K_pK_i^{-2}$ [see Eqs. (3) and (5)], and hence, the associated enthalpy changes ΔH_p and ΔH_n are interrelated as:

$$\Delta H_n=-\Delta H_{Re}=\Delta H_p-2\Delta H_i \tag{55}$$

Given that the oxidation proceeds in this way, $\Delta\bar{H}_O$ may be taken as a fractional sum of ΔH_p and ΔH_n such that:

$$\Delta\bar{H}_O = \frac{n}{n+p}\Delta H_n + \frac{p}{n+p}\Delta H_p \tag{56}$$

that takes the form, after some algebra using Eqs. (3) and (39),

$$\Delta\bar{H}_O = \frac{1}{2}(\Delta H_p + \Delta H_n) - \frac{\beta\delta}{\sqrt{(\beta\delta)^2 + K_i}} \cdot \frac{1}{2}(\Delta H_p - \Delta H_n) \tag{57}$$

This is essentially the same, due to Eq. (55), as Eq. (50) with $\chi^{-1} \rightarrow 0$. It is then possible immediately to identify $\Delta\bar{H}_O(\delta = 0)$ as:

$$\Delta\bar{H}_O(\delta = 0) = \frac{1}{2}(\Delta H_p + \Delta H_n) \tag{58}$$

If the holes are fully trapped or $\chi^{-1} \rightarrow \infty$, on the other hand, the oxygen incorporation may proceed via Eq. (55) and via, instead of Eq. (54),

$$\frac{1}{2}O_2(g) + V_O^{\bullet\bullet} + 2A'_{Ti} = O_O^{\times} + 2A_{Ti}^{x} \tag{59}$$

with the associated enthalpy $\Delta H_p - 2\Delta H_a$ [see Eqs (14) and (33)]. In the same line as in Eq. (56), one can immediately obtain Eq. (50). The same reasoning can also be applied to the case in which electrons are trapped, instead of holes.

Equation (56) indicates that when an oxygen atom is incorporated into the lattice, it picks up holes [Eq. (52)] and electrons [Eq. (53)] at random, depending on their availability. It is, therefore, quite natural that the partial molar enthalpy of the component oxygen, $\Delta\bar{H}_O$ is dependent on the relative amount of electrons and holes or the oxygen nonstoichiometry.

The relative partial molar entropy of component oxygen, $\Delta\bar{S}_O$ can also be obtained as a function of δ, either by using the thermodynamic identity $[\partial\Delta\bar{G}_O/\partial T]_\delta = -\Delta\bar{S}_O$ or by the same argument as in Eq. (56),

$$\Delta\bar{S}_O = \frac{n}{n+p}\Delta S_n + \frac{p}{n+p}\Delta S_p, \tag{60}$$

but it is not pursued any further. (Those readers interested in further details should refer to Ref. [19].)

10.3.2
Experimental Reality

Nonstoichiometry is normally measured by using either thermogravimetry or coulombmetric titrometry. In the former technique, the oxygen activity in the surrounding of a specimen oxide is changed stepwise, and the corresponding weight change is monitored. In the latter technique, a predetermined amount of oxygen is incorporated in the form of ionic current, and changes in equilibrium oxygen activity of the surroundings are measured. The experimental details of both methods are provided in Refs. [19] and [20], respectively. When using these techniques, a sufficient precision should be ensured so as to determine the nonstoichiometry variation, particularly in the near-stoichiometry region, because the nonstoichiometry variation versus oxygen activity in this region is the smallest [see Eqs. (45) and/or (49)]. The experimental data acquired are usually scarce in the near-stoichiometry region, but this is most likely due to insufficient precision. The stoichiometric point, $a^o_{O_2}$, in Eqs. (45) and (49), that is normally determined as in Eq. (47), is therefore less precisely known for most of the oxides, whether simple or complex. The experimental results on the nonstoichiometry of undoped $BaTiO_3$ will be detailed at this point.

For the system of undoped $BaTiO_{3-\delta}$, there were five different data sets [21–25] against oxygen activity, all of which were limited to the range of $\log a_{O_2} < -7$ over the temperature range of 1000 to 1340 °C. Furthermore, only relative changes in nonstoichiometry, and not the absolute values, were given [21, 24]. Indeed, even when the absolute values were reported [22, 23, 25], the stoichiometric points could not be located due to the poor precision. It has been only recently [19] that nonstoichiometry has been measured with the highest ever precision in the near-stoichiometry region, including the stoichiometric point $a^o_{O_2}$. These results are shown in Figure 10.9.

The solid curves Figure 10.9 are best fitted to Eq. (45), with K'_i and $a^o_{O_2}$ as fitting parameters. As can be seen, Eq. (45) satisfactorily describes the nonstoichiometry, even though the best-fitted values for the fitting parameters are subjected to rather large uncertainties. The latter is again attributed to the still poor precision of the measurement, particularly in the vicinity of the stoichiometric point.

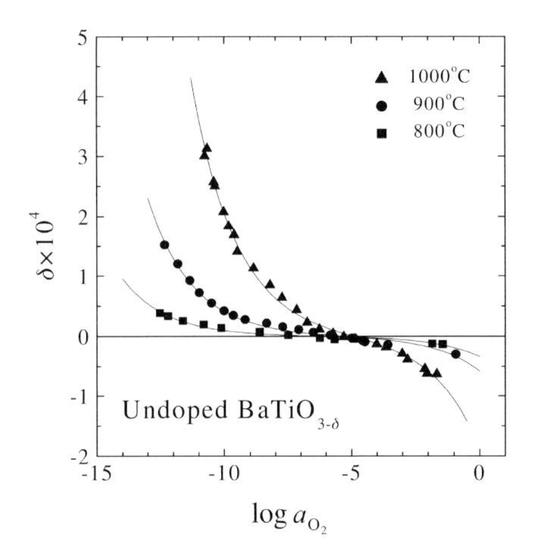

Figure 10.9 Oxygen nonstoichiometry versus oxygen activity at different temperatures, as measured on undoped $BaTiO_3$. The solid lines are best fitted to Eq. (45). Data from Ref. [19].

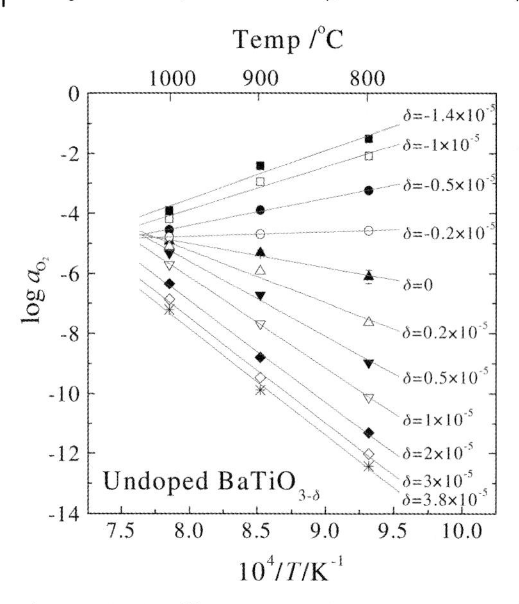

Figure 10.10 Equilibrium oxygen activity versus reciprocal temperature at different nonstoichiometry values; see Eq. (61). Data from Ref. [19].

The relative partial molar enthalpy of oxygen is evaluated at different nonstoichiometry values by using the thermodynamic identity

$$\frac{\Delta \overline{G}_{\mathrm{O}}}{RT} = \ln a_{\mathrm{O}_2} = \frac{2\Delta \overline{H}_{\mathrm{O}}}{RT} - \frac{2\Delta \overline{S}_{\mathrm{O}}}{R} \tag{61}$$

Figure 10.10 shows the plotted log a_{O_2} versus reciprocal temperature at different nonstoichiometries. As can be seen, this is generally linear for a fixed δ over the temperature range examined. The relative partial molar enthalpy and entropy of oxygen may then be evaluated from the slope and intercept, respectively. The results of $\Delta\overline{H}_{\mathrm{O}}$ are as shown in Figure 10.11. In agreement with Eq. (50), the partial molar enthalpy variation is best fitted as:

$$\Delta\overline{H}_{\mathrm{O}}(\delta)/\mathrm{kJ}\ mol^{-1} = -(76 \pm 5) - \frac{(272 \pm 38)\delta}{\sqrt{\delta^2 + (3.02 \pm 0.03) \times 10^{-11}}} \tag{62}$$

10.4
Nonstoichiometry Re-Equilibration

When the oxygen activity is changed in the surrounding of a binary oxide $AO_{1-\delta}$ that has previously been equilibrated with the surrounding, the nonstoichiometry of the oxide changes towards a new equilibrium value. The overall kinetics of this nonstoichiometry re-equilibration process typically consists of the surface reaction step

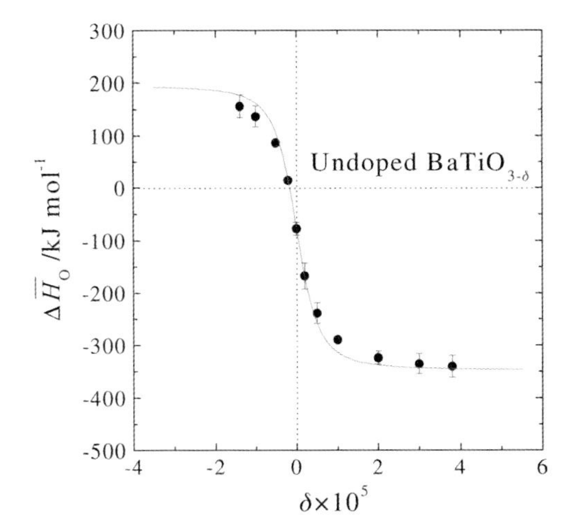

Figure 10.11 Relative partial molar enthalpy of component oxygen versus nonstoichiometry for undoped $BaTiO_3$. The solid curve is best fitted to Eq. (50). Data from Ref. [19].

and solid-state diffusion step, in series. The former, gas–solid reaction at the surface is usually regarded as a simple chemical reaction of the first order [26], whereas the latter diffusion refers to the chemical diffusion of the oxide. It is known that there should be one, only chemical, diffusion coefficient because there is only one composition variable for the binary oxide (although this is true only when the internal defect equilibrium prevails [34]).

For an infinite bar of $AO_{1-\delta}$ with the square cross-section of $2a \times 2a$, for example, the overall re-equilibration kinetics is normally described in terms of the two kinetic parameters – the surface reaction rate constant $\tilde{k}$, and chemical diffusivity $\tilde{D}$ as [12, 27]:

$$\frac{\bar{\delta}(t) - \delta(0)}{\delta(\infty) - \delta(0)} = 1 - \left[\sum_{n=1}^{\infty} \frac{2L^2 \exp\left(-\frac{\beta_n^2 \tilde{D} t}{a^2}\right)}{\beta_n^2(\beta_n^2 + L^2 + L)} \right]^2 \tag{63}$$

with β such that

$$\beta \tan \beta = L; \qquad L \equiv \frac{a\tilde{k}}{\tilde{D}} \tag{64}$$

Here, $\bar{\delta}(t)$, $\delta(0)$, and $\delta(\infty)$ are the mean (at time t), initial (at $t = 0$) and final (as $t \to \infty$) value of oxygen nonstoichiometry, respectively. By monitoring the temporal variation of the nonstoichiometry $\bar{\delta}(t)$ by either thermogravimetry or a δ-sensitive property (e.g., electrical conductivity), it is possible to determine the two kinetic parameters. With regards to binary systems, it is believed that the relaxation kinetics may be well understood. Chemical diffusion, in particular, has long been understood in the light of chemical diffusion theory [28], or in the light of the ambipolar diffusion theory [29].

With regards to the complex oxides, however, it is fair to say that the nonstoichiometry relaxation kinetics and chemical diffusion are, as yet, less well understood. In fact, it is only recently that the kinetics has been examined in a systematic manner [12, 13, 16, 30]. Here, the current understanding of the kinetics will be presented, and in particular the chemical diffusion process first for the undoped or acceptor-doped case, and then for the donor-doped case, in the order of complexity of the relaxation kinetics.

10.4.1 Undoped or Acceptor-Doped $BaTiO_3$

10.4.1.1 Relaxation Behavior and Chemical Diffusion

It has already been pointed out that the undoped and acceptor-doped $BaTiO_3$ have essentially the same defect structure (see Figures 10.3c and/or 10.4b), in the oxygen partial pressure range that is experimentally viable in practice, namely $-18 < \log a_{O_2} \leq 0$. Consequently, the equilibrium conductivities also have the same trend with oxygen activity, as shown in Figures 10.6a and b. Likewise, the relaxation kinetics appears also to be essentially the same [13, 16].

When the oxygen partial pressure in the surrounding of the system oxide is abruptly changed, either in its *n*-type branch of oxygen activity ($\sigma_{el} \propto n \propto a_{O_2}^{-1/4}$) or in the *p*-type branch ($\sigma_{el} \propto p \propto a_{O_2}^{1/4}$) (see Figure 10.6a or b), its electrical conductivity, as a direct measure of the oxygen nonstoichiometry ($\delta \approx n/2$ or $-p/2$, respectively), will relax in typical fashion, as shown in Figures 10.12(a) and (b).

These nonstoichiometry relaxations are satisfactorily described by Eq. (63), as depicted by the solid curves in Figure 10.12. The two kinetic parameters may subsequently be evaluated, as shown in Figure 10.13.

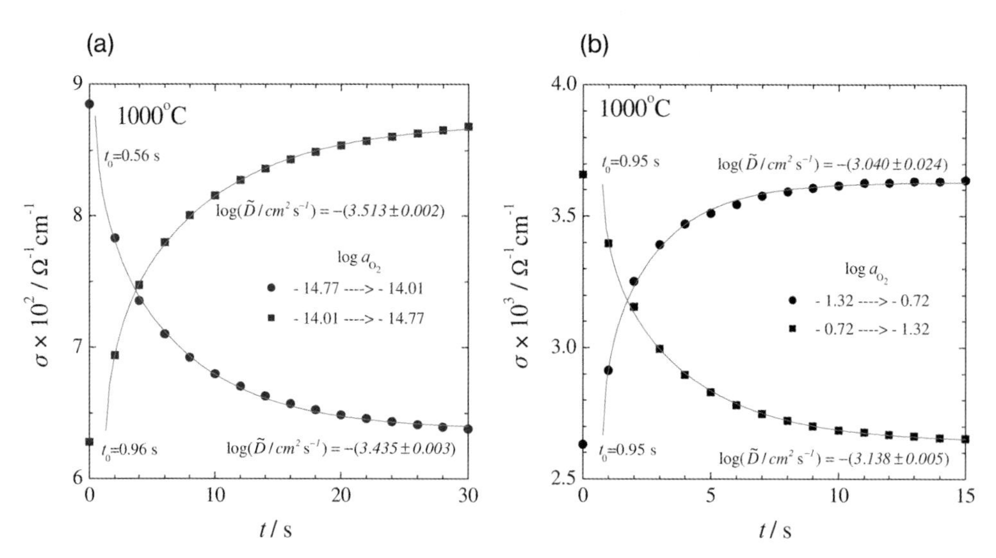

Figure 10.12 Typical nonstoichiometry relaxation of undoped $BaTiO_3$ in its *n*-type branch of oxygen activity (a) and p-type branch of oxygen activity (b) during reduction and oxidation at 1000 °C. The solid lines are best fitted to Eq. (63). Data from Ref. [12].

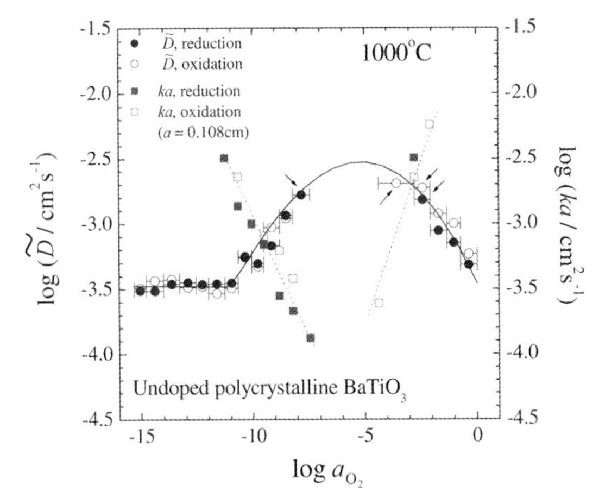

Figure 10.13 Chemical diffusivity ($\tilde{D}$) and surface-reaction-rate constant ($\tilde{k}$) versus oxygen activity of undoped $BaTiO_3$ at 1000 °C. The dotted and solid lines are for visual guidance only. Data from Ref. [12].

Upon comparison with the corresponding conductivity in Figure 10.6a or b, it can be seen that in the n–p mixed regime of oxygen activity – that is, in the near vicinity of the conductivity minimum in Figure 10.6a or b – the overall kinetics is governed mostly by the surface reaction step and otherwise, is controlled by the diffusion step. If the surface reaction step is totally ignored, or it is taken simply that $L \gg 1$ in Eq. (63), as was often practiced, then despite a somewhat larger uncertainty the diffusion coefficient obtained would be as shown in Figure 10.14.

Finally, the true chemical diffusivities that are responsible for the diffusion step in the nonstoichiometry relaxation kinetics have emerged to be as shown in

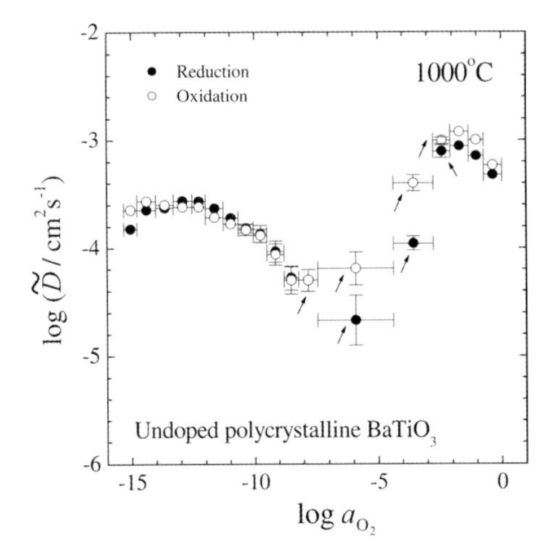

Figure 10.14 Apparent chemical diffusivity versus oxygen activity which would be obtained if the surface reaction step were to be ignored. Data from Ref. [12].

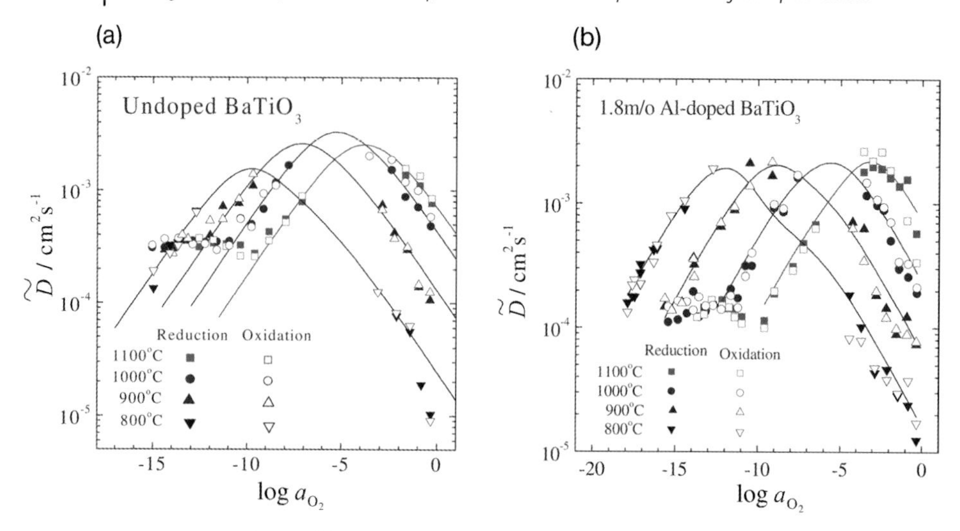

Figure 10.15 Chemical diffusivity isotherms for the (a) undoped and (b) acceptor (1.8 mol% Al)-doped $BaTiO_3$. Data from Refs [13] and [16].

Figures 10.15a and b for the undoped $BaTiO_3$ that is polycrystalline, and for the acceptor (Al)-doped $BaTiO_3$ that is single crystalline, respectively [12, 13]. It should be noted that the diffusivity takes on value up to the order of $10^{-3}\,cm^2\,s^{-1}$, depending on the oxygen activity and irrespective of the temperature in the range examined, and exhibits a maximum in the middle of the oxygen activity range examined. It may be of concern that the diffusivity for the polycrystal $BaTiO_3$ (Figure 10.15a) may have been subjected to the influence of grain boundaries and hence, the diffusivity would be somewhat enhanced. However, this seems not to be the case, because the results with polycrystals are quite similar to those for single crystals (Figure 10.15b), in both trend and magnitude.

10.4.1.2 **Defect-Chemical Interpretation**

When referring to the defect structures in Figure 10.3c or 10.4b and the corresponding conductivity isotherms in Figures 10.6a and b, for the undoped and acceptor-doped $BaTiO_{3-\delta}$, respectively, the defects in the majority must be $V_O^{\bullet\bullet}$ compensated by the acceptors A'_C, whether they are extrinsic (i.e., Al'_{Ti}) or not. Thus, oxide ions (O^{2-}) and electrons (e^-) may be taken as the most mobile charged components in this near-stoichiometry regime.

According to the chemical diffusion theory of Wagner [28], the chemical diffusion coefficient of component oxygen is given as:

$$\tilde{D}_O = \frac{RT}{8F^2}\sigma_{ion}t_{el}\left|\frac{\partial[V_O^{\bullet\bullet}]}{\partial \ln a_{O_2}}\right|^{-1} \tag{65}$$

where σ_{ion} denotes the partial ionic conductivity, t_{el} the electronic transference number, and F is the Faraday constant. The factor within the absolute-value signs is referred to the "thermodynamic factor."

When referring to the defect structure in the near-stoichiometry regime, $2[V_O^{\bullet\bullet}] \approx [A'_C]$ in Figure 10.3c or 10.4b, the ionic conductivity is due to $V_O^{\bullet\bullet}$, and essentially is independent of oxygen activity. The variation of the total conductivity in Figure 10.6a and b is, therefore, attributed to that of the partial electronic conductivity, σ_{el} that is due to electrons (e′) and holes, whether trapped (A_{Ti}^{x}) or not ($h^{\bullet}$) [17]. By using the solutions for n, p and $[Al_{Ti}^{x}]$ in Eq. (13) and Table 10.2, the total conductivities in Figures 10.6a and b can be written as [16, 17]:

$$\sigma = \sigma_{el} + \sigma_{ion} = \sigma_{el,m} \cosh\left[\frac{1}{4}\ln\frac{a_{O_2}}{a^*_{O_2}}\right] + \sigma_{ion} \tag{66}$$

where $\sigma_{el,m}$ denotes the minimum electronic conductivity that falls at the oxygen activity $a^*_{O_2}$ (see Figures 10.6a and b). The solid lines are best fitted to this equation. By using the values for $\sigma_{el,m}$, $a^*_{O_2}$ and σ_{ion} that are evaluated from σ as the fitting parameters, the electronic transference number t_{el} is calculated, as shown in Figure 10.16 for the case of, for example, the acceptor-doped $BaTiO_3$.

The thermodynamic factor can also be calculated from the defect structure. The oxygen nonstoichiometry is in general given as in Eq. (45) or Eq. (49). Differentiation leads to the thermodynamic factor as

$$\left|\frac{\partial[V_O^{\bullet\bullet}]}{\partial \ln a_{O_2}}\right|^{-1} = \frac{8}{n + p(1+\chi^{-1})} = \frac{4}{\sqrt{K'_i}\cosh\left(\frac{1}{4}\ln\frac{a_{O_2}}{a^o_{O_2}}\right)} \tag{67}$$

This is nothing but the inverse of the slope of the nonstoichiometry isotherm, as in Figure 10.9. If the system is pure enough, then it is possible to set $\chi^{-1} = 0$, although it has been actually found [17] that $\chi^{-1} \ll 1$ for the undoped, and $\chi^{-1} \gg 1$ for the Al-doped case.

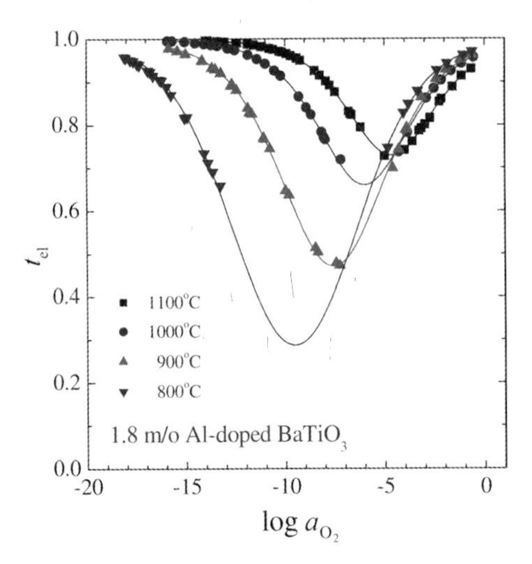

Figure 10.16 Electronic transference number for 1.8 mol% Al-doped $BaTiO_3$, as calculated from the total conductivity in Figure 10.6b. The solid lines are calculated from Eq. (66). Data from Ref. [13].

Finally, by substituting Eq. (67) into Eq. (65), one obtains

$$\tilde{D}_O = \frac{\tilde{D}_O^o t_{el}}{\cosh\left(\frac{1}{4}\ln\frac{a_{O_2}}{a_{O_2}^o}\right)} \tag{68}$$

with the factor that is independent of the oxygen activity

$$\tilde{D}_O^o = \frac{RT\sigma_{ion}}{2F^2\sqrt{K_i'}} \tag{69}$$

The solid lines in Figure 10.15 are best fitted to Eq. (68) by using the experimental values for σ_{ion} and t_{el} [13, 16]. As can be seen, Eq. (68) precisely explains the variation of the chemical diffusivity.

10.4.2 Donor-Doped $BaTiO_3$

10.4.2.1 Relaxation Behavior and Chemical Diffusion

The nonstoichiometry relaxation of the undoped or acceptor-doped $BaTiO_3$ is simple, being governed by the chemical diffusion of oxygen and the gas–solid oxygen exchange reaction. Whilst this is no different from ordinary binary oxides, it is no longer the case for the donor-doped case. In fact, a longstanding and notorious problem of $BaTiO_3$ (as well as of any other perovskite oxide) is that the kinetics of oxygen nonstoichiometry re-equilibration is usually very (perhaps even prohibitively) sluggish for donor-doped $BaTiO_3$, compared to undoped or acceptor-doped counterparts. As a consequence, it is difficult to ascertain whether a donor-doped specimen has been completely equilibrated upon re-equilibration. According to one report [31], it took more than four months to equilibrate a donor(Nb)-doped $BaTiO_3$ specimen measuring $4 \times 4 \times 12\,mm^3$, with an average grain size of a few microns, upon a change of temperature from 1308 to 1216 K in an atmosphere of fixed oxygen partial pressure of 0.0022 atm. If undoped or acceptor-doped, this process would have taken no more than a few hundreds of seconds (see Figure 10.12). This unusually sluggish kinetics may be termed the "kinetic anomaly" of donor-doped $BaTiO_3$. Not surprisingly, chemical diffusivity data are extremely sparse for the donor-doped case, with only two data sets having been identified: one from Wernicke [32] on La-doped $BaTiO_3$, and the other by Nowotny and Rekas [31] on Nb-doped $BaTiO_3$. In both cases the studies were conducted against temperature in a fixed Po_2 atmosphere (0.32 and 0.0022 atm, respectively).

The results of a recent study [30] showed that, compared to the usual relaxation behavior of the undoped counterpart (see Figure 10.12), the relaxation behavior of donor-doped $BaTiO_3$ appeared somewhat unusual, and depended on the oxygen activity.

It should be noted that in Figure 10.6c, the conductivity for the specimen denoted as V_{Ti} is the equilibrium conductivity of La-doped $BaTiO_3$ (nominal composition $Ba_{0.99}La_{0.01}Ti_{0.9975}O_3$), that has an average grain size of $0.86 \pm 0.03\,\mu m$, and a bulk

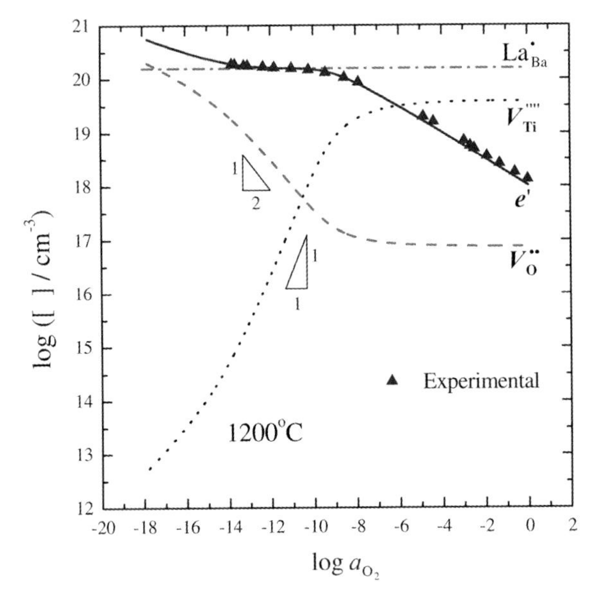

Figure 10.17 Equilibrium defect structure of 1 mol% La-doped $BaTiO_3$ at 1200 °C, as derived from the conductivity isotherm. Note that with decreasing oxygen activity, the majority disorder type shifts from $[La^{\bullet}_{Ba}] \approx 4[V''''_{Ti}]$ to $[La^{\bullet}_{Ba}] \approx n$ to $n \approx 2[V^{\bullet\bullet}_{O}]$. From Ref. [30].

density that was 96% of the theoretical value. The corresponding defect structure is basically as shown in Figure 10.3d, and in more detail for the system in present concern in Figure 10.17. Here, as the Po_2 decreases from 1 atm at the given temperature, the majority type of disorder shifts from $[La^{\bullet}_{Ba}] \approx 4V''''_{Ti}$, through $[La^{\bullet}_{Ba}] \approx n$, to $n \approx 2[V^{\bullet\bullet}_{O}]$.

Figures 10.18a–f show, sequentially, the as-measured conductivity relaxation curves for stepwise changes back (reduction) and forth (oxidation) between the pre-fixed oxygen activities specified, as the (geometric) mean oxygen activity decreases. Clearly, the relaxation behaviors differ depending on the oxygen activity windows imposed, or the mean oxygen activity. As the latter decreases from $\log a_{O_2} = 0$, the conductivity relaxes apparently with one relaxation time (Figures 10.18a and b), with two relaxation times (Figures 10.18c–e) and again with one relaxation time (Figures 10.18f). In other words, the kinetics varies from onefold to twofold to onefold with decreasing a_{O_2}. The onefold and twofold kinetics can be more clearly recognized from a plot of the relaxation curves $\sigma(t)$ against $\log t$, instead of t (see Figures 10.19). It is noted in twofold kinetics cases (Figures 10.18c–e) that the two relaxation times differ by orders of magnitude.

10.4.2.2 **Defect-Chemical Interpretation**

The twofold relaxation is understood as follows. The nonmolecularity η as well as the oxygen nonstoichiometry δ of the system is presumed, initially, to be spatially homogeneous. Then, as soon as a different oxygen activity is imposed upon such

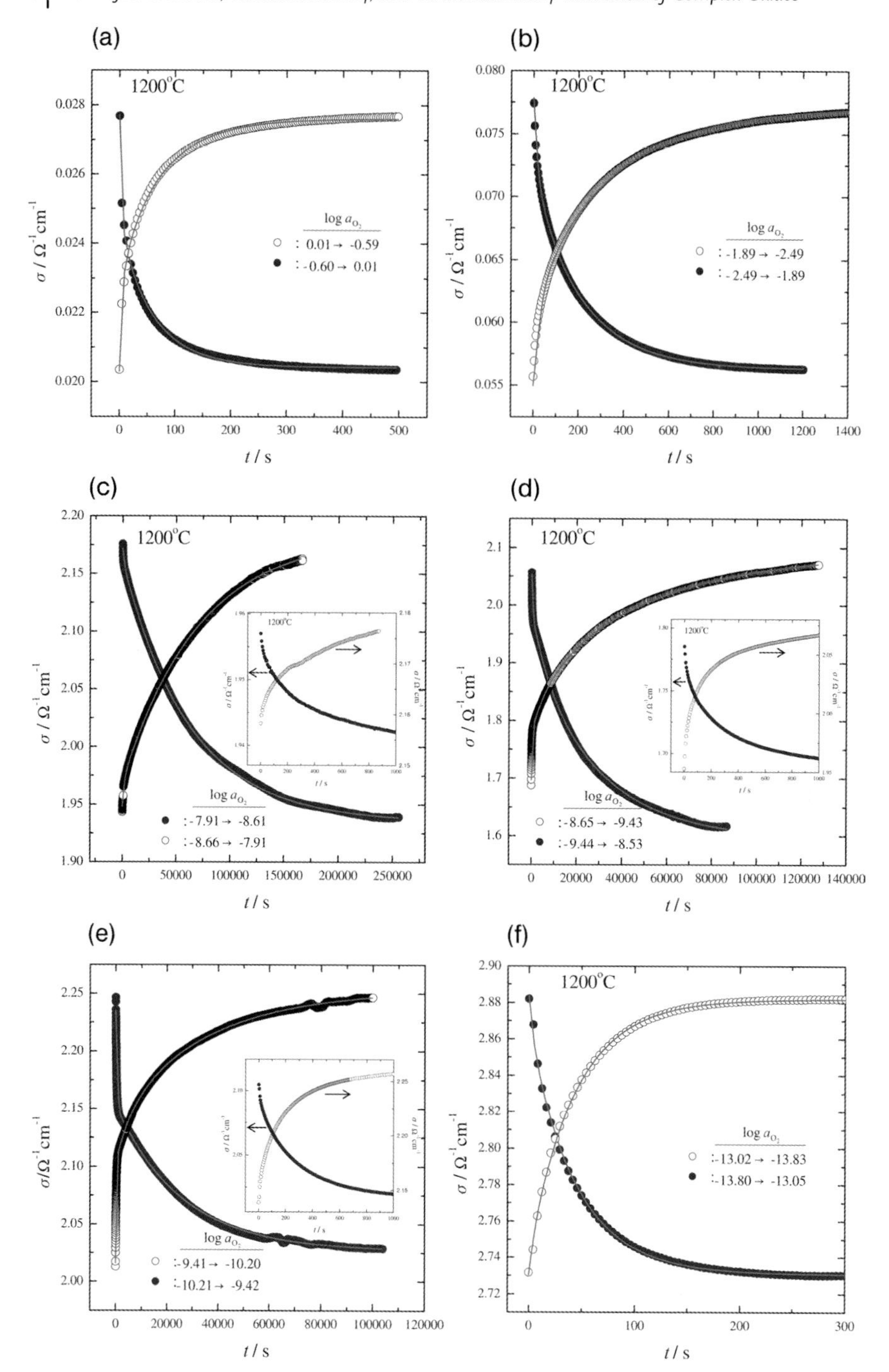

(a)
1200°C
log a_{O_2}
○ : 0.01 → -0.59
● : -0.60 → 0.01
σ / Ω⁻¹cm⁻¹
t / s
(b)
1200°C
log a_{O_2}
○ : -1.89 → -2.49
● : -2.49 → -1.89
σ / Ω⁻¹cm⁻¹
t / s
(c)
1200°C
log a_{O_2}
● : -7.91 → -8.61
○ : -8.66 → -7.91
σ / Ω⁻¹cm⁻¹
t / s
(d)
1200°C
log a_{O_2}
○ : -8.65 → -9.43
● : -9.44 → -8.53
σ / Ω⁻¹cm⁻¹
t / s
(e)
log a_{O_2}
○ : -9.41 → -10.20
● : -10.21 → -9.42
σ/Ω⁻¹cm⁻¹
t / s
(f)
1200°C
log a_{O_2}
○ : -13.02 → -13.83
● : -13.80 → -13.05
σ / Ω⁻¹cm⁻¹
t / s

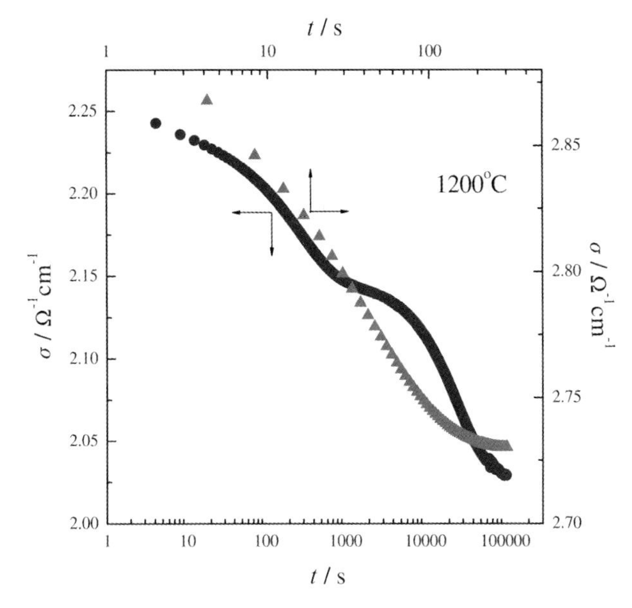

Figure 10.19 Conductivity versus time in a log scale. Note that the onefold and twofold kinetics can be clearly discerned, and if $\tau_{Ti} \gg \tau_{O}$, the twofold kinetics may turn like the onefold kinetics. Data from Ref. [30].

an homogeneous system, a new redox equilibrium will be immediately established at the surface of the specimen via the redox reaction of Eq. (5), assuming that the surface reaction is fast enough (of course, this is not always the case [26]), and proceeds inwards via the chemical diffusion of component oxygen, as is the case with the undoped or acceptor-doped $BaTiO_3$.

It can be further assumed that there also prevails initially an external equilibrium with respect to Ti-exchange (or Ba) between the system and a minority second phase [as in Eq. (6)]. The presence of a second phase has earlier been indicated upon shifting of the majority disorder types in the donor-doped case [11] (see Section 10.2). Such a second phase – however small it is – holds a_{TiO_2} fixed such that, when the oxygen activity is suddenly changed to a smaller value, a_{TiO_2} is fixed by the presence of the second phase. As a consequence, due to the equilibrium condition Eqs (4) and (6) (or $[V_{Ti}''''][V_O^{\bullet\bullet}]^2 = K_T a_{TiO_2}$), a gradient of $[V_{Ti}'''']$ is established at the surface, where the surface reaction is again assumed to be sufficiently fast. It is this gradient which

Figure 10.18 Conductivity relaxation for the stepwise changes back and forth between the oxygen activities (in log a_{O_2}): (a) $0.01 \leftrightarrow -0.59$; (b) $-1.89 \leftrightarrow -2.49$; (c) $-7.91 \rightarrow -8.61$ and $-8.66 \rightarrow -7.91$; (d) $-8.65 \rightarrow -9.43$ and $-9.44 \rightarrow -8.53$; (e) $-9.41 \rightarrow -10.20$ and $-10.21 \rightarrow -9.42$; (f) $-13.02 \rightarrow -13.83$ and $-13.80 \rightarrow -13.05$. Insets: exploded views of the first and faster relaxation. Note the change of kinetics from onefold (a,b) to twofold (c–e) to onefold (f), as the mean oxygen activity decreases. Solid curves are best fitted to Eq. (63), (73) or (74). Data from Ref. [30].

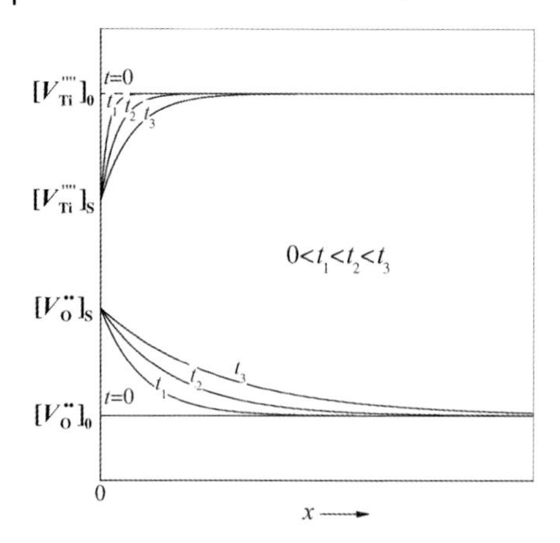

Figure 10.20 Thermodynamic situation and expected diffusion profiles of $[V_O^{\bullet\bullet}]$ and $[V_{Ti}^{''''}]$ with time upon an abrupt change of oxygen activity at the surface of a specimen $x = 0$. Note that the surface is assumed to be always in equilibrium, so that $[V_{Ti}^{''''}][V_O^{\bullet\bullet}]^2 = \text{const.}$ and $\tilde{D}_{Ti} \ll \tilde{D}_O$. Data from Ref. [30].

subsequently drives the chemical diffusion of, say, component Ti. This thermodynamic situation is depicted in Figure 10.20 [30].

It is well known that charge neutrality may break down at grain boundaries and surfaces. What is observed in the relaxation of Figure 10.18, however, is a spatial average property of the specimen, not a local property, and hence the space charge effect may be neglected (i.e., the Debye length is assumed to be sufficiently small compared to the grain size.) The overall charge neutrality condition may then be written for the present case [see Eq. (23)] as:

$$n = [La_{Ba}^{\bullet}] + 2[V_O^{\bullet\bullet}] - 4[V_{Ti}^{''''}] \tag{70}$$

Neglecting the mobility of the donor impurities $La_{Ba}^{\bullet}$, temporal variation of the electronic carrier density n may be written as:

$$\frac{\partial n}{\partial t} = 2\frac{\partial [V_O^{\bullet\bullet}]}{\partial t} - 4\frac{\partial [V_{Ti}^{''''}]}{\partial t} = -2\nabla J_{V_O^{\bullet\bullet}} + 4\nabla J_{V_{Ti}^{''''}} \tag{71}$$

As the present system is essentially an electronic conductor (electronic transference number, $t_{el} \approx 1$) over the entire a_{O_2} range of the present concern, as is clearly seen from Figure 10.17, the local electroneutrality field will be negligible. Hence, the diffusion of each type of ionic defect proceeds in the respective sublattices with no mutual coupling or

$$J_{V_O^{\bullet\bullet}} = -\tilde{D}_O \nabla [V_O^{\bullet\bullet}]; \qquad J_{V_{Ti}^{''''}} = -\tilde{D}_{Ti} \nabla [V_{Ti}^{''''}] \tag{72}$$

where $\tilde{D}_O$ and $\tilde{D}_{Ti}$ are the chemical diffusivity of component O and Ti, respectively.

Assuming these chemical diffusivities to be constant for small jumps of oxygen activity, it is possible to obtain the solution to Eq. (71) associated with Eq. (72) for the initial and boundary conditions, which are given in Figure 10.20 as:

$$\frac{\bar{\sigma}-\sigma_\infty}{\sigma_o-\sigma_\infty}=\frac{\bar{n}-n_\infty}{n_o-n_\infty}=\sum_{k=\mathrm{O,Ti}} A_k \frac{8}{\pi^2}\sum_{j=0}^{\infty}\frac{1}{(2j+1)^2}\exp\left[-\frac{(2j+1)^2\pi^2\tilde{D}_k t}{4a^2}\right] \quad (73)$$

with

$$A_\mathrm{O} \equiv \frac{2([V_\mathrm{O}^{\bullet\bullet}]_o-[V_\mathrm{O}^{\bullet\bullet}]_\infty)}{2([V_\mathrm{O}^{\bullet\bullet}]_o-[V_\mathrm{O}^{\bullet\bullet}]_\infty)-4([V_\mathrm{Ti}^{''''}]_o-[V_\mathrm{Ti}^{''''}]_\infty)}; \quad (74)$$

$$A_\mathrm{Ti} \equiv \frac{-4([V_\mathrm{Ti}^{''''}]_o-[V_\mathrm{Ti}^{''''}]_\infty)}{2([V_\mathrm{O}^{\bullet\bullet}]_o-[V_\mathrm{O}^{\bullet\bullet}]_\infty)-4([V_\mathrm{Ti}^{''''}]_o-[V_\mathrm{Ti}^{''''}]_\infty)} \quad (75)$$

where $\bar{\sigma}$, σ_o and σ_∞ are the mean conductivity at time t, the initial equilibrium conductivity at $t=0$, and the final equilibrium conductivity as $t\rightarrow\infty$, respectively, $\bar{n}$, n_o and n_∞ are the corresponding densities of the carrier electrons, and "$2a$" is the thickness of the present specimen that may be regarded as an infinite slab in the present case.

It is noted that $A_\mathrm{O}+A_\mathrm{Ti}=1$ and their ratio $R(=A_\mathrm{Ti}/A_\mathrm{O})$ may be regarded as a measure of the relative contribution of Ti-diffusion and O-diffusion processes to the overall re-equilibration kinetics, given the relaxation times ($\tau_k=4a^2/\pi^2\tilde{D}_k$): If $R\gg 1$ or $R\ll 1$, the conductivity relaxation kinetics may be represented essentially by either τ_Ti or τ_O, respectively; that is, onefold kinetics follows. Otherwise, the kinetics may have to be represented by both τ_Ti and τ_O; that is, twofold kinetics.

The equilibrium defect structure for the present case is calculated to be as shown in Figure 10.17 [30]. In oxidizing atmospheres (say, log $a_{\mathrm{O}_2}>-8$), where the majority type of disorder is $4[V_\mathrm{Ti}^{''''}]\approx[\mathrm{La}_\mathrm{Ba}^\bullet]\gg[V_\mathrm{O}^{\bullet\bullet}]$, $R=|\Delta[V_\mathrm{Ti}^{''''}]/\Delta[V_\mathrm{O}^{\bullet\bullet}]|=2[V_\mathrm{Ti}^{''''}]/[V_\mathrm{O}^{\bullet\bullet}](\approx 2[\mathrm{La}_\mathrm{Ba}^\bullet]/[V_\mathrm{O}^{\bullet\bullet}])\gg 1$ due to Eqs. (4) and (6) (or $[V_\mathrm{Ti}^{''''}][V_\mathrm{O}^{\bullet\bullet}]^2=K_T a_{\mathrm{TiO}_2}$); in reducing atmospheres (say, $\log a_{\mathrm{O}_2}<-13$) where $n\approx 2[V_\mathrm{O}^{\bullet\bullet}]\gg[V_\mathrm{Ti}^{''''}]$, $R=2[V_\mathrm{Ti}^{''''}]/[V_\mathrm{O}^{\bullet\bullet}]\ll 1$. In these $P\mathrm{o}_2$ regimes, the conductivity relaxation kinetics should appear to be onefold or with a single relaxation time, namely, τ_Ti and τ_O, respectively. In the intermediate oxygen activity region, then, the overall kinetics may appear twofold. Thus, the kinetics will have to shift from onefold (τ_Ti) to twofold (τ_Ti,τ_O) to onefold (τ_O) as the oxygen activity decreases from $a_{\mathrm{O}_2}=1$.

This is believed to be what has been observed in Figure 10.18. If it is the case, then the relaxation data may be fitted to Eq. (73) to evaluate the chemical diffusivities. In the onefold kinetics region of τ_Ti, all of the relaxation data could be sufficiently precisely fitted to Eq. (73) with $A_\mathrm{O}\approx 0$, as depicted by the solid curves in Figures 10.18a and b, and thus indicating that the rate of surface reactions, Eq. (63) and (64), are indeed fast enough compared with the chemical diffusion.

However, in the onefold kinetics region of τ_O in the reducing atmospheres, the surface reaction step must be taken into appropriate account, as in the undoped or acceptor-doped cases [12, 13, 26], for a sufficient precision of the fitting. The relaxations are, thus, fitted (solid lines in Figures 10.18a–e) to the conventional solution similar to Eq. (63), but in one dimension to evaluate $\tilde{D}_\mathrm{O}$ and the surface reaction rate constant $\tilde{k}$ simultaneously.

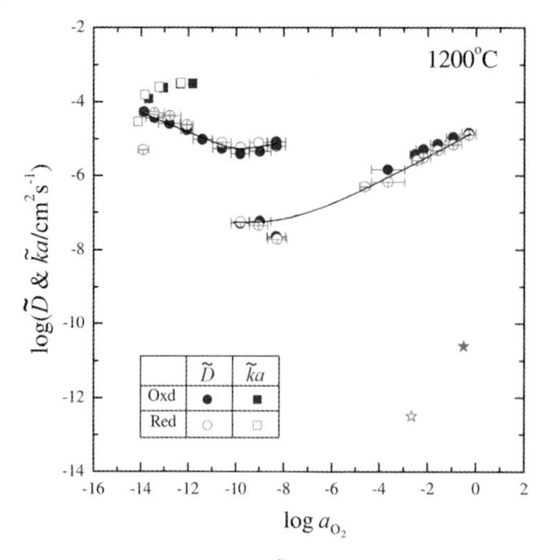

Figure 10.21 $\tilde{D}_{Ti}$, $\tilde{D}_O$, and $\tilde{k}a$(for oxygen exchange) versus oxygen activity at 1200 °C, as evaluated by using the specimen dimension as the decisive diffusion length. ★, $\tilde{D}_{Ba}$ from Ref. [32]; ☆, $\tilde{D}_{Ti}$ from Ref. [31].

All of the results, $\tilde{D}_{Ti}$, $\tilde{D}_O$ and $\tilde{k}a$ (for oxygen exchange), as obtained in this way are compiled in Figure 10.21. As can be seen, the $\tilde{D}_O$-values are on the order of magnitude of $-6 \sim -4$ (in cm^2 s^{-1}), depending on oxygen activity. Upon comparison with the $\tilde{D}_O$-values for the undoped and acceptor-doped in Figure 10.15, it can be recognized that the present $\tilde{D}_O$ values are quite reasonable as the chemical diffusivity of oxygen in $BaTiO_3$. In addition, the surface reaction rate constants "$\tilde{k}$" are also comparable, in both magnitude and trend, with those for the undoped or acceptor-doped systems [26] (see Figure 10.13).

Here, however, it should be noted that the values for $\tilde{D}_{Ti}$, being in the range of $10^{-8} \sim 10^{-5}$ cm^2 s^{-1}, seem too large in comparison with the two reported values for the chemical diffusivity of the cations in $BaTiO_3$ [31, 32] (see Figure 10.21). Remembering that, in the evaluation of $\tilde{D}_{Ti}$ via Eq. (73), the specimen thickness $2a$ (= 1.2 mm) has been employed as the decisive diffusion length, and indicates that the decisive diffusion length should rather be the grain size than the specimen thickness.

By simply assuming that the specimen consists of spherical grains with a mean radius α, Eq. (73) is modified to be [27]:

$$\frac{\bar{\sigma} - \sigma_\infty}{\sigma_o - \sigma_\infty} = A_O \frac{8}{\pi^2} \sum_{j=0}^{\infty} \frac{1}{(2j+1)^2} \exp\left[-\frac{(2j+1)^2 \pi^2 \tilde{D}_O t}{4a^2}\right] + A_{Ti} \frac{6}{\pi^2} \sum_{j=1}^{\infty} \frac{1}{j^2} \exp\left[-\frac{j^2 \pi^2 \tilde{D}_{Ti} t}{\alpha^2}\right] \tag{76}$$

On re-analyzing all of the experimental data for the twofold kinetics region and higher oxygen partial pressure region to this modified solution (solid lines in Figures 10.18a–e), the results are obtained as shown in Figure 10.22, along with all other literature-based data [31–33].

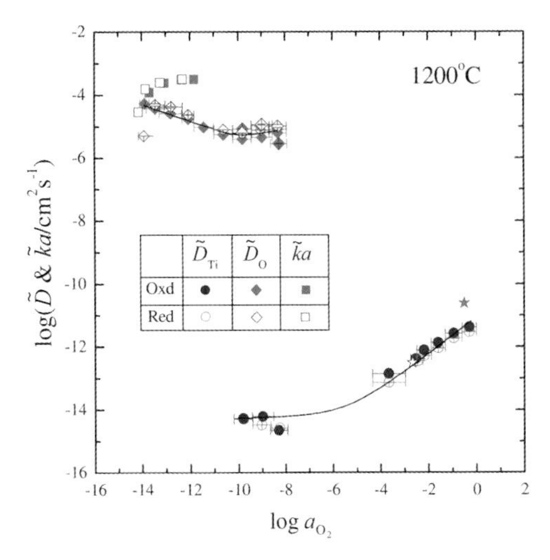

Figure 10.22 Chemical diffusivities re-evaluated by taking as the decisive diffusion length the sample dimension for $\tilde{D}_O$ and the mean grain size for $\tilde{D}_{Ti}$. ★, $\tilde{D}_{Ba}$ from Ref. [32]; ☆, $\tilde{D}_{Ti}$ from Ref. [31].

As can be seen, the re-evaluated values are now in a satisfactory agreement with the reported chemical diffusivity values [31, 32]. It may be concluded, therefore, that the twofold kinetics is due to the chemical diffusion of oxygen and cation (say Ti), the chemical diffusivities of which differ by about eight orders of magnitude from each other.

In summary, what happens in the donor-doped $BaTiO_3$ is that the oxygen sublattice has first been equilibrated with a decisive diffusion length of the overall specimen dimension, and the cation sublattice subsequently begins to re-equilibrate from the grain surfaces; that is, with the grain size as the decisive diffusion length. Whether or not the twofold kinetics is observed, seems to be determined by the combination of the relaxation times: If they are not much different (e.g., let $\tau_O \approx \tau_{Ti}$ or, in other words, $2a/\alpha \approx \sqrt{\tilde{D}_O/\tilde{D}_{Ti}}$), then it would not be possible to observe or discern the twofold kinetics, irrespective of the amplitude ratio R $(= A_{Ti}/A_O)$ in Eq. (74). For the specimen in question, $2a/\alpha \approx 10^3$ and $\sqrt{\tilde{D}_O/\tilde{D}_{Ti}} \approx 10^4$, thus, is observable [30]. If $\tau_O \ll \tau_{Ti}$, the twofold kinetics should be observed in principle, although no matter what the value of R, it would take too long to observe or there would be minimal symptoms of the cation sublattice relaxation in the experimentally viable time. If $R \ll 1$, on the other hand, the cation sublattice relaxation can hardly be detected experimentally, even if the relaxation does occur. This may be the case of undoped or acceptor-doped $BaTiO_3$ where the kinetics only appears onefold and very fast.

10.4.3
Defect Diffusivities

Again according to Wagner [28], the chemical diffusivities in Eq. (72) may be written for an electronic conductor system ($t_{el} \approx 1$) as:

$$\tilde{D}_{\mathrm{O}} = -D_{V_{\mathrm{O}}}\left(\frac{\partial\mu_{\mathrm{O}}}{RT\,\partial\ln[V_{\mathrm{O}}^{\bullet\bullet}]}\right) = -\frac{D_{V_{\mathrm{O}}}}{2}\left(\frac{\partial\ln a_{\mathrm{O}_2}}{\partial\ln[V_{\mathrm{O}}^{\bullet\bullet}]}\right) \tag{77}$$

and

$$\tilde{D}_{\mathrm{Ti}} = -D_{V_{\mathrm{Ti}}}\left(\frac{\partial\mu_{\mathrm{Ti}}}{RT\,\partial\ln[V_{\mathrm{Ti}}'''']}\right) = D_{V_{\mathrm{Ti}}}\left(\frac{\partial\ln a_{\mathrm{O}_2}}{\partial\ln[V_{\mathrm{Ti}}'''']}\right) \tag{78}$$

due to the identity $\nabla\mu_{\mathrm{Ti}} + 2\nabla\mu_{\mathrm{O}} = \nabla\mu_{\mathrm{TiO}_2} = 0$ in the presence of the second phase. The quantities within the parentheses, the thermodynamic factors, are nothing but the inverses of slopes of the curves for $[V_{\mathrm{O}}^{\bullet\bullet}]$ and $[V_{\mathrm{Ti}}'''']$, respectively, in Figure 10.17. These are evaluated graphically in Figure 10.23.

On the basis of Eqs. (77) and (78), it is then possible to evaluate the defect-diffusivities of $V_{\mathrm{O}}^{\bullet\bullet}$ and V_{Ti}'''', $D_{V_{\mathrm{O}}}$ and $D_{V_{\mathrm{Ti}}}$, respectively. The results are finally as shown in Figure 10.24. As can be seen, each defect diffusivity turns out to be fairly flat against oxygen activity, as expected. Their values at 1200 °C, $D_{V_{\mathrm{O}}} = 1.4 \times 10^{-5}$ cm^{2}s^{-1}; $D_{V_{\mathrm{Ti}}} = 1.4 \times 10^{-15}$ cm^{2}s^{-1} are quite reasonable in magnitude, upon comparison with published values (if available). It should be noted here that, whilst the defect diffusivity of the cation vacancy has been evaluated for the first time in $BaTiO_3$ and the like, it has yet to be elucidated which of V_{Ti}'''' or V_{Ba}'' is in the majority, and which is more mobile whether doped or undoped. In the past, it has generally been believed that Ti^{4+} is immobile relative to Ba^{2+}, for energetic reasons [33]. However, the results of a recent study [9] have shown that the impurity diffusivities of Sr (replacing Ba) and of Zr (replacing Ti) are comparable to each other.

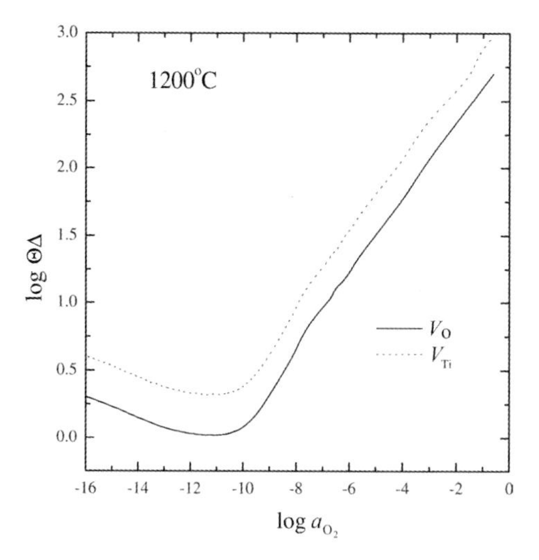

Figure 10.23 Thermodynamic factors (|ΘΔ|) for O (solid line) and Ti (dotted line) versus oxygen activity, as evaluated from the equilibrium defect structure of 1 mol% La-doped $BaTiO_3$ in Figure 10.17. Data from Ref. [30].

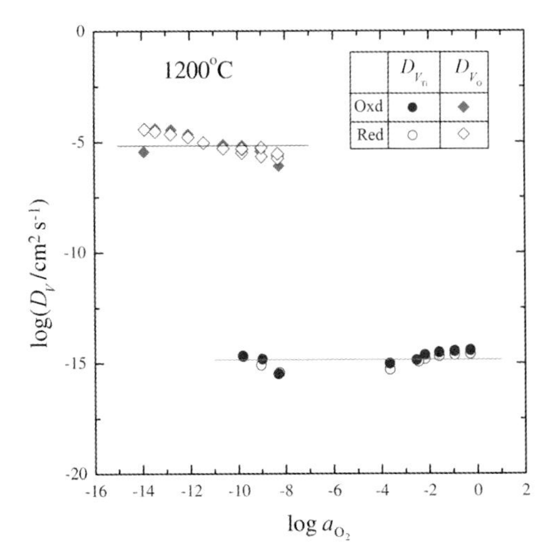

Figure 10.24 Defect diffusivities of O and Ti vacancies, D_{V_O} and $D_{V_{Ti}}$ versus oxygen activity at 1200 °C. The solid lines indicate the average values.

Acknowledgments

The author thanks C.E. Lee for producing the illustrations. These studies were supported financially by the Center for Advanced Materials Processing, under the "21C Frontier Program" of the Ministry of Commerce, Industry and Energy, Korea.

References

1 Wernicke, R. (1978) *Phys. Status Solidi A*, **47**, 139.
2 Schmalzried, H. and Wagner, C. (1962) *Z. Phys. Chem.*, **B31**, 198.
3 Schmalzried, H. (1964) Point defect in ternary ionic crystals, in *Progress in Solid State Chemistry*, vol. 2 (ed. H. Reiss), North-Holland Publishing Co., Amsterdam, pp. 265–303.
4 Kröger, F.A. (1974) in *The Chemistry of Imperfect Crystals*, 2nd edn, Ch. 20 and 21, North-Holland Publishing Co., Amsterdam.
5 Abelard, P. and Baumard, J.F. (1981) Nonstoichiometry in ternary oxides a new graphical representation of the defect configurations, in *Science of Ceramics*, vol. 11, Swedish Ceramic Society, Göteborg, pp. 143–148.
6 (a) Smyth, D.M. (1976) *J. Solid State Chem.*, **16**, 73;(b) Smyth, D.M. (1977) *J. Solid State Chem.*, **20**, 359.
7 Waser, R. (1991) *J. Am. Ceram. Soc.*, **74**, 1934.
8 Smyth, D.M. (2000) *The Defect Chemistry of Metal Oxides*, Ch. 14, Oxford University Press, New York.
9 Koerfer, S., De Souza, R.A., Yoo, H.-I., and Martin, M. (2008) *Solid State Sci.*, **10**, 725.
10 Yoo, H.-I., Song, C.-R., and Lee, D.-K. (2002) *J. Electroceram.*, **8**, 5.
11 Yoo, H.-I., Lee, S.-W., and Lee, C.-E. (2003) *J. Electroceram.*, **10**, 215.
12 Song, C.-R. and Yoo, H.-I. (1999) *Solid State Ionics*, **120**, 141.
13 Song, C.-R. and Yoo, H.-I. (2000) *J. Am. Ceram. Soc.*, **83**, 773.

14 Rase, D.E. and Roy, R. (1955) *J. Am. Ceram. Soc.*, **38** (3), 102.
15 Negas, T., Roth, R.S., Parker, H.S., and Minor, D. (1974) *J. Solid State Chem.*, **9**, 297.
16 Song, C.-R. and Yoo, H.-I. (2000) *Phys. Rev. B*, **61**, 3975.
17 Yoo, H.-I. and Becker, K.D. (2005) *Phys. Chem. Chem. Phys.*, **7**, 2068.
18 Wagner, C. (1971) *Prog. Solid State Chem.*, **6**, 1.
19 Lee, D.-K. and Yoo, H.-I. (2001) *Solid State Ionics*, **144**, 87.
20 Lee, D.-K., Jeon, J.-I., Kim, M.-H., Choi, W., and Yoo, H.-I. (2005) *J. Solid State Chem.*, **178**, 185.
21 Seuter, A.M.J.H. (1974) *Philos. Res. Rep. Suppl.*, **3**, 1.
22 Hagemann, H.-J. and Hennings, D. (1981) *J. Am. Ceram. Soc.*, **64**, 590.
23 Bois, G.V., Mikhailova, N.A., Prodavtsova, E.I., and Yusova, V.A. (1976) *News Acad. Sci. USSR, Inorg. Mater.*, **12**, 1302.
24 Panlener, R.J. and Blumenthal, R.N. (1971) *J. Am. Ceram. Soc.*, **54**, 610.
25 Hennings, D. (1976) *Philips Res. Rep.*, **31**, 516.
26 Yoo, H.-I., Song, C.-R., Lee, Y.-S., and Lee, D.-K. (2003) *Solid State Ionics*, **160** (3–4), 381.
27 Crank, J. (1975) *The Mathematics of Diffusion*, 2nd edn, Clarendon Press, Oxford, UK, pp. 60–61.
28 Wagner, C. (1933) *Z. Phys. Chem. B*, **21**, 25.
29 Van Roosbroeck, W. (1953) *Phys. Rev.*, **91**, 282.
30 Yoo, H.-I. and Lee, C.-E. (2005) *J. Am. Ceram. Soc.*, **88**, 617.
31 Nowotny, J. and Rekas, M. (1994) *Ceram. Int.*, **20**, 265.
32 Wernicke, R. (1976) *Philips Res. Rep.*, **31**, 526.
33 Lewis, G.V. and Catlow, C.R.A. (1983) *Radiat. Eff.*, **73**, 307.
34 Lee, D.-K. and Yoo, H.-I. (2006) *Solid State Ionics*, **177**, 1.

11
Interfaces and Microstructures in Materials

Wook Jo and Nong-Moon Hwang

11.1
Introduction

Since almost all materials of interest in this field of research are crystalline solids, a wide variety of important properties such as mechanical, thermal, dielectric, electronic, magnetic, optical, and so forth, are markedly dependent on both the crystallography and chemistry of each material. The effects of crystallography on the properties of a material are quite straightforward, as in the case of graphite and diamond. In this chapter, the term "chemistry" refers specifically to the type of constituting atom – that is, the chemical formula. For example, the properties of MgO and NaCl, the chemistries of which have nothing in common, differ significantly in many respects, yet their crystal structures are identical. In this regard, it can be said that properties originating from a combination of the chemistry and crystallography of a material are intrinsic, and consequently constitute a primary criterion for the correct choice of materials on demand.

It is also interesting to note that materials even of identical ingredients and crystal structure often exhibit significantly different properties when prepared via different processing routes. In most cases, these processing-dependent extrinsic properties are more important from a practical point of view. Hence, an understanding of the relationship between processing parameters and the consequent extrinsic properties has been of major concern to both materials scientists and engineers. One of the most important principles in the field of materials science and engineering is that a number of useful properties depend critically on the microstructure that can be engineered by adjusting processing parameters. Hence, a different processing route can result in a different microstructure.

The *microstructure* is defined as a network of one or more crystalline phases with various types of imperfection. The crystalline phase, which covers the greater part of the whole system, is the origin of the intrinsic properties of a specific material. *Imperfections* are defects to which most of the processing-dependent properties are attributed. Typical examples of imperfections include solid/vapor and solid/liquid interfaces, grain boundaries, phase boundaries, pores, secondary phases, and so on. Among those listed, the first three defects – that is, the solid/vapor and solid/liquid

Ceramics Science and Technology Volume 2: Properties. Edited by Ralf Riedel and I-Wei Chen

ISBN: 978-3-527-31156-9

interfaces and the grain boundaries – have been the subject of particular interest. This is not only because almost all engineering applications of materials are in service as a polycrystalline form through sintering, but also because the various useful properties of ceramic materials are closely related to the number density and the nature of those interfaces.

A good example where solid/liquid interfaces are important can be found in the fabrication of *cermets*, which are designed to have synergic properties from both hard ceramics and ductile metals. It should be noted that cermets are consolidated by a liquid-phase sintering (LPS), using the metallic constituent as a liquid matrix. The effect of grain boundaries is best seen in the Hall–Petch relationship, which relates the yield strength of materials to the grain size [1, 2]. The effect of these imperfections on the properties of a material is more pronounced, where the electric, magnetic, thermal, or optical properties are concerned, because these defects have their own structure and thereby disrupt the orderliness of the crystal structure of a given material. Note that in the case of ZnO polycrystalline ceramics, the well-known nonlinear current–voltage characteristics – which are a fundamental requirement for the varistor applications – are effective only in the presence of the resistive interfacial phases [3, 4].

The aim of this chapter is to develop and introduce a key concept denoted as *anisotropy in interfacial energies*. This is crucial in understanding microstructure evolution of materials, and how the interface anisotropy influences the microstructural evolution of materials. Particular attention is directed towards the interface-related topics, as many excellent reference materials are available that discuss the relationship between interface features and the resultant microstructures [5–7].

11.2 Interfaces in Materials

The aim of this section is to review the fundamental concepts needed to discuss the correlation between interfacial characteristics and the resulting microstructure evolution that develops during sintering. The discussion begins with a rather simple and well-understood type of interface that forms between a solid and its own vapor; this will hereafter be referred to as merely *surface*, following the convention. Because most concepts developed in the surface are also valid for the solid/liquid (S/L) interface, by their analogy in fluidity, these discussions of the S/L interface will be restricted to certain exclusive features derived from the presence of a liquid phase in contact with the surface. Subsequent discussions will lead to the concept of a *grain boundary*, which is the interface between two crystallographically distinctive solids.

11.2.1 Surface Fundamentals

11.2.1.1 Surface Energy

In order to introduce new surfaces into an object, a certain amount of energy must first be expended. In this regard, it can be said that surfaces – whatever they are – are always energetically less stable than their bulk counterparts, and that their "birth" always costs extra energy. This input energy, which is called *surface work*, is roughly twice as large as the energy of the surface created, because the surface work always involves the appearance of two identical surfaces. This means that the energy of a specific surface can be easily determined as long as the corresponding surface work can be measured. Unfortunately, however, this approach is not feasible in most cases from a practical point of view. Primarily, there is no way to determine the surface work other than via experiments, and these usually significantly underestimate such work in real crystals due to the inevitable presence of defects such as dislocations. Yet, even if the surface work is measured on a defect-free crystal, a perfect cleavage along a specific surface cannot be guaranteed. There is, therefore, a need to develop a more tractable way to quantify the surface work, so that the surface energies may be calculated.

The atoms that constitute objects are tightly bonded together and must be taken apart to introduce new surfaces. In this sense, the surface work can be said to be the energy required to break these atomic bonds. The surface energy is the total sum of the energy of all the broken bonds exposed on one of the two surfaces created. The calculation of surface energies based on this concept is commonly called the *broken bond model*. As the terminology implies, the energy of a surface in this model is determined merely by the number density of the broken bonds at the surface under consideration. Whilst this approach may appear too simplified, it is quite reliable in estimating the anisotropic nature of surface energies of materials. In fact, when it comes to the 4*d* transition metals the difference between the anisotropy among surfaces and that computed by the first-principles calculation is only 5%, at the most [8, 9]. Besides, it has been shown that there is a linear relationship between the energy of real metal surfaces and that estimated by the broken bond model [10–12]. Apart from these facts, the broken bond model represents one of the most powerful and widely used tools for understanding various surface-related phenomena. This is not only because it can estimate surface energies quite accurately, but also because it reflects most of the fundamental and important concepts regarding the surface energy. With this in mind, the principles and applications of the broken bond model are presented below in greater detail.

One of the most intuitive and easy ways to define surfaces is to regard them as a geometric boundary, where the periodicity of the crystallographic symmetry of the bulk crystal phase disappears discontinuously. The surface of this definition is often called the *Gibbs dividing surface* [13]. This definition implicitly implies that the surface is an independent phase which is distinguished from its bulk crystal. This is because the surface should not only be correlated crystallographically with its bulk phase, but also conceive the disorderly nature of the environment. The crystallographic correlation forces the surface to assume a two-dimensional (2-D) lattice that is a derivative of the symmetry of its bulk phase, and the conceived disorderliness is reflected by an excess energy – that is, the surface energy that makes it reactive. It follows that the first step for quantifying the surface energy is said to identify the crystallographic symmetry, which in turn allows the unit amount of excess energy conceived in the surface of interest to bc calculated.

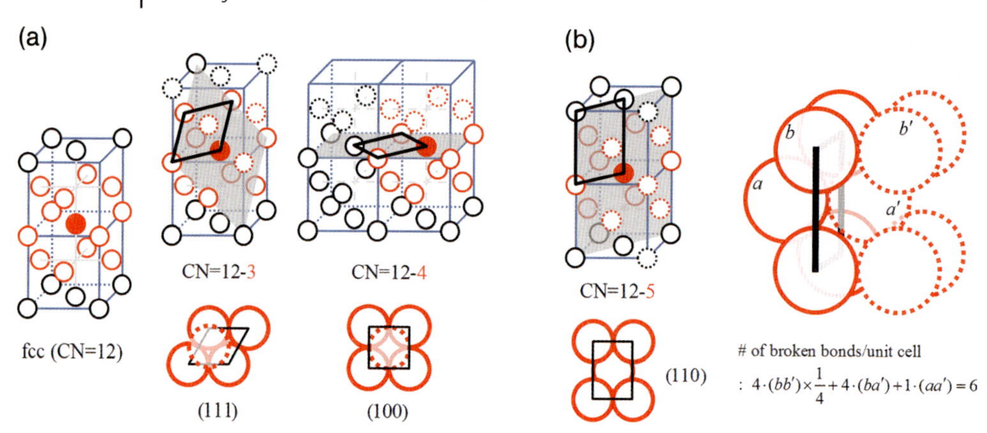

Figure 11.1 Schematic illustrations of the atomic configuration of (a) (100), (111), and (b) (110) surfaces of fcc symmetry, together with the corresponding unit cell of each surface. The surface energy per atom of a specific surface can be estimated by counting the total number of broken bonds belonging to the unit cell of the related surface. The atoms highlighted in red and those with a broken line refer to the 12 nearest neighbors of the atom filled with the red color and the atoms removed to introduce the surface, respectively.

An example of this is when calculating the surface energy of three principal surfaces of a pure metallic system[1] with the face-centered cubic (fcc) symmetry. Figure 11.1 shows the development of a 2-D lattice structure on the surfaces of a different crystallographic orientation. The loss of symmetry elements along the direction of a surface normal is best represented by the change in the coordination number (CN). The clinographic images constructed from the *hard sphere model* in Figure 11.1 clearly show that CN of 12 for the atoms in the bulk is reduced to 9, 8, and 7 for the atoms at (111), (100), and (110) surfaces, respectively. Moreover, along with the change in CN, the lattice type of each surface also changes to hexagonal (*p*6*mm*), square (*p*4*mm*), and rectangular (*p*2*mm*) systems, respectively. It is interesting to note that the symmetry of the 2-D lattice decreases with CN, which indicates that the surface energy also decreases with decreasing CN.

The number of broken bonds per unit cell of each surface is also easily counted to be three, four, and six, respectively. Attention needs to be paid to the fact that the number of broken bonds per atom at the (110) surface is not 5 but 6, since there is an extra broken bond related to the atoms on a sub-surface as is graphically manifested in Figure 11.1(b). In fact, in the case of simple fcc crystals, there are always additional broken bonds to be counted due to the atoms sitting on a subsurface, except for (100) and (111) surfaces. For example, although the number of broken bonds from the exact (210) surface is six, the total number of bonds to be broken in opening (210)

1) This choice of the system is made to simplify the discussion, because in this way each atom represented by a sphere from the hard sphere model corresponds directly to the crystallographic lattice points. In principle, however, the surface energy of any system can be calculated with the same scheme.

surface becomes 10 for each centered rectangular unit cell ($c2mm$), due to the presence of a subsurface.

Once the number of broken bonds per unit cell of a surface has been determined, the surface energy can be estimated from the following equation:

$$\gamma_{(\mathrm{hkl})} = N_S \frac{N_B}{2} \varepsilon_b, \tag{1}$$

where N_S, N_B, and ε_b refer to the number of surface atoms per unit area, the number of broken bonds per unit cell of surface, and the binding energy, respectively. As the heat of sublimation, ΔH_S, is equivalent to the energy required to break all the bonds in a crystal, the relationship between ΔH_S and ε_b in the case of fcc is given by

$$\Delta H_S = 12 \times N_A \times \frac{\varepsilon_b}{2}, \tag{2}$$

where N_A is Avogadro's number. From Eqs (1) and (2), the energy of (111), (100), and (110) surfaces of a pure fcc crystal, the lattice parameter of which is a, is given by:

$$\gamma_{111} = \frac{1}{\sqrt{3}} \cdot \frac{\Delta H_S}{a^2 N_A}, \quad \gamma_{100} = \frac{2}{3} \cdot \frac{\Delta H_S}{a^2 N_A} \quad \gamma_{110} = \frac{1}{\sqrt{2}} \cdot \frac{\Delta H_S}{a^2 N_A} \tag{3}$$

Therefore, the determination of the unit cell of a surface and the number of broken bonds graphically is quite straightforward, and intuitively easy to accept. However, when there is a need to consider more than the first nearest neighbor interactions (FNNs), or the surfaces with rather high Miller indices, this graphical method is no longer feasible, becoming highly subject to errors.

For this matter, Mackenzie *et al.* [14] developed a systematic way to calculate surface energies based on the broken bond model. The basic idea is that, since all the bonds to be broken for opening a surface should lie in the same direction of the surface normal, the total number of such bonding vectors,[2)] normalized by the area of the unit cell of the surface, is equal to the product, $NS \cdot NB$. For pure metals with the fcc symmetry, the energy of the (hkl) surface is given as:

$$\gamma_{(hkl)} = \frac{\Delta H_S}{3a^2 N_A} \cdot \frac{2h(1+\rho) + k(1+2\rho) + 2l\rho}{\sqrt{h^2+k^2+l^2}} \quad (h \geq k \geq l), \tag{4}$$

where ρ refers to the fraction of the contribution of the second nearest neighbor (SNN) interaction, which means that ρ equals 0 when only the FNN interaction is taken into account. The relative energies of the major surfaces in fcc with respect to that of the (111) surface, as well as the related parameters, are calculated from Eq. (4) and presented in Table 11.1.

Although the discussion on the broken bond model has been mainly focused on pure metallic surfaces, the extension can be easily made simply by replacing the binding energy term ε_b in Eq. (1) with the interaction energies among different atoms in the case of metallic alloys [15–17]. However, in the case of ceramic materials the

2) In the case of fcc, they refer to $1/2\ \langle 110 \rangle$ and $\langle 110 \rangle$ for FNN and SNN, respectively.

Table 11.1 A sample calculation of relative surface energies in a simple fcc crystal based on the broken bond model when the first and second nearest-neighbor interaction are considered.

(hkl)	$N_{B,FNN}$	$N_{B,SNN}$	N_S	$\gamma_{(hkl)}/\gamma_{(111)FNN}$ [‡]	$\gamma_{(hkl)}/\gamma_{(111)SNN}$ [†,‡]
111	3	4.2	$4/\sqrt{3}a^2$	1	1
100	4	4.8	$2/a^2$	1.155	0.990
110	6	7.6	$2/\sqrt{2}a^2$	1.225	1.108
311	7	9	$4/\sqrt{11}a^2$	1.219	1.116
331	9	11.8	$4/\sqrt{19}a^2$	1.192	1.119
210	10	12.4	$2/\sqrt{5}a^2$	1.291	1.142

†Contribution of SNN atoms is assumed to be 20% that of FNN.
‡FNN and SNN represent the first and the second nearest neighbor interactions.

same strategy often fails due to their inherent ionic bonding characteristics. As detailed discussions of the applicability of the broken bond model on ceramic materials are beyond the scope of this chapter, it is important to gain an idea how the ionic bonding characteristics influence the surface energy and the anisotropy, by using a rather simple example.

NaCl is one of the representative ceramic materials, which are featured by their strong ionic bonding characteristics. A clinographic view of the unit cells of NaCl is shown in Figure 11.2a.[3)] Although NaCl has an fcc symmetry, the space group of which is $Fm\bar{3}m$, the bonding vector is along $\langle 100 \rangle$, which connects an Na^+ and a Cl^- ion. Therefore, the CN of each ion becomes 6 in the bulk phase. Figure 11.2b shows an orthographic illustration of the (100) and (111) surfaces of NaCl on the (100) plane. By applying the same strategy as discussed above, the unit cell of each surface can be easily identified as square ($p4mm$) for (100), rectangular ($p2mm$) for (110), and hexagonal ($p6mm$) for (111). When the lattice constant is taken as a, the N_S for each surface is given as:

$$N_{S,100} = \frac{1 \cdot Na^+ + 1 \cdot Cl^-}{a^2/2}, \quad N_{S,110} = \frac{1 \cdot Na^+ + 1 \cdot Cl^-}{a^2/\sqrt{2}}, \quad N_{S,111} = \frac{1 \cdot Na^+}{\sqrt{3}a^2/4}. \tag{5}$$

It is readily noticed that the atomic structure of the (111) surface is completely different from that of the other two surfaces. This is because the (111) surface of NaCl-type ceramics is a polar surface where the charge neutrality fails, causing its energy to be extremely high, while the other two surfaces are nonpolar [18–20]. This argument becomes more clear by considering the number of broken bonds on each surface, as follows.

3) Though Na^+ ions were placed at the origin for the construction of NaCl crystal in this contribution, alternative constructions like placing Cl^- ions or the mean position of both ions are also possible. However, when Cl^- ions are assumed to be at the origin, the terminating ions on the (111) surface become Cl^- ions and those on the $(\bar{1}\bar{1}\bar{1})$ surface Na^+ ions.

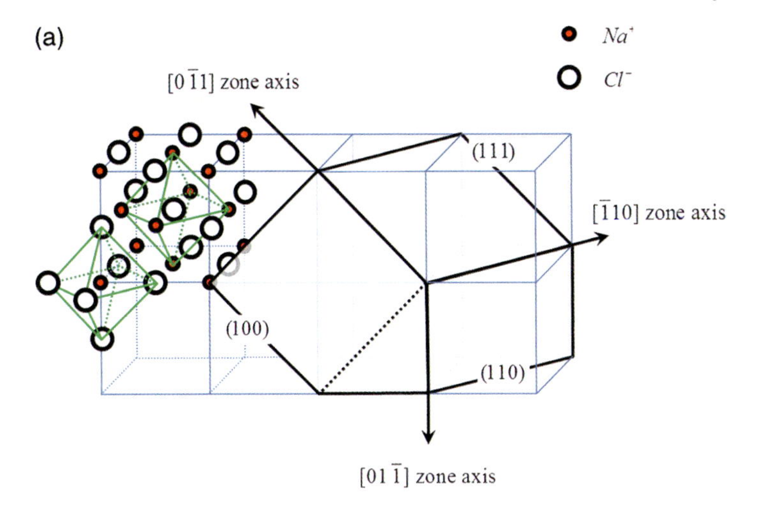

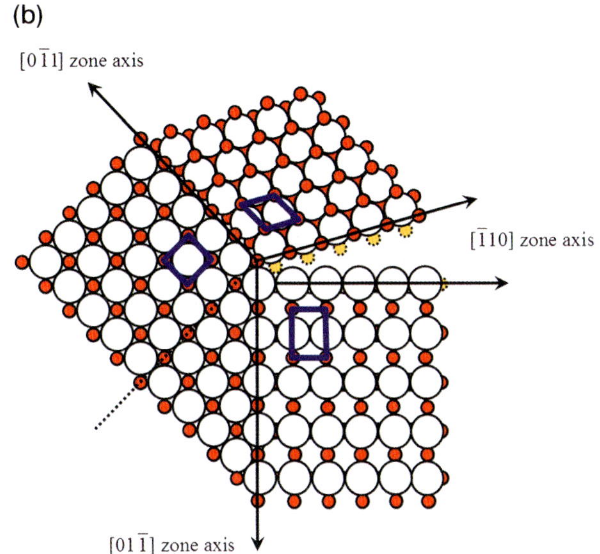

Figure 11.2 (a) A clinographic view of the unit cells of NaCl crystal. Note that the CN of both Na^+ and Cl^- ions in the bulk is 6; (b) An orthographic view of (100), (110), and (111) surfaces of NaCl crystal projected on (100) plane. Note the difference in the choice of the lattice point for the unit cell of each surface. Na^+ and Cl^- ions are not on the same level on (111) surface, while those are on the same level on (100) and (110) surfaces.

When only the FNN interaction is considered, the number of broken bonds per unit cell can be easily counted to be two, four, and three for the (100), (110), and (111) surface, respectively. However, great care must be taken here because the broken bonds generated on the (111) surface are completely different from those on the (100) and (110) surfaces. In fact, all three broken bonds created on the (111) surface are due to the loss of three Cl atoms. Therefore, effectively three single electrons are said to be at each Na atom on the surface. The broken bonds at the (100) and (110) surfaces consist of an equal number of both types of bonds – that is, one from the loss of Na, and the other from the loss of Cl. As these two different types of bonds are distributed in a checkerboard array, it can be expected that the energy of the broken bonds on these surfaces would be fairly low due to a coupling among the neighboring broken bonds. Together with Eq. (5), the surface energy of each surface of NaCl is expressed as follows:

$$\gamma_{100} = \frac{2}{a^2} \cdot [\varepsilon_{\mathrm{Na}^-} + \varepsilon_{\mathrm{Cl}^-}], \quad \gamma_{110} = \frac{4\sqrt{2}}{a^2} \cdot [\varepsilon_{\mathrm{Na}^-} + \varepsilon_{\mathrm{Cl}^-}], \quad \gamma_{111} = \frac{12}{\sqrt{3}a^2} \cdot [\varepsilon_{\mathrm{Na}^-}]. \tag{6}$$

When the ratio of $[\varepsilon_{\mathrm{Na}^-}]$ to $[\varepsilon_{\mathrm{Na}^-} + \varepsilon_{\mathrm{Cl}^-}]$ is denoted simply as $\varepsilon^{\mathrm{XS}}$, the relative surface energy is given as:

$$\gamma_{100} : \gamma_{110} : \gamma_{111} = 1 : 2.83 : 3.46\varepsilon^{XS}. \tag{7}$$

Equation (7) shows that in the case of ceramics with a NaCl structure, the energy of (111) surface is markedly higher than that of (100), even when $\varepsilon^{\mathrm{XS}}$ is assumed to be unity. In fact, $\varepsilon^{\mathrm{XS}}$ is known to be about 6 in the case of MgO [21], and on average is about 7 for various oxide ceramics with a NaCl structure [20]. Due to this high energy, (111) surfaces never arise naturally. Hebenstreit *et al.* [19] demonstrated this with an elegant experiment, by showing that the polar (111) surface of NaCl can be induced on (100) nano-islands supplied with an excessive amount of Na atoms, but that the extent of the conversion from (100) to (111) is too limited to produce a (111) surface, which is larger than several tens of square nanometers. However, polar surfaces are commonly observed in some ceramic systems belonging to the crystallographic polar classes such as ZnO with the wurtzite structure. In these systems, it is well known that preferentially existing polar surfaces in a crystal have a major influence on the crystal growth kinetics [22, 23], mechanical properties [24, 25], chemical etching characteristics [26, 27], and so on, due to their high surface energy.

At this point, it is worth emphasizing several important conclusions that can be drawn from these investigations:

- The origin of surface energies is the presence of broken bonds due to a termination of the crystallographic periodicity of their bulk phase.
- The number of broken bonds at a certain surface, equivalently the surface energy, is determined by correlating the crystallographic periodicity along the normal direction to the surface with the principal bonding vectors, determined by the symmetry of the bulk phase.
- As all the crystal systems have a finite number of the principal bonding vectors, it follows that the surface energies are, in principal, anisotropic.

11.2.1.2 **Wulff Plot**

No object on Earth is free from surfaces, because of its finiteness in size. The problem is that this unavoidable entity – that is, surface – is energetically unfavorable, as discussed in the previous section. This implies that all objects tend to take a specific shape by which they can minimize the excess energy due to the presence of the surfaces. This shape is unique for an object in a given thermodynamic condition, and is termed the *equilibrium shape*. Of course, whether an object can reach its equilibrium shape or not depends critically on how fast and easily the constituting atoms can transport in the object.

In the case of liquids such as water, there exists no perceptible long-range periodicity, and this in turn leads to an identical surface structure, which makes the surface energy isotropic. When the surface energy of an object is isotropic, its equilibrium shape is determined merely by the condition where the area of the entire surface is minimized for a given volume. This argument is equivalently described thermodynamically by Gibbs [13] with the Helmholtz free energy.

For example, imagine a liquid droplet, the volume of which is fixed at a given temperature, while the only way to lower the droplet's free energy is to change its shape. In this case, the Helmholtz free energy of the liquid droplet is merely given as

$$F = \oint_V \gamma dA = \gamma \oint_V dA. \tag{8}$$

In this specific case, the condition for the equilibrium is met when $dF = 0$, that is, $dA = 0$; this condition is fulfilled when the droplet shape becomes a sphere. Nevertheless, the equilibrium shape of solids is not determined merely by the condition $dA = 0$ (as in the case of liquids), because the surface energy γ varies as the crystallographic orientation changes. Curie [28] demonstrated that the equilibrium condition, $dF = 0$, as established by Gibbs [13], is satisfied when the shape of crystals[4)] changes in a way that the capillarity constant of each crystal surface becomes identical. To have a better idea on the capillarity constant concept, it is helpful to follow the argument proposed by Curie.

For this, suppose that a crystal with a tetragonal symmetry is in equilibrium with its vapor phase, as shown schematically in Figure 11.3. Then, Eq. (8) is expressed as

$$F = 2Ax^2 + 4Bxy, \tag{9}$$

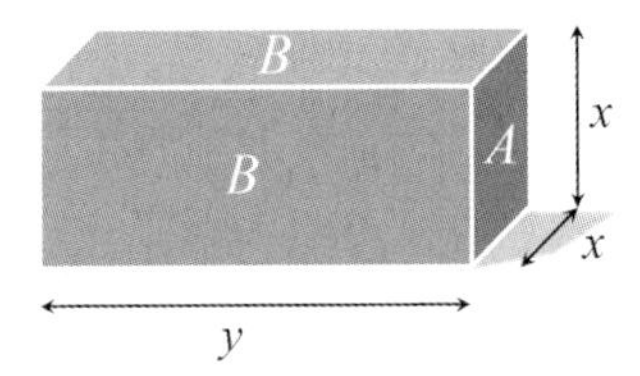

Figure 11.3 Outline of a crystal with the tetragonal symmetry at equilibrium. *A* and *B* represent the reciprocal reticular density of each surface. From Ref. [28].

4) From the physical point of view, the terminology of *solid* and *crystal* has no distinction and interchangeable.

where A and B refer to the reciprocal reticular density[5] of each type of surface. In order for a crystal to be at equilibrium, the chemical potential of the crystal phase (μ^C) and its vapor phase (μ^V) should be equal, which keeps the volume of the crystal constant. As the volume of the crystal is given as $V = x^2 y$, the condition for the equilibrium can be evaluated from Eq. (9) as follows:

$$F = 2Ax^2 + \frac{4BV}{x} \quad (10)$$

By differentiating Eq. (10), the condition for the equilibrium is given as:

$$\frac{A}{y} = \frac{B}{x}. \quad (11)$$

This ratio, which is constant for all of the surfaces of a crystal at equilibrium, is termed the *capillarity constant*, analogous to that defined in the well-known Laplace capillarity equation.

Inspired by the idea that the reticular density of a surface is inversely proportional to the distance of the surface from the origin where all surface normal vectors meet, Wulff [29] conducted a series of experiments seeking the experimental evidence for his hypothesis. With various inorganic crystals, he measured the growth rate of the surfaces for which the reticular density could be estimated. For this experiment, the degree of supersaturation was kept as low as possible so as to maintain the overall shape of crystals retained during the course of growth. In this way, the measured growth rate could be taken as a normalized distance of surfaces from the origin. A series of Wulff's experiments revealed an exquisite aspect of Nature. As shown in Table 11.2, the product between the experimentally determined growth rate and the reticular density was practically constant, regardless of the Miller indices of the surfaces. This observation led Wulff to propose a more generalized version of the Curie's capillarity constant concept such that, for an equilibrium crystal with N surfaces, the capillarity constant of a given crystal at equilibrium is:

$$\frac{\gamma_1}{h_1} = \frac{\gamma_2}{h_2} = \cdots = \frac{\gamma_n}{h_n} = \cdots = \frac{\gamma_N}{h_N}, \quad (12)$$

Table 11.2 Experimentally measured growth rate of various surfaces of a Mohr salt with cubic symmetry and the corresponding reticular density of each surface. Note that all the values are normalized by that of the (001) surface. After Ref. [29]..

	(001)	(111)	(201)	(310)	(112)
Growth rate	1.00	1.96	2.25	2.50	2.64
Reticular density	1.00	0.58	0.45	0.32	0.41

5) The *reticular density* has the identical meaning with the N_S that was introduced in Eq. (1) in the previous section, and its reciprocal is proportional to the energy of the corresponding surface.

where γ_n and h_n refer to the energy and the distance from the origin of the *n*th surface. The use of surface energies instead of reticular densities was based on the fact that the former was roughly inversely proportional to the latter. Equation (12) is commonly called the *Wulff theorem*, and the analytical description of the Wulff theorem was later given by Laue [30] as:

$$\frac{RT}{v_C M}\ln\frac{P}{P_\infty}=\frac{2\gamma_n}{h_n} \quad (n=1,2\ldots N), \tag{13}$$

where R, v_C, and M refer to the gas constant, crystal volume, and molecular weight, respectively.

The importance of the Wulff theorem is that it provides a graphical tool for constructing the equilibrium shape of any crystal, as long as the energy of each surface is known. This argument is illustrated schematically in Figure 11.4a. One thing to be noted here is that the surfaces with a very high energy (e.g., γ_2) are excluded from the equilibrium shape, but those surfaces whose energy is in an appropriate range (e.g., γ_3) are a part of the equilibrium shape.[6] Figure 11.4b is an example showing how surfaces with a relatively higher energy can appear in the equilibrium shape. Supposing that the energy of the {11} surface is 1.1 times as large as that of {10},

$$\frac{F^{\text{R.H.S.}}}{F^{\text{L.H.S.}}}=\frac{4\times 1.1\gamma_{10}\times 2.98+4\times\gamma_{10}\times 4.21}{4\times\gamma_{10}\times 10}\cong 0.75. \tag{14}$$

Note that the reduction in the free energy of the crystal with both {11} and {10} surfaces is no less than one-quarter.

Constructing the equilibrium shape of a crystal begins with checking the minimum segment that can represent the entire crystal by a symmetry operation. This

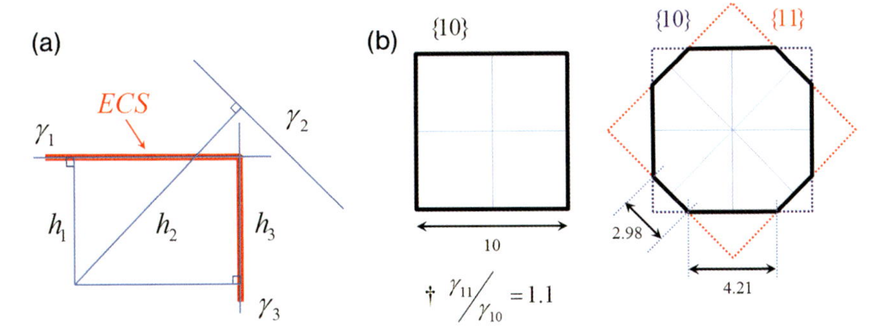

Figure 11.4 (a) Graphical interpretation of Wulff theorem. Note that the equilibrium shape is outlined by the envelopment of the innermost surfaces; (b) 2-D crystal enveloped only by {10} surfaces (L.H.S.) or by {10} and {11} surfaces (R.H.S.). Note that the area of each crystal is identical.

6) A rigorous and complete proof on this argument is tractable with the use of the Brunn-Minkowski inequality. Detailed discussion was presented by Gardener [*Bull. Am. Math. Soc.*, **39**, 355 (2002)].

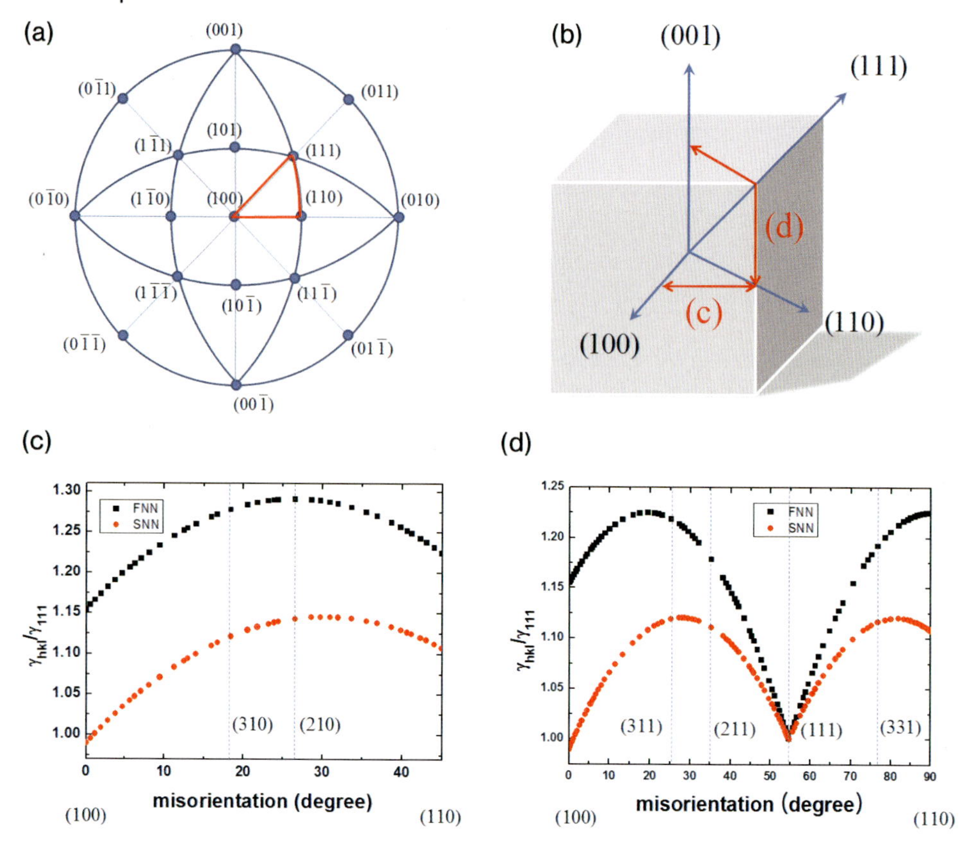

Figure 11.5 (a) The minimum symmetry segment for cubic crystals highlighted on a standard stereographic projection; (b) Contour lines which need to be defined to construct the equilibrium shape of a crystal with the cubic symmetry; (c, d) Calculated surface energies as a function of misorientation along the corresponding contour line.

minimum segment is determined by the number of the crystallographically equivalent general surfaces under the given symmetry; that is, the plane multiplicity factor for $\{hkl\}$ surfaces. For example, in the case of cubic systems, this minimum symmetry segment corresponds to a triangular region which connects three principal poles such as (100), (110), and (111), as highlighted on a stereographic projection in Figure 11.5a. The area of this segment is 1/48 of the entire space, since the plane multiplicity factor for $\{hkl\}$ surfaces in the cubic symmetry is 48. In principle, all the surface energies that belong to the areal fraction given by the reciprocal of the plane multiplicity factor need to be defined. As was discussed in the previous section, however, the lower the Miller indices are, the smaller the surface energies in the case of the simple metallic elements, except for the (111) surface [see Eq. (4)]. For a three-dimensional (3-D) visualization of the equilibrium shape of fcc metallic crystals, therefore, the energies of the surfaces lie on two perpendicular contour lines, as shown schematically in Figure 11.5b. Figure 11.5c and d show the surface energies

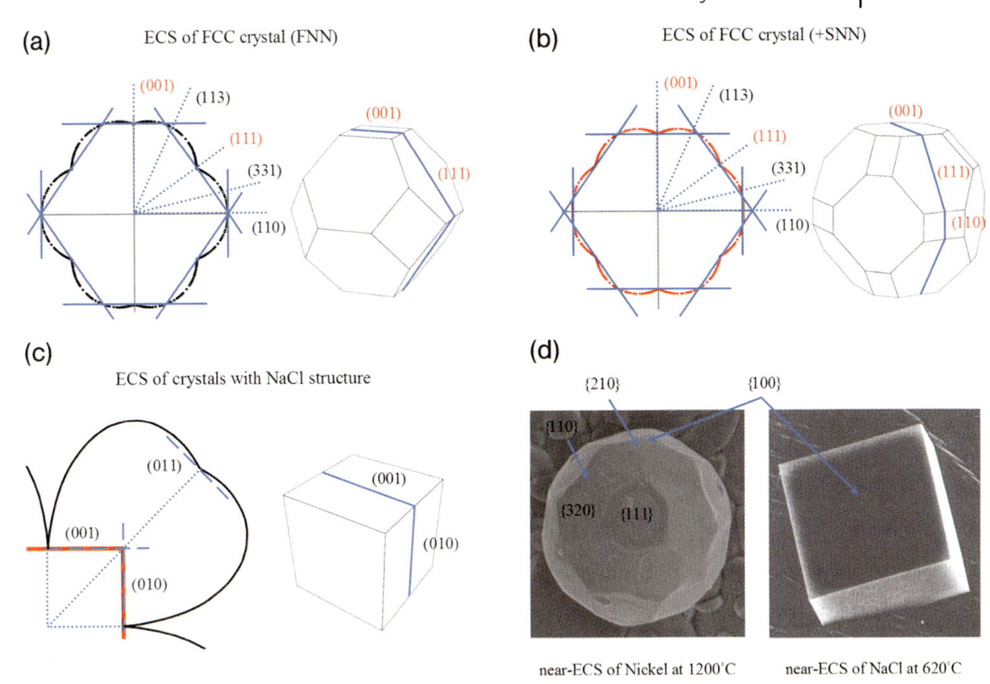

Figure 11.6 Wulff construction and related equilibrium shape of (a) a simple fcc crystal with only FNN broken bonds considered; (b) a simple fcc crystal with additional SNN broken bonds considered; and (c) a simple fcc crystal with ionic bonding characteristics; (d) A near-equilibrium shape of Ni at 1200 °C (left panel from Ref. [31]; right panel (NaCl at 620 °C) from Ref. [32]).

along each line, calculated from Eq. (4). Note that the consideration of the SNN interactions results in a marked difference in the γ-plot from that of the FNN interactions.

Figure 11.6a–c show how the equilibrium shape of crystals is constructed from a polar diagram of surface energies. It is noted that the equilibrium shape of fcc metals is rather difficult to determine, because even a small change in the calculation of surface energies may lead to a significant change in the equilibrium shape. For example, consideration of the SNN contribution not only changes the relative areal fraction between {100} and {111} surfaces, but also determines whether {110} surfaces are a part of the equilibrium shape, or not. However, in the case of crystals with the NaCl structure, surface energies are highly anisotropic due to the polar nature of most surfaces. This anisotropy is especially large in the vicinity of {100} orientations in the γ-plot. Hence, the equilibrium shape hardly changes even with a refined calculation on the surface energies. The argument can clearly be supported by the observation of real crystals near their equilibrium state. Figure 11.6d shows a near-equilibrium shape of a pure Ni crystal at 1200 °C (left) and sodium chloride at 620 °C (right).

It is well known that the real crystal surfaces usually deviate from the ideal ones predicted by the broken bond model. The deviation from the ideal is mostly attributed to two additional processes which decrease the energy of real crystal surfaces. One process is *relaxation*, and the other is *reconstruction*. The origin of the relaxation process comes from the different environment of atoms sitting at the surface. Due to a loss in their ordinary coordination number (for example, 12 for fcc crystals), the bonding distance that the surface atoms have is shorter than that of the bulk atoms. Since the displacement of atoms from their equilibrium position changes the potential energy of the atom, it can easily be expected that the relaxation process would also influence the surface energies as well as the equilibrium shapes. Moreover, considering that relaxation is mainly due to a smaller coordination number, this effect should be more pronounced in those surfaces with higher indices. In fact, Yu *et al.* [33] have shown that the relaxation effect in Pb reduces the anisotropy in surface energies, based on *ab initio* calculations as shown in Figure 11.7a. It is interesting to note that the degree of relaxation has also a linear dependence on the number of broken bonds.

The reconstruction process is a different type of energy-lowering process, in that it changes the surface crystallography. Figure 11.7b is one of the most famous reconstructions to be observed experimentally in the $\{110\}$ surfaces of Au [34] and Pt [35], and theoretically studied on various fcc metals [36]. As this reconstruction process involves removing an entire row of atoms out of every two row of surface atoms, it is called a (1×2) missing row reconstruction. Note that the unit cell formed after the reconstruction has the same lattice parameter along [110] direction, but a doubled parameter along [001]. Nonetheless, it is interesting to note that the surface energy of $\{110\}$ remains the same before and after this missing row reconstruction with respect to the broken bond model. In fact, the first-principles calculation also showed that the (1×2) missing row reconstruction could either decrease or increase the surface energy, depending on the system [37]. In contrast to metals, the reconstruction of ceramics or semiconductors – the bonding of which has a strong directionality and a rather open structure – is known to be a more common and important energy-lowering mechanism; a detailed discussion of this topic can be found elsewhere (see Ref. [38] and references therein).

As a practically important energy-minimizing process, *faceting* deserves special attention. In contrast to other processes such as relaxation and reconstruction, faceting has no influence either on the γ-plot or on the equilibrium shape. In this regard, faceting can be regarded as a process that meta-stabilizes thermodynamically unstable surfaces by replacing them with nearby thermodynamically stable surfaces. For example, if a non-equilibrium surface $\overline{BC}$, the energy of which corresponds to $\overline{OF}$, is introduced on a cubic crystal at equilibrium (as shown schematically in Figure 11.8a), the energy of the surface $\overline{BC}$ would be so high that it would be readily broken into step-like segments consisting of $\overline{AE}$ and $\overline{BE}$. This is shown schematically, and also by scanning electron microscopy, in Figure 11.8a.

In that faceting is an energy-reduction process, it can be estimated how much energy is reduced during the process. In the case presented in Figure 11.8a, the total

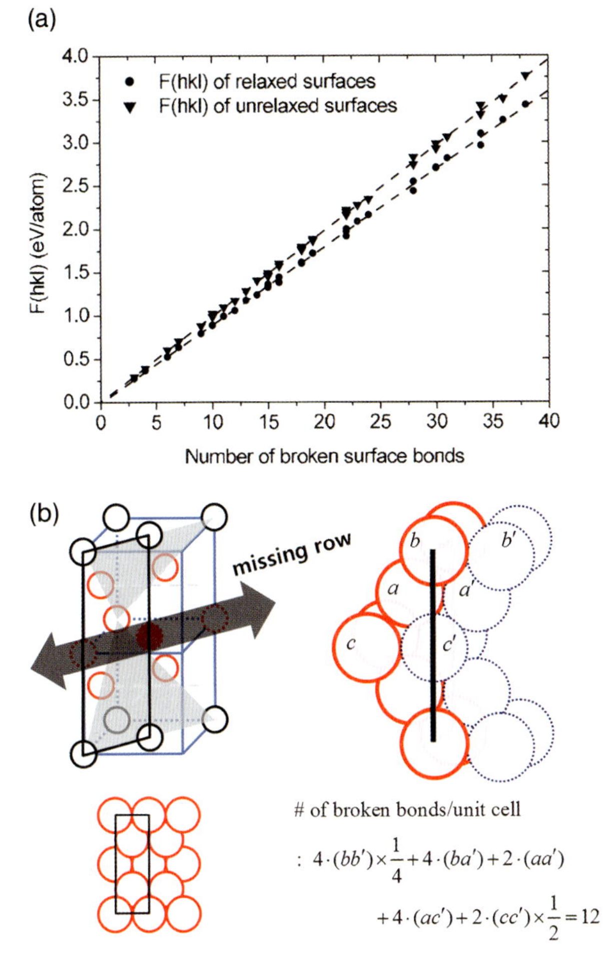

Figure 11.7 (a) Plot of theoretical surface energy F(*hkl*) in eV per surface atom versus the number of surface bonds which are broken in forming an (*hkl*) surface. First-principles calculation results are shown for unrelaxed and relaxed surfaces of Pb. The data are best fit by straight lines (From Ref. [9]); (b) Schematic illustrations showing (1×2) missing row reconstruction on {110} surfaces in fcc crystal. Comparison with Figure 11.1b will be helpful in understanding the meaning of a (1×2) missing row reconstruction.

energy of the initially introduced surface $\overline{BC}$ is given by

$$\gamma_{BC} = \overline{OF} \times \overline{BC} = \square OBFC, \tag{15}$$

and that of the faceted surface by

$$\gamma_{BE+CE} = \overline{OA} \times \overline{CE} + \overline{OB} \times \overline{BE} = \square OBEC = \square OBDC. \tag{16}$$

The comparison between Eqs (15) and (16) shows that faceting decreases the energy $\overline{OF}$ of the initially introduced surface to $\overline{OD}$ of the faceted surface. In fact, it should be noted that the energy of the faceted surface is not precisely $\overline{OD}$, but rather

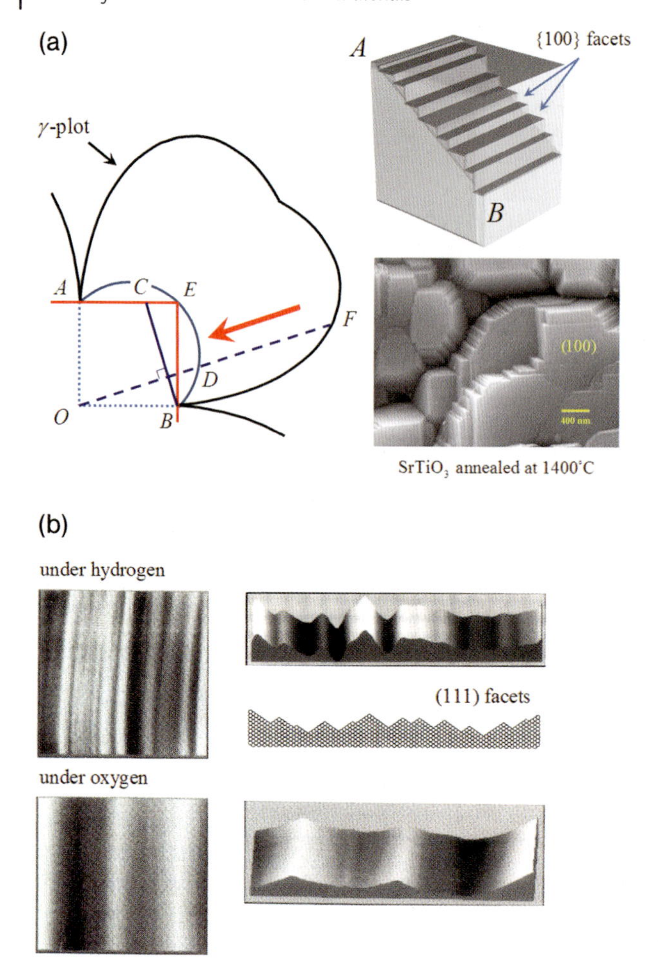

Figure 11.8 (a) Schematic illustration showing how much energy change takes place during faceting, and what the faceted morphology looks like. The faceted morphology of $SrTiO_3$ was observed by annealing slightly indented single crystals at 1400 °C for 200 h; (b) Adsorption-induced faceting of Pt(110) surface. From Ref. [39].

lies between $\overline{OD}$ and $\overline{OF}$. This deviation originates from the fact that the energy due to the presence of "edges" in the faceted surfaces is ignored in these calculations. This superfluous so-called "edge energy" drives facet coarsening until the entire energy decreases to $\overline{OD}$, that is the thermodynamically lowest. The circular energy contour $\frown AB$ that passes D and E is generally called the "convexified" γ-plot, because this γ-plot represents the convexified (i.e., the minimum free energy) contour line [40, 41].

It is well known that the adsorption of foreign atoms onto crystal surfaces always results in a reduction of the surface energies, and is orientation-dependent due to its preferential nature of orientation [38, 42]. This means that, in many cases, surface adsorption decreases the overall surface energy while simultaneously increasing the anisotropy in the surface energies of a given crystal. Consequently, faceting is usually promoted only when there are adsorbing substances in the atmosphere, as shown for example in Figure 11.8b [39, 43–46].

11.2.1.3 **Roughening Transition**

So far, for the purpose of simplicity, only the enthalpic aspect of the surface energy concept has been discussed. However, a precise description of various phenomena that occur on crystal surfaces requires a free energy concept, as the surface is a thermodynamic phase that is distinctive from its bulk phase, as discussed in the previous section. In this respect, it can easily be expected that an entropic effect which tends to make the surface phase disordered would drive it to undergo an *order–disorder* type of phase transformation at a sufficiently high temperature. Indeed, it has been observed – both theoretically and experimentally – that a surface phase transformation from (1×2) to (1×1) on an Au(110) surface takes place at around 650 K [47]. In addition, even a melting transformation – which is one of the most representative thermodynamic first-order phase transitions – has been identified both experimentally and theoretically at a surface level clearly below the corresponding bulk melting point (see Ref. [48] and references therein). Recently, all of these surface-specific phase transformations have become more important, both scientifically and practically, especially due to the advent of nanotechnology. However, as a detailed discussion of these topics does not comply with the main aim of this chapter, the discussion here will be confined to another surface-exclusive phase transition, namely *roughening transition*. One of the most practically important features of this transition lies in the fact that the mass transport mechanism along the crystal surfaces changes markedly before and after this transition.

For an effective discussion, it is necessary to introduce the Terrace-Step-Kink (TSK) model, which was proposed independently by Kossel [51] and Stranski [52] during the late 1920s. As might be expected from the name itself, this model simply regards the surface as a set of three physically distinctive entities such as terrace, step, and kink, as shown schematically in Figure 11.9a. The beauty of this model is that, despite its

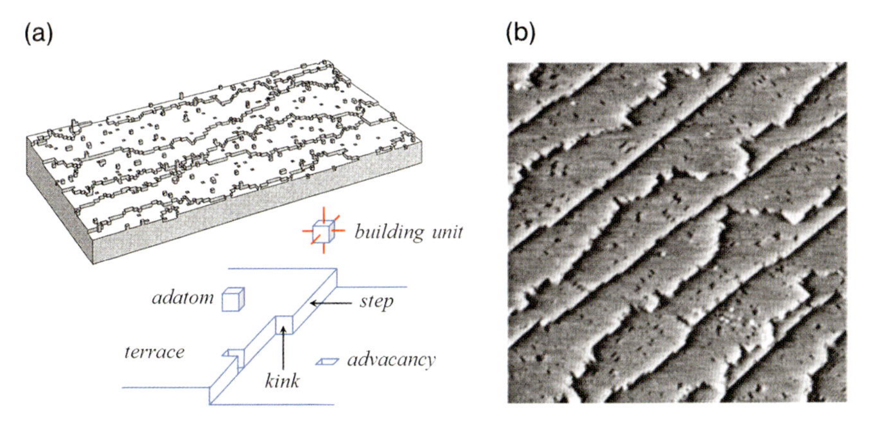

Figure 11.9 (a) A schematic illustration of a (001) vicinal surface of a silicon (001) [49] and a simplified view of a crystal surface in terms of TSK model; (b) Scanning tunneling microscopy image ($100 \times 100\,nm^2$) of a vicinal surface of Si(001). From Ref. [50].

simplicity, it enables an understanding of the most important physics behind surface-related phenomena. Another beauty of the model is that, although it is a product of speculation, the real crystal surfaces that are uncovered by the modern atomic resolution ultramicroscopy are not far away from those described by the model, as shown in Figure 11.9b. In this regard, all of the discussions – from the definition of each entity to the roughening transition given hereafter – are based on the TSK model.

The *terrace* refers to a flat surface, the energy of which has a local minimum value; that is, a cusp in γ-plot. This means that all surfaces existing in the equilibrium shape are terraces. Due to energetics, however, terraces mostly have low Miller indices, and the energetics and structural features of terraces conform to those of the surfaces.

The *kink* is defined as a site at steps where atoms can either attach or detach reversibly. This reversibility originates from the fact that neither attachment nor detachment changes the number of broken bonds on the surface [51, 52]. It follows that a kink site is always located at a place where the total number of broken bonds is equal to a half CN. For example, in the case of the Kossel crystal, where the atomic building block is assumed to be a cube with six broken bonds on each side, the corner where three perpendicular faces meet becomes the kink site, as shown in Figure 11.9a. As a result of this reversible nature of kink sites, it can be said that an atom sitting at a kink site belongs to the crystal and ambient vapor phase at the same time. For this reason, it was originally called a *half-crystal* position. Atoms in the ambient vapor phase deposit onto kink sites under a supersaturated ambient, and atoms depart from the kink sites under an unsaturated ambient. Hence, the number density of kinks at steps on a sufficiently large surface is roughly equal to the vapor pressure of the surface.

A *step* is an entity which compensates an angular deviation of a non-equilibrium surface from a nearby terrace. If the angular deviation of a vicinal surface from a terrace is θ, then this surface can be described by the terrace that is separated by periodically spaced steps. When the average mutual spacing of steps and the step height are l and h, respectively, the step density is given as $1/l$, equivalently $\tan\theta/h$ [53]. The free energy of this vicinal surface is then written as

$$f(\theta, T) = f^0(T) + \frac{\beta(\theta, T)}{h}|\tan\theta|, \tag{17}$$

where $f^0(T)$, $\beta(\theta, T)$ and h refer to the surface free energy of the unit area of terrace, the step free energy, and the step height, respectively. Note that the step free energy in Eq. (17) has an angular dependence; this is because the step–step interaction increases quadratically with the step density as the spacing between steps decreases (see Figure 11.10a). A statistical thermodynamic treatment[7] describes the angle-dependent step free energy function β as

$$\beta(\theta, T) = \beta_0(T) + \beta_2(T)|\tan^2\theta|, \tag{18}$$

7) Readers who wish to check the derivations and detailed explanation of each equation introduced, are encouraged to check Ref. [134].

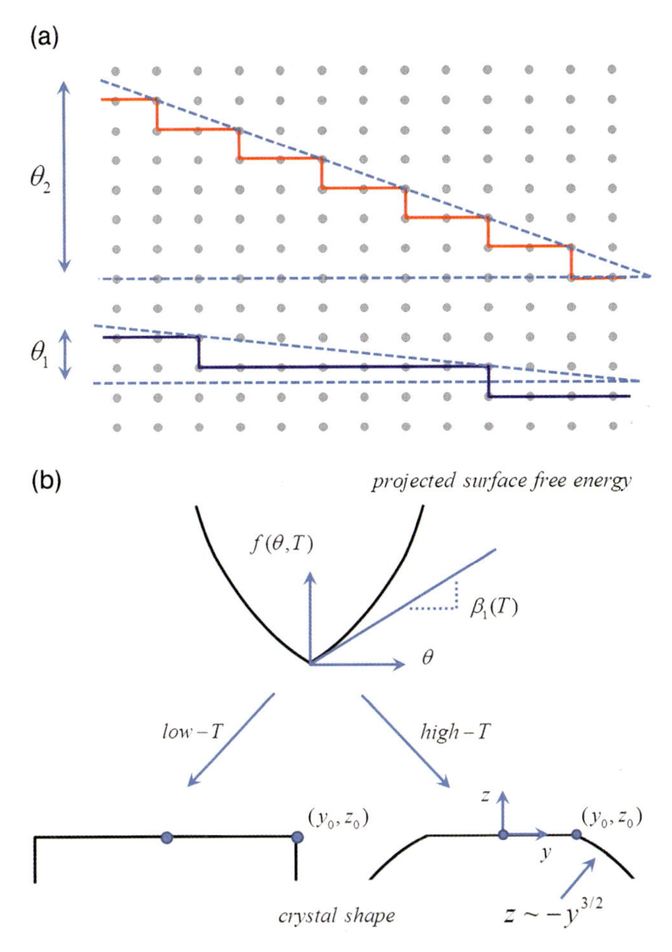

Figure 11.10 (a) Schematic illustration showing changes in the step density as a function of misorientation angle for a 2-D simple rectangular system. The step–step interaction increases quadratically with the step density; (b) Relationship between the projected surface free energy and the crystal shape as a function of temperature.

where $\beta_0(T)$ is the step free energy of an isolated step and $\beta_2(T)$ the step–step interaction. Combining Eqs (17) and (18) yields the following relationship:

$$f(\theta, T) = f^0(T) + \frac{\beta_0(T)}{h}|\tan\theta| + \frac{\beta_2(T)}{h}|\tan^3\theta|, \tag{19}$$

This is called the *projected surface free energy*, because it is the surface free energy projected along the surface normal direction, z. One of the most important meanings of the projected surface free energy is that it is equivalent to the crystal shape through the Legendre transformation. The conversion of a polar coordinates (f, θ) into a

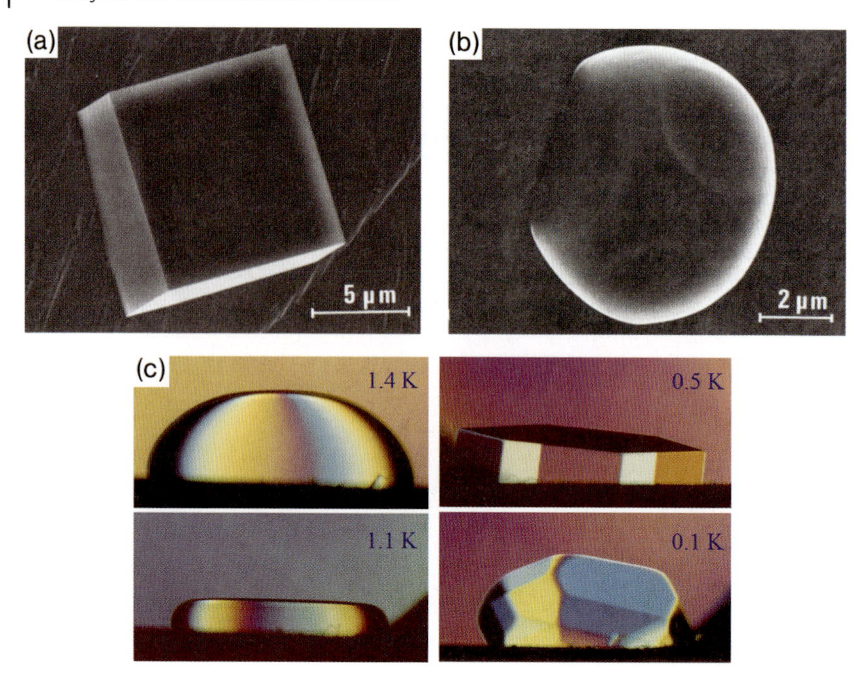

Figure 11.11 NaCl single crystal (a) at 620 °C and (b) 710 °C [32]; (c) An HCP ^{4}He crystal in a rapid growth as a function of temperature [54]. Note that a shape with perfect facets evolves into that with a mixture of facets and curved surfaces.

Cartesian coordinates (0, y, z) gives

$$z-z_0=-\frac{2\beta_2}{\lambda h}\left(\frac{\lambda h}{3\beta_2}\right)^{3/2}(y-y_0)^{3/2}, \tag{20}$$

where $\lambda = \Delta\mu/2\ \Omega$, with Ω being the atomic volume. Here, z_0 and y_0 refer to the vertical position of the terrace $(f^0(T)/\lambda)$ and a half of the facet size $(\beta_0(T)/\lambda h)$. In that the illustration of Eq. (20) visualizes the equilibrium shape of crystal, it can be said to be an analytical version of the Wulff theorem. The validity of the equilibrium shape described analytically by Eq. (20) was confirmed experimentally on various crystals [55–59]. Figure 11.11 shows, as an example, the experimentally observed thermal evolution of the equilibrium shape of NaCl and ^{4}He crystals. More precise experiments further confirmed that even the exponent 3/2 in Eq. (20) is also true for crystals with both rounded and faceted surfaces [59, 60]. An important observation to be noted here is that the facet size $2y_0$ in the equilibrium shape is proportional to the step free energy β_0, which indicates that the disappearance of the facet due to the roughening transition presupposes the step free energy to be zero. This is also obvious from the graphical representation of the step free energy in Figure 11.10b; that is, when $\theta = 0$, $f(\theta, T) = 0$.

The idea of the roughening transition was initially proposed by Burton *et al.* [61], who regarded steps and surfaces, respectively, as one- and two-dimensional entities in the Ising model. Based on this assumption, they proposed that surfaces should have a critical temperature – that is, a roughening transition temperature, where the order–disorder transition takes place. However, a rigorous analysis performed by Chui and Weeks [62] demonstrated that the roughening transition is not the order–disorder transition but rather the Kosterlitz–Thouless transition [63, 64], which designates an *infinite order transition*. This means that the roughening transition takes place continuously, like the ordinary thermodynamic second-order transition, but involves no disruption in symmetry. It follows that, even after the roughening transition, the crystal surfaces still hold their crystalline features. However, the correlation among atoms consisting of the surfaces disappears completely, so that atom movements along the surfaces have practically no energy barrier due to spontaneously generated steps and kinks. This was demonstrated by Chui and Weeks [65], who showed that the correlation length ξ, which is inversely proportional to the step free energy, follows the identical behavior with that of Kosterlitz–Thouless transition, regardless of the type of crystal:

$$\xi,\ \beta_0 \propto \exp\left[-c/\sqrt{T_R - T}\right], \tag{21}$$

where c and T_R denote a constant and the roughening transition temperature, respectively.

11.2.1.4 Kinetics of Surface Migration

Naturally grown crystals usually exhibit various faceted surfaces, and the interplanar angle measured between a given pair of the facets is constant, regardless of the overall shape. It has been of great interest to scientists why these faceted surfaces are always crystallographically specific, and how they can survive during crystal growth. As shown schematically in Figure 11.12a, it was speculated that the external morphology of crystals is a consequence of the internal periodicity, and that the growth of crystals takes place layer-by-layer so that the flat surface can survive throughout the growth process. Although this hypothesis was logical, at least two additional questions that are rather fundamental had to be answered:

- Why do only the surfaces with a higher atomic density survive?
- Why do atoms not sit on top of the layer before it completes, although the probability of sitting on top and flank is the same?

It was not until Gibbs [13] elaborated on the nucleation concept as a necessary prerequisite for phase transitions to take place that a clear answer to these questions was given.

For simplicity, consider a surface with a step, as shown in Figure 11.12b. Roughly speaking, there are five different positions for atomic attachment from the energetic point of view. When this surface is placed in an environment that is supersaturated with its own vapor, atoms tend to be deposited onto the surfaces. Note that there are always kink sites at steps at any finite temperature [61]. When a supersaturation is given to the system, atoms tend to nucleate onto the crystal surface. Atoms nucleated on the crystal surface then tend to migrate until they find the kink sites. The chance of

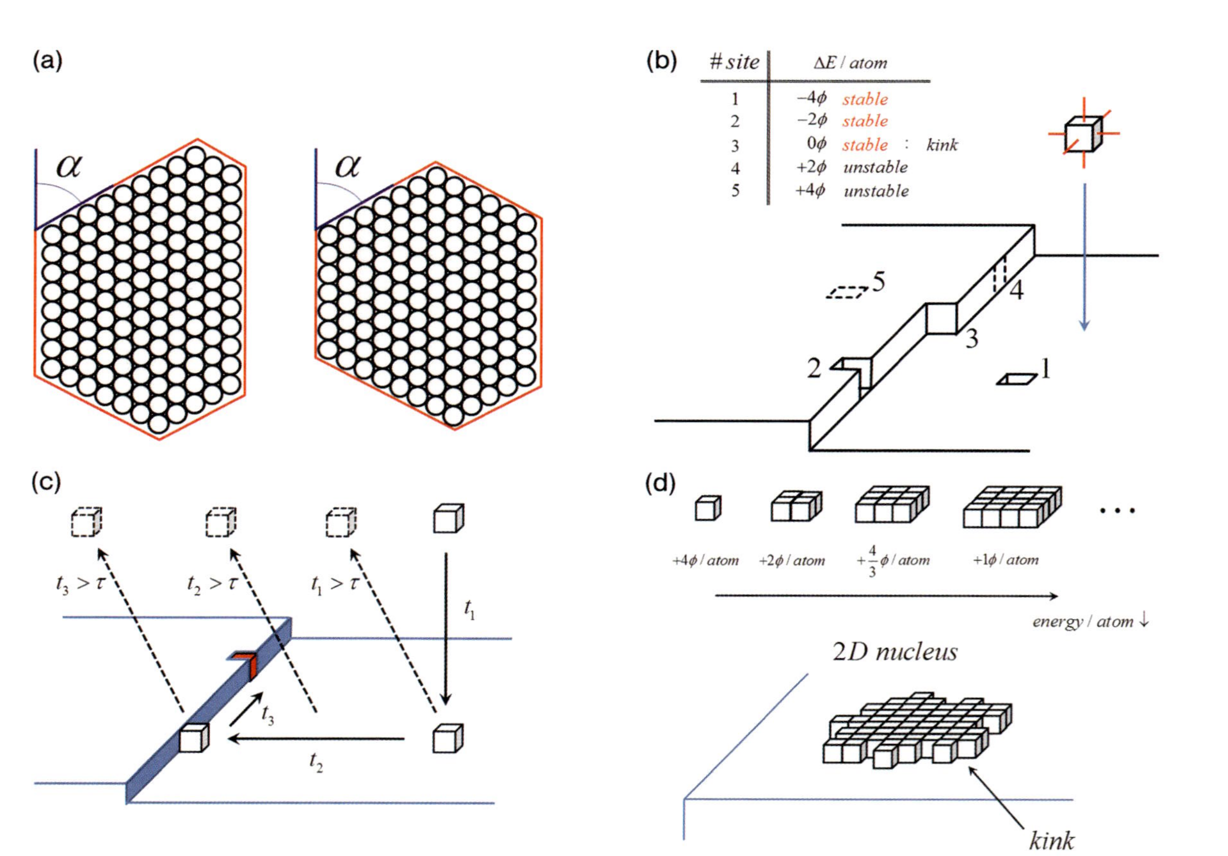

Figure 11.12 (a) A hard sphere model that explains the constancy of interplanar angle; (b) Possible sites for atomic attachments on the atomically flat crystal surface. Note that the promotion of atomic attachments in the presence of steps is due to the kink sites on them; (c) Stages for an atom to be incorporated into a crystal surface; (d) A schematic showing a 2-D nucleus. Note that due to thermal roughening, the edge of a 2-D nucleus consists of kinks.

atoms finding kink sites to be a part of the crystal is closely related to the number density of steps – that is, kinks. If atoms cannot find kink sites during the given relaxation time (τ), they are rejected from the crystal surface and will go back to the ambient due to energetics (see Figure 11.12c). This indicates that the higher the surface energy, the easier the atomic attachment, and explains why the surfaces with high energy – that is, a high step density – are usually not observed in macroscopically grown crystals.

Based on the discussion so far, it is easy to imagine that atoms deposited onto a terrace without any step cannot be a part of the crystal, and for this terrace to advance a step-generating source is needed. Figure 11.12d shows schematically the 2-D nucleation process which introduces steps to the terrace. Note that the number of broken bonds per atom for a group of atoms is always smaller than that for an individual atom when deposited onto a terrace. By supposing that the number of atoms is large enough, the shape of a 2-D nucleus can be assumed to be a disk due to the thermal fluctuation of monoatomic height steps. The energy cost required to produce such a nucleus can then be estimated as follows:

$$\Delta G = -\pi h r^2 \Delta G_V + 2\pi r \beta_0, \tag{22}$$

where ΔG_V denotes the free energy change for the phase transition from vapor to solid. From Eq. (22), the energy barrier for 2-D nucleation is given as

$$\Delta G^*_{2D} = \frac{\pi \beta_0^2}{h \Delta G_V}. \tag{23}$$

Since the nucleation rate is proportional to $\exp(-\Delta G^*_{2D}/kT)$, the growth rate as a consequence of the 2-D nucleation process has an exponential dependence on the driving force given by the supersaturation. Therefore, up to a certain level of driving force, practically no growth can be detected. In the meantime, it should be noted that the energy barrier for 2-D nucleation has a quadratic dependence on the step free energy, which vanishes with the roughening transition. In the case of the rough surface, 2-D nucleation is no longer needed due to spontaneously generated steps on the terrace. In this case, the growth rate has a linear relationship with the driving force.

Furthermore, the presence of defects – especially the screw dislocation emerging at the crystal surface – provides ledges containing a number of kink sites. These can serve as a perpetual source for the growth, with "no energy barrier" for the atomic attachment [61], as shown schematically in Figure 11.13a. In fact, it is well established experimentally that the growth of single crystals from a vapor or a melt is enhanced by screw dislocations [61, 66]. One of the interesting characteristics of this growth mode is that the ledge spacing of the growth spiral remains constant during the course of growth. Therefore, the growth rate in the presence of screw dislocation is given as [67, 68]:

$$v = \frac{C\sigma^2}{\sigma_c} \tanh\left(\frac{\sigma_c}{\sigma}\right) \cong \left\{ \begin{array}{l} \dfrac{C\sigma^2}{\sigma_c} (\sigma \ll \sigma_c) \\ C\sigma(\sigma \gg \sigma_c) \end{array} \right\}, \tag{24}$$

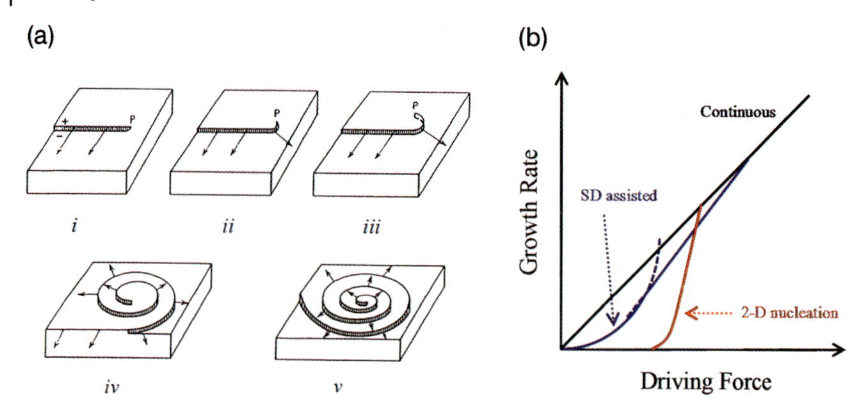

Figure 11.13 (a) Successive stages of step advancement originating from a screw dislocation (P) from *i* to *v* (after Ref. [38]); (b) Growth rate versus driving force diagram for three different types of growth mode.

where C and σ_c denote, respectively, a numerical parameter and a critical driving force where growth mode changes from quadratic to linear. As is evident from Eq. (24), the advance of the atomically flat surface is always possible in the presence of screw dislocation, even under an extremely small supersaturation.

Figure 11.13b summarizes the relationship between the rate of interface migration and the driving force. Note that there is a crossover in the growth rate between the screw dislocation and 2-D nucleation-assisted growths due to the exponential dependence of 2-D nucleation-assisted growth. This means that in the regime, where 2-D nucleation-assisted growth is active, it could dominate over the screw dislocation-assisted growth [69, 70].

11.2.2 Solid/Liquid Interfaces

In the case of S/L interfaces, two different types can be postulated. The first interface is between a solid and its own liquid phase, such as that between ice and water. This type of interface is relatively easy to handle, and provides an insight into the fundamental features of S/L interfaces. The second interface is between a solid and a liquid of different composition. In many practical situations, the S/L interfaces belonging to the latter type are particularly important. Note that the liquid phase sintering is carried out clearly below the melting point of the major solid phase. Here, the initial discussions relate to some of the essential features of S/L interfaces, derivable from the former type of S/L interface. The discussion can then easily be extended to the effect of a liquid phase of a different chemistry on the structures and energies of crystal surfaces.

The S/L interfaces, in general, have much in common with the surfaces from the structural point of view, in that both the vapor and liquid in contact with a crystal surface are a structureless fluid. However, the presence of a short-range structural order clearly distinguishes a liquid from a vapor. Bernal [72] noted that the short-range structural order derives from the fact that liquids tend to have as high a density as possible, without introducing a long-range translation symmetry, and proposed that all these requirements would be met with a dense random packing (DRP) model. The DRP model is featured by the presence of a predominant fivefold rotational symmetry among tetrahedron building units. It should be noted that the fivefold rotational symmetry disrupts the long-range translation order, because it cannot fill up the space in a regular way. Based on the DRP model, Spaepen [71] developed a model describing S/L interfaces, as shown in Figure 11.14.

Figure 11.15a and b show the crystal morphology of individual $Pb(Mg_{1/3}Nb_{2/3})O_3$–35 mol% $PbTiO_3$ (PMN–35PT) ceramics constantly in contact with vapor and with liquid during heat treatment, respectively. No noticeable trace of the development of any curved surfaces is detectable in (b), while curved surfaces – which are the

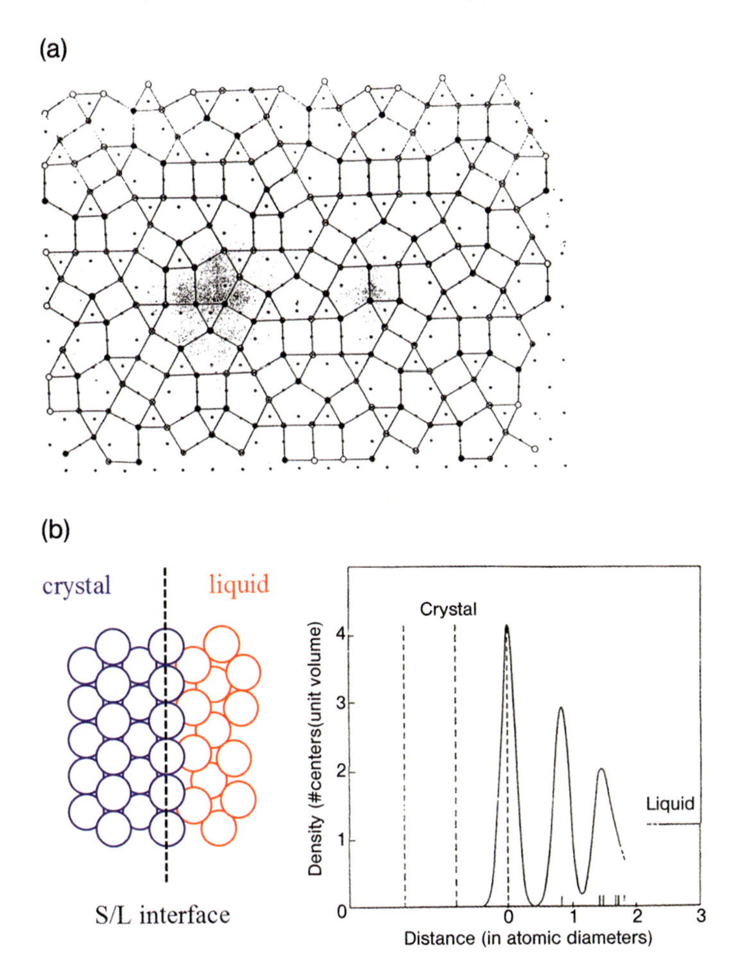

Figure 11.14 (a) Structure of the first layer of S/L interface on a {111} surface; (b) Schematic side view near the S/L interface showing how close-packing structure of a crystal decays in the liquid phase as the distance from the interface increases. After Ref. [71].

Figure 11.15 (a) As-sintered surface and (b) a fractured surface after leaching out the liquid phase in PMN–35PT ceramics sintered at 1200 °C for 100 h.

hallmark of a reduction of anisotropy in the surface energy – are quite common in (a). Because of a high volatility of PbO, the composition of both the vapor and liquid phases can be assumed to be more or less similar, in that they contain PbO as a major constituent. It follows that the effect of the presence of a liquid phase is basically the same as, but stronger than, the adsorption effect of a vapor phase.

Figure 11.16a and b show the typical microstructures of pure and VC-doped WC when heat-treated in a cobalt (Co) liquid under vacuum at 1500 °C, respectively. The addition of VC results in a drastic change in the overall shape of the WC crystals dispersed in a Co matrix. It is clear that the role of VC in this case is to increase the anisotropy in the S/L interfacial energy. The mechanism involved can be delineated from additional experimental facts presented in Figure 11.16c, d, and e. Yamamoto *et al.* [74] investigated the effect of VC on the grain growth behavior of WC by using high-resolution transmission electron microscopy (HRTEM), and showed that the added VC, regardless of the presence of a Co liquid phase, was located exclusively at the interface between WC/WC and WC/Co. Subsequent chemical analyses clearly showed the degree of adsorption at the interface to be anisotropic – that is, to be highest on a (0001) basal plane. It is of interest to note that this increase in anisotropy was significantly retarded by the addition of VC in the absence of a Co liquid phase. As summarized in Figure 11.16f, the morphological evolution from the truncated to perfect triangular prisms is attributed to the increased anisotropy due to the preferential adsorption of VC and Co on the WC surfaces.

This adsorption of VC appears to increase the step free energy of the WC/Co interface. The increase in the step free energy makes the 2-D nucleation barrier in Eq. (23) greater, which will in turn markedly retard the 2-D nucleation-assisted growth. In addition, it increases the step spacing of the screw dislocation and thereby reduces the screw dislocation-assisted growth rate. The effect of the addition of VC on the marked inhibition of the coarsening rate of WC grains, as shown in Figure 11.16a and b, could be attributed to the increase in the step free energy. The size ratio of large to small grains in Figure 11.16b is much greater than that in Figure 11.16a, indicating that the VC addition makes the abnormal grain growth behavior more pronounced. Choi *et al.* [75] demonstrated that the increase in the step free energy is correlated with

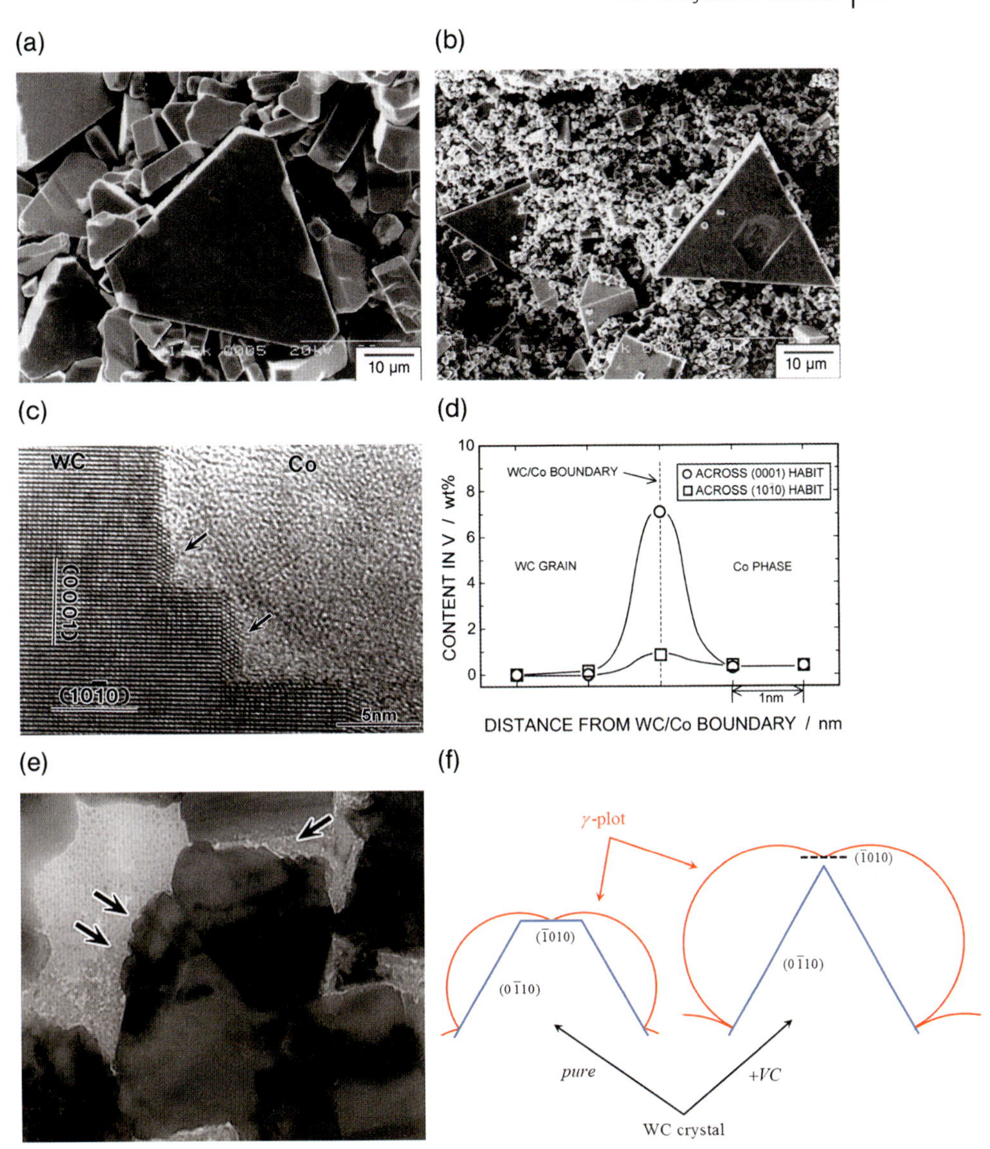

Figure 11.16 Scanning electron microscopy images of WC crystals after leaching out Co liquid phase from (a) WC-30Co and (b) WC-30Co-1VC (all in wt%). Both specimens were sintered at 1500 °C for 72 h (from Ref. [73]); (c) HRTEM image of the precipitates on (0001) habits in WC-12Co-0.5VC and (d) concentration profile of V across the S/L interface. Note the anisotropic nature of adsorption; (e) A VC-doped WC system without Co liquid phase. From the facets, the equilibrium shape is estimated to be a truncated triangular prism (from Ref. [74]); (f) The Wulff constructions, demonstrating the effect of VC addition.

the pronounced abnormal grain growth, which is attributed to the increase in the 2-D nucleation barrier.

11.2.3
Solid/Solid Interfaces

11.2.3.1 Fundamentals

Solid/solid interfaces, in general, include all types of interfaces in microstructures, such as grain boundary, stacking fault, twin boundary, both coherent and incoherent phase boundaries, and so forth. Special attention should be paid to the grain boundary, because it is the most frequent defect among those listed. This means that the microstructure evolution requires a clear understanding of the structure and energetics of grain boundaries. The grain boundary refers to an interface where the long-range translational symmetry of a crystal abruptly changes. Compared to the surfaces and S/L interfaces, grain boundaries are much more complicated in character with respect to their structure as well as their energetics. Such complexity mainly derives from the fact that no less than eight degrees-of-freedom (DOF) should be specified to describe even a single grain boundary completely – namely, five macroscopic DOF for the crystallographic characterization, and three microscopic or translational DOF for the determination of its energy [76]. Note that just a single degree of freedom – that is, surface normal – is good enough to define the solid/fluid interfaces.

Figure 11.17a shows, schematically, a general grain boundary formed by two randomly oriented crystals. When the normal vectors of crystallographic planes are designated with the same Miller indices of arbitrary choice, as $\hat{n}_1$ and $\hat{n}_2$, it is easy to

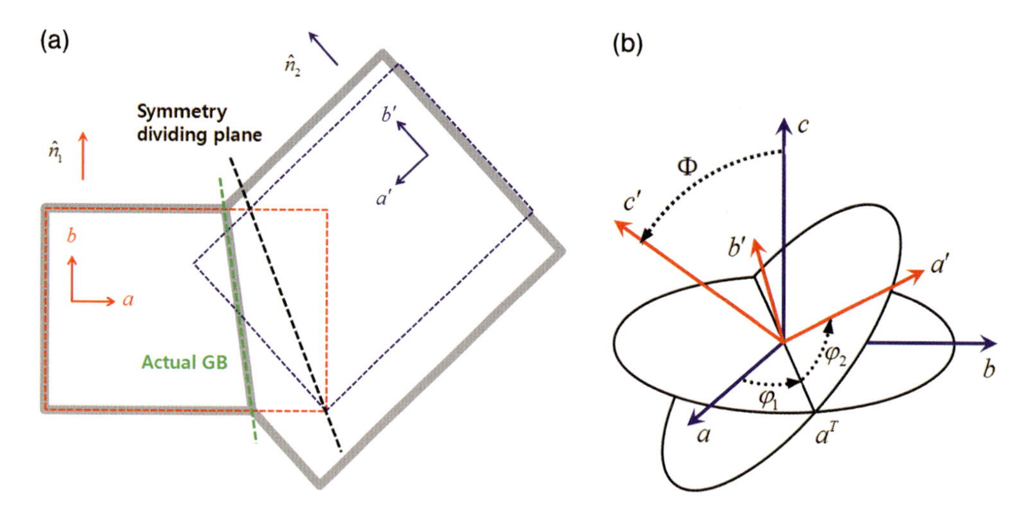

Figure 11.17 (a) Schematic illustration showing the meaning of macroscopic DOFs in two-dimensions, which is easily extendable to three-dimensions; (b) Illustration of the Euler's rotation theorem, which states that any arbitrary rotation can be described by three angular parameters, namely (φ_1,Φ, φ_2).

find the *common plane* that is parallel to $\hat{n}_1$ and $\hat{n}_2$ at the same time. The direction of the normal vector of this common plane is termed the *common axis*, and the plane which is perpendicular to the common plane is the *symmetry dividing plane*. Now, it is possible to bring two crystals into a single coordinate system by the use of three DOFs: $\hat{n}_1$, $\hat{n}_2$, and the misorientation angle μ that refers to the angle between $\hat{n}_1$ and $\hat{n}_2$. For this simple physical concept to be applied in practice, however, some mathematical treatments are needed. One of the most common procedures is to employ the Euler's rotation theorem, as illustrated in Figure 11.17b.

According to the Euler's rotation theorem, any arbitrary rotation in 3-D space can be described by three angular parameters, namely (φ_1,Φ, φ_2). For instance, to express the local coordinate system (a', b', c') with respect to a reference coordinate system (a, b,c), it is necessary only to apply three transformation matrices in the following order: the first rotation by an angle φ_1 about the c axis; the second rotation by an angle Φ about the transformed a axis (a^T axis); and the final rotation by an angle φ_2 about the transformed c axis (c' axis). Therefore, the symmetry-dividing plane can be determined unambiguously by the use of three angular parameters. In other words, three DOFs which are needed to define the misorientation relationship of two crystals can be said alternatively to be three Euler angles, (φ_1,Φ, φ_2).

As is evident from Figure 11.17a, however, this simple approach is insufficient for a complete description of grain boundaries. Note that the actual dividing plane (i.e., the grain boundary) generally deviates from the symmetry-dividing plane, except for the case of the symmetrical tilt grain boundaries (STGBs). This argument can be best visualized from the electron-beam back-scattered diffraction (EBSD) analysis on a penetration-twinned PMN–35PTceramic, as shown in Figure 11.18. Note that EBSD mapping only determines the three Euler angles which are required to make two crystals identical. As a consequence, two more DOFs – namely an appropriate amount of tilting and twisting of the symmetry-dividing plane – are required to adjust the deviation. These five DOFs, three Euler angles and tilt/twist angles, are often called the "macroscopic DOFs," as they characterize the crystallographic aspects of grain boundaries macroscopically [76]. Although the five macroscopic DOFs are usually good enough for most practical purposes, it should be noted that no information on the grain boundaries at the atomistic level can be specified merely by those macroscopic DOFs.

To stipulate the atomistic level geometry of grain boundaries, three additional DOFs, namely the translational vector $T = (T_x, T_y, T_z)$, are required. The first two components, T_x and T_y, account for a discontinuity in atomic arrangement or stacking, as they represent a translation parallel to the interface. T_z, which depicts a translation perpendicular to the interface, determines the grain boundary free volume that is closely related to the grain boundary energy [79, 80]. These three components of the space vector $T$are often called the "microscopic DOFs," as they do not affect the macroscopic crystallography of the grain boundaries.

When any deviation from the perfect crystal requires any of the macroscopic DOFs to be mediated, the related boundary can be termed the *macroscopic grain boundary*. Typical examples are general grain boundaries and special grain boundaries such as coincident site lattice (CSL) boundaries and twins. However, when only the micro-

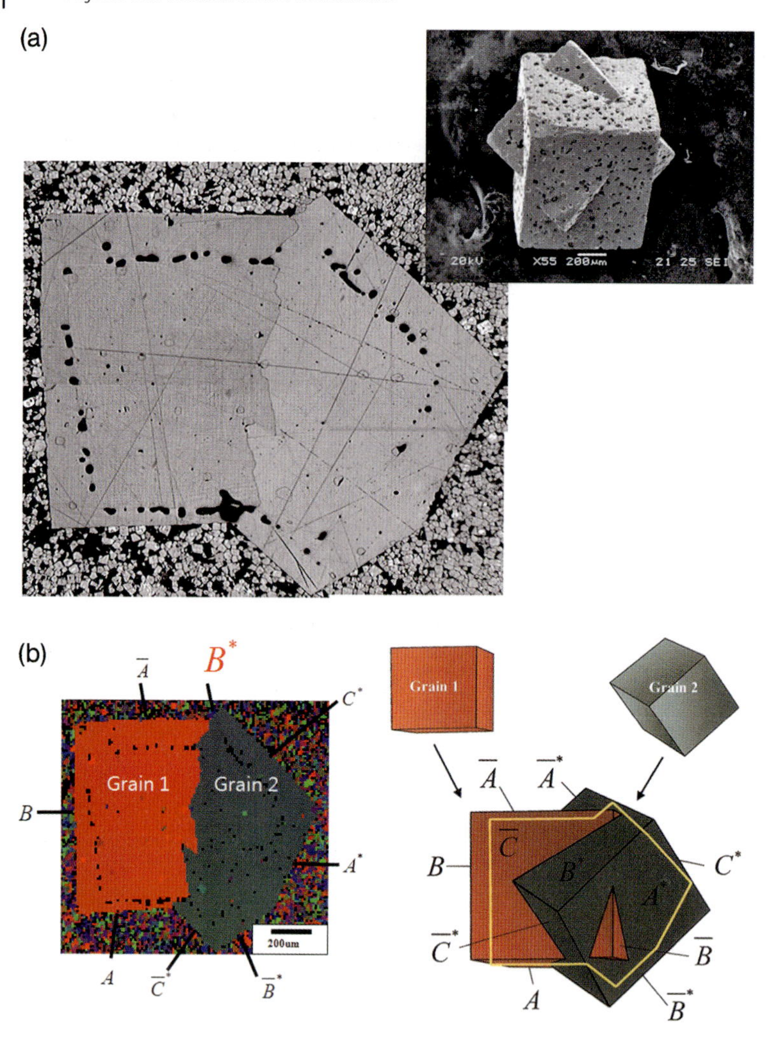

Figure 11.18 (a) Optical micrograph of an abnormally grown twinned PMN–35PT ceramic with its 3-D shape; (b) Electron-beam back-scattered diffraction analysis shows the crystallographic relationship between two domains in the twinned PMN–35PT ceramic. After Refs [77, 78].

scopic DOFs are good enough to bring two crystals into coincidence, the related boundary can be called the *microscopic grain boundary*; this is usually not treated as grain boundary, but rather as a special type of interfacial defect. The stacking fault is a typical example which belongs to this classification. Consequently, in this chapter the term *grain boundary* will be restricted to a reference to the macroscopic grain boundary.

11.2.3.2 **Structure and Energy**

The grain boundary is an interface where the regular periodicity of a crystalline phase undergoes an abrupt change. This change is featured by a slight or significant alteration in the crystallographic orientation between two neighboring grains on either side. Due to this structural discontinuity, it is inevitable that a certain level of disorderliness will result at the grain boundaries. In that such disorderliness derives directly from the unsaturated bonds at the grain boundaries, it is easy to expect that the degree of disorderliness increases as the misorientation between the neighboring grains in contact increases.

The first theoretical model quantifying disorderliness due to mismatched atomic bonds at the grain boundaries was developed by Read and Shockley [81]. Inspired by the dislocation model proposed by Burgers [82] and Bragg [83], these authors proposed that low-angle grain boundaries be interpreted as a planar defect composed of a periodic array of dislocations; that is, as parallel arrays of edge dislocations for tilt grain boundaries, and as hexagonal or square networks of screw dislocations for twist grain boundaries. This so-called "lattice dislocation model" was soon consolidated by clear experimental evidences [84–86], and has since become a standard model for the grain boundaries with a relatively small misorientation angle.

To date, high-angle grain boundaries have been far less understood. The increase in the misorientation between grains results in the overlap of dislocation cores when the misorientation exceeds about 10–15°. This implies that almost all high-angle grain boundaries are practically identical from the structural point of view. However, a number of experimental facts have shown that there are several high-angle grain boundaries, the atomistic structures of which are significantly simpler than that predicted by the lattice dislocation model [38, 84, 87, 88]. Several models have been proposed to account for the experimentally observed special high-angle grain boundaries, such as the coincidence site lattice (CSL) model, a "modified CSL model with structural units," a "near-coincidence model," and a "plane-matching model" [88]. Nevertheless, no general model conforming to all the experimental observations has yet been proposed. One point to be noted here is that, even though no model that perfectly explains the experimental observations is available, the grain boundary has at a certain misorientation a superlattice that makes it a thermodynamic phase distinctive from its bulk. This aspect can be intuitively grasped from the CSL model, which forms the basis for the other models.

Figure 11.19a shows a $\langle 110 \rangle$ symmetrical-tilt $\Sigma 3$ CSL boundary, which forms when an FCC lattice is rotated 70.53° along the $\langle 110 \rangle$ axis. As is evident from Figure 11.19a, every three lattice points in the original lattice constitute a superlattice (which is the reason why this superlattice is termed the $\Sigma 3$ CSL lattice). It should be noted that the CSL lattice develops along $\{111\}$ plane, which is identical to that forming when the lattice is rotated 60° along $\langle 111 \rangle$ axis. The terminology ΣN refers to the fraction of the CSL lattice points to the total number of the original lattice points. As the fraction is $1/N$, it is noted that the lower the N value, the smaller becomes the lattice mismatching between two crystals. However, a small CSL number does not necessarily mean that the corresponding boundary has a low energy with a good atomic matching, because a good lattice matching in the CSL model is relative. In other words, the degree of lattice matching depends on the reference plane. This argument becomes clearer with the following example.

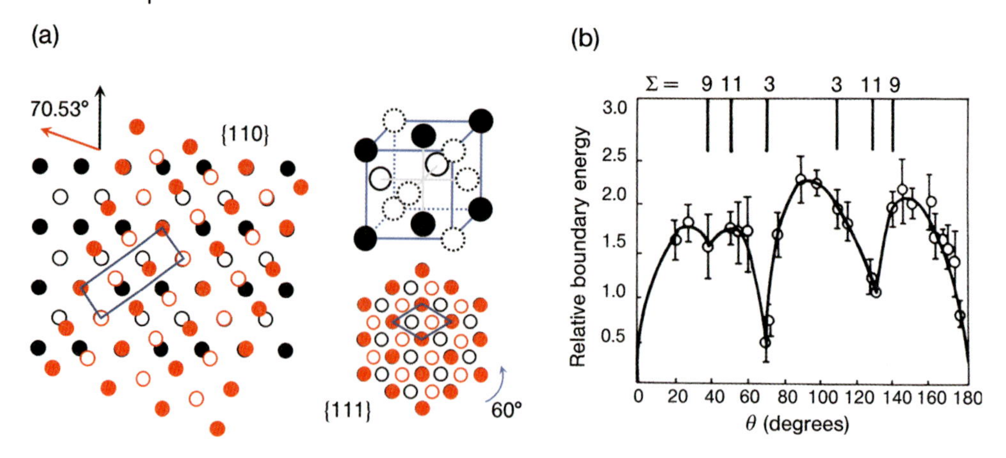

Figure 11.19 (a) Σ3 CSL boundary in FCC crystal. The solid and open circles represent the lattice points on {110} and {220} planes, respectively; (b) Experimentally determined grain boundary energies of aluminum as a function of misorientation. From Ref. [89].

Figure 11.19b shows the energies of experimentally determined ⟨110⟩ symmetrical-tilt grain boundaries of aluminum near its melting point. Note that even the grain boundaries with the same Σ values have a significantly different energy. For example, the energy of Σ3 CSL boundary at 70.53° is far smaller than that at 109.47°. The difference between these two Σ3 boundaries comes first from the fact that the crystallography of the boundary is different. The boundary plane for the former is {111} that is the densest in the pure metallic fcc crystals, while the boundary plane for the latter {112} has a significantly open structure. With respect to the first-nearest broken bond model in Eq. (5), the surface energy of {112} is about 1.18-fold as high as that of {111}. This means that the energy of Σ3, {112} should be at least 1.18-fold higher than that of Σ3, {111}, although everything else is identical.

Figure 11.20 shows, schematically, two Σ3 CSL lattices of current interest. Although the coincident site occurs in every three lattice points for both cases, the size and configuration of the CSL lattices are disparate. Furthermore, there is a significant difference in the degree of bonding mismatch, especially when more than the first nearest neighbor interactions are taken into account. This is quite evident from the different repeating unit at each grain boundary. Hence, it is no wonder that the energy of the CSL boundaries with a different composition plane should be different, even though the Σ values are the same. In fact, it is well known that there is very little correlation between grain boundary energy and the corresponding Σ value [90]. Nevertheless, the CSL boundaries are still worth paying attention to. As noted already in Figure 11.19b, any cusp in the energy diagram, which is a featured mark for the presence of anisotropy in the interfacial energies, corresponds to one of the CSL boundaries. This means that those grain boundaries orientated away from the cusps are also subject to faceting, just like the surfaces where the orientation is in between two cusped minima in the surface energy diagram.

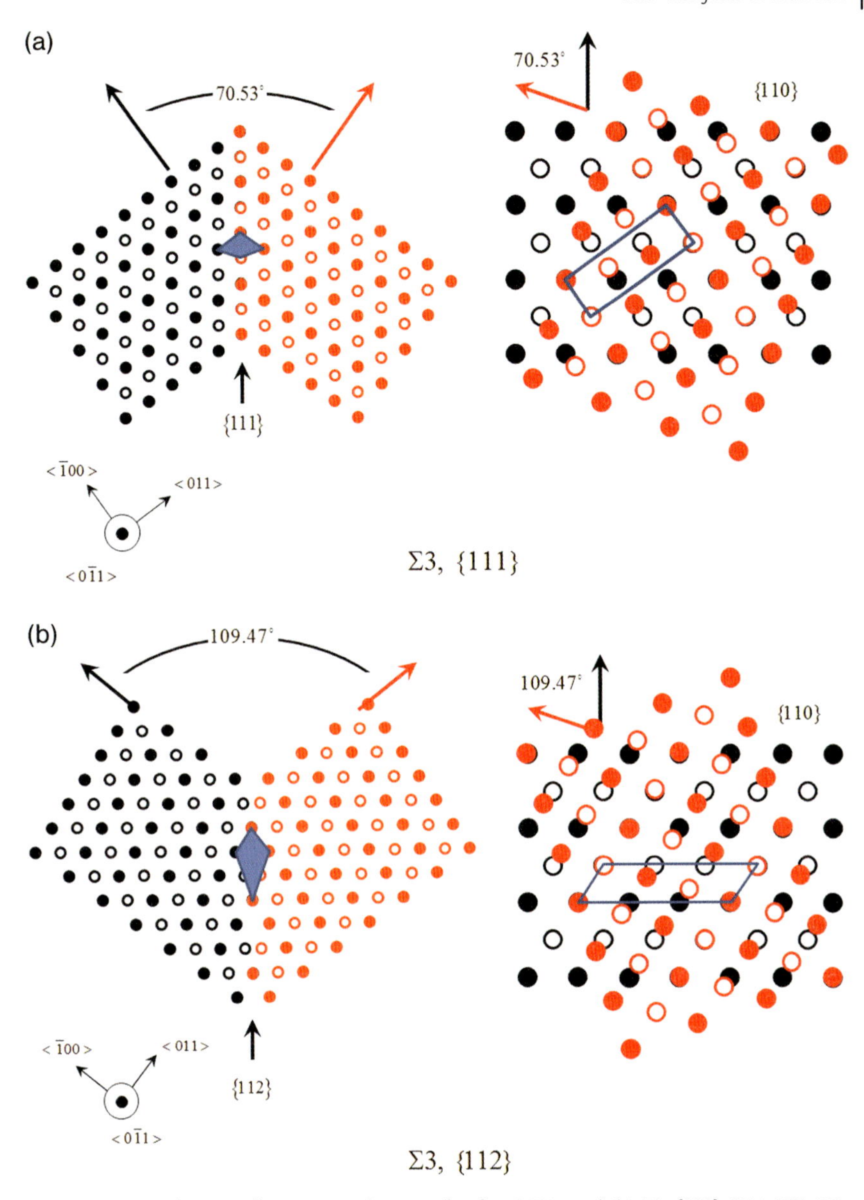

Figure 11.20 Schematic illustrations of (a) Σ3, {111} (70.53°) and (b) Σ3, {112} (109.47°) CSL boundaries occurring in the ⟨110⟩ symmetrical tilt (left side) and twist (right side) boundaries. The shaded region marked on the boundary region represents the repeating unit of each boundary.

The CSL model was elaborated initially for pure metallic crystals where lattice points can be considered as individual atoms, such that the coincidence in lattice points is equivalent to the formation of atomic bonds. However, it is interesting to note that the CSL model also holds true in ceramic systems. The energies and structures of various

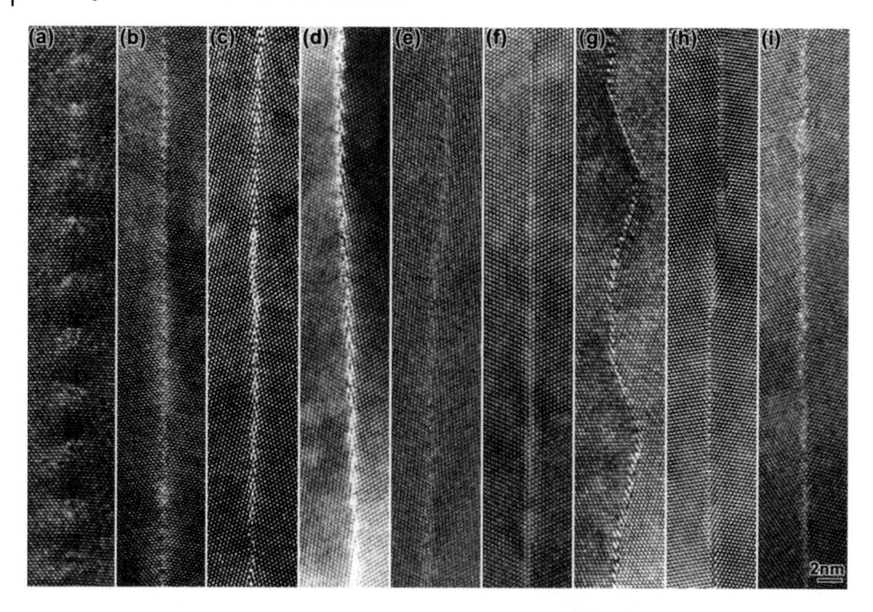

Figure 11.21 HRTEM images of symmetrical tilt CSL boundaries in yttria-stabilized zirconia. Misorientation angle is between {100} surfaces and the common axis {110}: (a) 5.0°; (b) 20.0°; (c) 39.0° (Σ9, {221}); (d) 50.4°, (Σ11, {332}); (e) 60.0°; (f) 70.6° (Σ3, {111}); (g) 109.4° (Σ3, {112}); (h) 129.6° (Σ11, {113}); (i) 141.0° (Σ9, {114}). From Ref. [91].

CSL boundaries in ceramic systems have been investigated intensively during the past decade with the aid of technological advancements in HRTEM [91–94]. Figure 11.21 shows a series of HRTEM images on the ⟨110⟩ symmetrical-tilt grain boundaries in an yttria-stabilized cubic zirconia bicrystal: (a) , a low-angle grain boundary; (b) and (c) , high-angle grain boundaries; (c, d, and f–i), CSL boundaries.

It is clear that the low- and high-angle grain boundaries, even in ceramics, are comparable to pure metals in that they are completely different form each other from the structural point of view. It must be remembered that from the low-angle grain boundaries are represented as a periodic array of edge dislocations, the grain boundaries for which the misorientation exceeds about 15° cannot be described in the same way [95]. Note the difference in the structure by comparing Figure 11.21a with Figure 11.21b or e. The high-angle grain boundaries consist of certain repeating units with their own structure, while the low-angle grain boundary has periodically spaced edge dislocations. It is also clear that there is a notable distinction in the grain boundary free volume and the degree of lattice matching between a pair of the CSL boundaries with the same Σ value. It is interesting to notice that the Σ3, {112} boundary in (Figure 11.21g) is faceted with {111} and {115}. This strongly indicates that {111} (Σ3) and {115} (Σ19*a*) in this system have a cusped minimum in the energy diagram [96].

Figure 11.22 is an example showing why the grain boundary faceting occurs, and what it looks like when there is a misorientation between two crystals in contact. Figure 11.22a shows a TEM image of the interface formed between two slightly misoriented sapphire single crystals. Roughly three distinctive regions, denoted by A, B, and C, are noted. As discussed above for the yttria-stabilized zirconia bicrystal in

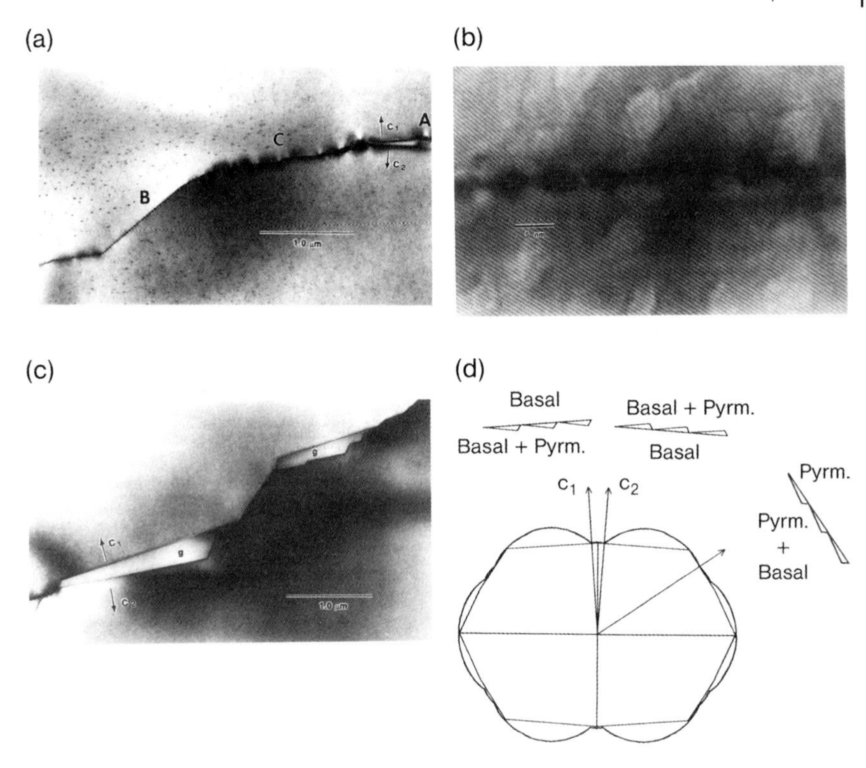

Figure 11.22 (a) An interface formed by joining two sapphire single crystals, the basal planes of which are misoriented at around 3.5° from the polished surface to be joined. Three distinctive regions are noted. A: fully wetted interface by anorthite glass, B: low-angle boundary, and C: a mixture of A and B; (b, c) Magnified images of regions B and C, respectively; (d) A corresponding Wulff construction of the bicrystal. From Ref. [92].

Figure 11.21a, a low-angle grain boundary of alumina is also composed of periodically spaced dislocations, as is evident from the HRTEM image in Figure 11.22b. Figure 11.22c shows a region where the grain boundaries are separated by liquid pockets; here, the grain boundary is faceted with (0001) and $\{1\bar{2}13\}$, which are known to be two major facets in the equilibrium shape [97]. It should be noted that the faceting behavior of the S/L interface and the grain boundaries is identical, which indicates strongly that the grain boundary faceting takes place for the same reason as does the surface and S/L interface faceting.

11.3
Practical Implications

Sintering is a process which converts a mass of powder into a dense product. This is achieved simply by introducing heat into the system, sufficiently below the melting temperature. The thermodynamic driving force for the process is a reduction in the interfacial free energy, caused by a replacement of the solid/vapor interfaces of powder compacts with the solid/solid or solid/liquid interfaces. The kinetic paths to reduce the excess free energy are *coarsening*, which reduces the total interfacial area, and *densification*, which replaces the interfaces of high energy with those of low energy. Since the birth of the powder metallurgy as a field of science during the mid 1940s by Frenkel [98], a number of research activities have been conducted in order to understand how the sintering process takes place. One of the most denoted and systematic analyses on the topic was provided by Kuczynski [99]. Since the real sintering process is too complicated to be modeled, Kuczynski formulated all of the possible diffusion processes occurring in an extremely simplified system – that is, two spherical particles of the same size in mutual contact. These studies were summarized in the very elegant and simple form of $x^n \propto t$, where n depends on the type of dominant diffusion mechanism.

Based on this two-particle model, Kingery [100] developed the densification theory of LPS, As the results based on this two-particle model are too simple to describe the microstructure evolution as a whole, Park and Yoon [101] and Kang *et al.* [102] reported the theoretical treatment of a more realistic model for LPS, where the stability of the pores inside the sintered body depends on whether or not the curvature of pores is larger than that of the liquid meniscus formed at the surface. According to the theory, with an increasing amount of the liquid phase the pores become unstable, which is in agreement with the observation that the densification during LPS is enhanced with an increasing amount of the liquid phase. Moreover, as the grains coarsen, the pores become unstable; this aspect is also in agreement with the observation that densification proceeds with coarsening.

By comparison, the invention in 1957 of a translucent alumina named as Lucalox™ by Coble [103–105] roused the importance of understanding the densification mechanism during sintering. Coble found that, if a small amount of magnesia was incorporated into pure polycrystalline alumina,[8)] then the coarsening of alumina grains was significantly retarded and the density of the specimen greatly enhanced at the same time. In other words, pure alumina doped with magnesia nearly reaches its theoretical density with a fine and uniform microstructure, eventually producing translucent alumina, whereas undoped pure alumina tends to exhibit an abnormal grain growth behavior with many pores trapped inside the abnormally grown large alumina grains, resulting in a poor density. Based on these experimental facts, it was considered that if the coarsening process were to be fast enough to trap the pores into the bulk interior, then no further densification would be practically possible because the elimination of pores from the bulk interior would require a relatively stagnant lattice diffusion process.

As a result, many of the later theoretical treatments of this topic were focused on how the coarsening process affected the densification process. In this respect, Burke [106] showed that the pores should remain at the grain boundary in order for polycrystalline materials to reach their full density. Later, Brook [107] suggested a condition whereby the pores would remain at the grain boundary, based on the Zener

8) Notice that the word "pure" in the context does not indicate the "pure" from the recent technological point of view.

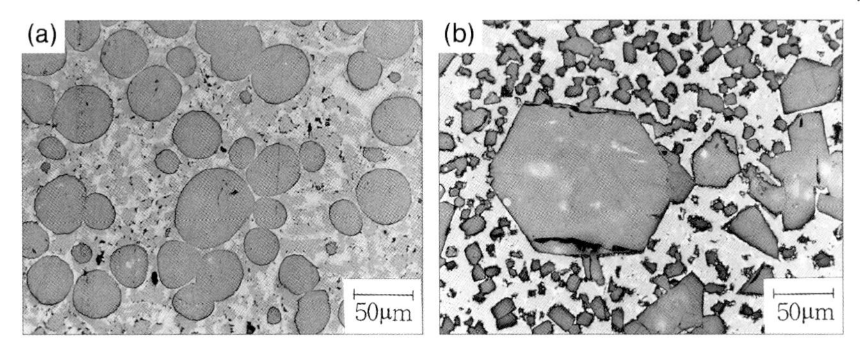

Figure 11.23 Microstructure of (a) dicalcium silicate and (b) tricalcium silicate dispersed in a clinker melt (1450 °C for 16 h). From Ref. [109].

drag effect. By assuming that the pores at the grain boundary would reduce the velocity of the grain boundary, Brook derived the condition for pores to migrate with the grain boundary. A more rigorous treatment on the so-called "pore–boundary separation" problem was elaborated by Hsueh *et al.* [108], this is generally considered to be one of the most acceptable works to date, as the model considers not only the mobility of pores and grain boundaries but also the dihedral angle of the system. Nonetheless, all of the theoretical models yet proposed have lacked any consideration of the structural transition of the interface and the frequent presence of the liquid phase.

Recently, increasing attention has been paid to the important role of interface structural transition on the microstructure evolution during sintering. This new approach was first applied to the coarsening behavior of ceramic materials dispersed in a liquid matrix [6, 7, 109, 110]. Figure 11.23 is a good example showing the effect of the interface structure on coarsening behavior. As discussed above, it is evident that the higher the interface anisotropy (or the higher the step free energy), the stronger the abnormal grain growth behavior becomes. This is because the advance of interfaces in the system with interface anisotropy is only possible when a layer-by-layer growth takes place, which is preceded by 2-D nucleation. It should be noted that, during LPS the driving force for an individual crystal to grow is given by the size advantage. Hence, only those grains that are large enough to induce 2-D nucleation can grow exclusively at the expense of smaller grains, and this results in a bimodal size distribution or abnormal grain growth in the final microstructure.[9)]

It should also be noted that, in order to induce 2-D nucleation by capillary driving force, the grain size should be in a nanometer range when the step free energy is approximated to the interface energy. To explain the abnormal grain growth by 2-D nucleation in real systems with grains of a few micrometers, the step free energy should be at most 20–30% of the interface energy [7, 112].

An enhanced understanding of the important role of the interface anisotropy on the microstructure evolution also raised a question on the conventional approach to the pore–boundary separation problem. It was noted that pores should remain at the

9) Details of this argument can be found in Ref. [7] and references therein.

grain boundary to be removed, as the grain boundary is an active source of pore removal. This means that if the pores are located in the intergranular liquid phase, they would be removed much faster, or at least stay at the advancing boundary, due to a fast mass transfer through the liquid phase. It should be noted that pore mobility is determined by the surface diffusivity of atoms from the front to rear side of the migrating pore. However, a number of experimental observations have shown that the trapping of pores into grains is far more pronounced in the presence of an intergranular liquid phase [78, 92, 111, 113–115]. It is clear from Figure 11.24 that the pore–boundary separation is closely related to the intergranular liquid phase. If a pore is located at the intergranular liquid film, there is no force balance and hence no barrier for the pore to be separated from the liquid film. This is in stark contrast to when a pore is located at the grain boundary, and suggests that the pore–boundary separation problem should also be reanalyzed since abnormal grain growth (as in the alumina system) is attributed to very thin intergranular liquid films, as indicated clearly by the elongated shape of abnormally growing grains [116].

MgO is known to increase the co-solubility of liquid, forming impurities in alumina such as CaO and SiO_2, and thereby acts as a scavenger of the liquid phase [117, 118]. This means that MgO-doped alumina tends to be free from thin intergranular liquid films. Consequently, the growth by 2-D nucleation is not operative in MgO-doped alumina, where abnormal grain growth is suppressed and

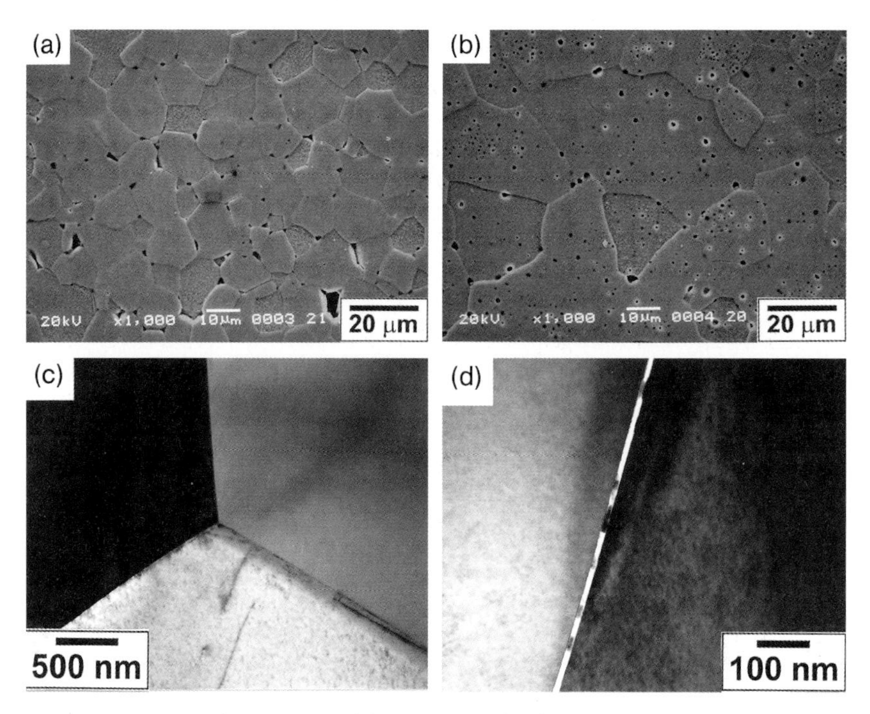

Figure 11.24 SEM and TEM images of the polished surface for (a and c) pure ZnO and (b and d) 0.05 mol% Bi_2O_3-doped ZnO specimens sintered at 1200 °C for 1 h. Note that Bi-doped ZnO exhibits a *wet* boundary, while pure ZnO exhibits a *dry* boundary. From Ref. [111].

there is a freedom from large, elongated alumina grains [112]. In fact Bae and Baik [119] showed that, even without MgO, alumina undergoes a normal grain growth as long as its purity is high enough so as not to induce a liquid phase during sintering. These authors proposed that the presence of the liquid phase in alumina is necessary for abnormal grain growth to occur. In this regard, it can be said that the role of MgO in suppressing the pore–boundary separation and abnormal grain growth is to remove a liquid phase.

However, even in the absence of thin intergranular liquid films, the tendency of pore–boundary separation during sintering is much higher in pure alumina than in MgO-doped alumina, which suggests that MgO has an additional role. It has been reported that MgO induces roughening of the interfaces in alumina such as surfaces, S/L interfaces, and grain boundaries [120–123]. This means that MgO-doped alumina should have much smaller step free energy than pure alumina, which was revealed by the surface structure [121]. Self-diffusion on the rough surface occurs much faster than that on the singular surface; similarly, self-diffusion on the surface with a smaller step free energy would be much faster than that with a larger step free energy. This indicates that the pore mobility of MgO-doped alumina would be much higher than that of pure alumina, resulting in a higher tendency for pore separation in pure alumina.

The dihedral angle of pores with grain boundaries has also been proposed to influence the pore–boundary separation behavior; notably, the higher the dihedral angle, the easier the separation [108]. Disordered grain boundaries in MgO-doped alumina, which can be assumed from the roughening effect by MgO, would have a narrow distribution of dihedral angles, whereas ordered grain boundaries in pure alumina would have a wider distribution of dihedral angles, as reported previously by Hendwerker *et al.* [124]. It is expected that pore–boundary separation would be easier in pure alumina with a wide distribution of dihedral angles, than in MgO-doped alumina.

The working mechanism for grain growth in the solid state, in the absence of a liquid phase, is quite different in many respects from LPS. In the case of LPS, the mass transfer occurs between individual grains and the liquid medium and, as long as there is a size distribution large enough to induce 2-D nucleation, the coarsening will take place. In contrast to the coarsening behavior during LPS (where coarsening occurs through atomic transfer, largely through the liquid medium), mass transfer during solid-state sintering is a purely local problem, being driven only by the chemical potential difference between two grains in contact. As the chemical potential difference is manifested by the curvature in the grain boundary, the grain boundaries will, in principle, move towards the center of curvature either until the boundary becomes flat, or until the smaller grain disappears. When only two grains of a different size are taken into account, the boundary migration would be completed only when the smaller grain disappears, because the difference in chemical potential becomes larger; that is, the boundary curvature becomes larger, as the boundary moves.

The intrinsic grain boundary mobility would depend on the grain boundary structure, with a disordered grain boundary migrating faster than an ordered boundary. When any impurity (such as solute atoms) is involved, the solute drag effect is higher in disordered grain boundaries than in ordered boundaries, as in the case of the effect of solute atoms on the mobility of random and special boundaries [126]. Grain boundary migration is also inhibited by the second phase or

Figure 11.25 Dense nanostructured Y_2O_3 induced by two-step sintering. From Ref. [125].

precipitates by the Zener drag effect [127]; indeed, fine precipitates are highly effective in inhibiting grain growth. The driving force for the precipitation of supersaturated solutes can easily be provided, normally by lowering the temperature of heat treatment from the high-temperature region. It is most likely that a dual heat treatment might explain the microstructure evolution of Figure 11.25, as reported by Chen *et al.* [125], who showed that a correctly chosen two-step heat treatment was highly effective in suppressing grain growth, and facilitating densification.

An important aspect to be noted here is that the migration of grain boundaries should be accompanied by the migration of their triple junctions, as shown schematically in Figure 11.26. Thus, the kinetics of grain boundary migration depends on that of triple junction migration, and *vice versa*. The migrations of grain boundaries are kinetically coupled to one another through triple junctions, and under

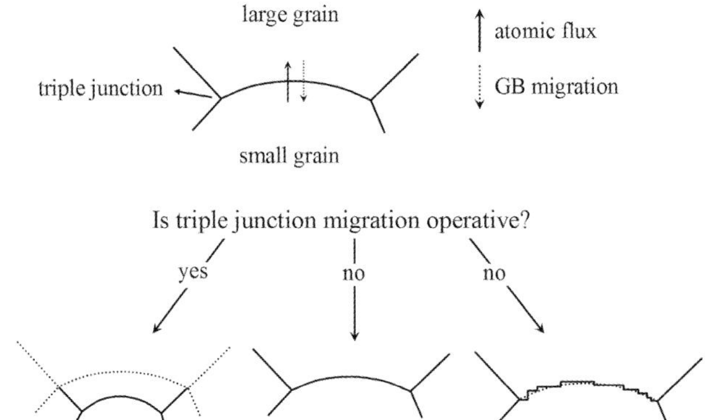

Figure 11.26 Schematic illustration showing the importance of the mobility of triple junction. Note that the rate of triple junction motions is a rate-determining step during curvature-driven grain boundary migration.

such a condition the grain growth tends to be uniform, resulting in a normal grain growth. However, under certain other conditions, a so-called "secondary recrystallization" occurs in metallic systems (this is also termed "abnormal grain growth"). The most famous example of this is the exclusive growth of Goss grains in Fe-3%Si steel. Steel sheets with a Goss texture of (110) [001] orientation are used in transformer cores because they provide an easy direction of magnetization.

It is very difficult to appreciate that some grains grow exclusively over other grains, leading to abnormal grain growth, when the migration of grain boundaries is kinetically coupled to that of triple junctions. Hwang *et al.* [128, 129] suggested that a kinetic decoupling would occur between the migrations of grain boundaries and triple junctions if solid-state wetting were to occur. If a grain has sub-grain boundaries with a very low energy, then the grain can drastically increase the probability of growing by solid-state wetting, when compared to other grains without sub-grain boundaries, as illustrated in Figure 11.27.

Figure 11.27 shows that the grains A and A^* share the sub-boundary with a low energy of 0.1, while the energies of the grain boundaries, AA^*, AB, and A^*B are 0.1, 2.0, and 2.2, respectively. Since $0.1 + 2.0 < 2.2$, the grain boundary A^*B should be replaced by two grain boundaries, AA^* and AB. In other words, grain A grows by penetrating the grain boundary A^*B by solid-state wetting. However, the energies of the grain boundaries, AA^*, AB, and A^*B are 0.1, 2.0, and 1.8, respectively. Since $0.1 + 1.8 < 2.0$, the grain boundary AB should be replaced by two grain boundaries AA^* and A^*B. Therefore, grains A and A^* with a sub-grain boundary, have the highest probability of growing by solid-state wetting.

Although, for simplicity, Figure 11.27 explains growth by solid-state wetting only with three grains in contact, the exclusive growth of grains with sub-grain boundaries can be tested in a more realistic 3-D polycrystalline structure by using computer simulations, such as Monte Carlo and phase field models [130, 132]. Figure 11.28 shows the cross-section of the 3-D Monte Carlo simulation of up to 4000 Monte Carlo steps (MCS); this indicates that white grains with sub-boundaries

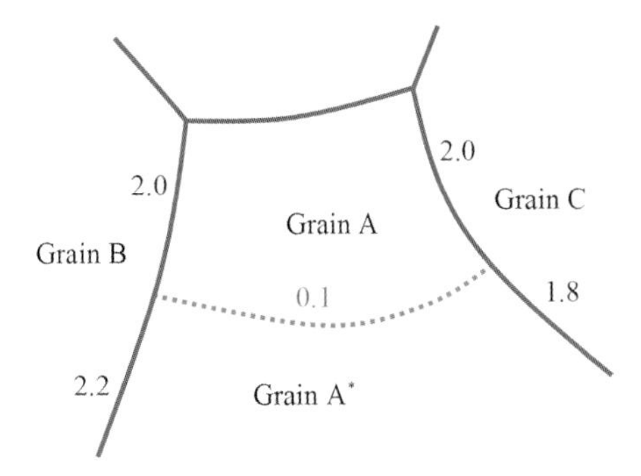

Figure 11.27 Schematic illustration showing that a sub-grain boundary with a low energy of 0.1 shared by grain denoted as A and A^* can wet high-energy boundaries. Since $0.1 + 2.0 < 2.2$, Grain A grows by wetting or penetrating the high-energy A^*–B boundary (energy 2.2), while a similar Grain A^* grows by penetrating the high-energy A–C boundary (energy 2.0).

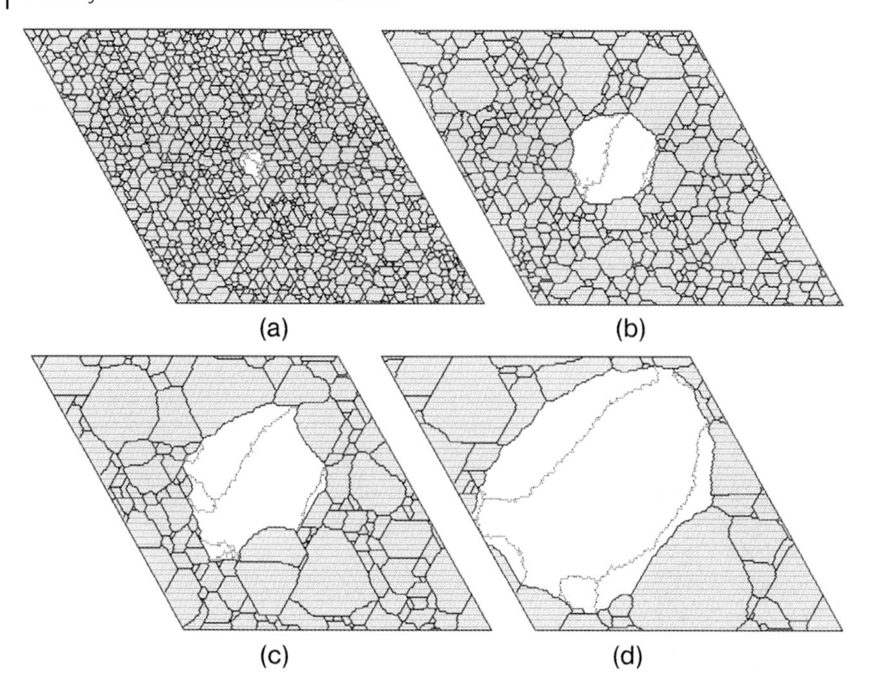

Figure 11.28 The cross-section of 3-D microstructure at z-axis = 80 after (a) 200 Monte Carlo steps (MCS), (b) 1000 MCS, (c) 2000 MCS, and (d) 4000 MCS under the condition where the sub-grain boundary energy inside the white grain is 0.05 and the energy of the other boundaries varies from 1 to 2. From Ref. [130].

of energy 0.05 grow abnormally over the other dark grains with energy randomly varying between 1 and 2 [130]. Dorner *et al.* [131] reported experimental evidence for the existence of sub-boundaries exclusively within Goss grains in Fe-3%Si steel, as shown in Figure 11.29.

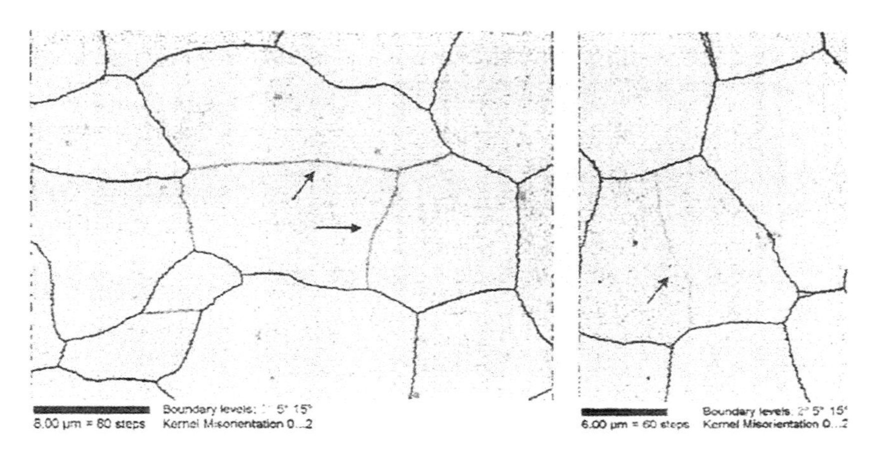

Figure 11.29 Sub-grain boundaries or very small-angle grain boundaries within recrystallized Goss grains with misorientations less than 1°. From Ref. [131].

Even in a 3-D polycrystalline structure, solid-state wetting along the triple junction can occur. Energetically, triple junction wetting is easier to be satisfied than grain boundary wetting for a given anisotropy of the grain boundary energies [133]. Therefore, growth by triple junction wetting seems to be the dominant mechanism of the abnormal grain growth in real systems. Figure 11.30 shows the four grains, A, B, C and D, where the grain A grows by triple junction wetting along the grain boundary shared by grains B, C, and D. Figure 11.30a and b show a 2-D section parallel to the direction of triple junction wetting and its schematic, respectively, whereas Figure 11.30c and d show a 2-D section vertical to the direction of the triple junction wetting and its schematic, respectively.

The microstructural evidence for Goss grains to grow by the triple junction wetting, as shown in Figure 11.30, can be found at the growth front of an abnormally

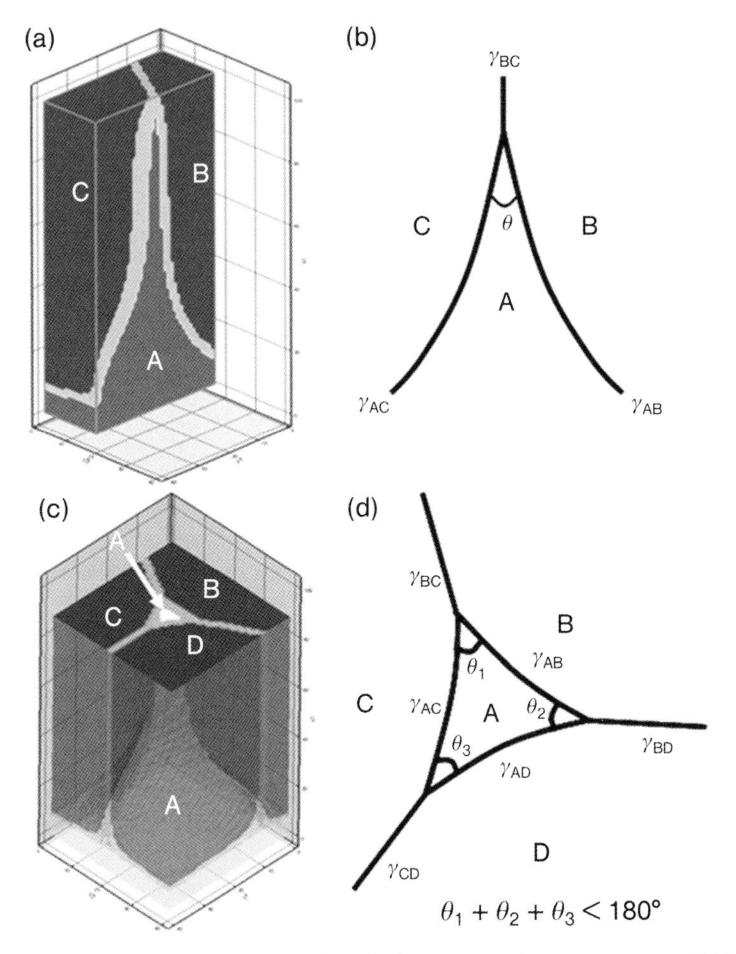

Figure 11.30 (a) 2-D section parallel to the direction of triple junction wetting; (b) Schematic of 2-D section image of panel (a); (c) 2-D section vertical to the direction of triple junction wetting; (d) Schematic of a 2-D section image of panel (c).

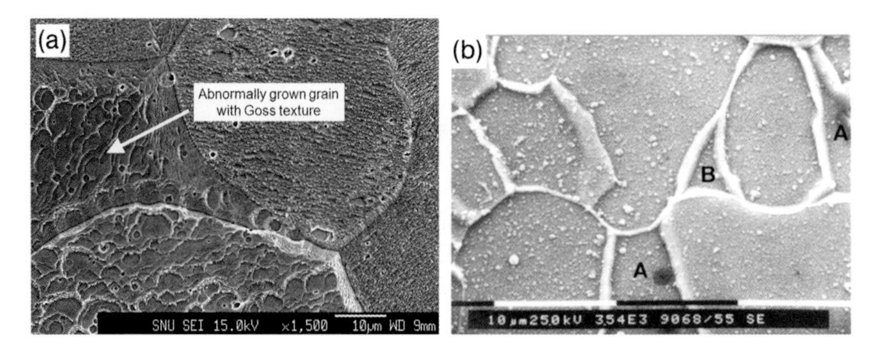

Figure 11.31 Microstructure evolution at the growth front of an abnormally growing Goss grain, showing that the Goss grain is undergoing the triple junction wetting. 2-D sections (a) parallel and (b) vertical to the wetting direction. In (b), grain 'A' is an abnormally growing Goss grain, and the three-sided grain 'B' with a negative curvature is not shrinking but growing, presumed to be an identical grain with the Goss grain 'A' being connected three-dimensionally with each other.

grown Goss grain (see Figure 11.31). Figure 11.31a and b correspond to the schematics in Figure 11.30b and d, respectively. Figure 11.32 shows serially sectioned images by optical microscopy near the growth front of an abnormally grown Goss grain. The grain at the central region within the black dotted-circle in Figure 11.32a

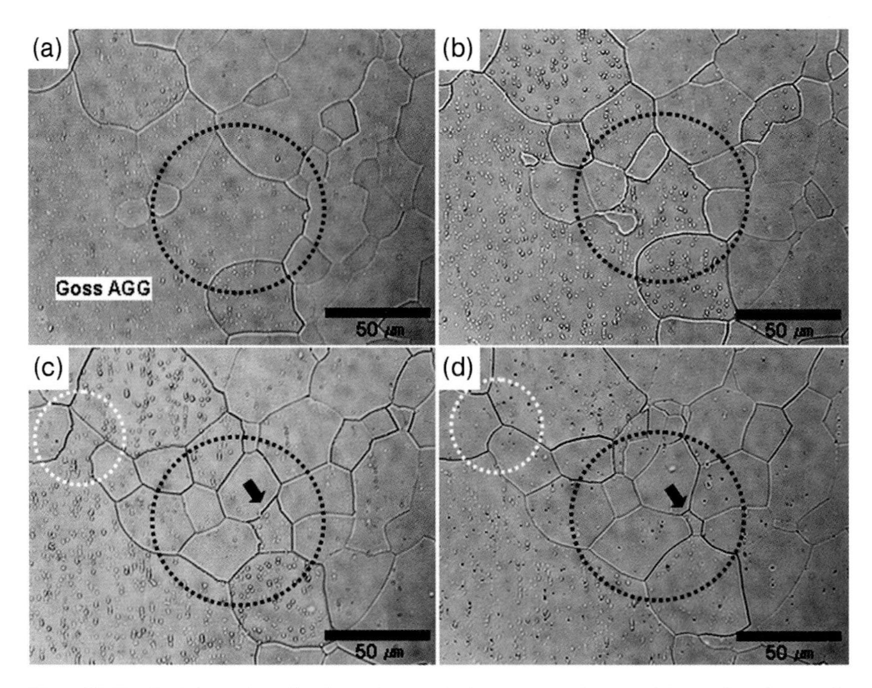

Figure 11.32 Serially sectioned optical microscopic images near the growth front of an abnormally grown grain in Fe-3% Si steel. The serial sections from (a) to (d) of the same sample were prepared by electrolytically polished and etched 8∼10 μm of the surface. See text for details.

and b is connected to the Goss grain. After polishing and etching, however, the same grain indicated by the black arrow in Figure 11.32c and d appears to be, respectively, isolated and separated from the Goss grain. In addition, there are other wetting morphologies showing in the microstructure within the white dotted circle in Figure 11.32c and d. The region within the white dotted circle in Figure 11.32c shows that the Goss grain is protruding into the regions among the matrix grains. After polishing and etching, however, the protruding morphology disappeared, as shown in Figure 11.32d. This indicates that the Goss grain is wetting along the triple junction at the center of the white-dotted circle of Figure 11.32d.

11.4 Summary and Outlook

Although, to date, interface anisotropy has received less consideration in the study of microstructure evolution, it is clear that it plays a critical role in determining the final microstructure of materials. It can be said that a desirable microstructure could only be obtained by a clear understanding of interface anisotropy, and making proper use of it. Today, most theoretical developments relating to interface anisotropy and microstructure evolution remain qualitative in nature, due mainly to the lack of a database on the interface anisotropy of the materials in use. Hence, an extensive accumulation of these data, and a better refinement of theory, should further enhance the present understanding of the microstructure evolution of materials.

References

1 Hall, E.O. (1951) The deformation and ageing of mild steel: III discussion of results. *Proc. Phys. Soc. London Sect. B*, **64**, 747–753.

2 Petch, N.J. (1953) The cleavage strength of polycrystals. *J. Iron Steel Inst.*, **174**, 25–28.

3 Blatter, G. and Greuter, F. (1986) Carrier transport through grain boundaries in semiconductors. *Phys. Rev. B*, **33** (6), 3952–3966.

4 Gupta, T.K. (1990) Application of zinc oxide varistors. *J. Am. Ceram. Soc.*, **73** (7), 1817–1840.

5 Kang, S.-J.L. (2005) *Sintering: Densification, Grain Growth & Microstructure*, Elsevier, Oxford.

6 Rohrer, G.S. (2005) Influence of interface anisotropy on grain growth and coarsening. *Annu. Rev. Mater. Res.*, **35**, 99–126.

7 Jo, W., Kim, D.-Y., and Hwang, N.-M. (2006) Effect of interface structure on the microstructural evolution of ceramics. *J. Am. Ceram. Soc.*, **89** (8), 2369–2380.

8 DaSilva, J.L.F., Barreteau, C., Schroeder, K., and Blügel, S. (2006) All-electron first-principles investigations of the energetics of vicinal Cu surfaces. *Phys. Rev. B*, **73**, 125402.

9 Yu, D., Bonzel, H.P., and Scheffler, M. (2006) Orientation-dependent surface and step energies of Pb from first principles. *Phys. Rev. B*, **74**, 115408.

10 Tyson, W.R. and Miller, W.A. (1977) Surface free energies of solid metal estimation from liquid surface tension measurements. *Surf. Sci.*, **62** (1), 267–276.

11 Galanakis, I., Bihlmayer, G., Bellini, V., Papanikolaou, N., Zeller, R., Blügel, S., and Dederichs, P.H. (2002) Broken-bond rule for the surface energies of noble metals. *Europhys. Lett.*, **58**, 751–757.

12 Galanakis, I., Papanikolaou, N., and Dederichs, P.H. (2002) Applicability of

broken-bond rule to surface energy of FCC metals. *Surf. Sci.*, **511** (1), 1–12.
13 Gibbs, J.W. (1906) On the equilibrium of heterogeneous substances. The scientific papers, in *Thermodynamics*, vol. 1 (ed. J.W. Gibbs), Longmans, Green & Co., pp. 55–349.
14 Mackenzie, J.K., Moore, A.J.W., and Nicholas, J.F. (1962) Bonds broken at atomically flat crystal surfaces-I. Face-centered and body-centered cubic crystals. *J. Phys. Chem. Solids*, **23**, 185–196.
15 Athenstaedt, W. and Leisch, M. (1996) The segregation behavior of a $Pt_{90}Rh_{10}$ alloy studied with a three dimensional atom-probe. *Appl. Surf. Sci.*, **94–95**, 403–408.
16 Yamamoto, M., Fukuda, T., and Nenno, S. (1987) Surface energy and equilibrium shape of L12-type A_3B ordering alloys. *J. de Physique*, **48**, C6-323–C6-328.
17 Liu, X.-Y., Ohotnicky, P.P., Adams, J.B., Rohrer, C.L., and Hyland, R.W. Jr (1997) Anisotropic surface segregation in Al-Mg alloys. *Surf. Sci.*, **373** (2–3), 357–370.
18 Tasker, P.W. (1979) The surface energies, surface tensions and surface structure of the alkali halide crystals. *Philos. Mag. A*, **39** (2), 119–136.
19 Hebenstreit, W., Schmid, M., Redinger, J., Podloucky, R., and Varga, P. (2000) Bulk terminated NaCl(111) on aluminum: a polar surface of an ionic crystal? *Phys. Rev. Lett.*, **25**, 5376–5379.
20 Liu, W., Liu, X., Zheng, W.T., and Jiang, Q. (2006) Surface energies of several ceramics with NaCl structure. *Surf. Sci.*, **600**, 257–264.
21 Goniakowski, J. and Noguera, C. (1999) Characteristics of Pd deposition on the MgO(111) surface. *Phys. Rev. B*, **60**, 16120–16128.
22 Li, W.-J., Shi, E.-W., Zhong, W.-Z., and Yin, Z.-W. (1999) Growth mechanism and growth habit of Oxide crystals. *J. Cryst. Growth*, **203** (1–2), 186–196.
23 Lee, J.-S. and Wiederhorn, S.M. (2004) Effects of polarity on grain-boundary migration in ZnO. *J. Am. Ceram. Soc.*, **87** (7), 1319–1323.
24 Wolff, G.A. and Broder, J.D. (1959) Microcleavage, bonding character and surface structure in materials with tetrahedral coordination. *Acta Crystallogr.*, **12**, 313–323.
25 Abrahams, M.S. and Ekstrom, L. (1960) Dislocations and brittle fracture in elemental and compound semiconductors. *Acta Metall.*, **8** (9), 654–662.
26 Mariano, A.N. and Hanneman, R.E. (1963) Crystallographic polarity of ZnO crystals. *J. Appl. Phys.*, **34** (2), 384–388.
27 Jo, W., Kim, S.-J., and Kim, D.-Y. (2005) Analysis of the etching behavior of ZnO ceramics. *Acta Mater.*, **53**, 4185–4188.
28 Curie, P. (1885) Sur la formation des crystaux et les constants de capillarite de leur different phase. *Bull. Soc. Franc. Minéral*, **8**, 145–150.
29 Wulff, G. (1901) Zur frage der geschwindigkeit des wachsthums und der aufl sung der krystall fächen. *Z. Krist.*, **34**, 449–530.
30 von Laue, M. (1943) Der wulffsche satz für die gleichgewichtsform von kristallen. *Z. Krist.*, **105**, 124–133.
31 Hong, J.-S. (2007) Effect of carbon on the equilibrium shape of nickel crystal. MS Thesis, Seoul National University, Seoul.
32 Heyraud, J.C. and Métois, J.J. (1987) Equilibrium shape of an ionic crystal in equilibrium with its vapor phase (NaCl). *J. Cryst. Growth*, **84**, 503–508.
33 Yu, D., Bonzel, H.P., and Scheffler, M. (2006) Orientation-dependent surface and step energies of Pb from first principles. *Phys. Rev. B*, **74**, 115408-1–115408-7.
34 Marks, L.D. (1983) Direct imaging of carbon-covered and clean gold (110) surfaces. *Phys. Rev. Lett.*, **51**, 1000–1002.
35 Kellogg, G.L. (1985) Direct observations of the (1×2) surface reconstruction on the Pt(110) plane. *Phys. Rev. Lett.*, **55**, 2168–2171.
36 Foiles, S.M. (1987) Reconstruction of FCC (110) surfaces. *Surf. Sci.*, **191**, L779–L786.
37 Zhang, J.-M., Li, H.-Y., and Xu, K.-W. (2007) Missing row reconstructed (110), (211), and (311) surfaces for FCC transition metals. *Surf. Interface Anal.*, **39**, 660–664.
38 Howe, J.M. (1997) *Interfaces in Materials*, John Wiley & Sons, Inc., New York.
39 McIntyre, B.J., Salmeron, M., and Somorjai, G.A. (1993) *In situ* scanning

tunneling microscopy study of platinum (110) in a reactor cell at high pressures and temperatures. *J. Vac. Sci. Technol. A*, **11** (4), 1964–1968.

40 Cahn, J.W. and Handwerker, C.A. (1993) Equilibrium geometries of anisotropic surfaces and interfaces. *Mater. Sci. Eng. A*, **162**, 83–95.

41 Cahn, J.W. and Carter, W.C. (1996) Crystal shapes and phase equilibria: a common mathematical basis. *Metall. Mater. Trans. A*, **27A**, 1431–1440.

42 Hondros, E.D. and McLean, M. (1969) Preferential chemisorption on crystalline surfaces – Cu/O system. *Nature*, **224**, 1296–1297.

43 Rieder, K.H. and Engel, T. (1980) Adatom configurations of H(2 × 6) and H(2 × 1) on Ni (110) analyzed using He diffraction. *Phys. Rev. Lett.*, **45**, 824–828.

44 Williams, E.D. and Bartelt, N.C. (1989) Surface Faceting and the equilibrium crystal shape. *Ultramicroscopy*, **31**, 36–48.

45 Somorjai, G.A. and van Hove, M.A. (1995) Restructuring of metal surfaces and adsorbed mono-layers during chemisorption and catalytic reaction. *Acta Crystallogr.*, **B51**, 502–512.

46 Phaneuf, R.J. and Schmid, A.K. (2003) Low energy electron microscopy: imaging surface dynamics. *Phys. Today*, **56**, 50–55.

47 Campuzano, J.C., Foster, M.S., Jennings, G., Willis, R.F., and Unertl, W. (1985) Au (110) (1 × 2)-to-(1 × 1) phase transition: a physical realization of the two-dimensional ising model. *Phys. Rev. Lett.*, **54** (25), 2684–2687.

48 Broughton, J.Q. and Gilmer, G.H. (1987) Interface melting: simulations of surfaces and grain boundaries at high temperatures. *J. Phys. Chem.*, **91**, 6347–6359.

49 Jeong, H.-C. and Williams, E.D. (1999) Steps on surfaces: experiment and theory. *Surf. Sci. Rep.*, **34**, 171–294.

50 Zandvliet, H.J.W. (2000) Energetics of Si (001). *Rev. Mod. Phys.*, **72** (2), 593–602.

51 Kossel, W. (1927) *Zur Theorie des Kristallwachstums*, Nachr. Ges. Wiss., Göttingen, pp. 135–143.

52 Stranski, I.N. (1928) Zur theorie des kristallwachstums. *Z. Phys. Chem.*, **142**, 259–278.

53 Gruber, E.E. and Mullins, W.W. (1967) On the theory of anisotropy of crystalline surface tension. *J. Phys. Chem. Solids*, **28**, 875–887.

54 Balibar, S., Alles, H., and Parshin, A.Y. (2005) The surface of helium crystals. *Rev. Mod. Phys.*, **77**, 317–370.

55 Heyraud, J.C. and Métois, J.J. (1980) Establishment of the equilibrium shape of metal crystallites on a foreign substrate: gold on graphite. *J. Cryst. Growth*, **50** (2), 571–671.

56 Heyraud, J.C. and Métois, J.J. (1983) Equilibrium shape and temperature: lead on graphite. *Surf. Sci.*, **128** (2–3), 334–342.

57 Balibar, S. and Castaing, B. (1985) Helium: solid-liquid interfaces. *Surf. Sci. Rep.*, **5** (3), 87–143.

58 Heyraud, J.C., Métois, J.J., and Bermond, J.M. (1989) Surface melting and equilibrium shape: the case of Pb on graphite. *J. Cryst. Growth*, **98** (3), 355–362.

59 Bermond, J.M., Métois, J.J., Heyraud, J.C., and Floret, F. (1998) Shape universality of equilibrated silicon crystals. *Surf. Sci.*, **416** (3), 430–447.

60 Rottman, C., Wortis, M., Heyraud, J.C., and Métois, J.J. (1984) Equilibrium shapes of small lead crystals: observation of Pokrovsky-Talapov critical behavior. *Phys. Rev. Lett.*, **52** (12), 1009–1012.

61 Burton, W.K., Cabrera, N., and Frank, F.C. (1951) The growth of crystals and the equilibrium structure of their surfaces. *Philos. Trans. R. Soc.*, **243A**, 299–358.

62 Chui, S.T. and Weeks, J.D. (1976) Phase transition in the two-dimensional coulomb gas, and the interfacial roughening transition. *Phys. Rev. B*, **14** (11), 4978–4982.

63 Kosterlitz, J.M. and Thouless, D.J. (1973) Ordering, metastability and phase transitions in two-dimensional systems. *J. Phys. C: Solid State Phys.*, **6**, 1181–1203.

64 Kosterlitz, J.M. (1974) The critical properties of the two-dimensional *xy* model. *J. Phys. C: Solid State Phys.*, **7**, 1046–1060.

65 Chui, S.T. and Weeks, J.D. (1978) Dynamics of the roughening transition. *Phys. Rev. Lett.*, **40** (12), 733–736.

66 Peteves, S.D. and Abbaschian, R. (1991) Growth kinetics of solid-liquid Ga

interfaces: part I. Experimental. *Metall. Trans. A*, **22A**, 1259–1270.

67 Elwell, D. and Scheel, H.J. (1975) *Crystal Growth from High-Temperature Solutions*, Academic Press, London.

68 Markov, I.V. (1995) *Crystal Growth for Beginners: Fundamentals of Nucleation, Crystal Growth, and Epitaxy*, World Scientific Publishing Co. Pte. Ltd.

69 Bennema, P. and van der Eerden, J.P. (1977) Crystal growth from solution: development in computer simulation. *J. Cryst. Growth*, **42**, 201–213.

70 Sunagawa, I. (1995) The distinction of natural from synthetic diamonds. *J. Gemmol.*, **24** (7), 485–499.

71 Spaepen, F. (1975) A structural model for the solid-liquid interface in monatomic systems. *Acta Metall.*, **23**, 729–743.

72 Bernal, J.D. (1964) The Bakerian Lecture, 1962. The structure of liquids. *Proc. R. Soc. Lond. A*, **280** (1382), 299–322.

73 Lee, H.R., Kim, D.J., Hwang, N.M., and Kim, D.-Y. (2003) Role of VC additive during sintering of WC-Co: mechanism of grain growth inhibition. *J. Am. Ceram. Soc.*, **86** (1), 152–154.

74 Yamamoto, T., Ikuhara, Y., and Sakuma, T. (2000) High resolution transmission electron Microscopy study in VC-doped WC-Co compound. *Sci. Technol. Adv. Mater.*, **1**, 97–104.

75 Choi, K., Hwang, N.-M., and Kim, D.-Y. (2002) Effect of Grain shape on abnormal grain growth in liquid-phase-sintered (Nb,Ti)C-Co alloys. *J. Am. Ceram. Soc.*, **85** (9), 2313–2318.

76 Wolf, D. (1992) Atomic-level geometry of crystalline interfaces, in *Materials Interfaces Atomic-Level Structure and Properties*, Chapman & Hall, London, pp. 1–57.

77 Chung, U.-J., Jo, W., Lee, J.-H., Hwang, N.M., and Kim, D.-Y. (2004) Coarsening process of the penetration twinned Pb$(Mg_{1/3}Nb_{2/3})O_3$-35 mol%$PbTiO_3$ grain. *J. Am. Ceram. Soc.*, **87** (1), 125–128.

78 Jo, W., Chung, U.-J., Hwang, N.M., and Kim, D.-Y. (2005) Effect of SiO_2 and TiO_2 addition on the morphology of abnormally grown large PMN-35mol/ grains. *J. Am. Ceram. Soc.*, **88** (7), 1992–1994.

79 Wolf, D. (1989) A Read-Shockley model for high-angle grain-boundaries. *Scripta Metall.*, **23** (10), 1713–1718.

80 Wolf, D. (1989) Correlation between energy and volume expansion for grain boundaries in FCC metals. *Scripta Metall.*, **23** (11), 1913–1918.

81 Read, W.T. and Shockley, W. (1950) Dislocation models of crystal grain boundaries. *Phys. Rev.*, **78** (3), 275–289.

82 Burgers, J.M. (1940) Geometrical considerations concerning the structural irregularities to be assumed in a crystal. *Proc. Phys. Soc.*, **52** (1), 23–33.

83 Bragg, W.L. (1940) The structure of a cold-worked metal. *Proc. Phys. Soc.*, **52** (1), 105–109.

84 Amelinckx, S. and Dekeyser, W. (1959) The structure and properties of grain boundaries. *Solid State Phys.*, **8**, 325–499.

85 Hirsch, P.B., Horne, R.W., and Whelan, J.J. (1956) Direct observation of the arrangement and motion of dislocations in aluminium. *Philos. Mag.*, **1**, 677–684.

86 Bollmann, W. (1956) Interference effects in the electron microscopy of thin crystal foils. *Phys. Rev.*, **103**, 1588–1597.

87 Gleiter, H. and Chalmers, B. (1972) Structure of grain boundaries. *Prog. Mater. Sci.*, **16**, 1–12.

88 Humphrey, P.H. (1976) Special high angle grain boundaries, in *Grain Boundary Structure & Properties*, Academic Press, London, pp. 139–200.

89 Hansson, G.C. and Goux, C. (1971) Interfacial energies of tilt boundaries in aluminum. Experimental and theoretical determinations. *Scripta Metall.*, **5**, 889–894.

90 Sutton, A.P. and Balluffi, R.W. (1997) *Interfaces in Crystalline Solids*, Oxford University Press, Oxford.

91 Shibata, N., Oba, F., Yamamoto, T., and Ikuhara, Y. (2004) Structure, energy and solute segregation behavior of [110] symmetric tilt grain boundaries in yttria-stabilized cubic zirconia. *Philos. Mag.*, **84** (23), 2381–2415.

92 Kim, D.-Y., Wiederhorn, S.M., Hockey, B.J., Handwerker, C.A., and Blendell, J.E. (1994) Stability of surface energies of wetted grain boundaries in aluminum oxide. *J. Am. Ceram. Soc.*, **77** (2), 444–453.

93 Matsunaga, K., Nishimura, H., Saito, T., Yamamoto, T., and Ikuhara, Y. (2003) High-resolution transmission electron microscopy and computational analyses of atomic structures of [0001] symmetric tilt grain boundaries of Al_2O_3 with equivalent grain-boundary planes. *Philos. Mag.*, **83** (36), 4071–4082.

94 Oba, F., Ohta, H., Sato, Y., Hosono, H., Yamamoto, T., and Ikuhara, Y. (2004) Atomic structure of [0001]-tilt grain boundaries in ZnO: A high-resolution TEM study of fiber- textured thin films. *Phys. Rev. B*, **70**, 125415–125420.

95 Li, J.C.M. (1961) High-angle tilt boundary – a dislocation core model. *J. Appl. Phys.*, **32** (3), 525–541.

96 Shibata, N., Oba, F., Yamamoto, T., Sakuma, T., and Ikuhara, Y. (2003) Grain-Boundary Faceting at a $\Sigma = 3$, [110]/{112} Grain Boundary in a Cubic Zirconia Bicrystal. *Philos. Mag*, **83** (19), 2221–2246.

97 Choi, J.H., Kim, D.-Y., Hockey, B.J., Wiederhorn, S.M., Handwerker, C.A., Blendell, J.E., Carter, W.C., and Roosen, A.R. (1997) Equilibrium shape of internal cavities in sapphire. *J. Am. Ceram. Soc.*, **80** (1), 62–68.

98 Frenkel, J. (1945) Viscous flow of crystalline bodies under the action of surface tension. *J. Phys.*, **9** (5), 385–391.

99 Kuczynski, G.C. (1949) Self-diffusion in sintering of metallic particles. *Trans. AIME*, **185** (2), 169–178.

100 Kingery, W.D. (1959) Densification during sintering in the presence of a liquid phase. I. Theory. *J. Appl. Phys.*, **30** (3), 301–306.

101 Park, H.-H. and Yoon, D.N. (1985) Effect of dihedral angle on the morphology of grains in a matrix phase. *Metall. Trans. A*, **16A**, 923–928.

102 Kang, S.-J.L., Kim, K.-H., and Yoon, D.N. (1991) Densification and shrinkage during liquid- phase sintering. *J. Am. Ceram. Soc.*, **74** (2), 425–427.

103 Coble, R.L. (1961) Sintering crystalline solids. I. Intermediate and final state diffusion models. *J. Appl. Phys.*, **32** (5), 787–792.

104 Coble, R.L. (1961) Sintering crystalline solids. II. Experimental test of diffusion models in powder compacts. *J. Appl. Phys.*, **32** (5), 793–799.

105 Coble, R.L.(March 20 1962) Transparent Alumina and Method of Preparation, U. S. Patent No. 3026210.

106 Burke, J.E. (1957) The role of grain boundaries in sintering. *J. Am. Ceram. Soc.*, **40** (3), 80–85.

107 Brook, R.J. (1969) Pore-grain boundary interactions and grain growth. *J. Am. Ceram. Soc.*, **52** (1), 56–57.

108 Hsueh, C.H., Evans, A.G., and Coble, R.L. (1982) Microstructure development during final/intermediate stage sintering-I. Pre/grain boundary separation. *Acta Metall.*, **30** (7), 1269–1279.

109 Kwon, S.-G., Hong, S.-H., Hwang, N.-M., and Kim, D.-Y. (2000) Coarsening behavior of growth of C_3S and C_2S grains dispersed in a clinker melt. *J. Am. Ceram. Soc.*, **83** (5), 1247–1252.

110 Park, Y.J., Hwang, N.M., and Yoon, D.Y. (1996) Abnormal growth of faceted (WC) grains in a (Co) liquid matrix. *Metall. Trans. A*, **27A** (9), 2809–2819.

111 Choi, J.-H., Hwang, N.-M., and Kim, D.-Y. (2001) Pore-boundary separation behavior during sintering of pure and Bi_2O_3-doped ZnO ceramics. *J. Am. Ceram. Soc.*, **84** (6), 1398–2300.

112 Jo, W., Kim, D.-Y., and Hwang, N.-M. (2007) Reply to the comment on effect of interface structure on the microstructural evolution of ceramics. *J. Am. Ceram. Soc.*, **90** (7), 2293–2295.

113 Kang, M.-G., Kim, D.-Y., Lee, H.-Y., and Hwang, N.M. (2000) Temperature dependence of the coarsening behavior of barium titanate grains. *J. Am. Ceram. Soc*, **83** (12), 3202–3204.

114 Kang, M.-K., Yoo, Y.-S., Kim, D.-Y., and Hwang, N.-M. (2000) Growth of $BaTiO_3$ seed grains by the twin plane re-entrant edge mechanism. *J. Am. Ceram. Soc.*, **83** (2), 385–390.

115 Jo, W., Chung, U.-J., Hwang, N.M., and Kim, D.-Y. (2003) Temperature dependence of the coarsening behavior of (Ba, Sr)TiO_3 grains dispersed in a silica-rich liquid matrix. *J. Eur. Ceram. Soc.*, **23** (10), 1565–1569.

116 Kwon, O.-S., Hong, S.-H., Lee, J.-H., Chung, U.-J., Kim, D.-Y., and Hwang, N.-M. (2002) Microstructural evolution during sintering of TiO_2/SiO_2-doped alumina: mechanism of anisotropic

abnormal grain growth. *Acta Mater.*, **50** (19), 4865–4872.

117 Gavrilov, K.L., Bennison, S.J., Mikeska, K.R., Chabala, J.M., and Setti, R.L. (1999) Silica and magnesia dopant distributions in alumina by high resolution scanning secondary ion mass spectrometry. *J. Am. Ceram. Soc.*, **82** (4), 1001–1008.

118 Handwerker, C.A., Morris, P.A., and Coble, R.L. (1996) Effect of chemical inhomogeneities on grain growth and microstructure in Al_2O_3. *J. Am. Ceram. Soc.*, **72** (1), 130–136.

119 Bae, S.I. and Baik, S. (1993) Sintering and grain growth of ultrapure alumina. *J. Mater. Sci.*, **28** (15), 4197–4204.

120 Park, C.W. and Yoon, D.Y. (2000) Effect of SiO_2, CaO, and MgO additions on the grain growth of alumina. *J. Am. Ceram. Soc.*, **83** (10), 2605–2609.

121 Park, C.W. and Yoon, D.Y. (2001) Effect of MgO on faceted vicinal (0001) surface of aluminum oxide. *J. Am. Ceram. Soc.*, **84** (2), 456–458.

122 Kim, M.-J. and Yoon, D.Y. (2003) Effect of MgO addition on surface roughening of alumina grains in anorthite liquid. *J. Am. Ceram. Soc.*, **86** (4), 630–633.

123 Park, C.W., Yoon, D.Y., Blendell, J.E., and Handwerker, C.A. (2003) Singular grain boundaries in alumina and their roughening transition. *J. Am. Ceram. Soc.*, **86** (4), 603–611.

124 Handwerker, C.A., Dynys, J.M., Cannon, R.M., and Coble, R.L. (1990) Dihedral angles in magnesia and alumina: distributions from surface thermal grooves. *J. Am. Ceram. Soc.*, **73** (5), 1371–1377.

125 Chen, I.-W. and Wang, X.-H. (2000) Sintering dense nanocrystalline ceramics without final-stage grain growth. *Nature*, **404**, 168–171.

126 Lin, P., Palumbo, G., Harase, J., and Aust, K.T. (1996) Coincidence site lattice (CSL) grain boundaries and Goss texture development in Fe-3% Si alloy. *Acta Mater.*, **44** (12), 4677–4683.

127 Manohar, P.A., Ferry, M., and Chandra, T. (1998) Five decades of the Zener equation. *ISIJ Int.*, **38** (9), 913–924.

128 Hwang, N.M., Lee, S.B., and Kim, D.-Y. (2001) Abnormal grain growth by solid-state wetting along grain boundary or triple junction. *Scripta Mater.*, **44**, 1153–1160.

129 Hwang, N.M., Lee, D.K., Ko, K.J., Lee, B.J., Park, J.T., Hong, B.D., Kim, J.K., and Kim, D.Y. (2005) Alternative mechanism of secondary recrystallization: Solid-state wetting along grain boundaries or triple junctions, in *Solid-to-Solid Phase Transformation in Inorganic Materials 2005, Volume 1: Diffusional Transformations* (eds J.M. Howe, D.E. Laughlin, J.K. Lee, U. Dahmen, and W.A. Soffa), TMS.

130 Lee, D.K., Ko, K.J., Lee, B.J., and Hwang, N.M. (2008) Monte Carlo simulations of abnormal grain growth by sub-boundary-enhanced solid-state wetting. *Scripta Mater.*, **58**, 683–686.

131 Dorner, D., Lahn, L., and Zaefferer, S. (2004) Investigation of the primary recrystallisation microstructure of cold rolled and annealed Fe 3% Si single crystals with Goss orientation. *Mater. Sci. Forum*, **467–470**, 129–134.

132 Ko, K.-J., Park, J.-T., Kim, J.-K., and Hwang, N.-M. (2008) Morphological evidence that Goss abnormally growing grains grow by triple junction wetting during secondary recrystallization of Fe-3% Si steel. *Scripta Mater.*, **59**, 764–767.

133 Hwang, N.M., Park, Y.J., Kim, D.-Y., and Yoon, D.Y. (2000) Activated sintering of Ni- doped tungsten: approach by grain boundary structural transition. *Scripta Metall.*, **42** (5), 421–425.

134 Saito, Y. (1996) Chapter II, Statistical mechanics of surface, *Statistical Physics of Crystal Growth*, World Scientific Publishing Co. Pte. Ltd.

III
Mechanical Properties

Ceramics Science and Technology Volume 2: Properties. Edited by Ralf Riedel and I-Wei Chen

ISBN: 978-3-527-31156-9

12
Fracture of Ceramics

Robert Danzer, Tanja Lube, Peter Supancic, and Rajiv Damani

12.1
Introduction

Ceramic materials have a large fraction of ionic or covalent bonds, and this results in some special behavior – and, consequently, some special problems – with their reliable use in engineering. Within the temperature range of technical interest, dislocations are relatively immobile, and dislocation-induced plasticity is almost completely absent in ceramics [1–3]. This is the basis for their extreme hardness and inherent brittleness, with typical values for the fracture toughness of ceramics ranging from 1 to 10 $MPa{\cdot}m^{1/2}$, and the total fracture strain being commonly less than a few parts per thousand [4].

In general, the plastic yield stress is a factor of 10 higher than the tensile strength [1–4], and local stress concentrations cannot therefore be relaxed by yield. This implies a need for the careful design of highly loaded ceramic parts, for more precise tolerances (compared to parts made of metals or polymers), and for the very careful handling of those parts, both in production and in application. It also entails the need for extremely careful stress analysis. In the region of load transfer into the component, a steep stress gradient may occur and dangerous tensile stresses even exist around compressive contact zones (a prominent example of the damaging action of these tensile stresses are Hertzian cone cracks, which can frequently be found in impacted surfaces [3, 5]). Other significant mechanical stresses, which are not accounted for by an overly simplistic mechanical analysis, may result from thermal mismatch strains.

Mechanical testing is more complicated for ceramics than for ductile materials, since problems arising from any misalignment of specimens can become severe. They cannot be balanced by small amounts of plastic deformation, and this makes specimens with complicated shapes or the use of very sophisticated testing jigs necessary. In the case of hard and brittle ceramics, the machining of specimens is both expensive and time-consuming.

The standardization of mechanical testing procedures began around 30 years ago at a national level. However, during the past ten years serious efforts on internationalization have also been made, and the first international standard (ISO 14704:

Ceramics Science and Technology Volume 2: Properties. Edited by Ralf Riedel and I-Wei Chen

ISBN: 978-3-527-31156-9

determination of bending strength at room temperature [6, 7]) was published in the year 2000. Yet, the process is far from complete. Even for such an important property such as fracture toughness, consensus on an appropriate testing procedure (i.e., an international standard) has not yet been achieved [7–12]. In light of this background, it is not surprising that good and comparable data – such as that needed for appropriate material selection – is missing for most ceramic materials [13].

In ceramics, a fracture starts in general from small flaws, which are discontinuities in the microstructure and which, for simplicity, can be assumed to be small cracks distributed in the surface or volume. The strength of a specimen then depends on the size of the largest (or critical) defect within it, and this varies from component to component [1–4, 14]. The strength of a ceramic material cannot, therefore, be described uniquely by a single number; rather, a strength distribution function is necessary, and a large number of specimens is required to characterize this. Design with ceramics must also be approached statistically. A structure is not safe or unsafe, but it does have a certain probability of failure or survival. A failure probability of zero is equivalent to the certainty that the component contains no defect larger than a given size. The critical defect size for typical design loads in technical ceramic components is around 100 μm or smaller, and is therefore too small to be reliably found by nondestructive testing techniques. Proof testing is usual for components needing high reliability.

Taking all these problems into account, it becomes clear that their high brittleness and insignificant ductility impedes the wider use of ceramics as advanced technical materials. A detailed understanding of the damage mechanisms and fracture processes is necessary to make the safe and reliable application of ceramic components possible.

12.2 Appearance of Failure and Typical Failure Modes

The fracture of ceramic components depends on many factors, including the chemistry and microstructure of the material and the resulting material's properties (e.g., toughness, *R*-curve behavior, etc.). The macroscopic appearance of fracture is, however, primarily influenced by the type of loading of the component and the resulting stress field.

At low and ambient temperatures, the fracture of ceramics is always brittle, and triggered by the normal tensile stresses in the component. An example of a typical fracture surface is provided in Figure 12.1, which shows the forced rupture of a silicon nitride valve tested in a tensile test [15]. The fracture path is almost perpendicular to the direction of loading (i.e., to the first principal stress), and no signs of plastic deformation can be detected. The fracture origin is surrounded by a smooth area, called the *fracture mirror*, and by rougher regions referred to as *mist* and *hackle* [16–19]. In this case, the fracture origin is an agglomerate related to a pore.

Such features are very common in brittle fracture of ceramic materials: the fracture origin is a critical flaw which behaves like a crack; the crack path is perpendicular to

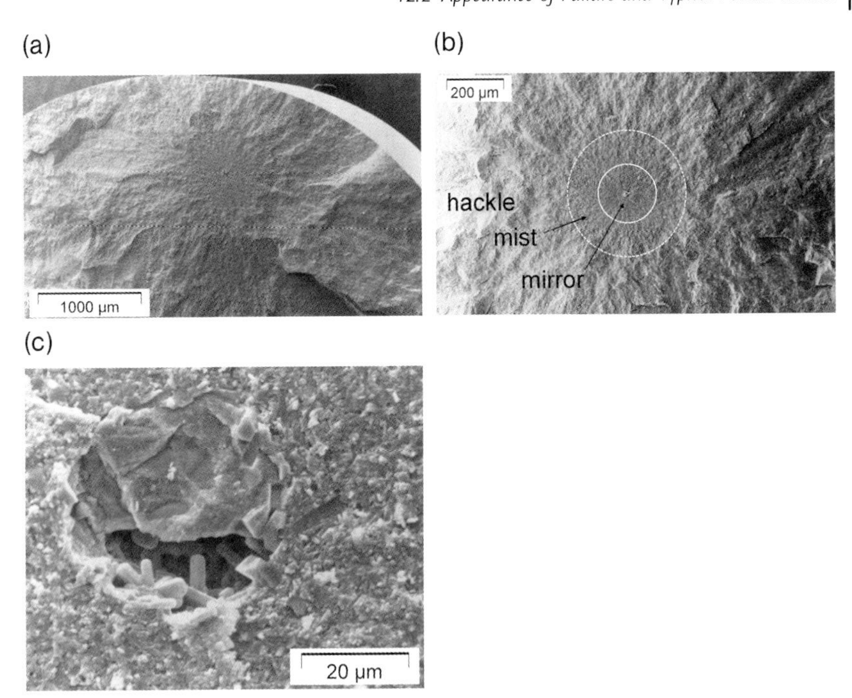

Figure 12.1 Fracture surface of a silicon nitride valve fractured in a rupture test. (a) Macroscopic view: the crack path is perpendicular to the loading direction; (b) Fracture mirror, mist and hackle can be recognized on the fracture surface; (c) The fracture origin (agglomerate related to a pore) can be found in the center of the mirror.

the first principal stress (at least at the beginning of crack extension); and the fracture origin is surrounded by mirror, mist, and hackle. The size of the fracture origin, mirror, and mist respectively are found to be proportional to σ^{-2}, where σ is the stress in an uncracked body at the position of the origin of fracture [16]. Values of mirror constants have been reported elsewhere [17, 18, 20].

This type of fracture can be found in both very large and very small systems, including fractured natural rocks, in tested specimens of advanced ceramic materials, and even in broken ceramic fibers. The fracture mirror size measurements can be used for post-mortem determination of the failure stress in components [21].

Compared to failure in a tensile test, the failure of components in operation can be much more complex, however. The stress field is generally not uniaxial, which may cause an intricate crack path, and in many cases secondary damage may confuse the initial picture. Figure 12.2 shows the remnants of a turbine rotor fractured in a spin test [22, 23], where the fracture origin was again identified as a flaw in the microstructure.

Many modes of failure in operation exist, and these are typically defined by the mode of loading, such that each results in a characteristic macroscopic fracture appearance. Nevertheless, a huge fraction of in-service failures can be traced back to just two groups of failure mode: (i) thermal shock; and (ii) contact loading, which are described in more detail in the following sections.

Figure 12.2 Turbine rotor made of silicon nitride ruptured in a spin test. (a) Fractured rotor (the location of the fracture origin is indicated by an arrow); (b) The fracture origin [22, 23].

12.2.1 Thermal Shock Failure

Thermal shock (or thermoshock) failure is a very prominent failure mode in ceramics. Indeed, based on the fractographic experience of the present authors, it is apparent that more than one-third of all rejections of ceramic components are caused by thermal shock.

Thermal shock occurs when rapid temperature changes cause temperature differences and thermal strains in the component [4, 19, 24–26]. If these strains are constrained, then thermal stresses which may cause crack propagation and failure can arise. The shape of the stress field depends on the boundary conditions. The most important example of thermal shock is cooling shock (e.g., due to the quenching of a hot part) since, during cooling, heat is transferred via the surface of a component into the environment and the surface cools down. As a result, the material in the surface region attempts to shrink but is constrained by the interior of the material, which is still at a higher temperature. Such constraint causes tensile stresses in the surface region, analogous to those in a hypothetical elastic layer stretched over the component, while the interior of the body is in compression. If the tensile stresses exceed certain critical values, damage will occur resulting in characteristic crack patterns. Some examples of thermal shock damage in samples and components [26, 27] are shown in Figures 12.3 and 12.4.

Thermal shock cracks also develop if the stresses become locally supercritical. The growth of such cracks reduces the tensile stress in their surroundings, and thus the driving force for further extension; consequently, the crack propagation may stop. The number of cracks increases with the severity of the shock. Whereas, thermoshock cracks typically run perpendicular to edges, at flat surfaces (where biaxial stress states exist) they tend to form networks, which closely resemble the mud-cracking patterns seen in dried-up marshes. In this case, the growth direction of the cracks is opposite to that of the heat flow. Due to their characteristic appearance, thermal shock cracks can generally be recognized by a simple optical inspection of the damaged component.

(a)

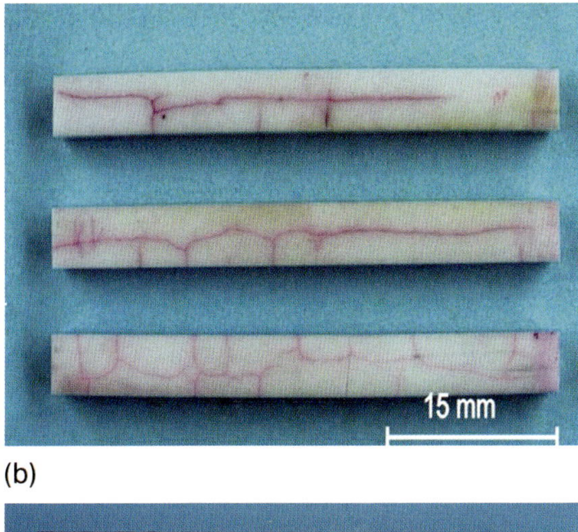

(b)

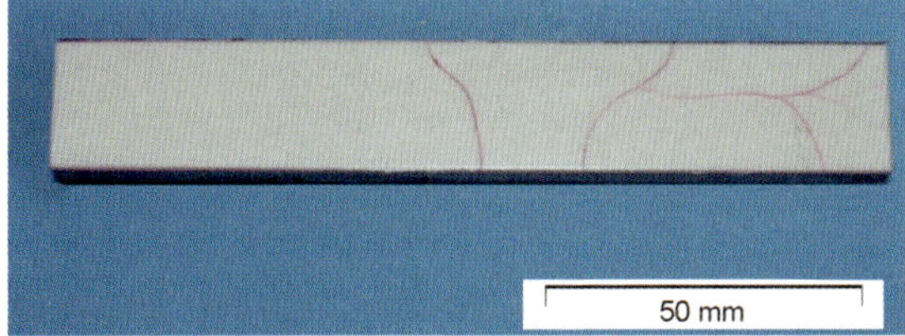

Figure 12.3 Thermal shock cracks in alumina ceramics. (a) In quenched specimens; (b) In a paper machine foil after severe operation. In this case the thermal shock was caused by water cooling after frictional heating. The cracks are highlighted by a dye penetrant.

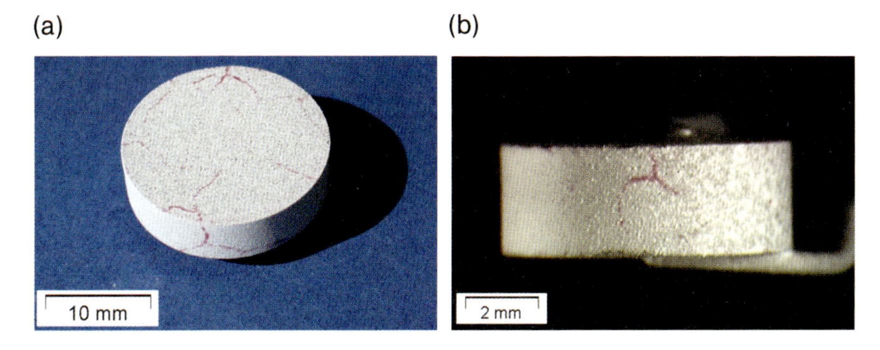

Figure 12.4 Thermal shock cracks in PTC-ceramics. (a) In a quenched specimen; (b) In a PTC resistor after soldering. The reason for the crack formation was an inappropriate handling of the component during soldering. The component has been taken out of the furnace by a cold pair of metal pliers. The cracks are marked by a dye penetrant.

12.2.2
Contact Failure

A loading contact of hard surfaces, for example, by static or dynamic impingement of bodies, can lead to cracking if the stresses exceed critical values. Such damage typically occurs in one of two modes: (i) local area loading or "blunt contact"; and point loading or "sharp contact."

Blunt contact is a classic scenario, and was comprehensively described by Hertz [5], who modeled it as the contact between a sphere and a flat surface. When the surfaces of two such bodies touch, they deform elastically and form a circular contact area. Under this contact area compressive stresses arise, the amplitude of which is a function of the elasticity of the two surfaces, and which increase with both the applied load and decreasing radius of the sphere. The contact zone is encircled by a narrow ring-shaped region where significant tensile stresses occur. These have a maximum amplitude of about one-quarter of the mean contact pressure [28]. If the tensile stresses locally exceed some critical value, a ring-shaped crack (Hertzian ring crack) is formed. Since the tensile stresses – and therefore the driving force for crack propagation – decrease steeply below a surface, Hertzian ring cracks generally do not penetrate deeply. (However, further increases in loading will cause such cracks to extend further at an oblique angle of about 22° from the indented surface to form a cone-shaped crack [28]). Such damage is typically caused and studied by means of blunt indentation techniques, such as Brinell hardness.

Hertzian cracks are often caused in service due to inappropriate handling, or to a localized overload. They may cause immediate (catastrophic) failure, or provide the origin for some sort of delayed failure. Figure 12.5a and b show a Hertzian ring crack in the surface of a silicon carbide specimen, and also cracks caused by contact damage in a silicon nitride roll for the rolling of high-alloyed steel wires [29]. Clearly, the reason for the cracks is similar in both cases. In the case of the roll, the shape of the working groove and the contact area of the wire with the groove had a significant influence on the shape of the crack [29, 30]. Hertzian contact damage is also likely to occur in ceramic balls in ball bearings.

Sharp contact is a (pseudo) point-loading situation, and is well modeled by sharp indentation techniques such as Vickers or Knoop hardness testing. Typically, any damage takes the form of severe plastic deformation beneath the point of contact, and the formation of radial cracks at any sharp features, such as the corners of the Vickers pyramid. The inelastic strains associated with plastic deformation produce high internal stresses which strongly influence the development of cracking and damage [28]. Lateral cracks may form below the plastically deformed region and extend parallel to the surface [31]. If these cracks intersect the radial cracks and the surface, then large fragments (chips) of the material may break away. In fact, this is one of the most prolific mechanisms of wear in ceramics [32].

Figure 12.5c and d show radial cracks and a break-out caused by lateral cracks originating from a Vickers indent in the surface of silicon nitride and silicon carbide ceramics.

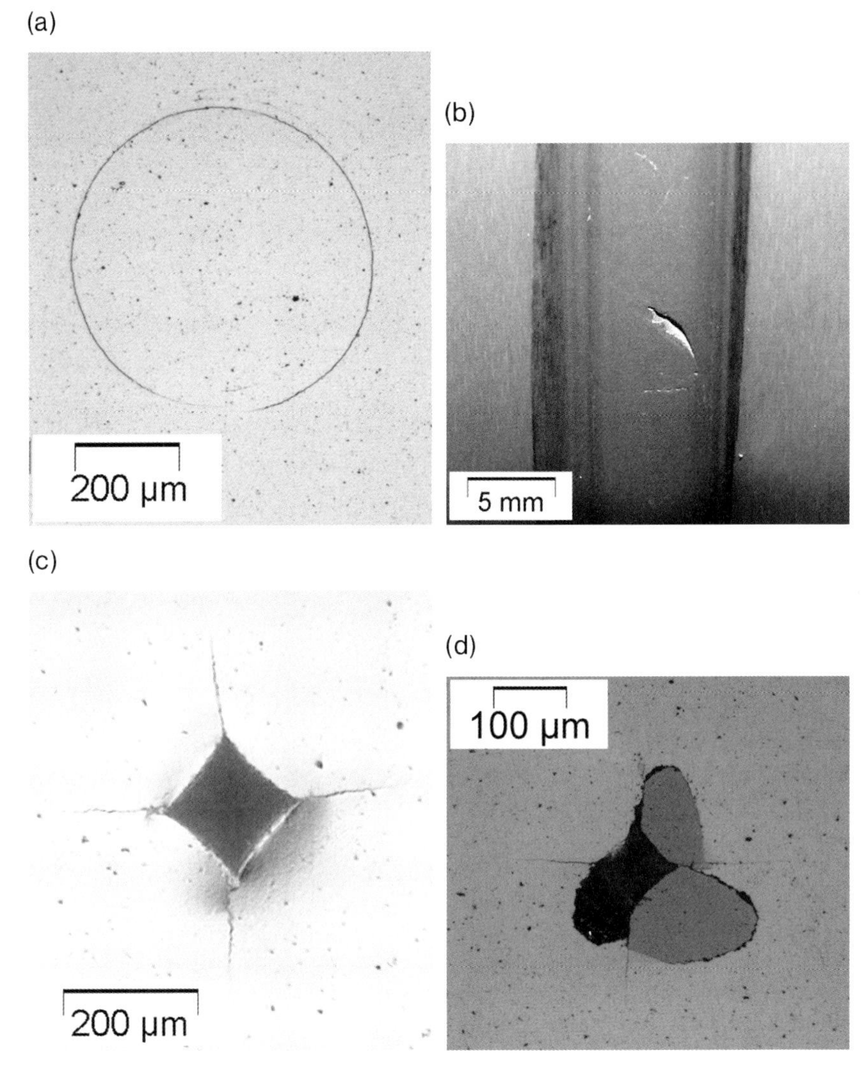

Figure 12.5 Contact damage (cracks) in the surface of ceramics. (a, b) Cracks caused by a blunt and (c, d) cracks caused by a sharp indenter. (a) Hertzian ring crack in silicon carbide; (b) Crack caused in operation by the rolling of high-alloyed steel wires [30]; (c) Radial cracks at the corners of a Vickers indent in silicon nitride; (d) Break out due to lateral cracks under a Vickers indent in silicon carbide.

An interesting variant of contact failure occurs if the contact is near an edge, as the crack may not stop there and an edge flake may break out [33–36]. Such edge flaking can often be observed in ceramic components used, for example, in classical ceramic products such as marble steps, tableware, and tiles [19]. However, it is also seen in technical ceramic products such as tool bits or guides for hot metal sheets in steel

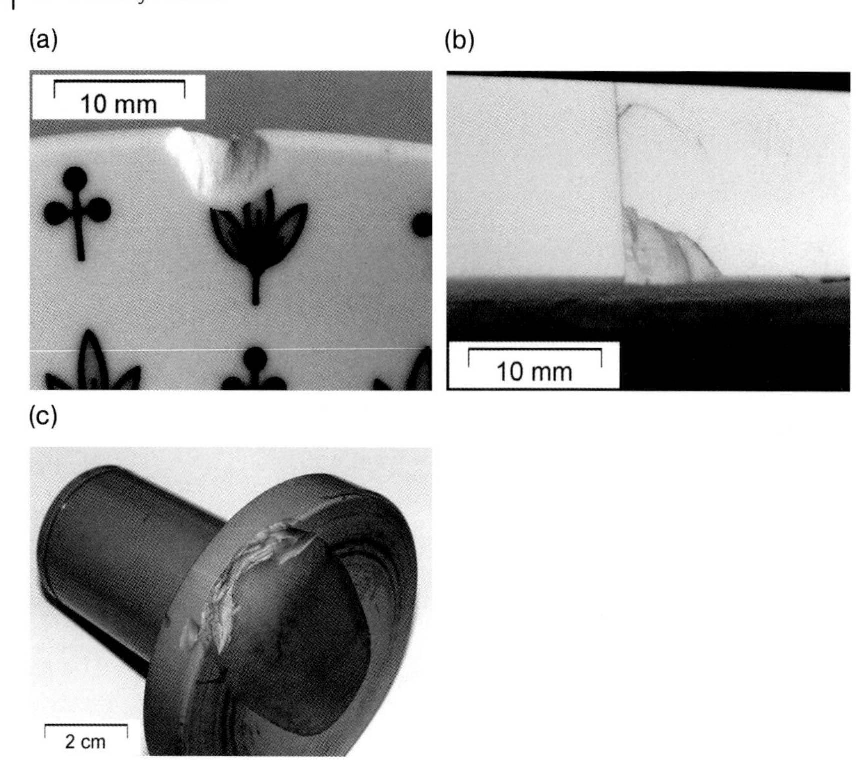

Figure 12.6 Examples of edge flakes in used ceramic components. (a) In a plate; (b) In a wear-resistant alumina foil in a paper machine; (c) In a silicon nitride guide for hot metal sheets in a steel mill.

works, and even in human teeth or tooth implants [37]. Some examples of edge flakes are shown in Figure 12.6.

Although the contact damage discussed so far is caused by some form of quasi-static loading, quite similar damage may occur from dynamic (impact) loading. For example, a strike with a hammer having a rounded head can cause a Hertzian-like ring crack at the surface of the component. A good everyday example is the chipping and cracking in the windscreens of cars, caused by the impact of grit and small stones.

Although frequently optically displeasing, contact damage does not always cause immediate failure, but may serve as the origin of some delayed failure caused by the subsequent steady or fast growth of cracks. The possible damage mechanisms will be discussed in Section 12.3.

12.3
A Short Overview of Damage Mechanisms

At room and ambient temperatures, almost any fracture in ceramics is brittle. It occurs without any significant plastic deformation, and the (elastic) failure strain is very small. When under load, cracks may grow steadily until they almost reach the speed of sound, at which point sudden catastrophic failure will occur. Several crack growth mechanisms exist, including sub-critical crack growth, fatigue, and corrosion. Ductile fracture only arises at very high temperatures (or at extremely slow deformation rates, though this is generally not relevant to technical applications) by creep. Several classes of damage mechanism have been identified [14], described briefly as:

- Sudden, catastrophic failure
- Sub-critical crack growth (SCCG)
- Fatigue
- Creep
- Corrosion or oxidation.

12.3.1 Sudden, Catastrophic Failure

Sudden, catastrophic failure is the most prominent damage mechanism in technical ceramics [1–4], and is inevitably the final active mechanism at failure. It is caused by the very rapid growth of a single crack (the critical crack). In fact, the growth rate is so high that crack arrest is almost impossible, and fracture of the component occurs almost immediately when a critical load is applied. Flaws, which may serve as critical cracks, may be generated during the production of a ceramic, by the machining of the component, by thermal shock, by contact damage or corrosion, and even by inadequate handling. Smaller cracks may grow to a critical size by SCCG, fatigue, creep, oxidation or corrosion (delayed failure). The time necessary for this growth determines the service life time of the component, after which sudden catastrophic failure occurs.

12.3.2 Sub-Critical Crack Growth

SCCG [38–41] is caused by the thermally activated breaking of bonds at the tip of a stressed crack [38]. It may be assisted by the corrosive action of some polar molecules (e.g., by water or water vapor), and is thus related to stress corrosion cracking as observed in metals [39, 40]. Some SCCG always precedes a final catastrophic fracture in brittle materials, but if the load is applied very quickly, the contribution can be rather small. It can cause delayed failure of components (long after the application of a load), without any plastic deformation of the ceramic material. Occasionally, SCCG is also referred to as fatigue (or static fatigue), although as it is not caused by cyclic loading (as in the case of fatigue in metals) this can be misleading. In order to avoid confusion, in the following sections the term “fatigue” will not be used in the context of SCCG.

12.3.3
Fatigue

For a long time, cyclic fatigue was thought not to occur in linear elastic ceramics. However, some thirty years ago evidence of damage due to the alternating application of loads was observed in advanced technical ceramics [42]. In the following, the term fatigue will be used exclusively to describe the damaging effects of cyclic loading.

In ceramics, fatigue crack growth is caused by repeated (cyclical) damage to microstructural elements, for example, by the breaking of crack bridges during the crack closure part of a loading cycle [43, 44]. As with SCCG, fatigue may also be a reason for delayed failure, and may precede the sudden catastrophic failure of a component.

12.3.4
Creep

Creep in ceramics is much less pronounced than in metals or polymers. The activation energy for creep of ceramics is much higher [45], and creep deformation and creep damage in ceramics therefore only occurs at very high temperatures [46]. (The activation energy approximately scales with the melting temperature [45], and the melting temperatures of most technical ceramics are high.) Creep damage is in general caused by the generation, growth, and coalescence of pores; it is not localized like crack growth, and may occur throughout a component [47]. The resulting final failure can be brittle as well as ductile, depending on the acting stresses and temperatures. In most applications of technical ceramics creep does not occur (exceptions include some applications of refractories, for example as components in kilns and furnaces, or as materials for lamps), and will therefore not be treated further at this point.

12.3.5
Corrosion

Corrosion, for example by oxidation, may cause serious damage in ceramics, and may represent an important wear mechanism [48]. If the corrosion products are gaseous, then a severe loss of material can result [49], and the oxidation of grain boundaries may even cause the total disintegration of a material [50]. Primarily, with respect to fracture, corrosion and oxidation damage acts as an initiation site for crack growth and consequent brittle fracture. For reasons of brevity, corrosion damage will also not be further discussed at this point.

In summary it can be stated that, at low and ambient temperatures, fracture always occurs by brittle failure, which starts from crack-like flaws in the microstructure. Brittle fracture can be preceded by some SCCG or fatigue crack growth [29], while the failure-initiating flaws can either result from the production process of the material or come into existence due to inappropriate machining, handling, or even by oxidation and corrosion. These mechanisms, and the consequences for mechanical design, will be treated in more detail in the following sections.

12.4
Brittle Fracture

In ceramics, brittle fracture is controlled by the extension of small flaws which are dispersed in a material or component's surface, and which behave like cracks. Flaws can arise not only from the production process, but also from handling and service. Some examples of critical flaws are shown in Figures 12.1 and 12.7.

In this chapter, brittle fracture is first discussed for the simple case of a uniaxial and homogeneous stress field (as in a tensile test), and for a single crack-like flaw which is oriented perpendicular to the stress field (in so-called mode I loading). These simple conditions are sufficient to explain the basic features of brittle fracture. Later, the analysis will be extended to more complex and more general situations, including multiple cracks, the arbitrary orientation of cracks (mode II and mode III loading), and multi-axial stress fields.

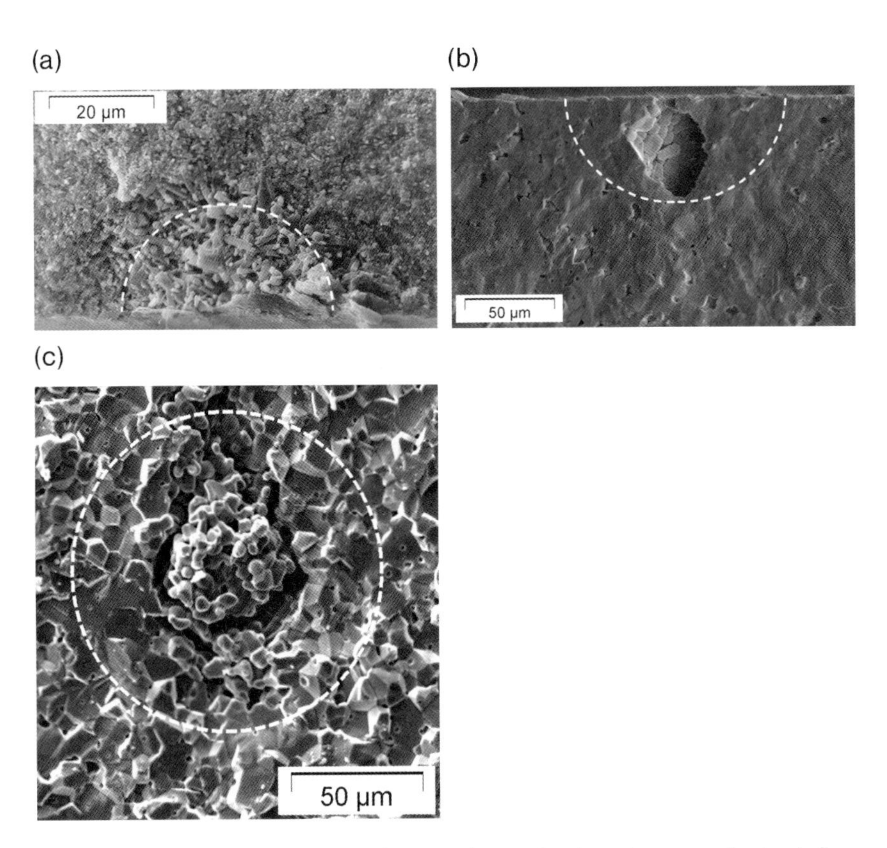

Figure 12.7 Fracture origins. (a) In a silicon nitride bending test specimen; (b) In a barium titanate (PTC ceramic) bending test specimen; (c) In an alumina bending test specimen. The origins are an agglomeration of coarse, elongated grains, a large pore due to a hollow agglomerate and a crescent-shaped pore surrounding a hard agglomerate, respectively. The Griffith crack size is indicated by white dotted lines.

12.4.1 Some Basics in Fracture Mechanics

Ultimately, fracture can only occur if all atomic bonds in an area are pulled apart and break. The stress necessary to break a bond (the theoretical strength) is between $E/5$ and $E/20$, where E is the elastic modulus of the material [51]. Typical tensile stresses applied to highly loaded components are in the order of $E/1000$ or even smaller, and yet fracture of components still occurs. To explain this discrepancy, it is necessary that certain strong local stress concentrations exist; these are termed *flaws*. The action of flaws can be discussed by using the simple example of an elliptical hole in a uniaxial tensile-loaded plate. At the tip of its major semi-axis (which is perpendicularly oriented to the stress direction), the stress concentration is [52]:

$$\sigma_n/\sigma = 1 + 2c/b. \tag{1a}$$

where σ is the applied far field stress and σ_n is stress at the tip of the major semi-axis c of the ellipse. The minor semi-axis is b. So, for any circular hole, independently of its diameter, the stress concentration is 3.

It is clear that the stress concentration may become very high, if the ellipse is extremely elongated; that is, if $c \gg b$. For that case, the equation can be approximated to give [51]:

$$\sigma_n/\sigma \approx 2\sqrt{c/\varrho}, \tag{1b}$$

where ϱ is the radius of curvature. It should be noted how the stress concentration depends on the shape rather than the size of the notch and that, for $\varrho \to 0$, the stress at the tip goes to infinity.

An elliptical hole must be extremely elongated to create the stress concentration necessary to reach the theoretical strength. To give a simple example, for an applied stress of $\sigma = E/1000$ the ratio of major to minor semi-axis must be $c/b \approx 50$ ($c/\varrho \geq 2500$) to create the stress concentration necessary to reach the theoretical strength at a notch tip: $\sigma_n \approx E/10$.

Now, consider the stress concentration produced by cracks. The stress field around a crack tip can be determined by linear elastic fracture mechanics (common solutions for this are available in standard text books [51, 53]). Here, a material is treated as a linear elastic continuum, and a crack is assumed to be a "mathematical" section through it (having a crack tip radius of zero). Under plane strain conditions, the components of the local stress field on a volume element, σ_{ij}, in a region near the crack tip is space dependent and can be expressed as (polar coordinates r and θ; origin at the crack tip) [51, 53]:

$$\sigma_{ij}(r,\theta) = (K/\sqrt{2\pi r}) \cdot f_{ij}(\theta). \tag{2}$$

The tensor f_{ij} is a function of the angle θ only, and values can be found in various text books [51, 53]. The stress field scales with the stress intensity factor, K, given by

$$K = \sigma Y \sqrt{\pi a}, \tag{3}$$

where σ is the nominal stress in the uncracked body, a is the crack length, and Y is a dimensionless geometric correction factor (for non-uniaxial tensile loadings, Y may also depend on the loading situation). Data for the geometric factor can also be found elsewhere [54–56]. For cracks, which are small compared to the component size (this is the general case in ceramics), Y is of the order of unity. Important examples of idealized crack geometry are "penny-shaped" cracks in the volume of a body, and straight-through edge cracks. The geometric correction factor is $Y = 2/\pi$ for small penny-shaped cracks, while in the latter case it is $Y = 1.12$.

The components of the stress tensor σ_{ij} have a square-root singularity at the crack tip. They therefore become infinite at the tip of a crack, and inevitably exceed the theoretical strength, even at very weak loads or small crack sizes. Note that this corresponds to the solution for an elliptical hole with a tip radius of zero [Eq. (1b)]. There exist cracks under load which do not grow; consequently, an additional condition for the growth of cracks must exist, and this was identified by Griffith, who analyzed the energy changes associated with the extension of a single crack in a brittle material [57].

According to Griffith, crack growth only occurs if the crack extension reduces the total energy in a body. There exists a "critical" crack length a_c: cracks of larger or equal size extend, but smaller cracks are stable. The Griffith analysis shows that the fracture stress (the strength) is related to the critical crack length according to the relationship: $\sigma_f \propto a_c^{-1/2}$.

Within the framework of linear elastic fracture mechanics, the total energy released by crack growth (work done and elastic energy released) under plane strain conditions is given by the product of the strain energy release rate [51, 53]:

$$G = \frac{K^2}{E}. \tag{4}$$

and the newly cracked area $(t \cdot \delta a)$. Fracture occurs if the released energy (per newly cracked area) reaches or exceeds the energy necessary to create the new crack:

$$G \geq G_c. \tag{5}$$

Inserting Eq. (12.4) into Eq. (12.5) gives the so-called Griffith–Irwin fracture criterion

$$K \geq K_c. \tag{6}$$

The critical stress intensity factor, K_c, is called the "fracture toughness," and is defined as:

$$K_c = \sqrt{E \cdot G_c}. \tag{7}$$

A theoretical lower bound for the fracture energy is the energy of the new surfaces (2γ). In reality, the fracture energy is at least one order of magnitude higher, because energy-dissipating processes ahead of the crack tip (in process zones [58]) or behind the crack tip (at crack bridges [59–61]) occur. A typical value for the surface energy of ceramics is in the order of $1\ \mathrm{J\ m^{-2}}$, while values for fracture energy range between 10 and $3000\ \mathrm{J\ m^{-2}}$.

12.4.2
Tensile Strength of Ceramic Components, and Critical Crack Size

The tensile strength of a component can be determined by inserting Eq. (3) into Eq. (6), and solving for the stress (at the moment of fracture):

$$\sigma_f = \frac{K_c}{Y\sqrt{\pi a_c}}. \tag{8}$$

The strength scales with the fracture toughness, and is inversely proportional to the square-root of the critical crack size a_c. This corresponds to the former results of the Griffith analysis.

Strength test results on ceramic specimens show, in general, a large scatter. This follows from the fact, that in each individual specimen, the size of the critical crack is a little different. Rearranging Eq. (8) provides a relationship for the critical crack size (Griffith crack size) in a specimen:

$$a_c = \frac{1}{\pi} \cdot \left(\frac{K_c}{Y\sigma_f}\right)^2. \tag{9}$$

Figure 12.7 shows several fracture origins in ceramic materials. It is clear that there exists a strong correlation between the occurrence of dangerous flaws and the processing of a material. Typical volume flaws, which may act as fracture origins, are agglomerates of second-phase, large pores, inclusions, large grains, or agglomerations of small pores [17, 19]. Typical fracture origins at the surface are grinding scratches and contact damage, and even grooves at the grain boundaries may act as fracture origins [62].

For many ceramic materials, the ratio K_c/σ_f is of the order $1/100\sqrt{m}$. If a fracture starts in the interior of the component, the geometric correction factor for a penny-shaped crack ($Y = 2/\pi$) can be used to calculate the corresponding Griffith crack size, giving a typical $a_c \approx 80\ \mu m$. The size of critical flaws reflects the state of the art in minimizing microstructural defects during the processing of ceramic specimens and components.

In Figure 12.7 it can be recognized that the Griffith crack size is often a little larger than the size of the critical flaw. Several reasons can be identified for such behavior, and some of these are discussed in the following sections.

In most materials, the resistance to crack propagation is not constant, but it rather increases at the start of a crack extension and reaches, after a degree of crack growth, a plateau value [3, 4]. In consequence, the toughness may depend on the testing conditions [4]. Under the conditions of typical fracture toughness tests, the toughness may be slightly higher than found under strength testing conditions. Consequently, in the evaluation of the Griffith crack size, a too-high value for toughness is occasionally used, but this may lead to an overestimation.

In most strength tests, some SCCG happens before the final fracture occurs, which means that the cracks has begun to grow at stress intensity factors less than the fracture toughness. The Griffith crack size will then correspond to the size of the flaw plus the crack extension due to SCCG.

Finally, it should also be remembered that flaws are not truly cracks, and that the microstructure of a material is not an ideal continuum; that is, the flaws and cracks have different stress fields. Whilst flaws cause a (finite) stress concentration [see Eq. (1)], cracks cause a stress singularity at their tips [see Eq. (2)]. As discussed above for elliptical holes, a major semi-axis to tip radius ratio of several thousand would be necessary to produce the stress concentration required to break the bonds at the tip. Whilst, at first glance, this seems improbable, it is not completely unreasonable as very sharp edges may occur on processing defects (an example is shown in Figure 12.7b). Here, the fracture origin is a large pore which, on a meso-scale, can be modeled by a spherical hole with a radius of about 25 μm. However, many sharp grooves can be detected at the grain boundaries that could be described as sharp notches distributed around the sphere [63, 64]. It is clear that the tip radius of these grooves in front of the pore is much less than 1 μm, giving a high aspect ratio.

In summary, the fracture of brittle materials is a complex topic, with the fracture process being largely influenced by the local microstructure, and which renders the use of suitable statistics significant. This topic will be examined more closely in Section 12.5.

12.5 Probabilistic Aspects of Brittle Fracture

In a series of fracture experiments on ceramic specimens, two important observations can be made, namely that the probability of failure increases with the load amplitude, and also with the size of the specimens [2–4, 14]. This strength–size effect is the most prominent and relevant consequence of the statistical behavior of the strength of brittle materials. However, these observations cannot be explained in a deterministic way by using a simple model of a single crack in an elastic body; rather, their interpretation requires an understanding of the behavior of many cracks distributed throughout a material.

In the following, it is assumed that many flaws (which behave like cracks) are distributed stochastically within a material. For convenience, it is assumed initially that the stress state is uniaxial and homogeneous, and that the cracks are perpendicularly oriented to the stress axis (more general situations will be discussed later).

It is further assumed that the cracks do not interact (i.e., their separation is large enough for their stress fields not to overlap), an assumption which is essential for the following arguments and which is equivalent to the weakest-link hypothesis: that the failure of a specimen is triggered by the weakest volume element or, in other words, by the largest flaw that it contains.

12.5.1 Fracture Statistics and Weibull Statistics

Using the assumptions made above, a cumulative probability of fracture F_S (the probability that fracture of a specimen occurs at a stress equal to or lower than σ) can be defined [65]:

$$F_S(\sigma) = 1 - \exp(-N_{c,S}(\sigma)). \tag{10}$$

where $N_{c,S}(\sigma)$ is the mean number of critical cracks in a large set of specimens (i.e., the value of expectation). The symbol S designates the size and shape of a specimen, and the number $N_{c,S}(\sigma)$ depends on the applied load via a fracture criterion, for example Eq. (5) or Eq. (6). At low loads, the number of critical cracks is small: $N_{c,S}(\sigma) \ll 1$. This is typical for the application of advanced ceramic components, as these are designed to be very reliable. In this case, the probability of fracture becomes equal to the number of critical cracks per specimen, $F_S(\sigma) \approx N_{c,S}(\sigma) \ll 1$, and at high loads this number may become high: $N_{c,S}(\sigma) \gg 1$. Some of the specimens may then contain even more critical cracks than the mean value $N_{c,S}$. Nevertheless, the (small) probability still exists that, in some individual specimens, no critical crack occurs. Consequently, there is always a nonvanishing probability of survival for an individual specimen, and the probability of fracture only asymptotically approaches one for very high value of $N_{c,S}(\sigma)$ [see Eq. (10)].

It should be noted that the fracture statistics given in Eq. (10) describe the experimental observations referred to at the start of this chapter correctly – that the probability of failure increases with load amplitude (because at higher loads more cracks become critical). In addition, the mean strength decreases with specimen volume (since the probability of finding large cracks in large specimens is higher than in small specimens).

Since, of course, a specimen can either fail or survive at any given load, it follows that $R_S(\sigma) + F_S(\sigma) = 1$, with $R_S(\sigma)$ (the reliability) being the probability of a specimen with size S surviving the stress σ.

In order to establish, on an analytical basis, the dependence of the probability of failure on the applied load, additional information relating to the crack population involved is required [66–70]. If a homogeneous crack-size frequency density function $g(a)$ exists, then the mean number of critical cracks per unit volume is

$$n_c(\sigma) = \int_{a_c(\sigma)}^{\infty} g(a)\, \mathrm{d}a, \tag{11}$$

and the mean number of critical cracks (Griffith cracks) per specimen is $N_{c,S}(\sigma) = V \cdot n_c(\sigma)$, where V is the volume of the specimen. (For an inhomogeneous crack size distribution, the number of critical flaws per specimen results from integration of the local critical crack density $n_c(\sigma, \vec{r}\,)$ over the volume, with $\vec{r}$ being the position vector.) The strength–size effect is a consequence of this relationship.

The stress dependence of the critical crack density results from the stress dependence of the Griffith crack size, which is the lower integration limit in Eq. (11).

In most materials, the size frequency density decreases with increasing crack size; that is, $g \propto a^{-p}$, where p is a material constant. For such cases, Eq. (11) can easily be integrated [67, 70, 71], and inserting the result into Eq. (10) produces the well-known relationship for Weibull statistics [72, 73]:

$$F(\sigma, V) = 1 - \exp\left[-\frac{V}{V_0}\left(\frac{\sigma}{\sigma_0}\right)^m\right]. \tag{12}$$

The Weibull modulus m describes the scatter of the strength data. The characteristic strength σ_0 is the stress at which, for specimens of volume $V = V_0$, the failure probability is: $F(\sigma_0, V_0) = 1 - \exp(-1) \approx 63\%$. The independent material parameters in Eq. (12) are m and $V_0\sigma_0^m$; the choice of the reference volume V_0 influences the value of the characteristic strength σ_0.

The material parameters in the Weibull distribution are related to the fracture toughness of the material, and also to parameters from the size frequency density of the cracks. (Details of this are available in Ref. [62].) For example, as shown in the noteworthy report of Jayatilaka *et al.* [67], the Weibull modulus depends only on the slope of the crack size frequency distribution: $m = 2(p-1)$.

The derivation indicated above shows clearly that the Weibull distribution is a special case among a class of more general distributions [65, 69, 70]. In particular, the type of flaw distribution may have an influence on the strength distribution function (this point will be discussed later). This notwithstanding, almost all sets of strength data determined on ceramics and reported to date can be fitted nicely by a Weibull distribution function, and consequently the Weibull distribution is used widely to describe strength data, and also for designing with brittle materials. Typically, advanced ceramics will have a Weibull modulus between 10 and 20, or even higher, whereas for classical ceramics the modulus will typically be between 5 and 10.

The Weibull distribution for several specialized situations is described in the following sections.

12.5.1.1 Weibull Distribution for Arbitrarily Oriented Cracks in a Homogeneous Uniaxial Stress Field

In general, the orientation of cracks will depend on the processing conditions of the specimens or components; furthermore, their distribution may be random, or there may be some preferred orientation. Although, until now, it has been assumed that cracks are perpendicularly oriented to a uniaxial tensile stress field, this assumption is not generally necessary and has only been made to simplify the above arguments.

With regards to the influence of random crack orientation in a homogeneous uniaxial tensile stress field, the most dangerous cracks are those oriented perpendicularly to the direction of stress. In this case, the crack boundaries are pulled apart by the applied stress (uniaxial tensile loading, opening, mode I). If the cracks are parallel to the uniaxial stress direction, the crack borders will not be opened by the stresses and they do not disturb the stress field; clearly, this type of loading is harmless. However, for any crack orientation between these extremes, some in-plane (sliding) or out-of-plane (tearing) shear loading (mode II and mode III, respectively) of the cracks occurs. The strain energy release rate is then $G = K_I^2/E' + K_{II}^2/E' + K_{III}^2(1-\mu)/E$, where K_I, K_{II}, and K_{III} are the stress intensity factors for each mode of loading, and $E' = E$ for plane stress conditions and $E' = E/(1-\mu^2)$ for plane strain conditions [51]. Here, μ is the Poisson ratio.

To describe the action of many cracks in a uniaxial stress field, it is important to know not only the crack-size distribution but also the distribution of crack orientations. Such an analysis shows that, whilst the results of the last section remain valid, the characteristic strength σ_0 is influenced by the fact that in general, the number of critical cracks depends not only on the length but also on the orientation of the crack [67].

12.5.1.2 Weibull Distribution for Arbitrarily Oriented Cracks in an Inhomogeneous Uniaxial Stress Field

For inhomogeneous but uniaxial stress fields, a generalization of Eq. (12) can be made (see Weibull [72, 73]):

$$F(\sigma_r, V_{\text{eff}}) = 1 - \exp\left[-\frac{V_{\text{eff}}}{V_0}\left(\frac{\sigma_r}{\sigma_0}\right)^m\right]. \tag{13}$$

where V_{eff}, the effective volume of a component, is given by the integration of the stress field over the volume:

$$V_{\text{eff}} = \int\limits_{\sigma>0} \left(\frac{\sigma(\vec{r})}{\sigma_r}\right)^m dV, \tag{14}$$

where σ_r is an arbitrary reference stress. The integration is conducted only over those volume elements where the (uniaxial) stress $\sigma(\vec{r})$ is tensile, such that any damaging action of compressive stresses is, therefore, neglected. However, as the compressive strength is generally around one order of magnitude higher than the tensile strength, this is in fact quite reasonable for any case where the compressive stress amplitude is no more than about threefold larger than the tensile stress amplitude.

Usually, the effective volume is defined as the volume of a tensile test specimen that would yield the same reliability as the component, if loaded with the reference stress. [(Note: Other definitions are also possible; for example, STAU uses the most damaging stress state (i.e., the isostatic–tri-axial stress state) for the definition of the equivalent stress [74].]

The effective volume is related to the reference stress via the equation: $\sigma_r^m V_{\text{eff}} = \sigma_0^m V_0$. In general, the maximum in the stress field is used as the reference stress (e.g., in bending tests, the outer fiber stress is commonly used). Since, for advanced ceramics, the Weibull modulus is a relatively high number (e.g., $m =$ 10–30), only the most highly loaded volume elements (stress more than 80% of the maximum) contribute significantly to the effective volume, and so the effective volume approximates to the "volume under high load." It should be noted that, if a stress lower than the maximum tensile stress is selected as the reference stress, then the effective volume may become large – possibly larger even than the real size of the component.

In simple cases – such as the stress field in a bending specimen – calculation of the integral in Eq. (14) can be made analytically, although in general – and especially in the case of components – a numerical solution is necessary. For each loading case the calculations must be performed only once, because the integral will scale with the load amplitude. As mentioned above, for modern ceramic materials the Weibull modulus can be high; this in turn makes a numerical determination of the effective volume very difficult, especially if high stress gradients are present. Today, however, commercial systems capable of handling the post-process data generated in finite element (FE) stress analyses are available (e.g., STAU [74] and CARES [75]). The details of effective volumes for standardized flexure bars and cylindrical rods in flexure have also been published [76, 77].

12.5.1.3 Weibull Distribution in a Multi-Axial Stress Field

Multi-axial stress states can be taken into account by replacing the stress σ in the above considerations by an equivalent stress σ_e:

$$\sigma \rightarrow \sigma_e. \tag{15}$$

The equivalent (uniaxial) stress is the uniaxial tensile stress which would have the same damaging action as an applied multi-axial stress state. The proper definition of the equivalent stress depends on the type of the fracture causing flaws, and how such flaws act as cracks (though this remains a matter of some debate). The definition also depends on the fracture criterion for multi-axial stress states, and this should also take into account the action of compressive stresses. Precise information on real flaw distributions and multi-axial strength data are rarely available, and the large scatter of strength values makes unambiguous interpretation of these data difficult.

Although, many different proposals to define equivalent stress in ceramics have been made [3, 4, 78–81], there is not yet a consensus on the best choice. For small data sets, of the type most commonly tested, the scatter obscures any differences between reasonable alternative fracture criteria. The most frequently used criteria are described below.

The first-principle stress (FPS) criterion simply assumes that only the first-principle stress σ_I is taken into account. The FPS criterion [3, 4] reads:

$$\sigma_{e,\mathrm{FPS}} = \sigma_I \quad \text{for} \quad \sigma_I > 0$$

and (16a)

$$\sigma_{e,\mathrm{FPS}} = 0 \quad \text{for} \quad \sigma_I \leq 0$$

The principle of independent action (PIA) accounts for the action of all principle stresses (σ_I,σ_{II} and σ_{III}), independently. The PIA criterion [78] is:

$$\sigma_{e,\mathrm{PIA}} = \sqrt[m]{\sigma_I{}^m \Theta(\sigma_I) + \sigma_{II}^m \Theta(\sigma_{II}) + \sigma_{III}^m \Theta(\sigma_{III})}$$

with (16b)

$$\Theta(x) = \begin{cases} 0 : x \leq 0 \\ 1 : x > 0 \end{cases}.$$

The action of compressive stresses is neglected in both cases, which is taken into account in Eq. (16b) by the use of the Heaviside function, Θ.

It is expected that the FPS criterion would apply nicely for flat, crack-like flaws, but for more spherical flaws, which have similar stress concentrations in all directions, the PIA criterion would seem to be more appropriate [63]. It should be noted, however, that the influence of the criterion on the equivalent stress is relatively limited: for a material with a Weibull modulus of $m = 20$ and in a biaxial stress state, the FPS equivalent stress is $\sigma_{e,\mathrm{FPS}} = \sigma_I$, whereas the PIA equivalent stress is $\sigma_{e,\mathrm{PIA}} = 2^{1/m}\sigma_I \approx 1.04\,\sigma_I$. In other words, the difference is only 4%, with such a small difference between equivalent stresses often being hidden by the scatter of the data. As an example, if the Weibull parameters are determined on a sample of 30 specimens, with $m = 20$, the 90% confidence interval of the characteristic strength would range from 98% to 102%, and that of the Weibull modulus from 79% to 139%. Therefore, a clear selection between the different criteria can barely be achieved.

Of course, the equivalent stress should also be used for the proper definition of the effective volume. This possibility (for several different definitions of the equivalent stress) is also accounted for in commercial FE programs [74, 75].

12.5.2 Application of the Weibull Distribution: Design Stress and Influence of Specimen Size

Until now, the Weibull distribution function and the ideas sketched above have formed the basis of the state-of-the-art mechanical design process for ceramic components [2–4]. Consequently, the strength testing [82–84] of ceramics and the determination of Weibull distributions have become standardized [6, 85–87].

In this section, examples of strength statistics and the resulting size effect are provided, using data acquired in a European research program organized by ESIS, that was conceived to determine a complete set of design-relevant data for a commercial silicon nitride ceramic. (Details on the ESIS reference material testing program can be found in Refs [13, 88, 89].)

An example of a strength distribution is provided in Figure 12.8, which shows the cumulative probability of failure versus four-point bending strength data. The sample consisted of 42 individual specimens, and was larger than in most other reported cases. The logarithmic scales were chosen in such a way that the Weibull distribution could be represented by a straight line. The distribution fitted nicely to the measured data, while the reference stress used was the maximum surface bending stress and the normalizing volume used was $V_{\mathrm{eff}} = V_0 = 7.7\ \mathrm{mm}^3$. The Weibull modulus was $m = 15.5 \pm 3$, and the characteristic strength of the sample resulting from the sampling procedure was $\sigma_0 = 844 \pm 15$ MPa. The indicated scatter related to the 90% confidence intervals.

The distribution function can be used to determine the relationship between reliability and the necessary equivalent design stress $\sigma_{e,r,R}(V_{\mathrm{eff}})$. For a component with the effective volume V_{eff} it holds:

$$\sigma_{e,r,R}(V_{\mathrm{eff}}) = \sqrt[m]{\frac{V_0\sigma_0^m}{V_{\mathrm{eff}}} \cdot \ln\frac{1}{R}}. \tag{17}$$

To give an example, for $V_{\mathrm{eff}} = V_0$, the first stage is to determine the maximum allowable stress for a reliability of $R = 99.9999\%$ in the bending tests shown in Figure 12.8. Following Eq. (17), the stress is $\sigma_{e,r,R}(V_{\mathrm{eff}}) = \sigma_0 \cdot (\ln 1/R)^{1/m}$. Inserting

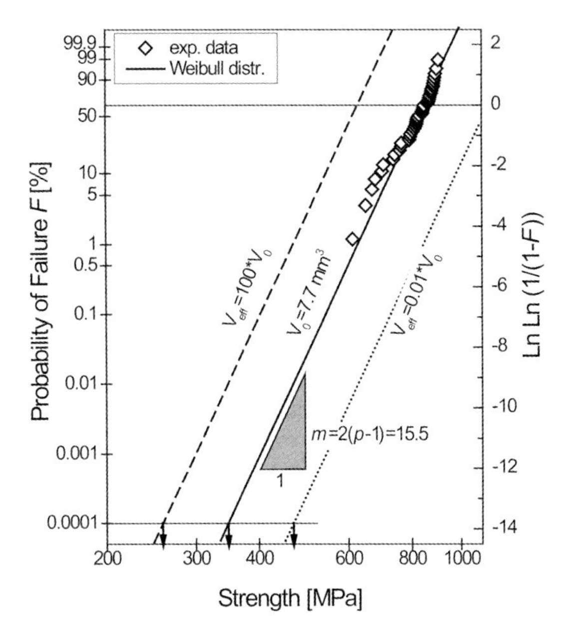

Figure 12.8 Bending strength test results for a silicon nitride ceramic. Plotted is the probability of failure versus the strength. The straight lines represent Weibull distributions. The measured Weibull distribution has a modulus of $m = 15.5$, and the characteristic strength is 844 MPa. Also shown are Weibull distributions for specimens with a 100-fold larger (left) and a 100-fold smaller (right) effective volume. Indicated are the 10^{-4}% failure probabilities and the design stresses for each set of specimens, respectively.

the numerical values for R, σ_0 and m gives $\sigma_{e,r,0.999999}(V_0) = 0.41 \cdot \sigma_0 = 348$ MPa. In other words, it is expected that one specimen out of a million would fail at a stress equal to or smaller than 348 MPa. This stress is also indicated in Figure 12.8. It should be remembered that the characteristic strength of the data set was 844 MPa.

It should be noted that, for the selected example, the design stress determined by probabilistic means corresponds – in a deterministic design approach – to a "safety factor" of about 844 MPa/348 MPa = 2.4. The advantage in the probabilistic approach is that a realistic reliability can be determined, and that total safety – which does not exist in reality – is not implied.

It is equally important to discuss the influence of the volume under load. For two sets of specimens of different volume (indicated by the indices 1 and 2) the (equivalent) stresses to achieve the same reliability, R must follow the condition:

$$\sigma_1(R)^m V_1 = \sigma_2(R)^m V_2 \quad \text{or} \quad \sigma_2(R) = \sigma_1(R) \cdot \left(\frac{V_1}{V_2}\right)^{1/m}, \tag{18}$$

where $\sigma_1(R)$ and $\sigma_2(R)$ are the corresponding equivalent reference stresses to achieve the desired reliability R, and V_1 and V_2 are the corresponding effective volumes of the specimens. If, in the above example, the effective volume is increased 100-fold, the design stress is decreased by a factor of $100^{1/15.5} = 0.72$ and becomes 0.72×348 MPa = 234 MPa (for a 100-fold smaller volume it becomes 451 MPa). These estimates assume a single flaw-type population initiating failure in all specimen sizes.

The Weibull distributions functions for the large and small specimens are also shown in Figure 12.8. It can be recognized that the Weibull line is shifted to lower strength values in the first case, and to higher strength values in the latter case.

The dependence of strength on specimen size is the most significant effect caused by its probabilistic nature since, as discussed above, this may have a dramatic influence on the design stress. To demonstrate such an influence, Figure 12.9 shows further experimental results from the ESIS reference materials testing program. Here, the results of three-point and four-point bending tests (uniaxial stress fields) and of ball-on-three balls tests (biaxial stress field) are reported (for details of the testing procedures, see Refs [88–92]). When the PIA criterion was used to calculate the equivalent stress, the effective volume of the specimens was found to range between about 10^{-3} mm^3 and 10^2 mm^3. Within this range the data followed the trend determined by the Weibull theory, with the dashed lines in Figure 12.8 describing the limits of the 90% confidence interval for a data prediction based on the four-point bending data. The characteristic strength of the individual samples varied from about 800 MPa (for the largest specimens) up to almost 1200 MPa (for the smallest specimens).

Of course, measurement uncertainties should also be considered in a proper design, and these points will be discussed in Section 12.5.3.

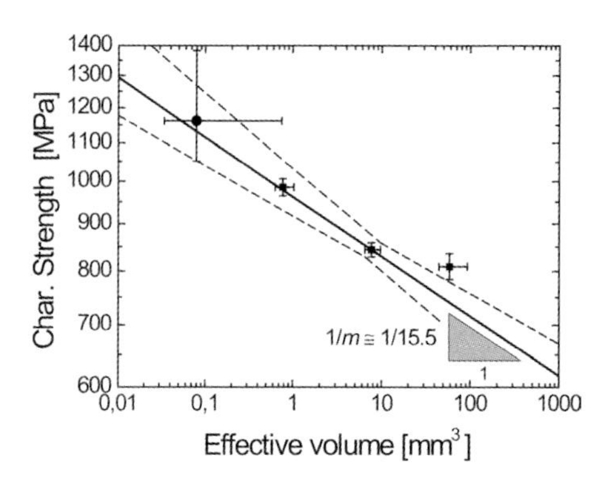

Figure 12.9 Strength test results of a silicon nitride ceramic. Shown are data from uniaxial and biaxial strength tests. To determine the equivalent stress, the PIA criterion was used. Plotted in log-log scale is the characteristic strength of sets of specimens versus their effective volume. The solid line indicates the trend predicted by the Weibull theory. The scatter bars and dashed lines refer to the 90% confidence intervals due to the sampling procedure. The relatively large scatter of the sample with the smallest specimens resulted from the small number of tested specimens.

12.5.3
Experimental and Sampling Uncertainties, the Inherent Scatter of Strength Data, and Can a Weibull Distribution be Distinguished from a Gaussian Distribution?

In most cases, the strength data of ceramics are determined in bending tests following EN 843-1 [82], ASTM 1161 [83], ISO 14704 [6], or JIS R 1601 [84]. The measurement uncertainty of an individual strength measurement should be less than 1% [93, 94]. The determination of the Weibull distribution is standardized in EN 843-5 [85]. When following this standard (and other similar American [86] and Japanese [87] standards), the Weibull distribution function must be measured on a sample of "at least" 30 specimens (due to high machining costs, it is not common practice to test larger samples).

Whilst the failure probability of an individual strength test is unknown, it can be estimated by using an estimation function. Many proposals exist for appropriate estimation functions, but the difference between them is – from a practical point of view – not very relevant. Typically, for large sample sizes the differences are very small, while for small sample sizes the scatter of data is larger than the difference between the different estimation functions. In standards, for example, in EN 843-5 [85], the function

$$F_i = \frac{2i-1}{2N} \tag{19}$$

is used, where F_i is the estimated failure probability of the ith specimen (the specimens are ranked with increasing strength), and the number of specimens in the sample is N (sample size). The range of failure probabilities which can be determined increases with sample size (from $F_1 = 1/2N$ to $F_N = (2N-1)/2N$). For a sample with $N = 30$ the range extends from $F_1 = 1/60$ to $F_{30} = 59/60$.

The inherent scatter of the Weibull parameters can be determined using Monte Carlo simulation techniques. Here, the first step is to generate (by "throwing dice") a random number between zero and one; this random number is defined as being the failure probability of a specimen. Then, for a material with a known Weibull distribution, the corresponding strength value can be generated using Eq. (12). By repeating this procedure N times, a virtual sample of size N is determined, which in turn allows millions of samples to be generated with limited effort. Moreover, this technique can be used to study the difference between sample and population [70].

It is clear that the difference between sample and population may become larger, the smaller a sample is. The minimum sample size of $N = 30$ as specified in standards represents a compromise between the large cost of specimen preparation and accuracy, although it should be noted that for $N = 30$ the uncertainty remains relatively large. Figure 12.10 shows the confidence intervals which result from the sampling procedure, for the Weibull modulus and the function σ_0^m (the scatter of the characteristic strength depends on the modulus) in dependence of the sample size, respectively. Of course, since in most practical situations the "sampling" is hypothetical, these values represent the uncertainty should the experiment be repeated.

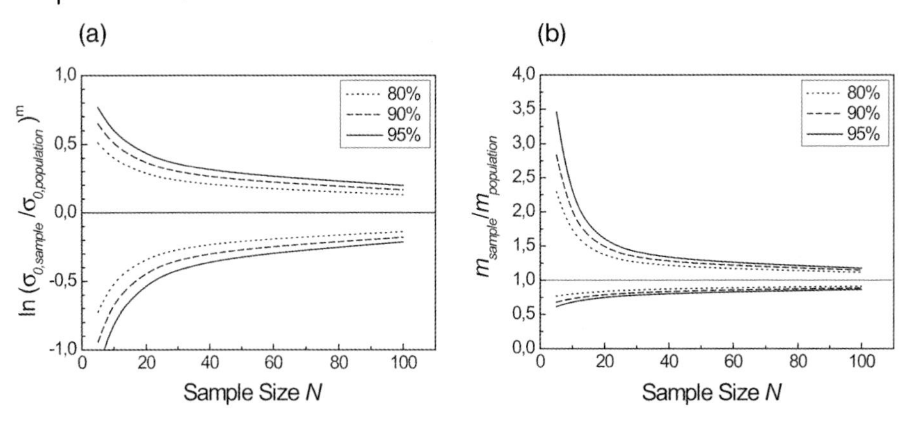

Figure 12.10 Confidence intervals in dependence of sample size (a) for σ_0^m and (b) for the Weibull modulus. The confidence intervals for the characteristic strength depend on the modulus.

It is important to realize that, on the basis of experiments, it is almost impossible to differentiate whether a population has a Weibull distribution or another type of distribution. In Figure 12.11 the data and the Weibull distribution of Figure 12.8 are plotted in a linear and in a log-log scale. Also shown is a Gaussian (Normal) distribution function, which fits to the experimental data.

Both functions describe the experimentally determined part of the distribution very well, but they show major differences at low failure probabilities, where no experimental data exist [see Eq. (10)]. This range of low probabilities, however, is exactly the range where distribution curves are applied to predict the reliability of components. For the above example, the $R = 99.9999\%$ design stress determined by Weibull analysis is 348 MPa, yet when determined on the basis of the Gaussian

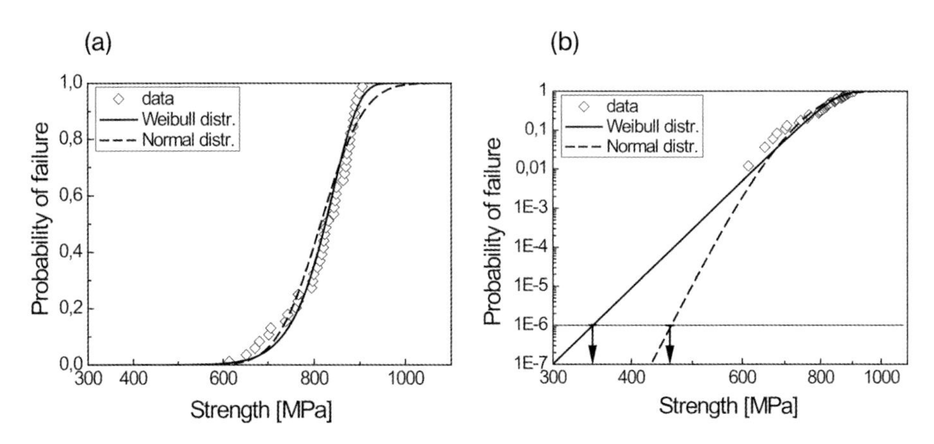

Figure 12.11 Strength distribution of a silicon nitride ceramic (same data as in Figure 12.8). A Weibull distribution function (solid curve) and a Gaussian distribution function (dashed curve) are fitted to the data. The data are plotted (a) in a linear and (b) in a log-log scale. Although both distributions fit the experimental data very well, the extrapolation to low probabilities of failure leads to quite different results.

function it is 460 MPa – that is, more than 30% larger! This large difference demonstrates the need for proper selection of the distribution function, although this can hardly be done on the basis of a small sample. From Eq. (19) and Figure 12.11 it can be seen that several thousand strength tests would be needed to detect, experimentally, any difference between a Weibull and a Gaussian distribution curve [62, 95]. Such an effort would exceed the possibilities of any research laboratory. However, the theoretical arguments provided at the start of Section 12.5 indicate that a Weibull distribution would be more appropriate to describe the strength data of brittle materials than would a Gaussian distribution. Weibull statistics also implies a size effect on strength (the Gaussian distributions does not), and such an effect has been demonstrated empirically (see Figure 12.10). It can be stated, therefore, that the strength of this material is Weibull-distributed.

12.5.4 Influence of Microstructure: Flaw Populations on Fracture Statistics

It follows from Eq. (10) that the strength distribution is strongly related to the flaw distribution. Previously in this chapter, it has been shown that the Weibull distribution occurs for volume flaws with a size distribution that follows an inverse power law: $g \propto a^{-p}$. It should be realized that the Weibull distribution is a direct function of the flaw-size distribution [see Eqs (10) and (11)]. By following this logic, any strength measurement can be interpreted as a measurement of the relative frequency distribution function of flaw sizes (albeit smeared by the scatter of data, as discussed above). Figure 12.12 shows the relative frequency of flaw sizes determined on the basis of the strength data shown in Figure 12.8. The frequency distribution follows the inverse power law (note the wide range of frequencies determined with this set of data), and the strength distribution is a Weibull distribution for that reason. A material with such behavior is termed a "Weibull material."

Other types of flaw distributions can also occur, with such examples being bimodal and multimodal flaw populations [69, 70, 96], the simultaneous occurrence of volume and surface defects [65, 70], or the occurrence of narrow peaked-flaw populations [70, 71]. In order to take into account the influence of several concurrently occurring flaw distributions, their mean numbers per specimen [see Eq. (12) and Ref. [65]] can simply be superposed:

$$N_{c,S}(\sigma) = \sum_i N_{c,S,i}(\sigma). \tag{20}$$

where $N_{c,S,i}(\sigma)$ is the mean number of critical defects of population i per specimen. The size S of the specimen can – depending of the type of flaw and the stress field in the specimen – relate to the (effective) volume (for volume flaws), the (effective) surface (for surface flaws), or the (effective) edge length (for edge flaws [70]). Multimodal flaw distributions cause perturbing structures in the strength distribution curves (bends or peaks in the straight "Weibull"-line, as shown in Figure 12.13), which cause a deviation from the simple analytic form of the Weibull distribution given in Eq. (12).

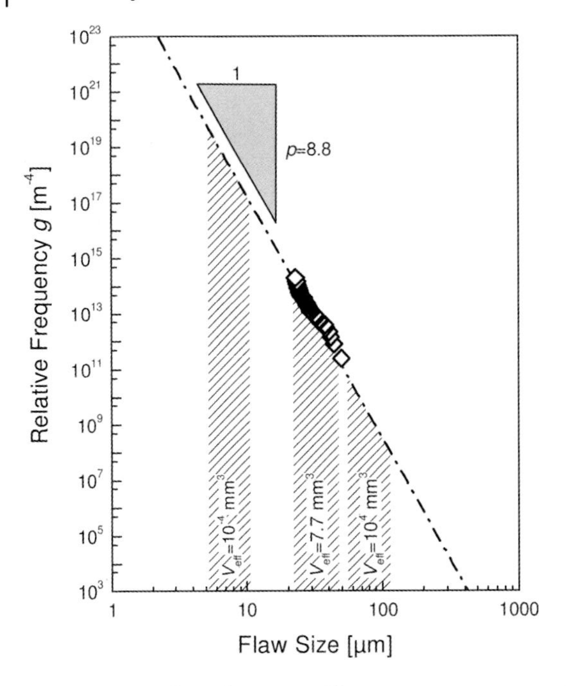

Figure 12.12 Relative frequency of flaw sizes versus their size. The distribution is plotted in a log-log scale. Shown are data of the same experiments as in Figure 12.8. The data are nicely arranged around a straight line, which indicates the Weibull behavior of the material (for details of the data evaluation, see Ref. [62]). The dashed areas correspond to the parameter ranges accessible with a sample of size 30 for specimens having the volume of the bending specimens (middle), a much larger volume (right), and a much smaller volume (left). It can be recognized that the size of fracture origins of specimens of different volumes is expected to be found in different ranges.

Figure 12.13 shows the example of a bimodal distribution, where a narrow peaked distribution is superimposed on to a wide flaw-size distribution. Figure 12.13a shows the relative frequency distribution of flaw sizes (top) and the density of critical flaws (below). The corresponding probabilities of failure are shown as a dashed line in Figure 12.13b. Also shown are results of Monte Carlo simulations, where the open symbols refer to a sample that represents the population very well, and the solid symbols refer to a sample that seems to behave like a Weibull distribution (dashed line). A small percentage of all samples ($N = 30$) have that behavior (it should be noted that this simulation exactly reflects the sampling procedure). Figure 12.13c shows the size effect of strength. Here, due to the relatively small scatter of the characteristic strength data (see 90% confidence intervals for samples of size 30), an investigation of the size effect on strengths seems to be an appropriate way to determine the fracture statistics of a material.

In summary, it must be stated that the strength distribution is a direct function of the size distribution of the flaws which act as fracture origins. The analytical Weibull equation [Eq. (12)] is a special case of a more general distribution which often – but

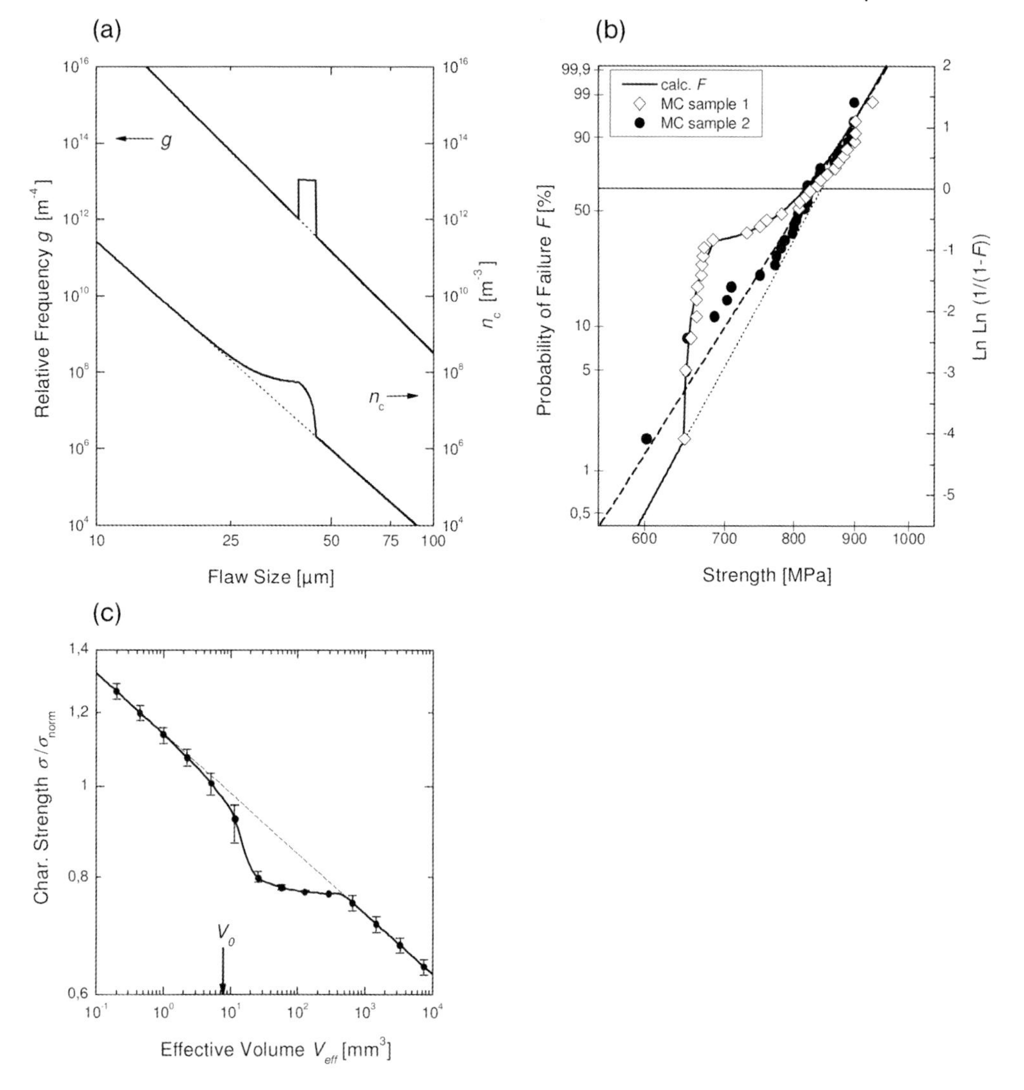

Figure 12.13 Bimodal flaw size distribution; a narrow peaked flaw population is superposed to a wide population. (a) Relative frequency of flaw sizes (bottom) and density of critical flaw sizes (top) versus the flaw size; (b) Weibull plot showing the probability function (through line) versus strength and two virtual samples (for details, see Ref. [70]) of that population; (c) Characteristic strength versus effective volume. The scatter bars refer to 90% confidence intervals for a sample of 30 specimens.

not always – occurs in brittle materials, and therefore, extrapolations from the experimentally explored parameter area should be considered with care. Both bimodal and multimodal flaw populations cause perturbations in the strength distribution, and can also cause major deviations from the behavior expected from a Weibull material. However, it should also be recognized that, on the basis of fitting distributions to the results of strength tests, the true type of distribution can hardly be determined, mainly because the data sets are too small to describe the tail ends of the distribution curve. The best way of determining distribution curves over a wide range is to test several subsets of specimens of different sizes [62, 71, 97].

12.5.5 Limits for the Application of Weibull Statistics in Brittle Materials

There are certain prerequisites for the occurrence of a Weibull distribution [65, 67, 71], the most important being: (i) that the structure fails if one single flaw becomes critical (weakest link hypothesis); and (ii) that dangerous flaws do not interact.

Notably, these prerequisites are not valid for ductile and fiber-reinforced materials. In the first case, stresses are accommodated by plastic deformation, whilst in the latter case the load can be transferred from the matrix into the fibers. Furthermore, a negligible interaction between flaws is only possible if the flaw density is low. Thus, Weibull theory does not apply to porous materials.

It has also been shown that Weibull theory does not apply to very small specimens [62], in which the fracture-causing flaws are also very small. For a Weibull material (i.e., for a material with relative flaw density corresponding to $g \propto a^{-p}$), the density of the flaws becomes so high for small flaws that an interaction between the flaws will undoubtedly occur.

In cases where the relative frequency of flaw sizes has a shape different to that of a Weibull material, the Weibull modulus becomes stress-dependent. The same happens for bimodal and multimodal flaw distributions [70, 71].

There are many other reasons for deviations from Weibull behavior [Eq. (12)], including internal stress fields [70], gradients in the mechanical properties [69], or if the toughness of the material increases with the crack extension (*R*-curve behavior) [71]. In these cases too, the modulus and the characteristic strength become stress- and volume-dependent, respectively.

In short, it can be stated that the Weibull distribution is often – but not always – the appropriate strength distribution function, and extrapolations must therefore be handled with care. It is helpful in many cases to determine strength data on specimens of different volume, as this procedure will enlarge the parameter ranges on which an extrapolation is based. Fractography is useful for ascertaining if a flaw distribution is either monomodal or multimodal.

12.6 Delayed Fracture

In the last section, attention was focused on fast, brittle fracture, in which a specimen under the action of a given load either failed immediately after the application of the load, or did not fail at all. However, it is well known that components also fail after some time in service. Of course, under a varying load, this may happen due to the incidental occurrence of an extra high peak in the load spectrum, but a delayed fracture may also be related to types of defects, which slowly grow to a critical size. The most relevant mechanisms for this behavior in advanced ceramics at low and intermediate temperatures are SCCG and, to a lesser extent, cyclic fatigue. The engineering aspects of these mechanisms will form the focus of this section (creep and corrosion will not be discussed at this point).

12.6.1 Lifetime and Influence of SCCG on Strength

In general, data relating to SCCG are scarce; examples of crack velocity versus stress intensity ($v-K$) curves, as determined by the method proposed by Fett and Munz [98], are shown in Figure 12.14. In this case, it can be recognized that the data follow a power law; that is, a Paris-type law for the crack growth rate is empirically observed (in fact several different crack growth mechanisms occur, which cause a more structured $v-K$ curve, but for the technically relevant velocity region, namely the range between 10^{-13} to $10^{-6}\,\mathrm{m\,s^{-1}}$, a Paris-type law can be used [39, 41, 88, 99]):

$$v = v_0 \left(\frac{K}{K_c} \right)^n, \tag{21}$$

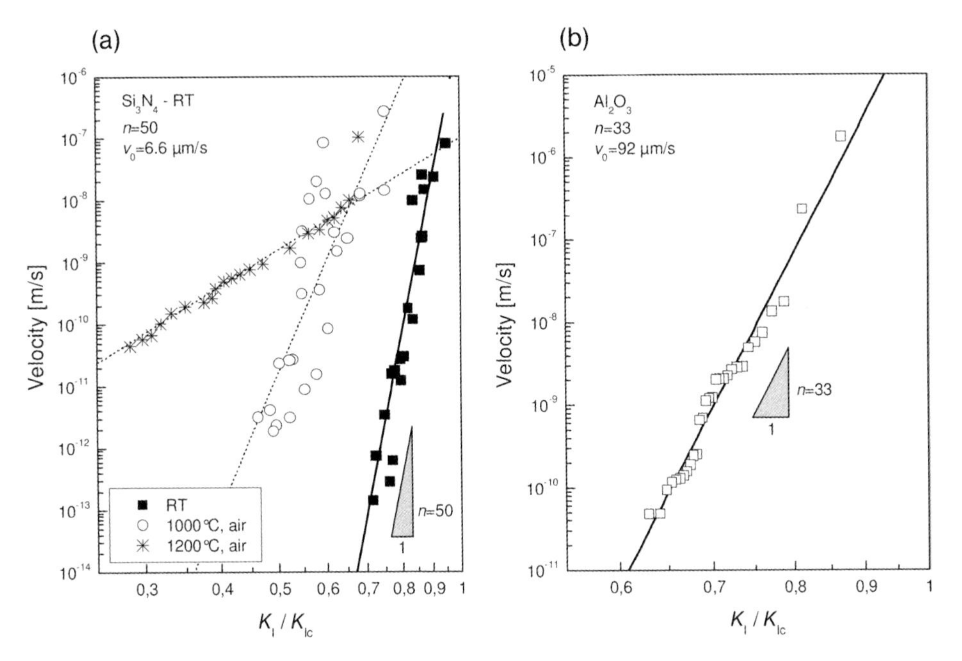

Figure 12.14 Crack growth rate versus stress intensity factor for (a) a silicon nitride [88] and (b) for an alumina ceramic [99] in a log-log representation. The data are arranged along a straight line, which indicates a power law dependence of the crack growth rate from on the stress intensity factor. At higher temperatures, the crack growth rate is temperature-dependent.

where v_0 and the SCCG exponent n are (temperature-dependent) material parameters. The data in Figure 12.14 were generated in laboratory air. The influence of humidity on the crack growth rate has been reported elsewhere [39–41].

The tests on silicon nitride (the ESIS reference testing material) (Figure 12.14a) were performed at several different temperatures (for details, see Ref. [88]). It can be seen that the crack growth rate significantly increased, if the temperature reached or exceeded the softening temperature of the glassy grain boundary phase. Also shown are data determined on a commercial alumina ceramic [99] (Figure 12.14b); here, the SCCG exponent at room temperature was $n \approx 33$ for the alumina ceramic and $n \approx 50$ for the silicon nitride, respectively. For ceramic materials which do not have a glassy grain boundary phase, the exponent may be even higher, with values of $n \approx 200$ being reported for silicon carbide ceramics [100]. The high number of the exponent suggests a loading situation of either negligible or very fast crack growth, causing failure within fractions of a second, but with an extremely small transition region in between.

A more detailed analysis, however, has shown that SCCG might have a significant influence on strength over a wide range of loading conditions. This point will be discussed in greater detail in the following sections.

For simplicity, in the following discussions a general uniaxial stress state is assumed. Of course, generalizations as made in the section above (equivalent stress, effective volume, etc.) are also possible in the case of delayed fracture.

The SCCG data shown in this section have been determined experimentally. Static and constant stress rate strength tests have been performed on standard bend specimens ($3 \times 4 \times 45\ \mathrm{mm}^3$) of a commercial 99.7% Al_2O_3 (Frialit-Degussit F99,7; Friatec AG, Mannheim, Germany) in four-point flexure [99]. All experiments were conducted at room temperature in laboratory air.

The growth of cracks under load caused a finite lifetime of loaded specimens or components: the crack grew from its initial length a_i to the critical (final) length $a_c = a_f$, after which fast failure occurred.

The lifetime can be determined in the following way. If the crack growth rate is the change of the crack length with time: $v = da/dt$, then the differential of the time is $dt = (1/v)$. Integration is possible by separation of variables. Using Eqs (21) and (3) gives:

$$\int_{t_i}^{t_f} \sigma^n \, dt = \frac{1}{v_0} \int_{a_i}^{a_c} \left(\frac{K_c}{Y\sqrt{\pi a}} \right)^n da. \tag{22}$$

Here, the time-dependent variables are written on the left-hand side and all crack length-dependent variables on the right-hand side of the equation. The lifetime, t_f, is composed of the incubation time, t_i, to originate a crack (of size a_i) and the time required for this crack to grow to its critical size $(t_f - t_i)$. In ceramic processing, defects exist which behave like cracks (see previous sections), and therefore the incubation time is, in general, negligible. Exceptions are cases where cracks come to existence during the operation of the component (e.g., by impact or by corrosion).

The geometric factor Y depends, in general, on the crack length, but if the cracks are very small compared to the dimensions of the specimen, and for slowly varying stress fields, then this dependence will be weak and can be neglected (this is generally the case for ceramics). An analytical integration of the right-hand side of Eq. (22) thus becomes possible. For $n > 2$ it holds:

$$\int_0^{t_f} \sigma(t)^n \, dt = \frac{2}{n-2} \cdot \frac{1}{v_0} \left(\frac{K_c}{Y\sqrt{\pi}} \right)^n \cdot \left[a_i^{(2-n)/2} - a_c^{(2-n)/2} \right]. \tag{23}$$

As the right-hand side of Eq. (23) is a constant value, the strength dependence of lifetime results from the left-hand side of Eq. (23). The lifetime depends heavily on the applied stress, because n is generally a high number. For $n \gg 1$, a rough approximation gives: $t_f \propto \sigma^{-n}$ (e.g., for $n = 33$ a stress reduction of 10% causes a 32-fold increase in lifetime), whilst for materials with a higher SCCG exponent this influence is even more pronounced. Therefore, a reduction of the applied stresses is a favorable method for increasing the lifetime in service.

Of course, the load spectrum [the function $\sigma(t)$ in Eq. (23)] can also depend on time. The lifetime t_f of the component is the upper integration limit of the integral on the right-hand side: the integration over the loading spectrum must be performed up to the time where the integral equates to the right-hand side. To understand the general consequences of SCCG on fracture, two simple cases will be discussed in the following.

12.6.1.1 Delayed Fracture under Constant Load

Delayed fracture under constant load may occur in service, for example, in components operating for a long time under stationary loading conditions. Delayed fracture occurs without any warning, as the growth of cracks causes only a very small, and therefore negligible, increase in a component's compliance. A thorough theoretical understanding of the consequences of SCCG is, therefore, of great significance for design.

At this point, it will be beneficial to analyze the simple case of a test under constant load (with homogeneous and uniaxial stress state σ_{stat}). In this case, and for $t_i = 0$, it can be shown that Eq. (23) becomes

$$t_f = \frac{2}{n-2} \cdot \frac{a_i}{v_i} \cdot \left[1 - \frac{a_c}{a_i} \cdot \frac{v_i}{v_f} \right], \tag{24}$$

with $v_i = v(K_i)$, $v_f = v(K_f) = v_0$, $K_i = \sigma_{stat} Y \sqrt{\pi a_i}$, and $K_f = \sigma_{stat} Y \sqrt{\pi a_c} = K_c$, respectively.

It should be noted that the second term in the square bracket can be neglected, if any significant crack growth occurs. This is a consequence of the high value of the SCCG exponent, n. For example, for $a_c \geq 1.2 \cdot a_i$ and $n = 33$, this term is smaller than $0.06 \ll 1$. Then, the time to failure depends only on crack length and crack velocity at the start of the experiment, and on the SCCG exponent.

With Eq. (21) the dependence of the lifetime on initial crack size can be evaluated. It is:

$$t_f \propto a_i^{-(n-2)/2} \cdot \sigma_{\text{stat}}^{-n}. \tag{25}$$

This underlines the high significance of the size of the initial cracks (and, of course, of the stress), which may grow in operation to a critical size. The lifetime of a specimen containing a crack 20% larger than a reference crack is decreased by a factor 0.06, 0.01, or 10^{-8}, for a SCCG exponent of $n = 33$, 50, or 200, respectively. To reduce the size of the flaws which can transform into growing cracks is, from the manufacturer's point of view, the most effective way of improving the lifetime of components by the optimization of microstructure.

The crack length at any given time under load (at constant stress) can be determined by integrating the crack growth rate over time. The result is:

$$a(t) = a_i \left[1 - \frac{n-2}{2} \cdot \frac{v_i}{a_i} \cdot t\right]^{2/(2-n)} \tag{26a}$$

or

$$a_i = a \left[1 + \frac{n-2}{2} \cdot \frac{v}{a} \cdot t\right]^{2/(2-n)} \tag{26b}$$

This function is plotted in Figure 12.15a for $n = 33$, 50, and 200 for alumina, silicon nitride, and silicon carbide, respectively, using the same material properties at a load level of $\sigma_{\text{stat}}/\sigma_i = 0.95$ with $\sigma_i = K_c/(Y\sqrt{\pi a_i})$ and $Y = 2/\pi$. The disproportionate influence of the SCCG exponent can easily be recognized. Due to the high value of this exponent, the crack needs a large part of its life to grow the first few percent. The crack growth rate then accelerates to such an extent that failure occurs in a very short fraction of the life time. Of course, the growth rate of cracks depends strictly on the initial crack size (see Figure 12.15b).

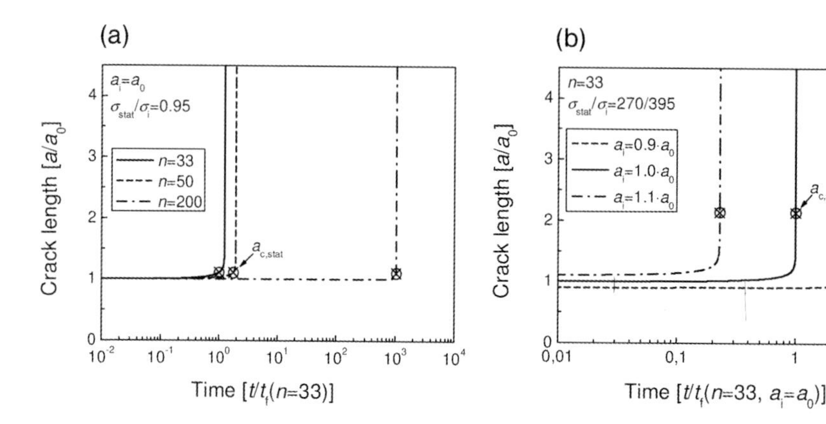

Figure 12.15 Crack length versus time for (a) materials having a different SCCG exponent n (used are data of alumina, silicon nitride and silicon carbide, respectively, $\sigma_{\text{stat}}/\sigma_i = 0.95$) and (b) for cracks of different initial size (used is the SCCG exponent of alumina). The cracks need a large fraction of specimen lifetime to grow the first few percent; they then accelerate, and fracture occurs very quickly. This behavior is promoted by high SCCG exponents.

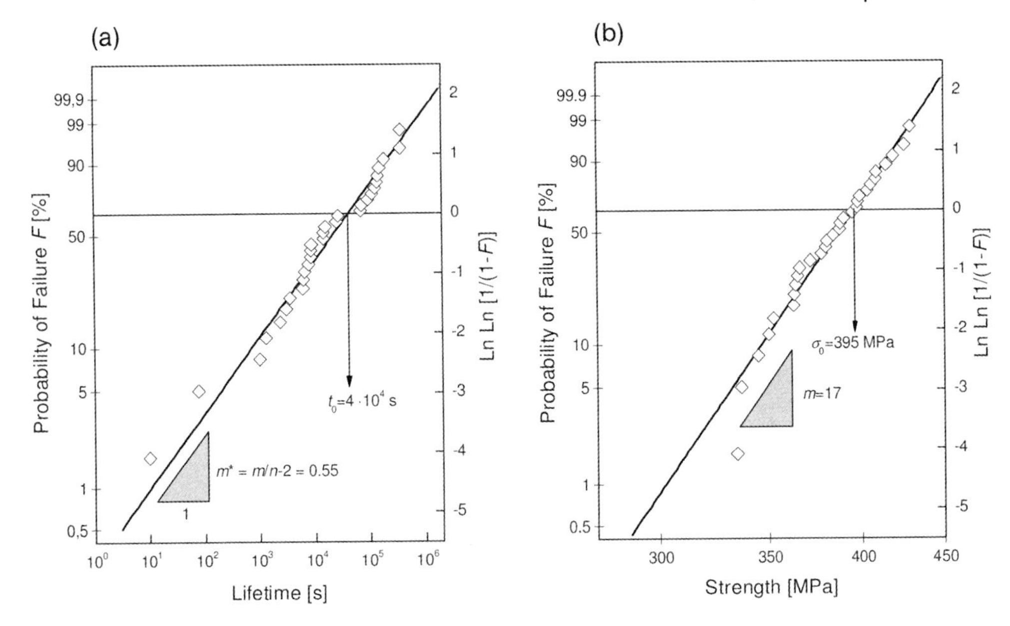

Figure 12.16 Bending tests at room temperature on an alumina ceramic [99]. Plotted is the cumulative frequency (a) versus time to failure for constant loading ($\sigma = 270$ MPa) and (b) versus strength for tests according to EN 843-1 [82] (i.e., inert strength). A comparison of both types of test makes the determination of the crack growth curve (v–K-curve) possible [98].

The large influence of the initial flaw size on life time can be studied with a series of constant stress tests. Figure 12.16a shows the test results of four-point bending tests at constant load ($\sigma = 395$ MPa, specimen geometry according to EN 843-1 [82]) performed on the alumina ceramic described above. The data (30 tests) are plotted as a Weibull graph (cumulative frequency versus life time). The measured lifetime ranged from around 10 s to almost 10^6 s. Also shown, in Figure 12.16b, is the Weibull distribution of the material, which has been determined in conventional four-point bending tests according to EN 843-1 [82].

Of course, the scatter of data in both distributions is a result of the specimen-to-specimen variation in the size of the critical flaw. Therefore, the "time Weibull distribution" can be deduced from the "strength Weibull distribution" [Eq. (12)] by taking SCCG into account.

Strength and crack size are related by the failure criterion [see, for example Eq. (9)], and the change of crack length with time is described by Eq. (26). By use of Eqs (26) and (9), the Weibull distribution becomes time-dependent [2, 4, 14]:

$$F(\sigma_{\rm stat}, V, t) = 1 - \exp\left[-\frac{V}{V_0}\left(\frac{\sigma_{\rm stat}}{\sigma_0}\right)^m \left(\frac{\sigma_{\rm stat}^2}{C}t + 1\right)^{m/(n-2)}\right], \tag{27}$$

where the constant $C = 2K_c^2/(\pi Y^2 v_0(n-2))$ is a material- and geometry-dependent parameter. The crack growth rate at fracture is $v_f = v_0$. If only little SCCG occurs,

$\sigma_{stat}^2 t/C \ll 1$ and Eq. (27) becomes equal to Eq. (12), and if SCCG is significant, $\sigma_{stat}^2 t/C \gg 1$, the time Weibull distribution is provided by:

$$F(\sigma_{stat}, V, t) \approx 1 - \exp\left[-\frac{V}{V_0} \cdot \left(\frac{t}{t_0(\sigma_{stat})}\right)^{m^*}\right] \tag{28}$$

with

$$t_0(\sigma) = \frac{C}{\sigma_{stat}^2} \left(\frac{\sigma_0}{\sigma_{stat}}\right)^{(n-2)}, \tag{29}$$

and

$$m^* = \frac{m}{n-2}. \tag{30}$$

The time Weibull modulus depends only on the slope of the flaw distribution (which is given by the Weibull modulus and the SCCG exponent).

It is clear that n can be determined by the measurement of m and m^*. In the case of the investigated alumina [99], $m = 17$ and $m^* = 0.55$, and the resultant SCCG exponent is $n \approx 33$.

One comprehensive way of representing design data to develop a strength–probability–time (SPT) diagram, which was first introduced some thirty years ago [2, 101]. In these diagrams, Eq. (28) is used to design Weibull diagrams (failure probability versus strength; for an example, see Figure 12.8), for a constant time to failure. Of course, the other design parameters (equivalent stress, effective volume) should also be properly accounted for.

A more detailed analysis of both types of Weibull distribution enables the determination of v–K curves for crack growth, without any special assumptions being made with regards to its analytical form [98]. In fact, the crack growth data shown in Figure 12.14b have been determined using such an analysis of the data shown in Figure 12.16a. The straight line in the diagram is a fit through the data giving $n \approx 33$.

12.6.1.2 **Delayed Fracture under Increasing Load: Constant Stress Rate Tests**

SCCG also influences loading situations in which the stress is not constant. A simple example of this is a test where the stress increases linearly with time (constant stress rate test: $\sigma = \dot{\sigma} \cdot t$). In this case, the left-hand side of Eq. (23) can also be integrated analytically and, after some rearrangement of the formulae, is obtained:

$$\sigma_f^{n+1} = \frac{2(n+1)}{n-2} \cdot \frac{a_i}{v_0} \cdot \sigma_i^n \cdot \dot{\sigma} \left[1 - \left(\frac{\sigma_f}{\sigma_i}\right)^{n-2}\right], \tag{31}$$

where $\sigma_i = K_c/(Y\sqrt{\pi a_i})$ is the inert strength (i.e., the strength of the specimen without any SCCG), σ_f is the strength of the specimen after some SCCG and, of course, $\sigma_f \leq \sigma_i$. Again, if any significant SCCG occurs, the second term in the square bracket is very small compared to unity, and can be neglected. The strength of

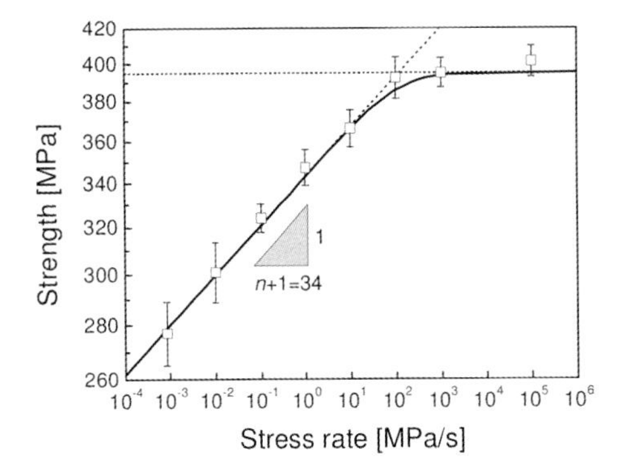

Figure 12.17 (Mean) strength versus stress rate for constant stress rate bending tests performed on alumina (same material as described above). For details, see Ref. [99]. The SCCG exponent determined using this data is $n = 33$. Each datum point results from seven (and in one case from 30) individual measurements. The scatter bars reflect the 90% uncertainties due to the sampling procedure, which are much smaller for the characteristic strength than for the Weibull modulus.

specimens tested with a constant stress rate will then depend on the stress rate: $\sigma_f \propto \dot{\sigma}^{1/(n+1)}$. This behavior can be recognized in Figure 12.17, where constant stress rate test results on alumina are shown. Presented here is the strength versus stress rate in a log-log plot. For "low" stress rates with $\sigma_f \propto \dot{\sigma}^{1/(n+1)}$, a fit to the data yields $n \approx 33$. Of special note here is the consistency of the SCCG-exponent determined with different methods. For stress rates higher than $\sim 10^2$ MPa s^{-1}, the testing time is too short for significant SCCG, and the strength is equal to the inert strength. It should be noted that for conventional bending tests on the analyzed alumina according to EN 843-1 [82], the testing velocity is high enough that significant SCCG does not occur. Thus, the determined strength is (almost) equal to the inert strength.

In summary, two loading regions can be recognized: (i) at low stress rates, a strong influence of SCCG on the strength exists; and (ii) at (very) high stress rates, SCCG has no significant influence on the strength. In both regions, the strength is controlled by the size and distribution of flaws. As discussed above, however, this situation must be described by statistical means.

12.6.1.3 **Delayed Fracture under General Loading Conditions**

In reality, multiaxial inhomogeneous stress fields, which vary with time, occur in components. Delayed fracture due to SCCG is, in general, accounted for on the basis of Eq. (22). Methods to consider the influence of inhomogeneous and multiaxial stress fields have been discussed in the previous sections. Thus, all necessary tools for a proper design against delayed fracture in brittle materials are available. A few examples for the application of such design methodology can be found elsewhere [101, 102].

12.6.2
Influence of Fatigue Crack Growth on Strength

Although cyclic fatigue has been recognized as a possible damage mechanism in ceramics, almost no cyclic fatigue data exist for engineering ceramics [43, 103–106]. A rare exception is the data of NGK Insulators Ltd, for their silicon nitride material [107]. It follows from the reported cyclic fatigue data that there also exists a Paris law for cyclic fatigue crack growth, where the crack growth per load cycle is

$$\frac{da}{dN} = \left(\frac{\Delta a}{\Delta N}\right)_0 \left(\frac{\Delta K}{K_c}\right)^{\tilde{n}}. \tag{32}$$

The material parameters, the factor $(\Delta a/\Delta N)_0$, and the SCCG exponent $\tilde{n}$, are each dependent on mean stress and temperature. Here, N is the number of load cycles and N_f is the numbers of cycles to failure.

Cyclic crack growth influences lifetime in similar manner to SCCG (see previous section). Under conditions where cyclic fatigue is the dominant damage mechanism, the cyclic lifetime can be determined in an analogous way to the procedures discussed above. For example, the equation analogous to Eq. (27) is [14]:

$$F(\sigma, V, N_f) = 1 - \exp\left[-\frac{V}{V_0}\left(\frac{\Delta\sigma}{\sigma_0}\right)^m \left(\frac{\Delta\sigma^2}{\Delta C} N_f + 1\right)^{m/(\tilde{n}-2)}\right], \tag{33}$$

where ΔC is a material parameter which is defined in an similar way to the parameter C.

It should be noted that, in the few cases reported, the SCCG exponent for cyclic fatigue is much larger than unity: $\tilde{n} \gg 1$ [105, 108, 109]. Thus, like SCCG, the cyclic fatigue life is heavily dependent on the applied stress range and the size of the initial flaws.

The action of cyclic fatigue and of SCCG can be assumed to be independent, and the crack advance due to both mechanisms can be simply added. However, as both growth rates depend heavily on the applied load, and since the SCCG exponents are, in general, different, it is quite likely that the growth rates are very different. The simplified model description can then be used whereby only one of the two mechanisms is dominant.

12.6.3
Proof Testing

Proof testing is an often (routinely) used technique to guarantee a component's minimum service life [2, 4, 102, 110]. In order to clarify the basic ideas of proof testing, reference is made again to a uniaxial tensile and homogeneous stress state.

In a proof test, the component is loaded with a stress σ_p, which is higher than the design service stress σ_a. The critical flaw size in a proof test is $a_c(\sigma_p) = (1/\pi) \cdot (K_c/Y\sigma_p)^2$. In a proof test, all components containing equal or larger sized flaws are destroyed; in other words, the frequency distribution of flaws is truncated at $a \leq a_c(\sigma_p)$.

The critical crack size in service is $a_c(\sigma_a) = (1/\pi) \cdot (K_c/Y\sigma_a)^2$ and $a_c(\sigma_p) < a_c(\sigma_a)$. If the crack growth law is known, the time can be determined that the crack needs to grow from $a_c(\sigma_p)$ to the critical size $a_c(\sigma_a)$.

If SCCG is the dominant damage mechanism, this can be calculated on the basis of Eq. (26) with $a_i = a_p$ and $a = a_a$; then, the time $t = t_{a,\min}$ is the shortest possible time for a crack in a component to grow to a critical size, and the time $t_{a,\min}$ is taken as the lifetime that can be guaranteed for a proof tested component. This procedure can be used to determine a suitable proof stress if the loading stress and the crack growth law are known, although in practice neither are known precisely. Consequently, proof testing is often carried out at much too-high stress values, which in turn results in an unnecessarily large proportion of rejections.

It has often been claimed that proof testing causes damage in the component, although of course any applied stress will cause some SCCG, thereby reducing the remaining strength. However, if the stress rate during proof testing is sufficiently high, then this loss of strength can be neglected. Consequently, in proof tests not only the loading rate but also (especially) the unloading rate should be high. If, for any reason, the application of a high loading rate in a proof test is not possible, the elongation of cracks during testing can be estimated and accounted for, so that the reliability of any surviving parts can still be guaranteed.

One often-occurring problem is that the in-service stress fields of components cannot easily be simulated in a test. For example, the stress fields caused by temperature changes or by contact loading can hardly be reproduced in simple mechanical tests. In such cases, large parts of a component are often subjected to a too-large proof stress, which increases the number of failures in the test. On the other hand, some parts of a component may be subjected to a too-small proof stress. Whilst the latter situation will reduce the probability of failure, some in-service failures will still be possible.

Yet, if cyclic fatigue is the dominant crack growth mechanism of cracks, the above-described arguments can be used in an analogous way.

12.7 Concluding Remarks

In this chapter, a close examination has been made of the phenomenon of fracture in ceramics. The macroscopic appearance of fracture and typical failure modes in ceramic materials has been analyzed, fracture mirrors and fracture origins have been identified, and the way in which fracture is intrinsically connected to the microstructure of a ceramic has been outlined. In particular, by detailing stress distributions it has been shown that fracture always starts at a single microstructural flaw, the stability of which can be described with simple linear elastic fracture mechanics. Notably, these features are responsible for the inherently statistical nature of failure in ceramic materials, an understanding of which can provide knowledge of the close correlation between defect populations and fracture statistics, and of how to develop materials parameters such as the characteristic strength, the Weibull modulus or effective volume, in the planning and design of appropriate material behaviors.

The influence of different damage mechanisms has been discussed in relation to materials design. Notably, the effects of both environment and time have been analyzed, with subcritical crack growth being recognized as the most significant damage mechanism in terms of the delayed failure of components. The huge influence of initial flaw (crack) size on a material's lifetime has also been demonstrated, and examples given. Finally, proof testing, which is aimed at overcoming these problems, has been discussed.

Of course, this short treatise cannot be claimed in any way to be comprehensive, and several important points have not been covered in the detail they deserve. This includes aspects of toughness [3, 4, 61, 111, 112] and increases in toughness with crack growth (*R*-curve behavior [3, 4, 113, 114]), the influence of internal stresses on toughness [115, 116] and strength [117–119], as well as concepts of layered materials [119–124], graded materials [125], and composites. Whilst each of these topics is deserving of a separate section, the basic concepts have nevertheless been presented, and the tools for handling them provided.

Today, the fracture mechanics of ceramics is used to a large extent in the analysis of fractured components [17]. Dangerous loading conditions can be recognized, and reasons for failing such as improper use [17], bad mechanical design [110], imperfect materials processing [19, 126], or inadequate materials selection identified. In some cases, local fracture stresses can be read from the fracture surface [16, 21, 127], and in this respect fractography represents a powerful tool to identify the weak points in a system and suggest possible improvements [19]. Although design tools for fracture statistical methods and for lifetime prediction are commercially available [74, 75], and are clearly required for design processes, they are rarely used in daily practice. The reasons for this are manifold, and include method complexity, an inadequate training of ceramic engineers in numerical mathematics, deficiencies in education, or simply a lack of appropriate material data [13, 89]. Yet, the increased technical demands, the use of technical ceramics in highly loaded applications, and the general acceptance of numerical simulation methods, will surely lead to an increased recognition and application of these methods in the near future.

Although the greatest – and indeed the fastest growing – area of ceramics applications relates to their functional (e.g., electronic or micromechanical) rather than to their structural properties [128–132], ultimately it will be their structural failure and poor mechanical properties that limits such applications. This is indeed a serious problem, with several orders of magnitude more fractures occurring in functional rather than in structural ceramic components, while studies of mechanical behavior remain in their infancy [26, 133–139]. Examples of such failure include thermal strains and stresses causing the fracture of microelectronic components, major strains and stresses in piezoelectric ceramics leading to failures in fuel injection systems, and positive temperature resistors suffering from major thermomechanical strains as a result of the phase transformation required for their functionality.

The concepts, mechanisms and tools discussed in this chapter are equally valid for these components as for structural parts. This is clearly demonstrated by two surface

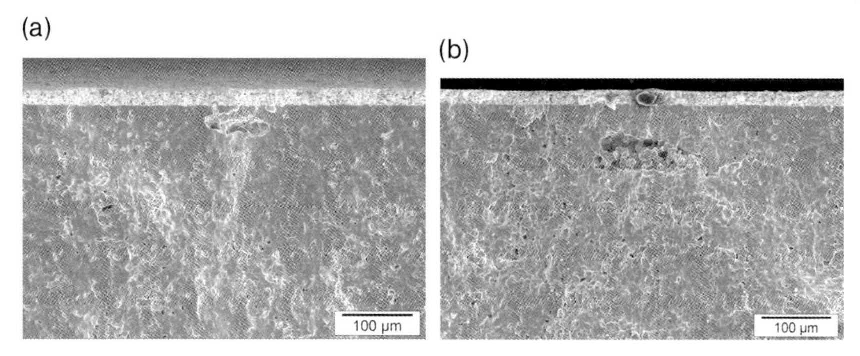

Figure 12.18 (a) Fracture surface in a thermistor component after over voltage loading; (b) Fracture surface in a bending test specimen made from the same barium titanate ceramic. Qualitative differences between both types of fracture surfaces are hardly recognizable.

fractures of a barium titanate positive thermal coefficient (PTC) component, first after a voltage overload of a thermistor (Figure 12.18a) and second, after bending a test specimen cut from the PTC (Figure 12.18b). Whilst no significant differences were detected on the fracture surfaces, one occurred due to a too-high *electrical* load, and the other to a too-high *mechanical* load.

The main challenge for the future will be not only the generation, acquisition, and collection of valid materials data for this economically increasingly important group of functional ceramic materials, but also the development of a clear understanding of the corresponding fracture mechanisms.

References

1 Kingery, D.W., Bowen, H.K., and Uhlmann, D.R. (1976) *Introduction to Ceramics*, John Wiley & Sons, New York.

2 Davidge, R.W. (1979) *Mechanical Behaviour of Ceramics*, Cambridge University Press, Cambridge.

3 Wachtman, J.B. (1996) *Mechanical Properties of Ceramics*, Wiley-Interscience, New York, Chichester.

4 Munz, D. and Fett, T. (1999) *Ceramics*, Springer, Berlin, Heidelberg.

5 Hertz, H. (1882) Über die Berührung fester elastischer Körper. *Crelles Journal*, **92**, 156–171.

6 ISO (2000) 14704. *Fine Ceramics (Advanced Ceramics, Advanced Technical Ceramics) - Test Method for Flexural Strength of Monolithic Ceramics at Room Temperature*, International Organization for Standardization, p. 20.

7 Danzer, R. and Lube, T. (2004) Werkstoffprüfung keramischer Werkstoffe - ein Überblick, in *Tagungsband Werkstoffprüfung* (ed. M. Pohl), Werkstoff-Informationsgesellschaft mbH, Frankfurt, pp. 245–254.

8 EN ISO (2003) 18756. *Fine ceramics (Advanced Ceramics, Advanced Technical Ceramics) - Determination of Fracture Toughness of Monolithic Ceramics at Room Temperature by the Surface Crack in Flexure (SCF) Method*, International Organization for Standardization.

9 ISO (2005) 24370. *Fine Ceramics (Advanced Ceramics, Advanced Technical Ceramics) – Determination of Fracture Toughness of Monolithic Ceramics at Room Temperature by the Chevron-notched Beam (CNB) Method*, International Organization for Standardization.

10 ISO (2007) DIS 23146. *Fine Ceramics (Advanced Ceramics, Advanced Technical Ceramics) – Determination of Fracture Toughness of Monolithic Ceramics at Room Temperature by the Single-edge Vee-notched Beam (SEVNB) Method,* International Organization for Standardization.

11 Primas, R.J. and Gstrein, R. (1997) ESIS TC round robin on fracture toughness. *Fatigue Fract. Eng. Mater. Struct.*, **20** (4), 513–532.

12 EN ISO (2003) 15732. *Fine ceramics (Advanced Ceramics, Advanced Technical Ceramics) - Determination of Fracture Toughness of Monolithic Ceramics at Room Temperature by the Single-edge Pre-cracked Beam (SEPB) Method,* International Organization for Standardization.

13 Lube, T., Danzer, R., and Steen, M. (2002) A Testing program for a silicon nitride reference material, in *Improved Ceramics Through New Measurements, Processing and Standards* (eds M. Matsui, S. Jahanmir, H. Mostgaci, M. Naito, K. Uematsu, R. Waesche, and R. Morrell), The American Ceramic Society, Westerville, Ohio, pp. 259–268.

14 Danzer, R. (1994) Ceramics: mechanical performance and lifetime prediction, in *The Encyclopedia of Advanced Materials* (eds D. Bloor, R.J. Brook, M.C. Flemings, S. Mahajan, and R.W. Cahn), Pergamon, pp. 385–398.

15 Lube, T. (1993) Entwicklung eines Zugversuches zur Zuverlässigkeitsprüfung von keramischen Ventilen. Diplomarbeit am Institut für Metallkunde und Werkstoffprüfung, Montanuniversität.

16 Mecholsky, J.J. (1991) Quantitative fractography: an assessment. *Ceram. Trans.*, **17**, 413–451.

17 Morrell, R. (1999) *Fractography of Brittle Materials,* National Physical Laboratory, Teddington.

18 European Committee for Standardization (2004) CEN/TS 843-6. *Advanced Technical Ceramics - Monolithic Ceramics. Mechanical Properties at Room Temperature - Part 6: Guidance for Fractographic Investigation,* European Committee for Standardization.

19 Danzer, R. (2002) Mechanical failure of advanced ceramics: the value of fractography. *Key Eng. Mater.*, **223**, 1–18.

20 ASTM (2005) C 1322-05b. *Standard Practice for Fractography and Characterization of Fracture Origins in Advanced Ceramics,* ASTM.

21 Morrell, R., Byrne, L., and Murray, M. (2001) Fractography of ceramic femoral heads, in *Fractography of Glasses and Ceramics IV* (eds J.R. Varner, G.D. Quinn, and V.D. Frèchette), The American Ceramic Society, Westerville, Ohio, pp. 253–266.

22 Mörgenthaler, K.D. and Tiefenbacher, E. (1983) Final report of project "Bauteile aus Keramik für Gasturbinen," BMFT-FB-T 86-027, ZK/NT/NTS1021 of the Bundesministerium fuer Forschung und Technik (BRD). Fachinformations-zentrum Karlsruhe, Karlsruhe.

23 Mörgenthaler, K.D. (1988) Keramikanwendung im Turbinenbau, *Technische Keramik Jahrbuch,* Vulkan-Verlag, Essen, pp. 285–297.

24 Hasselman, D.P.H. (1969) Unified theory of thermal shock fracture initiation and crack growth. *J. Am. Ceram. Soc.*, **52**, 600–604.

25 Schneider, G.A., Danzer, R., and Petzow, G. (1988) Zum Thermoschockverhalten spröder Werkstoffe. *Fortschr. Dtsch. Keram. Ges.*, **3** (3), 59–70.

26 Supancic, P. and Danzer, R. (2003) Thermal shock behavior of $BaTiO_3$ PTC-ceramics. *Ceram. Eng. Sci. Proc.*, **24** (4), 371–379.

27 Fellner, M. and Supancic, P. (2002) Thermal shock failure of brittle materials. *Key Eng. Mater.*, **223**, 97–106.

28 Lawn, B.R. (1998) Indentation of ceramics with spheres: a century after Hertz. *J. Am. Ceram. Soc.*, **81** (8), 1977–1994.

29 Lengauer, M. and Danzer, R. (2008) Silicon nitride tools for hot rolling of high alloyed steel and superalloy wires - crack growth and lifetime prediction. *J. Eur. Ceram. Soc.*, **28** (11), 2289–2298.

30 Lengauer, M., Danzer, R., and Harrer, W. (2004) Keramische Walzen für das Drahtwalzen - Simulation und Analyse der Werkzeugbeanspruchung, in *Walzen*

mit Keramik (eds A. Kailer and T. Hollstein), Fraunhofer IRB Verlag, Stuttgart, pp. 95–107.

31 Lube, T. (2001) Indentation crack profiles in silicon nitride. *J. Eur. Ceram. Soc.*, **21** (2), 211–218.

32 Jahanmir, S. (1994) *Friction and Wear of Ceramics*, Marcel Dekker, New York.

33 McCormick, N.J. and Almond, E.A. (1990) Edge flaking of brittle materials. *J. Hard Mater.*, **1** (1), 25–51.

34 Morrell, R. (2005) Edge flaking - similarity between quasistatic indentation and impact mechanisms for brittle materials. *Key Eng. Mater.*, **290**, 14–22.

35 Danzer, R., Hangl, M., and Paar, R. (2001) Edge chipping of brittle materials, in *Fractography of Glasses and Ceramics IV* (eds J.R. Varner, G.D. Quinn, and V.D. Frèchette), The American Ceramic Society, Westerville, Ohio, pp. 43–55.

36 Chai, H. and Lawn, B.R. (2007) Edge chipping of brittle materials: effects of side-wall inclination and loading angle. *Int. J. Fract.*, **145**, 159–165.

37 Quinn, J.B. (2007) The increasing role of fractography in the dental community, in *Fractography of Glasses and Ceramics V* (eds J.R. Varner, G.D. Quinn, and M. Wightman), Wiley-Interscience, Hoboken, pp. 253–270.

38 Schoeck, G. (1990) Thermally activated crack propagation in brittle materials. *Int. J. Fract.*, **44**, 1–44.

39 Wiederhorn, S.M. (1967) Influence of water vapor on crack propagation in soda-lime glass. *J. Am. Ceram. Soc.*, **50**, 407–414.

40 Michalske, T.A. and Freiman, S.M. (1982) A molecular interpretation of stress in silica. *Nature*, **95**, 511–512.

41 Danzer, R. (1994) Subcritical crack growth in ceramics, in *The Encyclopedia of Advanced Materials* (eds D. Bloor, R.J. Brook, M.C. Flemings, S. Mahajan, and R.W. Cahn), Pergamon, pp. 2693–2698.

42 Dauskardt, R.K., Yu, W., and Ritchie, R.O. (1997) Mechanism of cyclic fatigue crack propagation in a fine grained alumina ceramic: the role of crack closure. *Fatigue Fract. Eng. Mater. Struct.*, **20**, 1453–1466.

43 Gilbert, C.J. and Ritchie, R.O. (1987) Fatigue crack propagation in transformation toughened zirconia ceramic. *J. Am. Ceram. Soc.*, **70**, C248–C252.

44 Grathwohl, G. and Liu, T. (1991) Crack resistance and fatigue of transforming ceramics; 1. Materials in the $ZrO_2 - Y_2O_3 - Al_2O_3$ system. *J. Am. Ceram. Soc.*, **74**, 318–325.

45 Brown, A.M. and Ashby, M.F. (1980) Correlations for diffusion constants. *Acta Metall.*, **28**, 1085–1101.

46 Frost, H.J. and Ashby, M.F. (1982) *Deformation Mechanism Maps*, Pergamon Press, Oxford.

47 Riedel, H. (1987) *Fracture at High Temperatures*, Springer-Verlag, Berlin.

48 Nickel, K.G. and Gogotsi, Y.G. (2000) Corrosion of hard materials, in *Handbook of Ceramic Hard Materials* (ed. R. Riedel), VCH-Wiley, Weinheim, pp. 140–182.

49 Schneider, B., Guette, A., Naslain, R., Cataldi, M., and Costecalde, A. (1998) A theoretical and experimental approach to the active-to-passive transition in the oxidation of silicon carbide. *J. Mater. Sci.*, **33**, 535–547.

50 Alles, A.B. and Burdick, V.L. (1993) Grain boundary oxidation in PTCR barium titanate thermistors. *J. Am. Ceram. Soc.*, **76** (2), 401–408.

51 Lawn, B.R. (1993) *Fracture of Brittle Solids*, 2nd edn, Cambridge University Press, Cambridge.

52 Timoshenko, S.P. and Goodier, J.N. (1970) *Theory of Elasticity*, McGraw-Hill, New York.

53 Gross, D. and Seelig, T. (2006) *Fracture Mechanics*, Springer, Berlin.

54 Murakami, Y. (1986) *The Stress Intensity Factor Handbook*, Pergamon Press, New York.

55 Tada, H., Paris, P., and Irwin, G.R. (1985) *The Stress Analysis Handbook*, Del Research Corporation, St. Louis.

56 Newman, J.C. and Raju, I.S. (1981) An empirical stress-intensity factor equation for the surface crack. *Eng. Fract. Mech.*, **15** (1–2), 185–192.

57 Griffith, A.A. (1920) The phenomenon of rupture and flow in solids. *Philos. Trans. R. Soc. London*, **A221**, 163–198.

58 Evans, A.G. (1990) Perspective on the development of high-toughness ceramics. *J. Am. Ceram. Soc.*, **73** (2), 187–206.

59 Mai, Y.-M. and Lawn, B.R. (1987) Crack-interface grain bridging as a fracture resistance mechanism in ceramics: II theoretical fracture mechanics model. *J. Am. Ceram. Soc.*, **70** (4), 289–294.

60 Swanson, P.L., Fairbanks, C.J., Lawn, B.R., Mai, Y.-M., and Hockey, B.J. (1987) Crack-interface grain bridging as a fracture resistance mechanism in ceramics: II, experimental study on alumina. *J. Am. Ceram. Soc.*, **70**, 279–289.

61 Steinbrech, R.W. and Schenkel, O. (1988) Crack-resistance curves of surface cracks in alumina. *J. Am. Ceram. Soc.*, **71** (5), C271–C273.

62 Danzer, R. (2006) Some notes on the correlation between fracture and defect statistics: is the Weibull statistics valid for very small specimens? *J. Eur. Ceram. Soc.*, **26** (15), 3043–3049.

63 Zimmermann, A., Hoffmann, M., Flinn, B.D., Bordia, R.K., Chuang, T.-J., Fuller, E.R., and Rödel, J. (1998) Fracture of alumina with controlled pores. *J. Am. Ceram. Soc.*, **81** (9), 2449–2457.

64 Lu, C., Danzer, R., and Fischer, F.D. (2004) Scaling of fracture strength in ZnO: effects of pore/grain-size interaction and porosity. *J. Eur. Ceram. Soc.*, **24** (14), 3643–3651.

65 Danzer, R. (1992) A general strength distribution function for brittle materials. *J. Eur. Ceram. Soc.*, **10**, 461–472.

66 Hunt, R.A. and McCartney, L.N. (1979) A new approach to fracture. *Int. J. Fract.*, **15**, 365–375.

67 Jayatilaka, A.d.S. and Trustrum, K. (1977) Statistical approach to brittle fracture. *J. Mater. Sci.*, **12**, 1426–1430.

68 McCartney, L.N. (1979) Extensions of a statistical approach to fracture. *Int. J. Fract.*, **15**, 477–487.

69 Danzer, R., Reisner, G., and Schubert, H. (1992) Der Einfluß von Gradienten in der Defektdichte und Festigkeit auf die Bruchstatistik von spröden Werkstoffen. *Z. Metallkd.*, **83**, 508–517.

70 Danzer, R., Lube, T., and Supancic, P. (2001) Monte-Carlo simulations of strength distributions of brittle materials - type of distribution, specimen- and sample size. *Z. Metallkd.*, **92** (7), 773–783.

71 Danzer, R., Supancic, P., Pascual Herrero, J., and Lube, T. (2007) Fracture statistics of ceramics - Weibull statistics and deviations from Weibull statistics. *Eng. Fract. Mech.*, **74** (18), 2919–2932.

72 Weibull, W. (1939) *A Statistical Theory of the Strength of Materials*, Generalstabens Litografiska Anstalts Förlag, Stockholm.

73 Weibull, W. (1951) A statistical distribution function of wide applicability. *J. Appl. Mech.*, **18**, 293–298.

74 Brückner-Foit, A., Heger, A., and Munz, D. (1993) Evaluation of failure probability of multiaxially loaded components using the STAU postprocessor. *Ceram. Eng. Sci. Proc.*, **14** (7–8), 331.

75 Nemeth, N.N., Manderscheid, J.M., and Gyekenyesi, J.P. (1989) Designing ceramic components with the CARES computer program. *Am. Ceram. Soc. Bull.*, **68** (12), 2064–2072.

76 Quinn, G.D. (2003) Weibull strength scaling for standardized rectangular flexure specimens. *J. Am. Ceram. Soc.*, **86** (3), 508–510.

77 Quinn, G.D. (2003) Weibull effective volumes and surfaces for cylindrical rods loaded in flexure. *J. Am. Ceram. Soc.*, **86** (3), 475–479.

78 Freudenthal, A.M. (1968) Statistical approach to fracture, in *Fracture* (ed. H. Liebowitz), Academic Press, New York, London, pp. 591–619.

79 Lamon, J. (1990) Ceramics reliability: statistical analysis of multiaxial failure using the Weibull approach and the multiaxial elemental strength. *J. Am. Ceram. Soc.*, **73** (8), 2204–2212.

80 Thiemeier, T., Brückner-Foit, A., and Kölker, H. (1991) Influence of fracture criterion on the failure prediction of ceramics loaded in biaxial flexure. *J. Am. Ceram. Soc.*, **74** (1), 48–52.

81 Brückner-Foit, A., Fett, T., Munz, D., and Schirmer, K. (1997) Discrimination of multiaxiallity criteria with the Brazilian disc test. *J. Eur. Ceram. Soc.*, **17** (5), 689–696.

82 European Committee for Standardization (CEN) (1995) EN 843-1. *Advanced Technical Ceramics - Monolithic Ceramics - Mechanical Properties at Room Temperature: Part 1 - Determination of Flexural Strength.*
83 ASTM (2002) C 1161-02c. *Standard Test Methods for Flexural Strength of Advanced Ceramics at Ambient Temperature,* ASTM.
84 Japanese Standards Association (JSA) (1995) JIS R 1601. *Testing Method for Flexural Strength (Modulus of Rupture) of Fine Ceramics.*
85 European Committee for Standardization (CEN) (1997) ENV 843-5. *Advanced Technical Ceramics - Monolithic Ceramics - Mechanical Properties at Room Temperature: Part 5 - Statistical Evaluation*
86 ASTM (1995) C 1239-95. *Standard Practice for Reporting Uniaxial Strength Data and Estimating Weibull Distribution Parameters for Advanced Ceramics,* ASTM.
87 Japanese Standards Association (JSA) (1996) JIS R 1625. *Weibull Statistics of Strength Data for Fine Ceramics.*
88 Lube, T., Danzer, R., Kübler, J., Dusza, J., Erauw, J.-P., Klemm, H., and Sglavo, V.M. (2002) Strength and fracture toughness of the ESIS silicon nitride reference material, in *Fracture Beyond 2000* (eds A. Neimitz, I.V. Rokach, D. Kocanda, and K. Golos), EMAS Publications, Sheffield, pp. 409–416.
89 Lube, T. and Dusza, J. (2007) A silicon nitride reference material – a testing program of ESIS TC6. *J. Eur. Ceram. Soc.*, **27** (2–3), 1203–1209.
90 Börger, A., Supancic, P., and Danzer, R. (2002) The ball on three balls test for strength testing of brittle discs – stress distribution in the disc. *J. Eur. Ceram. Soc.*, **22** (8), 1425–1436.
91 Börger, A., Supancic, P., and Danzer, R. (2004) The ball on three balls test for strength testing of brittle discs – Part II: analysis of possible errors in the strength determination. *J. Eur. Ceram. Soc.*, **24** (10–11), 2917–2928.
92 Danzer, R., Harrer, W., Supancic, P., Lube, T., Wang, Z., and Börger, A. (2007) The ball on three balls test – strength and failure analysis of different materials. *J. Eur. Ceram. Soc.*, **27** (2–3), 1481–1485.
93 Baratta, F.I., Mathews, W.T., and Quinn, G.D. (03/ 1997) Errors Associated with Flexure Testing of Brittle Materials. MTL TR 87-35, U.S. Army Materials Technology Laboratory, Watertown.
94 Lube, T., Manner, M., and Danzer, R. (1997) The miniaturisation of the 4-point bend-test. *Fatigue Fract. Eng. Mater. Struct.*, **20** (11), 1605–1616.
95 Lu, C., Danzer, R., and Fischer, F.D. (2002) Fracture statistics of brittle materials: Weibull or normal distribution. *Phys. Rev.*, **E65** (6), Article no. 067102.
96 Sigl, L. (1992) Effects of the flaw distribution function on the failure probability of brittle materials. *Z. Metallkd.*, **83**, 518–523.
97 Danzer, R. and Lube, T. (1998) Fracture statistics of brittle materials: it does not always have to be Weibull statistics, in *Ceramic Materials, Components for Engines* (ed. K. Niihara), Japan Fine Ceramics Association, Tokyo, pp. 683–688.
98 Fett, T. and Munz, D. (1985) Determination of v-K curves by a modified evaluation of lifetime measurements in static bending tests. *J. Am. Ceram. Soc.*, **68** (8), C213–C215.
99 Baierl, R.G.A. (1999) Langsames Rißwachstum in Aluminiumoxid. Diplomarbeit am Institut für Struktur- und Funktionskeramik, Montanuniversität Leoben.
100 Richter, H., Kleer, G., Heider, W., and Röttenbacher, R. (1985) Comparative study of the strength properties of slip-cast and of extruded silicon-infiltrated SiC. *Mater. Sci. Eng.*, **71**, 203–208.
101 Hempel, H. and Wiest, H. (1986) Structural analysis and life prediction for ceramic gas turbine components for the Mercedes-Benz Research Car 2000. ASME Paper No. 86-GT-199.
102 Soma, T., Ishida, Y., Matsui, M., and Oda, I. (1987) Ceramic component design for assuring long-term durability. *Adv. Ceram. Mater.*, **2**, 809–812.
103 Kawakubo, T. and Komeya, K. (1987) Static and cyclic fatigue behavior of

a sintered silicon nitride at room temperature. *J. Am. Ceram. Soc.*, **70** (6), 400–405.

104 Reece, M.J., Guiu, F., and Sammur, M.F.R. (1989) Cyclic fatigue crack propagation in alumina under direct tension-compression loading. *J. Am. Ceram. Soc.*, **72** (2), 348–352.

105 Gilbert, C.J., Dauskardt, R.H., and Ritchie, R.O. (1995) Behavior of cyclic fatigue cracks in monolithic silicon nitride. *J. Am. Ceram. Soc.*, **78** (9), 2291–2300.

106 Gilbert, C.J., Cao, J.J., Moberlychan, W.J., DeJonghe, L.C., and Ritchie, R.O. (1996) Cyclic fatigue and resistance -curve behaviour of an *in situ* toughened silicon carbide with Al, B, C additions. *Acta Mater.*, **44** (8), 3199–3214.

107 Masuda, M., Soma, T., and Matsui, M. (1990) Cyclic fatigue of Si_3N_4 ceramics. *J. Eur. Ceram. Soc.*, **6** (4), 253–258.

108 Liu, S.-Y. and Chen, I.W. (1994) Plasticity-induced fatigue damage in ceria-stabilized tetragonal zirconia polycrystals. *J. Am. Ceram. Soc.*, **77** (8), 2025–2035.

109 Dauskardt, R.H., James, M.R., Porter, J.R., and Ritchie, R.O. (1992) Cyclic fatigue-crack growth in a SiC-whisker-reinforced alumina ceramic composite: long- and small-crack behavior. *J. Am. Ceram. Soc.*, **75** (4), 759–771.

110 Morrell, R. (1989) *Handbook of Properties of Technical & Engineering Ceramics, Part 1: An Introduction for the Engineer and Designer*, Her Majesty's Stationary Office, London.

111 McMeeking, R.M. and Evans, A.G. (1982) Mechanics of transformation-toughening in brittle materials. *J. Am. Ceram. Soc.*, **65**, 242–246.

112 Rödel, J. (1992) Crack closure forces in ceramics: characterization and formation. *J. Eur. Ceram. Soc.*, **9**, 325–334.

113 Shetty, D.K. and Wang, J.-S. (1989) Crack stability and strength distribution of ceramic that exhibits rising crack-growth-resistance (R-curve behavior). *J. Am. Ceram. Soc.*, **72**, 1158–1162.

114 Kendall, K., Alford, N.M., Tan, S.R., and Birchall, J.D. (1986) Influence of toughness on Weibull modulus of ceramic bending strength. *J. Mater. Res.*, **1**, 120–123.

115 Lugovy, M., Slyunyayev, V., Orlovskaya, N., Blugan, G., Kübler, J., and Lewis, M.H. (2005) Apparent fracture toughness of Si_3N_4-based laminates with residual compressive or tensile stress in surface layers. *Acta Mater.*, **53**, 289–296.

116 Lube, T., Pascual Herrero, J., Chalvet, F., and de Portu, G. (2007) Effective fracture toughness in Al_2O_3-Al_2O_3/ZrO_2 laminates. *J. Eur. Ceram. Soc.*, **27** (2–3), 1449–1453.

117 Virkar, A.V., Huang, J.L., and Cutler, R.A. (1987) Strengthening of oxide ceramics by transformation-induced stresses. *J. Am. Ceram. Soc.*, **70** (3), 164–170.

118 Pascual Herrero, J., Chalvet, F., Lube, T., and de Portu, G. (2005) Strength distributions in ceramic laminates. *Mater. Sci. Forum*, **492–493**, 581–586.

119 Bermejo, R., Pascual Herrero, J., Lube, T., and Danzer, R. (2008) Optimal strength and toughness of Al_2O_3-ZrO_2-laminates designed with external or internal compressive layers. *J. Eur. Ceram. Soc.*, **28** (8), 1575–1583.

120 Chan, H.M. (1997) Layered ceramics: processing and mechanical behaviour. *Annu. Rev. Mater. Sci.*, **27**, 249–282.

121 Rao, M.P., Sánchez-Herencia, A.J., Beltz, G.E., McMeeking, R.M., and Lange, F.F. (1999) Laminar ceramics that exhibit a threshold strength. *Science*, **286**, 102–105.

122 de Portu, G., Micele, L., and Pezzotti, G. (2006) Laminated ceramic structures from oxide systems. *Composites Part B*, **37** (6), 556–567.

123 Lube, T. (2007) Mechanical properties of ceramic laminates. *Key Eng. Mater.*, **333**, 87–96.

124 Pascual, J., Lube, T., and Danzer, R. (2008) Fracture statistics of ceramic laminates strengthened by compressive residual stresses. *J. Eur. Ceram. Soc.*, **28** (8), 1551–1556.

125 Erdogan, F. (1995) Fracture mechanics of functionally graded materials. *Compos. Eng.*, **5**, 753–770.

126 Danzer, R., Fellner, M., Börger, A., and Damani, M. (2003) Evolution von Gefügedefekten in Keramiken. *Prakt. Metallogr. Sonderband*, **34**, 451–458.

127 Rice, R.W. (1988) Perspective on fractography. *Adv. Ceram.*, **22**, 3–56.

128 Moulson, A.J. and Herbert, J.M. (1997) *Electroceramics*, Chapman & Hall, London.

129 Uchino, K. (1997) *Piezoelectric Actuators and Ultrasonic Motors*, Kluwer Academic Publishers, Boston/Dordrecht/London.

130 Uchino, K. (2000) *Ferroelectric Devices*, Marcel Dekker, Inc., New York, Basel.

131 Clarke, D.R. (1999) Varistor ceramics. *J. Am. Ceram. Soc.*, **82** (3), 485–502.

132 Pritchard, J., Bowen, C.R., and Lowrie, F. (2001) Multilayer actuators: review. *Br. Ceram. Trans.*, **100** (6), 265–273.

133 McMeeking, R.M. (2001) Towards a fracture mechanics for brittle piezoelectric and dielectric materials. *Int. J. Fract.*, **108**, 25–41.

134 Balzer, B., Hagemeister, M., Kocher, P., and Gauckler, L.J. (2004) Mechanical strength and microstructure of zinc oxide varistor ceramics. *J. Am. Ceram. Soc.*, **87** (10), 1932–1938.

135 Cao, H. and Evans, A.G. (1994) Electric-field-induced fatigue crack growth in piezoelectrics. *J. Am. Ceram. Soc.*, **77** (7), 1783–1786.

136 Fett, T., Munz, D., and Thun, G. (1999) Mechanical fatigue of a soft PZT ceramic under pulsating tensile loading. *J. Mater. Sci. Lett.*, **18**, 1895–1898.

137 Fett, T., Munz, D., and Thun, G. (1999) Tensile and bending strength of piezoelectric ceramics. *J. Mater. Sci. Lett.*, **18**, 1899–1902.

138 Supancic, P. (2000) Mechanical stability of $BaTiO_3$-based PTC thermistor components: experimental investigations and theoretical modelling. *J. Eur. Ceram. Soc.*, **20** (12), 2009–2024.

139 Supancic, P., Wang, Z., Harrer, W., and Danzer, R. (2005) Strength and fractography of piezoceramic multilayer stacks. *Key Eng. Mater.*, **290**, 46–53.

13
Creep Mechanisms in Commercial Grades of Silicon Nitride

František Lofaj and Sheldon M. Wiederhorn

13.1
Introduction

13.1.1
Motivation

Structural ceramics, such as silicon nitride, first attracted the attention of materials scientists and engineers during the period between 1980 and 2000, because of the promise of this material to operate at temperatures much higher than could be reached by high-temperature alloys [1–8]. These ceramics were to be used to produce more efficient gas turbines for trucks, automobiles and electric power-generation plants. Advanced grades of silicon nitride can operate in air at temperatures of at least 300 °C higher than the most advanced superalloys [7]. Ceramic components in small power gas turbines thus enable operating temperatures of up to 1400 °C, whereas alloys in the same turbines must operate at temperatures below 1000 °C. For example, tests of a Japanese 300 kW ceramic gas turbine, CGT 302, at 1420 °C demonstrated a significantly reduced pollutant emission and an increase in thermal efficiency to 42.1% [9], which was twice that of comparable metallic turbines. Although the turbine operated at 1400 °C for only 200 h [10], this was insufficient to prove its long-term reliability.

Additional problems arose as a consequence of the high sensitivity of ceramic blades to foreign object damage, and to surface recession due to corrosion by water vapor in exhaust gases [11–13]. The realization that corrosion and foreign object impact were serious problems led to a significant reduction in research and development activities during the late 1990s. Nevertheless, the substantial research that had already been carried out on ceramics for high-temperature structural applications has provided the materials engineering community with a deep insight how these materials perform under stress at high temperatures. Several problematic areas were identified, including failure due to creep and creep rupture, corrosion in aggressive environments, the lack of methods for lifetime prediction, and failure due to foreign object damage [10–14]. In this chapter, a review is presented of the creep behavior of silicon nitride, including details of the mechanisms of creep that

Ceramics Science and Technology Volume 2: Properties. Edited by Ralf Riedel and I-Wei Chen

ISBN: 978-3-527-31156-9

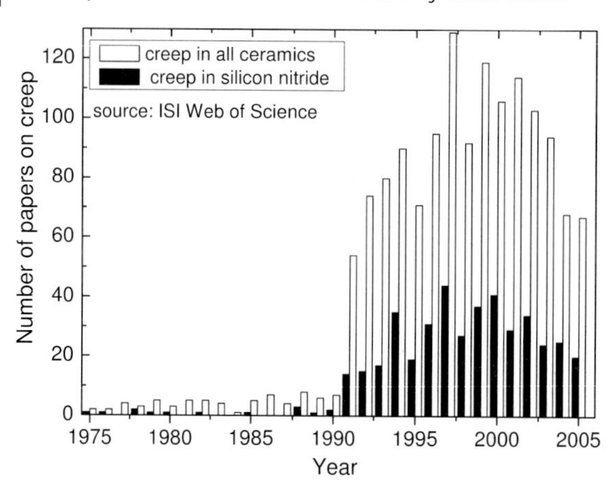

Figure 13.1 Creep research activities in ceramic materials and silicon nitride ceramics based on the number of scientific reports in the Institute for Scientific Information (ISI) database. The contribution of silicon nitride-related papers was approximately 30% of all research activities in this field [15].

result in significant differences in creep resistance of commercial grades of this material.

13.1.2
Creep of Silicon Nitride

The importance of creep to the structural reliability of silicon nitride was understood soon after the potential of its high-temperature mechanical properties was recognized [3–6]. Figure 13.1 compares three decades of research activities on the creep of silicon nitride with creep studies in all ceramic materials [15]. Almost 430 studies on silicon nitride were conducted between 1975 and 2005, most of which were carried out in flexure, which is difficult to analyze because of the mixed mode of loading and the nonstationary neutral axis. It is much easier to understand results from experiments carried out in a single mode of loading – that is, compression or tension. The first tensile creep study on silicon nitride was reported by Glenny and Taylor in 1961 [1], followed by Kossowsky, Miller and Diaz in 1975 [2]. Despite its relatively early beginning, the large expansion in this field of creep investigations occurred only after 1990 and peaked around 1997. The subsequent gradual decrease in creep activities paralleled the reduced interest in structural ceramics due to unresolved technical problems with blade surface recession in the gas turbine environment [9–13]. Despite this general decrease in activity, research on the creep of silicon nitride has continued, albeit at a slower rate. The ratio of creep studies on silicon nitride to those on all other ceramic materials over the past decade has been relatively steady, varying from about 25% to a maximum of 39% in 2000.

Creep behavior and creep mechanisms in advanced ceramics – particularly in silicon nitride – have been summarized in several reviews. Messier *et al.* [3] reviewed a number of reports, conference proceedings and journal contributions through 1974, while Quinn reviewed static fatigue data on silicon nitride up to 1982 [4]. Cannon and Langdon produced a large review of creep properties of early grades of numerous ceramics in 1983, and examined possible deformation mechanisms according to the models adopted from metals in 1988 [5, 6]. Most of the theoretical models applicable to creep in ceramics were described in an extensive review of Hynes and Doremus in 1996 [16]. Wilkinson summarized the effects of reinforcing phase and possible creep mechanisms in multiphase materials in 1998 [17, 18]. The last extensive review of creep in silicon nitride ceramics was published by Melendéz-Martínez and Domíguez-Rodríguez in 2004 [19]. These authors concluded that it is impossible to determine with accuracy the deformation mechanism under fixed experimental conditions, because none of the models fully accounts for all experimental observations. Apparently, the creep of silicon nitride must be explained in terms of several mechanisms that occur simultaneously. Hence, it was suggested that the main mechanism occurring during compressive creep is grain boundary sliding accommodated by solution–precipitation, whereas tensile creep is greatly influenced by cavitation.

Despite the difficulties in determining creep mechanisms in silicon nitride, progress in developing creep-resistant grades of silicon nitride has been tremendous over the past three decades. Figure 13.2 illustrates that the substitution of Y_2O_3 for MgO, and later the elimination of Al_2O_3, greatly improved the creep resistance of silicon nitride [20]. Figure 13.2 also shows that decreasing the amount of Y_2O_3, increases the creep resistance of silicon nitride.

As shown in Figure 13.2 [21], the creep resistance of NT 154 at 150 MPa is much greater than that of earlier commercial grades of silicon nitride.[1] Assuming a

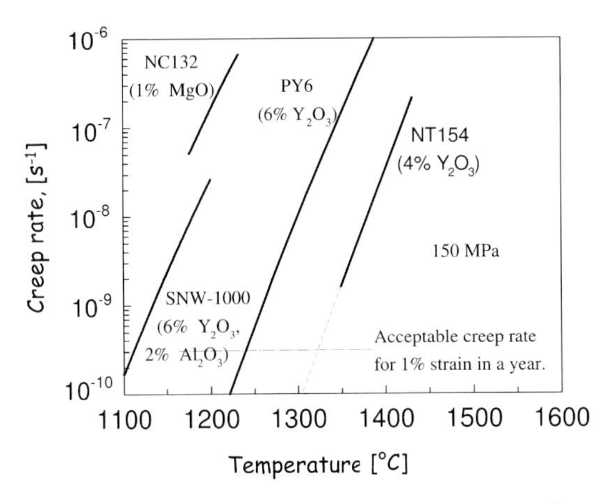

Figure 13.2 Comparison of the creep performance in different grades of silicon nitride ceramics at 150 MPa [21].

1) The use of commercial designations is for purposes of identification only, and does not imply any endorsement by the National Institute of Standards and Technology, nor does it imply that the items mentioned are the best ones for the intended application.

maximum allowable strain of $<1\%$ in 10 000 h to be a critical design limit for rotating blades in a gas turbine, the strain rate of the material must be $<3 \times 10^{-10}\,s^{-1}$ for long-term survival. Figure 13.2 thus suggests that the maximum temperature allowed for NT 154 at a stress of 150 MPa is more than 200 °C higher than that in NC 132 silicon nitride. The difference in creep rates of the two materials at the same temperature and stress exceeds four orders of magnitude. Such improvement was achieved by increasing the deformation resistance of the sintering aids, and reducing their volume fraction.

Detailed investigations of long-term creep behavior and accumulation of data for reliable creep life have been accomplished on a number of newer materials intended for use in ceramic turbine engines [9–12, 22]. The behavior of the best grades of silicon nitride at temperatures from 1200 to 1400 °C is often characterized by an exponential dependence of creep rate on stress, by a significant contribution of cavitation to the tensile strain, and by a high oxidation resistance. Density change measurements, small-angle X-ray scattering studies and microstructural studies using transmission electron microscopy (TEM) have shown that cavities produced in NT 154 [15] and in SN 88 [23–27] account for up to 90% of the tensile strain due to creep. Similar density changes were reported for siliconized/silicon carbide [28], and for vitreous-bonded aluminum oxide [29]. Cavitation is the creep-controlling process in NT 154 and SN 88, and probably also in other grades of vitreous-bonded ceramics. Based on these observations, Luecke and Wiederhorn suggested a new cavitation creep model [30], which is able to explain the gradual increase in stress dependence and the significant creep asymmetry typical for these materials.

Because of the importance of cavitation to the creep rate, the present chapter incorporates a review of creep in commercial grades of silicon nitride. During the chapter, a common approach to creep behavior of different grades of silicon nitride ceramics is suggested. By comparing the creep behavior of commercial grades of silicon nitride, and analyzing the contribution of cavitation to the tensile strain, it is possible to demonstrate that tensile creep in silicon nitride occurs via cavitation and the redistribution of secondary phases among multigrain junctions [30]. If the effective viscosity of secondary phases is low, the creep behavior is controlled by extensive cavitation at multigrain junctions [31]. In the opposite case, cavitation is suppressed and the creep resistance increases significantly. Another important consequence of these studies is that meso-mechanical models, which consider a redistribution of secondary phases among a large number of grains, are required to describe such creep processes. The need for these models is in contrast to conventional, micro-mechanical models, for which the redistribution of the material over only one grain is considered.

13.2
Material Characterization

In this chapter, three grades of technologically mature, commercially available silicon nitride are considered. Because of their high quality, their availability on the open market, and their excellent high-temperature properties, these materials have been studied extensively in many laboratories over the past two decades. Consequently, a large body of data exists which can be used for comparison with various models of creep.

NT 154: This is the oldest of the three grades; it is a hot isostatically pressed material with 0.04 mass fraction Y_2O_3 as a densification aid. Produced by Norton/TRW, NT 154 is an *in situ*-reinforced composite consisting of 10 μm × 1 μm reinforcing β-Si_3N_4 grains distributed among finer, equiaxed β-Si_3N_4 matrix grains[23, 32, 33]. The secondary phase in the interstitial regions is nearly completely crystallized, consisting of α-$Y_2Si_2O_7$ and $Y_5(SiO_4)_3N$. Residual glass remains on nonspecial boundaries as an amorphous boundary film about 1 nm thick [32].

SN 88: This is a commercial gas-pressure-sintered grade of silicon nitride produced by NGK Insulators, Ltd, Nagoya, Japan, for the hot-stage components of ceramic gas turbines. This grade of silicon nitride was extensively studied by Wiederhorn *et al.* [20, 24, 30], by Lofaj *et al.* [15, 25–27, 34–37], and by Wereszczak *et al.* [38]. Its microstructure consists of large β-silicon nitride grains 50–80 μm long with a diameter of 5–10 μm, embedded in a fine β-Si_3N_4 matrix. The mean size of the matrix grains is approximately 0.5–0.8 μm, with an aspect ratio of 3–5. Secondary phases contain Yb^{3+} and Y^{3+} cations. X-ray diffraction (XRD) analysis has shown that the major crystalline phases in the as-received materials involve β-Si_3N_4, ytterbium silicon oxynitride ($Yb_4Si_2N_2O_7$) and/or Y-based isotype, $Y_4Si_2N_2O_7$, and Y-apatite, $[Y_5(SiO_4)_3N]_x$. Trace phases identified as B-Yb_2SiO_5 and/or X_2-Y_2SiO_5 were also detected. Additional peaks correspond to $MoSi_2$, SiC and free Si, but their identification is uncertain.

SN 281: This is a gas-pressure-sintered silicon nitride subjected to additional hot isostatic pressing. Manufactured by Kyocera Corp., Kyoto, Japan, the material was designed for blades of the ceramic turbines operating for prolonged periods at temperatures of around 1400 °C and at relatively high stresses [39]. It consists of β-Si_3N_4 matrix grains with the mean diameter of about 1 μm, an aspect ratio of 2–4, and a small number of large grains of length up to 30 μm and diameter of 3–6 μm. XRD analysis revealed the presence of $Lu_2Si_2O_7$ and $Lu_4Si_2N_2O_7$ as the dominant crystalline secondary phases in the multigrain junctions [15, 40, 41].

These three grades represent a wide range of creep behavior in silicon nitride ceramics. Although, clearly, the family of the commercial grades of silicon nitride is much greater, such grades as GN 10, ST1, GR1 and others [19, 26, 32] have not been included in this chapter because they were short-lived, not widely available and, therefore, less well-studied. Tensile testing, density measurements and other procedures used to characterize the three grades of silicon nitride have been described in detail elsewhere [23, 42], and their details will not be repeated here.

13.3 Discussion of Experimental Data

In this section, the available data are discussed relating to the three materials chosen for analysis. First, it is shown that most data reported on creep rates do not fit the log–log plots that correspond to the classical theories of creep in metals. It is then demonstrated that NT 154 and SN 88 exhibit extensive cavitation, which can be quantitatively related to the creep strain, suggesting that creep in these materials is a consequence of cavity formation. This behavior is then rationalized by showing that the creep data for these materials is consistent with the cavitation creep model proposed by Luecke and Wiederhorn [30].

13.3.1 Creep Behavior

Most reported creep data on silicon nitride are expressed and analyzed in terms of classical creep mechanisms for which the creep rate, $\dot{\varepsilon}$, is expressed as a power function of the grain size, g, and the applied stress, σ, and an Arrhenius function of temperature, T:

$$\dot{\varepsilon} = \dot{\varepsilon} \cdot \left(\frac{g}{g_0}\right)^m \cdot \left(\frac{\sigma}{\sigma_0}\right)^n \cdot \exp\left(\frac{-Q}{RT}\right), \tag{1}$$

Here, the parameters, m, n, and Q, are constants of the fit of Eq. (1) to the data, and g_o and σ_o are normalization constants. In the discussion that follows, published data are discussed in terms of the grain size exponent, m, the stress exponent, n, and the activation energy, Q.

13.3.1.1 NT 154

Specimens tested in tension exhibited transient creep behavior. The strain rates slowly decreased until failure, and only a few samples showed steady-state creep (Figure 13.3). Typically, the initial creep rate was fourfold that of the creep rate at

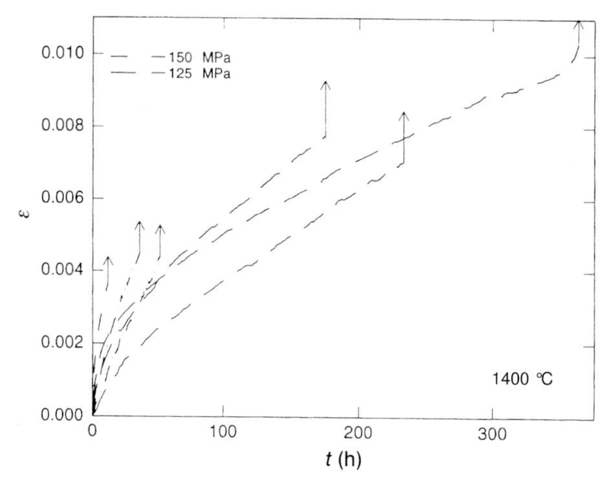

Figure 13.3 Tensile creep curves from repeated creep tests under stress of 125 MPa and 150 MPa at 1400 °C in NT 154 [23].

failure, and could not be correlated with temperature, or applied stress [23, 32]. Several specimens with more than 1% strain exhibited a roughly constant creep rate for at least half of their lifetime. Failure strains taken over temperatures from 1370 to 1430 °C were usually in the range from 0.5 to 1.0%. The maximum strain was 2% at 1430 °C under a stress of 75 MPa. Higher stresses produced smaller strains-to-failure, and some decrease in the strains-to-failure was observed when the temperature was increased.

Compressive creep behavior revealed signs of creep asymmetry [23]. The rates at the same strains in compression were at least one order of magnitude lower than in tension, and failure strains were not reached even after exposure times fivefold longer than in tension. Because of the absence of a clear steady-state creep rate, the rates measured at predetermined strains were used to plot creep rate as a function of applied stress. The stress exponents, n, in tension and compression are compared in Figure 13.4 [23]. The value of n in compression is close to unity for the whole range of stresses. Stress exponents in tension depend on stress and increase from $n \approx 1$ to $n > 5$ as the stress increased. Activation energies were also influenced by the lack of a steady-state creep rate, and depended on the strain used to determine strain rate. Typical values at 125 MPa ranged from 1000 to 1250 kJ mol^{-1}, while at 150 MPa they ranged from 1370 to 1615 kJ mol^{-1} [21].

13.3.1.2 **SN 88**

Creep curves at temperatures below 1300 °C, consisted of a primary and prolonged secondary creep rate, followed by an abrupt failure. At temperatures over 1350 °C, a prolonged, slightly accelerated tertiary stage appeared. Figure 13.5 shows tensile

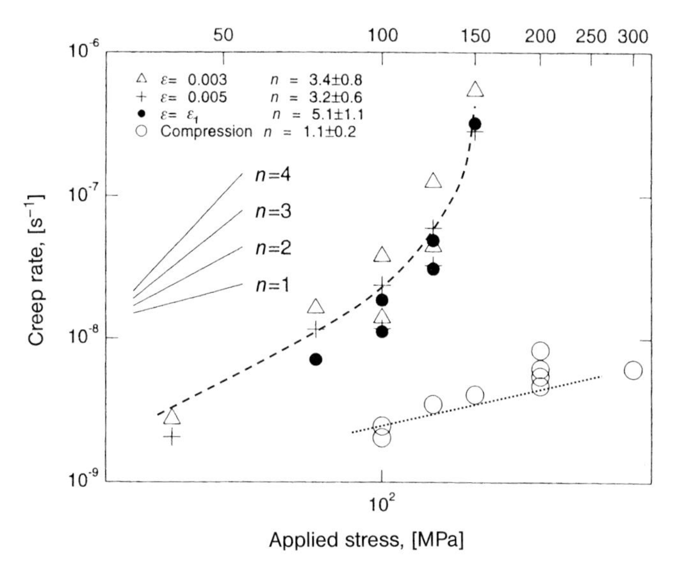

Figure 13.4 Stress dependence of the minimum creep rates in NT 154 in compression and tension at different strains and 1430 °C [23].

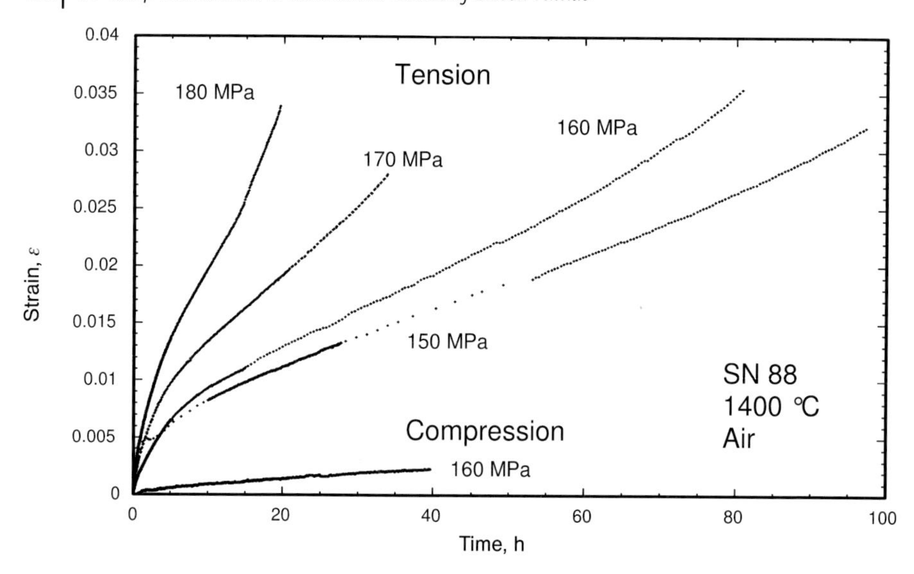

Figure 13.5 Tensile and compressive creep curves in SN 88 at 1400 °C under different stresses [27].

creep curves obtained at 1400 °C at different stresses, and one compressive creep curve at a stress of 160 MPa [25–27]. The maximum failure strain in tension for all of the stresses was almost 3.5%, which exceeds the values of ~0.4% obtained in NT 154 at the same conditions. Although, the minimum strain rates in SN 88 were slightly higher than those in NT 154, the lifetime was at least twofold longer at the same stress and temperature.

Figure 13.6 summarizes strain rates as a function of applied stress, for stresses ranging from about 10 to 400 MPa [43]. Over this stress range, stress exponents

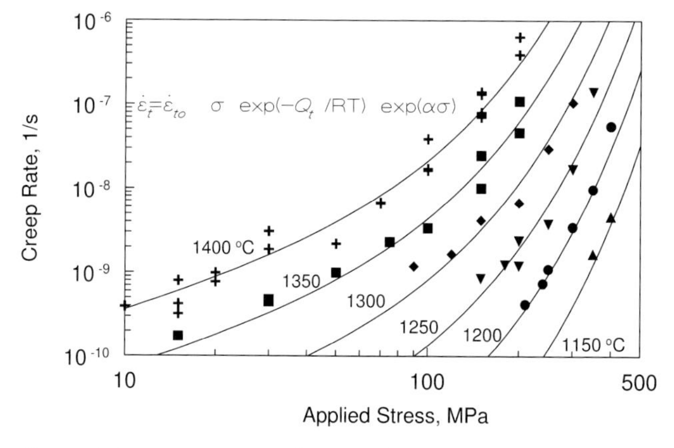

Figure 13.6 Strain rates as a function of stress in a wide range of stresses and temperatures in SN 88. The tests were performed on dog-bone samples. The curves represent the exponential Eq. (2) [43].

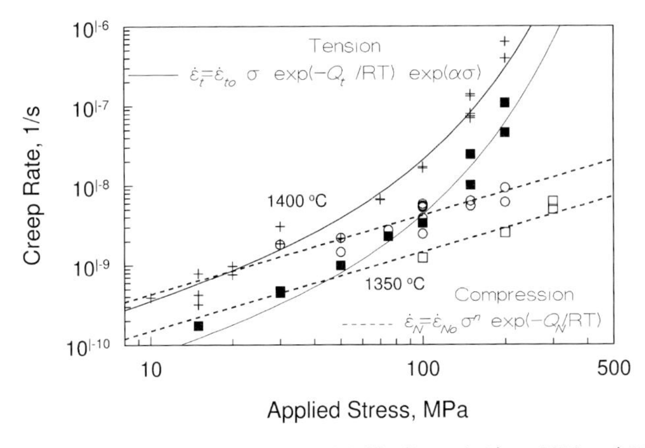

Figure 13.7 Creep asymmetry in SN 88 silicon nitride at 1350 and 1400 °C depends on stress. Tensile creep rates correspond to exponential dependence on stress, while compression creep follows the power law [43].

ranged from approximately 1 at 1400 °C and 10 MPa to approximately 12 at 1150 °C and 400 MPa. These data clearly do not fit a power-law function. Rather, the behavior is similar to that observed for the NT 154 (see Figure 13.4), and suggests an exponential dependence of creep rate on applied stress. Indeed, the exponential curve given in Figure 13.6 tracks the data over the entire range of test variables.

The observation of strong creep asymmetry in SN 88 was a subject of a special investigation (Figure 13.7) [43], in which compression and tension tests were made from the same billet. The results were similar to those obtained for NT 154 (see Figure 13.4). Creep asymmetry depends on stress; it appears to be greater at high stresses, and disappears at low stresses (see Figure 13.7). An even more important observation is that compressive creep follows conventional power-law dependence on stress, with n equal to about 1, whereas an exponential dependence of creep rate on applied stress better describes tensile creep. This difference in functional behavior suggests a significant difference in creep mechanism, as discussed below.

13.3.1.3 **SN 281**

Creep curves for this material often exhibit extended primary creep, prior to approaching steady state (Figure 13.8) [44]. The data in Figure 13.8 were collected at temperatures ranging from 1450 to 1550 °C, which were considerably higher than those used for any other grade of silicon nitride [19]. At 1400 °C and a stress of 200 MPa, the primary creep stage exceeded 4000 h and the lifetime exceeded 10 000 h [40, 41]. Depending on the stress, creep rates at 1400 °C are more than four orders of magnitude lower than those for SN 88.

In addition to its higher temperature capability, SN 281 exhibits a classical dependence of creep rate on stress and temperature. A log–log plot of the creep rate as a function of applied stress indicates no curvature in the creep plots over the range of stresses used for the study (Figure 13.9) [44]. The slope of the curves, $n = 1.87 \pm 0.48$ (95% confidence limits), is much lower than that obtained for either

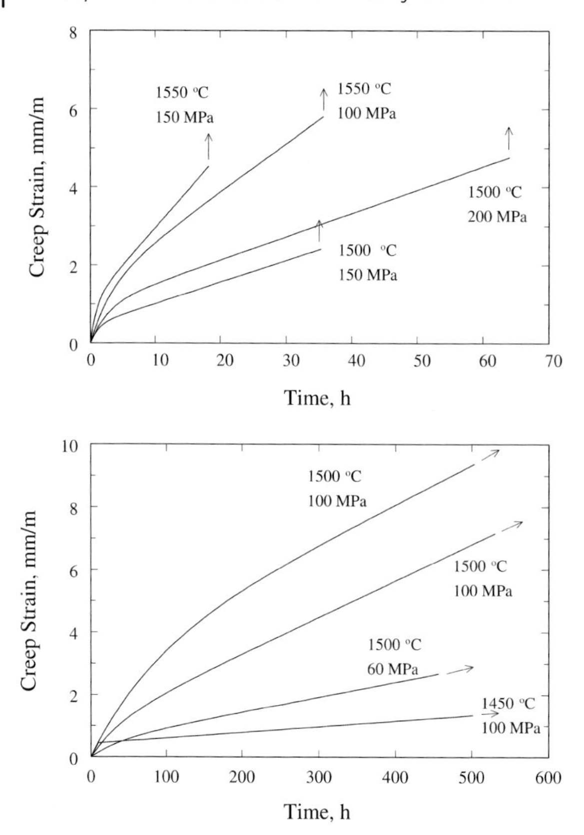

Figure 13.8 Creep behavior of SN 281 (billet #57) at temperatures, which are considerably higher than in the earlier generations of silicon nitride [44].

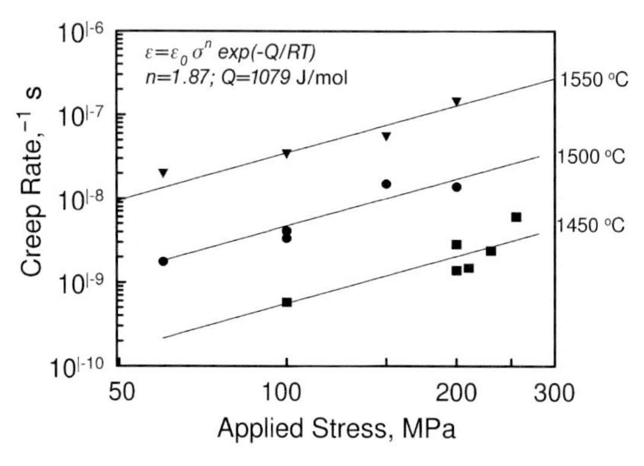

Figure 13.9 Stress–strain rate dependence in SN 281 [44]. The power-law behavior of this material over the entire range of applied stresses distinguishes the behavior of SN 281 from other commercial grades of silicon nitride.

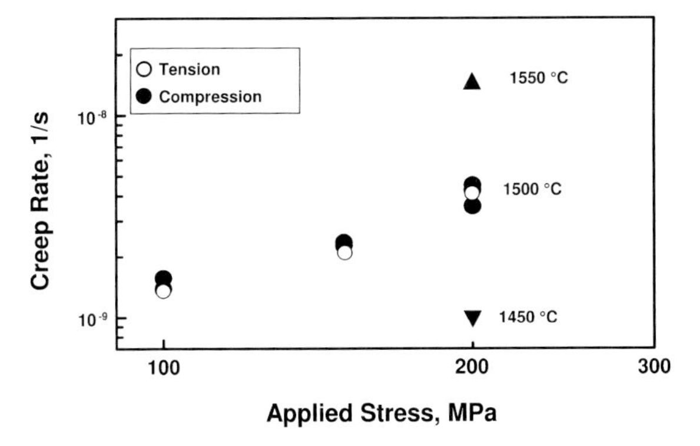

Figure 13.10 A comparison of tensile and compressive creep for SN 281 indicates that creep in tension and compression is equal. In addition, the slope of the creep curve is close to 1, suggesting a classical diffusional creep mechanism [45].

the SN 88 or the NT 154 over most of their creep range, and certainly at the higher stresses. The activation energy for the SN 281 was approximately 1079 ± 142 kJ mol^{-1} (also 95% confidence limits). Although high, this value was not atypical of activation energies obtained for the creep of silicon nitrides.

The results of recent studies on the creep of SN 281 have also indicated that the creep rate of this grade of material is the same in tension and compression (Figure 13.10) [45]. This finding is very different from that of the other commercial-grade silicon nitrides, and suggests that the mechanism of creep for the SN 281 might differ from that of other grades of silicon nitride. Except for its low strain to failure, the creep behavior of SN 281 is more like that of the high-temperature metals than it is of silicon nitride.

13.3.2 TEM Characterization of Cavitation

Extensive microstructural analysis on silicon nitride revealed that most of them form cavities as they creep in tension. In the case of SN 88 and NT 154, the creep damage can be divided into two groups, depending on the phase in which the damage forms (Table 13.1) [23, 24, 32, 36, 37]. The first group consists of the damage in the amorphous boundary phase, involving numerous and relatively large multi- and two-grain junction cavities (Figures 13.11 and 13.12), as well as microcracks and cracks, which form at the later stage of the lifetime. The two-grain junction cavities are rare relative to the cavities formed at multigrain junctions. Defects in the second group are formed in the silicon nitride grains themselves; these include broken large grains and crack-like intragranular cavities in long fiber-like grains, and small lens-shaped cavities [15, 37]. In contrast to NT 154 and SN 88, cavitation in SN 281 is rare, even after creep tests exceeding 10 000 h and strains in excess of 1% [41, 44].

Table 13.1 Classification of the creep cavities in the studied silicon nitride ceramics [23, 24, 28, 29].

	Type of defect	Quantity of cavities			Possible mechanisms
		NT154	SN88	SN281	
Amorphous secondary phase	Multigrain junction cavities	Many	Many	Few	GBS + VF + $S\text{-}P_{sec.phases}$
Amorphous secondary phase	Two-grain junction cavities	Few	Few	NO	GBS + VF
Si_3N_4 grains	Broken large grains	NO	Few	NO	GBS + VF
Si_3N_4 grains	Lenticular cavities	Some	Few	NO	GBS + $S\text{-}P_{SN}$
Si_3N_4 grains	Intragranular cavities	NO	Very few	NO	GBS + VF + $S\text{-}P_{SN}$

GBS = grain boundary sliding; VF = viscous flow; S-P = solution–precipitation.

13.3.3 Density Change and Volume Fraction Cavities

The role of cavity formation in the creep process has been elucidated by a number of experimental techniques, including density measurements [23, 43], measurements of ultrasonic velocities [34, 35], determination of elastic moduli by instrumented indentation [31], and anomalous ultra-small angle X-ray scattering (A-USAXS) [23, 46]. The results of density measurements are illustrated in Figure 13.13. The vertical axis of this figure was calculated from measured density changes in tensile or

Figure 13.11 Multigrain junction cavities of different shapes in SN 88 after creep for 1693 h at 1300 °C under stress of 155 MPa. [15, 37].

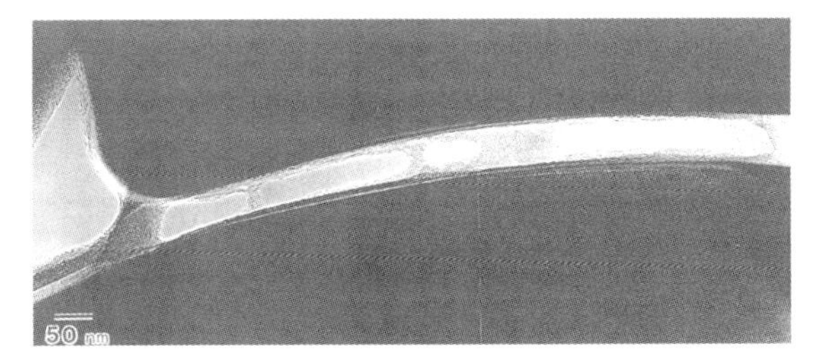

Figure 13.12 Two-grain junction cavities observed in SN 88 after creep at 1250 °C under stress of 180 MPa after 868 h [37].

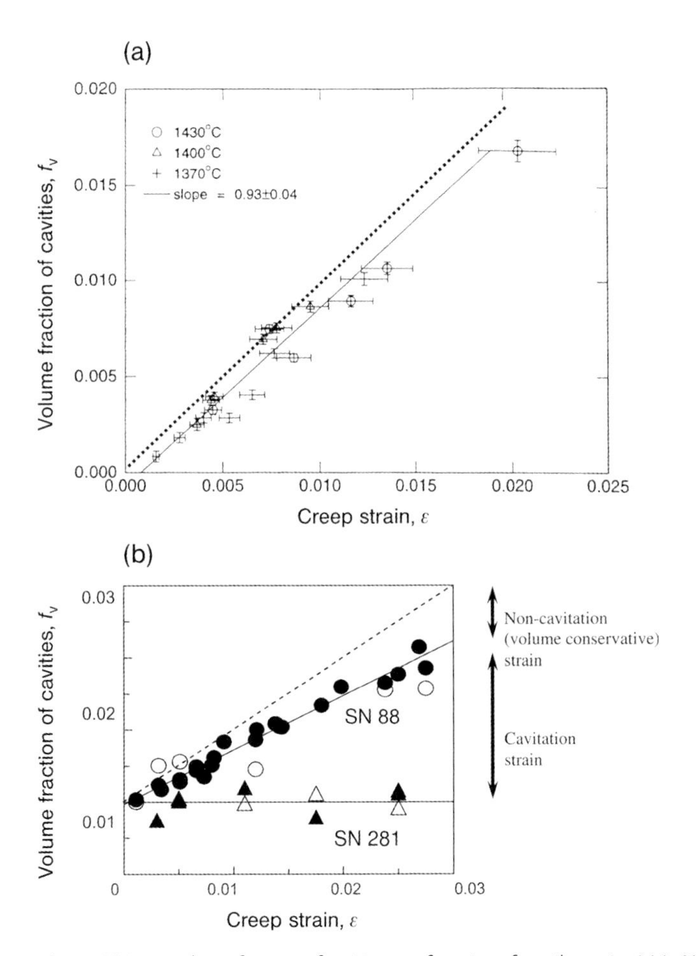

Figure 13.13 Volume fraction of cavities as a function of tensile strain. (a) In NT 154 [25]; (b) In SN 88 (full circle symbols) and SN 281 (full triangles); open symbols correspond to the volume fraction of cavities determined by A-USAXS (anomalous ultra-small-angle scattering) [40, 41].

compressive specimens of silicon nitride that had been deformed at high temperatures by creep [23]. The decrease in density is assumed to be entirely due to cavity formation, and hence can be converted easily into a volume fraction of cavities. The data in Figure 13.13 show that the volume fraction of cavities in NT 154 and SN 88 increases linearly with increasing tensile strain. This behavior contrasts sharply with that of SN 281, for which the volume fraction of cavities is zero, regardless of strain. For NT 154 and SN 88, the slopes of the lines in Figure 13.13 are 0.93 ± 0.04 [23] for NT154, and 0.75 ± 0.03 for SN 88 [25–27, 30, 35]. For SN 281, the slope is approximately 0 [40, 41]. The slopes determined from measurement of the change in density agree with those obtained by measurement of the elastic constants of the crept material. The methods used included ultrasonic measurements [35] and the instrumented indentation technique [34]. Both types of test were carried out on the same samples on which the density measurements in Figure 13.13 were made [34, 35].

The slopes of the lines in Figure 13.13, at > 0.75, are also typical for reaction-bonded silicon carbide [24, 28]. These data also indicate that over 75% of the measured axial tensile strain results from cavitation. The volumes generated by cavities are transferred primarily into axial tensile strain [25], which suggests that cavitation is the main creep mechanism of deformation in these ceramics. As the contribution of cavities to strain in SN 281 is close to zero, creep in this material is fundamentally different from that of other grades of silicon nitride [15, 40, 41, 44]. The suppression of cavitation in SN281 is most likely the reason for its increased creep resistance.

13.3.4
Size Distribution of Cavities Formed

Anomalous small-angle X-ray scattering is useful not only for determining the total volume of cavities (as shown above), but also for determining the size distribution of both the cavities and the secondary phase. Experiments to determine the evolution of the size distribution of cavities during the creep of NT 154 were carried out by Luecke *et al.* [23] as a function of deformation strain (Figure 13.14). The data in Figure 13.14 were obtained from both the gauge and the grips of the tensile specimens that had been subjected to a controlled amount of creep strain. Scattering data from the grips provide the background scattering that comes from the unstrained material and has gone through the same heat cycle as the gauge. The two sets of scattering data are subtracted; the difference gives that portion of the data that is due only to the scattering by cavities. The data in Figure 13.14 were collected on three individual specimens that had been strained by the specified amount prior to failure. An additional specimen was strained to failure. Besides collecting SAXS data, the densities of all four specimens were measured [23], from which volume fractions of cavities were calculated. These fell on a curve identical to that shown in Figure 13.13a, and indicated that the cavities had accumulated linearly with time. The total area under the curves in Figure 13.14 also fell on the curve shown in Figure 13.13a, indicating a consistency of the density data with the SAXS data.

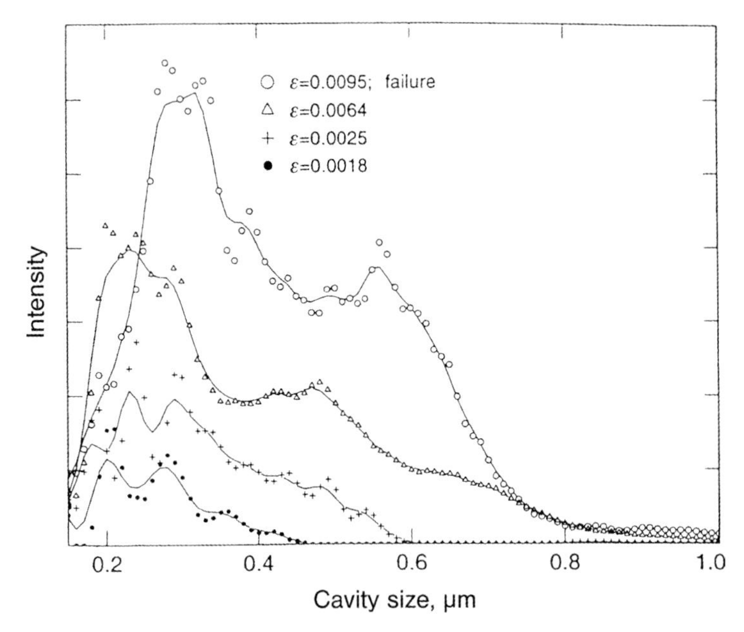

Figure 13.14 Cavity size distribution evolution as a function on strain for NT 154 sample tested at 1400 °C and 125 MPa [23].

The curves in Figure 13.14 consist of two peaks: one peak occurring at about 0.25 μm, and another appearing as a shoulder on the first peak at about 0.4–0.5 μm. For small strains (the first three curves in Figure 13.14), the peak at 0.25 μm does not change position with increasing strain, indicating that the addition of new cavities of the same size is the primary contributor to creep strain. As the strain increases (the largest curve in Figure 13.14), the peak that was at 0.25 μm has shifted to about 0.3 μm, indicating a growth of cavities within the gauge section. Over the same range of strain, the peak that developed as a shoulder shows both an increase in total volume of cavities and a movement to larger cavities. From these results it is reasonable to assume that, in the initial stages of deformation, cavities are simply added to the distribution, whereas at the later stage the cavity growth occurs perhaps as a consequence of the linkage of smaller cavities into larger ones.

The second SAXS study of creep-strained silicon nitride was carried out on SN 88, which employed a mixture of Yb_2O_3 and Y_2O_3 as a sintering aid [46]. Because Yb has a substantially higher atomic weight than Y, it exhibits a sufficiently large amount of X-ray scattering to be used when analyzing the distribution of $Yb_2Si_2O_7$ within the microstructure of the silicon nitride. In order to distinguish between the scattering due to cavitation and that due to the secondary phase, a series of eight wavelengths was selected for the collection of scattering data. The data were collected close to the Yb L_{III} absorption edge, where the amount of scattering from the $Yb_2Si_2O_7$ changes as a function of wavelength. As the cavities exhibited no wavelength dependence for the scattering, it was possible to separate the scattering due to the cavities from that due to the $Yb_2Si_2O_7$ [46].

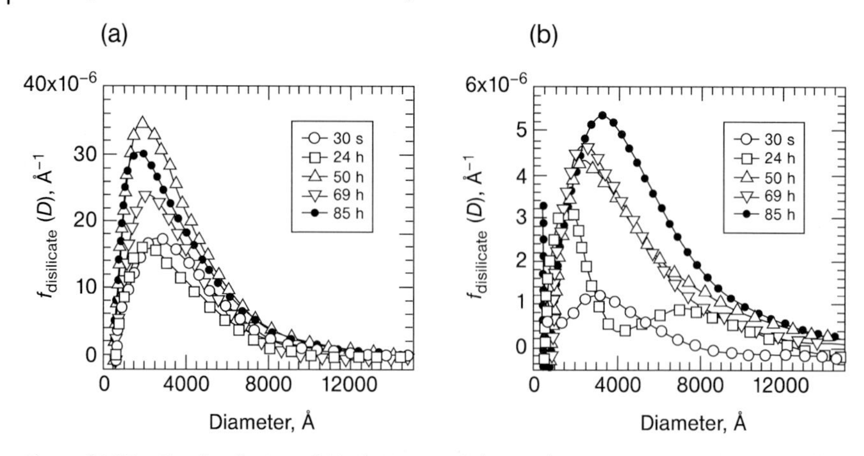

Figure 13.15 Size distribution of (a) $Yb_2Si_2O_7$ and (b) tensile creep cavities in SN 88 tested at 1400° C and 150 MPa [46]. The legend provides the creep times for each sample.

The study results are shown in Figure 13.15. The scattering due to the $Yb_2Si_2O_7$ resulted in a peak at about 0.2 μm, which increased with increasing creep time, but gave no indication of any change in the diameter of the second phase (see Figure 13.15a). Apparently, as the deformation strain increased, the number of small pockets of $Yb_2Si_2O_7$ increased, but not their size. Figure 13.15b shows how the volume fraction of cavities depends upon the cavity diameter. These results were similar to those obtained for the NT 154 (see Figure 13.14). The volume fraction of cavities increased with creep strain, as did the size of the cavities. Unlike NT 154, there seemed to be no range of strain over which the cavity diameter stayed constant; neither was a second peak observed in the distribution at larger diameters.[2)] Yet, by integrating the areas under the curves in Figure 13.15b, the total volume fraction of cavities could be obtained; these were in excellent agreement with the fraction of cavities obtained by density measurements.

13.4
Models of Creep in Silicon Nitride

In this section, the mechanisms of deformation that are active in two-phase materials such as silicon nitride are summarized. The grains are assumed to be completely surrounded by the secondary phase; conservative creep deformation of the material – that is, no change in volume – may then occur either as a consequence of deformation of the secondary phase, or by deformation of the silicon nitride particles themselves. If the volume is not conserved – as when cavities are formed – then other mechanisms can occur. Hence, the types of deformation that can occur in the above-mentioned commercial grades of silicon nitride will be briefly discussed.

2) For the 24 h strain, a peak was observed at about 0.8 μm, but this was swallowed up by the dominant peak, with further deformation.

13.4.1
Cavitation Creep Model in NT 154 and SN 88

The development of a model for cavitation-creep of materials with grain boundaries wetted by a secondary phase requires an understanding of why cavities form at multigrain junctions, and where the material that was formerly located in the cavity transfers to after cavitation. Luecke and Wiederhorn adapted ideas from soil mechanics theory to explain the source of the dilatational stresses in silicon nitride that lead to cavitation [30]. These arise because of the way in which fully dense materials made of a network of rigid grains deform. If the grains themselves cannot be deformed by the applied stresses, then the only way deformation can occur is by dilatation of the network of grains. Any sliding of grains along grain boundaries leads to a dilatation of the structure as a whole [47]. Dilatation in turn generates local hydrostatic tensile stresses within the secondary phase at grain boundaries and multigrain junctions. Cavities nucleate in the multigrain junctions where dilatational stresses are highest, provided that the stress exceeds the threshold stress for cavity nucleation [30].

Once a cavity forms within a multigrain junction, the stress in the junction relaxes to zero, and this results in a relatively large stress gradient from the cavity to other multigrain junctions that have no cavities. In this case, the stress gradient is the driving force for the redistribution of the secondary phase towards the cavity-free junctions, primarily along the tensile axis because the stress gradient is greatest in that direction. The formation of cavities and redistribution of secondary phase leads to an expansion of the silicon nitride – hence the decrease in density with increasing strain. Because cavitation relaxes the stresses within the junctions that cavitate, it also relaxes the hard contacts between the grains that surround the junctions. This process continues until specimen failure, either by the linking of cavities or by a combination of linkage and growth [28]. The gradual increase in the number of cavities generates additional volume, which results in a linear dependence of cavitation strain on tensile strain (Figure 13.13). The minimum strain rate of such cavitation creep according to the model of Luecke and Wiederhorn (L-W) is [30]:

$$\dot{\varepsilon} = \dot{\varepsilon}_o \cdot (\sigma/\sigma_o) \cdot \exp(-Q_c/RT) \cdot \exp(\beta_c \cdot \sigma) \tag{2}$$

where $\dot{\varepsilon}_o$, β_c, and Q_c are constants of the fit of the equation to the creep data. The parameter, σ_o, is a normalization-constant for the stress. This is the same equation that has been discussed with reference to Figure 13.6. Using the method of least squares, the constants for Eq. (2) were obtained from the data in Figure 13.6: $\ln(\dot{\varepsilon}_o) = 27.2 \pm 1.6$ ($\dot{\varepsilon}$ is in s^{-1}); $Q_c = (715.3 \pm 22.9)$ kJ mol^{-1} and $\beta_c = 0.0197 \pm 0.128$

For this mechanism of deformation to work, the stresses required for grain boundary sliding must be low relative to the stresses required to deform the grains. The main difference between this model of creep and the power-law model is the exponential dependence of the creep rate on stress. The increase of the stress exponent with stress, which often occurs in silicon nitride, is a natural consequence of Eq. (2). The model also successfully predicts the creep behavior of these two phase materials at low values of stress. As the stress, σ, approaches zero, the exponent term approaches 1, at which point the creep rate is determined by the linear stress term in Eq. (2). This explains why the creep curves approach a slope of 1 as the stresses become very small (Figures 13.4 and 13.7 [23, 43]). Equation (2) does not explain the apparent convergence of tensile and compressive creep rates at low stresses (see Figures 13.4 and 13.7), as cavities still form in the SN 88 at very low stresses, so that the mechanism of deformation in tension and compression are not the same at low stresses.

Possible mechanisms involved in cavitation include grain boundary sliding, (GBS), viscous flow (VF) of the glassy phase at the grain boundaries, solution–precipitation of the crystalline secondary phases ($S\text{-}P_{sec.\,ph.}$) between pockets, and solution–precipitation (S-P) of silicon nitride, either between grains or from one grain surface to another (see Table 13.1). A schematic description of the interactions between mechanisms during creep is shown in Figure 13.16 [15, 27]. Besides the continuous nucleation and growth of cavities at multigrain junctions, this scheme explains the development of other types of defects identified in Table 13.1.

Broken large grains form after primary grain boundary sliding because they hinder local grain boundary sliding and are, therefore, subjected to higher stresses than the surrounding matrix. The grains break when the stresses transferred to them exceed their strength. When those stresses are lower, large grains are subjected to loads for a prolonged time, which results in intragranular cavitation via S-P of silicon nitride [25]. Apparently, in the case of smaller matrix grains, when the local stress is not sufficient

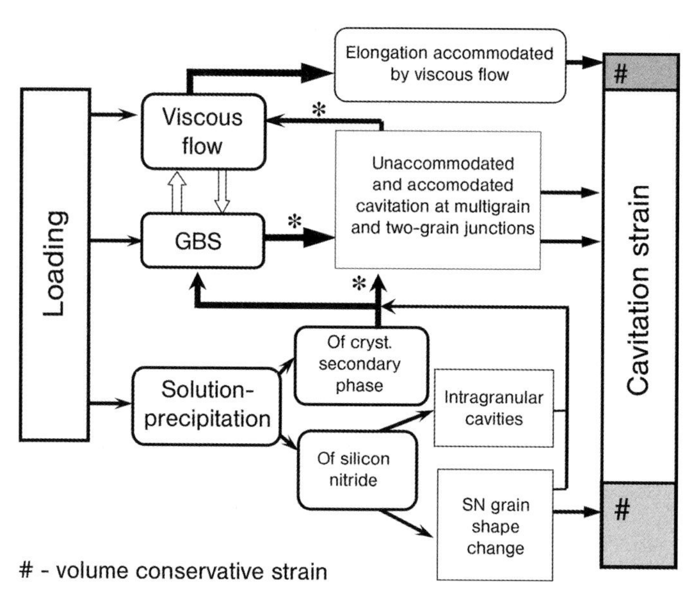

Figure 13.16 Phenomenological model of the cavitation processes during tensile creep deformation in silicon nitride [15, 27].

for intragranular cavitation, lenticular cavities develop at two-grain junctions [23, 32]. The scheme also assumes that the redistribution of the secondary phase changes the size distribution of cavities within the material, which is supported by experimental observations [23, 46]. Final proof of the cavitation creep model is the fact that Eq. (2) fits the experimental data over a wide range of stresses for both SN 88 and NT 154. Thus, the model shown in Figure 13.16 provides an understanding of the overall creep behavior of silicon nitride. Cavitation is the main creep mechanism in silicon nitride, and in other similarly bonded ceramics.

13.4.2 Noncavitation Creep

In the absence of cavitation, creep in vitreous-bonded materials would occur by S-P, wherein material dissolves from one side of the grain and deposits on another [48, 49]. No definitive studies have been made to date that support the dislocation creep models in which the grains of silicon nitride deform by dislocation motion. Studies of deformed silicon nitride grains have provided no evidence of the types of dislocation pileup that should be present in order for this type of mechanism to be active [50].

A minor degree of deformation can occur by plastic flow of the vitreous bonding phase that lies between all of the silicon nitride grains. When a mechanical load is applied to silicon nitride at high temperatures, the first response is transient creep in which deformation occurs primarily in the secondary phase that binds the grains of silicon nitride together. The thickness of the amorphous phase between the grains can either decrease or increase, depending on the sign of the local stress. That this mechanism of deformation actually occurs has been proven by Wilkinson and colleagues in some very careful studies of the thickness of grain boundaries during the deformation of silicon nitride. As expected from such a mechanism, some boundaries were thicker and some thinner than the starting average boundary thickness [18]. As deformation continues, the difficulty in forcing the amorphous phase from between the boundaries increases to the point that the deformation process appears to completely stop. For grain boundaries that are approximately 1 nm thick and grains that are 1 μm in dimension (typical of many grades of silicon nitride), the total strain allowable by such a mechanism is of the order of 0.1%. This strain is not sufficient to account for the fact that 1% strain is observed for many grades of silicon nitride; therefore, other mechanisms of deformation must be activated as this type of mechanism dies out.

The main difference between SN 281 and most other grades of silicon nitride (such as SN 88 and NT 154) is the suppression of cavity formation in SN 281. Without cavitation, the most likely deformation process in SN 281 is the S-P of silicon nitride and crystalline secondary phases [16, 19]. The glassy phase at the grain boundaries acts as a fast path for transport of the Si^{+4}, N^{-3} and other ions from one side of a grain to the other; the ions follow the chemical potential gradients along the grain boundaries. Because the grains can change their shape, they are no longer rigid and the types of stresses that build up in rigid-grained materials do not occur in deformable grains. With reference to Figure 13.16, the mechanism of deformation inferred is shown at the bottom of the scheme.

Cavitation can be suppressed by increasing the material's resistance to grain boundary sliding. Then, the dilatation necessary for cavity formation in silicon nitride can occur slowly enough that the stresses which normally would build up at multigrain junctions are relieved, either by S-P or by transport of the secondary phase towards the highly strained multigrain junctions. The most likely cause for the suppression of both grain boundary sliding and deformation of the secondary phase is an increase in the viscosity of the amorphous phase at the grain boundaries; this can be accomplished by changing the composition of the sintering additives used to densify silicon nitride.

13.4.3
Role of Lutetium in the Viscosity of Glass

The residual glass at grain boundaries of silicon nitride is an oxynitride glass that contains impurities and elements from sintering additives [51]. Its viscosity is controlled by the type and concentration of additives and impurities. The main difference between SN 281 and earlier grades of silicon nitride is the presence of Lu^{3+} instead of Yb^{3+} or Y^{3+} at the grain boundary. Lutetium, which has the smallest ionic radius of the lanthanides, significantly increases the viscosity of the oxynitride glassy phase compared to Yb^{3+} or Y^{3+} [52–55].

As the measurement of viscosity of the grain boundary is very difficult, bulk oxynitride glasses were used instead to quantify the role of rare-earth (RE) oxides on the viscosity of glass. A direct comparison of the viscosity changes in the bulk RE–Si–Mg–O–N glasses with various N contents revealed that the replacement of Yb^{3+} with Lu^{3+} increases the viscosity less than a factor of 20 [52–54]. However, the substitution of Lu^{3+} for Yb^{3+} also changed the amount of nitrogen in the glass, and nitrogen is much more effective than the Lu^{+} in enhancing the viscosity of the glass. Nitrogen can easily make up for approximately 2.5 orders of magnitude in viscosity. Thus, the suppression of cavitation may result from both the presence of the Lu^{+} and an increase in the concentration of N^{3-} in the glass. Lutetium, with its smaller ionic radius, "tightens" the glass network, while the incorporation of 3-coordinated N increases the network "crosslinking". Both effects increase the glass viscosity [54–56].

13.5
Conclusions

Tensile creep in silicon nitride ceramics are best described by meso-mechanical models based on the dilatation of granular solids. These models provide a rationale for the exponential dependence of creep rate on applied stress, creep asymmetry, and the role of cavitation in the creep process. Meso-mechanical models are based on the assumption that grains of silicon nitride are rigid during deformation, so that displacements between adjacent grains can only occur along the grain boundaries.

As a consequence, the network of grains must move apart, giving rise to dilatational stresses which are responsible for cavitation within the material. Cavitation occurs when the dilatational stress at a multigrain junction exceeds a critical stress for cavity nucleation. Subsequent cavity growth occurs via redistribution of the secondary phases from the cavity towards the uncavitated material. As the secondary phase flows away from the cavitated junction, the stress required for cavity nucleation relaxes, thus releasing the stressed contacts between silicon nitride grains. The rearrangement of dilatation stresses then results in a new round of cavity formation. The contribution of cavitation to tensile strain in vitreous-bonded silicon nitride exceeds 75–95%. Cavitation is the main creep mechanism in the tensile creep of silicon nitride.

Cavitation creep follows an exponential rather than power-law dependence on stress [30]. This dependence, together with the understanding that cavitation only contributes to axial tensile strain, explains creep asymmetry and the unusually high stress exponents measured in tension at high stresses.

If the dilatation stresses are not sufficient to nucleate cavities, the driving force for creep is then determined by the local gradient of stresses around individual grains. These stress gradients are responsible for S-P of the individual silicon nitride and secondary phase grains within the silicon nitride. At a given temperature and stress, creep by S-P occurs much more slowly than creep by cavitation. Creep by S-P is observed in compression for most grades of silicon nitride, and in compression and tension for SN281. Because creep by cavitation is the faster mechanism of deformation, the suppression of cavity formation is an effective way to increase the creep resistance of silicon nitride.

Increasing the viscosity of the grain boundary phase is one way to suppress cavitation, as it reduces grain boundary sliding, viscous flow and S-P. The replacement of Yb by Lu in the bulk oxynitride glass increases the viscosity of the grain boundary glass by about four orders of magnitude, partly due to a "tightening" of the glass structure by the smaller rare-earth ion and in part by an enhancement of N in the glass due to the presence of Lu. These two effects account for the more than four orders of magnitude increase of creep resistance of SN 281 compared to SN 88 or NT 154.

Acknowledgments

The contributions of W.E. Luecke of the National Institute of Standards and Technology to the development of some of the basic ideas used in this chapter, and G.G. Long and P. Jemian of Argonne National Laboratory for conducting the A-USAXS measurements, are gratefully acknowledged. The roles of the Science and Technology Agency, the Fulbright Foundation and the A. von Humboldt Foundation in the support of the Institute for Energy, JRC-EC, are appreciated. These studies were also supported by NANOSMART, Centre of Excellence, SAS, KMM-NoE EU 6FP Project and by the Slovak Grant Agency for Science, grants No. 2/0088/08 and 2/7194/27.

References

1 Glenny, E. and Taylor, T.A. (1961) *Powder Metall.*, **8**, 164–195.
2 Kossowsky, R., Miller, D.E., and Diaz, E.S. (1975) *J. Mater. Sci.*, **10**, 983–997.
3 Messier, D.R. and Murphy, M.M. (1975) Silicon nitride for structural applications – an annotated bibliography. Final Report AMMRC MS 75-1, Army Materials and Mechanics Research Center.
4 Quinn, G.D. (1982) *Ceram. Eng. Sci. Proc.*, **3** (1–2), 77–98.
5 Cannon, R.W. and Langdon, T.G. (1983) *J. Mater. Sci.*, **18**, 1–50.
6 Cannon, R.W. and Langdon, T.G. (1988) *J. Mater. Sci.*, **23**, 1–20.
7 Davies, J.R. (1998) *Heat Resistant Materials, ASM Specialty Handbook*, ASM International, USA.
8 Petzow, G. and Herrmann, M. (2002) *Struct. Bond.*, **102**, 47–167.
9 Tatsumi, T., Takehara, I., and Ichikawa, Y. (2001) in *Ceramic Materials and Components for Engines* (eds J.G. Heinrich and F. Aldinger), DKG-Wiley-VCH, Weinheim, Germany, pp. 45–50.
10 Fukudome, T., Tsuruzono, S., Karasawa, W., and Ichikawa, Y. (2002) ASME paper GT-2002-30627. American Society of Mechanical Engineers.
11 van Roode, M., Ferber, M.K., and Richerson, D.W. (2002) *Ceramic Gas Turbine Design and Test Experience: Progress in Ceramic Gas Turbine Development*, vol. 1, American Society of Mechanical Engineers. See also: http://www.eere.energy.gov/de/pdfs/microturbine_materials_technology_activities.pdf.
12 Lin, H.T., Ferber, M.K., Becher, P.F., Price, J.R., van Roode, M., Kimmel, J.M., and Jimenez, O.D. (2006) *J. Am. Ceram. Soc.*, **89**, 258–265.
13 Lin, H.T., Ferber, M.K., Westphal, W., and Macri, F. (2002) Proceedings ASME TURBO EXPO Land, Sea & Air, Paper GT2002-30629. ASME International Gas Turbine Institute.
14 Raj, R. (1993) *J. Am. Ceram. Soc.*, **76**, 2147–2174.
15 Lofaj, F. (2004) Creep behavior and mechanisms in the high-performance silicon nitride ceramics. DSc Thesis, IMR SAS, Košice, Slovakia.
16 Hynes, A. and Doremus, R. (1996) *Crit. Rev. Solid State Mater. Sci.*, **21**, 129–186.
17 Wilkinson, D.S. (1998) *J. Am. Ceram. Soc.*, **81**, 275–299.
18 Jin, Q., Wilkinson, D.S., and Weatherly, G.C. (1999) *J. Am. Ceram. Soc.*, **82**, 1492–1496.
19 Melendéz-Martínez, J.J. and Domíguez-Rodríguez, A. (2004) *Prog. Mater. Sci.*, **49**, 19–107.
20 Wiederhorn, S.M. (2000) *Z. Metallkd.*, **9**, 1053–1058.
21 Wiederhorn, S.M. and Ferber, M.K. (2001) *Curr. Opin. Solid State Mater. Sci.*, **5**, 311–316.
22 van Roode, M., Ferber, M.K., and Richerson, D.W. (2003) *Ceramic Gas Turbine Component Development and Characterization, Progress in Ceramic Gas Turbine Development, vol.* **2**, American Society of Mechanical Engineers.
23 Luecke, W.E., Wiederhorn, S.M., Hockey, B.J., Krause, R.F.., Jr., and Long, G.G. (1995) *J. Am. Ceram. Soc.*, **78**, 2085–2096.
24 Wiederhorn, S.M., Hockey, B.J., and French, J.D. (1999) *J. Eur. Ceram. Soc.*, **19**, 2273–2284.
25 Lofaj, F., Okada, A., and Kawamoto, H. (1997) *J. Am. Ceram. Soc.*, **78**, 1619–1623.
26 Lofaj, F., Okada, A., Ikeda, Y., and Kawamoto, H. (2000) *Key Eng. Mater.*, **171–174**, 747–754.
27 Lofaj, F. (2000) *Mater. Sci. Eng.*, **A279**, 61–72.
28 Fields, B.A. and Wiederhorn, S.M. (1996) *J. Am. Ceram. Soc.*, **79**, 977–986.
29 Wiederhorn, S.M., Hockey, B.J., and Chuang, T.J. (1991) Creep and creep rupture of structural ceramics, in *Toughening Mechanisms in Quasi-Brittle Materials* (ed. S.P. Shah), Kluwer Academic Publisher, The Netherlands, pp. 555–576.
30 Luecke, W.E. and Wiederhorn, S.M. (1999) *J. Am. Ceram. Soc.*, **82**, 2769–2778.
31 Wiederhorn, S.M., Krause, R.F., Jr, Lofaj, F., and Täffner, U. (2005) *Key Eng. Mater.*, **287**, 381–392.

32 Wiederhorn, S.M., Hockey, B.J., Cranmer, D.C., and Yeckley, R.Y. (1993) *J. Mater. Sci.*, **28**, 445–453.

33 Menon, M.N., Fang, H.T., Wu, D.C., Jenkins, M.G., Ferber, M.K., Moore, K.L., Hubbard, C.R., and Nolan, T.A. (1994) *J. Am. Ceram. Soc.*, **77**, 1217–1227.

34 Lofaj, F., Smith, D.T., Blessing, G.V., Luecke, W.E., and Wiederhorn, S.M. (2003) *J. Mater. Sci.*, **38**, 1403–1412.

35 Lofaj, F., Blessing, G.V., and Wiederhorn, S.M. (2003) *J. Am. Ceram. Soc.*, **86**, 817–822.

36 Lofaj, F., Usami, H., Okada, A., and Kawamoto, H. (1997) Long-term creep damage development in a self-reinforced silicon nitride, in *Engineering ceramics '96: Higher reliability through processing* (eds G.N. Babini, M. Haviar, and P. Šajgalík), NATO ASI Series, Kluwer Academic Publisher, The Netherlands, pp. 337–352.

37 Lofaj, F., Usami, H., Okada, A., and Kawamoto, H. (1999) *J. Am. Ceram. Soc.*, **82**, 1009–1019.

38 Wereszczak, A.A., Ferber, M.K., Barnes, A.S., and Kirkland, T.P. (1999) *Ceram. Eng. Sci. Proc.*, **20**, 535–544.

39 Carruthers, W.D., Kraft, E.H., Yoshida, M., Tanaka, K., Kubo, T., Terazono, H., and Tsuruzono, S. (2003) Technology development at Kyocera Corporation, in *Ceramic Gas Turbine Component Development and Characterization, Progress in Ceramic Gas Turbine Development*, vol. 2, (eds M. van Roode, M.K. Ferber, and D.W. Richerson), ASME Press, New York, pp. 125–134.

40 Lofaj, F., Wiederhorn, S.M., Long, G.G., and Jemian, P.R. (2001) *Ceram. Eng. Sci. Proc.*, **22**, 167–174.

41 Lofaj, F., Wiederhorn, S.M., Long, G.G., Hockey, B.J., Jemian, P.R., Bowder, L., Andreason, J., and Täffner, U. (2002) *J. Eur. Ceram. Soc.*, **22**, 2479–2487.

42 French, J.D. and Wiederhorn, S.M. (1996) *J. Am. Ceram. Soc.*, **79**, 550–552.

43 Yoon, K.J., Wiederhorn, S.M., and Luecke, W.E. (2000) *J. Am. Ceram. Soc.*, **83**, 2017–2022.

44 Wiederhorn, S.M., Krause, R.F., Jr, Lofaj, F., and Täffner, U. (2005) *Key Eng. Mater.*, **287**, 381–392.

45 Wiederhorn, S.M. and Krause, R.F. Jr (2009) *Ceram. Proc. Res.*, **10** (3), 269–277.

46 Jemian, P.R., Long, G.G., Lofaj, F., and Wiederhorn, S.M. (2000) Anomalous Ultra-Small-Angle X-ray Scattering from evolving microstructures during creep, in *MRS Symposium 590, Applications of Synchrotron Radiation Techniques to Materials Science V* (eds S.R. Stock S.M. Mini, and D.L. Perry), MRS, Warrendale, PA, pp. 131–136.

47 Reynolds, O. (1885) *Philos. Mag.*, **20** (127), 469–481.

48 Coble, R.L. (1963) *J. Appl. Phys.*, **34**, 1679–1682.

49 Wakai, F. (1994) *Acta Metall. Mater.*, **42**, 1163–1172.

50 Lange, F.F., Davis, B.I., and Clarke, D.R. (1980) *J. Mater. Sci.*, **15**, 601–610.

51 Gu, H., Pan, X., Cannon, R.M., and Rühle, M. (1998) *J. Am. Ceram. Soc.*, **81**, 3125–3135.

52 Becher, P., Waters, S.B., Westmoreland, C.G., and Riester, L. (2002) *J. Am. Ceram. Soc.*, **85**, 897–902.

53 Becher, P.F. and Ferber, M.K. (2004) *J. Am. Ceram. Soc.*, **87**, 1274–1279.

54 Lofaj, F., Dériano, S., LeFloch, M., Rouxel, T., and Hoffmann, M.J. (2004) *J. Non-Cryst. Solids*, **344**, 8–16.

55 Lofaj, F., Dorčáková, F., Dolekcekic, E., LeFloch, M., Rouxel, T., Hoffmann, M.J., and Hampshire, S. (2004) *Glastech. Ber. Glass Sci. Technol.*, **77C**, 273–279.

56 Lofaj, F., Dorčáková, F., and Hoffmann, M.J. (2005) *J. Mater. Sci.*, **40**, 47–52.

14
Fracture Resistance of Ceramics

Mark Hoffman

14.1
Introduction

Ceramic materials have many advantageous properties, including high corrosion and temperature resistance, special functional properties and high hardness and wear resistance. Due to the nature of the chemical bonding in ceramic materials, which usually is ionic or covalent, they have – in theory – very high strength in the order of over 3–5 GPa. However, they are generally considered poor materials for structural applications because of their low applied strength. The reason for this low strength is derived from their low fracture toughness or resistance to failure by crack propagation. A direct correlation exists between the strength of a ceramic material and its fracture resistance or fracture toughness [1] (see Figure 14.1). In this chapter, the theory behind fracture, and its application to the strength of ceramics, will initially be outlined, after which the various mechanisms employed to raise the fracture resistance of ceramics will be reviewed. This will be followed by a discussion of the significance of these mechanisms in terms of the applications of ceramics. Finally, methods for assessing the fracture resistance of various ceramic materials will be outlined.

14.2
Theory of Fracture

Fracture theory provides a link between the strength of a material, the size of defects within it, and a material property – fracture toughness. It has long been recognized that mechanical defects within a material, which result from either its intrinsic microstructure or manufacturing processes, may have a deleterious effect upon its strength. This issue was first addressed by considering stress concentration factors. Inglis developed a relationship which explained the stress at the end of an elliptical hole within a plate as a function of the shape of the ellipse and the applied stress [2].

Ceramics Science and Technology Volume 2: Properties. Edited by Ralf Riedel and I-Wei Chen

ISBN: 978-3-527-31156-9

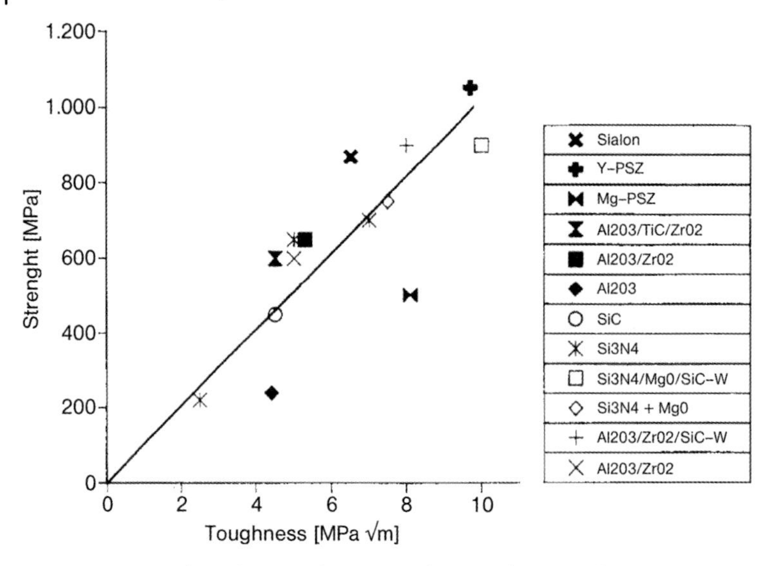

Figure 14.1 Correlation between fracture toughness and strength for ceramic materials. From Ref. [1].

Using the nomenclature shown in Figure 14.2, this relationship can be expressed as follows:

$$\sigma_C = \sigma\left(1 + \frac{2c}{b}\right) \tag{1a}$$

$$= \sigma\left[1 + 2\sqrt{\frac{c}{\varrho}}\right] \tag{1b}$$

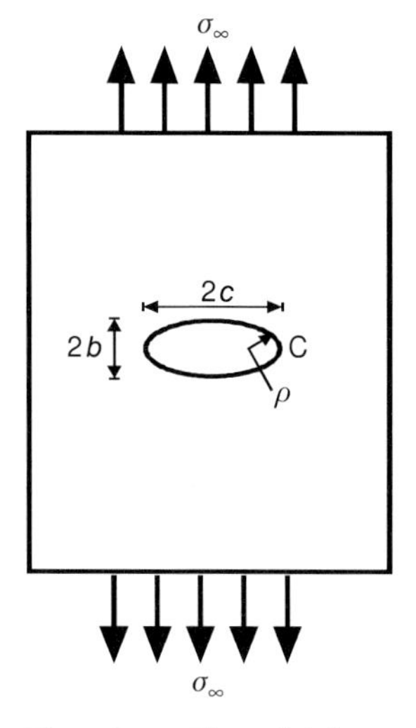

Figure 14.2 Elliptical defect, as modeled by Inglis.

where the radius of the ellipse, $\varrho = b^2/c$ for $b < c$. It can be seen that as the radius of the end of the ellipse at point C $\varrho \rightarrow 0$, equivalent to a sharp-tipped crack, then the stress at that point, $\sigma_C \rightarrow \infty$. A very thin, long, flat ellipse can be geometrically considered equivalent to the crack.

This stress concentration explanation, however, has a number of factors which are not consistent with the observation of the effect of cracks upon strength of materials. First, Eq. (1) states that the stress at point C is dependent upon the *shape* of the crack, and not its length; however, this is inconsistent with observations that long cracks are more likely to lead to premature failure than short cracks. Second, the insertion of a value ϱ consistent with the radius of a crack tip, which is believed to be in the order of a few atomic spacings, would result in failure stress values far lower than those which are observed in components containing defects. Hence, the concept of a stress concentration factor alone does not explain the deleterious effect of defects upon the strength of a material.

This issue was addressed by A.A. Griffith, whose theory is used today by engineers and scientists to determine the strength of materials containing defects [3]. This theory established an energy criterion for fracture which states that crack growth occurs when:

Energy deliverable by a system > Energy required to form an additional crack.

Indeed, this remains the crucial criterion for predicting failure in engineering, despite the fact that – interestingly – it was not used to any significant degree until more than thirty years after its development.

Griffith based his analysis on the following:

- A loaded plate contains stored elastic energy, U_0, as in Figure 14.3a.
- A crack of length $2a$ is placed in the plate. Stress is relieved around the crack in an area which is approximated as a circle for simplicity as in Figure 14.3b. This causes a reduction in the stored elastic energy, U_E.
- Energy is associated with forming two new surfaces (breaking atomic bonds, etc.), U_S.

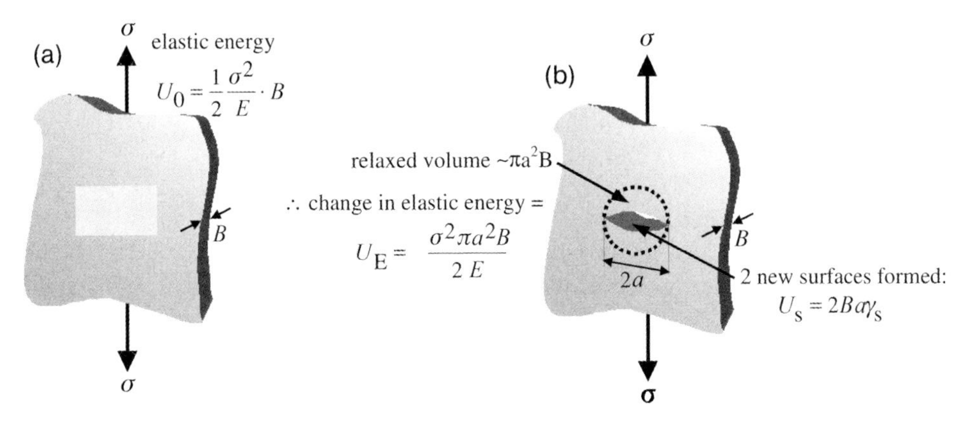

Figure 14.3 (a) Stressed plate with (b) through-thickness crack in a plate as used in the Griffith analysis.

The total potential energy of the system containing a crack is therefore:

$$U = \text{surface energy} - \text{elastic energy relieved (lost)} + \text{initial elastic energy} \quad (2a)$$

$$= U_s - U_E + U_0 \quad (2b)$$

$$= 2aB\gamma_s - \frac{\sigma^2 \pi a^2 B}{2E} + \frac{\sigma^2 B}{2E} \quad (2c)$$

where:

γ_s = specific surface energy
σ = applied stress
E' = Young's modulus = E (plane stress) or = $E/(1-\nu^2)$ (plane strain)

At this point, it should be noted that in his analysis, Griffith applied the stress analysis of Inglis, and found that the relaxation around the crack is underestimated by a factor of 2 by assuming a circle, as in the present case. Noting this and defining crack surface area $A = aB$, Eq. (2c) becomes:

$$U = 2A\gamma_s - \frac{\sigma^2 \pi a A}{E} + \frac{\sigma^2 B}{2E} \quad (3)$$

As shown in Figure 14.4, the system will be in equilibrium when the change in potential energy with crack growth (or fracture surface formation) is zero:

$$\frac{dU}{d2A} = 0 \quad (4)$$

The stress at this point will be the fracture stress, σ_f, because as soon as the stress increases then the system becomes unstable and fracture occurs. Differentiating Eq. (3) as per Eq. (4) gives:

$$\frac{dU}{dA} = 2\gamma_s - \frac{\sigma_f^2 \pi a}{2E} = 0 \quad (5)$$

or

$$\gamma_s = \frac{\sigma_f^2 \pi a}{2E} \quad (6)$$

which, upon rearranging, gives the fracture stress:

$$\sigma_f = \sqrt{\frac{2E\gamma_s}{\pi a}} \quad (7)$$

Griffith defined a term, the *mechanical energy release rate*, which is the amount of elastic energy released as a crack grows. In the case of the plate in Figure 14.3b, this would be the change in the energy in the "relaxed" zone as the crack grows,

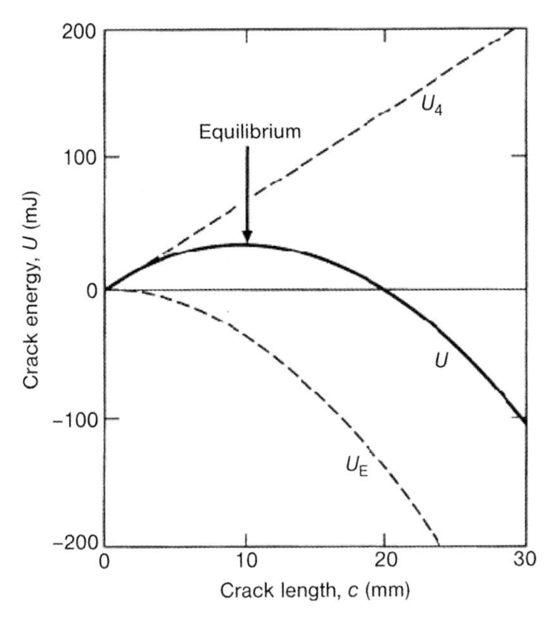

Figure 14.4 Energy associated with fracture of a crack uniformly loaded in far-field tension in glass. From Ref. [4].

that is:

$$G = \frac{d(U_E - U_0)}{dA} = \frac{\sigma^2 \pi a}{E'} \tag{8}$$

where G is energy per unit area, and therefore has units J m^{-2}.

The condition for crack growth [from Eqs (1b) and (3)] is:

$$\frac{d}{da}(U_s - U_E + U_0) = 0 \tag{9a}$$

or

$$\frac{d}{da}(U_E - U_0) = \frac{dU_s}{da} \tag{9b}$$

where

$\frac{d}{da}(U_E - U_0) = G$ is the mechanical energy release rate per unit area (driving force)

$\frac{dU_s}{da} = R$ is the intrinsic work rate per unit area (resistance)

By definition:

$$U_S = 2a\gamma_S \tag{10a}$$

$$\frac{dU_S}{da} = R = 2\gamma_S \tag{10b}$$

Returning to the system in Figure 14.3 of a through thickness crack in a large plate, Eq. (8) gives $G = \frac{\sigma^2 \pi a}{E'}$ and Eq. (10b) gives $R = 2\gamma_s$. Equation (9a) states that the critical condition for fracture is:

$$G = R \tag{11}$$

Therefore

$$\frac{\sigma^2 \pi a}{E'} = 2\gamma_s \tag{12}$$

which, on rearranging, gives the fracture stress, σ_f, defined in Eq. (7) as $\sigma_f = \sqrt{\frac{2E\gamma_s}{\pi a}}$. For this particular through-thickness crack configuration:

$$K = \sigma\sqrt{\pi a} \tag{13}$$

which when substituted in Eq. (8) gives:

$$G = \frac{K^2}{E'} \tag{14}$$

It can, therefore, be seen that the fracture strength of a material is proportional to the inverse of the square root of the size of a defect, and dependent upon the material properties of the stiffness of the material and the energy required to form new surfaces.

Subsequent to the analysis of Griffith, a solution was developed which defined the stresses at a crack tip as a function of an applied far-field stress, σ. A full solution will not be derived here, but can be found elsewhere [5, 6]. Using the nomenclature shown in Figure 14.5, the stresses may be defined in cartesian and polar coordinates as follows:

$$\sigma_{xx} = \frac{K_I}{\sqrt{2\pi r}} \cos\frac{\theta}{2}\left(1 - \sin\frac{\theta}{2}\sin\frac{3\theta}{2}\right) \tag{15a}$$

$$\sigma_{yy} = \frac{K_I}{\sqrt{2\pi r}} \cos\frac{\theta}{2}\left(1 + \sin\frac{\theta}{2}\sin\frac{3\theta}{2}\right) \tag{15b}$$

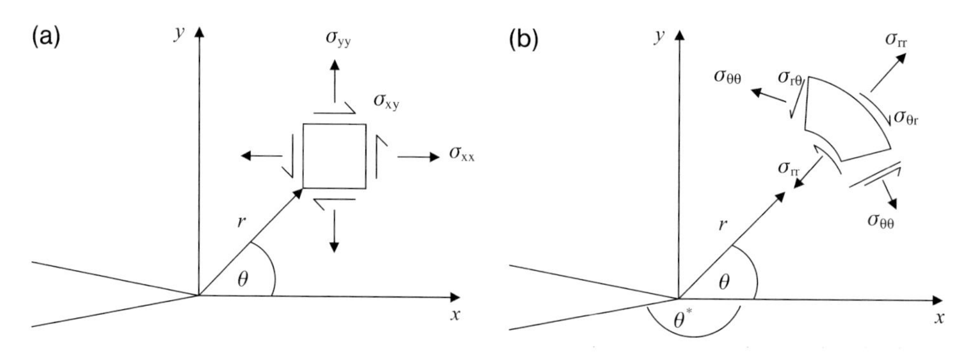

Figure 14.5 Stress element near the crack-tip in terms of (a) cartesian and (b) polar coordinates.

$$\sigma_{xy} = \frac{K_I}{\sqrt{2\pi r}} \sin\frac{\theta}{2} \cos\frac{\theta}{2} \sin\frac{3\theta}{2} \qquad (15c)$$

$$\sigma_{rr} = \frac{K_I}{\sqrt{2\pi r}} \left[\frac{5}{4}\cos\left(\frac{\theta}{2}\right) - \frac{1}{4}\cos\left(\frac{3\theta}{2}\right)\right] \qquad (16a)$$

$$\sigma_{\theta\theta} = \frac{K_I}{\sqrt{2\pi r}} \left[\frac{3}{4}\cos\left(\frac{\theta}{2}\right) + \frac{1}{4}\cos\left(\frac{3\theta}{2}\right)\right] \qquad (16b)$$

$$\sigma_{r\theta} = \frac{K_I}{\sqrt{2\pi r}} \left[\frac{1}{4}\cos\left(\frac{\theta}{2}\right) + \frac{1}{4}\cos\left(\frac{3\theta}{2}\right)\right] \qquad (16c)$$

Similarly, expressions for displacements in the *x*- and *y*-directions are defined, respectively, as:

$$u = 2(1+\nu)\frac{K_I}{E}\sqrt{\frac{r}{2\pi}}\cos\frac{\theta}{2}\left[1-2\nu+\sin^2\frac{\theta}{2}\right] \qquad (17a)$$

$$v = 2(1+\nu)\frac{K_I}{E}\sqrt{\frac{r}{2\pi}}\sin\frac{\theta}{2}\left[2-2\nu-\cos^2\frac{\theta}{2}\right] \qquad (17b)$$

K_I is known as the stress intensity factor and is, effectively, a measure of the magnitude of the singularity of the expression "at the crack tip" and may be defined as:

$$K_I = \sigma\sqrt{\pi a} \qquad (18)$$

By combining Eqs (7) and (17), a relationship can be developed between the stress intensity factor and the strain energy release rate as defined by Griffith:

$$G = \frac{K_I^2}{E'} \qquad (19)$$

Griffith stated that fracture will occur when the strain energy release rate G becomes equal to the fracture resistance of the material R. This expression has been refined for applications in engineering, namely, that fracture will occur when the stress intensity factor K_I becomes equal to a critical value K_c, the fracture toughness of the material.

At this stage, it should be noted that K_I refers to a crack loaded in uniform tension, or Mode I. Other crack loading modes exists as shown in Figure 14.6, and Eq. (19) may be more completely defined as:

$$G = G_{\mathrm{I}} + G_{\mathrm{II}} + G_{\mathrm{III}} = \frac{K_{\mathrm{I}}^2}{E'} + \frac{K_{\mathrm{II}}^2}{E'} + \frac{K_{\mathrm{III}}^2(1+\nu)}{E} \qquad (20)$$

and analogous expressions to Eqs (15) and (16) exist for stress in the crack-tip region.

The fracture toughness K_c, therefore, defines the fracture resistance of a material and, by Eq. (8), provides a material property which determines the strength of a component as a function of the size of a defect.

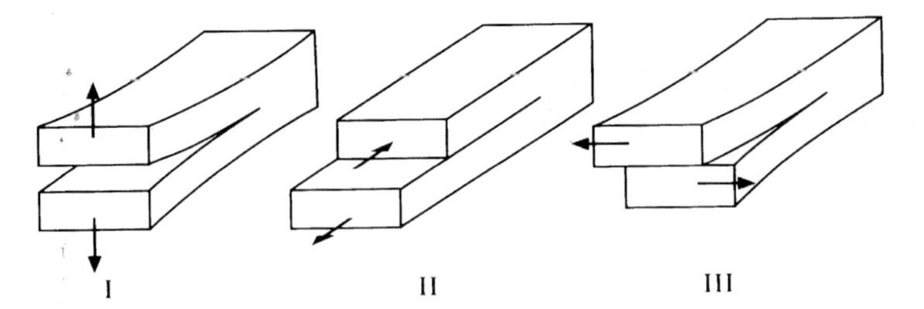

Figure 14.6 Crack loading modes. Mode I = tension; Mode II = shear; Mode III = torsion.

Figure 14.7 outlines, schematically, the two failure criteria for structural materials, general yielding and fast fracture. *General yielding* occurs when the applied stress is greater than the yield strength of a material, and is defect size independent. The *fast fracture* strength reduces with the inverse square root of the size of the defect.

Ceramics occupy a unique part of this curve. Compared to other structural materials, ceramics have very high yield strength and low fracture toughness, which means that they invariably operate in the regime where strength is determined by size of the defect populations within the component. This has a number of implications. First, defects have a size distribution within the material, which means that each component made from a ceramic will have a different strength from another. This leads to low levels of structural reliability and presents difficulties for designers. Furthermore, the strength of ceramic components becomes highly sensitive to factors such as the quality of surface finish and the quality of processing, which determines whether intrinsic defects such as pores and micro-cracks are present in the component.

14.2.1 Stress Concentration Factors

The theories outlined above are based upon a through-thickness crack in a large plate (see Figure 14.3). Defects in ceramics, however, are typically in the form of surface

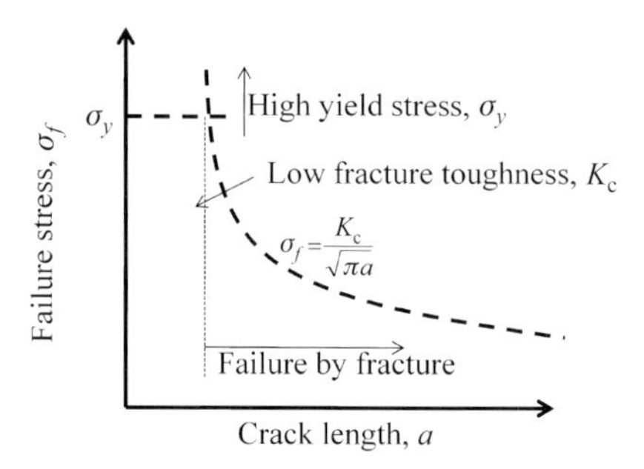

Figure 14.7 Loci of failure stress as a function of crack length for failure by general yielding and fracture.

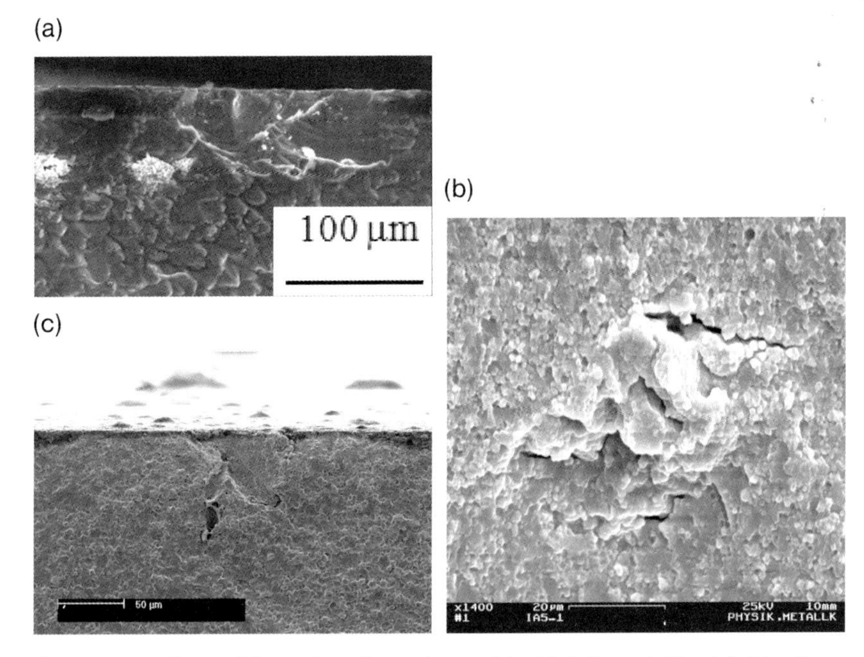

Figure 14.8 Defects at failure origin of ceramic materials. (a) Surface grinding defect in silicon (from O. Borerro-Lopez, unpublished results); (b) Internal pore in polycrystalline alumina; (c) Surface pore on alumina–silicon carbide nanocomposites.

cracks or pores or internal cracks within the component. Examples of these are shown in Figure 14.8. To account for these scenarios, adaptations of Eq. (18) have been developed using various analytical and numerical techniques; some common examples of these are shown in Table 14.1. The significance of choosing the appropriate expression for stress intensity factor can be seen in a study by Stech *et al.* [7]. Figure 14.9 shows a surface elliptical surface defect in a ceramic which was progressively loaded, adding dye at each step. From the relevant expression in Table 14.1 it can be seen that, for a flat ellipse, the stress concentration factor is larger at the base than it is at the surface; hence, the defect will propagate more at its base than at the surface, until such time as it becomes a perfect semicircle, when it will be stable (as seen in the figure).

The above examples outline the complexity of predicting the strength of ceramic materials, as the largest defect – or the one with the highest stress intensity factor – is rarely known *a priori*. However, it can be seen that the fracture toughness of a material remains a common factor in determining when fast fracture will occur.

14.2.2 Crack Closure Concept and Superposition

Stress intensity factors, provided that they are all of the same mode, may be either added or superimposed. One commonly used method to raise the fracture toughness

Table 14.1 Expressions for stress intensity factor for common ceramic defect types.

Internal penny-shaped defect:

$K_{\mathrm{I}} = \frac{2}{\pi}\,\sigma\sqrt{\pi a}$

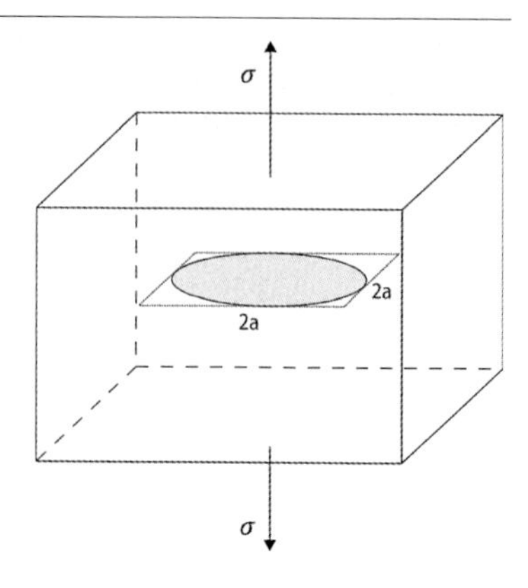

Half-elliptical surface crack:

$$K_{I(\phi=\pi/2)} = 1.12\frac{\sigma\sqrt{\pi a}}{\Phi} \quad K_{I(\phi=0)} = 1.12\frac{\sigma\sqrt{\pi\frac{a^2}{c}}}{\Phi}$$

where

$$\Phi = \int_0^{\pi/2}\left[1-\frac{c^2-a^2}{c^2}\sin^2\phi\right]^{1/2} d\phi$$

$$\approx \frac{3\pi}{8}-\frac{\pi}{8}\frac{a^2}{c^2}$$

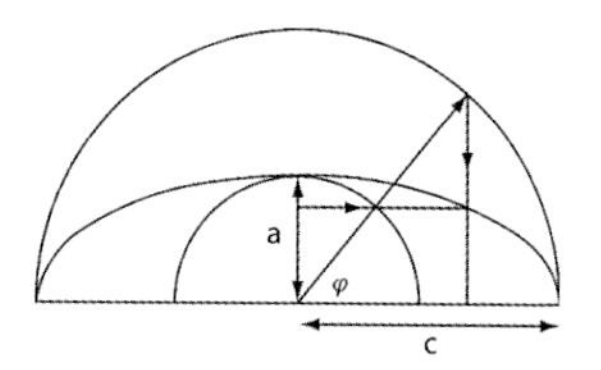

Circumferential crack around a spherical pore[a]:

$$\frac{K_{\mathrm{I}}}{\sigma\sqrt{\pi c}} = [n-k(\tan^{-1}c/R)^m]$$

$$\left\{\frac{1}{2(7-5\nu)}\left[(4-5\nu)\left(\frac{1}{c/R+1}\right)^3+9\left(\frac{1}{c/R+1}\right)^5\right]+1\right\}$$

where $n=1.12$, $m=2.748$ and $k=0.101$

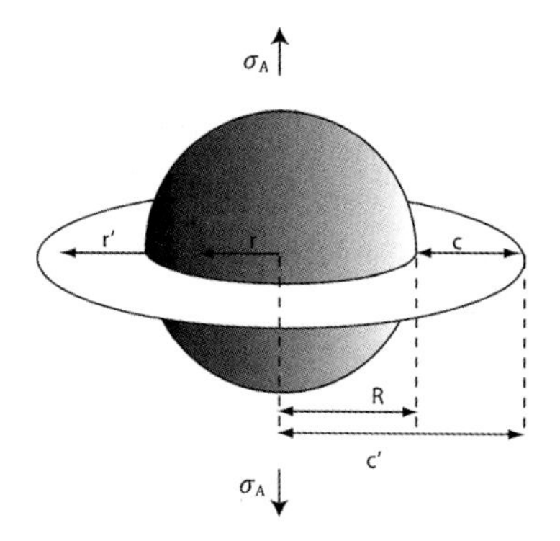

Table 14.1 (*Continued*)

Half-penny crack of radius a at the edge of a spherical pore of radius r with an applied stress σ perpendicular to crack direction[b)]:

$K = 2\sigma\sqrt{\pi a}[1 + 0.3(0.21 + a/r)^{-1}]$

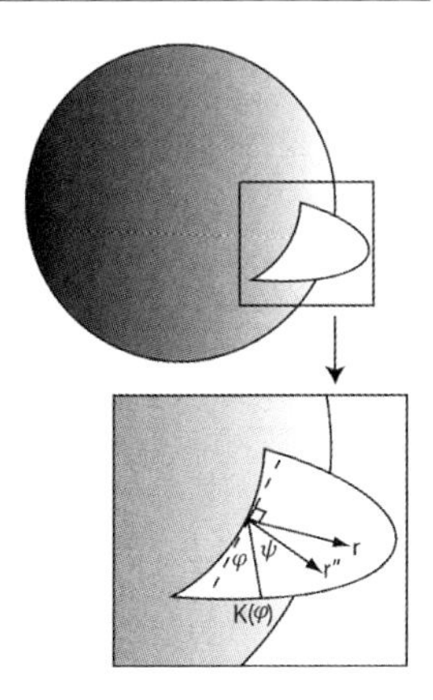

a) F. Baratta, *J. Am. Ceram. Soc.*, **61**, 490–493 (1978)
b) A.G. Evans *et al.*, *J. Am. Ceram. Soc.* **62** (1–2), 101–106 (1979)

of a ceramic is to develop microstructural mechanisms that apply a force or stress on the crack flanks, and which opposes the stress that is opening the crack, as shown schematically in Figure 14.10. The stress intensity factors associated with the crack closure force are given as:

$$K_A = \frac{P}{\sqrt{\pi a}}\sqrt{\frac{a+x}{a-x}} \quad \text{and} \quad K_B = \frac{P}{\sqrt{\pi a}}\sqrt{\frac{a-x}{a+x}} \tag{21}$$

for the point load P in Figure 14.10 at crack tips A and B. Should this point load be spread across the crack flanks a closure stress p results, giving:

$$K_{A,B} = \frac{p}{\sqrt{\pi a}}\int_n^m \left[\sqrt{\frac{a+x}{a-x}} + \sqrt{\frac{a-x}{a+x}}\right] dx \tag{22}$$

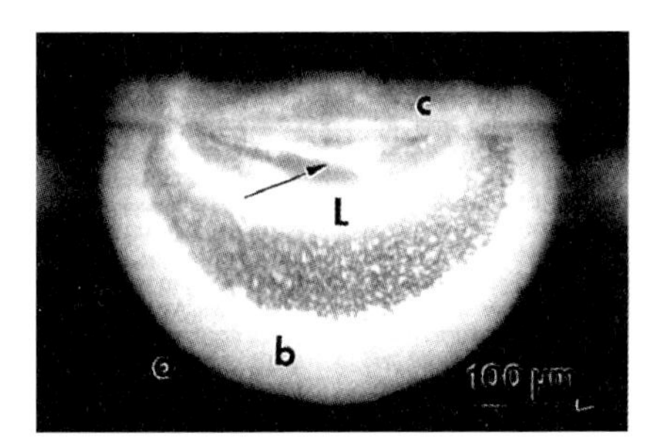

Figure 14.9 Fracture surface showing a progressively loaded elliptical defect in which dye penetrant was applied at each loading step, highlighting the greater growth of the crack in the vertical versus horizontal direction, consistent with stress intensity factor calculations. From Ref. [7].

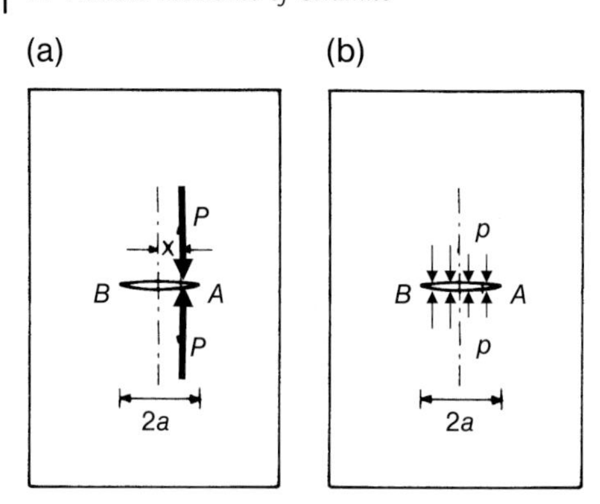

Figure 14.10 (a) Point load closure force, P, on crack flanks; (b) Closure pressure, p, on crack flanks.

where m and n represent the region of the crack surface over which the closure stress is applied, measured from the axis of symmetry,

Both of these represent a Mode I load, and hence can be added to Mode I stress intensity factors arising from other uniaxial stress perpendicular to the crack. The significance of this will subsequently become evident.

14.3 Toughened Ceramics

By rearranging Griffith's expression [Eq. (7)], $\sigma_f = \frac{K_c}{\sqrt{\pi a_c}}$, it can be shown that the structural deficiency of ceramics – that is, their low strength – may be redressed by either reducing the size of the largest defect (a_c) or raising the fracture toughness (K_c). The former method is generally effected by reducing the size of the microstructure, sintering under pressure, or providing high-quality surface finishes to components. Unfortunately, however, all of these methods usually involve high costs when implemented into mass production. The alternative approach is to develop microstructures which lead to a rise in the resistance to crack propagation or fracture resistance of the material.

Considerable research effort has been devoted over the past 25–30 years towards developing high-toughness ceramics, and these may be categorized into two approaches: (i) *bridged crack wake* methods; and (ii) *frontal wake* methods, as summarized in Figure 14.11. Bridged interface methods involve the development of a mechanism to induce forces behind the crack tip so as to hinder opening of the crack and, hence, to reduce the stress intensity factor at the crack tip. Frontal wake methods involve the introduction of a mechanism whereby energy is absorbed in a zone around the crack tip such that crack propagation is hindered. The implication from both methods is that the energy to cause the crack to propagate is not

Bridged wake methods

(a)

(b)

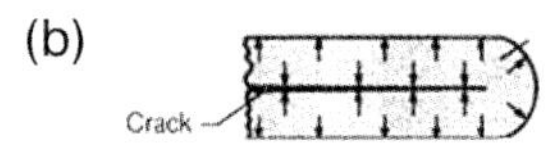

(c)

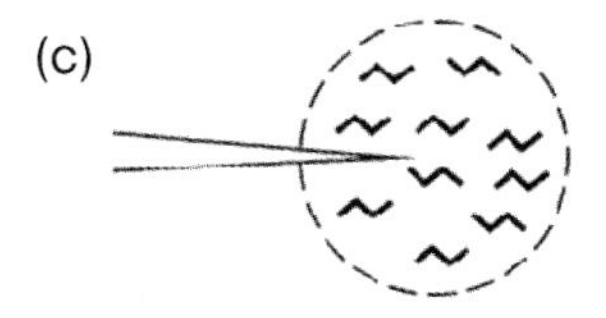

Figure 14.11 Microstructural design methods to increase fracture toughness in ceramics in the crack wake. (a) Grain bridging; (b) Crack-wake dilatants zones, (c) Microcracking. From M. Hoffman, PhD Thesis.

simply the energy of new surface formation γ_s, as outlined in Eq. (7), but also contains other components, which in many cases are significantly larger in magnitude than γ_s.

14.3.1 Bridged Interface Methods for Increasing Fracture Resistance

14.3.1.1 Grain Bridging

The principle here is to design a material in which the fracture resistance along the grain boundaries is less than that for transgranular fracture. This induces a meandering crack path which requires the grains to slide out of a gap as the crack opens (Figure 14.11a). Frictional forces between the grains then induce a closure stress on the crack flanks shielding the crack tip. Common examples of this phenomenon include polycrystalline alumina, silicon carbide and β-silicon nitride, whereby elongated grains induce closure stresses (Figure 14.12). The resistive stress exerted by the grains may be expressed as follows [10]:

$$p(\mathrm{v}) = p_M\left(1-\frac{\mathrm{v}}{\mathrm{v}*}\right) \tag{23}$$

where 2v is the crack opening, p_M is the maximum resistive stress which occurs just before the grain commences pull out, and $2\mathrm{v}^*$ is the crack opening at which the grain bridges disengage. These terms may furthermore be expressed as follows:

$$2\mathrm{v}* = \varepsilon_L l \tag{24a}$$

$$p_M = 4\varepsilon_B\mu\sigma_R(1-l^2/2d^2) \tag{24b}$$

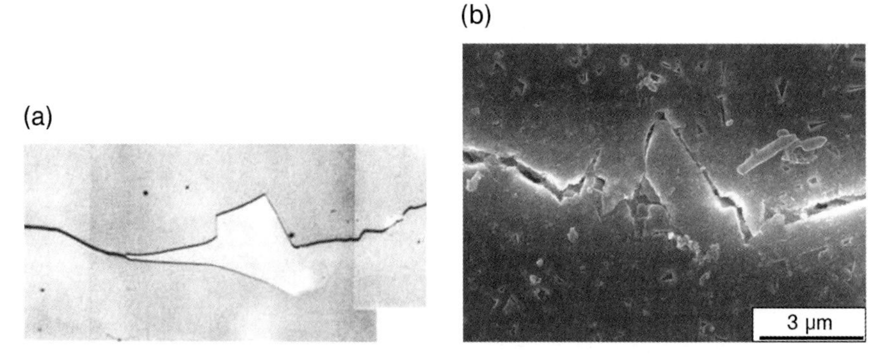

Figure 14.12 Examples of crack bridging by grains in (a) alumina (from Ref. [8]) and (b) sialon (from Ref. [9]).

where l is the grain size, ε_L is the grain bridge rupture strain, μ is the coefficient of friction, σ_R is the internal stress on the grains addressed momentarily, and d is the spacing of the bridging grains.

A number of points should be made regarding the equations above. First, the compressive stress resisting the grain pullout, σ_R, is influenced in alumina by thermal expansion anisotropy within alumina grains, resulting in a compressive stress on grain boundaries following sintering. It can also be seen that the length of the grain bridging zone [Eq. (24a)] and gain pullout stress [Eq. (24b)] increase with grain size, implying that larger grain size ceramics have higher toughness. It should also be noted that the same theory applies to toughening by a second bridging phase (this is discussed subsequently). First, however, the pertinent concept of crack growth resistance toughening will be introduced.

14.3.1.2 Crack Growth Resistance Toughening

Toughened structural ceramics, such as those outlined above, demonstrate a fracture behavior known as crack growth resistance (or *R*-curve) behavior, as demonstrated schematically in Figure 14.13. The process (note also Figure 14.11a) occurs as follows. Initially, a pre-existing defect of length a_0 does not have any bridging grains. Then, as the crack propagates the number of grains behind the crack tip (which bridge the crack) progressively increases, leading to an increase in the fracture resistance $R(= K_R^2/E)$ required to propagate the crack. At some point, the crack will propagate to a length such that the crack opening becomes larger than the length of the bridges ($2v^*$) at the point furthest from the crack tip. At this point the extent of crack tip shielding exerted by the grains reaches a steady state, and the crack growth resistance curve reaches a plateau.

The crack growth resistance, as a function of crack extension, may be expressed as follows:

$$K_R(Aa) = K_0 + K_s(Aa) \tag{25}$$

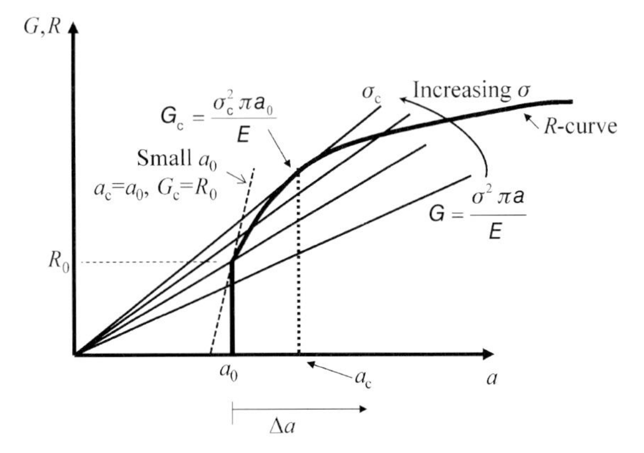

Figure 14.13 Mechanics of fracture of a material demonstrating crack-growth resistance behavior. The thick solid line represents the material's crack growth resistance, *R*, or *R*-curve from a crack of initial length a_0. The finer solid lines represent the strain energy release rate, *G*, as a function of crack length, *a*, for different values of applied stress, σ.

where K_0 is known as the intrinsic fracture toughness of the material, and is usually the stress intensity factor at which the crack will commence propagating. K_s is the crack tip shielding term; for toughening by crack bridging, this is defined as:

$$K_s = \sqrt{\frac{2}{\pi}\int_0^{\Delta a}\frac{p(X)}{\sqrt{X}}dX} \tag{26}$$

where *X* is the distance from the crack tip. *p*(*X*) is obtained by combining Eqs (23) and (17b), at $\theta = \pi$ where r becomes *X*, which is modified to account to account for bridging stresses[11]:

$$\mathrm{v}(X) = \frac{K_R}{E'}\sqrt{\frac{8x}{\pi} + \frac{2}{\pi E'}\int_0^l p(X')\ln\left|\frac{\sqrt{X'}+\sqrt{X}}{\sqrt{X'}-\sqrt{X}}\right| dX'} \tag{27}$$

Solving these relationships, however, is challenging and requires an iterative process.

Nevertheless, the overall implication is that materials which possess a fine microstructure and relatively short crack bridges have a short *R*-curve which rises to a plateau relatively quickly, whereas materials with a large grain size and, hence, large crack bridges, rise to reach a higher level of plateau toughness. This is exemplified in Figure 14.14, which shows the *R*-curve behavior of alumina of varying grain size.

14.3.1.3 Crack Bridging By a Second Phase

Ceramic designers have also incorporated a second phase into ceramic microstructures to provide crack bridging behind the crack tip. As seen in the schematic

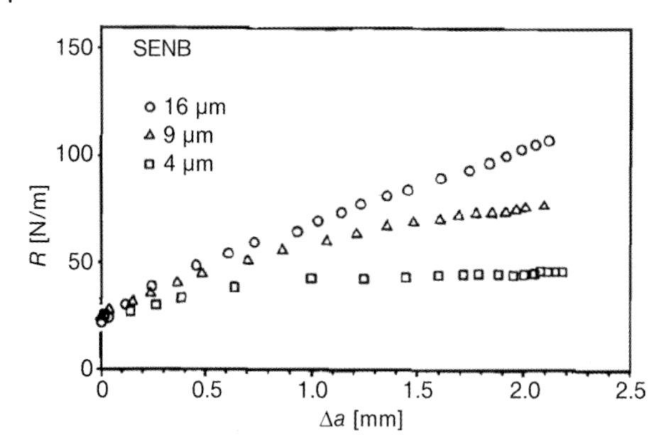

Figure 14.14 Crack growth resistance behavior as a function of grain size (4, 9, and 16 μm) for alumina using a single-edge notched beam (SENB) sample. From Ref. [12].

of Figure 14.15 this second phase can take the form of either a ductile or whisker phase. The greatest challenge in developing these microstructures is associated with their processing; notably, in the case of a whisker phase difficulties often arise with sinterability. Successful examples include the reinforcement of alumina with a silicon carbide fibers, an example of which is seen in Figure 14.15a. Here, the fibers break at a point which is not on the fracture plane; this then results in frictional fiber pull-out, with large crack openings required to completely disengage the fibers.

Ductile phase reinforcement is usually undertaken through a two-stage manufacturing process. One common method for this is the infiltration of a ductile metallic phase following the sintering of a preform with interconnecting porosity. Figure 14.15b shows a crack propagating through an alumina/copper system where the metal phase is bridging the crack behind the tip; usually, in the case of these materials, infiltration must be pressure-assisted [14]. Reactive processing has also

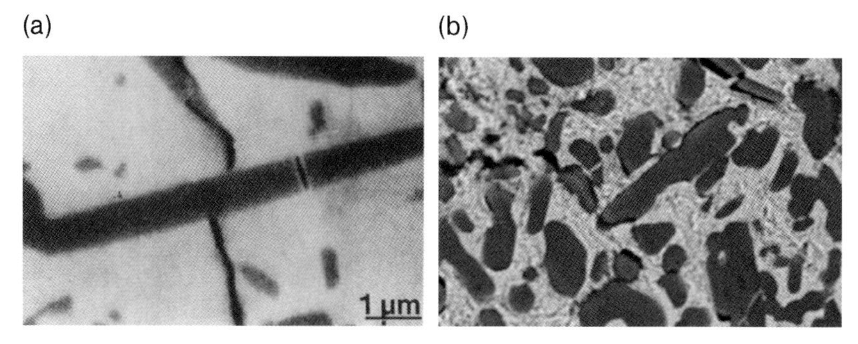

Figure 14.15 Crack propagation in (a) alumina and an alumina/SiC composite, showing bridging by a SiC fiber (from Ref. [13]) and (b) alumina–copper composites showing bridging of crack by metal phase (from O. Lott, unpublished results).

been successfully applied to these systems, but difficulties may arise due to large numbers of defects which occur within the material, thus reducing its strength [15].

14.3.1.4 **Phase Transformation or Dilatant Zone Toughening**

Phase transformation or dilatant zone toughening occurs in a number of zirconia-based ceramics, and utilizes a stress-induced phase transformation to increase toughness by an *R*-curve, as shown in Figure 14.11b. A metastable tetragonal phase is created through the judicious use of dopants, such as magnesia, ceria, or yttria. Upon the application of mechanical stress, the tetragonal phase transforms into a monoclinic phase with an approximate 4% dilation. In the region of a crack, the application of a stress intensity factor leads to high stress at the crack tip and a corresponding phase-transformation zone. As the crack then propagates, this dilated zone moves into the crack wake, inducing compressive stresses on the crack flanks and, consequently, crack growth resistance behavior [16].

The stability of the phase transformation depends on various factors, including the grain size; indeed, an increase in grain size has been found to facilitate phase transformation, leading to a larger zone [16]. A larger zone has consequently been found to lead to a greater extent of phase transformation toughening, as shown in Figure 14.16. Furthermore, the use of Raman spectroscopy has enabled mapping

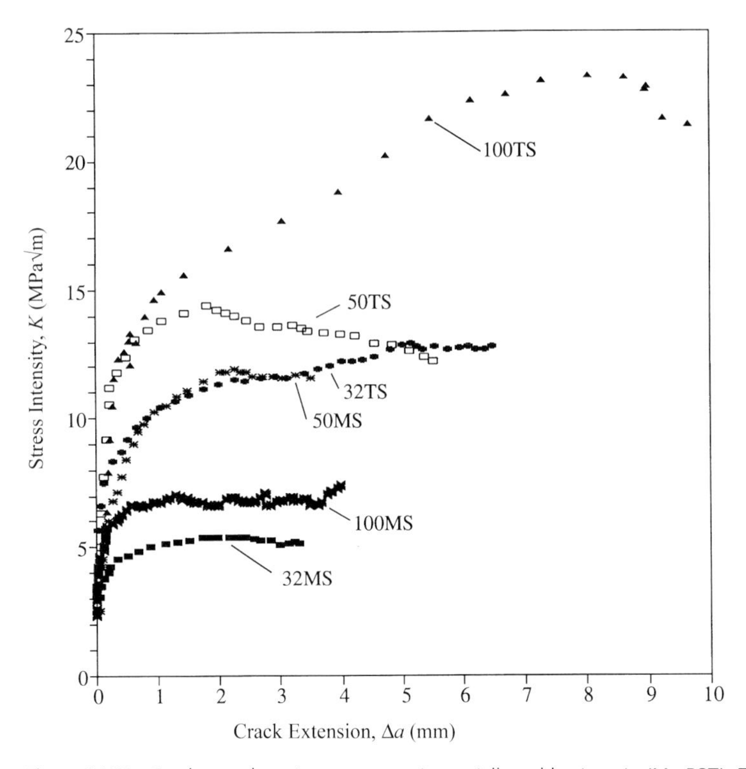

Figure 14.16 Crack growth resistance curves in partially stable zirconia (Mg-PSZ). The first two digits of the nomenclature show the average grain size, in microns. From Ref. [16].

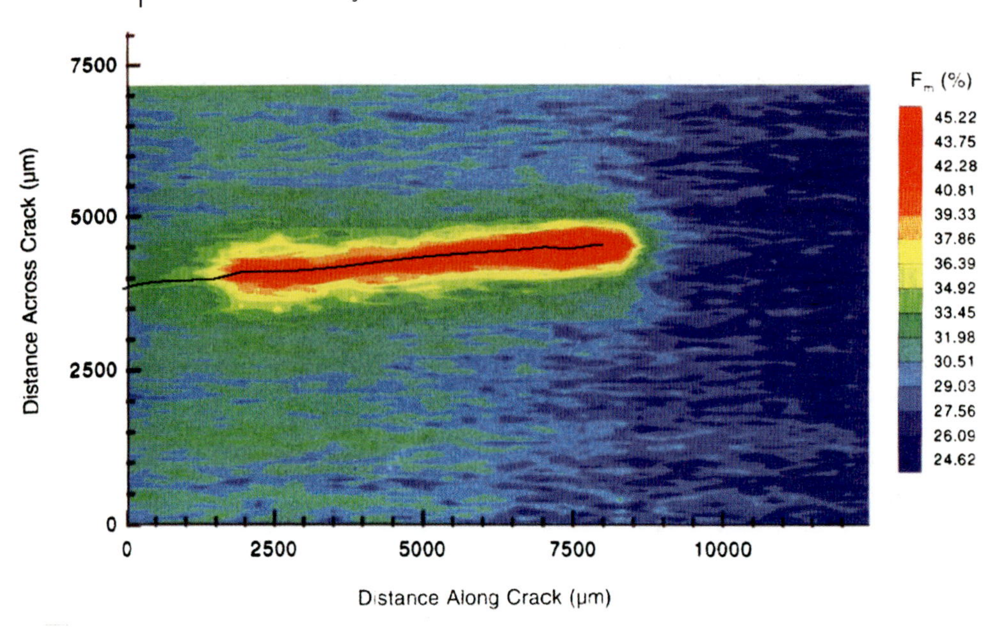

Figure 14.17 Monoclinic phase content as determined using Raman spectroscopy showing the phase transformation zone in the region of a crack in partially stabilized zirconia. From Ref. [16].

of the transformed phases around a crack, as shown in Figure 14.17. Interestingly, this phase transformation is reversible through annealing. Crack growth resistance behavior can be predicted as follows; the crack tip shielding may be expressed as:

$$K_s = \frac{E}{3\sqrt{2\pi}(1-\nu)} \int\int_A \varepsilon^T V R^{-3/2} \cos\left(\frac{3\theta}{2}\right) dA \tag{28}$$

where ε^T is the dilation strain, ν is Poisson's ratio, R and θ are polar coordinates around the crack tip, A is the area of the transformation zone, and V is the volume fraction of transformable material (metastable monoclinic phase). Assuming that phase transformation follows a hydrostatic stress profile, the boundary of the transformed zone at the crack tip may be expressed as:

$$R(\theta) = \frac{2}{9\pi}(1+\nu)^2 \left[\frac{K_{tip}}{\sigma_m^c}\right]^2 \cos^2\left(\frac{\theta}{2}\right) \tag{29}$$

where K_{tip} is the stress intensity factor at the crack tip and σ_m^c is the critical stress for phase transformation. Consequently, the width of the transformed zone at the crack flanks may be expressed as:

$$H = \frac{\sqrt{3}(1+\nu)^2}{12\pi} \cdot \left(\frac{K_{tip}}{\sigma_m^c}\right)^2 \tag{30}$$

Hence, the extent of phase transformation toughening and overall toughness is a function of the square root of the width of the transformed zone along the crack flanks.

In room-temperature applications, phase transformation-toughened zirconia ceramics are some of the toughest structural ceramics available. However, one drawback of these materials is that the phase-transformation process ceases to occur as the temperature rises above a few hundred degrees.

14.3.1.5 **Ferroelastic Toughening**

Ferroelectric ceramics, such as lead zirconate titanate (PZT) exhibit anisotropic crack growth resistance behavior in a process analogous to that demonstrated by phase transformation-toughened ceramics. Piezoelectric ceramics (a subset of ferroelectric ceramics) contain a perovskite structure in which the application of a mechanical stress results in the reorientation of a noncubic crystal. In the case of a crack tip, the application of a mechanical stress results in the orientation of domains, so as to exert a strain perpendicular to the crack plane, as shown schematically in Figure 14.18 [17]. Although, as the crack propagates, the majority of the reorientated domains partially return to their original orientation, some remain permanently reorientated and this leads to a compressive stress upon the crack flanks, and subsequent crack growth-resistance behavior [17].

It should be noted, however, that piezoelectric ceramics are often subjected to a process known as "poling" prior to application. This involves the application of a large electric field so as to cause a permanent orientation of the domains. Whether the material then subsequently exhibits crack growth resistance behavior depends upon the direction of the crack plane relative to the domain orientation (Figure 14.19) [18]. Should the domains be orientated perpendicular to the crack plane prior to crack propagation, then very little toughening will occur as the localized strain is unchanged. However, should they be orientated parallel to the crack plane and then become orientated perpendicular to it, then there will be a significant enhancement in toughness.

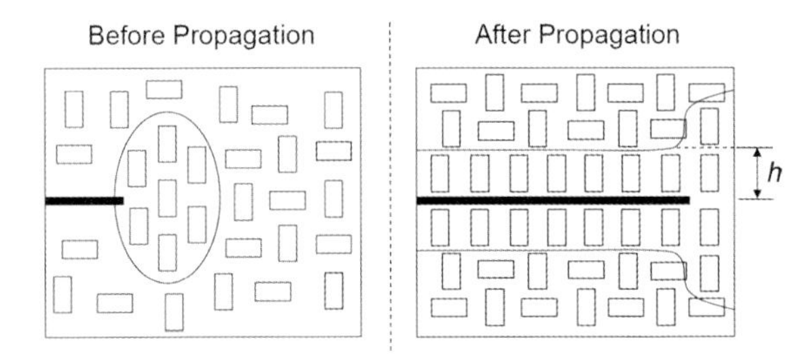

Figure 14.18 Schematic representation of the orientation of crystal domains in lead zirconate titanate (PZT) at a crack tip, following the application of stress perpendicular to the crack direction before propagation, and the wake zone of reorientated domains at the crack flanks after propagation resulting in a wake zone of width $2h$. From Ref. [17].

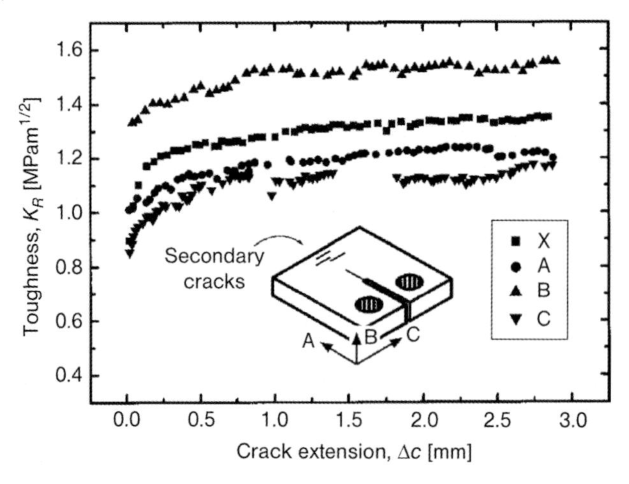

Figure 14.19 Crack growth resistance as a function of poling direction in PZT. From Ref. [18].

14.3.2 Toughening by Crack Tip Process Zone Effects

Figure 14.11g shows the development of micro-crack zones ahead of a crack tip following the application of stress. The principle behind this toughening mechanism is that the formation of micro-cracks requires the input of energy, and hence leads to an increase in toughness. Furthermore, micro-cracking will reduce the modulus of the material in the micro-crack zone, which may furthermore give an apparent increase in toughness due to the larger strains which may be accommodated [19]. However, it should also be noted that the formation of micro-cracks ahead of the crack tip may facilitate propagation of the crack, providing an easier crack path [20]. Systems which use this type of toughening included alumina containing zirconia particles, which may cause micro-cracking upon phase transformation, or leading to large zones around the crack [21], as seen in Figure 14.20.

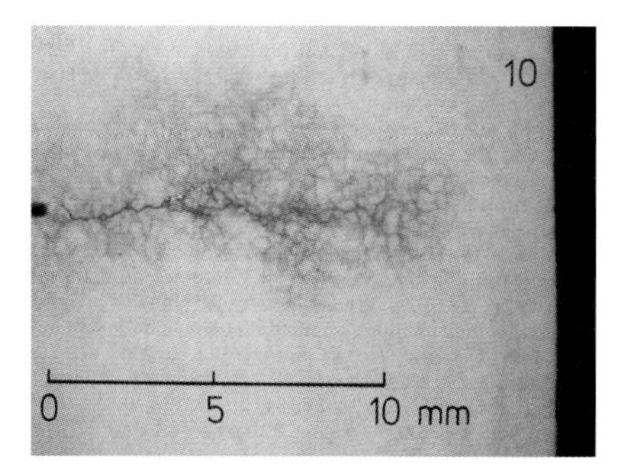

Figure 14.20 Microcrack process zone around a crack in an alumina–zirconia duplex ceramic, where the stress-induced phase transformation of zirconia particles induces microcracking. From Ref. [21].

14.4
Influence of Crack Growth Resistance Curve Upon Failure by Fracture

The presence of a crack growth resistance curve leads to a increasing fracture toughness with crack extension. The question then arises as to the appropriate value of fracture toughness, K_c, to be used when applying Eq. (13) to determine the fracture strength. Figure 14.13 shows a schematic crack growth resistance curve plotted as fracture resistance, R, versus crack extension, Δa. Also shown are lines for the strain energy release rate, G, as determined via Eq. (8), for various values of applied stress. It can be seen that as the stress increases, the crack will propagate and the effective toughness will consequently increase. However, when the strain energy release rate curve makes a tangent with the crack growth resistance curve, the failure strength is determined. This provides a second fracture criterion in addition to Eq. (11), namely that failure will occur when:

$$\frac{dG}{da} > \frac{dR}{da} \tag{31}$$

Furthermore, by considering differing initial defect sizes, a_0, it can be seen that the effective fracture toughness at failure will increase with the length of the initial defect, although the failure stress will decrease.

In the case of many structural ceramics, intrinsic defects (e.g., pre-existing micro-cracks or pores) often scale in the order of the microstructure – that is, grain size. Applying this to the R-curves of these materials reveals that the linear strain energy release rate curve makes a tangent with the crack growth resistance curve very close to its initiation point, $R_0 \cdot (= K_0^2/E)$, and unstable fracture occurs soon after crack propagation. This means that the effective fracture toughness at failure is close to the intrinsic toughness of the material, K_0. The implication of this is that the processes which have been developed to create tougher ceramics have, in many cases, a limited practical effect in improving the reliability and strength of ceramic components.

Lawn and coworkers have studied the effects of this dichotomy extensively through a process of inducing defects into ceramics through the use of sharp Vickers indenter at varying loads, and then ascertaining the failure strength [10]. Figure 14.21 shows an example of this for a range of alumina materials of varying grain size. (Note: as the grain size of the alumina increases, the R-curve will become less steep but then rise to a higher plateau toughness.) It can be seen that for very small flaws, or no extrinsically induced defects, the fine-grained material exhibits the highest strength; this is because intrinsic defects tend to scale with the grain size. However, fine-grained materials also exhibit the most dramatic reduction in strength, with an increasing indentation-induced defect size. This is because of their less pronounced R-curve behavior. The implication is that large-grained materials, or materials which have pronounced R-curves, are more damage-resistant than fine-grained materials, and their intrinsic strengths are lower. However, the strength in large-grained materials is also less affected by large extrinsically induced defects.

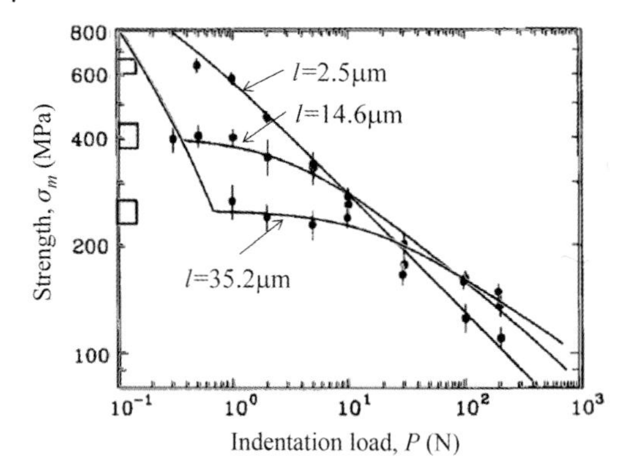

Figure 14.21 Strength of alumina of three different grain sizes following Vickers indentation. Adapted from Ref. [10].

14.5 Determination of Fracture Resistance

Given the significance of fracture resistance in determining the structural reliability of ceramic materials, measurement of fracture toughness as a mechanical property is critical. For most materials, such as metals and polymers, the determination of fracture resistance utilizes large samples which are machined, usually involving notches and holes for loading. In the case of ceramic materials, however, processes such as machining and polishing are very time-consuming. Furthermore, ceramic components are traditionally small in size and, hence, facilities for manufacturing large blocks of material are limited and of limited relevance. Considerable effort, therefore, has been invested in developing specialized methods to ascertain the fracture toughness of ceramics, including indentation fracture toughness, surface crack in flexure, and single-edged notched beams. Each of these is discussed in the following sections.

14.5.1 Indentation Fracture Toughness

The loading of a diamond-tipped Vickers indenter into most ceramic and glass materials results in the formation of cracks from the corners of the indentation (Figure 14.22). Anstis *et al.* [22] developed a method to correlate the length of these cracks with the fracture toughness of the material:

$$K_r = \chi \left(\frac{E}{H} \right)^{\frac{1}{2}} \frac{P}{c^{3/2}} \tag{32}$$

where E is the Young's modulus of the material, P is the indentation load, c is the average length of the cracks, and the hardness $H = P/2d^2$, where $2d$ is the length of

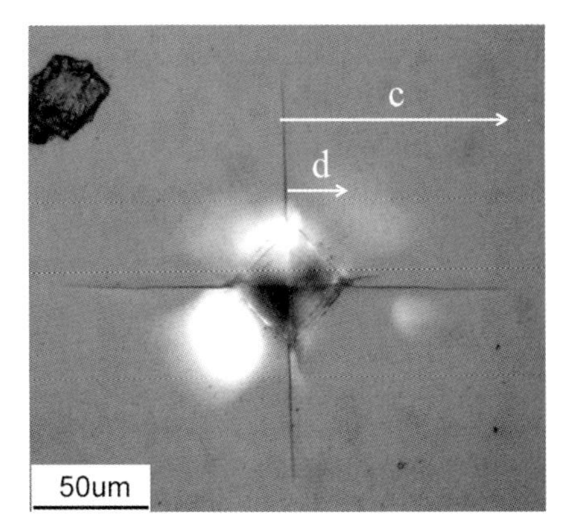

Figure 14.22 Vickers indent on glass, as used for measuring fracture toughness. (From Z.H. Luo, unpublished results.)

the diagonal across the indent. χ is a material-independent constant which, in the original study, was found to be equal to 0.016.

This method has a number of clear advantages, the primary benefit being the ease with which it may be undertaken; only a small volume of material is required and the necessary test equipment is commonly available. A flat, optical-quality polished surface is necessary to ensure the accurate measurement of crack lengths. The method does, however, require that the ceramic materials are "well behaved" – namely, that cracks propagate straight from the corners of the indent, and that the crack length is relatively large ($c > 2d$). It is also important that there is no chipping from the corners of the indent. Critically, as the crack develops, a well-defined medial crack system must form underneath the indent, although this is often difficult to ascertain without a further post-analysis investigation. Due to these issues, concerns have been raised regarding the accuracy of this test method, despite its simplicity [23].

14.5.2
Single-Edge Notched Beam

In order to overcome many of the difficulties associated with the indentation fracture toughness measurement as noted above, the single-edged notched beam (SENB) test – which often is used for metallic and polymeric materials – has been adapted to suit ceramics [24]. For this, a bar of ceramic, usually in the order of 3×4 mm cross-section and 25–45 mm length, has a notch introduced on one side. The bar is then loaded in four-point bending and the fracture toughness determined from the failure load, P_m, as per:

$$K_{IC} = \frac{P_m}{B\sqrt{W}} Y \qquad (33)$$

where $$Y = \frac{S-s}{W} \cdot \frac{3\sqrt{\alpha}}{2(1-\alpha)^{3/2}} \left[1.9887 - 1.1326\alpha - \frac{(3.49 - 0.68\alpha + \alpha^2)\alpha(1-\alpha)}{(1+\alpha)^2} \right],$$

$\alpha = \frac{a}{W}$, and a and W are the crack length and height of the sample, respectively, and S and s are the larger and smaller spans for the bending fixture.

The notch may be introduced using one of two methods. The first method involves making a row of Vickers indentations along the polished surface of one of the sides, and then loading the sample in a bridge compression rig, as shown in Figure 14.23. This provides a tensile stress on one side of the sample, causing the crack to propagate partially through the sample. For some materials this test process can be challenging, however, and consequently a technique that involved placing a sharpened sawn notch on one side of the sample was developed and subsequently tested [26]. The notch is sharpened by repeatedly sliding a sharp blade across the base of the notch after the insertion of some diamond paste, and has proved effective in creating notches of a radius in the order of 2–5 μm. The difficulty with this method is that the crack tip is blunter than that of a crack, and this may subsequently lead to overestimates of fracture toughness.

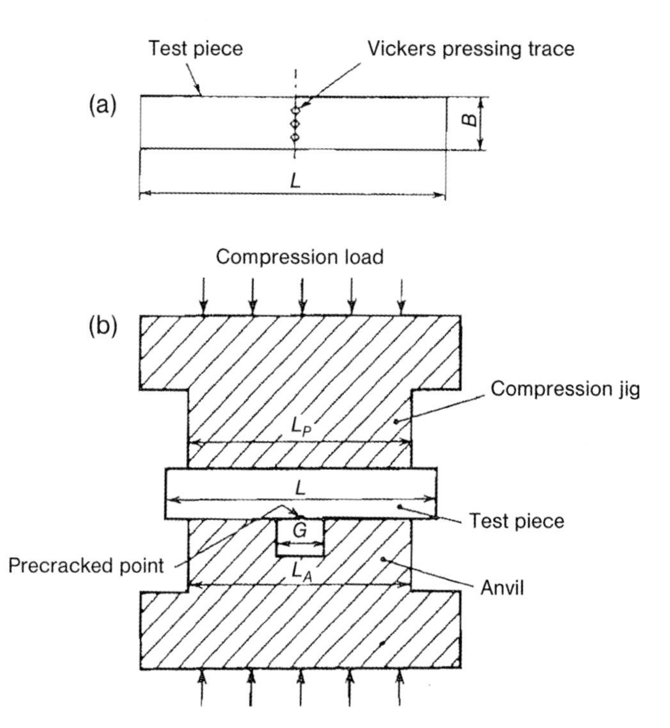

Figure 14.23 (a) Rows of Vickers indents and (b) compression loading used to create sharp initial crack for SENB testing. From Ref. [25].

14.5.3
Surface Crack in Flexure

The Vickers indentation and SENB methods involve the measurement of fracture toughness from cracks which are larger than those usually found in ceramic components. As discussed above, this may affect the measured fracture toughness value as a result of the crack growth resistance curve, whereby large initial cracks will provide a value of K_c higher up the crack growth resistance curve, while failure from larger initial cracks will also lead to higher values than may be the effective fracture toughness from failure from a microstructure-sized crack in a ceramic component.

The surface crack in flexure method seeks to address these issues [24]. This involves placing a Vickers indentation into the surface of a component, so as to induce a subsurface medial crack. The indentation is then polished away, leaving a small elliptical crack in the surface of the sample; the length of this crack is then measured and the sample loaded to failure, after which the fracture toughness is calculated using the appropriate equation in Table 14.1. While redressing a number of significant issues with other methods, the main difficulty with this approach is the high level of skill required to prepare the samples and to measure the crack length.

14.5.4
Determination of Intrinsic Toughness

As discussed in the Section 14.3.1.2, the intrinsic toughness – or the point at which crack propagation commences – is, in most practical cases, the most important measure for fracture resistance and subsequent failure of ceramic materials. Detecting the initiation of crack propagation during *R*-curve testing is experimentally challenging; furthermore, none of the methods discussed above enables its accurate determination. Even in the case of the surface crack in flexure method, it is not known whether the tangent point in Figure 14.13 is intersecting at the intrinsic toughness. The use of scanning electron microscopy (SEM) or atomic force microscopy (AFM), however, enables the accurate determination of the opening profile at the crack tip of a loaded crack within a ceramic. Assuming that the crack is loaded close to the critical point for propagation, the opening of the crack will follow the profile shown in Eq. (27), irrespective of the point on the *R*-curve to which the crack has propagated. The significance of this is seen, for example, in Figure 14.24, where the crack-opening profile following a Vickers indentation is plotted for a material which demonstrates a clear *R*-curve – polycrystalline alumina – and an alumina/silicon carbide nanocomposite where fracture is transgranular and no *R*-curve is present [27]. Both materials were shown to have the same intrinsic toughness as the crack-opening profiles close to the crack tip are the same, and hence the first term in Eq. (27) may be used to ascertain the intrinsic toughness.

Crack tip opening displacement (CTOD) measurements are complex to undertake due to the need to perform high-resolution scanning electron microscopy (HRSEM).

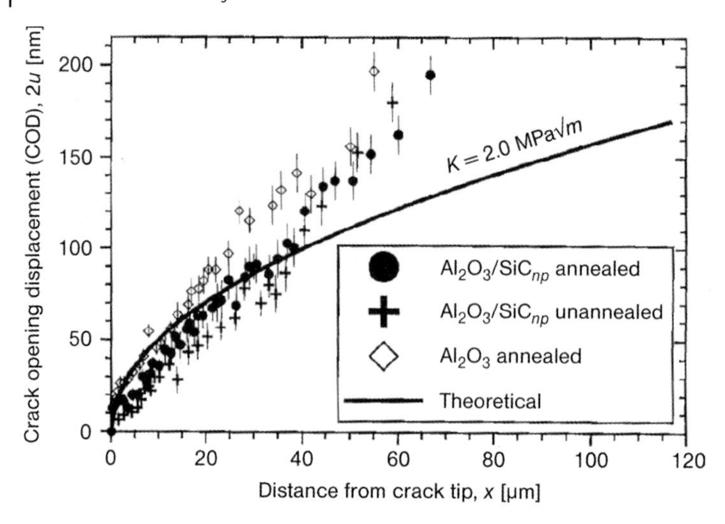

Figure 14.24 Crack-opening profiles for the critical load for crack propagation in polycrystalline alumina, which demonstrates a pronounced *R*-curve, and for an alumina–silicon carbide nanocomposite, which does not showing the divergence with distance from the crack tip. From Ref. [27].

However, the continual development of new bench-top instruments provides potential for the future use of this method.

14.6 Fatigue

14.6.1 Cyclic Fatigue Crack Propagation

The development of micromechanical mechanisms to raise the fracture resistance of structural ceramics has led to improved strength reliability and, in many cases, to higher strengths for ceramic materials that are used in some structural applications. Significant advantages have occurred, for example, in their application for biomedical uses, such as prosthetic implants and dental implants. A corollary, however, of the improvement in fracture resistance has been an increased susceptibility to the subcritical propagation of cracks under repeated cyclic loading, which has resulted in crack propagation well below the accepted toughness values for these materials.

The most common causes of fatigue are the frictional degradation of the bridging processes which lead to increased toughness [28]. Figure 14.25 shows, for example, the interfacial degradation between a cracked bridge where it is pulled out in an alumina ceramic during cyclic loading. This results in a reduction of the coefficient of friction between the two grains, and hence, the bridging stress [29]. Observations of cyclic fatigue degradation involving other toughening mechanisms, have also been

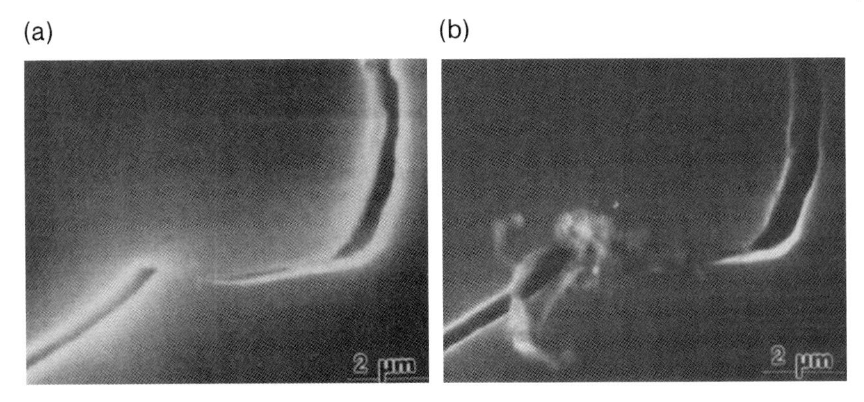

Figure 14.25 (a) Friction point of crack bridge in alumina and (b) following 45 000 load (opening) cycles, showing degradation leading to reduction in friction force. From Ref. [29].

made for phase transformation-toughened ceramics [30] and ferroelectric ceramics [31]. In the former case, this is believed to be a result of degradation of the intrinsic toughness of the material, whereas in the case of ferroelectric ceramics it is associated with degradation of the domain-switching processes.

14.6.2 Contact Fatigue

A common application of ceramic materials is on surfaces which are subjected to wear and cyclic contact loads. Being brittle materials, the primary degradation mechanism of the surfaces is microfracture of the ceramic material. It has been shown that, in the case of monotonic loading, a critical load must be reached before cracking occurs on a contact surface. On the surface of cyclic loading, repeated subcritical load may lead to progressive degradation. This is shown, for example, in Figure 14.26 in the case of silicon carbide. It is noticeable that the extent of degradation increases with microstructural size where a fine-grained material (which also demonstrates a far less-pronounced *R*-curve) shows markedly less susceptibility to increased damage on repeated contact, than does the coarse-grained silicon nitride material. It is clear that the hysteretic processes which result in improved *R*-curve behavior lead to an increased susceptibility to cyclic fatigue degradation.

This scenario is also relevant to degradation by sliding contact or wear. Figure 14.27a and b shows the ground surface of two microstructures [32]. One is a fine equiaxed-grained sialon material, while the other is an elongated-grained material. The former material shows significant pitting on the surface, while the latter is relatively undamaged. The reason for this is that the cracks in the elongated material propagate and then stop at relatively short distances due to the rising *R*-curve. However, in the equiaxed material the *R*-curve plateau is lower, and subsequently crack arrest does not occur. Nevertheless, it can be seen in Figure 14.27c and d

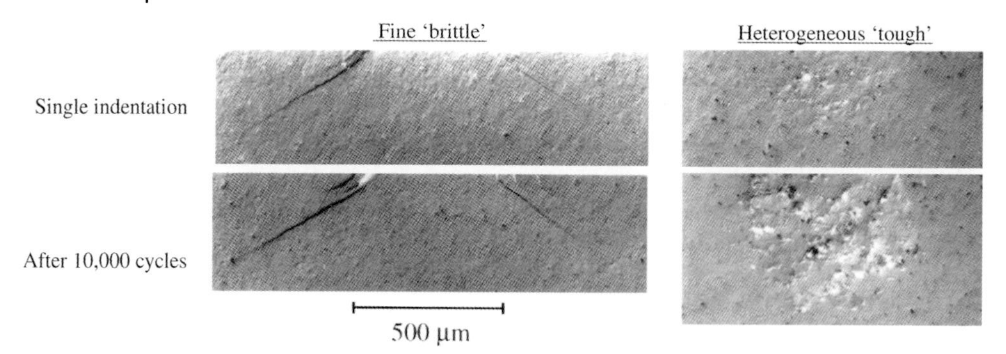

Figure 14.26 Cross-sectional images of Hertzian contact zone in fine microstructured silicon nitride which does not exhibit a pronounced *R*-curve, and a coarse heterogeneous structure which would be expected to have a pronounced *R*-curve. Comparison is seen between the effect of a single and multiple contacts. From Ref. [32].

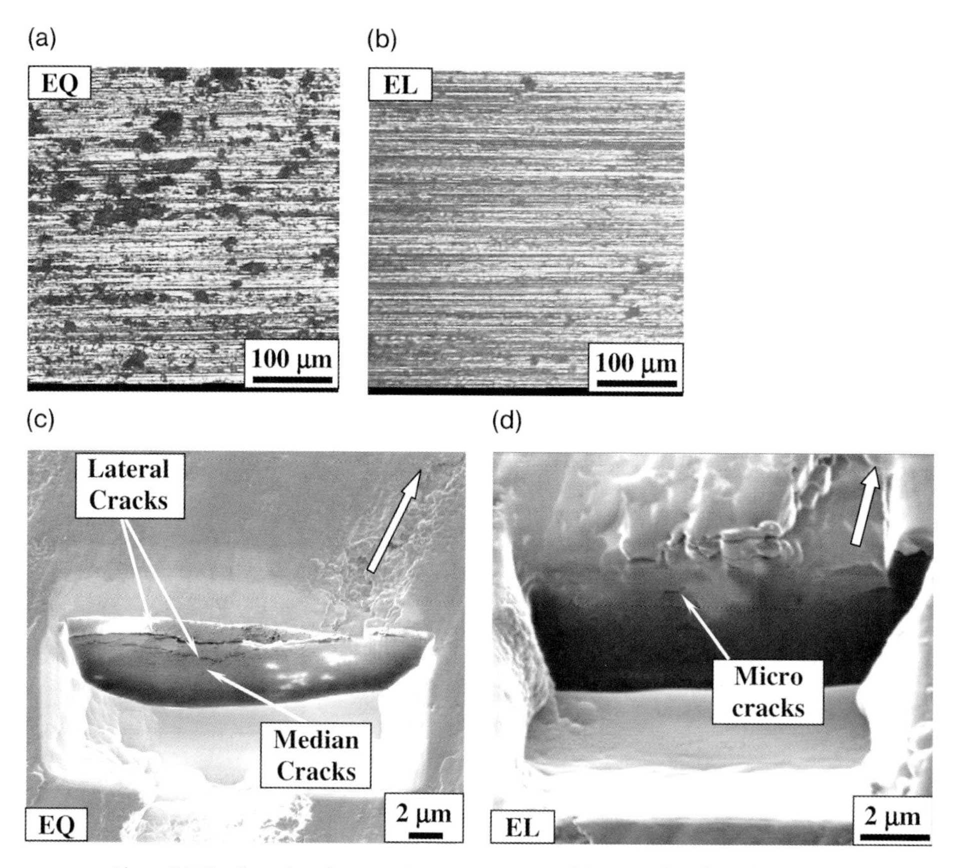

Figure 14.27 Scanning electron microscopy images of the ground surface of (a) an equiaxed (EQ) and (b) an elongated (EL) grained sialon, and cross-sections of (c) EQ and (d) EL. From Ref. [33].

that subsurface damage has occurred in the elongated grained material, consistent with the observations shown above for repeated contact loading. In the case of sliding wear, this behavior results in lower rates of wear for the elongated grained material, and also a delayed transition to the point where severe wear, associated with micro-cracking processes, occurs.

14.7
Concluding Remarks

Fracture resistance is the critical structural property for ceramic materials, notably because the strength of these materials is determined by Griffith-type fast fracture processes and the critical propagation of cracks. Hence, the ability to resist crack propagation determines the material's strength. The development of a range of methods for improving fracture toughness has led to significant enhancements in the fracture resistance of ceramic materials and crack growth resistance toughening. Nevertheless, the intrinsic toughness of most ceramic materials is essentially the same. The advantage, however, of materials with higher fracture resistance through *R*-curve behavior is that they become less susceptible to the presence of extrinsically induced flaws in determining their strength and, hence, have higher strength reliability. The effectiveness of crack growth resistance toughening in improving the effectiveness toughness of the materials is, however, limited due to the small size of the intrinsic defects that exist within ceramic materials. Hence, the homogeneity of microstructure and the presence of intrinsic or extrinsic processing and manufacturing defects is critical in determining the ultimate strength of ceramic materials. The presence of toughening mechanisms also makes ceramic materials more susceptible to cyclic fatigue degradation. However, these mechanisms may also lead to improved microfracture resistance, which is particularly important in sliding wear applications.

References

1 Steinbrech, R.W. (1992) Toughening mechanisms for ceramic materials. *J. Eur. Ceram. Soc.*, **10**, 131–142.

2 Inglis, C.E. (1920) Stresses in a plate due to the presence of cracks and sharp corners. *Trans. Inst. Naval Architects*, **221**, 219–241.

3 Griffith, A.A. (1920) The phenomena of rupture and flow in solids. *Philos. Trans. R. Soc.*, **221A**, 162–198.

4 Lawn, B.R. (1993) *Fracture of Brittle Solids*, 2nd edn, Cambridge University Press, Cambridge, UK.

5 Westergaard, H.M. (1939) Bearing pressures and cracks. *J. Appl. Mech.*, **6**, 49–53.

6 Williams, M.L. (1957) On the stress distribution at the base of a stationary crack. *J. Appl. Mech.*, **24**, 109–114.

7 Stech, M. and Roedel, J. (1996) Method for measuring short crack R-curves without calibration parameters: case studies on alumina and alumina/aluminum composites. *J. Am. Ceram. Soc.*, **79** (12), 297–397.

8 Swanson, P.L., Fairbanks, C.J., Lawn, B.R., Mai, Y.-W., and Hockey, B.J. (1987)

Crack-interface grain bridging as a fracture-resistance mechanism in ceramics. 1. Experimental-study on alumina. *J. Am. Ceram. Soc.*, **70** (4), 279–289.

9 Xie, Z.-H., Hoffman, M., and Cheng, Y.-B. (2002) Microstructural tailoring of a Ca α-SiAlON composition. *J. Am. Ceram. Soc.*, **85** (4), 812–818.

10 Chantikul, P., Bennison, S., and Lawn, B.R. (1990) Role of grain size in the strength and R-curve properties of alumina. *J. Am. Ceram. Soc.*, **73** (8), 2419–2427.

11 Barrenblatt, G.I. (1962) The mathematical theory of equilibrium of cracks in brittle fracture. *Adv. Appl. Mech.*, **7**, 55–129.

12 Deuerler, F., Knehans, R., and Steinbrech, R. (1986) Testing-methods and crack resistance behaviour of Al_2O_3. *J. Phys. Colloque C1*, **47** (Suppl. 20), C1–C617.

13 Roedel, J. (1992) Interaction between crack deflection and crack bridging. *J. Eur. Ceram. Soc.*, **10**, 143–150.

14 Prielipp, H., Knetchtel, M., Claussen, N., Streiffer, S.K., Mullejans, H., Ruhle, M., and Roedel, J. (1995) Strength and fracture toughness of aluminium/alumina composites with interpenetrating networks. *Mater. Sci. Eng. A*, **197**, 19–31.

15 Hoffman, M., Roedel, J., Skirl, S., Zimmermann, A., Fuller, E., and Mullejans, H. (1999) Tailoring of an interpenetrating network ceramic/metal microstructure to improve strength: Al_2O_3/Ni_3Al and Al_2O_3/Al. *Key Eng. Mater.*, **159–160**, 311–318.

16 Hoffman, M.J., Dauskardt, R.H., Ager, J., Mai, Y.-W., and Ritchie, R.O. (1995) Grain size effects on cyclic fatigue and crack-growth resistance behaviour of partially stabilised zirconia. *J. Mater. Sci.*, **30**, 3291.

17 Jones, J., Motahari, S.M., Varlioglu, M., Lienert, U., Bernier, J.V., Hoffman, M., and Üstündag, E. (2007) Crack tip process zone domain switching in a soft lead zirconate titanate ceramic. *Acta Mater.*, **55**, 5538–5548.

18 Lucato, S., Lupascu, D.C., and Rodel, J. (2000) Effect of poling direction on R-curve behavior in lead zirconate titanate. *J. Am. Ceram. Soc.*, **83** (2), 424–426.

19 Ruhle, M., Evans, A.G., McMeeking, R.M., Charalamibides, P.G., and Hutchinson, J.W. (1987) Microcrack toughening in alumina/zirconia. *Acta Metall.*, **35** (11), 2701–2710.

20 Rose, L.F. (1986) Effective fracture toughness of microcracked materials. *J. Am. Ceram. Soc.*, **69** (3), 212–214.

21 Lutz, E., Claussen, N., and Swain, M.V. (1991) K^R-curve behaviour of duplex ceramics. *J. Am. Ceram. Soc.*, **74** (1), 1–18.

22 Anstis, G.R., Chantikul, P., Lawn, B.R., and Marshall, D.B. (1981) A critical evaluation of indentation techniques for measuring fracture toughness: I. Direct Crack Measurements. *J. Am. Ceram. Soc.*, **64** (9), 533–538.

23 Quinn, G.D. and Bradt, R.C. (2007) On the Vickers indentation fracture toughness test. *J. Am. Ceram. Soc.*, **90** (3), 673–680.

24 ASTM (2007) C142101b. *Standard Test Methods for Determination of Fracture Toughness of Advanced Ceramics at Ambient Temperature*, ASTM International, PA, USA.

25 Japanese Industrial Standard (1990) JIS R 1607. *Testing Methods for Fracture Toughness of High Performance Ceramics*, Japanese Industrial Standard

26 Kübler, J. (1999) Fracture toughness of ceramics using the SEVNB method: Round Robin, ESIS Document D2-99, EMPA.

27 Hoffman, M. and Rödel, J. (1997) Suggestion for mechanism of strengthening in 'nanotoughened' ceramics. *J. Ceram. Soc. Jpn*, **105**, 1086–1090.

28 Chen, I.-W., Liu, S.-Y., Jacobs, D.S., and Engineer, M. (1996) Fracture mechanics of fatigue of structural ceramics, in *Fracture Mechanics of Ceramics*, vol. 12 (eds R.C. Bradt, *et al.*), Plenum Press, New York.

29 Lathabai, S., Rodel, J., and Lawn, B.R. (1991) Cyclic fatigue from frictional degradation at bridging grains in alumina. *J. Am. Ceram. Soc.*, **74** (6), 1340–1348.

30 Hoffman, M.J., Wakayama, S., Mai, Y.-W., Kishi, T., and Kawahara, M. (1995) Crack-tip degradation processes observed during

in situ cyclic fatigue of partially stabilised zirconia in the SEM. *J. Am. Ceram. Soc.*, **78** (10), 2801–2810.

31 Salz, C., Hoffman, M., Westram, I., and Roedel, J. (2005) Cyclic fatigue crack growth in PZT under mechanical loading. *J. Am. Ceram. Soc.*, **88** (5), 1331–1333.

32 Lawn, B.R. (1998) Indentation with spheres: a century after hertz. *J. Am. Ceram. Soc.*, **81** (8), 1977–1994.

33 Xie, Z.-H., Moon, R., Hoffman, M., Munroe, P., and Cheng, Y.-B. (2003) Role of microstructure in grinding and polishing of α-sialon ceramics. *J. Eur. Ceram. Soc.*, **23** (13), 2351–2360.

15
Superplasticity in Ceramics: Accommodation-Controlling Mechanisms Revisited

Arturo Domínguez-Rodríguez and Diego Gómez-García

15.1
Introduction

Superplasticity is defined macroscopically as the "ability of a polycrystalline material to exhibit large elongations at elevated temperatures and relatively low stresses." It is commonly found in a wide range of materials from metals to ceramics (bioceramics or high-temperature superconductors, among others) when the grain size is sufficiently small enough, such as a few micrometers for metals and less than one micron for ceramics.

Several strategies have been used to obtain fully dense ceramics with grain growth inhibitors, a necessity without which superplasticity would come to an abrupt end after a small elongation. In this chapter, the roles of dopants which can segregate or precipitate, and second phases with lower melting temperatures, are emphasized. In the first of these strategies, a drag force is created by the segregation species, which retards grain growth. In the second strategy, advantage is taken of the enhanced densification kinetics at lower temperatures when a low-melting-temperature second phase is added. Quite often, this low-temperature densification prevents grain growth.

Grain boundary sliding is commonly assumed to be the primary controlling mechanism in superplasticity. As the structure and the nature of the grain boundaries can be profoundly modified by segregation or glassy phases, the accommodation processes that control the strain rate of superplasticity are also affected by segregation and second phases. In the case of segregation, several important factors are recognized, including the ionic radius of the impurities, their charge, the binding energy between the host atoms, and the impurities. These factors commonly refer to cations, and become especially important in nanostructured materials. In the case of the secondary phase, it may act as a "lubricant" so that the viscous movement of the glassy phases constitutes an accommodation mechanism for grain boundary sliding. The grain boundaries may also become a fast diffusion pathway in the presence of a secondary glassy phase, when compared to the case when grain boundaries are "clean," thus improving the diffusional accommodation. Finally, the secondary phase

Ceramics Science and Technology Volume 2: Properties. Edited by Ralf Riedel and I-Wei Chen

ISBN: 978-3-527-31156-9

may provide sites for preferential nucleation and the growth of cavities during deformation, which is detrimental to superplasticity.

Depending on which of the above factors dominates during deformation, the accommodation mechanism may be regarded as viscous flow, solution–precipitation, or cavitation creep. In general, cavitation creep can be discarded as an accommodation mechanism for superplasticity, as the strain-to-failure afforded by this mechanism is rather small. Therefore, only viscous flow and solution–precipitation mechanisms are important. Obviously, too, whether these mechanisms apply depends on the presence or absence of a liquid phase. Solution–precipitation requires a liquid phase to envelop the grains, while viscous flow is facilitated by the fast diffusion path of the liquid, although it may also occur in a dry polycrystal via diffusional creep.

In this chapter, the macroscopic and microscopic aspects of superplasticity, the accommodation processes, the applications and the future prospects of ceramic superplasticity will be addressed.

15.2 Macroscopic and Microscopic Features of Superplasticity

The phenomenon of *superplasticity* was defined as the ability of a polycrystalline material to exhibit large elongation without necking, and was identified for the first time in metal for nominally low stresses and relatively high temperatures. Whilst it is not clear if the ancient swords of the Damascans were made by using superplastic deformation, the first report of a very high-tensile deformation prior to failure was in 1912 for a ($\alpha + \beta$) brass [1]. However it was only during the 1960s when the superplasticity of metals and metal alloys began to be studied intensively, from both scientific and commercial points of view.

Initially, superplasticity was restricted mainly to metallic alloys, as it required a small and stable grain size (<10 μm). However it was only after 1975, when Garvie *et al.* [2] published their article in *Nature*, "Ceramic steel?", which described the potential use of ceramics in structural applications, that a great effort was made to identify tough new ceramics, by means of processing and sintering ceramics. These techniques constituted the keystones of the investigations, which were aimed at obtaining fully dense ceramics with equiaxed grain sizes below 1 μm. As a consequence, Wakai *et al.* [3] reported the first observation of superplasticity in a 3 mol% yttria-stabilized tetragonal zirconia polycrystal (YTZP) ceramic with a grain size of 0.4 μm (Figure 15.1).

Nowadays, the list of ceramics and ceramics composites with superplastic behavior is very wide [4–9], and ceramics such as high-temperature superconductors YBaCuO behave superplastically (see Ref. [10] and the cover page of the *Journal of the American Ceramics Society*, volume 97, May 1996).

Nevertheless, it was not until superplasticity was defined from a microscopic point of view, as the deformation due to the grain boundary sliding, that new ceramics, intermetallics and metal and ceramic–matrix composite materials were considered

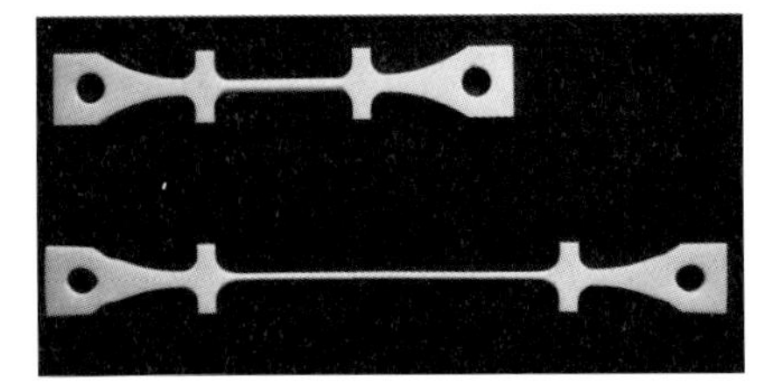

Figure 15.1 As-received and tensile-deformed specimens of superplastic yttria tetragonal zirconia polycrystals (YTZP), as displayed by Wakai *et al.*, in 1986. Illustration courtesy of Prof. Wakai.

as superplastic systems. This term was later extended to cover materials tested in compression and even in bending. This lack of a clear definition explains why materials deformed in compression by grain-boundary sliding – which, indeed, is at the origin of the superplasticity – were not classified as superplastic materials initially. Remarkable examples of this have been reported elsewhere [11–13].

From a microscopic point of view, when a polycrystalline material is deformed at high temperatures, grain boundary sliding (GBS) takes place in two different ways:

- The deformation is due to the flow of point defects: GBS then occurs to maintain grain coherency. This is termed as "Lifshitz grain boundary sliding" (Figure 15.2a), and is termed "diffusional creep," Nabarro–Herring if the diffusion takes place along the bulk, or Coble if it takes place along the grain

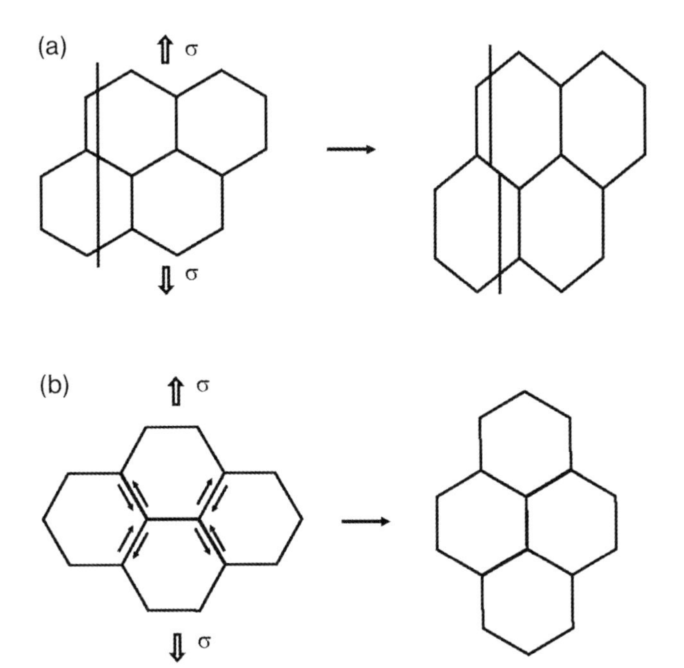

Figure 15.2 (a) A scheme of grain switching occurring in the Lipschitz-mechanism activated during diffusional creep; (b) The Rachinger mechanism.

boundaries. In these cases, each individual grain suffers almost the same deformation as the macroscopic grain, and the grains which are nearest neighbors remain nearest neighbors.

- A different picture arises when GBS is responsible for the deformation. In order to release stresses created during GBS, deformation is to be accompanied by intergranular slip throughout adjacent grains, by localized slip adjacent to the boundaries, or by diffusional processes involving point defect migration. Sometimes, the formation of a triple point follows, and the opening up of cracks at the triple points can also accommodate GBS; however, as soon as there is any coalescence of voids or cracks, the material fails, and that happens at not so-large deformations. This type of GBS, accommodated by the process described above, is termed "Rachinger grain boundary sliding" (Figure 15.2b). When this occurs, the average grain size and shape remain unchanged after large deformation values, and this is the main feature of this regime. A detailed analysis of the characteristic of the different types of grain boundary sliding can be found elsewhere [14].

The relative motion of two adjoining rigid grains has two components: one of these is parallel and the other perpendicular to their common grain boundary. In fact, GBS means a parallel motion to the grain boundary, and is responsible for 70–80% of the deformation in superplasticity of fine-grained ceramics [15–17]. This can be demonstrated by measurement of the grain aspect ratio, by scanning electron microscopy (SEM) and by atomic force microscopy (AFM), and displayed recently by Duclos [18] in YTZP.

Spectacular evidence of GBS during ceramic superplasticity can be observed in Figure 15.3, which shows the relative motion of grains during the deformation of YTZP [18]. It can observed how two small grains, initially separated by two larger grains in contact, gradually draw nearer such that finally the larger grains are separated. Several other examples of switching events are also shown in Ref. [18].

The perpendicular motion of the two adjoining grains creates a stress concentration at the grain boundaries which must be relaxed in order for the deformation to proceed. This is equivalent to a nonconservative motion, and it is associated with the diffusional fluxes of the components to/from or along the interfaces, or to the dislocation motion, these being the main accommodation processes and the rate-controlling process associated with GBS.

When GBS is accommodated by some of the mechanisms involving dislocation movement or the diffusion of point defects, the grains retain almost the original size and shape even after large deformations. Hence, GBS, as the primary mechanism for deformation, is indeed the basis for the high ductility exhibited for some materials at high temperatures, and therefore to their structural superplastic behavior.

Recently, cooperative GBS, which reveals itself through the sliding of grain groups as an entity, has been demonstrated in some superplastic alloys [19]. Such cooperative GBS has also been suggested to play an important role for fine-grained ceramics [20]. However, the direct observation of grain rearrangement during the superplasticity of YTZP revealed that the influence of cooperative GBS was very limited [18].

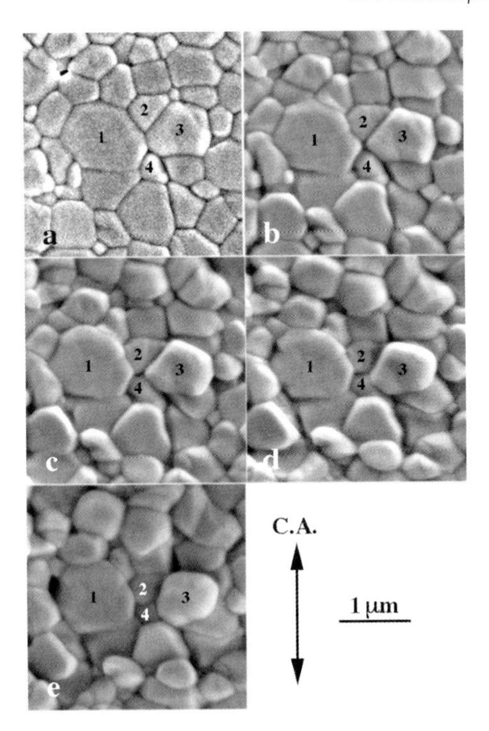

Figure 15.3 Experimental evidence of grain boundary sliding with microstructural invariance in superplastic YTZP. The arrow indicates the compression axis (C.A.). Illustration courtesy of Prof. Duclos.

Recently, Yasuda *et al.* have proved the existence of grain domains with long-range correlation of displacements that tend to disappear when deformation proceeds [21]. However, these authors pointed out that such long-range correlations were not involved with cooperative GBS.

The strain rate in the superplasticity of many metals and ceramics is often expressed by the following semi-empirical equation [7, 8, 22]:

$$\dot{\varepsilon} = \frac{AGb}{kT}\left(\frac{b}{d}\right)^p\left(\frac{\sigma-\sigma_0}{G}\right)^n D \tag{1}$$

where $\dot{\varepsilon}$ is the strain rate, b is the Burgers vector, G is the shear modulus, σ is the stress, σ_0 is a threshold stress, n is the stress exponent, D is an appropriate diffusion coefficient [$= D_0 \exp(-Q/kT)$, where D_0 is a frequency factor, Q is the activation energy and k is Boltzmann's constant], d is the grain size, and p is the grain-size exponent. The value of p is 2 when the rate-controlling process is lattice diffusion, and 3 when it is grain boundary diffusion. The threshold stress σ_0, which is zero in many cases, depends on the nature of the grain boundaries [8, 23].

The values of the creep parameters (p, n, and Q) identifying the superplastic behavior of ceramic-related materials are not unique, neither are they similar for the same types of material. Several factors can affect these parameters, among them the

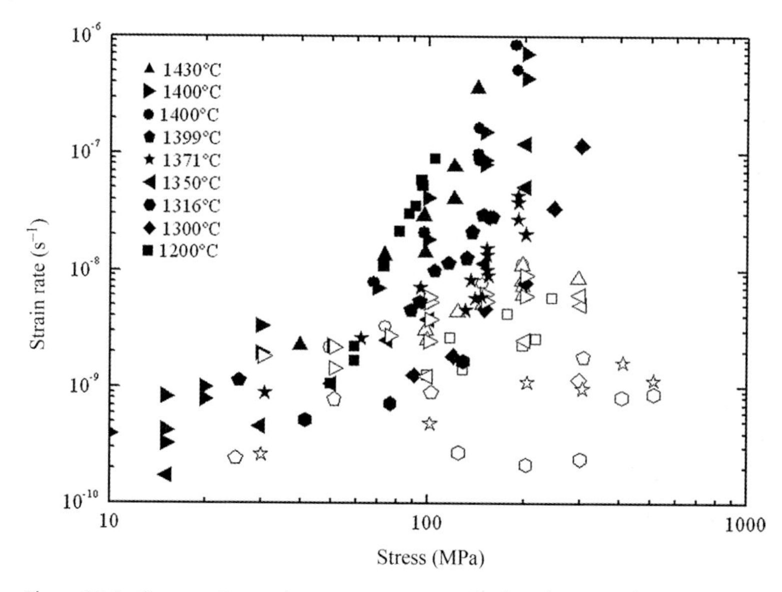

Figure 15.4 Comparative strain rate versus stress plot for silicon nitride ceramics crept under tension (solid symbols) and under compression (open symbols). The characteristic asymmetric behavior is clearly displayed. The experimental data are collected in Ref. [9].

purity of the ceramics, including the nature of the grain boundaries [6, 8, 9, 23] or the testing conditions, compression or tension when the aid-sintering phase is required during processing, as in silicon carbide and silicon nitride materials [9].

Taking the second example as the most important, the mechanical behavior will depend on the nature of the mechanical test, whether tension or compression [9] (Figures 15.4 and 15.5).

Even when concentrating on tensile tests only and analyzing them in terms of the classical creep equation, stress exponents ranging between only 2 and 13 are calculated. This fact is illustrated in Figure 15.6, where the strain rate values corrected for temperature dependence are plotted against stress, and a very wide scattering ranging from 10^{10} to 10^{40} is observed. The apparent activation energies for creep are also systematically higher under tension than under compression, ranging between 600 and 1645 kJ mol^{-1}. A more detailed analysis of the special features of silicon nitride ceramics is provided in Ref. [9].

The "asymmetry" between tensile and compressive creep is a typical feature of other systems, such as SiC-based and Al_2O_3-based ceramics. In fact, this is common in ceramic systems in which an easily deformable glassy phase located at the grain boundaries is present. In contrast, single-phase ceramics such as pure alumina polycrystals and YTZP are reported to be symmetric under both tension and compression.

Another important fact which must be taken carefully into account is the presence of impurities at the grain boundaries. In the case of YTZP ceramics, the stress exponent is reported to depend on the stress in high-purity specimens, whereas it is

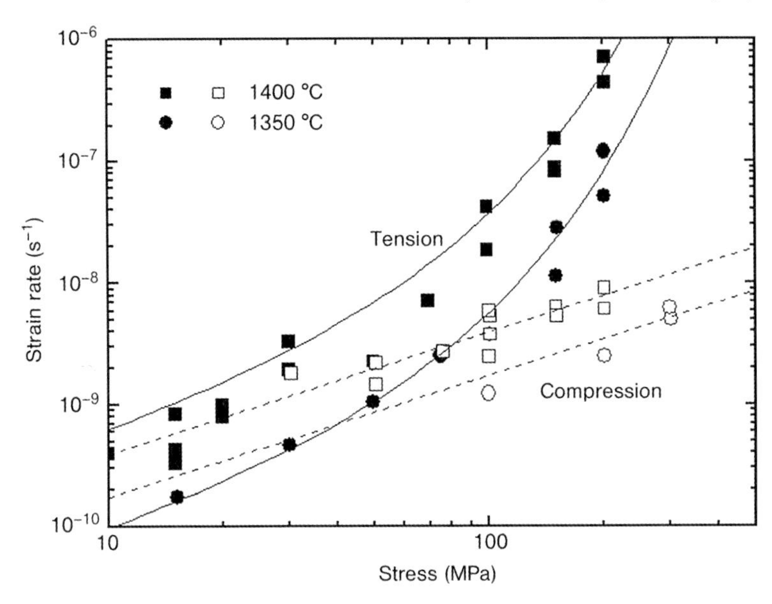

Figure 15.5 Tensile and compressive strain rates for silicon nitride ceramics showing that, in the limit of low stresses, the two strain rates approach each other. See Ref. [9] for further details of these data.

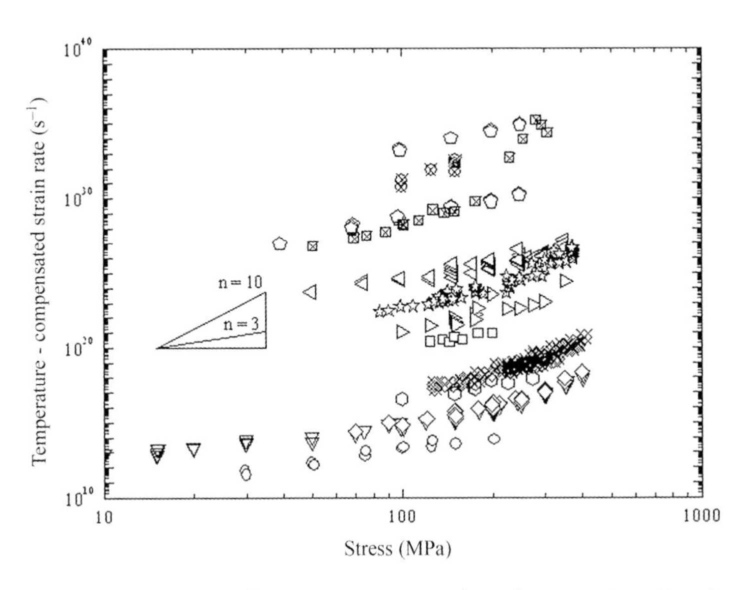

Figure 15.6 Experimental temperature-compensated strain rates versus stress plots for tensile creep of silicon nitride-based ceramics. The stress exponents $n = 3$ and $n = 10$ are shown for comparison. Note the extensive scattering exhibited by data, ranging 30 orders of magnitude. Data from Ref. [9].

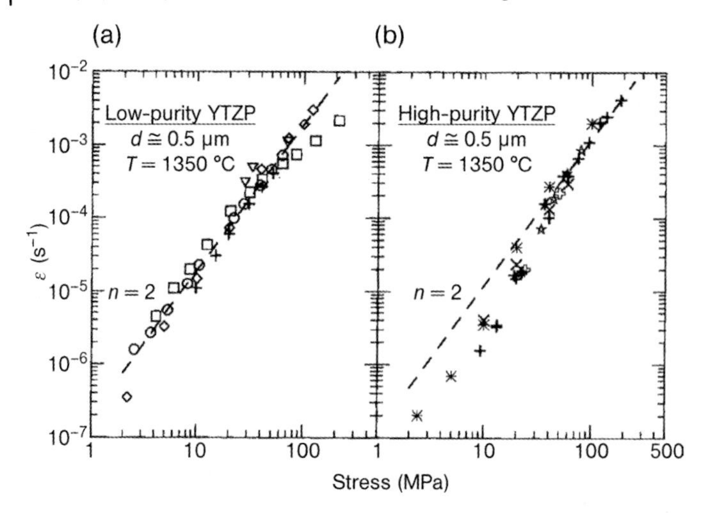

Figure 15.7 Strain rate versus stress for yttria tetragonal zirconia ceramics at 1350 °C. Note that the apparent stress exponent is fully dependent on the degree purity of the specimens. It is constant, equal to 2 in low-purity samples, whereas it is a function of the stress in high-purity samples. Data from Ref. [8].

constant with and equal to 2 in low-purity specimens (Figure 15.7). This phenomenon has been explained in terms of a "threshold stress," induced by cation segregation at the grain boundaries [8, 23]. The different values of the creep parameters are justified through the accommodation processes controlling the superplasticity, and these will be analyzed further in the following sections.

15.3 Nature of the Grain Boundaries

Before discussing the possible accommodation processes controlling superplasticity, it is important first to discuss how the nature of the grain boundaries can influence superplastic behavior.

It is well known that grain boundaries are sinks/sources of point defects, and that the addition of secondary phases, which normally are used as sintering aids, are distributed along the grain boundaries and triple-point junctions of the grains. Consequently, the nature of the grain boundary may either enhance or retard the diffusion coefficients of the process controlling superplasticity.

Many investigations have been conducted on the influence of grain boundary segregation on superplasticity in ceramics. Indeed, it has been shown successively that the yttrium in YTZP polycrystals segregates at the grain boundaries, and that this segregation was the possible cause of the threshold stress (σ_0). It also helped to explain (quantitatively) the dependence of σ_0 with temperature and grain size [23]. A detailed analysis is provided later in the chapter.

The segregation of yttrium atoms, the electric charge of which is different from that of the parent ions, induces a local density of negative charge produced by the Y'_{Zr} defects, accounting for a local electric field which is screened by the gradient of oxygen vacancies between the bulk and the boundaries. When the grain size of the polycrystal becomes a few screening lengths ($d \leq 100$ nm, nanoscale length), the electric field can influence the diffusional processes and the creep [Eq. (1)] will be multiplied by a factor α [24]:

$$\alpha = \frac{1}{1 + 4\frac{\lambda}{d}\left[\exp\left(\frac{-z_D e V(R)}{3\varepsilon_r kT}\frac{\lambda}{d}\right) - 1\right]} \tag{2}$$

where λ is the Debye attenuation (screening) length, z_D the valence of yttrium, $V(R)$ the electrical potential, and ε_r the relative dielectric constant.

A plot of α versus the average grain size from Eq. (2) is shown in Figure 15.8. From this plot, it can be seen that the effect of the nanostructured character of the YTZP specimen is increasingly pronounced the bigger λ is, or the smaller the grain size.

Whereas, an ample number of reports have described submicron-sized YTZP, the number decreases when going dropping to nanometer-size, due mainly to the fact that well-densified nanocrystalline YTZP have been available only recently. In fact, a recent improvement of creep resistance was reported in 50- to 70 nm-YTZP deformed at 1200 °C, in agreement with the predictions of that model [25, 26]. Few reports have been made [27–29] on nanocrystalline monoclinic ZrO_2, although among monoclinic materials there is no segregation at the grain boundaries, and the predictions of the model developed when segregation occurs cannot be

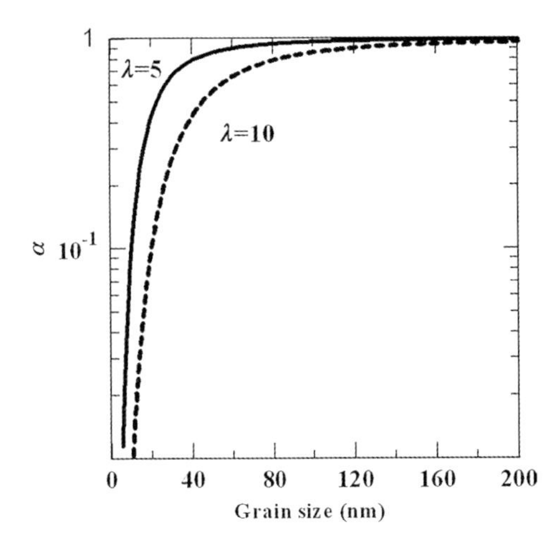

Figure 15.8 Dependence of alpha (the charge effect-induced hardening factor) versus grain size for different values of the Debye length, as displayed in Ref. [24]. Note that a value of $\lambda \cong 10$ nm is more realistic at high-temperature conditions.

tested. At this point, further systematic investigations are necessary on well-defined systems in order to verify the importance of an electric field in the diffusion process, when the grain size decreases to the Debye length scale. In this regard, it has been shown recently that segregation does not occur when the grain size is smaller than a critical value, which is around 50 nm [23], due to a thermodynamic restriction to segregation-induced phase changes. Indirect experimental evidence of this statement was reported by García-Gañán *et al.* [30], who sintered YTZP specimens by means of fast microwave heating which allowed fully dense nanostructured specimens to be obtained, with an average grain size of 40 nm claimed to have been achieved. With these a priori conditions, the plastic deformation of these specimens at 1150–1200 °C fits to a parabolic-type power law with the applied stress, regardless of its value. This is fully consistent with a nonthreshold stress regime, thus implying that no segregation had occurred at the grain boundaries.

Another aspect to be analyzed here is the fact that superplastic flow stress correlates with the ionic radius of the corresponding dopant [31, 32], as reported in 3 mol% YTP (3YTZP) specimens deformed at 1400 °C. Cations with smaller ionic sizes decrease the flow stress, whereas large ionic sizes increase the flow stress. The authors of these reports suggested that the flow stress was determined by the grain boundary diffusivity, which was affected by the dopant segregation. The same improvement in superplasticity was found respectively in 0.3 mol% SiO_2-doped 3YTZP with $d = 0.35\ \mu m$ and 0.18 mol% Al_2O_3-doped 3YTZP with $d = 0.4\ \mu m$ [32, 33]. This behavior also seemed to account for fine-grained Al_2O_3 with ZrO_2 as a dopant [35, 36].

However, various other explanations have been outlined. For example, in the case of Al_2O_3 doped with different impurities, the mechanical behavior can be explained in terms of the change in ionic bonding strength between Al and O and the covalent bonding between Al and the surrounding cations, which affects the grain boundary diffusivity [37–39]. This effect was recently elucidated at the atomic-scale structure using high-resolution transmission electron microscopy (TEM), where large Y-doping ions are energetically more stable at the expanded regions of the grain boundaries, thus increasing the bond strength by increasing the covalency of the Y–O bonds and, as a consequence, increasing the creep resistance despite the small amount of Y [40].

This explanation has been outlined by the same group in SiO_2–TZP doped with several types of metal oxide [39, 41, 42]. This change in bond strength has also been used to explain the superplasticity of SiC doped with small amounts of boron [43]. The doped boron was shown to segregate at the grain boundaries, removing silicon from its site, and then to form bonds in a local environment, similar to that in the B_4C structure [44]. This fact enhances the deformation by grain boundary diffusion.

When a secondary glassy phase is added as a sintering aid, it may act either as a lubricant for GBS, it can provide a preferential location for the nucleation and growth of cavities, or it can improve the diffusivity pathways along the grain boundary and, as a consequence, also the flow stress and elongation to failure. This fact has been reported in barium- and borosilicate-doped 3YTZP [8, 45]. In the case of nonoxide ceramics such as SiC and Si_3N_4, their densification is normally achieved with sintered additives which, by reacting with SiO_2 present in the surfaces of the powders, promote the formation of a higher-diffusivity phase, melting at a lower temperature and producing a densification of these ceramics. The two-phase composite formed was composed of a hard phase surrounded by a soft secondary glassy phase that was shown to control the mechanical behavior of the materials (this point will be referred to later when describing the accommodation processes controlling superplasticity).

In the case of Si_3N_4, glass pockets and thin glass films with thicknesses of about 1 μm often remain at the grain boundaries [46]. The viscosity of the intergranular glassy phase often controls plasticity, together with the solubility of the crystalline phase in the liquid and its diffusivity, as reported elsewhere [9, 47–49]. The solid solution of silicon nitride with some aluminum-based compounds or mixture form the so-called α' or β'-SiAlON compounds. These are made superplastic by the addition of a secondary glassy-phase; for example, Rosenflanz and Chen reported that Li-doped SiAlON deforms 10-fold faster than Si_3N_4 [50]. A revision of the mechanical properties of these compounds is available in Refs [9, 47]. The enhancement of superplastic deformation by an intergranular glass phase was also applicable to liquid-phase sintered SiC [51, 52].

15.4 Accommodation Processes in Superplasticity

The accommodation processes are responsible for the rate-control of superplasticity, and no single mechanism exists to accommodate GBS, even with regards to a particular ceramic system. As noted above, several factors can affect the different mechanisms, among which should be included the nature of the impurities present in the grain boundaries, the secondary phases, and the testing conditions. The different mechanisms for accommodation will be analyzed in the following sections.

15.4.1 GBS Accommodated by Diffusional Flow

This type of mechanism is due to a nonuniform flow of point defects. The mechanism's fundamental topological features are that the grains in the polycrystals change their neighbors, but show almost no change in shape after a large deformation. This fact was first observed by Rachinger in polycrystalline metals [53], and modeled for the first time by Ashby and Verrall (thereafter A-V) [54]. Notably, the latter authors employed a model which was two-dimensional (2-D) in nature, as they maintained that a three-dimensional (3-D) model would, most likely, be much more complicated to handle. Nonetheless, they claimed that only a numerical factor should appear in a rigorous 3-D treatment.

In the model employed, bulk diffusion through the grains and diffusion along the grain boundaries must be considered. These two processes are always dominant at low stresses, because the rates of diffusional flow are usually linear functions of stress, whereas those involving dislocation motion are nonlinear in nature [55].

In this situation, Ashby and Verrall considered a group of four grains deformed at constant pressure and stress. During this flow, four irreversible processes took place:

- A diffusive process by which the grains temporarily changed shape.
- The grain or phase boundaries were sinks and sources of point defects.
- Grain boundary sliding.
- Fluctuations of the boundary area.

By considering the entropy change in these four processes, the constitutive equation of the A-V model, when lattice and grain boundary diffusion are taken into account, can be written as:

$$\dot{\varepsilon} = \frac{100\,\Omega}{kTd^2}\left(\sigma - \frac{0.72\gamma}{d}\right)D_L\left(1 + \frac{3.3\delta D_{gb}}{dD_L}\right) \tag{3}$$

where Ω is the atomic volume of the diffusion controlling species, $\frac{0.72\gamma}{d}$ is a threshold stress for the grain-switching event and γ is the grain boundary free energy, δ is the thickness of the boundary, and D_L and D_{gb} are the lattice and grain boundary diffusion coefficients, respectively.

The principle of this model for grain rearrangement is shown in Figure 15.2b. This type of grain rearrangement retains the equiaxed grains after a large deformation, as shown in the typical microstructure features of superplastically deformed polycrystals [8, 18]. A modified A-V model which accounted for a more realistic symmetrical diffusion path has been developed by Spingarn and Nix, for when diffusional flow occurs only along the grain boundaries [56]. A number of modifications of the original A-V model have also been developed by several groups, and are reviewed elsewhere [14]; however, the main features of the original A-V model remain unchanged.

Two main features can be observed from Eq. (3): first, the strain rate is a linear function of the stress; and second, a threshold stress exists which is due to the fluctuation of the boundary area during the grain switching (although this threshold stress is generally too small). In fact, the value has been calculated in the case of NiO for which γ is $1\,\mathrm{J\,m^{-2}}$ [13], giving a threshold stress according to Eq. (3) equal to 0.08 MPa, which is far too small to be measured in a conventional experimental set-up. This suggests that, if a threshold stress is measured, then its origin must be quite different.

A linear dependence between the strain rate and the stress has been observed in ceramics such as NiO with grain sizes between 9 and 21 μm [13] and with UO_2 having grain sizes between 2 and 10 μm [57]. In the case of YTZP ceramics, the values of n is 1.5 ± 0.3 for a grain size of 1.8 μm [58]. However, for typical materials with a grain size <1 μm – such as many ceramics and ceramic composites, pure and impure metals, solid solution and fine-dispersed secondary phase-containing alloys – the dependence between strain rate and stress cannot be explained by the A-V model [6–8]. In the case of YTZP with grain size 0.3–0.5 μm, values of n between 2 and 5 have been found [8]. A detailed analysis of the n-values in YTZP will be provided later in the chapter.

The experimental data show that the stress exponent in superplasticity regime is 1 for polycrystalline materials with a coarse grain size, but tends towards 2 when the grain size decreases. This tendency has been found regardless of the nature of the grain boundaries, and was first identified by A-V [54] when analyzing diffusional creep. In order to explain this discrepancy, A-V modified their model in terms of an interface–reaction flow mechanism controlling plasticity, and this has been modeled in detail by Artz *et al.* [59]. It is based on the existence of boundary dislocations which have been observed using TEM. In these boundary dislocations, the Burgers vector is not a lattice vector, and therefore they are constrained to move along the grain boundaries. The model that was developed [59] can justify n equal to 2 for solute-drag-limited diffusional creep; however, the grain size dependence was equal to 1, a value not observed in many ceramics [6–8].

Although this explanation might seem adequate for metals, it cannot be applied to ceramics where dislocations at grain boundaries have never been observed, neither at the boundaries nor in the bulk during superplastic deformation.

Another problem arises from the fact that the A-V model is based on the movement of four equal hexagonal grains, and so may be altered for a distribution of smaller grains. A qualitative analysis of the different models accounting for the geometric aspects of superplastic flow has been provided [19], although a rigorous quantitative analysis is still to be conducted. This would be especially necessary to cover some of the weaknesses of the A-V model, the most important of which is the limited grain switching motion that accounts for the grain sliding and which imposes a very severe grain correlation to the system.

The mechanism of GBS accommodated by diffusional flow has been successfully used to explain the superplastic behavior of YTZP. In the case of YTZP, values of p between 1 and 3, of n between 2 and higher than 5, and of Q between 450 and 700 kJ mol^{-1} have been reported during creep [8].

Several explanations have been proposed to justify this scattering in experimental creep data, and a detailed analysis of the mechanisms involved is available [60]. Perhaps the most plausible explanation for the scatter of the creep parameters is based on a single mechanism involving GBS with a threshold stress (σ_0) [8, 60, 61].

When a threshold stress is introduced into the creep equation, all of the creep parameters in YTZP become $n = 2$, $p = 2$, and $Q = 460$ kJ mol^{-1} (the activation energy for cation lattice diffusion in a zirconia–yttria system [8]), whatever the stress or temperature of the test. Moreover, the data could be fitted to a constitutive equation which is identical to that found in metals, when the lattice diffusion is the rate-controlling mechanism [62]:

$$\dot{\varepsilon} = 2 \times 10^7 \frac{Gb}{kT} \left(\frac{\sigma - \sigma_0}{G} \right)^2 \left(\frac{b}{d} \right)^2 D_{lat}^{Zr} \tag{4}$$

The value of this σ_0 was found experimentally [61]:

$$\sigma_0 = 5 \times 10^{-4} \frac{\exp\left(\dfrac{120\ \text{kJ mol}^{-1}}{RT} \right)}{d} \tag{5}$$

where the units of d are μm.

However, as the discrepancy derives from the origin of the threshold stress, two different explanations have been proposed, and these are outlined critically as follows.

The first explanation for threshold stress was proposed recently by Morita and Hiraga [17, 63, 64], who claimed that n-values as high as 5 and Q-values of 680 kJ mol^{-1} can be related to the existence of a threshold stress for intragranular dislocation motion, with such motion being the accommodation process for GBS. This hypothesis was based on the observation of dislocation pile-ups, as reported elsewhere [64].

A dislocation pile-up is not a stable dislocation microstructure, as a strong dislocation repulsion of dislocating each other must occur; however, it will remain stable for as long as an external applied stress exists. Such stress can be calculated using the equation [64]:

$$\tau = \frac{Gb}{2L} N \tag{6}$$

where τ is the shear stress acting on the pile-up plane, L is the pile-up length, and N is the number of pile-up dislocations. Morita and Hiraga reported values of τ for the different pile-ups observed that ranged from 351 to 1260 MPa for nominal applied stresses between 15 and 50 MPa (see Table 15.1 in Ref. [64]). These values for the stress required for pile-up stability were in agreement with reported values of 400 MPa that were required for dislocation motion in yttria–tetragonal zirconia single crystals deformed at 1400 °C [65].

Domínguez-Rodríguez *et al.* have recently assessed the validity of a dislocation-driven model in yttria–tetragonal zirconia polycrystals [66], and concluded that such a mechanism would be quite unlikely to be operative in this system.

Taking into account that, for polycrystals, the applied stress is one-third of the shear stress acting on the pile-up plane, the relationship between the shear stress concentration (τ) and the applied stress σ (τ/σ) must lie in the range between 40 and 75, although this would be too large to be explained by the stress concentration induced at multiple-grain junctions during GBS. Clearly, the dislocations reported by

Table 15.1 Creep laws as predicted by Wakai's model [81], depending on the steps density and the controlling mechanism.

Controlling mechanism	**Density of steps**		
	Constant	**2-D-nucleation**	**Spiral step**
Diffusion in the liquid film	$\dot{\varepsilon} \propto \frac{\sigma}{kTd^3} D_L$ a)		
Diffusion in the adsorption layer	$\dot{\varepsilon} \propto \frac{\sigma}{kTd} D_S$ a)	$\dot{\varepsilon} \propto \frac{\sigma^n}{kTd} D_S$	$\dot{\varepsilon} \propto \frac{\sigma^2}{kTd} D_S$
Reaction at kink	$\dot{\varepsilon} \propto \frac{\sigma}{kTd} \frac{1}{\tau_K}$ b)	$\dot{\varepsilon} \propto \frac{\sigma^n}{kTd} \frac{1}{\tau_K}$	$\dot{\varepsilon} \propto \frac{\sigma^2}{kTd} \frac{1}{\tau_K}$

a) D_L and D_S are the diffusion coefficients in the liquid phase and the adsorption layer.
b) τ_K is a characteristic relaxation time for the integration of solute at kinks. $\tau_K \propto \exp\frac{\Delta G_K}{kT}$ where ΔG_K is the free enthalpy for the reaction at the kink.

Morita and Hiraga are more likely to be an artifact rather than an intrinsic mechanism, being created during cooling when the creep experiment has been completed.

Recently, another model was developed on the basis of segregation of the yttrium atoms at the grain boundaries, to account for σ_0. This model was capable of explaining, on a quantitative basis, the dependence of σ_0 with both temperature and grain size [23].

Nowadays, it is accepted that yttrium segregates at the grain boundaries or at the dislocation cores in YTZP. The origin of this segregation is a relaxation of the elastic energy around the yttrium atoms, as a consequence of the difference between the ionic radius of Y^{3+} and the Zr^{4+}, with Y^{3+} being about 20% larger than Zr^{4+}. This has been proven experimentally by using different techniques [67–72]. For example, yttrium segregation depends on both the temperature and bulk concentration (c_b) [73], and was shown recently also to depend on the grain size [72]. The space charge potential (V) due to the segregation has been measured, using impedance spectroscopy techniques, as a function of the grain size in a 3 mol% YTZP, and found to increase from 0.18 to 0.25 V when the grain size increased from 100 to 1000 nm [72]. Cation segregation to the grain boundaries alters the chemical composition as well as the electric space charge at the boundaries, creating a local electric field (E), that consequently affects the deformation mechanisms such as grain boundary sliding; this, in turn, may affect the mechanical behavior of the zirconia alloys, as shown recently in several reports. Details of how the segregation can influence the superplasticity in YTZP can be found in Refs [24, 74].

If it is assumed that the interfacial region between two grains is very narrow and is in electrostatic equilibrium, then the electric field can be determined as a function of the yttrium bulk concentration, the thickness of the segregation layer (Debye length λ), and of the reduction in energy per atom of dopant when one of these segregates to the grain boundaries (the absorption free energy ΔG) (for details, see Refs [24, 74]).

During deformation, one grain displaces in respect to each other, and the work done during such displacement can be written as [24, 74]:

$$\delta W = C\frac{1}{2}\varepsilon E^2 \delta V \tag{7}$$

where C is a geometric constant and δV is the infinitesimal change in volume between two grains upon sliding an infinitesimal distance $\delta\xi$. If ω is the thickness of the intermediate region between the grains and d is the grain size, then $\delta V = \delta\omega\delta\xi$.

As the work done during the displacement is also equal to the net force F per unit area times $\delta\xi$ (and this force can be considered as that needed to be applied for GBS to proceed), it can be considered as the threshold stress (σ_0) which, by substituting the different expression, can be written as:

$$\sigma_0 = \frac{A}{d}\left[\exp\left(-\frac{\Delta G}{kT}\right)-1\right]^2 \propto \frac{1}{d}\exp\left(-\frac{2\Delta G}{kT}\right) \tag{8}$$

where A is a constant with no temperature dependence.

In order to analyze Eq. (8), it is necessary to know the absorption free energy in YTZP. However, although it appears that no experimental data are available for this magnitude, in the case of the solid solution of antimony in copper this energy is $-0.68\,\text{eV}\,\text{atom}^{-1}$ [73]. In this system, the difference between the atomic radius of Sb ($r_1 = 1.53$ Å) and that of copper ($r_1 = 1.23$ Å), and the misfit between the two atoms is $+0.194$, which very close to that found in an yttria–zirconia solid solution [74]. On the other hand, the shear modulus of both systems is 41 GPa for Sb–Cu [73] and 45 GPa for YTZP [76]. Another similar system which can be compared to YTZP is the ceramic alloy Y_2O_3–TiO_2 [77], in which the Ti^{4+} cations have an ionic radius equal to 0.68 Å, whereas that for Y^{3+} is 1.015 Å. This creates an elastic energy equal to 0.60 eV at room temperature. As the misfit in this system is very close to that of the YTZP system, it is plausible to accept that a value of 0.60 eV would account for the segregation energy in YTZP ceramic alloys, in which the ionic radii are very close to those for Y_2O_3–TiO_2. In consequence, the experimental values from Eq. (8) are reasonable.

As can be observed in Eq. (8), the impurity segregation at the grain boundaries not only justifies the existence of the threshold stress, but also provides data relating to the grain size and temperature dependence found experimentally.

15.4.2 GBS Accommodated by Dislocation Motion

Another accommodation process widely explored in metals and metallic alloys is GBS accommodated by dislocation motion. When GBS is accommodated by the movement of dislocations, the strain rate can be deduced in a similar way as made in recovery creep: the dislocations generated at the grain boundaries move either along the grain boundaries or across the grain until they are piled up at an obstacle. This situation is then overcome when the stress at the head of the pile-up is high enough to promote the climb of the head dislocation (applied force). The climb velocity is governed by the rate of diffusion of point defects (chemical force). From this picture, the strain rate controlling GBS can be written as [78]:

$$\dot{\varepsilon} \approx \frac{2DL\sigma^2 b^3}{3\sqrt{3}h^2 GkTd} \tag{9}$$

with D being an appropriate diffusion coefficient, L the length of the pile-up, and h the climb distance.

Depending on the grain size, two limit cases have been analyzed [78]. At small grain sizes, it transpires that $L \sim \bar{l}$ and $h \sim 0.3\bar{l}$, where $\bar{l}$ is the mean linear intercept grain size ($d \cong 1.7\bar{l}$) and $D = D_{gb}$, such that Eq. (9) becomes:

$$\dot{\varepsilon} \approx \frac{AD_{gb}Gb}{kT}\left(\frac{b}{d}\right)^2\left(\frac{\sigma}{G}\right)^2 \tag{10}$$

where A is a dimensionless constant with a value of 10.

Conversely, when the grain size is large, $L \approx 20\frac{Gb}{\sigma}$, $h \approx \frac{bG}{16\sigma}$, and $D = D_l$ (for more details see Ref. [78]); then, by substituting these parameters Eq. (9) becomes:

$$\dot{\varepsilon} \approx \frac{AD_L Gb}{kT}\left(\frac{b}{d}\right)\left(\frac{\sigma}{G}\right)^3 \tag{11}$$

where A is a dimensionless constant having a value of the order of 10^3. The above analysis shows that, for materials with low grain sizes, the creep parameters are very similar to those found in superplastic ceramics. However, there are certain discrepancies, which are as follows:

- Equation (10) shows that the stress and grain size exponents are equal to 2; however, it is D_{gb} – the appropriate diffusion coefficient – that controls the accommodation process. Again, in the case of YTZP where a large number of experimental results exist [8, 60, 62], the activation energy has been always associated with the lattice diffusion of cation defects. On the other hand, for large grain sizes, the activation energy corresponds to that of the lattice diffusion coefficient; however, the stress exponent is 3 and the grain size exponent 1 [13, 57, 58], in disagreement with the reported experimental outputs.
- The above equations are deduced assuming the existence of dislocation activity. Although dislocation activity has been shown in YTZP [17, 63, 64], this is likely to be an artifact (for details, see Section 15.3.1). Recently, superplastically deformed nano-MgO with grain size of 37 nm has been reported to deform at temperatures between 700 and 800 °C, and the stress exponent results in a value of 2. Dislocation activation in this system with this grain size requires an applied compressive stress in excess of 3 GPa. Such calculated stresses are far higher than the yield stresses measured experimentally (i.e., 190 to 640 MPa) [79].

In conclusion, although this mechanism has been used to account for superplasticity in metals and metallic alloys, at present there is no evidence of its applicability in ceramics.

For second glassy-phase ceramics, the accommodation processes are governed by these phases, in spite of the diffusional processes occurring during creep. At high temperatures, the glassy phase may become less viscous and even liquid, and in consequence this may account for the plastic deformation. However, viscous flow creep is not regarded as a viable creep mechanism for superplasticity due to its limited deformation, which corresponds to the redistribution of the glassy phase and therefore to the squeeze of these secondary phase from grain boundaries subjected to compression [9]. In the next section, the two mechanisms that are considered to be controlled by secondary phase in ceramics superplasticity will be analyzed.

15.4.3 Solution–Precipitation Model for Creep

For ceramics with secondary glassy phases, the accommodation processes are governed by these phases in different ways:

- The glassy phases may act as a lubricant for GBS.

- They can improve the diffusivity pathways along the grain boundary.
- They can provide a preferential location for the nucleation and growth of cavities.

In this case, the secondary phases melt at temperatures lower than the matrix and, provided that the crystals are at least partially soluble in the glassy phase, creep may take place by:

- solution of the crystal in the liquid phase at grain boundaries under compression;
- diffusion along the liquid phase; and
- precipitation of the crystalline material at grain boundaries under traction.

The first solution–precipitation model was proposed by Raj and Chyung [80], using the original Nabarro diffusional creep model. The external stress induces a variation of the chemical potential at the solid–liquid interface. Depending on whether the solution of the crystal in the glassy phase or the diffusion along the glassy phase is the rate-controlling mechanism, two cases were modeled:

- The strain rate is controlled by the diffusion along the glassy phase. The case is similar to that modeled by Coble, and the strain rate is written as:

$$\dot{\varepsilon} = \frac{1}{\eta}\frac{1}{d^3}\sigma \tag{12}$$

where η is the viscosity of the secondary phase.

- The strain rate is controlled by the reaction of solution and precipitation at the interfaces. This is termed as interface-reaction-controlled creep, and the strain rate is written as:

$$\dot{\varepsilon} = \frac{1}{d}k_d\sigma \tag{13}$$

where k_d is a reaction constant.

This model postulates that the glassy phase in compression can support normal stresses because of the existence of islands that avoid the complete squeezing of the intergranular liquid. However, it has been shown that these islands are not necessary for the grain boundaries to support normal stresses. Several modifications of this first model have been created and revised [9].

An important modification of the Raj and Chyung [80] model was made by Wakai [81], who assumed that the solution and precipitation took place at steps (kinks) formed at the grain boundaries. It was proposed that the solution–precipitation process involved the movement of these steps, and the strain rate was therefore related to the step velocity and density, with an expression analogous to Orowan's equation for dislocation movement:

$$\dot{\varepsilon} = \frac{\varrho_S a v_S}{d} \tag{14}$$

where ϱ_S is the density of surface steps per unit length, a is the height of the steps, and v_S the velocity of the steps. This velocity depends on the process of integration into the

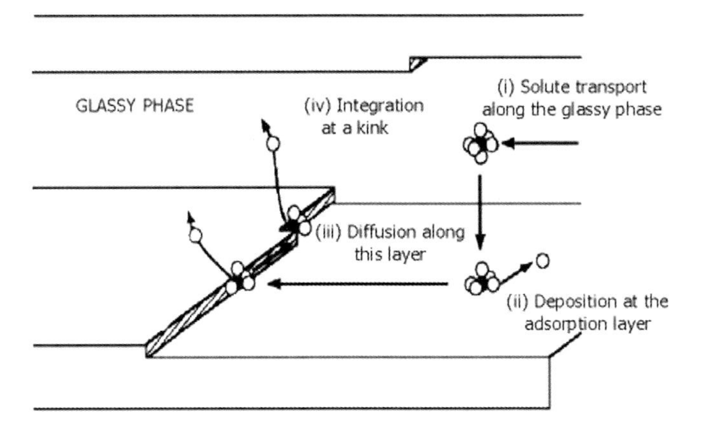

Figure 15.9 Schematic representation of the solution–precipitation mechanism. For further details, see Ref. [9].

crystal at a kink, the diffusion in the absorption layer, and the diffusion in the liquid film. The rate-controlling process will be the most resistant to the step movement.

The velocity of the steps is then written as:

$$v_S \propto \frac{\sigma}{kT}[R_1 + R_2 + R_3]^{-1} \qquad (15)$$

where R_1, R_2, and R_3 are the resistances to the step movement due to the processes of integration into the crystal at a kink, diffusion in the absorption layer, and diffusion in the liquid film, respectively. The most resistant of these would be rate-controlling in nature.

Figure 15.9 represents a scheme of the solution–precipitation step model.

Three different situations were analyzed by Wakai for the density of steps:

- The constant density of surface steps is independent of the applied stress.
- If the initial density of steps is very low, 2-D nucleation of the surface steps occurs.
- If the continuous source of steps is a screw dislocation, then a spiral step is generated.

The combination between the rate-controlling process and the density of steps give the different equations of the Wakai's model [81], which are summarized in Table 5 of Ref. [9] and displayed in a simplified way in Table 15.1.

The case of constant density of steps modeled by Wakai is equivalent to the diffusion-controlled creep modeled by Raj and Chyung [80], and it is also consistent with terms of the stress, temperature and grain size dependence of the strain rate for interface-reaction-controlled creep predicted by others [80]. However, in the two cases of bidimensional nucleation of step and spiral step, the creep parameters differ from those predicted by the authors cited above. In particular, for 2-D nucleation there is a divergence of the creep parameters which has been recently solved [81], considering in detail the precipitation or solution of the crystalline material at the step, which changes significantly the free enthalpy involved in the process.

From the Wakai's creep equation, the predicted n-values differ from 1, but this may lead to severe physical disagreement with the experimental evidence, as it yields an "apparent" stress exponent given by:

$$n^{app} = \left(\frac{\partial \ln \dot{\varepsilon}}{\partial \ln \sigma}\right)_T = \frac{2}{3} + \frac{\pi \gamma_S^2}{3a\sigma kT} \quad (16)$$

In the case of β-silicon nitride, and following Wakai's assumption – namely that the solution of silicon nitride in the glassy phase is ideal – the step energy per unit length γ_S can be evaluated from the enthalpy of fusion per molecule, Δh_f, in the form:

$$\gamma_S \cong \frac{N \Delta h_f}{\ell} \quad (17)$$

where ℓ is a characteristic unit cell length and the factor N is the number of molecules per unit cell; for β-silicon nitride, $N = 2$. The enthalpy Δh_f obviously depends on the particular solution or precipitation process for silicon nitride in the glassy phase. According to Raj and Morgan [82], its value can be estimated as $\Delta h_f \approx 10^{-19}$ J per molecule for most glassy phase systems. Thus, taking ℓ equal to the shortest lattice parameter for β-silicon nitride ($\ell \approx a = 3 \times 10^{-10}$ m [83]), the estimate for the step energy per unit length is $\gamma_S \approx 7 \times 10^{-10}$ J m^{-1}. Figure 15.10 represents the apparent stress exponent for several values of a and γ_S.

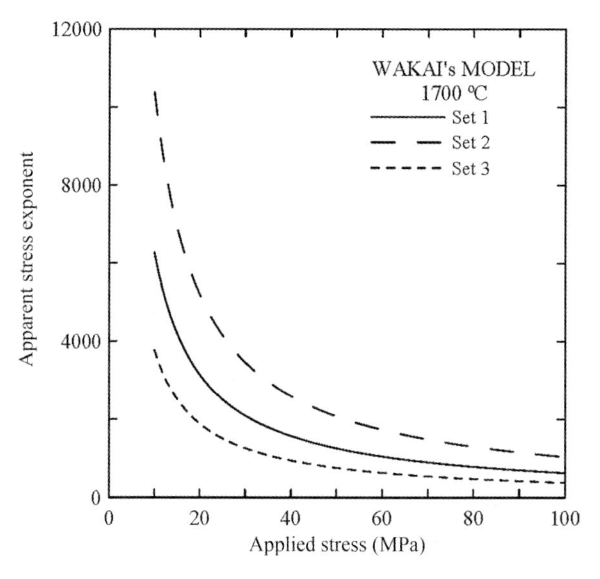

Figure 15.10 Dependence of the apparent stress exponent versus the applied stress, as predicted by Wakai's model [81]. The set of curves corresponds to different values of the surface step energy as shown: set 1, $\gamma_s = 7 \times 10^{-10}$ J m^{-1} and $a = 3 \times 10^{-10}$ m; set 2, $\gamma_s = 9 \times 10^{-10}$ J m^{-1} and $a = 3 \times 10^{-10}$ m; set 3: $\gamma_s = 7 \times 10^{-10}$ J m^{-1} and $a = 5 \times 10^{-10}$ m. Data from Ref. [49].

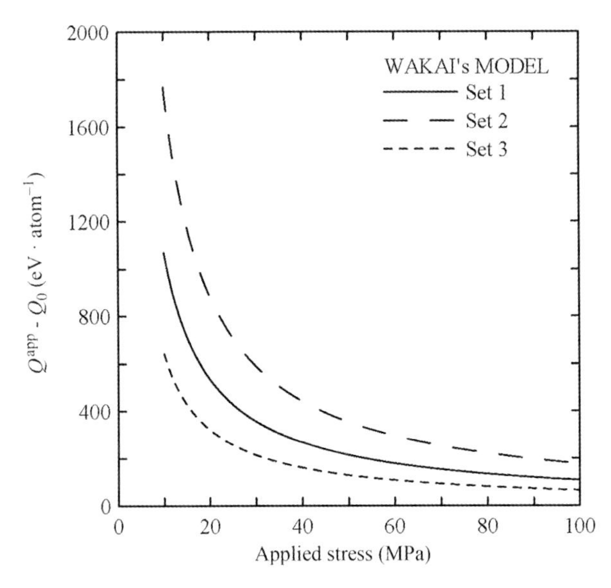

Figure 15.11 Dependence of the deviation of the apparent activation energy from the cation diffusion energy, as predicted by Wakai's model [81]. The three curves correspond to the three values of the surface step energy used for Figure 15.5. See Ref. [49] for further details.

Similar inconsistencies are found when analyzing the theoretical predictions of this model for the apparent activation energy. In the original model, this is given by:

$$Q^{\mathrm{app}}(\sigma) = Q_0 + \frac{\pi\gamma_S^2}{3a\sigma} \tag{18}$$

In Figure 15.11, it can be observed that the apparent activation energy is very high, whereas the reported values for the apparent activation energy for compressive creep in silicon nitride lie within the range 3.6–9.0 eV atom^{-1} [84]. Furthermore, there is a clear divergency towards the low-stress regime.

Recently, a modification of Wakai's model has been created [49], based on the fact that the detailed examination of a nucleation process requires the addition of two new terms to the Wakai's free energy [81]. One term is related to the free energy increases due to the molecular volume change when the solute (which initially was dissolved in the liquid phase) precipitates into the crystal at a step. The second term arises from the fact that an oversaturated solution of crystalline material in the glassy phase, giving rise to precipitation phenomena, is not in equilibrium with the crystal, at least locally in the surroundings of the step [85, 86] (also see Ref. [49] for further details).

With the change of the free energy, the apparent stress exponent and the apparent activation energy is now written as:

$$n^{\mathrm{app}} = \frac{\partial \ln\dot{\varepsilon}}{\partial \ln\sigma} = 1 - \frac{1}{3}\frac{\sigma_0}{\sigma_0 + \Delta g_v} + \frac{\pi}{3akT}\frac{\sigma_0\gamma_S^2}{(\sigma_0 + \Delta g_v)^2} \tag{19}$$

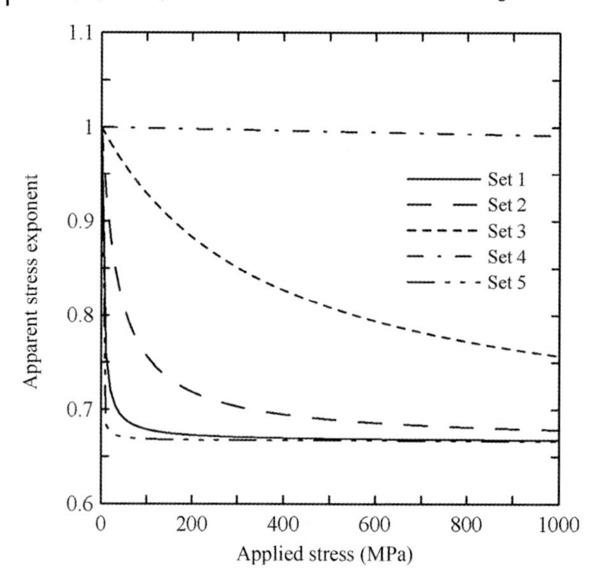

Figure 15.12 Apparent values of the stress exponent at 1700 °C, as predicted by the modified model developed by Meléndez-Martínez *et al.* [49]. The five curves correspond to the following values of $\Delta\mu^{\text{solid-sat}}$, respectively: set 1, $\Delta\mu^{\text{solid-sat}} = 3.7 \times 10^6\ \text{Jm}^{-3}$; set 2, $\Delta\mu^{\text{solid-sat}} = 3.7 \times 10^6\ \text{Jm}^{-3}$; set 3, $\Delta\mu^{\text{solid-sat}} = 3.7 \times 10^8\ \text{Jm}^{-3}$; set 4, $\Delta\mu^{\text{solid-sat}} = 3.7 \times 10^{10}\ \text{Jm}^{-3}$; set 5, $\Delta\mu^{\text{solid-sat}} = 3.7 \times 10^5\ \text{Jm}^{-3}$.

and

$$Q^{\text{app}}(\sigma) = Q_0 + \frac{\pi\gamma_S^2}{3a(\sigma + \Delta\mu^{\text{solid-sat}})} \tag{20}$$

Figures 15.12 and 15.13 display n^{app} and Q^{app} respectively when using the new approach.

In Figure 15.12, the different sets have been obtained for values of $\Delta\mu^{\text{solid-sat}}$ from $3.7 \times 10^5\ \text{J m}^{-3}$ to $3.7 \times 10^{10}\ \text{J m}^{-3}$, which correspond to a situation close to equilibrium, and one far from equilibrium, respectively. As can be observed in the figure, there is no longer any divergence in the limit $\sigma \to 0$, but the value tends to $n^{\text{app}} = 1$ as the applied stress decreases. The value decreases monotonically with increasing applied stress to an asymptotic value of $n \approx 0.7$ at high stresses. This behavior has been reported in Si_3N_4 deformed in compression [87].

$(Q^{\text{app}} - Q_0)$ as a function of the applied stress is displayed in Figure 15.13, and is obtained for $\gamma_S = 7 \times 10^{-10}\ \text{J m}^{-1}$, $a = 3 \times 10^{-10}\ \text{m}$. The different sets are obtained for the same values of $\Delta\mu^{\text{solid-sat}}$ used in that figure.

In the case of diffusion-controlled creep ($\beta \geq 1$), the n^{app} depends on $\frac{\partial \ln\sigma_0}{\partial \ln\sigma} = 1 + \frac{\sigma}{R_3}\left(\frac{\beta^2}{1+\beta}\right)\frac{\partial R_1}{\partial \sigma}$, and although this quantity is very difficult to estimate, a comparison with experimental data for silicon nitride ceramics, the main deforma-

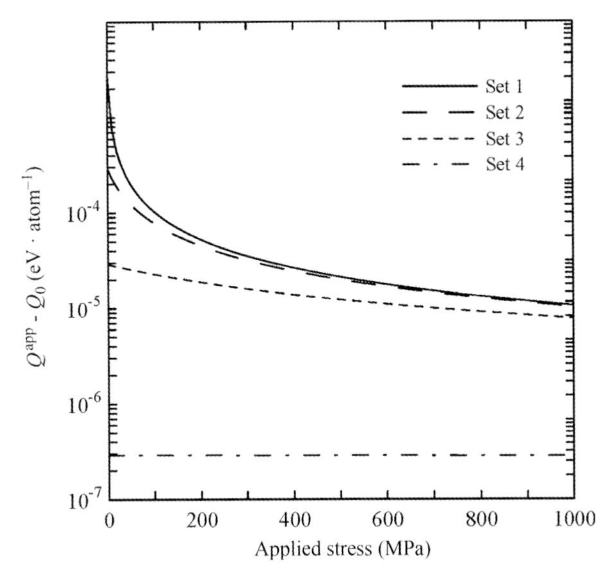

Figure 15.13 Dependence of the deviation of the apparent activation energy from the cation diffusion energy, as predicted by Meléndez-Martínez *et al.* [49]. The five curves correspond to the same values of $\Delta\mu^{solid-sat}$ as used for Figure 15.7.

tion features of which are consistent with the step model [81], suggests that $\frac{\partial \ln \sigma_0}{\partial \ln \sigma} \approx 0.8\text{-}0.9$.

For the activation energy, the same predictions made for reaction-controlled creep also apply here.

15.4.4
Shear-Thickening Creep

Shear-thickening creep was postulated to explain the compressive superplastic deformation of SiAlON, which undergoes a transition from $n = 1$ (Newtonian behavior) to $n = 0.5$ for a characteristic critical stress (σ_c), independent of both temperature and the composition of the secondary phase [88].

Figure 15.14 reports such a tendency, which is described below for the case of a SiAlON system.

The model is based on the idea that the glassy phase is composed of two layers: one layer is a normal glassy phase layer which behaves in a Newtonian way and is embedded into an overcondensed layer with a non-Newtonian behavior. Thus, for stresses lower than the critical stress, the creep is controlled by the normal glassy phase ($n = 1$). Then, when the stress overcomes a critical value, the squeeze of this phase forces the two overcondensed layers to come into contact with each other, such that the material will creep in a non-Newtonian manner ($n = 0.5$). Hence, the creep rate can be written as:

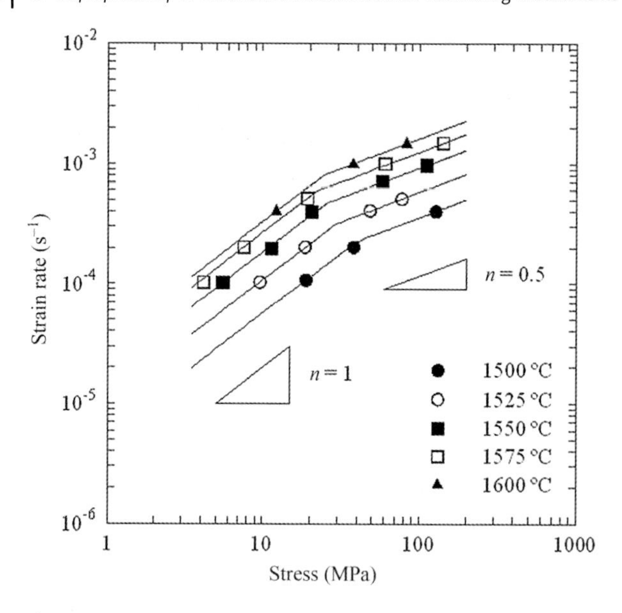

Figure 15.14 Strain rate versus stress plots, at temperatures between 1500 and 1600 °C, for a SiAlON (83.73% Si_3N_4, 7.77% AlN, 3.73% Al_2O_3, 3.70% Y_2O_3), showing the Newtonian ($n = 1$) → shear thickening ($n < 1$) transition. Note that the transition stress σ^* is approximately independent of the temperature. Data from Ref. [9].

$$\dot{\varepsilon} = (1-\nu_f)^{2.5}\frac{\sigma}{\eta'} \tag{21}$$

where η' is an apparent viscosity depending on the temperature, grain size, phase composition and liquid content, and ν_f is the volumetric fraction of the overcondensed rigid phase, which for a $\sigma \geq \frac{2}{3}\sigma_c$ yields a value:

$$\nu_f = 1-\sqrt{\frac{1}{4}+\frac{\sigma_c}{2\sigma}} \tag{22}$$

which explains the transition from $n = 1$ to $n = 0.5$. Again, a review of the model can be found elsewhere [9].

15.5 Applications of Superplasticity

The increasing applications of advanced ceramics in technical areas, including aerospace, automobile, sanitary, energy, electronics, and biology, often require complex shaped pieces that can be manufactured at low cost. The extensive potential applications of superplasticity, together with the possibility of processing dense ceramics and forming complex structures, have led to the synergistic appearance and rapid development of a large number of ceramic systems with superplastic capabilities. Several industrial processes within the metal and polymer industries have already made use of these highly ductile ceramics. The processing of dense ceramics includes sheet forming, blowing, stamping, forging, and joining. A good example of the superplastic forming of different ceramics is shown in Figure 1 of Ref. [4], where a flat, 1 mm-thick disk of Si_3N_4, 2Y-TZP:Al_2O_3, 2Y-TZP:Mullite and 2Y-TZP + 0.3% doped with Mn, Fe, Co, Cu, and Zn was stretched with a 6.5 mm radius punch at temperatures and forming times that depended on the ceramics used.

Sinter forging represents a very promising technique, mainly because the densification and net-shaping are achieved simultaneously [88]. Another advantage is that sinter forging avoids the creation of cavities and voids because it is produced by compression. A high strength and high fracture toughness of Si_3N_4 has been achieved by superplastic sinter forging, due not only to a reduction in flaw size but also to grain alignment [89].

When accommodated by some of the mechanisms involving dislocation movement or the diffusion of point defects, GBS forms the basis of the structural superplastic behavior of these materials (see Section 15.2). By taking advantage of the processes involved in superplasticity, it is possible to join ceramics superplastically. For example, when two pieces of the same ceramics in contact are deformed within a superplastic regime (i.e., as soon as GBS is activated), the grains of one part interpenetrate those of the other part. This produces a rapid and perfect junction of the two, in such a way that a shorter time and a lower temperature can be used than are commonly required in other conventional process for ceramics joining [90].

An example of this joining technique – and good proof again that GBS is the mechanism of superplasticity – is illustrated in Figure 15.15, which shows two 3Y-TZPs with different grain sizes joined at 1400 °C during a 15 min period. The use of nanoceramics as an interlayer can also lead to a drastic reduction in the temperature of joining. A good example of this was demonstrated for layers of YTZP which, in the presence an interlayer of 20 nm YTZP, could be joined at only 1150 °C – a temperature 200 °C below that required in the absence of the nanolayer [91].

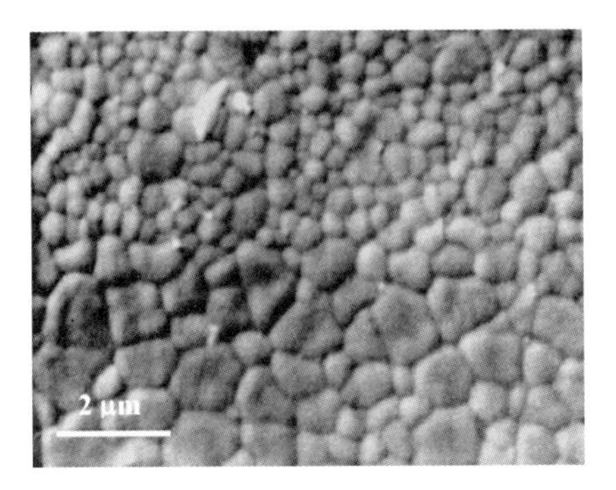

Figure 15.15 Micrograph of a ceramic joining of two ceramic parts. Note that the ceramic interface displays a perfect matching; that is, no voids or cracks can be found.

On the other hand, fracture behavior, wear resistance and superplasticity are grain size-dependent. Materials with coarse grains are more creep resistant than those with smaller grains [Eq. (1)]; a similar tendency is found for the toughness when the crack-bridging mechanism is operative [92]. In contrast, an opposed response is reported to occur for wear resistance [93, 94] and also for transformation toughness [95]. Based on this behavior, it is possible to join ceramic layers of different grain sizes so as to obtain a multilayer composite with clean and strong interfaces and heterogeneous mechanical behavior at both room and high-temperature, depending on the stress application. A good example of this can be found in Ref. [95]. Such junctions can be achieved by annealing the same as-received ceramics in order to obtain different grain sizes. Alternatively, the same behavior may be obtained when two layers of different compositions are joined, because in the case of yttria-stabilized zirconia (YSZ), the grain size is a function of the yttria content [96]. By using this technique, a multilayer made from four different layers 0.5 mm-thick 3YTZP with grain sizes of 0.3, 0.5, 0.8, and 1.0 μm, obtained by annealing the as-sintered ceramics (0.3 μm), have been joined together sequentially at 1400 °C; the result was a compound with anisotropic mechanical behavior, depending on the compression axis parallel or perpendicular to the interface [97]. When the compression axis was perpendicular to the interface, all of the layers were submitted to the same stress (thereby, the "isostress test"), and the layer with the smaller grain size controlled the superplasticity. However, when the compression axis was parallel to the interface, all of the layers suffered the same strain (thereby, the "isostrain test"), and the layer with the biggest grain size controlled the plasticity.

When using the creep model for duplex microstructure, as developed by French *et al.* [98], it is possible to fabricate a composite with a defined creep resistance that will control the grain size, the width of the layers, and the compression axis. In the case of isostrain, where the strain and strain rate are the same for each phase, and assuming the creep equation in a compact form as, $\dot{\varepsilon} = A_i \sigma^{n_i}$ the stress for this configuration, σ_c, is given as:

$$\sigma_c = \sum_{i=1}^{m} V_i \left(\frac{\dot{\varepsilon}_c}{A_i} \right)^{1/n_i} \tag{23}$$

where V_i is the volume fraction of each type of layer, n_i is the stress exponent (which is assumed to be the same for each type of layer), and m is the number of layers different. In this case, the mechanical behavior of the laminated compound will be controlled by the layer with the coarse grain size.

For the isostress case, the stress is the same for each layer and the strain rate for this configuration can be written as:

$$\dot{\varepsilon}_c = \frac{\sum_{i=m}^{m} A_i \sigma^n}{m} \tag{24}$$

In this case, the mechanical behavior of the laminated compound will be controlled by the softer layer deformed at the strain rate imposed by the testing machine.

In conclusion, with this technique, it is possible to obtain a defined functional gradient material (FGM).

15.6 Future Prospective in the Field

Ceramic superplasticity must still be regarded as a challenging issue from both scientific and technical points of view, and much endeavor remains to be undertaken in order to achieve a satisfactory understanding of the phenomenon. Indeed, this represents a necessary landmark to optimize the conditions for superplastic response according to industrial requirements.

From a basic point of view, two major goals must be fulfilled: the first goal comprises a deep analysis of the correlation between the nature of grain boundaries and their glide response under mechanical stresses. At this point, the role played by impurity segregation as agents that either enhance or inhibit (or even avoid) grain sliding must be fully defined. Recent valuable contributions have been reported by Ikuhara *et al.* [42], whose pioneering studies have described the sliding response of a bicrystal under shear stresses. A bicrystal is considered an ideal model for a grain boundary, as it can be doped with impurity cations under very controlled conditions. Notably, Ikuhara and coworkers' studies have highlighted the remarkable effects of small doping cations on the shear stresses for gliding at a fixed strain rate. Yet, this remains only a first step towards achieving a full correlation between grain boundary microstructure and plastic response.

A second major problem involves the consequences of the multiscale nature of superplasticity, notably the correlation between the individual behavior of one grain under shear and normal stresses, and the average collective behavior of these grains. This is a very difficult task, as plasticity is a nonequilibrium phenomenon, and grain motion cannot be completely treated as being thermal in nature. Rather, it implies that the usual methods of thermodynamics and statistical physics cannot be applied – or perhaps they can, but under certain restrictions.

Related to this problem is the fact that the topology of grain contact is likely to have a major influence on the shear capabilities of grains. Several clues have indicated, in particular, the origin of the stress exponent for superplasticity, and the dependence of the stress exponent on average grain size. More precisely, the reasons why the stress exponent in most ceramic systems seems to fit to a value 2, but in established models (such as the A-V model) it predicts a value equal to 1, remain unclear. Moreover, there remains no explanation for the transition from 2 to 1 with grain size. Clearly, a deep insight into the equations featuring superplasticity in monolithic ceramics is required.

When focusing on more technical aspects, the path towards high-strain rate superplasticity (HSRS) ($\dot{\varepsilon} \geq 10^{-2}\,s^{-1}$) and superplasticity at low temperatures has emerged as the main approach for industrial applications. In this regard, there are two good examples – one concerning the HSRS [98] and other the superplasticity at low temperature [78] – although obtaining these two conditions systematically is still a long way off.

To a large extent, research in superplasticity – whether from a scientific or a technical aspect – is today promoted by the new scientifically designed ceramic microstructures developed in ceramic systems. These are the logical result of the significant improvements in ceramic powder processing technology, with new high-purity ceramics with nanoscaled grain sizes benefiting from new techniques, such as hot-isostatic pressing, spark plasma sintering, or microwave sintering. Among these, one remarkable "intelligent" technique is that of a two-step temperature sintering process for nanocrystalline ceramics fabrication, developed by Chen and Wang [100]. This method exploits the fact that grain boundary migration and grain boundary diffusion exhibit different kinetics. Thus, a granular material can be heated to a sufficiently high temperature for a short time so as to achieve a minimum critical densification value (ca. 70–80% full density). The specimen is then cooled to a much lower temperature, at which grain growth is negligible as the grain boundary migration is inhibited but still remains a significant driving force, allowing the pores to shrink by capillarity forces. This technique has been used successfully for the fabrication of yttria and other ceramics [100, 101].

Clearly, the extensive use of ceramics will lead to the development of increasingly sophisticated structures, including extensively nanostructured materials. Moreover, the benefits of ceramic design will lead to new scientific challenges, creating new structural and functional applications. Despite the world of ceramics superplasticity being born over twenty years ago, its transition from childhood to maturity remains an ongoing process.

Acknowledgments

The authors acknowledge financial support from the Spanish 'Ministerio de Educación y Ciencia,' via research project MAT2009-14351-C02-01 and the regional government 'Junta de Andalucía' through the project of excellence FQM–337.

References

1 Bengough, G.D. (1912) *J. Inst. Metals*, **7**, 123.
2 Garvie, R.C., Hannink, R.C., and Pascoe, R.T. (1975) *Nature*, **258**, 703.
3 Wakai, F., Sakaguchi, S., and Matsuno, Y. (1986) *Adv. Ceram. Mater.*, **1**, 259.
4 Chen, I.-W. and Xue, L.A. (1990) *J. Am. Ceram. Soc.*, **73**, 2585.
5 Nieh, T.G., Wadsworth, J., and Wakai, F. (1991) *Int. Mater. Rev.*, **36**, 146.
6 Chokshi, A.H. (1993) *Mater. Sci. Eng.*, **A166**, 119.
7 Nieh, T.G., Wadsworth, J., and Sherby, O. (eds) (1997) *Superplasticity in Metals and Ceramics*, Cambridge University Press, London.
8 Jiménez-Melendo, M., Domínguez-Rodríguez, A., and Bravo-León, A. (1998) *J. Am. Ceram. Soc.*, **81**, 2761.
9 Meléndez-Martinez, J.J. and Domínguez-Rodríguez, A. (2004) *Prog. Mater. Sci.*, **49**, 19.
10 Albuquerque, J.M., Harmer, M.P., and Chou, Y.T. (2001) *Acta Mater.*, **49**, 2277.
11 Boullier, A.M. and Gueguen, Y. (1975) *Contrib. Mineral. Petrol.*, **50**, 93.
12 Schmid, S., Boland, J.N., and Peterson, M.S. (1977) *Tectonophysics*, **43**, 257.

13 Jiménez-Melendo, M., Domínguez-Rodríguez, A., Marquez, R., and Castaing, J. (1987) *Philos. Mag. A*, **56**, 767.
14 Langdon, T.G. (1981) *Metals Forum*, **4**, 14.
15 Ishihara, S., Tanizawa, T., Akashiro, K., Furushiro, N., and Hori, S. (1999) *Mater. Trans., JIM*, **40**, 1158.
16 Duclos, R., Crampon, J., and Carry, C. (2002) *Philos. Mag. Lett.*, **82**, 529.
17 Morita, K. and Hiraga, K. (2003) *Scripta Mater.*, **48**, 1403.
18 Duclos, R. (2004) *J. Eur. Ceram. Soc.*, **24**, 3103.
19 Zelin, M.G. and Mukherjee, A.K. (1996) *Mater. Sci. Eng.*, **A208**, 210.
20 Muto, H. and Sakai, M. (2000) *Acta Mater.*, **48**, 4161.
21 Yasuda, K., Okamoto, T., Shiota, T., and Matsuo, Y. (2006) *Mater. Sci. Eng. A*, **418**, 115.
22 Mukherjee, A.K. (2002) *Mater. Sci. Eng. A*, **322**, 1.
23 Gómez-García, D., Lorenzo-Martín, C., Muñoz, A., and Domínguez-Rodríguez, A. (2003) *Philos. Mag.*, **83**, 93.
24 Gómez-García, D., Lorenzo-Martín, C., Muñoz-Bernabé, A., and Domíguez-Rodríguez, A. (2003) *Phys. Rev. B*, **67** (14), 144101.
25 Gutierrez-Mora, F., Jimenez-Melendo, M., Domínguez-Rodriguez, A., and Chaim, R. (2000) *Key Eng. Mater.*, **171–74**, 787.
26 Gutierrez-Mora, F., Domínguez-Rodriguez, A., Jimenez-Melendo, M., Chaim, R., and Hefetz, M. (1999) *Nanostruct. Mater.*, **11**, 531.
27 Roddy, M.J., Cannon, W.R., Skandan, G., and Hahn, H. (2002) *J. Eur. Ceram. Soc.*, **22**, 2657–2662.
28 Yoshida, M., Shinoda, Y., Akatsu, T., and Wakai, F. (2002) *J. Am. Ceram. Soc.*, **85**, 2834.
29 Yoshida, M., Shinoda, Y., Akatsu, T., and Wakai, F. (2004) *J. Am. Ceram. Soc.*, **87**, 1122.
30 García-Gañán, C., Meléndez-Martínez, J.J., Gómez-García, D., and Domínguez-Rodríguez, A. (2006) *J. Mater. Sci.*, **41**, 5231.
31 Mimurada, J., Nakano, M., Sasaki, K., Ykuhara, Y., and Sakuma, T. (2001) *J. Am. Ceram. Soc.*, **84**, 1817.
32 Nakatani, K., Nagayama, H., Yoshida, H., Yamamoto, T., and Sakuma, T. (2003) *Scripta Mater.*, **49**, 791.
33 Morita, K., Hiraga, K., and Kim, B.N. (2004) *Acta Mater.*, **52**, 3355.
34 Sato, E., Morioka, H., Kuribayashi, K., and Sundararaman, D. (1999) *J. Mater. Sci.*, **34**, 4511.
35 Yoshida, H., Okada, K., Ikuhara, Y., and Sakuma, T. (1997) *Philos. Mag. Lett.*, **76**, 9.
36 Wakai, F., Nagano, T., and Iga, T. (1997) *J. Am. Ceram. Soc.*, **80**, 2361.
37 Yoshida, H., Yamamoto, T., Ikuhara, Y., and Sakuma, T. (2002) *Philos. Mag.*, **A82**, 511.
38 Yoshida, H., Ikuhara, Y., and Sakuma, T. (2002) *Acta Mater.*, **50**, 2955.
39 Ikuhara, Y., Yoshida, H., and Sakuma, T. (2001) *Mater. Sci. Eng.*, **A319–321**, 24.
40 Buban, J.P., Matsunaga, K., Chen, J., Shibata, N., Ching, W.Y., Yamamoto, T., and Ikuhara, Y. (2006) *Science*, **311**, 212.
41 Thavorniti, P., Ikuhara, Y., and Sakuma, T. (1998) *J. Am. Ceram. Soc.*, **81**, 2927.
42 Ikuhara, Y., Yamamoto, T., Kuwabara, A., Yoshida, H., and Sakuma, T. (2001) *Sci. Technol. Adv. Mater.*, **2**, 411.
43 Shinoda, Y., Nagano, T., Gu, H., and Wakai, F. (1999) *J. Am. Ceram. Soc.*, **82**, 2916.
44 Gu, H., Shinoda, Y., and Wakai, F. (1999) *J. Am. Ceram. Soc.*, **82**, 469.
45 Imamura, P.H., Evans, N.D., Sakuma, T., and Mecartney, M.L. (2000) *J. Am. Ceram. Soc.*, **83**, 3095.
46 Clarke, D.R. (1987) *J. Am. Ceram. Soc.*, **70**, 15.
47 Wilkinson, D.S. (1998) *J. Am. Ceram. Soc.*, **81**, 275.
48 Wakai, F., Kondo, N., and Shinoda, Y. (1999) *Curr. Opin. Solid State Mater. Sci.*, **4**, 461.
49 Melendez-Martinez, J.J., Gomez-Garcia, D., and Dominguez-Rodriguez, A. (2004) *Philos. Mag.*, **84**, 2305.
50 Rosenflanz, A. and Chen, I.W. (1998) *J. Am. Ceram. Soc.*, **81**, 713.
51 Wang, C.M., Mitomo, M., and Emoto, H. (1997) *J. Mater. Res.*, **12**, 3266.
52 Nagano, T., Gu, H., Shinoda, Y., Zhan, D., Mitomo, M., and Wakai, F. (1999) *Mater. Sci. Forum*, **304–306**, 507.
53 Rachinger, W.A. (1952) *J. Inst. Metals*, **81**, 33.

54 Ahsby, M.F. and Verrall, R.A. (1973) *Acta Metall.*, **21**, 149.
55 Farber, B., Chiarelli, A.S., and Heuer, A.H. (1995) *Philos. Mag. A*, **72** (1), 59.
56 Spingarn, J.R. and Nix, W.D. (1978) *Acta Metall.*, **26**, 1389.
57 Chung, T.E. and Davies, T.J. (1979) *Acta Metall.*, **27**, 635.
58 Bravo-León, A., Jiménez-Melendo, M., and Domínguez-Rodríguez, A. (1992) *Acta Metall. Mater.*, **40**, 2717.
59 Arzt, E., Ashby, M.F., and Verrall, R.A. (1983) *Acta Metall.*, **31**, 1977.
60 Jimenez-Melendo, M. and Dominguez-Rodriguez, A. (2000) *Acta Mater.*, **48**, 3201.
61 Domínguez-Rodríguez, A., Bravo-León, A., Ye, J.D., and Jiménez-Melendo, M. (1998) *Mater. Sci. Eng.*, **A247**, 97.
62 Jiménez-Melendo, M. and Domínguez-Rodríguez, A. (1999) *Philos. Mag.*, **A79**, 1591.
63 Morita, K. and Hiraga, K. (2002) *Acta Mater.*, **50**, 1075.
64 Morita, K. and Hiraga, K. (2001) *Philos. Mag. Lett.*, **81**, 311.
65 Muñoz, A., Gómez-García, D., Domínguez-Rodríguez, A., and Wakai, F. (2002) *J. Eur. Ceram. Soc.*, **22**, 2609.
66 Domínguez-Rodríguez, A., Gómez-García, D., and Castillo-Rodríguez, M. (2008) *J. Eur. Ceram. Soc.*, **28**, 571.
67 Theunissen, G.S.A.M., Winnubst, A.J.A., and Burggraaaf, A.J. (1992) *J. Mater. Sci.*, **27**, 5057.
68 Xu, C. (2005) *Ceram. Int.*, **31**, 537.
69 Stemmer, S., Vleugels, J., and Van der Biest, O. (1998) *J. Eur. Ceram. Soc.*, **18**, 1565.
70 Hines, J., Ikuhara, Y., Chokshi, A., and Sakuma, T. (1998) *Acta Mater.*, **46**, 5557.
71 Flewit, P. and Wild, R. (2001) *Grain Boundaries: Their Microstructure and Chemistry*, John Wiley & Sons, Ltd, Chichester, United Kingdom.
72 Guo, X. and Zhang, Z. (2003) *Acta Mater.*, **51**, 2539.
73 Cahn, R.W. and Haasen, P. (1983) *Physical Metallurgy, Part I*, North-Holland, Amsterdam.
74 Domínguez-Rodríguez, A., Gómez-García, D., Lorenzo-Martín, C., and Muñoz-Bernabé, A. (2003) *J. Eur. Ceram. Soc.*, **23**, 2969.
75 Domínguez-Ridríguez, A., Lagerlof, K.P.D., and Heuer, A.H. (1986) *J. Am. Ceram. Soc.*, **69**, 281.
76 Kandil, H.M., Greiner, J.D., and Smith, J.F. (1984) *J. Am. Ceram. Soc.*, **67**, 341.
77 Wang, Q., Lian, G., and Dickey, E.C. (2004) *Acta Mater.*, **52**, 809.
78 Langdon, T.G. (1994) *Acta Metall. Mater.*, **42**, 2437.
79 Domínguez-Rodríguez, A., Gómez-García, D., Zapata-Solvas, E., Shen, J.Z., and Chaim, R. (2007) *Scripta Mater.*, **56**, 89.
80 Raj, R. and Chung, C.K. (1981) *Acta Metall.*, **29**, 159.
81 Wakai, F. (1994) *Acta Mater.*, **42**, 1163.
82 Raj, R. and Morgan, P.E.D. (1981) *J. Am. Ceram. Soc.*, **65**, C–143.
83 Jennings, H.M., Edwards, J.O., and Richman, M.H. (1976) *Inorg. Chim. Acta*, **20**, 167.
84 Zhan, G.D., Mitomo, M., Xie, R.J., and Kurashima, K. (2000b) *Acta Mater.*, **48**, 2373.
85 Atkins, P.W. (1986) *Physical Chemistry*, 3rd edn, Oxford Univ. Press, UK, p. 153.
86 Gaskell, D.K. (2003) *Introduction to the Thermodynamics of Materials*, 4th edn, Taylor & Francis, New York, USA.
87 Melendez-Martinez, J.J., Gomez-Garcia, D., Jimenez-Melendo, M., and Dominguez-Rodriguez, A. (2004) *Philos. Mag.*, **84**, 3375 and 3387.
88 Venkatachari, K.R. and Raj, R. (1987) *J. Am. Ceram. Soc.*, **70**, 514.
89 Kondo, N., Suzuki, Y., and Ohji, T. (1999) *J. Am. Ceram. Soc.*, **82**, 1067.
90 Ye, J. and Dominguez-Rodriguez, A. (1995) *Scripta Metall. Mater.*, **33**, 441.
91 Gutiérrez-Mora, F., Domínguez-Rodriguez, A., Routbort, J.L., Chaim, R., and Guiberteau, F. (1999) *Scripta Mater.*, **41**, 455.
92 Chantikul, P., Bennison, S.J., and Lawn, B.R. (1990) *J. Am. Ceram. Soc.*, **73**, 2419.
93 Borrero-López, O., Ortiz, A.L., Guiberteau, F., and Padture, N.P. (2005) *J. Am. Ceram. Soc.*, **88**, 2159.
94 Cho, S.J., Hockey, B.J., Lawn, B.R., and Bennison, S.J. (1989) *J. Am. Ceram. Soc.*, **72**, 1249.
95 Domínguez-Rodriguez, A., Guiberteau, F., and Jiménez-Melendo, M. (1998) *J. Mater. Res.*, **13**, 1631.

96 Lee, I.G. and Chen, I.W. (1988) Sintering and grain growth in tetragonal-and cubic zirconia, in *Sintering 87*, vol. 1 (eds S. Somiya, M. Yoshimura, and R. Watanabe), Elsevier, London, pp. 340–345.

97 Domínguez-Rodríguez, A., Jiménez-Piqué, E., and Jiménez-Melendo, M. (1998) *Scripta Mater.*, **39**, 21.

98 French, J.D., Zhao, J., Harper, M.P., Chan, H.M., and Millar, G.A. (1994) *J. Am. Ceram. Soc.*, **77**, 2857.

99 Kim, B.N., Hiraga, K., Morita, K., and Sakka, Y. (2001) *Nature*, **413**, 288–291.

100 Chen, I.W. and Wang, X.H. (2000) *Nature*, **404**, 168.

101 Wang, X.-H., Deng, X.-Y., Bai, H.-L., Zhou, H., Qu, W.-G., Li, L.-T., and Chen, I.-W. (2006) Two-step sintering of ceramics with constant grain size. II. $BaTiO_3$ and Ni-Cu-Zn ferrite. *J. Am. Ceram. Soc.*, **89** (2), 438–443.

IV
Thermal, Electrical, and Magnetic Properties

Ceramics Science and Technology Volume 2: Properties. Edited by Ralf Riedel and I-Wei Chen

ISBN: 978-3-527-31156-9

16
Thermal Conductivity

Kiyoshi Hirao and You Zhou

16.1
Introduction

Most applications of polycrystalline ceramics take advantage of the materials' thermal-insulating or thermal-conducting behaviors. The use of the ceramics is subjected to the rate of heat transmitting through a ceramic at a given temperature gradient. The conduction of heat is proportional to the temperature difference, and given by the following relationship:

$$dq/dt = -k\,dT/dx \tag{1}$$

where dq/dt is the quantity of the heat, transmitting in a direction normal to a surface of a unit area in time dt, and dT/dx is the temperature gradient along the direction dx. The proportional factor, k, which is known as the *thermal conductivity*, is measured in Watts per meter-Kelvin ($W\,m^{-1}\,K^{-1}$).

From the viewpoint of industrial applications, a low thermal conductivity is needed for thermal insulation, while a high thermal conductivity is important for achieving high heat release, high thermal shock resistance and temperature homogeneity through a material [1]. In the semiconductor manufacturing industries, ceramic parts made from Al_2O_3, SiC, and AlN are used widely as guide stages, electrostatic wafer chucks, supporting parts in heating furnaces, and so on, because of their excellent electrical and thermal properties, as well as their high plasma resistance and chemical stability. In addition to engineering applications, high-thermal conductivity dielectric materials are required in electronic devices in order to dissipate the heat generated from the semiconductor, and today they are widely used as packages and substrates for IC products (e.g., accelerometers, gyroscopes, pressure sensors, probe cards, large-scale integration devices, radiofrequency modules, wireless communication devices, light-emitting diodes), optoelectronic products, semiconductor power devices, and so forth.

Alumina, a material commonly used for substrates, has a moderate thermal conductivity of about 20–30 $W\,m^{-1}\,K^{-1}$, which is about two orders higher than that of plastic substrates. In order to manage increasing heat with power upgrade in the semiconductor, and increasing density in the integrated circuit, AlN and SiC ceramic

Ceramics Science and Technology Volume 2: Properties. Edited by Ralf Riedel and I-Wei Chen

ISBN: 978-3-527-31156-9

materials with a high thermal conductivity of about 200 W m^{-1} K^{-1} have been developed. The applications of SiC ceramics are limited by their low electric resistance and high dielectric constant; hence, AlN has been mainly used for this purpose to date, because of its superior thermal conductivity and high electric resistivity.

With increasing demands for electric transportation systems and/or electric vehicles, semiconductor power modules such as electric power converters and DC-AC inverters will continue to expand in terms of their applications. In these systems, in order to transmit a high electric current, thick copper electrodes are often directly bonded to ceramic substrates, the structures of which may cause major residual stresses in ceramic parts. Thus, to avoid failure due to residual stress, ceramic materials are required to have a high strength and, in order to further improve the reliability of the systems, improvements in the mechanical properties of high-thermal conductivity materials are clearly required. Consequently, the electrical industries are continuing an active search for alternative materials with both high thermal conductivity and superior mechanical properties.

Silicon nitride ceramic (Si_3N_4), on the other hand, is well-known as a high-temperature structural ceramic, having high strength and high fracture toughness. The excellent mechanical properties of Si_3N_4 result from its unique microstructure, which is composed of hexagonal, rod-like grains, bonded together and reinforcing each other. Recently, Si_3N_4 has attracted much attention as a high-thermal conductivity material, with many research groups having predicted that β-Si_3N_4 crystal would have a high intrinsic thermal conductivity.

In this chapter, the intrinsic thermal conductivities of these nonoxide ceramic materials (AlN, Si_3N_4 and SiC) are described, and the development of high-thermal conductivity materials is reviewed. As many reports have been made on high-thermal conductivity AlN and SiC, attention in the chapter is focused on high-thermal conductivity Si_3N_4. In the case of thermal insulation, one major concern of engineers is that of pore control, because air is a highly thermal insulating material. In contrast, a high thermal conductivity is generally achieved only in a dense material composed of pure crystals, and consequently the development of high-thermal conductivity ceramics is closely related to microstructure control through sintering technologies that involve powder syntheses. The processing strategy for producing Si_3N_4 with both a high thermal conductivity and a high strength is discussed in terms of microstructure control in the final section of the chapter.

16.2 Thermal Conductivity of Dielectric Ceramics

16.2.1 Thermal Conductivity of Nonmetallic Crystals

Heat in solids is conducted in various carriers, depending on their nature of chemical bonding in structures: electrons, lattice vibrations (or phonons), magnetic

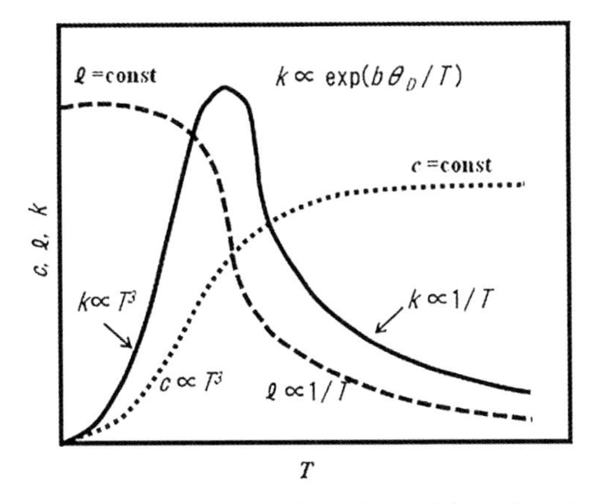

Figure 16.1 Temperature dependence of thermal conductivity, *k*, dominated by phonon transfer.

excitations, and so forth. In dielectric ceramic materials, the conduction of heat is dominated by phonon transfer and is, in general, expressed by the following equation [2]:

$$k = \frac{1}{3}\nu \int_{0}^{\nu_D} C(\nu)l(\nu)\mathrm{d}\nu \tag{2}$$

where ν_D is the Debye frequency, $C(\nu)$ is the phonon contribution of volumetric heat capacity, ν is the phonon group velocity, and $l(\nu)$ is phonon mean free path. $C(\nu)$ is proportional to T^3 at low temperatures, and becomes constant at high temperatures according to the Debye model. V corresponds to acoustic velocity, and is minimally affected by the temperature. $l(\nu)$ exhibits an inverse relationship with the temperature since at high temperatures it is limited by phonon–phonon interactions [2, 3].Thus, k exhibits a maximum as a function of temperature, as shown in Figure 16.1.

Anharmonicity in lattice vibrations prevents the propagation of lattice waves; in other words, they make $l(\nu)$ decrease. ν is higher in materials with light atoms and strong interatomic bonding, while there is a little difference in $C(\nu)$ among materials. Thus, the rules for identifying a nonmetallic crystal with a high thermal conductivity are: (i) a low atomic mass; (ii) strong interatomic bonding; (iii) a simple crystal structure; and (iv) a low anharmonicity [4, 5]. The materials with adamantine (diamond-like) structure such as diamond, SiC, Si, BeO, AlN, and GaP each satisfy these requirements, and exhibit a high thermal conductivity.

16.2.2
High Thermal Conductivity in Adamantine Compounds

In pure crystals the prime cause of thermal resistance at high temperatures is phonon–phonon scattering, whereas at sufficiently low temperatures the phonon mean free path is limited by the external boundaries, such as specimen size. In addition to these intrinsic resistance, phonons extensively scatter by almost all the imperfections in crystals such as point defects, interstitials, line defects, and planar defects at low and intermediate temperatures at which $l(\nu)$ is relatively long. Therefore, an estimate of thermal conductivity in pure crystals (intrinsic thermal conductivity) is strongly desirable in order to provide a guide for fabricating high-thermal conductivity materials, as well as seeking candidate materials for practical applications.

Thermal conductivity by phonon transportation in the temperature range equal to or higher than Debye temperature is given by:

$$k = B\bar{M}\delta\Theta_D^3/T\gamma^2 \tag{3}$$

where B is a constant, $\bar{M}$ is the mean atomic mass, δ is the cube root of the average volume occupied by one atom of the crystal, Θ_D is the Debye temperature, γ is the Grüneisen parameter, and T is the absolute temperature [4]. The equation is valid only at $T > \Theta_D$. From the viewpoint of applications, room temperature thermal conductivity is important, and all of the materials with a high thermal conductivity exhibit high values of Θ_D. An estimate of intrinsic thermal conductivity at room temperature was conducted by interpolation. Slack [5] indicated that the thermal conductivity of pure adamantine crystals at 300 K is proportional to the scaling parameter $\bar{M}\delta\Theta_D^3$, as shown in Figure 16.2. Based on this relationship, Slack estimated the intrinsic thermal conductivity of AlN at 300 K to be 320 W m^{-1} K^{-1}, and later measured the

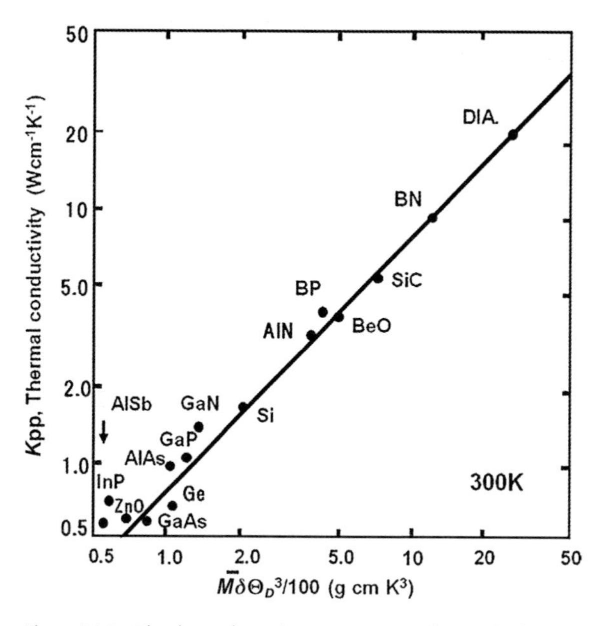

Figure 16.2 The thermal conductivity at 300 K of several adamantine structure crystals, as limited by intrinsic phonon–phonon scattering [5].

Table 16.1 Thermal conductivity at room temperature for several adamantine crystals.[a].

Crystal	Debye temperature Θ_D (K)	Thermal conductivity k (W m^{-1} K^{-1})
Diamond	2240	2000
BN	1750	760
SiC	1185	490
BeO	1280	370
BP	975	360
AlN	950	319
Si	648	156
GaN	525	130

a) Data for typical ceramic substances from Ref. [5].

thermal conductivity of high-purity single crystals of AlN, correcting the measured value for isotope scattering and obtained thermal conductivity of 319 W m^{-1} K^{-1} [5]. The thermal conductivity data at room temperature for several adamantine crystals, as reported by Slack, are summarized in Table 16.1 [5].

16.2.3
Estimate of Thermal Conductivity of β-Si_3N_4

Silicon nitride exists in the form of two crystal structures, namely α-Si_3N_4 and β-Si_3N_4 [6], and it is generally accepted that these are low- and high-temperature forms, respectively. The industrial synthesis routes lead mainly to α-Si_3N_4, which converts to the more stable β-Si_3N_4 phase during liquid-phase sintering (LPS) [7]. Silicon nitride is a highly covalent compound (70% covalence), composed of relatively light atoms of Si and N; hence, it is anticipated that Si_3N_4 would possess a high intrinsic thermal conductivity. When Haggerty and Lightfoot first estimated the thermal conductivity of Si_3N_4 crystal, they noted that the SiC and Si_3N_4 materials were almost identical except for the number of atoms (n) in each primitive cell, and reported predicted intrinsic values of 200 W m^{-1} K^{-1} for $n = 14$, and 320 W m^{-1} K^{-1} for $n = 7$ [8]. Similarly, Watari *et al.* estimated that the upper limit of thermal conductivity of β-Si_3N_4 would be 400 W m^{-1} K^{-1}, based on a proportional relationship between the intrinsic thermal conductivity and the scaling parameter of $\bar{M}\delta\Theta_D^3$ [9]. Very recently Hirosaki *et al.* estimated the ideal thermal conductivity of single-crystal α- and β-Si_3N_4, using the molecular dynamics method in conjunction with the Green–Kubo formulation, as a function of temperature. In Hirosaki's calculations, the estimated thermal conductivities in α- and β-Si_3N_4, along the a-axis and c-axis at room temperature were approximately 105 and 225 W m^{-1} K^{-1}, and 170 and 450 W m^{-1} K^{-1}, respectively [10]. The estimated thermal conductivities reported in the literature are summarized in Table 16.2. Besides these attempts to estimate the intrinsic thermal conductivity, direct measurements of thermal conductivity of well-developed β-Si_3N_4 grains in a sintered silicon nitride have been conducted by means of thermoreflectance microscopy [13]. The principal

Table 16.2 Theoretical estimates of the thermal conductivity of β-Si_3N_4 crystals.

Estimation method	Estimated value ($W\,m^{-1}\,K^{-1}$)	Reference
Slack equation	200 ($n=14$), 320 ($n=7$)	[8]
Scaling parameter, $M\delta\Theta_D^3$	<400	[9]
Molecular dynamics simulation	a-axis: 170, c-axis: 450	[10]
Slack formula	250	[11]
Temperature-dependent thermal diffusivity data	105	[12]
Thermoreflectance microscopy measurements	a-axis: 69, c-axis: 180	[13]

diffusivities obtained in individual grains were $0.32\,cm^2\,s^{-1}$ along the a-axis, and $0.84\,cm^2\,s^{-1}$ along the c-axis (the corresponding thermal conductivities were 69 and $180\,W\,m^{-1}\,K^{-1}$), respectively. The measured values were consistent with the estimated intrinsic values shown in Table 16.2.

16.3 High-Thermal Conductivity Nonoxide Ceramics

In general, the thermal conductivity of a sintered polycrystalline ceramic is lower than the estimated intrinsic thermal conductivity of a single crystal. The reported thermal conductivities of sintered silicon nitrides, fabricated using a variety of methods, are listed in Table 16.3. Most of the values reported were much lower than the estimated intrinsic values; even at the highest case, the thermal conductivity of the sintered specimen was about $150\,W\,m^{-1}\,K^{-1}$, and this was achieved after a long sintering time at high temperature. The sintered silicon nitrides have lower thermal conductivities for two main reasons: (i) the distribution of low-thermal conductivity secondary phases resulting from the addition of sintering additives; and (ii) the existence of compositional and crystallographic defects in the grains. These extrinsic factors, which affect the thermal conductivity of nonoxide ceramics (e.g., AlN, Si_3N_4 and SiC) are discussed in the following sections.

16.3.1 Thermal Conductivity of Composite Microstructures

16.3.1.1 Liquid-Phase Sintering of Nonoxide Ceramics

Due to the strong covalency of chemical bonding and low self-diffusion coefficient, the sintering of nonoxide ceramics such as AlN, Si_3N_4 and SiC to high densities normally requires the aid of sintering additives. In almost all cases, the densification of AlN [21–23] and Si_3N_4 ceramics [6, 24, 25] is achieved through liquid-state sintering, where rare-earth oxides and alkaline-earth oxides are typically used as sintering additives. In both cases, these oxide additives react with impurity oxides that mainly exist on the surface of nitride particles, alumina on AlN and silica on Si_3N_4,

Table 16.3 Thermal conductivity of sintered β-Si_3N_4 ceramics with various sintering additives at room temperature, as reported in the literature.

Year	Sintering additives	Sintering method[a)] and condition	Thermal conductivity ($W\,m^{-1}\,K^{-1}$)	Reference
1978	15 mol% MgO	HP: 2023 K, 0.5 h, 20 MPa	55	[14]
1981	Without additive	HHP: 2173 K, 1 h, 3 GPa	30	[15]
1989	6 mol% Al_2O_3	HIP: 2023 K, 60 MPa, 1 h	18	[16]
	6 mol% Y_2O_3	HIP: 2023 K, 60 MPa, 1 h	72	[16]
1992	Y_2O_3	GPS: 2173 K, 1 h, 100 MPa N_2	80	[17]
1996	4 mol% Y_2O_3-4 mol% Al_2O_3	GPS: 2273 K, 4 h, 100 MPa N_2	40	[1]
	0.5 mol% Y_2O_3-0.5 mol% Nd_2O_3	GPS: 2273 K, 4 h, 100 MPa N_2	120	[1]
1998	0.5 mol% Y_2O_3-0.5 mol% Nd_2O_3-4 mol% Al_2O_3	GPS: 2173 K, 4 h, 10 MPa N_2	40	[18]
	0.5 mol% Y_2O_3-0.5 mol% Nd_2O_3-2 mol% MgO	GPS: 2173 K, 4 h, 10 MPa N_2	128	[18]
2001	2 mol% Yb_2O_3-5 mol% $MgSiN_2$	GPS, 2173 K, 48 h, 0.9 MPa N_2	142	[45]
2003	8 mass% Y_2O_3-1 mass% HfO_2	GPS: 2173 K, 48 h, 0.9 MPa N_2	120	[19]
2003	10 mass% Yb_2O_3-2 mass% ZrO_2	GPS: 2223 K, 16 h, 0.9 MPa N_2	140	[20]
2003	10 mass% Yb_2O_3-2 mass% ZrO_2	GPS: 2173 K, 36 h, 0.9 MPa N_2 $\rightarrow$ Annealing: 1973 K, 100 h, 0.9 MPa N_2	150	[48]

a) Sintering method
HP: Hot-pressing; HHP: High-pressure hot-pressing; HIP: Hot-isostatic pressing; GPS: Gas-pressure sintering.

respectively, to form an eutectic liquid which enhances densification via particle rearrangement and dissolution–reprecipitation processes.

By contrast, the densification of SiC may proceed by either a solid-state sintering mechanism or a LPS mechanism, depending on the type of the sintering additive used. The typical sintering additives for the solid-state sintering of SiC are boron and carbon or boron carbide [26], while beryllium and beryllium oxide are also effective sintering aids under hot-pressing conditions [27]. Because the amounts of these additives needed to facilitate the solid-state sintering of SiC is very small (<1–2%), solid-state-sintered SiC is usually regarded as a single-phase ceramic. In contrast, for the LPS of SiC, which is similar to the sintering of AlN and Si_3N_4 ceramics, a relatively large amount of sintering additives is required. The most commonly used sintering additives include rare-earth or alkaline-earth oxides [28] which, during sintering, react with the impurity SiO_2 residing on the surface of the raw powder particles to form eutectic liquids which facilitate densification by LPS mechanisms.

Upon cooling, these liquid phases remain as glassy phases or as secondary crystalline phases in the sintered materials; consequently, these liquid-phase-sintered ceramics are actually composite materials consisting of a matrix of grains and dispersed secondary phases. The thermal conductivity of these composite materials will depend on the amount, distribution state and thermal conductivity of each constituent phase in the structure. The effects of secondary phases on the thermal conductivity of liquid-phase-sintered nonoxide ceramics are discussed in the following subsection.

16.3.1.2 **Effect of Secondary Phase on Thermal Conductivity of AlN Ceramic**

A number of models have been proposed to predict the thermal conductivity of a multiphase system. For the most simple case, in a two-phase system where the phases exhibit an alternative layer arrangement (this is well-known as Wiener's model), the thermal conductivity parallel to and perpendicular to the plane of the layers is given by the following equations [29]:

$$k_c = v_1 k_1 + v_2 k_2 \qquad \text{(Parallel arrangement)} \tag{4}$$

$$1/k_c = v_1/k_1 + v_2/k_2 \qquad \text{(Series arrangement)} \tag{5}$$

where k_1 and k_2 are the thermal conductivities, and v_1 and v_2 are the volume fractions of each phase. For the parallel arrangement, most of the heat flow is through the better conductor, such that the heat conduction process is dominated by the better conductor. In contrast, for the series arrangement, where heat flow is perpendicular to the plane of each layer, the amount of heat flow through each layer is equal, and so it is dominated by the poorer conductor. The thermal conductivity of an ordinary composite material lies between that of two extreme arrangements at a given composition. Equations (4) and (5) are reasonable approximations in cases where the amount of secondary phase is less than about 10% [30]. The secondary phase with a spherical shape, when discretely dispersed in a matrix, constitutes a better model for representing a two-phase system. If the spherical secondary phase with thermal

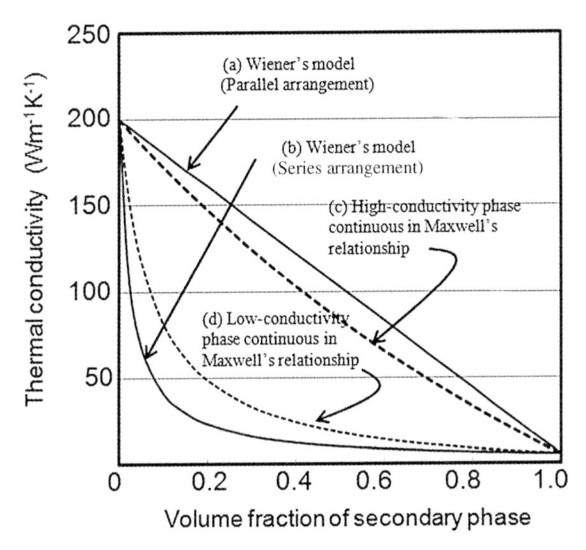

Figure 16.3 Calculated thermal conductivity in two-phase systems based on parallel arrangement in Wiener's model (solid line) and Maxwell's relationship (dashed line), provided that the thermal conductivity of each phase is 200 and $5\,\mathrm{W\,m^{-1}\,K^{-1}}$, respectively.

conductivity k_d is dispersed in a matrix phase with thermal conductivity k_m, then the thermal conductivity of the composite k_c is given by [31]:

$$k_c = k_m \left[\frac{1+2V_d\left[1-\dfrac{k_m}{k_d}\right] \Big/ \left[\dfrac{2k_m}{k_d}+1\right]}{1-V_d\left[1-\dfrac{k_m}{k_d}\right] \Big/ \left[2\dfrac{k_m}{(k_d)}+1\right]} \right] \tag{6}$$

where V_d and V_m are the volume fractions of the spherical dispersed phase and matrix phase, respectively. This model can be used for describing two extreme microstructures: one in which a concrete low-thermal conductivity phase is distributed in a contiguous high-thermal conductivity phase; and one in which a high-thermal conductivity phase is distributed in a contiguously low-thermal conductivity phase. The effects of the volume fraction of the dispersed phase in two different microstructures are compared in Figure 16.3 [30]. Clearly, the low-thermal conductivity secondary phase, with its high ability to wet the matrix grains, causes a significant disturbance to the high-thermal conductivity matrix.

As noted in the previous section, oxides are generally added as sintering additives for densifying AlN, and will react with any alumina impurity in an AlN powder. After sintering, the secondary phases of aluminate compounds will then remain in the microstructure. The thermal conductivity of the aluminates may be as low as a few $\mathrm{W\,m^{-1}\,K^{-1}}$, this value being about two orders lower than that of AlN [23]. If the secondary oxide phases are discretely distributed within the AlN matrix, the secondary phase is of minor importance, whereas if the AlN grains are covered with the low-thermal conductivity secondary phase the material's thermal conductivity will be significantly reduced. The AlN grains are generally equiaxed polyhedrons. The distribution of the secondary phases is reported to depend on processing parameters, including the type and amount of additives [23, 32], the sintering atmosphere [33], and the cooling rate [34]. In most cases, the grain boundary phases which are concentrated at the corner of the AlN grains exhibit very little penetration along the grain faces. For example, when Virkar *et al.* [23] systematically investigated the thermal conductivity of AlN with various amounts of Y_2O_3 additive, the thermal conductivity of the AlN ceramic was seen initially to increase in relation to the amount of Y_2O_3, because the latter had a high affinity for alumina. This, in turn, led to the purification of AlN grains via a dissolution–reprecipitation process (as discussed in Section 16.3.2). However, beyond a certain level of Y_2O_3 additive the thermal conductivity was decreased, due to the fact that Y_2O_3 plays a purification role in AlN grains during sintering, while simultaneously existing as yttrium aluminates with low thermal conductivities after sintering. It was also pointed out [32] that, after reaching a maximum value, the thermal conductivity of the specimens decreased in line with the amounts of secondary oxide phases, and that this was in accordance with the prediction of the dispersed-phase model [i.e., Eq. (6)]. This was provided that the aluminates with a low thermal conductivity of 13 $W\,m^{-1}\,K^{-1}$ were discretely dispersed in the AlN matrix grains with a high thermal conductivity of 230 $W\,m^{-1}\,K^{-1}$.

In some cases, thin grain boundary layers are observed between the AlN grains, but even in this case the thickness may be only a few nanometers. Thus, in the case of AlN the overall thermal conductivity is controlled to a lesser degree by the secondary phases.

16.3.1.3 Effect of Secondary Phase on Thermal Conductivity of Si_3N_4 Ceramic

In general, α-Si_3N_4 powders are used as starting raw powders, as the α-phase $\rightarrow$ β-phase transformation during LPS leads to the development of rod-like β-Si_3N_4 grains, on account of a preferential growth rate in the [001] direction of β-Si_3N_4 crystals [35]. Such grain growth behavior leads to the excellent mechanical properties for Si_3N_4, with the simultaneous achievement of high strength and high toughness. By contrast, a strong tendency towards the faceting of β-Si_3N_4 grains in liquid-phase-sintered Si_3N_4 ceramics causes the microstructure to be more complicated in terms of thermal conductivity. It is generally accepted that a thin-grain boundary amorphous layer exists between Si_3N_4 grains, the thickness of which – the so-called *equilibrium thickness* – is determined by the composition of the secondary glassy phase (1–2 nm) [36]. Although, the distance between grains is in equilibrium for some grains, the other parts should have greater inter-grain distances because the prismatic planes of β-Si_3N_4 grains are inclined to each other in most parts. Kitayama *et al.* assessed the effect of the grain boundary glassy phase on the overall thermal conductivity in Si_3N_4 by using a modified Wiener's model [37]. For this, they simplified the microstructure of Si_3N_4, as illustrated in Figure 16.4. In the idealized two-dimensional (2-D) microstructure, the (001) and (100) planes of the β-Si_3N_4 grains are perfectly oriented to the directions designated as "para" and "perp," respectively. Each grain is completely covered with the grain boundary film with an

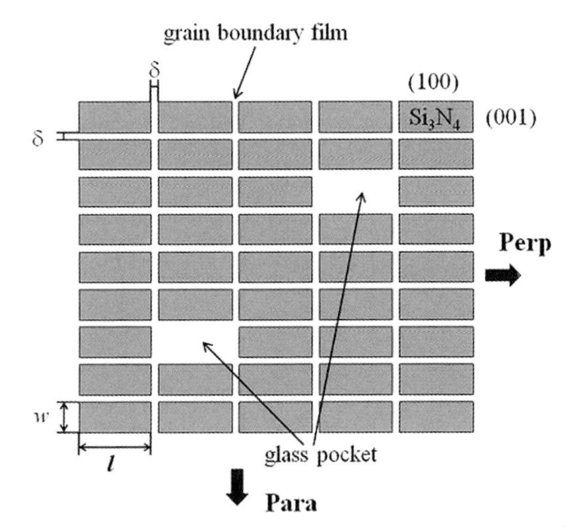

Figure 16.4 Illustration of idealized two-dimensional microstructure of β-Si_3N_4; the [210] and [001] directions of the β-grains are perfectly oriented to directions designated Para and Perp, respectively; the grain-boundary film thickness, δ, is independent of crystallographic orientations and completely covers the grain surfaces. If the amount of grain-boundary glassy phase exceeds that required for a complete covering of grains, the remainder is assumed to exist as glass pockets [37].

average thickness of *d*, and the remainder of the glassy phase is discretely dispersed in the microstructure as glassy pockets. In order to take the grain sizes and grain boundary film thickness, Wiener's parallel and serial formulae for the thermal conductivity of composite material were combined. In addition, the role of glassy pockets in the thermal conductivity was taken into account by using the well-known Maxwell's relationship for the thermal conductivity of the composite with a dispersed secondary phase. In order to simplify the model calculation, it was assumed that the thermal conductivity would be independent of the crystallographic orientation, although anisotropy in thermal conductivity might exist. A typical example of the calculation is shown in Figure 16.5, which illustrates the effect of the volume fraction of the glassy phase on the thermal conductivity of Si_3N_4 ceramics, with the aspect ratio of Si_3N_4 being 1. Here, the grain boundary thickness, *d*, was 1 and 10 nm, and the thermal conductivity of the β-Si_3N_4 crystal and the glassy phase were 180 and 1 W m^{-1} K^{-1}, respectively. The calculation indicated that the thermal conductivity of Si_3N_4 ceramic was heavily dependent on the average grain boundary film thickness, which was in the range of a few tenths of a nanometer. The thermal conductivity of Si_3N_4 initially increased steeply with increasing grain size, to reach constant values determined by the film thickness and the amount of glassy phase. These results indicated that grain growth is necessary to a certain extent for increasing the thermal conductivity of Si_3N_4 ceramics; however, decreasing the total amount of glassy phase resulted, of course, in a higher thermal conductivity, as shown in Figure 16.5.

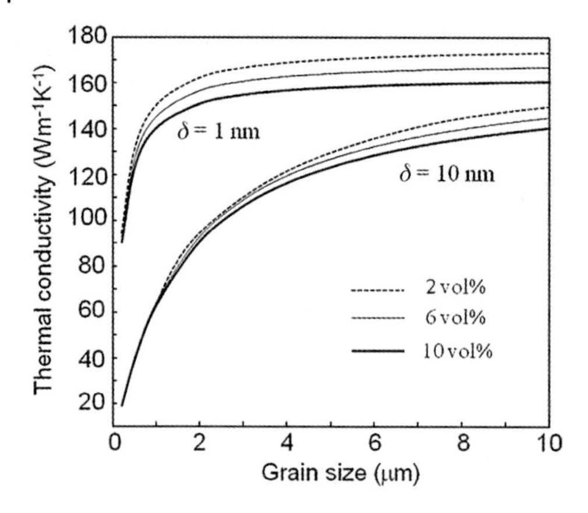

Figure 16.5 Effect of volume fraction of glassy phase on thermal conductivity of β-Si_3N_4 for R (aspect ratio) = 1 in β-grain and δ = 1 and 10 nm.

16.3.1.4 **Effect of Secondary Phase on Thermal Conductivity of SiC Ceramic**

Similar to the sintered AlN and Si_3N_4, the microstructure of liquid-phase-sintered SiC (LPS-SiC) consists of SiC grains that are usually surrounded by a film of amorphous grain boundary phase, with crystallized or partially crystallized secondary phases residing at the multijunctions between the SiC grains. When Sigl studied the role of microstructure on the thermal transport in a series of LPS-SiC doped with 3 to 30 vol% of YAG-AlN mixtures as the sintering additives, it was found that the amorphous grain boundary phase could act as a thermal resistance barrier and hence reduce the thermal conductivity of the LPS-SiC materials [38]. When the thermal conductivity of the LPS-SiC was calculated using Maxwell's model, it could be overestimated by about 20% if the presence of the grain boundary phase was neglected. High-resolution transmission electron microscopy (TEM) observations showed the typical width of the grain boundary film to be about 1.2 nm. The presence of an amorphous grain boundary film was also reported in LPS-SiC doped with other sintering additives, although their thickness might have varied when different sintering additives were used. For example, Zhan *et al.* reported that the grain boundary film was about 1 nm thick in a LPS-SiC doped with a mixture of Al_2O_3, Y_2O_3, and CaO [39]. Since the grain boundary film is known to have a detrimental effect on thermal conductivity, a reduction in the amount of grain boundary phase should result in an improved thermal conductivity. Indeed, this was the case, as revealed by Zhou *et al.*, when LPS-SiC was prepared by using a mixture of Y_2O_3 and La_2O_3 as the sintering additives [40, 41]. After undergoing an annealing treatment, the majority of grain boundary phases in the LPS-SiC were lost by their migration to the multijunctions. As a result, the contiguity between the SiC grains was increased, and the thermal conductivity of the annealed materials was shown to be 20% higher when compared to the as-sintered LPS-SiC. It should be noted, however, that although the microstructure exerted a major influence on the thermal conductivity, the factors that had the most substantial effects on the thermal conductivity of

LPS-SiC were the impurities and defects in the SiC grains. This subject is discussed in the following subsection.

16.3.2 Improvement of Thermal Conductivity via Purification of Grains During Sintering

As noted in Section 16.2, the thermal conductivity of crystals at room temperatures is lowered by lattice defects in crystals. In sintered ceramics, the grains always include various defects, the type and amount of which are related to the nature of the raw powder and the processing routes and that, in turn, govern the thermal conductivity of sintered ceramics. At this point, attention is focused on grain defects to demonstrate how the thermal conductivity of nonoxide ceramics might be improved through optimizing their processing.

16.3.2.1 Aluminum Nitride Ceramics

It is generally accepted that lattice oxygen – that is, impurity oxygen dissolved in the AlN lattice – is the dominant factor lowering the thermal conductivity of AlN ceramics, as described below. It was proposed by Slack that the incorporation of oxygen in AlN occurs by the dissolution of Al_2O_3, via the following plausible reaction [4, 5]:

$$Al_2O_3 \rightarrow 2Al_{Al} + 3O_N + V_{Al} \tag{7}$$

where O_N and V_{Al} denote oxygen in a nitrogen site and a vacancy in an aluminum site, respectively. Mass and strain misfits caused by the vacant aluminum site increase the scattering cross-section of the phonons, which in turn decreases the phonon mean free path, thereby lowering the thermal conductivity [23]. Thus, in order to achieve a high thermal conductivity it is essential to decrease lattice oxygen in the AlN grains.

Even high-purity AlN raw powder contains a small percentage of impurity oxygen, most of which is located inevitably on the surface of AlN particle as an oxidized layer, while the remainder exists as lattice oxygen. Therefore, the amount of impurity oxygen in the AlN raw powder will significantly affect the thermal conductivity. Miyashiro *et al.* fabricated hot-pressed- or pressureless-sintered AlN ceramics from AlN raw powder with various amounts of impurity oxygen using 3 mass% Y_2O_3 as a sintering additive, and showed a clear tendency for the thermal conductivity to increase with decreasing impurity oxygen in raw powder, for both fabrication methods [42].

Both alkaline-earth and rare-earth oxides are generally used as sintering additives for dual purpose. First, the additives react with the major oxide layer on the surface of AlN particles so as to form a liquid phase which enhances densification. The additives also play an important role in improving thermal conductivity, by reacting with alumina to form stable aluminates via which impurity oxygen in AlN will become trapped in the grain-boundary phases. It has been postulated that the higher the affinity of the sintering additive for reaction with Al_2O_3 to form aluminates, the higher the resultant thermal conductivity [23]. Watari pointed out that sintering additives such as rare-earth oxides (Y_2O_3, La_2O_3), and alkaline-earth oxides

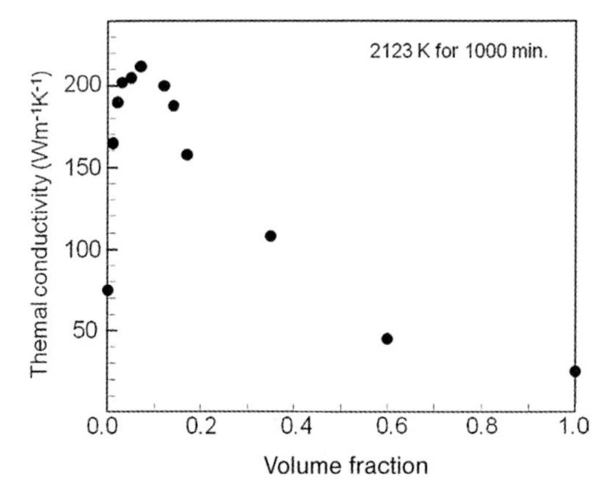

Figure 16.6 Thermal conductivity as a function of volume fraction of the system of the yttrium aluminate phases in AlN–Y_2O_3 compositions [32].

(CaO, MgO), which are effective additives for high thermal conductivity, exhibit a lower Gibbs free energy of formation for aluminates [43].

In addition to the type of sintering additives, the grain boundary phase composition is also important. Jackson *et al.* conducted a systematic investigation of the effect of phase composition in the Al_2O_3–Y_2O_3 additive system on the thermal conductivity of AlN ceramics by fabricating AlN without additives, doped with Y_2O_3, or doped with various yttrium aluminate compounds, notably $Y_3Al_5O_{12}$ (YAG), $YAlO_3$ (YAP), and $Y_4Al_2O_9$ (YAM). The study results showed the thermal conductivity of the fully dense specimen to increase in the order of AlN without additive, AlN doped with YAG, with YAP, and with Y_2O_3 [32]. These authors explained this phenomenon from a thermodynamic point of view, by suggesting that the thermodynamic driving force for oxygen removal would be increased with decreasing activity of Al_2O_3 in the phase field – that is, in the order of Al_2O_3–YAG, YAG–YAP, and YAM–Y_2O_3 phase fields. Their results also suggested that for a high thermal conductivity a certain amount of rare-earth oxide was required in order to shift the grain boundary phase field, from an Al_2O_3-rich composition towards a Y_2O_3-rich composition. Therefore, in general, the variation in thermal conductivity in relation to the amount of oxide additives first increased with higher amounts of additives, reached a maximum value, and then gradually decreased (see Figure 16.6).

16.3.2.2 Lattice Defects in β-Si_3N_4 Grains

Similar to AlN ceramics, dense Si_3N_4 ceramics are fabricated via LPS, with oxides being used as sintering additives, as these react with silica impurities located mainly on the Si_3N_4 particle surfaces as an oxidized layer, so as to form a liquid phase. Although Si_3N_4 is a highly covalent compound (70% covalence) [7], its decomposition temperature under atmospheric nitrogen pressure is rather low, at about 2100 K. Thus, compared to AlN ceramics, a larger quantity of sintering additive will be required, except when sintering via hot-pressing (HP) or hot isostatic pressing (HIP). The oxide sintering additives are, therefore, selected from among alkaline-earth oxides, rare-earth oxides, alumina, zirconia, hafnia, and so forth, in order to increase the volume fraction of the liquid phase and to decrease the eutectic temperatures.

It has been shown experimentally that the doping of alumina, an effective sintering additive for Si_3N_4, resulted in a substantial decrease in the thermal conductivity of Si_3N_4 due to the formation of a solid solution. The doped alumina first dissolves into β-Si_3N_4 lattice to form β-SiAlON by substituting Si for Al and N for O, after which the solid solution reduces the thermal conductivity by distorting the lattice and causing the difference between the mass of a normal site and a substituted site [3, 44]. As an example, Watari *et al.* fabricated fully dense Si_3N_4 doped with 6 mol% Y_2O_3 or 6 mol% Al_2O_3 by capsule-HIP sintering, and showed the thermal conductivities of the former and latter specimens to be 70 and 20 $W\,m^{-1}\,K^{-1}$, respectively [16]. In addition, the variation in thermal conductivity with temperature was assessed systematically for Si_3N_4 with Y_2O_3, or MgO, or Al_2O_3 as additive. The thermal conductivity of Si_3N_4 doped with Al_2O_3 had a near-constant value of approximately 20 $W\,m^{-1}\,K^{-1}$ between 300 and 1000 K, which typically indicates a suppression of the thermal conduction of a material by point defects [44]. In the case of gas-pressure-sintered Si_3N_4, Hirosaki *et al.* reported the room temperature thermal conductivity of Si_3N_4 doped with Y_2O_3–Al_2O_3 to be less than 80 $W\,m^{-1}\,K^{-1}$ at highest, and to decrease subject to the additive amount, whereas Si_3N_4 doped with Y_2O_3–Nd_2O_3 exhibited a higher thermal conductivity of over 120 $W\,m^{-1}\,K^{-1}$ [1]. Likewise, Okamoto *et al.* [18] reported that the thermal conductivity of Si_3N_4 doped with Y_2O_3–Nd_2O_3 was improved by the further addition of MgO, but was drastically decreased with the further addition of Al_2O_3. Consequently, Si_3N_4 ceramics with a higher thermal conductivity ($>100\,W\,m^{-1}\,K^{-1}$) have been successfully fabricated by carefully choosing an adequate sintering additive, excluding alumina. The additive systems identified were Y_2O_3–Nd_2O_3 [1], Y_2O_3–Nd_2O_3–MgO [18], Y_2O_3–HfO_2 [19], Yb_2O_3–ZrO_2 [20], Yb_2O_3–MgO [45], and so forth.

In addition to the formation of a solid solution (i.e., β-SiAlON), a variety of lattice defects for β-Si_3N_4 grains in sintered Si_3N_4 specimens have been reported, including the dissolution of oxygen in β-Si_3N_4 crystals [46, 47], dislocations [48], stacking faults [49], and precipitates inside β-Si_3N_4 crystals [50]. Each of these defects might have a negative effect on the thermal conductivity of Si_3N_4 ceramics although, when considering the similarity to AlN ceramics, the role of point defects – and in particular, of lattice oxygen – should be more significant than others. As a consequence, attention at this point is focused on the lattice oxygen.

Many reports have been made relating to the oxygen content in α-Si_3N_4 (as summarized in Table 16.4), with values ranging from 0.05 to 0.3 mass% for pure α-Si_3N_4 synthesized by chemical vapor deposition (CVD), and from 0.48 to 2 mass% for α-phase-rich Si_3N_4 (α-phase content $>80\%$) [46]. In contrast, no report had been made of the oxygen content in β-Si_3N_4 crystal lattices until Kitayama *et al.* [46] used a hot-gas extraction method that originally had been developed for measuring the lattice oxygen in an AlN crystal [51].

In these experiments, β-Si_3N_4 crystals were fabricated via the heat treatment of α-Si_3N_4 raw powder in which 15 mol% Y_2O_3–SiO_2 had been included as additives ($Y_2O_3 : SiO_2 = 1:2$ or $2:1$); this was followed by an acid rinse treatment to remove any secondary phases. The lattice oxygen contents in the β-Si_3N_4 crystal were reported as

Table 16.4 Oxygen content in Si_3N_4 crystal lattice prepared by various methods [46].

Fabrication method	Phase composition (%)		Oxygen content (mass%)
	α	β	
Unknown	90	10	2
Unknown	10	90	0.2
Si-metal nitridation	87	13	0.53 ± 0.05
Si-metal nitridation	81	19	0.48 ± 0.02
Si-metal nitridation	58	26	0.26 ± 0.03
Si-metal nitridation	6	83	0.31 ± 0.03
CVD	100	0	0.30 ± 0.05
CVD	100	0	0.05 ± 0.02
CVD	100	0	0.09 ± 0.02
LPS[a] (1: 2)	0	100	0.258 ± 0.006
LPS[a] (2: 1)	0	100	0.158 ± 0.003

a) Liquid-phase sintering (LPS) of α-Si_3N_4 powder doped with additives (ratios in the parenthesis indicate Y_2O_3: SiO_2 additive ratios).
CVD: chemical vapor deposition.

0.258 ± 0.006 mass% (in the case of $Y_2O_3 : SiO_2 = 1:2$) and 0.158 ± 0.003 mass% (in the case of $Y_2O_3 : SiO_2 = 2:1$), depending on the composition of the liquid phase. It is expected that the dissolution of oxygen into the β-Si_3N_4 lattice forms a vacancy in the Si site, as expressed by the following equation:

$$2\,SiO_2 \rightarrow 2\,Si_{Si} + 4O_N + V_{si} \tag{8}$$

where O_N and V_{Si} is a dissolved oxygen atom in a nitrogen site and a vacancy in the Si site, respectively. Accompanied by the substitution of O with N, a silicon vacancy is generated in order to maintain electrical neutrality, which was indirectly confirmed by measuring the concentration of nitrogen vacancy using electron spin resonance (ESR) analyses [46].

The dissolution of oxygen into the β-Si_3N_4 lattice would be expected to lower the thermal conductivity, due to the formation of a large number of lattice vacancies. Kitayama *et al.* investigated the relationship between thermal conductivity and lattice oxygen content for dense β-Si_3N_4 ceramics with various Y_2O_3/SiO_2 additive ratios fabricated by hot-pressing and subsequent annealing. In order to exclude microstructural factors such as grain size and the number of grain boundaries, the thermal conductivity of β-Si_3N_4 crystals in the sintered specimens was estimated using a modification of the Wiener method (see Section 16.3.1.3) [37]. The results showed a clear tendency for the thermal resistivity (the inverse of thermal conductivity) of the β-Si_3N_4 crystal to increase in line with increases in the lattice oxygen content (see Figure 16.7).

16.3.2.3 **Improvements in Thermal Conductivity for Silicon Nitride Ceramics**

When selecting the correct sintering additives and sintering conditions, the thermal conductivity will increase with increasing sintering time due not only to a reduction

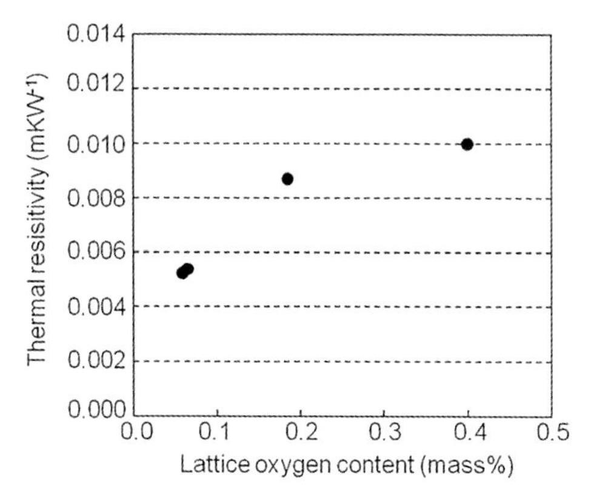

Figure 16.7 Relationship between the lattice oxygen contents determined by hot-gas extraction method and thermal resistivities (inverse of thermal conductivities) of β-Si_3N_4 crystal [47].

in the number of two-grain junctions [1, 37], but also to the purification of grains [45] as a result of grain growth via solution–reprecipitation processes. Figure 16.8 shows the variation of thermal conductivity with sintering time for Si_3N_4 specimens doped with 2 mol% MgO–5 mol% Yb_2O_3 [45], or with 1 mass% HfO_2–8 mass% Y_2O_3 [52] and sintered at 2273 K for between 2 and 48 h. In both experiments, β-Si_3N_4 raw powders were used in order to exclude the effect of any α-phase → β-phase transformation. Appreciable grain growth occurred during heating, as illustrated in Figure 16.9, and the thermal conductivities were increased with increasing sintering time. When the lattice oxygen content in β-Si_3N_4 grains (which had been extracted from Si_3N_4-sintered specimens by an etching treatment) was measured [45], it was clear that the increase in thermal conductivity was closely related to the decrease in lattice oxygen content in the β-Si_3N_4 grains (see Figure 16.10). More definitive experiments were conducted subsequently by Yokota *et al.* [19], in which

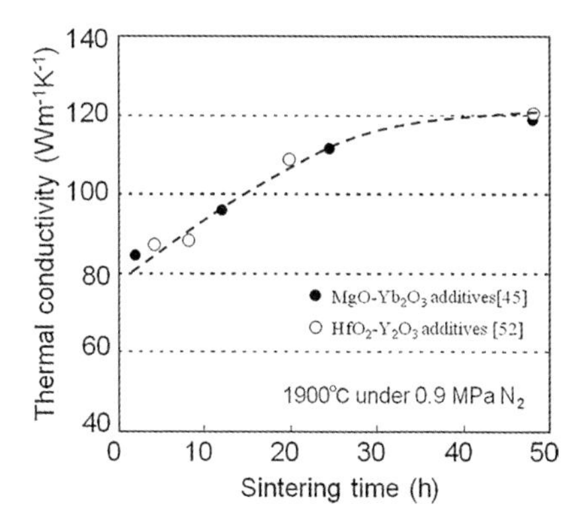

Figure 16.8 Variation of thermal conductivity with sintering time for the Si_3N_4 specimens doped with 2 mol% MgO–5 mol% Yb_2O_3 [45], or with 1 mass% HfO_2 –8 mass% Y_2O_3 [52] sintered at 2273 K for 2–48 h.

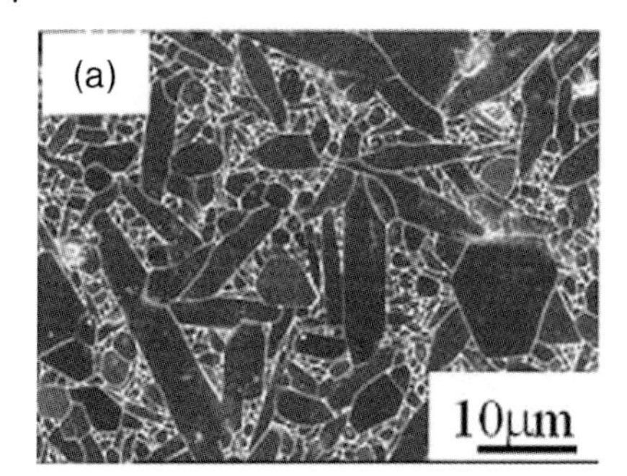

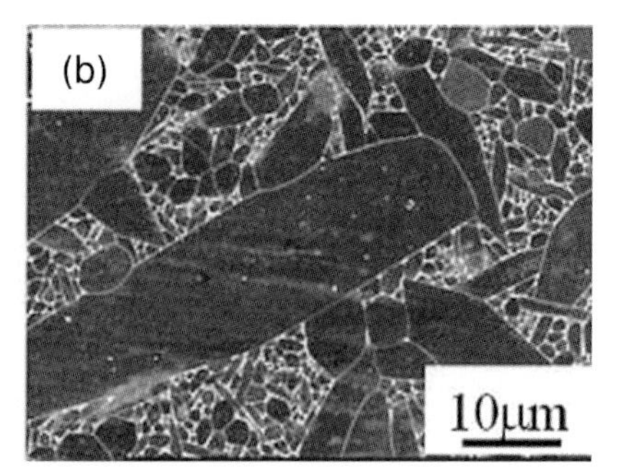

Figure 16.9 Microstructure of the Si_3N_4 specimens doped with 2 mol% MgO–5 mol% Yb_2O_3 sintered at 2173 K under 0.9 MPa N_2 for (a) 2 h and (b) 48 h [45].

Si_3N_4 doped with HfO_2–Y_2O_3 additives was sintered at 2273 K for 8 and 48 h, so as to obtain materials with thermal conductivities of 88 and 120 $W\,m^{-1}\,K^{-1}$, respectively. Both specimens exhibited a bimodal microstructure where larger elongated grains with a grain diameter in excess of 2 μm were dispersed in fine matrix grains. The volume fraction of the larger grains in the specimen sintered for 48 h was much higher than that of the specimen sintered for 8 h. When the amount of impurity oxygen in the smaller matrix grains and larger grains was carefully measured, there was a minimal change in lattice oxygen content of the smaller matrix grains between 8 and 48 h (1900 ppm for 8 h and 1920 ppm for 48 h, respectively). In contrast, there was a drastic decrease in the lattice oxygen content of larger, elongated grains (980 ppm for 8 h and 460 ppm for 48 h). The purification of β-Si_3N_4 grains occurs through a dissolution–reprecipitation process during long-term heating, due to a higher affinity for oxygen in Y_2O_3 containing oxynitride glassy phase. Thus, the thermal conductivity of Si_3N_4 ceramic was increased with increasing amounts of purified larger grains [52].

The relationship between the lattice oxygen content of β-Si_3N_4 and the thermal resistivity of the specimens, as reported in the literature, are summarized in

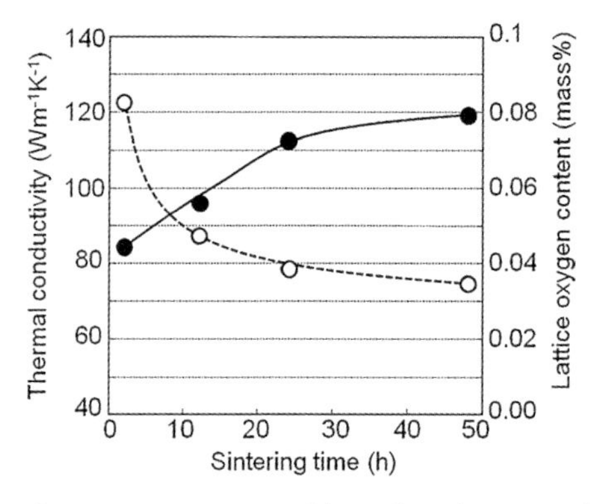

Figure 16.10 Variation of thermal conductivity and lattice oxygen content with sintering time for the Si_3N_4 specimens doped with 2 mol% MgO–5 mol% Yb_2O_3 sintered at 2173 K under 0.9 MPa N_2.

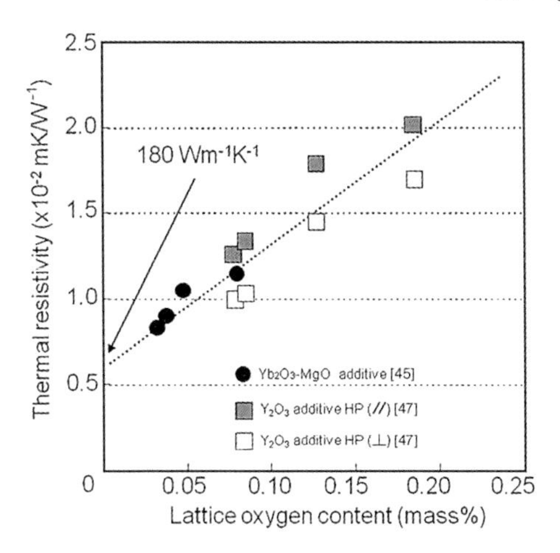

Figure 16.11 Effect of lattice oxygen content on the thermal resistivity of Si_3N_4 fabricated by gas-pressure sintering (filled circles) and hot-pressing (HP; squares). The thermal resistivities for the hot-pressed specimens were measured in two directions, parallel to and perpendicular to the HP direction [53].

Figure 16.11 [53]. Although the processing method, the type of raw Si_3N_4 powder and the additive system used were each quite different among these investigations, there was a clear tendency for the thermal resistivity to decrease with a decreasing lattice oxygen content in the β-Si_3N_4. It might be said that lattice oxygen is a dominant extrinsic factor governing the high thermal conductivity of Si_3N_4 ceramic, as has been demonstrated in AlN ceramic. Moreover, by extrapolation, the β-Si_3N_4 free of lattice oxygen should exhibit a thermal conductivity of at least $180\,W\,m^{-1}\,K^{-1}$.

Rare-earth oxides are well known as effective additives for preparing high-thermal conductivity Si_3N_4 [9]. Kitayama *et al.* investigated the effect of rare-earth oxides on the thermal conductivity of Si_3N_4 by hot-pressing α-Si_3N_4 powder doped with a series of rare-earth (La, Nd, Gd, Y, Yb, and Sc) oxides, followed by subsequent annealing [54]. With a decreasing ionic radius of the rare-earth element, the mean grain size was increased while the lattice oxygen was decreased, and hence the thermal conductivity was increased. Unfortunately, the results of these experiments could not confirm which factor – microstructure or lattice oxygen – was dominant, because there might have been a close connection between the grain growth and oxygen removal. Based on the results obtained, however, it was concluded that the type of rare-earth oxide additive had a significant influence on the thermal conductivity of β-Si_3N_4, unlike the case of AlN.

At this point, it might be worthwhile noting that, in addition to the types of sintering additive used, the ratio of additive oxide to SiO_2 existing in a raw Si_3N_4 powder would also affect the thermal conductivity of the sintered specimens. When Kitayama *et al.* fabricated Si_3N_4 with various Y_2O_3/SiO_2 additive ratios (0.289, 0.807, 1.267, and 2.029) by hot-pressing, the thermal conductivity of the specimens was shown to increase with increasing $Y_2O_3:SiO_2$ ratio, with a significant rise occurring when the ratio was close to 1 [47] (see Figure 16.12). This effect of grain-boundary

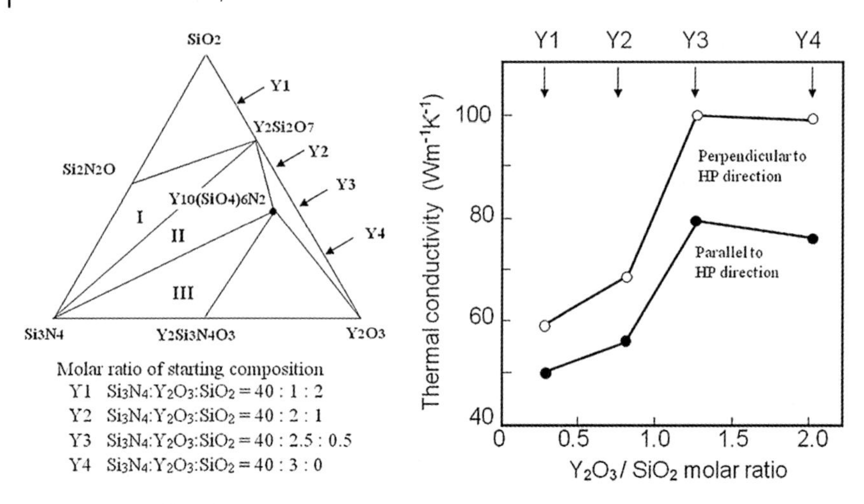

Figure 16.12 Effect of $Y_2O_3 : SiO_2$ ratio on the thermal conductivity of Si_3N_4 fabricated by hot pressing at 2073 K for 2 h under a pressure of 40 MPa [47]. The starting compositions in terms of the Si_3N_4–Y_2O_3–SiO_2 phase diagram are shown on the left-hand side.

composition on the thermal conductivity of Y_2O_3-doped Si_3N_4 was explained on the basis of arguments proposed by Jackson *et al.* [32] to explain the same effect occurring with Y_2O_3-doped AlN. With an increasing $Y_2O_3 : SiO_2$ ratio, the three-phase field was seen to shift from region I (Si_3N_4–Si_2N_2O–$Y_2Si_2O_7$), to region II (Si_3N_4–$Y_2Si_2O_7$–$Y_{20}N_4Si_{12}O_{48}$), and finally to region III (Si_3N_4–$Y_{20}N_4Si_{12}O_{48}$–$Y_2Si_3N_4O_3$). In corresponding with the variant of the three-phase field, the activity of SiO_2 was decreased in the order of regions I, II, and III. In this case, the grain-boundary phase dictated the lattice oxygen content of β-Si_3N_4, depending on the activity of SiO_2 in the three-phase field; notably, the lower the activity of SiO_2 in the three-phase field the lower was the solubility of oxygen in β-Si_3N_4. Thus, the highest thermal conductivity was achieved when both the $Y_{20}N_4Si_{12}O_{48}$ and $Y_2Si_3N_4O_3$ phases were present in the grain-boundary phase.

With regards to the amount of oxygen in the grain-boundary glassy phase, it has been suggested that the use of a nitride sintering additive might represent an effective way of improving the thermal conductivity of Si_3N_4. Hayashi *et al.* [45] sintered β-Si_3N_4 raw powder compacts doped with either MgO–Yb_2O_3 or $MgSiN_2$–Yb_2O_3 additives at 2273 K for 2 to 48 h under a 0.9 MPa nitrogen pressure, and then measured the lattice oxygen content of the β-Si_3N_4 grains and thermal conductivity in both specimens. When comparing the two types of Si_3N_4 sintered under the same conditions (i.e., the same sintering time), the specimen which had used $MgSiN_2$ as a magnesium supply source exhibited a lower lattice oxygen content, and thus a higher thermal conductivity. The thermal conductivity of the $MgSiN_2$-doped Si_3N_4 was about 20 $W m^{-1} K^{-1}$ higher than that of the MgO-doped Si_3N_4, and reached a maximum of over 140 $W m^{-1} K^{-1}$ following a 48 h period of sintering. The higher thermal conductivity in the $MgSiN_2$-doped Si_3N_4 was attributed not only to the purification of Si_3N_4 grains via a solution–reprecipitation process in the nitrogen-rich grain-boundary glassy phase, but also to an enhanced grain growth that had resulted in a microstructure where the larger grains were in contact with each other.

16.3.2.4 **Lattice Defects in SiC Grains and Improvement of Thermal Conductivity**

Slack reported that Al and N atoms could dissolve in SiC as electrically active impurities and cause phonon scattering, thereby substantially reducing the thermal conductivity of the SiC single crystal [55]. The commercial SiC powders may also contain such impurity atoms, and after sintering these would remain in the SiC grains of the polycrystalline materials. In contrast, the consolidation of SiC is usually accomplished with the aid of sintering additives. The elements Al, B, and Be (which are known as popular sintering aids for SiC) have been reported to dissolve in the SiC lattice and to become the most detrimental factor in affecting the thermal conductivity. The typical solid-state-sintered SiC ceramics doped with B_4C additive demonstrated a room-temperature thermal conductivity of about $120\,W\,m^{-1}\,K^{-1}$ [27], while the typical liquid-phase-sintered SiC with Al_2O_3–Y_2O_3 addition had a conductivity of 70–90 $W\,m^{-1}\,K^{-1}$ [56]. Yet, when sintered with the aid of BeO additives, the SiC could attain a room-temperature thermal conductivity of $270\,W\,m^{-1}\,K^{-1}$, which to date is the highest value for a sintered polycrystalline SiC material [27]. The much lower thermal conductivity of these polycrystalline SiC materials, when compared to a pure SiC single crystal ($490\,W\,m^{-1}\,K^{-1}$), was mainly attributed to the dissolution of the impurity atoms (Al, B, and Be) in the SiC lattice, while the difference between the SiCs doped with various sintering additives was related to the different concentrations of those atoms dissolved in the SiC grains. The solubility of Be in SiC was reported as $8 \times 10^{18}\,cm^{-3}$, which was much less than those of B and Al, at 1×10^{21} and $1 \times 10^{20}\,cm^{-3}$, respectively [9]. Notably, the higher the solubility of the impurity atoms in the SiC lattice, the lower was the thermal conductivity of the SiC ceramics.

Apart from the above-mentioned impurity atoms, oxygen may also dissolve in the lattice of SiC as an impurity, similar to the cases of AlN and Si_3N_4. Zhou *et al.* [57] have claimed that the lattice oxygen could be regarded as a substitutional impurity resulting from the dissolution of SiO_2 in SiC, according to the following defect reaction:

$$SiO_2 \rightarrow Si_{Si} + 2O_C + V_{Si} \tag{9}$$

where O_C and V_{Si} denote oxygen in a carbon site and a silicon vacancy, respectively. A single Si vacancy is created for each two dissolved oxygen atoms for charge compensation. As these point defects are detrimental to thermal conductivity by causing a scattering of phonons, reducing the lattice oxygen content should represent an effective way of improving the thermal conductivity of SiC. Following the thermodynamic arguments made for the lattice-purifying process of AlN and Si_3N_4 ceramics [4, 5, 23, 32, 47], Zhou *et al.* [57] proposed that the dissolved oxygen might be removed from the SiC lattice by utilizing the oxygen-gathering ability of some oxide additives during sintering; hence, they used a mixture of Y_2O_3 and La_2O_3 as sintering additives to prepare LPS-SiC ceramics. When using a hot-gas extraction method to measure the lattice oxygen contents of the SiC raw powder and the sintered SiC, the lattice oxygen content showed a drastic decrease after sintering, while the sintered SiC attained a thermal conductivity of up to 167 $W\,m^{-1}\,K^{-1}$, which was further improved (to over 200 $W\,m^{-1}\,K^{-1}$) after subsequent annealing. During both sintering and annealing, the driving force for the dissolved oxygen to diffuse out from the SiC lattice and react with Y_2O_3 and La_2O_3 to form the yttrium and lanthanum silicates ($Y_2Si_2O_7$ and La_2SiO_5) that existed as secondary phases in the materials, was considered to be the thermodynamic affinity that the rare-earth oxides had for SiO_2.

16.4 Mechanical Properties of High-Thermal Conductivity Si_3N_4 Ceramics

16.4.1 Harmonic Improvement of Thermal Conductivity and Mechanical Properties

As noted in Section 16.3, one key approach for improving the thermal conductivity of Si_3N_4 ceramic is to enhance the grain growth, thereby decreasing the lattice oxygen in β-Si_3N_4 grains via a dissolution–reprecipitation process. Of course, grain growth itself plays a role in reducing the number of grain boundary glassy phases that exist between two β-Si_3N_4 grains, so as to increase thermal conductivity of Si_3N_4 [37]. However, this approach is only effective below a critical value of grain size, equal to a few microns [37].

Almost all high-thermal conductivity (>100 $W\,m^{-1}\,K^{-1}$) Si_3N_4 ceramics have been fabricated via gas-pressure sintering at temperatures of about 2173 K for an unusually long period of time (1–2 days), or at temperatures in excess of 2373 K under high N_2 pressures (>10 MPa) for several hours, in order to achieve grain purification (see Table 16.3). Unfortunately, however, these sintering conditions provide a material with a very coarse microstructure that leads to drastic reductions in its mechanical properties. The data within the dotted elliptical curve in Figure 16.13 indicate the

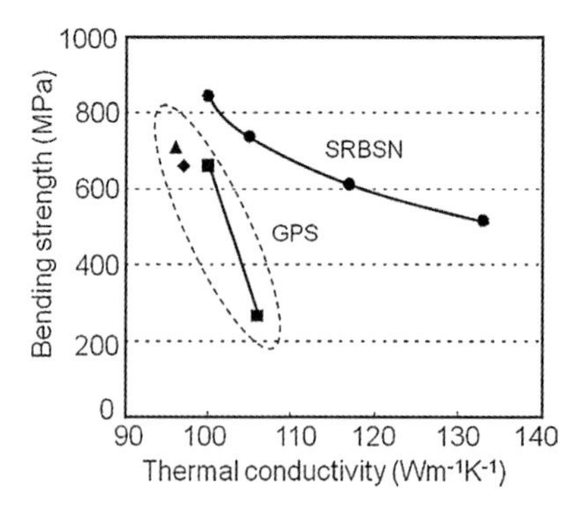

Figure 16.13 Relationship between thermal conductivity and four-point bending strength of various types of high-thermal conductivity Si_3N_4 fabricated by different routes. GPSN = gas-pressure-sintered silicon nitride; SRBSN = sintered reaction-bonded silicon nitride [62].

relationship between bending strength and thermal conductivity for Si_3N_4 ceramics doped with various sintering additives and densified by gas pressure sintering at 2173 K under a 0.9 MPa nitrogen pressure. Clearly, the improvement in thermal conductivity led to a sacrifice of the mechanical properties; hence, to achieve a high strength and a high thermal conductivity is antagonistic, and the fabrication of such materials with harmonic characteristics remains a major challenge.

One strategy by which this problem may be overcome is to apply a reaction bonding (RB) process, which is a well-known, classical method used to fabricate silicon nitride ceramics from a Si powder compact [58]. In 1995, Haggerty *et al.* noted that improvements in the thermal conductivity of reaction-bonded Si_3N_4 could be achieved by reducing the impurities in raw silicon powder to an extremely low level [8]. Recently, a research group at the National Institute of Advanced Industrial Science & Technology, in Japan, have investigated the preparation of high-thermal conductivity silicon nitrides via the route of the sintering of reaction-bonded silicon nitride (RBSN) [59–62]. In this scheme, when a Si powder compact is heated at about 1673 K in a nitrogen atmosphere, it will be nitrided so as to form a silicon nitride compact. Then, as the nitridation reaction proceeds (mainly via a gas–solid reaction system), the external dimensions of the compact are retained whilst each of the individual Si particles expands by about 22% [58]. It is, therefore, well-recognized that the reaction bonding process has the following benefits: (i) a cheaper Si powder can be used as the starting material, rather than the more expensive Si_3N_4 powder; and (ii) shrinkages accompanied by post-sintering can be minimized because of the higher density of the pre-sintered specimens.

In addition to the above-mentioned benefits, the SRBSN process has the potential to fabricate silicon nitrides with enhanced mechanical and thermal properties. A typical example of the morphological change before and after the nitridation of a Si powder compact is shown in Figure 16.14 where, clearly, the particle size of the nitrided compact is much finer than that of the starting Si powder. This phenomenon, combined with the higher density achieved in nitrided compacts, makes the RB process very attractive for fabricating high-performance silicon nitride ceramics,

Figure 16.14 Scanning electron microscopy images of microstructures (a) before and (b) after nitridation of a silicon compact. The nitridation reaction was conducted at 1673 K for 8 h under 0.1 MPa nitrogen pressure.

because the pre-sintered specimen with a finer microstructure and a higher density would be favorable for controlling the microstructure of the final product. Another benefit is that the entire process, from nitridation to post-sintering, can be carried out without exposing the compacts to the air, which would be especially favorable for controlling the oxygen content in Si_3N_4.

Zhu *et al.* investigated the effect of the impurity oxygen content of raw Si powder on the properties of a RBSN ceramic [60]. In these experiments, 2 mol% Y_2O_3 and 5 mol% $MgSiN_2$ as sintering additives were added into two types of raw Si powder, the properties of which are summarized in Table 16.5. The oxygen contents after milling, and the estimated oxygen contents in Si_3N_4 after complete nitridation of the Si powders, are also shown in Table 16.5. It should be noted that the estimated oxygen content after nitridation of the fine Si powder was almost identical to the impurity oxygen in the commercially available α-Si_3N_4 powder (UBE E-10 grade powder; Ube Industries Co. Ltd, Japan), while the estimated oxygen content of the coarse Si powder was about one-third that of the commercially available material. The thermal properties of the SRBSN fabricated by the post-sintering of reaction-bonded silicon nitride (RBSN) at 2273 K for 12 h is shown in the lowest line of Table 16.5, along with the data for Si_3N_4 obtained by the gas-pressure sintering of α-Si_3N_4 doped with the same additives. Although the microstructures of these three types of Si_3N_4 were roughly similar (see Figure 16.15), the thermal conductivity of the SRBSN prepared from a coarse Si powder was approximately $10\,W\,m^{-1}\,K^{-1}$ higher than that of the

Table 16.5 Characteristics of raw Si powders used for the preparation of sintered reaction-bonded silicon nitrides (SRBSN) and their thermal conductivities (for comparison, the properties of normal sintered Si_3N_4 are also shown) [60].

	Sintered reaction-bonded Si_3N_4 (SRBSN)		Sintered Si_3N_4 (for comparison)
Raw powder	**Fine Si powder**	**Coarse Si powder**	**Fine α-Si_3N_4 powder[a]**
Particle size[b]	<1	<10	
D 50[b]	0.7	7	0.1
Purity[b]	>99.9	>99.9	>99.9
Oxygen content (as received)	1.3 mass%	0.34 mass%	
Oxygen content (after milling)	1.8 mass%	0.7 mass%	
Estimated oxygen content in Si_3N_4 (after nitridation)	1.1 mass%[c]	0.42 mass%[c]	1.2 mass%[b]
Thermal conductivity of sintered specimen[d]	$112\,W\,m^{-1}\,K^{-1}$	$121\,W\,m^{-1}\,K^{-1}$	$111\,W\,m^{-1}\,K^{-1}$

a) α-Si_3N_4 powder, E-10 grade, UBE industries Co., Japan.
b) Data provided by the manufacturer.
c) Estimated oxygen content in Si_3N_4 after complete nitridation of Si powder.
d) Nitridation for SRBSN specimens was performed at 1673 K for 8 h under 0.1 MPa nitrogen pressure. All specimens were sintered at 2173 K for 12 h under 0.9 MPa N_2.

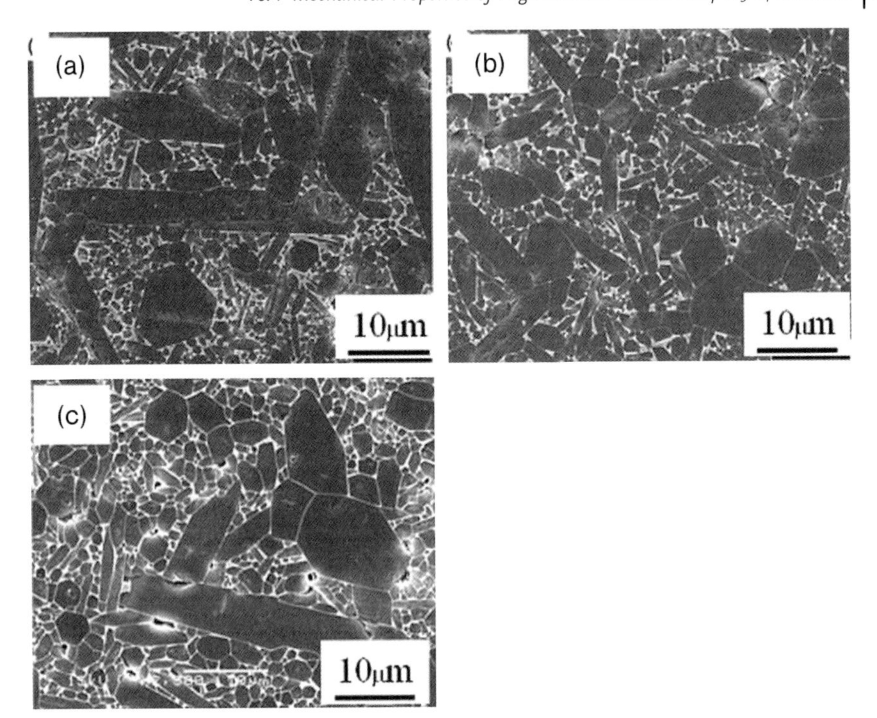

Figure 16.15 Scanning electron microscopy images of polished and etched surfaces for (a) sintered reaction-bonded silicon nitride (SRBSN) from a fine silicon powder; (b) SRBSN from a coarse Si powder; (c) Sintered silicon nitride from a commercial fine-grained α-Si_3N_4 powder. All of the specimens were doped with 2 mol% Y_2O_3 and 5 mol% $MgSiN_2$ as sintering additives. The nitridation reaction for specimens (a) and (b) was performed at 1673 K for 8 h. Post-sintering of nitrided specimens or sintering of specimen (c) was performed at 2173 K for 12 h under 0.9 MPa nitrogen pressure.

other silicon nitrides. These results confirmed that Si_3N_4 with an improved thermal conductivity could be fabricated by sintering a reaction-bonded Si_3N_4, using a coarse, high-purity Si powder.

Most recently, Zhou *et al.* [62] investigated the mechanical and thermal properties of SRBSN using a high-purity Si powder. In these experiments, a RBSN was prepared by nitriding a green compact composed of a high-purity Si powder (impurity oxygen 0.28 mass%) doped with Y_2O_3 and MgO additives, followed by post-sintering under a N_2 pressure of 0.9 MPa at 1900 °C for various times that ranged from 3 to 24 h, so as to prepare the SRBSN. The materials obtained by sintering for 3, 6, 12, and 24 h had thermal conductivities of 100, 105, 117, and 133 W m K^{-1}, respectively, and bending strengths of 843, 736, 612, and 516 MPa, respectively. By comparison, when a gas-pressure-sintered silicon nitride was prepared from a high-purity Si_3N_4 powder, the SRBSN material was shown to be superior to the sintered silicon nitride material in terms of a balance between thermal conductivity and bending strength (as shown in Figure 16.13). The study results showed that the SRBSN route does indeed represent

a promising method for preparing Si_3N_4 ceramics with both high thermal conductivity and high strength.

16.4.2
Anisotropic Thermal Conductivity in Textured Si_3N_4

Another approach for achieving a high thermal conductivity in Si_3N_4 ceramic is to develop a textured microstructure in which the elongated β-Si_3N_4 grains are oriented almost unidirectionally. In this way, a high thermal conductivity along the grain orientation would be expected, compared to a material with a random distribution of β-Si_3N_4 grains, because of the much higher thermal conductivity along the c-axis than along the a-axis in β-Si_3N_4 crystals [10, 13].

Such anisotropic microstructures can be fabricated by a combination of the seeding of rod-like β-Si_3N_4 nuclei and a forming process generating shear stress, such as tape-casting and extrusion [63–65]. This process is based on the grain-growth behavior of silicon nitride; during the α- to β-phase transformation, the preferential nucleation site of the newly formed β-Si_3N_4 phase is pre-existing β particles. However, after the transformation only a few large β grains will selectively, particularly along the c-axis direction, via a solution–reprecipitation reaction. The typical microstructure of seeded and extruded Si_3N_4 is shown in Figure 16.16 [66]. The

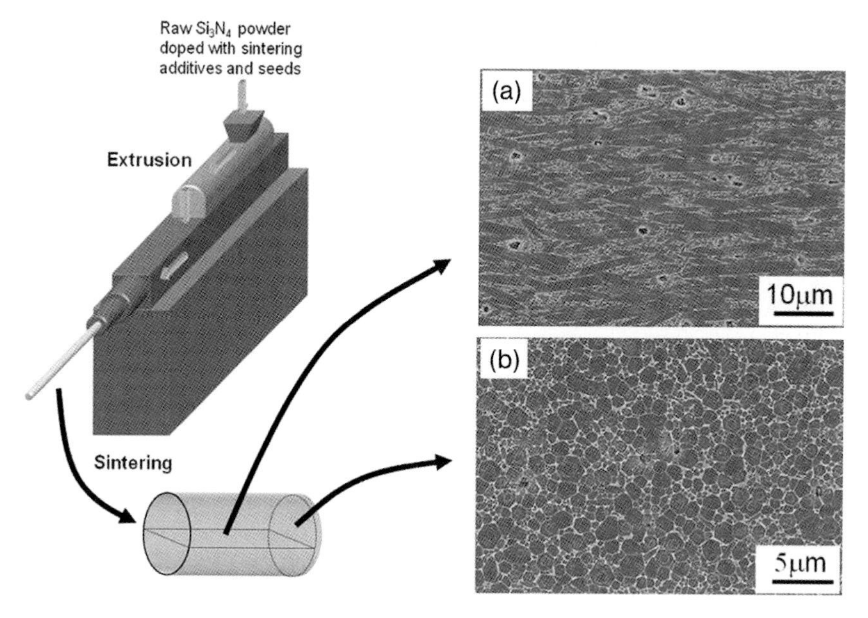

Figure 16.16 Scanning electron microscopy images of textured Si_3N_4 observed in the plane (a) parallel to and (b) perpendicular to the grain alignment. The specimen was fabricated by sintering the green body formed by extrusion of Si_3N_4 raw powder doped with Y_2O_3, Al_2O_3, and a small percentage of seed particles [66].

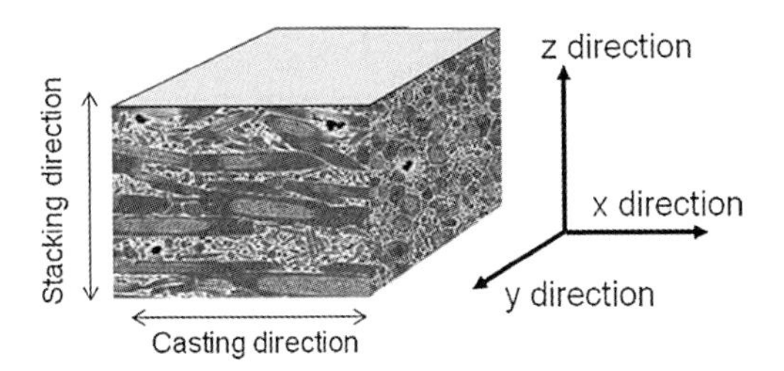

Figure 16.17 Microstructure of seeded and tape-cast Si_3N_4 (left-hand side) and three different directions for thermal conductivity measurements (right-hand side) [53].

Table 16.6 Thermal conductivity in three directions for textured Si_3N_4 fabricated by combined seeding and tape casting [53].

No.	Sintering additives	Annealing conditions	Thermal conductivity ($W m^{-1} K^{-1}$)			Reference
			x direction	*y* direction	*z* direction	
1	5 mass% Y_2O_3	2123 K, 66 h, 0.9 MPa N_2	121	75	60	[65]
2	5 mass% Y_2O_3	2773 K, 2 h, 200 MPa N_2	155	67	52	[67]
3	0.5 mol% Y_2O_3 - 0.5 mol% Nd_2O_3	2473 K, 4 h, 30 MPa N_2	138	97	72	[68]

a) Thermal conductivities were measured in three directions, as shown in Figure 16.17.

material exhibits a high anisotropy where the large elongated grains, which are grown from seeds, are almost unidirectionally oriented parallel to the extruding direction. The thermal conductivities, measured in three different directions (as shown in Figure 16.17) for the textured Si_3N_4 fabricated by this method with different sintering additives are summarized in Table 16.6 [53]. As expected, for each of these specimens the thermal conductivity was highest along the grain alignment. In particular, a highly anisotropic Si_3N_4 with very long grains and fabricated under extreme conditions, exhibited a high thermal conductivity of about $150\,W\,m^{-1}\,K^{-1}$ in the direction parallel to the grain alignment.

16.5 Concluding Remarks

In this chapter, the thermal conductivity of nonoxide ceramics (AlN, SiC, and Si_3N_4) has been reviewed from both theoretical and processing points of view. In particular, attention was focused on Si_3N_4 ceramics, which have recently attracted much attention as candidates for high-thermal conductivity substrates. Similar to the situation with AlN, the dominant factors that control the thermal conductivity of Si_3N_4 ceramic are the lattice defects of grains (particularly the lattice oxygen content), as well as the amount and distribution of grain boundary phases. Compared to high-thermal conductivity AlN and SiC, the Si_3N_4 ceramics have the prospect of achieving a balance between thermal conductivity and mechanical properties at a high level, although a more careful and thoughtful processing strategy is needed.

References

1 Hirosaki, N., Okamoto, Y., Ando, A., Munakata, F., and Akimune, Y. (1996) *J. Am. Ceram. Soc.*, **79**, 2878–2882.

2 Kingery, W.D., Bowen, H.K., and Uhlmann, D.R. (1960) *Introduction to Ceramics*, John Wiley & Sons Inc., New York, USA.

3 Touloukian, Y.S., Powell, R.W., Ho, C.Y., and Klemens, P.G. (1970) *Thermal Conductivity - Nonmetallic Solids - Thermophysical Properties of Matter*, vol. 2, IFI/Plenum, New York-Washington.

4 Slack, G.A. (1973) *J. Phys. Chem. Solids*, **34**, 321–335.

5 Slack, G.A., Tanzilli, R.A., Pohl, R.O., and Vandersande, J.W. (1987) *J. Phys. Chem. Solids*, **48**, 641–647.

6 Riley, F.L. (1983) *Progress in Nitrogen Ceramics*, Martinus Nijhoff Publishers, The Hague.

7 Dressler, W. and Riedel, R. (1997) *Int. J. Refract. Met. Hard Mater.*, **15**, 13–47.

8 Haggerty, J. S. and Lightfoot, A. (1995) Ceram. Eng. Sci. Proc. 16, in *19th Annual Conference on Composites & Advanced Ceramic Materials and Structures - A*, (ed. J.B. Watchman), The American Ceramics Society, Westerville, OH.

9 Watari, K. (2001) *J. Ceram. Soc. Japan*, **109**, S7–S16.

10 Hirosaki, N., Ogata, S., and Kocer, C. (2002) *Phys. Rev. B*, **65**, 134110-1–134110-11.

11 Morelli, D.T. and Heremans, J.P. (2002) *Appl. Phys. Lett.*, **81**, 5126.

12 Bruls, R.J., Hintzen, H.T., and Metselaar, R. (2005) *J. Eur. Ceram. Soc.*, **25**, 767–779.

13 Li, B., Pottier, L., Roger, J.P., Fournier, D., Watari, K., and Hirao, K. (1999) *J. Eur. Ceram. Soc.*, **19**, 1631–1639.

14 Kuriyama, M., Inomata, Y., Kijima, T., and Hasegawa, Y. (1978) *Ceram. Bull.*, **57**, 1119–1122.

15 Tsukuma, K., Shimada, M., and Koizumi, M. (1981) *Ceram. Bull.*, **60**, 910–913.

16 Watari, K., Seki, Y., and Ishizaki, K. (1989) *J. Ceram. Soc. Japan*, **97**, 56–62 [in Japanese].

17 Li, C.W., Yamanis, J., Whalen, P.J., Gasdaska, C.J., and Ballard, C.P. (1992) *Mater. Res. Soc. Symp. Proc.*, **251**, 103–111.

18 Okamoto, Y., Hirosaki, N., Ando, M., Munakata, F., and Akimune, Y. (1998) *J. Mater. Res.*, **13**, 3473–3477.

19 Yokota, H. and Ibukiyama, M. (2003) *J. Eur. Ceram. Soc.*, **23**, 55–60.

20 Yokota, H. and Ibukiyama, M. (2003) *J. Am. Ceram. Soc.*, **86**, 197–199.

21 Komeya, K. (1984) *Am. Ceram. Soc. Bull.*, **63**, 1158–1159.

22 Kurokawa, Y., Utsumi, K., and Takamizawa, H. (1988) *J. Am. Ceram. Soc.*, **71**, 588–594.

23 Virkar, A.V., Jackson, T.B., and Cutler, R.A. (1989) *J. Am. Ceram. Soc.*, **72**, 2031–2042.

24 Hoffmann, M.J., Becher, P.F., and Petzow, G. (1994) Silicon Nitride 93, in *Key Engineering Materials*, vol. **89–91**, Trans Tech Publications Ltd, Switzerland.

25 Kim, H.-D., Lin, H.-T., and Hoffmann, M.J. (2005) Advanced Si-Based Ceramics and Composites, in *Key Engineering Materials*, vol. **287**, Trans Tech Publications, Switzerland.

26 Prochazka, S. (1975) *Special Ceramics 6* (ed. P. Popper), British Ceramic Research Association, Stoke on Trent.

27 Takeda, Y. (1988) *Am. Ceram. Soc. Bull.*, **67**, 1961–1963.

28 Padture, N.P. (1994) *J. Am. Ceram. Soc.*, **77**, 519–523.

29 Wiener, O. (1904) *Phys. Z.*, **5**, 332–338.

30 Kingery, W.D. (1958) *J. Am. Ceram. Soc.*, **42**, 617–627.

31 Euckem, A. (1932) *Forsch. Geb. Ingenieurw.*, **B3** (353), 16.

32 Jackson, T.B., Virkar, A.V., More, K.L., Dinwiddie, R.B. Jr, and Cutler, R.A. (1997) *J. Am. Ceram. Soc.*, **80**, 1421–1435.

33 Buhr, H. and Muller, G. (1993) *J. Eur. Ceram. Soc.*, **12**, 271–277.

34 Kim, W.J., Kim, D.K., and Kim, C.H. (1996) *J. Am. Ceram. Soc.*, **79**, 1066–1072.

35 Lange, F.F. (1979) *J. Am. Ceram. Soc.*, **62**, 428–430.

36 Clarke, D.R. and Thomas, G. (1977) *J. Am. Ceram. Soc.*, **60**, 491–495.

37 Kitayama, M., Hirao, K., Tsuge, A., Watari, K., Toriyama, M., and Kanzaki, S. (1999) *J. Am. Ceram. Soc.*, **82**, 3105–3112.

38 Sigl, L.S. (2003) *J. Eur. Ceram. Soc.*, **23**, 1115–1122.

39 Zhan, G.D., Mitomo, M., and Mukherjee, A.K. (2002) *J. Mater. Res.*, **17**, 2327–2333.

40 Zhou, Y., Hirao, K., Yamauchi, Y., and Kanzaki, S. (2003) *J. Mater. Res.*, **18**, 1854–1862.

41 Zhou, Y., Hirao, K., Watari, K., Yamauchi, Y., and Kanzaki, S. (2004) *J. Eur. Ceram. Soc.*, **24**, 265–270.

42 Miyashiro, F., Iwase, N., Tsuge, A., Ueno, F., Nakahashi, M., and Takahashi, T. (1990) *IEEE Trans, Components, Hybrids, Manufact. Technol.*, **13**, 313–319.

43 Watari, K., Hwang, H.J., Toriyama, M., and Kanzaki, S. (1996) *J. Am. Ceram. Soc.*, **79**, 1979–1981.

44 Watari, K., Brito, M.E., Toriyama, M., Ishizaki, K., Cao, S., and Mori, K. (1999) *J. Mater. Sci. Lett.*, **18**, 865–867.

45 Hayashi, H., Hirao, K., Toriyama, M., Kanzaki, S., and Itatani, K. (2001) *J. Am. Ceram. Soc.*, **84**, 3060–3062.

46 Kitayama, M., Hirao, K., Tsuge, A., Toriyama, M., and Kanzaki, S. (1999) *J. Am. Ceram. Soc.*, **82**, 3263–3265.

47 Kitayama, M., Hirao, K., Tsuge, A., Watari, K., Toriyama, M., and Kanzaki, S. (2000) *J. Am. Ceram. Soc.*, **83**, 1985–1992.

48 Yokota, H., Abe, H., and Ibukiyama, M. (2003) *J. Eur. Ceram. Soc.*, **23**, 1751–1759.

49 Akimune, Y., Munakata, F., Matsuo, K., Hirosaki, N., Okamoto, Y., and Misono, K. (1999) *J. Ceram. Soc. Jpn*, **107**, 339–342.

50 Hirosaki, N., Saito, T., Munakata, F., Akimune, Y., and Ikuhara, Y. (1999) *J. Mater. Res.*, **14**, 2959–2965.

51 Thomas, A. and Muller, G. (1991) *J. Eur. Ceram. Soc.*, **8**, 11–19.

52 Yokota, H., Yamada, S., and Ibukiyama, M. (2003) *J. Eur. Ceram. Soc.*, **23**, 1175–1182.

53 Hirao, K., Watari, K., Hayashi, H., and Kitayama, M. (2001) *Mater. Res. Bull.*, **26**, 451–455.

54 Kitayama, M., Hirao, K., Watari, K., Toriyama, M., and Kanzaki, S. (2001) *J. Am. Ceram. Soc.*, **84**, 353–358.

55 Slack, G.A. (1964) *J. Appl. Phys.*, **35**, 3460–3466.

56 Liu, D.M. and Lin, B.W. (1996) *Ceram. Int.*, **22**, 407–414.

57 Zhou, Y., Hirao, K., Yamauchi, Y., and Kanzaki, S. (2003) *J. Mater. Res.*, **18**, 1854–1862.

58 Moulson, A.J. (1979) *J. Mater. Sci.*, **14**, 1017–1051.

59 Zhu, X.W., Zhou, Y., and Hirao, K. (2004) *J. Am. Ceram. Soc.*, **87**, 1398–1400.

60 Zhu, X.W., Zhou, Y., Hirao, K., and Lences, Z. (2006) *J. Am. Ceram. Soc.*, **89**, 3331–3339.

61 Zhu, X.W., Zhou, Y., Hirao, K., and Lences, Z. (2007) *J. Am. Ceram. Soc.*, **90**, 1684–1692.

62 Zhou, Y., Zhu, X.W., Hirao, K., and Lences, Z. (2008) *Int. J. Appl. Ceram. Technol.*, **5**, 119–126.

63 Hirao, K., Nagaoka, T., Brito, M.E., and Kanzaki, S. (1994) *J. Am. Ceram. Soc.*, **77**, 1857–1862.

64 Hirao, K., Ohashi, M., Brito, M.E., and Kanzaki, S. (1995) *J. Am. Ceram. Soc.*, **78**, 1687–1690.

65 Hirao, K., Watari, K., Brito, M.E., Toriyama, M., and Kanzaki, S. (1996) *J. Am. Ceram. Soc.*, **79**, 2485–2488.

66 Teshima, H., Hirao, K., Toriyama, M., and Kanzaki, S. (1999) *J. Ceram. Soc. Jpn*, **107**, 1216–1220 [in Japanese].

67 Watari, K., Hirao, K., Brito, M.E., Toriyama, S. and Kanzaki, S. (1999) *J. Mater. Res.* **14**, 1538–1541.

68 Hirosaki, N., Ando, M., Okamoto, Y., Munakata, F., Akimune, Y., Hirao, K., Watari, K., Brito, M.E., Toriyama, M., and Kanzaki, S. (1996) *J. Ceram. Soc. Jpn*, **104**, 1171–1173.

17
Electrical Conduction in Nanostructured Ceramics

Harry L. Tuller, Scott J. Litzelman, and George C. Whitfield

17.1
Introduction

Ceramics, initially viewed primarily from the perspective of their insulating or dielectric properties, are now utilized in many applications where their conductive properties are of prime interest. Electrically active ceramics, or *electroceramics*, exhibit the full spectrum of electrical properties spanning the gap from superconducting to insulating electronic conductors, and from fast ionic to insulating ionic conductors [1]. Commonly, it is the unique combination of electrically conductive and other desired properties that render electroceramics the obvious choice for many technological applications. For example, the use of indium tin oxide (ITO), as a transparent electrode, is central to the operation of liquid crystal displays (LCDs), semiconductor light sources and solar cells [2]. $La_{1-x}Sr_xMnO_3$ (LSM), as the cathode in high-temperature solid oxide fuel cells (SOFCs), satisfies the need for a metallic conductor capable of operating under a highly oxidizing environment, while yttria-stabilized zirconia (YSZ), as the electrolyte, supports high-oxygen ion conductivity [3]. Indeed, the ionic bonding of many ceramics allows, in some cases, for simultaneously high-electronic and -ionic conductivities, leading to *mixed ionic electronic conductivity* (MIEC). Solid electrolytes and MIEC represent the underpinnings of *solid-state ionics* and *solid-state electrochemistry*, a field of growing importance as society has become more acutely concerned with efficient and environmentally clean methods for energy conversion, conservation, and storage [4].

As is the case in many fields, for example, microelectronics, catalysis, and medicine, much interest has developed in understanding the role of nanodimensions in influencing and ideally optimizing properties. In this chapter, attention is focused on the impact of nanoscale dimensions on the properties of electrically conductive *electroceramics* and the implications that this may have on applications. This is particularly relevant since electroceramics, by their nature, are dominated by boundaries (grain boundaries, electrode interfaces, surfaces, etc.), with the notable exception of epitaxial thin films, which are optimized for dielectric and ferroelectric applications but are beyond the scope of this chapter [5]. Some boundaries are detrimental to

Ceramics Science and Technology Volume 2: Properties. Edited by Ralf Riedel and I-Wei Chen

ISBN: 978-3-527-31156-9

performance; for example, ion blocking in YSZ for SOFCs and automotive oxygen sensors [3], while others are advantageous, such as barriers in ZnO varistors, $BaTiO_3$ positive temperature coefficient (PTC) thermistors, and multilayer capacitors [6, 7]. The role of interfaces in controlling the electronic properties of electroceramics has been detailed previously [5]. The present studies were extended to include the impact of nanostructure on both ionic and semiconducting systems.

Nanoscale structures, as illustrated in Figure 17.1, imply the existence of high densities of surfaces and/or interfaces. For conductive materials this, in turn, results

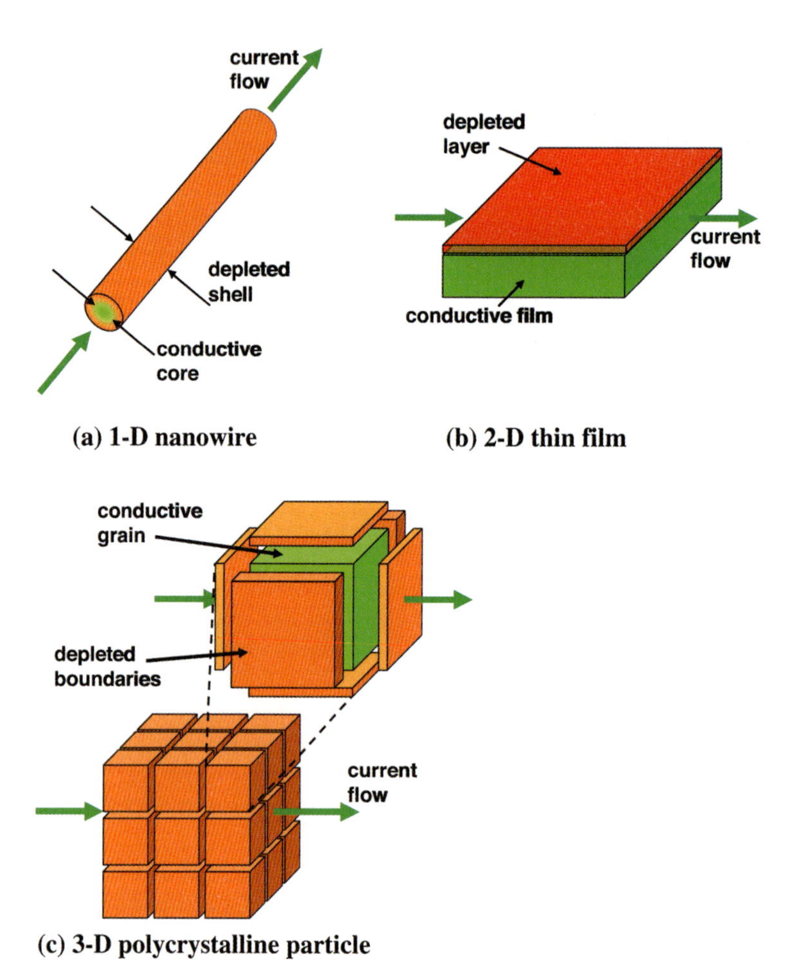

Figure 17.1 Schematic depiction of 1-D, 2-D, and 3-D nanostructures in which the diameter, thickness, or particle size of the respective structures are typically on the order of tens of nanometers or less in dimension. Modulation in the relative size and intensity of the depleted/accumulated regions (orange) with respect to the neutral (*bulk*) regions (green) can markedly influence the resistance of the overall structure. This effect is strongly multiplied as the dimensions of the space charge regions become comparable to the geometric dimensions of the depicted structures.

in the formation of adjacent space charge regions (as discussed in detail below). Nanostructures are classified by their dimensionality; that is, those that conduct in one, two, or three dimensions [one-, two- or three-dimensional (1-D, 2-D or 3-D) systems, respectively]. In contrast to transport in bulk materials, charge carriers in nanostructured materials are heavily influenced by interactions with surfaces and/or interfaces as they move along the conductive pathway. Canonical examples of 1-D, 2-D, and 3-D systems are shown in Figure 17.1, as a nanowire, a thin film, and a nanoporous/nanograined bulk material, respectively. In practice, nanostructured ceramic materials often contain a hierarchy of 1-D, 2-D, and 3-D structures, as illustrated later in the chapter.

The chapter begins with a review of space charge models, and how standard assumptions must be modified to address shrinkage in dimensions to the nanoscale. First, semiconducting materials are considered, with treatment later extended to include MIEC materials. These models are examined in the context of examples drawn from relevant technologies, including SOFCs, lithium batteries, solar cells, and gas sensors. Finally, promising areas for future research are suggested.

17.2
Space Charge Layers in Semiconducting Ceramic Materials

Space charge formation commonly occurs in semiconducting or ionic systems at or adjacent to surfaces and interfaces. The origins of these charges are several, and include the difference in work function at a heterojunction (e.g., a metal/semiconductor interface), the chemisorption of gaseous molecules at a semiconductor surface, or defect segregation to surfaces or interfaces driven either by intrinsic or extrinsic means. The end result of each of these phenomena is a redistribution of mobile species in response to a gradient in electrostatic and/or a chemical potential across an interface. This in turn impacts the operation of various electroceramic devices that are subject to these phenomena. In the following, the origins of space charge potential are first considered, after which resultant redistribution of charge carriers is derived, initially considering the case where ions are frozen in place and cannot redistribute. Next, the case of mixed ionic electronic conductors is described; at sufficiently high temperatures these permit the redistribution of both ionic and electronic charge carriers. Having derived expressions describing charge redistributions in the vicinity of interfaces in bulk systems, the consequences of shrinking dimensions to the nanoscale are then considered.

There are various driving forces that result in the formation of space charge across surfaces or interfaces. Perhaps the simplest example is the case of a Schottky diode that is formed from a metal–semiconductor heterojunction. For an ideal metal–semiconductor contact, the height of the potential barrier is given by

$$\Phi_B = \Phi_m - \chi_s \tag{1}$$

where Φ_m is the work function of the metal and χ_s is the electron affinity of the semiconductor [8].

Space charge potentials also arise due to inhomogeneities that occur within a single material, as is the case of *p–n* junctions in single crystalline materials. The so-called built-in potential ϕ_{bi} of a *p–n* junction is given by [9]:

$$\phi_{bi} = \frac{kT}{q} \ln\left(\frac{N_D N_A}{n_i^2}\right) \tag{2}$$

in which N_D, N_A and n_i are the donor density on the *n*-side, the acceptor density on the *p*-side, and the intrinsic carrier density, respectively.

A *grain boundary* is an interface between adjacent single crystalline grains, and in both ceramic and metallic polycrystalline materials, defects tend to segregate to such boundaries to lower their interfacial free energies, as first discussed by Gibbs [10]. In semiconducting oxides, these defects tend to form electron or hole traps within the band gap of the oxide, trapping free carriers from the adjoining grains. The barrier height of a back-to-back Schottky potential barrier formed at such electron-blocking grain boundaries between grains with donor density of N_D, is given by [11]:

$$\phi_B(0) = \frac{-Q^2}{8q\varepsilon N_D} \tag{3}$$

in which ε and Q are the dielectric constant and charge/area trapped at the grain boundary core, respectively. As mentioned above, such grain boundary barriers are, for example, essential to the operation of ZnO varistors and PTC thermistors. Both depend, in their operation, on the collapse of the space charge barrier; in the former case at a characteristic breakdown voltage, and in the latter case in the vicinity of the ferroelectric curie temperature [6, 7].

In the case of gases adsorbing onto the surface of a semiconductor (gas–semiconductor junction), the space charge potential arises from the charge transfer of electronic charge between the semiconducting oxide and the chemisorbed ions. Here, the magnitude of the potential is determined by charge per unit area of ions chemisorbed at the surface, Q_S, as given by Eq. (3), with a factor of 8 being replaced by 2 in the denominator. Q_S in turn is related to the partial pressure of gases, P, in the surrounding environment by:

$$Q_S = \frac{qN^* \beta P}{1 + \beta P} \tag{4}$$

where N^* is the number of chemisorption sites per unit area and β is an adsorption coefficient that depends on the electronic properties of the semiconductor, as derived by Vol'kenshtein [12]. This adsorption coefficient will be discussed in further detail later in the chapter, in the case study of chemical sensors.

Additional sources for potential barriers in ionic systems can be driven by intrinsic ionic processes. As first described by Frenkel, the formation of a net surface charge and a compensating space charge layer relates to the energy differences required to bring various ionic species to a surface [13]. Indeed, while ionic solids are macroscopically charge-neutral, local variations in both structure and chemistry lead to internal electrostatic potentials and electric fields. Space charge layers are formed to compensate this excess charge and to preserve electrochemical equilibrium. Subsequent studies by Lehovec [14] and Kliewer and Koehler [15] led to the refinement and widespread adoption of space charge theory in ionic systems.

As an example, the space charge properties of TiO_2 were analyzed by Ikeda and Chiang [16]. Assuming cation Frenkel disorder – that is, cation vacancies and interstitials are the predominant defects – the free energy change due to the introduction of such defects into a perfect crystal is given by

$$F = \int_0^\infty \left[n_{V_{Ti}}(x) g_{V_{Ti}} + n_{Ti_i}(x) g_{Ti_i} + \frac{1}{2} \varrho(x) \phi(x) \right] dx - TS_c \tag{5}$$

where n is the defect density, g the free energy per defect, ϱ the charge density per unit volume, T the absolute temperature, and S_c the configurational entropy. The authors derived the potential difference between the grain boundary core and the bulk in TiO_2 to be:

$$e\phi(0) = \frac{1}{8}(g_{Ti_i} - g_{V_{Ti}}), \tag{6}$$

thus relating the polarity and magnitude of the space charge potential to the difference in defect formation energies of titanium vacancies and interstitials.

The impact of space charge regions on combined electronic and ionic transport was presented by Maier [17]. The space charge potential, $\Delta\phi = [\phi(0) - \phi(\infty)]$, where $\phi(0)$ is the potential at the interface and $\phi(\infty)$ is the reference value in the bulk, determines the charge carrier profiles in the space charge layer. The equilibrium condition for species j is given by the constancy of the electrochemical potential:

$$\tilde{\mu}_j(x) = \mu_j(x) + z_j e\phi(x) \tag{7}$$

consisting of the chemical potential, μ_j, and the electrostatic potential $\phi(x)$ with $z_j e$ the net charge. For dilute defect concentrations, the chemical potential term in Eq. (3) can be expanded to include the standard chemical potential,μ_j^0, and concentration c_j (or *activity* for concentrated systems) of species j:

$$\mu_j = \mu_j^0 + kT \ln c_j. \tag{8}$$

By equating the electrochemical potential of species j at two different locations and solving, it is found that the ratio c_j at two locations, one at x and the other far from the interface ($x = \infty$), depends exponentially on the difference in potential between these two locations as described by:

$$\left(\frac{c_j(x)}{c_{j\infty}} \right)^{1/z_j} = \exp\left(-\frac{e}{k_B T} \Delta\phi(x) \right) \tag{9}$$

Equation (9) demonstrates that an electrical potential difference is compensated by nonuniform chemical profiles in order to preserve electrochemical equilibrium.

The importance of Eq. (9) is clear, as it demonstrates that carriers will either accumulate or deplete exponentially in the space charge layer in response to the space charge potential. In order to quantify the enhancement/depletion effects, it is first necessary to solve for the spatial variation of the electrical potential. Beginning with Gauss' law,

$$\nabla \cdot E = \frac{\varrho}{\varepsilon} \tag{10}$$

and the definition of an electric field:

$$E = -\nabla\phi \tag{11}$$

a partial differential equation is found as a specific example of Poisson's equation:

$$\nabla^2\phi = -\frac{\varrho}{\varepsilon_o\varepsilon_r} = -\frac{z_j e c_j}{\varepsilon_o\varepsilon_r} \tag{12}$$

Poisson's equation and Eq. (10) are combined to form the Poisson–Boltzmann differential equation that, for the case of electrostatic potential variation in one dimension and one predominant defect, results in:

$$\frac{d^2\phi}{dx^2} = -\frac{e z_j c_{j\infty}}{\varepsilon_o\varepsilon_r}\exp\left(\frac{-z_j e}{kT}\Delta\phi(x)\right) \tag{13}$$

Equation (13) represents an ordinary differential equation requiring two boundary conditions and a reference point for the potential, commonly set to zero in the bulk. If ionic defects are frozen in place and unable to move, the charge density in Poisson's equation is determined only by redistribution of electronic charge carriers that are ionized from the dopants. This is known as the Mott–Schottky approximation [18], which results in a simplification to Poisson's equation: by setting $c_j(x) = c_{j\infty}$:

$$\frac{\partial^2\phi}{dx^2} = -\frac{z_j e c_{j\infty}}{\varepsilon_o\varepsilon_r} \tag{14}$$

With boundary conditions $\phi(\lambda^*) = 0$ and $\phi(\lambda^*) = \phi_\infty = 0$, Eq. (14) can be integrated to yield (relative to the bulk reference potential, commonly set to zero):

$$\Delta\phi(x) = -\frac{z_j e c_{j\infty}}{\varepsilon_o\varepsilon_r}(x-\lambda^*)^2 \tag{15}$$

where λ^* is the depletion (space charge) width [19]:

$$\lambda^* = \sqrt{\frac{2\varepsilon_o\varepsilon_r\Delta\phi(0)}{z_j e c_{j\infty}}} \tag{16}$$

Note, that the Debye length of the material is defined as:

$$\lambda = \sqrt{\frac{\varepsilon_o\varepsilon_r kT}{2 z_j{}^2 e^2 c_{j\infty}}} \tag{17}$$

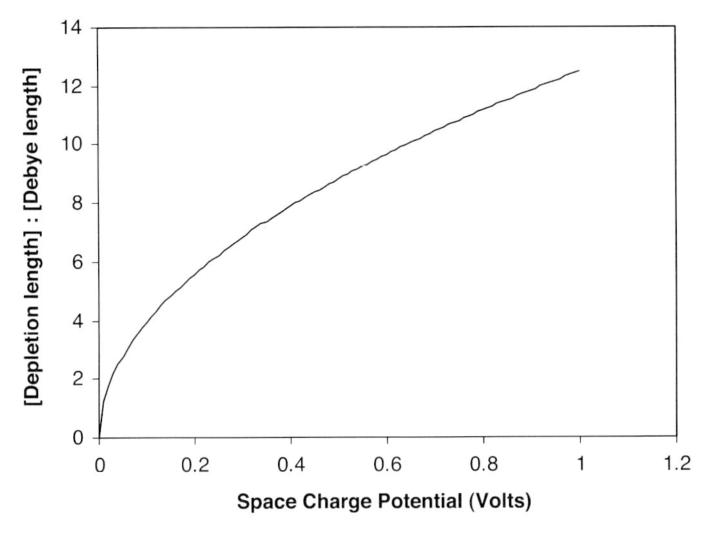

Figure 17.2 Ratio of depletion length to Debye length versus space charge potential at a depleted surface or interface. Note, $z_j = 1$ and $T = 300$ K.

Thus, λ and λ^* are directly related:

$$\lambda^* = \lambda\sqrt{\frac{4z_j e}{kT}\Delta\phi(0)} \tag{18}$$

It is important to note that the ratio of depletion length to Debye length is largely determined by the magnitude of the space charge potential, as illustrated in Figure 17.2 for the case of a $z_j = 1$ and $T = 300$ K. At zero potential, the space charge vanishes, and for moderate space charge potentials of 10 mV and above, the space charge length will be larger than the Debye length.

In order to find the spatial dependence of electronic charge carrier concentrations, the expression for the potential difference as a function of x, Eq. (15), can be substituted into the original equation for carrier enhancement/depletion, Eq. (9), yielding:

$$\frac{c_i(x)}{c_{i\infty}} = \exp\left[-\frac{z_i}{z_j}\left(\frac{x-\lambda^*}{2\lambda}\right)^2\right] \tag{19}$$

where defect i is enhanced or depleted, and defect j is the majority defect that determines the space charge width.

Considering the simple example of an n-type semiconductor ($c_{n\infty}$ being equal to the donor concentration, N_D), the electron concentration can be expressed as:

$$\frac{c_n(x)}{c_{n\infty}} = \exp\left[\left(\frac{x-\lambda^*}{2\lambda}\right)^2\right] \tag{20}$$

and the corresponding hole concentration is related to the electron concentration by:

$$\frac{c_n(x)}{c_{n\infty}} = \left(\frac{c_p(x)}{c_{p\infty}}\right)^{-1} \tag{21}$$

Note that, under static equilibrium, the electron and hole concentrations are additionally related by the familiar mass action law, $c_n\, c_p = n_i^2$, where n_i is the intrinsic charge carrier concentration of the material.

ZnO and TiO_2 are examples of wide-bandgap ($E_g \sim 3.2\,\text{eV}$) semiconducting ceramics that are commonly used in temperature ranges at which ionic defects are not mobile. At room temperature, ZnO is widely used in varistors and thin-film transistors, while TiO_2 has become heavily investigated as an anode material for dye-sensitized solar cells (see case study below). In addition, both materials can be used as chemical sensors that are operated from 200–400 °C; these are sufficiently high temperatures to promote reversible surface reactions, but are still low enough to avoid the motion of bulk defects during typical time scales of interest. Figure 17.3 illustrates the spatial dependence of electron and hole concentration in these two materials at various doping concentrations, under the Mott–Schottky approximation. It is interesting to note that, since the relative permittivity of TiO_2 is over 10-fold larger than that of ZnO, the depletion lengths will be significantly longer, according to Eq. (16). At constant space charge potential, materials with lower doping levels have longer depletion lengths and are easier to electronically invert, as illustrated by the cross-over in electron/hole concentration for $N_D = 10^{17}\,\text{cm}^{-3}$.

In certain materials, and at sufficiently high temperatures, ionic defects become mobile enough to redistribute in response to an electric field. If it is assumed that all charged defects in the space charge layer are able to move in this manner, then the spatial variation of the potential is given by [20]:

$$\phi(x) = \frac{2kT}{z_j e} \ln\left(\frac{1 + \Theta \exp\left(-\frac{x}{\lambda}\right)}{1 - \Theta \exp\left(-\frac{x}{\lambda}\right)}\right) \tag{22}$$

where Θ is the profile parameter:

$$\Theta = \tanh\left(\frac{z_j e \Delta\phi}{4kT}\right) \tag{23}$$

Finally, by combination of Eqs (9) and (23), the spatial profiles of defects in the space charge region are found:

$$\frac{c_j(x)}{c_{j\infty}} = \left(\frac{1 + \Theta \exp\left(-\frac{x}{\lambda}\right)}{1 - \Theta \exp\left(-\frac{x}{\lambda}\right)}\right)^{2z_j} \tag{24}$$

These boundary conditions, known as Gouy–Chapman, are valid when all defect species can redistribute in the space charge region.

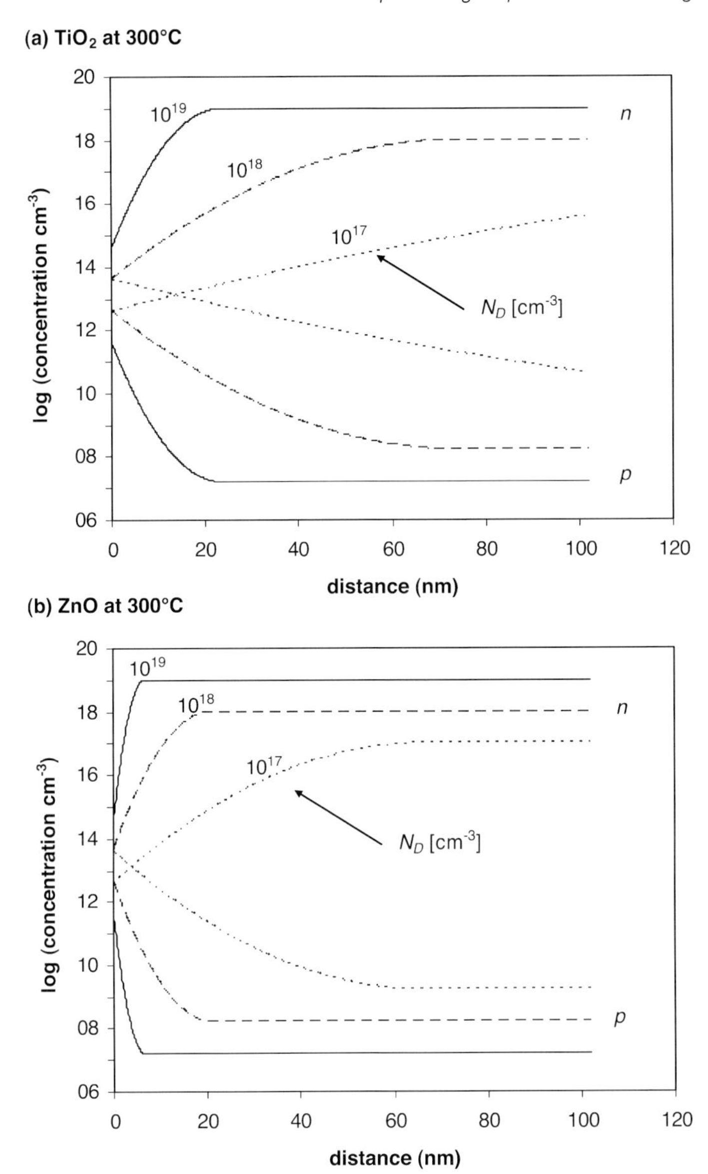

Figure 17.3 Space charge profiles of electrons and holes according to the Mott–Schottky model, in two common semiconducting ceramics, (a) TiO_2 and (b) ZnO, at 300 °C. Both materials have band gap $E_g \sim 3.2$ eV, but relative permittivity $\varepsilon_r \sim 100$ in TiO_2 versus $\varepsilon_r \sim 8$ in ZnO. Varying doping levels (10^{17}, 10^{18}, and 10^{19} cm^{-3}) are shown, for comparison at $\phi = 0.5$ volt space charge potential. Note that the density of the electronic states is assumed equal to that of silicon.

Two distinct differences can be seen in the relation for the space charge width in the Mott–Schottky compared to the Gouy–Chapman boundary conditions. When the majority defect cannot redistribute, the space charge width is dependent on the space charge potential, and the depletion width is greater in spatial extent due to a reduced charge screening ability.

For the case of oxygen vacancies depleted in the space charge regions, $z_i = 2$ and $z_j = 1$:

$$\frac{c_V(x)}{c_{V\infty}} = \exp\left[-\frac{1}{2}\left(\frac{x-\lambda^*}{\lambda}\right)^2\right] \tag{25}$$

and correspondingly for electrons:

$$\frac{c_n(x)}{c_{n\infty}} = \left(\frac{c_V(x)}{c_{V\infty}}\right)^{-\frac{1}{2}} \tag{26}$$

To facilitate visualization of the difference in defect profiles for the two models, schematic profiles are plotted in Figure 17.4 for the predominant defects in ceria (CeO_2) for a space charge potential of 0.44 V and an acceptor concentration of 1700 ppm at 500 °C. Under these conditions, a *depletion* of oxygen vacancies and an *accumulation* of electrons are predicted in both models in response to a positive potential in the grain boundary core. It is evident, however, that the natures of the profiles differ considerably. In the Gouy–Chapman case, there is a very large change in carrier concentration within only the first few nanometers from the grain boundary core. Thus, most of the enhancement/depletion effects occur approximately within one Debye length. Under Mott–Schottky conditions, however, the

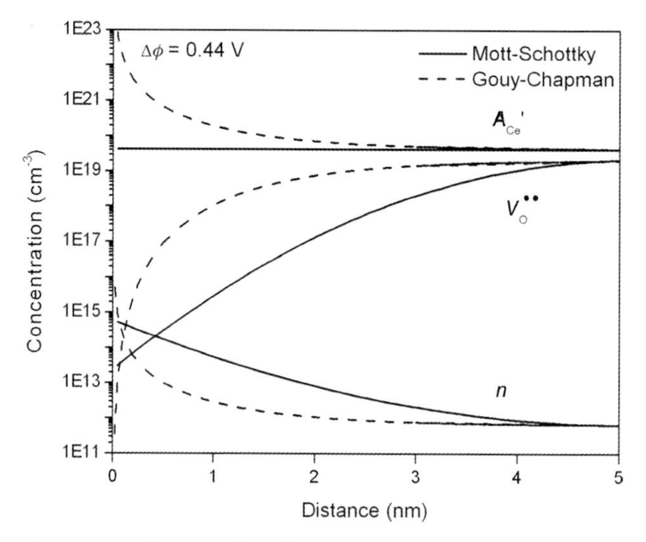

Figure 17.4 Charge profiles of acceptor dopants, oxygen vacancies, and electrons near a grain boundary interface with a space charge potential of +0.44 V, according to both the Gouy–Chapman (dotted lines) and Mott–Schottky (solid lines) models.

change in carrier concentration is less severe, but is still more than three orders of magnitude, depending on the charge of the defect. On the other hand, the extent of nonuniformity is much larger than the Debye length, due to reduced charge screening resulting from the immobile acceptor.

17.3
Effect of Space Charge Profiles on the Observed Conductivity

In order to examine the applicability of the space charge models to experimental data, relationships must be established between the carrier profile equations and the observed electrical conductivity. Maier presented a framework with which space charge contributions to the observed conductivity in ionic and mixed ionic–electronic conductors could be modeled under Gouy–Chapman conditions [21]. The effect of a grain boundary and the corresponding space charge regions can be understood in terms of the resistance perpendicular to the boundary, $Z^{\perp}$, and the conductance parallel to the space charge layers, $Y^{||}$, as well as the partial conductivities σ_m, as shown schematically in Figure 17.5. The widths of the grain boundary core, space charge layer, and grain interior are denoted as $2w$, 2λ, and d, respectively.

Approximate solutions for the Gouy–Chapman case were presented by Maier [21], while more recently analytical solutions for the Gouy–Chapman as well as the Mott–Schottky cases were presented by Litzelman *et al.* [22, 23]. The relevant expressions for each partial conductivity are summarized for the Gouy–Chapman case:

$$\sigma_{el}^{\perp,G-C} = \frac{\sigma_\infty}{\dfrac{2\Theta}{\Theta+\exp(2)}+1-\dfrac{2\Theta}{\Theta+1}} \tag{27}$$

$$\sigma_{el}^{||,G-C} = \sigma_\infty\left(1-\frac{2\Theta}{\exp(2)-\Theta}+\frac{2\Theta}{1-\Theta}\right) \tag{28}$$

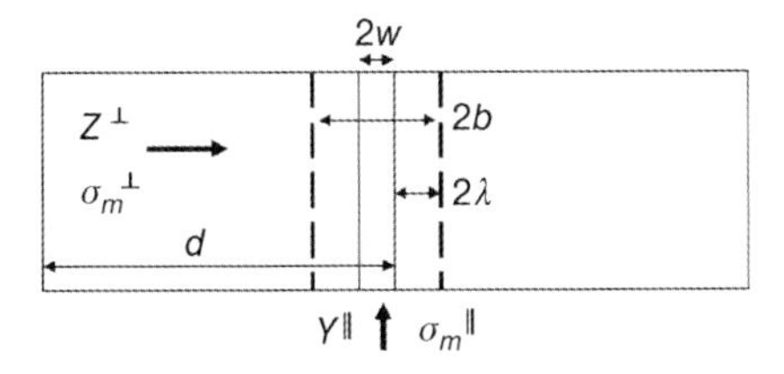

Figure 17.5 Schematic diagram of the resistance perpendicular to a boundary and the conductance parallel to it. Adapted from Ref. [21].

$$\sigma_{\text{ion}}^{\perp,G-C} = \sigma_\infty \left(\frac{\Theta\left[2\Theta^2 + 3\Theta\exp\left(\frac{8}{3}\right) + 3\exp\left(\frac{16}{3}\right)\right]}{\left(\Theta + \exp\left(\frac{8}{3}\right)\right)^3} + 1 - \frac{\Theta[2\Theta^2 + 3\Theta + 3]}{(\Theta + 1)^3} \right)^{-1} \tag{29}$$

$$\sigma_{\text{ion}}^{||,G-C} = \sigma_\infty * \left(1 - \frac{\Theta\left[2\Theta^2 - 3\exp\left(\frac{8}{3}\right)\Theta + 3\exp\left(\frac{16}{3}\right)\right]}{\left[\exp\left(\frac{8}{3}\right) - \Theta\right]^3} + \frac{\Theta[2\Theta^2 - 3\Theta + 3]}{[1 - \Theta]^3} \right) \tag{30}$$

where σ_∞ is equal to the partial conductivity in the bulk and Θ is the profile parameter, as defined in Eq. (23). The corresponding relations for the Mott–Schottky conditions are:

$$\sigma_{el}^{\perp,M-S} = \frac{\sigma_\infty \lambda^*}{\sqrt{\pi}(\lambda) erf\left(\frac{\lambda^*}{2\lambda}\right)} \tag{31}$$

$$\sigma_{el}^{||,M-S} = \frac{\sigma_\infty \sqrt{\pi}(\lambda) erfi\left(\frac{\lambda^*}{2\lambda}\right)}{\lambda^*} \tag{32}$$

$$\sigma_{\text{ion}}^{\perp,M-S} = \frac{\sigma_\infty \lambda^*}{1.25(\lambda) * erfi\left(0.707\frac{\lambda^*}{\lambda}\right)} \tag{33}$$

$$\sigma_{\text{ion}}^{||,M-S} = \frac{\sigma_\infty * 1.25(\lambda) * erf\left(0.707\frac{\lambda^*}{\lambda}\right)}{\lambda^*} \tag{34}$$

The overall partial conductivity is obtained as a weighted average of the bulk and boundary contributions, expressed by the boundary fraction φ_{gb} [21]:

$$\sigma_m = \frac{\sigma_\infty \sigma_{gb}^{\perp} + (2/3)\varphi_{gb}\, \sigma_{gb}^{||}\, \sigma_{gb}^{\perp}}{\sigma_{gb}^{\perp} + (1/3)\varphi_{gb}\, \sigma_\infty} \tag{35}$$

These expressions provide the means by which space charge accumulation and depletion can be related to the macroscopic conductivity of a material.

17.4 Influence of Nanostructure on Charge Carrier Distributions

As the characteristic dimension of a conducting nanostructure is decreased below the scale of the Debye length of the material, the semi-infinite approximation implicit in the carrier distributions shown in Figures 17.3 and 17.4, is no longer applicable. Instead, it becomes necessary to consider the influence of charge screening by multiple surfaces on the conductive properties. In the case of a material with no mobile ionic defects, the space charge density compensating the fixed surface/interface charge will be approximately constant in magnitude throughout the depleted region. Consider the example of a nanothin membrane with exposure at two free surfaces; as the membrane thickness approaches twice the Debye length, the membrane becomes nearly completely depleted. As the membrane thickness is further reduced, the total amount of compensating space charge within the material becomes geometrically constrained to the dimensions of the membrane. This, in turn, reduces the intensity of the electric fields that form perpendicular to the surface, and limits the total change in electrostatic potential across this region. This results in an energy band profile, as calculated by Leonard and Talin for nanowires of different dimensions, that approaches flat band conditions as the diameter of the wire approaches the space charge width [24] (Figure 17.6).

At even smaller dimensions – that is, as the characteristic feature sizes decrease below 5 nm – quantum confinement begins to have a strong impact on the conductive properties of semiconducting materials. First, the reduction of device dimensions results in bulk energy bands splitting into discrete energy levels, accounting for an increase in the HOMO–LUMO separation or, equivalently, an increase in the effective band gap. This quantization of states can be computed from a top-down approach by considering deviation from single-band conduction models [25], or from a bottom-up approach by using a tight-binding model [26].

Typically, at small conducting channel dimensions (e.g., nanowire diameter) and high surface-to-volume ratios, the conductivity would be expected to drop due to the heightened influence of surface scattering on carrier mobility, as is commonly observed at the channel/dielectric interface in field-effect transistors [9]. However,

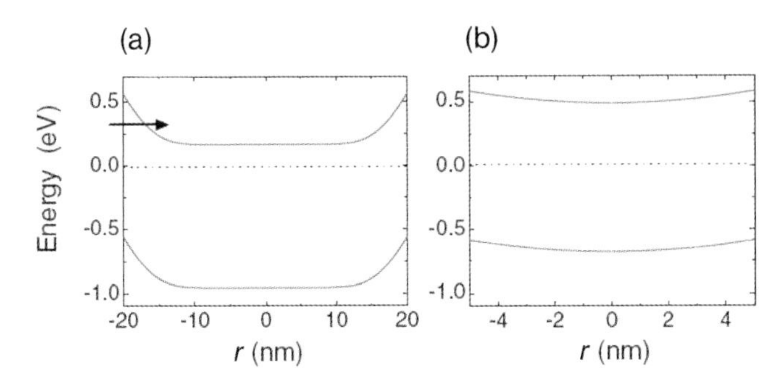

Figure 17.6 Numerical calculation of energy band bending across a metal/silicon nanowire Schottky barrier, a system that bears similarity to surface charging due to chemisorption. The calculation is for a nanowire of *n*-type doping density at $10^{19}\,cm^{-3}$ and diameter equal to (a) 40 nm (b) 10 nm. Note that at smaller nanowire diameter, band bending nearly disappears and the Fermi energy moves away from the conduction band towards mid gap. After Ref. [24].

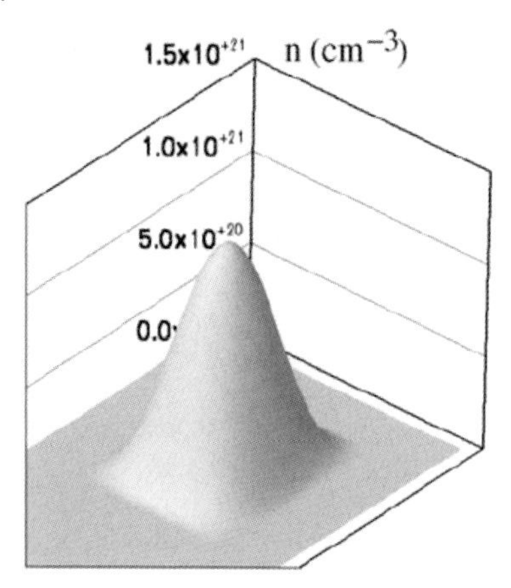

Figure 17.7 Electron density throughout the cross-section of a 2 nm-diameter silicon nanowire, determined by coupled numerical solution of the Schrödinger and Poisson equations. After Ref. [27].

for sufficiently small dimensions there appear to be several quantum effects that contribute towards a reduced surface scattering upon continued downscaling of, for example, nanowire diameter. Zheng *et al.* calculated that a restructuring of the energy bands decreases the charge carrier effective mass in the direction along the nanowire axis and increases the effective mass in directions transverse to the axis [25]. This contributes to an increased electron mobility along the nanowire axis and a reduced mobility normal to the axis. Based on numerical solutions to coupled Poisson–Schrödinger equations at and below 5 nm nanowire diameter, the spatial density of electronic states are reported to shift toward the center of the nanowire, further serving to remove charge carriers from the region of the surface, as illustrated in Figure 17.7 [27].

17.5 Case Studies

17.5.1 Case Study: Nanostructured Sensor Films

An important class of chemical sensors relies on the chemisorption of molecules onto the surfaces of semiconducting metal oxides (SMOs). Here, oxidizing or reducing gases chemisorb, by charge transfer of electronic charge to or from the adsorbed molecules, resulting in the charges being trapped at the surface. Due to the requirement of an overall charge neutrality, the surface charge is compensated by the formation of an adjacent space charge region within the semiconductor. This in turn, impacts the material's near-surface conductivity, $\sigma^{\|}$, that is, conductance parallel to the surface. Thus, it is possible to detect changes in the composition of the adsorbing gas by monitoring changes in the resistance of the material.

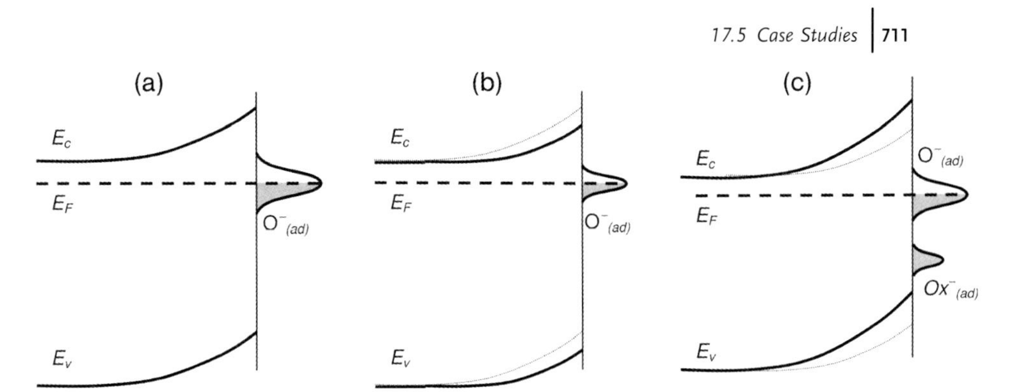

Figure 17.8 Schematic representations of the effect of the ambient gas atmosphere on the energy band diagram of: (a) an *n*-type metal-oxide in clean air; (b) reducing gases in air; and (c) oxidizing gases in air.

Figure 17.8 shows, schematically, the energy band diagram of an *n*-type semiconductor in the presence of several different gaseous environments. For example, adsorbed oxygen on an *n*-type SMO serves as an electron acceptor, resulting in a build-up of negative surface charge. This is compensated by electron depletion in the neighboring bulk of the material reflected in bending upwards of the conduction and valence bands (Figure 17.8a) [28]. The introduction of reducing gases (e.g., H_2) results in their reaction with the chemisorbed oxygen to form a product species (e.g., H_2O) and, in the process, this releases trapped electrons back into the bulk and reduces the degree of band bending (Figure 17.8b).

As discussed above, the concentration of chemisorbed ions at the surface is related to the partial pressure of the gaseous species according to Eq. (4), and increases with an adsorption coefficient, β. The relationship between β and the electronic properties of the semiconductor was originally derived by Vol'kenshtein and reproduced by Rothschild and Komem [29]:

$$\beta = \beta_0 \left(\frac{g_A + \exp\left(\frac{E_F - E_A^S}{kT}\right)}{g_A + \frac{\nu^-}{\nu^0} \exp\left(-\frac{E_C^S - E_F}{kT}\right)} \right) \tag{36}$$

where E_F is the Fermi energy, E_C^S is energy of the conduction band edge at the surface, E_A^S is the energy level of chemisorption-induced acceptor states at the surface, g_A is the degeneracy factor of chemisorption-induced states, υ^0 is the oscillation frequency of neutral species, ν^- is the oscillation frequency of charged chemisorbed species, and β_0 is the adsorption coefficient as given by the well-known Langmuir equation. Intuitively, the above equation demonstrates that the adsorption coefficient β is higher when the Fermi energy E_F is at higher positions above the energy level of surface acceptor states E_A^S. Similarly, as E_F increases and its separation from the conduction band edge at the surface E_C^S decreases, the adsorption coefficient β will increase. While E_C^S and E_A^S are properties of the semiconductor and its interaction with the chemisorbed gas, the position of E_F at the surface will depend on the dimensions of the space charge region, which in turn can be heavily dependent on the geometry of the sensor film, especially at the nanoscale.

The sensitivity of an SMO wire, film or porous layer to various gases depends on the fraction of the conductive channel modulated by the gas adsorption. It thus becomes obvious that sensitivity should depend on the ratio of the space charge to the conductive channel width. As discussed above, the penetration depth of the depletion region is closely related to the Debye length [see Eqs (16)–(18)], and ranges from tens to hundreds of nanometers for typical semiconducting metal oxides of interest (see Figure 17.3). Thus, it is on the scale of nanopatterned features that enhancements to sensor performance become significant. Currently, many research groups are exploiting novel processing techniques to achieve desired porous, hierarchical nanostructures that enable a strong interaction of the sensor material with the surrounding environment. Several examples are briefly discussed below, illustrating the anticipated trends with scale.

One common method for creating nanograined porous structures is by sintering a ceramic powder at a reduced temperature, and allowing the individual grains to begin forming sinter necks with neighboring grains, while maintaining the overall porosity. The average grain size and porosity of the final product will depend on the initial grain size of the starting powders and the sintering parameters such as temperature, pressure, and atmosphere. Xu *et al.* prepared SnO_2-based chemical sensors with an average grain size ranging from 5 to 32 nm, and reported significant improvements in sensitivity to both H_2 and CO for grain sizes below 10 nm [30]. Later, Rothschild and Komem confirmed that Xu's results were in quantitative agreement with a sensor response model based on modulation of surface depletion layers [31]. As predicted by the model, the trend in sensitivity is proportional to the surface-to-volume ratio; that is, proportional to $1/D$ in the case of spherical grains, where D is the characteristic grain diameter. Model predictions are shown plotted together with experimental data in Figure 17.9.

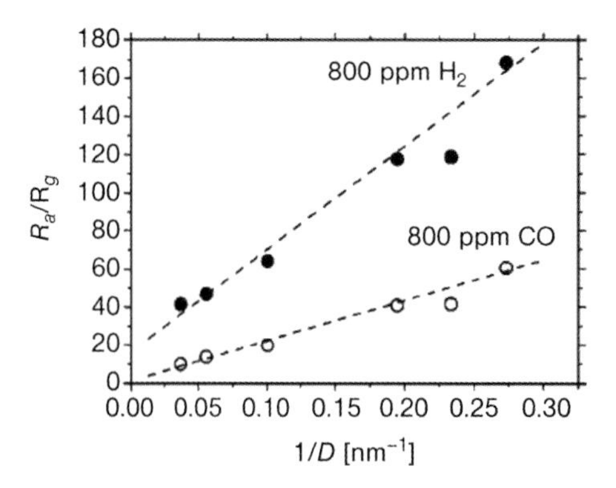

Figure 17.9 Response of nanoporous SnO_2 elements to 800 ppm H_2 or CO in air at an operating temperature of 300 °C as a function of the reciprocal diameter of the grains. Reproduced with permission from Ref. [31]; © 2004, Springer.

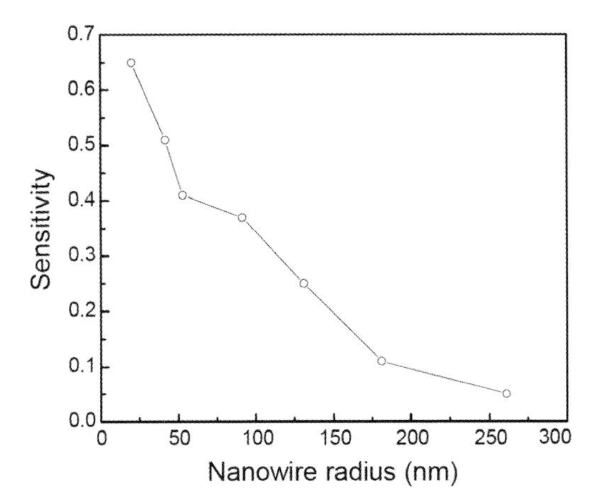

Figure 17.10 ZnO nanowire sensitivity to chemisorption of 20% oxygen, as a function of nanowire radius. Sensitivity is defined as $\Delta G/G_0$, where G_0 is conductance of the nanowire in an environment of in 10^{-2} Torr vacuum and ΔG is change in nanowire conductance when it is exposed to 20% oxygen at atmospheric pressure. Reproduced with permission from Ref. [42]; © 2005, IEEE.

The popularity of carbon nanotubes (CNTs) has inspired, in turn, numerous means for preparing *quasi-1-D metal oxide nanostructures* [32]. This has included the preparation of nanofibers, wires, belts, tubes, and springs [33, 34] with the ability, quite often, to control the size, composition, and electronic properties of the metal oxide nanostructures during growth [35]. Synthesis methods for metal oxide nanofibers have focused on electrospinning [36, 37] or vapor–liquid–solid and vapor–solid growth [38] with an ability, in some cases, to incorporate longitudinal and coaxial heterostructures. Examples of recent efforts directed specifically towards metal oxide nanoscaled devices for gas sensing have been summarized elsewhere [39–41]. Similar to the case of nanograin chemical sensors, the device sensitivity increases as the nanowire diameter decreases, as illustrated in Figure 17.10 for oxygen detection using ZnO nanowires [42]. Although such nanowires are highly sensitive, they are difficult to grow and contact reproducibly, and consequently alternative methods for achieving wire or fiber-like structures have been investigated. One such approach has been the preparation of nanofiber mats by electrospinning [43, 44]. These tend to be of a larger diameter and are more complex microstructurally, being polycrystalline and often porous (Figure 17.11). Nevertheless, they can also exhibit high sensitivity while remaining structurally robust and easy to contact [44].

Next moving to 2-D structures, it might be imagined that very thin films should be exceptionally sensitive, given that the width of the space charge layer becomes comparable to the thickness of the film. Surprisingly, however, this is often not the case. It is suspected that this might be due to the space charge layer emanating from the substrate–film interface, which largely *pins* the carrier density in the film. Consistent with this hypothesis, very thin films grown onto microsphere templates, which served to *lift* the films off of the substrate (see Figure 17.12), indeed showed

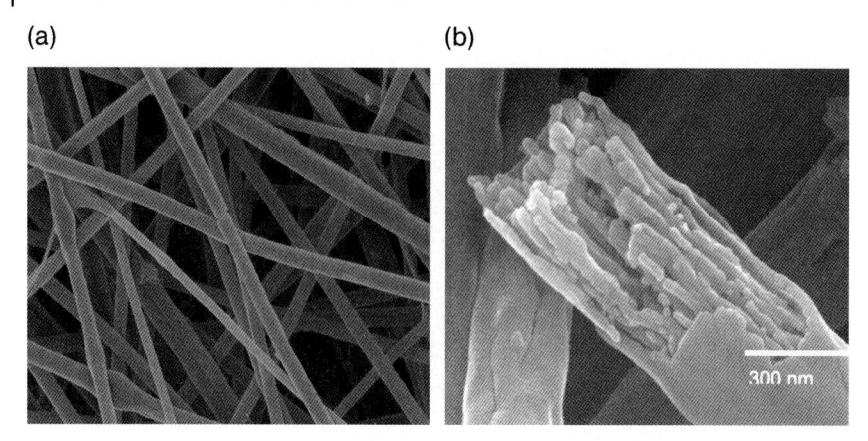

Figure 17.11 (a) As-spun TiO_2/PVAc 200–500 nm-diameter fibers; (b) Following calcination at 450 °C, 30 min, bundle structure of anatase TiO_2 composed of sheaths of 200–500 nm, cores filled with 10 nm fibrils. Reproduced with permission from Ref. [44]; © 2006, American Chemical Society.

a markedly enhanced sensitivity to hydrogen as compared to the identical film deposited onto a flat substrate [45].

17.5.2
Case Study: Interfaces in Ionic and Mixed Conducting Materials

Ionic conduction in the solid state is essential to the operation of electrochemical energy storage and conversion devices such as the lithium ion battery and the SOFC. In the solid state ionics arena, there is much interest, for example, in reducing the operating temperatures of SOFCs by shifting from bulk to thin-film electrolytes [e.g., yttria-stabilized ZrO_2 (YSZ) or gadolinium-doped ceria (GDC); $Gd_xCe_{1-x}O_{2-\delta}$]. Indeed, solid electrolyte membranes of ~150 nm thickness would provide an adequate electrolyte conductance down to temperatures of ~200 and 400 °C with GDC and YSZ, respectively [46–48]. The possibility thereby exists to embed miniaturized SOFC structures together with microelectromechanical systems (MEMS) components, and other active electronics, into the same silicon wafer for portable power generation in devices such as laptop computers and mobile telephones [49–51]. Because such thin films are commonly characterized by features with nanodimensions, other factors come into play regarding performance, leading to either improved or degraded properties. As a consequence, nanosize effects, connected with heterointerfaces and grain boundaries, are considered in some detail in this case study.

The electrical conductivity of an ionic solid, like that of a semiconducting solid, is normally controlled and enhanced by *homogeneous* doping, in which a donor and/or an acceptor species is substituted on specific lattice sites.[1] Such doping occurs

1) Exceptions exist such as Ag ion conduction in αAgI in which the Ag sublattice is intrinsically disordered.

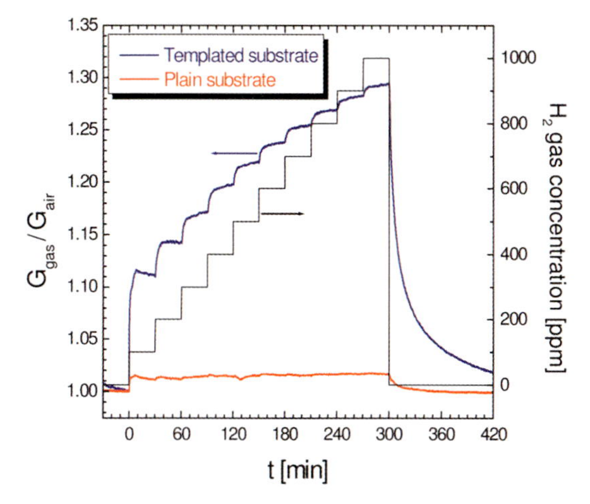

Figure 17.12 (a) Hollow microsphere-templated thin films of macroporous $CaCu_3Ti_4O_{12}$ made up of grains on the order of 30 nm; (b) Relative conductance response to H_2 gas of $CaCu_3Ti_4O_{12}$ films grown on flat versus a microsphere-templated surface. Reproduced with permission from Ref. [45]; © 2006, American Chemical Society.

uniformly throughout the material, and the ionized defect must be compensated by the formation of an additional defect, either electronic or ionic, in order to preserve charge neutrality. There is, however, another method with which the electrical conductivity can be controlled, namely *heterogeneous* doping. This concept was first demonstrated in dramatic fashion by Liang, who discovered that the addition of an insulating second phase (Al_2O_3) to a lithium ion conductor (LiI) increased its Li ion conductivity by as much as 50-fold [52].

While the improvement of ionic conductivity via an insulating phase may appear counterintuitive, the concept was first discussed with respect to the space charge model by Wagner [53]. A thorough review and analysis of this effect was later presented by Maier [17]. The conductivity enhancement in systems such as $LiI{:}Al_2O_3$

have been attributed to fast conduction along interfacial space charge regions close to the LiI–Al_2O_3 interface. Analogous to the discussion above, defects can be either enhanced or depleted in space charge layers, and the spatially varying conductivity will depend on the relative mobility of the accumulated charge-carrying defects. The conductivity could increase or decrease monotonically as the boundary is approached, or potentially reach a local minimum. In the LiI:Al_2O_3 system, the space charge layer induced by the LiI–Al_2O_3 interface is enhanced in the high-mobility I^- species, resulting in an enhanced ionic conductivity relative to the single phase [17].

Enhanced ionic conduction at heterointerfaces has more recently been demonstrated in more highly controlled structures. A notable example was described by Sata *et al.*, who fabricated CaF_2/BaF_2 nanoscaled *superlattices* by using molecular beam epitaxy (MBE) [54]. In this case, the period of the CaF_2/BaF_2 layers was varied from 16 to 430 nm. The conductivity enhancement of the multilayer relative to the two end-members, as well the increasing magnitude of the effect with decreasing period, are depicted in Figure 17.13. The authors proposed that the conductivity enhancement had resulted from a transfer of F^- ions from the BaF_2 to CaF_2 layers, resulting in an "artificial ion conductor."

In a follow-up report, Guo and coworkers proposed a model in which F'_i defects are depleted and $V_F^\bullet$ enhanced in the space charge region of the BaF_2 layers, based on the observed activation energies and the large difference in mobility between the enhanced and depleted ionic defects [55]. In the same report, these authors also concluded that the properties of the ionic heterostructure were best described by the

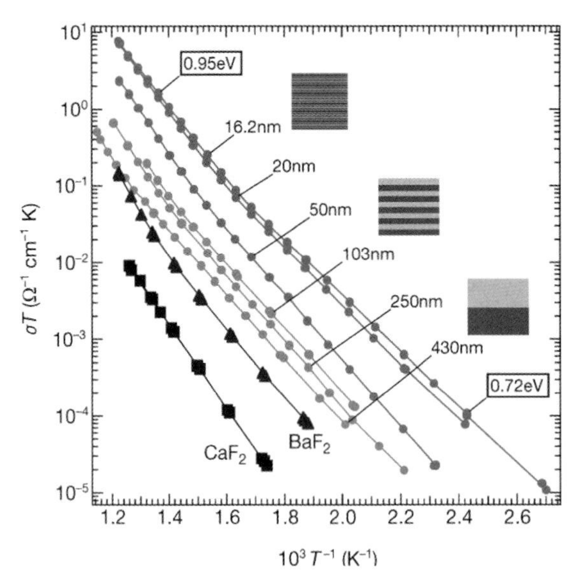

Figure 17.13 Arrhenius plot of the ionic conductivity of CaF_2/BaF_2 multilayers for period spacings of 16 to 430 nm. Also plotted are the bulk conductivity values for CaF_2 and BaF_2. Reproduced with permission from Ref. [54]; © 2000, Nature Publishing Group.

Mott–Schottky model, with a modification to include an impurity gradient near the layer interface.

A recent and even more exceptional example of enhanced ionic conductivity due to the introduction of heterointerfaces was reported by Garcia-Barriocanal *et al.*, who prepared YSZ/$SrTiO_3$ multilayers by using MBE, and recorded a conductivity increase by up to eight orders of magnitude that was ascribed to increases in the oxygen ion conductivity of the YSZ layers [56]. Confirmation of these results await either electron-blocking experiments [23] or open-circuit Nernst potential studies [57].

While these results are both intriguing, and potentially of great importance, the very thin cross-sectional area of the multilayer ionic conductors results in a high resistance, even for structures with hundreds of layers. This implies a low power output from such a structure when utilized as the solid electrolyte in a device such as a fuel cell. An alternative approach would be to seek a solution which does not require artificially prepared interfaces. Polycrystalline films with nanometer-sized grains form during film deposition, particularly if low substrate temperatures and high deposition rates are utilized, and this leads to very high interface area-to-volume ratios. However, if the grain boundaries could be engineered to induce an accumulation of mobile ionic defects, then an enhanced ionic conductivity would be expected. Indeed, this research team reported that the electrical conductivity of bulk nanocrystalline CeO_2 was increased substantially over that of a coarsened microcrystalline CeO_2 [58, 59]. The source of increased conductivity was, however, tied to an increase in electronic conductivity.

Subsequently, Tschöpe [60] and Kim and Maier [61] attributed the increase in electronic and decrease in ionic conductivity to space charge effects resulting from the segregation of positive charges to the grain boundary cores. Similar effects were reported more recently in vapor phase-deposited ceria thin films with nanosized grains [62]. The logical question emerges: can the grain boundary be re-engineered so that it becomes negatively charged? Litzelman and coworkers recently reported such an attempt via heterogeneous doping in nanocrystalline CeO_2 by grain boundary diffusion [63].

It was postulated that if negatively charged defects were to be introduced into the grain boundary core, then the space charge potential would shift to less-positive values and oxygen vacancies, initially depleted, would be reintroduced into the space charge region. This could be accomplished, for example, by the introduction of Ni onto Ce sites within the grain boundary core. A schematic decrease in $\Delta\Phi$ from $+0.5$ to $+0.3$ V, as shown in Figure 17.14, would result in a reduced depletion of oxygen vacancies near the interface. The decrease in potential barrier height should also reduce the blocking of ionic charge transport across the boundaries, and this effect alone may significantly improve ionic conduction in materials. At the isoelectric point, where the space charge potential is zero (flat band condition), the bulk equilibrium defect concentration would remain uniform throughout the bulk and boundary regions. If sufficient negative charge were to be introduced at the boundary core, however, then oxygen vacancies would accumulate in the space charge region, leading to further enhancements in ionic conductivity along the boundaries.

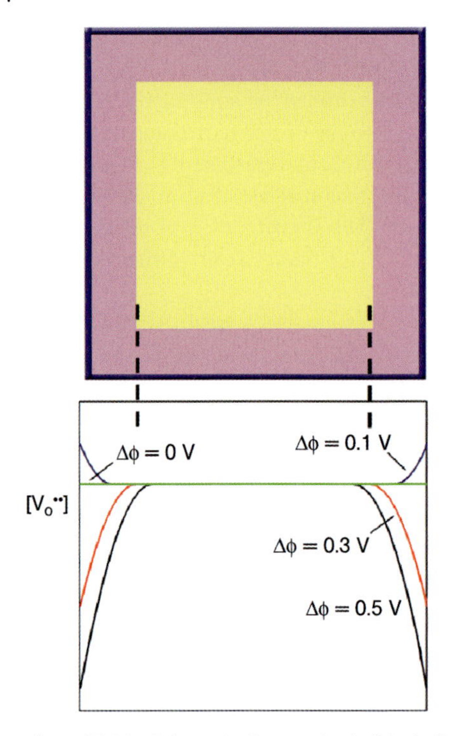

Figure 17.14 Schematic diagram (top) of the bulk (center), space charge layers (frame), and grain boundaries (outer dark lines) in a polycrystalline material. Shown below are Mott–Schottky oxygen vacancy profiles for $\Delta\Phi$ values of +0.5, +0.3, 0, and −0.1 V in the space charge layer.

Surprisingly, electron-blocking measurements performed on Ni in-diffused ceria thin films (which would be expected at least to reduce the magnitude of the positive charge trapped at the grain boundary cores) revealed that the ionic conductivity, rather than being enhanced, was depressed along with the electronic conductivity by the treatments [64]. This was inconsistent with a shift in the space charge potential alone. However, if the acceptor impurities were assumed to be sufficiently mobile so as to redistribute at in-diffusion temperatures, the Gouy–Chapman model would predict a decrease in both partial conductivities [64]. These results demonstrated the complex interplay that occurs between dopants and defects in the vicinity of grain boundaries. Yet, as research groups become more adept at grain boundary engineering, improvements in ionic conduction in nanocrystalline oxides can be expected.

17.5.3
Case study: Lithium-Ion Battery Materials

Solid-state batteries promise to improve upon existing liquid electrolyte technology by increasing the energy density and shelf life, and offering a greater potential for miniaturization. Much of the research effort in this field has been directed toward Li-ion cells because of their low weight and high energy density [65]. As a commercial system, Li-ion batteries have been affected more directly by reduced-dimension geometries than materials for fuel cells. There are several ways in which nanostructuring can lead to improvements in battery technology [66]. First, the time to charge or discharge a homogeneous particle will be limited by the diffusivity of the slower lithium ion (as compared to electrons). In simplified form, the solution of the diffusion equation for a fixed, single boundary into a semi-infinite medium is given by [67]:

$$c(x,t) = c_o erfc\left(\frac{x}{\sqrt{4Dt}}\right) \tag{37}$$

where $\sqrt{4Dt}$ is the "diffusion length," L_D, the characteristic length scale of the system. This implies the potential for a tremendous decrease in response time for a nanocrystalline system, as the time scales as the square of the length. Thus, given two particles sizes of 50 nm and 50 μm, the response time for the nanocrystalline sample should be six orders of magnitude faster than its microcrystalline counterpart, simply because of the geometric realization of a reduced diffusion length scale. Second, electron transport has been reported to increase in nanostructured positive electrodes [68]. Third, nanostructured electrodes may enable the interfacial storage of Li, in addition to the traditional insertion/removal (intercalation) and reactive (Li alloys) mechanisms [69]. This model was presented as part of a theoretical three-stage form of lithium storage in which insertion occurs for small amounts of Li. Phase separation then occurs if the solubility limit is exceeded. If the two phases are of different character – such as one conducting only ions and the other phase conducting only electrons – then additional amounts of the two species may be stored at the interface between the two phases. Fourth, the solid solution range may increase over the composition space, and degradation due to strain resulting from the intercalation process may decrease [70].

These effects are related primarily to the positive electrode. In a recent review on nanostructured materials for Li-ion batteries, possible advancements for the electrolyte and negative electrodes were also considered [71]. Much research has been devoted to solid-state electrolyte layers, although crystallization often leads to a reduced conductivity at temperatures below 70 °C. In this vein, nanocomposite electrolytes are being explored, in which nanoscaled ceramic materials are added to inhibit the formation of crystalline chains. In addition, crystalline polymer materials have recently begun to be examined as possible electrolytes. Key to increasing the conductivity of these layers is nanoscale-control of the chain length and facilitation of polydisperse chain lengths. Many structures have also been explored for the negative electrode, including nanoparticles, nanowires, and nanoalloys of Li and a second metal [71].

17.5.4
Case Study: Dye-Sensitized Solar Cells

Increasing energy demands and growing environmental concerns are driving the search for new, sustainable sources of energy. While solar energy is by far the most abundant clean energy source available, only a tiny fraction of these needs are presently met by photovoltaics. Large-scale solar cell deployment, however, only becomes feasible when it is possible simultaneously to satisfy the needs for low cost, nontoxic and easily recyclable materials, coupled with reasonable energy conversion efficiencies and operating lifetimes. An important milestone was the invention of excitonic solar cells, commonly called dye-sensitized solar cells (DSSCs) [72], which are currently the most efficient third-generation solar technology (as high as 11%), while promising a low cost and ease of manufacture. Instead of the potentially costly and/or toxic semiconductors that are used in other thin-film technologies, DSSC utilizes nanocrystalline semiconducting metal oxides such as TiO_2 – the low-cost, inert pigment which is used in paint.

In a conventional solar cell, *light absorption* takes place by means of the band-to-band generation of electron hole pairs across the energy band gap of a semiconducting material such as amorphous or crystalline silicon. Next, *charge separation* occurs in response to the influence of electric fields present within space charge regions typically formed at internal *p–n* junctions. *Charge collection* then takes place as excess electrons and holes diffuse in opposite directions, away from the space charge region, towards electrical contacts that connect the solar cell to an external circuit.

In DSSCs, light active synthetic dyes, bound to nanostructured wide band gap TiO_2, are used as photosensitizers to harvest the solar energy and generate excitons (see Figure 17.15) [73]. Operationally, photon excitation of the dye results in the transfer of electrons from the dye to the semiconductor, with the holes remaining in the dye. Although it is energetically possible for electrons to recombine with holes

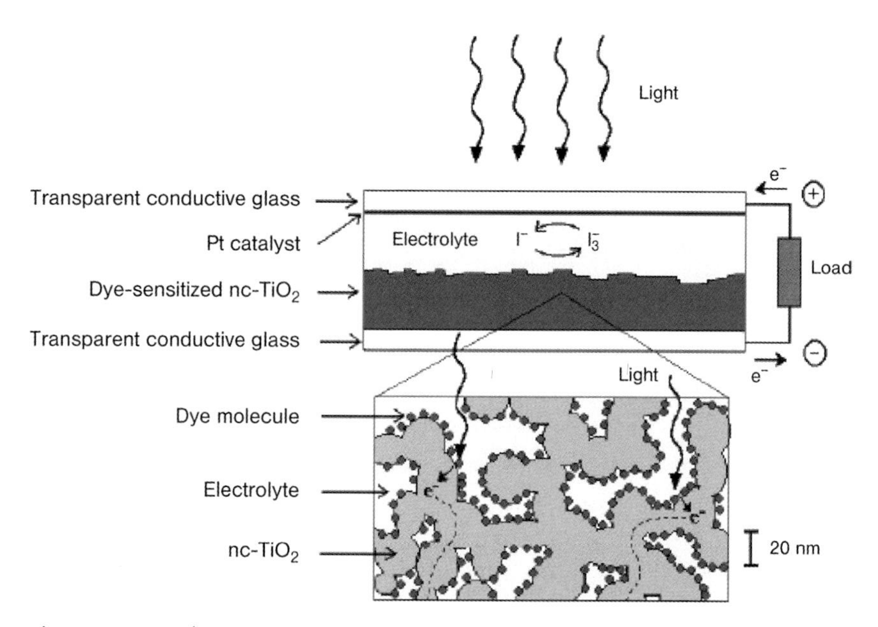

Figure 17.15 A schematic representation of the structure and components of the dye-sensitized solar cell. Reproduced with permission from Ref. [73].

in the dye, the rate at which this occurs is very slow compared to the rate that the dye regains an electron from the surrounding electrolyte. The oxidized species in the electrolyte I_3^- are then reduced at the counterelectrode with electrons supplied, via the outer circuit, from the photoanode.

The primary role of the nanostructure in this system is to increase the efficiency of the light-absorption step by increasing the total surface area of the TiO_2 covered by the dye. The TiO_2 mesoporous structure consists of a 3-D nanocrystalline network of grains, each of which is on the order of tens to hundreds of nanometers in diameter; this serves to scatter and trap the incident light as it enters the nanocrystalline network and is absorbed by the dye. Graetzel's studies have shown that a 10 μm-thick layer of mesoporous dye-sensitized TiO_2 is able to increase the active surface area by three orders of magnitude beyond the area of a flat surface, resulting in an increase in incident photon-to-electron conversion efficiency (IPCE) from 0.13% to 88% at a wavelength of 530 nm [74]. It is important to note that, whilst the IPCE is very high at this particular wavelength, the total efficiency of photovoltaic energy conversion remains at approximately 11%, and is primarily limited by the absorption spectrum of the dye as compared to the spectrum of incident solar radiation.

The introduction of nanostructure also bears a significant impact on electron transport within the TiO_2, the consequences of which have been studied extensively in the context of charge collection for the DSSC [75]. Charge screening occurs across the TiO_2/dye/electrolyte interface, and since the concentration of ions in the electrolyte is typically much greater than the concentration of electrons in TiO_2 ($10^{21}\,cm^{-3}$ ion concentration versus $10^{18}\,cm^{-3}$ electron concentration), depletion lengths on the order of hundreds of nanometers would arise if the TiO_2 layer were sufficiently thick. As the TiO_2 nanoparticle diameter is typically on the order of 20 nm, the entire structure can be considered to be completely depleted, with a negligible electric field along the cross-section. As a result, electrons that are injected into the mesoporous TiO_2 from the dye are subsequently transported to the underlying collection electrode via random-walk diffusive processes. It is important to note that due also to the effect of charge screening from electrolyte to TiO_2, the diffusion that occurs is not strictly electronic but is in fact ambipolar, due to electrostatic coupling of injected charge carriers with mobile ions in the electrolyte. However, as the concentration of electrons is much lower than the concentration of ions, the observed diffusion coefficient will be approximately determined by that of the electrons.

Independent of the dye, it is desirable to improve the efficiency of the SMO conductive pathway. Under operating conditions, the electrons must diffuse several microns in the SMO surrounded by electron acceptors (dye, electrolyte) only nanometers away. Fortuitously, the recombination is slow, but the overall efficiency remains limited by the slow effective diffusivity of electrons within the TiO_2-conductive pathways. These diffusivities of $\sim 10^{-5}$ to $10^{-4}\,cm^2\,s^{-1}$ are at least several orders of magnitude lower than those reported for single crystal anatase [76]. The low D-values are attributed to multiple trapping of carriers with trap states exhibiting an exponential energy distribution [75]. While the traps are suspected to be due to charged defects, grain boundaries, and/or surface states, only limited evidence supports these choices. Because the SMO networks are composed of arrays of nanosized particles, connected either three-dimensionally or in the form of pseudo 1-D wires, each electron must traverse large numbers of grain boundaries before reaching the transparent collection electrode. Small changes in grain boundary chemistry and/or structure can, therefore, have a major impact on the density and distribution of traps at the boundaries.

Electron transport dynamics modeled using simulated mesoporous *random* nanoparticle TiO_2 films by random-walk diffusion find that: (i) the average number of particles visited by electrons increases and, as a result, the pathway increases significantly with higher porosity; and (ii) films composed of ordered particles with similar porosity show a reduced path length. Alternatively, the use of ordered arrays of nanofibers or nanowires has been proposed as a means for *directed* electron transport medium as well as enhanced light scattering. Figure 17.16 depicts one such DSSC system that was created using ZnO nanowires that were formed in aqueous solution using a seeded growth process [77]. While transport efficiency is found to be improved in the case of the ordered nanowire array, the surface area of ceramic material that is covered by dye is geometrically reduced, resulting in a lower photovoltaic conversion efficiency of 1.5%, as observed in this study.

17.6 Conclusions and Observations

Given the polycrystalline nature of most ceramics, it is not surprising that boundaries and interfaces play a key role in determining their electrical and electrochemical properties. As the dimensions of ceramic materials and structures shrink towards the nano domain, this connection becomes even more evident and dramatic. In this chapter, attention has been focused on the source and distribution of local electric

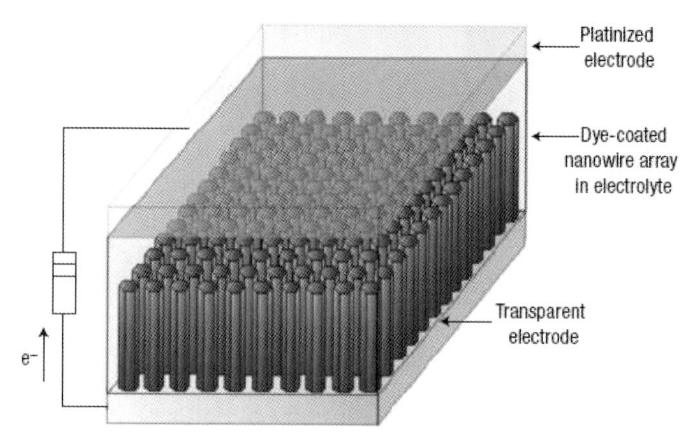

Figure 17.16 Schematic diagram of a DSSC that utilizes an ordered nanowire array in place of the mesoporous nanoparticle network. Reproduced with permission from Ref. [77]; © 2005, Nature Publishing Group.

fields in the vicinity of boundaries, and the resultant spatial redistribution of charged ionic and electronic species.

Through case studies, an examination has been conducted as to how such carrier redistributions could impact properties and device performance. Some typical boundary controlled devices were not examined (e.g., varistors, PTC resistors), as their desired properties are expected to markedly degrade as the grains become fully depleted due to space charge overlap from opposite boundaries. In contrast, sensors generally benefit greatly from scaling down to nanodimensions, due to an increased ability to modulate the conducting channel by changes in space charge potential. For solid electrolytes or mixed conductors used in fuel cells or batteries, the increased role of space charge in controlling conduction in nanoscale ceramics can result in the enhancement or degradation of desired properties, depending on the polarity of the space charge field. Heterogeneous doping was introduced as one important potential means for *engineering* grain boundaries so as to achieve desired properties.

Nanodimensioned ceramics are sometimes prepared to take advantage of other features which do not relate directly to conductivity control. This was the case for DSSCs, where the mesoporous structure of the TiO_2 was needed to enable a high dye loading to insure high solar absorptivity and a high *triple-phase* boundary length between the electrode, electrolyte, and dye. In the case of the lithium battery, short diffusion lengths insure more rapid charge and discharge cycles. Nevertheless, the transport properties of these materials are also impacted by changes in scale that must be considered when examining the overall performance of the materials or devices.

Today, experimental data relating to the physical and chemical properties of nanodimensioned ceramics remains limited. However, as the routes to prepare such materials in a reproducible manner and with controlled chemistries and morphologies become more refined, and as high-resolution measurements of their properties are extended, interesting new insights and phenomena can be expected. Much progress has been made since Tuller's original article entitled *Solid State Electrochemical Systems: Opportunities for Nanofabricated or Nanostructured Materials* in 1997 [66]. Even greater strides can be expected in the coming decade.

Acknowledgments

Funding for this research was provided by NSF (DMR-0243993) under a Materials World Network collaboration with Profs E. Ivers-Tiffee and D. Gerthsen, Universität Karlsruhe, The Korean Institute for Science and Technology, and partial support for G. Whitfield via an Intel fellowship. RWTH Aachen University is gratefully acknowledged for providing a Charlemagne scholarship for a student visitation by S.J. Litzelman, and to Prof. M. Martin and Dr. R. De Souza of RWTH Aachen University for their collaboration on the grain boundary diffusion studies. Helpful discussions with, and assistance from, present or previous group members, including W.C. Jung, I.D. Kim, A. Rothschild, and K. Sahner, are greatly appreciated.

References

1 Tuller, H.L. (2004) Highly conducting ceramics, *Ceramic Materials for Electronics*, 3rd edn (eds R.C. Buchanan), Marcel Dekker, New York, pp. 87–140.

2 Ginley, D.S. and Bright, C. (2000) Transparent conducting oxides. *MRS Bull.*, **25**, 15–22.

3 Minh, N.Q. and Takahashi, T. (1995) *Science and Technology of Ceramic Fuel Cells*, Elsevier Science, Amsterdam, p. 118.

4 Knauth, P. and Tuller, H.L. (2002) Solid-state ionics: roots, status, and future prospects. *J. Am. Ceram. Soc.*, **85**, 1654–1680.

5 van de Krol, R. and Tuller, H.L. (2002) Electroceramics – the role of interfaces. *Solid State Ionics*, **150**, 167–179.

6 Kulwicki, B.M., Amin, A., Lukasiewicz, S.J., Subramanyam, S., and Tuller, H.L. (2004) Ceramic sensors, *Ceramic Materials for Electronics*, 3rd edn (eds R.C. Buchanan), Marcel Dekker, New York, pp. 377–429.

7 Levinson, L.W. (2004) ZnO varistor technology, *Ceramic Materials for Electronics*, 3rd edn (eds R.C. Buchanan), Marcel Dekker, New York, pp. 431–464.

8 Rhoderick, E.H. and Williams, R.H. (1988) *Metal-Semiconductor Contacts*, Oxford University Press, New York.

9 Sze, S.M. (1981) *Physics of Semiconductor Devices*, John Wiley & Sons, New York.

10 Gibbs, J.W. (1906) *The Scientific Papers of J. Willard Gibbs*, vol. 1, Dover Publications, New York.

11 Blatter, G. and Greuter, F. (1986) Electrical breakdown at semiconductor grain boundaries. *Phys. Rev. B*, **34**, 8555–8572.

12 Vol'kenshtein, F.F. (1991) *Electronic Processes on Semiconductor Surfaces During Chemisorption* (ed. R. Morrison), Consultants Bureau, New York, USA.

13 Frenkel, J. (1946) *Kinetic Theory of Liquids*, Oxford University Press, New York.

14 Lehovec, K. (1953) Space-charge layer and distribution of lattice defects at the surface of ionic crystals. *J. Chem. Phys.*, **21**, 1123–1128.

15 Kliewer, K.L. and Koehler, J.S. (1965) Space charge in ionic crystals. I. General approach with application to NaCl. *Phys. Rev.*, **140**, A1226–A1240.

16 Ikeda, J.A.S. and Chiang, Y.-M. (1993) Space charge segregation at grain boundaries in titanium dioxide: I, Relationship between lattice defect chemistry and space charge potential. *J. Am. Ceram. Soc.*, **76**, 2437–2446.

17 Maier, J. (1995) Ionic conduction in space charge regions. *Prog. Solid State Chem.*, **95**, 171–263.

18 (a) Mott, N.F. (1939) The theory of crystal rectifiers. *Proc. R. Soc. London*, **171**, 27–38; (b) Schottky, W. (1939) Zur halbleitertheorie der sperrschicht- und spitzengleichrichte. *Z. Phys.*, **113**, 367–414.

19 Sze, S.M. (1985) *Semiconductor Devices*, John Wiley & Sons, New York.

20 Evans, D.F. and Wennerström, H. (1999) *The Colloidal Domain*, Wiley-VCH, New York.

21 Maier, J. (1986) On the conductivity of polycrystalline materials. *Ber. Bunsen. Phys. Chem.*, **90**, 26–33.

22 Litzelman, S.J. (2008) Modification of space charge transport in nanocrystalline cerium oxide by heterogeneous doping. Ph.D. thesis, Massachusetts Institute of Technology.

23 Litzelman, S.J. and Tuller, H.L. (2008) Modulation of mixed conductivity in nanocrystalline electrolytes by heterogeneous doping, in Electrochemical Society Transactions, Vol. 16, Solid State Ionic Devices 6 - Nano-Ionics; Electrochemical Society Meeting, 12–17 October 2008, Honolulu, HI (eds E. Wachsman, J. Weidner, K. Abraham, E. Traversa, S. Yamaguchi, K. Zaghib, R. Mukundan, and S. Minteer), Electrochemical Society, pp. 3–12.

24 Leonard, F. and Talin, A.A. (2006) Size-dependent effects on electrical contacts to nanotubes and nanowires. *Phys. Rev. Lett.*, **97**, 026804.

25 Zheng, Y., Rivas, C., Lake, R., Boykin, T.B. and Klimeck, G. (2005) Electronic properties of silicon nanowires. *IEEE Trans. Electron. Devices*, **52**, 1097–1103.

26 Boykin, T.B., Klimeck, G., and Oyafuso, F. (2004) Valence band effective-mass expressions in the sp ^{3}d^{5}s* empirical tight-binding model applied to a Si and Ge parametrization. *Phys. Rev. B*, **69**, 115201–115210.

27 Gnani, E., Marchi, A., Reggiani, S., Rudan, M., and Baccarani, G. (2006) Quantum-mechanical analysis of the electrostatics in silicon-nanowire and carbon nanotube FETs. *Solid-State Electron.*, **50**, 709–715.

28 Vol'kenshtein, F.F. (1963) *The Electronic Theory of Catalysis on Semiconductors*, Pergamon Press, New York.

29 Rothschild, A. and Komem, Y. (2002) Quantitative evaluation of chemisorption processes on semiconductors. *J. Appl. Phys.*, **92**, 7090–7097.

30 Xu, C., Tamaki, J., Miura, N., and Yamazoe, N. (1990) Grain size effects on gas sensitivity of porous SnO_2-based elements. *J. Electrochem. Soc. Jpn*, **58**, 1143–1148.(1991) *Sens. Actuators B*, **3**, 147–155.

31 Rothschild, A. and Komem, Y. (2004) On the relationship between the grain size and gas-sensitivity of chemo-resistive metal-oxide gas sensors with nanosized grains. *J. Electroceram.*, **13**, 697–701.

32 Wang, J., Xie, S., and Zhou, W. (2007) Growth of binary oxide nanowires. *MRS Bull.*, **32**, 123–126.

33 Law, M., Goldberger, J., and Yang, P. (2004) Semiconductor nanowires and nanotubes. *Annu. Rev. Mater. Sci.*, **34**, 83–122.

34 Baratto, C., Comini, F., Faglia, G., Sberveglieri, G., Zha, M., and Zappettini, A. (2005) Metal oxide nanocrystals for gas sensing. *Sens. Actuators B*, **109**, 2–6.

35 Lieber, C.M. and Wang, Z.L. (2007) Functional nanowires. *MRS Bull.*, **32**, 99–104.

36 Chronakis, I.S. (2005) Novel nanocomposites and nanoceramics based on polymer nanofibers using electrospinning process – a review. *J. Mater. Process. Technol.*, **167**, 283–293.

37 Li, D. and Xia, Y. (2004) Electrospinning of nanofibers: reinventing the wheel? *Adv. Mater.*, **16**, 1151–1170.

38 Fan, H.J., Werner, P., and Zacharias, M. (2006) Semiconductor nanowires: from self-organization to patterned growth. *Small*, **2**, 700–717.

39 Patolsky, F., Timko, B.P., Zheng, G.F., and Lieber, C.M. (2007) Nanowire-based nanoelectronic devices in the life sciences. *MRS Bull.*, **32**, 142–149.

40 Baratto, C., Comini, E., Faglia, G., Sberveglieri, G., Zha, M., and Zappettini, A. (2005) Metal oxide nanocrystals for gas sensing. *Sens. Actuators B*, **109**, 2–6.

41 Law, M., Kind, H., Messer, B., Kim, F., and Yang, P. (2002) Photochemical sensing of NO_2 with SnO_2 nanoribbon nanosensors at room temperature. *Angew. Chem. Int. Ed. Engl.*, **41**, 2405–2408.

42 Fan, Z. and Lu, J.G. (2005) Chemical sensing with ZnO nanowires. *IEEE Sensors*, **3**, 834–836.

43 Teo, W.E. and Ramakrishna, S. (2006) A review on electrospinning design and nanofibre assemblies. *Nanotechnology*, **17**, R89–R106.

44 Kim, I.-D., Rothschild, A., Lee, B.H., Kim, D.Y., Jo, S.M., and Tuller, H.L. (2006) Ultrasensitive chemiresistors based on electrospun TiO_2 nanofibers. *Nano Lett.*, **6**, 2009–2013.

45 Kim, I.-D., Rothschild, A., Hyodo, T., and Tuller, H.L. (2006) Microsphere templating as means of enhancing surface activity and gas sensitivity of $CaCu_3Ti_4O_{12}$ thin films. *Nano Lett.*, **6**, 193–198.

46 Bieberle-Hütter, A., Beckel, D., Infortuna, A., Muecke, U.P., Rupp, J.L.M., Gauckler, L.J., Rey-Mermet, S., Muralt, P., Bieri, N.R., Hotz, N., Stutz, M.J., Poulikakos, D., Heeb, P., Müller, P., Bernard, A., Gmür, R., and Hocker, T. (2008) A micro-solid oxide fuel cell system as battery replacement. *J. Power Sources*, **177**, 123–130.

47 Bieberle-Hütter, A., Hertz, J.L., and Tuller, H.L. (2008) Fabrication and electrochemical characterization of planar Pt-CGO microstructures. *Acta Mater.*, **56**, 177–187.

48 Huang, H., Nakamura, M., Su, P.C., Fasching, R., Saito, Y., and Prinz, F.B. (2007) High-performance ultrathin solid oxide fuel cells for low-temperature operation. *J. Electrochem. Soc.*, **154**, B20–B24.

49 Baertsch, C.D., Jensen, K.F., Hertz, J.L., Tuller, H.L., Vengallatore, S.T., Spearing, S.M., and Schmidt, M.A. (2004) Fabrication and structural characterization of self-supporting electrolyte membranes for a micro solid-oxide fuel cell. *J. Mater. Res.*, **19**, 2604–2615.

50 Beckel, D., Bieberle-Hütter, A., Harvey, A., Infortuna, A., Müecke, U.P., Prestat, M., Rupp, J.L.M., and Gauckler, L.J. (2007) Thin films for micro solid oxide fuel cells. *J. Power Sources*, **173**, 325–345.

51 Hertz, J.L. and Tuller, H.L. (2009) *Microfabricated Power Generation Devices: Design and Technology* (eds A. Mitsos and E.I. Barton), Wiley-VCH, Weinheim, p. 45.

52 Liang, C.C. (1973) Conduction characteristics of the lithium iodide-aluminum oxide solid electrolytes. *J. Electrochem. Soc.*, **120**, 1289–1292.

53 Wagner, J.B. (1972) The electrical conductivity of semi-conductors involving inclusions of another phase. *J. Phys. Chem. Solids*, **33**, 1051–1059.

54 Sata, N., Eberman, K., Eberl, K., and Maier, J. (2000) Mesoscopic fast ion conduction in nanometre-scale planar heterostructures. *Nature*, **408**, 946–949.

55 Guo, X., Mantei, I., Jamnik, J., Lee, J.-S., and Maier, J. (2007) Defect chemical modeling of mesoscopic ion conduction in nanosized CaF_2/BaF_2 multilayer heterostructures. *Phys. Rev. B*, **76**, 125429.

56 Garcia-Barriocanal, J., Rivera-Calzada, A., Varela, M., Sefrioui, Z., Iborra, E., Leon, C., Pennycook, S.J., and Santamaria, J. (2008) Colossal ionic conductivity at interfaces of epitaxial ZrO_2:Y_2O_3/$SrTiO_3$ heterostructures. *Science*, **321**, 676–680.

57 Tuller, H.L. (1981) Mixed conduction in nonstoichiometric oxides, *Non-Stoichiometric Oxides* (ed. O.T. Sorensen), Academic Press, New York, pp. 271–335.

58 Chiang, Y.M., Lavik, E.B., Kosacki, I., Tuller, H.L., and Ying, J.Y. (1996) Defect and transport properties of nanocrystalline CeO_{2-x}. *Appl. Phys. Lett.*, **69**, 185–187.

59 Chiang, Y.M., Lavik, E.B., Kosacki, I., Tuller, H.L., and Ying, J.Y. (1997) Nonstoichiometry and electrical conductivity of nanocrystalline CeO_{2-x}. *J. Electroceram.*, **1**, 7–14.

60 Tschöpe, A. (2001) Grain size-dependent electrical conductivity of polycrystalline cerium oxide. II: Space charge model. *Solid State Ionics*, **139**, 267–280.

61 Kim, S. and Maier, J. (2002) On the conductivity mechanism of nanocrystalline ceria. *J. Electrochem. Soc.*, **149**, J73–J83.

62 Kek-Merl, D., Lappalainen, J., and Tuller, H.L. (2005) Electrical properties of nanocrystalline CeO_2 thin films deposited by *in situ* pulsed laser deposition. *J. Electrochem. Soc.*, **153**, J15–J20.

63 Litzelman, S.J., De Souza, R.A., Butz, B., Tuller, H.L., Martin, M., and Gerthsen, D. (2008) Heterogeneously doped nanocrystalline ceria films by grain boundary diffusion: Impact on transport properties. *J. Electroceram.*, **22**, 405–415.

64 Tuller, H.L., Litzelman, S.J., and Jung, W.C. (2009) Micro-ionics: next generation power sources. *Phys. Chem. Chem. Phys.*, **11**, 3023–3034.

65 Julien, C. and Nazri, G.-A. (1994) *Solid State Batteries: Materials Design and Optimization*, Kluwer Publishers, Boston.

66 Tuller, H.L. (1997) Solid state electrochemical systems: opportunities for nanofabricated or nanostructured materials. *J. Electroceram.*, **1**, 211–218.

67 Crank, J. (1980) *The Mathematics of Diffusion*, Oxford University Press, New York, USA.

68 Arico, A.S., Bruce, P., Scrosati, B., Tarascon, J.-M., and Van Schalkwijc, W. (2005) Nanostructured materials for advanced energy conversion and storage devices. *Nat. Mater.*, **4**, 366–377.

69 Jamnik, J. and Maier, J. (2003) Nanocrystallinity effects in lithium battery materials: aspects of nano-ionics, Part IV. *Phys. Chem. Chem. Phys.*, **5**, 5215–5220.

70 Meethong, N., Huang, H.-Y.S., Carter, W.C., and Chiang, Y.-M. (2007) Size-dependent lithium miscibility gap in nanoscale $Li_{1-x}FePO_4$. *Electrochem. Solid State Lett.*, **10**, A134–A138.

71 Bruce, P.G., Scrosati, B., and Tarascon, J.-M. (2008) Nanomaterials for

rechargeable lithium batteries. *Angew. Chem., Int. Ed.*, **47**, 2930–2946.

72 O'Regan, B. and Graetzel, M. (1991) A low-cost, high-efficiency solar cell based on dye-sensitized colloidal titanium dioxide films. *Nat. Nanotechnol.*, **353**, 737–740.

73 Halme, J. (2002) Dye-sensitized nanostructured and organic photovoltaic cells: technical review and preliminary tests, M.S. Thesis, Helsinki University.

74 Graetzel, M. (2005) Mesoscopic solar cells for electricity and hydrogen production from sunlight. *Chem. Lett.*, **34**, 8–13.

75 Frank, A., Kopidakis, N., and Lagemaat, J. (2004) Electrons in nanostructured TiO_2 solar cells: transport, recombination and photovoltaic properties. *Coord. Chem. Rev.*, **248**, 1165–1179.

76 Solbrand, A., Lindstrom, H., Rensmo, H., Hagfeldt, A., Lindquist, S., and Sodergren, S. (1997) Electron transport in the nanostructured TiO_2–electrolyte system studied with time-resolved photocurrents. *J. Phys. Chem. B*, **101**, 2514–2518.

77 Law, M., Greene, L., Johnson, J., Saykally, R., and Yang, P. (2005) Nanowire dye-sensitized solar cells. *Nat. Mater.*, **4**, 455–459.

18
Ferroelectric Properties

Doru C. Lupascu and Maxim I. Morozov

18.1
Introduction

Ferroelectrics are a subgroup of pyroelectric materials with reversible spontaneous polarization. According to their crystal symmetry, all ferroelectrics possess piezoelectric properties. Besides, these materials usually reveal nonlinear dielectric and semiconductor properties, and some of them also demonstrate many other pronounced effects such as the photovoltaic effect, electrostriction, or a positive temperature coefficient of resistance (PTCR). Due to such multifunctional variety, ferroelectric materials have found numerous applications in areas such as high-dielectric-constant capacitors, nonvolatile memories, and integrated optical devices. Ferroelectric ceramics, in general, are technologically easier to manufacture than single-crystal piezoelectrics, and also have a greater capacity in the technological variety of their formulations, forms (bulk, films), and fabrication. Comprehensive reviews of ferroelectric ceramic applications have been provided by Haertling [1] and Waser [2]. The latest achievements in the applications of ferroelectric materials of a new generation have been reviewed by Ahn *et al.* [3] and Scott [4]. Applications of ferroelectric ceramics can be classified by their forms of realization (bulk or films), and by their functional effects responsible for device operation.

The first area of ferroelectric ceramic application was that of capacitor engineering, where the dielectric effect is exploited. Most ceramic capacitors are, in reality, high-dielectric-constant ferroelectric compositions in which the ferroelectric properties (hysteresis loop) are suppressed with suitable chemical dopants while retaining a high dielectric constant over a broad temperature range. Historically, the first composition used for such capacitors was $BaTiO_3$ and its modifications, but today lead-containing relaxors and other compositions are also included.

Ferroelectrics are the prime material for piezoelectric applications in ceramic form, due to the discovery of the electrical poling process that aligns the internal polarization of the crystallites within the ceramic, and causes the ceramic to act in similar fashion to a piezoelectric single crystal. Three categories of applications have

Ceramics Science and Technology Volume 2: Properties. Edited by Ralf Riedel and I-Wei Chen

ISBN: 978-3-527-31156-9

been identified depending on whether the direct or converse piezoelectric effects, or their combinations, are employed:

- **Charge or voltage generators (direct piezoelectric effect)**: This class of devices includes various sensors, such as microphones, phonograph cartridges, gas igniters, accelerometers, photoflash actuators, pressure sensors, impact fuses and so on. The operating characteristic varies from millivolts to kilovolts.
- **Displacement actuators (converse piezoelectric effect)**: This class contains various actuators – loudspeakers, camera shutters, buzzers, ink-jet printers, microrobots, relays, pumps, fuel injection systems, and others.
- **High-frequency transformers and resonant devices**: These include various transducers, sonar devices, ultrasonic cleaners, ultrasonic welders, filters, transformers, delay lines, and test-heads for nondestructive testing. Such devices operate at working frequencies ranging from kilohertz to megahertz.

Transparent ferroelectric single crystals are traditionally used for electro-optic applications. Since the optical transparency was first discovered in lead-lanthanum-zirconate-titanate (PLZT), ferroelectric ceramics have been investigated in great depth such that, today, their characteristics allow them to compete with single crystals for certain electro-optic applications. The electro-optic properties of PLZT compositions are intimately related to their ferroelectric properties. Variations in ferroelectric polarization with an electric field, such as in a hysteresis loop, also affect the optical properties of the material.

The applications of ferroelectric materials in film form are growing in number due to the need for miniaturization and the integration of electronic components and also to the developments in microtechnology, materials fabrication, and engineering. Aside from the obvious advantages, such as a smaller size, less weight, and easier incorporation into integrated circuit technology, ferroelectric films offer additional benefits such as a lower operating voltage, a higher speed, and the ability to fabricate unique microlevel structures. In addition, the sintering temperatures of the films are usually hundreds of degrees Celsius lower than that of the bulk, and this often can be the decisive factor in a successful design for application [4–7]. Among the most common ferroelectric film applications are nonvolatile memories, integrated optics, electro-optic displays, microactuators, microtransducers, and capacitors fabricated in film form.

Most of the important ferroelectric properties are closely linked to the variation of spontaneous polarization under the influence of external factors. In this chapter, the important effects are first reviewed related to the key properties of ferroelectrics, such as anisotropy, dielectric hysteresis, and nonlinearity. Some recent trends in the fabrication and application of ferroelectric materials in various forms, structures, and compositions are then highlighted.

The first section details the purely intrinsic response of ferroelectrics and discusses their anisotropic properties, the useful application of which can be controlled by the correct orientation of a single crystal or by texturing a polycrystalline material. Attention is then focused on one of the most significant extrinsic contributions to the polarization response of ferroelectrics, namely the motion of domain walls. The effect of the domain wall contribution can be controlled by the hardening–softening of a ferroelectric system; in this relationship, a phenomenon termed "ferroelectric aging" will be discussed. Following an outline of the relationships between ferroelectricity and magnetism, a final discussion will include the effect of ferroelectric fatigue that manifest itself in a similar manner to aging, but originates from different mechanisms.

18.2
Intrinsic Properties: The Anisotropy of Properties

This section of the chapter details the features of polarization response in ferroelectrics that are solely due to contributions of the crystal lattice, and which usually are referred to as *intrinsic*. In the majority of applications, the polarization response of real materials is often dominated by extrinsic properties, such as displacement of the domain walls, charge carriers, effects of grain boundaries, and other nonlattice contributions. The extrinsic effects – especially a partially irreversible displacement of the domain walls – may induce a high degree of nonlinearity and hysteresis of the material response, and are not always desirable. For example, piezoelectric devices based on "soft" lead zirconate titanate (PZT) ceramics possess a large electromechanical response, but demonstrate poor accuracy, low precision, and high heat generation during operation. Such undesirable effects can be avoided by using hardening dopants that depress the hysteretic motion of the domain walls (see Section 18.3), albeit at the expense of efficiency, or by exploiting the high intrinsic response in electrostrictive ceramics or in actuators based on so-called "domain-engineered" single crystals that have a special domain configuration excluding the domain wall contribution in electromechanical response in certain crystallographic directions [8, 9]. Examples of such behavior for various ferroelectric materials are shown in Figure 18.1.

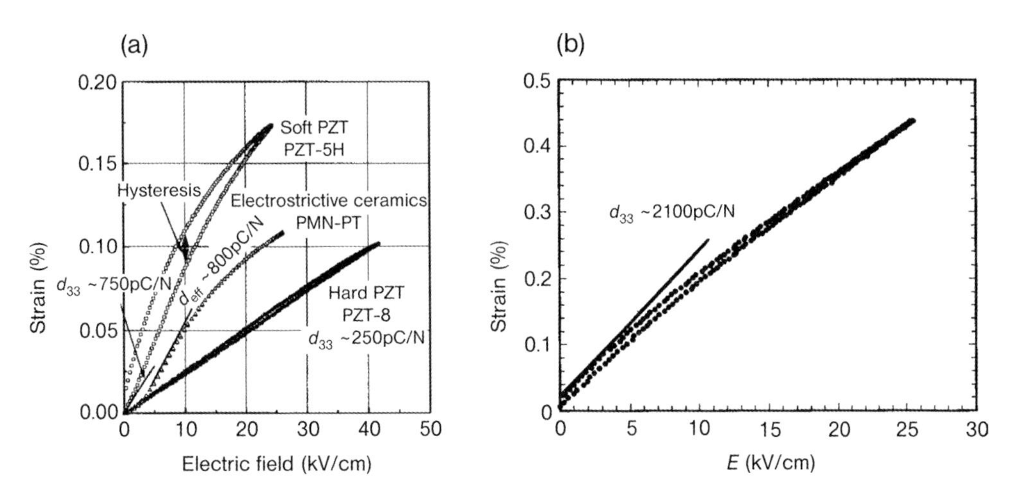

Figure 18.1 Electromechanical responses for various materials. (a) Electromechanical ceramics [8]; (b) Domain-engineered ⟨001⟩ single crystal of $Pb(Zn_{1/3}Nb_{2/3})_{0.955}Ti_{0.045}O_3$ [9].

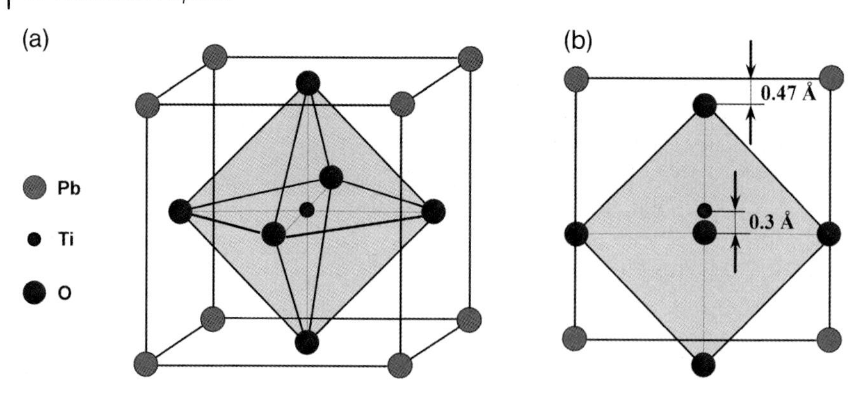

Figure 18.2 Perovskite crystal structure. (a) The nonpolar paraelectric cubic phase; (b) A polar ferroelectric phase due to tetragonal distortion. Data for $PbTiO_3$ stem from Ref. [10].

The main intrinsic property of ferroelectrics is the existence of an effective dipole moment, which forms the polar state that can be switched by an applied electric field. The most technologically important ferroelectrics are an ABO_3 oxide with perovskite structure, in which the spontaneous macroscopic polarization appears as a net dipole moment per unit volume. This is due to a shift of cations relative to anions in a distorted lattice of lower symmetry with respect to the cubic perovskite structure of the paraelectric phase (Figure 18.2) [10]. In a large crystalline volume, a very high electroelastic energy of the polar state may be naturally reduced by the formation of a multidomain structure with different orientations of spontaneous polarization in the domains. A contribution to the polarization response of a ferroelectric due to displacement of the domain interfaces (domain walls) is extrinsic in nature, and will not be considered in this section. The intrinsic dielectric, mechanical, and electromechanical properties are structure-related and depend on the crystal symmetry. The type of perovskite distortion, and in turn the number of phase transitions, depend on the nature of the chemical bonds forming the crystal structure of the compound. As shown by Cohen [11], different ground-state symmetries occur in $BaTiO_3$ and $PbTiO_3$. In both ferroelectrics, the hybridization between the titanium *3d* states and the oxygen *2p* states results in a softening Ti–O repulsion and allows ferroelectric instabilities that favor the formation of a low-symmetry phase. In $PbTiO_3$, an additional hybridization occurs between the lead and oxygen states, leading to a large strain that stabilizes the tetragonal phase. In $BaTiO_3$, the interaction between barium and oxygen is completely ionic, and the influence of tetragonal strain is much lower. Calculations of the energy states for both rhombohedral and tetragonal phases, without accounting for strain, demonstrate that both $BaTiO_3$ and $PbTiO_3$ energetically favor the rhombohedral structure [11]. The rhombohedral strain is energetically small, but the tetragonal strain has a considerable effect on $BaTiO_3$; in $PbTiO_3$, it stabilizes the tetragonal structure over the rhombohedral phase. Thus, the electron configuration of the A-cation can be a decisive factor in the establishment of the ground state of a perovskite ferroelectric. Following this concept, the nature of the phase transition can be interpreted as an increase of the ferroelastic strain that couples to the ferroelectric distortions.

In a variety of ferroelectrics with the cubic perovskite prototype structure of the paraelectric phase, possible lattice distortions in the ferroelectric ground state include the tetragonal (T), orthorhombic (O), rhombohedral (R), and monoclinic (M) phases. Interesting effects of intrinsic ferroelectric softening are often observed in the vicinity of the phase transitions, especially in those ferroelectric systems with a morphotropic phase boundary (MPB). The concept of a MPB was introduced by Jaffe *et al.* [12] for the phase diagram of the $PbZrO_3$–$PbTiO_3$ solid solutions (PZT), as a near-vertical phase boundary between the rhombohedral and tetragonal phases. A rough structural transformation in these phases is sensitive to the variation of composition rather than to temperature. The region near the morphotropic phase boundary in PZT and other ferroelectric systems has attracted the profound interest of many research groups, as they show excellent properties and possess large dielectric, piezoelectric, and electromechanical coupling constants in the vicinity of the MPB [12, 13]. Recently, a breakthrough in understanding the origins of these properties has emerged, due to both theoretical and careful experimental investigations of the MPB phenomenon. Ishibashi and Iwata explained, on a theoretical basis, the increase in the dielectric susceptibility and piezoelectric constant in the vicinity of the MPB in relation to the Landau–Ginsburg–Devonshire (LGD) theory, assuming a possible composition dependence of β-parameters in the polynomial expansion of the free energy [14–16]. Experimentally, very minute and accurate structural studies [17–19] of the region of PZT compositions near to the MPB have revealed the presence of a monoclinic phase separating the tetragonal and rhombohedral phases of $Pb(Zr,TiO)_3$ over a wide range of temperatures. Shortly thereafter, similarly narrow regions of the intermediate phases were found near the MPB of other ferroelectric systems with distorted perovskite structures: in $(1-x)Pb(Mg_{1/3}Nb_{2/3})O_3$–$xPbTiO_3$ (PMN-PT) [20] and $(1-x)Pb(Zn_{1/3}Nb_{2/3})O_3$–$xPbTiO_3$ (PZN-PT) [21, 22]. In the framework of the LGD theory, the ferroelectric monoclinic phase first appears on the phase diagram if the energy expansion is considered higher than a sixth-order term. Vanderbilt and Cohen [23] showed that an eighth-order expansion could account for three types of equilibrium monoclinic phases, in which the polarization is confined not to a symmetry axis but rather to a symmetry plane. These are a monoclinic of type C (M_C), in which the polarization P is anywhere along [0uv], and two monoclinic phases M_A and M_B, in which P lies along [uuv] directions, with $u < v$ and $u > v$, respectively (Figure 18.3). The M_A phase is observed in PZT [24, 25], M_C in PMN-PT, and the structure of an intermediate phase in PZN-PT is either a monoclinic M_C [21] or orthorhombic [22] that can easily turn to M_A or M_C if a very small electric field is applied in the [001] direction [26]. These three phase diagrams with intermediate phases are shown in Figure 18.4, as presented in a recent review by Noheda [25]. More recently, the monoclinic M_B phase has been established between the rhombohedral R and M_C phases in PMN-PT solid solutions, and a R–M_B–O–M_C–T pathway for the polarization rotation over the transition region has been suggested for both PZN-PT and PMN-PT systems [27–29]. Another example of intermediate monoclinic phases existing near the MPB has recently been found in the $(1-x)Pb(Sc_{1/2}Nb_{1/2})O_3$–$xPbTiO_3$ (PSN-PT)

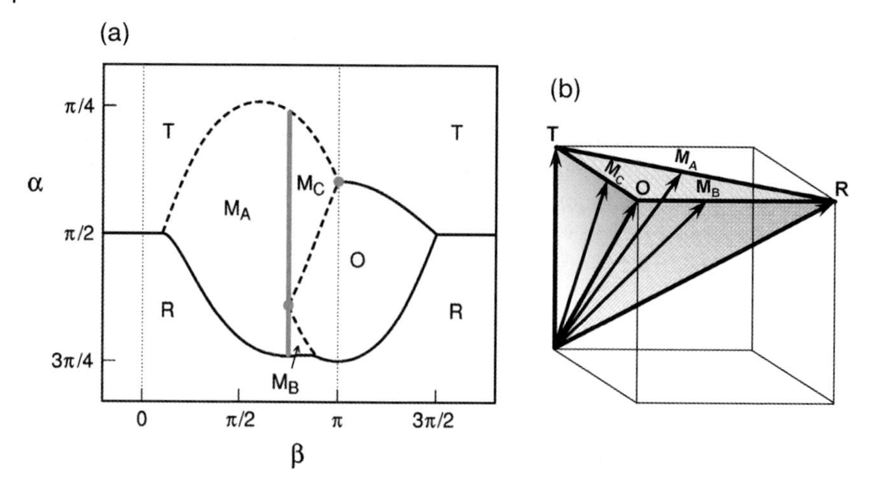

Figure 18.3 (a) Phase diagram for ferroelectric perovskites in the space of the dimensionless parameters α and β derived by Vanderbilt and Cohen [23] in the framework of the LGD theory using an eighth-order expansion of the free energy. (b) Schematic of the corresponding polar axes in the rhombohedral (R), orthorhombic (O), tetragonal (T), and monoclinic (M_A, M_B, M_c) phases. The polar axes of the monoclinic phases are not fixed, but can rotate within the shaded planes.

system [30, 31]. Further analysis of the systems with MPB, including a refinement of their structure, has been a subject of many recent and present studies that will hopefully establish the phase diagrams of many other important ferroelectric systems.

As the polarization is not confined to a fixed direction in a monoclinic phase, its unconstrained rotation can explain the highly valued properties of ferroelectric systems near the MPB [25]. A relationship between the polarization rotation and a strong enhancement of the electromechanical response in perovskite ferroelectrics was demonstrated by Fu and Cohen [32], from a first principles study of the internal

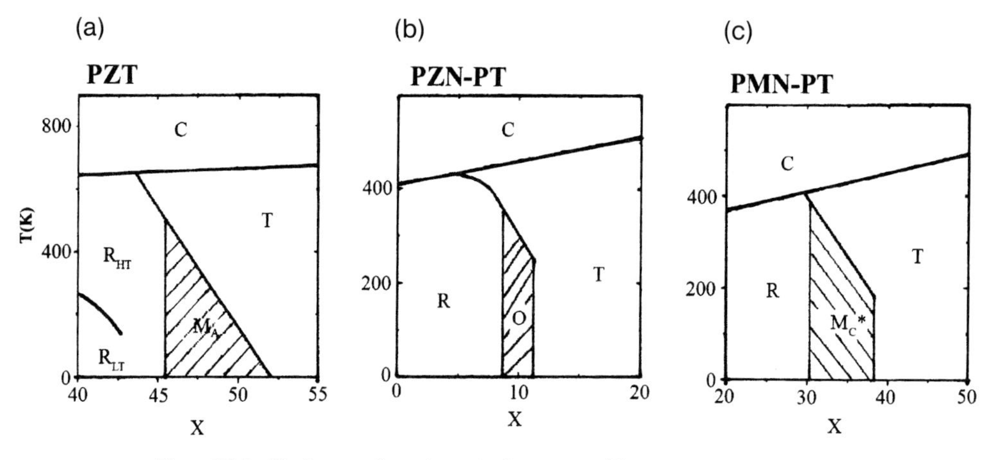

Figure 18.4 The intermediate phases in the vicinity of the MPB in perovskite ferroelectric systems. (a) PZT [19]; (b) PZN-PT [22]; (c) PMN-PT [27]. From Ref. [25].

energies for different polarization directions in $BaTiO_3$ within three crystallographic planes, as indicated in Figure 18.3b. The direct R–M_A–T path was found to be advantageous in energy for polarization rotation between the initial direction of spontaneous polarization [111] and the direction of applied field [001]. Fu and Cohen's analysis suggested that high electromechanical properties could be obtained due to a field-induced rotation of polarization, from the rhombohedral to tetragonal direction, if the material possessed a high level of strain in the tetragonal phase and a flat energy surface near the rhombohedral phase.[1)] The calculated magnitude of the electric field needed to induce the rhombohedral to tetragonal transition in $BaTiO_3$ was very large (300 kV cm^{-1}), but a similar process was observed experimentally on a PZN–8% PT single crystal at ~40 kV cm^{-1}.

The mechanism of polarization rotation causing an enhanced piezoelectric response along the nonpolar direction is intrinsic by nature, and is usefully exploited in actuators based on the domain-engineered crystals with almost linear and anhysteretic characteristics (Figure 18.1b). A domain-engineered crystal is a ferroelectric crystal of a rhombohedral phase poled along the tetragonal [001] direction, as proposed in Ref. [8]. Such poling reduces the number of possible domain orientation states from eight to four, as shown in Figure 18.5. The subsequent application of an electric field to this domain-engineered structure along [001] direction will not favor any further domain motion; rather, the polarization will rotate towards the applied field until the tetragonal phase is established [8, 32]. The absence of any domain wall motion provides an anhysteretic response for a wide range of applied fields. Nevertheless, passing the induced-phase transition may cause a certain hysteresis.

An important aspect of the enhanced intrinsic response of ferroelectrics is anisotropy, the direction dependence of properties. By symmetry, ferroelectric crystals are anisotropic with respect to dielectric, mechanical, and electromechanical properties, and this issue is essential when designing devices that exploit the highest material responses by correctly orienting single crystals [8, 33–35]. The crystal axes, along which the highest material response occurs, may not coincide with the polar directions of spontaneous polarization for a given ferroelectric phase. This is the case

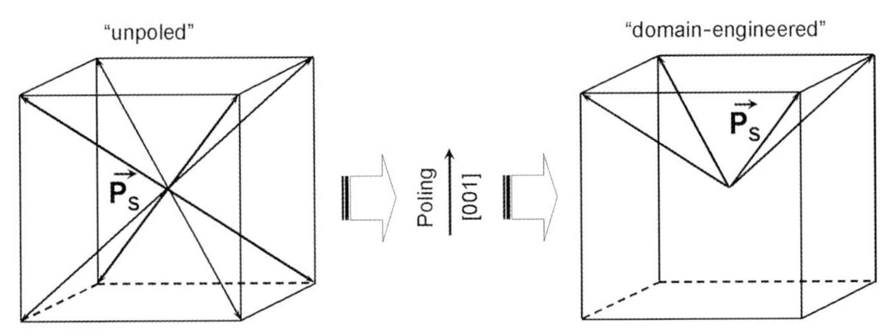

Figure 18.5 Possible domain orientation states in an unpoled rhombohedral crystal and in the domain-engineered crystal poled along the [001] direction.

1) The rotation of polarization at the standard resolution of XRD can also be interpreted as the average over nano-domain patterns. For a detailed discussion see Ref. 288.

in the above-mentioned domain-engineered PMN-PT and PZN-PT crystals, the excellent electromechanical properties of which are observed in the rhombohedral state, but along the tetragonal nonpolar axis, as is typical of that in other materials with the same symmetry. By using a phenomenological approach, Du *et al.* [35, 36] showed that the maximum longitudinal piezoelectric coefficient d_{33} and the electromechanical coupling factor k_{33} coincide with the polar direction in the tetragonal phase, whilst in the rhombohedral phase they are canted to the polar direction at $\sim$57 and $\sim$51°, respectively. The canting angles of maximal d_{33} vary with respect to composition, but remain close to the perovskite [001] direction (Figure 18.6).

As demonstrated in Figure 18.7, both the intrinsic response of PZT and the anisotropy response increase when approaching the MPB region. Although a PZT single crystal has not yet been grown and investigated, the direction-dependence of lattice deformation has been examined in this material using high-resolution synchrotron X-ray diffraction [37], while the piezoelectric anisotropy has been studied on oriented PZT thin films [38], which both justified the predictions of the phenomenological theory.

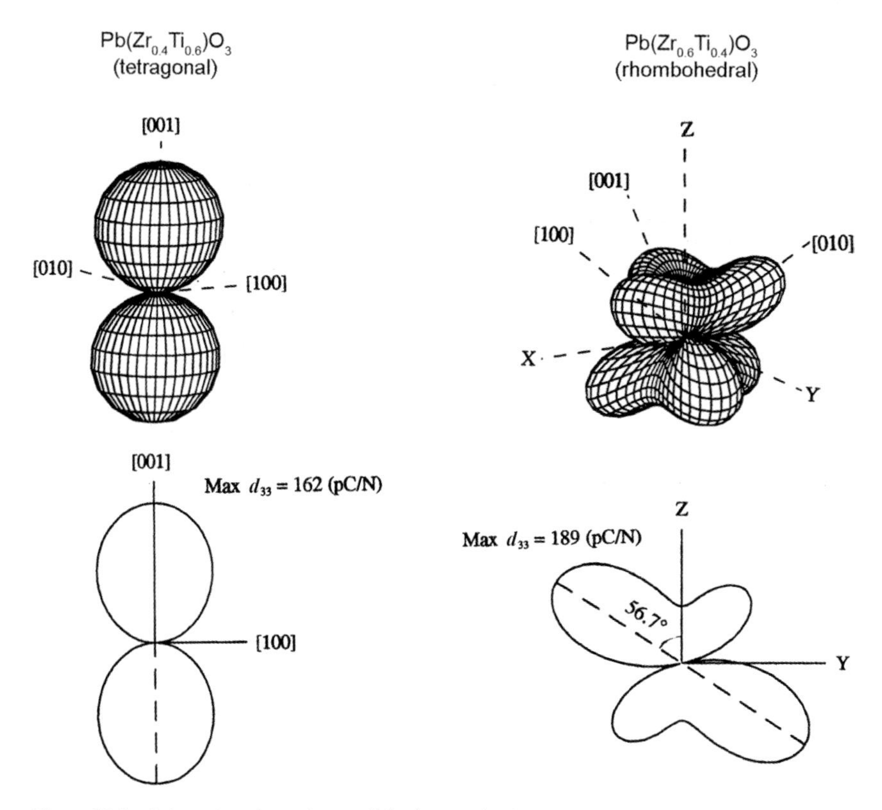

Figure 18.6 Orientation dependence of the longitudinal piezoelectric coefficient d_{33} in the tetragonal and rhombohedral phases of PZT. From Ref. [35].

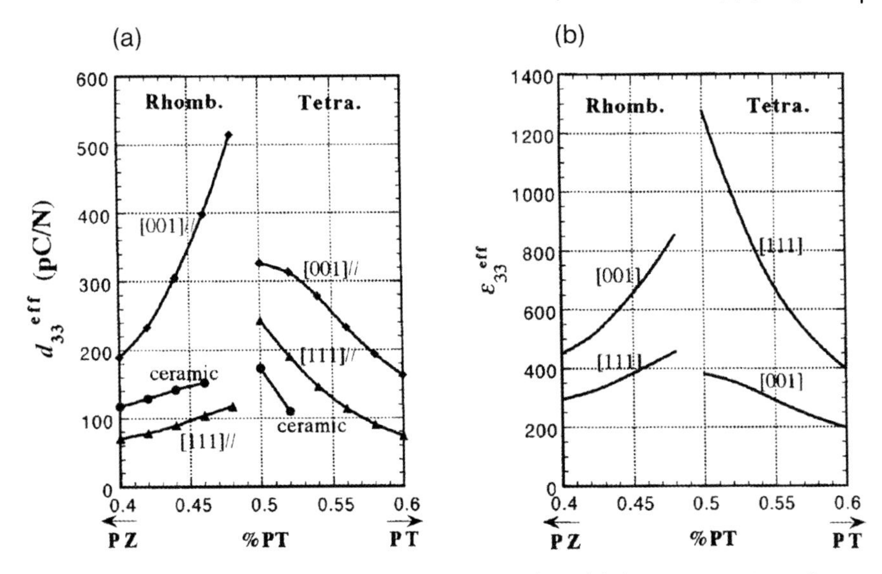

Figure 18.7 Effective longitudinal piezoelectric coefficient d_{33} and dielectric permittivity ε_{33} for various PZT compositions. The curves are calculated for a single crystal using a phenomenological approach, and shown in comparison with experimental data for a PZT ceramic sample. From Ref. [36].

An enhanced anisotropic response in the vicinity of a ferroelectric–ferroelectric phase transition exists in other perovskites. Furthermore, it can be associated not only to the transitional processes near the MPB, but also to similar effects of instabilities in crystals, either near temperature-induced polymorphic phase transitions or under external fields [39–43]. Budimir *et al.* [40] showed that the fundamental thermodynamic process behind the piezoelectric enhancement and anisotropy can be explicitly described by examining a flattening of the Gibbs free energy function at "critical" conditions, such as the degeneration of two or more ferroelectric phases or a very high level of applied electric field or stress near to the thermodynamic coercive limit. This thermodynamic approach suggests that the anisotropy of the free-energy flattening is the origin of the anisotropic enhancement of the piezoelectric response [40, 41].

The anisotropy of dielectric and piezoelectric properties can be expressed in terms of tensor components of the relative dielectric susceptibility χ_{ij}, and piezoelectric coefficients d_{ij}. For the tetragonal, orthorhombic, and rhombohedral phases of perovskite ferroelectrics, the longitudinal coefficients χ_{33}, and d_{33} are related to the collinear polarization extension along the polar direction, while polarization rotation in response to an applied electric field along the nonpolar direction is related to the transverse dielectric susceptibilities χ_{11}, χ_{22}, and the piezoelectric shear coefficients d_{15} and d_{24} [42]. The mutual relations between χ_{33}, and d_{33}, as well as between χ_{11}, χ_{22}, and d_{15}, d_{24}, are established via corresponding electrostrictive coefficients [44, 45]. Thus, for an arbitrary nonpolar direction of the applied field, the anisotropy factors such as d_{15}/d_{33} and χ_{11}/χ_{33}, or d_{24}/d_{33} and χ_{22}/χ_{33}, correspond to the relative effects of polarization rotation and polarization extension. Considering in detail this issue,

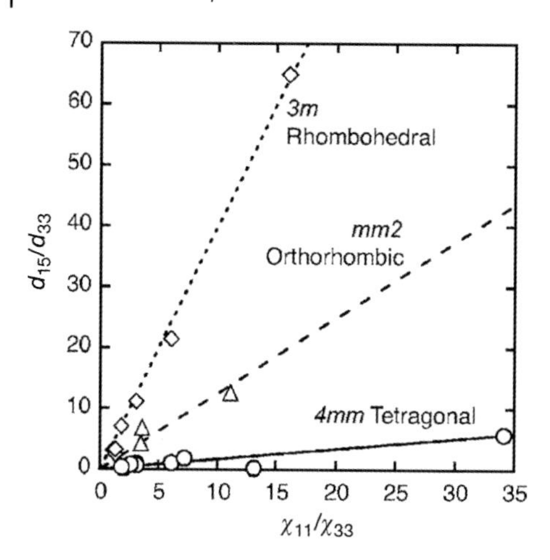

Figure 18.8 Piezoelectric anisotropy versus dielectric anisotropy for various oxygen octahedra ferroelectrics with 3*m*, *mm*2, and 4*mm* point group symmetries. From Ref. [42].

Davis *et al.* [42] have recently introduced the terms "rotator" and "extender" for a variety of ferroelectrics based on oxygen octahedra, in order to classify them with respect to whether the shear or the collinear effect dominates in the piezoelectric response. In extenders, the dominant polarization extension is directly related to the collinear piezoelectric effect, whereas in rotators the dominant contribution to the piezoelectric effect is the polarization rotation, that is directly related to the shear piezoelectric effect. Thus, extenders are ferroelectrics with a large longitudinal piezoelectric coefficient d_{33} that is related to a large relative dielectric susceptibility χ_{33}, while rotators are ferroelectrics with large shear coefficients d_{15} and d_{24}, which are related to transverse susceptibilities χ_{11} and χ_{22}, correspondingly. Electrostrictive correlations between dielectric and piezoelectric anisotropies depend on the point group of the crystal (Figure 18.8), and result from different orientations of the polar axis with respect to the component oxygen octahedra [42]. As shown in Figure 18.8, for a group of rotators with a high dielectric anisotropy χ_{11}/χ_{33}, an enhanced polarization response along the nonpolar direction is expected to be largest in *3m* crystals and smallest in *4mm* crystals, due to their inherent symmetry-related electrostrictive anisotropy. A comprehensive discussion of the mechanisms that can contribute to the enhanced electromechanical anisotropy in relation to a variety of intrinsic and extrinsic processes in perovskite ferroelectrics is presented in a recent topical report from Damjanovic *et al.* [43].

In concluding this section, it is important to address the issue of application aspects of intrinsic properties. Such properties of ferroelectric crystals are well pronounced but highly anisotropic in a variety of advanced materials, such as "rotator" ferroelectrics, including systems with the morphotropic phase boundary or experiencing ferroelectric–ferroelectric phase transitions. An application of these materials requires the correct orientation of the crystal. Although not all advanced materials are available in single-crystal form (e.g., PZT), it is possible to take

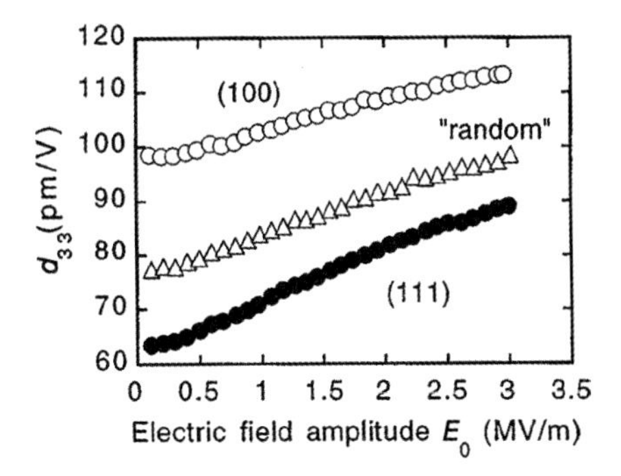

Figure 18.9 Longitudinal d_{33} piezoelectric coefficients for the PZT 60/40 thin films with (100), (111), and "random" orientation as a function of the driving field amplitude. From Ref. [38].

advantage of the intrinsic anisotropies in their physical properties by texturing polycrystalline ceramics. For example, the desired texture of PZT can be achieved in thin films by using an appropriate substrate, the pattern of which governs an interface with PZT controlling its nucleation and crystallographic orientation during growth [38, 46]. As expected from theoretical predictions (Figure 18.7a), the highest piezoelectric response of rhombohedral $Pb(Zr_{0.6}Ti_{0.4})O_3$ is achieved along the [100] nonpolar direction, and is minimal along the polar [111] direction, as shown in Figure 18.9 [38]. The nonlinearity along the [100] direction is lowest due to the domain-engineering effect. The observed nonlinearity and discrepancy between the values of measured and predicted piezoelectric coefficients (10–50%) are most likely a result of imperfect film orientation that remains a technological issue.

The texturing of PMN-PT ceramics by using a templated grain growth technique leads to an increase in the piezoelectric response by about a factor of 2 with respect to their randomly oriented counterparts [47–49]. This permits the production of relatively inexpensive materials with high piezoelectric properties, using a ceramic processing route. Although the longitudinal piezoelectric coefficient of textured PMN-PT ceramics is very high ($d_{33} \sim 1600$ at low fields $<5\,\mathrm{kV\,cm^{-1}}$), the useful combination of the properties of this material is still not competitive with the single crystals (see Figure 18.10), due to an apparent hysteresis and nonlinearity, plausibly resulting from the presence of tetragonal domain states, porosity, and internal stresses [48].

Wider analyses of utilizing the advantages of textured materials for piezoelectric applications have been provided recently by Messing *et al.* [49] and Kimura [50]; these will be discussed in more detail later in the chapter.

18.3 Extrinsic Properties: Hard and Soft Ferroelectrics

The reversibility of the spontaneous polarization in ferroelectrics is attributed to their basic property that permits the switching of spontaneous polarization by the

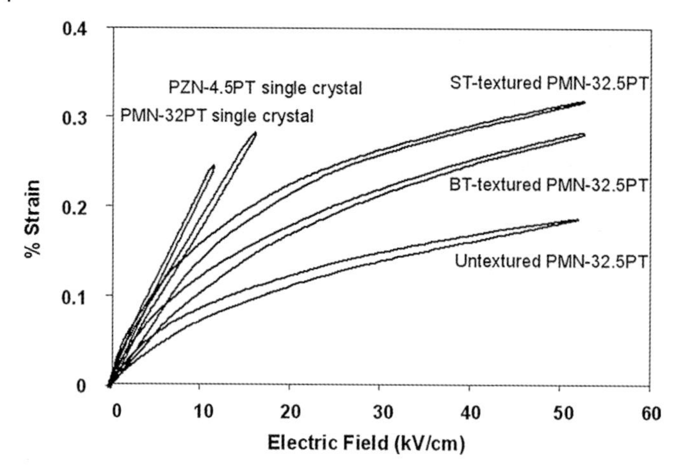

Figure 18.10 Comparison of the high-field strain response of textured (001) PMN-PT with single crystals of similar composition. From Ref. [49].

application of an external electric field. With respect to this feature, ferroelectrics demonstrate a hysteresis phenomenon that manifests itself in a hysteretic dependence of the material polarization (P) in response to the applied electric field (E). Hysteresis loops may appear in many different forms, depending on the composition and prehistory of the ferroelectric material. Figure 18.11a shows two types of hysteresis loop peculiar to "hard" and "soft" materials. As for ferromagnets, these terms first of all characterize the material's suitability to switching. "Soft" materials allow an easy switching of spontaneous polarization by applying fields under normal conditions, whereas "hard" ferroelectrics usually require special thermo-temporal conditions for reliable poling. Nevertheless, a set of other distinguishing

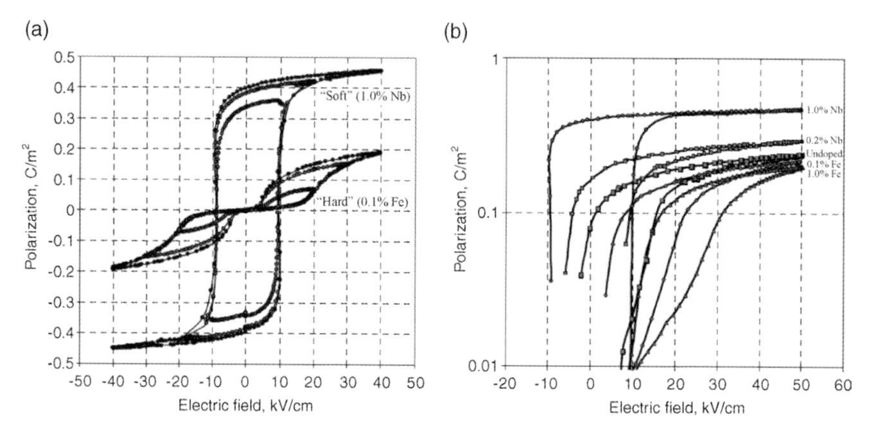

Figure 18.11 Ferroelectric hysteresis loops for hard and soft $Pb_{0.58}Zr_{0.42}TiO_3$ ceramics modified by various amounts of Nb and Fe dopants: (a) on linear scales; (b) on semi-logarithmic scale (the first quadrant).

properties can be attributed to soft and hard materials [13]. In general, soft materials are characterized by square hysteresis loops, a high dielectric loss, high mechanical compliance, and reduced aging. In contrast, hard ferroelectrics are characterized by poorly developed hysteresis loops, a lower dielectric constant, low dielectric losses, low compliances, and a higher aging rate. The term "hard" is usually associated with an aged state of ferroelectrics (with reduced polarization response), and "soft" for materials that do not undergo aging (they have stable properties in the course of time). As an aged hard material can be "deaged" and become soft by thermal quenching or other techniques [51–56], the "hardening–softening" concept entirely covers the sense of "aging–deaging" terminology.

One well-known way to achieve hardening or softening in ABO_3 ferroelectric materials is to dope them with aliovalent additives. The donor-type dopants causing cation-site vacancies lead to softening, while acceptor-type additives result in a decrease in the polarization response and thus a hardening of the ferroelectric switching. As shown in Figure 18.11b, a gradual variation of dopant type and concentration results in a gradual hardening–softening transition in PZT ceramics.

Besides this effect of hardening–softening by dopants, a similar gradual transformation of hysteresis loops can be achieved by several other external influences, such as a very long time (aging) [51, 52], electric field cycling [53, 54], thermal quenching [54, 55], or light illumination [56]. A recent systematic study of such effects on the hardening–softening of PZT ceramics in both switching and subswitching conditions has revealed strong similarities among them, and suggested that these effects are of a common origin and related to the ordering–disordering processes in the charged defect environment of the domain wall [57, 58].

An understanding of hardening–softening properties can be achieved through the analysis of the domain wall contribution to the polarization response of ferroelectrics. It should be noted here that this is not the only contribution to the polarization response; rather, the intrinsic polarization response as well as surface, boundary, and interface effects may also contribute significantly to the total polarization of a ferroelectric material, especially in thin films. However, the dominant contribution to the dielectric, elastic, and piezoelectric properties in ferroelectric materials is extrinsic, and typically originates from displacement of the domain walls [59].

The spontaneous polarization of each domain contributes to the total polarization response of ferroelectrics. In equilibrium, the domain configuration of multidomain ferroelectrics with the absence of any external field yields a net zero polarization state. When the external electric field is applied, the domain configuration changes favoring growth of those domains of which spontaneous polarization conforms with the direction of the external field such that the domain walls are displaced. The displacement of a domain wall depends on its interaction with crystal lattice defects, which act as pinning centers for the domain wall and cause its nonuniform motion [60]. With respect to this interaction, two types of domain wall motion can be considered, namely *reversible* (a bowing of the domain wall pinned by defects) and *irreversible* ("jumps" between pinning centers).

A direct observation of the pinning and bowing of a single ferroelectric domain wall under a weak electric field has been described and analyzed by Yang *et al.* [61]. By

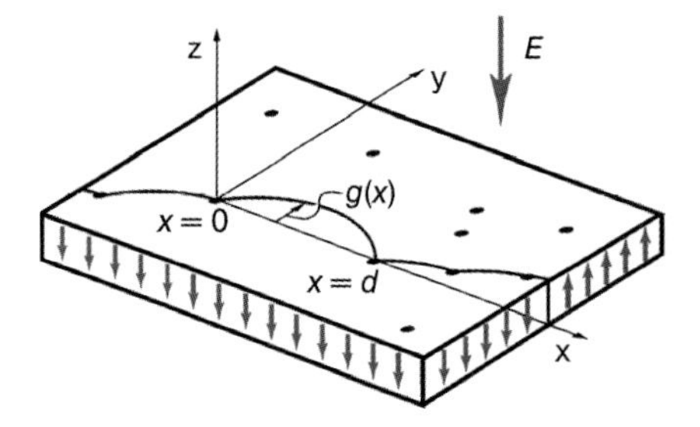

Figure 18.12 Schematic of bowing of a pinned domain wall under an applied electric field. The dots denote defects. From Ref. [61].

using a simple two-dimensional (2-D) model (as shown in Figure 18.12), these authors showed that the domain wall profile $g(x)$ is a circle segment of radius $R = \sigma_w/(2P_sE)$, where σ_w is the domain wall energy per unit area, E is the strength of the static field applied, and P_s is the spontaneous polarization. Thus, the domain wall curvature is unique for a given material at a fixed electric field E, and independent of the distance between the pinning sites. The movement of the domain wall was revealed to be reversible and unipolar with respect to the applied field.

It is most likely that local Barkhausen jumps observed in ferroelectrics [62–64] at switching field conditions are the next step of the domain wall motion under an external electric field after bending between pinning centers.

The domain reconfiguration process during electric field switching for non-aged multidomain ferroelectrics has been well known for a long time, whereas for a hard material an *in situ* observation of the domain pattern transformation during a switching cycle has been presented only very recently for both multidomain [55] and single domain [65] cases of Mn-doped $BaTiO_3$ single crystals. Such direct observations of the domain structure during P–E loop measurements give a very clear image of the domain wall contribution to the polarization response. It had been shown that in a hard, aged material the pattern of the multidomain configuration is evidently restored after a switching cycle by the application of an electric field. In the same deaged material, the multidomain–single domain–multidomain process occurs, but after each cycle the domain pattern changes. These experiments strongly suggest the presence of domain wall-pinning centers distributed in the bulk of an aged hard material, in a special way that leads to a restoration of the domain pattern after removal of the electric field. Charged point defects are the first candidates for such pinning centers.

Several models have been proposed that consider the various interactions of the domain wall with electrostatically ordered defect pinning centers. These can be conditionally classified into three different scenarios with respect to the charged defects destinations adopted during ordering:

- The first scenario suggests an electrostatic alignment of the charged defect or their associates such as $\{V_O^{\bullet\bullet}-B_{B^{4+}}^{3+}\}^+$ in ABO_3 along the direction of spontaneous polarization within a domain (Figure 18.13a), thus revealing the symmetry-conforming property of point defects [66]. This scenario is the most frequently used to

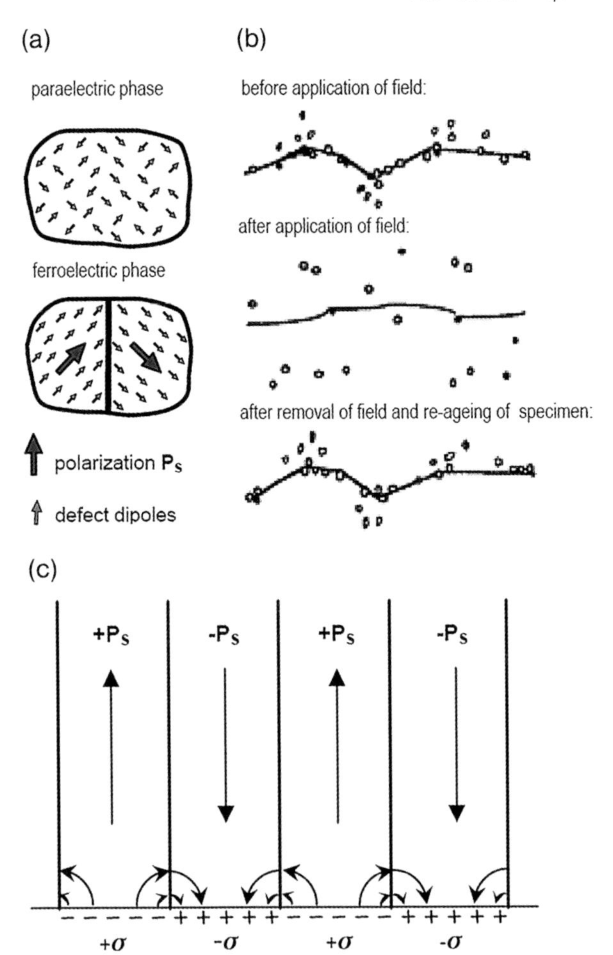

Figure 18.13 Stabilization of the domain wall position through interaction with charged defects. (a) The "volume effect" of charged dipole ordering along the spontaneous polarization (P_s); (b) The "domain wall effect" due to diffusion of defects to domain boundaries resulting in pinning (after Ref. [54]); (c) The "grain boundary" effect; an example of a simplified 2-D model of 180° domain walls crossing the grain boundary. See Ref. [77] for details.

interpret biased or pinched hysteresis loops in bulk ferroelectrics in general [67], in perovskites [68–70], as well as in particular cases of triglycine sulfate (TGS) [71], Rochelle salt [72], and barium titanate [55, 65, 73, 74] single crystals, when these were doped with impurities or subjected to either X-ray or gamma-irradiation.

- The second scenario assumes a longer distance in the arrangement of charge defects (or their complexes) considering their drift towards the domain boundaries (to the domain walls) [54, 75, 76], which results in pinning, as shown in Figure 18.13b.

- The third type of domain wall-clamping effect is assumed due to charge drift towards grain boundaries, resulting in the creation of local fields (Figure 18.13c) which then exert a clamping pressure on the domain walls [77]. This mechanism also requires longer distances in the electrostatic rearrangement of charged defects in comparison with the first scenario, where charged defects are assumed to move within a unit cell.

The discussion concerning the domination of the domain wall-clamping mechanisms in bulk ferroelectrics has been the subject of several reports [51–55, 65, 66, 73, 78] since the discovery of ferroelectric aging, and still remains an open question. Recently, it has been shown by calculations [77] that the clamping pressure exerted by oriented defect dipoles in a 1-D model is two to three orders of magnitude weaker than that exerted by the same amount of charges accumulated on grain boundaries.

The arguments for the volume effect are based on a direct observation of dipole alignment [68–70] and some experimental studies, such as observing the biased hysteresis loop on Mn-doped $BaTiO_3$ single crystals, both before and after the surface layer removal, by etching to half of the initial thickness within a time much shorter than the relaxation time of the stabilization process [78]. Direct evidence of domain pattern conservation during the switching process by *in situ* microscopy observation of 90° ferroelectric domains of Mn-doped $BaTiO_3$ during electric field cycling [55] also suggests an electrostatic rearrangement of charged defects in the material bulk.

The models mentioned above have been developed for aging mechanisms as an effect of "hard" (or acceptor) doping, assuming that it originates from either the short- or long-range electrostatic arrangement of mobile charge carriers, for example, oxygen vacancies. However, the effect of softening is not understood entirely. Notably, the question remains open as to whether soft doping serves only to compensate the naturally occurring oxygen vacancies, or whether it causes an independent effect. The smooth transition of material behavior with respect to the gradual change in type and concentration of dopants (Figure 18.11b) suggests that hardening and softening are of the same origin. Nevertheless, softening may still occur for much higher concentrations than necessary for full compensation of naturally occurring hard dopants. Irrespective of the exact dominant mechanism of domain wall clamping, a phenomenological approach can be applied by linking the degree of electrostatic arrangement of pinning centers for a domain wall with the randomness, roughness, and steepness of the domain wall energy profile with respect to its displacement [57, 58]. Such a phenomenological model considers the hardening–softening transition as a continuous variety of intermediate states for the domain wall potential energy profile between two limiting cases, as shown in Figure 18.14.

As for a rate-independent process, the hysteresis results from an irreversibility of the domain wall motion, with the two limiting cases for the hardening–softening process being defined with respect to the number of trapping sites for the domain wall in its potential energy landscape:

- **The "Very Hard" Limit**: The V-potential profile was proposed by Robels and Arlt [79] and its nonlinear extension was proposed by Li, Cao, and Cross [80]. This

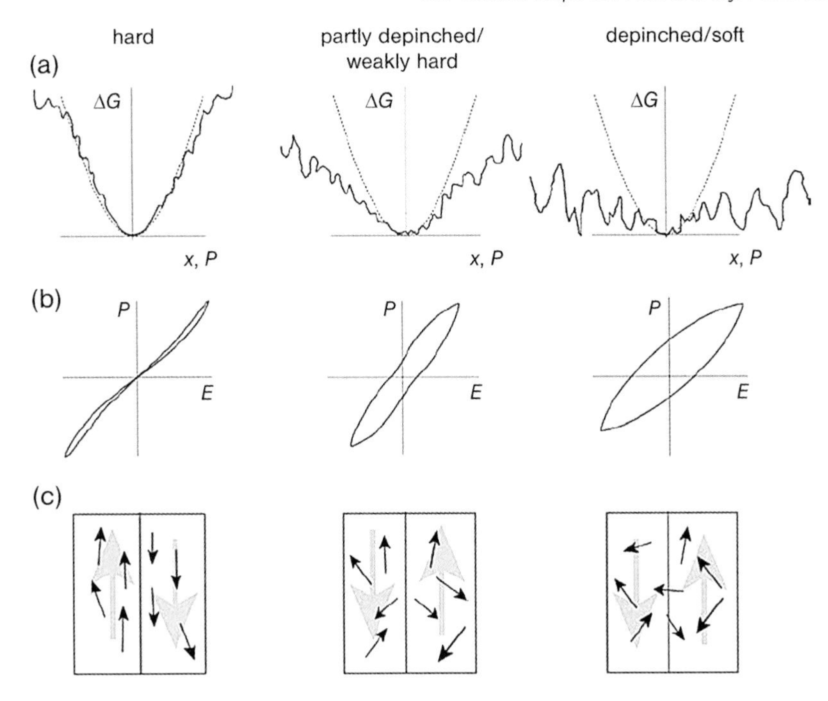

Figure 18.14 V-potential, with introduced roughness for a very hard, a partially de-pinched or a weakly hard ferroelectric material, and for a field-relaxed or soft sample. (a) Energy profile of a domain wall; (b) Polarization-field hysteresis; (c) Schematic example of possible mechanism for electrostatic ordering of charged defects: case of the "volume effect" [58].

is the case of a very hard ferroelectric experiencing strong domain wall clamping by an electrostatically ordered distribution of charged defects, according to any of the above-mentioned scenarios. A motion of the domain wall in such an energy landscape is purely reversible, and gives only anhysteretic contributions to the polarization response.

- **The "Soft" Limit**: A random energy landscape corresponding to a random distribution of pinning centers (disordered state) and attributed to a soft ferroelectric. This is the case where the Rayleigh model [81] can be applied. This model was originally proposed to explain the magnetization process in ferromagnets, and later a statistical theory was developed to justify this model for material response due to domain wall motion in a media with a random distribution of the pinning centers in ferromagnets [82, 83] and ferroelectrics [84]. According to the Rayleigh model, the domain wall contribution to the polarization response consists of two components – the linear term and a hysteretic term – depending on whether the field is ascending or descending. Hysteresis and nonlinearity are entirely linked to each other in the Rayleigh model.

All other possible varieties of material response are accounted for by intermediate states of the energy landscape, as shown in Figure 18.14. The hysteresis first appears when the initial perturbation in the electrostatic arrangement of charged defects occurs. Any further disordering of the pinning center distribution increases the randomness of the energy landscape, thus increasing the number of domain wall trapping sites and increasing the irreversible (hysteretic) contribution to the polarization response. This transitional process may last until the total randomization of pinning centers towards the stochastic energy landscape corresponding to the Rayleigh model is reached. A good correlation between experimental and calculated P–E dependences has been established qualitatively for a sample of Fe-doped PZT ceramic in the aged state, and quantitatively (using the Rayleigh formalism) for a de-aged state of the same sample in which the distribution of pinning centers had been disordered by thermal quenching (Figure 18.15). The quenching assures the freezing of the random distribution of pinning centers that exist in the paraelectric phase, where neither spontaneous polarization nor intrinsic fields occur in the material.

As mentioned above, a hard doped material can be softened in several different ways. Besides thermal quenching, the disorder in the distribution of pinning centers can be introduced by periodical cycling of the spontaneous polarization by an external field at switching amplitude strength (Figure 18.16). A clear dependence of the hysteresis loop constriction in a hard material on the cooling rate is shown in Figure 18.16a. However, the rectangular form of the hysteresis loop peculiar to a "soft" material has not been observed, even after fast quenching into warm water (quenching into cold water often causes destruction of the samples). Another well-known method to achieve hysteresis relaxation is to cycle a sample by a switching field [53]. The higher the temperature of the sample, the faster the relaxation occurs. The thermal activation of the oxygen vacancy motion facilitates the disordering effect caused by electric field cycling (Figure 18.16b and c). As shown in Figure 18.17, a softening treatment of PZT ceramics doped with 1% Fe leads to qualitatively identical response as of PZT doped with 1% Nb.

These experimental observations suggest that a smooth and gradual transformation of the polarization response with respect to the concentration and type of dopants occurs in the same way as being caused by various external influences that affect the degree of electrostatic arrangement of the charged defects. This allows the consideration of hardening and softening as a sole transitional process, that can be described in terms of a model linking the domain wall energy landscape with the degree of electrostatic arrangement of mobile charged defects [57].

It has been shown recently [85] that hardening by aging also occurs in hybrid-doped $BaTiO_3$ ceramics, even when the concentration of the soft dopant (Nb) is twice that of the hard dopant (Mn). This case is different from soft monodoped ceramics, where donor-type dopants serve to compensate naturally occurring acceptor centers originating from impurities in the raw materials, as well as from technological processes. It is possible that the presence of "hardening" acceptor centers in the crystal lattice favors the formation of oxygen vacancies in close vicinity, and the creation of complex defect associates. An analysis of charge transport processes in typical ferroelectric perovskites such as PZT and $BaTiO_3$ suggests that the oxygen vacancies are the only lattice defects that may have significant mobility in these materials [86].

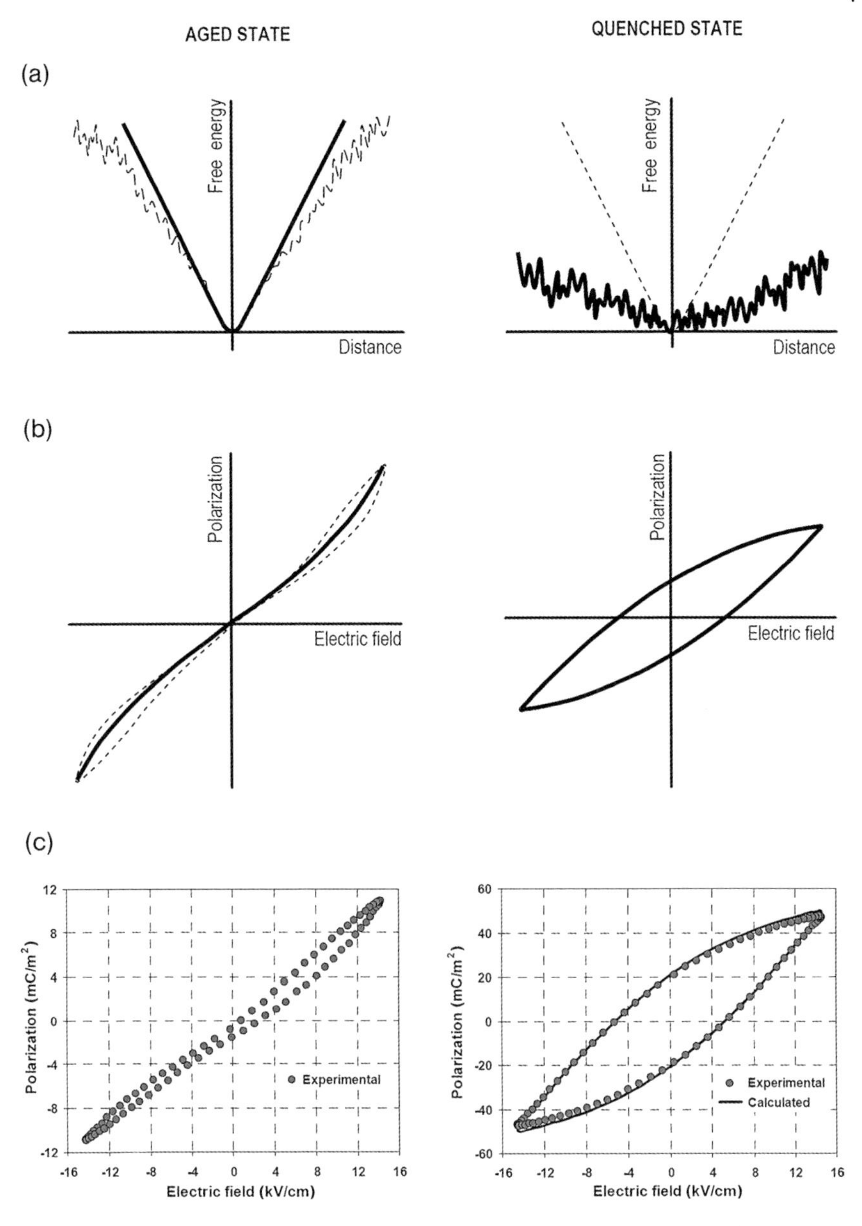

Figure 18.15 Application of the model of hardening–softening transitions to the case of an aged and quenched state of a ferroelectric system with alternating distribution of charged defects. (a) Domain wall free energy versus distance; (b) Calculated *P-E* dependences using a model of smeared V-potential (from Ref. [58]) and the Rayleigh relationships; (c) Experimental hysteresis loops obtained at 1 kHz driving field on $Pb(Zr_{0.58}Ti_{0.42})_{0.99}Fe_{0.01}O_3$ ceramics.

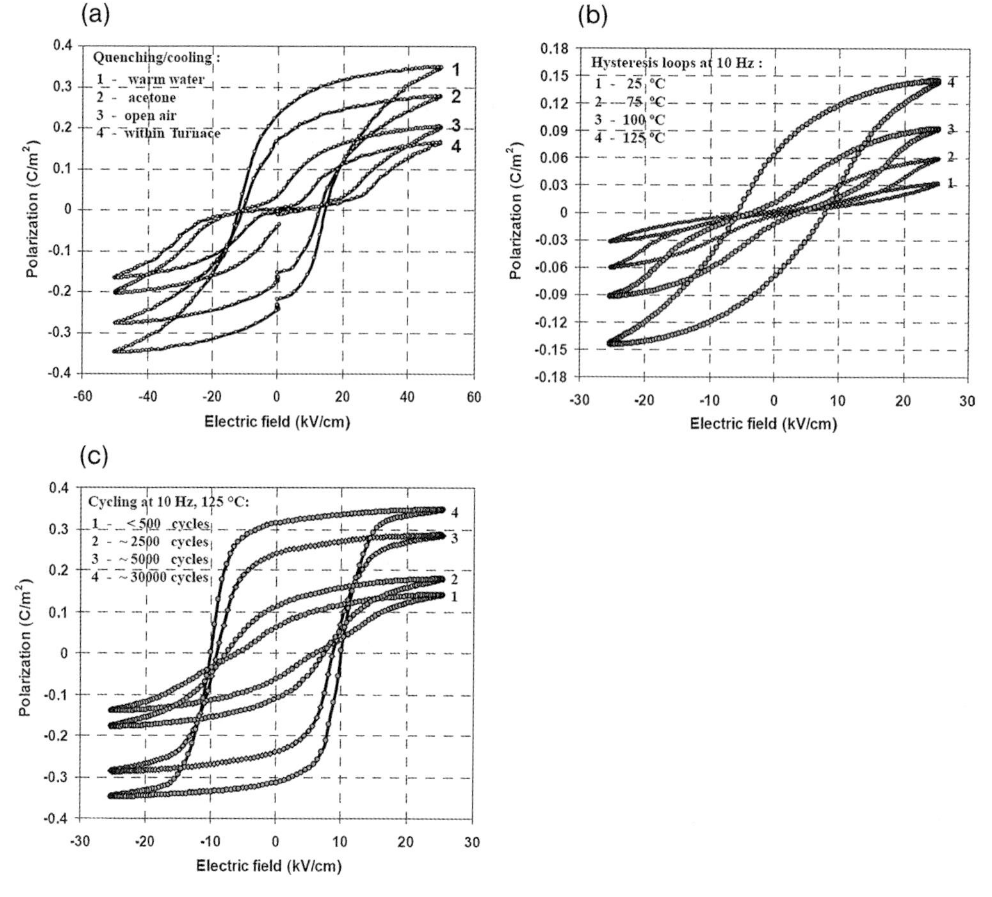

Figure 18.16 Relaxation of the ferroelectric hysteresis loops in hard (doped with 1.0 atom% Fe) PZT (58/42) ceramics. (a) For different cooling rates from the paraelectric state; (b) The thermal activation of hysteresis relaxation (characterized at 10 Hz ac-field); (c) Switching cycling by an ac-field (10 Hz) at 125 °C. From Ref. [289]

Nevertheless, softening cannot be entirely explained in the framework of the model considering randomization of the domain wall energy landscape as a result of ordering–disordering transitions in the distribution of pinning centers. Indeed, for the limiting case of an "ideal" soft material, such a model predicts a Rayleigh-type polarization response characterized by a linear dependence of the dielectric permittivity on the field amplitude, the slope of which is uniquely related to the hysteresis according to the Rayleigh relationships:

$$D = (\varepsilon_{\text{init}} + \alpha_\varepsilon E_0)E \mp \frac{\alpha_\varepsilon}{2}(E_0^2 - E^2)$$

$$\varepsilon^{\text{eff}} = (\varepsilon_{\text{init}} + \alpha_\varepsilon E_0),$$

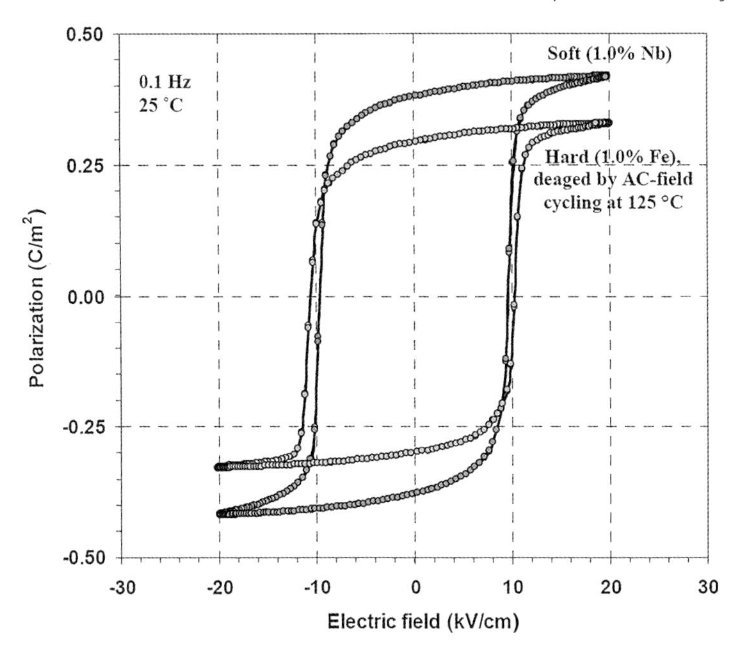

Figure 18.17 Comparison of ferroelectric hysteresis loops. Soft (1.0 atom% Nb-doped) PZT (58/42) ceramics and hard (1.0 atom% Fe-doped) PZT (58/42) ceramics relaxed by ac-field cycling at 125 °C. The measurements are carried out under the same conditions.

where D is the total dielectric displacement, the $\mp$sign represents the ascending/descending branches of hysteresis, $\varepsilon_{\text{init}}$ is a field-independent dielectric permittivity related to the intrinsic contribution to the linear dielectric displacement, E_0 is the amplitude of the electric field E, ε^{eff} is the effective field-dependent dielectric permittivity of a nonlinear dielectric material, and α_ε is the Rayleigh coefficient linking dielectric nonlinearity and hysteresis. It is possible to show that the above-mentioned model predicts either the linear (for the soft limiting case) or sublinear dependence of the effective dielectric permittivity on electric field amplitude. Besides, the area of the Rayleigh-type polarization hysteresis loop is a top limit in this model. The experimental results, however, demonstrate a slightly stronger than linear dependence of dielectric permittivity and a more developed hysteresis loop for soft compositions than that calculated using the Rayleigh formalism (Figure 18.18). This means that the experimentally observed response of a real soft material is softer, in practice, than the model limit can assume. Nevertheless, these deviations are relatively small and may result from other contributions that are not accounted for by the present model.

It should be noted that only the A- or B-site donor dopants cause softening of ABO_3 ferroelectric materials. As shown in Ref. [87], the fluorine doping of PZT leads to a hardening of the dielectric and electromechanical response. This effect can be explained by ordering of point defects or their associates involving the charged

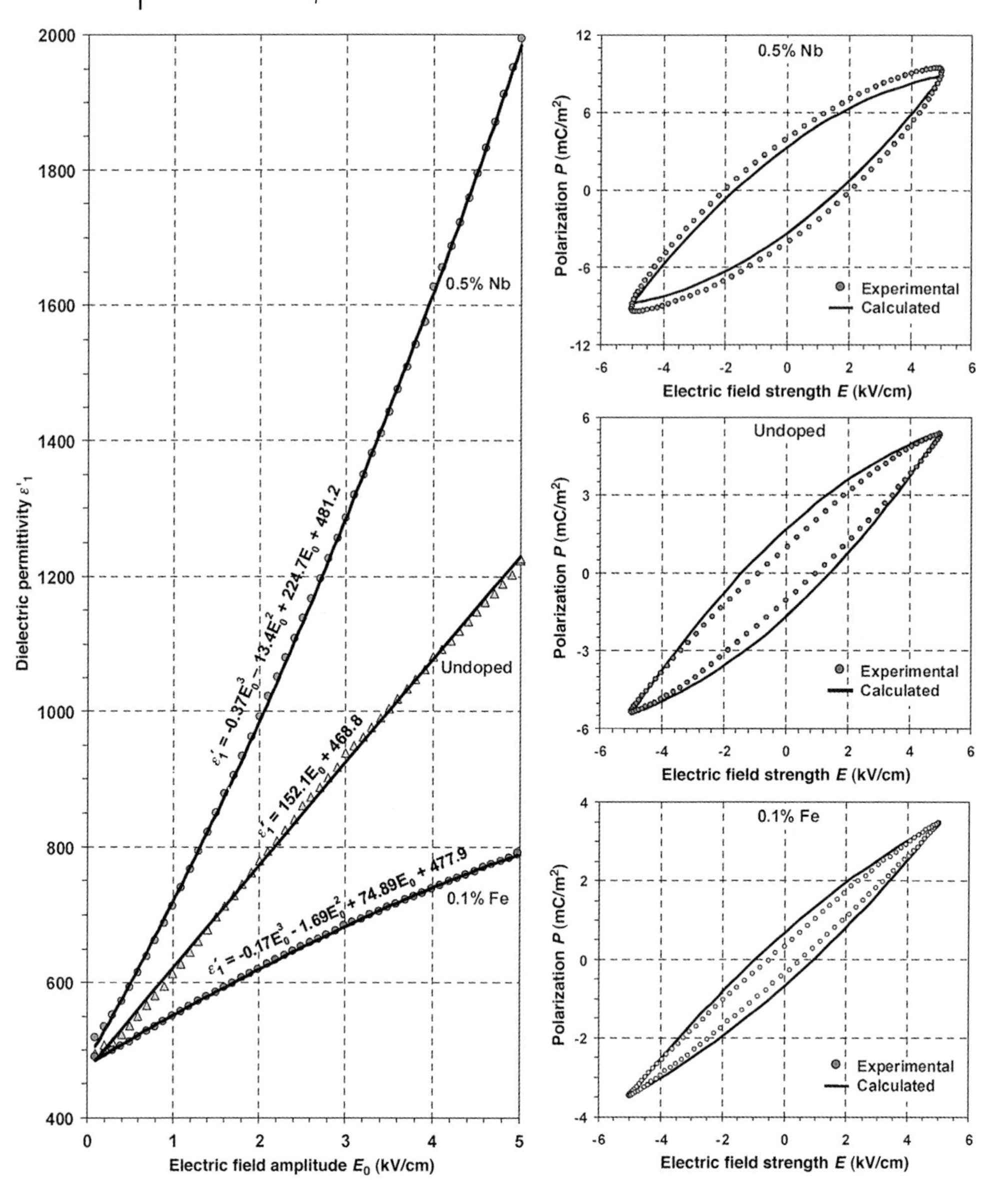

Figure 18.18 Real part of the dielectric permittivity as a function of electric field amplitude (left) and corresponding P–E hysteresis loops (right). The dots correspond to experimental data; the solid lines are calculated curves using Rayleigh's formalism modified with near-linear polynomial interpolations of ε^{eff} and the Rayleigh coefficient α^*. The dopant concentrations for the hard and soft $Pb(Zr_{0.58}Ti_{0.42})O_3$ ceramics are indicated in atom%. All measurements were performed at 1 kHz.

fluorine at the oxygen site. The perovskite ferroelectrics (co)doped with fluorine may have interesting properties, as they represent a rare case of hard ferroelectric material with *n*-type conductivity caused by the donor doping effect [13].

18.4 Textured Ferroelectric Materials

Texturing is a well-known method used to control the functionality of polycrystalline materials by attaining a nonrandom distribution of grain orientations, thus enhancing the resultant anisotropy of their properties. This method is widely used in modern ceramic technology including texturing ferroelectric ceramics for piezoelectric applications. Texture engineering is concerned with two important aspects that will be considered here: (i) texture development by control of the microstructure; and (ii) the evaluation of texture using appropriate methods and tools. When considering the dielectric or piezoelectric response of polycrystalline materials, there are two contributions of texture that affect the properties of ferroelectric ceramics: (i) the crystallographic texture formed by the crystallographic orientation of grains; and (ii) the domain texture formed by the distribution of spontaneous polarization within the grains. The domain texture can be introduced by poling ferroelectric ceramics. Usually, the ferroelastic (non-180°) domains restrict the attainable domain texture due to switching strains when poling. Furthermore, for poling along the direction of crystallographic texture, the attainable domain texture may also depend on the chosen direction [88, 89]. The crystallographic texture can be developed during the ceramic processing cycle, with modern texture engineering employing a variety of preparation methods for texturing piezoelectric ceramics that can be allocated to four groups: oriented consolidation of anisometric particles (OCAP); templated grain growth (TGG); heterotemplated grain growth (HTGG); and reactive templated grain growth (RTGG) [50].

18.4.1 OCAP

The OCAP technique deals only with naturally anisometric particles of plate-like or needle-like shapes that are typical for compounds with a pronounced anisometric unit cell, such as bismuth-based layered perovskites and tungsten bronzes. Technologically, the shape anisotropy of sintered particles can be controlled by the presence of fused salts during synthesis [90–92]. The anisometric particles can be consolidated using various mechanical processes, such as hot-forging [93] or tape-casting [94]. Hot-forging was the earliest method used to develop a crystallographic texture in bismuth titanate ceramics, in which the plate-like grains become oriented with (001) planes normal to the uniaxial force applied to the ceramic at or near the sintering temperature. For this method, the intergranular sliding is the main mechanism of texture development. Tape-casting and extrusion are used for particle alignment in a green compact, due to the shear stresses applied during forming. In the

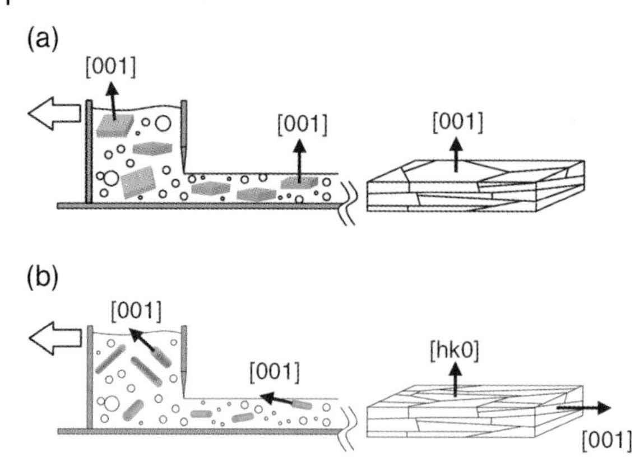

Figure 18.19 Schematic of the tape-casting process and the resulting texture in the sintered ceramic for (a) 001-oriented, platelet-shaped $Bi_4Ti_3O_{12}$ templates, and (b) 001-oriented, needle-shaped $PbNb_2O_6$ templates. From Ref. [89].

tape-casting process (shown schematically in Figure 18.19), a film of ceramic slurry is spread over a flat surface with the desired thickness being controlled by a doctor blade located above the moving carrier surface. When the solvent contained in the slurry evaporates, this results in the formation of a ceramic green sheet that can be stripped off from the supporting surface. The process optimization for increasing texture in the tape-casting system requires accounts to be made for the particle's shape and concentration, as well as the slurry viscosity and velocity [95–97].

18.4.2
TGG

The texture development of equiaxed particles can be achieved using the TGG method, in which a minority of larger anisometric templated particles is dispersed in a matrix of relatively finer and equiaxed particles. The template particles in the mixture are aligned using the above-mentioned extrusion or tape-casting methods; the tape-cast film is then thermally treated to initiate the densification process. The microstructure evolves with heating as shown schematically in Figure 18.20. When possible, template and matrix materials of the same composition are used, but

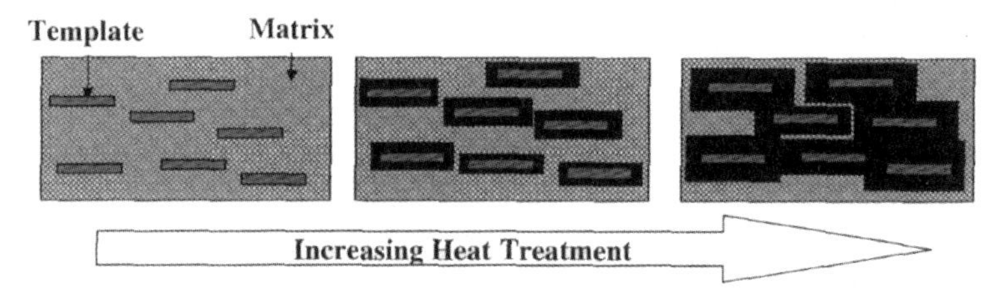

Figure 18.20 Schematic of the microstructure evolution with heating for the templated grain growth process. From Ref. [49].

when the template particles and matrix material are of different composition two types of TGG are possible: (i) heterotemplated grain growth; and (ii) reactive grain growth, in which the chemical reaction between the template and the matrix material is used with the purpose of obtaining a textured material of a new phase.

18.4.3 HTGG

For the HTGG process the templates represent an essential component, because they act as a substrate for epitaxy and as a seed for exaggerated grain growth [49]; consequently, a number of requirements must be fulfilled by the templates. As the templating phase is supposed to nucleate and grow from the oriented template particle, the latter must possess a similar crystal structure and a low mismatch with the desired phase to be templated. In addition, the template particle must be thermodynamically stable with the environment in order to avoid any chemical reaction or dissolution into the matrix material before the stable and oriented nuclei form and grow. Single crystals and thin-film-coated substrates of many compositions have been proposed to template the growth of textured Pb-based ferroelectric thin films such as PMN, PT, PMN-PT, PZT. These can be grown on MgO, $SrTiO_3$, $MgAl_2O_4$, $LaAlO_3$, $LaNiO_3$, $YBa_2Cu_3O_7$, (La,Sr)CoO_3, $SrRuO_3$, and Pt [49]. In some cases, an ultra-thin TiO_2 film is used as a very efficient seed layer for the nucleation of PZT(111). It has been shown that the growth conditions leading $PbTiO_3$ (100) on bare platinum are changed to a (111) orientation, if a TiO_2 seed layer is used [98].

18.4.4 RTGG

The RTGG process was first proposed for texturing ferroelectric materials with a perovskite-like structure by Tani [99]. It subsequently became a key processing technique for textured lead-free piezoelectric ceramics with properties comparable to typical actuator-grade PZT [100]. In this method, the precursor anisometric particles of a material with a simpler composition and an easier fabrication route than the target compound are used as a starting material. This is then aligned and converted by chemical reaction into the target material while preserving the crystallographic orientation of the template. The process scheme is as follows: First, the reactive templates are mixed with the complementary reactants in the slurry; after which, the slurry is tape-cast in order to align the templates. A thermal treatment is applied in the next steps to activate a topotactic chemical conversion, followed by epitaxial grain growth and densification. Bismuth-based layered perovskites are often used as a precursor material as these are available in anisometric shape. For example, plate-like $Bi_4Ti_3O_{12}$ particles can be used as a reactive template to form a textured perovskite $Bi_{0.5}Na_{0.5}TiO_3$ as a target material [99]. If necessary, several chemical reactions can also be involved in the process. In order to obtain anisometric templates for the alkaline niobate-based perovskite materials, a special technique to synthesize

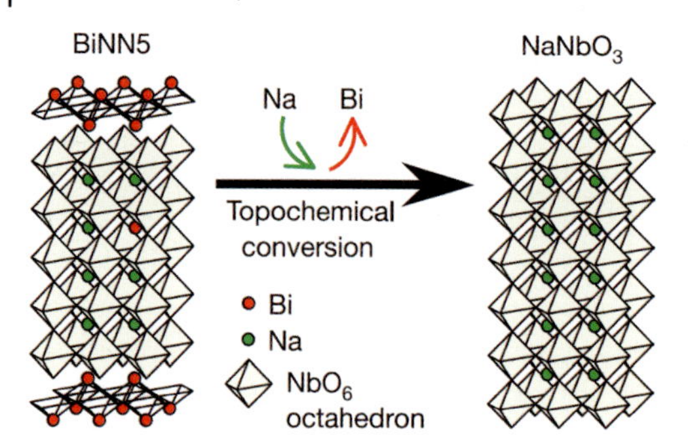

Figure 18.21 Schematic diagram of topochemical conversion from bismuth layer-structured $Bi_{2.5}Na_{3.5}Nb_5O_{18}$ particles to plate-like $NaNbO_3$ particles. From Ref. [100].

plate-like $NaNbO_3$ particles has recently been proposed by Saito *et al.* [100]. These authors synthesized $Bi_{2.5}Na_{3.5}Nb5O_{18}$ (BiNN5) platelets, initially by a molten salt method; in a second molten salt reaction, the BiNN5 platelets were then converted into NaNbO3 platelets to be used as a reactive template for textured $K_{0.5}Na_{0.5}NbO_3$ ceramics. The topochemical reaction, in which $NaNbO_3$ is formed from BiNN5 by ion-exchange of Na for Bi, preserves the particle morphology with a developed (001) plane (Figure 18.21).

At present, the technologies for texturing ferroelectric materials are undergoing a rapid development, with the latest achievements in developing texture, as well as the application issues of textured materials having been the subjects of many recent reviews [49, 50, 89, 101].

The evaluation of texture is an important issue for the engineering of ferroelectric ceramics. Both, domain and crystallographic texture can be evaluated qualitatively or quantitatively using either microscopy or neutron and X-ray diffraction (XRD). The Lotgering factor, *f*, a commonly used assessment of texture using XRD analysis, relates the relative intensities of (00*l*) reflections to all observed reflections in a coupled θ–2θ powder XRD spectrum [102]:

$$f = \frac{p - p_0}{1 - p_0}; \qquad p = \frac{\sum I_{00l}}{\sum I_{00l} + \sum I_{non-00l}}$$

where I_{hkl} is the integrated intensity of the (*hkl*) reflection, and p_0 is the *p*-value of a sample with random crystallographic orientation. In the case where a sample with truly random crystallographic orientation is not attainable, a simplified formula that compares intensity of the (00*l*) reflection to all or several reflections can be used. For example, the following parameter is proposed for $Bi_4Ti_3O_{12}$ ceramics [103]:

$$f' = \frac{I_{006}}{I_{006} + I_{119}}.$$

Although the Lotgering factor is often used for texture estimation due to its simplicity, it can be very imprecise as a qualitative assessment of texture since, in some cases, it depends heavily on the reflections included in the calculations [104]. A more careful analysis of texture can be provided using an orientation distribution function (ODF) which quantifies, in units termed multiples of random distribution (mrd), the probability of given crystalline orientations with respect to the reference sample coordinate axes [105]. For example, the ODF density of 2 mrd describes a crystalline orientation that is twice as probable in textured material as in a random material. The 3-D ODF is a probability density function $f(\varphi_1,\Phi,\varphi_2)$, where the orientations φ_1, Φ, and φ_2 are the angles describing a coordinate transformation [88, 106]. The ODF is difficult to represent linearly in two dimensions, and "pole figures" are typically used; these are integrations of the ODF in two dimensions, and represent either the densities of a single crystallographic direction in sample orientation directions, or the densities of all crystallographic directions in a single sample direction (the "inverse pole figure"). Inverse pole figures are usually plotted on a stereographic projection. The ODF can be calculated using experimental data obtained with various techniques of pole figure measurements, such as an X-ray area detector diffractometry, synchrotron XRD, and time-of-flight (TOF) neutron diffraction [107]. The results of quantifying texture in ferroelectric ceramics by using the ODF or the Lotgering factor indicate an advantage in quantitative adequacy of the former over the latter [104, 107].

Another commonly used model equation to fit and quantify the measured texture distribution is the March–Dollax function [49, 108–110]:

$$F(\nu,r,\theta) = \nu\left(r^2\cos^2\theta + \frac{\sin^2\theta}{r}\right)^{-\frac{3}{2}} + (1-\nu),$$

where θ is the angle between the texture (orientation) axis and the scattering vector, ν corresponds to the volume fraction of oriented material, and r is a texture factor that characterizes the width of the orientation distribution. For a random sample $r = 1$, while for a perfectly textured sample $r = 0$. In this representation of the March– Dollase function, the probability distribution of the orientation of the textured axis in the polycrystalline material is quantified with the use of two fitting parameters, r and ν.

The domain texture can be evaluated analyzing the intensity interchanges in symmetry-dependent reflections of XRD spectra. Jones *et al.* [88, 111, 112] proposed the following formulae to determine the degree of domain preference:

- $f_{001}(\text{mrd}) = 3\frac{I_{00h}/I^R_{00h}}{I_{00h}/I^R_{00h} + 2I_{h00}/I^R_{h00}}$ – the 001 pole density for tetragonal symmetries; and
- $f_{100}(\text{mrd}) = 2\frac{I_{h00}/I^R_{h00}}{I_{h00}/I^R_{h00} + 2I_{0h0}/I^R_{0h0}}$ – the 100 pole density orthorhombic symmetries, with two possible orthogonal ferroelastic variants such as $Bi_4Ti_3O_{12}$ or $PbNb_2O_6$.

18.5 Ferroelectricity and Magnetism

Ferroelectrics may find application in magnetoelectric devices, either in combination with magnetic materials or intrinsically as a multiferroic material. The latter represents a rare class of materials that are both ferroelectric and ferromagnetic, and often also ferroelastic, in the same phase [113]. This means that, simultaneously, they can undergo spontaneous polarization, magnetization, and deformation that can be reoriented by the respective application of an electric or magnetic field, or by mechanical stress. The magnetization and polarization effects are both used extensively in data storage applications and a possible coupling between the two may permit data to be written electrically and read magnetically; potentially, this represents an exceptional future storage medium. Interest in materials that demonstrate a correlation between the electric and magnetic properties was triggered when the effect of an electric-field-induced magnetization and a magnetic-field-induced polarization in antiferromagnetic Cr_2O_3 was first predicted [114] and subsequently observed [115–118] during the early 1960s. As a consequence, during the past few decades many magnetoelectric materials have undergone extensive investigation, despite there being a limited number of appropriate compounds and the weakness of their magnetoelectric effect (ME) often hampering possible applications. Nonetheless, a recent revival of interest in the ME was observed [119] following the successful investigation and development of new materials capable of overcoming the inherent weakness of the microscopic ME. Two major sources for large MEs led to this revival: (i) the development of composite materials yielding a product property of a magnetostrictive and a piezoelectric compound with a well-pronounced mechanical coupling; and (ii) the use of multiferroic materials that possessed a giant ME due to the presence of multiple long-range orderings that could enhance the internal magnetic or electric fields. To date, although a high magnetoelectric response in multiferroics has been achieved only at very low temperatures, the intense pursuit of convenient multiferroics with high magnetoelectric properties at room temperature is ongoing. The rapid expansion of research into magnetoelectric phenomena during the past decade has been reflected by the surge in published reports on the topic [119, 120], especially in the field of multiferroics where the number has grown exponentially (Figure 18.22). An understanding of the ME in multiferroics, as well as of

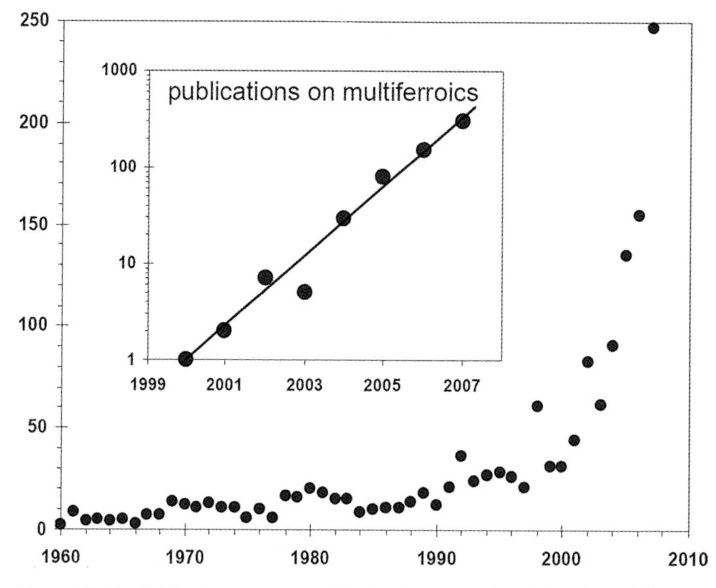

Figure 18.22 Publications per year with the "magnetoelectric" and "multiferroic" (inset) as keywords. According to the Web of Science [119, 120].

progress in the development of new materials and their applications, have formed the focal points of recent reviews summarizing the state of art in this field [119, 121–125].

In composite materials, the ME can be achieved as a result of the elastic interaction between piezoelectric or electrostrictive and piezomagnetic or magnetostrictive phases which, independently, may have no ME. Thus, as a product property of the composite, the ME can be described as [126]:

$$\mathrm{ME} = \frac{\text{electrical}}{\text{mechanical}} \cdot \frac{\text{mechanical}}{\text{magnetical}}$$

Similarly, the converse effect or a coupled ME by thermal interaction in a pyroelectric–pyromagnetic composite can be obtained. A thorough theoretical investigation of the ME in composites and polymers was presented by Nan *et al.* [126–128], where a clear consideration of all six different variables describing the electrical, magnetic, and mechanical properties in composite materials was treated in terms of rigorous theoretical modeling, based on Green's function technique and perturbation theory. The numerical results of this analysis forecast a possible giant ME in composite materials and polymers [127, 128].

In practice, the ME in composites is usually characterized for a weak (up to 10 Oe) ac magnetic field (H) applied in the presence of a large (up to 10 kOe) dc bias, with the ac field frequency range of 0.1 to 1000 kHz [119]. It should be noted that, as a recent trend, new magnetoelectric composites have demonstrated a pronounced effect at a lower dc bias of a few Oersteds, or even without any dc bias. The voltage induced by the ac electric field (E) is proportional to the ac field amplitude. The magnetoelectric voltage coefficient $\mathrm{d}E/\mathrm{d}H$ is typically specified in units of mV $(\mathrm{cm} \cdot \mathrm{Oe})^{-1}$, although a more relevant parameter to characterize the ME would be the magnetoelectric susceptibility $\delta M/\delta E$ or $\delta P/\delta H$, which has been shown to be a fundamental parameter of a complex quantity [129]. All magnetoelectric coupling parameters are usually denoted in the literature as α or α_{ME}; in some sources, the magnetoelectric voltage coefficient is denoted as V_{ME}.

The magnetoelectric response depends on the magnetic and dielectric permeabilities of the constituents, as well as on the stoichiometry, microstructure, and the interphase interface. A possible intimate contact between the electromechanical and magnetomechanical phases in composites can be organized in various forms, such as grain mixtures, laminates, sandwiched multilayer structures, or nanopillars embedded on a substrate [130–138], as shown schematically in Figure 18.23. Due mainly to the low resistivity of the magnetic phase, and chemical reactions occurring between phases during sintering, the magnetoelectric product property of the grain mixture composites is generally low. In laminated structures, this problem may be overcome by the macroscopic separation of the piezomagnetic and piezoelectric constituents. As shown in Figure 18.24, a comparison of the two best results achieved with a granular composite and with a PZT sandwiched between Terfenol-D ($Tb_{1-x}Dy_xFe_{2-y}$), revealed a difference in the magnetoelectric voltage coefficient of one order of magnitude [130, 131].

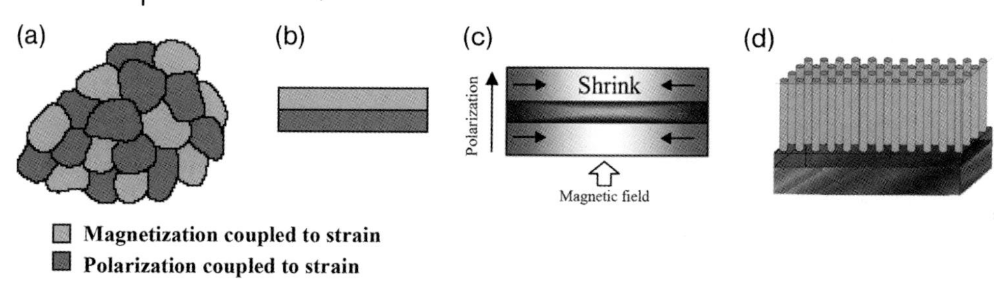

Figure 18.23 Mechanically coupled magnetoelectric composites. (a) Grain mixture; (b) Thin-film heterostructure (from Ref. [123]); (c) Sandwich structure (from Ref. [130]); (d) Spinel ferro/ferrimagnetic nanopillars embedded on a perovskite ferroelectric matrix (from Ref. [137].)

The continuous search for composites with a higher magnetoelectric coupling has resulted in the identification of a best magnetoactive and electroactive material, in combination with an optimal design in the form of laminated structures. The general trend of the search has been towards increasing the magnetoelectric coefficients and decreasing the dc magnetic bias over a wide range of frequencies. A dramatic increase in magnetoelectric coefficient can be achieved at the resonant frequency. Depending on the material and the sample geometry used, the magnetoelectric resonance occurs in the range of tens or hundreds of kHz [132–136], coinciding with the electromechanical resonance. The effect of ferromagnetic resonance on magnetoelectric interactions in the microwave frequency range has also been reported [139]. Recently, a very high ME was measured in a composite of a piezofiber layer laminated between two magnetostrictive FeBSiC alloy layers of a high magnetic permeability [132]. As reported, the magnetoelectric voltage coefficient in this composite was up to 22 V $(cm \cdot Oe)^{-1}$ at nonresonant frequencies, albeit under a small biases of 5 Oe that was approaching predicted theoretical limits; in contrast, at the resonant frequency this value was strongly enhanced up to 500 V $(cm\,Oe)^{-1}$ (Figure 18.25). Another very recent report [133] described the magnetoelectric effect in three-phase composites of magnet-metal-cap-piezoceramic, which require no external dc bias magnetic field to enable an obvious magnetoelectric voltage coefficient.

Recently, rapidly developing nanotechnologies have made a challenging impact on the area of ferroelectric random access memory (FeRAM), the memory density of which is currently approaching 1 Terabit per inch [4]. The application of a ME in data storage devices is very attractive, if it could be used to exploit the best of FeRAM and magnetic data storage, and possibly avoid such undesirable effects as fatigue, a destructive data read, and a need for a strong local magnetic field when data writing [4, 123, 140]. Magnetoelectric coupling in a nanostructured composite has recently been demonstrated [137, 138]. These composites have a structure as shown schematically in Figure 18.23d. A sufficiently high magnetoelectric susceptibility (such as $10^{-2}\,G \cdot cm\,V^{-1}$) and a magnetization reversal, that can be controlled by an electric field and observed in the $BiFeO_3$–$CoFe_2O_4$ nanostructured composite at room temperature [138], proved to be key properties for a possible magnetoelectric memory application.

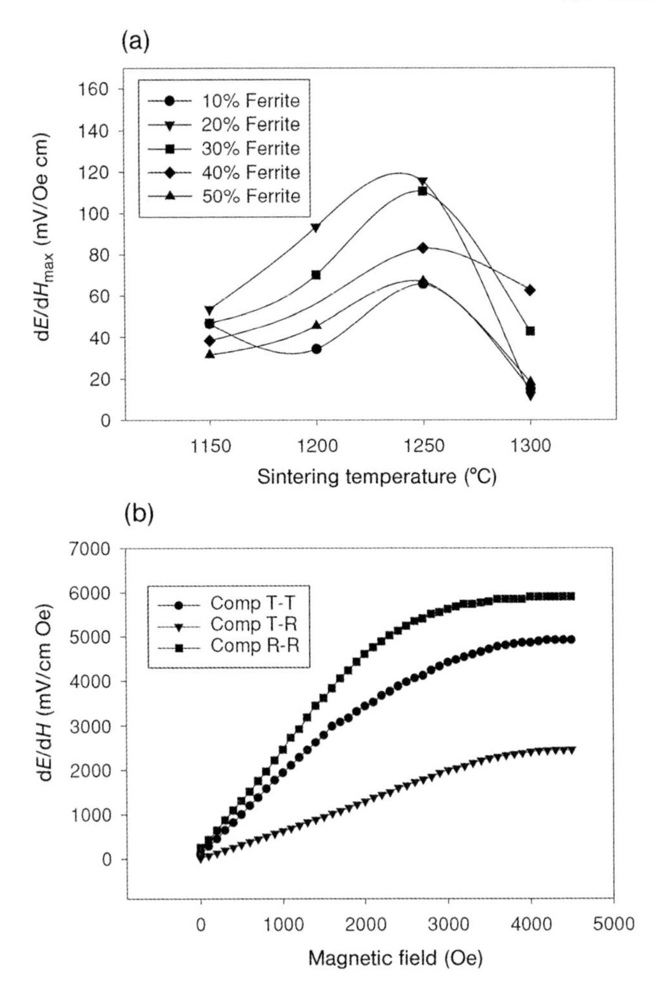

Figure 18.24 Comparison of the magnetoelectric effect in granular and laminated composites. (a) maximum ME voltage coefficient of the PZT and Ni-ferrite particulate composites as functions of sintering temperature and Ni-ferrite particle content; (b) ME voltage coefficient as a function of applied DC magnetic bias for PZT sandwiched between two Terfenol-D disks with different direction of the magnetostriction (T and R denote correspondingly the thickness and the radial directions of the magnetostriction in the Terfenol-D disks). From Ref. [131].

The major challenge to identify a sufficiently strong ME in a single-phase material has given rise to intensive investigations of multiferroics (Figure 18.22, inset). Nevertheless, the number of known prospective multiferroics is not large, most likely because of the mutually exclusive requirements that magnetism and ferroelectricity place on the material. This issue has been explicitly investigated by Hill ("Why are there so few magnetic ferroelectrics?" [125]), and touched upon in other topical reviews dedicated to multiferroics [121–124]. Basically, among a large group of

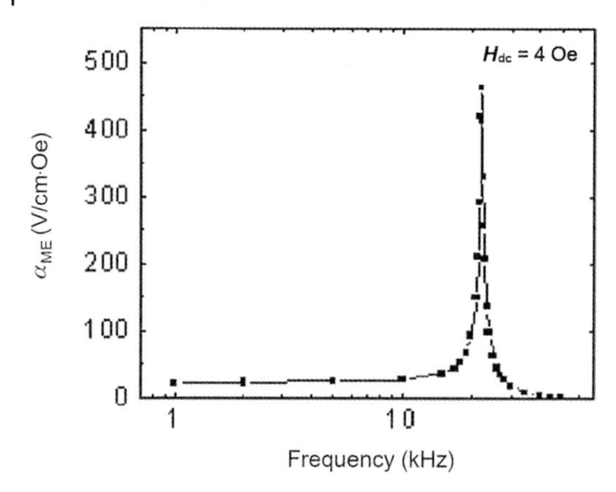

Figure 18.25 The magnetoelectric voltage coefficient as a function of frequency, illustrating a strong enhancement at the electromechanical resonance frequency for FeBSiC/piezofiber laminates. (From Ref. [132].)

materials the contraindications between ferroelectricity and magnetism arise from the electron configuration of the transition metal ions that are mainly involved in both effects. Most of the ferroelectric perovskites ABO_3 contain off-centered B-site cations in a d^0 state (such as Ti^{4+}, Zr^{4+}, Nb^{5+}, Ta^{5+}, W^{6+}), which is plausibly essential for ferroelectricity [11], whereas for magnetic perovskites only a partial occupancy of the d-orbital is required to establish the magnetic order. Another important issue here is the insulating properties of materials. Strong ferromagnets are often metals with a high density of states at the Fermi level, and thus are conductive by nature. However, most ferrimagnets – as well as some weak ferromagnets – are insulators, and so only relatively weak magnetic properties can be revealed by ferroelectrics.

Among perovskites there are two examples that combine ferroelectric and magnetic properties in one phase, namely $BiMnO_3$ and $BiFeO_3$. Although, both of these materials contain a non-empty d-shell of the B-site cation, it is noteworthy that the ferroelectric ordering occurs in these perovskites due rather to the A-site than to the B-site cation off-shift [141].

Aside from perovskites, many other structural types exist where multiferroism can be found. An explicit systematic classification of symmetries and related ferroic properties has been presented by Aizu [142], although among real materials a strong coupling between ferroic properties is generally rare [113, 125, 143]. At present, the most extensively studied multiferroic systems have included hexagonal manganites $RMnO_3$ (R = Sc, Y, In, Ho, Er, Tm, Yb, Lu), boracites MB_7O_{13} (M = Cr, Mn, Fe, Co, Ni, Cu; X = Cl, Br, I), compounds with the general formula $BaMF_4$ (M = Mg, Mn, Fe, Co, Ni, Zn), and many others [119].

Multiferroism in the recently discovered "frustrated" magnets, such as $RMnO_3$ (with perovskite structure; R = Gd, Tb) [144, 145], RMn_2O_5 (R = Tb, Y) [146], $Ni_3V_2O_8$ [147], hubnerite $MnWO_4$ [148], delafossite $CuFeO_2$ [149], and chromite spinels MCr_2O_4 (M = Mn, Fe, Co) [150], now attract special attention as their

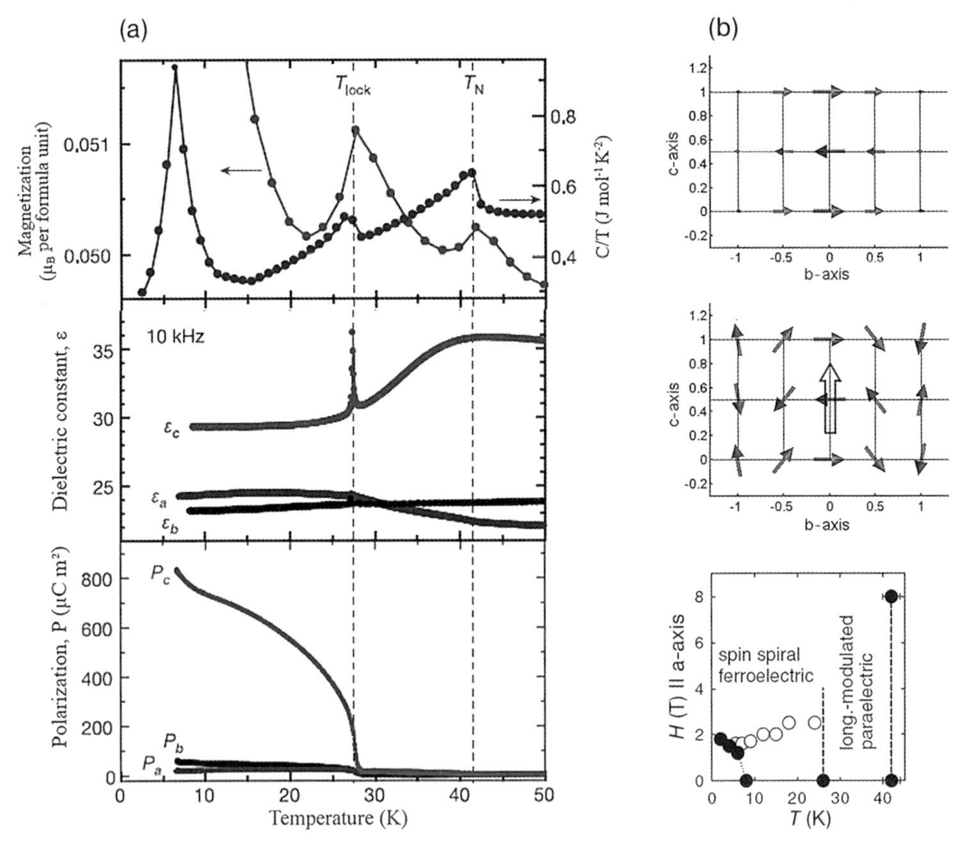

Figure 18.26 (a) Magnetic and dielectric properties of $TbMnO_3$ in the low-temperature region (0–50 K) (from Ref. [144]); (b) Schematic of the magnetic structure at the temperatures $T = 35$ and $T = 15$ K, projected onto the b–c plane. The filled arrows indicate the direction and magnitude of Mn moments, the unfilled arrow indicates the electric polarization. The phase diagram shows a series of phase transitions indicated by solid circles. From Ref. [152].

ferroelectricity appears only in a certain phase with a special (e.g., spiral) magnetic order [121, 124]. The electric dipoles induced by magnetic ordering represent a source of high dielectric tunability by the applied magnetic field, which makes this group of frustrated magnets prospective for useful multiferroics. An example of such a spin frustration effect in $TbMnO_3$ is shown in Figure 18.26, where the sequence of anomalies in the temperature dependence of magnetic and dielectric properties, as well as of the specific heat divided by temperature (Figure 18.26a), indicates phase transitions occurring in the material at 7 K, 27 K, and 41 K [144]. Below the Néel temperature ($T_N = 41$ K), the Mn^{3+} spin develops a long-range order, while the Tb^{3+} moment ordering occurs below $T = 7$ K [151]. The anomaly at 27 K is related to the transition between the paraelectric, magnetically incommensurate phase with sinusoidally modulated collinear magnetic order and the ferroelectric phase, with noncollinear incommensurate magnetic order [152]. The low-temperature fragment of the phase diagram and a schematic of magnetic structures above and below $T = 27$ K are shown in Figure 18.26b. The longitudinally modulated phase respects inversion symmetry along the c-axis, but the spiral phase violates it, thus allowing for an electric polarization.

An inhomogeneous magnetic order in which the magnetization varies over the crystal is often a crucial point for developing the ME because, by symmetry, it allows for the third-order coupling between polarization and magnetization [121]. The spatial variation of magnetization in a frustrated system can be of a different modulation, and often is incommensurate with the crystal lattice period. However, the type of magnetic order is important for establishing the symmetries that govern coupling between electric polarization and magnetization. For example, when the magnetic spiral order sets in, it conforms to two necessary conditions: (i) it spontaneously breaks the time-reversal symmetry; and (ii) it breaks the spatial inversion symmetry: there may be either a left-turning or a right-turning spiral. Thus, the symmetry of the spiral state allows for the simultaneous presence of electric polarization, the sign of which is coupled to the direction of spin rotation [121]. A more accurate consideration shows that ferroelectricity can be induced only if the spin rotation axis does not coincide with the wave vector of a spiral [153]. The detailed microscopic mechanism of inducing ferroelectricity by magnetic ordering is not yet understood. Plausible concepts consider this effect in connection with the spin supercurrent in noncollinear magnets [154], the Dzyaloshinskii–Moria interaction that helps to stabilize spiral magnetic structure at low temperature [155], the magnetic Jahn–Teller effect [156], or magnetoelastic coupling [157, 158], which allows an explanation of the appearance of ferroelectricity without need to invoke a spiral magnetic structure as an essential feature.

The magnetoelectric properties of multiferroics with magnetically induced polarization make these materials potential prospects for a variety of applications, and also as a partial substitute for composites. The electric polarization appearing with a spiral magnetic order, and the tunability of dielectric permittivity by magnetic field in $TbMnO_3$, have been well studied, but the magnitudes of these effects are relatively weak (Figures 18.26a and 18.28a). For comparison, the dielectric permittivity in $DyMnO_3$ is increased by up to 500% in an applied magnetic field, as shown in Figure 18.28b [159], and a highly reproducible 180° polarization reversal by magnetic field (Figure 18.27) has been observed in $TbMn_2O_5$. Nevertheless, the very low temperatures at which these interesting phenomena occur remain, to date, the main obstacles in the technological application of these multiferroics. There is a general trend that the magnetic-to-paramagnetic phase transition in frustrated magnets is shifted to lower temperatures when compared to the Curie–Weis temperatures, as obtained by the asymptotical extrapolation of high-temperature magnetic susceptibility [121]. A hexagonal ferrite $Ba_{0.5}Sr_{1.5}Zn_2Fe_{12}O_{22}$ demonstrated a magnetically induced polarization in an intermediate phase that persisted up to room temperature, although the low resistivity of the crystal hampered the dielectric characterization at temperatures above $\sim$130 K [160].

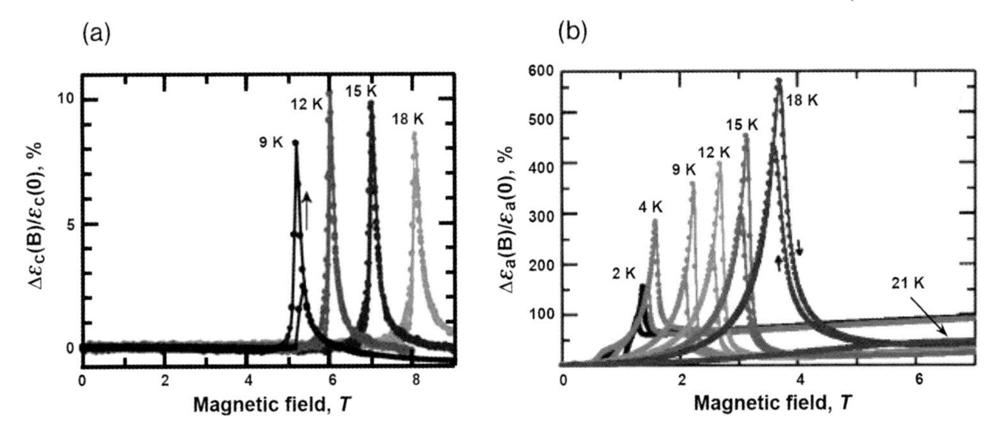

Figure 18.27 (a) Polarization reversal by magnetic fields in $TbMn_2O_5$ at the temperatures $T = 28$ and $T = 3$ K. The arrows indicate directions of applied magnetic field and the induced polarization; (b) Polarization flipping at 3 K by linearly varying the magnetic field from 0 to 2 T. From Ref. [146].

A remarkable combination of dielectric and magnetic properties, as well as their coupling at room temperature, has been discovered in thin films of $BiFeO_3$ [161]. This compound in the bulk form has long been known as an antiferromagnet with a Néel temperature $T_N = 643$ K [162], and simultaneously as a weak ferroelectric [163] with a Curie temperature $T_C > 1100$ K [143, 164] that is near to the melting point [13, 165]. The early investigations of dielectric properties in bulk and film forms of $BiFeO_3$ showed a large spread of values that were mostly very low [166]. Recently, however, large ferroelectric polarizations – exceeding those of the prototypical ferroelectrics $BaTiO_3$ and $PbTiO_3$ – have been reported in high-quality thin films of $BiFeO_3$ [161, 167, 168]. Moreover, the magnetic properties of the $BiFeO_3$ thin films were seen to differ from those of the bulk. As reported in [161], the films demonstrated an enhanced thickness-dependent magnetism and a strong magnetoelectric coupling (Figure 18.29).

In concluding this section, it is important to note that investigations of the correlation between ferroelectricity and magnetism represents a rapidly developing

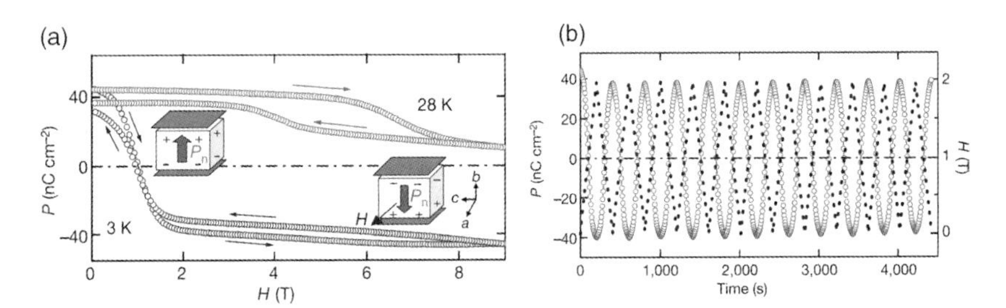

Figure 18.28 Effect of magnetocapacitance in (a) $TbMnO_3$ (from Ref. [144]) and (b) $DyMnO_3$ (from Ref. [159]) as a change in dielectric permittivity induced by magnetic field at various temperatures.

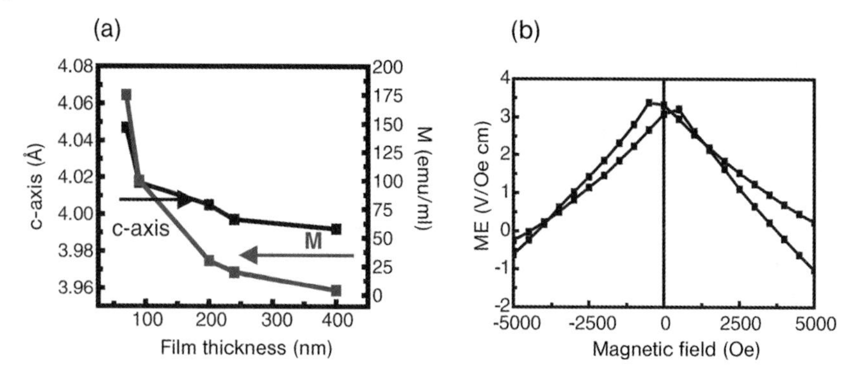

Figure 18.29 (a) Thickness dependence of saturated magnetization; (b) Magnetoelectric effect in a 70 nm thin film of $BiFeO_3$. From Ref. [161].

area of physics, chemistry, and materials science, with a promising capacity for the application engineering of a new generation of memory devices.

18.6 Fatigue in Ferroelectric Materials

One major impediment to the widespread use of ferroelectrics in certain applications has been their susceptibility to fatigue and, indeed, the very first attempts to produce ferroelectric memories have recognized this problem [169]. In fact, the problem relating to ferroelectric memory thin films has remained until recently, when layered ferroelectrics and certain electrode systems were developed [170, 171] although, under certain loading regimes these problems persist for actuator materials. Initially, this fatigue (which still remains a topic of controversy) was termed "dynamic aging," because the ferroelectric hysteresis was decreased in polarization amplitude and the dielectric response was diminished [169, 172]. In recent years, however, it has become common practice to use the term "aging" to mean shelf-aging, and "fatigue" to mean any material deterioration due to the cyclic application of electric fields, mechanical stresses, or both [170, 173–178]. A consistent overall terminology has still to be agreed on, however. Although, initially, it was considered that a detailed investigation of the fatigue would lead to the identification of a single microscopic mechanism responsible for the observed effects, it has by now become clear that several mechanisms are involved in the material modification. In thin films, the proximity of the electrodes and the partly very harsh environments encountered during processing each have different effects on the fatigue properties for the set of electrodes and materials chosen [170]. Moreover, breakdown is a significant issue [179], with many fatal effects being related to microscopic or macroscopic cracking in bulk ferroelectrics.

18.6.1
Macrocracking

One very straightforward effect that is mostly encountered in multilayer actuator devices is that of macroscopic cracking parallel to the electrode faces [180–183], which leads to the introduction of an air gap in series with the ferroelectric material [184]. Due to a continuity of the dielectric displacement [185], this air gap reduces the electric fields experienced by the ferroelectric material, and thus the device's external response. The cracking is due initially to a strain mismatch between the active and inactive volume elements in the multilayer design of the standard comb-like electrode structure [186]. For certain materials, this initial crack formation can be suppressed either by static mechanical preload externally applied to the entire actuator [175, 187], or by a moderate loading regime during poling. Prior texturing by transverse compression has also been attempted [188], with any subsequent crack growth being, again, a cyclic process [189–191]. Currently, it is clear that the fully nonlinear hysteretic material law must be considered to provide a satisfactory description of the crack opening forces exerted by the crack process zones ahead of the crack tip [186, 192, 193]. Likewise, time effects and creep are also encountered [194], and extended unloading times will reduce the initial stress intensity factor, K_{Ii} [195]. Initially, various theoretical approaches based on mechanical modeling assumed that the electric field would become zero at the crack face (an impermeable crack), although this assumption stemmed from the mechanical case where, for open cracks without crack bridging, no forces are conveyed across the crack [196–200]. As the dielectric displacement is continuous, however, such an assumption proved to be invalid and insufficient to describe the crack intensity factors in cracks that were additionally subject to electric loads [201, 202]. A description of the monotonic loading of nonlinear stress–strain relationships in ferroelectrics has become feasible, starting from different constitutive models and being applied to crack-processing zones [183, 186, 199, 201, 203]. Moreover, a good match of the experimental data has been achieved. Whilst cyclic crack propagation requires a knowledge of all stress and strain components in the plastic zone [189], these have only recently become experimentally accessible, as displayed by direct liquid crystal images (see Figure 18.30) [192, 193].

It has been shown experimentally that crack advance occurs in an electrically loaded double cantilever beam specimen only when the entire material reverses its polarization orientation [204]. Any previously used small-scale yielding models are, therefore, insufficient to provide an experimental explanation [205–207], because the major effects only occur when the *entire* material reverses polarity, and not just a small switching zone at the crack tip. In the case of a polycrystalline material, crack advance occurs during polarization reversal in each cycle [204], whereas in a single-crystal $BaTiO_3$ crack advance occurs as more of a "jump" at certain cycle numbers [191]. For purely electric cyclic loading, it was shown recently (on a theoretical basis) that crack propagation depends critically on the ratio of the coercive field to the field of full ferroelectric saturation, again under a small-scale assumption [208]. The data

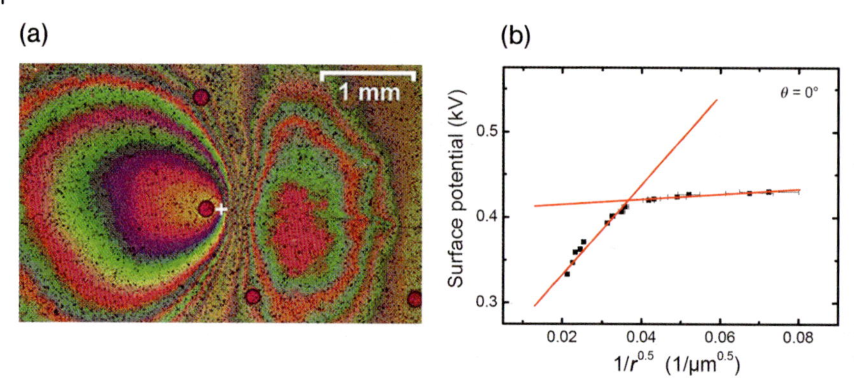

Figure 18.30 (a) Color image of the stresses in a thickness-poled compact tension specimen of bulk ferroelectric lead-zirconate-titanate (PIC151, $Pb_{0.99}[Zr_{0.45}Ti_{0.47}(Ni_{0.33}Sb_{0.67})_{0.08}]O_3$). Depending on the electrical potential induced at the sample surface, the liquid crystal molecules rotate to a different degree, thus modifying the refractive index which yields the different colors Ref. [193]. Reproduced with permission from © The American Ceramic Society; (b) Quantitative analysis of the stress values in the process zone near the crack tip at distance *r* from the tip. A detailed analysis yields a hardening exponent of 0.6 for the soft ferroelectric PZT (PIC 151). From Ref. [192].

acquired using a model based on the concept of a cohesive zone at the crack faces have also approached the experimentally acquired data [209]. Direct experiments on a propagating crack, using transmission electron microscopy (TEM), have revealed that the pores at the grain boundary triple junctions will tend to open under an electric load, ahead of an existing crack tip [210]. In conclusion, it is clear that in multilayer actuators, even a simple unipolar loading will, ultimately, lead to a delamination fracture along the electrodes [198, 211, 212]. However, this normally is not an instantaneous occurrence; rather, it occurs as a cyclic crack propagation process. Although the mechanical properties of the interface to the electrode for simple crack propagation have been widely investigated [213], this remains a theoretical and experimental task for cyclic loading.

18.6.2
Microcracking

Microcracking is, in general, a phenomenon that is not fully understood in ceramics [214–216]. It often occurs at grain boundaries, which are potentially weaker than the neighboring grains. In ferroelectric PZT, a large portion of the local stresses may occur due to the interaction of domain walls with the grain boundaries. As has been shown theoretically, the stresses at an intersection point of a wall with the grain boundary are not much enhanced for a noncracked sample [217]. The singularity is of order $1/r^{0.1}$, which is comparatively weak compared to the $1/\sqrt{r}$ dependence encountered for macroscopic cracks. If other microcracks already exist, then the local stress singularity will become more pronounced than for long cracks; indeed, a $1/r^{0.6}$ dependence has been identified, which means that any prior microcracking

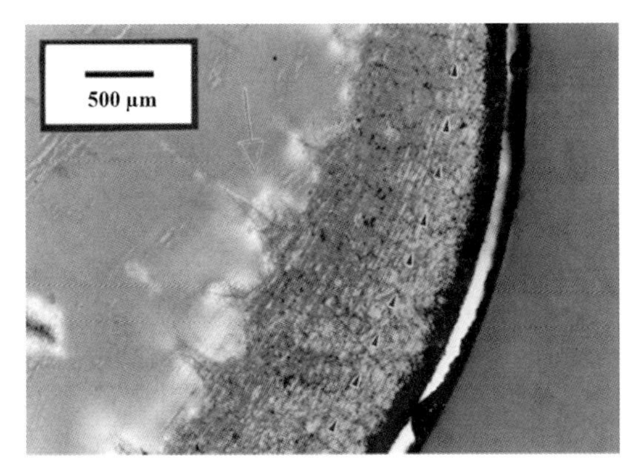

Figure 18.31 Microcracks emanating from an electrode edge after 3×10^6 switching cycles at $2\,kV\,mm^{-1}$ [219].

would enhance the susceptibility of a material towards further microcracking. Microcracking often starts at extrinsic singularities, such as an electrode edge where high local stresses arise due to macroscopic constraints [218]. This type of behavior is best seen in an electrostrictive transparent material, such as 9.5/65/35 PLZT (Figure 18.31).

Microcracking is induced strictly mechanically, because purely electrostrictive 9.5/65/35 PLZT – which does not develop a domain structure or a ferroelectric hysteresis – yields similar crack patterns [219]. This proposal was also supported by XRD data on fatiguing PZT, which exhibited microcracking after a certain cycle number. Zhang *et al.* observed a decreased XRD intensity that reflected a reduced domain switching and a consequent increase in mechanical stresses [220]. For sufficiently large grains in polycrystalline PZT, poling alone can induce sufficient internal stresses so as to drive (anisotropic) microcracking [221, 222]. The strain mismatch between differently oriented adjacent grains induces the stresses necessary for microcrack formation. When a sample was fatigued, the distribution of microcracks was homogeneous throughout its interior after 10^8 cycles, and a weak anisotropy was observed [174].

Cracking may even occur within grains for intersecting domains on their own. Microcracking at the intersection point of needle domains in macroscopic single-crystal $BaTiO_3$ has been observed directly, using optical microscopy. The intersecting domains in real ceramic microstructures may induce stresses of up to 1 GPa, unless the elastic strains can be partly accommodated by point defects or dislocations. This effect was demonstrated in a detailed TEM study on $Pb_{0.965}La_{0.01}Sr_{0.02}(Zr_{0.45}Ti_{0.55})O_3$ [223], whereby the microcracking altered the impedance response of the material such that an altered effective medium was generated [224, 225].

18.6.3
Breakdown

Breakdown has long been a topic associated with ferroelectric memory devices, and is recognized as being a second major failure mechanism [226]. Breakdown can be initiated at cracks that offer an open breakdown path due to the low breakdown strength of air, and the open surfaces bearing electronically sensitive surface states [227]. In thin films, breakdown is an electrochemical process from the very start. Although the initial step has not been identified, once a location for a breakdown path develops, then propagation as a dendritic tree into the ferroelectric is established [226]. As the breakdown paths are generally much more conducting than the surrounding material, the local electric field between the dendritic trees of opposite polarity increases as the trees grow towards each other, enhancing dendrite growth even further. Then, when the dendrites touch, a run-away of electric current causes a thermal breakdown and final failure. Recently, it was shown that local temperatures and/or enhanced electric fields were sufficient to induce a chemical disintegration of the ferroelectric material into simpler oxides of the constituent metals, in this case β-PbO_x (which is not the stable lead oxide phase at ambient conditions), Rutile TiO_2, and ZrO_2 in the case of PZT [179]. Because each of these simpler oxides is a better conductor than the ferroelectric composition, the breakdown current is enhanced during formation of the breakdown phases, and this constitutes the final short circuit. Surprisingly, the disintegration was shown to be inhomogeneous across $80\,\mu m \times 80\,\mu m$-wide electrode pads. Similar to the edge effects in the microcracking of bulk samples, breakdown occurs more frequently near the electrode edges, although the similarity of the effect is also partly accidental. Among thin films investigated for breakdown, the electrode pads were applied using shadow masks, and yielded an inhomogeneous electrode thickness and, potentially, an adhesion to the underlying ferroelectric material. As a precursor mechanism to final breakdown, Wang *et al.* attributed the reduced time to breakdown of polycrystalline (Pb,Sr)TiO_3-films processed at lower oxygen partial pressures to the increased density of oxygen vacancies. These generate shallow electron states in the band-gap which in turn strongly contribute to conduction under moderate electric fields and temperatures and thus become paths for breakdown [228]. This led to the conclusion that there was, indeed, a correlation between fatigue and breakdown.

Multiferroic $BiFeO_3$ is difficult to process in such a way that the ohmic conductivity is low and ferroelectric effects can be observed [229]. Although Lebeugle obtained high-purity single crystals that yielded a low leakage current [230], a rapid degradation due to cyclic loading and subsequent leakage were identified. Subsequently, these authors applied the external field along the cubic (010), which corresponded to the easy (012) hexagonal axis, this being the direction in which polarization switched with the most ease. Thin films, on the other hand, were reported to be strictly fatigue-free [231].

18.6.4
Frequency Effect

Fatigue studies beyond 10^{10} cycles are rare, even though at high repetition rates such as 1 MHz a memory would acquire 10^{10} cycles within only a few hours. Jiang *et al.* observed that Pt/PZT/Pt degraded under cyclic loading [232], although beyond 10^{10} cycles of a near-rectangular shape external voltage of rise-time 8 ns, the degradation essentially depended on the residence time at maximum voltage. For hold times at maximum voltage in excess of 40 ns, the degradation was more rapid. Jiang *et al.* ascribed this effect to charge injection since, during polarization reversal all of the charge current compensated the polarization and no injection took place. Subsequently, the high fields drove a charge injection that generated a space charge layer, in turn causing a fatigue effect. On a completely different time/frequency scale, Colla *et al.* arrived at a similar conclusion [233], after having cycled their materials in the mHz range to produce complete material fatigue within a few cycles. The peculiarity of these experiments was that the system was loaded not at maximum voltage, but rather for different times near the coercive field. With this loading regime, the number of domains could be maximized and their arrangement kept mainly random while the electric field was applied. The authors argued that a maximum number of domain walls could be electrically compensated by free charge carriers. In full saturation, the ferroelectric would contain only a few domain walls, and the clamping effect would be small. An influence of frequency was also observed in $Bi_{3.15}Nd_{0.85}$-$Ti_{3-x}W_xO_{12}$, where a similarly higher loading frequency produced less fatigue [234].

The rate dependencies of the ferroelectric material properties are also reflected in the dynamics "after" fatigue. Initially, most of the domain system will be switched almost instantaneously [235], and only a small amount of polarization will creep for longer time periods [194]. A highly retarded stretched exponential relaxation was observed after bipolar fatigue treatment [235], and these observations correlated well with the thermally activated domain dynamics. If the overall materials response was represented in a rate-dependent constitutive material law [236], however, then a growing defect cluster size would retard the domain dynamics considerably. Hard and soft material behaviors were also representable as different barrier heights to a thermally activated domain wall motion, as demonstrated by the theoretical studies of Belov and Kreher [236].

18.6.5
Defect Agglomeration

Electrochemical models may consider different effects, including ionic segregation, ionic conduction, or interface reactions at the electrode. In the following, only transport will be considered. Ionic motion can remain localized, as in the cage motion of, for example, oxygen vacancies around acceptor defects in ferroelectrics (this was discussed above as an aging model in ferroelectric materials). The itinerant motion of defects, and particularly of oxygen vacancies, has been considered as a fatigue mechanism, as it only involves transport and does not refer to any further chemical reaction. There are several approaches to this. One approach considers the direct diffusion of defects to the domain wall, and this then can also serve as an aging mechanism [64]. In order to have a significant effect on domain wall mobility, the resultant configuration must contain a wrinkled domain wall which, at each charged defect, will contain an atomic wide step [237]. The fatigue mechanism then must involve the enhancement of defect concentration due to domain wall motion (this point has not been covered in the literature in the context of this mechanism).

A second approach is that the oxygen defects agglomerate to localized clusters, due to an enhanced defect migration in the vicinity of the passing domain walls. The fundamental driving force to be considered here is the electric field, which becomes very large at the very moment a domain reverses polarity in the vicinity of an already existing defect [174, 238]. Considering that the resulting planar defects are intrinsically stabilized at the atomic scale, the model of cyclic defect accumulation yields a set of reasonable time, cycle number, and external field dependencies. This approach does not require the existence of other defects such as external electrodes, grain boundaries, or similar. The observation of chemical etch grooves in fatigued lead zirconate titanate, without any microcracking, appeared to offer a reasonable justification of this approach. Based on the same argument, Geng and Yang have recently developed a continuum mechanical approach in a similar sense, but on a significantly larger length scale [239], which discussed the migration of pores due to the cyclic passing of the domain walls. At the moment when the domain wall touches the pore, a local instability at the triple point generates an effective driving force for particle drift within the pore. An energetic argument then yields the effective driving forces on the entire pore. Geng and Yang's finite element calculations yielded a pore accumulation and finally an extended microvoid formation at the length scale of macroscopic samples. The effect of real microstructures and, for example, grain boundaries, was not considered. The motion of oxygen vacancies appeared not to be that of isolated vacancies, but rather of larger units. This was at least interpreted from the relaxor-like behavior of $Bi_4Ti_3O_{12}$ (BiT) treated under different oxygen vacancies. The Cole–Cole relaxation in these measurements indicated an increasing degree of collective motion of several defects with treatment under decreasing oxygen partial pressure [240].

One approach taken in thin films is that of vacancy accumulation underneath metallic electrodes, due to the much reduced diffusivity of oxygen in the metal. The charged vacancies generate a space charge cloud underneath the electrode, which in turn hampers the domain wall motion [241, 242]. For some time, very vivid discussions have been conducted as to whether the domain wall motion or the formation of new domains is the limiting step in polarization reversal [171, 241, 242]. In fact, at one point the accumulation of oxygen vacancies was considered to be the origin of a percolative phase transition [243]. Experimental findings very much in favor of the oxygen vacancy agglomeration model underneath electrodes have included the successful use of oxide electrodes to suppress fatigue (in some systems, entirely). $SrRuO_3$ [244], Ir/IrO_x [245], RuO_x [246, 247], and $(Bi,La)_4Ti_3O_{12}$ [248] each constitute fatigue-free systems, and even PtO_x as an electrode material yields a reduced susceptibility to fatigue [249]. Recently, Cao *et al.* directly determined the oxygen vacancy content in the subelectrode layer of PZT using *in situ* high-resolution grazing incidence X-ray specular reflectivity (XRSR) [250]. In this case, it was found that the vacancy content did not increase for samples with platinum electrodes that were strongly fatigued after 10^7 bipolar cycles, and the upper limit for a vacancy concentration in the top layer of PZT was 5.5%. The authors concluded that this was not sufficient for assigning fatigue to an ionic modification of the interface layer to the metallic electrode. The data of oxygen tracer diffusivity for an IrO_x-electrode system on PZT also suggested that the relevance of IrO_x as an oxygen ion source was less than initially discussed [251]. Nevertheless, it remains an open question as to what extent the annealing step employed (5 min at 500 °C, rapid thermal annealing) might lead to the rearrangement of any ionic defects [252]. In a similar vein was the finding by Kingon and Srinivasan, that copper electrodes would yield fatigue-free thin film devices if processed under a certain external oxygen partial pressure [171]. Thus, a vacancy-clustering model beneath the electrodes is only possible in accordance with these data if the vacancy concentration is reduced to zero by the annealing procedure at that particular oxygen partial pressure. Yet, these data suggest a more complex mechanism. An ionic picture is supported by the fact that 2% calcium-doped PZT becomes almost fatigue-free [253], which means in turn that the loss of lead from the PZT might be the most critical effect that is sufficiently compensated by a 2% Ca additive (and 20% excess lead before firing the sol–gel films). This somewhat electronic origin of the fatigue effect was concluded from measurements of all-oxide electrode/ferroelectric/electrode systems of different majority carriers in the electrodes [254]. An *n*-type oxide electrode ($La_{0.07}Sr_{0.93}SnO_3$) yielded a strong fatigue, whereas a *p*-type $La_{0.7}Sr_{0.3}MnO_3$ yielded a fatigue-free system. Chen *et al.* interpreted their data by suggesting the presence of a depletion layer at the interface between the ferroelectric and the electrode, which in turn induced a charge injection for *n*-conducting electrodes. The electronic nature of fatigue-like material behavior was illustrated by using X-ray irradiation to stabilize an internal bias in an otherwise fatigue-free $SrBi_2Ta_2O_9$ thin films. However, this bias could be recovered if X-ray irradiation and bipolar fields were applied simultaneously. Electronic traps were thus generated by the irradiation, and later rebalanced under the bipolar fields [255]. Another way of reducing fatigue due to oxygen vacancy motion is to reduce their number in the ferroelectric itself, and for this strontium proved to be a suitable substitute for lead [256]. In fact, a concentration of only 20% strontium led to a 20-fold reduction in the number of oxygen vacancies with respect to the unmodified PZT phase, and provided fatigue resistance for up to 10^{10} cycles with Pt electrodes. A similarly beneficial influence of reducing the oxygen vacancy content in the ferroelectric was shown for lanthanum-substituted $Bi_{4-x}La_xTi_3O_{12}$, where a substitution of $x = 0.5$ was sufficient to generate a fatigue-free system [257], again using ruthenium electrodes [258]. Barium proved to be equally beneficial in PZT, where a 10% substitution of lead was sufficient to suppress fatigue [259]. The oxygen vacancies appeared to play a double role; it was not their migration but rather their change of charge state that was proposed as the origin of the fatigue effect, using a 2-D, four-state Potts model [260], with the simulation data shadowing the experimental data very impressively. Thus, the nature of the traps in fatigue appears to be electronic, with the ionic vacancy as a favored site for charge trapping.

One very impressive experimental argument for the role of point defects in fatigue is the strong temperature-dependence of the effect [261–263]. $BaTiO_3$ demonstrates the greatest degree of fatigue at room temperature, but this decreases at higher temperatures [261], due either to a reduced spontaneous polarization or to a higher mobility of defects above room temperature, which yields a partial dissociation of clustered defects as a result of thermal fluctuations. At temperatures well below room temperature, fatigue can be avoided entirely in PZT and $Pb_{0.75}La_{0.25}TiO_3$ [262, 263]. Besides domain wall motion, a thermal activation is needed to yield an effective fatigue effect. However, whether cluster formation is suppressed, or whether the thermionic charge injection from the electrodes is the crucial process, has not yet been fully determined.

Electronic carriers appear to be at least partly involved in the fatigue process. The fatigue rates were seen to be lowest for the minimum conductivity of a material, and this was observed independent of whether the minimum was achieved by varying the oxygen partial pressure [264], or by doping. When the latter approach was taken, using tungsten-doped PLZT as an example [265], the calculations suggested that the Ti 3d states were preferentially occupied by additional electrons. Kim *et al.* suggested, from their calculations, that the PZT compositions richer in Ti should thus exhibit higher fatigue rates [266]. However, any experimental proof was obscured by the stronger effects of strain mismatch in the Ti-richer tetragonal PZT compositions, and by electrode effects. Also, otherwise fatigue-free systems such as $SrBi_2TaNbO_9$ become susceptible to fatigue when electronic carriers are introduced, in this case by optical illumination [267].

18.6.6 Electrode Effects

The interaction of the electrode metal with the ferroelectric has been discussed on many occasions (see Refs [169, 173] and references therein). Platinum electrodes have been the major obstacle to ferroelectric thin film use as a memory material (as discussed above). The entire interplay of point defect equilibrium, metal work function and ferroelectric was clarified by the studies of Kingon and Srinivasan, who determined the necessary oxygen partial pressure for processing nearly fatigue-free PZT films using copper electrodes [171]. These authors used an external partial pressure that maintained the metallic copper and did not reduce the lead oxide. In difference to similarly processed samples using platinum electrodes, the switchable polarization in the copper system showed an increase during cyclic loading, rather than a deterioration. Heteroepitaxial films produced by a hydrothermal process on niobium-doped conductive strontium-titanate single crystals at low temperatures, when the material was already in its ferroelectric state, yielded fatigue-free PZT films, whereas pure lead titanate still suffered from fatigue [268]. The direct synthesis of the material in the ferroelectric state appears, therefore, not to be a crucial prerequisite to a fatigue-free system.

18.6.7 Domain Nucleation or Wall Motion Inhibition

In order to reverse the polarization of a perfect single crystal, the reverse domains must be nucleated and then grown by forwards and sideways motions of the domain wall (this is a classic nucleation and growth mechanism in phase transformations). In polycrystalline ferroelectric materials, no poling process can annihilate all of the reverse domains, and consequently in bulk samples reverse nuclei will always exist, with the majority of them still being sizeable domains. Fatigue is, therefore, predominantly an issue of reduced domain wall mobility (as discussed above in the context of aging). In polycrystalline or epitaxial thin films this may be different due to the high degree of texture obtained in many cases; the inhibition of domain nucleation might then become the rate-determining step of switching. Tagantsev *et al.* discussed this mechanism with respect to several other models for fatigue [173], and concluded that either a near-electrode charge injection into low-mobility electronic defect states, or a chemical modification of the near-electrode volume might be responsible for the apparent fatigue in thin film memory devices. The fact that external surfaces – and particularly the inhomogeneities therein – are necessary for domain nucleation has been demonstrated indirectly by Shin *et al.* [269]. These authors used *ab initio* calculations to determine the energy barriers for nucleation and growth of domain walls, and found reverse domain nucleation within existing domains to be a process associated with a very large energy barrier. Thus, extrinsic nucleation is dominant.

18.6.8 Clamping in Thin Films

In 111-oriented PZT thin films, the internal strain increases constantly with cycle number as an increasing number of a/c-domain systems become clamped due to the cycling process. In single crystals of tetragonal perovskites, c-domains have a polarization perpendicular to the external crystal face, while a-domains are in plane. This means that the larger components of polarization will interfere with the grain boundary and thus contribute to clamping much more than would mechanically and electrically adapted structures. Even though a large number of fixed-domain systems initially exists in 115-oriented SBT, this material showed no fatigue up to 10^{10} cycles [270]. Fatigue was accompanied by the appearance of ferroelastic domains that were formed to relieve any arising mechanical stresses, as was investigated using piezoresponse force microscopy (PFM) on commercial lead zirconate titanate ceramics (PZT) [271]. The strongest effect was observed in the regions adjacent to the electrode. However, annealing to temperatures above the phase transition promoted the partial recovery of the initial domain structure.

18.6.9 Domain Splitting and Crystal Orientation Dependence

An observation by Takemura *et al.* was an enhanced density of the domain walls in PZN–4.5%PT after bipolar fatigue along $[111]_{qc}$ (qc = quasi cubic indexing), while fatigue along $[001]_{qc}$ reduced the number of domain walls [272]. These authors explained such an effect in the $[111]_{qc}$-oriented case by a total pinning of some domains at defects. For any subsequent switching to occur, the new domain walls must be generated at the same nucleation sites as previous domains, with the driving forces being slightly higher due to the already-fixed domains. Although the nature of the pinning centers was not discussed, the domain wall density was shown also to increase for PZT due to fatigue. Menou *et al.* arrived at a similar result, when they interpreted their synchrotron X-ray data before and after fatigue as fatigue-induced transitions between the tetragonal and rhombohedral ferroelectric phases in $PbZr_{0.45}Ti_{0.55}O_3$ (Pt-electrode) thin films [273].

Chou *et al.* were able to show, by using TEM, that zigzag domain patterns are rapidly generated from an initially simple domain pattern during the fatigue process [274]. Such patterns emanate from simple straight 90° domain walls, and then grow in size and lateral extension until they intersect with the 90° domain walls of the initial domain structure. The zigzag domain wall structure appears to be generated during single polarization reversal processes, such that the increasing stress in the microstructure ultimately yields a high dislocation density. The fatigue effect was considered to be due to cumulated mechanical stresses in the microstructure, increasingly limiting the domain wall mobility.

In thin films, the crystallite orientation was shown to be important for fatigue behavior. Rhombohedral 111-oriented PZT fatigues more than the 100-oriented rhombohedral counterpart (on the $Pt/TiO_2/SiO_2/Si$-electrode). Tetragonal 001-oriented PZT on Pt/MgO was shown to fatigue to the greatest degree [275]. Oriented PZT on oriented buffer layers of $BaPbO_3$ on ruthenium metal electrode displayed a similar tendency for improved fatigue resistance for non-001-oriented films [276], although in neither study was any conclusive explanation provided for this orientation dependence. Likewise, Yoon *et al.* failed to identify any difference in the fatigue behavior of 00l- and 117-oriented BLT ($(Bi,La)_4Ti_3O_{12}$), but rather observed a threshold loading pulse length of 20–50 ns, beneath which no fatigue was observed, irrespective of the orientation of the film [277].

18.6.10 Combined Loading

Practically all ferroelectrics also show a ferroelastic response, with the purely ferroelastic response being constituted by non-180° domain switching. Electrically driven switching yields an apparently higher strain than a purely mechanically driven hysteretic material deformation, because each 180° domain is also piezoelectric. Upon polarization reversal, the piezoelectric strain also reverses under the non-zero electric field applied. In practice, all multilayer actuators are housed in an elastic environment or a mechanical spring, in order to provide some compressive stress to the ceramic suppressing delamination [211] and interface cracking [213], which also partly increases the strain output [175]. Under dynamic loading, the inertia of the ceramic part and/or the housing may induce additional electromechanical coupling effects. Cyclic mechanical loading may also stem from external sources without any actuator drive. Three loading conditions can be differentiated, according to the degree of switching involved, namely elastically clamped, cyclic purely mechanical, and mixed (out of phase) loading [175–177]:

- In elastically clamped loading, a more or less static elastic load is applied to the ceramic or stack actuator; this affords an initially negative strain of the device and/or material. The applied electric field then suffices to drive the actuator to almost the same saturation strain value as without any mechanical preload, and thus permits a higher total strain output than its uncompressed counterpart. The material fatigue does not differ considerably from the purely electrical unipolar case.
- Cyclic, purely mechanical loading yields strain rates that are an order of magnitude higher than static loads, and which generate simple creep [176]. The strain accumulates in a similar fashion as in metals or under electrical fatigue in ferroelectrics, generating a distinctly different microstate after loading than before loading.
- Out-of-phase (mixed) loading may occur for example, in badly designed control circuits in damping units or under dynamic loading in resonance of the device or the multilayer actuator itself. Out-of-phase loading clearly yields a material fatigue that is significantly higher than for a purely static compressive load. It arises in the bulk material as well as for the entire device, and is thus a material and not a device effect.

Sesquipolar (partly negative fields but less than the coercive field) loading is the next, more severe, degree of material fatigue [175, 278]. In this case, the material fatigue increases as soon as the hysteresis opens up, indicating significant internal losses within the material [278]. The heating-up of the material or device simultaneously becomes relevant to the fatigue effect. Sesquipolar loading not only allows for larger strains [175] but also entails a stronger defect agglomeration due to an increased, thermally activated mobility of the responsible defects [174, 178]. Overall, a hierarchy of increasing fatigue rates can be determined, namely unipolar, unipolar under compressive preload, mixed, sesquipolar, and bipolar [178]. It is thus crucial in device output and circuit design to avoid resonances of the system, to keep the internal losses as low as possible, and to refrain from the temptation of larger strain output at the expense of a more hysteretic material or device system.

18.6.11
Antiferroelectrics

In terms of their crystal structure, antiferroelectrics are constituted by adjacent unit cells of opposite polarity. As such, they provide an interesting comparison with regards to fatigue, because their initial and stable crystal state is nonpolar, despite each unit cell being polar. Unless they are active on a single unit cell scale, none of the mechanisms that involve charged defects will affect the neutral status of such a sample; rather, they will only influence the behavior at higher fields and (potentially) also the phase-transition boundary. Bulk ceramic samples of $Pb_{0.97}La_{0.02}(Zr_{0.55}Sn_{0.33}Ti_{0.12})O_3$ exhibit a certain degree of degradation after 10^8 bipolar fatigue cycles, exceeding the saturation voltage of the antiferroelectric by $1.5\,kV\,mm^{-1}$. A large acoustic emission activity during the antiferroelectric to ferroelectric and reverse-phase transitions is observed [279], whilst a similar treatment in a ferroelectric bulk sample will produce a strong fatigue degradation [280]. Both, the ferroelectric and antiferroelectric materials develop bulk etch-grooves that might be interpreted as agglomerates of point defects. If the antiferroelectric to ferroelectric transition is sharp, then microcracks will develop; however, agglomerates will develop in both cases. Hence, whilst the overall fatigue scenario is very similar for both antiferroelectrics and ferroelectrics, the extent of fatigue is much less in antiferroelectric compositions.

18.6.12
Fatigue-Free Systems

For ferroelectric memory applications, two practical solutions to the fatigue problem have been found. The first replaces metallic electrodes by oxide interfaces; in this case, iridium-containing electrodes which generate an Ir/IrO_x-equilibrium [245], ruthenium oxide [246, 247], and strontium ruthenate [244] have each proven to be successful. The pile-up of vacancies considered responsible for the fatigue effect in thin films with metallic electrodes [242] can be avoided with oxide electrodes. For Ir, the formation of an IrO_x-layer is assumed to allow for the diffusion of oxygen vacancies into the electrode material. The processing of the other oxide electrodes requires the application of several metals and/or their subsequent oxidation; hence, most manufacturers rely on the Ir-system. Ferroelectric-layered bismuth compositions [$SrBi_2Ta_2O_9$ (SBT) and $Bi_3Ta_4O_{12}$ (BTO)] represent a material class that is not susceptible to fatigue, even for metallic electrodes [170], except for pure $Bi_4Ti_3O_{12}$ [281]. Pure $Bi_4Ti_3O_{12}$ contains unstable layers of oxygen octahedra which can be stabilized by lanthanum substitution to yield fatigue-free $Bi_{3.25}La_{0.75}Ti_3O_{12}$ (BLT). Pure $Bi_4Ti_3O_{12}$ shows a delayed and reduced switching due to fatigue which is assigned to a domain wall-pinning mechanism, as in the lead-containing perovskites [282]. The layered bismuth ferroelectrics do not contain 90° domain walls. Large mechanical stresses during switching are avoided, and this is considered to be a reason for their fatigue-resistant character [283]. Dimos *et al.* made the large band-gap (and thus the lower mobility of deep electronic trap states) responsible for the fatigue resistance particularly of SBT [284]. In their TEM studies of BLT, Ding *et al.* [285] observed a large number of antiphase boundaries that were not present in pure $Bi_4Ti_3O_{12}$ [286], but were considered to facilitate the 90° switching. Hence, this argument differs from the electrochemical argument of oxide vacancy pile-up beneath electrodes in the perovskite ferroelectrics. The pile-up is considered to be suppressed by the Bi_2O_2-layers blocking the diffusion of oxygen vacancies towards the electrodes [287].

Acknowledgments

The authors thank Dr. Dragan Damjanovic for valuable discussions of recent achievements in the field of ferroelectricity. This manuscript was compiled at Technische Universität Dresden, Institut für Werkstoffwissenschaft, 01062 Dresden, Germany.

References

1 Haertling, G.H. (1999) Ferroelectric ceramics: history and technology. *J. Am. Ceram. Soc.*, **82** (4), 797–818.

2 Waser, R. (1999) Modeling of electroceramics – applications and prospects. *J. Eur. Ceram. Soc.*, **19**, 655–664.

3 Ahn, C.H., Rabe, K.M., and Triscone, J.M. (2004) Ferroelectricity at the nanoscale: Local polarization in oxide thin films and heterostructures. *Science*, **303** (5657), 488–491.

4 Scott, J. (2007) Applications of modern ferroelectrics. *Science*, **315** (5814), 954–959.

5 Setter, N., Damjanovic, D., Eng, L., Fox, G., Gevorgian, S., Hong, S., Kingon, A., Kohlstedt, H., Park, N.Y., Stephenson, G.B., Stolitchnov, I., Tagantsev, A.K., Taylor, D.V., Yamada, T., and Streiffer, S. (2006) Ferroelectric thin films: Review of materials, properties, and applications. *J. Appl. Phys.*, **100** (5), 051606.

6 Sheppard, L.M. (1992) Advances in processing of ferroelectric thin films. *Am. Ceram. Soc. Bull.*, **71** (1), 85–95.

7 Schwartz, R.W., Boyle, T.J., Lockwood, S.J., Sinclair, M.B., Dimos, D., and Buccheit, C.D. (1995) Sol-gel processing PZT thin films: A. Review of the state of the art and process optimization strategies. *Integr. Ferroelectr.*, 7, 259–277.

8 Park, S.E. and Shrout, T.R. (1997) Ultrahigh strain and piezoelectric behavior in relaxor based ferroelectric single crystals. *J. Appl. Phys.*, **82** (4), 1804–1811. Reprinted with permission from © American Institute of Physics.

9 Liu, S.F., Park, S.E., Shrout, T.R., and Cross, L.E. (1999) Electric field dependence of piezoelectric properties for rhombohedral 0.955 $Pb(Zn_{1/3}Nb_{2/3})O_3$ – 0.045 $PbTiO_3$ single crystals. *J. Appl. Phys.*, **85** (5), 2810–2814. Reprinted with permission from © American Institute of Physics.

10 Shirane, G., Pepinsky, R., and Frazer, B.C. (1956) X-ray and neutron diffraction study of ferroelectric $PbTiO_3$. *Acta Crystallogr.*, **9** (2), 131–140.

11 Cohen, R.E. (1992) Origin of ferroelectricity in perovskite oxides. *Nature*, **358** (6382), 136–138.

12 Jaffe, B., Roth, R.S., and Marzullo, S. (1955) Properties of piezoelectric ceramics in the solid-solution series lead titanate-lead zirconate-lead oxide-tin oxide and lead titanate-lead hafnate. *J. Res. Nat. Bur. Stds*, **55** (5), 239–254.

13 Jaffe, B., Cook, W.R., and Jaffe, H. (1971) *Piezoelectric Ceramics*, Academic Press, New York.

14 Ishibashi, Y. and Iwata, M. (1998) Morphotropic phase boundary in solid solution systems of perovskite type oxide ferroelectrics. *Jpn. J. Appl. Phys.*, **37** (8B), L985–L987.

15 Ishibashi, Y. and Iwata, M. (1999) A theory of morphotropic phase boundary in solid-solution systems of perovskite-type oxide ferroelectrics. *Jpn. J. Appl. Phys.*, **38** (2A), 800–804.

16 Iwata, M. and Ishibashi, Y. (2005) Analysis of ferroelectricity and enhanced piezoelectricity near the morphotropic phase boundary. *Topics Appl. Phys.*, **98**, 127–148.

17 Noheda, B., Cox, D.E., Shirane, G., Gonzalo, J.A., Cross, L.E., and Park, S.E. (1999) A monoclinic ferroelectric phase in the $Pb(Zr_{1-x}Ti_x)O_3$ solid solution. *Appl. Phys. Lett.*, **74** (14), 2059–2061.

18 Noheda, B., Gonzalo, J.A., Cross, L.E., Guo, R., Park, S.E., Cox, D.E., and Shirane, G. (2000) Tetragonal-to-monoclinic phase transition in a ferroelectric perovskite: The structure of $PbZr_{0.52}Ti_{0.48}O_3$. *Phys. Rev.*, **61** (13), 8687–8695.

19 Noheda, B., Cox, D.E., Shirane, G., Guo, R., Jones, B., and Cross, L.E. (2001) Stability of the monoclinic phase in the ferroelectric perovskite $PbZr_{1-x}Ti_xO_3$. *Phys. Rev. B*, **63** (1), 014103.

20 Uesu, Y., Yamada, Y., Fujishiro, K., Tazawa, H., Enokido, S., Kiat, J.-M., and Dkhil, B. (1998) Structural and optical studies of development of the long-range order in ferroelectric relaxor $Pb(Zn_{1/3}Nb_{2/3})O_3$–$PbTiO_3$. *Ferroelectrics*, **217**, 319–325.

21 Kiat, J.-M., Uesu, Y., Dkhil, B., Matsuda, M., Malibert, C., and Calvarin, G. (2002) Monoclinic structure of unpoled PMN-PTand PZN-PTcompounds. *Phys. Rev. B*, **65** (6), 064106.
22 La-Orauttapong, D., Noheda, B., Ye, Z.-G., Gehring, P.M., Toulouse, J., Cox, D.E., and Shirane, G. (2002) New phase diagram of the relaxor ferroelectric $(1-x)Pb(Zn_{1/3}Nb_{2/3})O_3$–$xPbTiO_3$. *Phys. Rev. B*, **65** (14), 144101.
23 Vanderbilt, D. and Cohen, M.H. (2001) Monoclinic and triclinic phases in higher-order Devonshire theory. *Phys. Rev. B*, **63** (9), 094108.
24 Ragini, R.R., Mishra, S.K., Pandey, D. (2002) Room temperature structure of Pb $(Zr_xTi_{1-x})O_3$ around the morphotropic phase boundary region: A Rietveld study. *J. Appl. Phys.*, **92** (6), 3266–3274.
25 Noheda, B. (2002) Structure and high-piezoelectricity in lead oxide solid solutions. *Curr. Opin. Solid State Mater. Sci.*, **6** (1), 27–34.
26 Noheda, B., Zhong, Z., Cox, D.E., Shirane, G., Park, S.E., and Rehrig, P. (2002) Electric-field-induced phase transitions in rhombohedral $Pb(Zn_{1/3}Nb_{2/3})_{(1(x)}Ti_xO_3$. *Phys. Rev. B*, **65** (22), 224101.
27 Noheda, B., Cox, D.E., Shirane, G., Gao, J., and Ye, Z.G. (2002) Phase diagram of the ferroelectric relaxor $(1-x)PbMg_{1/3}Nb_{2/3}O_3$–$xPbTiO_3$. *Phys. Rev. B*, **66** (5), 054104.
28 Singh, A.K. and Pandey, D. (2003) Evidence for M_B and M_C phases in the morphotropic phase boundary region of $(1-x)[Pb(Mg_{1/3}Nb_{2/3})O_3]$-$xPbTiO_3$: A Rietveld study. *Phys. Rev. B*, **67** (6), 064102.
29 Singh, A.K. and Pandey, D. (2003) Confirmation of M_B-type monoclinic phase in $Pb[(Mg_{1/3}Nb_{2/3})_{0.71}Ti_{0.29}]O_3$: A powder neutron diffraction study. *Phys. Rev. B*, **68** (17), 172103.
30 Haumont, R., Dkhil, B., Kiat, J.M., Al-Barakaty, A., Dammak, H., and Bellaiche, L. (2003) Cationic-competition-induced monoclinic phase in high piezoelectric $(Pb(Sc_{1/2}Nb_{1/2})O_3)_{1-x}$–$(PbTiO_3)_x$ compounds. *Phys. Rev. B*, **68** (1), 014114.
31 Haumont, R., Al-Barakaty, A., Dkhil, B., Kiat, J.M., and Bellaiche, L. (2005) Morphotropic phase boundary of heterovalent perovskite solid solutions: Experimental and theoretical investigation of $PbSc_{1/2}Nb_{1/2}O_3$–$PbTiO_3$. *Phys. Rev. B*, **71** (10), 104106.
32 Fu, H.X. and Cohen, R.E. (2000) Polarization rotation mechanism for ultrahigh electromechanical response in single-crystal piezoelectrics. *Nature*, **403** (6767), 281–283.
33 Nakamura, K. and Kawamura, Y. (2000) Orientation dependence of electromechanical coupling factors in $KNbO_3$. *IEEE Trans. Ultrason. Ferroelectr. Freq. Control*, **47** (3), 750–755.
34 Service, R.F. (1997) Materials science: shape-changing crystals get shiftier. *Science*, **275** (5308), 1878.
35 Du, X.H., Belegundu, U., and Uchino, K. (1997) Crystal orientation dependence of piezoelectric properties in lead zirconate titanate: theoretical expectations for thin films. *Jpn. J. Appl. Phys.*, **36** (9A), 5580–5587. Reprinted with permission from © American Institute of Physics.
36 Du, X.H., Zheng, J.H., Belegundu, U., and Uchino, K. (1998) Crystal orientation dependence of piezoelectric properties of lead zirconate titanate near the morphotropic phase boundary. *Appl. Phys. Lett.*, **72** (19), 2421–2423.
37 Reszat, J.T., Glazounov, A.E., and Hoffmann, M.J. (2001) Analysis of intrinsic lattice deformation in PZT-ceramics of different compositions. *J. Eur. Ceram. Soc.*, **21** (10–11), 1349–1352.
38 Taylor, D. and Damjanovic, D. (2000) Piezoelectric properties of rhombohedral $Pb(Zr, Ti)O_3$ thin films with (100), (111), and "random" crystallographic orientation. *Appl. Phys. Lett.*, **76** (12), 1615–1617. Reprinted with permission from © American Institute of Physics.
39 Budimir, M., Damjanovic, D., and Setter, N. (2003) Piezoelectric anisotropy–phase transition relations in perovskite single crystals. *J. Appl. Phys.*, **94** (10), 6753–6761.
40 Budimir, M., Damjanovic, D., and Setter, N. (2006) Piezoelectric response and free-energy instability in the perovskite

crystals $BaTiO_3$, $PbTiO_3$, and $Pb(Zr,Ti)O_3$. *Phys. Rev.*, **73** (17), 174106.

41 Damjanovic, D. (2005) Contributions to the piezoelectric effect in ferroelectric single crystals and ceramics. *J. Am. Ceram. Soc.*, **88** (10), 2663–2676.

42 Davis, M., Budimir, M., Damjanovic, D., and Setter, N. (2007) Rotator and extender ferroelectrics: Importance of the shear coefficient to the piezoelectric properties of domain-engineered crystals and ceramics. *J. Appl. Phys.*, **101** (5), 054112. Reprinted with permission from © American Institute of Physics.

43 Damjanovic, D., Budimir, M., Davis, M., and Setter, N. (2006) Piezoelectric anisotropy: Enhanced piezoelectric response along nonpolar directions in perovskite crystals. *J. Mater. Sci.*, **41** (1), 65–76.

44 Haun, M.J., Furman, E., Halemane, T.R., Zhuang, Z.Q., Jang, S.J., and Cross, L.E. (1989) Thermodynamic theory of the lead zirconate titanate solid solution system. Parts I-V. *Ferroelectrics*, **99**, 13–86.

45 Sundar, V. and Newnham, R.E. (1992) Electrostriction and polarization. *Ferroelectrics*, **135** (1–4), 431–446.

46 Muralt, P., Maeder, T., Sagalowicz, L., Hiboux, S., Scalese, S., Naumovic, D., Agostino, R.G., Xanthopoulos, N., Mathieu, H.J., Patthey, L., and Bullock, E.L. (1998) Texture control of $PbTiO_3$ and $Pb(Zr,Ti)O_3$ thin films with TiO_2 seeding. *J. Appl. Phys.*, **83** (7), 3835–3841.

47 Sabolsky, E.M., Trolier-McKinstry, S., and Messing, G.L. (2003) Dielectric and piezoelectric properties of ⟨001⟩ fiber-textured 0.675 $Pb(Mg_{1/3}Nb_{2/3})O_3$ −0.325 $PbTiO_3$ ceramics. *J. Appl. Phys.*, **93** (7), 4072–4080.

48 Kwon, S., Sabolsky, E.M., Messing, G.L., and Trolier-McKinstry, S. (2005) High strain, ⟨001⟩ textured 0.675 $Pb(Mg_{1/3}Nb_{2/3})O_3$–0.325 $PbTiO_3$ ceramics: Templated grain growth and piezoelectric properties. *J. Am. Ceram. Soc.*, **88** (2), 312–317.

49 Messing, G.L., Trolier-McKinstry, S., Sabolsky, E.M., Duran, C., Kwon, S., Brahmaroutu, B., Park, P., Yilmaz, H., Rehrig, P.W., Eitel, K.B., Suvaci, E., Seabaugh, M., and Oh, K.S. (2004) Templated grain growth of textured piezoelectric ceramics. *Crit. Rev. Solid State Mater. Sci.*, **29** (2), 45–96.

50 Kimura, T. (2006) Application of texture engineering to piezoelectric ceramics – A review. *J. Ceram. Soc. Jpn*, **114** (1325), 15–25.

51 Jonker, G.H. (1972) Nature of aging in ferroelectric ceramics. *J. Am. Ceram. Soc*, **55**, 57–58.

52 Schulze, W.A. and Ogino, K. (1988) Review of literature on aging of dielectrics. *Ferroelectrics.*, **87**, 361–377.

53 Carl, K. and Härdtl, K.H. (1978) Electrical after-effects in $Pb(Ti,Zr)O_3$ ceramics. *Ferroelectrics*, **17**, 473–486.

54 Tan, B.Q., Li, J.-F., and Viehland, D. (1997) Ferroelectric behaviours dominated by mobile and randomly quenched impurities in modified lead zirconate titanate ceramics. *Philos. Mag. B*, **76** (1), 59–74.

55 Zhang, L.X. and Ren, X. (2005) *In situ* observation of reversible domain switching in aged Mn-doped $BaTiO_3$ single crystals. *Phys. Rev.*, **B71**, 174108.

56 Dimos, D., Warren, W.L., Sinclair, M.B., Tuttle, B.A., and Schwartz, R.W. (1994) Photoinduced hysteresis changes and optical storage in $(Pb,La)(Zr,Ti)O_3$ thin films and ceramics. *J. Appl. Phys.*, **76** (7), 4305–4315.

57 Morozov, M. (2005) Softening and hardening transitions in ferroelectric Pb $(Zr,Ti)O_3$ ceramics. Thesis #3368. Swiss Federal Institute of Technology-EPFL.

58 Damjanovic, D. (2005) Hysteresis in piezoelectric and ferroelectric materials, in *Science of Hysteresis* (eds G. Bertotti and I. Mayergoyz), Elsevier, pp. 337–465.

59 Berlincourt, D. and Krueger, H.H.A. (1959) Domain processes in lead titanate zirconate and barium titanate ceramics. *J. Appl. Phys.*, **30** (11), 1804–1810.

60 Shur, V., Rumyantsev, E., Batchko, R., Miller, G., Fejer, M., and Byerb, R. (1999) Physical basis of the domain engineering in the bulk ferroelectrics. *Ferroelectrics*, **221**, 157–167.

61 Yang, T.J., Gopalan, V., Swart, P.J., and Mohadeen, U. (1999) Direct observation of pinning and bowing of a single

ferroelectric domain wall. *Phys. Rev. Lett.*, **82** (20), 4106–4109.

62 Ma, H.Z., Kim, W.J., Horwitz, J.S., Kirchoefer, S.W., and Levy, J. (2003) Lattice-scale domain wall dynamics in ferroelectrics. *Phys. Rev, Lett.*, **91** (21), 217601.

63 Chynoweth, A.G. (1958) Barkhausen pulses in barium titanate. *Phys. Rev.*, **110** (6), 1316–1332.

64 Little, E.A. (1955) Dynamic behavior of domain walls in barium titanate. *Phys. Rev.*, **98** (4), 978–984.

65 Zhang, L.X. and Ren, X. (2006) Aging behavior in single-domain Mn-doped $BaTiO_3$ crystals: Implication for a unified microscopic explanation of ferroelectric aging. *Phys. Rev.*, **B73**, 094121.

66 Ren, X. (2004) Large electric-field-induced strain in ferroelectric crystals by point-defect-mediated reversible domain switching. *Nat. Mater.*, **3**, 91–94.

67 Lines, M.E. and Glass, A.M. (1979) *Principles and Applications of Ferroelectrics and Related Materials*, Clarendon Press, Oxford.

68 Warren, W.L., Dimos, D., Pike, G.E., Vanheusden, K., and Ramesh, R. (1995) Alignment of defect dipoles in polycrystalline ferroelectrics. *Appl. Phys. Lett.*, **67** (12), 1689–1691.

69 Warren, W.L., Pike, G.E., Vanheusden, K., Dimos, D., Tuttle, B.A., and Robertson, J. (1996) Defect-dipole alignment and tetragonal strain in ferroelectrics. *J. Appl. Phys.*, **79** (12), 9250–9257.

70 Warren, W.L., Vanheusden, K., Dimos, D., Pike, G.E., and Tuttle, B.A. (1996) Oxygen vacancy motion in perovskite oxides. *J. Am. Ceram. Soc.*, **79** (2), 536–538.

71 Keve, E.T., Bye, K.L., Annis, A.D., and Whipps, P.W. (1971) Structural inhibition of ferroelectric switching in triglycine sulfate. 1. Additives. *Ferroelectrics*, **3** (1), 39.

72 Okada, K. (1961) Ferroelectric properties of X-ray-damaged Rochelle salt. *J. Phys. Soc. Jpn*, **16** (3), 414–423.

73 Lambeck, P.V. and Jonker, G.H. (1978) Ferroelectric domain stabilization in $BaTiO_3$ by bulk ordering of defects. *Ferroelectrics*, **22**, 729–731.

74 Misarova, A. (1960) Aging of barium titanate single crystals. *Solid State Phys.*, **2** (6), 1160–1165.

75 Postnikov, V.S., Pavlov, V.S., and Turkov, S.K. (1968) Interaction between 90° domain walls and point defects of the crystal lattice in ferroelectric ceramics. *Sov. Phys. - Solid State*, **10** (6), 1267–1270.

76 Postnikov, V.S., Pavlov, V.S., and Turkov, S.K. (1970) Internal friction in ferroelectrics due to interaction of domain boundaries and point defects. *J. Phys. Chem. Solids*, **31**, 1785–1791.

77 Lupascu, D., Genenko, Yu., and Balke, N. (2006) Aging in ferroelectrics. *J. Am. Ceram. Soc.*, **89** (1), 224–229.

78 Lambeck, P.V. and Jonker, G.H. (1986) The nature of domain wall stabilization in ferroelectric perovskites. *J. Phys. Chem. Solids*, **47** (5), 453–461.

79 Robels, U. and Arlt, G. (1993) Domain wall clamping in ferroelectrics by orientation of defects. *J. Appl. Phys.*, **73** (7), 3454–3460.

80 Li, S., Cao, W., and Cross, L.E. (1991) The extrinsic nature of nonlinear behavior observed in lead zirconate titanate ferroelectric ceramic. *J. Appl. Phys.*, **69** (10), 7219–7224.

81 Lord Rayleigh, R.S. XXV (1887) Notes on Electricity and Magnetism - III. On the behaviour of iron and steel under the operation of feeble magnetic forces. *Philos. Mag.*, **23**, 225–245.

82 Néel, L. (1942) Théories des lois d'aimantation de Lord Rayleigh. *Cahiers Phys.*, **12**, 1–20.

83 Kronmüller, H. (1970) Statistical theory of Rayleigh's law. *Z. Angew. Phys.*, **30**, 9–13.

84 Boser, O. (1987) Statistical theory of hysteresis in ferroelectric materials. *J. Appl. Phys.*, **62** (4), 1344–1348.

85 Liu, W., Chen, L.Y., Zhang, L., Wang, Y., Zhou, C., Li, C., and Ren, X. (2006) Ferroelectric aging effect in hybrid-doped $BaTiO_3$ ceramics and the associated large recoverable electrostrain. *Appl. Phys. Lett.*, **89**, 172908.

86 Raymond, M.V. and Smyth, D.M. (1996) Defects and charge transport in

perovskites ferroelectrics. *J. Phys. Chem. Solids*, **57** (10), 1507–1511.

87 Guiffard, B., Audigier, D., Lebrum, L., Troccaz, M., and Pleska, E. (1999) Effects of fluorine-oxygen substitution on the dielectric and electromechanical properties of lead zirconate titanate ceramics. *J. Appl. Phys.*, **86** (10), 5747–5752.

88 Key, T.S., Jones, J.L., Shelley, W.F., Iverson, B.J., Li, H.Y., Slamovich, E.B., King, A.H., and Bowman, K.J. (2005) Texture and symmetry relationships in piezoelectric materials. *Mater. Sci. Forum*, **495–497**, 13–22 (Texture of Materials – ICOTOM 14, Parts 1 & 2).

89 Jones, J.L., Iverson, B.J., and Bowman, K.J. (2007) Texture and anisotropy of polycrystalline piezoelectrics. *J. Am. Ceram. Soc.*, **90** (8), 2297–2314.

90 Arendt, R.H. (1973) The molten salt synthesis of single magnetic domain $BaFe_{12}O_{19}$ and $SrFe_{12}O_{19}$ crystals. *J. Solid State Chem.*, **8** (4), 339–347.

91 Duran, C., Messing, G.L., and Trolier-McKinstry, S. (2004) Molten salt synthesis of anisometric particles in the $SrO\text{-}Nb_2O_5\text{-}BaO$ system. *Mater. Res. Bull.*, **39** (11), 1679–1689.

92 Holmes, M., Newnham, R.E., and Cross, L.E. (1979) Grain-oriented ferroelectric ceramics. *Am. Ceram. Soc. Bull.*, **58** (9), 872.

93 Takenaka, T. and Sakata, K. (1980) Grain orientation and electrical properties of hot-forged $Bi_4Ti_3O_{12}$ ceramics. *Jpn. J. Appl. Phys.*, **19** (1), 31–39.

94 Watanabe, H., Kimura, T., and Yamaguchi, T. (1989) Particle orientation during tape casting in the fabrication of grain-oriented bismuth titanate. *J. Am. Ceram. Soc.*, **72** (2), 289–293.

95 Seabaugh, M.M., Cheney, G.L., Hasinska, K., Azad, A.M., Sabolsky, E.M., Swartz, S.L., and Dawson, W.J. (2004) Development of a templated grain growth system for texturing piezoelectric ceramics. *J. Intell. Mater. Syst. Struct.*, **14** (3), 209–214.

96 Descamps, M., Ringuet, G., Leger, D., and Thierry, B. (1995) Tape casting: relationship between organic constituents and the physical and mechanical properties of the tape. *J. Eur. Ceram. Soc.*, **15** (4), 357–362.

97 Kim, H.J., Krane, M.J.M., Trumble, K.P., and Bowman, K.J. (2006) Analytical fluid flow models for tape casting. *J. Am. Ceram. Soc.*, **89** (9), 2769–2775.

98 Muralt, P., Maeder, T., Sagalowicz, L., Hiboux, S., Scalese, S., Naumovic, D., Agostino, R.G., Xanthopoulos, N., Mathieu, H.J., Patthey, L., and Bullock, E.L. (1998) Texture control of $PbTiO_3$ and $Pb(Zr,Ti)O_3$ thin films with TiO_2 seeding. *J. Appl. Phys.*, **83** (7), 3835–3841.

99 Tani, T. (1998) Crystalline-oriented piezoelectric bulk ceramics with a perovskite-type structure. *J. Korean Phys. Soc.*, **32**, S1217–S1220.

100 Saito, Y., Takao, H., Tani, T., Nonoyama, T., Takatori, K., Homma, T., Nagaya, T., and Nakamura, M. (2004) Lead-free piezoceramics. *Nature*, **432** (7013), 84–87.

101 Tani, T. and Kimura, T. (2006) Reactive-templated grain growth processing for lead free piezoelectric ceramics. *Adv. Appl. Ceram.*, **105** (1), 55–63.

102 Lotgering, F.K. (1959) Topotactical reactions with ferromagnetic oxides having hexagonal crystal structures – I. *J. Inorg. Nucl. Chem.*, **9**, 113–123.

103 Watanabe, H., Kimura, T., and Yamaguchi, T. (1989) Particle orientation during tape casting in the fabrication of grain-oriented bismuth titanate. *J. Am. Ceram. Soc.*, **72** (2), 289–293.

104 Jones, J.L., Slamovich, E.B., and Bowman, K.J. (2004) Critical evaluation of the Lotgering degree of orientation texture indicator. *J. Mater. Res.*, **19** (11), 3414–3422.

105 Bunge, H.J. (1982) *Texture Analysis in Material Science: Mathematical Methods*, Butterworths, Boston.

106 Bowman, K.J. (2004) *Mechanical Behavior of Materials*, John Wiley & Sons, New York.

107 Jones, J.L., Vogel, S.C., Slamovich, E.B., and Bowman, K.J. (2004) Quantifying texture in ferroelectric bismuth titanate ceramics. *Scripta Mater.*, **51**, 1123–1127.

108 Dollase, W.A. (1986) Correction of intensities for preferred orientation in powder diffractometry – application of

the March model. *J. Appl. Crystallogr.*, **19**, 267–272.
109 Amorin, H., Kholkin, A.L., and Costa, M.E.V. (2006) Texture-property relationships in Bi-layered ferroelectric ceramics: a case study of $SrBi_2Ta_2O_9$. *Mater. Sci. Forum*, **514–516**, 170–174.
110 Garcia, R.E., Carter, W.C., and Langer, S.A. (2005) The effect of texture and microstructure on the macroscopic properties of polycrystalline piezoelectrics: Application to barium titanate and PZN-PT. *J. Am. Ceram. Soc.*, **88** (3), 750–757.
111 Jones, J.L., Hoffman, M., and Bowman, K.J. (2005) Saturated domain switching textures and strains in ferroelastic ceramics. *J. Appl. Phys.*, **98** (2), 024115.
112 Jones, J.L., Slamovich, E.B., Bowman, K.J., and Lupascu, D.C. (2005) Domain switching anisotropy in textured bismuth titanate ceramics. *J. Appl. Phys.*, **98** (10), 104102.
113 Schmid, H. (1994) Multi-ferroic magnetoelectrics. *Ferroelectrics*, **162**, 317–338.
114 Dzyaloshinskii, I.E. (1960) On the magneto-electrical effect in antiferromagnets. *Sov. Phys. JETP*, **10** (3), 628–629.
115 Astrov, D.N. (1960) The magnetoelectric effect in antiferromagnetics. *Sov. Phys. JETP*, **11** (3), 708–709.
116 Astrov, D.N. (1961) Magnetoelectric effect in chromium oxide. *Sov. Phys. JETP*, **13** (4), 729–733.
117 Rado, G.T. and Folen, V.G. (1961) Observation of magnetically induced magnetoelectric effect and evidence for antiferromagnetic domains. *Phys. Rev. Lett.*, **7** (8), 310–311.
118 Folen, V.J., Rado, G.T., and Stalder, E.W. (1961) Anisotropy of magnetoelectric effect in Cr_2O_3. *Phys. Rev.*, **6** (11), 607–608.
119 Fiebig, M. (2005) Revival of the magnetoelectric effect. *J. Phys. D*, **38**, R123–R151.
120 Web of knowledge http://portal.isiknowledge.com/.
121 Cheong, S.-W. and Mostovoy, M. (2007) Multiferroics: a magnetic twist for ferroelectricity. *Nature Mater.*, **6**, 13–20.
122 Ramesh, R. and Spaldin, N.A. (2007) Multiferroics: progress and prospects in thin films. *Nature Mater.*, **6**, 21–29.
123 Eerenstein, W., Mathur, N.D., and Scott, J.F. (2006) Multiferroic and magnetoelectric materials. *Nature*, **442**, 759–765.
124 Khomskii, D.I. (2006) Multiferroics: different ways to combine magnetism and ferroelectricity. *J. Magn. Magn. Mater.*, **306**, 1–8.
125 Hill, N.A. (2000) Why are there so few magnetic ferroelectrics? *J. Phys. Chem. B*, **104**, 6694–6709.
126 Nan, C.W. (1994) Magnetoelectric effect in composites of piezoelectric and piezomagnetic phases. *Phys. Rev. B*, **50** (9), 6082–6088.
127 Nan, C.W., Li, M., and Huang, J.H. (2001) Calculations of giant magnetoelectric effects in ferroic composites of rare-earth-iron alloys and ferroelectric polymers. *Phys. Rev. B*, **63**, 144415.
128 Nan, C.W., Li, M., Feng, X.Q., and Yu, S.W. (2001) Possible giant magnetoelectric effect of ferromagnetic rare-earth-iron-alloys-filled ferroelectric polymers. *Appl. Phys. Lett.*, **78**, 2527–2529.
129 Zhai, J.Y., Li, J.F., Viehland, D., and Bichurin, M.I. (2007) Large magnetoelectric susceptibility: The fundamental property of piezoelectric and magnetostrictive laminated composites. *J. Appl. Phys.*, **101**, 014102.
130 Ryu, J., Carazo, A.V., Uchino, K., and Kim, H.E. (2001) Magnetoelectric properties in piezoelectric and magnetostrictive laminate composites. *Jpn. J. Appl. Phys.*, **40**, 4948–4951.
131 Ryu, J., Priya, S., Uchino, K., and Kim, H.E. (2002) Magnetoelectric effect in composites of magnetostrictive and piezoelectric materials. *J. Electroceram.*, **8**, 107–119.
132 Dong, S., Zhai, J., Li, J., and Viehland, D. (2006) Near-ideal magnetoelectricity in high-permeability magnetostrictive piezofiber laminates with a (2-1) connectivity. *Appl. Phys. Lett.*, **89**, 252904. Reprinted with permission from © American Institute of Physics.

133 Jia, Y.M., Or, S.W., Lam, K.H., Chan, H.L.W., Tang, Y.X., Zhao, X.Y., and Luo, H.S. (2007) Magnetoelectric effect in composites of magnet, metal-cap, and piezoceramic. *Appl. Phys. A*, **86** (4), 525–528.

134 Xing, Z.P., Dong, S.X., Zhai, J.Y., Yan, L., Li, J.F., and Viehland, D. (2006) Resonant bending mode of Terfenol-D/steel/Pb(Zr, Ti)O_3 magnetoelectric laminate composites. *Appl. Phys. Lett.*, **89** (11), 112911.

135 Zhai, J.Y., Dong, S.X., Xing, Z.P., Li, J.F., and Viehland, D. (2006) Giant magnetoelectric effect in Metglas/polyvinylidene-fluoride laminates. *Appl. Phys. Lett.*, **89** (8), 083507.

136 Zhang, N., Ke, W., Schneider, T., and Srinivasan, G. (2006) Dependence of the magnetoelectric coupling in NZFO–PZT laminate composites on ferrite compactness. *J. Phys.: Condens. Matter*, **18**, 11013–11019.

137 Zheng, H., Wang, J., Lofland, S.E., Ma, Z., Mohaddes-Ardabili, L., Zhao, T., Salamanca-Riba, L., Shinde, S.R., Ogale, S.B., Bai, F., Viehland, D., Jia, Y., Schlom, D.G., Wuttig, M., Roytburd, A., and Ramesh, R. (2004) Multiferroic $BaTiO_3$-$CoFe_2O_4$ nanostructures. *Science*, **303**, 661–663.

138 Zavaliche, F., Zheng, H., Mohaddes-Ardabili, L., Yang, S.Y., Zhan, Q., Shafer, P., Reilly, E., Chopdekar, R., Jia, Y., Wright, P., Schlom, D.G., Suzuki, Y., and Ramesh, R. (2005) Electric field-induced magnetization switching in epitaxial columnar nanostructures. *Nano Lett.*, **5** (9), 1793–1796.

139 Shastry, S., Srinivasan, G., Bichurin, M.I., Petrov, V.M., and Tatarenko, A.S. (2004) Microwave magnetoelectric effects in single crystal bilayers of yttrium iron garnet and lead magnesium niobate-lead titanate. *Phys. Rev. B*, **70** (6), 064416.

140 Scott, J.F. (2007) Data storage – Multiferroic memories. *Nat. Mater.*, **6** (4), 256–257.

141 Seshadri, R. and Hill, N.A. (2001) Visualizing the role of Bi 6s “lone pairs” in the off-center distortion in ferromagnetic $BiMnO_3$. *Chem. Mater.*, **13**, 2892–2899.

142 Aizu, K. (1970) Possible species of ferromagnetic, ferroelectric, and ferroelastic crystals. *Phys. Rev. B*, **2** (3), 754–772.

143 Smolenskii, G.A. and Chupis, I.E. (1982) Ferroelectromagnets. *Sov. Phys. Uspekhi*, **25** (7), 475–493.

144 Kimura, T., Goto, T., Shintani, H., Ishizaka, K., Arima, T., and Tokura, Y. (2003) Magnetic control of ferroelectric polarization. *Nature*, **426** (6962), 55–58.

145 Kimura, T., Ishihara, S., Shintani, H., Arima, T., Takahashi, K.T., Ishizaka, K., and Tokura, Y. (2003) Distorted perovskite with e(g)(1) configuration as a frustrated spin system. *Phys. Rev. B*, **68** (6), 060403.

146 Hur, N., Park, S., Sharma, P.A., Ahn, J.S., Guha, S., and Cheong, S.W. (2004) Electric polarization reversal and memory in a multiferroic material induced by magnetic fields. *Nature*, **429** (6990), 392–395.

147 Lawes, G., Harris, A.B., Kimura, T., Rogado, N., Cava, R.J., Aharony, A., Entin-Wohlman, O., Yildirim, T., Kenzelmann, M., Broholm, C., and Ramirez, A.P. (2005) Magnetically driven ferroelectric order in $Ni_3V_2O_8$. *Phys. Rev. Lett.*, **95** (8), 087205.

148 Hyer, O., Hollmann, N., Klassen, I., Jodlauk, S., Bohaty, L., Becker, P., Mydosh, J.A., Lorenz, T., and Khomskii, D. (2006) A new multiferroic material: $MnWO_4$. *J. Phys.: Condens. Matter*, **18** (39), L471–L475.

149 Kimura, T., Lashley, J.C., and Ramirez, A.P. (2006) Inversion-symmetry breaking in the noncollinear magnetic phase of the triangular-lattice antiferromagnet $CuFeO_2$. *Phys. Rev. B*, **73** (22), 220401.

150 Yamasaki, Y., Miyasaka, S., Kaneko, Y., He, J.P., Arima, T., and Tokura, Y. (2006) Magnetic reversal of the ferroelectric polarization in a multiferroic spinel oxide. *Phys. Rev. Lett.*, **96** (20), 207204.

151 Quezel, S., Tcheou, F., Rossatmignod, J., Quezel, G., and Roudaut, E. (1977) Magnetic structure of the perovskite-like compound $TbMnO_3$. *Physica B + C*, **86–88** (2), 916–918.

152 Kenzelmann, M., Harris, A.B., Jonas, S., Broholm, C., Schefer, J., Kim, S.B., Zhang, C.L., Cheong, S.W., Vajk, O.P.,

and Lynn, J.W. (2005) Magnetic inversion symmetry breaking and ferroelectricity in $TbMnO_3$. *Phys. Rev. Lett.*, **95** (8), 087206.

153 Mostovoy, M. (2006) Ferroelectricity in spiral magnets. *Phys. Rev. Lett.*, **96** (6), 067601.

154 Katsura, H., Nagaosa, N., and Balatsky, A.V. (2005) Spin current and magnetoelectric effect in noncollinear magnets. *Phys. Rev. Lett.*, **95** (5), 057205.

155 Sergienko, I.A. and Dagotto, E. (2006) Role of the Dzyaloshinskii-Moriya interaction in multiferroic perovskites. *Phys. Rev. B*, **73**, 094434.

156 Chapon, L.C., Blake, G.R., Gutmann, M.J., Park, S., Hur, N., Radaelli, P.G., and Cheong, S.W. (2004) Structural anomalies and multiferroic behavior in magnetically frustrated $TbMn_2O_5$. *Phys. Rev. Lett.*, **93** (17), 177402.

157 Chapon, L.C., Radaelli, P.G., Blake, G.R., Park, S., and Cheong, S.W. (2006) Ferroelectricity induced by acentric spin-density waves in YMn_2O_5. *Phys. Rev. Lett.*, **96** (9), 097601.

158 Aliouane, N., Argyriou, D.N., Strempfer, J., Zegkinoglou, I., Landsgesell, S., and Zimmermann, M.V. (2006) Field-induced linear magnetoelastic coupling in multiferroic $TbMnO_3$. *Phys. Rev. B*, **73** (2), 020102.

159 Goto, T., Kimura, T., Lawes, G., Ramirez, A.P., and Tokura, Y. (2004) Ferroelectricity and giant magnetocapacitance in perovskite rare-earth manganites. *Phys. Rev. Lett.*, **92** (25), 257201.

160 Kimura, T., Lawes, G., and Ramirez, A.P. (2005) Electric polarization rotation in a hexaferrite with long-wavelength magnetic structures. *Phys. Rev. Lett.*, **94** (13), 137201.

161 Wang, J., Neaton, J.B., Zheng, H., Nagarajan, V., Ogale, S.B., Liu, B., Viehland, D., Vaithyanathan, V., Schlom, D.G., Waghmare, U.V., Spaldin, N.A., Rabe, K.M., Wuttig, M., and Ramesh, R. (2003) Epitaxial $BiFeO_3$ multiferroic thin film heterostructures. *Science*, **299** (5613), 1719–1722.

162 Kiselev, S.V., Zhdanov, G.S., and Ozerov, R.P. (1962) Detection of magnetic arrangement in $BiFeO_3$ ferroelectric by means of neutron diffraction study. *Dokl. Akad. Nauk SSSR*, **145** (6), 1255–1257.

163 Teague, J.R., Gerson, R., and James, W.J. (1970) Dielectric hysteresis in single crystal $BiFeO_3$. *Solid State Commun.*, **8**, 1073–1074.

164 Fedulov, S.A. (1961) Determination of Curie temperature for $BiFeO_3$ ferroelectric. *Dokl. Akad. Nauk SSSR*, **139** (6), 1345–1346.

165 Kozumi, H., Niizeki, N., and Ikeda, T. (1964) An X-ray study on Bi_2O_3-Fe_2O_3 system. *Jpn. J. Appl. Phys.*, **3**, 495–496.

166 Neaton, J.B., Ederer, C., Waghmare, U.V., Spaldin, N.A., and Rabe, K.M. (2005) First-principles study of spontaneous polarization in multiferroic $BiFeO_3$. *Phys. Rev. B*, **71** (1), 014113.

167 Li, J.F., Wang, J., Wang, N., Bai, F., Ruette, B., Pyatakov, A.P., Wuttig, M., Ramesh, R., Zvezdin, A.K., and Viehland, D. (2004) Dramatically enhanced polarization in (001), (101), and (111) $BiFeO_3$ thin films due to epitaxial-induced transitions. *Appl. Phys. Lett.*, **84** (25), 5261–5263.

168 Yun, K.Y., Ricinschi, D., Kanashima, T., Noda, M., and Okuyama, M. (2004) Giant ferroelectric polarization beyond 150 $\mu C/cm^2$ in $BiFeO_3$ thin film. *Jpn. J. Appl. Phys.*, **43** (5a), L647–L648.

169 Merz, W.J. and Anderson, J.R. (1955) Ferroelectric storage device. *Bell Lab. Record*, **33**, 335–342.

170 Scott, J.F. (2000) *Ferroelectric Memories*, Springer, Berlin.

171 Kingon, A.I. and Srinivasan, S. (2005) Lead zirconate titanate thin films directly on copper electrodes for ferroelectric, dielectric and piezoelectric applications. *Nat. Mater.*, **4** (3), 233–237.

172 Mason, W.P. (1955) Aging of the properties of $BaTiO_3$ and related ferroelectric ceramics. *J. Acoust. Soc. Am.*, **27**, 73–85.

173 Tagantsev, A.K., Stolichnov, I., Colla, E.L., and Setter, N. (2001) Polarization fatigue in ferroelectric films: Basic experimental findings, phenomenological scenarios, and microscopic features. *J. Appl. Phys.*, **90**, 1387–1402.

174 Lupascu, D.C. (2004) *Fatigue in Ferroelectric Ceramics and Related Issues*, Springer, Heidelberg.

175 Chaplya, P.M., Mitrovic, M., Carman, G.P., and Straub, F.K. (2006) Durability properties of piezoelectric stack actuators under combined electromechanical loading. *J. Appl. Phys.*, **100**, 124111.

176 Jones, J.L., Salz, C.R.J., and Hoffman, M. (2005) Ferroelastic fatigue of a soft PZT ceramic. *J. Am. Ceram. Soc.*, **88**, 2788–2792.

177 Lupascu, D.C., Aulbach, E., and Rödel, J. (2003) Mixed electromechanical fatigue of lead zirconate titanate. *J. Appl. Phys.*, **93** (9), 5551–5556.

178 Lupascu, D.C. and Rödel, J. (2005) Fatigue in bulk lead zirconate titanate actuator materials. *Adv. Eng. Mater.*, **7** (10), 882–898.

179 Lou, X.J., Hu, X.B., Zhang, M., Morrison, F.D., Redfern, S.A.T., and Scott, J.F. (2006) Phase separation in lead zirconate titanate and bismuth titanate during electrical shorting and fatigue. *J. Appl. Phys.*, **99** (4), 044101.

180 Winzer, S.R., Shankar, N., and Ritter, A.P. (1989) Designing cofired multilayer electrostrictive actuators for reliability. *J. Am. Ceram. Soc.*, **72**, 2246–2257.

181 Schneider, G.A., Rosteck, A., Zickgraf, B., and Aldinger, F. (1994) Crack growth in ferroelectric ceramics under mechanical and electrical loading, in *Electroceramics IV*, vol. **II** (eds R. Waser*et al.*), Augustinus Buchhandlung, Aachen.

182 Parton, V.Z. (1976) Fracture behavior of piezoelectric materials. *Acta Astronaut.*, **3**, 671–683.

183 Yang, W. (2002) *Mechatronic Reliability*, Springer, Berlin.

184 Dunn, M.L. (1994) The effects of crack face boundary conditions on the fracture mechanics of piezoelectric solids. *Eng. Fract. Mech.*, **48**, 25–39.

185 Jackson, J.D. (1975) *Classical Electrodynamics*, John Wiley & Sons, New York.

186 Kamlah, M. and Böhle, U. (2001) Finite element analysis of piezoceramic components taking into account ferroelectric hysteresis behavior. *Int. J. Solids Struct.*, **38**, 605–633.

187 Schwartz, R.W. and Narayanan, M. (2002) Development of high performance stress-biased actuators through the incorporation of mechanical pre-loads. *Sens. Actuators A, Physical*, **101**, 322–331.

188 Granzow, T., Leist, T., Kounga, A.B., Aulbach, E., and Rödel, J. (2007) Ferroelectric properties of lead zirconate titanate under radial load. *Appl. Phys. Lett.*, **91**, 142904.

189 Lynch, C.S., Chen, L., Suo, Z., McMeeking, R.M., and Yang, W. (1995) Crack growth in ferroelectric ceramics driven by cyclic polarization switching. *J. Intell. Mater. Syst. Struct.*, **6** (2), 191–198.

190 Cao, H.C. and Evans, A.G. (1994) Electric-field-induced fatigue crack growth in piezoelectrics. *J. Am. Ceram. Soc.*, **77**, 1783–1786.

191 Fang, F., Yang, W., Zhang, F.C., and Luo, H.S. (2005) Fatigue crack growth for $BaTiO_3$ ferroelectric single crystals under cyclic electric loading. *J. Am. Ceram. Soc.*, **88** (9), 2491–2497.

192 Kounga-Njiwa, A.B., Lupascu, D.C., and Rödel, J. (2004) Crack tip switching zone in ferroelectric ferroelastic materials. *Acta Mater.*, **52** (16), 4919–4927.

193 Lupascu, D.C., Kreuzer, M., Lucato, S.L.S., Rödel, J., and Lynch, C.S. (2001) A liquid crystal display of stress fields in ferroelectrics. *Appl. Phys. Lett.*, **78** (17), 2554–2556.

194 Fett, T. and Thun, G. (1998) Determination of room-temperature tensile creep of PZT. *J. Mater. Sci. Lett.*, **17**, 1929–1932.

195 Salz, C.R.J., Hoffman, M., Westram, I., and Rödel, J. (2005) Cyclic fatigue crack growth in PZTunder mechanical loading. *J. Am. Ceram. Soc.*, **88**, 1331–1333.

196 Freiman, S.W. and Pohanka, R.C. (1989) Review of mechanically related failures of ceramic capacitors and capacitor materials. *J. Am. Ceram. Soc.*, **72**, 2258–2263.

197 Sosa, H. (1992) On the fracture mechanics of piezoelectric solids. *Int. J. Solids Struct.*, **29**, 2613–2622.

198 Suo, Z., Kuo, C.-M., Barnett, D.M., and Willis, J.R. (1992) Fracture mechanics for piezoelectric ceramics. *J. Mech. Phys. Solids*, **40**, 739–765.

199 Lynch, C.S. (1998) Fracture of ferroelectric and relaxor electro-ceramics:

influence of electric field. *Acta Mater.*, **46**, 599–608.

200 Park, S. and Sun, C.T. (1996) Fracture criteria for piezoelectric ceramics. *J. Am. Ceram. Soc.*, **78**, 1475–1480.

201 Landis, C. (2003) On the fracture toughness of ferroelastic materials. *J. Mech. Phys. Solids.*, **50**, 1347–1369.

202 McMeeking, R.M. (2004) The energy release rate for a Griffith crack in a piezoelectric material. *Eng. Fract. Mech.*, **71**, 1149–1163.

203 Kuna, M. (2006) Finite element analyses of cracks in piezoelectric structures – a survey. *Arch. Appl. Mech.*, **76**, 725–745.

204 Westram, I., Oates, W.S., Lupascu, D.C., Rödel, J., and Lynch, C.S. (2007) Mechanism of electric fatigue crack growth in lead zirconate titanate. *Acta Mater.*, **55**, 301–312.

205 Wang, B.L. and Han, J.C. (2007) An accumulation damage model for fatigue fracture of ferroelectric ceramics. *Eng. Fract. Mech.*, **74** (9), 1456–1467.

206 Fang, F., Yang, W., and Zhu, T. (1999) Crack tip 90° domain switching in tetragonal lanthanum-modified lead zirconate titanate under an electric field. *J. Mater. Res.*, **14**, 2940–2944.

207 Zhu, T. and Yang, W. (1999) Fatigue crack growth in ferroelectrics driven by cyclic electric loading. *J. Mech. Phys. Solids*, **47**, 81–97.

208 Beom, H.G. and Jeong, K.M. (2005) Crack growth in ferroelectric ceramics under electric loading. *Acta Mech.*, **177**, 43–60.

209 Arias, I., Serebrinsky, S., and Ortiz, M. (2006) A phenomenological cohesive model of ferroelectric fatigue. *Acta Mater.*, **54**, 975–984.

210 Tan, X. and Shangy, J.K. (2002) In-situ transmission electron microscopy study of electric-field-induced grain-boundary cracking in lead zirconate titanate. *Philos. Mag. A*, **82**, 1463–1478.

211 Dewitte, C., Elst, R., and Delannay, F. (1994) On the mechanism of delamination fracture of $BaTiO_3$-based PTC thermistors. *J. Eur. Ceram. Soc.*, **14**, 481–492.

212 Furuta, A. and Uchino, K. (1993) Dynamic observation of crack propagation in piezoelectric multilayer actuators. *J. Am. Ceram. Soc.*, **76**, 1615–1617.

213 Ru, C. (2000) Electrode-ceramic interfacial cracks in piezoelectric multilayer materials. *J. Appl. Mech.*, **67**, 255–261.

214 Kingery, W.D., Bowen, H.K., and Uhlmann, D.R. (1976) *Introductions to Ceramics*, John Wiley & Sons, New York.

215 Vedula, V.R., Glass, S.J., Saylor, D.M., Rohrer, G.S., Carter, W.C., Langer, S.A., and Fuller, E.R. Jr, (2001) Residual stress prediction in polycrystalline alumina. *J. Am. Ceram. Soc.*, **84**, 2947–2954.

216 Zimmermann, A., Carter, W.A., and Fuller, E.R. Jr, (2001) Damage evolution during microcracking in brittle solids. *Acta Mater.*, **49**, 127–137.

217 Zhang, Y. and Jiang, Q. (1995) Twinning-induced internal stress in ferroelectric ceramics. *Proc. SPIE*, **2442**, 11–22.

218 Lucato, S.L.S., Lupascu, D.C., Kamlah, M., Rödel, J., and Lynch, C.S. (2001) Constraint induced crack initiation at electrode edges. *Acta Mater.*, **49**, 2751–2759.

219 Nuffer, J., Lupascu, D.C., and Rödel, J. (2001) Microcrack clouds in fatigued electrostrictive 9.5/65/35 PLZT. *J. Eur. Ceram. Soc.*, **21**, 1421–1423.

220 Zhang, Y., Chen, Z.W., Cheng, X., and Zhang, S. (2004) *In situ* XRD investigation of domain switching in ferroelectric ceramics PLZT during an electric fatigue process. *Acta Metall. Sinica*, **40** (12), 1299–1304.

221 Kroupa, F., Nejezchleb, K., and Saxl, I. (1988) Anisotropy of internal stresses in poled PZT ceramics. *Ferroelectrics*, **88**, 123–137.

222 Kroupa, F., Nejezchleb, K., Rataj, J., and Saxl, I. (1989) Non-homogeneous distribution of internal stresses and cracks in poled PZT ceramics. *Ferroelectrics*, **100**, 281–290.

223 MacLaren, I., Schmitt, L.A., Fuess, H., Kungl, H., and Hoffmann, M.J. (2005) Experimental measurement of stress at a four-domain junction in lead zirconate titanate. *J. Appl. Phys.*, **97**, 094102.

224 Verdier, C., Morrison, F.D., Lupascu, D.C., and Scott, J.F. (2005) Fatigue studies

in compensated bulk lead zirconate titanate. *J. Appl. Phys.*, **97**, 024107.

225 Prabakar, K. and Mallikarjun Rao, S.P. (2007) Complex impedance spectroscopy studies on fatigued soft and hard PZT ceramics. *J. Alloys Compd.*, **437**, 302–310.

226 Duiker, H.M., Beale, P.D., Scott, J.F., Paz de Araujo, C.A., Melnick, B.M., Cuchiaro, J.D., and McMillan, L.D. (1990) Fatigue and switching in ferroelectric memories: Theory and experiment. *J. Appl. Phys.*, **68** (11), 5783–5791.

227 Mayoux, C. (2000) Degradation of insulating materials under electrical stress. *IEEE Trans. Dielectr. Electr. Insul.*, **7**, 590–601.

228 Wang, J.-L., Lai, Y.-S., Chiou, B.-S., Tseng, H.Y., Tsai, C.-C., Juan, C.-P., Jan, C.-K., and Cheng, H.-C. (2006) Study on fatigue and breakdown properties of Pt/(Pb,Sr)TiO_3/Pt capacitors. *J. Phys.: Condens. Matter*, **18**, 10457–10467.

229 MacChesney, J.B., Jetzt, J.J., Potter, J.F., Williams, H.J., and Sherwood, R.C. (1966) Electrical and magnetic properties of the system $SrFeO_3$–$BiFeO_3$. *J. Am. Ceram. Soc.*, **49** (12), 644–647.

230 Lebeugle, D., Colson, D., Forget, A., and Viret, M. (2007) Very large spontaneous electric polarization in $BiFeO_3$ single crystals at room temperature and its evolution under cycling fields. *Appl. Phys. Lett.*, **91**, 022907.

231 Lee, D., Kim, M.G., Ryu, S., Jang, H.M., and Lee, S.G. (2005) Epitaxially grown La-modified $BiFeO_3$ magnetoferroelectric thin films. *Appl. Phys. Lett.*, **86**, 222903.

232 Jiang, A.Q., Lin, Y.Y., and Tang, T.A. (2007) Unsaturated charge injection at high frequency of Pt/Pb(Zr,Ti)O_3/Pt thin-film capacitors. *Appl. Phys. Lett.*, **91**, 082901.

233 Colla, E.L., Taylor, D.V., Tagantsev, A.K., and Setter, N. (1998) Discrimination between bulk and interface scenarios for the suppression of the switchable polarization (fatigue) in Pb(Zr,Ti)O_3 thin film capacitors with Pt electrodes. *Appl. Phys. Lett.*, **72** (19), 2478–2480.

234 Li, W., Yin, Y., Su, D., and Zhu, J. (2005) Ferroelectric properties of polycrystalline bismuth titanate films by Nd^{3+}/W^{6+} cosubstitution. *J. Appl. Phys.*, **97**, 084102.

235 Lupascu, D.C., Fedosov, S., Verdier, C., von Seggern, H., and Rödel, J. (2004) Stretched exponential relaxation in fatigued lead-zirconate-titanate. *J. Appl. Phys.*, **95**, 1386–1390.

236 Belov, A.Yu. and Kreher, W.S. (2006) Simulation of microstructure evolution in polycrystalline ferroelectrics–ferroelastics. *Acta Mater.*, **54**, 3463–3469.

237 Mueller, V. and Shchur, Ya. (2004) Aging, rejuvenation and memory due to domain-wall contributions in RbH_2PO_4 single crystals. *Europhys. Lett.*, **65** (1), 1 37–143.

238 Lupascu, D.C. (2006) Fatigue in ferroelectric ceramics due to cluster growth. *Solid State Ionics*, **177**, 3161–3170.

239 Geng, L. and Yang, W. (2006) Agglomeration of point defects in ferroelectric ceramics under cyclic electric field. *Modell. Simul. Mater. Sci. Eng.*, **14**, 137–155.

240 Li, W., Chen, A., Lu, X., and Zhu, J. (2005) Collective domain-wall pinning of oxygen vacancies in bismuth titanate ceramics. *J. Appl. Phys.*, **98**, 024109.

241 Dawber, M. and Scott, J.F. (2000) A model for fatigue in ferroelectric perovskite thin films. *Appl. Phys. Lett.*, **76**, 1060–1062 and 3655.

242 Dawber, M. and Scott, J.F. (2001) Fatigue and oxygen vacancy ordering in thin-film and bulk single crystal ferroelectrics. *Integr. Ferroelectr.*, **32**, 259–266.

243 Scott, J.F. (2001) Fatigue as a phase transition. *Integr. Ferroelectr.*, **38**, 125–133.

244 Tsukada, M., Cross, J.S., Fujiki, M., Tomotani, M., and Kotaka, Y. (2000) Evaluation of Pb(Zr,Ti)O_3 capacitors with top $SrRuO_3$ electrodes. *Key Eng. Mater.*, **181–182**, 69–72.

245 Hase, T., Noguchi, T., Takemura, K., and Miyasaka, Y. (1998) Fatigue characteristics of PZT capacitors with Ir/IrO_x electrodes. Proceedings 11th IEEE International Symposium on Applied Ferroelectrics, Montreux, Switzerland, pp. 7–10.

246 Bernstein, S.D., Wong, T.Y., Kisler, Y., and Tustison, R.W. (1993) Fatigue of ferro-electric $PbZr_xTi_yO_3$ capacitors with Ru and RuO_x electrodes. *J. Mater. Res.*, **8**, 12–13.

247 Vijay, D.P. and Desu, S.B. (1993) Electrodes for $PbZr_xTi_{1-x}O_3$ ferroelectric thin films. *J. Electrochem. Soc.*, **140**, 2640–2645.

248 Bao, D., Zhu, X., Alexe, M., and Hesse, D. (2005) Microstructure and ferroelectric properties of low-fatigue epitaxial, all (001)-oriented $(Bi,La)_4Ti_3O_{12}/Pb(Zr_{0.4}Ti_{0.6})O_3/(Bi,La)_4Ti_3O_{12}$ trilayered thin films on (001) $SrTiO_3$ substrates. *J. Appl. Phys.*, **98**, 014101.

249 Huang, C.-K., Chiou, Y.-K., Chu, Y.-C., Wu, T.-B., and Tsai, C.-J. (2006) Enhancement in ferroelectric properties of $Pb(Zr_{0.4}Ti_{0.6})O_3$ thin-film capacitors with PtO_x electrodes. *J. Electrochem. Soc.*, **153** (6), F115–F119.

250 Cao, J.-L., Solbach, A., Klemradt, U., Weirich, T., Mayer, J., Schorn, P.J., and Böttger, U. (2007) Probing fatigue in ferroelectric thin films with subnanometer depth resolution. *Appl. Phys. Lett.*, **91**, 072905.

251 Jo, J.Y., Yoon, J.-G., Lee, J.K., Koo, J.M., Won, J.Y., Kim, S.P., and Noh, T.W. (2004) Role of IrO_2 electrode in reducing the retention loss of $Ir/IrO_2/Pb(Zr,Ti)O_3/Ir$ capacitors. *Integr. Ferroelectr.*, **67**, 143–149.

252 Larsen, P.K., Dormans, J.M., Taylor, D.J., and van Veldhoven, P.J. (1994) Ferroelectric properties and fatigue of $PbZr_{0.51}Ti_{0.49}O_3$ thin films of varying thickness: Blocking layer model. *J. Appl. Phys.*, **76**, 2405–2413.

253 Ezhilvalavana, S. and Samper, V.D. (2005) Ferroelectric properties of sol-gel derived Ca modified $PbZr_{0.52}Ti_{0.48}O_3$ films. *Appl. Phys. Lett.*, **87**, 132902.

254 Chen, F., Liu, Q.Z., Wang, H.F., Zhang, F.H., and Wu, W. (2007) Polarization switching and fatigue in $Pb(Zr_{0.52}Ti_{0.48})O_3$ films sandwiched by oxide electrodes with different carrier types. *Appl. Phys. Lett.*, **90**, 192907.

255 Menou, N., Castagnos, A.-M., Muller, Ch., Goguenheim, D., Goux, L., Wouters, D.J., Hodeau, J.-L., Dooryhee, E., and Barrett, R. (2005) Degradation and recovery of polarization under synchrotron X-rays in $SrBi_2Ta_2O_9$ ferroelectric capacitors. *J. Appl. Phys.*, **97**, 044106.

256 Wang, Y., Shao, Q.Y., and Liu, J.-M. (2006) Enhanced fatigue endurance of ferroelectric $Pb_{1-x}Sr_x(Zr_{0.52}Ti_{0.48})O_3$ thin films prepared by sol-gel method. *Appl. Phys. Lett.*, **88**, 122902.

257 Simões, A.Z., Riccardi, C.S., Cavalcante, L.S., Longo, E., Varela, J.A., Mizaikoff, B., and Hess, D.W. (2007) Ferroelectric fatigue endurance of $Bi_{4-x}La_xTi_3O_{12}$ thin films explained in terms of X-ray photoelectron spectroscopy. *J. Appl. Phys.*, **101**, 084112.

258 Furukawa, T., Kuroiwa, T., Fujisaki, Y., Sato, T., and Ishiwara, H. (2005) Fatigueless ferroelectric capacitors with ruthenium bottom and top electrodes formed by metalorganic chemical vapor deposition. *Jpn. J. Appl. Phys.*, **44**, L378–L380.

259 Wang, Y., Wang, K.F., Zhu, C., Wei, T., Zhu, J.S., and Liu, J.-M. (2007) Fatigue suppression of ferroelectric $Pb_{1-x}Ba_x(Zr_{0.52}Ti_{0.48})O_3$ thin films prepared by sol-gel method. *J. Appl. Phys.*, **101**, 046104.

260 Li, K.T. and Lo, V.C. (2005) Simulation of oxygen vacancy induced phenomena in ferroelectric thin films. *J. Appl. Phys.*, **97**, 034107.

261 Kudzin, A.Yu., Panchenko, T.V., and Yudin, S.P. (1975) Behavior of 180° domain walls of barium titanate single crystal during the "fatigue" and recovery of switching polarization. *Sov. Phys. Solid State*, **175**, 1589–1590.

262 Paton, E., Brazier, M., Mansour, S., and Bement, A. (1997) A critical study of defect migration and ferroelectric fatigue in lead zirconate titanate thin film capacitors under extreme temperatures. *Integr. Ferroelectr.*, **18**, 29–37.

263 Liu, J.-M., Wang, Y., Zhu, C., Yuan, G.L., and Zhang, S.T. (2005) Temperature-dependent fatigue behaviors of ferroelectric $Pb(Zr_{0.52}Ti_{0.48})O_3$ and $Pb_{0.75}La_{0.25}TiO_3$ thin films. *Appl. Phys. Lett.*, **87**, 042904.

264 Brazier, M., Mansour, S., and McElfresh, M. (1999) Ferroelectric fatigue of $Pb(Zr,Ti)O_3$ thin films measured in atmospheres of varying oxygen concentration. *Appl. Phys. Lett.*, **74**, 4032–4033.

265 Shannigrahi, S. and Yao, K. (2005) Effects of WO_3 dopant on the structure and electrical properties of $Pb_{0.97}La_{0.03}Zr_{0.52}Ti_{0.48}O_3$ thin films. *Appl. Phys. Lett.*, **86**, 092901.

266 Kim, Y.S., Kim, Y.S., Kim, S., and No, K. (2004) Electronic structure and chemical bonding of Zr substitution of Ti site on $Pb(Zr_{1-x}Ti_x)O_3$ using DV-Xα method. *Integr. Ferroelectr.*, **64**, 297–303.

267 Yang, P., Deng, H., Shi, M., Tong, Z., and Qin, S. (2007) Growth and properties of $SrBi_2TaNbO_9$ ferroelectric thin films using pulsed laser deposition. *Mater. Sci. Eng. B*, **137**, 99–102.

268 Han, S.H., Ahn, W.S., Lee, H.C., and Choi, S.K. (2007) Ferroelectric properties of heteroepitaxial $PbTiO_3$ and $PbZr_{1-x}Ti_xO_3$ films on Nb-doped $SrTiO_3$ fabricated by hydrothermal epitaxy below Curie temperature. *J. Mater. Res.*, **22**, 1037–1042.

269 Shin, Y.-H., Grinberg, I., Chen, I.-W., and Rappe, A.M. (2007) Nucleation and growth mechanism of ferroelectric domain-wall motion. *Nature*, **449**, 881–884.

270 Liu, J.S., Zhang, S.R., Dai, L.S., and Yuan, Y. (2005) Domain evolution in ferroelectric thin films during fatigue process. *J. Appl. Phys.*, **97**, 104102.

271 Shvartsman, V.V., Kholkin, A.L., Verdier, C., Yong, Z., and Lupascu, D.C. (2005) Investigation of fatigue mechanism in ferroelectric ceramic via piezoresponse force microscopy. *J. Eur. Ceram. Soc.*, **25** (12), 2559–2561.

272 Takemura, K., Ozgul, M., Bornard, V., and Trolier-McKinstry, S. (2000) Fatigue anisotropy in single crystal $Pb(Zn_{1/3}Nb_{2/3})O_3$-$PbTiO_3$. *J. Appl. Phys.*, **88**, 7272–7277.

273 Menou, N., Muller, Ch., Baturin, I.S., Shur, V.Ya., and Hodeau, J.-L. (2005) Polarization fatigue in $PbZr_{0.45}Ti_{0.55}O_3$-based capacitors studied from high resolution synchrotron X-ray diffraction. *J. Appl. Phys.*, **97**, 064108.

274 Chou, C.-C., Hou, C.-S., and Yeh, T.-H. (2005) Domain pinning behavior of ferroelectric $Pb_{(1-x)}Sr_xTiO_3$ ceramics. *J. Eur. Ceram. Soc.*, **25**, 2505–2508.

275 Le Rhun, G., Poullain, G., Bouregba, R., and Leclerc, G. (2005) Fatigue properties of oriented PZT ferroelectric thin films. *J. Eur. Ceram. Soc.*, **25**, 2281–2284.

276 Liang, C.-S. and Wu, J.-M. (2005) Characterization of highly (110) and (111)-oriented $Pb(Zr,Ti)O_3$ films on $BaPbO_3$ electrode using Ru conducting barrier. *Appl. Phys. Lett.*, **87**, 022906.

277 Yoon, S.-M., Lee, N.-Y., Ryu, S.-O., Shin, W.-C., You, I.-K., and Yu, B.-G. (2005) Effect of ferroelectric switching time on fatigue behaviors of (117)- and (00l)-oriented $(Bi,La)_4Ti_3O_{12}$ thin films. *Thin Solid Films*, **484**, 374–378.

278 Balke, N., Lupascu, D.C., Granzow, T., and Rödel, J. (2007) Fatigue of lead zirconate titanate ceramics II: sesquipolar loading. *J. Am. Ceram. Soc.*, **90** (4), 1088–1093.

279 Zhou, L., Zuo, R.Z., Rixecker, G., Zimmermann, A., Utschig, T., and Aldinger, F. (2006) Electric fatigue in antiferroelectric ceramics induced by bipolar electric fatigue. *J. Appl. Phys.*, **99**, 044102.

280 Nuffer, J., Lupascu, D.C., and Rödel, J. (2000) Damage evolution in ferroelectric PZT induced by bipolar electric cycling. *Acta Mater.*, **48**, 3783–3794.

281 Noh, T.W., Park, B.H., Kang, B.S., Bu, S.D., and Lee, J. A new ferroelectric material for use in FERAM: Lanthanum-substituted bismuth titanate. Proceedings IEEE, 12th International Symposium on the Applications of Ferroelectrics, ISAF 2000, Honolulu, HI, USA, pp. 237–242.

282 Li, W., Chen, A., Lu, X., Zhu, J., and Wang, Y. (2005) Priority of domain wall pinning during the fatigue period in bismuth titanate ferroelectric thin films. *Appl. Phys. Lett.*, **86**, 192908.

283 Liu, J.S., Zhang, S.R., Dai, L.S., and Yuan, Y. (2005) Domain evolution in ferroelectric thin films during fatigue process. *J. Appl. Phys.*, **97**, 104102.

284 Dimos, D., Al-Shareef, H.N., Warren, W.L., and Tuttle, B.A. (1996) Photoinduced changes in the fatigue behaviour of $SrBi_2Ta_2O_9$ and $Pb(Zr,Ti)O_3$ thin films. *J. Appl. Phys.*, **80**, 1682–1687.

285 Ding, Y., Liu, J.S., Qin, H.X., Zhu, J.S., and Wang, Y.N. (2001) Why lanthanum-substituted bismuth-titanate becomes fatigue free in a ferroelectric capacitor with platinum electrodes. *Appl. Phys. Lett.*, **78**, 4175–4177.

286 Ding, Y., Liu, J.S., Maclaren, I., Wang, Y.N., and Kuo, K.H. (2001) Study of domain walls and their effect on switching property in $Pb(Zr,Ti)O_3$, $SrBi_2Ta_2O_9$, and $Bi_4Ti_3O_{12}$. *Ferroelectrics*, **262**, 37–46.

287 Paz de Araujo, C.A., Cuchiaro, J.D., McMillan, L.D., Scott, M.C., and Scott, J.F. (1995) Fatigue-free ferroelectric capacitors with platinum electrodes. *Nature*, **374**, 627–629.

288 Theissmann, R., Schmitt, L.A., Kling, J., Schierholz, R., Schonau, K.A., Fuess, H., Knapp, M., Kungl, H., Hoffmann, M.J. (2007) Nanodomains in morphotropic lead zirconate titanate ceramics: On the origin of the strong piezoelectric effect, *J. Appl. Phys.*, **102** 024111.

289 Morozov, M.I. and Damjanovic, D. (2008) Hardening-softening transition in Fe-doped $Pb(Zr,Ti)O_3$ ceramics and evolution of the third harmonic of the polarization response, *J. Appl. Phys.*, **104** 034107. Reprinted with permission © American Institute of Physics.

19
Magnetic Properties of Transition-Metal Oxides: From Bulk to Nano

Polona Umek, Andrej Zorko, and Denis Arčon

19.1
Introduction

Magnetism in transition-metal oxides (TMOs) has attracted human attention since ancient times. One of the first technological applications exploring the magnetic properties of these materials dates back to the twelfth century, with the use of a magnetic compass needle made from lodestone, an iron ore mineral known already by the ancient Greeks. During the past two centuries, however, the science of magnetic TMO materials has grown into a vast and rapidly developing field that continues to surprise. More recently, the discovery of high-temperature superconductivity in copper oxide phases, of colossal magnetoresistance in manganese oxides, and of half-metallic behavior of CrO_2 or unusual quantum states in low-dimensional copper and vanadium oxide phases, has had a major influence on research in areas of physics and chemistry, and also on the applications of TMOs. The richness of TMOs stems from the incomplete filling and characteristics of the transition metal $3d$ orbitals. Correlation effects and/or coupling to the lattice frequently bring TMOs close to various electronic or magnetic instabilities, which constantly challenges existing theories.

In recent years, TMOs have become associated with a very wide variety of applications, and the possibility of preparing them in the form of nanoparticles has made the choice even wider. Magnetic TMO particles and nanoparticles are of special interest for applications in magnetic fluids, data storage, catalysis, and bioapplications, to name but a few. Nanosized magnetic particles demonstrate magnetic properties that are totally different from those of their bulk counterparts; that is, they frequently display superparamagnetic behavior, a higher coercivity, a lower Curie temperature, enhanced magnetic susceptibility, and so on. Moreover, the constant requirements due to the ever-increasing demands of users have already led to the development of patterned media, in which data are stored in an array of single-domain TMO magnetic (nano)particles, close to the physical limitation of sub-50 nm magnetic nanoparticles that enables recording densities of up to 150 $Gbit\,cm^{-2}$ (1 $Tbit\,in^{-2}$) [1]. Understanding and tailoring the magnetic properties of sub-50 nm

Ceramics Science and Technology Volume 2: Properties. Edited by Ralf Riedel and I-Wei Chen

ISBN: 978-3-527-31156-9

magnetic nanoparticles and their assembly is currently one of the major challenges of condensed matter physics, chemistry, and the materials sciences. The current biomedical applications of magnetic TMO particles and nanoparticles include cellular therapy such as cell labeling and targeting, and as a cell biology research tool to separate and purify cell populations, for tissue repair, drug delivery, and for diagnostic methods using magnetic resonance imaging (MRI) [2–4]. Today, TMO magnetic systems are also believed to show the greatest promise as materials for *spintronic devices*, where it is not the electron charge but rather the electron spin that carries the information [5, 6].

It is almost impossible to cover the entire range of magnetic TMOs in one chapter, and for this reason the aim here will be to describe the structural and magnetic characteristics of only a few iron-oxide phases, ferrites, CrO_2 and, in more detail, a family of manganese oxides. This choice is purely that of the authors, and does not prejudice the (non)importance of other TMO magnetic phases. Later in the chapter, the correlation between the structural and magnetic properties of different manganese oxide phases will be addressed, after which the effects observed when these systems are prepared in nanoparticle form will be described.

19.2 Properties of Transition Metal 3d Orbitals

The basic building blocks of most of TMO phases are $(TM)O_6$ octahedra (where TM = transition metal). The blocks can form a large variety of structures by sharing edges and/or corners, with most such structures belonging to either one-dimensional (1-D) or two-dimensional (2-D) layered structures. In the case of 1-D structures, typically single, double or triple chains of edge-sharing $(TM)O_6$ octahedra create tunnels, which can accommodate different species (e.g., H_3O^+) or various cations (e.g., alkali or alkaline earth metal ions) that balance the negative charge on the TMO framework. In contrast, 2-D structures are formed as stacks of sheets of edge-sharing $(TM)O_6$ octahedra, with the interlayer space again filled with different cations or small molecules (H_2O, . . .).

The local environment will influence the electronic state of the transition metal ion within the basic building block, the $(TM)O_6$ octahedron. Its effect on the TM d-energy levels is usually treated within a crystal-field theory (or a ligand-field theory as an improved approximation). The valence-band orbitals of the $(TM)O_6$ octahedra, that mainly have TM $3d$ character in the octahedral (O_h) symmetry, split into two classes. The three t_{2g} orbitals (d_{xy}, d_{xz}, and d_{yz}) have smaller overlap with the ligand oxygen $2p$ atomic orbitals, and are therefore lower in energy than the two e_g orbitals ($d_{3z^2-r^2}$ and $d_{x^2-y^2}$), which overlap with the ligand p-orbitals to a more pronounced degree. For more than half-filled orbitals, the electron representation is usually replaced with a hole representation, which inverts the t_{2g} and e_g levels. A schematic splitting of the TM $3d$-electron levels in an octahedral crystal field is shown in Figure 19.1. It must be stressed here that in known TMO structures the symmetries of $(TM)O_6$ octahedra are often lowered from the octahedral symmetry. Tetrahedral distortions, for instance,

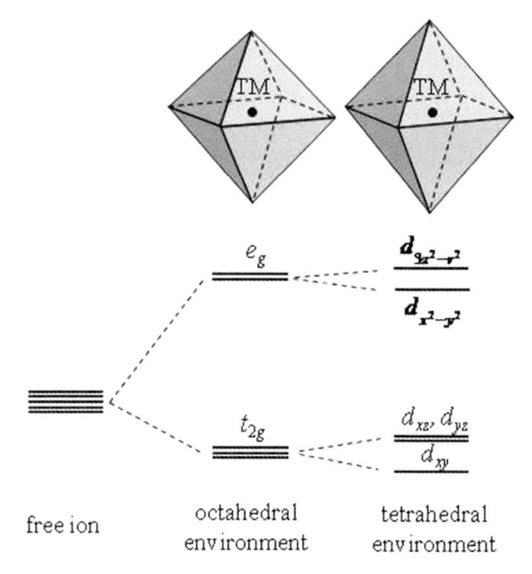

Figure 19.1 Crystal-field splitting of $3d$ one-electron levels in octahedral and tetrahedral environment within the $(TM)O_6$ octahedron.

split the e_g and the t_{2g} levels, as shown in Figure 19.1. However, when such distortions are small enough – as is usually the case – they will not significantly alter the energies of the orbitals, and the above division of TM d-orbitals will still be a very good approximation.

When these five d-orbitals are filled with electrons, Hund's rules must be obeyed. It is the competition between the energy of the crystal-field splitting and the Hund's correlation energy that defines the magnetic state of the transition-metal ion. In Table 19.1 are listed the electronic configurations and the ground-state effective spins of transition-metal ions for the case of weak and strong octahedral crystal fields (i.e., compared to the Hund's correlation energy), leading to a high- and a low-spin state, respectively.

19.3 Iron Oxides

19.3.1 Iron Oxide Structures

Iron oxides and iron hydroxides generally consist of arrays of iron ions and O^{2-} or OH^- ions. Iron appears in Fe^{2+} ($3d^6$) and Fe^{3+} ($3d^5$) valence states (Table 19.1), and is octahedrally coordinated to negatively charged O^{2-}/OH^-. FeO_6 (or $FeO_3(OH)_3$) octahedra are linked by corners, edges, or faces so as to form a number of different structures (a summary of known structures is given in Table 19.2). For hydroxide compounds – that is, goethite, lepidocrocite, akaganéite, δ-FeOOH and δ′-FeOOH –

Table 19.1 Electronic configurations and ground-state effective spins of 3*d* transition-metal ions in weak (high-spin state) and strong (low-spin state) octahedral crystal field.

Transition metal	Ionic form	Electronic configuration	Effective spin	
			High-spin	Low-spin
Titanium	Ti^{3+}	$[Ar]3d^14s^0$	1/2	1/2
Vanadium	V^{2+}	$[Ar]3d^34s^0$	3/2	3/2
	V^{3+}	$[Ar]3d^24s^0$	1	1
	V^{4+}	$[Ar]3d^14s^0$	1/2	1/2
Chromium	Cr^{2+}	$[Ar]3d^44s^0$	2	1
	Cr^{3+}	$[Ar]3d^34s^0$	3/2	3/2
Manganese	Mn^{2+}	$[Ar]3d^54s^0$	5/2	1/2
	Mn^{3+}	$[Ar]3d^44s^0$	2	1
	Mn^{4+}	$[Ar]3d^34s^0$	3/2	3/2
Iron	Fe^{1+}	$[Ar]3d^74s^0$	3/2	1/2
	Fe^{2+}	$[Ar]3d^64s^0$	2	0
	Fe^{3+}	$[Ar]3d^54s^0$	5/2	1/2
Cobalt	Co^{1+}	$[Ar]3d^84s^0$	1	1
	Co^{2+}	$[Ar]3d^74s^0$	3/2	1/2
	Co^{3+}	$[Ar]3d^64s^0$	2	0
Nickel	Ni^{2+}	$[Ar]3d^84s^0$	1	1
	Ni^{3+}	$[Ar]3d^74s^0$	3/2	1/2
Copper	Cu^{2+}	$[Ar]3d^94s^0$	1/2	1/2

the basic building block is the $FeO_3(OH)_3$ octahedron. In the case of goethite, which is one of the thermodynamically most stable iron oxides at ambient temperature, double chains of octahedra formed by edge sharing run along the [100] direction. The double chains are linked by corner-sharing to a nearest-neighboring pair, being displaced by $b/2$ and resulting in a tunnel-like structure. Among iron oxides, hematite (α-Fe_2O_3) and magnetite (Fe_3O_4) have the most interesting structures:

- *Hematite* is isostructural with corundum, and can be described as consisting of *hcp* arrays of oxygen ions stacked along the [001] direction. Two-thirds of the voids are then filled with Fe^{3+} ions.
- *Magnetite* has a face-centered cubic unit cell where tetrahedral sites are occupied by Fe^{3+} ions, while both Fe^{2+} and Fe^{3+} ions can be found at octahedral sites.

19.3.2 Magnetic Properties of Iron Oxides

Magnetic moments in iron oxides are associated with the Fe^{2+} and Fe^{3+} ions. Fe^{3+} ($3d^5$) is almost always in the high-spin state ($S = 5/2$); that is, the electronic configuration can be written as $(t_g)^3(e_g)^2$. The effective magnetic moment of Fe^{3+} is $\mu_{eff} = 5.9\,\mu_B$ because of the quenching of the orbital angular momentum. For the Fe^{2+} ion, the measured magnetic moment $\mu_{eff} = 5.1$–$5.5\,\mu_B$ is slightly larger than the

Table 19.2 Structural and magnetic properties of the different iron oxide phases.

Compound	Space group	Lattice constants	Magnetic properties	Reference(s)
Goethite [α-FeO(OH)]	Orthorhombic (*Pnma*)	$a = 9.96$ Å, $b = 3.02$ Å, $c = 4.61$ Å	Antiferromagnetic, $T_N = 400$ K	[7, 8]
Lepidocrocite [γ-FeO(OH)]	Orthorhombic (*Bbmm*)	$a = 3.07$ Å, $b = 12.5$ Å, $c = 3.87$ Å	Antiferromagnetic, $T_N = 77$ K	[9]
Akaganéite [β-FeO(OH)]	Monoclinic (*I2/m*)	$a = 10.6$ Å, $b = 3.03$ Å, $c = 10.5$ Å, $\beta = 90.63°$	Antiferromagnetic, $T_N = 290$ K	[10, 11]
Feroxyhyte [δ′-FeO(OH)]	Hexagonal (*P3ml*)	$a = 2.93$ Å, $c = 4.56$ Å	Ferrimagnetic, $T_C = 455$ K	[12]
δ-FeO(OH)	Hexagonal (*P3ml*)	$a = 2.93$ Å, $c = 4.49$ Å	Ferrimagnetic, $T_N = 440$–460 K	[12]
Ferrihydrite	Hexagonal (*P31c*; *P3*)	$a = 2.96$ Å, $c = 9.37$ Å	Speromagnetic, $T_N \sim 350$ K	[13, 14]
Hematite (α-Fe_2O_3)	Hexagonal (Rhombohedral)	$a = 5.03$ Å, $c = 13.8$ Å	Weakly ferromagnetic, $T_C = 956$ K; antiferromagnetic, $T_M = 260$ K,	[15–17]
Magnetite (Fe_3O_4)	Cubic (*Fd3m*)	$a = 8.40$ Å	Ferrimagnetic, $T_C = 850$ K	[14, 18]
Maghemite (γ-Fe_2O_3)	Cubic (*P4₃32*) Tetragonal (*P41212*)	$a = 8.35$ Å, $c = 25.0$ Å	Ferrimagnetic, $T_C = 820-986$ K	[11, 19]
Wüstite ($Fe_{1-x}O$)	Cubic (*Fm3m*)	$a = 4.30$ Å (high-Fe), $a = 4.28$ Å (low-Fe)	Antiferromagnetic, $T_N = 203-211$ K	[20]
(ε-Fe_2O_3)	Orthorhombic (*Pna2₁*)	$a = 5.10$ Å, $b = 8.79$ Å, $c = 9.44$ Å	Antiferromagnetic, $T_N = 1026$ K	[21]
$Fe(OH)_2$	Hexagonal (*P3ml*)	$a = 3.26$ Å, $c = 4.60$ Å	Planar antiferromagnetic, $T_N = 34$ K	[22, 23]
Bernalite	Orthorhombic (*Immm*)	$a = 7.54$ Å, $b = 7.56$ Å, $c = 7.56$ Å	Weakly ferromagnetic, $T_N \sim 427$ K	[24, 25]

calculated value of $\mu_{eff} = 4.9\,\mu_B$, due mostly to the non-zero orbital contribution. The main superexchange Fe–Fe interaction is of an antiferromagnetic nature. Antiferromagnetic superexchange interactions are strong, when the Fe^{3+}–O–Fe^{3+} or Fe^{2+}–O–Fe^{2+} angles are in the range between 120° and 180°, but become much weaker when the angles approach 90° (typically when the FeO_6 octahedra share faces) [26]. Another complication arises when Fe^{3+} and Fe^{2+} are nearest neighbors (as is the case in magnetite). In such cases, electron delocalization between Fe^{2+} and Fe^{3+} can be expected and the effective exchange interaction can be changed dramatically, due to a double exchange mechanism. The sensitivity of exchange interactions on the structural details of the iron ion's local environment also explains why the magnetic properties of the different iron oxide phases vary so widely (see Table 19.2).

Goethite orders antiferromagnetically below $T_N = 400$ K [7], with iron spins in each chain aligned along the crystal *b*-axis. The orientation alternates between two neighboring chains. The dominant antiferromagnetic exchange interaction is between corner-sharing octahedra in continuous chains, where the Fe–O–Fe angle has an almost optimal value of 124°.

Hematite (α-Fe_2O_3) undergoes a transition to a weakly ferromagnetic state at $T_C = 956$ K. The Fe^{3+} moments are aligned ferromagnetically within each of the basal layers, while the spins between layers are ordered antiferromagnetically along the *c*-axis. [17]. However, the magnetic moments are canted with a very small canting angle (<0.1°), leading to a weakly ferromagnetic state. The dominant exchange interaction is antiferromagnetic coupling between Fe^{3+} across the shared octahedral faces along the crystal *c*-axis. At the so-called Morin transition ($T_M = 260$ K), the competition between the weak anisotropy of the Fe^{3+} ions and the dipolar anisotropy causes the magnetic moments to reorient from the basal plane to an angle of 7° with respect to the *c*-axis [15], whereas below T_M the magnetic moments become collinear such that the ground state is that of collinear antiferromagnet [16].

Magnetite (Fe_3O_4) is a fascinating material, as it is the only mixed-valence iron oxide, and is metallic ferrimagnetic below $T_C = 850$ K. The ferrimagnetic state arises as iron ions occupy two different sites in the crystal structure: the tetrahedral sites are occupied by Fe^{3+} cations, while the octahedral sites are occupied by both Fe^{3+} and Fe^{2+}. As a result, two interpenetrating and inequivalent sublattices lead to a ferrimagnetic response. The electrons are thermally delocalized over Fe^{2+} and Fe^{3+} ions at the octahedral sites; this delocalization explains the high conductivity of magnetite. At the Verwey transition ($T_V = 118$ K), a discontinuous drop in conductance is due to charge ordering between the Fe^{2+} and Fe^{3+} cations. The electronic properties of Fe_3O_4 seem to be that of so-called half-metallic ferromagnets (this subject will be revised later in the chapter, when the properties of CrO_2 are discussed).

It is well known, and also well documented, that iron oxides can be prepared in the form of nanoparticles. The majority of such studies [26–39] have concentrated on the size effects on the magnetic properties in different iron-oxide phases. Typically, it has been found that transition temperatures decrease with decreasing particle size. For example, in hematite the Morin transition shifts from $T_M = 263$ K in bulk to temperatures below 4 K in particles smaller than 8–20 nm [33]. It is interesting to note that magnetic anisotropy rapidly increases for particles with diameters less than

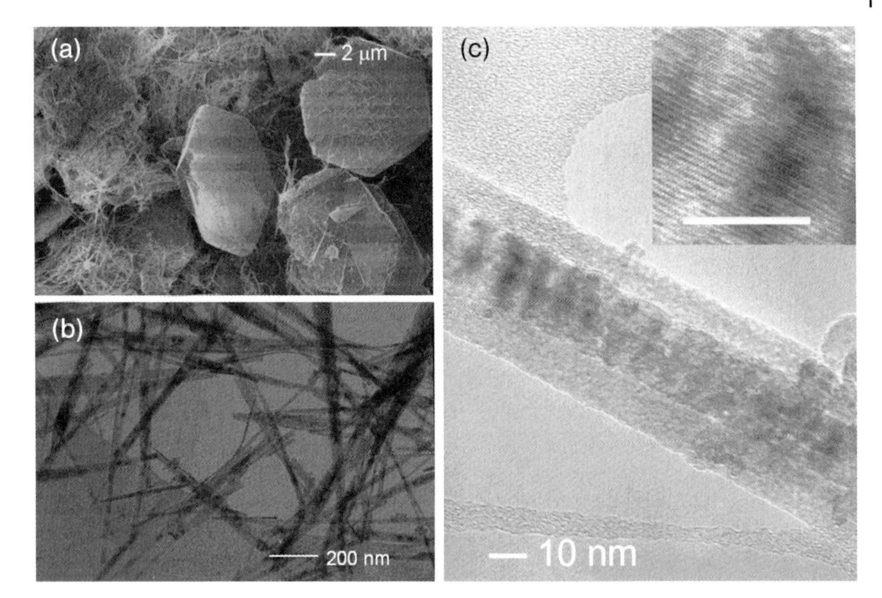

Figure 19.2 (a) Scanning electron microscopy image of the as-prepared sample. In addition to goethite α-FeOOH nanowires, large micrometer-sized crystallites of α-Fe_2O_3 (hematite) were found; (b) Transmission electron microscopy (TEM) image of the sample after sonication where hematite crystallites were removed; (c) High-resolution TEM image of the individual nanowire. Note that the nanowire substructure is composed of units with a 10–15 nm diameter. The inset shows an enlarged image of the crystalline core, where the layered structure is clear (scale bar = 10 nm). (Reproduced with permission from Ref. [40].)

~15 nm [39]. However, a common belief here is that strains, low-crystallinity, stoichiometry deviations and surface effects are responsible for such behavior. Recently, Pregelj *et al.* [40] studied the effect of nanoparticle shape on the magnetic properties of goethite by preparing goethite α-FeOOH nanowires with a typical diameter of about 10–15 nm and lengths of up to 600 nm (Figure 19.2). A superparamagnetic behavior of goethite nanowires was found from the susceptibility measurements, and also from the field-dependent measurements magnetization measurements at 4 K. The superparamagnetic blocking temperature was 34 K, and superparamagnetic resonance was observed below 375 K with a characteristic temperature dependence of the linewidth and signal center. The reported values for the remanent magnetization were $M_R = 22 \pm 1$ emu mol^{-1} (Fe), and for the coercive field $H_C = 490 \pm 10$ Oe. Annealing the goethite nanowires to temperatures above 530 K led to an irreversible transformation into the hematite phase.

19.4
Ferrites

Ferrites represent a general class of magnetic ceramic compounds with the chemical formula AB_2O_4, where A and B represent various metal cations, one of which is the

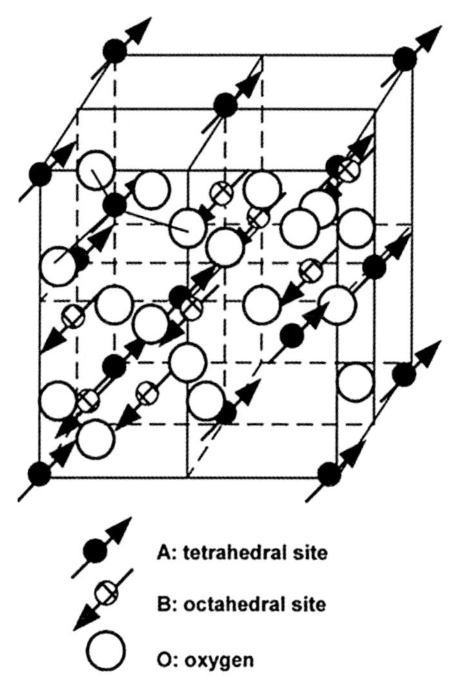

Figure 19.3 Schematic of a partial unit cell and ferrimagnetic ordering of spinel ferrite structure. Reproduced with permission from Ref. [44].

iron ion. O^{2-} ions form a cubic close-packed arrangement, while *A* cations occupy one-eighth and *B* cations one-half of the octahedral holes (Figure 19.3). In recent years, ferrites have attracted considerable attention for their wide range of technological applications, such as transformer cores, recording heads, antenna rods, memory, ferrofluids, biomedical application, and sensors [41, 42].

At this point, the spinel ferrite structure MFe_2O_4, where *M* refers to a metal, will be briefly discussed. As a typical representative, reference is made to the $ZnFe_2O_4$ structure, with Fe^{3+} ions occupying the octahedral sites and half of the tetrahedral sites. The remaining tetrahedral sites in this spinel are occupied by Zn^{2+} [43]. Depending on the distribution of the M^{2+} and Fe^{3+} cations between the tetrahedral and octahedral sites, the spinel structures are termed either "normal," "inverse," or "mixed." In the case of a normal spinel structure, all M^{2+} ions occupy tetrahedral sites; this is the case for $ZnFe_2O_4$. In the case of an inverse spinel structure, all M^{2+} are in octahedral positions and the Fe^{3+} cations are distributed equally between the tetrahedral and octahedral sites. The ferrites $NiFe_2O_4$ and $CoFe_2O_4$ have an inversed spinel structure (magnetite, Fe_3O_4, can also be classified as this structure). Finally, a mixed spinel structure is realized when the cations M^{2+} and Fe^{3+} occupy both tetrahedral and octahedral positions; $MnFe_2O_4$ is a typical example of this structure.

The basic magnetic properties of ferrites have been described previously by Néel [45]. The magnetic properties of ferrites are due to interactions between the two sublattice magnetizations – the sublattice of magnetic moments of cations in the tetrahedral sites, and the sublattice magnetization of cations in the octahedral site. Usually, the exchange interaction between magnetic ions of different sublattices is the strongest. with exchange interactions between tetrahedral nearest neighbors being almost an order of magnitude weaker. Interactions between the octahedral magnetic moments are the weakest of the three.

In $ZnFe_2O_4$ (normal spinel structure) the Zn^{2+} ions are diamagnetic, and therefore only the superexchange coupling between octahedrally coordinated Fe^{3+} magnetic moments will define the magnetic properties. The exchange pathway passes through the tetrahedral sites. As noted above, such coupling would be expected to be rather weak, and it is therefore not surprising that $ZnFe_2O_4$ orders antiferromagnetically below $T_N = 9–11$ K [46–48]. However, the detailed low-temperature magnetic behavior seems to be considerably more complex than just a simple transition into an antiferromagnetic state. The results of a recent muon-spin relaxation study showed that a magnetically short-range ordered (SRO) fraction appeared already at about 60 K and then coexisted with a long-range antiferromagnetically ordered fraction even well below T_N. Increasing the degree of inversion – that is, in structures where the Zn^{2+} cations partially occupy also octahedral sites – has a dramatic effect on the magnetic properties. For example, the behavior of a quenched sample is more typical for a highly frustrated system which cannot support long–range order and deviates into a spin glass-like magnetism [49]. This would mean that an originally present weak frustration in the antiferromagnetic couplings would be enhanced by a higher degree of inversion. In nanostructured samples, the non-stoichiometry and the degree of inversion may be rather high, while coupling between the tetrahedral and octahedral sites is increasingly important and may lead ultimately even to ferromagnetic exchange couplings. Clearly, $ZnFe_2O_4$ nanoparticles can behave in a quite different manner compared to their bulk counterparts, and hence are frequently described as superparamagnetic particles [47, 49].

Cobalt ferrite ($CoFe_2O_4$) has, in recent years, become one of the most important and most abundant magnetic materials. It is a ferromagnetic oxide with a magnetic moment of about 3.7 μ_B per formula unit, and with a Curie temperature of about 793 K. At low temperatures, cobalt ferrite undergoes a similar magnetic transition as magnetite at 90 K, such that it is no longer cubic below its magnetic transition temperature [50]. $CoFe_2O_4$ is well known to have a large magnetic anisotropy, a moderate saturation magnetization, remarkable chemical stability, and a mechanical hardness, all of which make it a good candidate for recording media. Cobalt ultrafine powders [51] and films [52, 53] are applied in a wide range of technological applications, but mainly as transformer cores, recording heads, antenna rods, and as memory and ferrofluids [41, 42]. Numerous methods for preparing cobalt ferrite nanoparticles have been proposed, with conventional techniques including sol–gel processing, hot-spraying, evaporation–condensation, matrix isolation, laser-induced vapor-phase reactions, and aerosols. In order to protect the oxidation of these nanoparticles from atmospheric oxygen, and also to block their agglomeration, the particles are usually coated and dispersed in a medium such as sodium dodecyl sulfate [54]. High-coercivity cobalt-ferrite nanoparticles that are <50 nm in size and which have a high coercivity (1440 Oe) and saturation magnetization (65 emu g^{-1})

were prepared from water and oil microemulsions [55]. Even smaller nanoparticles, with sizes ranging from 15 to 150 nm, were produced using a reverse micellar system of AOT (sodium bis(kethylhexy1) sulfosuccinate):isooctane [56, 57]. A microemulsion technique was reported to produce $CoFe_2O_4$ nanoparticles with a mean size of 12 nm [58] which, typically, behaved as superparamagnetic particles. Both, the blocking temperature and coercive field were typically increased with the increasing size of the nanoparticles [59].

Manganese-zinc ferrites ($Zn_xMn_{1-x}Fe_2O_4$) and nickel–zinc ferrites are industrially important magnetic materials. The former are used when constructing the cores of intermediate-frequency transformers, inductors, loudspeakers and other electromagnetic devices, while the latter have found application in the modern electronics industry, due primarily to their high electrical resistivity.

19.5 Chromium Dioxide

Chromium dioxide (CrO_2) shows unique magnetic properties, and has long been an important material in magnetic recording technology. Lately, CrO_2 has attracted renewed attention due not only to its practical importance in the technology of spintronics, tunneling magnetic resonance devices, magnetic heads and magnetic field sensors [60–62], but also to its distinct transport, optical, electronic, and structural properties. Currently, CrO_2 is a potential candidate for the fabrication of magnetic tunneling junction devices with a desirable low-field magnetoresistance. Nanoparticles of CrO_2 have the potential for use as various types of sensor, such as high-density magnetic recording medium in magnetic tapes, read sensors, random-access memory devices, optical data storage systems, biological sensory, and land-mine detectors [63–67].

CrO_2 crystallizes with a tetragonal rutile-type structure (space group: $P4_2/mnm$) where the chromium atoms form a tetragonal unit cell and the chromium sites are octahedrally coordinated by the oxygen atoms. The lattice unit cells are $a = 4.421$ Å and $c = 2.916$ Å [68]. At ambient pressure, but above 400 K, CrO_2 transforms into a thermodynamically more stable structure, Cr_2O_3, with lattice parameters $a = 4.951$ Å and $c = 13.566$ Å, and space group $R\bar{3}c$ [69].

Magnetically, Cr_2O_3 is an insulator where the Cr^{3+} ion has three well-localized $3d$ electrons that form a spin $S = 3/2$ antiferromagnet below 308 K. In contrast, CrO_2 is a metal, although with an unusually high room-temperature resistivity [70]. CrO_2 orders into a ferromagnetic ground state at $T_C = 392$ K, with a magnetic moment of 2 μ_B per chromium atom [71]. Although, ferromagnetic coupling has been interpreted in terms of a double-exchange mechanism [72, 73], from a simple electronic scheme it is difficult to anticipate that a double-exchange mechanism could be activated in CrO_2, since in the naïve picture only Cr^{4+} ions occupy the lattice. One of the possibilities investigated during recent years has been that the mixed-valence state of chromium ions arises from the oxygen nonstoichiometry. The results of recent theoretical studies have suggested that self-doping in CrO_2 generates a mixed-valence state of chromium ions and an oxygen-mediated double-exchange interaction [73, 74]. In these models, the chromium t_{2g} orbitals split into xy localized and $xz \pm yz$ itinerant states while, at the same time, the Cr(3d)–O(2p) hybridization induces a mixed-valence state of chromium ions. The existence of chromium ions in two different valence states has been confirmed using ^{53}Crnuclear magnetic resonance (NMR) measurements [75]. The zero-field ^{53}Cr NMR spectra measured on CrO_2 nanorods showed two clearly separated peaks with similar intensities, in contrast to the expected single peak corresponding to a uniform single-valence Cr^{4+} state. Self-doping opens the possibility for a double-exchange mechanism via the following scheme. Instead of the single-valence state of $Cr^{(4+)+\delta}O_2^{(2-)-\delta/2}$, electron hopping creates two chromium sites with different noninteger valence states, namely $Cr_{0.5}^{(4+)+x}Cr_{0.5}^{(4+)-x}O_2^{(2-)}$. Here, x is a self-doping ratio for $\delta \approx 0$. In the zero-field NMR, the position of the ^{53}Cr resonance is proportional to the magnetic moment of the chromium site. From the observed resonance frequencies (37.1 and 26.4 MHz), it was estimated that the two chromium sites had magnetic moments of 2.34 μ_B and 1.66 μ_B; in other words, their valence states corresponded to $+4.34$ and $+3.66$ ($x = 0.34$) [75]. This result was also consistent with the Hall resistivity measurements [76, 77].

Ultrasonic treatment may have a major influence on the saturation magnetization, resistivity, and magnetoresistance (MR) of the CrO_2 powder [78]. Both, the particle size and shape also affect the magnetic properties, with nanoparticles characteristically displaying a considerably lower saturation magnetization [79]. Nanoparticles in the form of thin platelets with a fairly sharp size distribution and an average crystallite size of 35 nm were prepared from a Cr^{4+}-polyvinyl alcohol (PVA) polymer precursor [80]. In this case, the reduced saturation magnetization in fine particles was most likely due to the existence of "magnetically dead" layer, which was estimated as being a few nanometers thick.

The CrO_2 electronic structure was solved only in 1987 by Kämper *et al.* [63]. CrO_2 is a "half-metallic" ferromagnet; that is, it is a metal for the majority (spin up) electrons, but exhibits a semiconductor-type gap for the majority (spin down) electrons (Figure 19.4). This picture was confirmed by a spin-resolved photoemission from polycrystalline CrO_2 films, which showed a spin polarization of almost 100% for

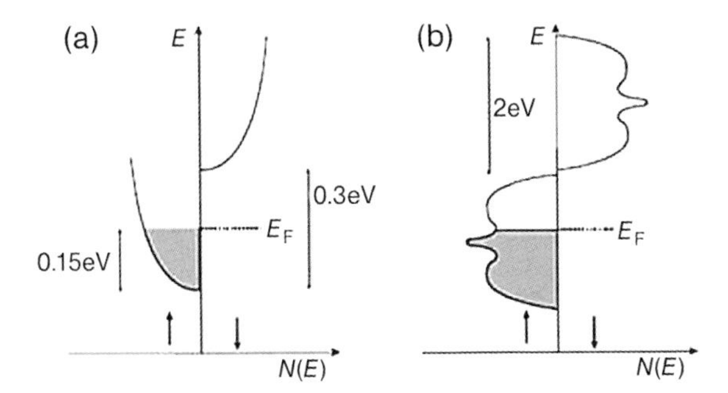

Figure 19.4 (a) Schematic densities of states $N(E)$ for a concentrated magnetic semiconductor below T_c; (b) Schematic densities of states $N(E)$ for the half-metallic ferromagnet CrO_2. Note that the energy scale is almost 10-fold larger in panel (b). Reproduced with permission from Ref. [81].)

binding energies close to 2 eV below the Fermi level (E_F). The d-band width was estimated to be about 1.6 eV, and did not cut the Fermi level. Electronic correlations are also important, as the Coulombic repulsion is estimated to be $U = 0.9$ eV. As a result, magnetism cannot be treated within the conventional Slater–Stoner–Wolfarth band-type, but rather must be associated with the localized moments aligned by some type of interatomic exchange interaction. Strong electronic correlations are seen experimentally as a characteristic T^2-dependence of the resistivity [82]. The MR was relatively small at low temperatures (<5% at 4.2 K) because the Jahn–Teller effect was not present [83]. On the other hand, in films or polycrystalline samples the MR was much larger (up to 50%) due to the half-metallic, spin-polarized electronic structure of CrO_2 [84, 85]. Tunneling MR most likely depends on the surface state, and the size and orientation of the acicular CrO_2 particles, and many experiments have appeared to corroborate with this interpretation [86, 87]. By introducing an interface barrier of Cr_2O_3, a dramatically enhanced MR effect is found in cold-pressed powder and nanoparticles [85]. There is, however, some confusion regarding the sign of the MR, with both Gupta *et al.* [88] and Watts *et al.* [77] having reported a positive 25% transverse MR with current parallel to the c-axis, whereas a negative transverse MR was reported by Suzuki *et al.* [83]. In order to resolve this ambiguity, Yuan *et al.* [89] prepared highly grain-oriented CrO_2 (100) films deposited onto TiO_2 (100) substrates; the low-temperature negative MR observed in some of these films was attributed to defects in the CrO_2 crystal, with the defects most likely occurring during the deposition process. A large positive MR was measured with the current parallel to the c-axis.

19.6
Manganese Oxide Phases

Manganese is the tenth most abundant element in Earth's crust, with more than thirty different manganese oxide minerals being identified among a wide range of geological settings. At the same time, a wide variety of synthesis routes has been developed to prepare these materials in different forms and in a controlled manner. For many centuries, manganese oxides have been used as pigments and also as an essential component in steel-making, although more recently they have found use in alkaline batteries as a cathodic material. In addition, manganese oxides are involved in many other different applications, including roles as catalysts to remove volatile organic compounds, as magnetic particles in magnetic resonance imaging, and in certain biogenic and bioscience applications. One of the most important uses of manganese oxide phases is related to the discovery of the colossal MR in certain manganites with perovskite structures, $Re_{1-x}A_xMnO_3$ (Re = rare earth; A = alkaline earth) [90]. The use of magnetoresistive materials in the read heads of hard disk drives allows the data not only to be stored in a more dense manner but also to be read at a faster rate.

The magnetism of manganese oxides is particularly strong, with the diversity of their magnetic properties being based on the electronic flexibility of the manganese ions. Manganese exists in range of different oxidations states, from Mn^{2+} to Mn^{7+}, although among the structures of most interest the valence varies between Mn^{2+} and Mn^{4+}. Strong local interactions involving Hund's rule exchange coupling, interaction to lattice via Jahn–Teller phonons, as well as onsite Coulombic correlations, are each believed to be responsible for the complex magnetic behavior of manganese ions in bulk. However, along with a decreasing size of the magnetic particles (down to nanometer range), finite-size effects will become increasingly relevant and eventually dominate the magnetic properties of the monodomain nanoparticles. On a nanoparticle surface, the magnetic properties would be expected to differ from those of the bulk phase, due to the different atomic coordination, and to the presence also of structural defects and strains. However, in addition to affecting the nanoparticles' magnetic properties, their transport properties may also be affected, leading in turn to dramatically different ground states compared to the bulk materials.

At this point, the main interest will concern Mn^{2+} ($3d^5$), Mn^{3+} ($3d^4$), and Mn^{4+} ($3d^3$) ions in octahedral environments, because these all occur in great abundance in compounds. In the simplest case of Mn^{4+} there are only three *d*-electrons, so that the three t_{2g} orbitals are all singly occupied; thus, by applying Hund's rule, their spins will be coupled to form a total spin $S = 3/2$. The three electrons in half-filled t_{2g} orbitals have only one possible arrangement, and consequently the $3d^3$ ground state is an orbital singlet. Due to its large splitting to excited orbital states, the expectation value of the orbital moment operators will be efficiently quenched in a cubic environment. As a result, the measured g-factor values will therefore be close to the free-electron value ($g_e = 2.0023$), and the magnetic anisotropies will be small. Mn^{3+} has one extra electron compared to Mn^{4+}. Although transition-metal ions with more than three electrons can exist in both high- and low-spin states, in the case of the octahedrally coordinated Mn^{3+} the high-spin state with a ${t_{2g}}^3{e_g}^1$ electronic configuration and a total spin of $S = 2$ is the ground state, this being an orbital doublet. By having a single electron in the doubly degenerate e_g orbitals, Mn^{3+} will be Jahn–Teller active, which means that the MnO_6 octahedron will be spontaneously distorted because the energy gain due to an increased elastic energy of deformation will be smaller than the energy loss due to the reduced electronic energy of one of the e_g orbitals. This is always possible at zero temperature, because the magnitude of the electronic splitting is linear in the ionic displacement *d*, while the potential energy is quadratic in *d*. At higher temperatures, the Jahn–Teller deformation can be a function of temperature, as higher-energy orbitals may also become populated. It should be mentioned at this point that the same distortion for Mn^{4+} ($3d^3$) would not be effective, because there is no net lowering of the electronic energy; Mn^{4+} is Jahn–Teller inactive. For Mn^{2+} ($3d^5$), each of the five $3d$ orbitals is singly occupied ($t_{2g}^3 e_g^2$) and all spins are ferromagnetically correlated (Hund's coupling) to give a total spin of $S = 5/2$ in the orbital singlet ground state. Again, the high-spin configuration ($t_{2g}^3 e_g^2$) is much more stable than the low-spin configuration ($t_{2g}^5 e_g^0$), while the exchange splitting of the Mn^{2+} crystal field orbitals is found to be quite large (4.5 eV) [91].

The above-described overview of the properties of manganese ions shows that, in manganese oxides, there will be a strong correlation between electronic systems where the interplay between spin, charge, orbital and lattice degrees of freedom plays a crucial role in determining the physical properties. As will be shown later, manganese oxides exhibit a plethora of magnetic, electronic and structural phase transitions. Moreover, the competition between the different phases makes these materials highly sensitive to lattice dimensionality and particle size, when grown in the form of nanoparticles, and also extremely vulnerable to external perturbations such as external fields and defects.

19.6.1 Manganese Oxide Structures

Binary metal oxides frequently show polymorphism, and manganese dioxide (MnO_2) is in this respect exceptionally rich, with at least fourteen polymorphs having been reported [92]. Such structural diversity is believed to originate from the small ionic radius of the Mn^{4+} ion (0.53 Å) and the $3d^3$ electronic configuration of Mn^{4+}, which prefers a octahedral over a tetrahedral coordination by about 2.79 eV [91]. MnO_2 structures also show intriguing tunnel motifs, where tunnels of different sizes (1×1, 1×2, 2×2, ...) result from a particular packing of edge- and corner-sharing MnO_6 octahedra. The structural characteristics of the most important forms of manganese oxides and the related hydroxides, will be briefly reviewed in the following sections (see also Table 19.3 and Figure 19.5).

19.6.1.1 One-Dimensional Structures

Pyrolusite (β-MnO_2) β-MnO_2 (mineral pyrolusite) is, thermodynamically, the most stable structure of all MnO_2 polymorphs. It has a structure analogous to the rutile structure, with a tetragonal symmetry ($P4_2/mnm$, JCPDS 24-0735) [93], and where single chains formed by edge-sharing MnO_6 propagate along the crystallographic *c*-axis. These chains link to four neighboring chains via corner-sharing. Such a MnO_6 network defines tunnels with square cross-section; that is, one octahedron by one octahedron, also referred as 1×1 tunnels (Figure 19.5a). The cross-section of the tunnels is too small to accommodate other chemical species, except for Li^+ ions, although even the intercalation of Li^+ ions is possible only up to 0.2 Li^+ per formula unit at room temperature [94]. It should be also noted, that minerals grown in nature frequently exhibit structural defects, due mostly to the oxidation of manganese [95], and that the same observation holds for samples prepared at low temperatures [96].

Ramsdellite (R-MnO_2) The ramsdellite (R-MnO_2) structure is closely related to that of β-MnO_2, except that the single-strand chains are replaced by double chains; that is, two adjacent chains also share their octahedral edges. The resultant structure again consists of tunnels [97] which are, in this case, of rectangular shape and determined with 1×2 octahedra sides (Figure 19.5b). In natural minerals this increased size of tunnels permits the intercalation of water or different cations, such as Na^+ or Ca^{2+}. Ramsdellite, which is isostructural with goethite (FeOOH), usually occurs as a rare mineral, but can be prepared synthetically from the spinel $LiMn_2O_4$ [98].

Table 19.3 Crystallographic summary of tunnel and layered manganese oxide structures with corresponding bulk magnetic properties.

Crystal. phase	Space group	Lattice constants	Magnetic properties	Reference
β-MnO_2 (pyrolusite)	Tetragonal ($P4_2/mnm$)	$a = 4.39$ Å, $c = 2.87$ Å	Helical AFM structure; $T_N = 92$ K	[93]
α-MnO_2 (hollandite) $K_xMn_8O_{16}\cdot yH_2O$ $Ba_xMn_8O_{16}\cdot yH_2O$	Tetragonal ($I4/m$) or monoclinic ($I2/m$)	$a = 4.39$ Å, $c = 2.87$ Å or $a = 10.03$ Å, $b = 5.76$ Å, $c = 9.90$ Å, $\beta = 90.42°$	AFM below $T_N = 40$ K ($Ba_{1.2}$) or $T_N = 52$ K ($K_{1.5}$)	[111]
R-MnO_2 (ramsdellite)	Orthorhombic ($Pbnm$)	$a = 4.53$ Å, $b = 9.27$ Å, $c = 2.87$ Å		[97]
γ-MnO_2 (nsutite)	Hexagonal	$a = 9.65$ Å, $b = 4.43$ Å		[112]
$Ba_{0.66}Mn_5O_{10}\cdot xH_2O$ (romanechite)	Monoclinic ($C2/m$)	$a = 13.929$ Å, $b = 2.8459$ Å, $c = 9.678$ Å		[113]
λ-MnO_2	Cubic ($Fd3m$)	$a = 8.225$ Å	3-D pyrochlore-type array with AFM LRO $T_N = 32$ K	[96]
$Mn_7O_{13}\cdot 5H_2O$ (birnessite)	Rhombohedral ($P3m1$)	$a = 2.84$ Å, $c = 7.27$ Å, $\gamma = 120°$		[114]

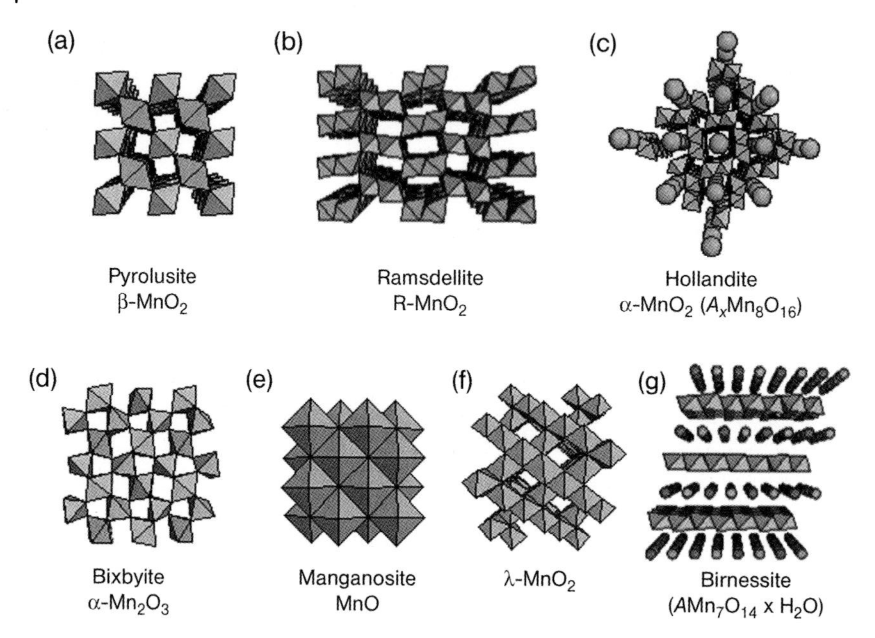

Figure 19.5 Polyhedral presentations of the crystal structures of some manganese oxide phases. (a) Pyrolusite; (b) Ramsdellite; (c) Hollandite; (d) Bixbyite; (e) Manganosite; (f) λ-MnO_2; (g) Birnessite.

It should be noted here that the γ-MnO_2 phase (Nsutite minerals), which is derived from pyrolusite (β-MnO_2) and ramsdellite (R-MnO_2), consists of domains with 1×1 and 1×2 tunnels, but usually is of poor crystallinity. The intergrowth of the two phases is described in terms of the de Wolff disorder [99]; that is, as a random growth of pyrolusite phases in the ramsdellite matrix. It should be stressed here that, among all MnO_2 structures, γ-MnO_2 is the best-known electrode material for batteries.

Hollandite (α-MnO_2) The next MnO_2 structure with an even larger size of tunnels is α-MnO_2, and the corresponding minerals found in nature are hollandite ($Ba_xMn_8O_{16}$) and cryptomelane ($K_xMn_8O_{16}$). α-MnO_2 grows in the tetragonal phase (space group $I4/m$, JCPDS No. 44-0141) with $a = b = 9.8776$ Å and $c = 2.8654$ Å. Cation intercalation distorts the structure so that the hollandite minerals usually adopt the monoclinic $I2/m$ symmetry. Hollandite structures consist of double chains of edge-sharing MnO_6 octahedra, resulting in 2×2 tunnels that extend along the c-axis of the tetragonal unit cell (Figure 19.5c). The α-MnO_2 tunnel structure is stabilized by Ba^{2+} or K^+ intercalation, or by water molecules located in the center of the 2×2 tunnels. Without large stabilizing cations, α-MnO_2 can be synthesized in a highly crystalline form [100, 101] by heat treatment at 300 °C; this removes the intercalated water without causing any collapse of the α-MnO_2 structure. Even though the instability of the 2×2 tunnel structure seems to limit the applicability of α-MnO_2, hollandite MnO_2 structures have in recent years been examined intensively for different applications, including ionic conductors, ion and molecular sieves, positive electrodes in Li-batteries, and oxidation catalysts, all of which take advantage of the material's porous structure.

MnO_2 frameworks with even larger tunnels can be found in Nature, with two of the best-known examples being romanechite $A_2Mn_5O_{10}{\cdot}xH_2O$ ($A = Ba^{2+}$, K^+, ...; tunnel size 2×3) and todorokite (tunnel size 3×3) [102, 103]. The detailed description of these materials is beyond the scope of this chapter.

19.6.1.2 Layered Structures

Birnessite (δ-MnO_2) Layered forms of MnO_2 are often termed "birnessite-type" compounds, after the mineral birnessite. These compounds have a characteristic layered structure (Figure 19.5g), with an interlayer distance of approximately 7 Å and intercalated stabilizing cations such as Na^+ or K^+ and a significant water content. In general, it has been proved to be quite difficult to remove the water from the structure without causing structural degradation. The structure [104] of the synthetic Na-, Mg-, and K-rich birnessites (e.g., $Na_4Mn_{14}O_{27}{\cdot}9H_2O$) is closely related to that of ternary oxides such as chalcophanite $Na_2Mn_3O_7$ [105]. Na-rich birnessite consists of almost vacancy-free layers composed of MnO_6 octahedra. The distortion of the hexagonal symmetry of the layers is caused by the Jahn–Teller distortion, associated with the substitution of Mn^{3+} for Mn^{4+}. The supercell $A = 3a$ parameter ($a = 5.172$ Å) arises from the ordered distribution of Mn^{3+}-rich rows parallel to [010] and separated from each other along [100] by two Mn^{4+} rows [106].

19.6.1.3 Three-Dimensional Structures

Spinel λ-MnO_2 λ-MnO_2 is a metastable phase of manganese dioxide, which is obtained by extracting lithium from the $LiMn_2O_4$ compound and retaining its cubic spinel structure (Figure 19.5f) [107]. This can be achieved by either acid leaching [107, 108] or by electrochemical delithiation [108, 109]. At optimal conditions, the former method leads to samples with the composition $Li_{0.03}MnO_2$ [96]. It remains unclear as to whether the λ-MnO_2 phase can be stabilized without any lithium. The 3-D spinel structure with the above symmetry has a general formula $A_x[B_2]O_4$, where the B cations reside on the octahedral 16d sites, the oxygen anions on the 32e sites, and the A cations occupy the tetrahedral 8a sites. In $LiMn_2O_4$, a cubic close-packed array of oxide ions incorporates the MnO_6 octahedron sharing two opposing corners with LiO_4 tetrahedra. The room-temperature cubic lattice unit cell parameter a varies with the lithium content, and increases from $a = 8.155(1)$ Å in $Li_{0.27}Mn_2O_4$ to $a = 8.225(1)$ Å in $Li_1Mn_2O_4$ [110].

19.6.2 Magnetic Properties of Selected Manganese Oxide Phases: The Manifestation of Magnetic Frustration

Magnetic interactions between manganese ions in manganese oxides can originate from both the direct Mn–Mn exchange, due to a direct overlap of manganese orbitals, and also from Mn–O–Mn exchange. The latter requires bridging oxygen orbitals and involves charge transfer, real or virtual. *Real charge transfers* are treated in first-order perturbation theory and lead to ferromagnetic double-exchange interactions in metals [115, 116]. *Virtual charge transfers* are treated in higher-order perturbation theory and result in superexchange interactions in insulators [117], which can be either ferromagnetic of antiferromagnetic. Their sign depends on the number of d electrons [117], and is closely related to the symmetry relationship between occupied cation orbital state and the overlapping anion orbitals [118]. For Jahn–Teller-active ions, lattice vibrations are capable of correlating electronic configurations on neighboring cations, and thus determining the configurations cooperating in the superexchange [119]. The dominance of either the direct or the superexchange mechanisms in a particular insulating manganese oxide depends on the distances between the manganese ions, the distances to the bringing ions, and the bridging Mn–O–Mn angles.

The magnetic behavior of manganese oxides is determined by the dominant magnetic interactions and the symmetry of the underlying spin lattice. This can be particularly complex and unpredictable on geometrically frustrated lattices, where not all of the pair-wise correlations can be simultaneously satisfied. The most common examples of this are triangular-based spin arrangements with antiferromagnetic nearest-neighbor interaction.

19.6.2.1 **Helical Order in Pyrolusite (β-MnO_2)**

Among the edge-shared manganese oxides, the pyrolusite β-MnO_2, featuring 1×1 tunnels, is a prototype compound because of its very simple crystal structure. From the perspective of magnetism, however, pyrulosite is extremely important as it represents one of the first known examples of nontrivial magnetic order. Its archetypal role is reflected in the fact that it has been presented as a model example of helical magnetism in several textbooks on magnetism.

Magnetism in β-MnO_2 is due to Mn^{4+} spins $S = 3/2$, residing on two nonequivalent magnetic sites, $(0, 0, 0)$ and $(1/2, 1/2, 1/2)$. The three $3d$ electrons of Mn^{4+} occupy localized t_{2g} orbitals (see Figure 19.1). As reviewed by Ohama *et al.* [120], although susceptibility measurements demonstrated significant discrepancies between different samples, an antiferromagnetic long-range ordering (LRO) transition was regularly detected in the susceptibility curve around $T_N = 92$ K [120]. This phase-transition was also observed from complementary heat-capacity [120, 121] and neutron diffraction measurements [122]. Recent single-crystal susceptibility measurements yielded an isotropic curve which, as expected, split below T_N [123]. As shown in Figure 19.6a, below T_N the susceptibility continued to increase with decreasing temperature for the c crystal axis, but started to decrease for the perpendicular directions; this makes the c-axis a hard magnetic axis.

Many years ago, in 1959, to account for the magnetic structure in the ordered state, a helical screw-type structure was proposed by Yoshimori [124], who suggested a spin structure in which the manganese spins screwed along the c-axis with a pitch of $7c/2$ (Figure 19.6b). In this way, the magnetic unit cell would be enlarged by a factor of seven in the c-direction with respect to the chemical unit cell. To stabilize this phase,

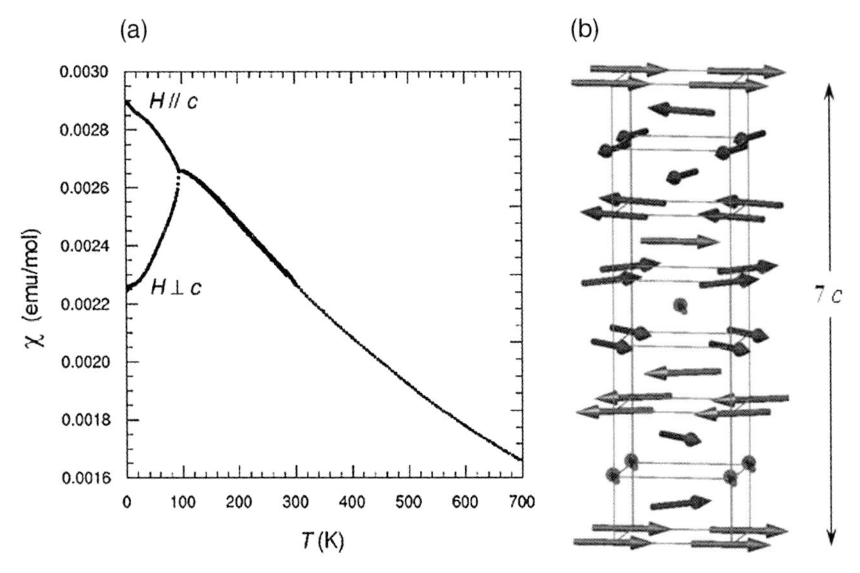

Figure 19.6 (a) Magnetic susceptibility of β-MnO_2 measured in the magnetic field of 1 T. Reproduced with permission from Ref. [123]; (b) Helical magnetic structure of β-MnO_2, as proposed by Yoshimori [124].

the three dominant exchange interactions – J_1, J_2, and J_3, between nearest neighbors in the (001), (111) and (100)/(010) directions, respectively [120] – must satisfy Yoshimori's constraints [32]. The magnetic propagation vector $\mathbf{k} = (0, 0, k_z)$ was strongly influenced by the ratio J_1/J_2, while the $k_z = 2/7$ value was stabilized by a crystal-field anisotropy only in the close vicinity of $J_1/J_2 = 1.6$, when $J_3 < J_2^2/2J_1$. For this reason, Yoshimori suggested that the pitch could be off from the precise 7/2 value. Taking into account the above constraints, the exchange constant $J_1 = 27$ K was determined from the susceptibility value at T_N; however, such a value caused a significant overestimate of T_N in a mean-field picture. This implied the existence of a delocalization effect of t_{2g} electrons, and an important role of conducting e_g electrons in the magnetism of β-MnO_2 [123]. Since J_1 and J_2 are similar in magnitude and both are antiferromagnetic, a strong geometric frustration was seen to be present between the body-centered site and its eight neighbors.

The helical order, commensurate with the underlying crystal structure, was consistent with early neutron diffraction measurements, within experimental accuracy [122, 124]. Recently, much effort has been devoted to refining the magnetic order, ultimately in an attempt to determine how "accurately" the magnetic helix is commensurate to the crystal lattice. A recent X-ray magnetic scattering study conducted by Sato *et al.* [125] revealed that the magnetic helix is, in fact, incommensurate to the crystal lattice, which was later confirmed by neutron diffraction [126]. The propagation vector length $k_z = 2/7 + \varepsilon$ was slightly larger than 2/7 and exhibited a local maximum near T_N, $\varepsilon = 0.0120(1)$ [127]. The results of the neutron diffraction study also showed that the magnetic order could be well described by a single-order parameter [126].

In addition to magnetic neutron Bragg peaks due to LRO, broad diffuse magnetic scattering was observed over a wide temperature range below and above T_N [127]. The width of the peaks did not vary between 50 K and 300 K, indicating a constant correlation length $\xi = 9.5$ Å within a SRO phase. Interestingly, this distance was close to the modulation period $\sim 7c/2$ of the LRO state [127]. The region where SRO was observed above T_N was unusually wide, and this proved to be one of the "fingerprints" of frustrated antiferromagnetism. In contrast, the coexistence of SRO and LRO below T_N was of fundamental importance.

β-MnO_2 is also important from another fundamental viewpoint, namely, that of phase transitions. Helical antiferromagnets belong to a new universality class because of their chiral degeneracy – that is, degeneracy between left- and right-handed helices. The critical exponent of the order parameter was found experimentally to be anomalously small, $\beta = 0.25(5)$, for a 3-D XY spin system on which β-MnO_2 can be mapped, whilst for a collinear case this value was 0.346 [128]. However, according to Monte-Carlo simulations, β would be reduced to 0.253(1) [129] if the chiral degree of freedom were to be considered, and which in turn would indicate major effects due to chiral degeneracy.

19.6.2.2 **Magnetic Properties of the Mixed-Valence Hollandite (α-MnO_2)**

Although, hollandite-type manganese oxides $A_xMn_8O_{16}$ (A = K, Ba, Pb, etc.) with 2 × 2, 1-D tunnels have been intensively studied for various possible applications, much less attention – quite surprisingly – has been paid to their magnetic properties. The tunnels are bordered by double chains of edge-sharing MnO_6 octahedra, linked through corners to each other. The double chains consist of manganese isosceles triangles (Figure 19.7a), which promote the effects of geometric frustration. It is important to stress here that the Mn–O–Mn angles in the double chains range

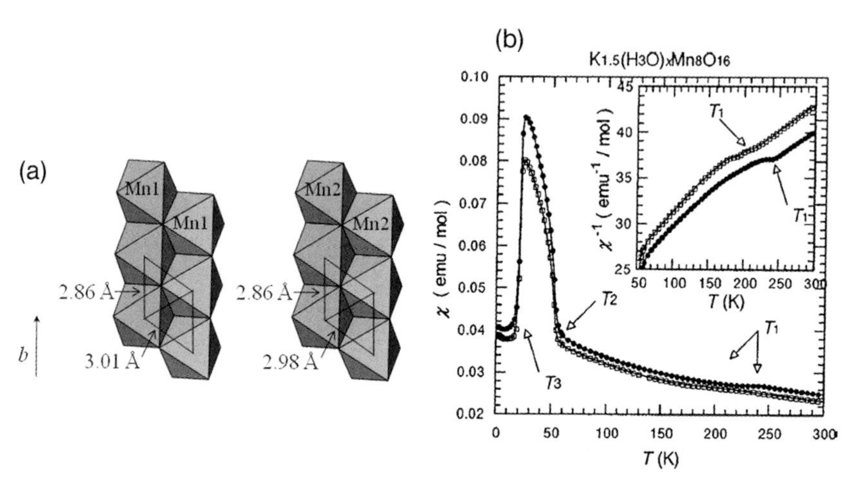

Figure 19.7 (a) Topology of the two different double chains and selected distances in $Ba_{1.2}Mn_8O_{16}$; (b) Magnetic susceptibility for two batches of $K_{1.5}(H_3O)_xMn_8O_{16}$ in the field of 1 T. Reproduced with permission from Ref. [131]).

between 90° and 100°, while the Mn–O–Mn angles for manganese belonging to neighboring double chains fall into the range of 125–129°. By applying Kanamori–Goodenough rules for the Mn–O–Mn superexchange, it can be deduced that the coupling between the double chains should be rather weak, and that the magnetic properties can be approximated with the behavior of independent double chains.

Another complication with the $A_xMn_8O_{16}$ hollandites is the role of intercalated cations in their electronic and magnetic properties; notably, the charge of the large cations is compensated by the oxidation state of manganese in the framework. Although it is speculated that Mn^{2+} (and Mn^{4+}) ions are more stable than Mn^{3+}, structural refinements of hollandites speak in favor of Mn^{3+} [111]. The presence of Mn^{3+} ions introduces delocalized e_g electrons and triggers interesting conducting and magnetic properties of the hollandite structures. The magnetism in hollandites, which is due to both the Mn^{3+} and Mn^{4+} ions (mixed-valence compounds), is therefore expected to change considerably with the A content.

In $Ba_{1.2}Mn_8O_{16}$ the average manganese valence is $+3.7$, and both types of ion are found in high-spin states [130]. This compound shows magnetic ordering at $T_N =$ 40 K, which is about tenfold lower than the Weiss constant, thus implying strong frustration effects. Although the magnetic state below T_N is fundamentally antiferromagnetic, there is a small ferromagnetic component which causes magnetic hysteresis and the splitting of field-cooled (FC) and zero-field-cooled (ZFC) susceptibility curves [130]. Magnetic neutron diffraction peaks suggest a complex spin arrangement with a doubled unit cell in the plane perpendicular to the double chains [130].

Another example of the hollandite-type manganese oxides is $K_{1.5}(H_3O)_x Mn_8O_{16}$ [131, 132], where the H_3O^+ groups occupy part of the 1/4 vacant sites in the tunnels, not occupied by potassium ions. These groups intercalate into the structure during a hydrothermal synthesis. From susceptibility measurements, three phase transitions were observed [131, 132]; at $T_1 = 180$–250 K, $T_2 = 52$ K, and $T_3 =$ 20 K (Figure 19.7b). The first transition corresponded to a charge localization of the Mn^{3+} and Mn^{4+} moments, and was very sensitive to the synthesis conditions. The Mn^{3+} and Mn^{4+} ions, which were in the averaged valence state $S \sim 1.63$ above T_1, underwent a charge separation below T_1 [131]. Between T_2 and T_3, a weak ferromagnetism appeared, with the spontaneous magnetization perpendicular to the direction of the double chains [131]. The spontaneous magnetization (0.011 μ_B per manganese atom) corresponded to only 0.3% of saturated magnetization, and completely disappeared below T_3. The susceptibility was almost temperature-independent below T_3, and its anisotropy was interpreted as an evidence of a proper-type helical structure with the screw axis parallel to chains [131].

19.6.2.3 Magneto-Elastic Coupling in the Layered α-$NaMnO_2$ Compounds

Among layered manganese oxides, the alpha polymorph (α-$NaMnO_2$) has recently attracted much attention because of its intriguing magnetic and elastic properties [133, 134]. Its crystal structure is of the α-$NaFeO_2$ type, with cubic close packing (*ABCABC*) of the oxygen ions, while the sodium and manganese ions occupy octahedral sites in consecutive oxygen interlayers [135]. The original rhombohedral structure is, however, distorted due to the Jahn–Teller distortion of MnO_6 octahedra [135]. As a direct result, the energy of one of the e_g states is lowered below the Fermi level, so that the high-spin state ($S = 2$) is stabilized for the Mn^{3+} magnetic moments. This was confirmed experimentally from the values of the high-temperature magnetic susceptibility [134].

The Mn^{3+} spin system can be mapped on a spatially anisotropic 2-D triangular lattice (Figure 19.8a). Depending on the ratio between the nearest-neighbor intrachain J_1 and interchain J_2 exchange interaction, different ground states could be realized, including LRO on two or three sublattices, an incommensurate magnetic order, or various exotic spin-liquid phases. The determination of these couplings was, therefore, highly desirable for further theoretical investigations of the ground state of α-$NaMnO_2$. The exchange couplings $J_1 = 65$ K and $J_2/J_1 = 0.44$ were determined from numerical fits of the susceptibility curve (Figure 19.8b) by Zorko *et al.*, using a finite-temperature Lanczos method (FTLM) [134]. These authors also reported that magnetic anisotropy was significant in this system. This magnetocrystalline contribution is of the single-ion type DS_z^2 ($D = -4.1$ K), and sets a magnetic easy axis in the direction of the Jahn–Teller elongation [134].

The system shows an antiferromagnetic LRO below $T_N = 45$ K [133]. This transition was hardly expressed in the magnetic susceptibility curve, and only its derivative showed a clear anomaly at that temperature (inset to Figure 19.8b). However, the setting of the order below T_N was unambiguously confirmed by magnetic Bragg peaks in high-resolution neutron powder diffraction measurements [133]. The proposed magnetic structure (Figure 19.8a) with the propagation vector $k = (1/2, 1/2, 0)$ was collinear, which corroborated with the strong magnetic easy axis in this material. The collinear structure was highly frustrated in terms of the

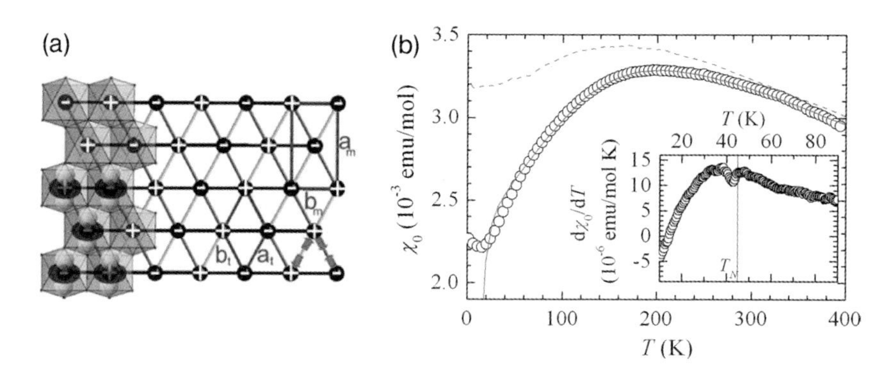

Figure 19.8 (a) Triangular lattice with nonequivalent exchange along the chains, J_1 (b_m direction), and perpendicular to them, J_2. The + and − signs indicate the direction of the ordered moments along the $d_{3r^2-z^2}$ orbitals (shown in gray) below $T_N = 45$ K. Reproduced with permission from Ref. [41]; (b) Magnetic susceptibility in α-$NaMnO_2$ at 0.1 T. The solid and broken lines correspond to fits with the Lanczos method and Monte-Carlo simulations, respectively. The temperature derivative of the susceptibility is shown in the inset. Reproduced with permission from Ref. [134].

interchain exchange, as each spin was coupled to one parallel and one antiparallel spin on each of the neighboring chains.

The frustration of the interchain exchange was, however, reported to be at least partially lifted below T_N through a structural phase transition which represented an interesting example of a coupling between magnetic and elastic degrees of freedom [133]. Namely, the magnetoelastic coupling caused a strong anisotropic broadening of the diffraction profiles, and ultimately led to a structural phase transition at T_N, lowering the symmetry from monoclinic to triclinic. Although the contraction/elongation of the interchain distances was only of the order of 0.1%, it seemed sufficient to lift the frustration to the level which, if supported by the strong magnetic anisotropy [134], would allow the stabilization of LRO. A subtile balance of degeneracy was proposed to effectively reduce dimensionality of spin excitations to 1D and lead to cancellation of interchain interaction.

The importance of the geometric frustration in α-$NaMnO_2$ is reflected in a broad cooperative paramagnetic region, $T_N \leq T < \theta$. Namely, the discrepancies between the theoretical predictions of the magnetic susceptibility between the classical Monte-Carlo simulations and quantum Lanczos simulations were observed up to room temperature (Figure 19.5b) [134]. Hence, quantum corrections should be important even at high temperatures, despite the relatively large value of the manganese spins, $S = 2$. The frustration was seen to be further responsible for a wide temperature region above T_N where the SRO phase was present. Broad asymmetric magnetic peaks showed up in the neutron diffraction patterns at rather high temperatures (i.e., below 200 K) [134]. This magnetic Warren-type scattering was attributed to 2-D magnetic ordering with the correlation length $\xi = 10$ Å at high temperatures. The presence of SRO was confirmed up to 260 K by a broadening of the electron spin resonance (ESR) spectra with decreasing temperature [134]. Interestingly, below T_N, α-$NaMnO_2$ exhibited a magnetic phase segregation, as 3- and 2-D ordered magnetic domains were shown to coexist down to 4 K [133].

19.6.2.4 Frustrated Magnetism of the "Defect" Spinel λ-MnO_2

Spinel λ-MnO_2 is obtained from the $LiMn_2O_4$ compound by removing Li. As no reports of a perfect delithiation have been reported to date, some of the observed magnetic properties in this material might not be inherent but rather provoked by the remaining Li^+ ions and/or induced Mn^{3+} ions. The Mn^{4+} ions occupy 16*d* octahedral sites. The magnetic sublattice is comprised of 3-D pyrochlore-type array of corner-shared manganese tetrahedra (Figure 19.9a). As such, by assuming dominant nearest-neighbor antiferromagnetic exchange interactions, the lattice may become highly geometrically frustrated because it cannot simultaneously satisfy the antiferromagnetic arrangement of all nearest neighbors on the tetrahedron [136]. In the classical Heisenberg model, with only nearest-neighbor exchange, the magnetic ground state is macroscopically degenerate [137]. The difficulty in selecting a unique ground state then results in an absence of any magnetic order, illustrating the most extreme consequence of the geometric frustration. The ground state should remain disordered down to the lowest temperatures, even if thermal fluctuations are considered.

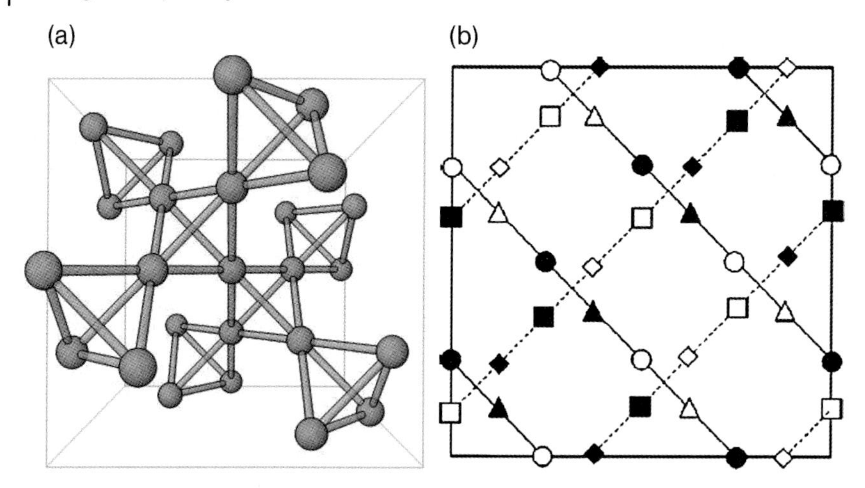

Figure 19.9 (a) Three-dimensional corner-sharing tetrahedral lattice found on the B-site of spinels; (b) The magnetic ground state of λ-MnO_2, drawn in a doubled conventional spinel unit cell, with four different sublattices (circles, triangles, diamonds, and squares). The filled and empty triangles represent up and down spins, respectively. Reproduced with permission from Ref. [138].

Magnetic exchange interactions in λ-MnO_2 have both antiferromagnetic and ferromagnetic contributions. For nearest neighbors, the former are due to direct exchange, while the latter result from the superexchange mechanism [139]. By performing first-principles calculations and parametrizing the obtained energies with the Heisenberg Hamiltonian, Morgan *et al.* [138] recently argued that the Heisenberg model is extremely accurate if interactions up to third-nearest neighbor are taken into account. According to local-density-approximation (LDA) calculations, the dominant exchange should then be of the order of 60 K [138].

Experimentally, λ-MnO_2 shows evidences of LRO [108], with a maximum in the susceptibility curve being observed around $T_N = 32$ K [108] and corresponding to long-range antiferromagnetic ordering. Below T_N, a FC/ZFC splitting was found [108, 109], which was interpreted as a spin-glass behavior by Jang *et al.* [109]. These authors proposed two scenarios to explain the suggested coexistence of LRO and spin glass. According to the first possible explanation, a cluster-spin-glass phase would be formed, in which case the fundamental freezing entities would be antiferromagnetic spin clusters with a finite correlation length. In the second explanation, a spatial segregation of antiferromagnetically ordered and spin-glass ordered regions might be present. Such a complex magnetic structure could be attributed to the geometric frustration and/or magnetic disorder arising from the small fraction of residual Li^+ ions. This picture of cluster formation was reinforced by measurements on single-crystal films of λ-MnO_2, grown on MgO substrates by plasma-assisted molecular beam epitaxy [140]. Subsequent susceptibility measurements implied an additional, weak, long-range ferromagnetic ordering among magnetization of clusters with a short-range antiferromagnetic order around $2T_N$. This transition would not be seen in polycrystalline samples due to averaging out of the field-induced magnetic anisotropy [140].

The presence of the LRO state below T_N was confirmed by neutron diffraction, where several magnetic Bragg peaks appeared at positions that required indexation with a doubled chemical cell, suggesting the ordering wave vector $\mathbf{k} = (1/2, 1/2, 1/2)$ [20]. In addition, diffuse magnetic scattering due to the presence of short-range order was detected above T_N up to at least 50 K. This was found not to be commensurate with the LRO magnetic unit cell, and its correlation length gradually decreased with temperature [108].

Both, the observation of the complex LRO and the presence of SRO over a wide temperature range can be attributed to a strong geometric frustration. The complexity of LRO is reflected in 128 manganese moments that constitute the magnetic unit cell. The originally proposed magnetic structure by Greedan *et al.* [108] was recently refined by Morgan *et al.* [138], who realized that there were other candidate ground states which met the condition of the face-centered magnetic ordering ($\mathbf{k} = (1/2, 1/2, 1/2)$) but possessed significantly lower energies than the state initially proposed. By using the same neutron diffraction data, these authors concluded that the magnetic structure had four types of sublattice (Figure 19.9b), connected through the third-nearest neighbor interaction. Within the isotropic Heisenberg model, there is an infinite class of degenerate ground states with the same energy as the collinear state shown in Figure 19.9b. The reason for this is that all relative orientations of the four sublattices with respect to one another, and remaining collinear within themselves, are degenerate in energy. In contrast, in the real system, long-range interactions or magnetic anisotropy might cause the sublattices to align.

19.6.3
Synthesis of MnO_2 Nanostructures

During the past few years, a significant effort has been directed towards the synthesis of low-dimensional MnO_2 nanostructures with controlled morphologies. For instance, α-, β-, γ-, δ-, λ-, and ε-MnO_2 polymorphs were synthesized in the shape of nanorods [141–145], nanowires [144, 146–148], nanofibers [149, 150], nanoneedles [151, 152], and nanotubes [153, 154]. Lately, the attention of a considerable number of chemists and materials scientists has also been oriented towards the self-assembly of 1-D nanostructured manganese dioxides into 2- and 3-D ordered microstructures [144, 155–158].

Currently, several established synthetic approaches have been reported for the synthesis of MnO_2 nanostructures, based on the redox reactions of MnO_4^- and/or Mn^{2+}; these include the hydrothermal method [143, 153, 156, 157, 159–161], as well as reflux [150, 162, 163], electrochemical [164], sonochemical [152] and solid-state chemistry [145, 165–168] approaches. The reaction conditions, crystallographic phases, precursors, and morphologies of the synthesized MnO_2 nanostructures are summarized in Table 19.4.

19.6.3.1 Synthesis of α-MnO_2 Nanostructures

A combination of hydrothermal conditions, an acidic environment, and a temperature within the range of 60 to 180 °C, tends to be favorable for the synthesis of α-MnO_2 nanostructures of different morphologies (Table 19.4). It seems that the

Table 19.4 Crystalline phases, precursors, synthesis conditions and morphologies overview for the synthesized MnO_2 nanostructures.

Crystalline phase	Precursors	Acid/base	Reaction conditions	Morphology	Reference(s)
α	$KMnO_4$	HCl	HR, 140 °C, 12 h	Nanowires/nanotubes	[146, 154]
α	$KMnO_4$	H_2SO_4	HR, 150 °C, 12 h	Nanorods	[169, 170]
α	$KMnO_4$	H_2SO_4	70–95 °C, 0.5 h	Nanorods and nanoneedles	[151]
α	$KMnO_4$, graphite	H_2SO_4	0 °C, 24 h	Nanorods	[171]
α	$KMnO_4$, fumaric acid	H_2SO_4, NH_4OH	RT, 30 min, pH = 4.5–5	Nanofibers	[149]
α	$KMnO_4$, Cu foil	H_2SO_4	HR, 60 °C, 8 h	Sea-urchin-like structures composed of nanorods	[157]
α	$KMnO_4$, $MnSO_4$		HR, 140 °C, 12 h	Nanowires	[144]
α and β[78]	$KMnO_4$, $MnSO_4 \cdot H_2O$		HR, 120–180 °C, 1–18 h	Nanorods/nanowires	[141, 142, 172]
α and β	$KMnO_4$, $MnCl_2$		HR, 180 °C, 48 h	Nanowires and nanorods	[161]
α	$KMnO_4$, $Mn(CH_3COO)_2$		Reflux, 1–10 h	Nanofibers	[150]
α	$MnSO_4 \cdot H_2O$, $(NH_4)_2S_2O_8$, $AgNO_3$ (or Ag foil)	H_2SO_4	RT, 1–2 days	Spherically aligned nanorods	[158]
α and β	$MnSO_4 \cdot H_2O$, $(NH_4)_2S_2O_8$, $AgNO_3$		HR, RT–80 °C	Sea-urchin-like structures composed of nanowires; nanorods	[173]
α and β	$MnSO_4 \cdot H_2O$, $(NH_4)_2S_2O_8$, $((NH_4)_2SO_4)$		HR, 120 °C, 12 h	Nanowires	[160]
α	$MnSO_4$, $K_2S_2O_8$, CuSO4,	H_2SO_4	RT, 3 days	Nanospheres	[174]
α	$Mn(CH_3COO)_3$	LiOH	SR, 30 °C, 3 h, pH = 7.2, Ar	Nanoneedles	[152]

β	$Mn(NO_3)_2$	(HNO_3)	HR, 160–180 °C, 3–72 h	Nanowires, microrods (dendrite-like nanostructures)	[144]
β	$Mn(NO_3)_2$, H_2O_2,	NH_4OH	HR, 250 °C, 12 h	Nanorods	[175]
β	$Mn(NO_3)_2 \cdot H_2O$, PEG-*b*-PPG	Citric acid, LiOH	HR, 100 °C, 48 h	Nanorods	[143]
β	$MnSO_4 \cdot H_2O$, $NaClO_3$, (PVP)		HR, 160 °C, 10 h	Nanorods (nanotubes)	[153, 176]
β	$MnCl_2 \cdot H_2O$, SDBS	NaOH	RT, 15 min	Spherical nanoparticles, nanorods	[177]
β	MnOOH nanorods		350 °C, 4 h	Nanorods	[145]
β	MnOOH		300 °C, 1 h	Nanowhiskers	[168]
β	γ-MnOOH		250–300 °C, 1–2 h	Nanorods	[167, 168]
γ	γ-MnO_2 powder	NH_4OH	HR, 120–160 °C, 24 h	Nanowires	[147]
γ	$MnSO_4$, $(NH_4)_2S_2O_8$		HR, 90 °C, 24 h	Nanorods and nanowires	[144]
γ	$MnSO_4$, $KBrO_3$, PEG or PVP		HR, 110–130 °C, 16 h	Sea-urchin-like nanostructures composed of nanorods and nanowires	[156]
γ	$[\{Mn(SO_4)(4,4'\text{-bpy})(H_2O)_2\}_n]$,	NaOH	Reflux, MeOH	Nanowires	[163]
γ	γ-MnO_2 powder	NH_4OH	HR, 120–160 °C, 24 h	Nanowires	[147]
δ	$MnSO_4$, $KMnO_7$,		HR, 160 °C,	Nanorods	[172]
λ	$Mn(Ac)_2 \cdot 4H_2O$, $N_2H_4 \cdot H_2O$, PVP		Reflux in DMSO and EtOH, 1–24 h	Nanodiscs	[162]
ε	$MnCl_2 \cdot 4H_2O$, $NaClO_4$		HR, 140–180 °C, 24 h	3-D hierarchical nanostructures	[155]
Birnessite	Mn_2O_3	NaOH	HR, 170 °C, 1 week	Nanobelts	[159]
Cryptomelane	$MnSO_4$, $KMnO_4$	H_2SO_4/KOH	Aging at 25, 45, and 95 °C	Nanowires	[148]
	$Mn(CH_3COO)_2$, Na_2SO_4		EC, RT	Nanotubes	[164]

Abbreviations: HR: hydrothermal reaction; SP: sonochemical process; EC: electrochemical reaction; RT: room temperature; PEG-*b*-PPG: poly(ethylene glycol)block-poly (propylene glycol); PEG: poly(ethylene glycol); PVP: polyvinylpyrrolidone.

Figure 19.10 (a) Transmission electron microscopy image of α-MnO_2 nanoparticles; (b) Scanning electron microscopy (SEM) image of α-MnO_2 nanotubes; (c) SEM image of sea-urchin-like structures composed of α-MnO_2 nanorods. Reproduced with permission from Refs [165], [154], and [157], respectively.).

reaction temperature [157, 158, 178] and time [141] have an impact on the secondary structure. Typically, nanorods/nanowires that were synthesized at temperatures below 100 °C and for reaction times between 1 and 3 h form round microstructures (Figure 19.10c). Under higher magnification it can be seen that, in some cases, these sea-urchin-like microstructures with diameters 1–2 μm have hollow cores [157].

Beside nanorods and nanowires, the existence of nanotubes and nanoparticles (Figure 19.10) with the α-MnO_2 crystallographic phase have also recently been reported. Nanoparticles (Figure 19.10a) with diameters of between 10 and 50 nm were synthesized [165] via a solid-state reaction between $KMnO_4$ and $MnCO_3$, while nanotubes (Figure 19.10b) were synthesized under hydrothermal conditions by a reduction of $KMnO_4$ in a solution of $HCl_{(aq)}$ [154]. The diameter of the nanotubes thus produced ranged between 80 and 150 nm (Figure 19.10b). When the reaction took place under similar conditions, with only difference being that $HCl_{(aq)}$ was exchanged with $H_2SO_{4(aq)}$, α-MnO_2 was grown in the nanorod morphology [151, 157, 169, 170]. In addition, MnO_2 nanorods in the α-phase were also prepared by the sonochemical hydrolysis of $Mn(CH_3COO)_3$ at pH 7.2 [152].

19.6.3.2 **Synthesis of β-MnO_2 Nanostructures**

In general, β-MnO_2 nanostructures were synthesized hydrothermally by the oxidation of a Mn^{2+} precursor in either neutral [153, 160, 161, 173, 176] or alkaline [143, 175] media, or by thermal decomposition of a MnOOH precursor [145, 166–168] that yielded only nanorod and nanowhisker morphologies. Products obtained via the thermal decomposition of MnOOH typically retained the dimensions of the MnOOH precursor. The diameters of the β-MnO_2 nanorods were found to be within the range of 30 to 400 nm, with lengths of up to several micrometers.

Hydrothermal synthesis, on the contrary, resulted in nanotube [153] and nanoparticle [177] morphologies, as well as in hexagonal star- and dendrite-like 2-D nanostructures [144] (Figure 19.11a and b). In this case, $Mn(NO_3)_2$ and $Mn(SO_4)_2$ salts were used as the most common sources of Mn^{2+}, with reaction temperatures ranging between 100 and 250 °C. Kim *et al.* [143] reported that the addition of a polyvinylpyrolidone (PVP) polymer to the reaction mixture, in combination with the post-treatment (calcination), resulted in a thin layer of α-Mn_2O_3 phase covering

Figure 19.11 Scanning electron microscopy images of (a) 2-D hexagonal star-like β-MnO_2 crystals synthesized from $Mn(NO_3)_2$ solution; and (b) dendrite-like hierarchical β-MnO_2 nanostructures synthesized from $Mn(NO_3)_2$ solution with the addition of nitric acid. Reproduced with permission from Ref. [144].

the β-MnO_2 nanorod surface (Figure 19.12). Clearly, a carbon coating during calcination in air reduced the manganese ions on the surface of the β-MnO_2 nanorods, from oxidation state 4+ to 3+.

19.6.3.3 Synthesis of γ-MnO_2 Nanostructures

Both, nanowires [144, 147, 156, 163] and nanorods [144, 156] of γ-MnO_2 were synthesized from different precursors. For instance, nanowires were synthesized by boiling a coordination polymer $[\{Mn(SO_4)(4,4'\text{-bpy})(H_2O)_2\}_n]$ in NaOH/MeOH solution [163], and also by the hydrothermal treatment of commercial γ-MnO_2 in water or NH_4OH solution [147] (Figure 19.13). In both cases, the diameters of the synthesized nanowires were found to range between 20 and 40 nm [163] and between 5 and 50 nm [147], with lengths of several micrometers. Nanorods were also

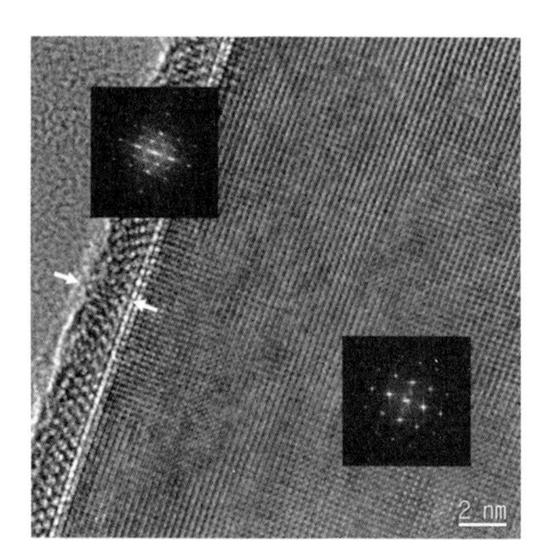

Figure 19.12 Atomic-resolved high-resolution transmission electron microscopy image of β-MnO_2 nanorod, showing that the surface of single crystalline β-MnO_2 nanorod is covered with a 2.5 nm-thin surface layer of α-Mn_2O_3 nanorod (located between the two arrows). The electron diffraction patterns of the two phases are also shown. Reproduced with permission from Ref. [143].

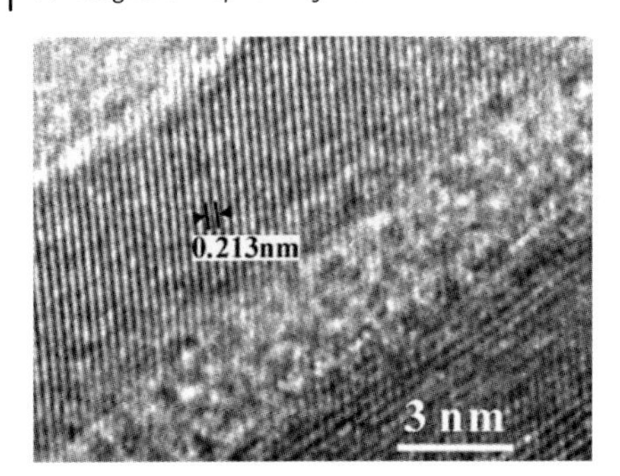

Figure 19.13 High-resolution transmission electron microscopy image of a single γ-MnO_2 nanowire. Reproduced with permission from Ref. [147].

synthesized by the oxidation of a Mn^{2+} precursor with $KBrO_3$ [156] or $(NH_4)_2S_2O_8$ [144] in a thermal bomb. When the polymers poly(ethylene glycol) (PEG) or PVP [156] were added to the reaction mixture of Mn^{2+} precursor and $KBrO_3$, the nanorods and nanowires formed uniform, sea-urchin-like 3-D γ-MnO_2 nanostructures.

19.6.3.4 **Synthesis of λ-MnO_2 Nanodiscs**

A solution-phase synthesis was used for the growth of λ-MnO_2 nanodiscs [162], using PVP as the modifying and protecting agent (Figure 19.14). In this case, a one-pot procedure combined the formation of nanoparticle precursors from manganese acetate, a 2-D assembly, phase transformation, and disk-shaping. The diameters of the nanostructures produced were 1–2 μm, while their thicknesses ranged from 50 to 100 nm.

Figure 19.14 Transmission electron microscopy image of the as prepared λ-MnO_2 nanodiscs. Reproduced with permission from Ref. [162].

19.6.3.5 Synthesis of MnO_2 Nanostructures in other Crystallographic Phases

A phase-controlled synthesis, using a comproportionation method, was employed to synthesize δ-MnO_2 nanorods. In this method, $KMnO_4$ was used as the oxidizing reagent for $MnSO_4$. In addition, according to Wang *et al.* [172], δ-MnO_2 served as an important intermediate in the synthesis of other crystallographic phases of 1-D MnO_2 nanostructures.

Cryptomelane-type MnO_2 nanowires [148] were synthesized from $MnSO_4$ and $KMnO_4$ in water between 60 and 95 °C. In this case, the reaction pH (which was either adjusted initially or fixed during the process of product formation) had a major influence on the diameters, lengths, and specific surface areas of the synthesized nanowires.

Birnessite-related MnO_2 nanobelts with a narrow width distribution (5–15 nm) and of high purity were synthesized under hydrothermal conditions from commercial Mn_2O_3 powders in NaOH solution [159].

19.6.4 Magnetic Properties of Manganese Dioxide Nanoparticles

As mentioned above, the magnetism in manganese oxide polymorphs is extremely rich due to the structural and electronic flexibility of the MnO_2 phases. Yet, when nanosize is introduced the behavior becomes even more complex. Notably, the nanoparticles frequently show dramatic structural inhomogeneities where different structures compete at the nanoscale. In the case of nanoparticles, it is not unusual for their core to grow in one phase, while their surface is formed in another polymorph (sometimes it is even amorphous). Therefore, the relatively complicated magnetic responses of investigated samples have been regularly interpreted as the coexistence of different phases. In magnetic nanoparticles, a large surface-to-volume ratio will increase the importance of shape and magnetocrystalline anisotropy, which may in turn alter their magnetic properties compared to the bulk analogous. Decreasing the particle size below a few nanometers would be expected to lead to monodomain behavior and associated superparamagnetic properties. Finally, as stoichiometry is usually poorly controlled during the nanoparticle synthesis, in the case of manganese oxides there may be a need to consider the existence of manganese ions with different valence states.

19.6.4.1 β-MnO_2 (Pyrolusite)

β-MnO_2 frequently grows in the form of nanorods. Wang *et al.* [175] reported the synthesis of β-MnO_2 nanorods with diameters of 25–40 nm and lengths of 240–440 nm. A kink in the magnetic susceptibility was noted at $T_N = 98$ K –that is, about 6 K higher than in the bulk β-MnO_2 [120]. This increase in T_N was attributed to the special morphology and effects of surface spins in as-prepared nanorods. It should be noted, that an increase in T_N to ~100 K was also reported for single-crystalline, dandelion-like β-MnO_2 3-D nanostructures [179]. For β-MnO_2 nanorods, another minor anomaly in the magnetic susceptibility in the temperature range 30–50 K was observed [175], though this might have been due to the presence of

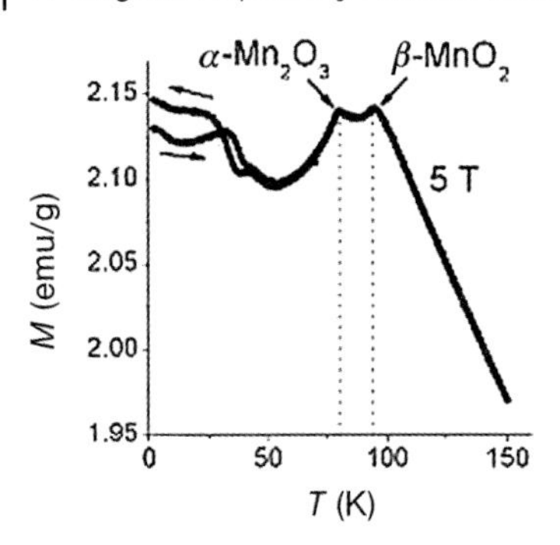

Figure 19.15 Magnetization of β-MnO_2 nanorods as a function of temperature measured in a magnetic field of 5 T. Both zero-field-cooled and field-cooled runs are shown. Reproduced with permission from Ref. [143].

another phase on the surface of the nanorods. In particular, Kim *et al.* [143] showed that their β-MnO_2 nanorods, with an outer diameter of 50–100 nm and a length of several micrometers, were covered with a thin surface layer (∼2.5 nm) of α-Mn_2O_3. This resulted in a very complicated temperature-dependence of magnetic susceptibility, where the core β-MnO_2 and surface α-Mn_2O_3 each added their own magnetic signal to the total measured magnetization, with two maxima in $M(T)$ at 93 K and 80 K being observed (Figure 19.15), both of which were independent of the magnetic field. The upper value coincided with the transition into a incommensurately modulated helical state of β-MnO_2 (see Section 19.6.2), and thus it was concluded that the magnetic transition was not affected by the size of nanorods with dimensions of 50 nm or more. The lower value was considered due to a magnetic ordering of the α-Mn_2O_3 surface layer, although despite the fact that the surface layer thickness was only ∼2.5 nm, the transition temperature did not deviate from that of the bulk phase. Interestingly, another anomaly was noted below about 45 K (Figure 19.15), which was close to the transition temperature of the γ-Mn_2O_3 nanoparticles [180]. As the electron microscopy data did not show any evidence of the presence of the latter phase, it was suggested that this anomaly reflected magnetic ordering within the mismatch layer at the interface between the β-MnO_2 and γ-Mn_2O_3.

In an attempt to study the role of nanoparticle shape, Jana *et al.* [181] reported the magnetic properties of exclusively single-crystalline β-MnO_2 nanospheres and nanorods. Typically, spherical nanoparticles have diameters of 5 ± 2 nm, whereas uniform nanorods grow to about 200 nm in length and about 8 nm in diameter. The magnetic properties of these particles differ dramatically from those of the bulk, and depend heavily on the nanoparticle shape. The field-dependent magnetization curve at 77 K is consistent with the superparamagnetic state, whilst at 2 K a small hysteresis, with a coercive field of 435 Oe and remnant magnetization of 0.84 emu g^{-1}, was measured. The temperature where a maximum in the ZFC $M(T)$ susceptibility curve occurs (Figure 19.16) has been interpreted as the "blocking temperature" (T_B) for the superparamagnetic phase, and this is dependent on the particle shape. Typically, T_B would be 4 K for nanorods, and about 40 K for nanospheres.

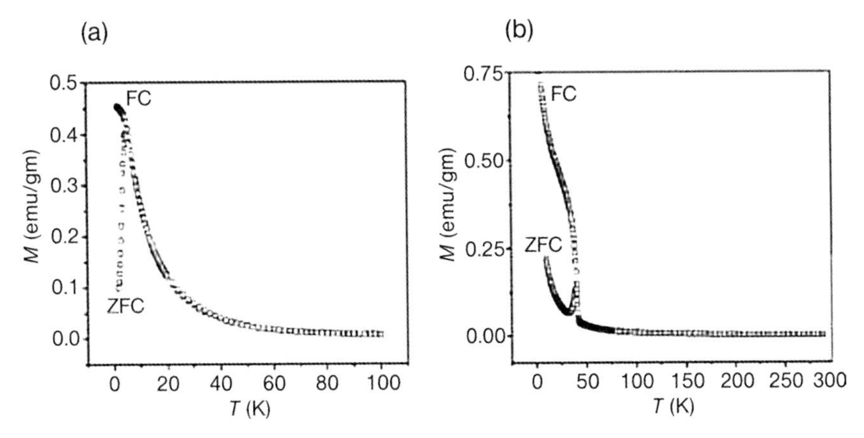

Figure 19.16 Temperature-dependence of the magnetization measured in an applied field of 100 Oe for (a) rod-shaped and (b) spherical β-MnO_2 nanoparticles. Reproduced with permission from Ref. [181].

To summarize, it appears that the magnetic properties of β-MnO_2 nanoparticles vary dramatically from sample to sample. For typical particles with a size exceeding 25–50 nm, a small increase in T_N is usually found, but no transition analogous to bulk β-MnO_2 is found in very small nanoparticles with dimensions below 10 nm; rather, a superparamagnetic-like behavior is encountered. The main driving force for a dramatic change in the sub-10 nm range has still not yet been resolved; consequently, it remains unclear whether it is the finite size effect and the magnetic anisotropy, the mismatch between a propagation vector and particle dimension, or a lack of perfect stoichiometry that makes these samples more closely resemble β-MnO_{2-x}. Notably, the presence of extra delocalized charges (e_g states) can have dramatic effects on the magnetism and conductivity of the studied systems; however, conductivity measurements should shed some additional light on these aspects.

Finally, as β-MnO_2 seems to compete with different manganese oxide phases, it might be instructive to refer to a recent study of the magnetic properties of Mn_3O_4 and MnO nanoparticles [182]. Depending on the precursor used, the composition of the final product can be varied in such a way that mainly Mn_3O_4 or MnO is formed. With regards to the shape of the Mn_3O_4 nanoparticles, a variety of morphologies was observed, including plate-like forms, spherules, ellipsoids, cubes, and rhombohedra. On the other hand, the shapes in the case of MnO particles were more uniform, with nanoplatelets as the main species. In the case of Mn_3O_4, the average crystallite size was 25.4 nm, while the MnO particle diameters ranged from 12 to 88 nm, with an average value of 37 nm. Especially in the case of MnO, an ordered manganese vacancy cubic superstructure has been noted.

The magnetic properties of both MnO and Mn_3O_4 were also investigated (Figure 19.17). For the MnO sample, a Curie–Weiss dependence of magnetic susceptibility was found at high temperatures, although the Curie temperature ($\theta = -462(17)$ K) was reduced compared to the value reported for bulk MnO ($\theta = -548$ K) [183]. The anomaly (bump) in the magnetic susceptibility at $T \sim 130$ K

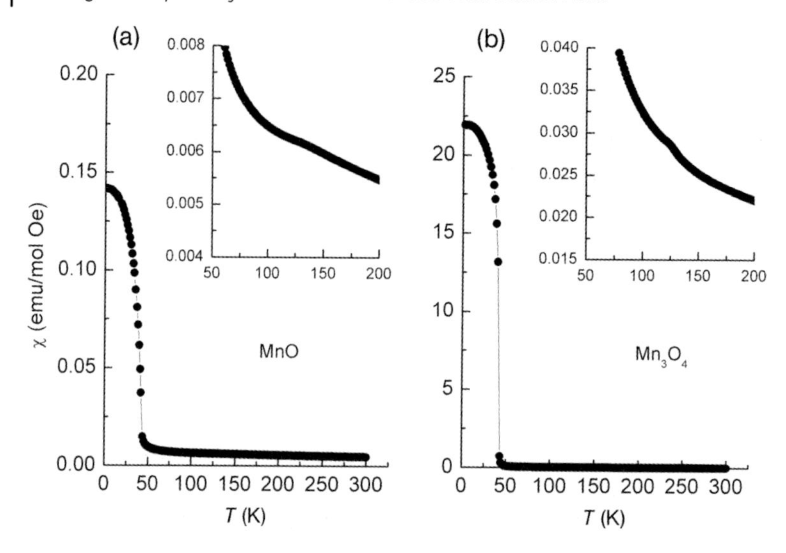

Figure 19.17 Magnetic susceptibility of (a) MnO sample (average nanoparticle size 37 nm) and (b) Mn_3O_4 sample (average nanoparticle size 25.4 nm). The insets show an enlargement of the region around 120 K. Note the differences in scales for the magnetic susceptibility. The applied magnetic field was 100 Oe in both cases. Reproduced with permission from Ref. [182].

(inset in Figure 19.7a) seemed to coincide well with the Neel transition temperature for bulk MnO ($T_N = 122$ K) [183], although the true LRO never really developed. At low temperatures, a weak ferromagnetic response was noted, due most likely to the coexistence of a small amount of Mn_3O_4 particles in the sample. For the Mn_3O_4 nanoparticles, a drastic reduction in the effective magnetic moment has been reported, but the transition temperature still matched very well with that found in the bulk Mn_3O_4. The ferrimagnetic transition temperature in bulk was $T_N = 42$ K, but the canted-spin order of the Yafet–Kittel type was realized below $T' = 33$ K, and a spin spiral structure along the [010] axis was observed in the temperature range of $33\text{ K} < T < 39\text{ K}$ [184]. No such anomalies were noted for these studied nanoparticles.

19.6.4.2 α-MnO_2 (Hollandite)

To date, the magnetic properties of α-MnO_2 nanoparticles have been rarely reported. Zhu *et al.* [185] prepared 1-D α-MnO_2 nanowires with a controlled width of 10–20 nm by means of ultrasonic waves from mesoporous carbon, using $KMnO_4$ as the precursor. For α-MnO_2 grown in hexagonal mesoporous carbon, the ZFC/FC splitting below 75–100 K indicated the transition to a low-temperature ferromagnetic state. The existence of ferromagnetism was also supported by the observation of a magnetic hysteresis loop, although the remanent magnetization and coercive field were both very small. At the same time, when a peak in the ac susceptibility was observed at 100 K, this was interpreted as evidence for a transition to the antiferromagnetic or spin-glass state coexisting with the ferromagnetic state. α-MnO_2 nanowires showed a kink in magnetization at about 50 K in the FC measurements, indicating a transition to an antiferromagnetic state. It should be noted that this temperature coincides roughly with the reported transition temperatures for the bulk α-MnO_2 phases [130, 131].

The details of highly crystalline α-MnO_2 nanowires were also reported by Li *et al.* [186], who showed the average diameter of α-MnO_2 nanorods to increase from 9.2 to 16.5 nm when the hydrothermal reaction temperature was raised from 140 to 220 °C. The effective magnetic moments calculated from the high-temperature magnetic susceptibility data indicated a mixed-valence state, Mn^{3+}/Mn^{4+} (the average valence states of Mn ions are in the range 3.93–3.83). At low temperatures, all of the α-MnO_2 nanorods showed two magnetic transitions that were characterized by a slightly suppressed Néel temperature (from 24.5 K to 22.8 K, judged from the peak in $d\chi/dT$) $T_N = 24.5$ K, and with a rod diameter reduction. A similar observation was also reported for needle-like α-MnO_2 [187]. In contrast, α-MnO_{2-x} nanowires with a diameter of between 20 and 50 nm were shown to undergo a magnetic transition at $T_N = 13$ K [188]. It should be emphasized that the concentration of oxygen vacancies as estimated from the X-ray photoemission spectrum was rather large [188] (i.e., $x = 0.3$); indeed, it has been proposed that these vacancies can have a dramatic effect on the magnetic properties.

A very recent report detailed the first synthesis of hollandite manganese oxide nanotubes, using the hydrothermal technique [189]. In this case, the outer diameter of the nanotubes was about 30–40 nm, the inner diameter 6–8 nm, and the length about 300 nm (Figure 19.18a). Subsequent high-resolution transmission electron microscopy (TEM) and X-ray diffraction (XRD) analyses confirmed that the nanotubes had the expected hollandite structure. A chemical analysis also signaled the presence of intercalated K^+ ions that most likely resided in the 2×2 tunnels. The temperature dependence of the magnetic susceptibility is shown in Figure 19.18b. In this case, a strong Curie-like dependence was found, consistent with the manganese charge localization identified also in the low-temperature conductivity measurements. However, in order to fit the temperature dependence, a rather strong temperature-independent contribution to the magnetic susceptibility, χ_0, must be taken into account, which probably reflects the charge doping due to the K^+ intercalation in the channels. The Curie–Weiss temperature was found to be −58 K, which indicated antiferromagnetic interactions between the Mn moments. Below ~70 K, deviations from the linear dependence of $1/(\chi - \chi_0)$ were taken as an indication for the development of SRO effects. Another anomaly in the magnetic susceptibility was observed at $T_C = 13.6$ K (inset to Figure 19.18b). Below this temperature, a difference between the FC and ZFC measurements was noted, and the magnetization in a 100 Oe applied field converged to the value of ~2.6 emu mol^{-1} Mn at 2 K. This was taken as a clear indication for a transition to a weak ferromagnetic state. The magnetic hysteresis loop was also found for $T < T_N$, while the corresponding coercive field was about 400 Oe at $T = 5$ K. This magnetization was still increased even for a 50 kOe applied magnetic field, such that the ground state was essentially antiferromagnetic in character.

The results of more detailed structural, transport and magnetic properties studies [190] have suggested that hollandite nanotubes are, in fact, structurally inhomo-

(a)

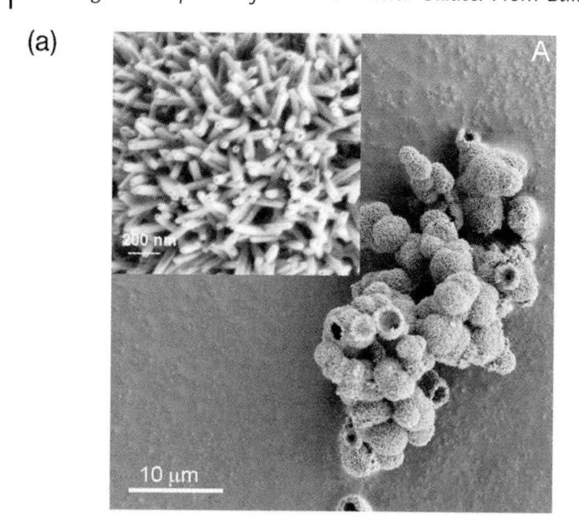

(b)

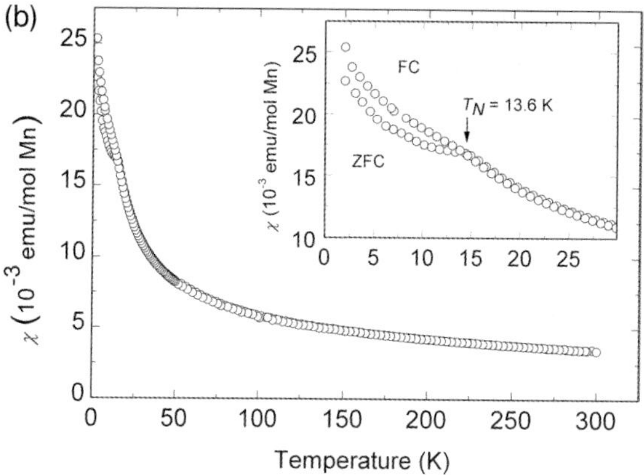

Figure 19.18 (a) Scanning electron microscopy image of as-prepared manganese oxide samples, showing the "sea-urchin"-type of nanotube organization. The inset shows an enlargement of an individual structure, with the nanotubular form of individual structures visible; (b) Temperature-dependence of dc magnetic susceptibility measured in the applied magnetic field 100 Oe. The inset shows an enlargement of the region around the magnetic ordering temperature $T_N = 13.6$ K, showing the difference between ZFC and FC measurements [190].

geneous. In other words, at least two regions in the sample must be distinguished: (i) a K^+-depleted region, where the manganese ions are predominantly in the Mn^{4+} valence state; and (ii) a K^+-rich region with a mixed Mn^{3+}–Mn^{4+} valence state. In the Mn^{4+} region, electron paramagnetic resonance (EPR) measurements suggested powerful antiferromagnetic interactions between the Mn^{4+} ($S = 3/2$) moments ($\theta = -52$ K), although it is not clear at present whether this phase is really ordered antiferromagnetically, or not. It should also be stressed at this point that Mn^{4+} is not a Jahn–Teller active ion, so that any geometric frustration deriving from the Mn arrangement in the double chain might be very important and suppress the long-range magnetic ordering. The other phase is due to regions with a mixed

Mn^{3+}–Mn^{4+} state, which demonstrates a small polaron-type conductivity. One very likely reason for such behavior is an ordering of the K^+ ions inside the tunnel structure. Notably, due to the electrostatic interactions, it is expected that K^+ ions will accumulate around less electronegative Mn^{3+} ions. Such K^+ ion ordering would create a Coulombic potential barrier for e_g hopping, even in the mixed Mn^{3+}–Mn^{4+} valence state. It is therefore not surprising that an activated type of behavior would dominate both the conductivity as well as the EPR linewidths. It should also be mentioned that anomalies between 250 and 300 K may be seen in the conductivity, susceptibility and EPR data, but this might be associated with a freezing out of the K^+ ion motion in the tunnel.

Nanoparticles with the hollandite structure can be prepared also from other transition-metal oxide phases, and it may be instructive at this point to compare these different systems. A recent report was made [191] on the synthesis and

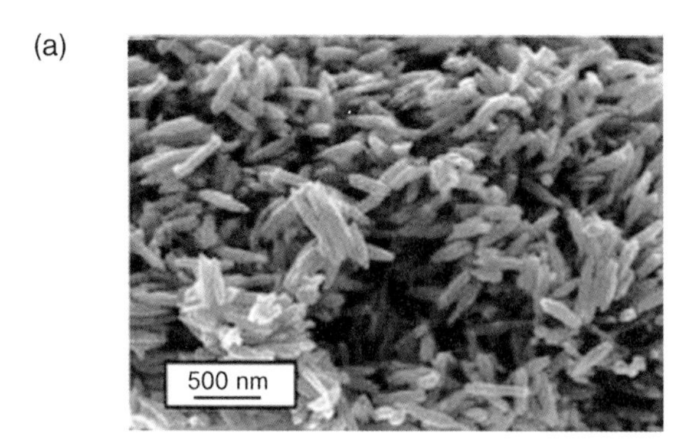

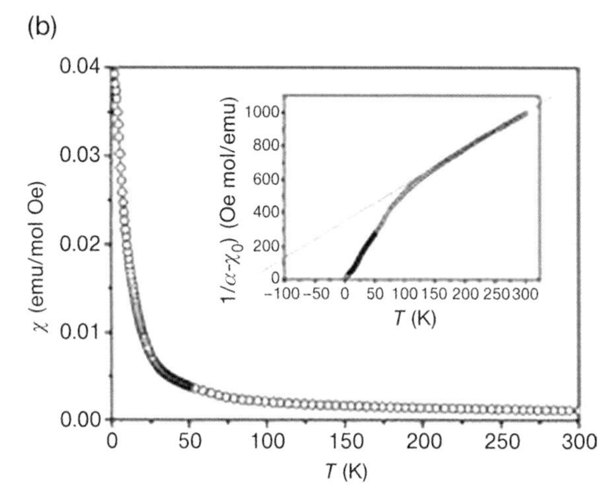

Figure 19.19 (a) Scanning electron microscopy overview images of $VO_{1.52}(OH)_{0.77}$ nanorods; (b) Temperature-dependence of the magnetic susceptibility measured in DC field $H = 1000$ Oe. The inset shows the reciprocal susceptibility as a function of the temperature. The straight line is the best fit to the Curie–Weiss law at high temperatures (150–300 K). Reproduced with permission from Ref. [191].

magnetization of vanadium oxyhydroxide ($VO_{1.52}(OH)_{0.77}$) nanorods, which typically had an ellipsoidal morphology and were up to 500 nm long and about 100 nm in diameter (Figure 19.19a). According to structural probes and magnetic measurements (Figure 19.19b), the effective moment was $1.94\,\mu_B\,V^{-1}$, consistent with the formal valence of vanadium of $+3.81$ (V^{4+}/V^{3+}, atomic ratio ~4). The close analogy with the manganese nanotubes described above should be stressed at this point. The Curie–Weiss temperature ($\theta = -157(2)$ K) indicated the presence of powerful antiferromagnetic interactions between the vanadium moments, although the magnetic order appeared only below 25 K. One likely reason for such a high ratio between the Curie–Weiss and transition temperatures was the magnetically frustrated lattice characteristic for the hollandite structure.

19.7 Concluding Remarks

Even after several centuries of research, transition-metal oxides continue to spring surprises! Their extreme richness in structural motifs is reflected in their diverse and intriguing magnetic properties. Moreover, the preparation of such samples in the form of nanoparticles has generated another option to tailor the magnetic response of these materials via an enhanced magnetic anisotropy or off-stoichiometry. Based on their large surface-to-volume ratio, special electronic and magnetic properties, biocompatibility, and low costs, transition-metal oxide nanoparticles are clearly expected to become even more widely used in everyday life in the near future.

References

1 Ross, C.A. (2001) *Annu. Rev. Mater. Res.*, **31**, 203–235.
2 Arbab, A.S., Bashaw, L.A., Miller, B.R., Jordan, E.K., Lewis, B.K., Kalish, H., and Frank, J.A. (2003) *Radiology*, **229** (3), 838–846.
3 Reimer, P. and Weissleder, R. (1996) *Radiology*, **36**, 153–163.
4 Pankhurst, Q.A., Connolly, J., Jones, S.K., and Dobson, J. (2003) *J. Phys. D: Appl. Phys.*, **36**, 167–181.
5 Wolf, S.A., Awschalom, D.D., Buhrman, R.A., Daughton, J.M., Von Molnár, S., Roukes, M.L., Chtchelkanova, A.Y., and Treger, D.M. (2001) *Science*, **294**, 1488–1495.
6 Soulen, R.J. Jr, Byers, J.M., Osofsky, M.S., Nadgorny, B., Ambrose, T., Cheng, S.F., Broussard, P.R., Tanaka, C.T., Nowak, J., Moodera, J.S., Barry, A., and Coey, J.M.D. (1998) *Science*, **282**, 85–88.
7 De Grave, E. and Vandenberge, R.E. (1986) *Hyperfine Interact.*, **28**, 643–646.
8 Sampson, C.F. (1969) *Acta Crystallogr. B*, **25**, 1683–1685.
9 Oles, A., Szytula, A., and Wanic, A. (1970) *Phys. Status Solidi*, **41**, 173–177.
10 Post, J.E. and Buchwald, V.F. (1991) *Am. Mineral.*, **76**, 272–277.
11 Murad, E. (1988) in *Iron in Soils and Clay Minerals* (eds J.W. Stucki, B.A. Goodman and U. Schwertmann), D. Riedel Publishing Co., Dordrecht, Holland, NATO ASI Ser. 217, pp. 309–350.
12 Chukhrov, F.V., Zvyagin, B.B., Gorshkov, A.I., Ermilova, L.P., Korovushkin, V.V., Rudnitskaya, E.S., and Yakubovsakaya, N.Yu. (1976) *Ser. Geol*, **5**, 5–24.

13 Chukhrov, F.V., Zvyagin, B.B., Gorshkov, A.I., Ermilova, L.P., and Balashova, V.V. (1973) *Izv. Akad. Nauk. SSSR, Ser. Geol.*, **4**, 23–33.

14 Murad, E. and Johnston, J.H. (1987) *Mössbauer Spectroscopy Applied to Inorganic Chemistry*, vol. **2**, Modern Inorganic Chemistry (ed. G.J. Long), Springer, Heidelberg, pp. 507–582.

15 Morrish, A.H., Johnston, G.B., and Curry, N.A. (1963) *Phys. Lett.*, **7**, 177.

16 Artman, J.O., Murphy, J.C., and Foner, S. (1965) *Phys. Rev.*, **138**, A912–A917.

17 Shull, C.G., Strauser, W.A., and Wollan, E.O. (1951) *Phys. Rev.*, **83**, 333–345.

18 Hill, R.J., Craig, J.R., and Gibbs, G.V. (1979) *Phys. Chem. Miner.*, **4**, 317–339.

19 Hägg, G. (1935) *Chem. Abstracts B*, **29**, 95–103.

20 Battle, P.B. and Cheetham, A.K. (1979) *J. Phys. C: Solid State Phys.*, **12**, 337–345.

21 Tronc, E., Chanéac, C., and Jolivet, J.P. (1998) *J. Solid State Chem.*, **139**, 93–104.

22 Bernal, J.D., Dasgupta, D.R., and Mackay, A.L. (1959) *Clay Miner. Bull.*, **4**, 15–29.

23 Miyamoto, H. (1976) *Mater. Res. Bull.*, **11**, 329–336.

24 Birch, W.D., Pring, A., Reller, A., and Schmalle, H. (1992) *Naturwissenschaft*, **79**, 509–511.

25 McCammon, C.A., Pring, A., Keppler, H., and Sharp, T. (1995) *Phys. Chem. Miner.*, **22**, 11–20.

26 Cornell, R.M. and Schwertmann, U. (2003) *The Iron Oxides*, Wiley-VCH Verlag GmbH & Co. KGaA, Weinheim.

27 Carbone, C., Di Bendetto, F., Marescotti, P., Sangregorio, C., Sorace, L., Lima, N., Romanelli, M., Lucchetti, G., and Cipriani, C. (2005) *Miner. Petrol.*, **85**, 19.

28 Kliava, J. and Berger, R. (1999) *J. Magn. Magn. Mater.*, **205**, 328–342.

29 Berger, R., Bissey, J., Kliava, J., Daubric, H., and Estournés, C. (2001) *J. Magn. Magn. Mater.*, **234**, 535–544.

30 Ayyub, P., Multani, M., Barma, M., Palkar, V.R., and Vijayaraghavan, R. (1988) *J. Phys. C: Solid State Phys.*, **21**, 2229–2245.

31 Schroeert, D. and Nininger, R.C. Jr (1967) *Phys. Rev. Lett.*, **19**, 632–634.

32 Yamamoto, N. (1968) *J. Phys. Soc. Jpn*, **24**, 23–28.

33 Amin, N. and Arajs, S. (1987) *Phys. Rev. B*, **35**, 4810–4811.

34 Muench, G.J., Arajs, S., and Matijevič, E. (1981) *J. Appl. Phys.*, **52**, 2493–2495.

35 Dormann, J.L., Cui, J.R., and Sella, C. (1985) *J. Appl. Phys.*, **57**, 4283–4285.

36 Kündig, W., Bömmel, H., Constabaris, G., and Linquist, R.H. (1966) *Phys. Rev.*, **142**, 327–333.

37 Mansilla, M.V., Zysler, R., Fiorani, D., and Suber, L. (2002) *Physica B*, **320**, 206–209.

38 Balasubramanian, R., Cook, D.C., and Yamashita, M. (2002) *Hyperfine Interact.*, **139/140**, 167–173.

39 Bødker, F. and Morup, S. (2000) *Europhys. Lett.*, **52**, 217–223.

40 Pregelj, M., Umek, P., Drolc, B., Jančar, B., Jagličić, Z., Dominko, R., and Arčon, D. (2006) *J. Mater. Res.*, **21**, 2955–2962.

41 Bodkar, F., Morup, S., and Linderoth, S. (1994) *Phys. Rev. Lett.*, **72**, 282–285.

42 Gopal Reddy, C.V., Manorama, S.V., and Rao, V.J. (2000) *J. Mater. Sci. Lett.*, **19**, 775–778.

43 Shriver, D.F., Atkins, P.W., Overton, T.L., Rourke, J.P., Weller, M.T., and Armstrong, F.A. (2006) *Inorganic Chemistry*, 4th ed, Oxford University Press, Oxford, p. 151.

44 Mathew, D.S. and Juang, R.-S. (2007) *Chem. Eng. J.*, **129**, 51–65.

45 Neel, L. (1948) *Ann. Phys. Paris*, **3**, 137–198.

46 Schiessl, W., Potzel, W., Karzel, H., Steiner, M., Kalvius, G.M., Martin, A., Krause, M.K., Halevy, I., Gal, J., Schäfer, W., Will, G., Hillberg, M., and Wäppling, R. (1996) *Phys. Rev. B*, **53**, 9143–9152.

47 Burghart, F., Potzel, W., Kalvius, G., Schreier, E., Grosse, G., Noakes, D., Schafer, W., Kockelmann, W., Campbell, S., Kaczmarek, W., Martin, A., and Krause, M. (2000) *Physica B*, **289–290**, 286–290.

48 Ho, J.C., Hamdeh, H., Chen, Y., Lin, S., Yao, Y., Willey, R., and Oliver, S. (1995) *Phys. Rev. B*, **52**, 10122–10126.

49 (a) Gaulin, B.D. (1994) *Hyperfine Interact.*, **85**, 159–171;(b) Anantharaman, M., Jagathesan, S., Malini, K., Sindhu, S., Narayanaswamy, A., Chinnasamy, C.,

Jacobs, J., Reijne, S., Seshan, K., Smits, R., and Brongerma, H.J. (1998) *J. Magn. Magn. Mater.*, **189**, 83–88.
50 Rooksby, H.P. and Willis, B.T.M. (1953) *Nature*, **172**, 1054–1055.
51 Charles, S.W., Davies, K.J., Wells, S., Upadhyay, R.V., Grady, K.O., El Hilo, M., Meaz, T., and Morup, S. (1995) *J. Magn. Magn. Mater.*, **149**, 14–18.
52 Martens, J.W.D., Eeters, W.L., van Noort, H.M., and Errnan, M. (1985) *J. Phys. Chem. Solids*, **46**, 411–418.
53 Okuno, S.N., Hashimoto, S., and Inomata, K. (1992) *J. Appl. Phys.*, **71**, 5926–5929.
54 Feried, T., Shemer, G., and Markovich, G. (2001) *Adv. Mater.*, **13**, 1158–1161.
55 Pillai, V. and Shah, D.O. (1996) *J. Magn. Magn. Mater.*, **163**, 243.
56 Lee, Y., Lee, J., Bae, C.J., Park, J.-G., Noh, H.-J., Park, J.-H., and Hyeon, T. (2005) *Adv. Funct. Mater.*, **15**, 503–509.
57 Doker, O., Bayraktar, E., Mehmetoğlu, U., and Calimli, A. (2003) *Rev. Adv. Mater. Sci.*, **5**, 498–500.
58 Rondinone, A.J., Samia, A.C.S., and John Zhang, Z. (1999) *J. Phys. Chem. B*, **103**, 6876–6880.
59 Liu, C., Rondinone, A.J., and Zhang, Z.J. (2000) *Pure Appl. Chem.*, **72**, 37–45.
60 Moodera, J.S., Kinder, L.R., Wong, T.M., and Meservey, R. (1995) *Phys. Rev. Lett.*, **74**, 3273–3276.
61 Gallagher, W.J., Parkin, S.S.P., Lu, Y., Bian, X.P., Marley, A., Roche, K.P., Altman, R.A., Rishton, S.A., Jahnes, C., Shaw, T.M., and Xiao, G. (1997) *J. Appl. Phys.*, **81**, 3741–3746.
62 Boeve, H., De Boeck, J., and Borghs, G. (2001) *J. Appl. Phys.*, **89**, 482–487.
63 Kämper, K.P., Schmitt, W., Güntherodt, G., Gambino, R.J., and Ruf, R. (1987) *Phys. Rev. Lett.*, **59**, 2788–2791.
64 Suzuki, K. and Tedrow, M.P. (1999) *Appl. Phys. Lett*, **74**, 428–429.
65 Ivanov, P.G., Watts, S.M., and Lind, D.M. (2001) *J. Appl. Phys.*, **89**, 1035–1040.
66 Biswas, S. and Ram, S. (2004) *Chem. Phys.*, **306**, 163–169.
67 Kuznetsov, A.Y., de Almeida, J.S., Dubrovinsky, L., Ahuja, R., Kwon, S.K., Kantor, I., Kantor, A., and Guignot, N. (2006) *J. Appl. Phys.*, **99**, 053909.
68 Thamer, B.J. (1957) *J. Am. Chem. Soc.*, **79**, 4298–4305.
69 Finger, L.W. and Hazen, R.M. (1980) *J. Appl. Phys.*, **51**, 5362–5367.
70 Rodbell, D.S., Lommel, J.M., and DeVries, R.C. (1966) *J. Phys. Soc. Jpn*, **21**, 2430.
71 Li, X.W., Gupta, A., and Xiao, G. (1999) *Appl. Phys. Lett.*, **75**, 713–715.
72 Rüdiger, U., Rabe, M., Samm, K., Özyilmaz, B., Pommer, J., Fraune, M., Güntherodt, G., Senz, S., and Hesse, D. (2001) *J. Appl. Phys.*, **89**, 7699–7701.
73 Korotin, M.A., Anisimov, V.I., Khomskii, D.I., and Sawatzky, G.A. (1998) *Phys. Rev. Lett.*, **80**, 4305–4308.
74 Schlottmann, P. (2003) *Phys. Rev. B*, **67**, 174419.
75 Shim, J.H., Lee, S., Dho, J., and Kim, D.-H. (2007) *Phys. Rev. Lett.*, **99**, 057209.
76 Li, X.W., Gupta, A., McGuire, T.R., Duncombe, P.R., and Xiao, G. (1999) *J. Appl. Phys.*, **85**, 5585–5587.
77 Watts, S.M., Wirth, S., von Molnár, S., Barry, A., and Coey, J.M. (2000) *Phys. Rev. B*, **61**, 9621–9628.
78 Fan, L.N., Chen, Y.J., Zhang, X.Y., and Yao, D.L. (2005) *J. Magn. Magn. Mater.*, **285**, 353–358.
79 Singh, G.P., Biswas, B., Ram, S., and Biswas, K. (2008) *Mater. Sci. Eng. Struct. Mater.*, **498**, 125–128.
80 Biswas, S., and Ram, S. (2004) *Chem. Phys.*, **306**, 163–169.
81 Wolf, S.A., Awschalom, D.D., Buhrman, R.A., Daughton, J.M., von Molnar, S., Roukes, M.L., Chtchelkanova, A.Y., and Treger, D.M. (2001) *Science*, **294**, 1488–1495.
82 Ranno, L., Barry, A., and Coey, J.M.D. (1997) *J. Appl. Phys.*, **81**, 5774–5776.
83 Suzuki, K., and Tedrow, P.M. (1998) *Phys. Rev. B*, **58**, 11597–11602.
84 Hwang, H.Y. and Cheong, S.-W. (1997) *Science*, **278**, 1607–1609.
85 Coey, J.M., Berkowitz, A.E., Balcells, L., Putris, F.F., and Barry, A. (1998) *Phys. Rev. Lett.*, **80**, 3815–3818.
86 Wang, K.-Y., Spinu, L., He, J., Zhou, W., Wang, W., and Tang, J. (2002) *J. Appl. Phys.*, **91**, 8204–8206.

87 Dai, J. and Tang, J. (2001) *Phys. Rev. B*, **63**, 054434.

88 Gupta, A., Li, X.W., and Xiao, G. (2000) *J. Appl. Phys.*, **87**, 6073–6078.

89 Yuan, L., Ovchenkov, Y., Sokolov, A., Yang, C.-S., Doudin, B., and Liou, S.H. (2003) *J. Appl. Phys.*, **93**, 6850–6852.

90 For reviews, see: (a) Rao, C.N.R., Cheetham, A.K., and Mahesh, R., (1996) *Chem. Mater.*, **8**, 2421 (b) Raveau, B., Maignan, A., Martin, C., and Hervieu, M. (1998) *Chem. Mater.*, **10**, 2641–2652.

91 Sherman, D.M. (1984) *Am. Mineral.*, **69**, 788–799.

92 Giovanoli, R. (1969) *Chimia*, **2**, 470–472.

93 Baur, W.H. (1976) *Acta Crystallogr. B*, **32**, 2200–2204.

94 Murphy, D.W., Di Salvo, F.J., Carides, J.N., and Waszczak, J.V. (1978) *Mater. Res. Bull.*, **13**, 1395–1402.

95 Yamada, N., Ohmasa, M., and Horiuchi, S. (1986) *Acta Crystallogr. B*, **42**, 58–61.

96 Mosbah, A., Verbaere, A., and Tournoux, M. (1983) *Mater. Res. Bull.*, **18**, 1375–1381.

97 Byström, A.M. (1949) *Acta Chem. Scand.*, **3**, 163.

98 Thackeray, M.M., Rossouw, M.H., Gummow, R.J., Liles, D.C., Pearce, K., de Kock, A., David, W.I.F., and Hull, S. (1993) *Electrochim. Acta*, **38**, 1259–1267.

99 Chabre, Y. and Pannetier, J. (1995) *Prog. Solid State Chem.*, **23**, 1–130.

100 Rossouw, M.H., Liles, D.C., Thackeray, M.M., David, W.I.F., and Hull, S. (1992) *Mater. Res. Bull.*, **27**, 221–230.

101 Muraoka, Y., Chiba, H., Atou, T., Kikuchi, M., Hiraga, K., Syono, Y., Sugiyama, S., Yamamoto, S., and Grenier, J.-C. (1999) *J. Solid State Chem.*, **144**, 136–142.

102 Turner, S. and Buseck, P.R. (1981) *Science*, **212**, 1024–1027.

103 Turner, S., Siegel, M.D., and Buseck, P.R. (1982) *Nature*, **296**, 841–842.

104 Post, J.E., and Veblen, D.R. (1990) *Am. Mineral.*, **75**, 477–489.

105 Chang, F.M., and Jansen, M. (1985) *Z. Anorg. Allgem. Chem.*, **531**, 177–182.

106 Drits, V.A., Silvester, E., Gorshkov, A.I., and Manceau, A. (1997) *Am. Mineral.*, **82**, 946–961.

107 Hunter, J.C. (1981) *J. Solid State Chem.*, **39**, 142–147.

108 Greedan, J.E., Raju, N.P., Wills, A.S., Morin, C., Shaw, S.M., and Reimers, J.N. (1998) *Chem. Mater.*, **10**, 3058–3067.

109 Jang, Y.-I., Huang, B., Chou, F.C., Sadoway, D.R., and Chiang, Y.-M. (2000) *J. Appl. Phys.*, **87**, 7382–7388.

110 Berg, H., Rundlöv, H., and Thomas, J.O. (2001) *Solid State Ionics*, **144**, 65–69.

111 Post, J.E., Von Dreele, R.B., and Buseck, P.R. (1982) *Acta Crystallogr. B*, **38**, 1056–1065.

112 Zwicker, W.K., Meijer, W.O.J.G., and Jaffe, H.W. (1962) *Am. Mineral.*, **47**, 246–266.

113 Turner, S. and Post, J.E. (1988) *Am. Mineral.*, **73**, 1155–1161.

114 Giovanoli, R. and Leuenberger, U. (1969) *Helv. Chim. Acta*, **52**, 2333.

115 Zener, C. (1951) *Phys. Rev.*, **82**, 403–405.

116 de Gennes, P.-G. (1960) *Phys. Rev.*, **118**, 141–154.

117 Anderson, P.W. (1950) *Phys. Rev.*, **79**, 350–356.

118 Goodenough, J.B., Wold, A., Arnott, R.J., and Menyuk, N. (1961) *Phys. Rev.*, **124**, 373–384.

119 Kanamori, J. (1959) *J. Phys. Chem. Solids*, **10**, 87–98.

120 Ohama, N. and Hamaguchi, Y. (1971) *J. Phys. Soc. Jpn*, **30**, 1311.

121 Kelley, K.K. and Moore, G.E. (1943) *J. Am. Chem. Soc.*, **65**, 782.

122 Gonzalo, J.A., and Cox, D. (1970) *An. Fis.*, **66**, 407.

123 Sato, H., Enoki, T., Isobe, M., and Ueda, Y. (2000) *Phys. Rev. B*, **61**, 3563–3569.

124 Yoshimori, A. (1959) *J. Phys. Soc. Jpn*, **14**, 807.

125 Sato, H., Wakiya, K., Enoki, T., Kiyama, T., Wakabayashi, Y., Nakao, H., and Murakami, Y. (2001) *J. Phys. Soc. Jpn*, **70**, 37–40.

126 Regulski, M., Przeniosøo, R., Sosnowska, I., and Hoffmann, J.-U. (2003) *Phys. Rev. B*, **68**, 172–401.

127 Regulski, M., Przeniosøo, R., Sosnowska, I., and Hoffmann, J.-U. (2004) *J. Phys. Soc. Jpn*, **73**, 3444–3447.

128 Sato, H., Kawamura, Y., Ogawa, T., Murakami, Y., Ohsumi, H., Mizumaki, M., and Ikeda, N. (2003) *Physica B*, **329**, 757–758.

129 Kawamura, H. (1992) *J. Phys. Soc. Jpn*, **61**, 3062–3066.

130 Ishiwata, S., Bos, J.W.G., Huang, Q., and Cava, R.J. (2006) *J. Phys. Condens. Matter*, **18**, 3745–3752.

131 Sato, H., Enoki, T., Yamaura, J.-I., and Yamamoto, N. (1999) *Phys. Rev. B*, **59**, 12836.

132 Sato, H., Yamaura, J.-I., Enoki, T., and Yamamoto, N. (1997) *Physica B*, **262–263**, 443.

133 Giot, M., Chapon, L.C., Androulakis, J., Green, M.A., Radaelli, P.G., and Lappas, A. (2007) *Phys. Rev. Lett.*, **99**, 247211. Stock, C., Chapon, L.C., Adamopoulo, O., Lappas, A., Giot, M., Taylor, S.W., Green, M.A., Brown, C.M., Radaelli, P.G. (2009) *Phys. Rev. Lett.*, **103**, 077202.

134 Zorko, A., El Shawish, S., Arčon, D., Jagličić, Z., Lappas, A., van Tol, H., and Brunel, L.C. (2008) *Phys. Rev. B*, **77**, 024412.

135 Parant, J.-P., Olazcuaga, R., Devalette, M., Fouassier, C., and Hagenmuller, P. (1971) *J. Solid State Chem.*, **3**, 1–11.

136 Villain, J. (1978) *Z. Phys. B*, **33**, 31–42.

137 Reimers, J.N. (1992) *Phys. Rev. B*, **45**, 7287–7294.

138 Morgan, D., Wang, B., Ceder, G., and van der Walle, A. (2003) *Phys. Rev. B*, **67**, 134404.

139 Goodenough, J.B., Anthiram, A.M., James, A.C.W.P., and Strobel, P. (1989) Solid State Ionics, in Materials Research Society Symposium Proceedings, No. 135 (eds G. Nazri, R.A. Higgins, and D.H. Shriver), Materials Research Society, Warrendale, PA, p. 391.

140 Guo, L.W., Peng, D.L., Makino, H., Hanada, T., Hong, S.K., Sumiyama, K., Yao, T., and Inaba, K. (2001) *J. Appl. Phys.*, **90**, 351–354.

141 Subramanian, V., Zhu, H., Vajtal, R., Ajayan, P.M., and Wei, B. (2005) *J. Phys. Chem. B*, **109**, 20207–20214.

142 Wang, X. and Li, Y. (2002) *Chem. Commun.*, 764–765.

143 Kim, H.J., Lee, J.B., Kim, Y.-M., Jung, M.-H., Jagličić, Z., Umek, P., and Dolinšek, J. (2007) *Nanoscale Res. Lett.*, **2**, 81–86.

144 Cheng, F., Zhao, J., Song, W., Li, C., Ma, H., Chen, J., and Shen, P. (2006) *Inorg. Chem.*, **45**, 2038–2044.

145 Zhang, W., Yang, Z., Wang, X., Zhang, Y., Wen, X., and Yang, S. (2006) *Catal. Commun.*, **7**, 408–412.

146 Chen, X., Li, X., Jiang, Y., and Shi, C., Li, X. (2005) *Solid State Commun.*, **136**, 94–96.

147 Yuan, Z.-Y., Zhang, Z., Du, G., Ren, T.-Z., and Su, B.-L. (2003) *Chem. Phys. Lett.*, **378**, 349–353.

148 Portehault, D., Cassaignon, S., Baudrin, E., and Jolivet, J.-P. (2007) *Chem. Mater.*, **19**, 5410–5417.

149 Sungantha, M., Ramakrishnan, P.A., Hermann, A.M., Warmisingh, C.P., and Ginley, D.S. (2003) *Int. J. Hydrogen Energy*, **28**, 597–600.

150 Zheng, Y., Cheng, Y., Bao, F., Wang, Y., and Qin, Y. (2006) *J. Crystal Growth*, **286**, 156–161.

151 Cheng, Y., Feng, L., and Cheng, H.-M. (2005) *J. Alloys Compd.*, **397**, 282–285.

152 Kumar, V.G. and Kim, K.B. (2006) *Ultrason. Sonochem.*, **13**, 549–556.

153 Zheng, D., Sun, S., Fan, W., Yu, H., Chunhua, F., Cao, G., Yin, Z., and Song, X. (2005) *J. Phys. Chem. B*, **109**, 16439–16443.

154 Luo, J., Zhu, H.T., Fan, H.M., Liang, J.K., Shi, H.L., Rao, G.H., Li, J.B., Du, Z.M., and Shen, Z.X. (2008) *J. Phys. Chem. C*, **112**, 1498–1506.

155 Ding, Y.-S., Shen, X.-F., Gomez, S., Luo, H., Aindow, M., and Suib, S.L. (2006) *Adv. Funct. Mater.*, **16**, 549–555.

156 Wu, C., Xie, Y., Wang, D., Yang, J., and Li, T. (2003) *J. Phys. Chem. B*, **107**, 13583–13587.

157 Li, B., Rong, G., Huang, L., and Feng, C. (2006) *Inorg. Chem.*, **45**, 6404–6410.

158 Li, Z., Ding, Y., Xiong, Y., Yang, Q., and Xie, Y. (2005) *Chem. Commun.*, 918–920.

159 Ma, R., Band, Y., Zhang, L., and Sasaki, T. (2004) *Adv. Mater.*, **16**, 918–922.

160 Wang, X. and Li, Y. (2002) *J. Am. Chem. Soc.*, **124**, 2880–2881.

161 Huang, X., Lv, D., Yue, H., Attia, A., and Yang, Y. (2008) *Nanotechnology*, **19**, 225606.

162 Wang, N., Cao, X., He, L., Zhang, W., Guo, L., Chen, C., Wang, R., and Yang, S. (2008) *J. Phys. Chem. C.*, **112**, 365–369.

163 Xiong, Y., Xie, Y., Li, Z., and Wu, C. (2003) *Chem. Eur. J.*, **9**, 1645–1651.

164 Wu, M.-S., Lee, J.-T., Wang, Y.-Y., and Wan, C.-C. (2004) *J. Phys. Chem. B*, **108**, 16331–16333.

165 Li, Q., Luo, G., Li, J., and Xia, X. (2003) *J. Mater. Process. Technol.*, **137**, 25–29.

166 Sun, X., Ma, C., Wang, Y., and Li, H. (2002) *Inorg. Chem. Commun.*, **5**, 747–750.

167 Xi, G., Peng, Y., Zhu, Y., Ku, L., Whang, Z., Yu, W., and Qian, Y. (2004) *Mater. Res. Bull.*, **39**, 1641–1648.

168 Fang, Z., Tang, K., Gao, L., Wang, D., Zeng, S., and Liu, Q. (2007) *Mater. Res. Bull.*, **42**, 1761–1768.

169 Wang, H., Lu, Z., Qian, D., Li, Y., and Zhang, W. (2007) *Nanotechnology*, **18**, 115616.

170 Wang, H.-E. Qian, D., Lu, Z., Li, Y., Cheng, R., and Zhang, W. (2007) *J. Crystal Growth*. doi: 10./016/j.jcrysgro.2007.01.034

171 Yang, R., Wang, Z., Dai, L., and Chen, L. (2005) *Mater. Chem. Phys.*, **93**, 149–153.

172 Wang, X. and Li, Y. (2003) *Chem. Eur. J.*, **9**, 300–306.

173 Li, Z., Ding, Y., Xiong, Y., and Xie, Y. (2005) *Cryst. Growth Des.*, **5**, 1953–1958.

174 Wang, H.-E., Lu, Z., Qian, D., Fang, S., and Zhang, J. (2008) *J. Alloys Compd.*, **466**, 250–257.

175 Wang, G., Tang, B., Zhuo, L., Ge, J., and Xue, M. (2006) *Eur. J. Inorg. Chem.*, (11), 2313–2317.

176 Jia, Y., Xu, J., Zhou, L., Liu, H., and Hu, Y. (2008) *Mater. Lett.*, **62**, 1336–1338.

177 Jana, S., Basu, S., Pande, S., Ghosh, S.K., and Pal, T. (2007) *J. Phys. Chem. C*, **111**, 16272–16277.

178 Wang, H.-E. and Qian, D. (2008) *Mater. Chem. Phys.*, **109**, 399–403.

179 Tang, B., Wang, G., Zhuo, L., and Ge, J. (2006) *Nanotechnology*, **17**, 947–951.

180 Kim, S.H., Choi, B.J., Lee, G.H., Oh, S.J., Kim, B., Choi, H.C., Park, J., and Chang, Y. (2005) *J. Korean Phys. Soc.*, **46**, 941–944.

181 Jana, S., Basu, S., Pande, S., Ghosh, S.K., and Pal, T. (2007) *J. Phys. Chem. C*, **111**, 16272–16277.

182 Djerdj, I., Arčon, D., Jagličić, Z., and Niederberger, M. (2007) *J. Phys. Chem. C*, **111**, 3614–3623.

183 Tyler, R.W. (1933) *Phys. Rev.*, **44**, 776–777.

184 Srinivasan, G. and Seehra, M.S. (1983) *Phys. Rev. B*, **28**, 1–7.

185 Zhu, S., Wang, X., Huang, W., Yan, D., Wang, H., and Zhang, D. (2006) *J. Mater. Res.*, **21**, 2847–2854.

186 Li, L., Pan, Y., Chen, L., and Li, G. (2007) *J. Solid State Chem.*, **180**, 2896–2904.

187 Yamamoto, N., Endo, T., Shimada, M., and Takada, T. (1974) *Jpn. J. Appl. Phys.*, **13**, 723–724.

188 Yang, J.B., Zhou, X.D., James, W.J., Malik, S.K., and Wang, C.S. (2004) *Appl. Phys. Lett.*, **85**, 3160–3162.

189 Arčon, D., Umek, P., Pregelj, M., Dominko, R., Gloter, A., Jagličić, Z., Cvetko, D., and Lappas, A. (2008) α-K_xMnO_2 nanotubes: magnetic and transport properties study, in International Conference on Superconductivity and Magnetism, 25–29 August 2008, Side, Antalya, Turkey, Abstract & program book, p. 34.

190 Umek, P., Gloter, A., Pregelj, M., Dominko, R., Jagodic, M., Jagličić, Z., Zimina, A., Brzhezinskaya, M., Potočnik, A., Filipič, C., Levstik, A., and Arčon, D., *J. Phys. Chem. C*, **113**, 14798–14803.

191 Djerdj, I., Sheptyakov, D., Gozzo, F., Arčon, D., Nesper, R., and Niederberger, M. (2008) *J. Am. Chem. Soc.*, **130**, 11364–11375.

Index

a

$A_2BB'O_6$ 269
– cation arrangements, perovskite-type structures 269
– double perovskites 269
ab initio calculations 492, 773
abnormal grain growth (AGG) 9
– triggering impurities 11
abnormal plate like alumina grains 9
ABO_3 perovskite
– crystal structure 259, 260
– polytypes, $BO_{6/2}$ octahedra network 264
accommodation processes 634, 640, 643, 648
Acheson furnaces 138
– SiC furnaces 137
adamantine structure crystals 670
– thermal conductivity 670, 671
adjacent anion sites
– oxygen vacancy migration, curved path 284
aging-deaging terminology 740
alkaline earth metals 393
– cations 40
Al_2O_3-Al_4C_3 system 169
Al_2O_3-AlN system 91, 107
Al_2O_3-Y_2O_3 additive system 680
– thermal conductivity 680
Al-SiC composites alloys, development 202
alum, calcination 6
alumina 513, 615, 622, 667
– based composites 4
– based materials 4
– calcia-silica system 9
– carbothermal reduction 74
– ceramics 535, 563
–– alumina-zirconia duplex, microcrack process zone 620
–– bending tests 563
–– strength vs. stress rate 565
–– thermal shock cracks 535
–– Weibull graph 563
– crack
–– bridge, friction point 626
–– opening profiles 626
–– propagation 616
– crystals, crystallographic faces 10
– doping 681
–– co-doping 9
– electric properties 13
– grain boundaries
–– high-resolution transmission electron microscopy images 9
– grain, solid-liquid interphase energies of 11
– impurities in 516
– interfaces 517
– low-angle grain boundary 513
– low-loss dielectric material 13
– polycrystalline alumina 514
– R-curve behavior 615
– single crystal 13
– strength 622
– translucent alumina, see Lucalox™
aluminum
– compounds 155
– grain boundary energies 510
– metal-organic compounds 6
aluminum nitride (AlN) 74
– band-gap semiconductors 75
– ceramics 75, 679
–– alumina impurity 675
–– effect of secondary phase 674
–– lattice-purifying process 687
–– sintering 674
–– thermal conductivity 674, 676, 679

Ceramics Science and Technology Volume 2: Properties. Edited by Ralf Riedel and I-Wie Chen

ISBN: 978-3-527-31156-9

– GaN/InN
–– hexagonal wurtzite-structured lattice parameters 102
– properties 75
– β-SiC powder mixtures, CVD process 167
– SiC solid solution powders 167
– structure 74
– synthesis 74
aluminum oxide
– crystal structure 4
– liquid-phase sintered (LPS) alumina 8–11
– natural sources 5
– polycrystalline alumina, properties 11–13
– powders, preparation 5
–– chemical methods 6
–– high-temperature/flame/laser synthesis 6
–– mechanically assisted synthesis 6
– solid-state sintered alumina
–– submicrometer/transparent alumina 7
ammonia gas atmosphere
– molecular precursor 120
– structural transformation 116
anion-deficient perovskites 273
– ABO_{3-x}-type perovskites 272
– copper-containing complex compounds 274
anomalous ultra-small angle X-ray scattering (A-USAXS) 588
antiferroelectrics 279, 776
– saturation voltage 776
antiferromagnetic (AFM) 278
$AO(ABO_3)_n$ Ruddlesden–Popper phase 266
applied moment double cantilever beam (AMDCB) techniques 359, 365
– R-curve behaviors 366
– specimen 359
archetypal perovskite 257
– $CaTiO_3$, composition 257
argon
– atmosphere 148
– gas 160
Ashby and Verrall (A-V) model 643, 644
– features 644
– interface-reaction flow mechanism 645
– principle 644
atmospheric plasma spraying (APS) 38
atomic force microscopy (AFM) 625, 636
atomic resolution ultramicroscopy 496
atomic-resolved high-resolution transmission electron microscopy 819
Aurivillius bismuth compounds 266
Avogadro's number 440, 483
A_xWO_3 bronzes, structures 271

b

$BaBiO_3$, semiconductive properties 276
Ba-doped titania varistors
– nonlinear exponent, variation 26
$Ba_2In_2O_5$
– O^{2-}-ion conductivity, Arrhenius plot 284
– oxygen-deficient brownmillerite-structure 285
barium titanate ($BaTiO_3$) 261, 437, 438, 447, 454, 735
– acceptor-doped 443, 457
–– chemical diffusivity isotherms 466
–– defect-chemical interpretation 466
–– defect structure 453
–– relaxation behavior and chemical diffusion 464
– Al-doped 454, 455
–– electronic transference number 467
– chemical diffusivity of oxygen 474
– defect concentration 441
–– two-dimensional representations 445
– defect structures 439
– donor-doped 445, 450, 468
–– conductivity vs. time 471
–– defect-chemical interpretation 469
–– kinetic anomaly 468
–– relaxation behavior and chemical diffusion 468
– equilibrium total electrical conductivity vs. log a_{O_2} 454
– La-doped 468
–– equilibrium conductivity 468
–– equilibrium defect structure 469
– log a_{O_2} vs. log a_{TiO_2}, configuration 444, 445
– log [S] 446–449
–– vs. log a_{O_2} 448, 449, 451
–– vs. log a_{TiO_2} 446, 447
– majority disorder types 441, 442
–– allocation rules 443
–– numerical values 442
–– regimes 449
–– type of region 443
– Mn-doped 744
– multilayer ceramic capacitors (MLCC) 437
– phase transformations 261
– polarization directions 735
– polycrystal 466
– pure 439
– tetragonal 262
–– transition 735
– undoped 461
–– chemical diffusivity isotherms 466
–– defect-chemical interpretation 466

-- relaxation behavior and chemical diffusion 464
-- surface-reaction-rate constant vs. oxygen activity 465
basal plane slip
- log(CRSS) vs. temperature, plots 383
$Ba_2Si_5N_8$, atomic arrangement 243
$Ba_{1-x}Sr_xRuO_3$
- room-temperature pressure vs. composition phase diagram 265
$Ba_5Ta_4O_{15}$ structure, schematic projection 272
Bayer process 6
B_4C polytypes
- Gibbs free-energy analysis 200
- SiC particulate composites 179
- transition metal boride composite materials, development 181
beryllium oxide (BeO) 396
- wurtzite structure of 396
$BiFeO_3$ thin films 764
- magnetic properties 764
- magnetoelectric effect 764
$Bi_{3.25}La_{0.75}Ti_3O_{12}$ (BLT), TEM studies 777
Bi_2MoO_6, ideal structure 268
$Bi_4Ti_3O_{12}$ (BiT) 771
- pseudo-tetragonal unit cell of 267
- relaxor-like behavior 771
binary phase diagram, thermodynamic assessment 135
binary ZrO_2-Ln_2O_3 systems, maximum conductivity 40
biomorphous silicon carbide composites 171
biomorphous SiSiC ceramics, SEM image 172
bismuth
- based layered perovskites, schematic diagram 754
- rich phase 17
BO_6 octahedrons 260, 261
boron carbide (B_4C) 131
- acids/alkali liquids 200
- application 207–210
- based composites 179, 180
- ceramics, physical properties 209
- comparison 196
- crystal structure 136
- densification 177
- electrical conductivity 195
- gaseous hydrogen 200
- magnesium-containing impurities 150
- microstructure 176
- nanostructures 151
- oxidation 201
- oxygen/water vapor 201
- Poisson's ratio 200
- powder 150, 176
- properties 200
-- calorific properties 198
-- electrical properties 195–198
-- mechanical properties 198–200
-- neutron-absorbing capability 198
-- optical properties 194
-- thermophysical properties 198
- room temperature isotropic elastic moduli 198
- shock compression data analysis 200
- sintering 174
- submicron B_4C powders 150
- systematic doping 197
- technical-scale production 149
- titanium carbide, reaction 182
boron nitride (BN) 71
- crystallographic structures
-- cubic boron nitride 72
-- hexagonal boron nitride 71
-- wurtzitic boron nitride (w-BN) 72
- industrial synthesis 72
- properties 73
-- c-BN vs. w-BN 73
-- h-BN 73
- thin films 72
brittle materials 545, 555, 557, 627
- application of Weibull statistics, limits for 558
- strength 545
-- data 555
brittle-to-plastic transition (BPT) 336
broken bond model 481, 483, 484, 510
broken bonds 482, 483, 486
brookite, synthesis 23
- crystal structures 21
Brouwer approximation 441
Burgers vector 321, 384, 386, 418, 422
- cubic rare-earth sesquioxides 391
- dissociation reaction 419
- partial dislocations 387
- for $SrTiO_3$ 411

c

$CaAlSiN_3$, atomic arrangement 245
calcium aluminum silicate (CAS) intergranular films 11
calcium silicate 30
- intergranular films 10
$(Ca,La)_{0.3}(Si,Al)_{12}(O,N)_{16}$ α-sialon, M cation coordination 240
$Ca_2Mn_2O_5$

– ab plane, vacancy ordering 273
– oxygen vacancies 273
capacitance transient techniques 123
capillarity constant concept 487, 488
carbide-based hard materials
– boron carbide 134
–– phase diagram 135
–– preparation 134
–– structural aspects 135–137
– silicon carbide (SiC) 131–134
–– carborundum 131
–– phase relations 132
–– structural aspects 132–134
carbon
– bearing materials 140
– carbon black 176
– containing compounds 175
– fiber-reinforced plastic 170
carbon nanotubes (CNTs) 713
carbothermal reduction and nitridation (CRN) 79
Carnot machine, theoretical efficiency 197
Cartesian coordinates 498
Ca-Si-Al-O-N system, glass-forming region 248
cation diffusion energy 653, 655
cavitation 587, 595–597
– creep model 595
– density change/volume fraction 588
– role 596
– size distribution 590
– TEM characterization 587
ceramic material(s) 515, 531, 567, 578, 601, 608, 633, 657, 722, 775
– application 568, 627
– blades, high sensitivity 577
– brittle fracture 541–545
–– application of Weibull statistics, limits in brittle materials 58
–– ceramic components/critical crack size, tensile strength 544
–– experimental and sampling uncertainties 553–555
–– fracture mechanics, basics 542
–– fracture statistics/Weibull statistics 545–550
–– influence of microstructure 555–558
–– probabilistic aspects 545–558
–– Weibull distribution, application 550–552
– coarsening behavior 515
– components 544
–– critical crack size 544
–– Griffith crack size 544
–– tensile strength 544
– contact damage 537
– crack growth resistance curve, influence 621
– creep research activities 578
– critical structural property 629
– crystal structures 388–390
–– dislocation dissociations 391
–– slip, crystallography 388
–– slip systems 390
– damage mechanisms, overview 538–540
–– corrosion 540
–– creep 540
–– fatigue 540
–– sub-critical crack growth 539
–– sudden, catastrophic failure 539
– defects at failure origin 609
– delayed fracture 558–567
–– fatigue crack growth on strength, influence 566
–– proof testing 566
–– SCCG on strength, lifetime and influence 559–565
– dimensions 722
– edge flakes, examples 538
– failure modes/appearance 532–538
–– contact failure 536–538
–– thermal shock failure 534–536
– fatigue 626–629
–– contact fatigue 627
–– cyclic fatigue crack propagation 626
– fracture theory 531, 601–612
–– crack closure concept/superposition 609
–– fracture mirror 532
–– mechanics 531, 568
–– resistance, determination 622–626
–– stress concentration factors 608
–– surface 569
–– toughness 531, 532
–– type 533
– functional properties 3
– inertia 775
– joining, micrograph 657
– mechanical testing, standardization 531
– penny-shaped cracks 543
– products 537
– statistical nature of failure 567
– strength 532
– microstructural mechanisms 611
– powder processing technology 660
– strength 609
– structural deficiency 612
– structural elements 352
– structural reliability 622
– superplasticity 633

-- accommodation-controlling mechanisms 633
-- accommodation processes 643–656
-- applications 656–659
-- future prospective in field 659
-- goals 659
-- grain boundaries, nature 640–643
-- macroscopic/microscopic features 634–640
- systems 486, 512
-- CSL boundaries 512
- toughened ceramics 612–620
-- bridged interface methods for increasing fracture resistance 613–620
-- toughening by crack tip process zone effects 620
- use 660, 667
ceramic matrix composite (CMC) systems 145, 205, 634
ceria (CeO_2) 43, 44, 47, 705, 717
- based ceramics
-- applications of 44
-- mechanical properties 45
-- room temperature, dependence 46
-- stabilized ZrO_2 ceramic, scanning electron microscopy image 37
- based solid electrolytes, properties 47
- nanocrystalline, electrical conductivity 717
- polishing powders 46
- predominant defects 705
cerium oxide, see ceria (CeO_2)
cermets, fabrication 480
charged transport processes 749
charge neutrality equation 452
- constraint 457
chemical diffusion theory 463, 466
chemical diffusivity 463, 472, 475
chemical mechanical planarization (CMP) 43
chemical vapor deposition (CVD) 60, 142, 330, 681
- molecular beam epitaxy (MBE) approaches 124
- SiC, coatings 160
- single crystals 330
chemical vapor infiltration (CVI) 169
chevron-notched beam (CNB) technique 369, 371
chromium dioxide (CrO_2) 791, 800–802
- ^{53}Cr NMR spectra 801
- half-metallic behavior 791
- magnetic moment 800
- magnetoresistance (MR) 801
chromium nitride (CrN) 77
coarse-grained samples 338
- aluminas 13
- crack-growth resistance 335
- microstructure 191
- SiC powder fractions 152
- Ti_3SiC_2 microstructures
- Ti_3SiC_2, tensile stress-strain curves 338
coarsening process 514
cobalt ferrite ($CoFe_2O_4$) 799
- high coercivity 799
- nanoparticles 800
Cohen's analysis 735
coincidence site lattice (CSL) model 509, 511
- boundaries 510
-- HRTEM images 512
-- schematic illustrations 511
- lattice 509, 510
Cole–Cole relaxation 771
colloidal methods 6
colossal magnetoresistance (CMR) effect 277
- manganites 275
- phenomenon 278
commercial anatase powders, morphologies 22
complex, definition 229
complex oxides 437
- applications 437
- $BaTiO_3$, multilayer ceramic capacitors (MLCC) 437
- defect-chemical analyses 455
- defect structure 437–455
-- acceptor-doped case 443
-- complication, hole trapping 452
-- defect concentrations, two-dimensional representations 445–451
-- donor-doped case 445
-- kinetic aspects 438
-- pure case 439–443
-- and reality 453–455
-- sensitive properties 453
- molecular ratio 439
- nonstoichiometry 437
-- oxygen nonstoichiometry 456–462
-- re-equilibration 462–476
-- relaxation 437
- TiO_2 438, 439
complex Si-Al-O-N
- lanthanum new phase 242
- $M(Si,Al)_3(O,N)_5$ phases 243
- $M_2(Si,Al)_5(O,N)_8$ oxynitrides 242
- $M_3(Si,Al)_6(O,N)_{11}$ phases 244
- M-Si-Al-O-N oxynitrides 238
-- JEM phase 239
-- sialon S-phase 240–242
- $MSi_2O_2N_2$ oxynitrides 245

– sialon polytypoid phases 232
– sialon X-phase 231
– wurtzite oxynitrides 245
– Y-Si-O-N oxynitrides 233–238
composite microstructures, thermal conductivity 672
– secondary phase effect
–– AlN ceramic 674–676
–– liquid-phase sintering of nonoxide ceramics 672–674
–– SiC ceramic 678
–– Si_3N_4 ceramic 676–678
contact 536
– blunt contact 536
– failure 536–538
– sharp contact 536
continuous fiber-reinforced SiC matrix composites 169
coordination number (CN) 482, 492
Coulombic correlations 803
Coulombic potential barrier 827
Cr_2AlC
– elastic properties 305
– oxidation resistance 319
crack 603
– bridging mechanism 163, 353, 359, 658
– closure concept/superposition 609
– closure force 611
–– point load closure force 612
– deflection 146, 159, 163
– flanks 611, 618
– fracture, energy associated 605
– function of length 608
– growth
–– condition for 605
–– resistance behavior 618
–– resistance curve 621
–– resistance, effectiveness 629
– interlinking 13
– loading modes 607, 608
– micro-cracking 620
– pertinent concept of crack growth resistance 614
– potential energy 604
– shape 603
– subcritical propagation 626
– tip opening displacement measurements 625
– wake interaction 163
– wake toughening mechanisms 352
creep curves approach 594
creep mechanisms 577
– resistance 658
critical resolved shear stress (CRSS) 380
– deformation, kink mechanism 382–384
– experimental observations 380–382
– kink pair nucleation, modification 384–386
– theory vs. experiment
–– nonstoichiometric spinel 388
–– sapphire and stoichiometric spinel 387
cross-head displacement (CHD) rate 340
cryptomelane-type MnO_2 nanowires 821
crystal 487, 489
– equilibrium shape 491
– fcc metallic crystals 490
–– Wulff construction 491
– free energy 489
– growth 499
– kink, density 496
– minimum symmetry segment 490
– structures, of oxides 389
– surfaces 496, 501
– terrace, definition 496
– tetragonal symmetry 487
crystal-field theory 792
crystalline boron carbide nanoparticles 151
crystalline oxide materials, applications 437
crystallographically complex 229
cubic perovskite structure 273
– ABO_3 273
– transformation 264
cubic $SrTiO_3$, crystal structure 411
cubic tungsten bronzes 270
cubic ZrO_2 matrix
– tetragonal phase nucleated, oblate spheroids 33
Curie's capillarity constant concept 488
Curie–Weiss temperature 763, 828
cyclic crack propagation process 767
cyclic stress-strain curves
– linear elastic solids 326
– typical compressive 326

d

damage mechanism 538
– classes 539
– corrosion 540
– creep 540
– fatigue 540
– overview 538–540
– sub-critical crack growth 539
– sudden, catastrophic failure 539
Debye frequency 669
Debye length 472, 641, 702, 707, 708
– vs. space charge potential 703
Debye model 669
Debye temperatures 315
defect agglomeration 770

deformation mechanisms 579, 592, 647
– degree of deformation 595
de Gennes double exchange 275
degrees-of-freedom (DOF) 505, 506
– macroscopic, schematic illustration 505, 507
delayed fracture 558
– loading regions 565
– under constant load 561
– under general loading conditions 565
– under increasing load 564
dense B_4C-TiB_2 composite materials 180
dense boron carbide
– boron carbide-based composites
–– B_4C-based MMCs 182
–– B_4C-SiC 179
–– B_4C-TiB_2 180–182
– evaporation-condensation processes 174
– sintered boron carbide 175
–– hot isostatic-pressed boron carbide 178
–– pressureless sintered boron carbide 175–178
–– spark-plasma-sintered boron carbide 179
dense random packing (DRP) model 503
– five fold rotational symmetry 503
dense silicon carbide
– bodies, fabrication 162
– ceramically bonded silicon carbide 152
– platelet-reinforced 145
–– alumina 146
– properties 79
– reaction-bonded silicon carbide 154
– recrystallized silicon carbide 152–154
– sintered silicon carbide
–– Al-B-C-based additives 159
–– AlN 167
–– Al_2OC 168
–– alloying 163–172
–– B_4C 165–167
–– biomorphous composites 171
–– chemical/physical vapor-deposited silicon carbide 160
–– CVI, fabrication 169
–– hot isostatic-pressed SiC 160
–– hot-pressed silicon carbide 159
–– liquid-phase-sintered SiC (LPS-SiC) 156
–– LPI, fabrication 170
–– LSI, fabrication 170
–– metal matrix composites (MMCs) 172–174
–– oxide additives 157–159
–– pressure less solid-state sintered silicon carbide 154
–– silicon carbide nanoceramics 162
–– silicon carbide wafers 161
–– spark plasma sintering (SPS) 155
–– TiB_2 164
–– TiC 164
density functional theory (DFT) 109
deterministic design approach 551
dicalcium/tricalcium silicate, microstructure 515
dielectric ceramic materials 669
diesel engines 204
diesel particulate filters (DPFs) 204
diffusion process 463, 514, 642
– coefficient 649
– controlled model 395
dilatation stresses 597
discontinuously reinforced aluminum (DRA) 202
dislocation theory
– application 379
– interactions 321
– in oxides 392
– particular oxides 393
– spinel, Hornstra 392
dissolution-reprecipitation processes 673
2-D nucleation process 501, 515, 651
– energy barrier 501
domain-engineered crystals 731, 735
– characteristics 735
– domain orientation states 735
domain-multidomain process 743
domain reconfiguration process 742
domain wall 746
– clamping effect 743
–– stabilization 744
– pinning mechanism 777
– trapping sites 746
dopant 445
– mass-conservation equation 445
– roles 633
doping
– heterogeneous doping 715
– homogeneous doping 714
2-D simple rectangular system 497
– step density, changes in 497
3d transition-metal ions 794
– electronic configurations 794
– ground-state effective spins 794
Duralcan process 173
DWA-aluminum composites manufactures 202
dye-sensitized solar cells (DSSCs) 704, 719–722
– light active synthetic dyes 720

– schematic diagram 722
– structure and components 720
Dzyaloshinskii–Moria interaction 762

e
earth-doped oxynitride glasses 66
edge energy 494
EKasic silicon carbide laser-structure 204
elasticity theory 408
elastic modulus 374
electrically active ceramics, see electroceramics
electrical poling process 729
electric field 765, 770
– cyclic application 765
– intensity 709
electric transportation systems 668
– parameters, summary 310
electroceramics 3, 697
– devices 699
– electrically conductive, properties 697
– electronic properties, role of interfaces 698
electrochemical models 770
electromechanical anisotropy 738
electron-beam back-scattered diffraction (EBSD) analysis 507
electron beam curing method 147
electron beam physical vapor deposition (EB-PVD) 38
electron energy loss spectroscopy (EELS) 62, 306
electronic charge carriers, redistribution 702
electronic conductor system 475
electron paramagnetic resonance (EPR) measurements 826
electron spin resonance (ESR) analysis 682
– spectra 813
electron transport dynamics 722
electrostrictive ceramics 731
electrostrictive effects 280
electrostrictive transparent material 767
Elektroschmelzwerk Kempten (ESK) process 138
elliptical defect 602
elliptic cylinder, schematic representation 327
emulsion techniques 44
energy-dispersive X-ray (EDX) analysis 252
energy-dissipating processes 543
energy-lowering process, see reconstruction processes
energy-minimizing process, faceting 492
energy-reduction process, faceting 492
environmental barrier coatings (EBCs) 205
equilibrium conductivity 473
equivalent stress, definition 549
Er-Si-O-N system 237
ESIS reference materials 552
ethylenediaminetetra-acetic acid (EDTA) 31
Euler angles 507
Euler's rotation theorem 506
explosion boron nitride (e-BN) 71
extreme-UV (EUV) spectral region
– complex refractive index 195
– space astronomy/earth sciences 194

f
fabrication method 150, 180
face-centered cubic (fcc)
– crystal, CSL boundary 510
– lattice 102
– symmetry 482
faceting 492
– energy change 494
Faraday constant 466
fatigue-free systems 776
– thin film devices 771
fatigue process 770, 772
Fermi energy 711
Fermi level 113
ferrites 797–800
– ferrimagnetic ordering 798
– magnetic properties 798
– partial unit cell 798
ferroelectric ceramics 619, 729, 750, 755
– applications 729
– engineering 755
– lead zirconate titanate (PZT) 619
– – crystal domains orientation, schematic representation 619
– – poling direction 620
– properties 750
ferroelectricity 258, 762
ferroelectric system 729, 735, 747, 765, 770, 775
– acceptor defects 770
– aging phenomenon 731
– crystals, properties 738
– enhanced intrinsic response 735
– fatigue in ferroelectric materials 765–777
– – antiferroelectrics 776
– – breakdown 768
– – clamping in thin films 774
– – combined loading 775
– – defect agglomeration 770
– – domain nucleation or wall motion inhibition 773
– – domain splitting and crystal orientation dependence 774

-- electrode effects 773
-- fatigue-free systems 776
-- frequency effect 769
-- macrocracking 765–767
-- microcracking 767
- ferroelectricity and magnetism 756–764
- hardening-softening transitions 747
- hysteresis loops 741, 748
-- comparison 749
-- relaxation 748
- lead-zirconate-titanate, image of stresses 766
- materials 730, 742, 745
-- applications 730
-- piezoelectric properties 742
-- properties, rate dependencies 730, 769
-- V-potential 745
- memory devices 765
- perovskites 734, 749, 750, 760
-- PZT/$BaTiO_3$ 749
- phase transition 281
- polarizations 764
- properties 729
-- extrinsic properties, hard and soft ferroelectrics 739–750
-- intrinsic properties, anisotropy of properties 731–739
- random access memory 760
- single crystals 730
- spontaneous polarization 739
- stress-strain relationships 766
- textured ferroelectric materials 750–756
-- HTGG 753
-- OCAP 752
-- RTGG 754–756
-- TGG 753
- use 765
- variety 733
ferromagnetism 258
fiber-reinforced SiC matrix composites
- mechanical and physical data 158
fibrous grains 358, 373
- schematic presentation 370
fibrous β-reinforcing grains 359
fibrous β-seed crystals, use 353
fibrous β-silicon nitride grains 365
field-cooled (FC)/zero-field-cooled (ZFC) susceptibility curves 811, 825
field effect transistor (FET) devices 92, 203
fine-grained material 621
- microstructure 162
-- yields materials 351
- Ti_3SiC_2 samples, cyclic-fatigue study 334
finite element (FE) stress analyses 549
finite-temperature Lanczos method (FTLM) 812
first law of thermodynamics 438
first nearest neighbor interactions (FNNs) 483, 485
- interactions 491
first-principle stress (FPS) criterion 549
flexure method, surface crack 625
forsterite, Mg_2SiO_4
- crystal structure of 417
- high-temperature creep deformation 418
Fourier transform infrared (FTIR) spectroscopy 115
fractography 558
fracture 546, 615
- energy strength, comparison of 375
- probability 546
- resistance, determination 622–626
-- advantages 623
-- indentation fracture toughness 622
-- intrinsic toughness, determination 625
-- significance 622
-- single-edge notched beam 623
-- surface crack in flexure 625
- statistics 546
- strain 376
- strength 373
-- porosity dependence 372
-- dependencies 373
- theory 601–612
-- crack closure concept/superposition 609
-- fracture toughness vs. strength for ceramic materials 602
-- stress concentration factors 608
free carbon content
- four-point bending strength vs. sintering temperature 199
French process 16
Frenkel disorder 701
Friedel–Fleischer statistics 428, 430
functional gradient material (FGM) 659

g

gadolinium-doped ceria (GDC) 714
gallium arsenide 112
gallium nitrate, thermal treatment of 115
gallium nitride (GaN) 93, 112
- anion-substituted 108
- bulk single crystals 100
- films
-- crystallinity 99
-- oxidation 108
- (Ga,Al,In)N solid solutions, properties 101
- GaN-Ga_2O_3 system 109

-- phase separation 118
- metal-oxide-semiconductor (MOS) capacitors 123
- phase 93
-- description 94–98
- powder 96
- solubility of oxygen 111
- substrates, optoelectronic materials 91
-- molecular precursors, pyrolysis of 117
-- thermally activated transformation 115–117
- synthesis routes 99
-- bulk gallium nitride, synthesis 99
-- chemical precursor routes, synthesis 101
-- thin film, synthesis of 100
- system 94
gallium oxide nitride 111
- films 123
-- sensor properties 114
- materials 113
- phases, synthesis 117
gallium oxides
- based materials 104
- β-gallium oxide displays 94
- GaN solid solutions 121
- GaN system 108, 110
- β-Ga_2O_3 phase
-- Raman spectra 105
-- X-ray diffraction patterns 105
- β-Ga_2O_3 polymorph 94
- ε-Ga_2O_3 polymorph 107
- growth techniques, synthesis 106
- phase description and properties 102–106
- system 104
-- phase transformations/synthesis temperatures 103
- thin films 94
gallium oxonitrides 107
- compounds, syntheses 114, 116
- gallium oxide nitride phases 110–114
- Ga-O-N materials, nomenclature issues 108
- growth techniques, synthesis 114
-- crystalline phases 118–121
-- precursor approach 115–118
- photoluminescence measurement 119
- potential applications 122–124
- spinel
-- phase 118
-- properties of 109
-- structure 118
- theoretical predictions 108–110
gallium tris(tert-butoxide) dimethylamine adduct 117
Ga/N/O atoms
- secondary ion mass spectroscopy profiles 113
$Ga_{2.8}N_{0.64}O_{3.24}$
- spinel-type structured, Rietveld refinement 119
Ga-O-N films, useful production 92
Gaussian distribution 553, 554
- curve 555
$Ga_xO_yN_z$ phase
- diffraction pattern 121
- solid-state compounds 108
Gibbs dividing surface 481
Gibbs free energy 458
- function 737
Gibbs–Helmholtz equation 458
glass
- grain interfacial energy 10
- pockets, triple grain junctions
-- energy-dispersive X-ray spectra 366
- transition temperatures 66
glassy-phase ceramics 649
- accommodation processes 649
- layers 655
global positioning system (GPS) 202
goethite α-FeOOH nanowires 797
- scanning electron microscopy image 797
- superparamagnetic behavior 797
Goss grains 519, 520, 523
- exclusive growth 519
- microstructural evidence 521, 522
- optical microscopic images 522
- sub-grain boundaries 520
Gouy–Chapman model 718
- boundary conditions 704, 707
-- vs. Mott–Schottky 706, 707
grain 614
- alignment 364
- crack bridging, examples 614
- crack growth resistance behavior 616
- morphologies 365
grain boundary(ies) 7, 11, 47, 368, 480, 508–510, 516, 519, 595, 633, 700
- coincident site lattice (CSL) boundaries 507
- crystal phase 351
- energies, anisotropy 521
- glass 11
-- corrosion resistance 13
- glassy phase 686, 688
- microscopic grain boundary 507, 508
- mobility 517
-- schematic illustration 518
- molecular dynamics (MD) simulations of 10

– nature 640–643
– sub-grain boundary, schematic illustration 519
– thickness 595
– viscosity 596, 597
grain boundary sliding (GBS) 594, 633, 635, 636
– accommodation process 646
– advantage 657
– mechanism 645
– point defects 635, 636
– strain rate controlling 648
grained sialon, equiaxed/elongated 628
– scanning electron microscopy images 628
Green's function technique 757
Griffith analysis 543, 544, 603, 606
– mechanical energy release rate 604
– stressed plate/through-thickness crack 603
– stress element, cartesian/polar coordinate 606
Griffith–Irwin fracture criterion 543

h

H^+-ions
– binding force 286
– conductivity, masses 287
Hall coefficient 310
hardening mechanisms 423
hardening-softening concept 740
hard sphere model 482, 500
heat-exchanger systems 204
Heisenberg model 813, 815
Helmholtz free energy 487
Hertzian cone cracks 531
Hertzian ring crack 536, 538
heterotemplated grain growth (HTGG) 752, 753
hexagonal BN (h-BN) 71
– functions 71
hexagonal gallium nitride 120
hexagonal/rhombohedral polytypes 421
hexagonal wurtzite structure 96
high-energy milling 14
high-pressure nitrogen solution growth (HPNSG) method 99
high-resolution electron energy loss spectroscopy (HREELS) 112
high-resolution scanning electron microscopy (HRSEM) 625
high-resolution synchrotron X-ray diffraction 736
high-resolution transmission electron microscopy (HRTEM) 106, 134, 504, 642, 678, 820, 825
– images, technological advancements 512
high-strain rate superplasticity (HSRS) 659
hole trapping, trapping factor 452
hollow microsphere templated thin films 715
homogeneous uniaxial stress field 547, 548
– arbitrarily oriented cracks, Weibull distribution 547, 548
HOMO–LUMO separation 709
hot isostatic pressing (HIP) 8, 160, 178
hot-pressed silicon carbide 159, 203
hot-wall horizontal quartz tube reactor 142
Hugoniot elastic limit (HEL) 200
Hund's correlation energy 793
Hund's rule 803
hydrocarbon combustion 47
hysteresis 745, 746
– ascending/descending branches 750
– loops 740

i

ideal perovskite ABO_3 structure, presentations 258
IKB
– critical resolved shear stress (CRSS) 328
– formation 327
incident photon-to-electron conversion efficiency (IPCE) 721
indium tin oxide (ITO) 697
injection laser diode, synthesis processes 100
in-situ scanning electron microscopy (SEM) 365
insulating gallium oxide nitride layers 92
interface anisotropy, role 515
intergranular amorphous phase 365
intermediate-temperature solid oxide fuel cells (IT-SOFCs) 41
intra-atomic electron-electron coulomb energies 275
intrinsic electronic excitation, pseudo-equilibrium constant 458
ion-beam-assisted deposition 60
ion beam-deposited boron carbide 194
ionic and mixed conducting materials 707, 714
– interfaces in 714–718
ionic conductivity 47, 285, 716
– Arrhenius plot 716
ionic solid, electrical conductivity 714
ionic systems 699, 700
– potential barriers 700
– space charge formation 699
iron oxides 793–797
– arrays 793
– hematite (α-Fe_2O_3) 794, 796

– magnetic moment 794
– magnetic properties 794–797
– magnetite (Fe_3O_4) 794, 796
– structural properties 795
– structures 793
Ising model 499

j
Jahn–Teller-active ions 808
Jahn–Teller effect 763, 802
Jahn–Teller phonons 803
JEM phase, crystal structure 240
J-phase
– monoclinic cells 236
– structure 236
– unit cell volume vs. ionic radius 237

k
Kagomé layers 389
– basal dislocations 391
– Shockley-like quarter partials 391
kink
– band formation, scheme 325
– behavior, analysis of 421
– boundaries 337
– formation energy 384
– nucleation model 384, 385, 388
–– modification 385
– pair nucleation, diagrams 386, 413, 419
Knoop hardness anisotropy 199, 400, 401
Kosterlitz–Thouless transition 499

l
$LaMnO_3$ 276
– Mn vacancy migration 288
Landau–Ginsburg–Devonshire (LGD) theory 733
Landolt–Börnstein tables 379
Langmuir equation 711
lanxide, see Primex process
large fibrous grains
– alignment 371, 374
– crack interacting, in situ observations 367
large-scale solar cell 720
large β-seed crystals, incorporation of 360
laser-heated diamond anvil cell (LH-DAC) 118
laser-heated diamond cell methods 118
laser scanning systems 207
$La_4Si_2O_7N_2$
– crystal structures, [0 1 0] projections of 237
lattice dislocation model 509
lattice molecule 439, 443, 445, 450
lattice wave propagation 669
lead-lanthanum zirconate-titanate (PLZT) 730
– electro-optic properties 730
lead zirconate titanate (PZT) ceramics 731, 774, 781
least squares method 593
Le Châtlier's principle 450
Legendre transformation 497
ligand-field theory 792
light-emitting diodes (LEDs) 161
light-emitting nitride 80
linear elastic fracture mechanics framework 543
Lipschitz-mechanism, grain switching, scheme 635
liquid crystal displays (LCDs) 697
liquid phase
– advantage of 157
– sintered aluminas 4
– sintered EKasic T 193
– sintered materials 11
– sintered silicon carbide (LPS-SiC) 156
–– tailored properties of 203
liquid-phase sintering (LPS) mechanism 157, 246, 480, 514, 517, 671
lithium-ion
– battery materials 718
– conductor 715
–– Al_2O_3 system 716
– densification theory 514
– realistic model 514
lithium sialon system
– α-sialon stability, region 239
local-density-approximation (LDA) calculations 109, 814
Lodgering factor 755
log(CRSS)-T law 382
longitudinal piezoelectric coefficient 736, 737
– orientation dependence 736
long-range ordering (LRO) transition 808, 810
– antiferromagnetic 812
low-energy electron diffraction (LEED) methods 112
low-Z rare earth sialon systems 249
Lu_2O_3-SiO_2 sintering additives
– seeded 369
– tape-cast silicon 369
Lucalox™ 514
lutetium, role 596

m
macrocracking 765
magic angle spinning nuclear magnetic resonance (MAS-NMR) spectroscopy 116, 231

magnesia, see magnesium oxide (MgO)
magnesium acetate 14
magnesium oxide (MgO) 13, 14, 516
– addition 11
– crystal structure 14
– crystals, work hardening 423–424
– – spinel 424–426
– Fe^{3+} dopant, yield stress 430
– MgO-doped alumina 516, 517
– – pore mobility 517
– natural sources and production 14
– polycrystalline magnesia 15
– powders, sinterability 14
– role 517
– single-crystal, physical properties 14
– slip systems, log(CRSS) vs. temperature 380
magnetic anisotropy 796
magnetic frustration, manifestation 807–815
– defect spinel λ-MnO_2, frustrated magnetism 813–815
– layered α-$NaMnO_2$ compounds, magneto-elastic coupling 811–813
– – geometric frustration 813
– – magnetic susceptibility 812
– mixed-valence hollandite (α-MnO_2), magnetic properties 810
– pyrolusite (β-MnO_2), helical order 808–810
magnetic nanoparticles 821
magnetic recording technology 800
magnetic semiconductor 801
– schematic densities 801
magnetic spiral order sets, conditions 762
magnetic susceptibility 823, 824
– Curie–Weiss dependence 823
magnetization 822
– temperature-dependence 823
magnetoelectric devices 756
magnetoelectric effect (ME) 756, 757
– challenge 760
– comparison 759
– product property 757
– sources 756
magnetoelectric voltage coefficient 758, 759, 760
manganese oxide 802, 804
– crystallographic summary 805
– frameworks 807
– hollandite-type 811
– layered forms 807
– layered structures 807
– – birnessite (δ-MnO_2) 807
– magnetic behavior 808
– magnetism 802
– one-dimensional structures 804–807
– – hollandite (α-MnO_2) 806
– – pyrolusite (β-MnO_2) 804
– – ramsdellite (R-MnO_2) 804–806
– phases 802–821
– – crystal structures, polyhedral presentations 806
– – magnetic properties 807–815
– – manganese dioxide nanoparticles, magnetic properties 821
– – manganese oxide structures 804–807
– – MnO_2 nanostructures, synthesis 815–821
– samples, scanning electron microscopy image 826
– three-dimensional structures 807
– – spinel λ-MnO_2 807
manganese dioxide nanostructures 815, 821
– crystalline phases/precursors/synthesis conditions/morphologies 816, 817
– magnetic properties 821–828
– – α-MnO_2 (hollandite) 824–828
– – β-MnO_2 (pyrolusite) 821–824
– synthesis 815–821
– – in crystallographic phases 821
– – α-MnO_2 nanostructures, synthesis 815–818
– – β-MnO_2 nanostructures, synthesis 818
– – γ-MnO_2 nanostructures, synthesis 819
– – λ-MnO_2 nanodiscs, synthesis 820
March–Dollas function 755, 756
mass-action law 440
mass conservation constraint 452
materials 731
– chemistry 479
– electromechanical responses 731
– fatigue, degree 776
– imperfections 479
– interfaces 479–513
– – anisotropy 523
– – energies, anisotropy in 480
– – kinetics of surface migration 499–502
– – roughening transition 495–499
– – solid/liquid interfaces 502–506
– – solid/solid interfaces 506–513
– – surface energy 480–486
– – Wulff plot 486–495
– microstructures 479
– practical implications 513–523
– properties 479
MAX phases 300
– compressive stresses 329
– cyclic stress-strain loops 328
– discovery of 304
– electrical conductivities 307

– electronic properties 312
– experimental vs. theoretical bulk, comparison 306
– kink bands (KBs) 321
– mechanical properties 320
– MX carbides 329
– nanolaminate nature of 324
– partial DOS 304
– Raman modes 307
– rutile-forming 319
– Si/Ge-containing 313
– Sn-containing 305
– TECs, summary 317
– temperature-dependence 309
– thermal conductivities 314, 315
– thermal expansion coefficients (TECs) 316
– thermal shock-resistant 333
– thin films 299
– Ti_3SiC_2 320
– transport parameters 312
– unit cells 301
– Vickers indentation 333
– Young's modulus/shear modulus/Poisson ratio 305
mechanically coupled magnetoelectric composites 758
mesoporous random nanoparticle TiO_2 films 722
metal/silicon nanowire Schottky barrier 709
– electron density 710
– energy band bending 709
metallic systems, secondary recrystallization 519
metal matrix composites (MMCs) 172
– advantage of 182
– discontinuously reinforced 173
– whiskers, use 145
metal-organic chemical vapor deposition (MOCVD) 101
metal oxide 642
– nanostructures, electronic properties 713
metal-oxygen binding energies 283
methyltrimethoxysilane (MTMS) 140
MgOċnAl_2O_3 385
– mechanical properties 413
– spinel 385, 411
– stacking fault energy of 416
$MgSiN_2$, crystal structure of 245
microcracking 31, 767, 768
microcrystalline ceria 45
microelectromechanical systems (MEMS) 714
micromechanical mechanisms, development 626
micro-mechanical models 580
microstructural elements, types 351
microwave hydrothermal (MH) synthesis 22
Miller indices 488, 490, 496, 505
mixed ionic electronic conductivity (MIEC) 697
$M_{n+1}AX_n$ phases 299
– bonding and structure 300–303
– elastic properties 303–307
– electronic transport 307–313
– list of 302
– machinability 342
– mechanical properties 299
–– compressive properties 336–338
–– creep 338–341
–– dislocations and arrangements 320
–– hardness and damage tolerance 329–333
–– high-temperature properties 336
–– incipient kink band microscale model 325–329
–– plastic anisotropy, internal stresses, and deformation mechanisms 321–325
–– quasi-single crystals/polycrystals, compression behavior of 329
–– R-curve behavior and fatigue 334
–– solid-solution hardening and softening 341
–– tensile properties 338
–– thermal shock resistance 333
– properties 299
– thermal properties
–– chemical reactivity and oxidation resistance 318
–– thermal conductivities 313–316
–– thermal expansion 316–318
–– thermal stability 318
– tribological properties 342
mobile dislocation walls (MDWs) 322, 326
molecular beam epitaxy (MBE) 716
monoclinic β-Ga_2O_3, crystal structure of 103
monolithic SiC-plate heat exchanger 205
Monte-Carlo simulations techniques 553, 810
– 3-D microstructure, cross-section 519, 520
– steps 519
morphotropic phase boundary(MPB) 282, 733, 734
– vicinity, intermediate phases 734
Mott–Schottky model 705, 717
– approximation 702
– conditions 708
– electrons/holes, space charge profiles 705
$MSi_2O_2N_2$ oxynitrides 245
M-Si-Al-O-N systems 248

– oxynitrides 230
M-Si-O-N liquid phases 247
multianvil pressure apparatus (MAP)
 synthesis 61
multi-axial stress field, Weibull
 distribution 549
multiferroics 763
– magnetoelectric properties 763
multilayer ceramic capacitors (MLCC) 437
multilayer ionic conductors 717
multiwalled carbon nanotubes (MWCNTs)
 149

n

NaCl 484
– single crystal, equilibrium shape 498
– structure 486, 491
– surface energy 486
– type ceramics 484
– unit cells, clinographic view 484, 485
nanocrystallines 78, 719
– anatase 22
– B_4C 151
– gallium oxide nitride ceramic 120
– MgO 15
– YSZ ceramics 40
nanodimensioned ceramics 723
– physical and chemical properties 723
nanofiber, preparation 713
nanometer-thin glass grain boundary films
– crystallization of 11
nanometric boron carbide particles
– CVD 151
nanometric carbon black 151
nanoporous SnO_2 elements 712
nanostructured ceramics 697
– case study 710–722
– – dye-sensitized solar cells 719–722
– – ionic/mixed conducting materials,
 interfaces 714–718
– – lithium-ion battery materials 718
– – nanostructured sensor films 710–714
– charge carrier distributions, nanostructure
 influence 708–710
– electrical conduction, introduction 697
– introduction 697–699
– observed conductivity, space charge profiles
 effect 707
– semiconducting Ceramic Materials, Space
 Charge Layers 699–707
nanostructures 698
– 1-D/2-D/ 3-D systems 698, 699
– – hierarchy 699
– – schematic depiction 698
– sensor films 710–714
– Y_2O_3 518
– – microstructure evolution 518
– – by two-step sintering 518
nano-TiO_2 coating
– hydroxyapatite layer formation, cross-
 sectional view of 27
nanowires, synthesis 819
National Institute of Advanced Industrial
 Science & Technology 689
natural plants
– morphology 171
Nd_2CuO_4, T′-tetragonal structure 268
Nd_2O_3-Yb_2O_3
– thermal conductivity 38, 39
neutron diffraction 274
– spectra 316
nierite (Si_3N_4) 62
– unit-cell parameters 62
nitric oxide (NO) reduction 104
nitride-based structures 230
nitride compounds 91
– development 229
nitrogen atmosphere 152, 167
nitrogen pyroxene $MgYSi_2O_5N$
– atomic arrangement 252
nitrogen-rich M-Si-Al-O-N oxynitrides 230
noise equivalent power (NEP) 123
nonagglomerated powders 47
nonmetallic crystal 669
– high thermal conductivity 669
nonmolecularity 439, 441, 444, 445, 447,
 449–451, 455, 469
nonoxide ceramic materials 668
– intrinsic thermal conductivities 668
– sintering 672
– thermal conductivity 679, 693
nonoxide ceramics 642
nonstoichiometric spinel 388
nonstoichiometry 437, 438, 442, 444
– re-equilibration 462–476
– – defect diffusivities 475
– – donor-doped $BaTiO_3$ 468–475
– – undoped/acceptor-doped $BaTiO_3$
 464–468
– relaxations 464
nonstoichiometry re-equilibration
 process 462
no poling process 773
novel γ-Si_3N_4 phase, synthesis 61
NT 154 579–582
– cavitation creep model 593
– cavity size distribution evolution 591
– creep-controlling process 580

– creep rates 583
– creep resistance 579
– tensile creep curves 582
– volume fraction of cavities 589
n-type semiconductor 437, 703
nuclear/electron configurations
– features 261
nuclear magnetic resonance (NMR) spectra 137
– measurements 801
nucleation process 653

o
on-site electron-electron coulomb energies 276
orientation distribution function (ODF) 755
– integrations 755
oriented consolidation of anisometric particles (OCAP) technique 752
Orowan's equation 650
orthorhombic $CaZrO_3$
– inter-octahedra proton, dynamical simulations 286
Ostwald's rule 132
oxalate coprecipitation method 45
oxide-ion conductors
– BIMEVOX family 267
oxide perovskites 271, 292
oxonitride compounds 91
oxygen 461
– activity 462
–– conductivity relaxation 471
–– thermodynamic situation/expected diffusion profiles 472
–– vs. defect diffusivities 477
–– vs. reciprocal temperature 462
–– vs. thermodynamic factors 476
– deficient compounds 267
– nitrogen exchange 290
– nitrogen disorder 110
– partial molar enthalpy 462
–– vs. nonstoichiometry for undoped $BaTiO_3$ 463
oxygen nonstoichiometry 456–462, 469
– vs. $BaTiO_3$ 456
– chemical diffusivity vs. oxygen activity 465
– control 456
– experimental reality 460–462
– in general 457–460
– vs. oxygen activity 461
– vs. partial molar enthalpy of component 459
oxygen vacancies 771
– accumulation 771
oxygen vacancy agglomeration model 771
oxynitride 108, 231
– field 231
– glass ceramics 248, 249
–– B-Phase 249, 250
–– Iw Phase 250
–– Nitrogen Pyroxenes 251, 252
–– U-Phase 250, 251
–– W-Phase 251
– properties 290
– structures 230
oxynitride glasses 66, 246–248
– feature 247
oxynitride phosphors 80, 81

p
paraelectric state 279
partially stabilized zirconia (PSZ) 32
– Ca-PSZ 34
– Mg-PSZ 32–34
– plasma-sprayed zirconia 38
– Y-PSZ 34
particle-matrix interface 174
particular oxides, dislocations 392
– aluminum oxide
–– basal slip 406
–– Castaing law 406
–– deformation twinning 408–410
–– dipoles and climb dissociation 407, 408
–– dislocation-dissociation 406
–– prism-plane slip 406, 407
–– stacking fault energy 408
–– Verneuil crystals 405
– climb vs. glide dissociation 422, 423
– fluorite structure, oxides
–– dislocation-dissociation 401, 402
–– uranium oxide 400, 401
–– zirconia 398–400
– forsterite 417
–– crystal structure 417
–– water-weakening/dislocation-dissociation 418, 419
– $MgO\text{-}Al_2O_3$ spinel single crystals
–– dislocation-dissociations 415–417
–– mechanical properties 413
–– slip planes 414, 415
– non-oxide ceramics
–– silicon carbide 421
–– silicon nitride 422
– oxides
–– cubic rare-earth sesquioxide structure 420
–– garnet structure 420, 421
–– slip systems 419
–– YAG 420, 421

– perovskite structure, oxides 410
– – $SrTiO_3$, inverse brittle-to-ductile transition (BDT) 410, 411
– – strontium titanate, dislocation-dissociation 411–413
– rock-salt structure, oxides
– – magnesium oxide 393
– – stoichiometry, effect 393–396
– – transition metal oxides 393–396
– rutile structure, oxides
– – titanium oxide 402
– silicon oxide
– – dislocation-dissociation 404, 405
– – hydrolytic weakening 403, 404
– – polymorph 402
– Wurtzite structure, oxides
– – BeO, dislocation dissociation 397, 398
– – beryllium oxide 396, 397
– – zinc oxide 398
particulate/liquid hybrid mixtures 140
Pauling's rules 235, 251
Pb-based ferroelectric thin films 753
Pb-containing perovskites 292
Peierls barrier 383
Peierls stress 382, 393
perfect single crystal 773
– polarization 773
perovskiteABO_3 structure 262
perovskite catalysts
– designing strategy 291
Perovskite crystal structure 732
perovskite families
– characterization 289
perovskite oxides
– B-site vacancies 271
– electrostatic Madelung energy, ionic compounds 274
– metallic conductivity 275
– thermal stability 291
perovskites 761
– chemical/catalytic properties 289–292
– – synthesis 292
– crystal chemistry 259
– ionic transport, computer modeling 288
– physical properties 281
– spectroscopic studies 282
– structural family 264
perovskites, crystal structure
– ABO_3 perovskite 259
– AO_3 layers, polytypes 264, 265
– cation ordering 269, 270
– ideal perovskite structure 259, 260
– nonstoichiometry
– – anion-deficient perovskites 272–274
– – anion-excess nonstoichiometry 274
– – A-site vacancies 270, 271
– – B-site vacancies 271, 272
– – vacancy-ordered structures 272–274
– perovskite intergrowth structures
– – aurivillius phases 266, 267
– – ruddlesden-popper (RP) phases 266
– perovskite-related copper oxide structures 267–269
– structural distortions and phase transitions 260–264
perovskites, physical properties 274
– ABO_3 ionic transport
– – computer modeling 288
– – thermal stability 290
– applications 289
– cation transport 287, 288
– ferroelectricity 279, 280
– ion conductivity
– – oxide-ion conductivity 283–285
– – proton conductivity 285–287
– morphotropic phase boundary (MPB) compositions 282
– optical properties 282, 283
– physical properties 274–277
– – A-site/B-site cations 275
– – colossal magnetoresistance (CMR) phenomenon 277–279
– relaxor ferroelectrics 280–282
perovskites containing 276
perovskites demonstrate relaxational properties 280
perovskites exhibit 280
perovskite structure 257, 259
– BO_3 array 270
– direct cation-cation interaction 274
perovskite-type oxide structure
– periodic table, metallic elements 260
persistent photoconductivity (PPC) 123
phase-controlled synthesis 821
ε phase transforms 77
phenolic resin, pyrolysis 176
phenomenological approach 745
photocatalysts, redox catalysis 290
photoelectrochemical water splitting 282
physical vapor deposition (PVD) methods 92
– SiC
– – novel applications 161
– – techniques 107
piezoelectric anisotropy vs. dielectric anisotropy 738
piezoelectric ceramics 619, 752
– groups 752
– poling 619

piezoelectric effects 258, 730
– converse piezoelectric effect 730
– direct piezoelectric effect 730
piezoresponse force microscopy (PFM) 774
plane-plane interactions 420
plane-strain condition 373
plasma-assisted reaction process 151
plasma-enhanced CVD (PECVD) 151
– boron carbide nanostructures, morphologies 152
plastic anisotropy 321
– brittle-to-plastic transition 321
plastic anisotropy 331
plastic deformation 162, 403, 421, 422, 538
plastic-to-brittle transition 336
plastic yield stress 531
plate-like α-SiC particles 145
platinum electrodes 773
PMN-PT ceramics 504, 507, 739
– as-sintered surface 504
– fractured surface 504
– high-field strain response 740
– longitudinal piezoelectric coefficient 739
– optical micrograph 508
– texturing 739
Pnma perovskite structure 283
Pnma phase 263
point defects 437, 772
– role 772
Poisson's equation 702
Poisson's ratio 618
Poisson–Boltzmann differential equation 702
Poisson–Schrödinger equations 710
poly(ethylene glycol) (PEG) 820
polycarbosilane (PCS) 141
polycrystalline (Pb,Sr)TiO_3-films 769
– breakdown 769
polycrystalline α-SiC fibers 148
polycrystalline alumina 4, 7, 8, 11, 12, 13, 50
– characteristic properties of 12
– corrosion of 13
polycrystalline boron carbide
– hardness 199
polycrystalline ceramics 667
– applications 667
– preparation of 25
– thermal conductivity 672
polycrystalline ferroelectric materials 773
polycrystalline films 141, 717
Polycrystalline materials 750
– functionality 750
polycrystalline samples
– Raman spectra 308
polycrystalline silicon carbide 133
polycrystalline silicon nitride
– materials, properties 67
– mechanical properties 67
– microstructure 65
polycrystalline titanium dioxide ceramics 24
– physical properties of 25
polycrystalline yttria ceramics 49
– fracture toughness of 49
polymer-pyrolysis-derived fibers (PP-fibers) 146
– silicon carbide fibers 147
polytypes, proportions 134
poly vinyl pyrolidone (PVP) polymer 818
pore-boundary separation 515, 516
– tendency 517
porous biocarbon monoliths
– preparation 171
porous silicon nitride 370, 373
– fracture energy 372, 374
–– porosity dependence 374
– fracture strength, porosity dependence 373
– fracture surface 372, 375
– microstructures 371
positive-temperature coefficient resistor (PTCR) 437, 729
– performance 437
positive thermal coefficient (PTC) component 569
– thermal shock cracks 535
– thermistors 698
post-HIPed composites, TEM images 182
post-HIP process 177
– boron carbide materials, mechanical properties 198
Potts model 772
power-law model 593
Pr_6O_{11}-doped varistors 19
Preform infiltration 173
preparation of submicrometer-grained aluminas 6
pressureless sintered
– flexural strength 166
pressureless sintered boron carbide 175
pressure-pulsed-CVI (P-CVI) 170
Primex process 173
principle of independent action (PIA) criterion 549, 552
prism plane slip
– log(CRSS) vs. temperature, plots 383
probability 545
projected surface free energy 497
0001 projection, schematic representation 323

p-type conductivity 455
pure metallic system, principal surfaces 482
pure rutile, synthesis of 23
pyrocarbon (PyC) 170
pyrochlore phase 18
pyroelectric behavior 282
pyroelectric materials 729
– ferroelectrics 729
PZT 733, 736, 758, 768, 771, 772. see also lead zirconate titanate ceramics
– compositions 772
– Fe-doped 746
– fluorine doping 750
– hardening-softening 740
– IrO_x-electrode system 771
– longitudinal piezoelectric coefficients 739
– nucleation 753
– softening treatment 748

q
quartz 137. see also silicon oxide
quartz deformation 402
– stress-strain curves 404
β-quartz phase field 405
quasi-1-D metal oxide nanostructures 713

r
Rachinger grain boundary sliding 636
Rachinger mechanism 635
Raman active phonon spectra 136
Raman microspectroscopy 137
Raman spectroscopy 118, 180, 617
Raman spectrum 200
rare-earth (RE) oxides 596, 685
– role 596
rate controlling mechanism 637, 645, 650, 651
rate-independent process 745
Rayleigh coefficient 750
Rayleigh model 746
Rayleigh-type polarization 749
– hysteresis loop 750
R-curve measurements 352, 353
– self-reinforced silicon nitrides 353
reaction-bonded silicon nitride (RBSN) 689, 690
– post-sintering 690
– sintering 689
reaction bonding (RB) process 689
– benefits 689
reactive templated grain growth (RTGG) 752, 754–756
reciprocal reticular density 488
reconstruction processes 492
– missing row reconstruction 492
– type 492
reference coordinate system 506
relaxation processes, origin 492
relaxation time 501
relaxor ferroelectrics 280
relaxor materials 281
ReO_3-type framework 259
residual porosity 8
resistance, schematic diagram 707
rock-salt structure 393, 394
rutile powder, micrograph 23

s
Sapphire (α-Al2O3) 405. see also aluminum oxide
– critical resolved shear stress 429
– crystal structure of 5
– deformation 405
– deformation twinning 408
– – basal twinning 410
– – rhombohedral twinning 408
– dislocation-dissociation 406, 407
– experimental and theoretical fault energies 409
– faulted dipole, formation 408
– gemstone 405
– prism-plane slip 406, 407
– stacking fault energy 408
– stress-strain curves 426, 427
– structure 390
– work hardening 426
– – basal plane slip 426, 427
– – prism plane slip 427, 428
sapphire single crystals 513
SAXS data 590
Scandia-stabilized zirconia (ScSZ) 41
– electrolyte sheets 41
– possesses 35
scanning electron microscopy (SEM) 94, 625, 636, 819, 827
Schottky barriers 17
Schottky diode 699
screw dislocation 502
– successive stages 502
$ScYSi_2O_4N_2$ glass ceramic
– fine-grained microstructure 252
secondary ion mass spectroscopy (SIMS) 113
second nearest neighbor (SNN) interactions 491
Seebeck coefficients
– dependence 197
– values 309
Seebeck effects 197

seeded silicon nitride
– fracture toughness and strength 355
– fracture toughness and strength 357
– microstructure 356
– R-curve behaviors 358, 359
selected area diffraction (SAD) techniques 109
selected area electron diffraction (SAED) 62
self-propagating high-temperature synthesis (SHS) 181
semiconducting ceramics 704
semiconducting metal oxides (SMOs) 710
– n-type 711
– sensitivity 712
semiconducting systems, space charge formation 699
semiconductor 711
– doping 195
– electronic properties 711
sensor response model 712
sensors, class 730
shear stress 646
Shockley-type partials 402
short-circuit diffusion 422
short-range ordered (SRO) fraction 799
[(Si,Al)(O,N)$_4$] tetrahedral 3-D interlinked arrangement 240
[(Si,Al)$_2$(O,N)$_7$]-type units 242
Si-(Al)-Y(Ln)-O-N oxynitride glasses 367
Si_2N_2O, potential applications 71
Si_3N_4-4(AlN)-4(YN)-2(Y_2O_3)-2(Al_2O_3)-3(SiO_2) system
– Jänecke prism 68
α-sialon 238
– forming area 69
– fracture surface 69
– JEM phase, typical microstructure 241
– materials 70
– mechanical properties 69
– solid solutions 69
sialon ceramics 655
– compressive superplastic deformation 655
– mechanical properties 70
– strain rate vs. stress plots 656
sialon polytypoid phases
– base-shared tetrahedral, undulating layers 234
sialon S-phase 240
Si-Al-O-N system
– aluminum nitride 232
– phase relationships 232, 234
sialon U-phase
– [1 1 0] projection 250
sialon W-phase
– acicular microstructure 251
sialon X-phase
– crystal structure, projection 233
Si-based ternary nitrides
– luminescent properties 80
$SiBN_3C$, fibers produced 147
SiCNTs, potential applications 149
silane (SiH_4)
– gas-phase reactions 60
– NH_3, laser-induced CVD 60
silica
– carbon mixtures 140
– carbothermal reduction 138
– heat-treating a mixture 139
– oxynitride 81
–– principles 230, 231
– polymorphs 230
silicon 495
– infiltration 154
– scanning tunneling microscopy image 495
– vicinal surface, schematic illustration 495
silicon atoms, nitrogen linked 244
silicon-based ternary nitrides 78
silicon carbide (SiC) 131, 138, 187
– Al_2OC solid solutions 168
– alloying 163
– AlN materials 167, 168
– AlN system 168
– anisotropic tribological behavior 188
– application 202–207
– B_4C composite materials 165
– chemical properties 186, 187
– continuous silicon carbide fibers 146
–– polymer pyrolysis-derived silicon carbide fibers (PP-Fibers) 147, 148
–– sintered powder-derived α-silicon carbide fibers (SP-Fibers) 148
–– substrate-based fibers (CVD) 146, 147
– densification 159
– iron pair 188
– Ohm's law 184
– oxidation 164, 186
– physical properties
–– electrical properties 184
–– mechanical properties 184–186
–– optical properties 183
–– thermal and calorific properties 184
– post-densification 160
– reaction bonding 154
– β-SiC filaments 146
– β-SiC nanoceramics
–– preparation 162
– α-SiC whisker 144
–– properties, Comparison 144

– silicon carbide nanofibers 148
– silicon carbide nanotubes (SiCNTs) 149
– silicon carbide platelets 145, 146
– β-silicon carbide powder
–– silica, carbothermal reduction 138–141
–– synthesized submicron β-SiC powder 141, 142
–– vapor phase, deposition 142
– α-silicon carbide, technical-scale production 137, 138
– silicon carbide whiskers 142
–– vapor-liquid-solid process 143
–– vapor phase formation and condensation 144
–– vapor solid reaction 144, 145
– sintered 193, 205
–– SEM image 193
– submicron/nanosized powders 140, 142
– substrate material 194
– synthesis 141
– thin films 161
– TiB_2 composite 164, 165
–– types 165
– TiC system 164
– toughened alumina 146
– transition metal diboride 164
– tribological properties 187
–– sintered α-SiC ceramics 187–190
–– sliding wear application, material development 190–194
– unlubricated friction 189
– whisker growth 142
–– routes 143
–– VLS process 143
silicon carbide ceramics 184, 668
– applications 668
– composite materials 163
–– fabrication 163
– densification 674
– microstructure 153, 192
– physical properties 185, 206
– solid-state sintering 674
– synthetic polymers, advantage 141
silicon carbide fibers
– mechanical characteristics 146
– preparation 141
silicon carbide nanoceramics, fabrication
– liquid-phase sintering 162, 163
– solid-state sintering 162
silicon carbide nanofibers 148, 155
– synthesis 148
silicon carbide nanotubes (SiCNTs) 149
silicon carbide polytypes
– crystal structures 133
– transformation 155, 168
silicon carbide powders, formation 160
silicon carbide roll, ESD facility 138
silicon carbide shapes 152
silicon carbide single crystal 687
– lattice 687
– thermal conductivity 687
silicon-carbon system 133
– binary phase 132
– phase diagram 132
silicon compact 689
– microstructures, scanning electron microscopy images 689
silicon nitride 60, 62, 67, 356, 541, 577, 628, 692
– AMDCB techniques, R-curve behaviors 361
– amorphous silicon nitride (α-Si_3N_4)
–– crystal structures 61
–– molecular dynamics (MD) calculation 63
–– tensile strength 63
– β-silicon nitride
–– hexagonal prisms 353
–– seed crystals 371
– cavitation processes, Phenomenological model 594
– compressive specimens 590
– creep cavities, classification 588
– creep data 582
– creep mechanisms in commercial grades 577
–– creep behavior 582
–– creep of silicon nitride 578–580
–– experimental data, discussion 581–592
–– material characterization 580
–– motivation 577
–– NT 154 582
– creep performance 579
– creep research activities 578
– creep-resistant grades 579
– creep-strained, SAXS study 591
– crystallographic modifications 60
– deformation 592
– extensive microstructural analysis 587
– fabrication procedure, schematic illustration 356
– fracture origin 533, 541
– fracture toughness and strength 364
– γ-Si_3N_4
–– hardness 64
–– thermal expansion coefficient 64
– grades 577
– grain boundaries 596
– grain-growth behavior 692

– Hertzian contact zone 628
– mechanical behavior 65
– microstructures 354, 360
– models of creep 592–596
–– cavitation creep model in NT 154/SN 88 593–595
–– lutetium, role in viscosity of glass 596
–– noncavitation creep 595
– NC 132 silicon nitride 580
– properties 62
– single crystal grains 66
– solution-precipitation (S-P) 594, 597
– structural phases 60
– structural reliability 578
– superplastic deformation 361
silicon nitride body, original formed 364
silicon nitride ceramic 59, 60, 352, 551, 596, 638, 668
– bending strength test 551
– β-Si_3N_4 crystal 668, 692
–– lattice defects in grains 680–682
–– lattice oxygen contents vs. thermal resistivities 683
–– thermal conductivity 672, 673, 677, 678
–– two dimensional microstructure 677
– comparative strain rate vs. stress plot 638
– crack velocity vs. stress intensity curve 559
– experimental data 654
– experimental temperature compensated strain rates vs. stress plots 639
– lattice defects in SiC grains 687
– lattice-purifying process 687
– mechanical behavior 67
– microstructure 684
– oxygen content 682
– polycrystalline ceramic materials 66
– processing strategy 668
– seeded/tape-cast, microstructure 693
– strength distribution 554
– strength test results 552
– tensile/compressive strain rates 639
– tensile creep 596
– tests on 560
– textured, Scanning electron microscopy images 692
– thermal conductivity 682, 688, 693
–– improvements in 682–688
–– of $MgSiN_2$-doped 586
–– variation 684
–– with sintering time 683
– thermal resistivity, lattice oxygen content effect 685
silicon nitride fabricated
– microstructure 368
silicon nitride grains 248
silicon nitride seeds
– fibrous crystals 354
silicon nitride single crystals
– mechanical properties
–– γ-Si_3N_4 63, 64
silicon oxide
– activity 686
– chemical sensors 712
– polymorph 402
silicon oxynitride 70
silicon tetrachloride 142
Si-melt infiltration 171
Si-N bondlength 230
SiN_4 tetrahedra 61
single crystal α-alumina, properties of 5
single-crystal MgO, physical properties 14
single-crystal seeding 355
single-crystal ZnO, physical properties of 16
single-crystal ZrO_2, physical properties 29
single-edged notched beam (SENB) test method 623, 625
single-edge precracked-beam (SEPB) method 357
single-edge-V-notched-beam (SEVNB) fracture toughness 363
single-walled carbon nanotubes (SWCNTs) 149
sintered powder-derived α-silicon carbide fibers (SP-Fibers) 148
sintered reaction-bonded silicon nitrides (SRBSN) 690
– preparation 690
– scanning electron microscopy images 691
sinter forging 371, 657
sinter-HIP techniques 8
sintering process 513, 514
sintering techniques
– development of 3
SiO_4 tetrahedra chains 251
Si-Si bonds 421
SiSiC
– seal rings 203
– slip-cast, combustion tubes 204
Slater–Stoner–Wolfarth band-type 802
{111}⟨110⟩ slip
– dissociated screw dislocations cross-slip 381
– stoichiometric spinel, Log(CRSS) vs. temperature 382
– stoichiometries, Log(CRSS) vs. temperature 381
slip casting 178
– SiSiC, combustion tubes 204

slip planes
– Kagomé layer 415
– oxides, spinel structure 414
slip systems
– in oxides 391
$Sm_3Si_6N_{11}$
– atomic arrangement 244
small-angle X-ray scattering 590
SMO wire 712
– conductive pathway 721
SN 281 581, 585
SN 88 581, 583
– cavitation creep model 593
– cavity formation 595
– creep asymmetry 585
– creep behavior 586, 587
– multigrain junction cavities 588
– strain rates 584
– stress-strain rate dependence 586
– tensile and compressive creep curves 583, 584, 287
– tensile creep cavities 592
– two-grain junction cavities 589
– volume fraction of cavities 589
sodium silicate
– water, leaching 30
sole transitional process 748
sol-gel method, see colloidal methods
sol-gel synthesis routes 102
– nanostructured SiC, morphologies 148
solid/liquid interfaces 502
– features 502
– first layer, structure 503
– type 502
solid/solid interfaces 505
– fundamentals 505
– structure and energy 508
– symmetry dividing plane 506
solid oxide electrolyzer cells (SOECs) 283
solid oxide fuel cells (SOFCs) 258, 697
– oxide electrodes 288
solid-state diffusion 438
solid-state electrochemistry 697
solid-state-sintered 186
– aluminas 4
solution hardening
– oxides
–– aliovalent cations 429, 430
–– isovalent cations 428, 429
solution-precipitation process 650, 652
– schematic representation 651
solution-reprecipitation processes 683
space charge barrier 699
space charge density 709
space charge layers, schematic diagram 718
space charge models 699, 707
– applicability 707
space charge potential 700, 701
spark-plasma-sintered boron carbide 179
spark plasma sintering (SPS) 15, 74, 155
sphalerite structure 396
S-phase, orthorhombic structure
– atomic arrangement 241
– $[(Si,Al)_{10}(O,N)_{25}]$ structural unit 242
spinel crystals
– stress-strain curves 425
– work hardening 424–426
– work softening 424–426
spinel grains 18
spinel-structured gallium oxonitride 120, 121
spin-spin interactions 275
spray deposition process 173
spray gun, aluminum alloy 173
$Sr_3Ln_{10}Si_{18}Al_{12}O_{18}N_{36}$
– schematic representation 246
sintered reaction-bonded silicon nitrides (SRBSN) 690
– thermal conductivity 690
short-range ordered (SRO) 810
$SrTiO_3$ crystals
– perovskite structure 259
– yield stress vs. temperature 412
stacking of {111} planes
– in spinel viewed 389
stoichiometric boron carbide
– densification 175
stoichiometric spinel 381
strain rate
– vs. stress, summary log-log plot 340
strain transient dip tests (STDTs) 340
stress intensity factor 542, 607, 612, 615, 617
– expression 609–611
– fracture toughness, defined 543
stress-strain behavior 172
stress tensor, components 543
structural ceramics 621. see also silicon nitride
structural elements
– ceramics 352
– fibrous grain-aligned porous silicon nitrides
–– dense silicon nitrides 375, 376
–– porous silicon nitrides 370
–– sinter-forging technique 374, 375
–– tape-cast porous material 371–374
– fibrous grain-aligned silicon nitrides
–– large grains 355–361
–– small grains 361–364

– grain boundary phase control
–– fracture resistance 365–367
–– heat resistance 368–370
– self-reinforced silicon nitrides 352–355
– well-organized control 351
sub-critical crack growth (SCCG) 544, 560, 567
– consequences on fracture 561
– crack length vs. time 562
– crack velocity vs. stress intensity curve 559
– exponent 560, 561, 562, 565
– lifetime and influence 559
submicrometer aluminas 7
submicrometer-grained aluminas 7
submicron boron carbide powder 178, 181
submicron-sized microstructure 155
superhard compounds 131
superplastically deformed silicon nitride 362, 363
superplastically sinter-forged body 364
superplastic forging
– schematic illustration 362
superplasticity 633, 649, 659
– accommodation processes 643–656
–– GBS accommodated by diffusional flow 643–648
–– GBS accommodated by dislocation motion 648
–– shear-thickening creep 655
–– solution-precipitation model for creep 649–655
– applications 656–659
– creep mechanism 649
– definition 633, 634
– macroscopic/microscopic features 634–640
– multiscale nature 659
– research in 660
– strain rate 633
– stress exponent 659
surface energy 480–486, 493, 501
– atomic configuration, Schematic illustrations 482
– calculation 484
– enthalpic aspect 495
– origin 486
– plot 493
– reduction of anisotropy 504
surface-related phenomena 481
surface work 481
symmetrical tilt grain boundaries (STGBs) 507
synchrotron X-ray data 774

t
tantalum mononitride (TaN) 77
tape-casting process 752, 753
– schematic 752
tape-cast silicon nitride 357, 368
– fracture toughness and strength 357
– microstructure 356
– R-curve behaviors 358
$TbMnO_3$ 762
– magnetic and dielectric properties 762
– magnetic field 763
–– polarization reversal 764
– magnetocapacitance effect 763
– spin frustration effect 761
templated grain growth (TGG) 752, 753
– heterotemplated grain growth 753
– microstructure evolution, Schematic 753
– reactive grain growth 753
tensile stresses 542
– flaws 542
– stress concentration 542
tentative SiC-AlN phase diagram 168
ternary nitrides 78
– schematic drawing 92
– ternary silicon nitrides
–– alkaline earth silicon nitrides 79
–– $LaSi_3N_5$ 79
–– lithium silicon nitride ($LiSi_2N_3$) 80
–– $MgSiN_2$ 78, 79
terrace-step-kink (TSK) model 495
TETRABOR boron carbide
– venturi blast nozzle, cross-sectional area 208
tetragonal zirconia polycrystals (TZP)
– typical properties of 36
– ceramics 35
tetrahedral 3-D interlinking 242
textured materials, advantages 739
texture engineering, aspects 750
theoretical density (TD) 145
thermal barrier coatings (TBC) 35, 39
thermal conductivity 351, 667
– of dielectric ceramics 668–671
–– in adamantine compounds 669–671
–– of nonmetallic crystals 668
–– β-Si_3N_4, thermal conductivity, estimate 671
– high-thermal conductivity nonoxide ceramics 672–688
–– of composite microstructures 672–679
–– improvement via purification of grains during sintering 679–688
– introduction 667

– Si_3N_4 ceramics, mechanical properties 688–693
–– anisotropic thermal conductivity in textured Si_3N_4 692
–– thermal conductivity/mechanical properties, harmonic improvement 688–692
– temperature dependence 669
thermal conversion 117
thermal expansion anisotropy, correlation 318
thermal expansion coefficient 198
thermal shock 534
– cracks 534
– failure 534–536
thermal stresses 534
thermodynamically stable phase 94
thermodynamic approach 737
thermodynamic factor 466, 467
thermoelectric devices
– materials, suitability 197
thermoreflectance microscopy 671
thin film 774
– clamping in 774
– glassy film 7
– PVD deposition 76
third-generation solar technology 720
third-order Birch–Murnaghan equation 110
– parameters 110
three-dimensional corner-sharing tetrahedral lattice 814
threshold stress 640, 646
Ti_2AlC compounds 304
– cyclic compressive stress-strain curves 325
Ti_3GeC_2 samples
– post-quench four-point flexural strength 333
Ti_3SiC_2 339, 341
– basal planes 342
– bulk moduli 341
– creep parameters, Summary 339
– cycle, fine-grain sample 337
– engineering stress-strain curves 330
– fine-grained microstructures 338
– grain, load-displacement curve 332
– loaded tension/compression 338
– oxidation 319
– polycrystalline samples, crack fatigue behavior 334
– response
– stress-strain curves 324
– tensile stresses 338
time-of-flight (ToF) neutron diffraction 755
titania single crystals
– physical properties of 24
titanium, oxychlorides 23
titanium dioxide (TiO_2) 20, 701, 721
– anatase 26
– applications of 25–27
– biocompatibility of 27
– concentration of electrons 721
– crystal structure 20, 21
– mesoporous dye-sensitized 721
– natural sources and production 22
–– anatase, synthesis of 22, 23
–– brookite, synthesis of 23, 24
–– rutile, synthesis of 23
– n-type semiconductor 26
– polycrystalline titania 24, 25
– polymorphs 20
–– anatase 20
–– brookite 20
–– properties of 24
–– rutile 20
– relative permittivity 704
– single-crystal, properties of 20, 21
– space charge properties 701
– surface area 721
titanium nitride (TiN) 75
– characteristic properties 76
– powders 76
– properties 76
– structure 75
– synthesis 75, 76
– thin film 75
titanium-oxygen interstitial defect 429
$(TM)O_6$ octahedron 792
– octahedral and tetrahedral environment 793
–– 3d one-electron levels, Crystal-field splitting 793
– symmetries 792
– valence-band orbitals 792
toughened ceramics 612–620
– bridged crack wake methods 612
– bridged interface methods for increasing fracture resistance 613–620
–– crack bridging by second phase 616
–– crack growth resistance toughening 614–616
–– ferroelastic toughening 619
–– grain bridging 613
–– phase transformation/dilatant zone toughening 617–619
– frontal wake methods 612
– microstructural design methods 613
– toughening by crack tip process zone effects 620

toughening mechanisms 365
transition metal nitrides 77
transition-metal oxides (TMOs) 791
– applications 791, 792
– chromium dioxide 800–802
– 3d orbitals, properties 792
– ferrites 797–800
– iron oxides 793–797
– – magnetic properties 791, 794–797
– – structures 793
– manganese oxide phases 802–821
– – magnetic properties 807
– – manganese dioxide nanoparticles, magnetic properties 821
– – manganese oxide structures 804–807
– – MnO_2 nanostructures, synthesis 815–821
– oxygen defects 394
– single crystals, Plastic deformation 393
transmission electron microscopy (TEM) 62, 181, 320, 580, 767, 818, 820
tribological applications
– SiC ceramics, physical properties 192
triglycine sulfate (TGS) 743
trimethylgallium 101
triple junction wetting 521
– 2-D section parallel 521, 742

u
ultrafine-grained titania powders
– preparation of 22
ultraviolet (UV) emitters 93
– wavelengths 102
UO_{2+x}, dislocation structure 401

v
vanadium on a magnesium oxide support (V/MgO) 14
vanadium oxyhydroxide 828
– magnetization 828
van der Waals forces 71
vanVechten predicted phase transitions 95
vapor-liquid-solid (VLS) process 143
– whisker 143
vapor phase formation 144
varistor-forming oxides (VFOs) 19
VC 504
– adsorption 504
– effect 504
– role 504
Verneuil method 106
Verwey transition 796
Vickers hardness 162, 182, 331
Vickers indentations 358, 624, 625
– loads 331
Vickers indenter 621
– diamond-tipped 622
– on glass 623
– rows 624
virtual charge transfers 808
volatile organic compounds (VOC) 20
voltage-current characteristics (VCC) 20

w
Wakai's model 646, 651
– cation diffusion energy 653
– creep equation 652
– creep laws 646
– equations 651
– free energy 653
– modification 653
– stress exponent vs. applied stress 652
– – apparent values 654
WC crysal 504
– coarsening rate 504
– grain growth behavior 504
– scanning electron microscopy images 505
weakest-link hypothesis 545
wedge effects 370
Weibull analysis 554
Weibull distribution function 550, 552
– application 550
Weibull equation 556
Weibull material 555, 558
– bimodal flaw size distribution 557
– flaw sizes, relative frequency 556
Weibull modulus 357, 359, 547, 548, 550, 553, 564, 568
– confidence intervals in 554
Weibull statistics 357, 545, 546
Weibull theory 558
white light-emitting diode (LED)
– schematic function 80
Wiedmann–Franz law 313
Wiener's model 674, 676
– two-phase systems 675
Wiener method 682
Willis defects 401
Wulff plot 486
– equilibrium shape 487
Wulff theorem 489
– analytical version 498
– graphical interpretation 489
wurtzite crystal structure 16, 95, 167
– stacking fault energy 397, 398
wurtzite-structured GaN 123
– O/N substitution 123, 124
wurtzitic boron nitride (w-BN) 72

x

X-ray absorption spectroscopy (XRAS) 106
X-ray diffraction (XRD) analysis 133, 581, 755, 825
– amorphous phase 319
– intensity 768
– symmetry-dependent reflections 756
X-ray photoelectron spectroscopy (XPS) 179
– spectra 112
X-ray specular reflectivity (XRSR) 771

y

$Yb_2Si_2O_7$ 591
– distribution 591
– scattering 591
– size distribution 592
YN-Si_3N_4 system 236
Young's modulus 12, 373, 622
Y-Si-Al-O-N system 249, 250
– B-phase produced 249
$Y_2Si_4N_6C$, atomic arrangement 238
$Y_2Si_3O_3N_4$, ^{29}Si MAS-NMR spectrum of 235
yttria
– ceramics, use 49
– Poisson ratio 50
– powder
–– morphology 49
–– ultra-fine-grained 49
yttria-stabilized tetragonal zirconia polycrystal (YTZP) ceramic 634, 638, 640, 644
– absorption free energy 648
– as-received/tensile-deformed specimens 635
– borosilicate-doped 642
– deformation 636
– dislocation cores 647
– grain boundary sliding, microstructural invariance 637
– layers 657
– segregation energy 648
– submicron-sized 641
– superplastic behavior 645
– superplasticity 636
yttria-stabilized zirconia (YSZ) 35, 658, 697, 714
– bicrystal 512
– disadvantages 39
– layer 39
– nanostructured thin films 41
yttria tetragonal zirconia ceramics 640
– strain rate vs. stress 640
– polycrystals, dislocation driven model 646
yttrium-aluminum-garnet (YAG) 48, 420
– powder 157
yttrium-aluminum ratios 365
yttrium atoms 641
– segregation 641, 647
yttrium-iron-garnets 48
yttrium oxide 48
– ceramic material 48
– microstructure 50
– optical ceramic 50
– thin economical layer 51
– SiO_2 additive ratios 685
– stabilized ZrO_2
–– plastic deformation 399
–– single crystals 399

z

Zener drag effect 518
Zener polarons 278
zincblende (zb)
– gallium nitride crystallizes 96
– phase 97
– structured GaN, experimentally/theoretical investigations 97
– structures 95, 96
– type phase 112
zinc oxalate dihydrate, thermal decomposition 16
zinc oxide (ZnO) 15, 516
– applications
–– ceramics 19
–– varistors 17–19
– based diluted magnetic semiconductors (DMS) 20
– crystals
–– properties 16
–– structure 16
– nanocolumns, scanning electron microscopy images of 20
– nanocrystals 16
– nanowire sensitivity 713
– natural sources and production 16
– n-type semiconductor 17
– pure/ Bi_2O_3-doped 516
–– SEM/TEM images 516
zirconia 617, 619
– aging-related nucleation 42
– based ceramics 27
– based materials 34
–– surface monoclinic phase fraction vs. exposure time 36
– biomedical-grade 42
– chlorination 29
– crack growth resistance curves 617
– electrolytes 40, 258
–– fuel cell, schematic diagram of 41

– melting temperature 32
– metastability 42
– orthorhombic 33
– particles, microcrack formation 31, 32
– phase transformation 31, 42, 619
– polymorphs, lattice parameters of 28
– Raman spectroscopy, monoclinic phase content 618
– single crystals, physical properties of 29
– thermal decomposition 29
– yttria 34
zirconia-toughened alumina (ZTA) 4, 31, 37
zirconium oxide 27–28
ZrO_2 polymorphs, crystal structures 28
ZrO_2-Y_2O_3 single crystals, yield stresses 400